THE ATLAS OF BREEDING BIRDS OF ALBERTA: A SECOND LOOK

The Federation of Alberta Naturalists

2007

PRINTED APRIL 2007

Published by the Federation of Alberta Naturalists

Additional copies of this book may be obtained by contacting the Federation of Alberta Naturalists at:

11759 Groat Road
Edmonton, Alberta
T3M 0K6
(780) 427-8124
www.fanweb.ca

Printed and bound at Friesens Printers
Altona, Manitoba, Canada

COVER PHOTOGRAPHS:
Front Cover
  Common Merganser: Randy Jensen
Back Cover
  Song Sparrow: Randy Kimura
  Whooping Crane: Gordon Court
  Northern Pygmy-Owl: Alan MacKeigan
  Sharp-tailed Grouse: Gerald Romanchuk
  Mallard: Gerald Romanchuk
  Mountain Bluebirds: Alan MacKeigan

CANADIAN CATALOGING IN PUBLICATION DATA
The Atlas of breeding birds of Alberta : a second look / Federation of Alberta Naturalists.

Includes bibliographical references and index.
ISBN 978-0-9696134-9-7
Printed in Canada

1. Birds–Alberta. 2. Birds–Alberta–Geographical distribution. 3. Birds–Alberta–Geographical distribution–Maps. 4. Birds–Alberta– Nests. I. Federation of Alberta Naturalists

QL685.5.A86A84 2007 598.097123 C2007-900865-8

Mixed Sources
Product group from well-managed forests and other controlled sources
www.fsc.org Cert no. SW-COC-1271
© 1996 Forest Stewardship Council

Contents

List of Figures

List of Tables

Acknowledgements

The publishing of *The Atlas of Breeding Birds of Alberta: A Second Look* was possible through the combined efforts of organizations, corporations and over 1500 individuals who contributed time, expertise, and resources to the Alberta Bird Atlas Update Project. This complex project involved a major field survey component that relied on atlassers, compilers, regional coordinators, regional rare bird committees, and office support staff, who gathered, reported, and verified data from all across Alberta. Many of our province's fields, sloughs, and woodlots would not have been explored for their rich and varied bird life were it not for the cooperation of Alberta's landowners. The atlas also benefitted from the naturalists, photographers, and biologists who allowed access to their notebooks, records, and expertise, and the agencies and companies that provided access to their data.

The continued support of volunteers, organizations, and corporations in Alberta allowed this publication to become a reality. Where recorded, their names are listed. If not recorded, their contributions to the project are still greatly appreciated.

This book is dedicated to all the people who contributed in some fashion to the success of this project, and to those individuals and groups who will continue to contribute to the knowledge of Alberta Birds through various volunteer programs such as the Breeding Bird Survey, the Alberta Nest Record Scheme, and the Alberta Birdlist Program.

Financial Support

Sponsoring Organizations

Alberta Conservation Association
Alberta Ecotrust Foundation
Alberta Sustainable Resource Development
George Cedric Metcalf Foundation
North American Waterfowl Management Plan
Slave Lake Pulp
Suncor Energy Foundation
Sundance Forest Industries Ltd.
Weyerhaeuser Ltd.

Major Donors

Ainsworth Lumber Ltd.
Alberta Sport, Recreation, Parks, and Wildlife Foundation
Baillie Bird Fund
Canadian Wildlife Service
Student Career Placement Program

Supporters

Bennett, Christine
Burkinshaw, Lois
Cartwright, John
Charles Ivey Foundation
Dolman, Teresa
Dunn, Robert
Gauthier, B.J.
Griffiths, Deirdre
Hall, Gerald
Imperial Oil Ltd.
Klauke, Richard
Mountain Equipment Coop
Nexen Inc.
Nordstrom, Stanley
O'Shea, Michael
Shell Environmental Fund
Smith, Jeanne
Student Temporary Employment Program
Szabo, Vincent
Weldwood of Canada Ltd / West Fraser Mills
Windsteig, Romana
Winter, Pres
Woolgar, Janet
Worona, Robert

Project Management and Coordination

Project Coordinators

Penner, Philip
Wagner, Greg

Assistant Project Coordinators

Boulet, Fiona
Hood, Fenella
Koch, Michelle
Moes, Cory

Management Committee

Clayton, Pat
Court, Gordon
Dickson, Loney
Kennedy, Bruce
Quinn, Mike
McGillivray, Bruce
McKeating, Gerry
Newton, George
Penner, Philip
Semenchuk, Glen
Stelfox, Harry
Wagner, Greg

Technical Committee

Boyce, Mark
Court, Gordon
Dale, Brenda
Duncan, Dave
Hudon, Jocelyn
Lein, M. Ross
Pilny, Jerry
Prescott, Dave
Schmiegelow, Fiona

Technical Advisors

Baresco, Dennis
Cumming, Steve
Lapalme, Monique
Lele, Subhash
Mazerolle, Dan
McGillivray, Bruce
McKeating, Gerry
Newton, George
St. Clair, Colleen Cassidy

Alberta Bird Record Committee

Carroll, Bob
Hudon, Jocelyn
Klauke, Richard
Knapton, Richard
Lein, M. Ross
Riddell, John
Ritchie, Brian
Slater, Andrew
Thormin, Terry
Wershler, Ray

Regional Rare Bird Committees

Arbuckle, Reg
Barclay, Reid
Baresco, Dennis
Bennett, Lloyd
Cerney, Linda
Collister, Doug
Craig, Gavin
Dolman, Teresa
Hervieux, Margot
Hindmarch, Ted
Kerr, Joan
Klauke, Richard
Knapton, Richard
Lumbis, Ken
McIntyre, David
Park, Jack
Ritchie, Brian
Romanchuk, Gerald
Wagner, Greg
Whiley, Fred

Regional Coordinators

Barclay, Ried
Beckmann, Christa
Bennett, Lloyd
Biener, Tanja
Boulet, Fiona
Boyd, Judy
Burkinshaw, Lois
Cerney, Linda
Dacyk, Kimberley
Davies, Iris
Grey, Robert
Henry, Grant
Hervieux, Margot
Hood, Fenella
Horch, Phil
Kerr, Joan
Klauke, Richard
Legris, Andre
Lohr, Lloyd
MacCallum, Beth
Moes, Corey
Okrainec, Jennifer
O'Shea, Michael
Penner, Philip
Prescott, Dave
Priestley, Chuck
Shier, Jack
Smith, Janice
Stiles, Don
Storms, Bob
Wagner, Greg

Publication and Communications

Managing Editor

Semenchuk, Glen

Technical Editor

Wiens, Trevor

Copy Editor

Burns, James / Woolly Mammoth Services, Edmonton

Authors

Dale, Brenda
Glasgow, Bill
Hudon, Jocelyn
MacCallum, Beth
McGillivray, Bruce
Penner, Philip
Priestley, Charles
Priestley, Lisa
Semenchuk, Glen
Wagner, Greg
Wiens, Trevor

Cartography and Graphs

Hanneman, Matthew
Wiens, Trevor

Data Analysis

Wiens, Trevor

Data Management

Bijelic, Vid
Penner, Philip
Semenchuk, Michael
Wiens, Trevor

Illustrations

de Jong, Anne

Photo Selection

Penner, Philip

Layout and Design

Fushtey, Judy / Broken Arrow Solutions Inc.
Wiens, Trevor

Newsletter Editor

Legris, Andre
Penner, Philip

Office Staff / Data Entry

Arnold, John
Baptiste, Shirley
Bruinsmas, David
Bykowski, Alana
Coates, Tricia
Dacyk, Kimberley
Foshay, Jennifer
Gordey, Joan
Holtz, Michael
Laforce, Mike
Monk, Dana
Pirk, Wendy
Semenchuk, Michael
Shannon, Jolene
Sneyd, Erin

Field Programs

Data Contributors

Alberta Biodiveristy Monitoring Program
Alberta Natural History Information Centre
Biodiversity/Species Observation Database
Bird Studies Canada
Breeding Bird Survey
Canadian Wildlife Service
Ducks Unlimited Canada
Eastern Irrigation District
Fish and Widlife Management Information System
Golder and Associates Ltd.
Nocturnal Owl Monitoring Program
Project Sapsucker
Westworth Associates Environmental Ltd.
Wood Buffalo National Park

Registered Atlassers

Acey, Tim
Albers, Frankie
Aldred, Karen
Allair, Jody
Allan, Ray
Allen, Jim
Allen, Peter
Allen, Rory
Alsterlund, Enola
Ancelin, Ryan
Anderson, Colleen
Anderson, Paul

Anderson, Philip
Arnold, John
Backen, Sandra
Baden, Dale
Bailey, Linda
Baker, Brian
Baker, Terence
Ball, Linda
Bargholz, Stanley (Bud)
Barr, Michael
Bartlett, Catherine
Bartlett, David
Baumbach, Melvin
Bayne, Eric
Benford, William
Bennett, Lloyd
Bennett, Christine
Biggs, Brian
Bird, Ann
Bird, Charles
Bissell, Cheryl
Bjorge, Myron
Bjorge, Ron
Blackwell, Wendy
Bonar, Rick
Bovell, J.D. (Des)
Boyd, Judy
Braun, Nicole
Briscoe, Diane
Brown, Anne
Brown, Bill
Brown, Dick
Bruns, Eldon
Buffalo Lake Naturalists
Burkinshaw, Lois
Burkinshaw, Phil
Burns, Rod
Busat, Vivian
Byers, Pat
Cabelka, Fred
Cabelka, Ruth
Calverley, Pat
Calverley, Ross
Campbell, Ingrid
Campbell, June
Campbell, Marilyn
Campbell, Wayne
Cannady, Steve
Cartwright, John
Chandler, Laurence
Cheesebrough, Jane
Chemago, Richard
Christensen, Eileen
Christensen, Paul
Christiansen, David
Christiansen, Marilyn
Chruszcz, Bryan
Cole, Allan
Collister, Doug
Coogan, Sean
Cooke, Doreen
Cooke, Moira
Cookson-Hills, Heather
Cormier, Kathy
Coughlin, Judith
Cowie, Ralph
Craig, Jill
Cram, Phil
Crane, Helene
Creaser, Jeanette
Cryderman, Alf
Csaky, Marg
Csaky, Tony
Cunningham, Jade
Davies, Iris
Deamer, Terry
Deneka, Shirley
Depner, Jane
Dickinson, Dawn
Diggle, Harry
Dobson, Brenda
Dolman, Doug
Dolman, Teresa
Downey, Dean
Downie, Ernest
Dubitz, Larry
Dubitz, Sharon
Dudragne, Mary Ann
Dunn, Robert
Dwornik, Wayne
Ellis, Carmel B.
Elser, Jane
Ewald, Mike
Feidler, Randy
Feleski, Val
Ferguson, Lorne
Fitch, Janet
Flynn, Dick
Flynn, Lenora
Ford, Eileen
Fraser, Joyce
Frisky, Dennis
Garrison, Lynn
Garrison, Troy
Gauthier, Bernie
Gosche, Stan
Gosche, Tim
Graham, Karen
Graupe, Shel
Greening, Glen
Grenier, Karen
Grenier, Wayne
Grey, Robert
Griffiths, Deirdre
Groom, Ronda
Groves, Jon
Guenette, Simone
Habib, Lucas
Hachey, Carole
Haddock, Peter
Hall, Dawn
Hall, Gerald
Hall, Robert
Halmazna, George
Hamilton, Jim
Harpe, Illo
Harris, Brian
Harris, Caroline
Heard, Anne
Heard, Stuart
Hervieux, Margot
Higgins, Sue
Hindmarch, (Ted) Edmond
Hofman, Ed
Holmes, Dale
Homister, Chris
Homister, Rosemary
Horch, Phil
Hudon, Jocelyn
Hudson, Velma
Huget, Del
Hummel, Ray
Hummerstone, Carol
Hunwick, Anne
Hvenegaard, Glen
Isaac, Alf
Isaac, Bryan
Jensen, Patti
Jensen, Randy
Johner, Brent
Johnson, Eunice
Johnson, Gwen
Johnston, Nicole
Joosten, Francis
Kelly, Carol
Kelly, Shane
Kenyon, Al
Kerr, Joan
Kilmury, Amy
Kilmury, Tyler
Kinley, Jessie
Kinsella, Wayne
Klauke, Richard
Klein, Tom
Kline, Keith
Knapik, Dwight P.
Knorr, Jack
Knorr, Mildred
Korth, Judy
Kraft, Carol
Kraft, Mel
Kullman, Linda
Lambert, Donna
Lane, Steve
Lange, Jim
Larsen, Ken
Larsen, Linda
Lavender, Darlene
Legris, Andre
Lehman, Aaron
Lemon, Steve
Leriger, Ralph
Livingston, Sharon
Logan, Keith
Logan, Sandy
Lubbers, Gerry
Lund, Mark
Lussier, Ruth
MacCallum, Beth
MacCulloch, Lynnette
MacIntosh, Carol
MacKay, Bill
MacKeigan, Alan
MacSween, April
MacSween, Blaise
Mandrusiak, Kristy
Manley, Lucille
Mann, Joyce
Marshall, Ely
Marshall, Lloyd
Matacio, Doug
McCallum, Cindy
McCarty, Marilyn
McCarty, Michael
McCarty, Rolland
McCracken, Andy
McCracken, Anne
McDonald, Robert

McDonell, Lorna
McKinnon, John
McKown, Dave
McKown, Donna
McQuid, Florence
Medicine River Wildlife Centre
Metz, Karen
Metz, Ken
Meyer, Greg
Miller, Carolin
Miller, Doug
Miller, Jeanne
Mills, Chris
Mills, Don
Mitchell, Patricia
Moker, Jeff
Moore, Dolores
Moore, Don
Moore, Joyce
Morgan, Bob
Morgan, Linda
Morigeau, Howard
Moysa, Alex
Murphy, Andy
Nelson, Wayne
Newton, George
Nissen, Clinton
Nissen, Juanita
Nnabuihe, Rita
Nordstrom, Stan
Oakes, Ken
O'Connor, Laura
Okrainec, Jennifer
Olsen, Lloyd
Olson, Chet
Olson, Lloyd
Osborne, Mark
O'Shea, Michael
Ottenbreit, Kim
Pearman, Myrna
Peden, Mary Jane
Pelzer, Aileen
Penner, David
Peterson, El
Petrie, Reny
Pfeiffer, Angela
Pilny, Jerry
Popp, Sabrina
Potter, Jim
Prescott, Dave
Priestley, Chuck
Priestley, Lisa
Putney, Natalie
Quinlan, Richard
Quist, Wendy
Red Deer River Naturalists
Reddecliff, Eleanor
Rippin, Blair
Rivet, Anne
Robertson, Ron
Rogers, Jason
Rowan, Glen
Rutter, Laura
Schultz, Murray
Schultz, Sharon
Scott, Barbara
Scott, Dean
Sepos, Blain
Shyry, Darcey
Simpson, Gary
Simpson, Nell
Sinton, Helga
Skrepnek, Shelley
Slater, Andrew
Smith, Cyndi
Smith, Jeanne
Smith, Lorna
Snyder, Carol
Snyder, Mira
Sommerhalder, Reno
Sommers, Jack
Soveran, Marilylle
Spitzer, Milton
Spytz, Chris
Stavne, Merril
Stavne, Robb
Steele, Donna
Stepnisky, Dave
Stiles, Don
Storm, Pat
Storm, Svend
Strand, Elizabeth
Strand, Keith
Stroebel, Karen
Sunpine Forest Products
Sutherland, Fred
Szabo, Sandy
Taylor, Alan
Taylor, Deborah
Teeter, Barb
Thomson, Jim
Thurston, John
Toews, Marilee
Toma, Michael
Turney, Donna
Turney, Jim
Tuvidale, John
Veitch, Arthur
Vesak, Tessa
Vos, Anne
Wagner, Greg
Walcheske, Art
Wallis, Ian
Wescott, Haze
Wild, Monty
Wilkinson, Lisa
Windsteig, Romana
Wingert, Kevin
Winter, Pres
Witskey, James
Wojnowski, Jul
Wood, Gwen
Woods, Ray
Woolgar, Janet
Worona, Rob
Wotton, Amy
Wotton, Tyler
Wright, Jack
Yaki, Gus
Yee, Karen
Young, Colin
Young, Jan
Zombori, Joseph
Zombori, Shirley
Zroback, Rick

Other Contributors

Achuff, P
Acorn, J
Albers, F
Alberta Fish and Wildlife
Albus, N
Alderson, J
Alderson, M
Aldridge, C
Alexander, W
Allarie, E
Allen, D
Allen, F
Allen, L
Allen, V
Alliance Pipeline
Allsebrook, K
Ament, L
Amy, R
Ancelin, R
Anderson, B
Anderson, C
Anderson, D
Anderson, E
Anderson, G
Anderson, J
Anderson, L
Anderson, N
Anderson, S
Anderson, W
Andrew, G
Anweiler, G
Applewhaite, C
Archibald, J
Archibald, M
Armstrong, P
Arnold, J
Arsenault, R
Arseneault, S
Ashbee, C
Attia, Y
Attwell, J
Attwell, R
Atwell, J
Auffray, A
Ayer, S
Baden, D
Bahmiller, D
Bahmiller, J
Bahnmiller, D
Balagus, P
Ball, G
Ball, L
Ball, T
Bank, J
Barbar, S
Barclay, R
Baresco, D
Bargman, J
Barkwell, M
Barnes, A
Barnes, L
Barnetson, B
Barrett, L
Barr, M
Barry, M
Bartlett, C
Barttett, D
Bast, P
Baumgarter, A

Baumgarter, D
Bayne, E
Beattie, L
Beaubien, E
Beazer, H
Beck, B
Beck, J
Beck, R
Beggs, J
Behm, J
Belesky, A
Beleyme, J
Bell, J
Benford, B
Benford, G
Benford, S
Bennett, B
Bennett, C
Bennett, J
Bennett, K
Bennett, L
Bennett, M
Bennett, Y
Benoit, S
Benson, M
Berger, S
Berg, M
Berg, R
Beriault, R
Berry, J
Bertram, H
Bevans, M
Beyersbergen, G
Bickle, L
Bielman, V
Biggs, B
Bijelic, V
Bileau, W
Binder, H
Bird, A
Bird, C
Bjorge, K
Bjorge, M
Bjorge, R
Black, J
Blake, A
Blake, T
Bogden, L
Bohnert, J
Bohnert, P
Bohnert, T
Boissey, H
Boisvert, G
Boisvert, P
Bolietz, C
Bolli, C
Bonar, J
Bonar, R
Bonar, S
Boonstra, E
Boonstra, M
Booth, C
Booth, G
Booth, J
Borduas, J
Borgardt, A
Boucher, B
Boukall, B
Boulian, J
Bourdages, S
Bourdin, D
Bourouiba, M
Bovell, J
Bowtell, B
Boyd, A
Boyd, J
Boyd, L
Boylen, D
Bradley, C
Brauner, K
Bray, S
Bregar, J
Brilling, M
Brimacombe, B
Brimacombe, E
Briscoe, C
Briscoe, D
Brissette, V
Bristow, C
Brooke, S
Brook, H
Brown, A
Brown, B
Brown, D
Brown, E
Brown, K
Brown, L
Brown, M
Brown, N
Brown, R
Brown, W
Bruinsma, D
Brundy, V
Bruns, E
Bull, M
Bulman, P
Burger, J
Burkinshaw, L
Burkinshaw, P
Burns, B
Burt, L
Burton, B
Burton, J
Burton, R
Burton, S
Busat, V
Bywater, D
Caddy, J
Cadieux, E
Cairns, A
Calverley, R
Cameron, C
Cameron, K
Camey, L
Campbell, D
Campbell, H
Campbell, J
Campbell, M
Campbell, S
Campbell, W
Canadian Natural Resources Ltd.
Canadian Wildlife Service
Cannady, S
Carpenter, J
Carrier, R
Carroll, B
Cartwright, J
Cartwright, R
Cass, P
Cathy, S
Cennie, N
Cerney, L
Cerney, M
Cesco, S
Chaboylo, R
Chandler, L
Chang, E
Charest, L
Charles, H
Charles, N
Cherriere, C
Cheeseborough, J
Chilton, G
Chisholm, R
Choquette, T
Chouinard, B
Choy, D
Christensen, B
Christensen, G
Christensen, H
Christensen, I
Christensen, M
Christensen, R
Christiansen, A
Christiansen, D
Christiansen, J
Christiansen, M
Christopher, K
Chruszcz, B
Clark, D
Clark, P
Clark, S
Claussen, J
Clawson, F
Cleanwater, L
Clements, F
Clements, J
Clibbon, B
Coates, P
Coccioloni/Amatto, D
Cole, A
Cole, J
Cole, L
Cole, M
Coleman, M
Coleman, P
Coleman, R
Collier, E
Collister, D
Comstock, J
Constable, M
Conway, H
Cooke, D
Cooke, H
Cooke, J
Cooney, S
Cooper, D
Cooper, P
Cooper, V
Coppock, M
Cordeiro, E
Cordeiro, J
Corkery, K
Corlaine, G
Cornejo, C
Corrigan, R
Costello, R
Costello, S
Cotter, G
Cotter, N
Coughlin, M
Coulter, C
Court, G
Coutts, J
Couture, N
Cowan, L
Cowie, R
Cowley, T
Cowtan, E
Cox, M
Craig, E
Craig, G
Craig, J
Cram, P
Crane, B
Crawford, M
Creaghan, C
Crescent, A
Crichton, J
Crighton, J
Cromie, R
Crone, B
Crosley, B
Crutchley, I
Cryderman, A
Cullum, J
Cunningham, D
Dagenois, L
Dahl, R
Dainard, B
Dale, B
Daniel, E
Dann, E
Darlington, J
Davidson, C
Davies, I
Davies, R
Davies, W
Davis, K
Dawson, D
Dawson, F
Deagle, J
Dean, L
Dean, N
deGraaf, D
deGraaf, N
Dekker, D
De La Mare, C

Demulder, P
Deneka, S
Depeel, M
Deroucher, N
de Sahurigen, W
deRaadt, T
Detomasi, D
Dickinson, D
Dickson, D
Dickson, R
Diepenbroek, J
Dittrich, E
Dober, J
Dobson, B
Doe, J
Dolman, D
Dolman, J
Dolman, T
Domes, A
Donaldson, A
Donaldson, G
Donnelly, B
Doohan, P
Dorado, M
Dougherty, L
Douglas, C
Downey, D
Droppo, O
Dubitz, L
Dubitz, S
Ducks Unlimited Canada
Dudragne, M
Dudragne, R
Dudra, M
Duffy, G
Duke, E
Dulley, S
Duncan, J
Dunlop, B
Dunn, B
Dunn, J
Dunn, M
Dunn, R
Dupres, J
Dutcher, R
Duthie, J
Dutoit, B
Du Toit, L
du Toit, S
Duxbury, J
Dwornik, W

Eastern, A
Ebel, R
Eddy, L
Ektvedt, I
Ellenwood, D
Ellenwood, M
Elliott, J
Ellis, M
Ellison, J
Elser, J
EnCana Ltd.
Eng, C
English, I
English, W
Ergezinger, C
Ergezinger, D
Ergezinger, N
Ergezinger, R
Ergezinger, W
Erickson, G
Erickson, P
Erven, D
Erven, M
Esler, J
Etue, C
Evans, P
Ewasiuk, L
Failler, M
Fairless, D
Fairweather, J
Fairweather, R
Falcione, J
Faulder, D
Feddema, H
Feesty, T
Feidler, R
Felesky, V
Felsberg, G
Felt, D
Fenton, H
Fenton, T
Fenwick, L
Ferguson, S
Ficht, B
Ficht, J
Fields, S
Fietz, C
Figiel, C
Findlay, B
Fisher, B
Fisher, C
Fishkin, A

Fitch, L
Fitzgerald, K
Fitzpatrick, H
Flemming, W
Fletcher, C
Flockhart, T
Flynn, D
Flynn, L
Flynn, R
Foley, J
Ford, B
Ford, C
Ford, E
Ford, E
Ford, S
Ford, W
Forget, L
Forrest, A
Foster, J
Foster, L
Foster, S
Frank, A
Fraser, B
Fraser-Hrynyk, H
Fraser-Hrynyk, N
Fraser, J
Fraser, P
Fraser, S
Fraser, W
Fratton, G
Fredeen, G
Freeman, C
Frenette, J
Frew, B
Friedt, T
Friends of Jasper
Frisky, D
Froggart, K
Frusel, P
Fry, K
Fujikawa, C
Fujikawa, J
Fujimori, T
Fuller, E
Fyfe, R
Gad, B
Gamble, H
Gammon, J
Gardner, G
Gardner, R
Garrett, B
Garrett, C

Gasser, E
Gauthier, B
Gauthier, C
Gehlert, R
George, D
Gibbard, A
Gibson, D
Gierulski, A
Gierulski, W
Giesbrecht, P
Giesbrecht, B
Giesbrecht, C
Gilbert, R
Gillard, L
Gill, I
Gill, J
Glass, S
Glendinning, S
Glenn, G
Glenn, Y
Godkin, D
Godsalve, A
Godsalve, B
Godwin, W
Goodall, C
Gooding, O
Goodwin, W
Gorda, M
Gorda, T
Gordon, B
Gordon, D
Gordon, E
Gorrie, S
Gorten, J
Gosche, C
Gosche, S
Gosche, T
Gotceitas, V
Gottfred, A
Goulet, B
Goulet, T
Gourley, B
Grafe, G
Grafe, T
Graham, K
Gratto-Trevor, C
Graves, C
Green, C
Green, J
Green, M
Green, P
Greening, A

Greenlee, G
Greenlee, P
Gregoire, J
Gregoire, P
Gregory, R
Greig, J
Grenier, K
Grenier, W
Grey, B
Grey, N
Grey, R
Griffiths, D
Groenveld, J
Groenveld, K
Groenveld, R
Grosfield, C
Gross, A
Gross, B
Gross, E
Gross, L
Gross, M
Gross, R
Grothman, H
Grout, W
Grove, C
Groves, J
Groves, K
Gulley, Z
Gurski, K
Gussie, G
Gustavson, A
Habib, L
Haddock, P
Haesler, N
Hafner, B
Hagen, J
Haig, B
Haley, C
Halladay, I
Hall, B
Hall, D
Hall, G
Hall, R
Halmazna, G
Hamilton, G
Hamilton, J
Hamilton, M
Hammell, T
Handy, D
Hanes, G
Hannah, K
Hanneman, M

Hannesschlager, M
Hanrahan, T
Hansen, C
Hargrove, J
Haring, C
Harpe, I
Harris, H
Harrison, L
Harrison, M
Harrison, R
Harrison, S
Harvey, P
Hastings, I
Hatch, L
Havard, C
Havard, K
Hay, M
Hayward, M
Haywood, M
Hazlett, D
Hazlett, G
Heard, A
Heard, S
Heinsen, B
Heinsen, L
Helmer, A
Hennessy, F
Henry, J
Henson, H
Henwick, A
Herbut, M
Herbut, S
Hervieux, M
Hess, L
Hess, T
Hewitt, W
Hewlett, W
Hider, J
Hider, L
Higgins, S
Hill, B
Hill, R
Hillworth, A
Hindle, M
Hindmarch, D
Hindmarch, J
Hindmarch, T
Hingston, A
Hitchon, C
Hitchon, R
Hiyet, D
Hoare, B
Hoare, M
Hockstadt, J
Hodgetts, R
Hoffman, W
Hogg, J
Holder, A
Holloway, K
Holmes, D
Holroyd, G
Homister, C
Homister, R
Honsaker, L
Hopkins, D
Hopkins, E
Horch, P
Horlick, M
Hornby, B
Horn, J
Hornsby, D
Hoscheit, R
Hostvedt, J
Hourabielle, C
Howard, L
Howland, R
Howle, M
Hrynyk, N
Hudon, J
Hudson, H
Hudson, M
Hudson, V
Huget, D
Hughes, A
Hughes, G
Hughes, R
Huibers, K
Hull, K
Humber, S
Hummel, R
Hummerstone, C
Hunka, S
Hunt, C
Hurabielle, J
Hurly, A
Husky Energy Ltd.
Huston, L
Huston, M
Hutchinson, B
Hutchison, L
Hvenegaard, G
Illo, H
Jackson, A
Jacobsen, T
Jacobson, K
Jacobs, R
James, W
Jamieson, J
Jamieson, R
Janke, J
Jasinsky, B
Jasinsky, E
Jelfer, E
Jensen, J
Jeschke, T
Jetime, J
Jim, S
Joan, D
Jodi, K
Joehnke, L
Johner, B
Johnson, F
Johnson, J
Johnson, L
Johnson, T
Johnston, D
Johnstone, M
Joly, R
Jones, A
Jones, E
Jones, J
Jullyan, B
Jullyan, D
Junior Forest Rangers
Kadey, G
Kadey, P
Kagume, K
Kamin, J
Karpuk, E
Kassai, S
Kasteel, G
Kaye, R
Keane, A
Keeping, D
Kendrick, J
Kendrick, P
Kennedy, D
Kennedy, W
Kenyon, A
Kenzie, L
Kerby, J
Kerr, J
Kibsey, C
Kilcullen, K
Kiliaan, H
Kilmury, T
Kimberley, A
King, H
Kinley, J
Kinnee, K
Kinsella, W
Kinsmen, E
Kitagawa, L
Kitchingham, F
Kitigawa, L
Klauke, R
Klein, T
Kline, K
Kloepfer, K
Knapik, D
Knapp, E
Knapp, L
Knapton, R
Knaus, C
Knight, S
Knight, U
Koivula, M
Kondrackyj, M
Konward, K
Korbut, F
Korda, Z
Korolyk, T
Korpola, S
Korrula, M
Korth, J
Koskie, N
Kottmann, J
Kovacs, J
Kozyra, S
Kraft, C
Kraft, M
Krikun, R
Kruger, M
Kruger, R
Kruse, A
Kruse, V
Kublik, L
Kudras, S
Kuhn, G
Kulstead, M
Kurtz, G
Kusiek, R
Kuyt, E
Labour, S
Laffin, T
Landry, J
Lane, B
Lange, J
Lange, R
La Rose, A
Larson, G
Larson, J
Larson, L
Lavallee, B
Lavallee, D
Lawrence, M
Lawrence, S
Lawrence, V
Leary , G
Leary , S
Leavitt, C
Lee, C
Lee, D
Lee, I
Lee, J
Lee, P
Legris, A
Lehman, A
Lehmann, A
Lehmann, R
Lein, R
Lennie, N
Leong, G
Lepage, D
Lepard, H
Leroux, R
Lewis, S
Lieske, D
Liger, B
Lim, E
Lindbloom, D
Lind, V
Linowski, R
Lipsett, K
Liske, K
Litke, G
Litke, L
Livingstone, D
Livingstone, S
Lloyd, B
Lloyd, K
Lockerbie, C
Lockhart, B
Lockwood, A
Loewen, J
Logan, K
Lohr, M
Lois, B
Lomow, M
Lord, D

Lord, M
Lorenzen, A
Lorenzen, B
Lozeman, D
Lubbers, G
Lucas, J
Lumley, L
Lussier, R
Luterbach, B
Lyndon, G
Lysons, K
Macaulay, D
Maccagno, A
Maccagno, T
MacCallum, B
MacCulloch, L
MacDonald, J
Macfarlane, K
MacIntyre, D
Mack, G
MacKinnon, T
Mackney, L
MacLean, D
Macura, J
Magnusson, D
Magoose, C
Mah-Lim, E
Mahoney, D
Mahoney, J
Manchak, J
Mandbley, J
Manley, C
Manley, L
Manley, M
Mann, J
Marklevitz, P
Marler, S
Marlowe, P
Marshall, J
Marshall, L
Marsh, R
Martin, F
Martin, J
Martin, R
Matacio, D
Matechuk, K
Mather, J
Mather, L
Mathison, S
Matthews, G
Matthews, R
McBride, R
McCallum, C
McCarty, B
McCarty, M
McCarty, R
McCauley, D
McClaskey, M
McClung, M
McCormick, J
McCrae, F
McCulloch, B
McCulloch, A
McCullough, L
McDonald, B
McDonald, M
McDonald, R
McDonell, L
McElhaney, R
McFarland, J
McGinnis, C
McGrath, A
McGrath, F
McGrath, T
McGregor, D
McIlroy, M
McIntosh, J
McIvor, D
McIvor, M
McJunkin, A
McKay, J
McKeage, M
McKeating, G
McKeigan, A
McKenzie-Brown, P
McKenzie, D
McKibbon, D
McKillop, L
McKinnon, J
McKinnon, K
McKown, D
McLean, G
McLean, L
McLean, S
McLeod, D
McLeod, S
Mcllvaim, D
McLure, F
McNabb, B
McParland, C
McPike, S
McQuid, C
McQuid, D
McQuid, F
Megasse, C
Melnyk, R
Merrlowe, P
Metz, K
Meyer, C
Meyer, G
Meyers, S
Middleton, E
Middleton, R
Midgley, G
Mike, N
Millar, W
Miller, C
Miller, D
Miller, J
Miller, S
Miller, V
Miller, W
Mills, C
Mills, D
Mills, P
Milner, J
Mitchell, P
Moes, C
Moker, J
Mokoski, B
Mondea, B
Mone, M
Moore, C
Moore, D
Moore, J
Moore, K
Moreira, N
Moreira, R
Morgan, B
Morgan, L
Morgan, R
Morgan, V
Morrison, B
Morrison, G
Morrison, J
Moysa, A
Mozak, L
Mueller, M
Muhlbach, B
Muhlbach, M
Muhlenfeld, E
Mulligan, M
Munro, M
Munz-Gue, M
Murphy, A
Myers, N
Nadeau, D
Nadeau, G
Nadeau, T
Naton, E
Neill, G
Nelson, R
Nelson, V
Ness, D
Neufeld, S
Newman, J
Newton, G
Ng, J
Nickolson, J
Niederleitner, J
Nietfeld, M
Nightengale, J
Nims, N
Nissen, C
Noon, M
Nordstrom, S
Nordstrom, W
Norgard, D
Norgard, G
Norgard, J
Norgard, K
Norgard, M
Norstrom, K
Norton, M
Noton, D
Noyes, C
Noyes, M
Oakley, D
Oakwood, A
O'Connell, K
O'Connor, L
Ogilvy, A
Ogilvy, J
O'Hara, C
O'Hara, E
O'Hara, R
Ohlsen, J
Okrainec, J
Okrainec, P
Oliver, D
Oliver, M
Olsen, R
Olson, C
Olson, M
Ora, L
Oro, L
Osborne, M
O'Shea, L
O'Shea, M
O'Shea, N
Osterud, S
Oullet, D
Overly, K
Oxamitny, M
Ozawa, Y
Page, N
Palichuk, S
Palindat, R
Pankratz, H
Pankratz-Smith, E
Pardieck, K
Parker, S
Park, J
Parks, J
Parr, J
Parson, C
Parsons, B
Parsons, E
Parsons, J
Patterson, Y
Patti, D
Peacock, J
Peacock, S
Pearce, K
Pearman, M
Pearson, K
Peckham, A
Peddie, K
Peddy, K
Pedersen, W
Pederson, D
Pederson, S
Pelandat, R
Peleshok, L
Pelzer, A
Pennell, R
Penner, B
Penner, D
Penner, P
Penner, R
Pernarowski, C
Pernarowski, R
Peter, E
Peters, J
Peterson, B
Peterson, G
Petherbridge, S
Phal, D
Pierce, B
Pike, T

Pilny, J
Pilny, R
Pimm, D
Pinel, H
Piorecky, M
Plant, B
Platt, C
Player, D
Pletz, H
Poland, G
Ponomar, W
Poole, P
Poriz, D
Porter, C
Posey, J
Potter, J
Poulton, T
Prescott, D
Preuss, E
Preuss, S
Price, G
Price, J
Priestley, C
Priestley, J
Priestley, L
Prior, H
Prirgnitz, J
Proudfoot, J
Pruess, E
Pryor, P
Pucci, B
Putney, N
Quade, J
Quinlan, L
Quinlan, P
Quinlan, R
Quist, D
Quist, W
Radke, V
Raeburn, F
Raeburn, J
Rae, R
Raffan, J
Ralph, J
Ramcharita, R
Ramel, M
Ramsey, M
Ramsey, R
Raniseth, A
Ranson, S
Rappel, R
Redcliff, D

Reddecliff, E
Reeves, R
Regnier, J
Reid, M
Reinhart, J
Rempel, J
Reutter, L
Reynolds, J
Reynolds, L
Richard, Q
Richardson, J
Richards, R
Ries, M
Rindero, D
Rinn, C
Rippin, B
Ritchie, B
Rivet, A
Rocher, E
Rockwell, R
Rocky Mountain
Eagle Research
Foundation
Rodrigues, C
Roe, K
Roe, R
Rogers, J
Roman, K
Roman, W
Romanchuk, G
Romanchuk, S
Romanow, T
Romeril, M
Romeril, V
Rose, J
Rose, M
Routledge, M
Rowan, G
Rowbotham, S
Rowell, P
Rowledge, G
Rowledge, M
Roxburgh, P
Roy, J
Roy, K
Roy, L
Ruddy, D
Ruddy, G
Rudiak, A
Rud, R
Rue, M
Rusnick, P

Russell, A
Russell, C
Russell, F
Russell, K
Ruth, L
Rutter, L
Ruttle, P
Rutton, B
Ryder, H
Saba, D
Sacker, L
Saker, C
Saker, G
Saker, K
Sallee, J
Salmon, W
Sanders, B
Sardstrom, C
Saunders, L
Savage, S
Savignac, K
Savignal, C
Schelhas, V
Schell, B
Schiebelbein, D
Schiebelbein, M
Schieck, A
Schieck, J
Schieck, K
Schmelzeisen, R
Schowalter, T
Schubert, H
Schuler, M
Schultz, A
Schultz, F
Schultz, M
Schultz, S
Scotland, S
Scott, A
Scott, G
Scott, I
Scott, J
Scott, K
Seaton, B
Seaton, J
Semenchuk, G
Semenchuk, M
Seneviratne, N
Serian, P
Setters, M
Sharp, B
Sharun, D

Shaw, A
Shaw, B
Shaw, D
Shaw, R
Shaw, W
Sheibelbein, M
Sherman, J
Sherrington, B
Sherrington, P
Sherrin, P
Shesterniak, M
Shier, J
Shilman, D
Shimizu, K
Shukster, I
Shuya, D
Sidwell, C
Siga, V
Sillito, S
Sillito, W
Simons, I
Simpson, B
Simpson, G
Simpson, N
Sinclair, C
Singbeil, L
Sinton, E
Sinton, H
Skilnich, J
Skrepnek, R
Skrepnek, S
Skretting, S
Slater, A
Slater, D
Slater, J
Slater, S
Slatter, G
Sloan, B
Sloan, L
Slot, D
Slot, K
Smith, B
Smith, C
Smith, F
Smith, G
Smith, J
Smith, K
Smith, L
Smith, M
Smith, P
Smith, S
Smithson, D

Smiton, H
Sneath, M
Snyder, A
Snyder, C
Snyder, M
Snyder, R
Soloman, W
Sommerhalder, R
Sommers, J
Song, S
Sorenson, D
Soveran, M
Spalding, E
Sparling, C
Spencer, M
Sperling, F
Spitzer, E
Spitzer, M
Stambaugh, C
Stanley, E
Starke, B
Starke, G
Staudzs, E
Stavne, M
Stavne, R
Steckley, J
Steele, D
Steingarten, L
Stein, M
Stein, S
Stephens, E
Stepnisky, D
Stevens, B
Stewart, L
Stiles, A
Stiles, D
Stiles, M
Stiles, P
Stinson, A
St. Laurent, J
St. Laurent, K
Stobart, T
Stoker, D
Storey Smith, K
Storms, B
Strandquist, D
Strauss, E
Stromsmoe, E
Stromsmoe, L
Strong, W
Struik, C
Stryde, S

Sturdy, J
Sturgess, E
Sturney, D
Sturney, R
Susut, J
Sutherland, F
Swallow, N
Swan, J
Swan, V
Swift, T
Swystun, M
Sykes, J
Sylvester, E
Symington, K
Szabo, S
Szabo, V
Szkarupa, T
Tailfeathers, D
Takahashi, A
Tamboulian, J
Tarrant, H
Taylor, A
Taylor, B
Taylor, C
Taylor, D
Taylor, J
Taylor, K
Taylor, M
Tedder, W
Temofychuk, A
Ternoster, M
Tetherington, G
Theberge, J
Theerabelle, J
Thibault, M
Thibault, P
Thiessen, S
Thomas, L
Thomas, M
Thomas, R
Thompkins, B
Thompson, C
Thompson, E
Thompson, J
Thomsen, W
Thomson, G
Thomson, J
Thormin, T
Thorpe, C
Tidman, L
Tidy, B
Tidy, R
Tietz, G
Timmons, T
Toews, A
Toma, M
Tomasson, G
Tomlinson, E
Totino, N
Tourangeau, B
Townsend, B
Tremblay, M
Tremblay, V
Trip, K
Tugult, C
Tull, E
Turcotte, G
Turnbull, G
Turney, D
Turney, J
Twiss, I
Twitchell, C
Two Feathers
Truckstop
Tyzuk, D
Ulfsten, S
Umeris, S
Unruh, D
Ursus Ecosystem
Management Ltd.
Van den Dolder, L
Vanderpol, H
Vanderwal, F
Vanes, B
Van Kamer, N
Van Peteghen, J
Van Rosendaal, M
Van Staten, J
VanStraten, J
Van Wageningen, A
VanWilgenburg, S
Vaxwick, L
Velner, B
Velner, D
Velner, M
Ver Beek, D
Vesak, R
Vesak, T
Vinsome, K
Vogt, L
Voon, V
Vos, A
Vujnovic, D
Wagner, E
Wagner, G
Walcheska, A
Walcheske, A
Walentowitz, M
Walker, A
Walker, B
Wallace, E
Walsh, L
Wapple, R
Waring, D
Warkentin, L
Warkheulin, L
Warren, R
Warren, W
Washbrook, J
Watkinson, J
Watson, M
Watt, R
Weakes, L
Weaver, T
Webber, W
Weber, F
Wege, R
Weidl, D
Weingartshofer, M
Welke, D
Welke, J
Wellier, B
Wells, G
Wendall, D
Wendell, G
Wenzel, W
Werezuk, R
Wershier, R
Wesbrook, M
Wescott, H
Weston, J
Westover, W
Wetsch, M
Whiley, F
Whitbread, J
White, R
Whitson, A
Whittingham, E
Whittington, J
Whitworth, J
Wiacek, R
Wicker, G
Widmer-Carson, L
Wiedeman, B
Wiens, T
Wild, M
Wild Store Junior
Birders
Wilkinson, J
Wilkinson, R
Williams, A
Willms, T
Wilson, B
Wilson, D
Wilson, J
Wilson, K
Wilson, M
Windsteig, R
Winter, P
Wishart, R
Wojnowski, J
Wong, B
Woods, A
Woods, R
Woodward, C
Woolgar, B
Woolgar, J
Worobetz, T
Worona, R
Worsely, E
Wotton, A
Wright, J
Wright, K
Wudel, G
Wyatt, B
Wyatt, H
Yaki, G
Yamazaki, K
Yanke, J
Yaremko, C
Yeoman, S
Yeoman, Z
Yeudall, B
Young, C
Young, I
Young, J
Zalesky, P
Zanewich, L
Zeilsdorf, D
Zerr, F
Zidek, E
Zimmerman, C
Zoeteman, B
Zoeteman, F
Zombori, J
Zombori, S
Zroback, R
Zurfluh, W
Zwick, L
Zwick, W

Figure 1: Natural Regions of Alberta

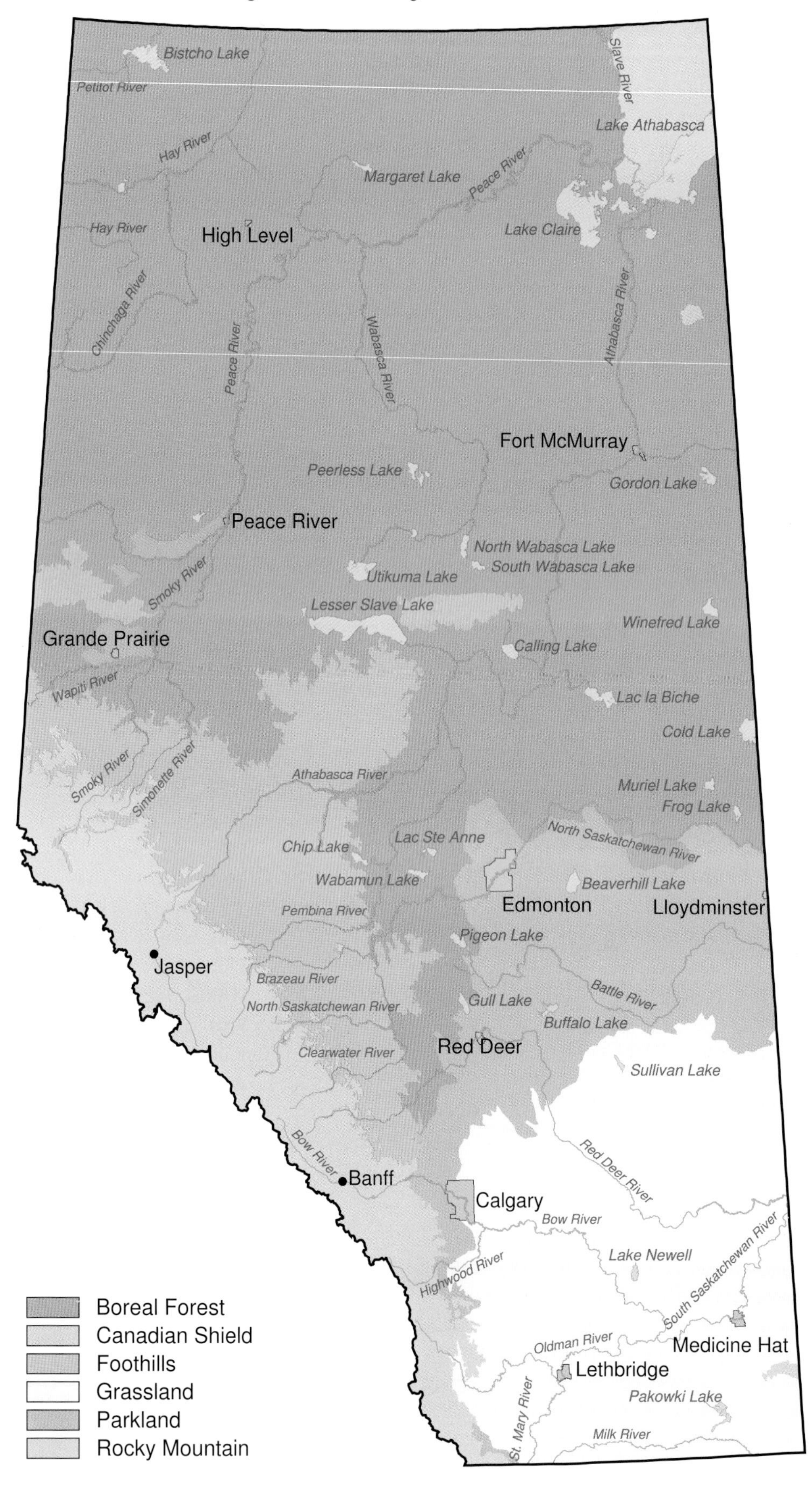

Natural Regions of Alberta

Introduction

The following text concerning the Natural Regions of Alberta has been abstracted from Downing and Pettapiece (2006).

The Province of Alberta includes some of the most diverse terrain in North America. Mountains, foothills, and plains temper regional climates, the intensity of solar radiation decreases markedly from the 49th to the 60th parallel, and regional landscapes transform solar and climatic influences to produce an intricate ecological complex.

Plant communities and soil patterns provide, in part, the evidence for delineating the climatic and physiographic patterns which control vegetation and soil distribution. Both the plant communities and soils develop in response to abiotic factors and biotic factors.

The relative influence of each factor at any place in the landscape is determined by the interaction of atmospheric and landscape attributes—climate, topography, parent materials and biotic elements—all acting over time, as described by Major (1951) and Jenny (1941) for vegetation and soils, respectively. These attributes can be delineated and represented as abstract ecological map units, and may be described at various scales.

At the global scale, the Biome or Vegetation Zone is recognized (Scott, 1995; Walter, 1979). At the national scale in Canada, Ecozones, Ecoregions and Ecodistricts are described Group (1995). In Alberta, Natural Regions and Natural Subregions constitute the broadest levels of ecologically based landscape classification.

The value of regional ecological land classification systems as a foundation for sustainable resource management has been recognized for at least four decades in Canada. These systems provide a means of presenting and understanding biophysical patterns in a geographic context, and a common basis for communication.

In Alberta, Natural Regions and Natural Subregions have supplied the provincial ecological context within which resource management activities have been planned and implemented since the 1970s. Examples of these activities include regional to local integrated resource plans, protected areas program plans based on Natural Subregion themes, numerous forest and range inventory and analysis systems nested within Natural Subregions, and provincial state of the environment reports at the Natural Region level.

Boreal Forest Natural Region

The Boreal Forest Natural Region includes the Dry Mixedwood, Central Mixedwood, Lower Boreal Highlands, Upper Boreal Highlands, Athabasca Plain, Peace-Athabasca Delta, Northern Mixedwood, and Boreal Subarctic Natural Subregions. It occupies most of northern Alberta, and extends south almost to Calgary.

Elevations range from about 150 m in the Northern Mixedwood Natural Subregion near the Alberta-Northwest Territories border to over 1100 m in the Upper Boreal Highlands Natural Subregion near the Alberta-British Columbia border. Level to undulating till and lacustrine plains interspersed with extensive wetlands are the dominant landforms; hummocky landscapes, high-elevation plateaus and extensive dune fields also occur.

This Natural Region is vegetated by deciduous, mixedwood and coniferous forests. Cultivation is limited to those areas that have a sufficiently long growing season. Aspen and balsam poplar are the most common deciduous species; white spruce, black spruce and jack pine are the dominant conifers. Wetlands are dominantly black spruce, shrub or sedge fens. The dominant soils are Luvisols on uplands and Mesisols in wetlands.

The boundaries between some boreal Natural Subregions are relatively well defined, such as those defined by elevation criteria or slope breaks (e.g., the boundary between the Lower Boreal Highlands Natural Subregion and the adjacent Boreal Subarctic or Upper Boreal Highlands Natural Subregions). The boundaries for other boreal Natural Subregions are more gradual (e.g., the boundary between the Dry Mixedwood and

Central Mixedwood Natural Subregions), and differences in vegetation and soils may not be evident for several kilometres on either side of the mapped boundary.

Climate

The Boreal Forest Natural Region is comparable to the Rocky Mountain Natural Region in terms of overall climatic variability. However, landscape changes across the Boreal Forest Natural Region are more gradual and produce less abrupt climatic differences between Natural Subregions. whereas the Rocky Mountain Natural Region is highly variable over short distances because of pronounced elevation changes. There is a nearly 5°C difference in mean annual temperature between the warmest and most southerly Natural Subregion (Dry Mixedwood) and the coldest and most northerly Natural Subregion (Boreal Subarctic). The differences in degree-day accumulations and mean annual precipitation are almost as pronounced.

Summers are short, with only one or two months in which the average daily temperature exceeds 15°C. Winters are long and very cold, with average daily temperatures below -10°C for four months or more in most Natural Subregions, and below -20°C for two months or more in the most northerly Natural Subregions.

Precipitation follows a summer-high continental pattern, with peak rainfalls occurring in July and about 60 to 70 percent of the annual precipitation received between April and August.

Vegetation

The Boreal Forest Natural Region is a mosaic of upland forests and extensive wetlands in low-lying areas. Both forest composition and the prevalence of wetlands changes from south to north moving through the Region. In the south, aspen forests are dominant, with white spruce occurring as scattered trees or small stands in moister locales. Wetlands are often shrubby or sedge fens and marshes. Cultivation is widespread where the growing season is sufficiently long to allow crop growth.

Moving northward, mixed forests of aspen, balsam poplar and white spruce are the dominant upland vegetation, with a mix of treed fens, shrubby fens and sedge fens occupying a larger proportion of the landscape. Undulating plains and elevated plateaus in the far north are vegetated mainly by black spruce fens and bogs, shrubby fens and sedge meadows. Upland vegetation is associated with slightly higher landscape positions, and is a mixture of deciduous, mixedwood and coniferous forests.

Jack pine stands occur throughout the Natural Region on dry, well drained sandy soils and are extensive in some areas. Lodgepole pine or lodgepole pine-jack pine hybrids form pure or mixed stands with black spruce and deciduous species on the slopes and tops of northern highlands

Canadian Shield Natural Region

The Canadian Shield Natural Region has only one Natural Subregion, the Kazan Upland. The following discussion of the Kazan Upland Natural Subregion applies to the Canadian Shield Natural Region as well. The Kazan Upland Natural Subregion occurs in the far northeastern corner of Alberta. The main area lies north of Lake Athabasca. It is bordered on the east and north by the Alberta-Saskatchewan and Alberta-Northwest Territories boundaries, respectively, on the west by the Slave River, and on the south by Lake Athabasca. There is a small outlier east of the Athabasca River between Fort McMurray and Fort Chipewyan embedded within the Athabasca Plain Natural Subregion. Elevations range from about 150 m to over 400 m. Extensive outcrops of Precambrian bedrock, the westernmost edge of the Canadian Shield, define the limits of the Kazan Upland Natural Subregion; on average, 60 percent of the landscape is exposed bedrock. Topography is hummocky to rolling, with local relief of up to 50 m. Parent materials are ice-scoured bedrock and coarse textured glacial drift. Bedrock barrens are interspersed with "pocket" communities vegetated by lichens, mosses and drought-tolerant ferns.

Open jack pine, aspen and birch stands occur where the soil is sufficiently deep. Acidic bogs and poor fens occur adjacent to the many small lakes and in low spots on the more subdued terrain in the western part of the Natural Subregion. The conventional reference site concept of deep, medium textured, well drained soils and associated vegetation does not adequately fit most of the Kazan Upland Natural Subregion. Characteristic sites are rocky exposures or dry, rapidly drained coarse glacial deposits.

Climate

The Kazan Upland Natural Subregion is characterized by short summers in which July is the warmest month, and the coldest winters of any Natural Subregion in Alberta, reflecting the influence of continental polar and continental arctic weather systems. July is the month of maximum precipitation, and winter snowfalls account for about 40 percent of the total annual precipitation.

The Alberta Climate Model indicates the Kazan Upland Natural Subregion receives about the same annual precipitation as the Dry Mixedgrass Natural Subregion, and has a potential summer moisture deficit comparable to that of the Northern Fescue Natural Subregion. The prevalence of rock barrens and well to rapidly drained glacial deposits, together with low rainfalls, produce conditions that are favourable for nonvascular and vascular plants that are adapted to dry conditions.

Vegetation

Vegetation in the Kazan Upland Natural Subregion is strongly influenced by the distribution and acidic characteristics of granitic bedrock exposures and well to rapidly drained glacial drift, and by frequent fires. Beckingham and Archibald (1996) produced a general description of vegetation communities based on limited plot data that included both the Athabasca Plain and the Kazan Upland Natural Subregion. Wallis and Wershler (1984) described five upland and three wetland types from the east portion of the Natural Subregion.

Communities associated with dry rock barrens are widespread. Although species diversity is low in any given locale, the variety of habitat types results in high species diversity across the barrens. Various lichen communities occupy south-facing and steep rock faces and slopes, and most are found only in the Kazan Upland. "Pocket" communities grow in rock crevices and in sheltered locations where mineral soil has accumulated and moisture conditions are better. Stunted jack pine and Alaska birch form open stands with a sparse understorey of bearberry, ground juniper, bog cranberry, and a variety of drought-tolerant ferns and other herbs, mosses and lichens.

Coarse textured, rapidly drained and dry sandy or gravelly soils support more vigorous jack pine growth. The driest sites are vegetated by open jack pine stands with a patchy carpet of lichens on the forest floor. Moister sites support more diverse understoreys of green alder, common blueberry, bearberry, common Labrador tea, Canada buffaloberry, bunchberry, other herbs and feathermosses. Aspen, Alaska birch and black spruce are locally common in places. Brunisols are common soils. Moist communities of aspen, balsam poplar, Alaska birch, white spruce and a diverse and lush shrub and forb understorey develop in bands adjacent to wetlands and along lakes.

Bog communities are the dominant wetland type. Black spruce forms open-to-dense stands with an understorey of common Labrador tea, leatherleaf, bog cranberry, cloudberry and peat moss on Organic soils. Permafrost is discontinuous but widespread. Nutrient-rich wetlands typically have open forests of tamarack, willow, dwarf birch, sedges and richsite mosses. Marshes can be locally extensive in sheltered lake bays or along creek channels, and are dominated by water and small bottle sedge, bulrushes, and in deeper water, pondweeds.

Grassland Natural Region

The Grassland Natural Region includes the level to rolling part of Alberta that is sometimes called prairie. Although in its natural state, prairie is thought of as an expanse of grasses, shrublands are found in moister areas. Even forests occur, but they are restricted to coulees and river valleys.

The Grassland Natural Region includes the Dry Mixedgrass, Mixedgrass, Foothills Fescue and Northern Fescue Natural Subregions arrayed in concentric bands from the Alberta-Montana border north to the Grassland-Central Parkland Natural Subregion and west to its boundary with the Foothills Parkland and Montane Natural Subregions.

Chernozemic soils are characteristic of the Grassland Natural Region. Elevations range from about 550 m in the Dry Mixedgrass Natural Subregion near the Alberta-Saskatchewan border, to over 1500 m in the Foothills Fescue Natural Subregion and about 1450 m at the highest elevations of the Mixedgrass Natural Subregion on the Cypress Hills.

Undulating plains are characteristic of much of the Grassland Natural Region, with hummocky uplands also occurring in northern portions along with rolling terrain that is more characteristic of higher elevation areas to the west. Much of the Region has been cultivated, but the remaining native prairies and their associated soils reflect the interactions of dry, warm climates and topography.

Variations in the types of Chernozemic soils are used to differentiate Natural Subregions within the Grassland Natural Region, because extensive cultivation has removed much of the native vegetation and detailed maps of the remaining native plant communities are not available for the entire area. The boundaries between Natural Subregions in the Grassland Natural Region are not as clearly defined as those in the Rocky Mountain and Foothills Natural Regions, reflecting the more gradual influence of latitudinal changes on climate in the comparatively gentle prairie terrain.

Climate

The Grassland Natural Region is the warmest, driest Region in Alberta. The mean annual precipitation of the driest Natural Subregion in Alberta, the Dry Mixedgrass, is only a third of that received by the wettest Natural Subregion, the Alpine. Summers are very warm and the growing season is longer in the Grassland Natural Region than in any other Region.

Precipitation follows a typical continental summer-high pattern, with a maximum in June. There is a pronounced moisture deficit during the latter part of the growing season. Both temperature and precipitation vary with latitude and proximity to the Front Ranges. In the north part of the Region, summer and winter temperatures are slightly lower than in the south and precipitation is higher. A similar increase in precipitation and decrease in summer temperatures occurs in the westernmost part of the Region, but winters are milder due to a higher incidence of Chinooks adjacent to the Front Ranges.

Climate characteristics for the Foothills Fescue Natural Subregion are more similar to the Montane Natural Subregion than to any of the Grassland Natural Subregions.

Vegetation

The Grassland Natural Region includes some of the most productive croplands in Alberta, and much of the Region has been cultivated. The characteristic native vegetation form is prairie, with shrublands in areas receiving water and on the cooler north and east aspects. Narrow forests parallel rivers where groundwater provides sufficient moisture for tree growth.

Many prairie species are adapted through dormancy, physiology or anatomy to survive the severe moisture deficits that occur during mid-to-late summer. In the driest areas (the Dry Mixedgrass Natural Subregion), extensive areas of native prairie remain, typically with a mixture of drought-tolerant, mid-height (e.g., needle-and-thread) and short (e.g., blue grama) grasses. Moister climates to the north and west in the Mixedgrass Natural Subregion support more widespread cultivation.

Where native grasslands remain, they are a mix of taller needle-and-thread, porcupine grass, and northern and western wheatgrasses. Plains rough fescue is the dominant grass on native range in the northernmost Northern Fescue Natural Subregion. Mountain rough fescue, bluebunch fescue and Parry oatgrass grasslands characterize the westernmost Foothills Fescue Natural Subregion, and many species that occur in the Montane and Foothills Parkland Natural Subregion also occur in the Foothills Fescue.

Foothills Natural Region

The Foothills Natural Region includes the Lower and Upper Foothills Natural Subregions. The Region extends along the eastern flank of the Rocky Mountains north from the Bow River valley to just south of Grande Prairie. It also includes the Swan Hills and Pelican Mountain outliers to the east and the Saddle Hills outlier north of Grande Prairie.

The topography is highly variable, ranging from sharp, bedrock-controlled ridges near the mountains to rolling and undulating terrain in the north and east. Elevations within the Foothills Natural Region range from a low of 700 m in the most northerly areas to a maximum of about 1700 m in the south.

Mixed forests of aspen, lodgepole pine, white spruce and balsam poplar with variable understoreys on Gray Luvisolic soils are dominant on average sites at lower elevations. Lodgepole pine forests with less diverse understoreys and well developed feathermoss layers on Brunisolic Gray Luvisols are typical of higher elevations.

Climate

Distinctive vegetation and soils patterns reflect subregional climate changes and characterize the Lower Foothills and Upper Foothills Natural Subregions. Both receive relatively high annual precipitation, and only the Alpine and Subalpine Natural Subregions are wetter. Average July precipitation is higher in the Lower and Upper Foothills Natural Subregions than in any others.

The Lower Foothills Natural Subregion has somewhat warmer summers and colder winters than the Upper Foothills Natural Subregion. The growing season is longer and total precipitation is lower especially in the winter months, indicating a stronger continental climate influence. Higher elevations and proximity to the mountains produce cooler summers, warmer winters and more precipitation than is characteristic of the adjacent Central Mixedwood Natural Subregion to the north and east.

Vegetation

Forests are the dominant vegetation cover in both Natural Subregions, and lodgepole pine stands are considered a good indicator of the Foothills Natural Region-Boreal Forest Natural Region boundary. Forests on upland sites within the Lower Foothills Natural Subregion are typically deciduous or mixedwood with aspen, balsam poplar, white birch, lodgepole pine, white spruce and black spruce as common associates. Wetlands are mainly vegetated by stunted black spruce and tamarack or shrub-graminoid communities.

The boundary between the Lower and Upper Foothills Natural Subregions is reasonably well defined by a change in dominance from mixedwood and deciduous stands on all aspects in the Lower Foothills Natural Subregion to conifer-dominated forests in the Upper Foothills Natural Subregion.

Parkland Natural Region

The Parkland Natural Region includes the Foothills Parkland, Central Parkland and Peace River Parkland Natural Subregions. Of these, the Central Parkland Natural Subregion is the most extensive, occurring in a broad arc from 200-250 km wide in central Alberta, narrowing to about 50 km where it joins the Foothills Parkland Natural Subregion in west-central Alberta.

The Foothills Parkland Natural Subregion occupies a discontinuous and narrow band along the foothills, extending south to the Alberta-Montana border. The Peace River Parkland occurs in three isolated patches in northwestern Alberta. Elevations range from 300 m in the Peace River Parkland Natural Subregion to about 1600 m in the Foothills Parkland Natural Subregion.

Undulating till plains and hummocky uplands are characteristic of the Central and Peace River Parklands. Rougher foothills terrain and steep, slumping river valley slopes are attributes of the Foothills and Peace River Parkland Natural Subregions, respectively.

The Parkland Natural Region is the most densely populated Natural Region in Alberta, and has been extensively cultivated since the late 1800s. The Natural Region has been strongly influenced by agriculture, and soil types have been used to define its boundaries.

Climate

The Parkland Natural Region represents a climatic transition between the Grassland and Cordilleran ecoclimatic provinces to the south and west (i.e., the Foothills Parkland Natural Subregion) and the Grassland and Boreal ecoclimatic provinces to the north (i.e., the Central Parkland Natural Subregion). Climate and unique site conditions together define the present extent of the Peace River Parkland Natural Region.

Mean annual temperature, growing season and mean annual precipitation values are intermediate between the Grasslands Natural Region to the south and the adjacent Rocky Mountain or Boreal Natural Regions to the west and north. Winters are cooler in the more northerly Peace River Natural Subregion because of more pronounced continental polar influences, and warmer in the Foothills Parkland Natural Subregion because of Chinooks.

Vegetation

The Parkland Natural Region includes highly productive croplands, and most of the Region has been cultivated. The characteristic remaining native vegetation is usually an aspen-grassland mosaic; an area in the Foothills Parkland Natural Subregion has a willow-grassland mosaic.

Grasslands are the dominant native vegetation in the southern part of the Region, and islands of aspen forest or patches of willow shrubland occur in moist depressions or on northerly aspects. To the north, aspen stands occupy a wider range of habitats, and closed aspen and balsam poplar stands with grassland inclusions are typical of the northern Central Parklands Natural Subregion.

The Peace River Parkland Natural Subregion is a complex of closed aspen stands and grasslands on uplands, the latter occurring mainly on Solonetzic soils, and on steep southfacing river valley slopes. Remnant Parkland vegetation also includes wet, low-lying areas unsuitable for agriculture. These are most common in the northern Central Parkland and upland Peace River Parkland Natural Subregions, and support cattail marshes, willow-sedge shrublands or treed fens.

Rocky Mountain Natural Region

The Rocky Mountain Natural Region includes the Alpine, Subalpine and Montane Natural Subregions within and adjacent to the Front Ranges and Main Ranges of the Rocky Mountains. It spans the widest elevational range in Alberta, from a low of about 825 m in northern Montane Natural Subregion valleys to a high of over 3600 m (Mount Columbia) in the Alpine Natural Subregion. Rapid aspect changes and extreme slopes are characteristic of the Alpine Natural Subregion, with less pronounced ridged and rolling landscapes in the Subalpine and Montane Natural Subregions.

Climate

The Rocky Mountain Natural Region has on average the coolest summers, shortest growing season, highest mean annual precipitation and snowiest winters of any Region. Within the Natural Region, climates are highly variable; the Montane Natural Subregion receives much less annual precipitation than the Alpine and Subalpine Natural Subregions, and has milder winters than most other Natural Subregions. The high elevation Alpine and Subalpine Natural Subregions generally receive more annual precipitation and have snowier winters than any other Subregion, and the lower elevation Montane Subregion tends to be drier.

Vegetation

All three Natural Subregions are strongly influenced by Cordilleran climates and by elevation, aspect and substrate. Local regional vegetation and soil patterns clearly reflect these differences, and a complex mosaic of plant communities and soil types has developed in response. The growing season is too short for tree growth in the Alpine Natural Subregion, and plant communities are found on microsites protected from wind and temperature extremes. The highest elevations are essentially barren, with permanent snowfields in some locations.

Open coniferous stands and herbaceous meadows at higher elevations, and closed coniferous stands at lower elevations, characterize the Subalpine Natural Subregion. The Montane Natural Subregion is a mix of grasslands and deciduous-coniferous forests on southerly and westerly aspects, and predominantly coniferous forests on northerly aspects and at higher elevations.

Alberta's Changing Landscape

Introduction

Since the initiation of the *Atlas of Breeding Birds of Alberta* project in 1986, the landscapes of this province have experienced substantive changes. The same forces of landscape change operate in all parts of the globe, but what makes the Alberta situation unique is the cumulative effect of particular meteorological, geological, and anthropogenic processes on this provinces respective land base. In the last two decades, however, important physical changes to the landscape in the province have been associated notably with changes in land-use practices, mainly oil and gas development, forestry, agriculture, and urban development. At the same time, there are natural factors that have influenced local habitats within these landscapes. Obvious natural factors include annual temperatures, precipitation, hours of sunlight, fire, and pests. A basic understanding of the capability of these factors to alter landscapes and habitats—particularly since the mid-1980s—may make it easier to understand the changes that Alberta's bird species have had to deal with in that same time frame.

Several studies have been initiated in the last twenty years to document these changes in Alberta. Following is a summary—though hardly comprehensive—of some of these research efforts, which provides a background to help in understanding the nature of the causes as well as the extent to which they have modified Alberta's landscapes and habitats. The summary can then be related to the information provided in the individual bird species accounts, which document changes observed in the distribution and abundance of these species between publication of the first Atlas (1992) and the current volume, *The Atlas of Breeding Birds of Alberta: A Second Look.*

Cumulative Effects Modelling

Dr. Brad Stelfox is the architect and developer of the ALCES (A Landscape Cumulative Effects Simulator) model, a program gaining rapid acceptance by governments, industry, the scientific community, and non-government organizations (NGOs) to explore issues that deal with landscapes, land uses (agriculture, forestry, oil and gas, mining, human populations, tourism, and transportation sectors), and ecological and economic integrity. The ALCES model was developed over a period of 7 years for the purpose of tracking industrial footprints and ecological processes under alternative management scenarios.

To run this model, Dr. Stelfox has populated it with historical data, from government and industry sources that document the change of many parameters from 1905 to the present. He has made the following information available to the Atlas project to enable us to clearly show some of the anthropogenic change in Alberta that may have contributed to the changes to bird populations documented in this Atlas.

Table 1: Human population

Type	1986	2006
Urban	1,877,760	2,699,581
Rural	488,070	590,499
Total	2,365,825	3,290,350

Figure 2: Road Network 1990 and 2000

The road network maps reflect all roads, public and private, constructed in Alberta in this period; whereas, table 2 reflects public roads constructed by some level of government. Resource extraction companies constructed the majority of private roads shown in figure 2.

Table 2: Roads (km) Note: this does not include privately built resource roads

Type	1987	1999
Highways	13,587	13,761
Minor Roads	139,744	167,477
Total	153,331	181,238

The consolidation to larger farms and the implementation of more intensive farming practices have been documented in several species accounts as being an influence on bird populations and abundance. The following tables quantify some of the changes to agriculture in Alberta since Atlas 1.

The intensification of activities in the oil and gas industry is a reality in Alberta. The magnitude of that intensification is documented in the following tables and maps. Loss of habitat and the fragmentation of habitat are the result of this activity in the Foothills, Boreal Forest, Parkland and Grassland Natural Regions.

Table 3: Livestock populations

Animal	1986	2001
Cattle	3,746,000	6,500,000
Chickens	8,852,415	12,175,246
Farmed Bison	0	79,731
Farmed Elk	0	42,021
Horses	135,025	159,962
Pigs	1,508,000	2,029,400
Sheep	183,600	287,000

Table 4: Crop area (ha)

Type	1986	2005
Cereals	6,945,455	5,696,364
Oilseeds	1,272,727	1,954,545
Forage	1,697,324	2,429,545
Specialty	44,069	301,818

Figure 3: Natural Gas Wells 1985 and 2005

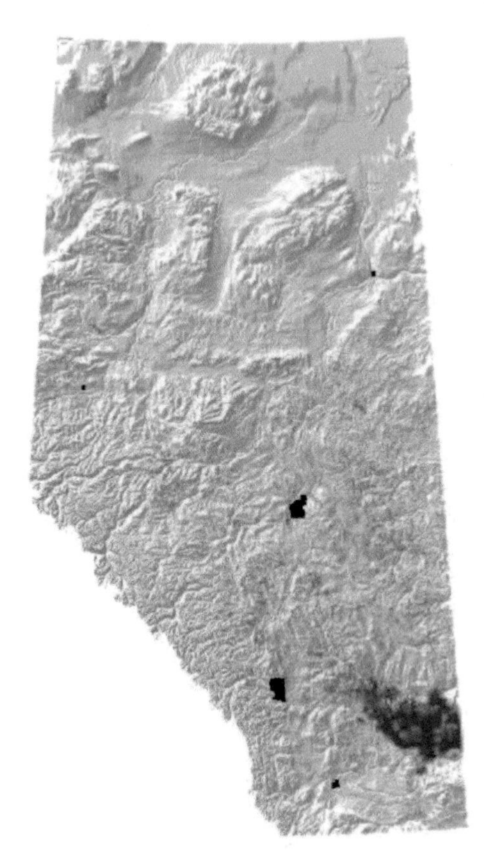

Figure 4: Conventional Wells 1985 and 2005

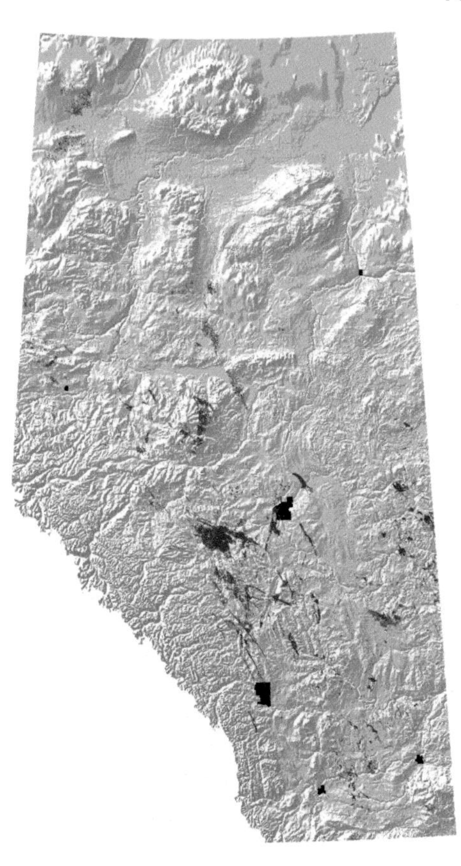

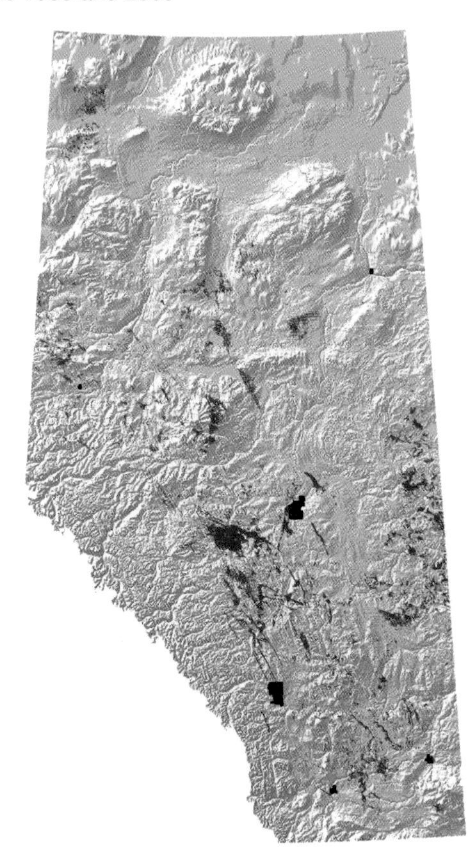

Table 5: Oil and gas wells completed

Type	1987	2005
Gas Wells	39,864	131,085
Oil Wells	38,169	85,865
Total	110,347	299,839

Table 6: Petroleum crown land sales (ha)

Year	Area	Year	Area
1987	2,274,942	1997	5,149,841
1988	2,630,286	1998	3,014,636
1989	2,263,664	1999	2,996,164
1990	2,410,126	2000	3,850,464
1991	2,217,421	2001	3,908,169
1992	1,765,598	2002	2,775,834
1993	2,969,132	2003	3,150,078
1994	4,738,879	2004	3,123,162
1995	4,368,470	2005	3,243,520
1996	4,682,768		

Figure 6: Forest Management Areas 1995-1999

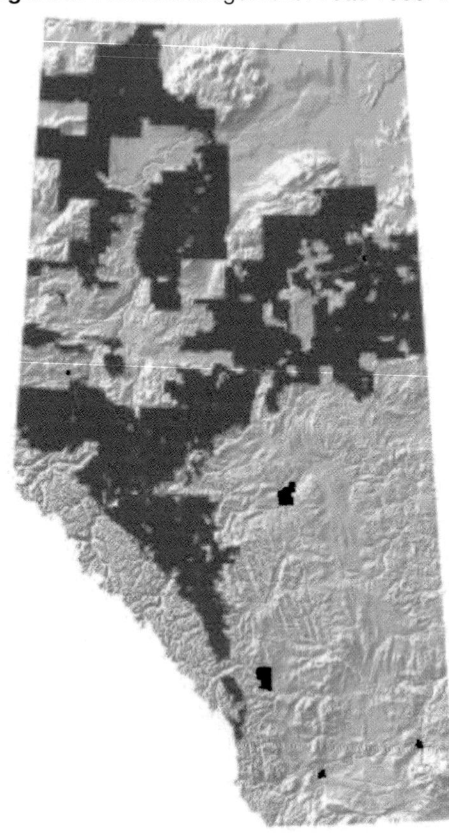

Figure 5: Pipeline Network 2005

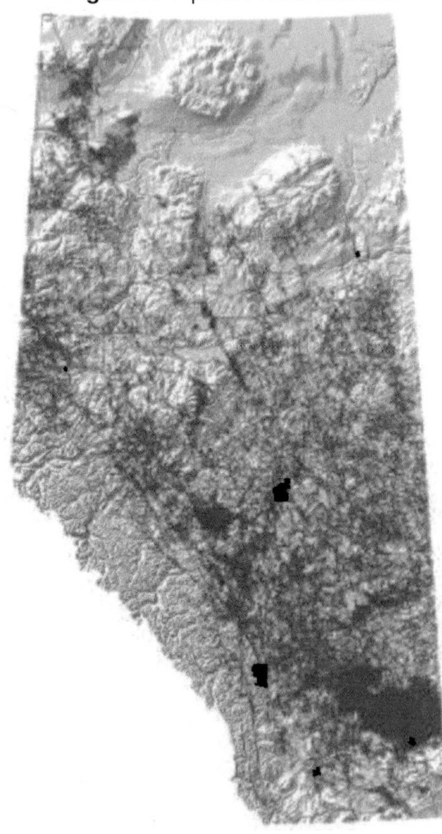

On the Alberta landscape the activities demonstrated herein are overlain with activities related to power generation and transmission, mining, urbanization, and the creation of country residential sub-divisions.

Prairie Habitat Joint Venture (PHJV)

The PHJV is the largest Canadian program under the North American Waterfowl Management Plan, signed in 1986 by Canada and the United States. The area of the PHJV in Alberta is comprised mainly of the agricultural areas of the province. Its vision is of healthy prairie, parkland, and boreal landscapes that support sustainable bird populations and provide ecological and economic benefits for society. The PHJV mission is to provide leadership to achieve healthy and diverse waterfowl and other bird populations through conservation partnerships. These partnerships strive for sustainable and responsible management of the landscape taking into account social, economic and environmental factors.

One of the recent reports from Environment Canada's Prairie & Northern Habitat Monitoring Program documents recent habitat trends in the PHJV area (Watmough and Schmoll, 2007). The following is taken from that report.

The Canadian Federal government strives to promote the

conservation of Canada's wetlands resources, and to sustain the ecological and socio-economic functions of this wetland resource for current and future generations. One of the goals of wetland policy in Canada is to promote "no net loss" of wetland functions. The most applicable measure of wetland function is often the quantitative measurement of the extent of wetland area.

Canada does not have a national wetland inventory or a national wetland status and trends monitoring program. The status and trends of wetlands in Canada have been estimated, over time, in several independent studies, most of which provide information at local scales and occasionally regional scales. Canada contains an estimated 127 million ha of wetlands or one-quarter of the world's total wetland area. Nationally, wetland loss has been estimated at 20 million ha since the 1800s and these losses degraded the wetland resource base in all areas of the country.

Expansion and development in the areas of agriculture, urbanization, transportation networks, resource extraction, recreational properties, forestry, etc. continue to result in losses to the wetland resource base in Canada. In many areas this continued loss has resulted in significant alterations to entire ecoregions, thus compromising the overall ecosystem function of these landscapes. One such landscape is the Prairie province region of Canada.

Numerous studies have estimated total wetland loss for the Prairies to be somewhere around 40–70 percent since settlement. Prairie wetlands are an important resource especially to the migratory bird populations of North, Central and South America as they provide habitat for the important breeding component of many migratory bird species life cycles. The lack of a consistent wetland status and trends program in Canada makes it difficult for conservation planners to piece together a holistic vision of the problem. Being the lead in wetland conservation in Canada, the Canadian Wildlife Service (CWS) has implemented a wetland habitat monitoring program, which periodically samples the Prairie ecozone of Canada to determine wetland status and trends. Data from this monitoring program are helping to direct the wetland conservation efforts of the CWS and its conservation partners.

Results for Alberta

Percentages are used to report relative and proportional change/composition. Percent relative change between baseline and update values is calculated by the division of the absolute change (area or counts) by the absolute baseline value. For example assume that the baseline wetland value for a group of transects was 10 ha and the updated value for the same group of transects was 5 ha, the result is a -5 ha net change in wetland area. The relative change is calculated by dividing the net change number, -5, into the 10 ha baseline value, thus, a -50% relative change in wetland area. Unless otherwise stated, the values presented here are direct measurements obtained though the examination of lands mapped by habitat monitoring transects, and are not estimates of change at the landscape scale.

Wetland Loss

The definition of a wetland loss is: A measurable, anthropogenically created wetland basin alteration sufficient in magnitude to impose permanent effects to a wetland's capacity to 1) hold water and/or 2) function as a wetland habitat.

Net Wetland Change

Measured wetland loss data represent the area of wetland removed from the landscape. Losses recorded are considered permanent, whereby the area was no longer considered as wetland habitat.

Native Grassland Change

For this analysis any grassland polygon that showed signs of past or current seeding or cultivation was not included as native grassland.

Table 7: Wetland and Grassland Change Summary

Landscape Feature	Type of Change	Percent Change
Wetland Area	Loss	-6.28
	Net Change	-5.42
Number of Wetlands	Gross Loss	-7.24
	Gross Gain	1.35
	Net Change	-5.89
Native Grass	Loss	-11.20

Global Forest Watch Canada

Global Forest Watch is an independent organization that monitors and maps logging, mining, road-building and other forest development within major forested regions of Canada to provide access to better information about development activities in forests and their environmental impact. To this end, GFW: tracks existing and planned development activities; identifies the actors—including companies, individuals, government agencies, and others—engaged in this development; monitors the implementation of laws and regulations established in the interest of forest stewardship;

and provides data on forest ecosystems to highlight the environmental and economic tradeoffs that development options entail.

Alberta's forests are being rapidly opened up for extraction of timber, energy, and mineral resources. The most diverse and productive forest ecosystems have undergone widespread fragmentation by roads and other access routes, and the bulk of their area is under logging tenures. Cumulative impacts with industrial uses are escalating. For example, most of the forests of the Western Canadian Sedimentary Basin have been severely fragmented by linear disturbances, usually caused by the oil and gas industry.

Figure 7: Change in Foothills Natural Region

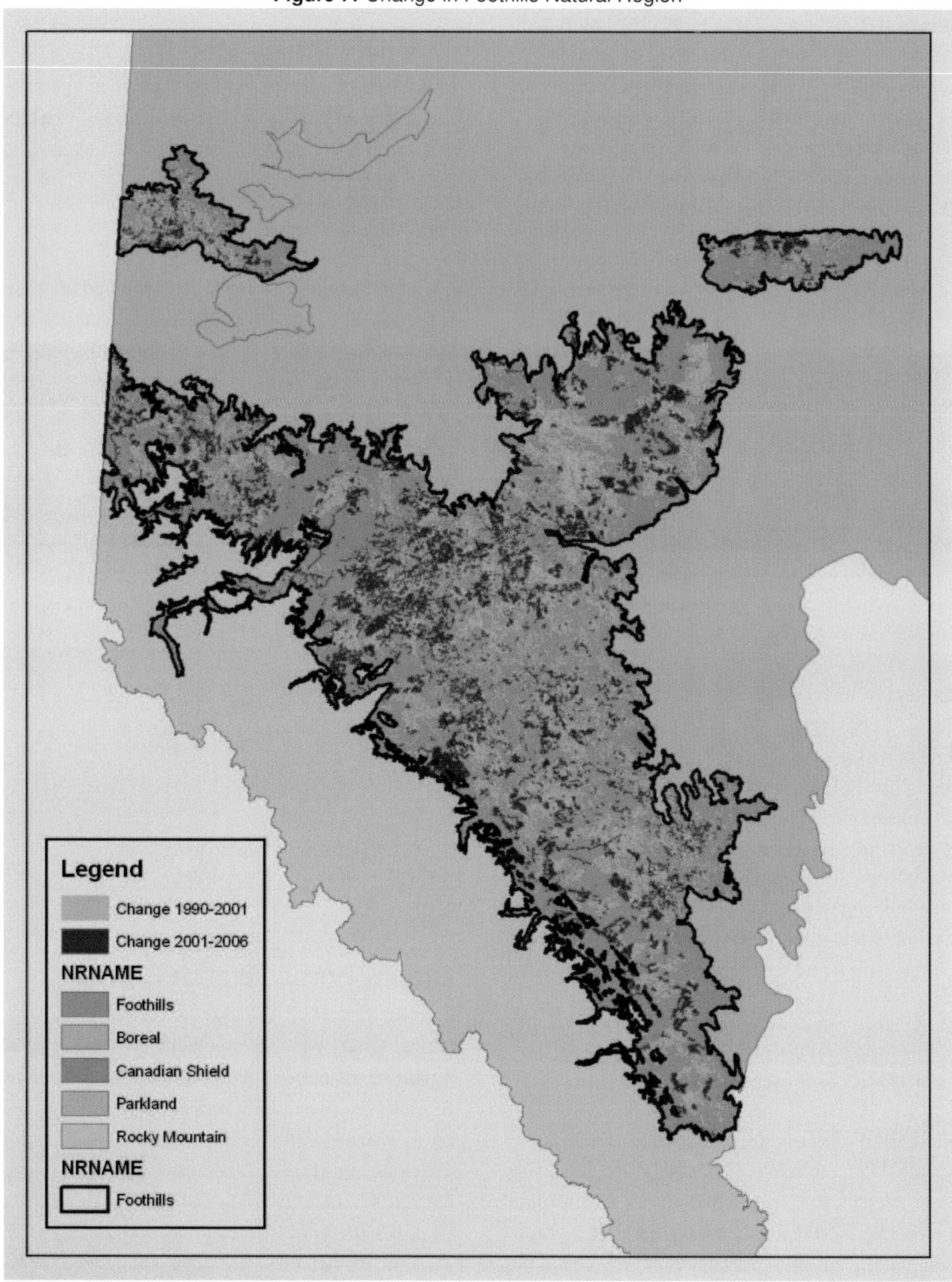

Analysis of the Atlas 2 data has indicated that there is a decrease in abundance for a number of species in the Foothills Natural Region. One of the suspected causes is the change in habitat and fragmentation of existing habitats. Fire and pests, such as the Pine Beetle, contribute to this but the largest impact comes as a result of activities associated with forestry and oil and gas development. A 1996 Alberta government study determined that, of 740 townships (areas of 100 sq. km or 36 sq. mi.) comprising the Rocky Mountain Foothills, only five were "intact", i.e., lacking well sites, logging, or significant linear disturbance. With less than 5% of this Natural Region under some form of protection and almost the entire area under a forestry or oil and gas disposition, the rate of change is not expected to lessen.

The following map, generated by Global Forest Watch Canada, shows the extent of change in the Foothills Natural Region of the province from 1990–2006. This represents both natural and man-made disturbances.

Canadian Wildlife Service Waterfowl Committee

The Canadian Wildlife Service Waterfowl Committee's 2005 report "Population Status of Migratory Game Birds in Canada" states:

Canadian Prairies and Western Boreal Canada

Breeding waterfowl populations are monitored annually through the Waterfowl Breeding Population and Habitat Survey. The traditional area of the survey encompasses the Canadian Prairies and Western Boreal Canada (northwestern Ontario to Old Crow Flats in the Yukon), as well as the north central U.S. (U.S. Prairies) and parts of Alaska. The U.S. Fish and Wildlife Service (USFWS) and CWS have been conducting this survey, using fixed-wing aircraft in combination with ground counts, since 1955. Breeding population estimates have been corrected for visibility bias since 1961. The southern portion of the survey area is typically covered again later in the summer to provide indices of overall waterfowl production (conducted by the USFWS, known as the July (Brood) Production Survey).

In the prairie pothole region (Canadian and U.S. Prairies), weather has a strong influence on waterfowl breeding habitat conditions and, consequently, on the abundance of waterfowl populations. Drought in the late 1980s and early 1990s created particularly difficult breeding conditions for ducks. Spring habitat conditions (as measured by the number of ponds in May) improved into the late 1990s from the low levels during the drought of the late '80s and early '90s. The May 2005 estimate of 5.4 million ponds in the prairie pothole region represented an increase of 37% from 2004. The increase was particularly striking in the Canadian Prairies where the number of ponds increased by 56% as compared to 2004. The 2005 value for the prairie pothole region was 12% above the long-term average. Analysis of trends showed significant increases ($P<0.05$) in the number of ponds for the prairie pothole region over the long-term.

Climate Data

Chetner et al. (2003) created a new climate normals database for Alberta using climate station data provided by the Weather Service of Canada. We selected climate stations with complete precipitation records for the years of Atlas 1 and Atlas 2 and calculated the mean precipitation values for each Natural Region in both Atlases. The differences from the mean values are presented in Figure 8. The chart clearly shows that the Boreal Forest and Parkland were drier in Atlas 2 than in Atlas 1. Also, the Grassland was notably wetter in Atlas 2 than in Atlas 1. There were differences in other areas, but the variability across those regions was too great to make a generalized statement.

Figure 8: Mean annual precipitation deviations from normal (Error bars display standard error)

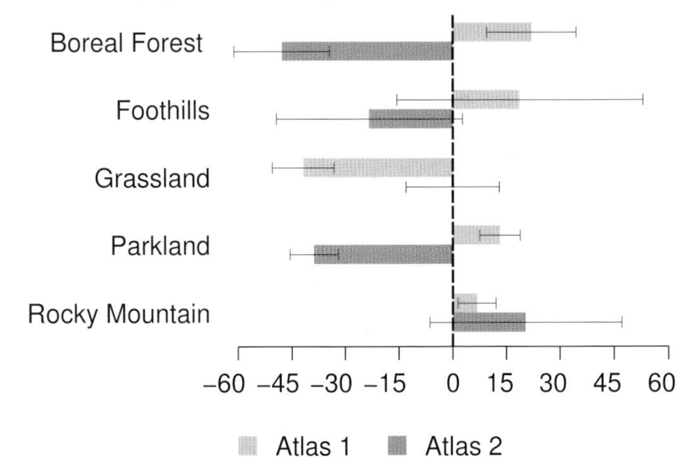

Methods

It is easy to lie with statistics.
It is hard to tell the truth without it.
Andrejs Dunkels

Things should be made as simple as possible —
but no simpler.
Albert Einstein

Introduction

In many scientific disciplines, amateurs are limited to reading about the work of professionals in magazines and popular books. There are, however, a few exceptions in which amateurs are active participants and contributors to scientific knowledge. These projects are commonly referred to as citizen science. The monitoring of plants and animals is one area in which amateurs can make an important contribution, applying some of their effort within the framework of a number of formalized schemes.

For birds in particular, the large population of interested and skilled observers has contributed to many successful projects, with the well known collaboration between amateurs and professionals in Britain (Greenwood, 2003) being a frequently cited example. In North America as well, there are several long-running programs, such as the Christmas Bird Count (CBC) and the Breeding Bird Survey (BBS) that have been enjoyable for participants, and have also contributed significantly to the knowledge of the status of North American birds (Bart and Klosiewski, 1989; Root, 1988).

For the observer who is conservation-minded, the results of these projects can often have significant impacts. For example, monitoring programs and atlas projects in Britain identified a widespread decline in farmland birds (Greenwood, 2003). As a result, changes were made to agricultural policies which shifted emphasis toward supporting rural communities and wildlife and away from production.

Like the CBC, atlas projects are popular with birders. The primary objective of these projects is to gain a snapshot in time of the distribution of the birds within a region. With standardized methods and either even or stratified coverage, grid-based atlases provide the best available estimate of the size, shape, and location of bird ranges (Donald and Fuller, 1998). The resulting databases contain information on the species compositions of surveyed locations, and are commonly queried by both professionals and amateurs. These projects have used a variety of field methods. For example, the first atlas projects in Alberta, Ontario, and the Maritimes employed the traditional method in which one card is filled out per survey plot per year (Cadman et al., 1990; Erskine, 1992; Semenchuk, 1992). However, projects in South Africa, Australia, and Québec, as well as this Atlas, have employed single-trip data recording methods (Barrett et al., 2003; Cyr and Larivée, 1995; Harrison et al., 1997).

There are benefits to using a single-trip format, also known as a day-list or checklist format, over a traditional atlas design. Perhaps the foremost benefit is that once the infrastructure is put into place and the volunteers are trained, the checklist program can be continued after the atlas project is finished and form an important component to a complete set of bird monitoring programs (Droege et al., 1998; Roberts et al., 2005). Of course, if there is not easy access to the information gathered in ongoing programs, it can be demoralizing for volunteers (Greenwood, 2003) and cause programs to falter. To address this, the Federation of Alberta Naturalists (FAN) is currently in discussions with Bird Studies Canada to ensure cooperation and data sharing with the eBird program, launched by Bird Life International. Checklist-based programs can provide information on timing of breeding and migration, as well as act as repositories for unusual sightings. These factors, and the success of the checklist program in Québec demonstrating the value of the data for monitoring population trends (Cyr and Larivée, 1993; Dunn et al., 1996), were the primary influence on the choice of a checklist format for the updated Alberta atlas (Atlas 2).

Conduct of the Survey

Field Protocols

Initiated in 1999, the Atlas Update Project (Atlas 2) launched its pilot field season in 2000. The main field work was conducted from 2001 until 2005. Prior to the start of the update project, FAN had initiated the Alberta Birdlist Program, which is a checklist program in which observers are asked to record all the birds they observe on their birding excursions at one time and at one location, along with weather, habitat, and duration. Following the success of the checklist programs in Québec and South Africa, it was decided that Alberta should use this checklist method for its update project. There were valid concerns raised that changing field protocol in this way would make comparisons with the first atlas more difficult, but it was concluded by the Atlas Technical Committee that the benefits of the checklist field protocol for ongoing monitoring of Alberta birds outweighed those concerns. Some volunteers found the change of the protocol confusing, but the majority had little difficulty.

Regarding the field program, the Technical Committee identified squares that would provide a representative sample of the different landscapes, but accessibility played a role in prescribing this coverage. Data swapping with other agencies proved to be important to gaining better coverage in these areas.

Project Scope

The goal of the original work (Atlas 1) and the updated work (Atlas 2) was to record the breeding status and the relative abundance of all breeding bird species in Alberta within a five-year period. For Atlas 1, data collection was conducted between 1987 and 1991, while field surveys for Atlas 2 were conducted between 2000 and 2005.

Based on a review of the results of the first atlas effort, an assessment of available resources, an evaluation of atlas survey methods (Smith, 1990) and other bird census techniques, the Atlas Technical Committee identified the following guiding principles for Atlas 2, listed in descending order of priority:

1. The primary focus of the atlas update is to document the distribution of breeding birds in Alberta. First presence/absence, then breeding status.

2. The survey intensity for each square should be such that at least 75% of the bird species expected to breed in a square are recorded. The amount of effort to reach this level of detection will vary depending on observer experience, habitat types present and terrain, but as a rule of thumb about 75% of breeding species should be detected after 16 to 20 hours of atlassing in a square.

3. Once a square has been adequately surveyed, atlassers should undertake surveys in a new square.

4. Atlassers should record the maximum number of individuals of each species seen or heard in a square during each visit to that square.

This approach was similar to that of Atlas 1, in which atlassers were encouraged to survey all habitat types in a square and consult expected species lists compiled by Regional Coordinators. A minimum of 20 hours of atlassing effort per survey square was required. During this time atlassers were likely to record at least 75% of expected species, and to confirm breeding for at least 50% of this number, while placing 35% and 15% in the probable and possible categories, respectively.

Figure 9: Atlassing regions

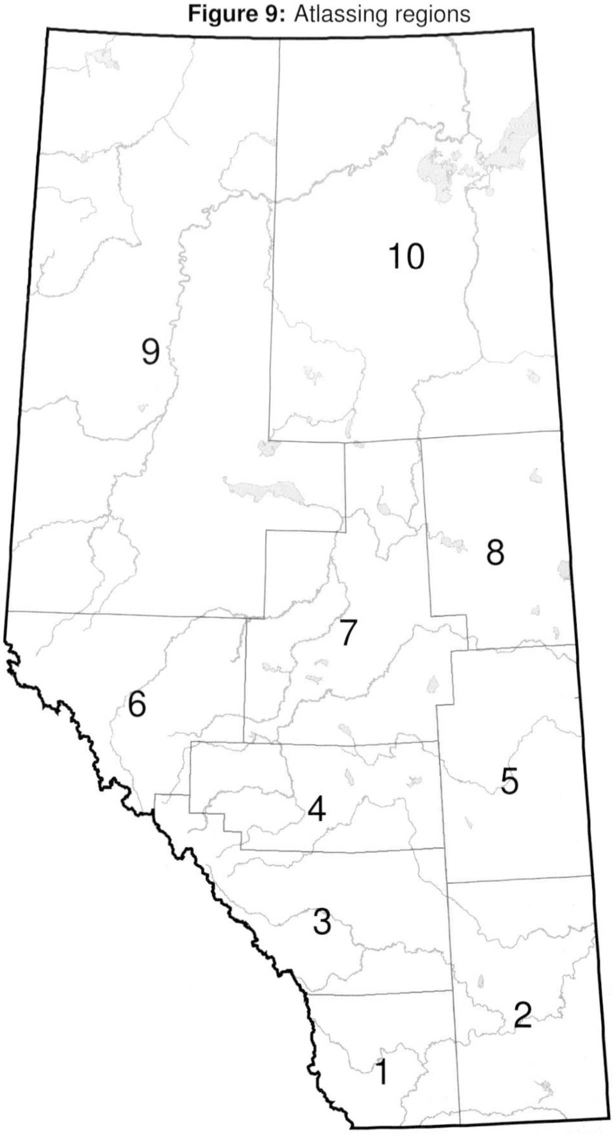

Figure 10: Atlas Grid UTM NAD27 block and square designations. Zone 11U is west of 114°. Zone 12U is east of 114°. Example square 11U PH 22 is labelled in red.

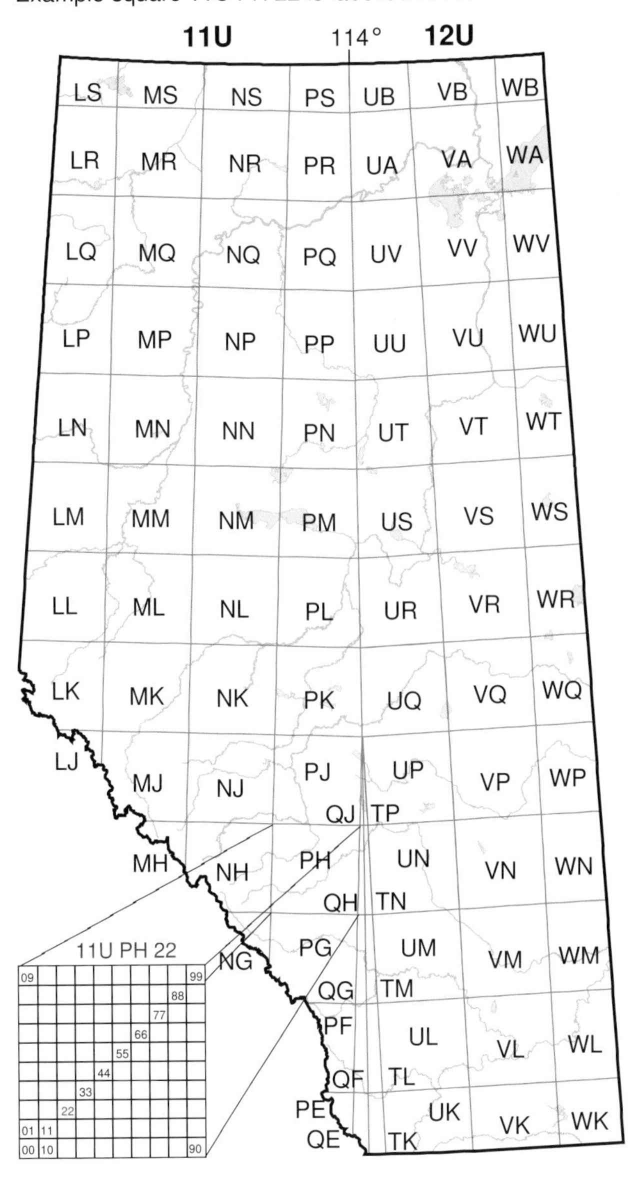

The Grid System

Atlas 2 used the same Universal Transverse Mercator (UTM) grid system as Atlas 1. This has been established as the standard for defining sampling units for breeding bird atlas projects in Canada (Robbins and Geissler, 1990) and facilitates comparisons to other North American atlas projects.

In both atlases, the province was divided into 10 regions (Figure 9). Each region was assigned a Regional Coordinator (RC) who was required to provide project materials and expertise to volunteers, and collate results. These RCs assigned atlassers to 10 km x 10 km UTM squares. These squares are marked on all Canadian topographic maps (1:50,000 and 1:250,000 scales).

A complete square designation is composed of a zone, block, and square code (Figure 10). The zone and block codes are given in an inset box in the margin of all NTS maps. (The area to the east of 114° West longitude is zone 12U: west of 114° West longitude is zone 11U. The two-letter block code can be read from the NTS map.) Squares are numbered from 00 in the southwest corner of the block to 99 in the northeast corner.

To correct for the inaccuracies of a two-dimensional grid lying over the earth's curved surface, adjustments are made along zone lines. In Alberta, the zone line is the 114° West line of longitude. On either side of this line and along provincial boundaries, partial squares result. For purposes of both atlases, data for squares smaller then 50 km² were added to neighbouring full squares for ease of surveying and mapping.

Coverage Goals

The goal of bird atlas projects is to achieve the best possible coverage of the mapping area; the ideal would be to achieve complete coverage of the entire study area. In Alberta, there are more than 6,000 10 km x 10 km UTM squares, many in remote and nearly inaccessible locations. To census all the breeding birds of these squares was beyond the scope of both atlas projects. Therefore, Regional Coordinators were instructed to deploy survey volunteers strategically.

Regarding Atlas 1, it was decided that one square from each block of four (i.e., from a 20 km x 20 km block) should be termed a priority square. Regional Coordinators in southern Alberta were asked to choose priority squares as those having the greatest habitat diversity within the 20 km x 20 km block. In northern Alberta, coverage was greatly hindered by limited access and the low atlasser population; hence, a minimum of one square for each 100 km x 100 km block (a 1:100 ratio) was assigned priority. However, squares were chosen to ensure sampling from all northern areas and, in most cases, several squares per block were surveyed.

For Atlas 2, the Technical Committee produced guidelines for setting priority squares. For atlas regions 1 through 8 (Figure 9), the target was to survey 50% of all squares in each 100 x 100 km UTM block, taking into account a fair representation of all the Natural Regions in the UTM block. For atlas regions 9 and 10, the target was to survey 10-20% of all squares in each block, taking into account accessibility. Atlas squares were also selected as priorities to provide an even distribution between previously surveyed squares (Atlas 1) and squares being surveyed for the first time. Regional Coordinators were then supplied with regional maps that showed the location of these priority squares.

Data Collection

Data Collected in Remote Areas

A major objective of Atlas 2 was to increase survey coverage in the more remote areas of the province over that of Atlas 1. With this in mind, a remote areas program was launched.

In addition to the remote survey work of the atlas update project, the University of Alberta (U of A) also conducted surveys. Led by Dr. Fiona Schmiegelow and Dr. Steve Cumming, this U of A program used 10 two-person field crews annually to collect information on breeding bird populations in remote sections of the boreal forest of Alberta. The purpose of the U of A program was to test and refine bird-monitoring methods for widespread application throughout the forested regions of Alberta. Field sampling was initiated in May of 2001 in north-eastern Alberta.

Within each landscape, several survey methods were used during the breeding season (May-July). The location of sampling areas within landscapes was based on reclassified forest inventory data, which was reduced to twelve major habitat classes. Triangular line transects were established to sample forested habitat types roughly in proportion to their abundances, adapting a protocol widely used for wildlife surveys in Finland and Sweden.

Point count surveys were conducted in one or more large patches of older commercial forests. In the first year the sampling focussed on patches of older deciduous- or deciduous-dominated forest. In 2002, a modified protocol and increased sampling effort placed equal emphasis on patches of older coniferous-dominated forest, in order to better capture species associated with this habitat type. Point counts were augmented by playback methods to increase detections of breeding activity. Each landscape was visited once during the breeding season, in the year of sampling. Winter surveys for resident bird communities were also conducted in many of the sample landscapes.

Collecting and Recording Field Data

During both atlas projects, atlassers were provided with handbooks, paper checklists, and rare bird forms. Atlassing techniques were standardized variously through discussions in quarterly newsletters, contact with the Regional Coordinators, and regional atlasser meetings. Paper checklists included fields for recording location, hours, date, species, breeding, and abundance. Rare and hard-to-identify species were indicated on the checklists to alert the atlassers to file a rare bird report or to be thorough with respect to these species. After bird observations were recorded, checklists were then mailed to the FAN office in Edmonton, or entered electronically using the Personal Birdlist software, and e-mailed to the FAN office.

In Atlas 1, project atlassers were asked to record their bird observations on one checklist for the entire year. For Atlas 2, atlassers were requested to file a separate checklist or electronic checklist for each survey (the minimum requirement was one checklist per square per day). Atlas 2 protocol required submission of single-day checklists with the knowledge that this survey method enhances the data and allows for a better understanding of breeding bird populations. At the same time it provides information at a greater level of detail for future use during assessment and research studies.

Atlassers were also encouraged to fill out separate checklists to record birds found in different parts of the survey square on a given day (i.e., per visit). For example, if atlassers were to visit a wetland, a forest and a grassland habitat in different parts of the square, they could fill out a separate checklist for each of the areas visited in the square, as well as an additional checklist for the birds observed between these locations. By recording data in this manner, important habitat relationships within each survey square are more easily identified.

Breeding Evidence

The North American Ornithological Atlas Committee (NORAC) has developed recommended breeding codes for North American atlas projects. The Alberta Breeding Bird Atlas has adapted NORAC's breeding code definitions for its own surveys (www.bsc-eoc.org/norac/atlascodes.htm). The objective of gathering breeding information for the atlas was to obtain the strongest possible evidence of breeding for each species within a survey square. There are four lines of breeding evidence: observed, possible breeding, probable breeding, and confirmed breeding.

Within each of these levels, there are categories of breeding evidence denoted by a letter code. These codes represent behavioural and empirical evidence that indicates breeding activity. These codes are to be entered on each checklist for each visit to each square (see Table 8 for breeding codes used in Atlas 2).

For the most part, breeding codes in both atlases are identical. One slight change was made for the probable breeding code of Territory (T) from Atlas 1 to Atlas 2. The original definition of the probable breeding code for Territory (T) read: "Territory presumed through territorial nesting behaviour in the same location on at least two occasions a week or more apart". It was changed to read: "Territory presumed through territorial nesting behaviour or the presence of a singing male in the same area over a four-day time interval".

The original breeding code for Territory was designed to avoid false positives that are associated with birds that sing while on migration, by spacing the observations a week or more apart. This would apply to birds like the American Tree Sparrow, which is known to breed in the northern region of the province but is also commonly reported singing in the southern part of the province while on migration. Some species of warblers also sing while on migration through to northern Alberta but do not breed while on migration.

Table 8: Atlas breeding codes

Breeding Code	Description
Observed	
X	Species identified but no indication of breeding
Possible Breeding	
H	Species observed, or breeding calls heard in suitable nesting habitat
Probable Breeding	
P	Pair observed in suitable nesting habitat.
T	Territory presumed through territorial nesting behaviour, or the presence of a singing male in the same area over a four-day interval
C	Courtship behaviour between a male and a female
V	Visiting probable nest site, but no further evidence obtained
N	Nest building or excavation of a nest hole by wrens or woodpeckers
Confirmed Breeding	
NB	Nest Building or adult carrying nest material; used for all species except wrens and woodpeckers
DD	Distraction Display or injury feigning
UN	Used Nest or eggshells found
FL	recently Fledged young or downy young
ON	Occupied Nest indicated by adult entering or leaving nest-site or adult seen incubating
CF	Carrying Food; adult seen carrying food or fecal sac for young
NE	nest with Eggs
NY	nest with Young

The duration of many remote area surveys was often less than one week; so the Technical Committee decided that the breeding code definition for Territory in Atlas 1 might exclude many legitimate Territory breeding observations in Atlas 2. Reducing the time required made it more likely that birders would still be in the area to make the required observations to confirm a Territory breeding code of a singing male. The Technical Committee also knew that, by using play-back tapes or other observation cues, a breeding code for Territory could be confirmed for birds; whereas, the original Territory breeding code caught only territorial breeding evidence observed in the same location on at least two occasions a week or more apart. To avoid these problems, the breeding code definition was altered to include any bird exhibiting territorial nesting behaviour or the presence of a singing male in the same area during a four-day interval. This allowed birders to identify a breeding code of Territory by observing territorial behaviour only, or by observing a singing male during a reduced required time period.

Abundance Data

Atlas 1 abundance data were collected only in the first year of the project using an abundance code system. An analysis of the first year's data showed that abundance data were not collected consistently, thereby reducing their usefulness. Subsequently a new approach for gathering abundance data was used. Abundance data for Atlas 2 recorded the actual number of individuals observed, and did not include extrapolated estimates of the total breeding population for the square. Though the primary focus of the atlas project was to document distribution and breeding status of species, it was understood that valuable information could be gained by keeping accurate counts of birds encountered, for example, deriving indices of relative abundance of species across the province. Atlassers were asked to record the maximum number of individuals observed in each square each time the square was surveyed.

Data Review and Records Confirmation

Atlas 2 used state of the art database and GIS software for data review and analysis; these powerful technologies were unavailable to Atlas 1 researchers. All bird observations forwarded to the Atlas office were entered into a database. Rare, unusual, or breeding records that fell outside the normal range of a species were flagged and forwarded to Regional Rare Bird Committees.

For both atlases, field observers filled out rare bird documentation forms for all rare species. Four committees were formed to adjudicate these and other records of regionally rare and hard-to-identify species. The Alberta Bird Atlas Regional Rare Bird Committees (RRBC) consisted of a Chairman and several voting members chosen for their ability to contribute on the basis of their respective areas of ornithological expertise.

Atlassers were required to fill out a rare bird form for any rare bird seen while conducting atlas surveys. A rare bird was defined as:

- any species not listed on paper Alberta Birdlist checklists or the electronic checklist in the Personal Birdlist software;
- a species listed as a rarity on the Alberta Birdlist checklists or the electronic checklist in the Personal Birdlist software; and
- a selected breeding species observed in prescribed areas in the province as outlined in Appendix 1 of the Atlassers' Handbook.

Data Sources

Several data collection sources were used for both atlases, including: the traditional volunteer effort; a campaign that solicited data from researchers; a Remote Areas Program that assisted volunteers in accessing northern squares; species-specific studies such as owl-monitoring programs; and data collected by the U of A remote areas program.

Collecting Owl Data

Many owl species are nocturnal and highly secretive and are, therefore, difficult to detect during the day. In both atlas projects, a special effort was made to document Alberta's varied owl life. For Atlas 2, guidelines for monitoring nocturnal owls (Takats et al., 2001) were provided to atlassers. Atlas 2 also incorporated data collected by the Alberta Nocturnal Owl Monitoring Program. During both atlas projects, atlassers were provided with owl tapes obtained from the Library of Natural Sounds, Cornell University, to help identify owl calls.

Breeding Bird Surveys

Breeding Bird Surveys were developed in Laurel, Maryland in the mid-1960s as a means of monitoring continental breeding bird populations (Sauer et al., 2005a). Breeding bird surveys are road-side surveys conducted along 24.5-mile-long routes, with 50 sampling points located at 0.5-mile intervals (Figure 11). A three-minute point count is conducted at each stop, during which the observer records all birds heard or seen within a 0.25-mile radius of the stop. The routes are randomly selected and use a consistent sampling methodology. They can, therefore, be used to document changes in relative bird abundance through time.

The Technical Committee for Atlas 2 agreed that Breeding Bird Surveys are one of the best tools available for detecting bird population changes through time because they utilize highly standardized methods. When coupled with atlas results they provided a more complete understanding of changes in the distribution and abundance of breeding birds. Breeding Bird Survey results are incorporated into the discussion of bird population changes in this atlas. Atlas participants were encouraged to conduct annual Breeding Bird Surveys in addition to their atlas surveys, particularly in northern Alberta where a limited number of surveys is currently conducted.

Figure 11: Alberta Breeding Bird Survey routes (historical and current courtesy of the Canadian Wildlife Service)

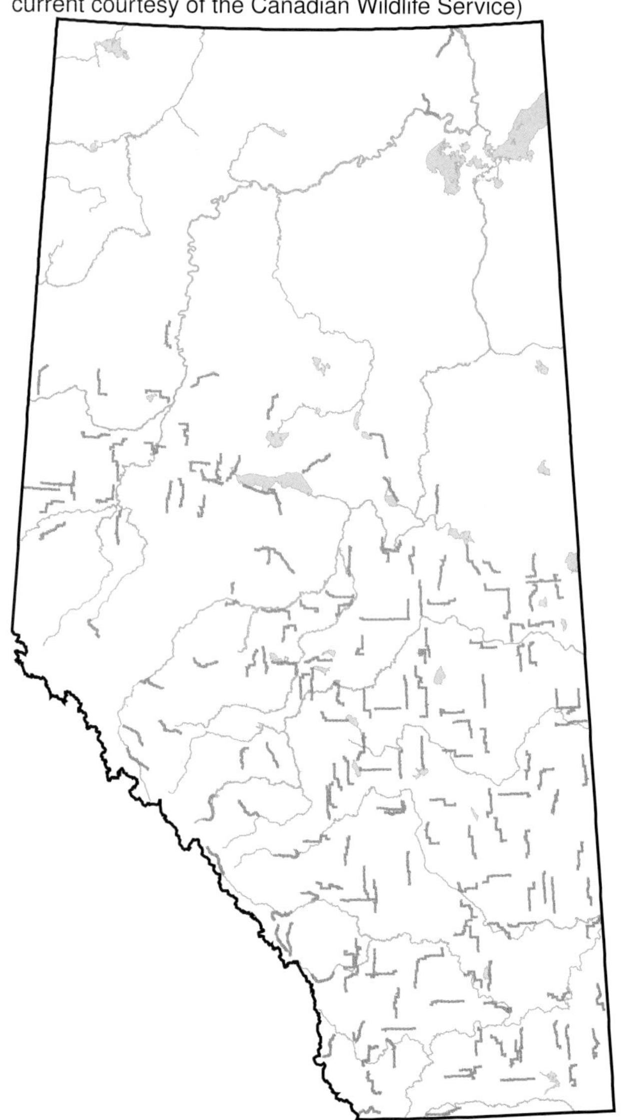

Data Analysis

Overview

The contributions of the volunteers can never be overstated. In that light it was felt that all possible effort should be made to make the most use of this effort. The following sections not only detail the statistical methods used to create the information presented in the species accounts but also detail the efforts made to gain maximum value of the data generated. Methods tried and rejected are also included, along with the rationale, in order to assist those who are planning these types of projects in the future.

Over the course of the project the Technical Committee had discussions about the analytical approach to be taken with the atlas data sets. However, in the spring of 2006, once the majority of the data were entered and could be examined, serious research into what was possible was initiated. The analytical methods used in this publication were not finalised until late in 2006 after all possible avenues were fully explored and tested.

One benefit of using a single time and date format for gathering data in the current atlas project was that it allowed use of many other data sources with similar field methods. Business and government partners were contacted to gain access to data that were already being gathered, allowing atlas efforts to be focussed elsewhere to make the best use of the volunteers. Using volunteer data and the data from other agencies ensured a much more extensive coverage in the update project than was possible in Atlas 1 (Table 9).

Examination of the proportion of data from different sources shows that more than half of the species records came from outside sources (Table 10). One problem with using these other data is that some of the data fields from the checklist form were not recorded, so their application was limited (Table 11). Although the vast majority of atlassers recorded both effort and habitat, others were less faithful in recording hours, and for the most part did not provide any habitat information.

External Sources

The external sources used for Atlas 2 were numerous. These data sets were carefully reviewed and included in two ways. For data sources that had very similar field protocols to the day-list methods in which all species detected were recorded, these were treated as day lists and used for calculating reporting rate maps and graphs, and in the change analysis. Data that were species-specific, or in some way incompatible with our field methods, were treated as incidental records and used in a limited manner for the reporting rate calculation (as outlined in the section on hybrid reporting rate) and for the production of distribution maps.

Data Limitations

In most cases, citizen science projects have less rigorous field protocols than those of professional scientists because extremely strict field protocols are less enjoyable and often daunting to volunteers. The challenge for project and program planners lies in designing field programs that provide a balance between the dual need for theoretical robustness and practicality (Greenwood, 2003).

Recent reviews of citizen science projects have identified a series of common concerns about study designs and ways in which these projects could be modified to improve data quality. Statistical methods for analyzing data depend upon standardized field methods and random samples. When the source data are not standardized, artifactual variability can be introduced into the data; this is commonly referred to as noise. Improved data quality increases the signal-to-noise ratio which in turn allows for detection of more subtle phenomena. When aspects of the data collection are not standardized, correction factors can sometimes be applied, but when source data are beyond repair, potentially valuable information is lost.

Analyzing the CBC data, Dunn et al. (2005) identify variance in effort as the most significant problem with the data gathered in these counts. They point out that the relationship between effort and the number of birds seen is not linear, and is not the same for all species. This means that if x (number of) species are seen in the first hour, it is unlikely that as many new species will be detected in the second hour. Moreover, in subsequent hours still fewer new species will be acquired. To add to this, the time of day, weather, and observer skill are factors that can affect the rate at which new species are acquired. For survey projects like atlases or checklist programs, time of year must also be considered. Therefore, simple correction factors that use birds per unit effort are potentially seriously biased and can introduce more bias than using unadjusted counts. Large-scale analysis of the CBC data reveals two additional problems. First, the density of count circles varies from region to region. Second, the distribution of CBC count locations is not random and may not be representative of the regions for which they are used. Dunn et al. (2005) cite examples of studies in which successful adjustments were made for effort and uneven density of survey locations, but they recommend improved field protocol standardization and recording of effort.

The BBS uses a field protocol that is more strictly standardized than the one used in the CBC or in the atlas projects in Alberta and Ontario; however, there is still considerable concern raised in the literature about biases in the data (Bart et al., 2004; Francis et al., 2005; Sauer et al., 2005b, 2004). The concerns call into question the faithfulness of data representation for the entire area, and also observer bias. Francis et al. (2005) note that although corrections can be made to account for some of

Table 9: Atlas coverage (S - squares with survey data, I - squares with incidental data, % - percent coverage)

Natural Region	Squares	Atlas 1			Atlas 2			Overlap		
		S	I	%	S	I	%	S	I	%
Boreal	3799	560	180	14.74	758	589	19.95	263	44	6.92
Canadian Shield	100	9	3	9.00	6	21	6.00	1	2	1.00
Foothills	668	180	51	26.95	174	95	26.05	55	19	8.23
Grassland	956	377	281	39.44	594	273	62.13	257	83	26.88
Parkland	610	270	96	44.26	274	126	44.92	152	24	24.92
Rocky Mountain	491	160	31	32.59	145	69	29.53	86	11	17.52
Total	6624	1556	642	23.49	1951	1173	29.45	814	183	12.29

Table 10: Composition of data from atlassers and others

Natural Region	Atlassers		Others		All	
	Trips	Records	Trips	Records	Trips	Records
Boreal	6963	78527	7851	80375	14973	161088
Canadian Shield	7	333	131	155	138	488
Foothills	536	9473	623	13019	1159	22492
Grassland	2929	57963	10394	100893	13391	160188
Parkland	3211	41966	1892	38048	5204	80985
Rocky Mountain	1962	20392	892	14333	2916	36034
Total	15608	208654	21783	246823	37781	461275

the bias, it is unlikely that the bias can ever be fully eliminated. Dunn et al. (1996), in their analysis of trends using checklist data from Québec, did not attempt to correct for weather effects or uneven distribution of observers, as these factors were not likely to introduce systematic bias into the data. Furthermore, the most likely sources of systematic bias—changes in observer behaviour or skill—cannot be mitigated post-hoc in most cases.

The last major issue related to bias is representative sampling. For surveys like the BBS, which sample only along roadsides, recommendations are now being considered to start an off-road portion of the survey to help gain a more representative sample of the landscape (Sauer et al., 2005b). In this updated atlas, the number of kilometres of roads per square was calculated, excluding cut lines and other smaller features, and grouped into a series of categories (Table 12). Examination of the table makes it clear that the atlas project, like the majority of surveys without remote areas helicopter support, are dependent on roads for access, which creates a bias.

During analysis of the results of citizen science surveys, remedies may be required to correct for systematic survey biases, where possible. The results derived from less well standardized studies need more qualification and can provide strong evidence of probable problems, but cannot give definitive conclusions. The issues discussed above apply to this atlas and, therefore, the results must be interpreted in the same light.

As standard methods for conducting change analysis were not possible with our data sets, we sought both internal and external sources of corroboration. For internal corroboration we chose to calculate our change analysis using two different methods, each of which would be more suitable for some species than for others. This approach allowed us to say something meaningful about the change for more species than relying upon a single method. For external corroboration we used the BBS as an independent abundance-based method of change. As such, this work will serve as a signpost to researchers by suggesting areas for further research, and to managers with protocol and analytical concerns.

Table 12: Relationship between roads and survey effort (Note: Surveyed denotes squares with one or more general surveys.)

km of roads	Total	Surveyed	Coverage
0	2219	169	7.62
<10	626	139	22.20
10–25	912	261	28.62
25–50	920	417	45.33
50–75	783	367	46.87
75–100	873	425	48.68
>100	291	173	59.45

Table 11: Completeness of supplemental information from atlassers and others

Natural Region	Atlassers		Others		All	
	Effort	Habitat	Effort	Habitat	Effort	Habitat
Boreal	79.4	75.7	9.8	12.0	42.1	41.5
Canadian Shield	57.1	100.0	0.0	20.6	2.9	24.6
Foothills	72.2	92.2	28.4	21.2	48.7	54.0
Grassland	83.7	89.0	8.7	1.2	25.1	20.4
Parkland	75.3	94.1	23.9	14.9	55.1	63.5
Rocky Mountain	67.0	96.7	37.4	10.2	56.5	68.2
Total	77.5	85.2	12.1	7.3	39.0	39.4

Abundance Data

Possibly the most common question asked during the project was if abundance data really needed to be gathered and whether it would be used. This was related to the fact that during Atlas 1, abundance data were gathered inconsistently and couldn't be used. When the Alberta Birdlist program was initiated after the completion of Atlas 1, the potentially confusing abundance codes were abandoned in favour of actual numbers of individuals observed. This same data format was used with Atlas 2 and most atlassers recorded abundance in a consistent fashion. However, again in this publication, there are no abundance numbers reported.

The reason there are no abundance numbers reported in this publication relates to the previous discussion about standardized methods. As we have many different observers with different skill levels, making observations in a non-standardized way, for non-standardized lengths of time, direct use of abundance values to estimate population sizes would yield erroneous results. The work of Dunn et al. (2001) demonstrated that although trends calculated from checklist-style abundance data often mirrored those from standardized surveys, those trends were less reliable than those calculated using frequency or reporting rate data. To be as scientifically robust as possible, we chose to use reporting rate as a proxy for relative abundance. Of equal importance, the choice of reporting rate as a proxy for abundance allowed for meaningful comparisons between the Atlas 1 and Atlas 2 data sets.

Although the abundance data were not used directly in this atlas, there are several compelling reasons to have recorded, and to continue to record, abundance information:

- Rare Species. In cases of rare species identifications, the number of individuals observed is often an indication of the reliability of the identification. Further, should a new or rare species be identified, knowing if one bird or ten were observed is important information for both the scientific community and recreational birders.
- Future Use. The statistical methods available today are much more sophisticated than were available in 1991 when Atlas 1 was published. Right now, it is problematic to use non-standardized abundance data, but methods are being actively researched to deal with this problem.
- Research Use. Related to the previous item, having a large database with abundance data gathered in a non-standard way, provides a valuable resource for professional researchers to test methods of using these data.
- Practical Use. For environmental consultants or interveners in Environmental Impact Assessments, knowing as much as possible about an area before planning a field program is valuable information. Having abundance data, along with other extras such as habitat, time of day, number of observers, and self-assessed skill, in addition to a species list, provides valuable information and makes the Atlas 2 data set more valuable to the scientific and land-management communities.
- Reliability. If observers must identify the same species multiple times, their chances of missing species, misidentifying species, or even recording them in the wrong row on the checklist form are reduced.
- Education. Recording abundance provides good training for more advanced citizen science projects such as the BBS where recording abundance is the primary objective of the survey.

Reporting Relative Abundance

Having chosen to use reporting rate as our measure of relative abundance, the next decision to be considered was how to report this information in a way that would be accessible and useful to readers of the book. Since recreational birders and scientists alike are interested in when, in what habitat, and where the species were most abundant, approaching our reporting in the same manner seemed an appropriate approach.

Variance Adjusted Reporting Rate (*varr*)

When there are few surveys conducted in an area, any sightings will have a large effect on the reporting rate (e.g., 1 in 2 vs. 1 in 10). As more surveys are conducted, the calculated reporting rate will more accurately approximate the true reporting rate and the variance will decrease. We applied this fact by grouping squares together in small clusters within the same natural subregion for mapping or over an entire Natural Region for general habitat use. Within these groups we calculated an average weighted by the inverse of the variance. Thus observations from areas with the most survey effort were weighted the strongest and observations from areas with the least survey effort were weighted the weakest, giving the best possible estimate of reporting rate possible from the source data. The calculation of the *varr* is outlined in detail on page 39.

Temporal

Use of the checklist data format provided time of year information, not available using an old style atlas data format. To simplify the display and reporting, a monthly bar chart was used discriminating between probable and confirmed breeding and other records. Each month was split into four periods and montly *varr* values were calculated. Because some months had considerably fewer surveys than others, we set a minimum of 500 surveys per month. For months with less than 500 general surveys, species *varr* values in those months were scaled by the number of surveys conducted in those months divided by 500. This way, records from months with few surveys were prevented from creating potentially erroneous spikes. An important point to consider when looking at these charts is that the species behaviour may affect detectability and create false impressions of relative abundance. However, the month in which the reporting rate is the highest is when one is most likely to observe the species, even though it may not be the month when the species was most abundant.

Habitat

Choosing an appropriate habitat classification for reporting habitat selection was difficult. Although many observers did record habitat information with their checklists, some did not. Using these data would have forced the exclusion of many records. We considered the utility of extracting habitat classifications based on satellite habitat classifications such as the Land Cover Map of Canada (Canada Centre for Remote Sensing, 2000), but in the end, there were too many uncertainties, creating the possibility of reporting false information. Use of Natural Subregions was considered, but this was limited by not having sufficient numbers of surveys in each subregion to provide accurate assessments. In the end, Natural Regions were chosen, excluding the Canadian Shield which had insufficient numbers of surveys to make reliable estimates of reporting rates.

Data from each species' expected breeding period were used to create a horizontal bar chart. To make this chart representative of a species' habitat preferences, bars for specific Natural Regions were excluded if less than 3% of observation records or less than 1% of breeding records were from that Natural Region. As with the temporal reporting rate, the Natural Regions with the highest reporting rates may not be the Natural Regions with the highest relative abundance, because some species are more easily detected in some habitats than others. However, for individuals who want to know where they are most likely to find a species, this information is provided in an easily comprehended fashion.

Spatial

Choosing an appropriate spatial display for reporting rate was not trivial. One method of displaying this type of information is to treat each data point as a virtual tent peg and drape a mathematical canvas (kriging, spline interpolation, etc.) across them to create a continuous surface. The results of these types of manipulations are very attractive and have merit in some situations, but are inappropriate for this type of data. The underlying assumption in interpolating a surface from points in this way is that the area is homogeneous in reference to what is being reported. In our case, we are reporting relative abundance which is a function of habitat selection. Thus we need to be able to assert that the habitat is homogeneous across a species range; this is clearly false. Further, for interpolation to be robust, the point grid should be regularly spaced across the range, which again was a requirement we could not fulfill. If these two assumptions could not be met, then the resulting surface would contain many errors and the locations of those errors would not have been obvious or correctable.

The results might have been reported on a square by square basis, but the limitation is that only squares with a minimum number of surveys (such as 10) could have been used thus avoiding the overestimation of relative abundance in areas with low survey effort. The end result would have been a fairly limited number of points. A second option was to identify squares that could be considered fully surveyed and use them as centroids for building a Voronoi diagram for the province. In this case, areas closest to the completed squares would be contained in the areas created by the Voronoi diagram algorithm. This option was ruled out because again it would assume a homogeneous habitat layer across the province. The third option was to create two levels of clustered squares within the Natural Subregion boundaries. The smaller clusters were used to calculate the *varr* for the larger clusters. In this way each area displayed contained a homogeneous general habitat classification (Natural Subregion) so that projection across those regions to squares that were not surveyed was reasonable. Further, this allowed for clear identification of areas with insufficient data to generate a reasonable estimate, which would not have been possible using surface interpolation. Although this method was manually intensive it generated the most complete and accurate relative abundance maps possible from the Atlas data. As with the habitat use graph, the reporting rate map was calculated using records from the documented breeding period for each species.

Change Analysis Methods

The field protocols of Atlas 1 and Atlas 2 were not the same. Thus the first task in conducting any change analysis was to create two compatible data sets. In Atlas 1, data were gathered on a yearly basis. In Atlas 2 data were gathered on a daily basis. Since the Atlas 1 data couldn't be split, the Atlas 2 data were merged. However, in some cases during Atlas 1, more than one atlas card was submitted for a single square in one year. We therefore created a summary of both data sets by year and square with the highest breeding evidence. This summary was limited to only those squares where general surveys had been conducted. It was clear that the differences in field methods between Atlas 1 and 2 were not fully eliminated by merging the data in this way. In Atlas 1, there was generally higher breeding evidence recorded in most squares, as can be seen in the differences between the two observed distribution maps for most species. We therefore could not use breeding evidence in our change analysis and expect to obtain a reliable result. The size of the merged data sets were similar and the seasonal proportions were similar except for fall and early winter; in Atlas 2 a slightly larger proportion of records had been submitted from September through December than in Atlas 1. Despite this difference, the data were similar enough to expect reasonable results if interpreted responsibly.

Traditionally change analysis is conducted by examining the percent of completed squares (POCS) where the species in question occurred in two atlas projects, and increases or decreases in occurrence are considered indicative of increases or decreases in abundance. Ideally this comparison should be limited to squares that were fully surveyed in both atlases. This ideal is difficult to achieve where the area is large and the population base is relatively small. In many cases this would mean using non-matched squares with similar habitat types in order to have enough records to make meaningful comparisons.

Level of effort measured in hours is the standard method for classifying square completeness. The underlying assumption in this determination is that one hour of field effort by individual A in place X is equivalent to one hour of effort by individual B in place Y. This assumption is questionable when varying habitats, skill levels, field behaviour, weather conditions, time of day, and time of year are considered. However, as this has been and remains the standard method, we wanted to test the viability of this method with our data.

First, we tested this directly using matched as well as unmatched squares using 20 hours of effort as a determinant of completeness. In these cases it is common to correct for more or fewer hours in individual squares by estimating, by means of a linear regression, the level of evidence or number of species likely to have been detected in that square at the specified level of effort. We had calculated species accumulation curves by Natural Region and found that noise in the data overwhelmed the signal in all regions between 15–20 hours of survey effort. We therefore decided that effort correction was not feasible with our data set and simply used squares with 20 or more hours of survey effort.

Second, we made this comparison indirectly as we lacked hours information for many surveys and wanted to use as many of the contributed data as possible. We used documented species range data in Alberta to generate theoretical species counts for each square. Then surveys with time data were used to calculate the percentage of a theoretical species count after 20 or more hours of survey effort. The mean values for Atlas 1 and Atlas 2 were used to define a percentage of theoretical species for each square that was needed for those squares to be considered complete. As previous, both matched and unmatched square comparisons were made.

In order to test the reliability of these measures we selected a number of species that were well represented in the Atlas and the BBS. We compared change analysis results from these methods with BBS results. Of these, the hour based methods produced confusing results. Use of the theoretical species count was more reliable. However, to have a sufficient number of records to conduct statistical significance tests, use of unmatched squares was required. Despite this allowance, this method was still based on a relatively small sub-sample of our data. Thus, we sought other methods that would give larger sample sizes and increase the power of our statistical testing.

We calculated the *varr* for each Atlas using square summary data as the source for our Natural Region calculation. By using the *varr* method we could employ the vast majority of our data and have it weighted by a defensible measure of confidence. Further the selection of data for inclusion was not dependent upon an artificial definition of completeness. The potential fault in this method is that it would report erroneous change in cases of systematic bias in one survey but not the other. The duty then fell upon authors to consider the differences between the Atlas 1 and Atlas 2 surveys for potential sources of systematic bias.

To assist authors in identifying systematic bias in Atlas 1 versus Atlas 2, we calculated an occurrence index in which the reporting rate within the survey was compared to that of all other species. This value provided a within-survey corrected measure of how often a bird was observed and was reported in the species accounts as such. This occurrence or rarity index could then be used to identify possible biases in the data and analytical results.

Other Methods

In addition to these methods we considered developing habitat selection models for certain species and using these models to predict species ranges. Upon careful consideration of the scale of the data and the limited data available to us, this approach was abandoned due to its likelihood of failure.

Analytical Details

Data Limitations

As previously discussed, there are several sources of bias to be considered when analyzing the Atlas data. These are:

1. Different field behaviour on some surveys (e.g., were the observers making an incidental observation or were they attempting to identify all the birds in the area);

2. Variable observer skill, timing (day, year), and environmental conditions;

3. Varying levels of habitat representation in different regions;

4. Variable effort on each survey.

These issues were addressed as follows:

1. Surveys were classified as either incidental or general surveys based on information provided or number of species. If species counts were four or less, the surveys were classified as incidental. Similarly, surveys that were species- or group-specific were classified as incidental, as well as surveys extending beyond a single day or not adhering to day-trip or day-trip-like field protocols.

2. For some analytical methods we used data only from the birds' breeding season. Other factors such as observer skill or time of day, as discussed above, were not likely causes of systematic bias. As such, we did not attempt to adjust for these factors.

3. Reporting rate was calculated for both of the spatial groupings and used to plot the reporting rate map and for the change analysis on a Natural Region basis. In the first case, a coverage index was applied to lower the values in areas (see section Coverage Index below) with small survey counts to prevent overestimation of reporting rate in areas with limited data. In the second case, the areas and numbers of squares involved were of sufficient number to be considered representative of those Natural Regions.

4. We did not have consistent and reliable effort information for much of the data in Atlas 1 or Atlas 2, so various methods described below were used to account for variable effort.

Understanding these limitations, we endeavoured to use statistical methods dependent on the fewest assumptions and methods that could use as many of the data as possible to allow the valuable information gathered in the project to be extracted.

Basic Approach

In analyzing Atlas data for the purpose of change analysis, numbers are usually reduced to presence / absence data. This type of data is referred to as binomial, and well-known methods exist to analyze these data. Comparisons were made using methods for examining the differences between binomial proportions.

Reporting Rate

One of the reasons that the checklist format was selected for the Atlas update project was that it provides the means to do within-data-set comparisons using reporting rate, or frequency of occurrence, as a proxy for abundance. The probability of a species being detected is a function of its abundance, size, behaviour, and distribution and it is independent of survey effort. We assumed [as did Temple and Cary (1990)] that reporting rate is influenced more by changes in abundance than by other factors such as observer skill or behaviour.

Dunn et al. (1996) compared abundance-based models and frequency-of-occurrence models based on checklist data from Québec against BBS trends. The abundance-based results were more variable than the frequency-based results, and the frequency-based results, although less sensitive, produced fewer false negatives. In addition to this study, knowledge of various sources of variance within our data from non-standardized field protocols led us to take a cautious approach and to use the frequency-based method to report abundance.

Although this method is not as statistically robust as a model-based approach, the problems of non-standardized field data and the lack of sufficient information to create successful models for common species make its use with day-list data an obvious choice. For this Atlas, this metric was used to construct a map indicating the relative abundance of individual species across Alberta during their breeding seasons. It is intended that the reporting rate map be used in combination with the habitat chart to provide a good indication of where to find a species of interest.

Reporting rate—the proportion of surveys in which a species was detected—is an indicator of relative abundance for a species across a study area. For example, if a species was observed in four of twenty surveys, it had a reporting rate of 0.2, or 20%. If another area had a reporting rate of 0.7, or 70%, it is reasonable to assume that the species was more abundant in the second area than in the first. Fundamental to this statistical measure is the assumption that the surveys were representative of all the habitats in their respective areas.

The basic calculation of reporting rate is a binomial proportion.

$$rr = \frac{o}{n} \tag{1}$$

- rr: reporting rate
- o: number of occurrences
- n: number of trips

Hybrid Reporting Rate (hrr)

As noted above, it is inappropriate to include incidental data in the same fashion as general surveys, as incidental records are, by definition, not surveys of all the species in the area. However, if these data were not included in some fashion, valuable information would be lost. Therefore, we devised a hybrid method in which incidental records were used to augment the basic data. In areas where a species occurred in the general surveys, the incidental data were ignored. However, in areas where general surveys were conducted but the species in question was not detected, the incidental data were consulted and if one or more records were found, then both o and n were incremented by one and the reporting rate was calculated using the basic reporting rate formula. In this way, incidental records were assigned the lowest possible importance in calculating reporting rate, but areas of species occurrence were not ignored.

Variance Adjusted Reporting Rate (*varr*)

We derived the following method based on *The New Atlas of Australian Birds* in which (Barrett et al., 2003) used a spatially adjusted reporting rate based on simple average of areal subunits within larger survey areas. The *varr* method accounts for spatial heterogeneity in reporting and, additionally, weights those values by their relative accuracy as expressed by the inverse of the variance.

If the values of n are small, then each positive occurrence has a large influence on reporting rate. In these areas the confidence is low and the variance is high. In areas with many surveys, the confidence is high and the variance is low. Considering this, it is rational to weight the values by the inverse of their variances. In this manner, the minimal number of surveys needed is greatly reduced. Theoretically, areas with two general surveys could be used. The problem arises in cases of 0% or 100% occurrence; the variance is zero, so the inverse is infinite, which is clearly wrong. To account for this, a single positive or negative case was added to these occurrences for the purpose of calculating variance. In cases with 100% occurrence, we also used this modified reporting rate for the calculation of the weighted average. This was done to penalize areas with few surveys more heavily. The combined effect of these conditional treatments allowed the use of all squares where general surveys had been conducted.

For mapping purposes, we aggregated Atlas squares into clusters of approximately nine squares. As noted by Donald and Fuller (1998), the problem with large regular-shaped survey grids is that their boundaries are usually different from those of blocks of contiguous habitat. Therefore, in all cases these clusters were limited by the natural subregion boundary to ensure that the coarse-scale habitat was similar. These were then grouped into larger areas of three to eight subgroups, again within a single natural subregion. These areas formed the reporting rate base map shown in Figure 12. Since the size and shape of these boundaries were somewhat irregular, we considered whether or not this irregularity would introduce bias (Bibby et al., 2000). However, since all the resulting spatial units were large and similar in size to other regions within the same subregion, and also of consistent basic habitat classification, we reasonably assumed that boundary irregularities would not introduce systematic bias if proportional metrics were used.

Within each areal subgroup the hybrid reporting rate was calculated. The hybrid reporting rate was used to estimate the probability of occurrence of that species in the subgroup area ($\hat{p}$). The $\hat{p}$ values for each subgroup were then used to calculate the variances ($\hat{v}_i$) for their respective subgroup. The subgroup hybrid reporting rate values were then used to calculate a weighted average using the inverse of the variance.

The formula for these calculations are shown below.

$$\hat{p} = \frac{o}{n} \quad (2)$$

$$\hat{v}_i = \frac{\hat{p}_i(1-\hat{p}_i)}{n_i} \quad (3)$$

$$w_i = \frac{\frac{1}{v_i}}{\sum_{j=1}^{n} \frac{1}{v_j}} \quad (4)$$

- $\hat{p}$: sample proportion
- o: number of occurrences
- n: number of trips
- $\hat{v}$: sample variance
- w: weight

As mentioned above, for cases in which the hybrid reporting rate was 100%, a single negative instance was added to enable the calculation of inverse of the variance. If this was not done the weight would then be infinite and the entire group would be rated at 100%. This was clearly inappropriate; so by adding a single negative case, areas with many surveys (high confidence areas) were only minimally affected and areas with few surveys (low confidence areas) had their impact appropriately reduced. In this case the alternative formula below was used.

$$\hat{p}' = \frac{o+0}{n+1} \quad (5)$$

In the opposite case where the reporting rate was zero, no variance could be calculated. For the purposes of calculating the variance and weights, a single positive case was added. Again in cases with high confidence (many surveys) the impact would be negligible and in areas with low confidence (few surveys) the influence would be greater. The formula in this case would be as follows.

$$\hat{p}'' = \frac{o+1}{n+1} \quad (6)$$

With reporting rates of 100% and 0%, the sample variance was calculated using $\hat{p}'$ or $\hat{p}''$ and dividing by $n_i + 1$ as appropriate. Together these adjustments lead to the final formula in which *hrr* is the hybrid reporting rate.

$$varr = \sum_{j=1}^{n} w_j hrr \quad (7)$$

The result of this rather complex calculation is that we were able to use the vast majority of the data, including incidental records, to construct a map of a species' relative abundance, taking into consideration the level of confidence with which values can be reported.

Coverage Index

The *varr* was used to account for differential coverage in calculating group reporting rates or in Natural Region reporting rates. For the reporting rate map, however, we needed a simple method to adjust for differential effort between groups. In planning the coverage goals for Atlas 2, we set a target of 10–20% coverage for northern areas. We thus applied a simple filter to all reporting rate values for mapping purposes based on a coverage goal of 15%. This was used primarily to filter extreme values from northern reporting rate group areas with extremely sparse data. In the south, our square coverage values were predominantly in the 25–50% range. Thus, the 15% coverage target had no impact on those areas. In areas with square coverage below 15%, the square coverage was divided by 15 as an indication of our confidence in the *varr* calculated for that group. This confidence factor was then multiplied by the species' *varr* values for that group prior to mapping to prevent over-reporting in areas of low survey coverage.

As mentioned in the description of the coverage index, reporting rate is a proxy for abundance based on occupancy. As such, it will be less sensitive to subtle changes in abundance change than a true abundance measure. However, because abundance data were not gathered in standardized fashion, reporting rate is more reliable, and less likely to produce false negatives. The resulting reporting rate map will show the relative abundance of species corrected by the measure of effort. This map will underestimate areas with low survey effort because a reliable measure cannot be calculated in those areas.

This same technique was applied to the production of the monthly reporting rate graphs. A value of 500 surveys per month (100 surveys / month / year) was used as the monthly coverage index analogue for scaling low survey effort month *varr* values.

Change Analysis

One of the objectives of Atlas 2 was to determine if the relative abundance and observed distribution of bird species had changed in the intervening 20 years (1985–2005) between Atlases. All change analysis methods for comparing one Atlas data set to another are based on differences in occurrence. In statistical terms the measures employed are binomial (having two possible values: observed or not observed) and the measures of occurrence are sample proportions. Therefore, tests between the two are based on binomial sampling and testing the difference between two independent proportions. We must remember, however, that relying on occupancy as a measure of abundance is likely to be most effective for species that vary significantly in both occupancy and abundance (Gaston, 1999). Therefore, changes observed for species not fitting these criteria must be more carefully considered.

Figure 12: Reporting rate base map showing numbers of general surveys during Atlas 2. Incidental data are excluded. Black outlines are reporting rate groups and grey outlines are reporting rate subgroups.

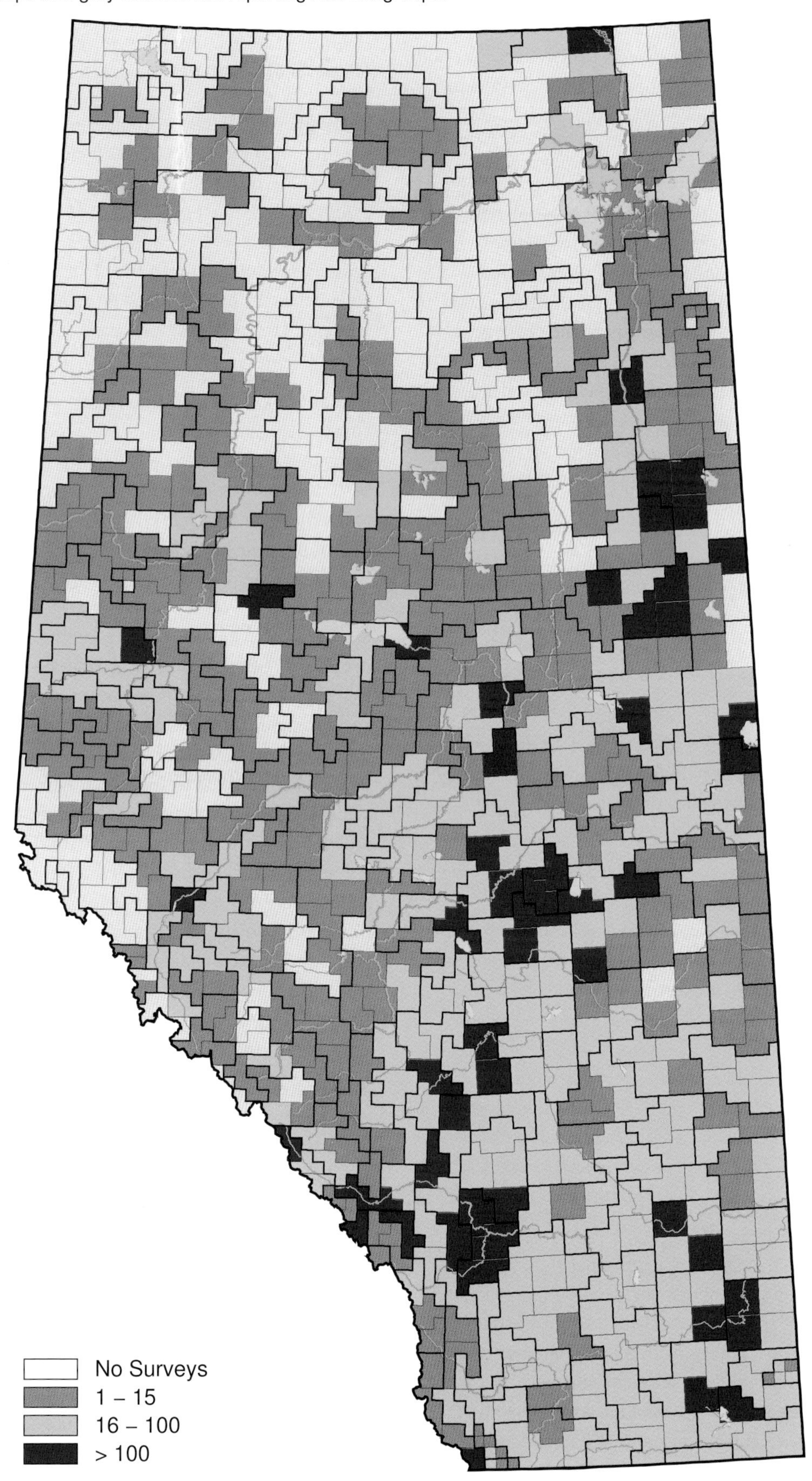

Traditionally this comparison has been made using only squares classified as completely surveyed, where completeness is defined by a certain number of survey hours. Statistical methods are often employed to correct for differential levels of effort and, usually, only matched squares (fully covered in both projects) are used. Unfortunately, this inevitably eliminates large proportions of data from consideration.

The underlying assumption of change analysis using complete squares is that broad scale occupancy is an indicator of higher abundance than limited occupancy. The relationship between abundance and occupancy is generally positive for most species (Gaston, 1999; Gaston et al., 2000; McGeoch and Gaston, 2002). Using the BBS dataset, Bart and Klosiewski (1989) showed a strong positive correspondence between occupancy and abundance trends. Dunn et al. (1996) demonstrated the same relationship using checklist data.

For Atlas 2, there were two additional factors to consider. First, the study designs of Atlas 1 and Atlas 2 were not the same. Second, with the incorporation of compatible external data sources, the coverage of the landscape in Atlas 2 was considerably greater.

One of the external sources used to provide better habitat selection information and more complete distribution maps was the BBS. The BBS database was excluded from the change analysis calculations to allow it to be used as an external corroborating data source.

In Atlas 1 observers were asked to keep track of observed birds by year and square. Data from both Atlases were combined into yearly square lists to enable comparisons between comparable data sets.

Change analysis based on occurrence has two weaknesses. First, as noted by Gaston et al. (2000), although there is a positive occupancy-abundance relationship for many species, this is not the case with all bird species. Second, if there is a positive relationship, then changes in occupancy identified using this method will likely indicate relatively large changes in abundance, not subtle ones. Therefore, the direction of the occupancy, and presumably abundance, changes detected may be accurate, but the magnitude of the change, may be quite different from that detected by abundance-based measures. As a result, in the species accounts, no reference is made to the magnitude of the detected change, only the direction of the change.

The underlying mechanisms driving occupancy-abundance relationships are not well known, but there is support for habitat and environmental heterogeneity being important (McGeoch and Gaston, 2002). In all cases, conclusions drawn from this analysis must be corroborated using an abundance-based measure. This message was echoed by Dunn et al. (1996) using checklist data and by Chamberlain and Fuller (2001) in their studies of farmland birds in Britain. We relied on the BBS trends and provincial status reports as independent measures of change during the last 20 years.

Methods of Comparison

The biggest challenge to comparing the two Atlas projects was to ensure that the comparisons were made between like objects. For example, if the coverage of habitats were more extensive in the first project than in the second, the species representations would not be equal, and observed increases or decreases would be functions of survey completeness, not actual changes in bird abundance and distribution. Various methods exist to adjust for these sources of variation.

Percent of Completed Squares. A common type of comparative analysis requires that the landscape units being compared must have received complete survey coverage in both projects. However, the central issue in the analysis is how to define what complete coverage is. As stated above, the traditional measure is hours of survey effort. Yet, it is important to realize that due to variation in observer skill, time of day, time of year, coverage of all the habitats in the square, and field behaviour, an hour spent on one survey will not necessarily be equivalent to an hour on another survey. Because we know that the relationship between effort and species identification is a case of diminishing returns, the logical approach would be to identify level of effort, either by general habitat class or overall, at which the rate of species accumulation approaches zero. Additional to this measure, the amount of "noise" involved must also be considered. Although it can be overly conservative (van Belle, 2002), overlapping standard error is commonly employed to indicate when the difference between the "signal" and the "noise" can no longer be discriminated. The rate of species accumulation approaches zero as the "noise" overwhelms the "signal", and we calculated that the "signal" was extinguished after 15–20 hours of observer effort.

Since we lacked effort information for a significant number of records in Atlas 1 and Atlas 2, this measure was used subsequently to define a percent of theoretical maximum number of species found in surveys with 20 hours of effort. Using known species ranges, a theoretical count was generated for each square. It was recognized that this number would be inflated in most cases due to habitat heterogeneity. An average for each Natural Region was calculated for Atlas 1 and Atlas 2 and the mean of those values by Natural Region was used to classify squares as either complete or incomplete.

Although this approach allowed for a common definition of completeness to be calculated in both Atlas projects, it also created a situation in which many changes were not likely to be detected. If there had been a broad-scale decline of species occurrence between Atlas 1 and Atlas 2, it is unlikely to have been detected by defining completeness in this way, because some squares with equal effective effort, but showing a genuine decline in species abundance, would not be assessed. With that limitation recognized, declines identified by this measure had a high probability of being real and not artifactual.

Variance Adjusted Reporting Rate. The second measure of change we used was the *varr* based on yearly square summary species lists from both Atlases. The *varr* was calculated for both Atlases for each species on a Natural Region basis. These regional values were then treated as two independent proportions and the significance of the differences between them was tested. In this way the majority of the data from both Atlas projects could be used. As noted by Dunn et al. (1996), reporting rate trend analysis results tend to have a positive bias, so we could be confident that few false negatives would be reported.

The results of the change analysis and changes in the observed distributions are the central focus of the species account text as opposed to graphical presentation that would be more susceptible to misinterpretation. As stated above, in all cases results were compared to external data sources for corroboration.

Occurrence Index. In an attempt to better understand the different data sets, we calculated the relative rarity of each species in both Atlas 1 and Atlas 2.

Radford and Bennett (2005) outlined an index of species occurrence which combined reporting rate and breadth of distribution to identify rare, common, widespread, and restricted species. Because their index is an indicator of a species' relative abundance in relation to the species assemblage with which it is found, it is a good measure of the relative rarity within different habitat classes. This quality makes it a valuable measure of possible survey bias differences between the Atlas 1 and Atlas 2.

Radford and Bennett (2005) corrected for extreme variance in survey coverage by randomly sampling a subset of their spatial units. Our approach was to use the variance adjusted reporting rate described on page 39 to adjust for variance in survey effort.

Consider a simplified example. Bird "A" was found in two squares. In the first square it was detected in 3 of 10 visits, or 0.3. On those ten lists the sum of the reporting rates for all other species forms the denominator. If the number of species in a list is large, the sum of their reporting rates will be similarly larger, resulting in fairly low occurrence index scores in landscape unit 1. If in landscape unit 2, Bird "A" is detected in 3 of 10 surveys again, but the overall species abundance is lower, then the occurrence index score of Bird "A" in landscape unit 2 will be higher. This variable ranking is by design, because species detected in areas with low survey effort are likely to be more locally abundant than those that are detected only with high effort. The formula is detailed below.

$$oi = \sum_{i=1}^{c} \frac{varr_{ji}}{\sum_{j=1}^{k} varr_{ji}} \tag{8}$$

- oi: occurrence index
- $varr$: variance adjusted reporting rate per landscape unit
- c: number of landscape units
- i: landscape unit
- k: number of species within landscape unit i
- j: species

The occurrence index was calculated for each Natural Region. After the occurrence index was calculated for all species, individual species scores were then scaled from 0 to 100.

This calculation provided a single metric which described both density and distribution. This index would not have been an effective measure in areas of very low effort, where not all of the commonly occurring species are detected. However, since it was calculated using the variance adjusted reporting rate across an entire Natural Region, this limitation was not a concern. The index is incapable of distinguishing between rare and difficult-to-detect species.

The occurrence index indicates the relative abundance and distribution of a species in relation to all other species found in that area. These values were used to produce a table referenced by authors in writing species accounts and reported in the accounts as the frequency of observation in Atlas 1 and 2. These figures were used to determine whether a given species' relative detectability had changed from Atlas 1 to Atlas 2. The index would help to identify potential sources of bias to be considered in examining the change analysis results.

Testing for Significance

All change analysis methods are based on binomial sampling and testing the difference between two independent proportions. For the tests, we assumed that the sample probability of occurrence $\hat{p}$ is an indication of changes in abundance. Therefore, the standard deviation of the sampling distribution is based upon a weighted average of the two proportions under the null hypothesis that they are equal (Kachigan, 1986).

$$\hat{p}_w = \frac{n_1\hat{p}_1 + n_2\hat{p}_2}{n_1 + n_2} \quad (9)$$

$$\sigma_{\hat{p}_1-\hat{p}_2} = \sqrt{\hat{p}_w\hat{q}_w(\frac{1}{n_1} + \frac{1}{n_2})} \quad (10)$$

- $\hat{p}_w$: the weighted average of the sample estimates
- $\hat{q}_w$: $1 - \hat{p}_w$ the probability of failure
- $\hat{p}_1$: the probability estimate for Atlas 1
- n_1: the sample size for Atlas 1
- $\hat{p}_2$: the probability estimate for Atlas 2
- n_2: the sample size for Atlas 2
- $\sigma_{\hat{p}_1-\hat{p}_2}$: the standard deviation of the difference between the proportions

Since we are interested in knowing not only if the difference between the two is significant, but also what the direction of the difference is (increase or decrease), a single-tailed test was applied (Wagner, 1992).

$$z = \frac{\hat{p}_1 - \hat{p}_2}{\sigma_{\hat{p}_1-\hat{p}_2}} \quad (11)$$

To be able to assume that the sample distribution is normal, the general guideline is that the values of $n\hat{p}$ and $n(1-\hat{p})$ must be greater than, or equal to, 5.0. When this requirement was not met, the results was treated as not significant. In cases where this requirement was met, we treated the results as significant for $\alpha = 0.05$ and which has the corresponding z value of 1.645 or greater for increases and -1.645 or less for decreases.

Data Processing / Software

Prior to the start of Atlas 2, FAN had initiated the Alberta Birdlist program and was distributing, free of charge, the Personal Birdlist Software. Many atlassers entered their observations at home and e-mailed their data to FAN. Others used conventional data cards and mailed them. Once the data were received at the FAN office, they were imported or entered into the central database for data checking and verification. After the completion of the field program, these data were then imported into PostgreSQL 8.1.3 (www.postgresql.org) and spatial layers were imported into PostGIS 1.1.2 (postgis.refractions.net).

Double checking of the data was conducted and imports of swapped data were conducted. Duplicate locations, observer names, and other potential sources of corruption were carefully examined to ensure the integrity of the data. After checking of the data was complete, SQL views and derivative tables were created to speed the analytical and book-production processes.

Because of the large number of species, the analytical methods, map and graph production, as well as the layout were automated to both save time and reduce the likelihood of human error. These processes were scripted using Python 2.4 (www.python.org) and the PyGreSQL library (www.pygresql.org). Maps were produced using Generic Map Tools 4.1 (gmt.soest.hawaii.edu) and graphs were produced using Ploticus software (ploticus.sf.net). LaTeX 2_ε was used to layout this publication.

Atlas Highlights

Introduction

Atlas 2 benefitted from the efforts of almost 1700 observers, most of whom were volunteers. Atlas 2 also was the recipient of large volumes of shared data from both government agencies and private companies. This data sharing enabled us to better focus our volunteer efforts. In Atlas 1, 23.5% of the 6,624 squares in Alberta had at least some survey coverage and in Atlas 2, this figure increased to 29.5% (see Table 9 in Methods chapter for details). Of all Alberta's Atlas squares, 12% were surveyed in both Atlasses. The end result was two complementary data sets for assessing changes in distribution and relative abundance over the period of 1987–2005.

This chapter provides some interesting highlights from the Atlas data as well as a summary of the Atlas results. Lastly, we include a list of unusual sightings that occurred during the period of Atlas 2.

Table 13: Atlas effort by year

Year	Trips	Records	Observers	Hours
2000	5535	54452	320	4213
2001	7490	61406	520	6569
2002	4606	76169	546	8205
2003	6398	89874	634	9393
2004	6390	88293	641	10209
2005	7362	91081	386	7266
Total	37781	461275	1695	45858

Figure 13: Effort over time (percent of total trips by month)

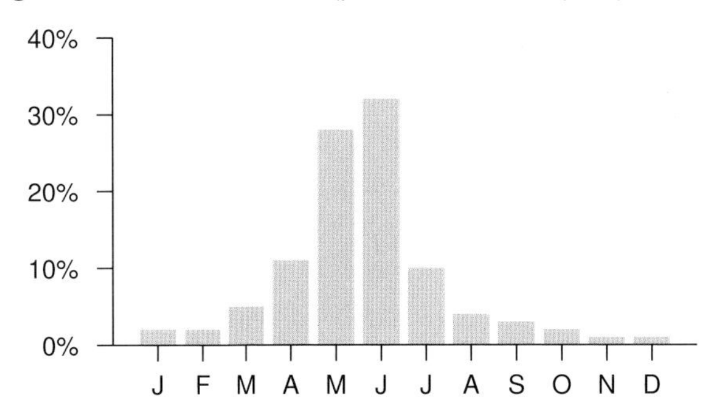

Table 14: Squares with the most species detected

Square Name (NAD27)	Location	Number of species
QG05	Calgary	250
VL11	Taber	221
WR53	Cold Lake	220
VR37	Lac La Biche	218
VL22	NE of Taber	207
PG34	Elbow-Sheep Wildland Park	203
VM30	Brooks	200
QG06	Calgary	198
VL38	Lac La Biche	197
UL60	Lethbridge	195

Table 15: Squares with the most species breeding detected

Square Name (NAD27)	Location	Number of species
VR37	Lac La Biche	134
WR53	Cold Lake	125
WR50	Sours Lake (South of Cold Lake)	124
PH76	SW of Dickson	122
VL22	NE of Taber	122
PG82	NW of Black Diamond	119
VL11	Taber	119
WR23	East of LaCorey	118
QG05	Calgary	116
UR38	Crooked Lake	116

Table 16: Species detected in the most squares

Species Name	Number of squares
American Robin	916
Mallard	824
Red-winged Blackbird	788
Chipping Sparrow	760
Red-tailed Hawk	674
Canada Goose	661
Yellow Warbler	644
Blue-winged Teal	640
Clay-colored Sparrow	635
American Crow	627

Figure 14: Species Diversity (red: > 100, yellow: 51–100, orange: 26–50, beige: 1–25, grey: no data)

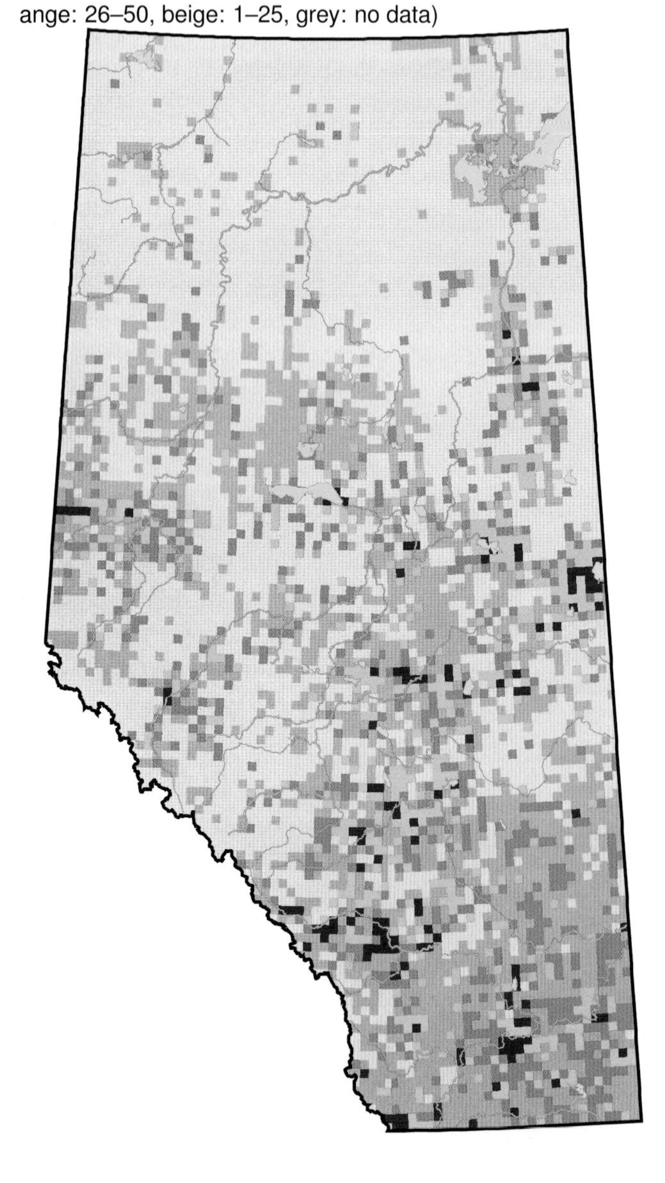

Distribution and Abundance

The National Audubon Society released a report in June, 2007 that listed twenty common bird species that have lost more than half their populations since 1967. The report is based on the analysis of forty year's of bird population data collected by Audubon's annual Breeding Bird Surveys and Christmas Bird Counts. Of the 20 species listed, 12 are found in Alberta and the Alberta Bird Atlas has confirmed a similar decline in abundance for each of these species in at least one Natural Region in the province. The Alberta species include the American Bittern, Boreal Chickadee, Common Grackle, Common Tern, Evening Grosbeak, Grasshopper Sparrow, Horned Lark, Lark Sparrow, Loggerhead Shrike, Northern Pintail, Ruffed Grouse, and Rufous Hummingbird.

Results from comparisons between Atlas 1 and Atlas 2 are summarized in the tables below.

Table 17: Changes in Observed Distribution

Type of Change	Number of Species
Expanding	26
Possibly Expanding	10
Contracting	10
Possibly Contracting	13
Mixed	1
Possibly Mixed	2
Unchanged	199
Possibly Unchanged	11

Table 18: Changes in Relative Abundance

Type of Change	Number of Species
Increasing	51
Possibly Increasing	18
Decreasing	46
Possibly Decreasing	20
Mixed	56
Possibly Mixed	26
Unchanged	27
Possibly Unchanged	25

What follows is the list of each species in taxonomic order with a summary of the Atlas results given as observed distribution / relative abundance. Observed distribution changes are classified as unchanged, expanding, contracting, or mixed. Mixed changes in observed distribution refers to cases where a species' distribution is expanding in some areas and contracting in others. Relative abundance is classified as stable, increasing, decreasing, or mixed. Mixed in terms of relative abundance refers to situations where increases were detected in one or more Natural Regions and decreases were detected in one or more different Natural Regions. In all cases, where the results are unclear due to lack of corroborative evidence, uncertain biological causes, etc., a question mark is added to indicate the lack of certainty associated with this result.

Common Name . . Observed Distribution / Relative Abundance

Ducks, Geese, and Swans

Canada Goose . . . . unchanged / increasing
Trumpeter Swan . . . . unchanged / increasing
Wood Duck . . . . .expanding / increasing
Gadwall . . . . .expanding / increasing
American Wigeon . . . . unchanged / mixed
Mallard . . . . .unchanged / mixed
Blue-winged Teal . . . . .unchanged / mixed
Cinnamon Teal . . . . .unchanged / mixed
Northern Shoveler . . . . unchanged / mixed
Northern Pintail . . . . unchanged / mixed
Green-winged Teal . . . . unchanged / decreasing
Canvasback . . . . unchanged / mixed
Redhead . . . . unchanged / increasing
Ring-necked Duck . . . . expanding / mixed
Lesser Scaup . . . . .unchanged / mixed
Harlequin Duck . . . . unchanged / stable
Surf Scoter . . . . expanding / stable?
White-winged Scoter . . . . unchanged / decreasing
Bufflehead . . . . .unchanged / mixed
Common Goldeneye . . . . expanding? / mixed
Barrow's Goldeneye . . . . unchanged / increasing
Hooded Merganser . . . . .expanding / increasing?
Common Merganser . . . . unchanged / mixed
Red-breasted Merganser . . . . expanding / increasing
Ruddy Duck . . . . unchanged / mixed?

Pheasants, Grouse, Ptarmigan, and Turkey

Gray Partridge . . . . unchanged / decreasing?
Ring-necked Pheasant . . . . expanding / decreasing
Ruffed Grouse . . . . unchanged / decreasing?
Greater Sage-Grouse . . . . unchanged / stable
Spruce Grouse . . . . unchanged / decreasing
Willow Ptarmigan . . . . unchanged? / stable?
White-tailed Ptarmigan . . . . unchanged / stable
Dusky Grouse . . . . unchanged / stable
Sharp-tailed Grouse . . . . unchanged / mixed?
Wild Turkey . . . . .unchanged / stable

Loons and Grebes

Red-throated Loon . . . . .unchanged / stable?
Pacific Loon . . . . unchanged? / stable?
Common Loon . . . . unchanged / increasing
Pied-billed Grebe . . . . .unchanged / mixed
Horned Grebe . . . . contracting / mixed
Red-necked Grebe . . . . unchanged / mixed
Eared Grebe . . . . .expanding / mixed
Western Grebe . . . . unchanged / mixed
Clark's Grebe . . . . .expanding / stable

Pelicans, Cormorants, Herons, Egrets, Ibises, and New World Vultures

American White Pelican . . . . .unchanged / increasing
Double-crested Cormorant . . . . .unchanged / increasing
American Bittern . . . . .unchanged? / decreasing
Great Blue Heron . . . . .unchanged / mixed?
Black-crowned Night-Heron . . . . .unchanged / decreasing
White-faced Ibis . . . . expanding / stable
Turkey Vulture . . . . .expanding / increasing

Hawks, Eagles, and Falcons

Osprey . . . . .unchanged / increasing
Bald Eagle . . . . .unchanged / increasing
Northern Harrier . . . . unchanged / mixed
Sharp-shinned Hawk .unchanged / mixed?
Cooper's Hawk . . . . unchanged / decreasing?
Northern Goshawk . . . . unchanged / mixed
Broad-winged Hawk .unchanged / mixed?
Swainson's Hawk . . . . unchanged / decreasing
Red-tailed Hawk . . . . unchanged / mixed
Ferruginous Hawk . . . . unchanged / increasing
Golden Eagle . . . . .unchanged / increasing
American Kestrel .unchanged / decreasing
Merlin . . . . .unchanged / mixed?
Peregrine Falcon . . . . unchanged / mixed?
Prairie Falcon . . . . unchanged / increasing

Rails, Coots, and Cranes

Yellow Rail . . . . .expanding? / stable?
Virginia Rail . . . . .unchanged / decreasing?
Sora . . . . .unchanged / mixed?
American Coot . . . . contracting? / mixed
Sandhill Crane . . . . unchanged / mixed?
Whooping Crane . . . . unchanged / stable?

Plovers, Sandpipers, Phalaropes, Jaegers, Gulls, and Terns

Semipalmated Plover . . . . unchanged / mixed?
Piping Plover . . . . .unchanged / stable?
Killdeer . . . . .contracting / decreasing
Mountain Plover . . . . .unchanged / stable?
Black-necked Stilt . . . . unchanged / increasing
American Avocet . . . . unchanged / stable
Spotted Sandpiper . . . . unchanged / decreasing
Solitary Sandpiper . . . . unchanged / stable?
Greater Yellowlegs . . . . .unchanged / increasing?
Willet . . . . unchanged / decreasing?

Lesser Yellowlegs .... unchanged / decreasing
Upland Sandpiper .... unchanged / decreasing
Long-billed Curlew .... unchanged / increasing
Marbled Godwit .... unchanged / decreasing
Least Sandpiper .... unchanged / mixed?
Short-billed Dowitcher .... unchanged / decreasing?
Wilson's Snipe .... unchanged / mixed?
Wilson's Phalarope .... contracting? / mixed
Red-necked Phalarope .... unchanged / mixed?
Franklin's Gull .... unchanged / mixed
Bonaparte's Gull .... unchanged / mixed
Mew Gull .... unchanged / decreasing?
Ring-billed Gull .... expanding? / mixed?
California Gull .... expanding? / mixed?
Herring Gull .... expanding / decreasing
Caspian Tern .... mixed? / increasing?
Black Tern .... unchanged / decreasing
Common Tern .... unchanged / mixed
Forster's Tern .... unchanged / mixed

Pigeons, Doves, and Cuckoos

Rock Pigeon .... unchanged / mixed?
Mourning Dove .... mixed / mixed
Black-billed Cuckoo .... contracting / decreasing?

Owls

Great Horned Owl .... unchanged / decreasing?
Northern Hawk Owl .... unchanged / decreasing
Northern Pygmy-Owl .... expanding? / stable
Burrowing Owl .... contracting / stable?
Barred Owl .... unchanged / increasing
Great Gray Owl .... unchanged / mixed?
Long-eared Owl .... expanding / decreasing?
Short-eared Owl .... unchanged / decreasing
Boreal Owl .... unchanged / increasing
Northern Saw-whet Owl .... unchanged / mixed?

Nightjars, Swifts, and Hummingbirds

Common Nighthawk .... unchanged / decreasing
Common Poorwill .... unchanged / stable?
Black Swift .... unchanged? / stable
Ruby-throated Hummingbird .... unchanged / decreasing?
Calliope Hummingbird .... unchanged / stable
Rufous Hummingbird .... unchanged / increasing

Kingfishers and Woodpeckers

Belted Kingfisher .... unchanged / decreasing
Yellow-bellied Sapsucker .... unchanged / mixed
Red-naped Sapsucker .... expanding? / increasing
Downy Woodpecker .... unchanged / mixed
Hairy Woodpecker .... unchanged / mixed
American Three-toed Woodpecker .... unchanged / increasing
Black-backed Woodpecker .... unchanged / increasing
Northern Flicker .... unchanged / decreasing
Pileated Woodpecker .... unchanged / mixed?

Flycatchers

Olive-sided Flycatcher .... unchanged / decreasing?
Western Wood-Pewee .... unchanged / decreasing
Yellow-bellied Flycatcher .... unchanged / increasing
Alder Flycatcher .... unchanged / mixed
Willow Flycatcher .... unchanged / stable
Least Flycatcher .... unchanged / mixed
Hammond's Flycatcher .... unchanged / stable
Dusky Flycatcher .... unchanged / increasing
Cordilleran Flycatcher .... unchanged? / decreasing?
Eastern Phoebe .... unchanged / decreasing
Say's Phoebe .... unchanged / stable?
Great Crested Flycatcher .... unchanged / stable?
Western Kingbird .... unchanged / decreasing
Eastern Kingbird .... unchanged / decreasing

Shrikes and Vireos

Loggerhead Shrike .... contracting / decreasing?
Northern Shrike .... unchanged? / mixed?
Cassin's Vireo .... unchanged? / increasing
Blue-headed Vireo .... unchanged / increasing
Warbling Vireo .... unchanged / mixed
Philadelphia Vireo .... unchanged / stable?
Red-eyed Vireo .... unchanged / decreasing

Jays, Crows, and Allies

Gray Jay .... unchanged / increasing?
Steller's Jay .... expanding / increasing
Blue Jay .... expanding / mixed
Clark's Nutcracker .... unchanged / increasing?
Black-billed Magpie .... unchanged / mixed
American Crow .... unchanged / mixed
Common Raven .... expanding / mixed

Larks, Swallows, and Chickadees

Horned Lark .... unchanged / mixed
Purple Martin .... expanding / decreasing
Tree Swallow .... unchanged / mixed
Violet-green Swallow .... unchanged / increasing
Northern Rough-winged Swallow .... unchanged / mixed?
Bank Swallow .... contracting? / decreasing
Cliff Swallow .... unchanged / decreasing
Barn Swallow .... unchanged / decreasing
Black-capped Chickadee .... unchanged / mixed
Mountain Chickadee .... unchanged / increasing
Boreal Chickadee .... unchanged / increasing?

Nuthatches, Creepers, Wrens, and Dippers

Red-breasted Nuthatch .... contracting? / increasing
White-breasted Nuthatch .... unchanged / increasing

Brown Creeper . . . . unchanged / increasing
Rock Wren . . . . contracting? / stable
House Wren . . . . unchanged / mixed
Winter Wren . . . . unchanged / increasing?
Sedge Wren . . . . unchanged / decreasing
Marsh Wren . . . . unchanged / mixed
American Dipper . . . . unchanged / stable

Kinglets, Bluebirds, and Thrushes

Golden-crowned Kinglet . . . . expanding? / increasing?
Ruby-crowned Kinglet . . . . unchanged / decreasing?
Eastern Bluebird . . . . unchanged / stable?
Western Bluebird . . . . unchanged? / stable?
Mountain Bluebird . . . . unchanged / decreasing?
Townsend's Solitaire . . . . unchanged / mixed?
Veery . . . . unchanged? / mixed?
Gray-cheeked Thrush . . . . unchanged / stable?
Swainson's Thrush . . . . unchanged / increasing?
Hermit Thrush . . . . expanding / mixed?
American Robin . . . . unchanged / decreasing
Varied Thrush . . . . expanding / stable

Mockingbirds and Thrashers

Gray Catbird . . . . unchanged / mixed
Northern Mockingbird . . . . contracting? / stable
Sage Thrasher . . . . unchanged / stable
Brown Thrasher . . . . unchanged / decreasing

Starlings, Pipits, and Waxwings

European Starling . . . . contracting / decreasing
American Pipit . . . . unchanged / mixed
Sprague's Pipit . . . . unchanged / mixed
Bohemian Waxwing . . . . unchanged / mixed
Cedar Waxwing . . . . unchanged / decreasing

Wood-Warblers, Tanagers, Sparrows, Grosbeaks, and Buntings

Tennessee Warbler . . . . unchanged / increasing
Orange-crowned Warbler . . . . unchanged / increasing
Nashville Warbler . . . . expanding / stable
Yellow Warbler . . . . unchanged / mixed
Chestnut-sided Warbler . . . . expanding / increasing
Magnolia Warbler . . . . unchanged / increasing?
Cape May Warbler . . . . unchanged / increasing
Yellow-rumped Warbler . . . . unchanged / increasing
Black-throated Green Warbler . . . . unchanged / increasing?
Townsend's Warbler . . . . expanding / increasing
Blackburnian Warbler . . . . expanding / stable
Palm Warbler . . . . unchanged / increasing
Bay-breasted Warbler . . . . expanding / increasing
Blackpoll Warbler . . . . contracting? / increasing?
Black-and-white Warbler . . . . unchanged / increasing
American Redstart . . . . expanding? / increasing
Ovenbird . . . . unchanged / increasing
Northern Waterthrush . . . . unchanged / increasing
Connecticut Warbler . . . . unchanged / stable
Mourning Warbler . . . . expanding? / increasing
MacGillivray's Warbler . . . . expanding / stable
Common Yellowthroat . . . . unchanged / decreasing
Wilson's Warbler . . . . contracting? / increasing?
Canada Warbler . . . . unchanged / increasing
Yellow-breasted Chat . . . . unchanged / stable?
Western Tanager . . . . unchanged / mixed?
Spotted Towhee . . . . unchanged / stable
American Tree Sparrow . . . . unchanged / mixed
Chipping Sparrow . . . . unchanged / decreasing?
Clay-colored Sparrow . . . . unchanged / decreasing
Brewer's Sparrow . . . . unchanged / stable?
Vesper Sparrow . . . . unchanged / decreasing
Lark Sparrow . . . . unchanged / decreasing?
Lark Bunting . . . . contracting? / decreasing?
Savannah Sparrow . . . . unchanged / mixed
Grasshopper Sparrow . . . . unchanged / increasing?
Baird's Sparrow . . . . unchanged / decreasing
Le Conte's Sparrow . . . . unchanged / decreasing
Nelson's Sharp-tailed Sparrow . . . . contracting? / stable
Fox Sparrow . . . . unchanged / decreasing?
Song Sparrow . . . . unchanged / mixed?
Lincoln's Sparrow . . . . unchanged / mixed
Swamp Sparrow . . . . unchanged / increasing?
White-throated Sparrow . . . . unchanged / mixed
White-crowned Sparrow . . . . unchanged? / mixed?
Golden-crowned Sparrow . . . . unchanged / stable?
Dark-eyed Junco . . . . contracting? / decreasing
McCown's Longspur . . . . unchanged / increasing
Chestnut-collared Longspur . . . . contracting / stable?
Rose-breasted Grosbeak . . . . unchanged / mixed?
Black-headed Grosbeak . . . . unchanged / stable
Lazuli Bunting . . . . contracting? / increasing

Blackbirds and Allies, Finches and Weaver Finches

Bobolink . . . . contracting / stable
Red-winged Blackbird . . . . unchanged / mixed
Western Meadowlark . . . . unchanged / mixed?
Yellow-headed Blackbird . . . . unchanged / mixed
Rusty Blackbird . . . . unchanged / decreasing
Brewer's Blackbird . . . . contracting / mixed
Common Grackle . . . . unchanged / mixed
Brown-headed Cowbird . . . . expanding? / decreasing
Baltimore Oriole . . . . mixed? / decreasing
Bullock's Oriole . . . . unchanged? / stable?
Gray-crowned Rosy-Finch . . . . unchanged / stable?
Pine Grosbeak . . . . contracting? / increasing?
Purple Finch . . . . unchanged / decreasing

Cassin's Finch .... unchanged / stable?
House Finch .... expanding / stable?
Red Crossbill .... unchanged / stable
White-winged Crossbill .... unchanged / increasing?
Common Redpoll .... unchanged / increasing
Pine Siskin .... unchanged / increasing?
American Goldfinch .... unchanged / decreasing
Evening Grosbeak .... unchanged / decreasing
House Sparrow .... contracting / mixed

Rarities

These are the noteworthy sightings of rarities that took place during Atlas 2 (2000-2005) as compiled by the Alberta Bird Record Committee.

Provincial Firsts (br — breeding)

- Little Stint (*Calidris minuta*) — Bonnyville, September 2000
- Slaty-backed Gull (*Larus schistisagus*) — Calgary, September 2000
- Eurasian Collared-Dove (*Streptopelia decaocto*) (br) — Red Deer, June - July 2002
- Flammulated Owl (*Otus flammeolus*) — Edmonton, Fall 2000
- Yellow-throated Vireo (*Vireo flavifrons*) — Calgary, August 2003
- Siberian Accentor (*Prunella montanella*) — Calgary, March 2002
- Black-throated Gray Warbler (*Dendroica nigrescens*) — Jenner, May 2001
- Hermit Warbler (*Dendroica occidentalis*) — Elk Island National Park, May 2002
- Field Sparrow (*Spizella pusilla*) — Kananaskis, June 2000

Species with less than five sightings (first sighting during Atlas)

- King Eider (*Somateria spectabilis*) — Banff National Park, December 2005
- Common Eider (*Somateria mollissima*) — Cold Lake, December 2001
- Little Blue Heron (*Egretta caerulea*) — Kininvie Marsh, May 2004
- Great Black-backed Gull (*Larus marinus*) — Grande Prairie, March 2005
- Ivory Gull (*Pagophila eburnea*) — McLennan, June 2004
- Long-billed Murrelet (*Brachyramphus perdix*) — Hinton–Jasper, August 2005
- White-winged Dove (*Zenaida asiatica*) — Lacombe, September - October 2003
- Yellow-billed Cuckoo (*Coccyzus americanus*) — Edmonton, April 2000
- Barn Owl (*Tyto alba*) — Bashaw, December - January 2000
- Red-breasted Sapsucker (*Sphyrapicus ruber*) Bottrel, April 2000
- Gray Flycatcher (*Empidonax wrightii*) — Mount Lorette, July 2005
- Wood Thrush (*Hylocichla mustelina*) — Ralston, May 2004
- Blue-winged Warbler (*Vermivora pinus*) — Vegreville, May 2000
- Hooded Warbler (*Wilsonia citrina*) — Banff, September - October 2000
- Summer Tanager (*Piranga rubra*) — Brown-Lowery Provincial Park, June 2001
- Green-tailed Towhee (*Pipilo chlorurus*) — Brooks, May 2000
- Painted Bunting (*Passerina ciris*) — Bowden, July 2000
- Dickcissel (*Spiza americana*) — Drayton Valley, May 2000

Species with less than twelve sightings (first sighting during Atlas)

- Green Heron (*Butorides virescens*) — Lethbridge, June 2000
- Long-tailed Jaeger (*Stercorarius longicaudus*) — Beaverlodge, August 2000
- Little Gull (*Larus minutus*) — Irricana, May 2002

- Lesser Black-backed Gull (*Larus fuscus*) — Calgary, April 2000
- Anna's Hummingbird (*Calypte anna*) — Lethbridge, October 2002
- Northern Parula (*Parula americana*) — Calgary, September 2000

First recorded breeding in Alberta

- Great Egret (*Ardea alba*) — St. Albert, June 2005
- Red-headed Woodpecker (*Melanerpes erythrocephalus*) — Red Deer, November 2004

How to Use this Book

Overview

As with all bird Atlas projects, the core material of this book is the species accounts. The chapters preceeding the species accounts provide background in methods, Alberta's Natural Regions, and changes that have occurred in Alberta since the start of Atlas 1 until the completion of Atlas 2. These materials are supplemental and are provided for readers who wish to understand the reasoning behind the presentation of the individual species material, as well as how Alberta has changed since 1987 and how this might have affected Alberta birds.

In Atlas 1, there was no baseline for comparison. So, like many other Atlasses, each species account provided a great deal of natural history information. The impetus to initiate Atlas 2, after only 15 years had passed, was due in part to the rapid rate of industrial development in Alberta. As an organization of naturalists and convervationists, FAN was concerned about how these changes might affect Alberta birds and decided that initiating a second Atlas project would be an important exercise in assessing whether Alberta's bird populations were indeed changing, as was believed. As a result the information presented for each species is designed to address this question.

Species Account Elements

Natural History information

It was decided to not reiterate the natural history information provided in the first Atlas, but instead provide a summary of this information in a table providing nesting details and a series of symbols on the top right page for each species account. Natural history details for each species were assembled referencing Atlas 1 (Semenchuk, 1992) and *The Birds of North America* (Poole et al., 1992). These were supplemented by other literature as needed. Information gathered on clutch size, incubation, fledging, and nest height is presented in tabular form for each species. Information gathered on habitat, nest location, nest type, and diet were then clustered by similarity and ease of representation. This process was difficult as having many symbols would have been accurate, but not practical, and having few, would have been too coarse. The final set is shown on the following pages and the details of each of these symbols are provided at the end of this chapter.

Distribution and range

Observed distribution maps from both Atlas 1 and Atlas 2 are provided. This enables the reader to make direct comparisons and reference these against the breeding range map provided for each species. The distribution maps show both coverage and breeding evidence to assist readers in evaluating whether the observed changes were a function of actual change or simply changes in coverage.

Reporting rate

Reporting rate is a measure of how often a bird is detected. For example, if a bird is observed in 3 out of 10 surveys, its reporting rate is 30%. Atlas 2 used a day-trip format for recording data, enabling us to use reporting rate to assess relative abundance. As outlined in the Methods, this is a more reliable measure of abundance than using the numbers directly because each survey card has different levels of effort associated with it. Reporting rate is displayed in three ways for each species.

- Monthly reporting rate, for both observations and probable and confirmed breeding, is shown in a vertical bar chart. Under the x-axis of this chart, the expected breeding period is displayed as a horizontal orange bar.
- Natural Region reporting rate, for both observations and probable and confirmed breeding, is shown in a horizontal bar chart. The data set used to create this chart was limited to the expected breeding period of this species to reflect its breeding habitat preference. The Canadian Shield is excluded from this chart due to insufficient data.

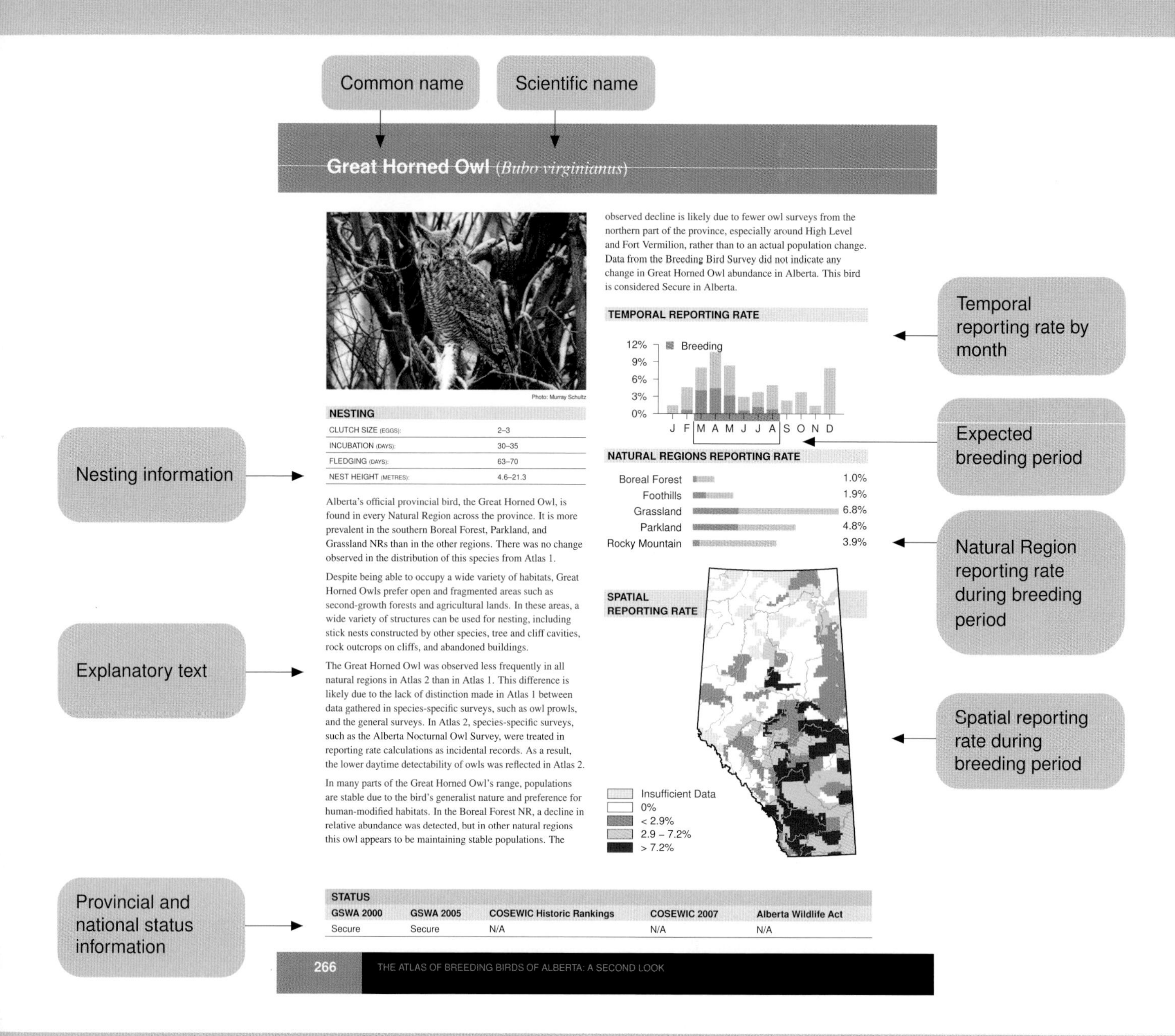

Natural History Symbols (Habitat and Nest Type)

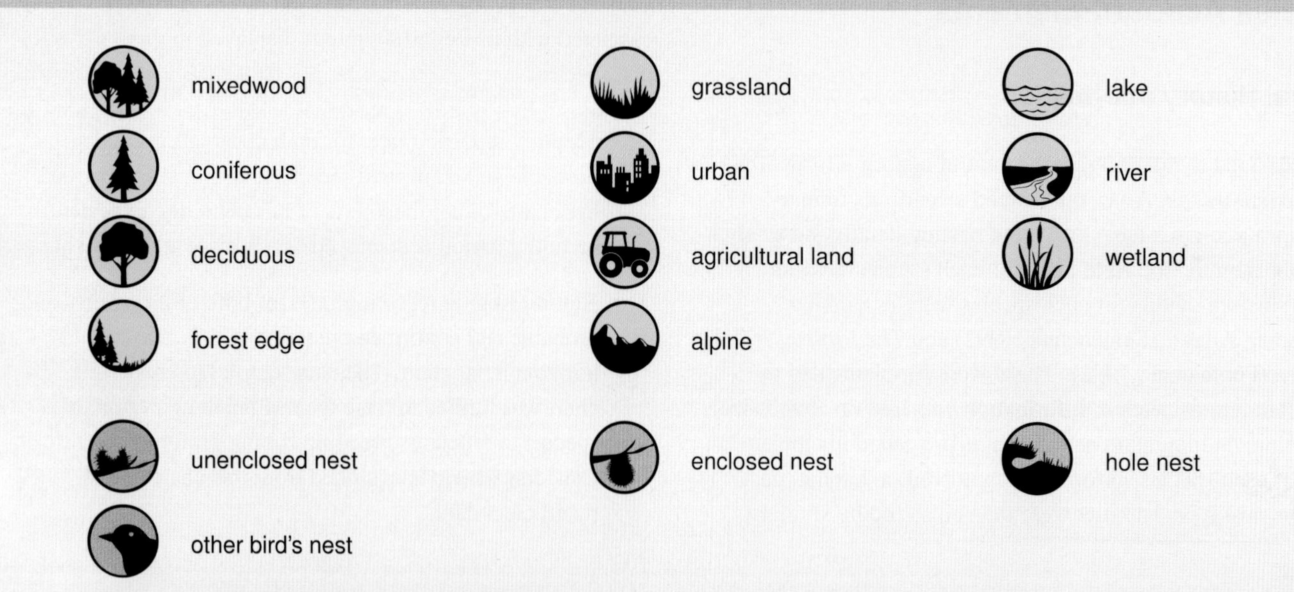

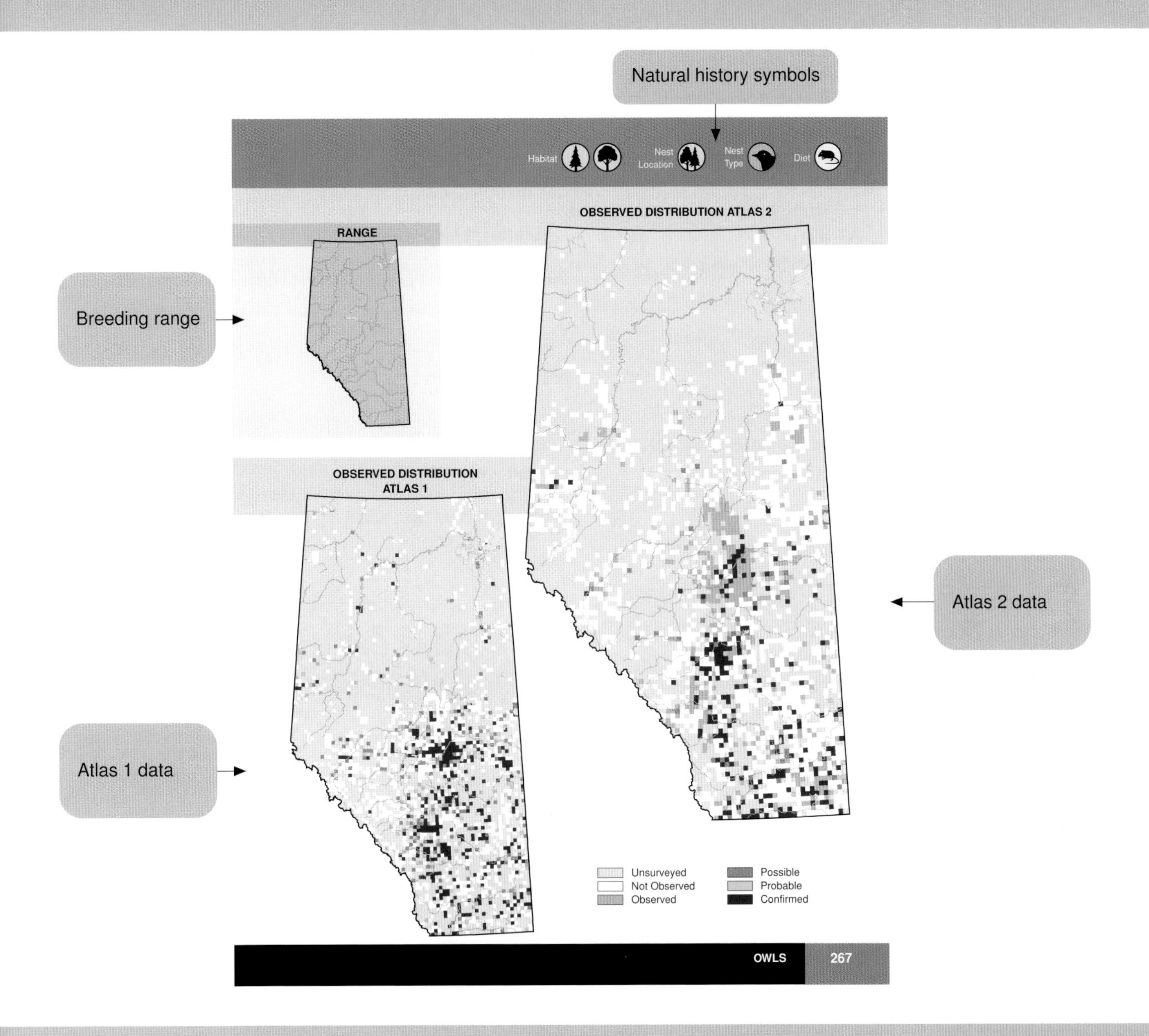

Natural History Symbols (Nest Location and Diet)

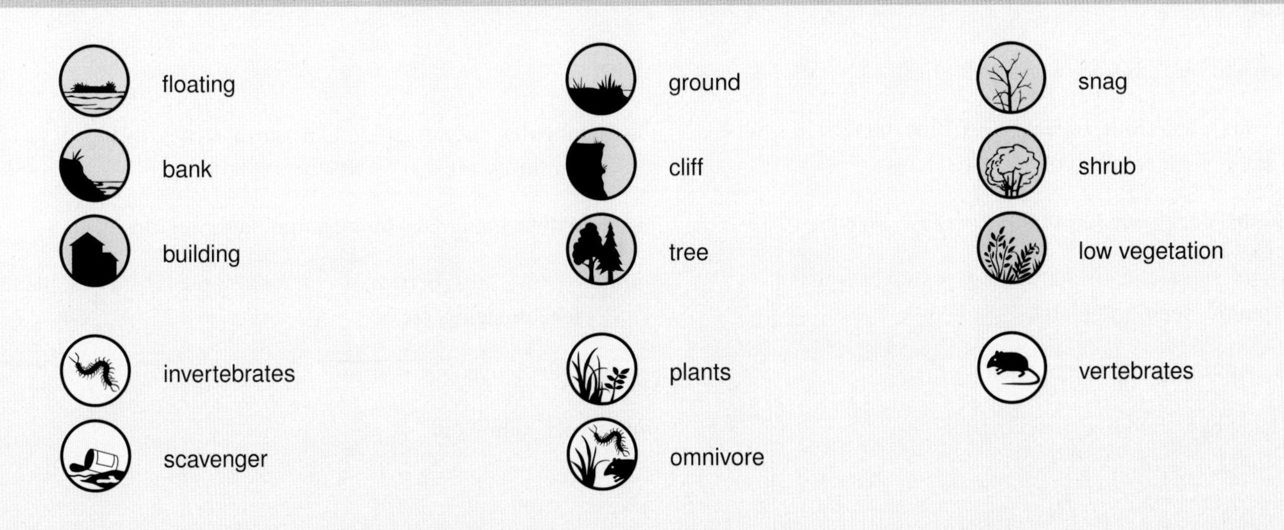

- Spatial reporting rate, with clusters of squares comprised of the same natural subregion, is used to display locations where the bird was detected most often. As with the Natural Regions chart, the data set used to generate this map is restricted to each species' expected breeding period. Although not true in all cases, it can often be assumed that areas where a bird was detected most often are areas of highest relative abundance.

Provincial and national status

For quick reference, the information on the provincial status (General Status of Wildlife in Alberta, Alberta Wildlife Act) and the national status (Committee on the Status of Endangered Wildlife in Canada) is provided in a table for each species.

Changes in distribution and relative abundance

The text for each species account is designed to focus on what changes, if any, were detected, the certainty of those changes, and their likely causes. In this section, natural history information, such as breeding habitat requirements, are included where pertinent to understanding the detected changes. Where available, Breeding Bird Survey (BBS) results are cited as a source of corroboration or contrast to the Atlas' detected changes.

Natural history symbol details

In each section, symbol names and examples of descriptions used in the literature are provided. In many cases, more than one symbol was used in the species accounts to provide a more complete summary of the species' natural history.

Habitat

mixedwood mixedwood, woodland, forest

coniferous coniferous, boreal forest, pine, tamarack, spruce, fir, jack pine, lodgepole pine

deciduous deciduous forest, alder, willow, aspen

grassland grassland, meadow, field, pasture, savannah, prairie, semi-arid plains

alpine alpine, montane, mountain

lake lake, bay, shoreline, pond, island

river river, stream, creek

wetland wetland, muskeg, bog, lagoon, fen, marsh, sedge swamp, reed bed, mudflat, slough, sedge meadow

urban urban, human structure

agricultural land cultivated field, farmland

forest edge forest clearing, forest edge, edge, woodland clearing

Nest Location

floating floating, above water

ground ground

snag snag, dead tree, decaying tree, stump

bank bank

cliff cliff

shrub shrub, bush or bushes

building building, ledge, bridge, rafters, eaves

tree tree, coniferous, deciduous

low vegetation low vegetation, reeds, sedge, grasses

Nest Type

unenclosed nest unenclosed, cup, platform, saucer, scrape, no construction

enclosed nest pendant, sphere, oven, gourd-shaped

hole nest cavity, crevice, burrow, nest box

other bird's nest abandoned nest, host species

Diet

invertebrates insects, spiders, worms, leeches, millipedes, molluscs, etc.

vertebrates mammals, birds, amphibians, reptiles, fish

plants fruits, nuts, berries, nectar, cones, seeds, grasses, plants leaves, stems, etc.

scavenger scavenger, carrion

omnivore omnivore

Species Accounts

Canada Goose (*Branta canadensis*)

Photo: Randy Jensen

NESTING

CLUTCH SIZE (EGGS):	4–6
INCUBATION (DAYS):	25–28
FLEDGING (DAYS):	63
NEST HEIGHT (METRES):	0

Many races of Canada Goose pass through Alberta but it appears that only the "large" race (*B. c. moffitti*) breeds here. Atlases 1 and 2 record them breeding throughout Alberta with concentrations in the central and southern parts of the province.

Highly adaptable, Canada Geese usually congregate and nest near lakes, ponds, sloughs, rivers, and irrigation reservoirs (Semenchuk, 1992). This goose has adapted well to human presence and is also found nesting near sewage lagoons, city lakes and parks, golf courses, urban subdivisions, highway medians, and on tops of city buildings (Mowbray et al., 2002).

Increases in relative abundance were detected in the Boreal Forest, Grassland, Parkland, and Rocky Mountain Natural Regions. No changes in relative abundance were detected in the Foothills NR. The Breeding Bird Survey (1985–2005) detected an increase in abundance in Alberta and Canada-wide.

During the 1930s, breeding Canada Geese all but disappeared from Alberta (Semenchuk, 1992). Now rated a true success story in wildlife management (Mowbray et al., 2002), they occur in abundance throughout North America. Mowbray et al. (2002) reports that, increasingly, Canada Goose managers are shifting from careful conservation of the species to programs designed to limit further population growth and distribution; mounting numbers of complaints about nuisance and damage associated with growing populations emphasize the conflict with agricultural and suburban landowner interests. The concerns in urban areas arise from locally breeding geese that concentrate in relatively large numbers and come into conflict with human land-use practices. Problems include: fecal deposition on public and private property; contamination of drinking-water sources and swimming pools; occasional aggressive behaviour toward humans and their pets; and the threat to aircraft safety (Mowbray et al., 2002). In Alberta the Canada Goose is listed as Secure.

TEMPORAL REPORTING RATE

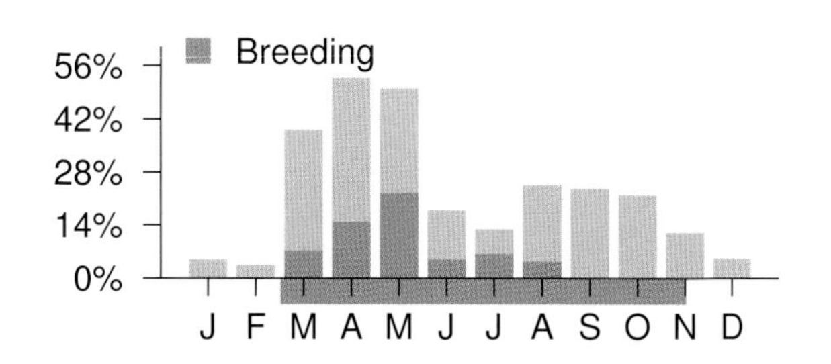

NATURAL REGIONS REPORTING RATE

SPATIAL REPORTING RATE

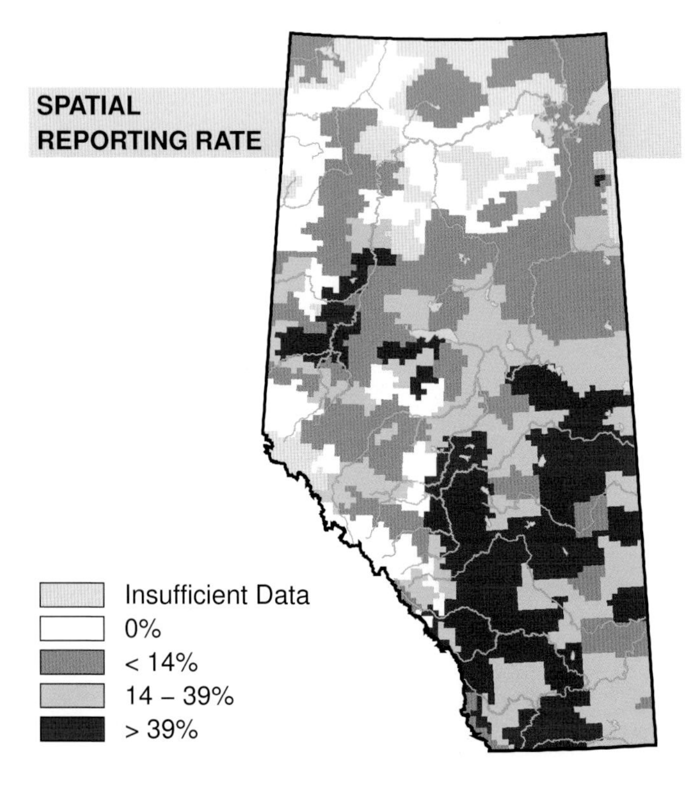

STATUS

GSWA 2000	GSWA 2005	COSEWIC Historic Rankings	COSEWIC 2007	Alberta Wildlife Act
Secure	Secure	N/A	N/A	N/A

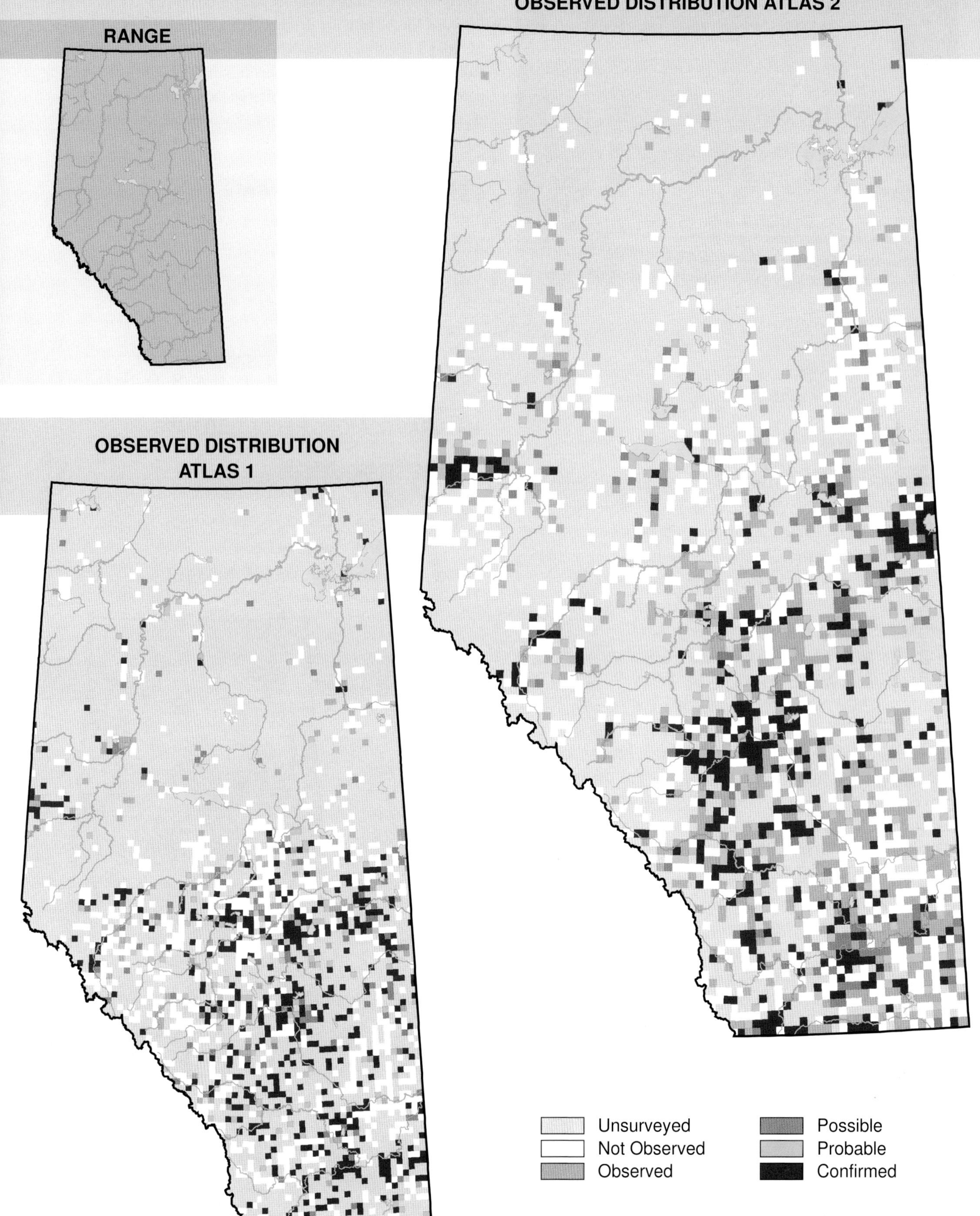
OBSERVED DISTRIBUTION ATLAS 2
RANGE
OBSERVED DISTRIBUTION
ATLAS 1
Unsurveyed
Not Observed
Observed
Possible
Probable
Confirmed

Trumpeter Swan (*Cygnus buccinator*)

Photo: Randy Jensen

NESTING

CLUTCH SIZE (EGGS):	2–9
INCUBATION (DAYS):	32
FLEDGING (DAYS):	100
NEST HEIGHT (METRES):	0

The majority of Alberta's Trumpeter Swans are found in the Grande Prairie area. In Atlas 2, the number of breeding observations increased in the northeast, between Lesser Slave Lake and Peerless Lake. Confirmed breeding was recorded in Elk Island and Waterton Lakes national parks, as well as in some new areas including Lac La Biche, Cold lake, Buck Lake, and south of Calgary.

Trumpeter Swans use a variety of breeding habitats that provide open water, access to food, and security from disturbance. Migratory behaviour differs widely among various flocks. This, along with specific patterns of habitat use and demography, make it mandatory to manage flocks on an individual basis (Mitchell, 1994). In Alberta this species is found on isolated, small- to medium-sized, shallow lakes that have well-developed emergent and submergent plant communities (Semenchuk, 1992).

Increases in relative abundance were detected in the Boreal Forest, Grassland, and Parkland Natural Regions. No changes were detected in other NRs. The Breeding Bird Survey (1985–2005) lacked sufficient data to permit the estimation of trends for this species. Birdwatching, photography, boating, and other activities in or near nesting areas may cause nest failures, as well as unintentional and malicious shooting, which remains a problem (Mitchell, 1994). The Committee on the Status of Endangered Wildlife in Canada (COSEWIC) has listed this species as Not at Risk because the North American population now numbers close to 20,000 birds and continues to increase. In Alberta the Trumpeter Swan is listed as At Risk. In Atlas 1 it was reported that there were about 50 breeding pairs. The 2005 provincial status report estimated 166 breeding pairs in Alberta. Breeding habitat in the province is relatively secure but a critical shortage of key winter habitat limits population growth.

TEMPORAL REPORTING RATE

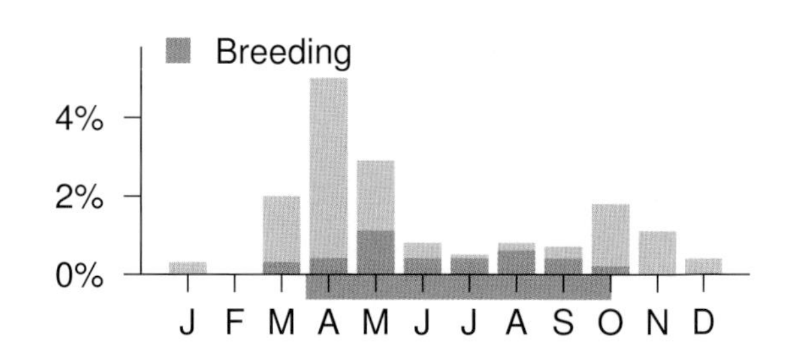

NATURAL REGIONS REPORTING RATE

SPATIAL REPORTING RATE

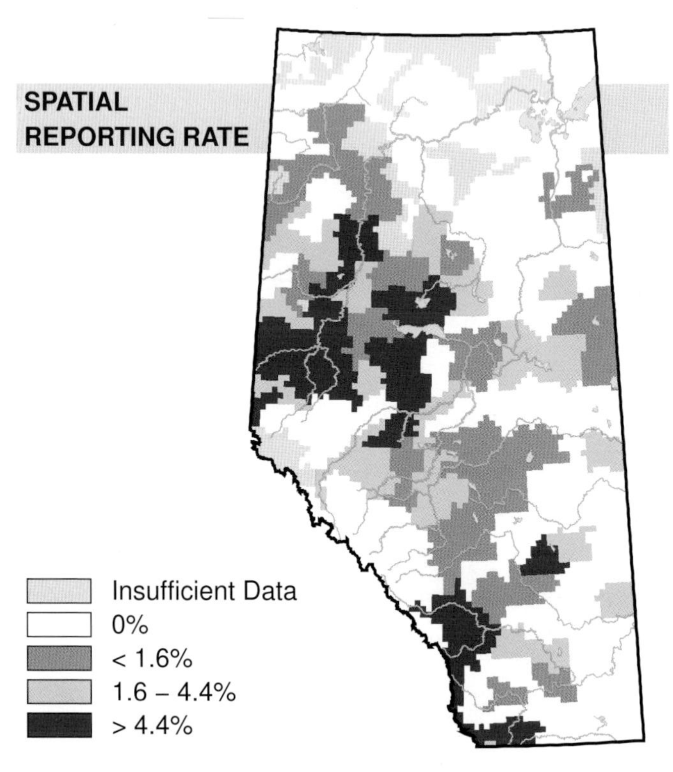

STATUS

GSWA 2000	GSWA 2005	COSEWIC Historic Rankings	COSEWIC 2007	Alberta Wildlife Act
At Risk	At Risk	Not At Risk	Not At Risk	Threatened

OBSERVED DISTRIBUTION ATLAS 2

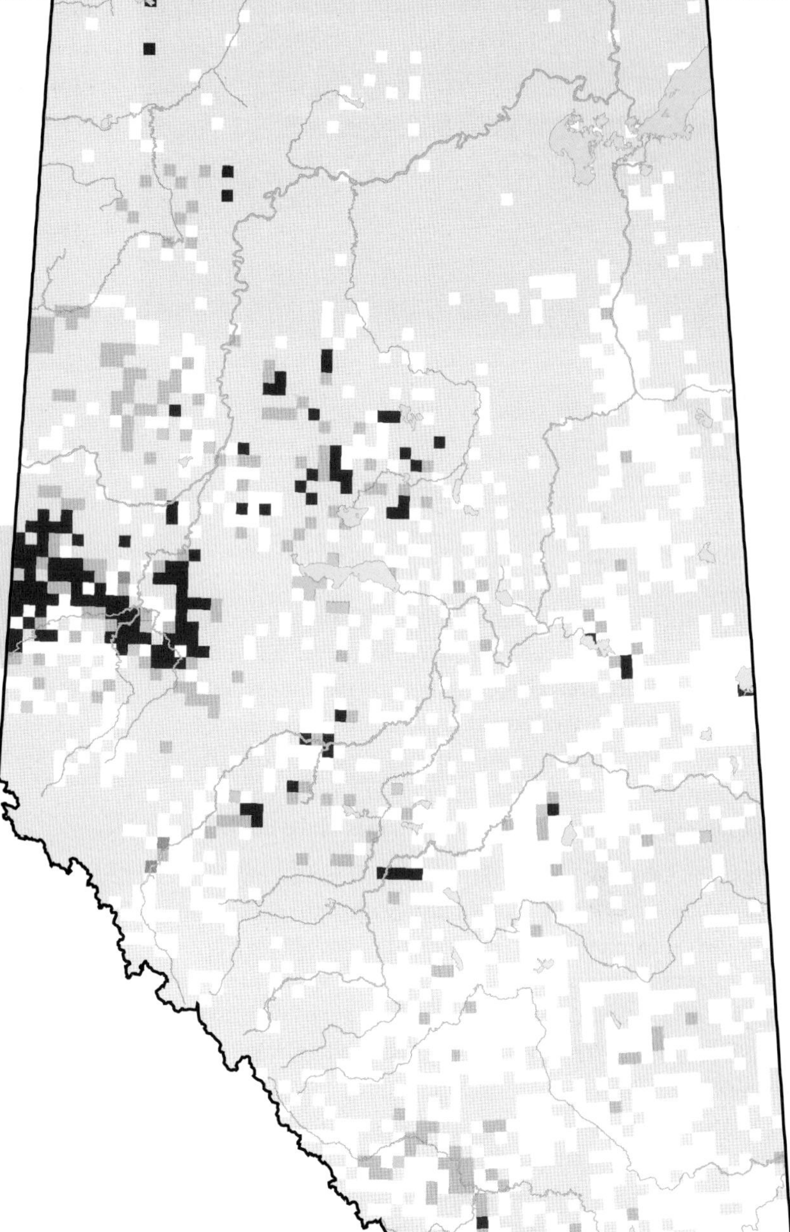

RANGE

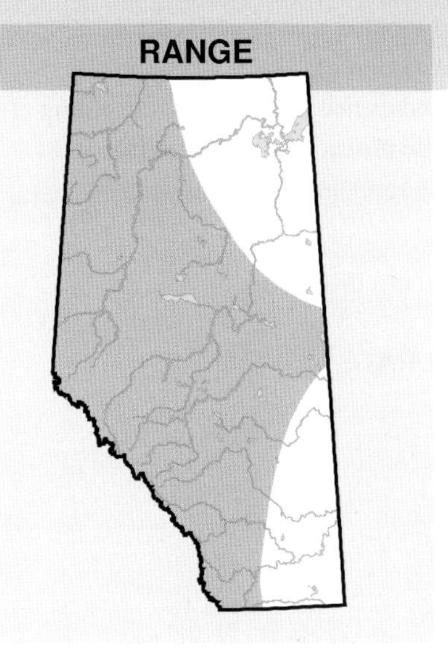

OBSERVED DISTRIBUTION ATLAS 1

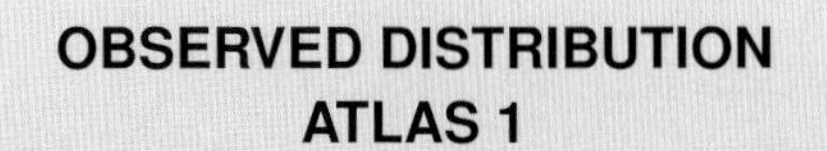

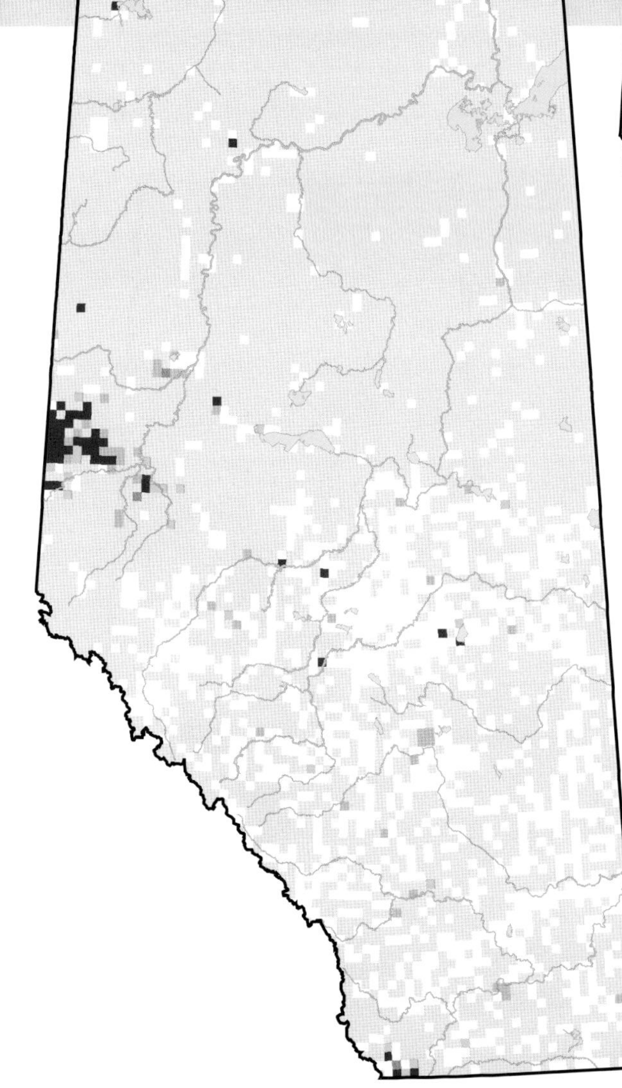

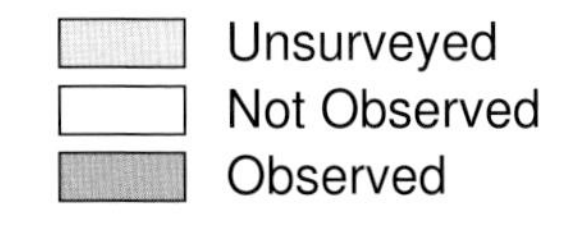

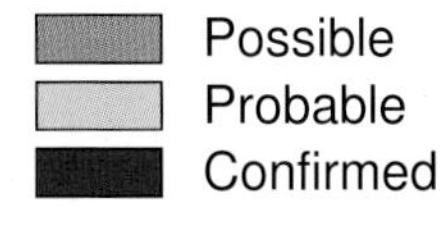

Wood Duck (*Aix sponsa*)

Photo: Gerald Romanchuk

NESTING

CLUTCH SIZE (EGGS):	13–15
INCUBATION (DAYS):	28–30
FLEDGING (DAYS):	60
NEST HEIGHT (METRES):	0.6–19.8

The Wood Duck occurs sporadically in Alberta where it is at the northern limits of its North American range. It is found mainly in the Grassland Natural Region with some records in the Parkland, southern Boreal Forest, and southern Rocky Mountain NRs. The distribution of this species has expanded northward since Atlas 1. There were new records for the Lac La Biche, Cold Lake, and Bonnyville areas in the northeast.

Wood Ducks utilize a wide variety of habitats that include wooded lakes and ponds, and streams with large willows and cottonwoods present. Interspersion of flooded shrubs, water-tolerant trees, and small areas of open water that result in about 50-75% cover are favoured. Beaver ponds with abundant vegetative cover typically provide good habitat (Hepp and Bellrose, 1995).

Increases in relative abundance were detected in the Grassland NR. This may be attributed to the increase in precipitation in this region in the latter part of Atlas 2. No changes in relative abundance were detected in the Boreal Forest, Parkland, and Rocky Mountain NRs. The Breeding Bird Survey (1985–2005) detected no change in this species' abundance across Canada, although an increase was detected for the period 1968–2005. No BBS data were available for Alberta. It is the most successful of the seven species of North American ducks which regularly nest in natural cavities (Hepp and Bellrose, 1995).

In order to promote an increase in Wood Duck presence, Hepp and Bellrose (1995) variously prescribe: the elimination of stream channelization; the establishment of greenways of timber and shrubs along stream banks that would reduce erosion and provide food, cover, and nest sites; the reduction of the drainage of wooded wetlands; encouragement to develop beaver and farm ponds; and, the provision of predator-resistant nest houses in situations where food and cover requirments warrant such efforts.

The Wood Duck is rated as Secure in Alberta.

TEMPORAL REPORTING RATE

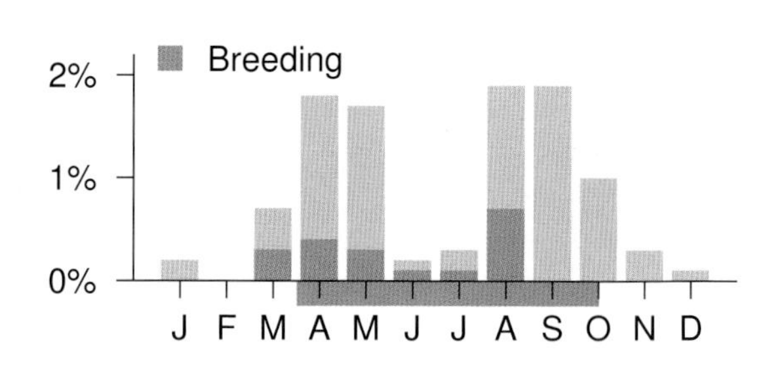

NATURAL REGIONS REPORTING RATE

SPATIAL REPORTING RATE

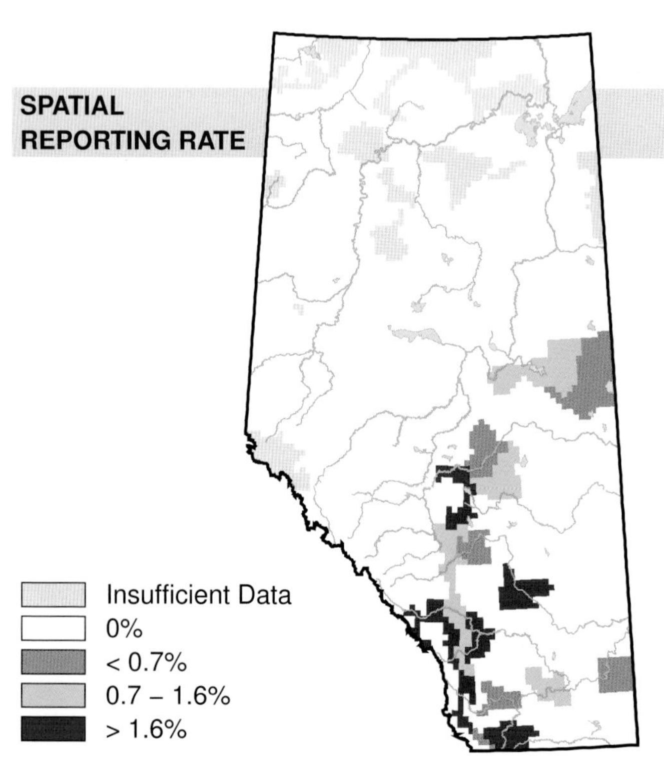

STATUS

GSWA 2000	GSWA 2005	COSEWIC Historic Rankings	COSEWIC 2007	Alberta Wildlife Act
Secure	Secure	N/A	N/A	N/A

RANGE

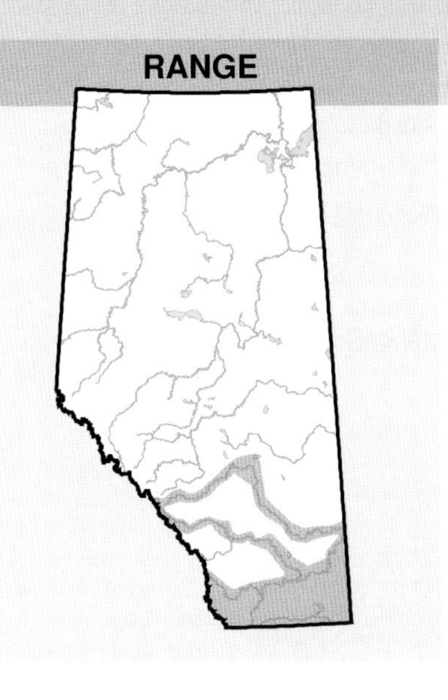

OBSERVED DISTRIBUTION ATLAS 1

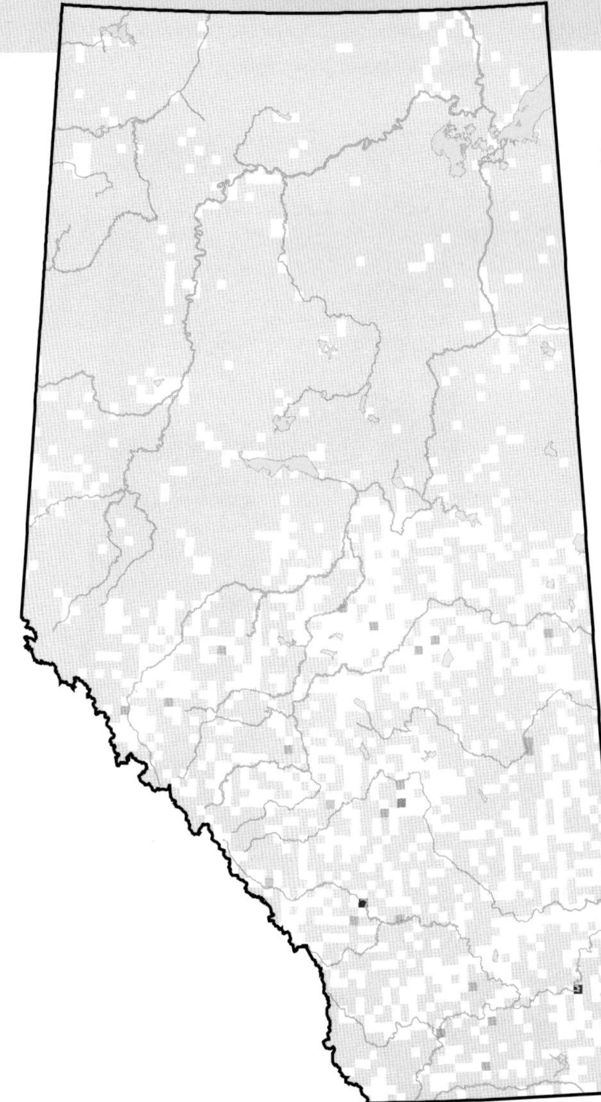

OBSERVED DISTRIBUTION ATLAS 2

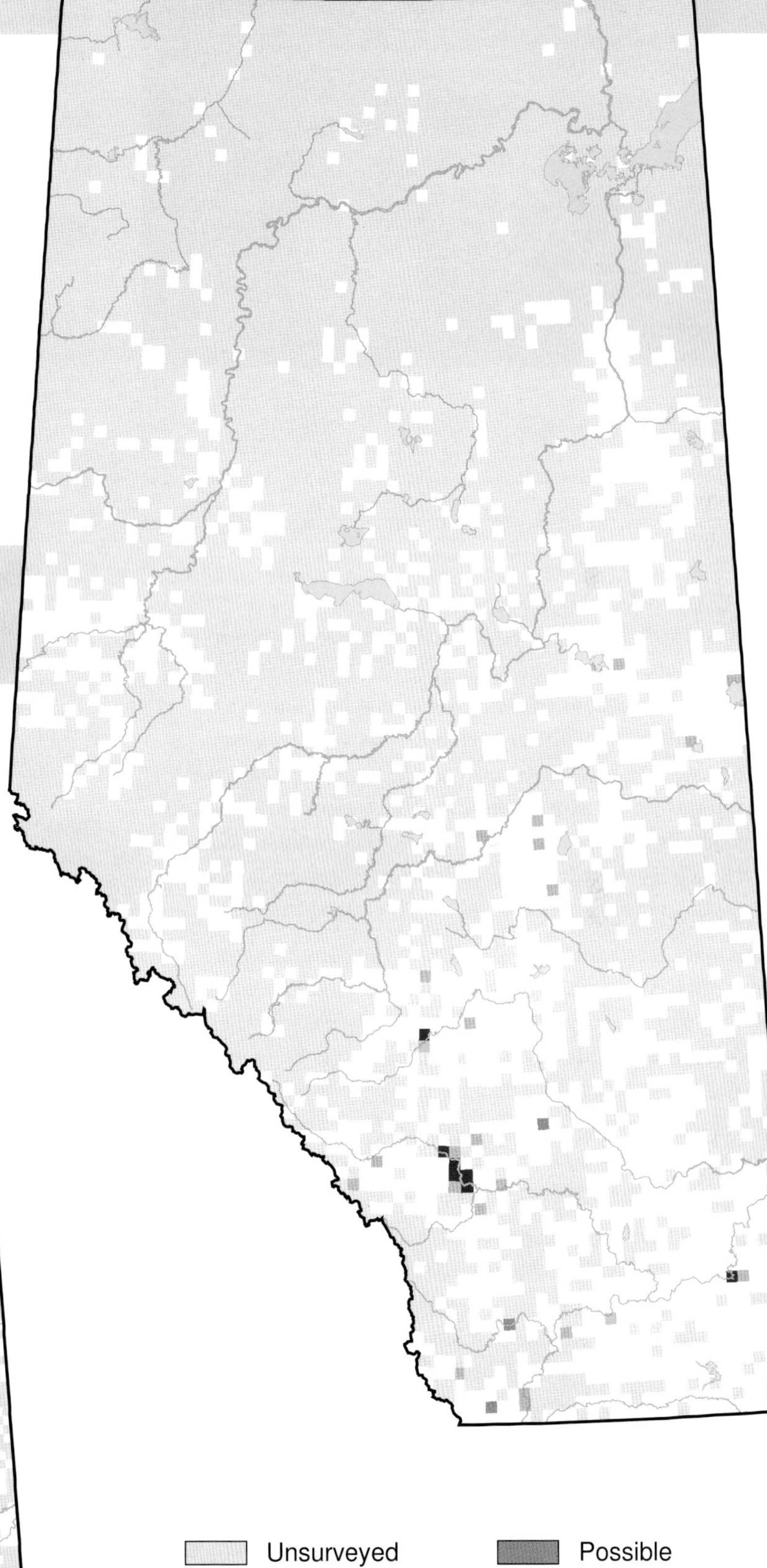

Gadwall (*Anas strepera*)

Photo: Gerald Romanchuk

NESTING

CLUTCH SIZE (EGGS):	8–12
INCUBATION (DAYS):	25–28
FLEDGING (DAYS):	49
NEST HEIGHT (METRES):	0

The Gadwall was found in all Natural Regions of the province, but it breeds mainly in the Grassland and Parkland NRs and in the southern portions of the Boreal Forest NR. Surveys during Atlas 2 indicate a range extension for this species in the northwestern corner of the province as far as the border with the Northwest Territories. In Atlas 1 it was reported that some of the highest nest densities on the continent occur on Jesse Lake near Bonnyville. In Atlas 2, this northeastern area again had many breeding records and very high reporting rates.

Gadwalls utilize marshes, sloughs, and shallow lake margins with good cover extending back from the shoreline. Grassy islands within this habitat are favourite nesting sites. In areas under intensive agriculture pressure, Gadwalls use untilled upland habitat almost exclusively (LeSchack et al., 1997). These seasonal and semi-permanent water bodies that Gadwalls prefer also make them vulnerable to year-to-year changes in precipitation.

Increases in relative abundance were detected in the Grassland and Parkland NRs. The increase in the Grassland NR is likely related to the wetter conditions in this NR during Atlas 2 than during Atlas 1. The increase in the Parkland NR is surprising as, overall, this NR was drier in Atlas 2 than in Atlas 1. Without probable biological cause or corroboration, further research is needed to evaluate this change. No changes in relative abundance were detected in the Boreal Forest NR and similarly in the Foothills and Rocky Mountain NRs where this species is only sporadically recorded. The Breeding Bird Survey (1985–2005) detected no change in the abundance of this species either in Alberta or across Canada.

The Gadwall is rated as Secure in Alberta.

TEMPORAL REPORTING RATE

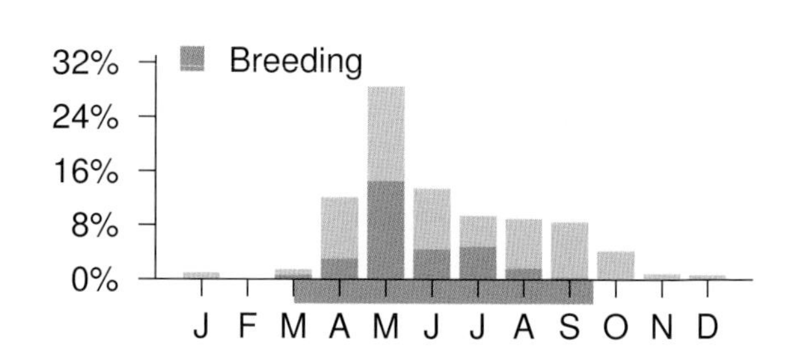

NATURAL REGIONS REPORTING RATE

SPATIAL REPORTING RATE

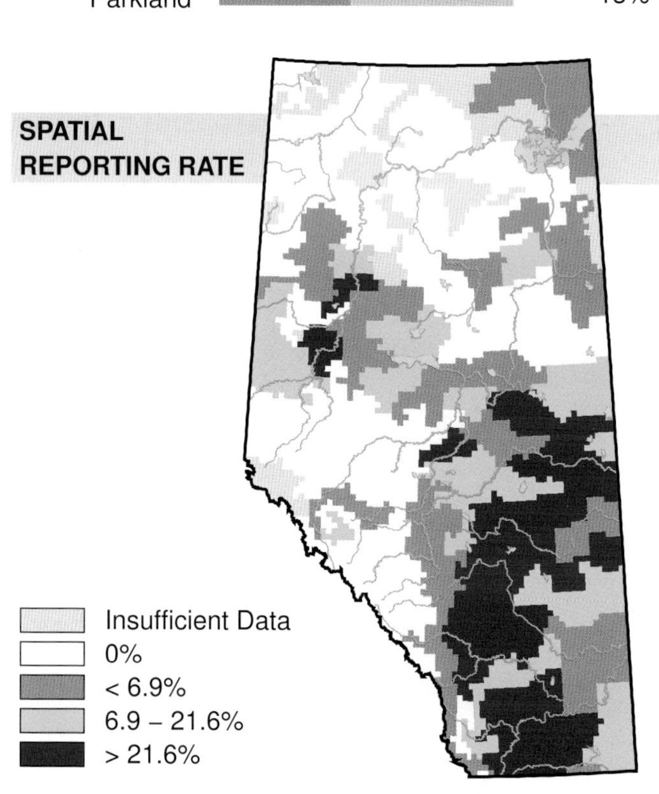

STATUS

GSWA 2000	GSWA 2005	COSEWIC Historic Rankings	COSEWIC 2007	Alberta Wildlife Act
Secure	Secure	N/A	N/A	N/A

RANGE

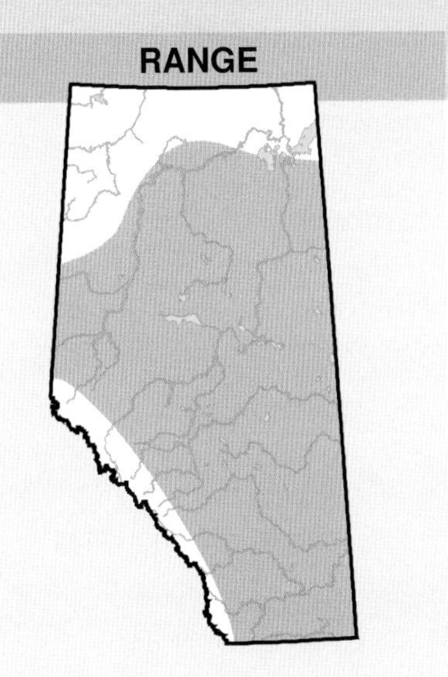

OBSERVED DISTRIBUTION ATLAS 1

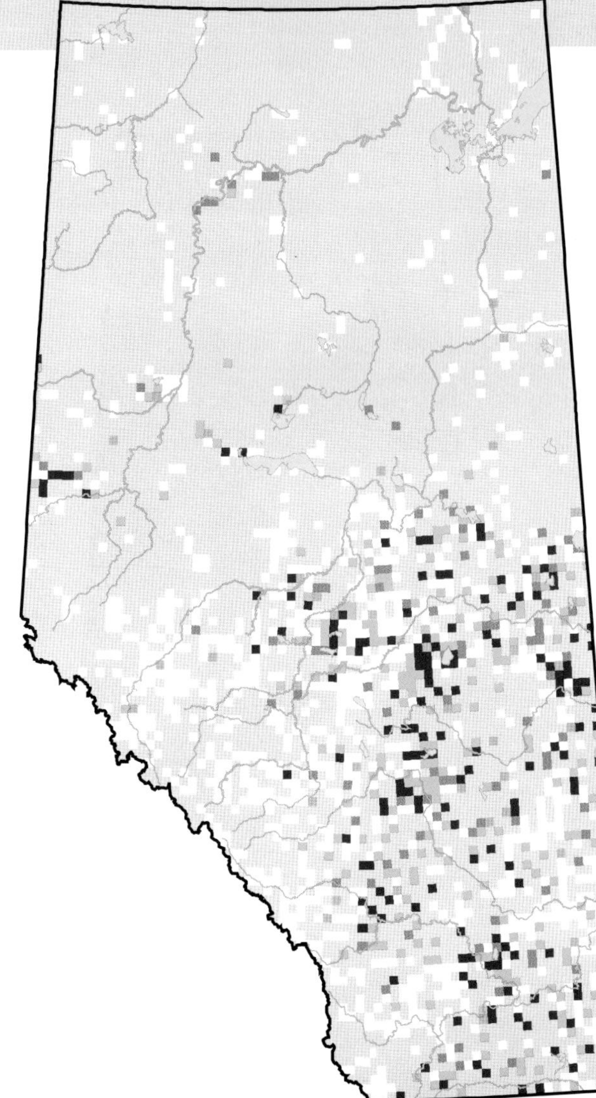

OBSERVED DISTRIBUTION ATLAS 2

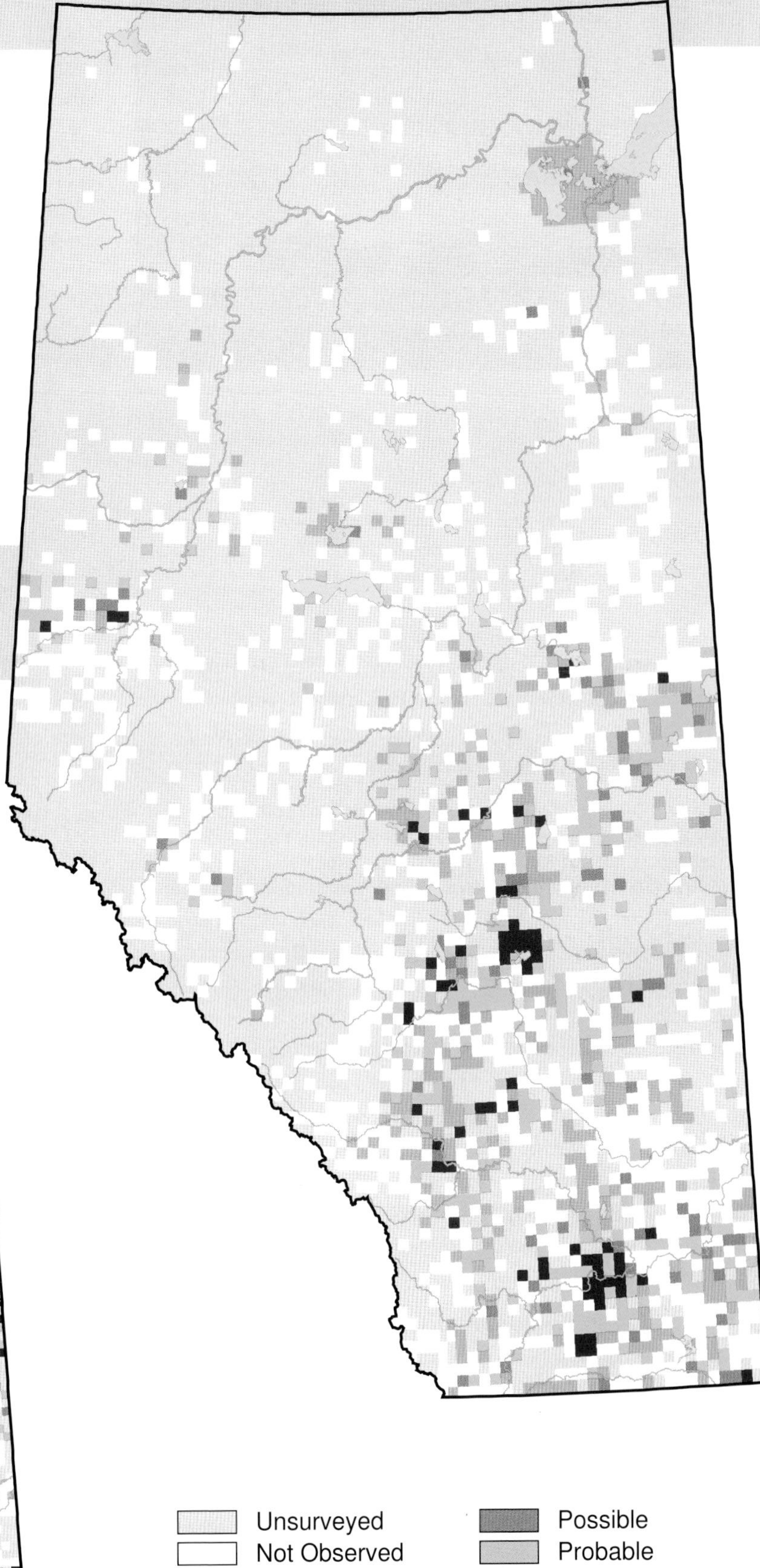

Unsurveyed	Possible
Not Observed	Probable
Observed	Confirmed

American Wigeon (*Anas americana*)

Photo: Gerald Romanchuk

NESTING

CLUTCH SIZE (EGGS):	9–11
INCUBATION (DAYS):	23–25
FLEDGING (DAYS):	45–48
NEST HEIGHT (METRES):	0

The American Wigeon, commonly known as the Baldpate, nests in appropriate habitat in every Natural Region in the province, with a preference for the Parkland and southern Boreal Forest NRs. There were no changes in the breeding distribution observed between Atlas 1 and Atlas 2.

This duck breeds near shallow, freshwater wetlands where it nests in areas with an upland cover of brush and grass.

Increases in relative abundance were detected in the Grassland, Parkland, and Rocky Mountain NRs. Increased precipitation in the southern part of the province during Atlas 2 may account in part for the observed changes. Decreases in relative abundance were detected in the Boreal Forest NR. These changes may be explained by Mowbray (1999) who stated that breeding densities are high in prairie and parkland habitats of the southern Prairie Provinces in wet years when ponds are numerous; in dry years, when ponds tend to dry up, birds homing to these habitats overfly the region and travel farther north. No changes in relative abundance were detected in the Foothills NR. The Breeding Bird Survey (1985–2005) found no change in abundance in Alberta or across Canada.

Mowbray (1999) cautions that in the Canadian Prairie-Parklands, where agricultural expansion has significantly reduced the available habitat for breeding waterfowl, habitat restoration programs should focus on areas where the highest quality waterfowl habitat and the lowest quality agricultural lands overlap. These areas offer the greatest potential for effecting the recovery of the breeding duck populations.

This species is rated as Secure in Alberta.

TEMPORAL REPORTING RATE

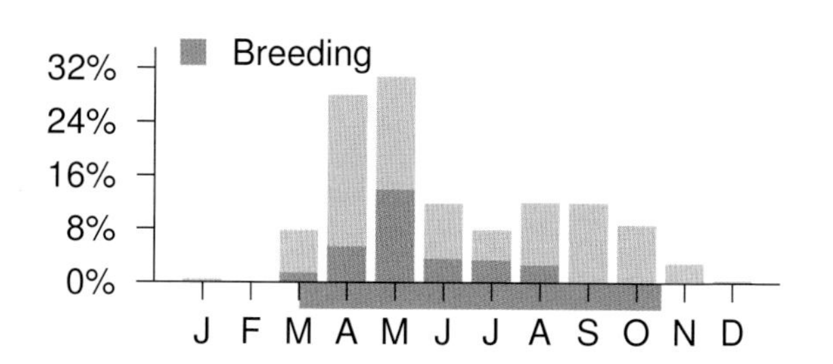

NATURAL REGIONS REPORTING RATE

SPATIAL REPORTING RATE

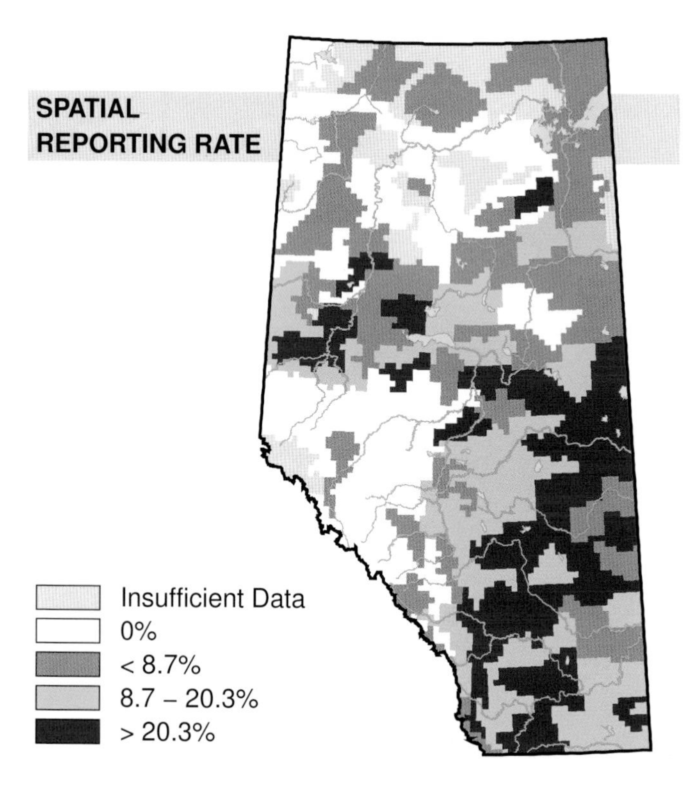

STATUS

GSWA 2000	GSWA 2005	COSEWIC Historic Rankings	COSEWIC 2007	Alberta Wildlife Act
Secure	Secure	N/A	N/A	N/A

RANGE

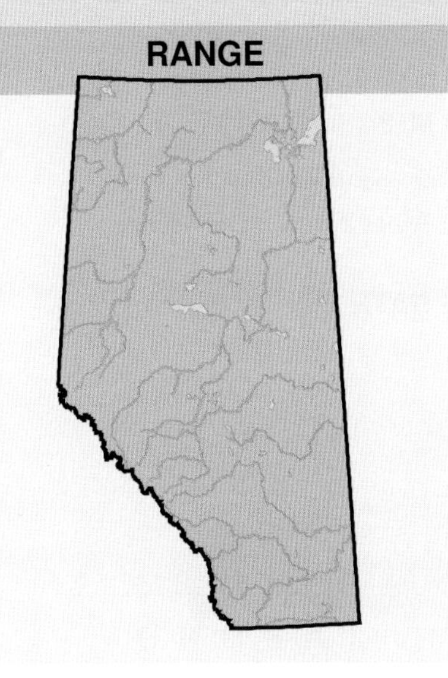

OBSERVED DISTRIBUTION ATLAS 1

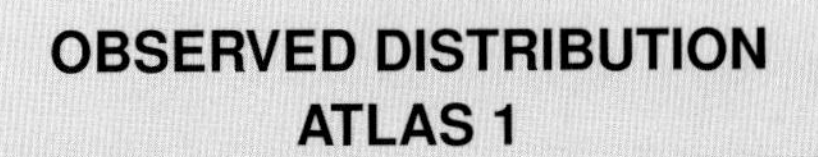

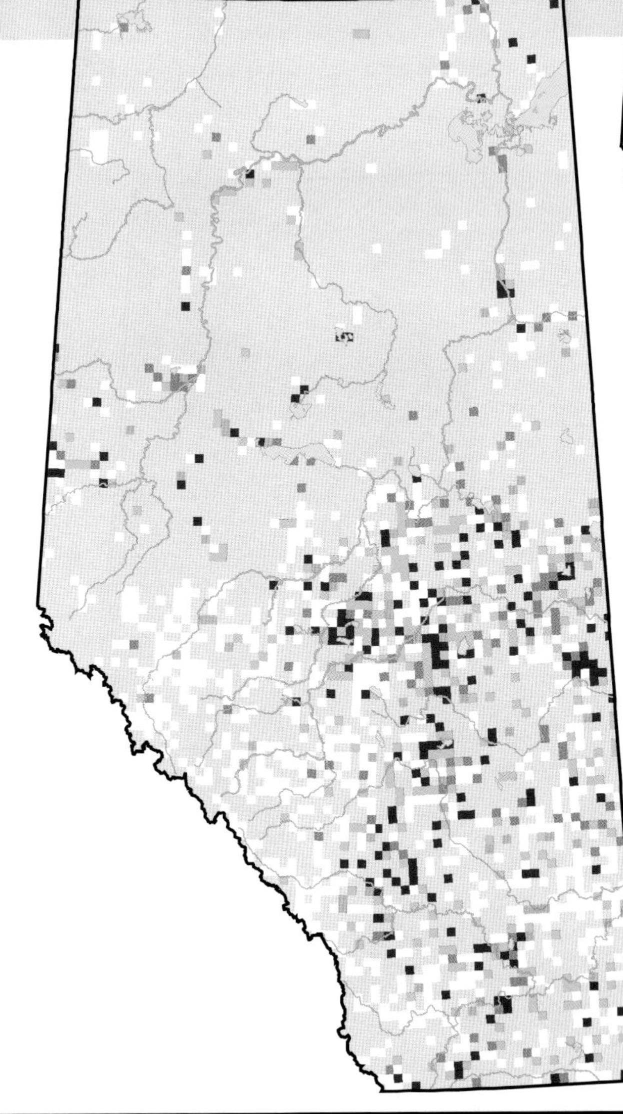

OBSERVED DISTRIBUTION ATLAS 2

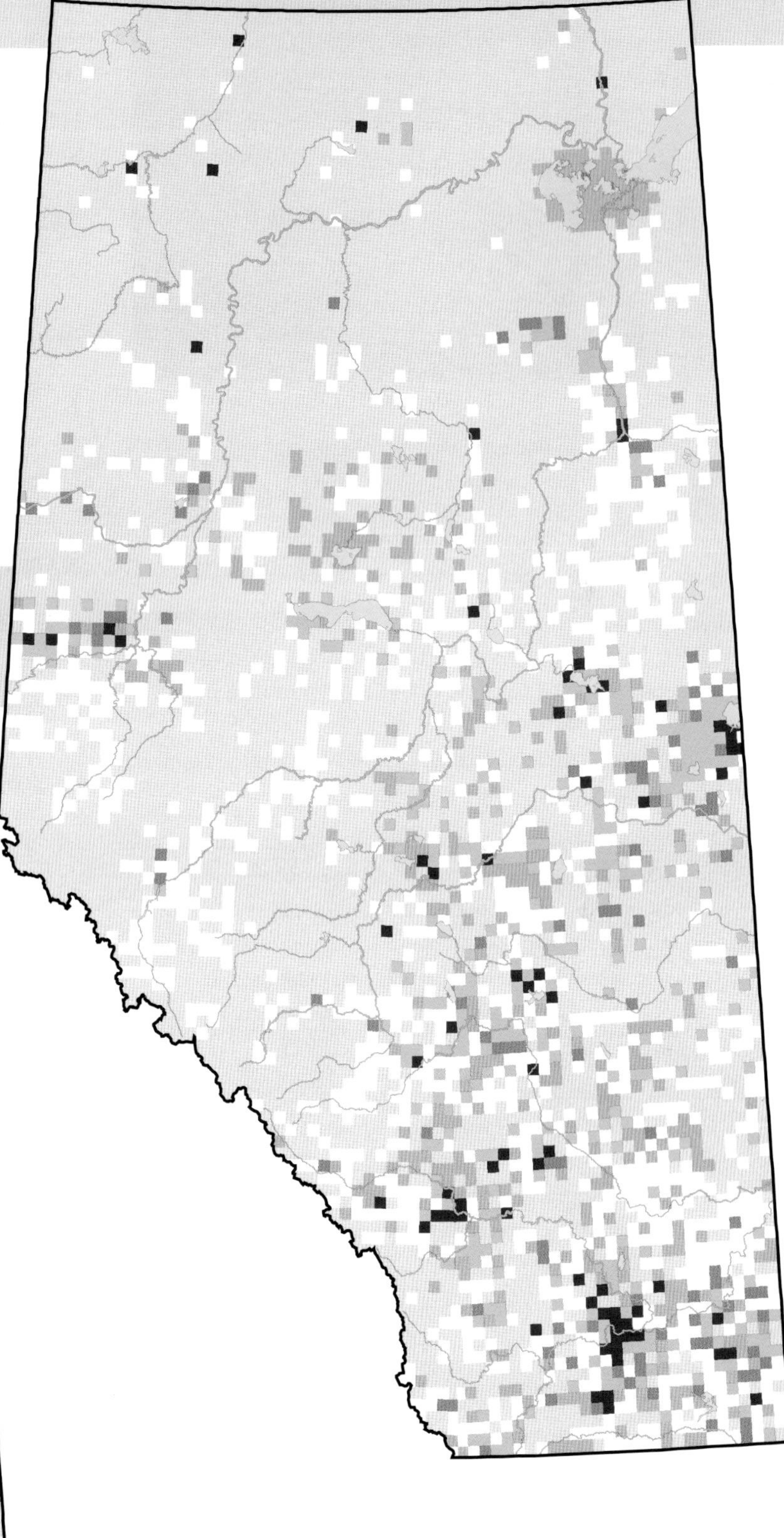

Mallard (*Anas platyrhynchos*)

Photo: Gerald Romanchuk

NESTING

CLUTCH SIZE (EGGS):	8–10
INCUBATION (DAYS):	26–29
FLEDGING (DAYS):	56
NEST HEIGHT (METRES):	0

The Mallard breeds in all Natural Regions across the province where suitable habitat is available. It is most prevalent in the Parkland, Grassland, Foothills, and southern Boreal Forest NRs. There was no change in the observed distribution from Atlas 1 to Atlas 2.

This highly adaptable duck uses various habitats including marshes, ponds, margins of lakes, quiet waters of streams and rivers, and flooded land in both forested and treeless country. It nests on fairly dry ground where there is vegetation for cover. The presence of shallow-water feeding areas and the availability of suitable nest sites appear to be the only critical limiting factors for breeding (Semenchuk, 1992).

Increases in relative abundance were detected in the Grassland NR. This area had higher levels of precipitation in the latter part of Atlas 2. Positive correlations occur between cumulative winter-spring precipitation and reproductive success of Mallards because they use breeding habitat in proportion to the availability of water (Drilling et al., 2002). Decreases in relative abundance were detected in the Boreal Forest and Foothills NRs. Most puddle ducks, including Mallards, fly farther north in dry years to find suitable wetland conditions. The decrease in these two regions in Atlas 2 could be attributed to a population shift to more preferred habitats in the Grassland NR. No changes in relative abundance were detected in the Parkland and Rocky Mountain NRs. The Breeding Bird Survey (1985–2005) reported no change in abundance in Alberta or across Canada.

Large-scale alterations of wetlands and grasslands by agriculture, urbanization, and industrial activities have affected distribution and abundance of Mallards (Drilling et al., 2002).

The Mallard is not threatened or endangered in any part of its North American range, and in Alberta is listed as Secure.

TEMPORAL REPORTING RATE

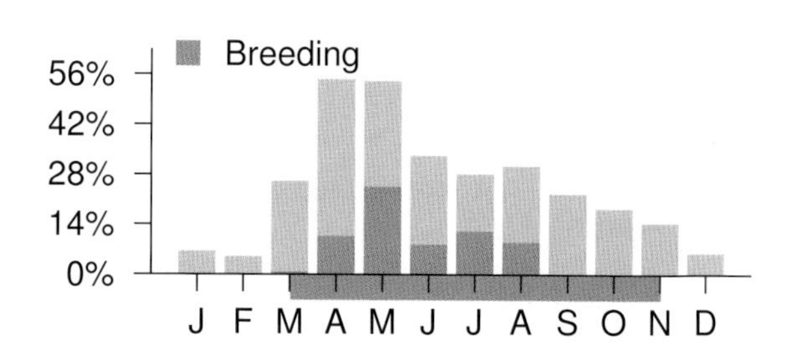

NATURAL REGIONS REPORTING RATE

Boreal Forest	8%
Foothills	21%
Grassland	49%
Parkland	61%
Rocky Mountain	19%

SPATIAL REPORTING RATE

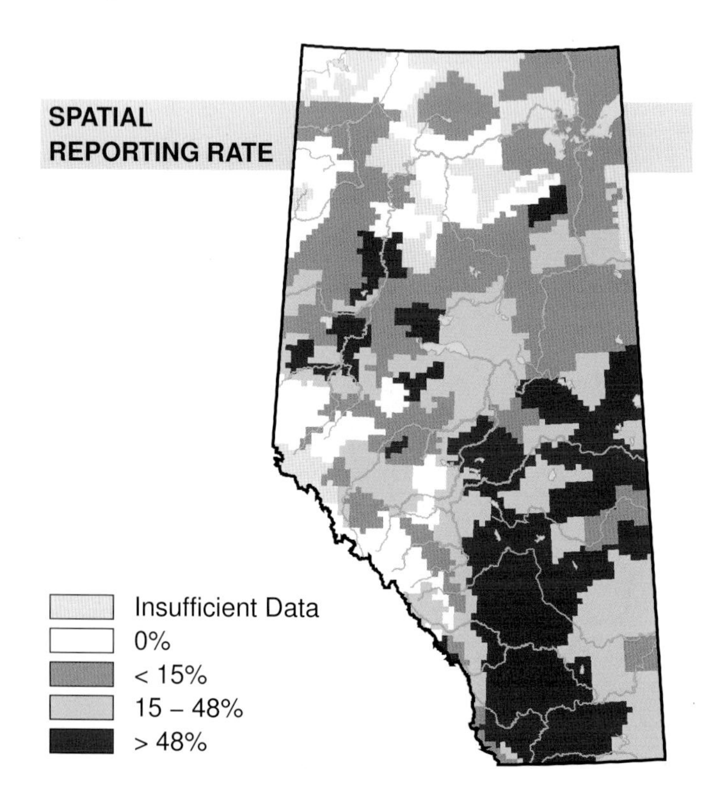

STATUS

GSWA 2000	GSWA 2005	COSEWIC Historic Rankings	COSEWIC 2007	Alberta Wildlife Act
Secure	Secure	N/A	N/A	N/A

RANGE

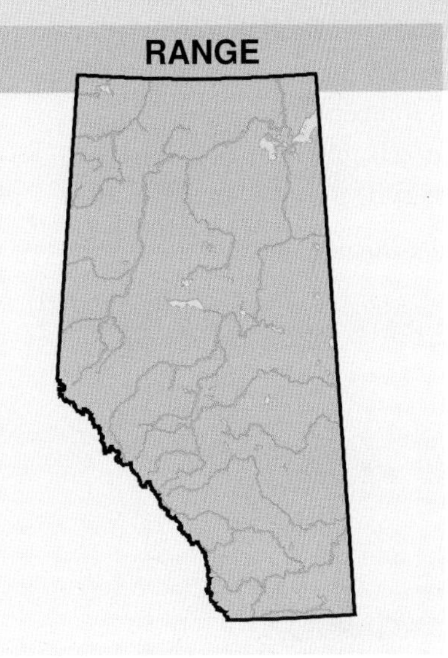

OBSERVED DISTRIBUTION ATLAS 1

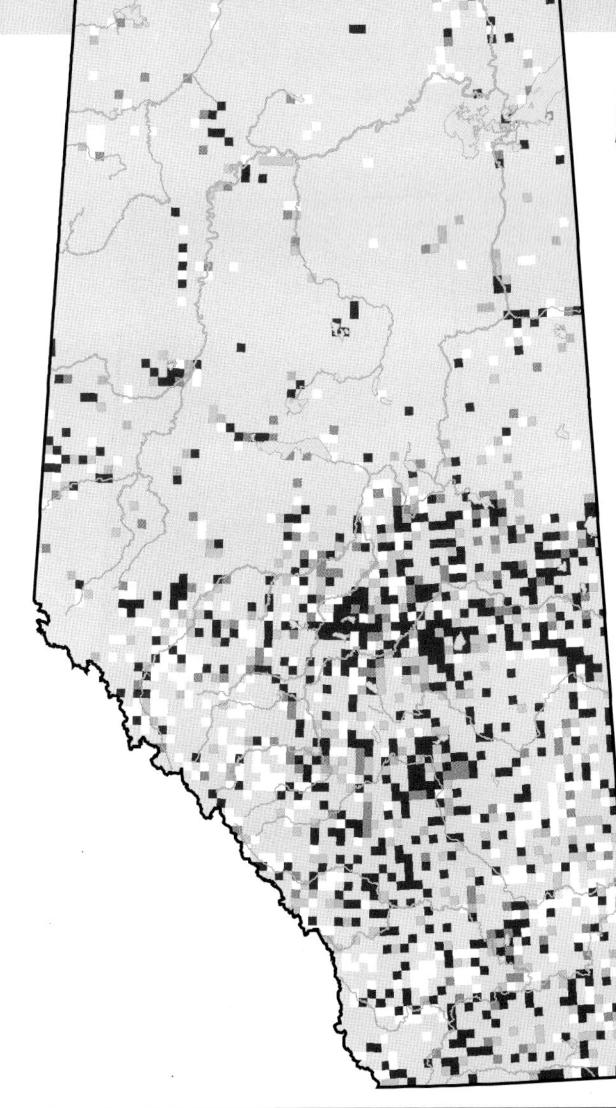

OBSERVED DISTRIBUTION ATLAS 2

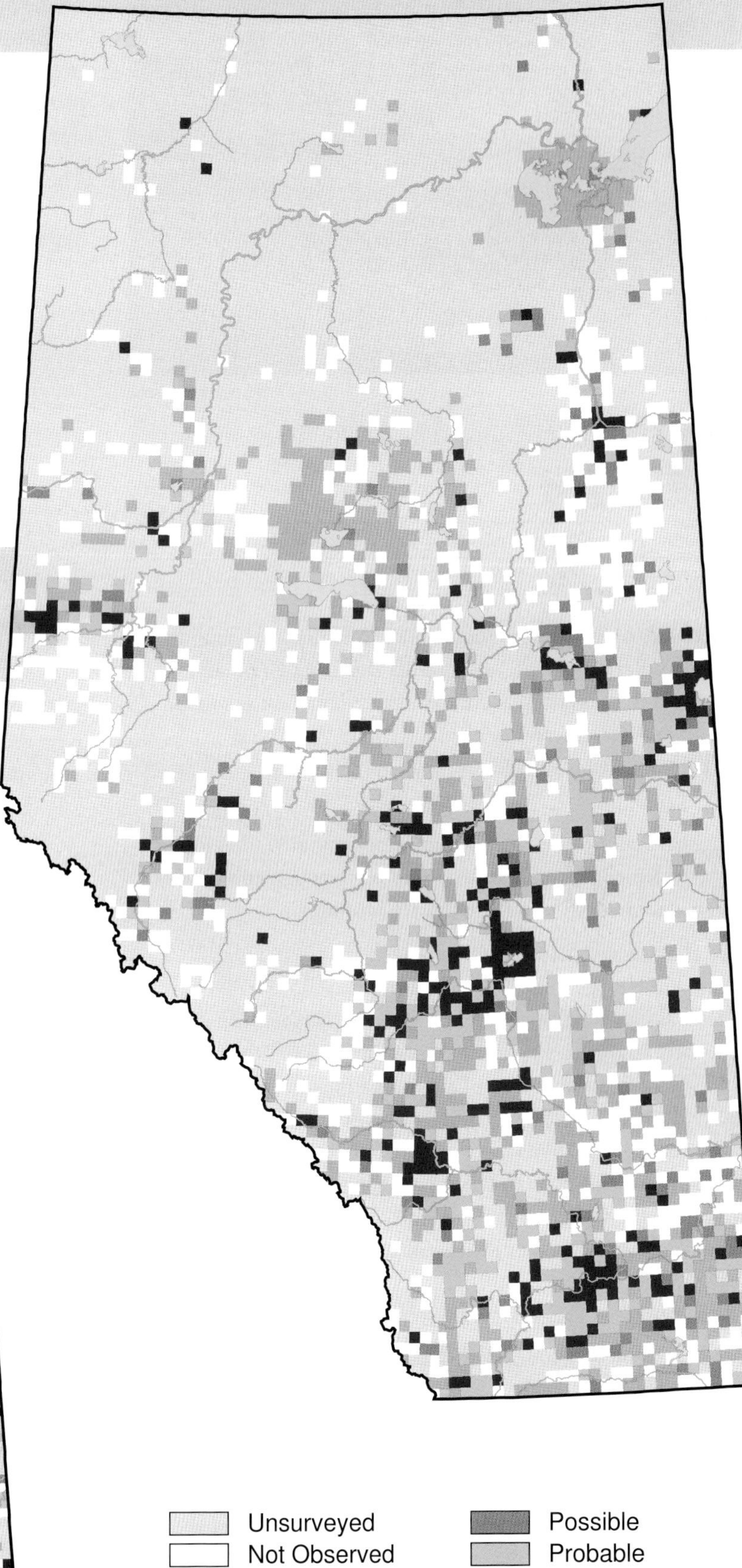

Blue-winged Teal (*Anas discors*)

Photo: Gerald Romanchuk

NESTING

CLUTCH SIZE (EGGS):	8–12
INCUBATION (DAYS):	22–24
FLEDGING (DAYS):	40–44
NEST HEIGHT (METRES):	0

The Blue-winged Teal is found in all Natural Regions of the province but is most prevalent in the Parkland, Grassland, southern Boreal Forest and the eastern edge of the southern Foothills NRs. There was no change in the observed distribution of this species between the two atlases.

This duck prefers shallow ponds, marshes, sloughs, and dugouts that provide abundant invertebrates. It will be found more often near shorelines than on open water, and it utilizes vegetated uplands near the water for nesting.

No changes in relative abundance were detected in the Parkland NR. Increases in relative abundance were detected in the Grassland and Rocky Mountain NRs. This may be due to the increased precipitation in the southern part of the province in the latter part of Atlas 2. Decreases in relative abundance were detected in the Boreal Forest and Foothills NRs. This may be attributed to a shift in population due to the more favourable conditions that exist in its preferred habitat regions as the Breeding Bird Survey (1985–2005) found no change in abundance in Alberta or across Canada. Frequency of observation relative to other species mirrored the Atlas-observed changes in relative abundance.

The population status of the Blue-winged Teal mirrors wetland conditions on its prairie breeding grounds. Populations dropped to a 40-year low in 1990 after several dry years but, in the decade following, numbers more than doubled (Rowher et al., 2002). This suggests that long-term wetland degradation on the prairies had not irreversibly damaged this teal's breeding habitat. The largest remaining expanses of grasslands in the Canadian Prairies are in southern Alberta and their proper management directly affects the future of a number of duck species in Western Canada.

The Blue-winged Teal is rated as Secure in Alberta.

TEMPORAL REPORTING RATE

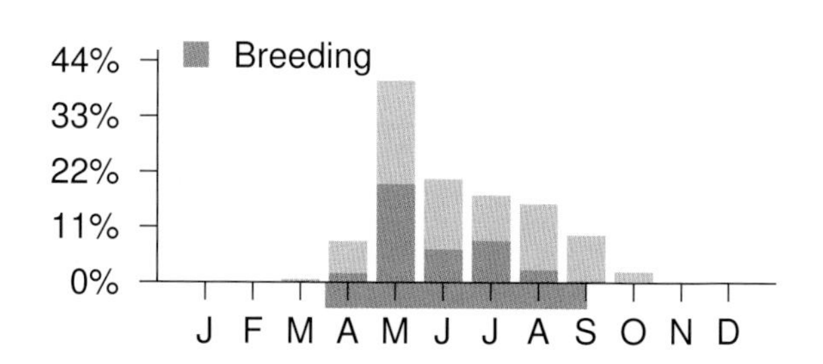

NATURAL REGIONS REPORTING RATE

SPATIAL REPORTING RATE

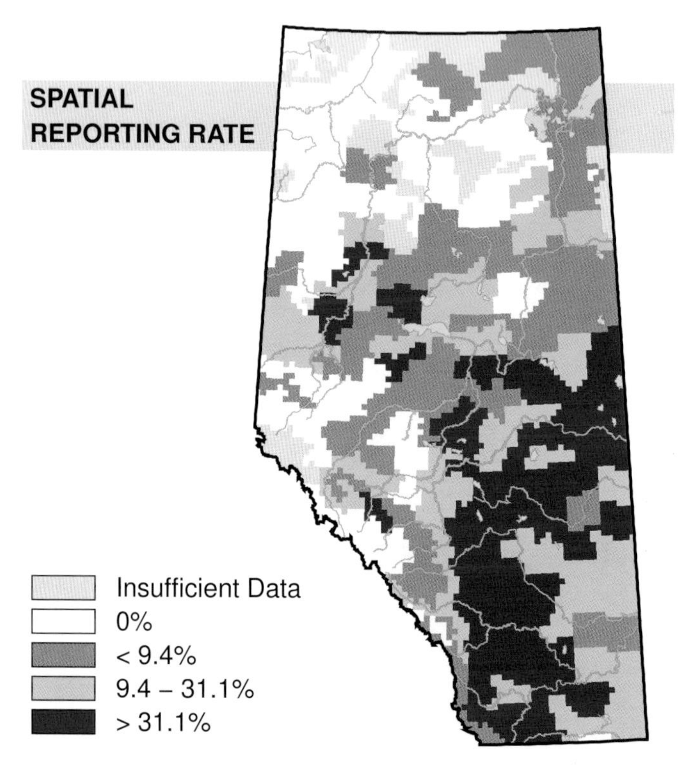

STATUS

GSWA 2000	GSWA 2005	COSEWIC Historic Rankings	COSEWIC 2007	Alberta Wildlife Act
Secure	Secure	N/A	N/A	N/A

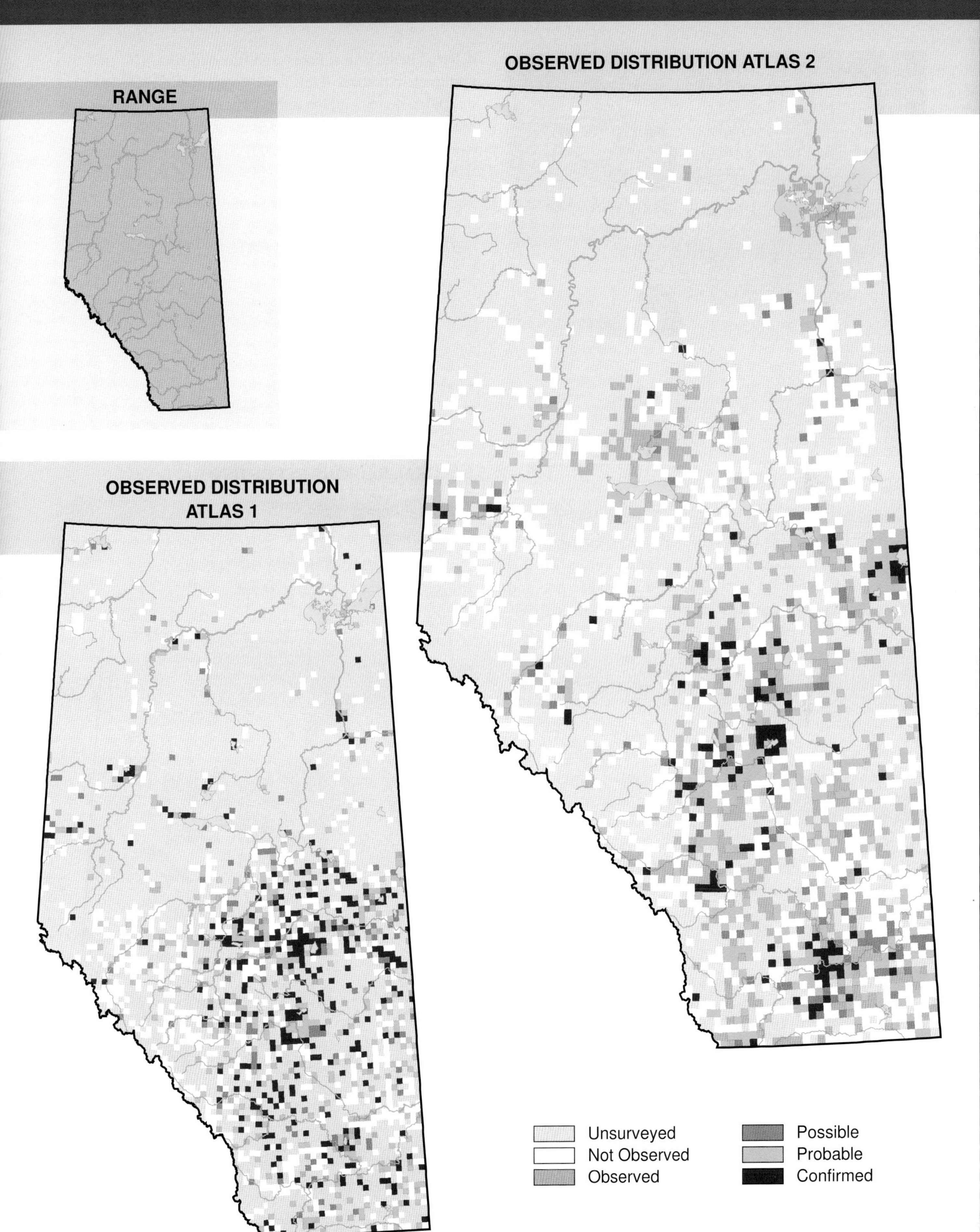
RANGE
OBSERVED DISTRIBUTION ATLAS 2
OBSERVED DISTRIBUTION ATLAS 1
Unsurveyed
Not Observed
Observed
Possible
Probable
Confirmed

Cinnamon Teal (*Anas cyanoptera*)

Photo: Alan MacKeigan

NESTING

CLUTCH SIZE (EGGS):	8–12
INCUBATION (DAYS):	22–25
FLEDGING (DAYS):	49
NEST HEIGHT (METRES):	0

Cinnamon Teal distribution in Alberta runs north from the Grassland Natural Region through the Parkland Natural Region and into the southern part of the Boreal Forest NR. There are also sporadic records from the southern Foothills and southern Rocky Mountain NRs. A record north of Peace River in Atlas 1 was not duplicated in Atlas 2. There was no change in the observed distribution of this species between Atlas 1 and Atlas 2, but consistent in both atlases were the observations of this species in the Grande Prairie area, as disjunct records from the main distribution to the south-east.

This duck uses shallow lake margins, freshwater wetlands both seasonal and semi-permanent, dugouts and ditches that have emergent vegetation, and muddy shorelines. In larger bodies of water it is never found far from shore. It appears to prefer basins with well-developed stands of emergent vegetation where it uses these emergent zones to a greater extent than open-water portions of basins. It usually nests near water in low, dense perennial vegetation. The availability and quality of wetlands and surrounding upland nesting habitats in the arid West may provide the most important limitation on North American populations (Gammonley, 1996).

Increases in relative abundance were detected in the Grassland and Rocky Mountain NRs. This could be attributed to the increase in precipitation in the southern part of Alberta in the latter part of Atlas 2. Decreases in relative abundance were detected in the Boreal Forest and Parkland NRs. The northeastern portion of this duck's range experienced lower than normal precipitation during Atlas 2. No changes in relative abundance were detected in the Foothills NR. The Breeding Bird Survey (1985–2005) reported no change in Alberta but a deline was detected across Canada.

The Cinnamon Teal is rated as Secure in Alberta.

TEMPORAL REPORTING RATE

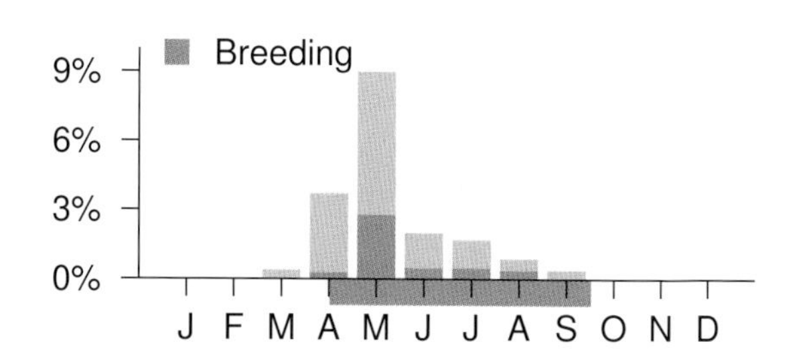

NATURAL REGIONS REPORTING RATE

SPATIAL REPORTING RATE

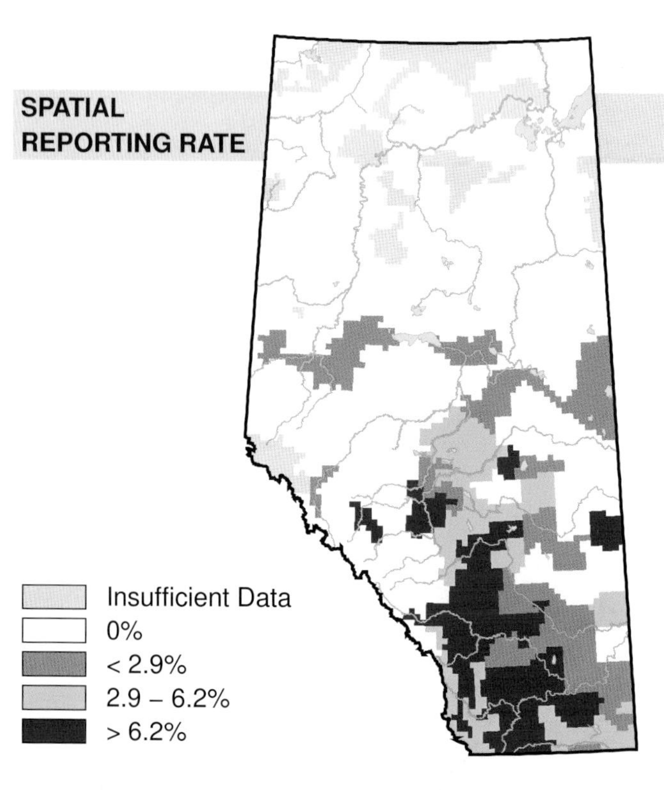

STATUS

GSWA 2000	GSWA 2005	COSEWIC Historic Rankings	COSEWIC 2007	Alberta Wildlife Act
Secure	Secure	N/A	N/A	N/A

OBSERVED DISTRIBUTION ATLAS 2

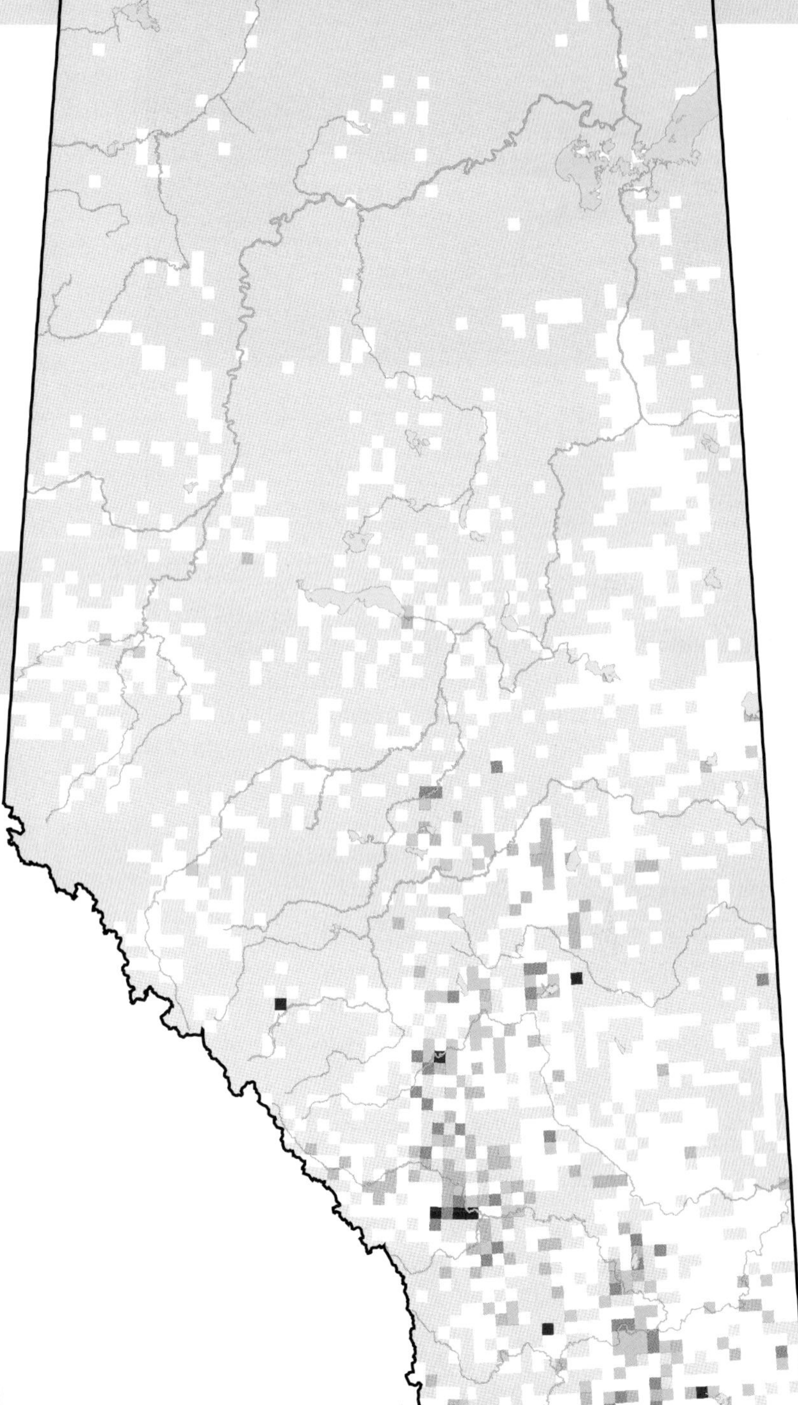

RANGE

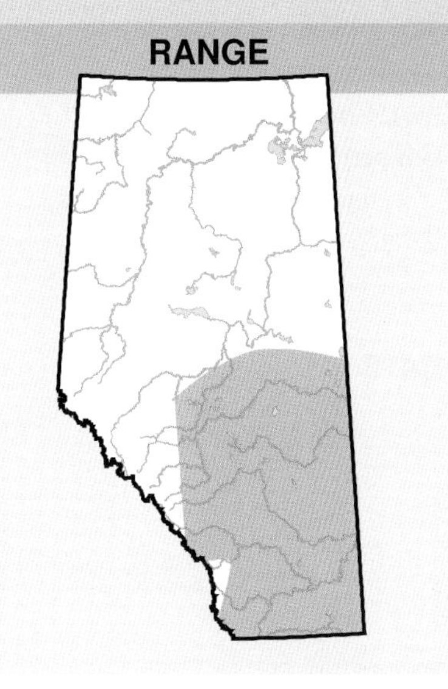

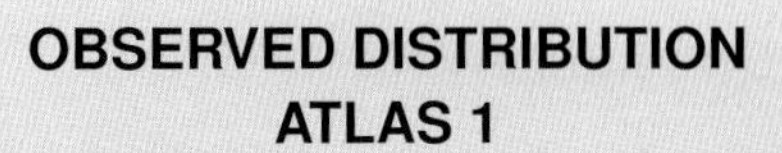

OBSERVED DISTRIBUTION ATLAS 1

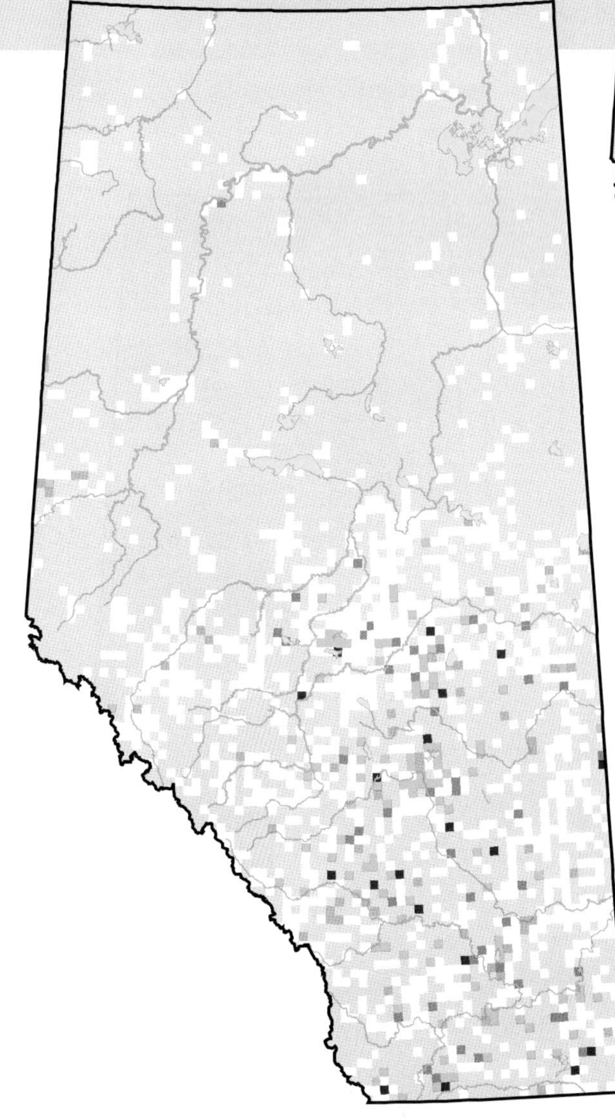

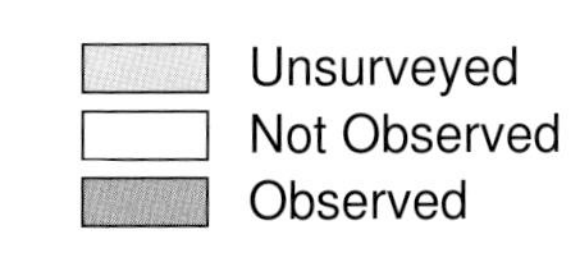

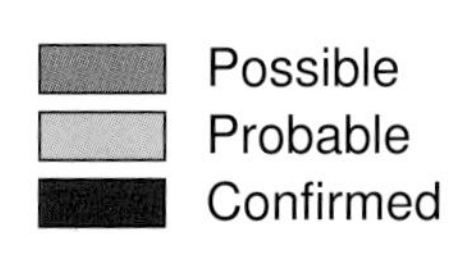

Northern Shoveler (*Anas clypeata*)

Photo: Mark Williams

NESTING

CLUTCH SIZE (EGGS):	8–12
INCUBATION (DAYS):	22–25
FLEDGING (DAYS):	56
NEST HEIGHT (METRES):	0

The Northern Shoveler breeds in all parts of the province with the preferred areas being the Grassland, Parkland, and southern Boreal Forest Natural Regions. It is somewhat rarer in the northern Boreal Forest, Foothills, and Rocky Mountain NRs due to a lack of suitable habitat. There was no change in the observed distribution between the two Atlases, although in Atlas 2 the Northern Shoveler was not recorded in the northwestern corner of the province.

The Northern Shoveler is found in a variety of wetland habitats. It uses open marshy areas with shallow waterways, muddy freshwater lakes, and sloughs with good cover and abundant submergent vegetation. Dry, sheltered, upland sites for nesting are nearby.

Increases in relative abundance were detected in the Grassland, Parkland, and Rocky Mountain NRs. Increased precipitation in the Grasslands and southern Rocky Mountain NRs in the latter part of Atlas 2 may have contributed to the increases in relative abundance in these two regions. Decreases in relative abundance were detected in the Boreal Forest NR. Drier conditions in this Natural Region in Atlas 2 compared to Atlas 1 may have contributed to this decline. No changes in relative abundance were detected in the Foothills NR. The Breeding Bird Survey (1985–2005) reported no change in the abundance of Northern Shovelers in Alberta or across Canada.

The largest factor affecting Northern Shoveler populations is the degradation of nesting grounds due to expansion and intensification of agriculture. The loss of plant cover leaves nests and young more vulnerable to predation, while farming operations (e.g., haying) increase mortality rates of current residents (Beauchamp et al., 1996). Changes in climate have also affected populations, as when periods of drought have reduced the number of habitable wetlands (Bethke and Nudds, 1995). The Northern Shoveler is rated as Secure in the province.

TEMPORAL REPORTING RATE

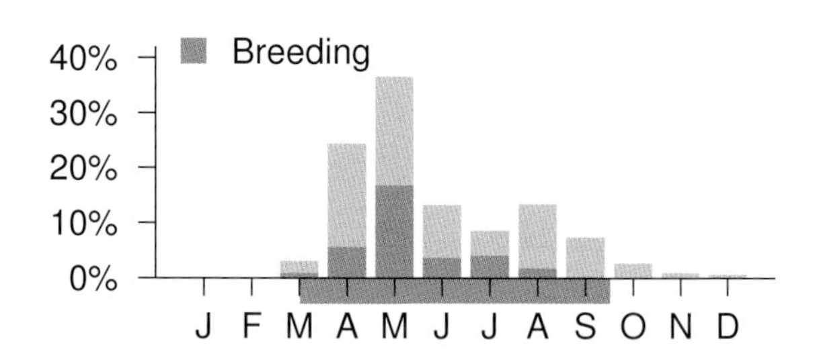

NATURAL REGIONS REPORTING RATE

Boreal Forest	3%
Grassland	25%
Parkland	20%
Rocky Mountain	4%

SPATIAL REPORTING RATE

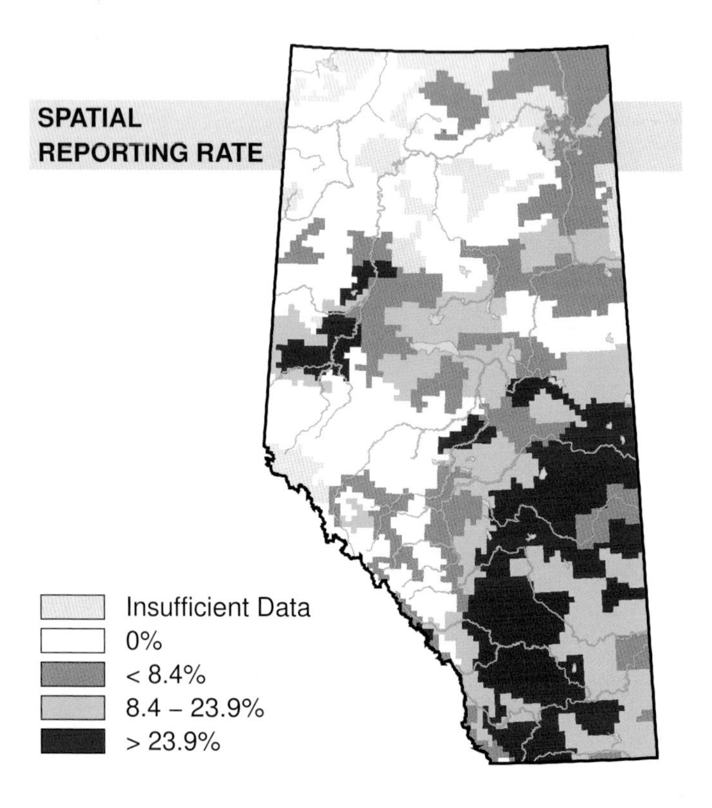

STATUS

GSWA 2000	GSWA 2005	COSEWIC Historic Rankings	COSEWIC 2007	Alberta Wildlife Act
Secure	Secure	N/A	N/A	N/A

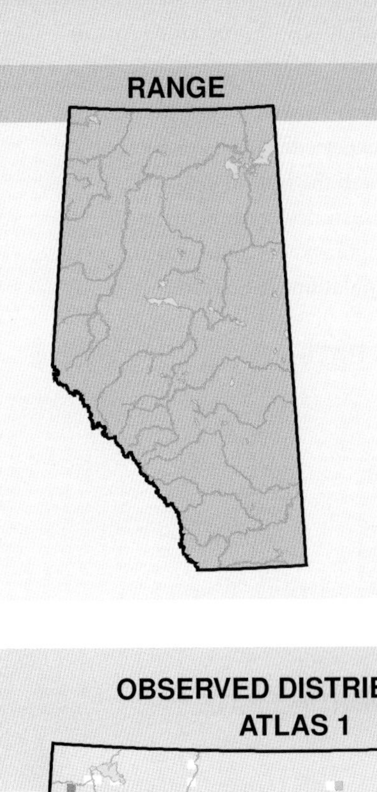
RANGE

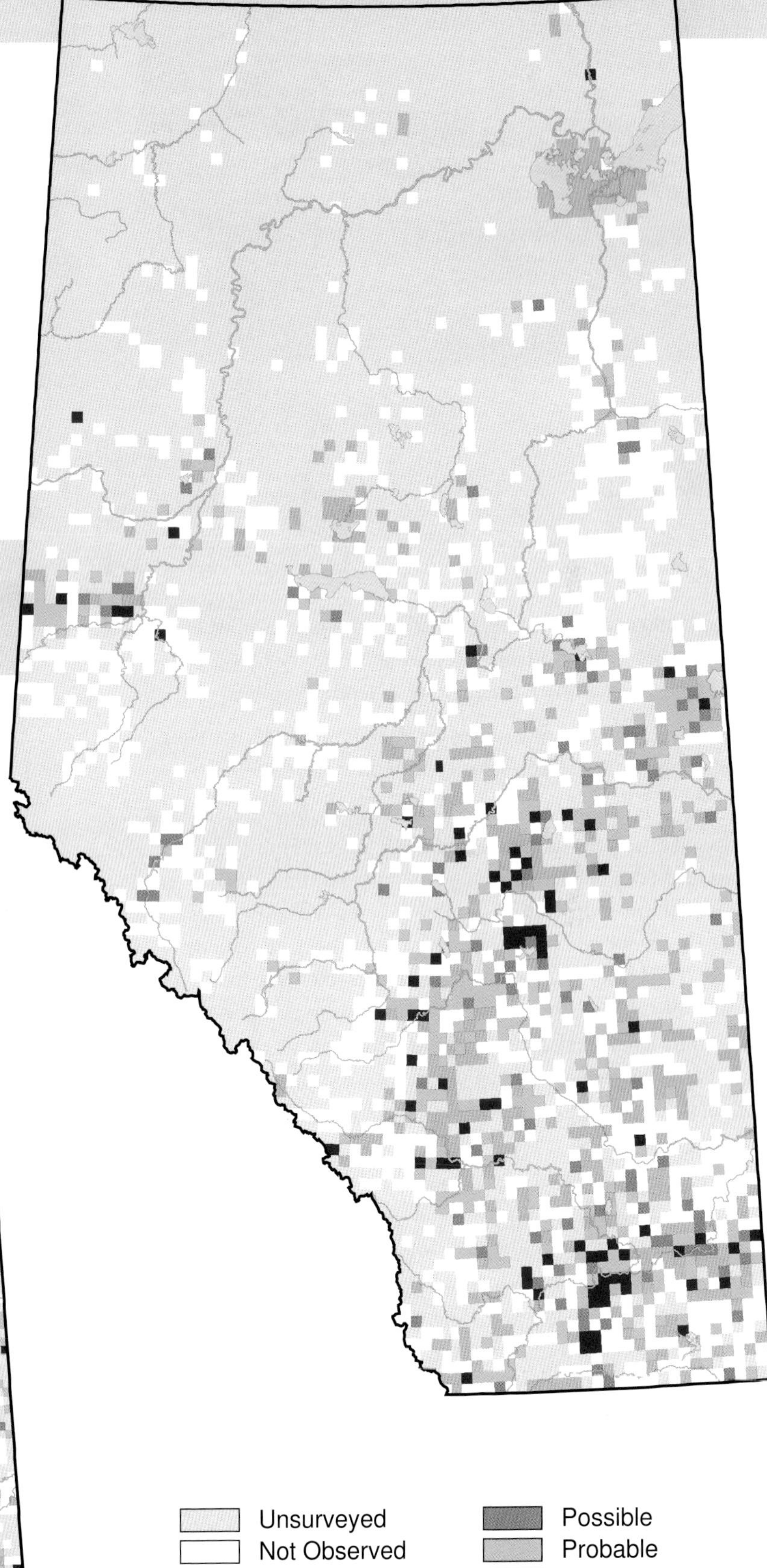
OBSERVED DISTRIBUTION ATLAS 2
Unsurveyed
Not Observed
Observed
Possible
Probable
Confirmed

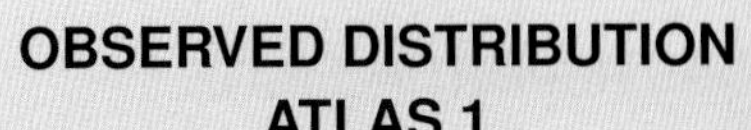
OBSERVED DISTRIBUTION ATLAS 1

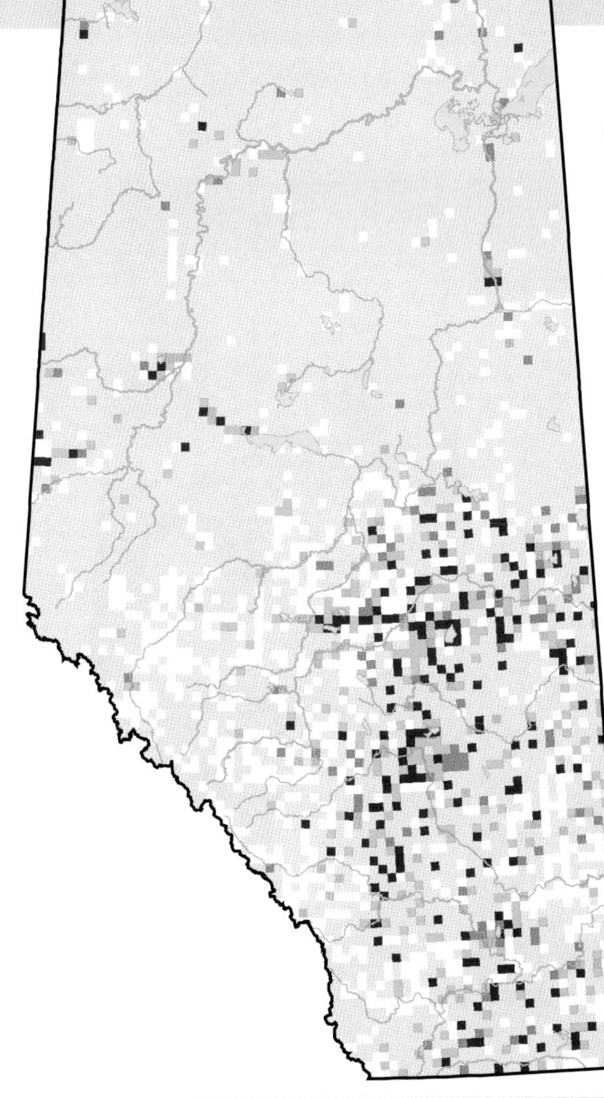

Northern Pintail (*Anas acuta*)

Photo: Mark Williams

NESTING

CLUTCH SIZE (EGGS):	7–10
INCUBATION (DAYS):	23–26
FLEDGING (DAYS):	42–46
NEST HEIGHT (METRES):	0

The Northern Pintail breeds throughout the province, where suitable habitat is available, predominantly in the Grassland, Parkland, and southern Boreal Forest Natural Regions.

It favours open terrain with seasonal shallow ponds, marshes, and reedy shallow lakes, usually with drier margins.

Increases in relative abundance were detected in the Grassland NR and decreases were detected in the Boreal Forest and Parkland NRs. This is attributed mainly to differences in precipitation, with the Grassland NR being wetter in Atlas 2 than in Atlas 1 and the Boreal Forest and Parkland NRs being drier. This difference provided additional suitable habitat in the south. No changes in relative abundance were detected in the Foothills and Rocky Mountain NRs. The Breeding Bird Survey (1985–2005), which rolls up all provincial data, reported no abundance change for Alberta or across Canada.

Annual nest success and productivity vary with water conditions, predation, and weather. Periods of extended drought in prairie nesting regions cause dramatic population declines, usually followed by periods of recovery (Austin and Miller, 1995). While populations of other prairie ducks have increased dramatically following the droughts of the late 1980s and early 2000s, Northern Pintail populations remain below their long-term average. The continued conversion of grasslands to cropland, and a reduction in farm acreage lying fallow each season, results in the destruction of many nests by agricultural machinery and high levels of predation (Austin and Miller, 1995).

The status of Northern Pintails in Alberta is Sensitive, as this widespread species has been experiencing severe population declines across North America in the last 40 years. Its preferred wetland habitat is threatened by drought and drainage. Conservation of temporary wetlands in native habitats is essential for the populations to stabilize.

TEMPORAL REPORTING RATE

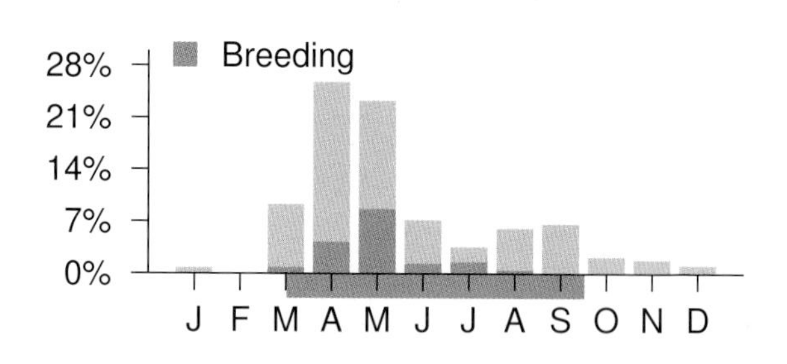

NATURAL REGIONS REPORTING RATE

SPATIAL REPORTING RATE

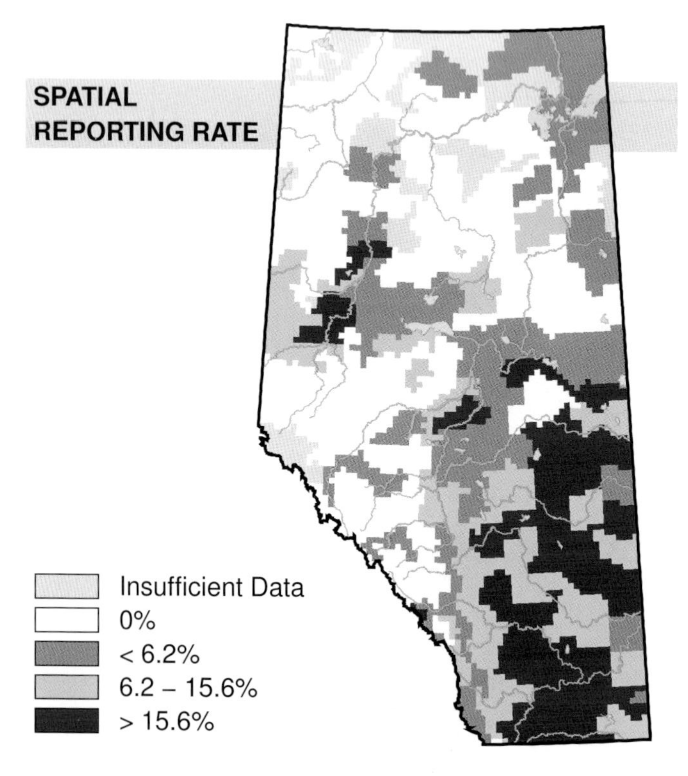

STATUS

GSWA 2000	GSWA 2005	COSEWIC Historic Rankings	COSEWIC 2007	Alberta Wildlife Act
Secure	Sensitive	N/A	N/A	N/A

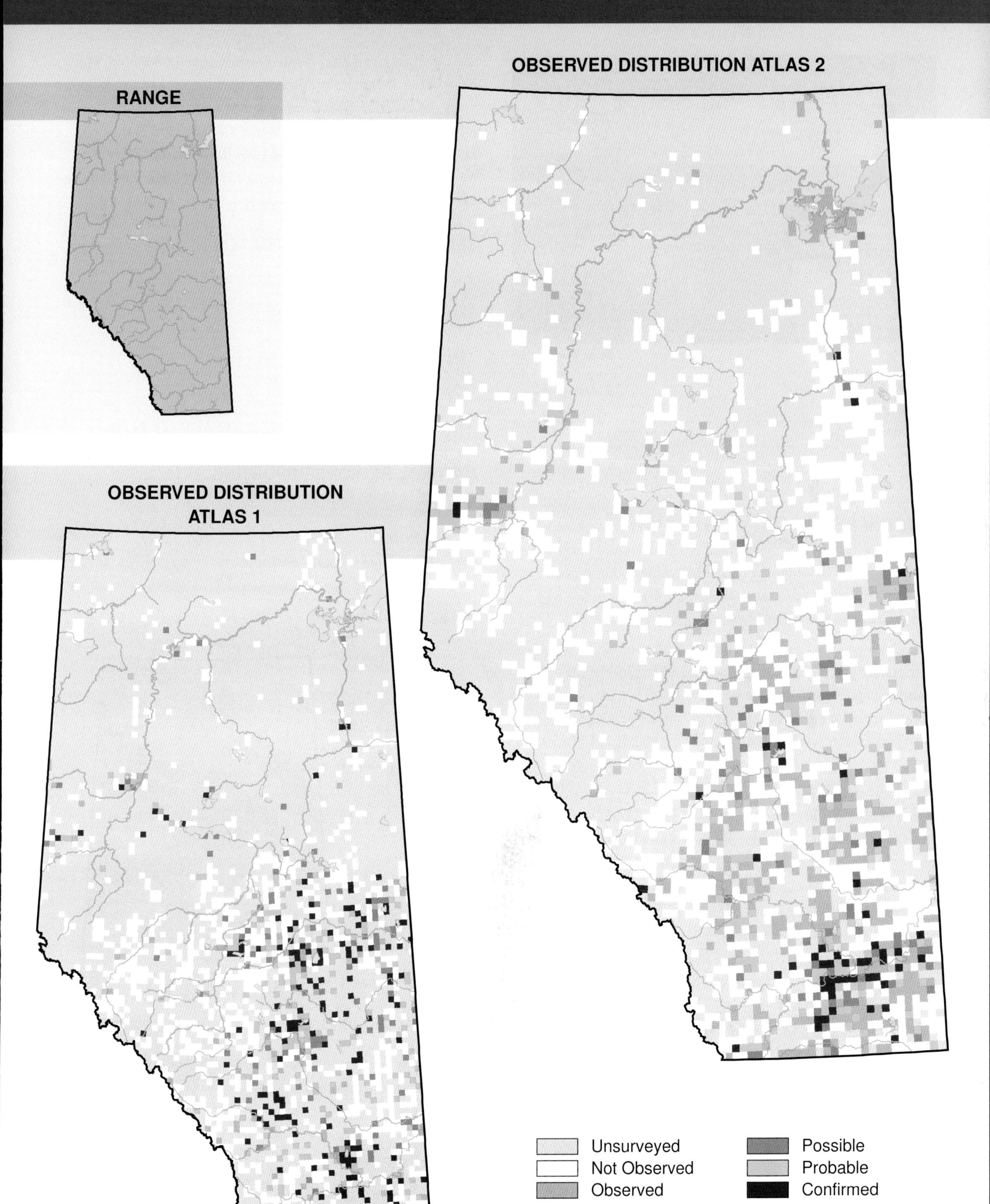

Habitat
Nest Location
Nest Type
Diet
OBSERVED DISTRIBUTION ATLAS 2
RANGE
OBSERVED DISTRIBUTION ATLAS 1
Unsurveyed
Not Observed
Observed
Possible
Probable
Confirmed

Green-winged Teal (*Anas crecca*)

Photo: Alan MacKeigan

NESTING

CLUTCH SIZE (EGGS):	8–10
INCUBATION (DAYS):	21–23
FLEDGING (DAYS):	44
NEST HEIGHT (METRES):	0

The Green-winged Teal is found breeding in all natural areas of Alberta. It is found predominantly in the central part of the province in the Grassland, Parkland, central Foothills, and southern Boreal Forest Natural Regions. The distribution of this species did not change between Atlas 1 and Atlas 2.

Green-winged Teal are unlike most dabblers in North America in that they prefer the wooded ponds and streams of the deciduous parklands and the boreal areas. The nests are in upland areas, in dense cover, often in shrubs or sedges. Breeding Green-winged Teals in the Grasslands NR place their nests mostly in sedge cover on low ground near sloughs.

Decreases in relative abundance were detected in the Boreal Forest, Foothills, and Parkland NRs. In the Boreal Forest NR this may be due in part to the drier conditions in that region during Atlas 2. No changes in relative abundance were detected in the Grassland and Rocky Mountain NRs. The Breeding Bird Survey (1995–2005) reported a decrease in Alberta and across Canada.

In the past, populations appeared to be relatively stable, and because its breeding habitat is relatively inaccessible, little attention has been given to the management of Green-winged Teal. Unfortunately, increasing exploration and development in the boreal forest and its fringes may have a negative effect on this species (Johnson, 1995). The rapid expansion of forestry, oil and gas production, mining, and agriculture impacts the boreal forest every year.

The Green-winged Teal is rated as Sensitive in Alberta. The 2005 status document prepared by the Government of Alberta states, “A common, widespread species with no known threats but is rapidly decreasing in Alberta, Canada, and North America." However, preliminary results from the 2006 Waterfowl Breeding Population and Habitat Survey indicate an increase in the population of this species in Alberta.

TEMPORAL REPORTING RATE

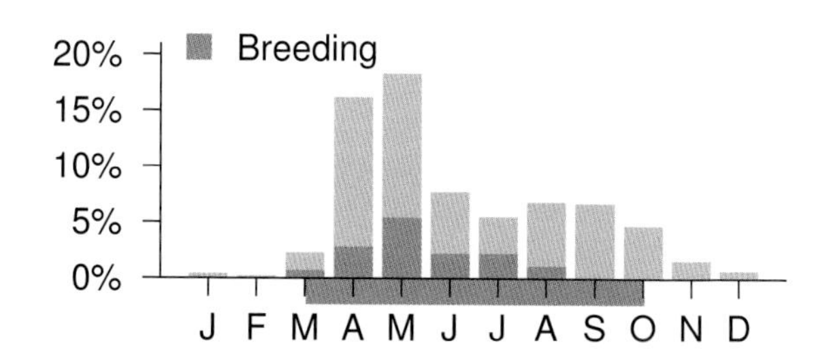

NATURAL REGIONS REPORTING RATE

SPATIAL REPORTING RATE

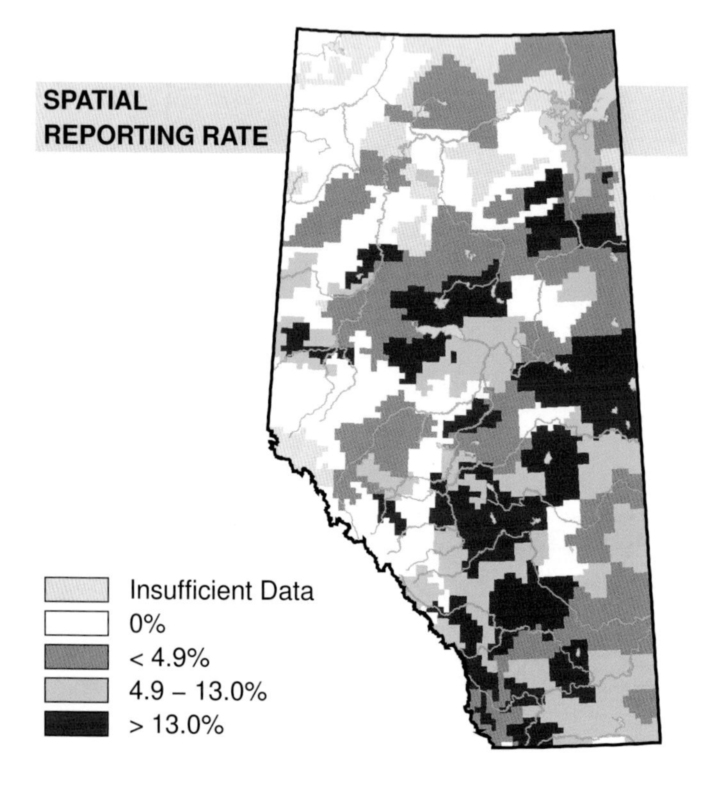

STATUS

GSWA 2000	GSWA 2005	COSEWIC Historic Rankings	COSEWIC 2007	Alberta Wildlife Act
Secure	Sensitive	N/A	N/A	N/A

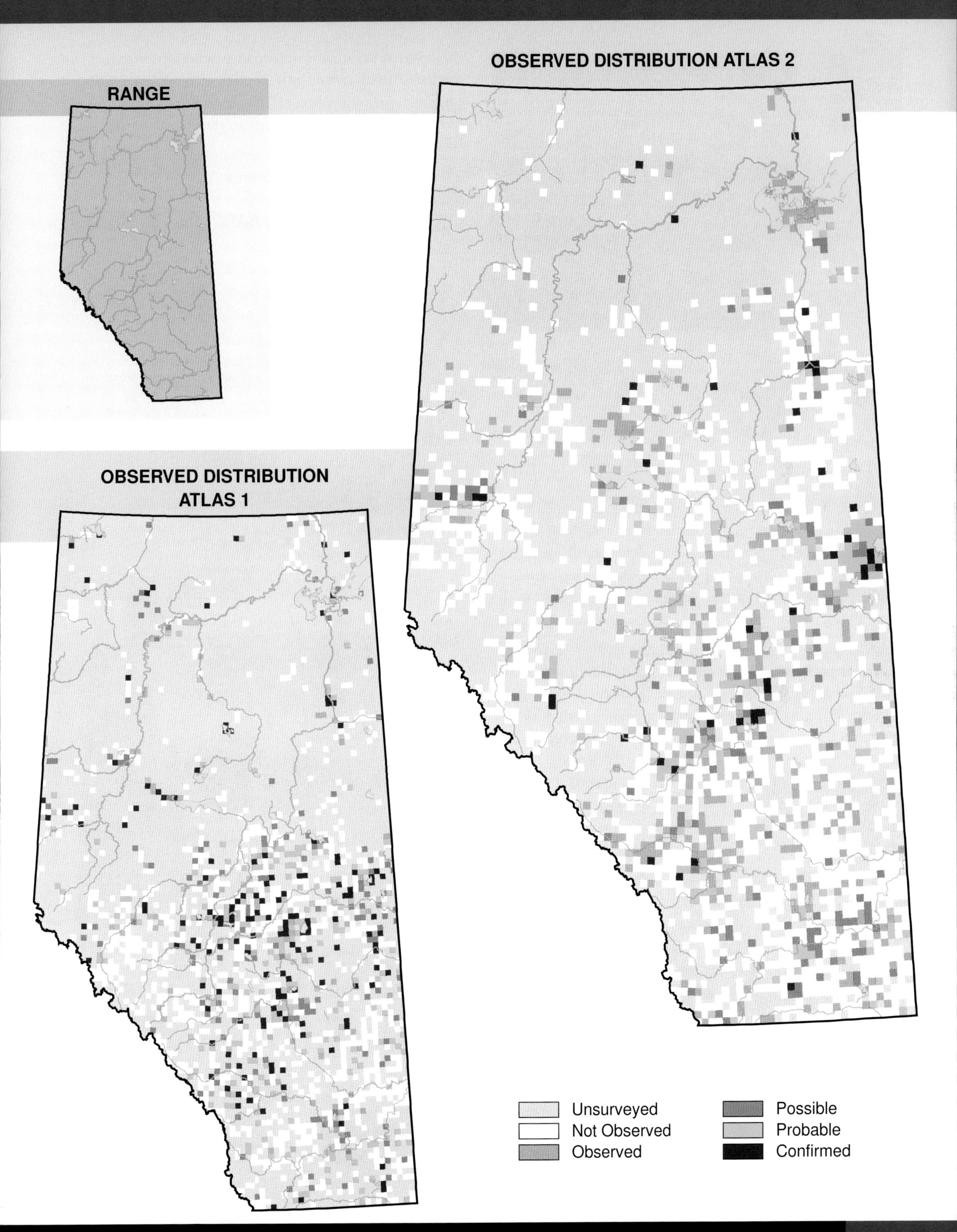
OBSERVED DISTRIBUTION ATLAS 2
RANGE
OBSERVED DISTRIBUTION
ATLAS 1
Unsurveyed
Not Observed
Observed
Possible
Probable
Confirmed

Canvasback (*Aythya valisineria*)

Photo: Mark Williams

NESTING

CLUTCH SIZE (EGGS):	7–10
INCUBATION (DAYS):	24–27
FLEDGING (DAYS):	63
NEST HEIGHT (METRES):	0

The Canvasback predominantly utilizes the Parkland, Grassland and southern Boreal Forest Natural Regions for breeding habitat. Data from Atlas 2 confirm that this species rarely occurs in the Rocky Mountain NR and in the extreme northern portion of the Boreal Forest NR. There was no change in the observed distribution of this species between Atlas 1 and Atlas 2.

This duck prefers deep, stable ponds, small lakes, marshes, and shallow river impoundments with emergent vegetation and vegetated margins as it nests over water. During periods of severe drought, it responds promptly by delaying breeding or failing to breed at all, accounting in part for its small population size. In wet years, increased availability of wetlands results in increased breeding densities, improved nest success, and a greater number of Canvasbacks (Mowbray, 2002).

No changes in relative abundance were detected in the Foothills and Rocky Mountain NRs. Increases in relative abundance were detected in the Grassland NR. The wetter conditions in this NR in the latter part of Atlas 2 may have been responsible for this increase. Decreases in relative abundance were detected in the Boreal Forest and Parkland NRs. This may be attributed to drier conditions in these NRs and to population shifts as the Canvasback utilized preferred habitat to the south. Mowbray (2002) documents that increases in cultivated land in Alberta take place at the expense of unimproved land that is important as waterfowl breeding habitat. There has been an expansion of agricultural lands in these two NRs since Atlas 1. The Breeding Bird Survey (1985–2005) detected no change in abundance for this species in Alberta or across Canada.

The Canvasback is rated as Secure in Alberta.

TEMPORAL REPORTING RATE

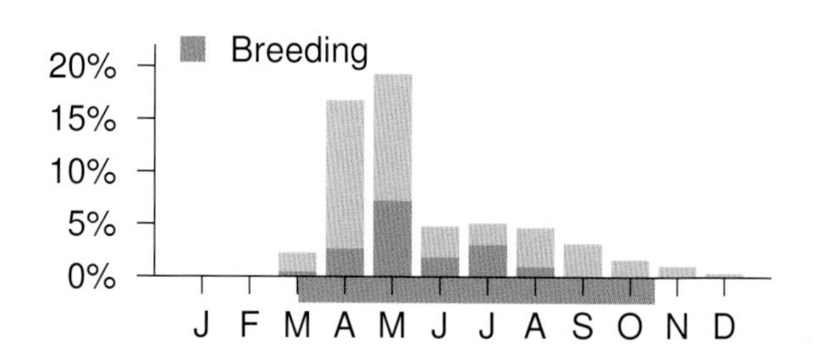

NATURAL REGIONS REPORTING RATE

SPATIAL REPORTING RATE

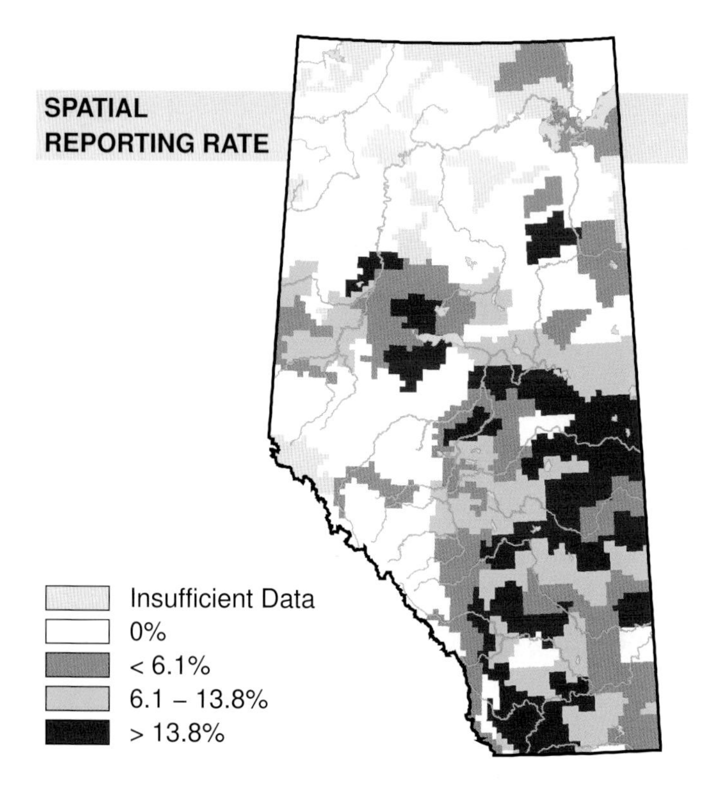

STATUS

GSWA 2000	GSWA 2005	COSEWIC Historic Rankings	COSEWIC 2007	Alberta Wildlife Act
Secure	Secure	N/A	N/A	N/A

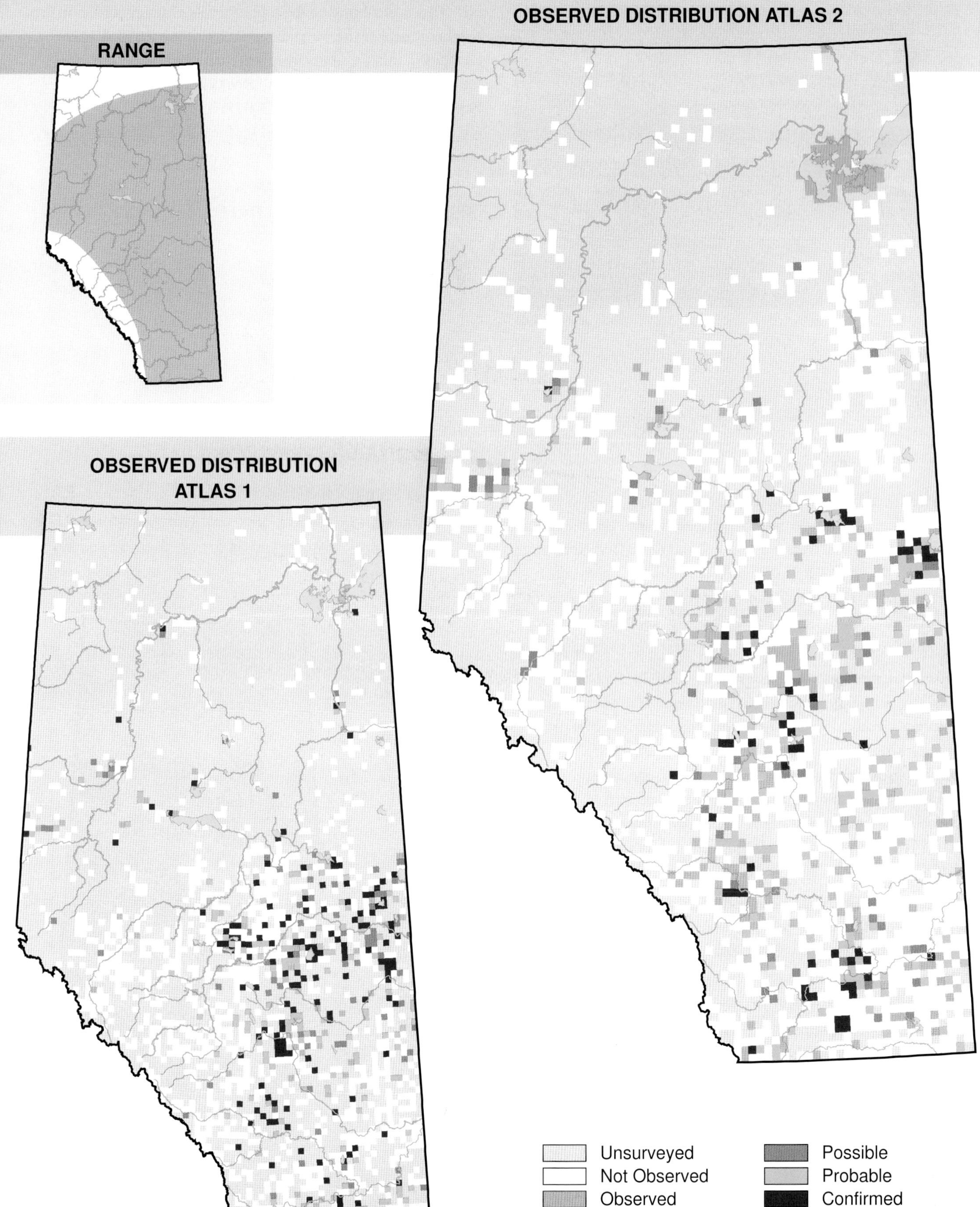
OBSERVED DISTRIBUTION ATLAS 2
RANGE
OBSERVED DISTRIBUTION
ATLAS 1
Unsurveyed
Not Observed
Observed
Possible
Probable
Confirmed

Redhead (*Aythya americana*)

Photo: Gerald Romanchuk

NESTING

CLUTCH SIZE (EGGS):	9–14
INCUBATION (DAYS):	24–28
FLEDGING (DAYS):	56–63
NEST HEIGHT (METRES):	0

The Redhead breeds locally throughout Alberta (Semenchuk, 1992) but prefers the central part of the province encompassing the southern Boreal Forest, Parkland, and Grassland Natural Regions. No change in distribution was observed between Atlas 1 and 2. However, Redheads were detected less often in the northern part of the Boreal Forest NR, likely due to more favourable conditions in its preferred habitat to the south.

This species exhibits a high degree of flexibility in habitat, food use, and reproductive behaviour; yet, paradoxically, its numbers are consistently surpassed by populations of most other prairie-nesting ducks (Woodin and Michot, 2002). It nests in predominantly non-forested areas supporting seasonally and semipermanently flooded wetlands, with fairly dense emergent vegetation for cover. This preference resulted in the Redhead's being more commonly observed in the Parkland and Grassland NRs in both Atlas 1 and 2.

Woodin and Michot (2002) report, "Redhead numbers have fluctuated wildly in the past ... Drainage of wetlands in the Prairie Pothole Region is a continuing concern in breeding range. Wetland creation and enhancement projects undertaken by North American Waterfowl Management Plan for waterfowl breeding in Prairie Pothole Region and for ducks wintering along the west coast of the Gulf of Mexico benefit this species."

Increases in relative abundance were detected in the Grassland NR. This is attributed to the higher levels of precipitation in the southern parts of the province in Atlas 2 than in Atlas 1. No changes in relative abundance were detected in the Boreal Forest, Foothills, Parkland, and Rocky Mountain NRs. The Breeding Bird Survey (1985–2005) reported no change in abundance in its provincial roll up for Alberta. This species is considered Secure in Alberta.

TEMPORAL REPORTING RATE

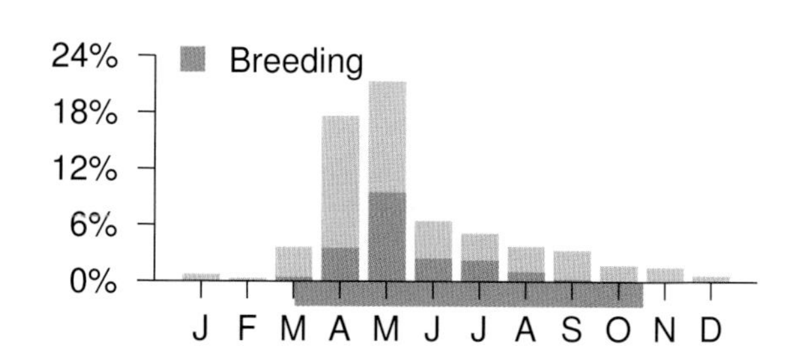

NATURAL REGIONS REPORTING RATE

SPATIAL REPORTING RATE

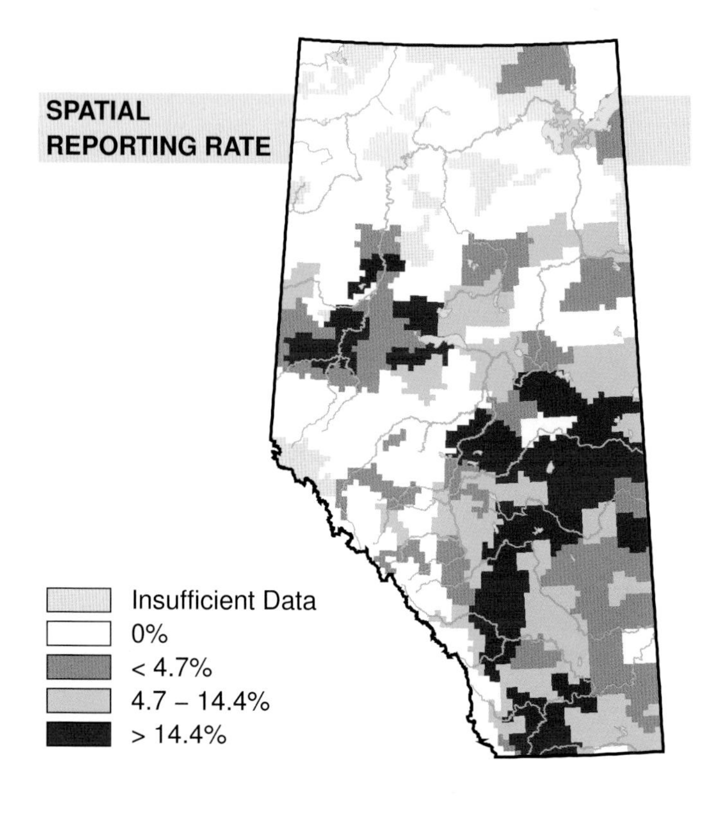

STATUS

GSWA 2000	GSWA 2005	COSEWIC Historic Rankings	COSEWIC 2007	Alberta Wildlife Act
Secure	Secure	N/A	N/A	N/A

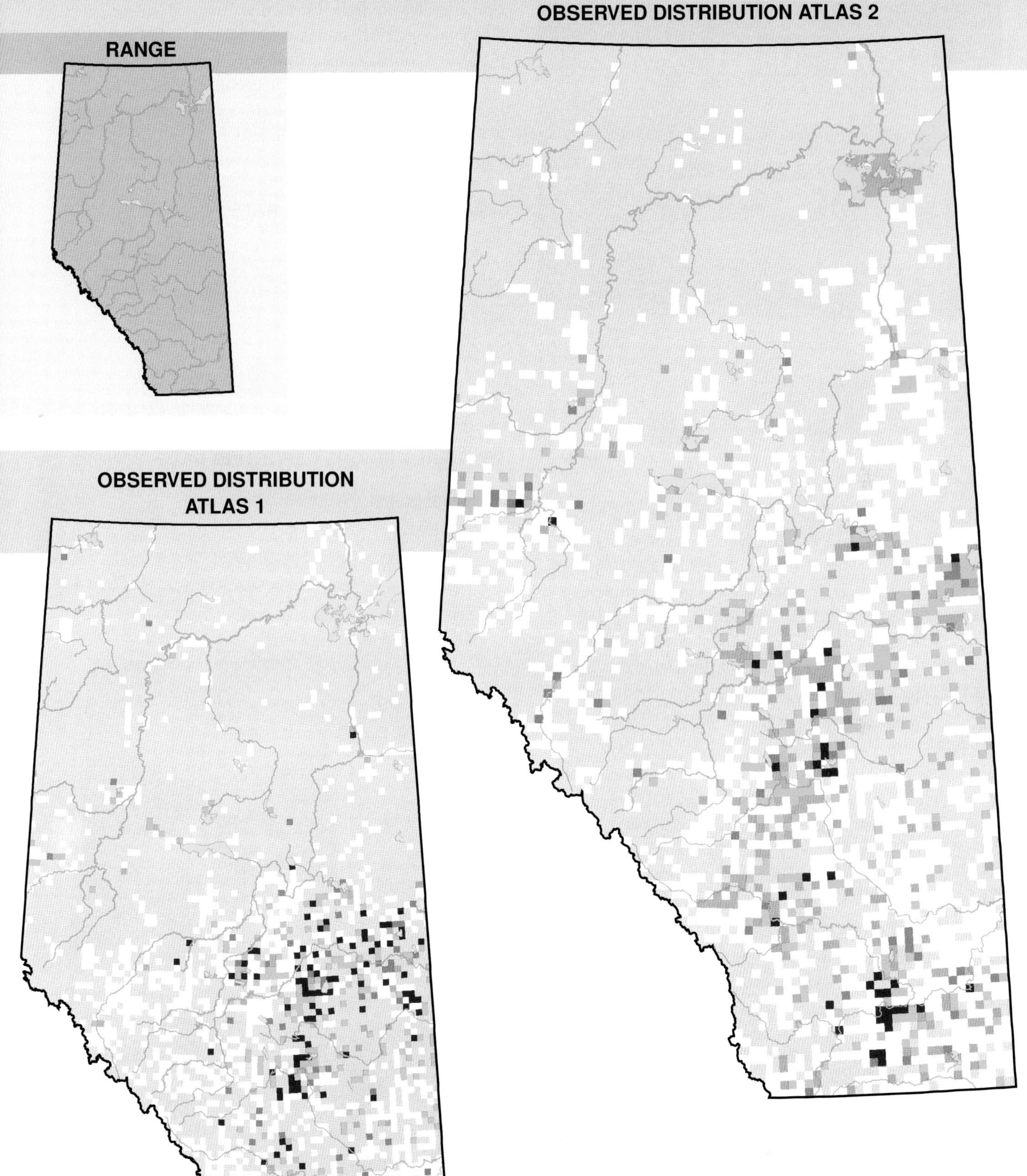
OBSERVED DISTRIBUTION ATLAS 2
RANGE
OBSERVED DISTRIBUTION ATLAS 1
Unsurveyed
Not Observed
Observed
Possible
Probable
Confirmed

Ring-necked Duck (*Aythya collaris*)

Photo: Raymond Toal

NESTING

CLUTCH SIZE (EGGS):	8–12
INCUBATION (DAYS):	25–28
FLEDGING (DAYS):	49–56
NEST HEIGHT (METRES):	0

The Ring-necked Duck is misleadingly named for the chestnut-coloured ring around the breeding male's black neck, the ring is barely visible on birds in the field. This duck breeds in northern and central Alberta south to Red Deer, with localized breeding in the Grassland Natural Region and the southern Rocky Mountain NR. In Atlas 2, this species was recorded breeding farther southeast into the Grasslands NR and south of the Cypress Hills. The increased precipitation in this region in the latter part of Atlas 2 may have provided additional suitable breeding habitat.

These ducks prefer marshes, swamps and bogs where the nest is built in damp situations and not on dry ground (Semenchuk, 1992). The brackish or alkaline water of the Grassland NR is not a preferred breeding habitat; still, the bird's scarcity in these habitats is puzzling because food resources and nesting substrate clearly are adequate for this species (Hohman and Eberhardt, 1998).

Increases in relative abundance were detected in the Grassland, Parkland, and Rocky Mountain NRs. Decreases in relative abundance were detected in the Boreal Forest and Foothills NRs. As this species breeds in permanently flooded wetlands, the increased industrial development in these regions and the associated removal of tree cover may be affecting the preferred breeding habitat. The Breeding Bird Survey (1985–2005) detected no change in abundance for Alberta or across Canada.

Curiously, the breeding distribution expanded and the population increased during the 1980s and early 1990s, when populations of most other North American ducks—especially prairie-nesting species—were in decline (Hohman and Eberhardt, 1998). Population surveys can be complicated due to the Ring-necked Duck's similarity in appearance to scaups. The Ring-necked Duck is rated as Secure in Alberta.

TEMPORAL REPORTING RATE

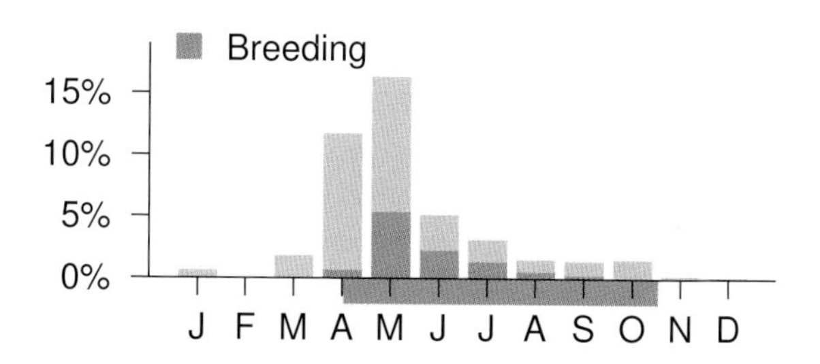

NATURAL REGIONS REPORTING RATE

SPATIAL REPORTING RATE

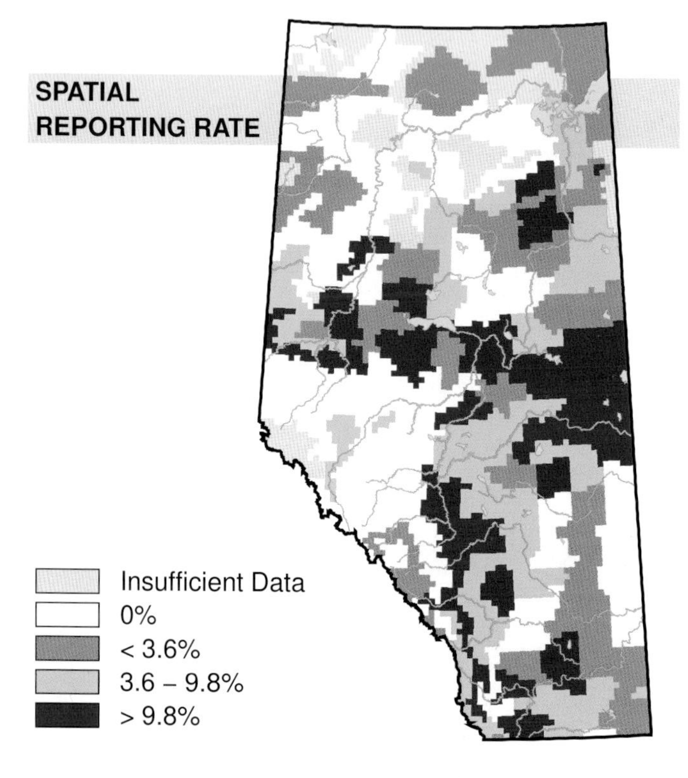

STATUS

GSWA 2000	GSWA 2005	COSEWIC Historic Rankings	COSEWIC 2007	Alberta Wildlife Act
Secure	Secure	N/A	N/A	N/A

Habitat

Nest
Location

Nest
Type

Diet

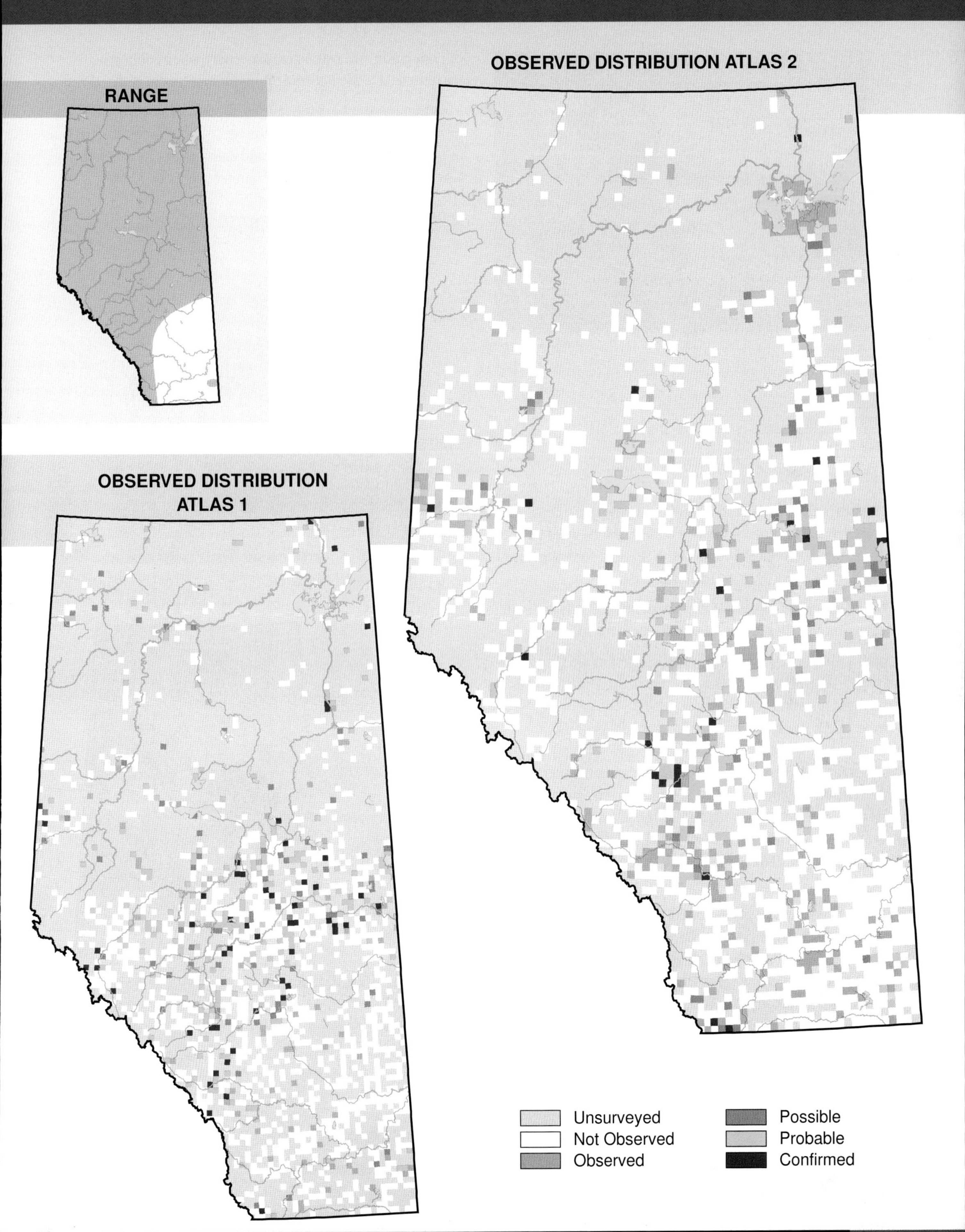
OBSERVED DISTRIBUTION ATLAS 2
RANGE
OBSERVED DISTRIBUTION
ATLAS 1
Unsurveyed
Not Observed
Observed
Possible
Probable
Confirmed

Lesser Scaup (*Aythya affinis*)

Photo: Raymond Toal

NESTING

CLUTCH SIZE (EGGS):	9–12
INCUBATION (DAYS):	23–26
FLEDGING (DAYS):	49
NEST HEIGHT (METRES):	0

The Lesser Scaup is found in all Natural Regions in the province but predominantly in the Parkland, Grassland, eastern Foothills, and southern Boreal Forest NRs. It is one of the most widely distributed of North American ducks.

They prefer permanent and semi-permanent wetlands with tall, dense herbaceous vegetation nearby for nest cover.

Increases in relative abundance were detected in the Grassland and Rocky Mountain NRs. The increase in the Grassland NR may be due to wetter conditions in Atlas 2. The cause of increase in the Rocky Mountain NR is not understood. Decreases were detected in the Boreal Forest NR. New logging activities in the boreal forests of the Canadian provinces and territories may threaten breeding habitat (Austin et al., 1998). Conservation efforts have focussed on the prairie and parkland areas of Canada with little activity in the Boreal Forest NR. No changes in relative abundance were detected in the Foothills NR.

The Breeding Bird Survey (1985–2005) reported no change in Alberta, but a decline in abundance was detected across Canada. Because of the timing of spring migration (May) and of the Breeding Waterfowl Population Survey (extending from May in southern regions to mid-June in the north), the potential exists for double-counting of birds or missing many individuals south of the survey limits; these factors may contribute to the high annual variation in counts and subsequently low precision of population size estimates, especially when cool spring weather delays the northward migration (Austin et al., 1998).

The rating of the Lesser Scaup in Alberta moved from Secure in 2000 to Sensitive in 2005. Surveys show a long-term decline in populations within Alberta and surrounding jurisdictions. Alteration and loss of suitable habitat may pose threats.

TEMPORAL REPORTING RATE

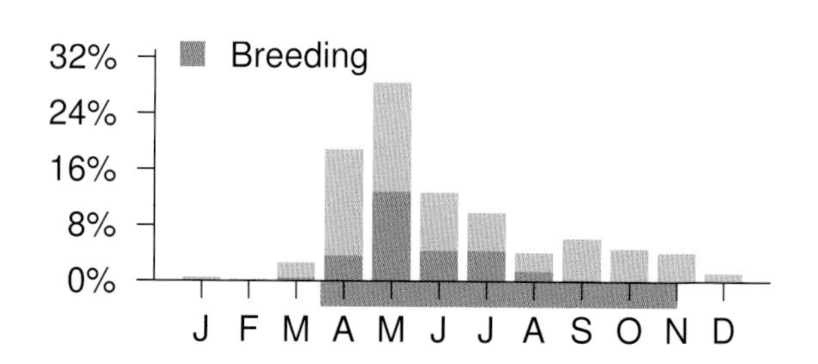

NATURAL REGIONS REPORTING RATE

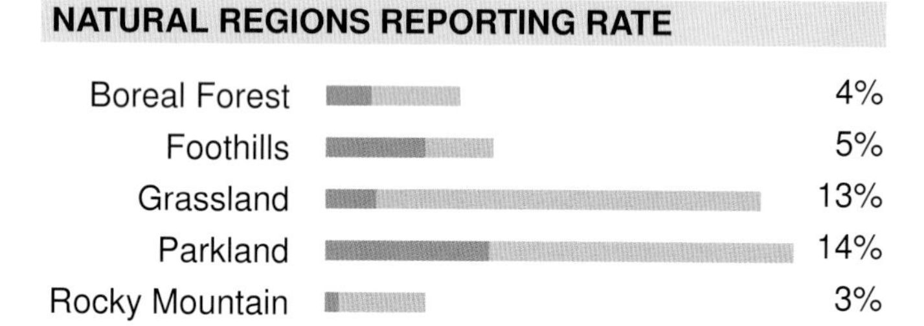

SPATIAL REPORTING RATE

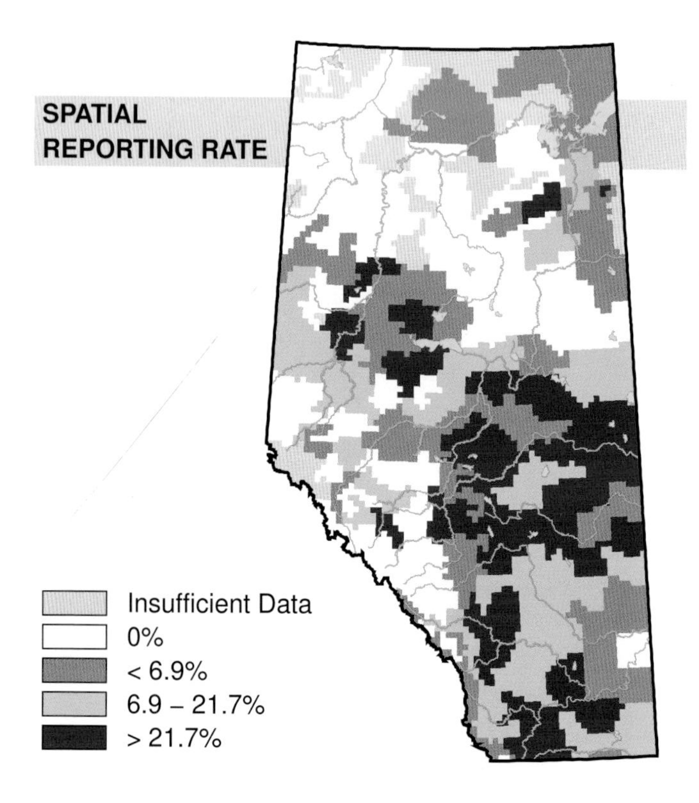

STATUS

GSWA 2000	GSWA 2005	COSEWIC Historic Rankings	COSEWIC 2007	Alberta Wildlife Act
Secure	Sensitive	N/A	N/A	N/A

RANGE

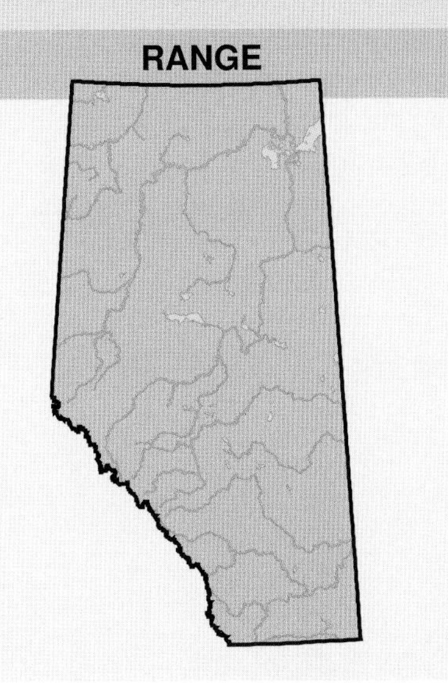

OBSERVED DISTRIBUTION ATLAS 2

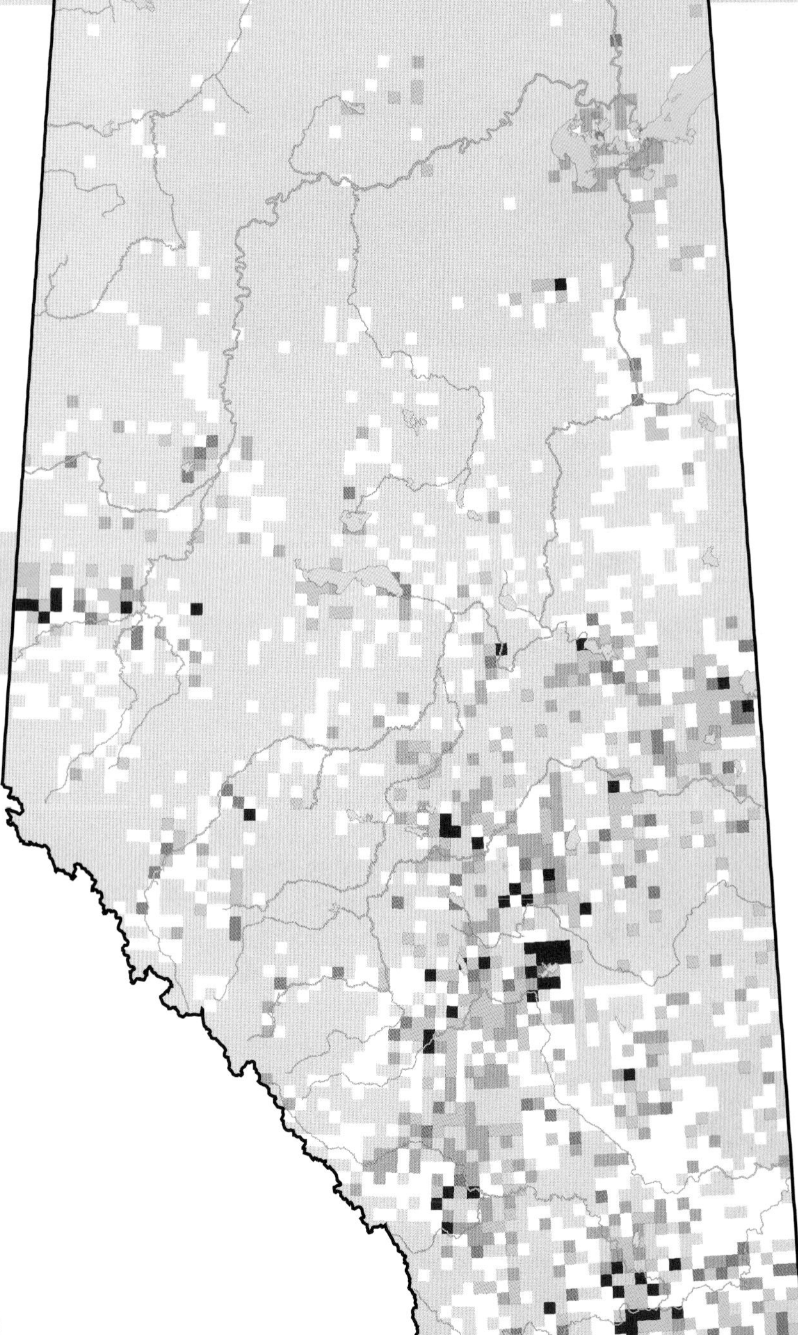

OBSERVED DISTRIBUTION ATLAS 1

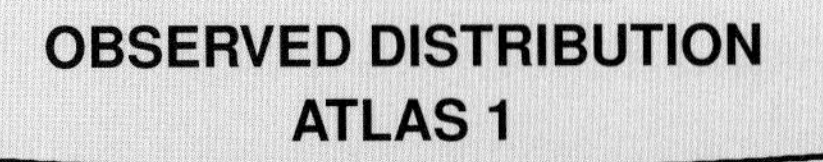

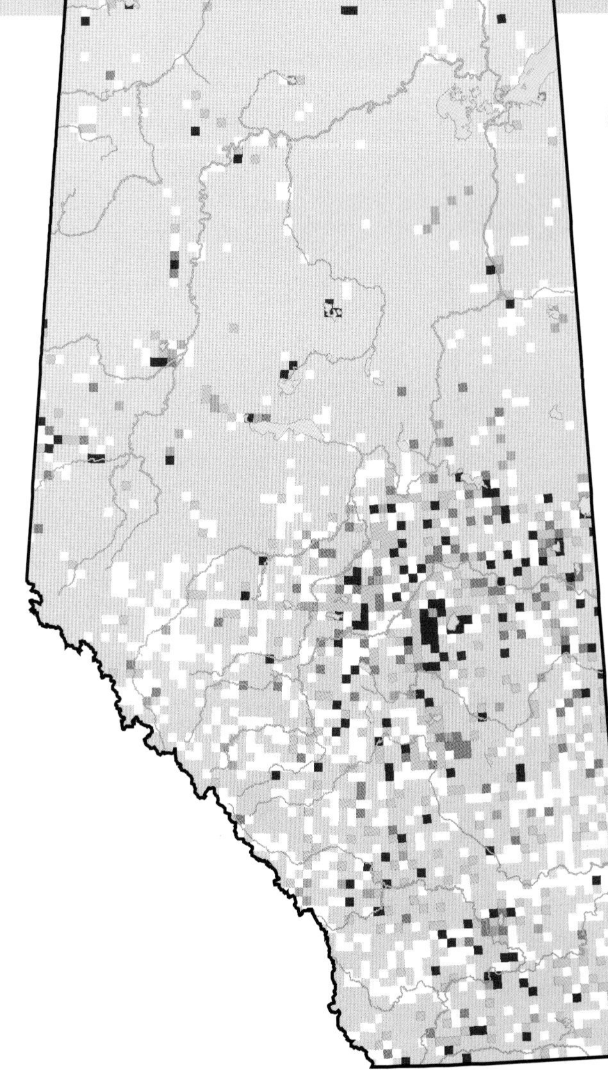

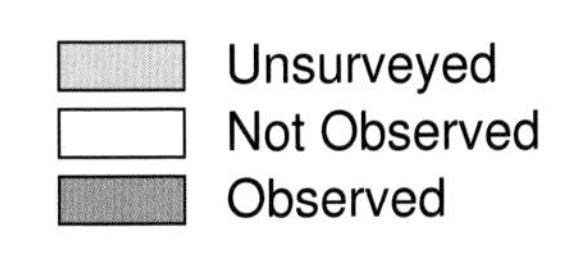

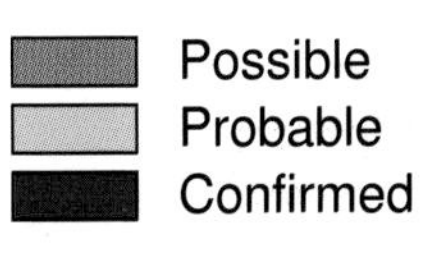

Harlequin Duck (*Histrionicus histrionicus*)

Photo: Gerald Romanchuk

NESTING

CLUTCH SIZE (EGGS):	6–8
INCUBATION (DAYS):	27–30
FLEDGING (DAYS):	40
NEST HEIGHT (METRES):	0

The Harlequin Duck is a small diving duck whose distribution is restricted to the foothills and mountainous part of southwestern Alberta. Its range extends from Waterton Lakes National Park in the south to north of the Kakwa River in the north. No change in distribution was observed between Atlas 1 and Atlas 2. An apparent gap in distribution between the North Saskatchewan and Clearwater rivers in Atlas 2 is due to lack of coverage. Spring pair surveys conducted by the Canadian Wildlife Service in May 1999 confirmed the presence of breeding pairs in this area on the North Ram and Clearwater rivers. Likewise, substantial numbers of breeding pairs were observed in several streams and rivers in the Willmore Wilderness Park area during surveys conducted in 1998 and 1999 by the Canadian Wildlife Service (Gregoire, 2000).

A higher reporting rate was recorded in the headwaters of the Red Deer, Bow, and Oldman watersheds than in the North Saskatchewan, Athabasca and Smoky watersheds. This may be due to completion of a higher proportion of squares, and may not actually indicate higher relative abundance. Access to breeding streams frequented by Harlequin Ducks is generally more difficult in the northern part of this species' range than in the southern part. Observations in the Boreal Forest and Parkland Natural Regions were made during the non-breeding periods or were not breeding observations. This was also true of the Grassland NR with the exception of a few records in the transition zone from the Foothills NR.

No changes in relative abundance were detected between Atlas 1 and Atlas 2 for this species. The male Harlequin Duck migrates to the west coast once females begin incubation and females with broods are very cryptic in the breeding streams; it is likely that these ducks are underestimated by the Atlas. The provincial status of this species is Sensitive due to threats to its breeding habitat from industrial development, grazing, and recreational activities.

TEMPORAL REPORTING RATE

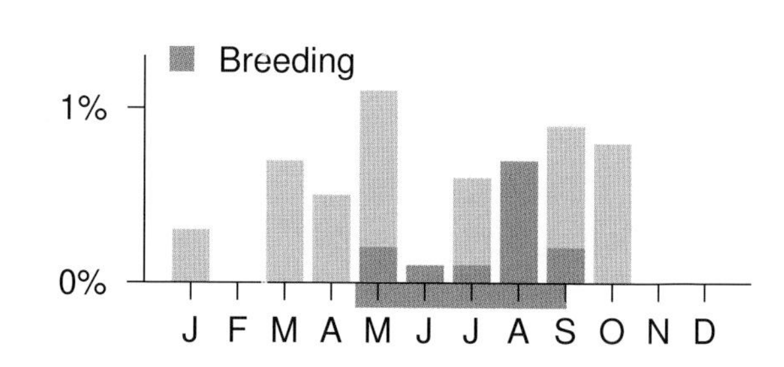

NATURAL REGIONS REPORTING RATE

SPATIAL REPORTING RATE

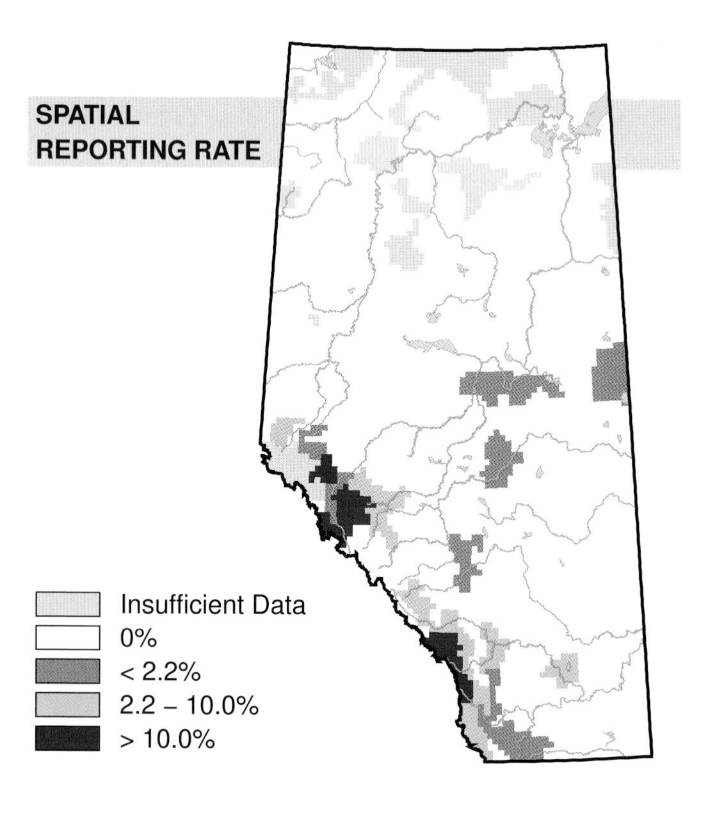

STATUS

GSWA 2000	GSWA 2005	COSEWIC Historic Rankings	COSEWIC 2007	Alberta Wildlife Act
Sensitive	Sensitive	N/A	N/A	N/A

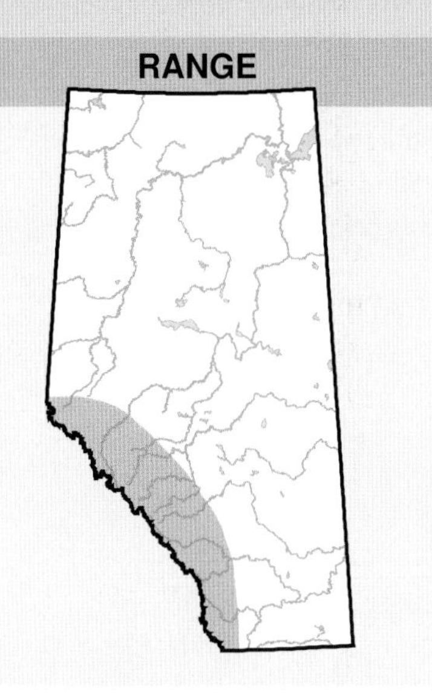

RANGE

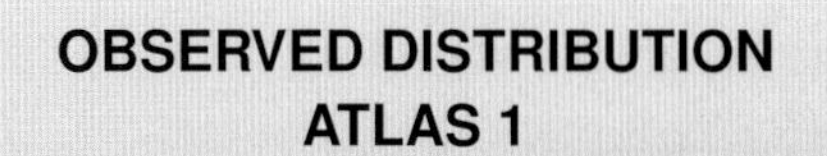

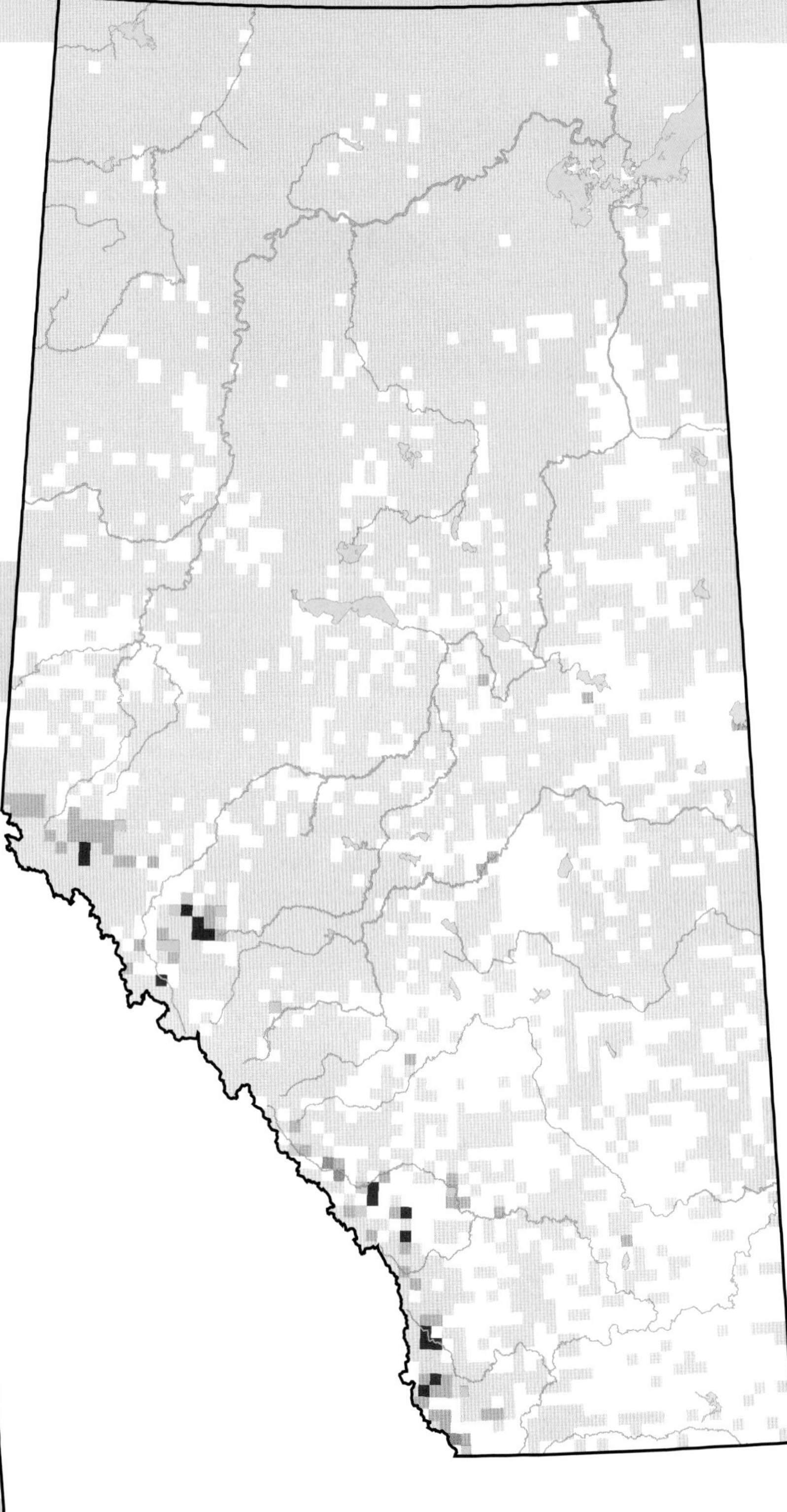

OBSERVED DISTRIBUTION ATLAS 2

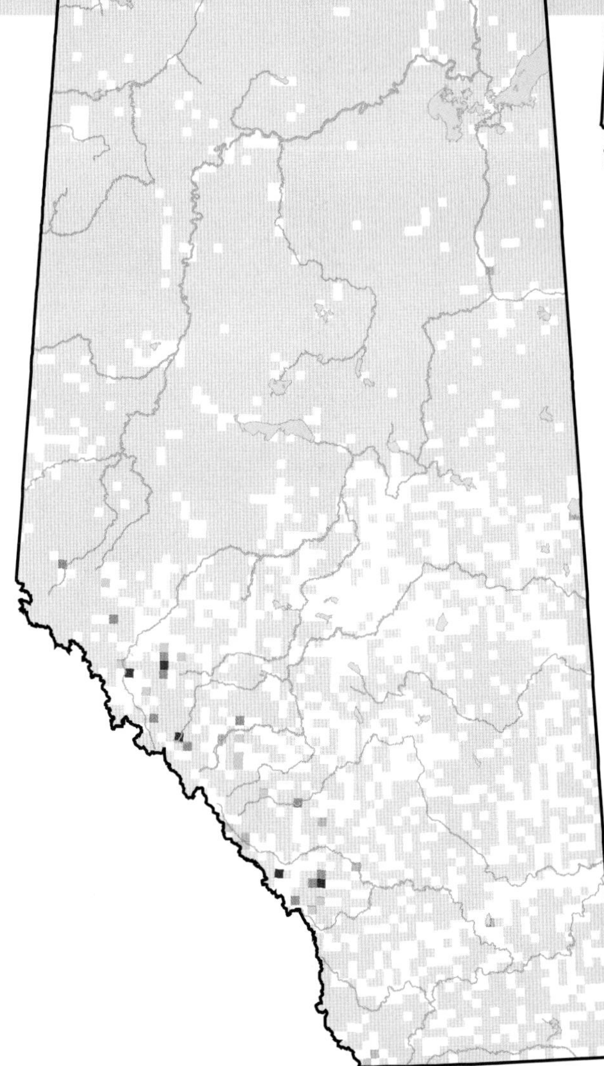

OBSERVED DISTRIBUTION ATLAS 1

	Unsurveyed		Possible
	Not Observed		Probable
	Observed		Confirmed

Surf Scoter (*Melanitta perspicillata*)

Photo: Royal Alberta Museum

NESTING

CLUTCH SIZE (EGGS):	5–9
INCUBATION (DAYS):	28–30
FLEDGING (DAYS):	55
NEST HEIGHT (METRES):	0

Alberta is at the southern limits of this duck's continental breeding range where it is a rare breeder. Nesting records are few, quite sporadic and widespread throughout the central and northern parts of Alberta, although on migration this scoter is observed in all Natural Regions. Data from Atlas 2 indicate that the western portion of the Surf Scoter's known breeding range has extended southward into the upper Kakwa River. In fact, this area had one of the highest reporting rates for this species during Atlas 2. In Atlas 1, only six breeding records were noted and only one of these was a confirmed breeding record. In Atlas 2, there were 17 breeding records, none of which was a confirmed breeding record.

The Surf Scoter prefers quiet and slow-moving waters including ponds, bogs, and streams with adjacent shrubs and woody cover. Several sources have indicated that the nests of this species are difficult to find as they are widely scattered and in inaccessible places.

No changes in relative abundance were detected in the Boreal Forest, Parkland, and Rocky Mountain NRs. There is no Breeding Bird Survey information available for the Surf Scoter. Caution is required in interpreting trend data, since surveys are not well adapted for estimating scoter numbers (Savard et al., 1998). Better estimates clearly are needed. Difficulties in detecting this species include the following: few studies in most of its wide probable distribution; sparse distribution of observers; difficulty in distinguishing between female Surf and White-winged scoters during surveys; and the secretive breeding behaviour of the species.

This species is highly sensitive to adult mortality, so hunting and accidental deaths due to fishing nets or oiling can have a significant impact on the population (Savard et al., 1998). In Alberta the Surf Scoter is rated as Secure.

TEMPORAL REPORTING RATE

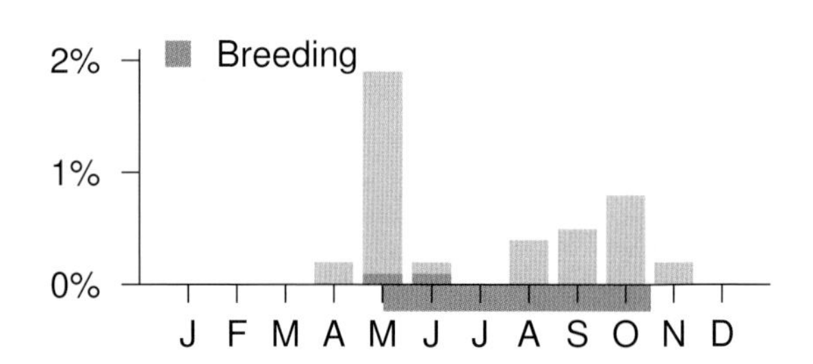

NATURAL REGIONS REPORTING RATE

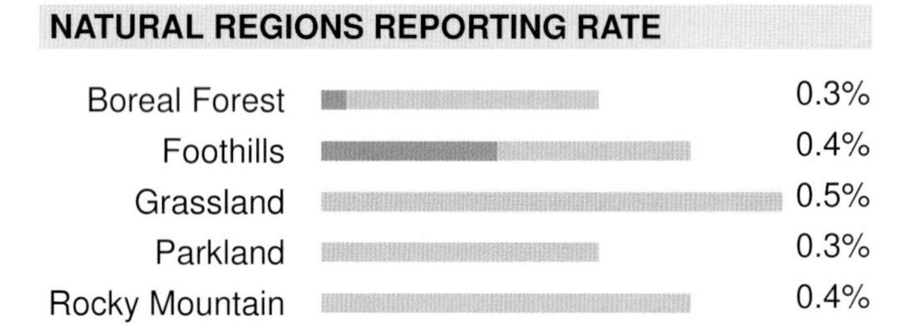

SPATIAL REPORTING RATE

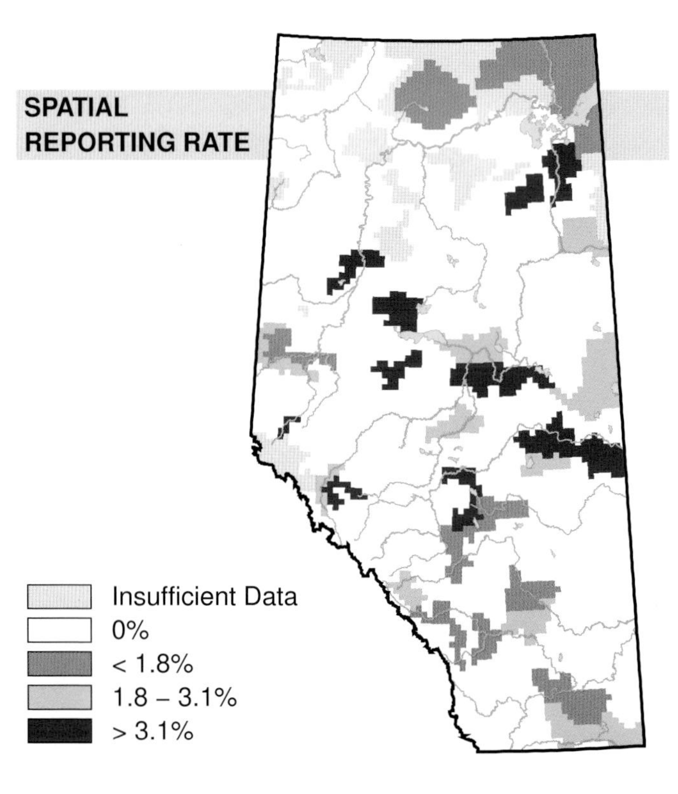

STATUS

GSWA 2000	GSWA 2005	COSEWIC Historic Rankings	COSEWIC 2007	Alberta Wildlife Act
Secure	Secure	N/A	N/A	N/A

Habitat Nest Location Nest Type Diet  

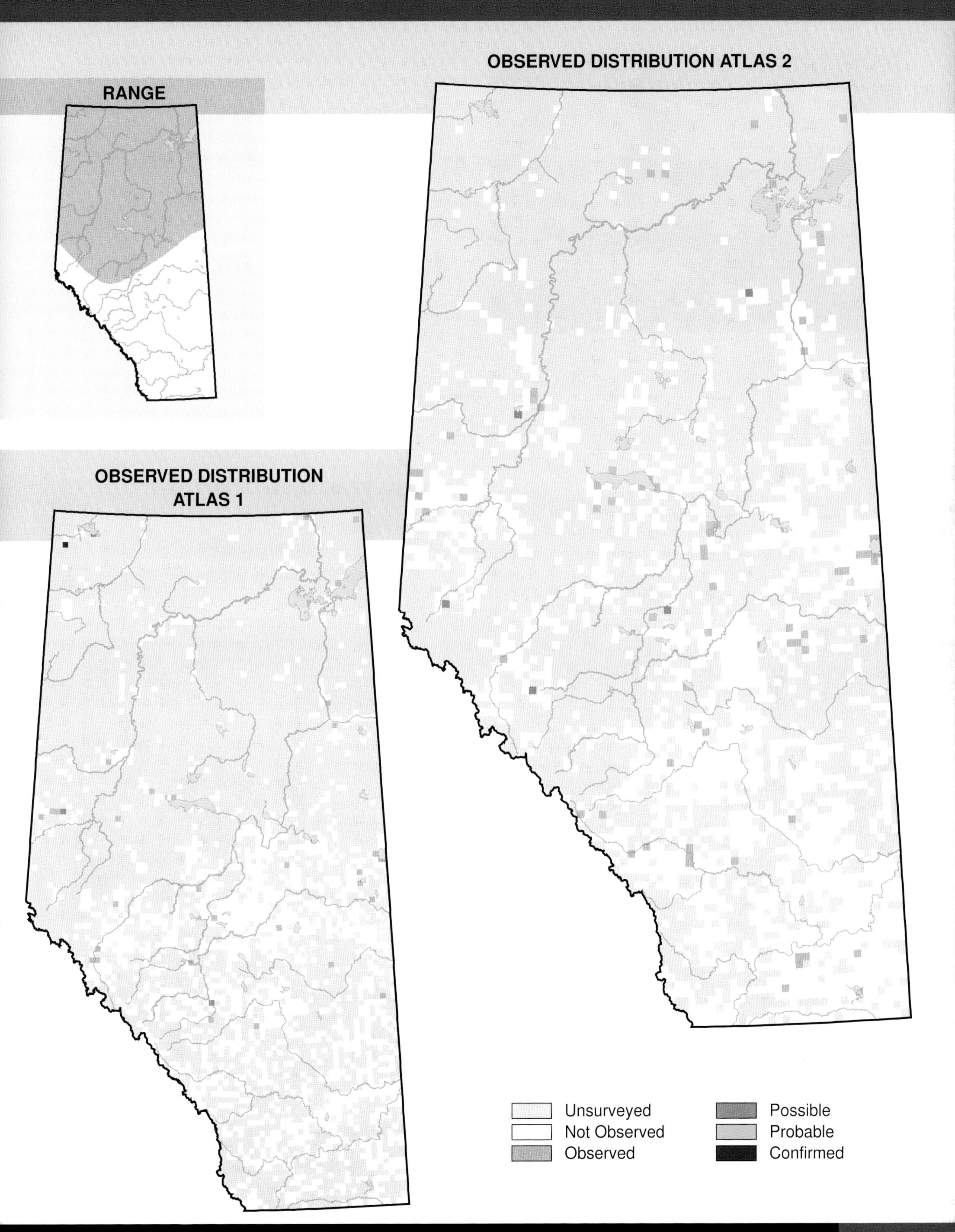

White-winged Scoter (*Melanitta fusca*)

Photo: Gerry Beyersbergen

NESTING

CLUTCH SIZE (EGGS):	9–14
INCUBATION (DAYS):	27–28
FLEDGING (DAYS):	28–35
NEST HEIGHT (METRES):	0

In Alberta, White-winged Scoters breed in all Natural Regions of the province except the Rocky Mountain NR, where the species is recorded only on migration. It prefers the Grassland, Parkland, and Boreal NRs. Although widely dispersed, it is not an abundant breeder anywhere in its Albertan range.

The White-winged Scoter breeds near lakes, ponds, and slow-moving streams in relatively open country. There must be dense and low ground-cover in association with these areas. The characteristic of nesting long distances from water in dense, often thorny, vegetation puts this species in jeopardy because this preferred nesting habitat is among the first to be eliminated (Brown and Fredrickson, 1997). In some parts of the province the preferred breeding habitat is now usually associated with undisturbed islands on deepwater lakes.

Decreases in relative abundance were detected in the Boreal Forest and Parkland NRs. No changes in relative abundance were detected in the Foothills, Grassland, or Rocky Mountain NRs. The Breeding Bird Survey (1985–2005) did not report on the abundance of this species in Alberta but reported a decrease across Canada. Brown and Fredrickson (1997) report, "Because of this species' low rate of recruitment and strong philopatry to nesting areas, disturbance during the nesting season and hunting on breeding areas have the potential to eliminate local populations. Reduction in breeding range on the prairies probably occurred because of changes in land use associated with agriculture ... Throughout northern prairies and aspen parkland areas of North America, intensification of agriculture has eliminated important nesting habitats."

The White-winged Scoter is listed as Sensitive in Alberta as waterfowl breeding population surveys indicate a long-term decline. This species is vulnerable to nesting disturbance, heavy-metal contamination of winter food supply, and—because of the bird's gregarious nature during non-breeding season—oil spills.

TEMPORAL REPORTING RATE

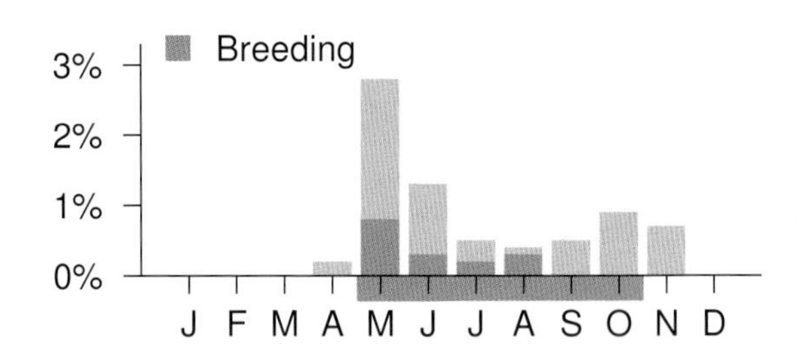

NATURAL REGIONS REPORTING RATE

SPATIAL REPORTING RATE

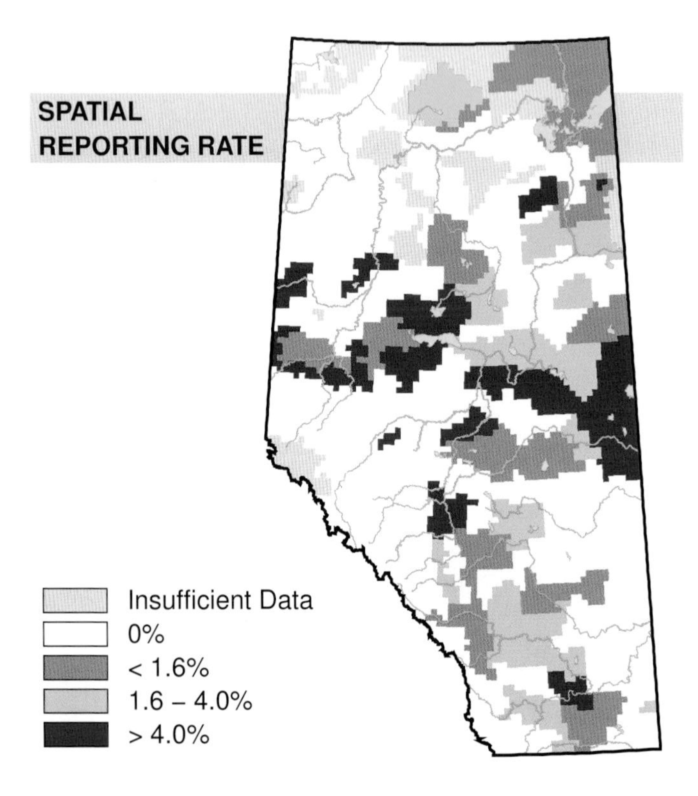

STATUS

GSWA 2000	GSWA 2005	COSEWIC Historic Rankings	COSEWIC 2007	Alberta Wildlife Act
Sensitive	Sensitive	N/A	N/A	N/A

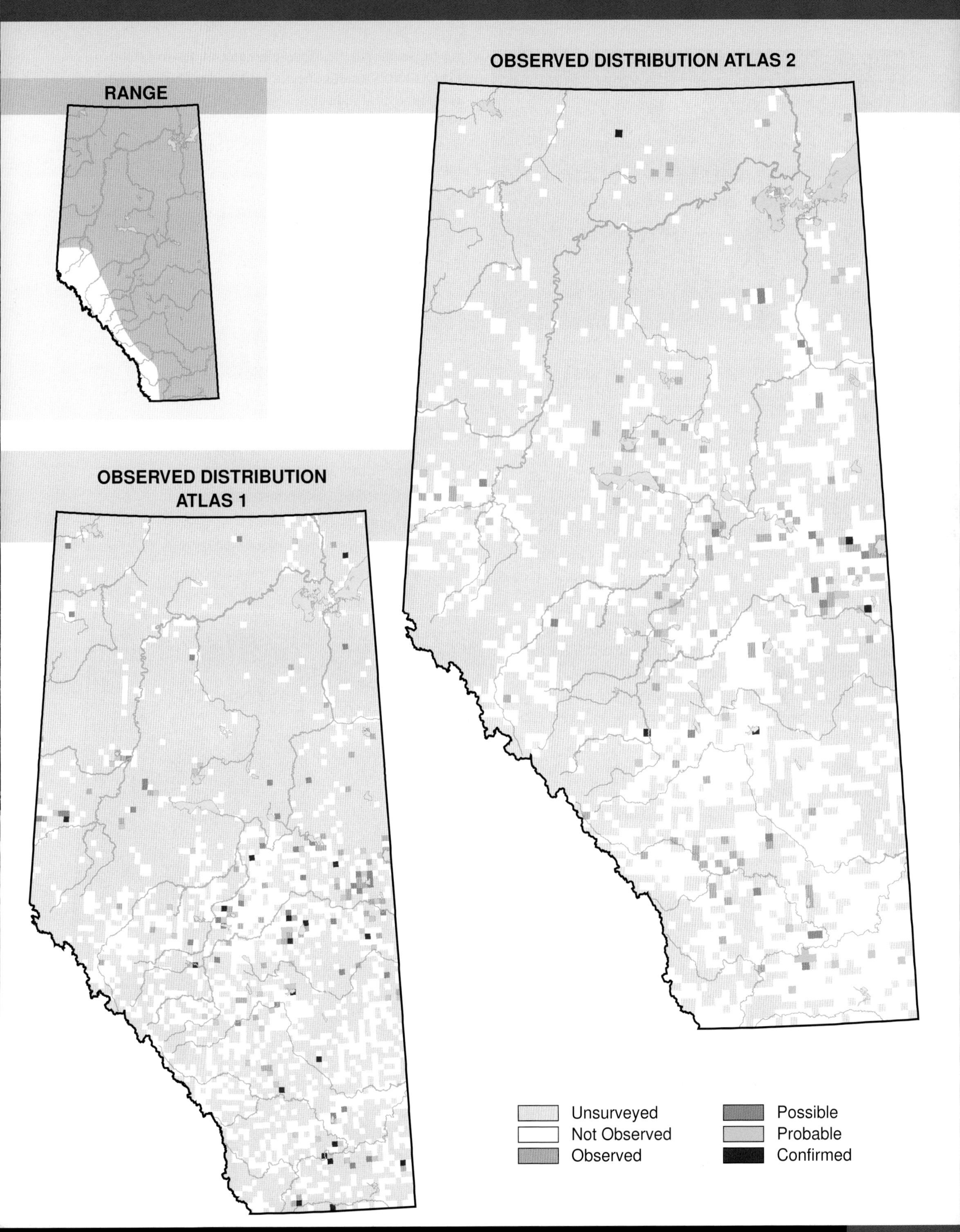
RANGE
OBSERVED DISTRIBUTION ATLAS 2
OBSERVED DISTRIBUTION ATLAS 1
Unsurveyed
Not Observed
Observed
Possible
Probable
Confirmed

Bufflehead (*Bucephala albeola*)

Photo: Gerald Romanchuk

NESTING

CLUTCH SIZE (EGGS):	7–11
INCUBATION (DAYS):	30
FLEDGING (DAYS):	50–55
NEST HEIGHT (METRES):	0–15.2

The Bufflehead is found breeding in all Natural Regions of the province but is concentrated in central Alberta in the Foothills, Parkland, and southern Boreal Forest and Rocky Mountain Natural Regions. There was no change in the distribution of this species between Atlas 1 and Atlas 2.

Because of its cavity-nesting habit, the breeding distribution is limited by the distribution of woodlands and of the Northern Flicker, the main source of nesting cavities. The lakes and ponds preferred for breeding in central Alberta are positioned in areas where poplar communities dominate and there is no emergent or floating vegetation.

Increases in relative abundance were detected in the Grassland, Parkland, and Rocky Mountain NRs. Additional suitable habitat may have been available as there was an increase in precipitation in the southern parts of the province in the latter years of Atlas 2. No changes in relative abundance were detected in the Boreal Forest NR. Decreases in relative abundance were detected in the Foothills NR. As these birds are dependent on large-diameter mature trees for nesting and on nearby water for feeding and brood-rearing, forest removal has the potential to influence Bufflehead habitat (Gauthier, 1993). The Foothills NR has seen an increase in the removal of forest cover due to both forestry and petroleum activities since Atlas 1. The Breeding Bird Survey which constitutes a provincial roll-up of individual surveys, detected no change in recent years, but did report a long-term positive trend (1968–2005) for this species in Alberta. The loss of habitat from increased industrial activity in the boreal forest could result in reduced adult and duckling survival, consequently causing the population to decline. The Province of Alberta has rated this duck as Secure.

TEMPORAL REPORTING RATE

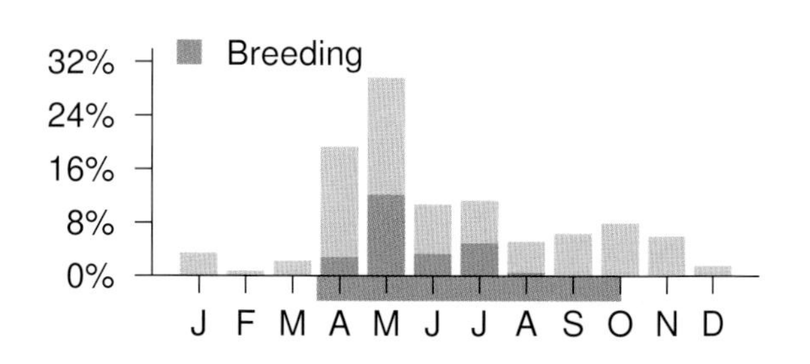

NATURAL REGIONS REPORTING RATE

SPATIAL REPORTING RATE

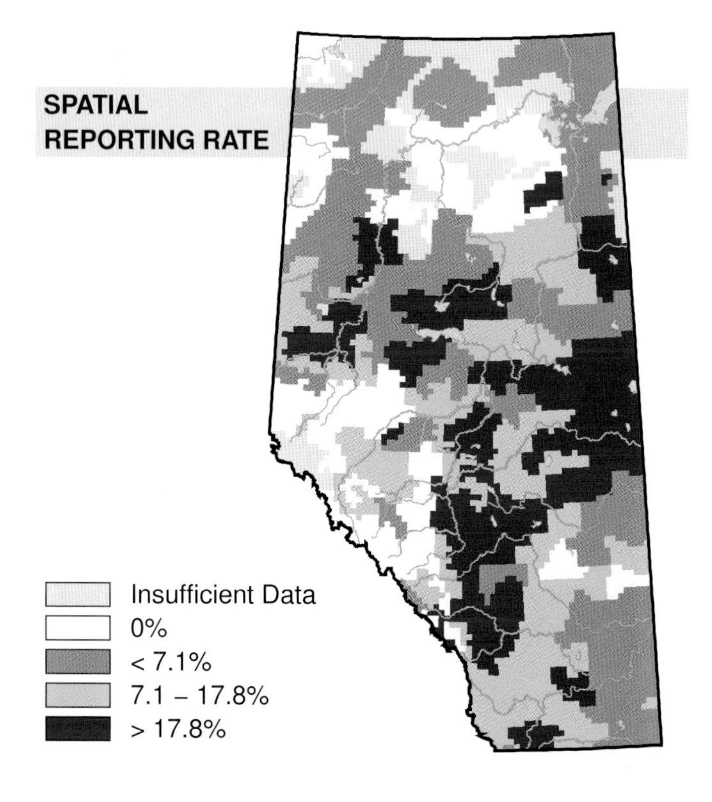

STATUS

GSWA 2000	GSWA 2005	COSEWIC Historic Rankings	COSEWIC 2007	Alberta Wildlife Act
Secure	Secure	N/A	N/A	N/A

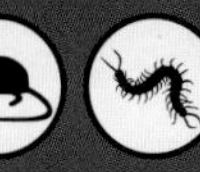

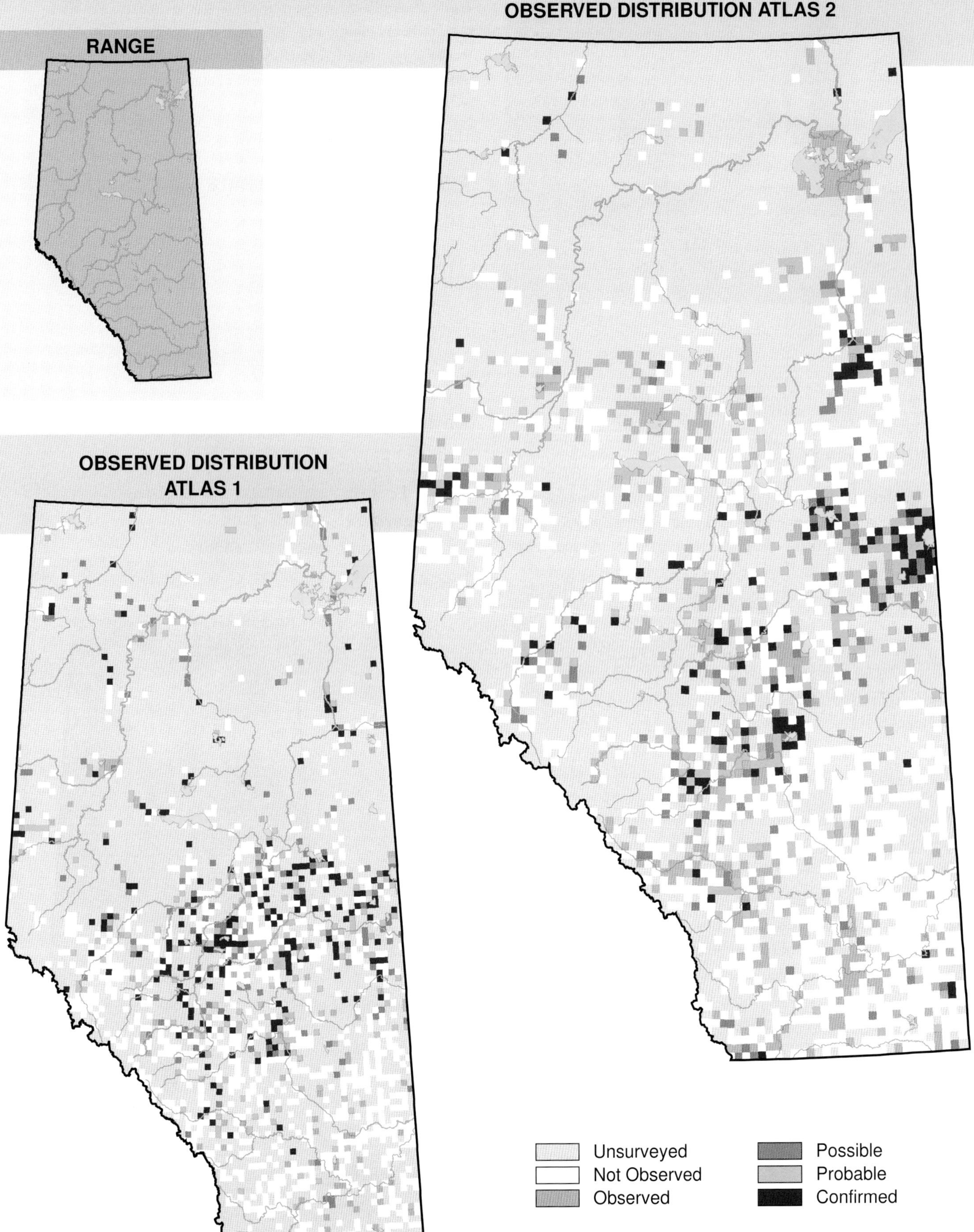
OBSERVED DISTRIBUTION ATLAS 2
RANGE
OBSERVED DISTRIBUTION ATLAS 1
Unsurveyed
Not Observed
Observed
Possible
Probable
Confirmed

Common Goldeneye (*Bucephala clangula*)

Photo: Gerald Romanchuk

NESTING

CLUTCH SIZE (EGGS):	7–12
INCUBATION (DAYS):	27–30
FLEDGING (DAYS):	55–60
NEST HEIGHT (METRES):	2–15

The Common Goldeneye breeds in all parts of the province except the southeastern corner. It is found in the central part of the province comprising the Foothills, Parkland, and southern Boreal Forest Natural Regions. The observed distribution of this species did not change between Atlas 1 and Atlas 2 with the exception of a slight expansion to the south-east. This may be related to increased precipitation in this area in the latter part of Atlas 2 contributing to more suitable breeding habitat.

The breeding habitat of this species is primarily woodland wetlands and lakes, shallow stretches of rivers, and muskeg ponds bordered by forests mature enough to provide suitable tree cavities. The most important factor limiting populations is probably nest-cavity availability, particularly in recently or historically logged regions. In breeding areas, availability of trees providing suitable cavities has been reduced by forestry practices (Eadie et al., 1995).

Increases in relative abundance were detected in the Grassland, Parkland, and Rocky Mountain NRs. The cause for these increases is unknown and should thus be interpreted with caution. Decreases in relative abundance were detected in the Boreal Forest and Foothills NRs. This may be related to the rate of industrial development in these regions that is removing mature tree cover critical for the nesting success of this species. The Breeding Bird Survey (1985–2005) reported no change in abundance for this species in Alberta or across Canada.

Eadie et al. (1995) suggest that this species' nesting habitat requirements, along with its sensitivity to prey quality and availability, make it a species potentially suitable for monitoring the effects of environmental perturbation on boreal wildlife.

The Common Goldeneye is rated as Secure in Alberta.

TEMPORAL REPORTING RATE

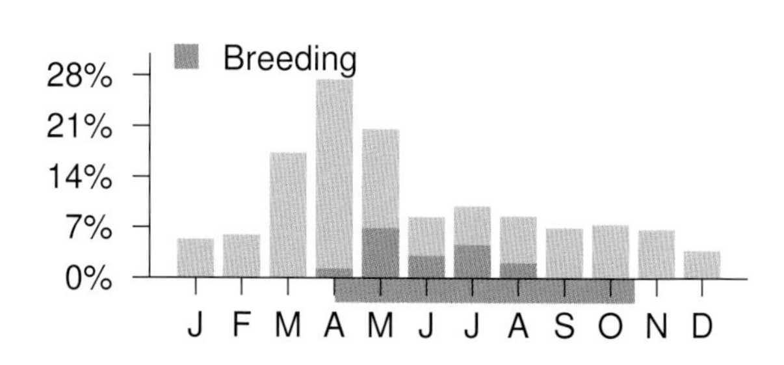

NATURAL REGIONS REPORTING RATE

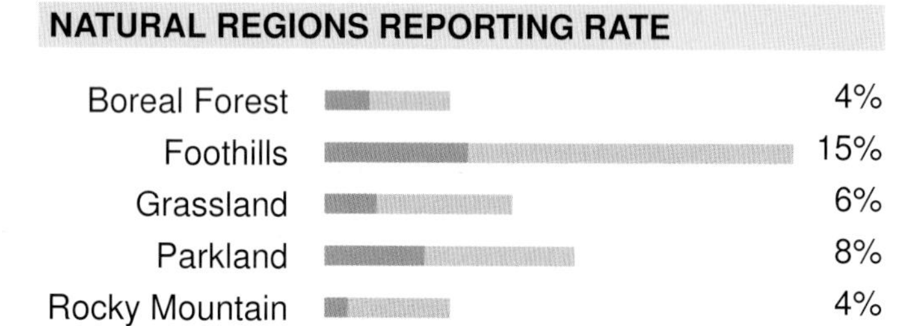

SPATIAL REPORTING RATE

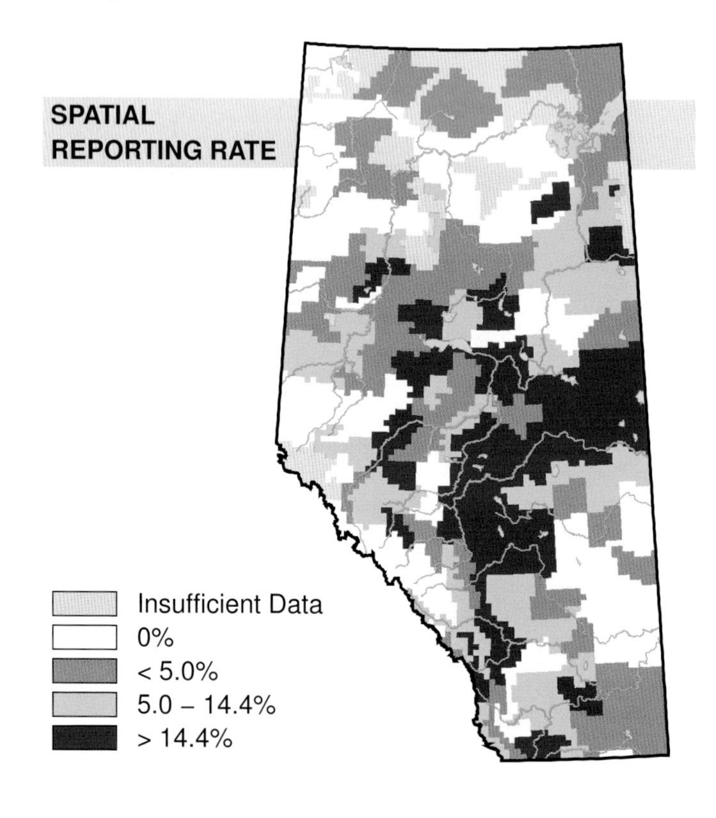

STATUS

GSWA 2000	GSWA 2005	COSEWIC Historic Rankings	COSEWIC 2007	Alberta Wildlife Act
Secure	Secure	N/A	N/A	N/A

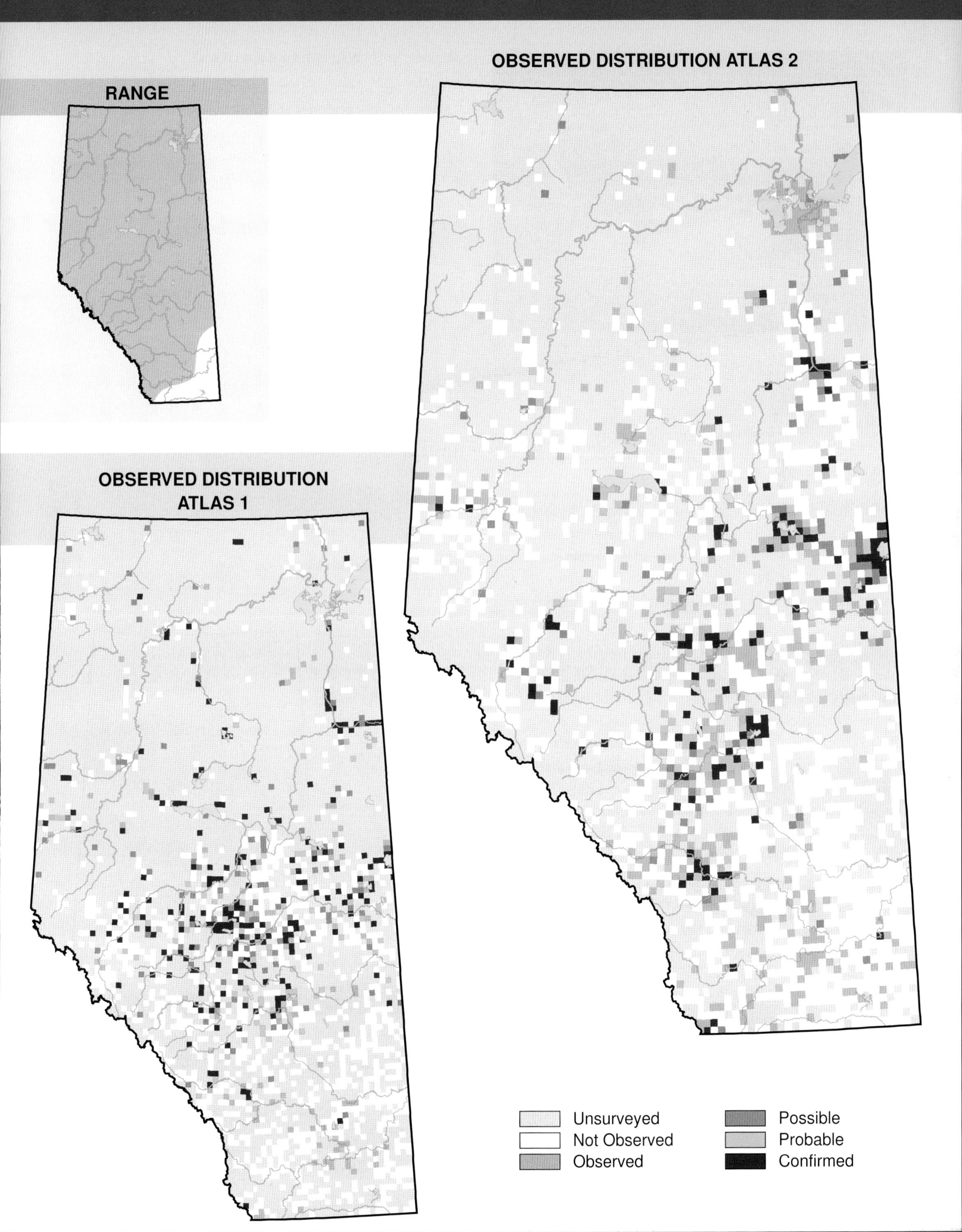
OBSERVED DISTRIBUTION ATLAS 2
RANGE
OBSERVED DISTRIBUTION
ATLAS 1
Unsurveyed
Not Observed
Observed
Possible
Probable
Confirmed

Barrow's Goldeneye (*Bucephala islandica*)

Photo: Raymond Toal

NESTING

CLUTCH SIZE (EGGS):	6–15
INCUBATION (DAYS):	30–34
FLEDGING (DAYS):	56
NEST HEIGHT (METRES):	0–15.2

Barrow's Goldeneye is found mainly in the Rocky Mountain and Foothills Natural Regions with some records in the Parkland and Grassland NRs. As with Atlas 1, some extralimital records were observed in Atlas 2. However, there was no change in the observed distribution of this species in Atlas 2.

In its alpine and subalpine habitat, this diving duck is found on lakes, ponds, and sloughs that have wooded margins, as it is a secondary cavity nester that uses the abandoned holes of Pileated Woodpeckers and flickers. Barrow's Goldeneye can be affected negatively by logging operations that reduce the availability of nesting cavities and possibly increase predation rates at remaining cavities (Eadie and Savard, 2000). The preferred breeding habitat in the Parkland and Grassland NRs comprises water bodies at least 1 metre deep bordered by forest or open range and with no dense emergent vegetation.

Increases in relative abundance were detected in the Parkland NR where, relative to other species, Barrow's Goldeneye was observed more often in Atlas 2 than in Atlas 1. The cause for this increase is not known. No changes in relative abundance were detected in the Boreal Forest, Foothills, Grassland, and Rocky Mountain NRs. The Breeding Bird Survey (1985–2005) did not have sufficient data to report on the abundance of this species for Alberta. The western population is believed to be stable or decreasing slightly although data are sparse outside of British Columbia. Surveys in the same period did record a decrease in abundance on a national scale.

An epidemic of mountain pine beetle in the BC interior has spurred logging activities resulting in removal of all large trees from extensive areas. Similar practices in Alberta could negatively impact Barrow's Goldeneye.

In Alberta this species is currently rated as Secure.

TEMPORAL REPORTING RATE

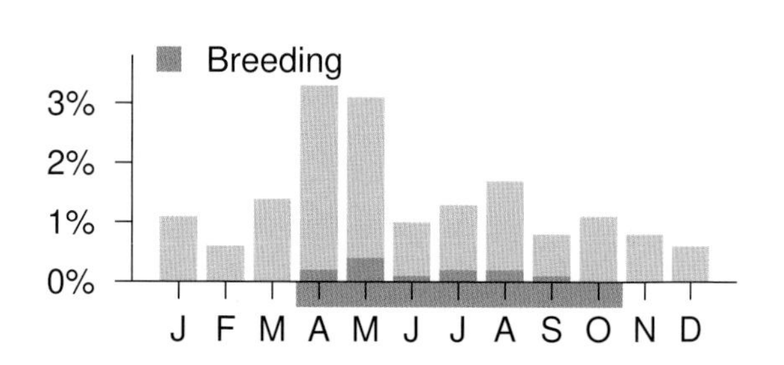

NATURAL REGIONS REPORTING RATE

Boreal Forest	0.3%
Foothills	2.5%
Grassland	1.3%
Parkland	1.0%
Rocky Mountain	9.3%

SPATIAL REPORTING RATE

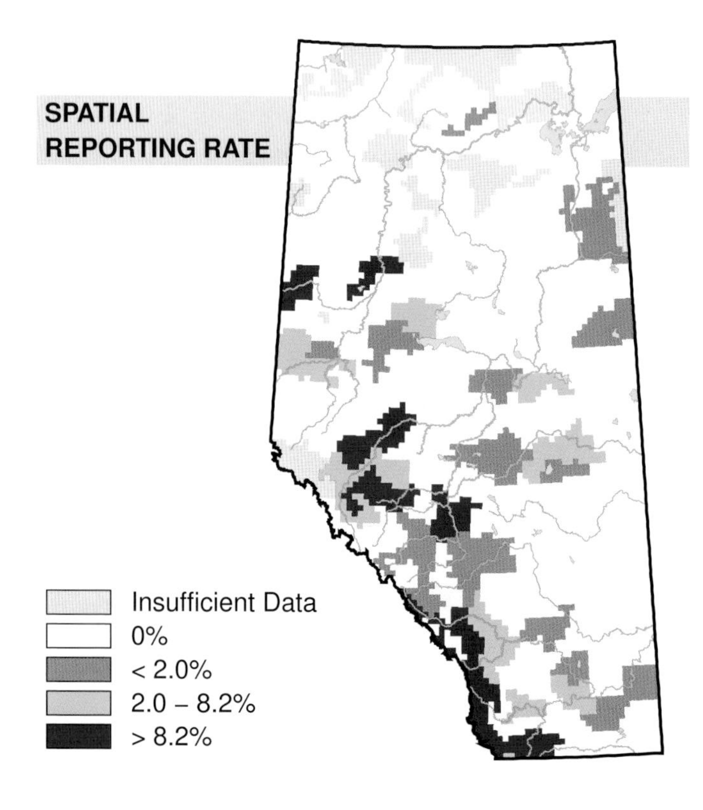

STATUS

GSWA 2000	GSWA 2005	COSEWIC Historic Rankings	COSEWIC 2007	Alberta Wildlife Act
Secure	Secure	N/A	N/A	N/A

RANGE

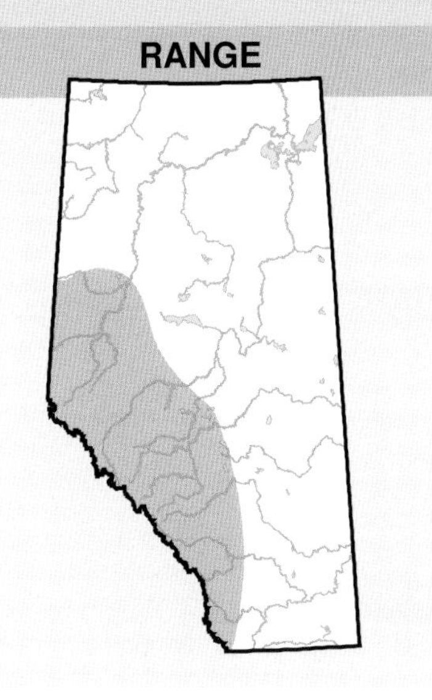

OBSERVED DISTRIBUTION ATLAS 2

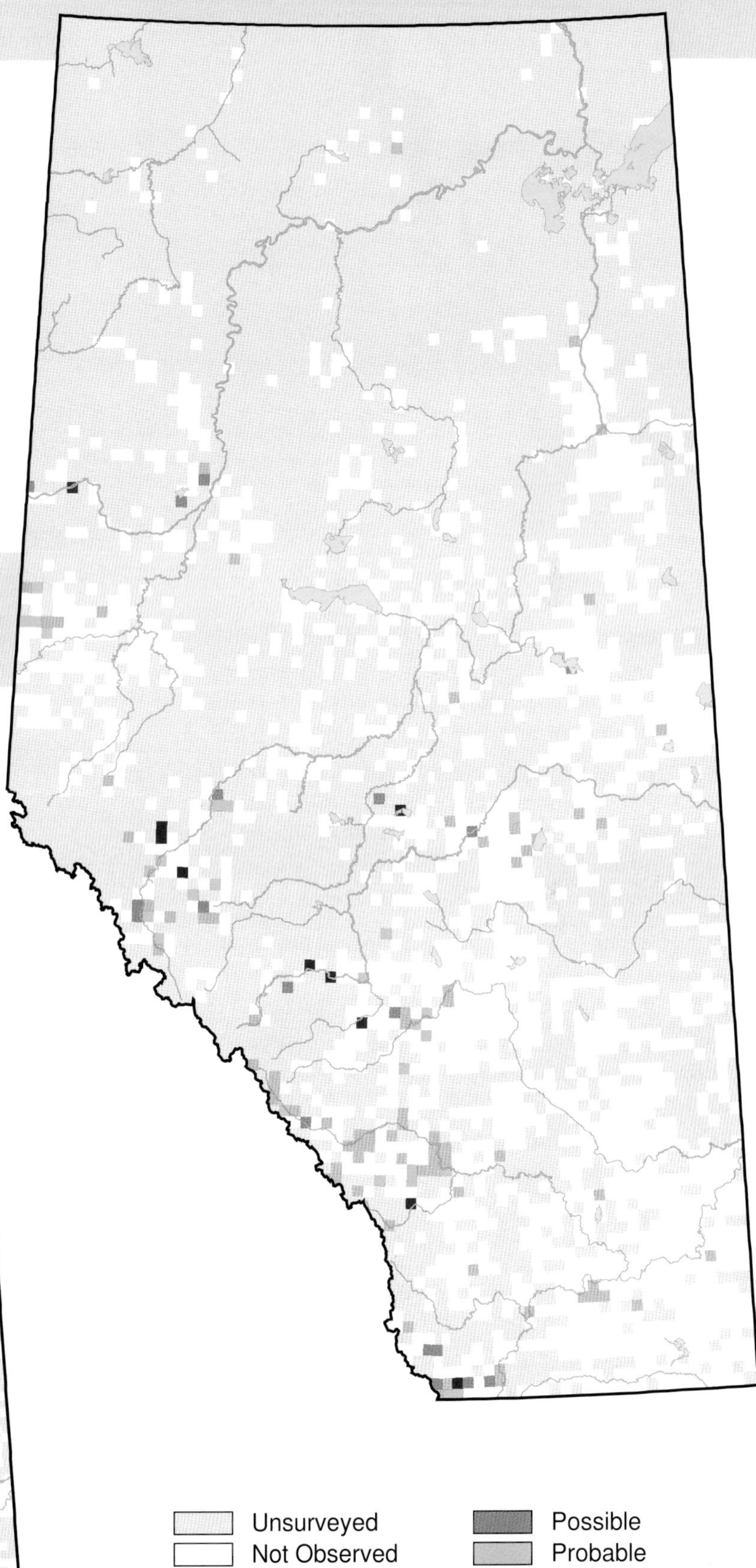

OBSERVED DISTRIBUTION ATLAS 1

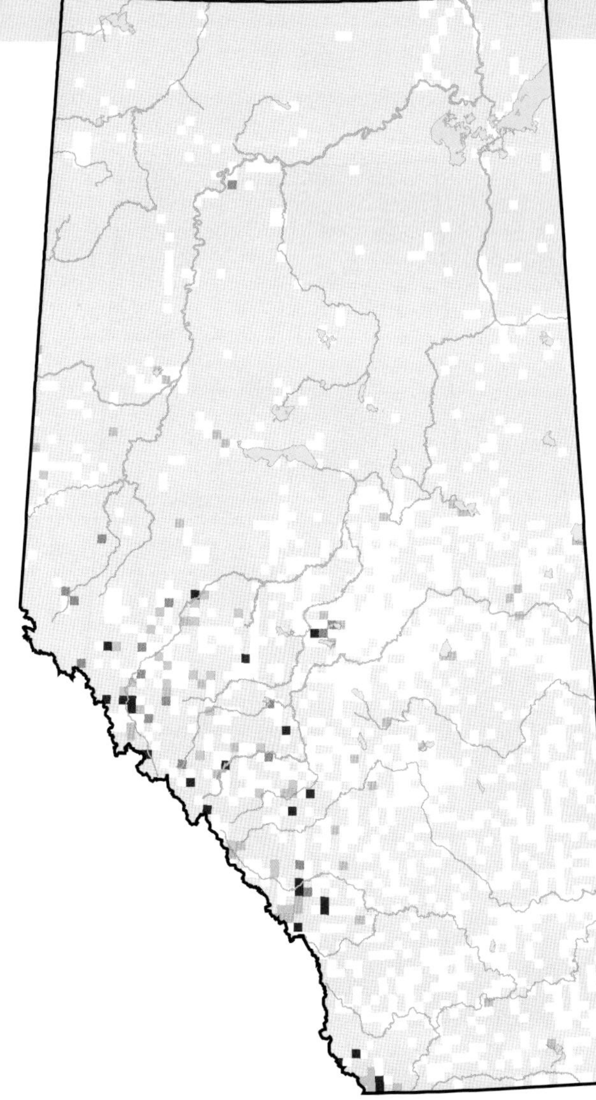

Hooded Merganser (*Lophodytes cucullatus*)

Photo: Gerald Romanchuk

NESTING

CLUTCH SIZE (EGGS):	5–12
INCUBATION (DAYS):	29–37
FLEDGING (DAYS):	71
NEST HEIGHT (METRES):	3–6

The Hooded Merganser is recorded breeding intermittently at several locations in Alberta but it is observed more frequently during migration. The range of this species has been reported previously as two disjunct areas. Data from Atlas 2 indicate that the two areas are probably contiguous, and that the northern limits in the central part of the province have extended north to Utikuma Lake and in the northeast to north of Lake Athabasca. The breeding range has also expanded eastward into the Grassland Natural Region.

Because of this species' secretive nature and dispersed population, less is known about it relative to other merganser species. It favours ponds, lakes, and rivers that have a ready supply of food including aquatic insects, fish, snails, earthworms, and amphibians. This habitat must include a wooded border of mature trees to provide suitable nest sites, making the Hooded Merganser susceptible to activities that reduce mature tree cover around preferred aquatic habitat. Dugger et al. (1994) indicate that forest management goals in this species' range include the conservation of trees that provide nesting cavities (i.e., are more than 100 years old and more than 30 cm in diameter) and the maintenance of riparian forested corridors and forests located within 1 km of suitable brood habitat.

Increases in relative abundance were detected in the Boreal Forest, Parkland, and Rocky Mountain NRs. No changes in relative abundance were detected in the Foothills and Grassland NRs. Increases must be interpreted cautiously because this species is secretive and dispersed, which makes reliable tracking of changes in abundance very difficult. The Breeding Bird Survey (1985–2005) detected no changes in the abundance of this species either in Alberta or across Canada.

The Hooded Merganser is rated as Secure in Alberta.

TEMPORAL REPORTING RATE

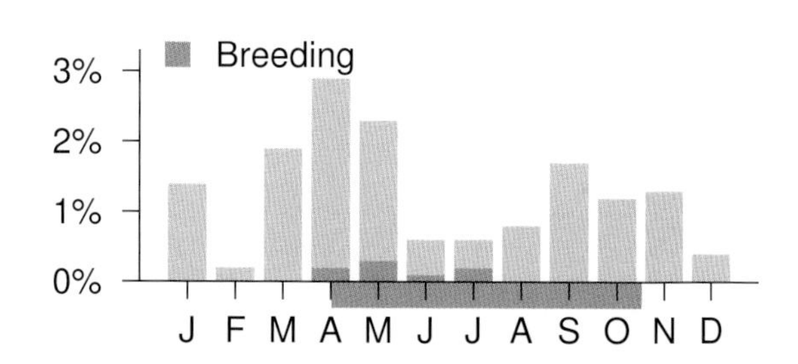

NATURAL REGIONS REPORTING RATE

SPATIAL REPORTING RATE

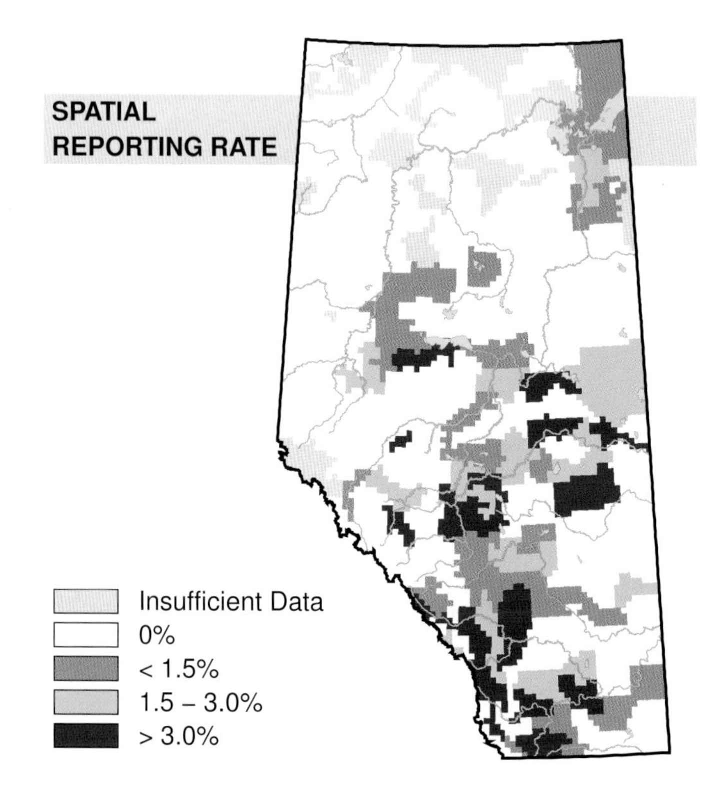

STATUS

GSWA 2000	GSWA 2005	COSEWIC Historic Rankings	COSEWIC 2007	Alberta Wildlife Act
Secure	Secure	N/A	N/A	N/A

RANGE

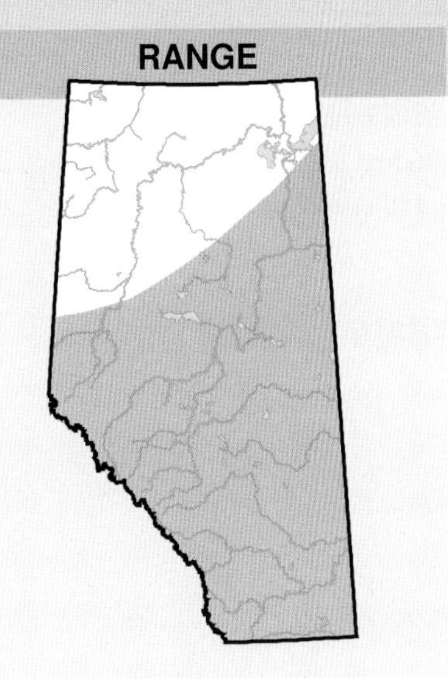

OBSERVED DISTRIBUTION ATLAS 2

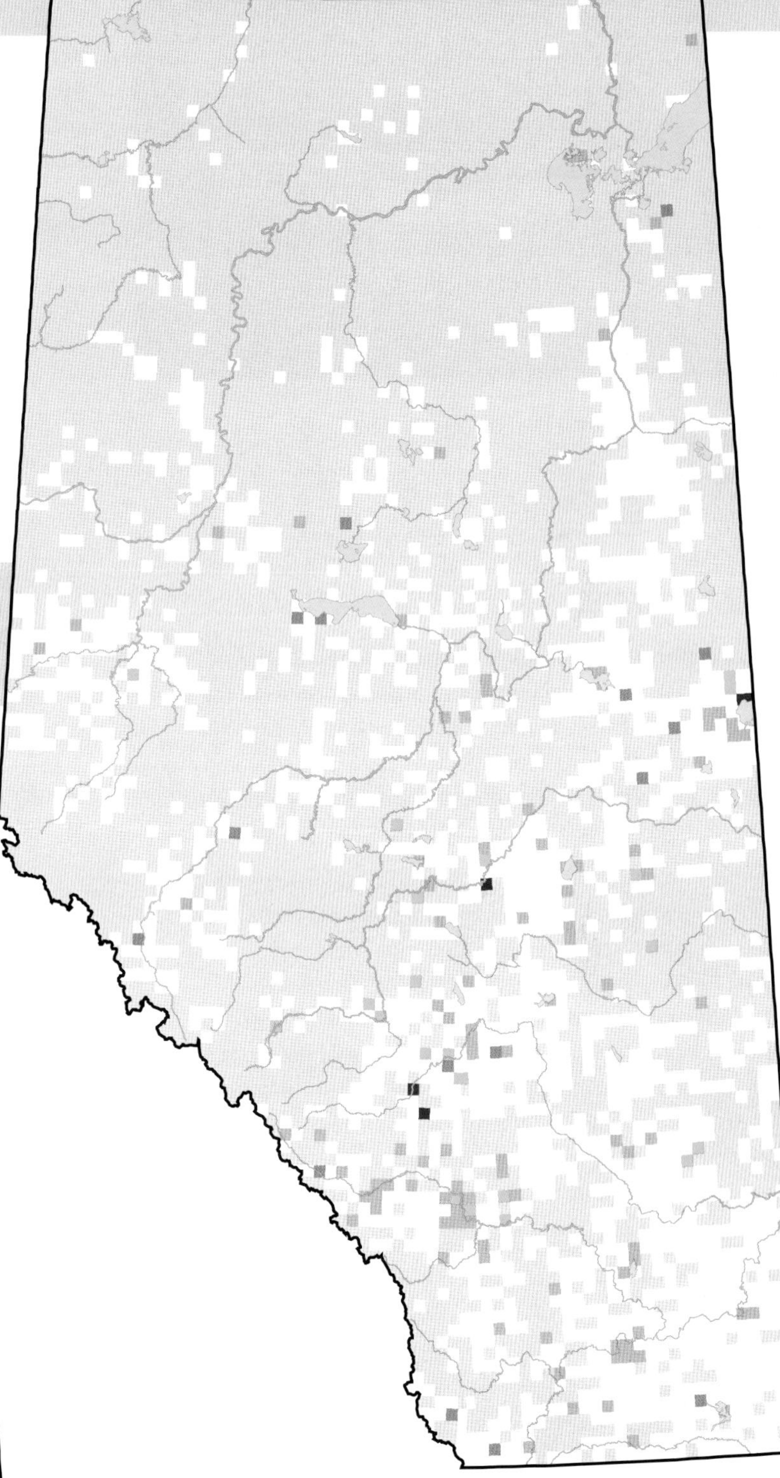

OBSERVED DISTRIBUTION ATLAS 1

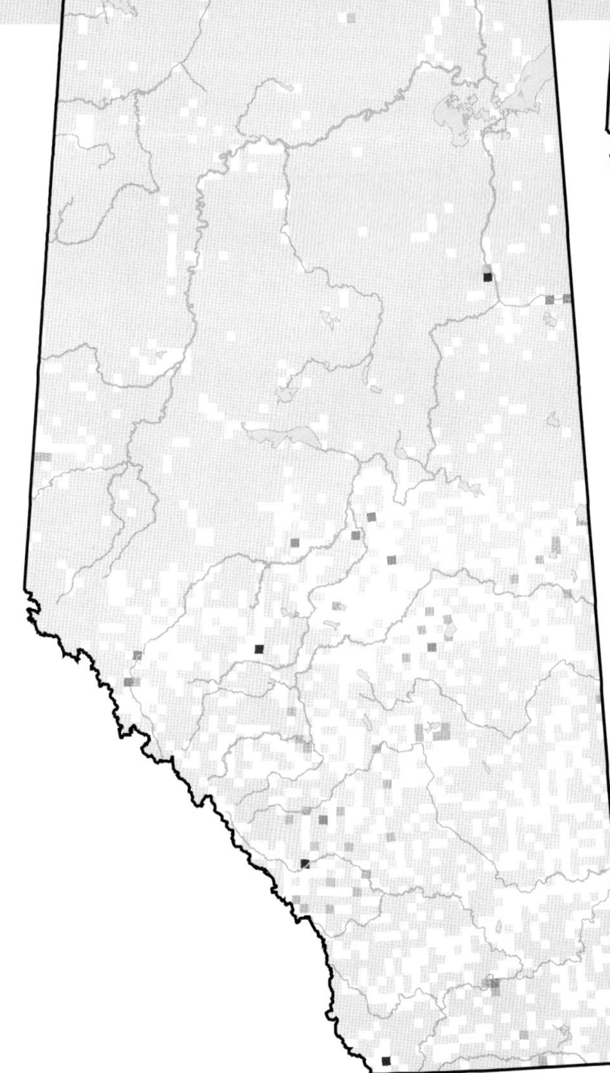

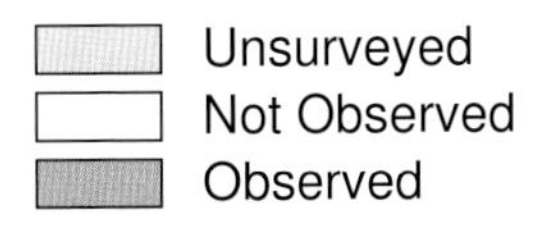

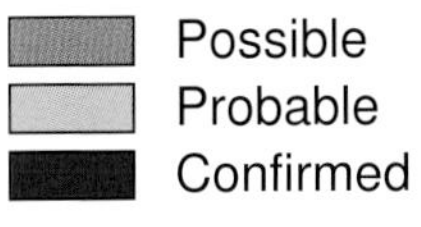

Common Merganser (*Mergus merganser*)

Photo: Gerald Romanchuk

NESTING

CLUTCH SIZE (EGGS):	9–16
INCUBATION (DAYS):	28–35
FLEDGING (DAYS):	65–85
NEST HEIGHT (METRES):	5–16

The Common Merganser breeds in all Natural Regions of the province with the exception of the eastern portions of the Grassland and Parkland NRs. There was no change in the observed distribution of this species between Atlas 1 and Atlas 2, but it was observed less often in the northern portions of the Foothills and Rocky Mountain NRs.

As this bird's preferred prey is fish, it favours lakes with clear water and wooded shores for breeding habitat. Perhaps the most important factor limiting Common Merganser populations is availability of nest cavities in trees. Consequently, activities (e.g., forestry and petroleum) that remove mature trees suitable for nesting from the landscape can have a detrimental effect. As a top predator in aquatic food chains, this species has served as an indicator of environmental health with respect to contaminants (pesticides, toxic metals) and lake acidification (Mallory and Metz, 1999).

Increases in relative abundance were detected in the Parkland and Grassland NRs. This may be related to the increase in precipitation in southern Alberta in the latter part of Atlas 2. In the same period, decreases in relative abundance were detected in the Foothills NR. Since the completion of Atlas 1, there has been a marked increase in industrial development, forestry, and oil and gas in this natural region. The amount of forest cover has been reduced, possibly affecting the amount of suitable breeding habitat available. No changes in relative abundance were detected in the Boreal Forest and Rocky Mountain NRs.

The BBS detected a decrease for this species across Canada from 1985–2005, and a steeper decline occurred in Alberta specifically during this same period. This species is considered Secure in Alberta. Several sources indicate that the Common Merganser is one of the least studied ducks in North America and much remains to be learned to enable effective management.

TEMPORAL REPORTING RATE

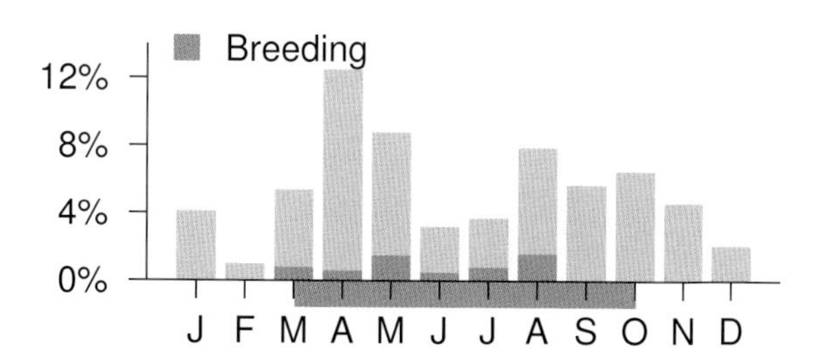

NATURAL REGIONS REPORTING RATE

Region	Rate
Boreal Forest	1.2%
Foothills	3.5%
Grassland	4.6%
Parkland	1.9%
Rocky Mountain	9.8%

SPATIAL REPORTING RATE

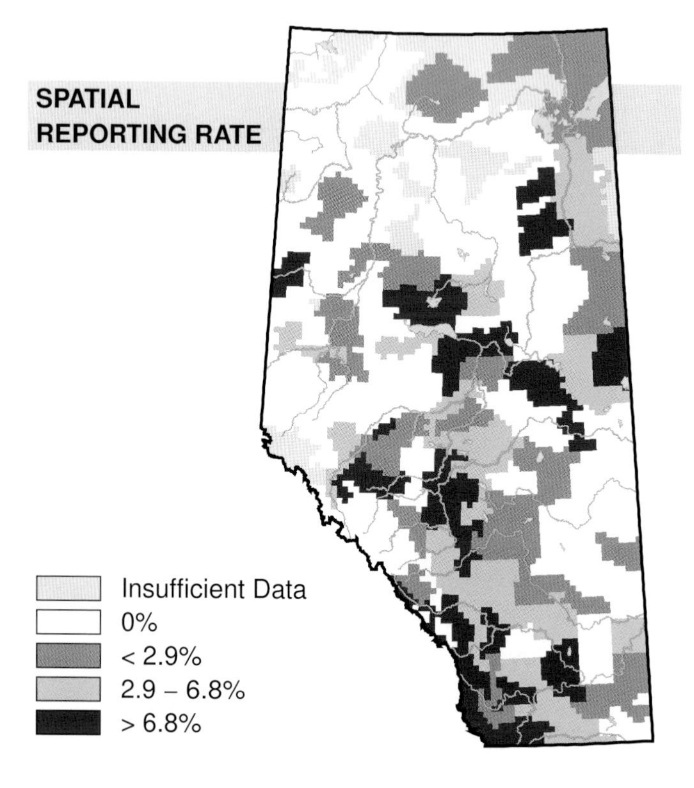

STATUS

GSWA 2000	GSWA 2005	COSEWIC Historic Rankings	COSEWIC 2007	Alberta Wildlife Act
Secure	Secure	N/A	N/A	N/A

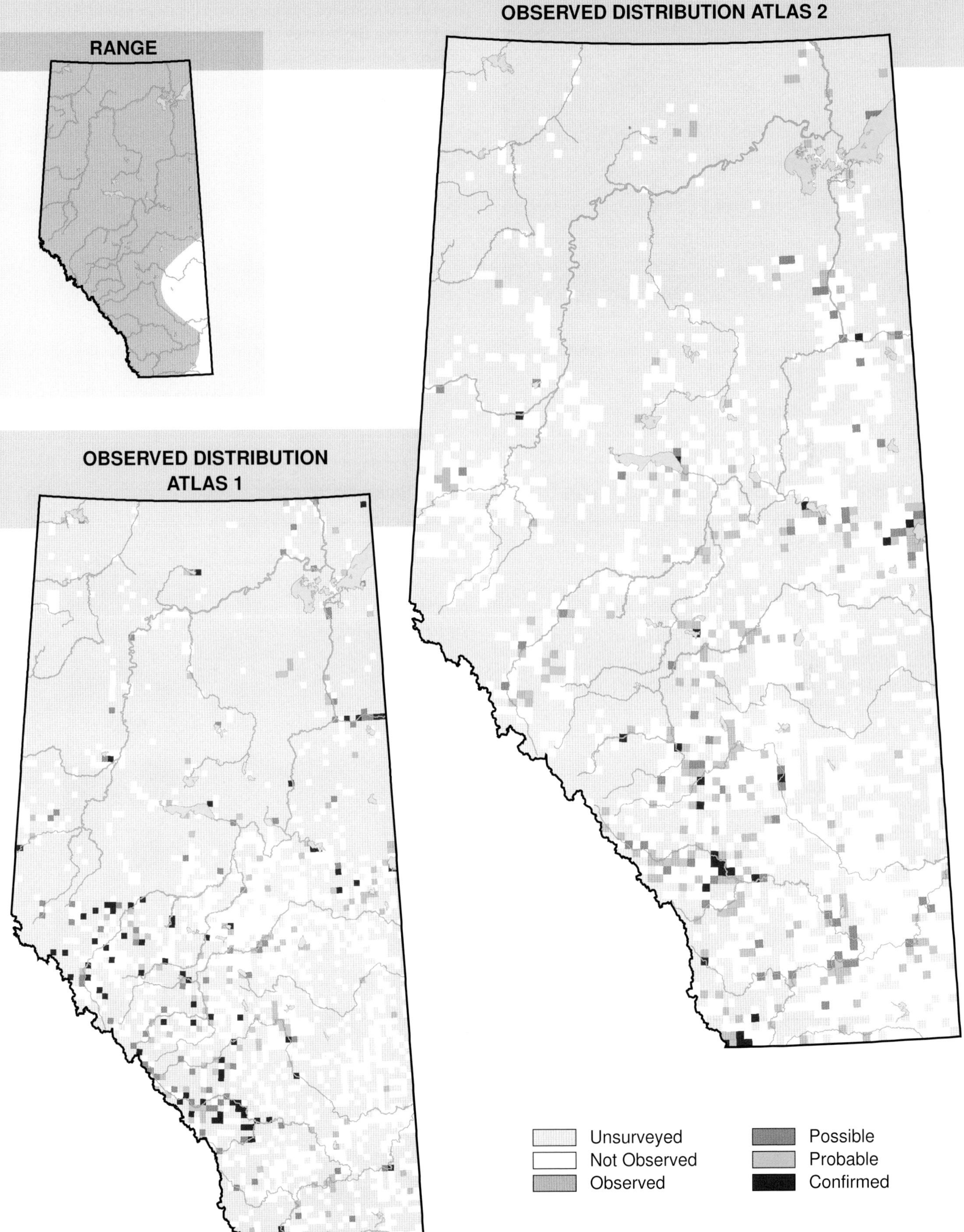
Habitat
Nest Location
Nest Type
Diet
OBSERVED DISTRIBUTION ATLAS 2
RANGE
OBSERVED DISTRIBUTION ATLAS 1
Unsurveyed
Not Observed
Observed
Possible
Probable
Confirmed

Red-breasted Merganser (*Mergus serrator*)

Photo: Gerald Romanchuk

NESTING

CLUTCH SIZE (EGGS):	7–12
INCUBATION (DAYS):	29–35
FLEDGING (DAYS):	60
NEST HEIGHT (METRES):	0

The Red-breasted Merganser is reported throughout the province while on its protracted migration that begins in April and moves north as water bodies thaw. Historically, it was reported breeding only north of Lesser Slave Lake. Atlas 1 records extended the breeding range south to Rocky Mountain House and the Red Deer River. Atlas 2 records have extended the breeding range even farther south into the southern portion of the Parkland Natural Region and the northeastern portion of the Grassland NR. However, the only confirmed breeding records were in its historical, forested range. Eighty-four percent of the records for this species during Atlas 2 were those of non-breeders.

It was noted in Atlas 1 that the breeding status of Red-breasted Mergansers may have been underrepresented because of the remoteness of their distribution, their well concealed nests, and the fact that the female resembles the Common Merganser.

This bird is usually associated with tundra and boreal forest zones where it prefers larger, deeper and more open lakes and rivers with abundant small fish to sustain them.

No changes in relative abundance were detected in the Boreal Forest, Foothills, Parkland, and Rocky Mountain NRs. Increases in relative abundance were detected in the Grassland NR. This may be the result of higher water levels in southern lakes due to increased precipitation in the latter part of Atlas 2. The Breeding Bird Survey (1985–2005) did not report on abundance for Alberta, and no change in abundance was detected across Canada.

In Alberta the Red-breasted Merganser is listed as Secure.

TEMPORAL REPORTING RATE

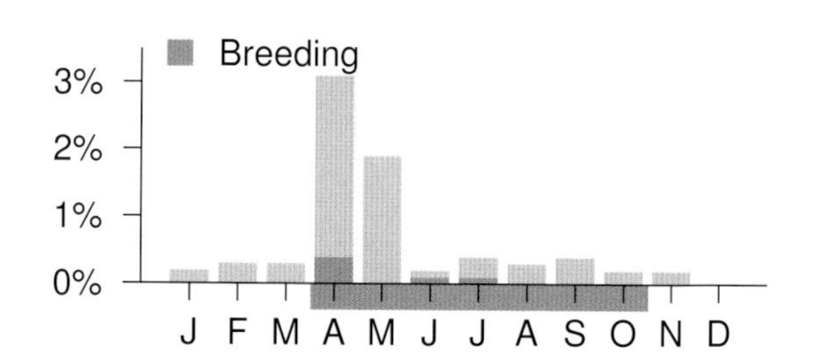

NATURAL REGIONS REPORTING RATE

SPATIAL REPORTING RATE

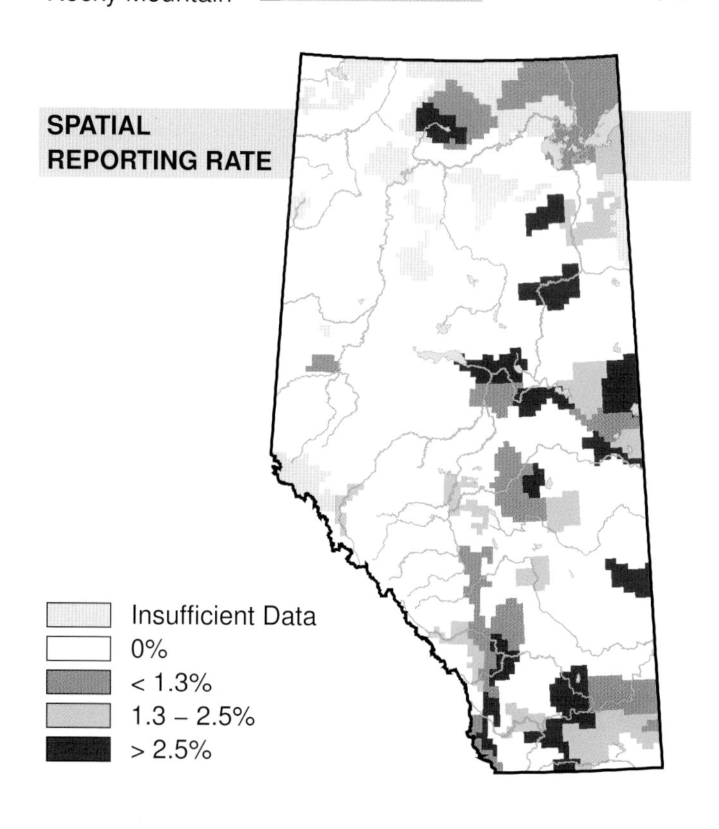

STATUS

GSWA 2000	GSWA 2005	COSEWIC Historic Rankings	COSEWIC 2007	Alberta Wildlife Act
Secure	Secure	N/A	N/A	N/A

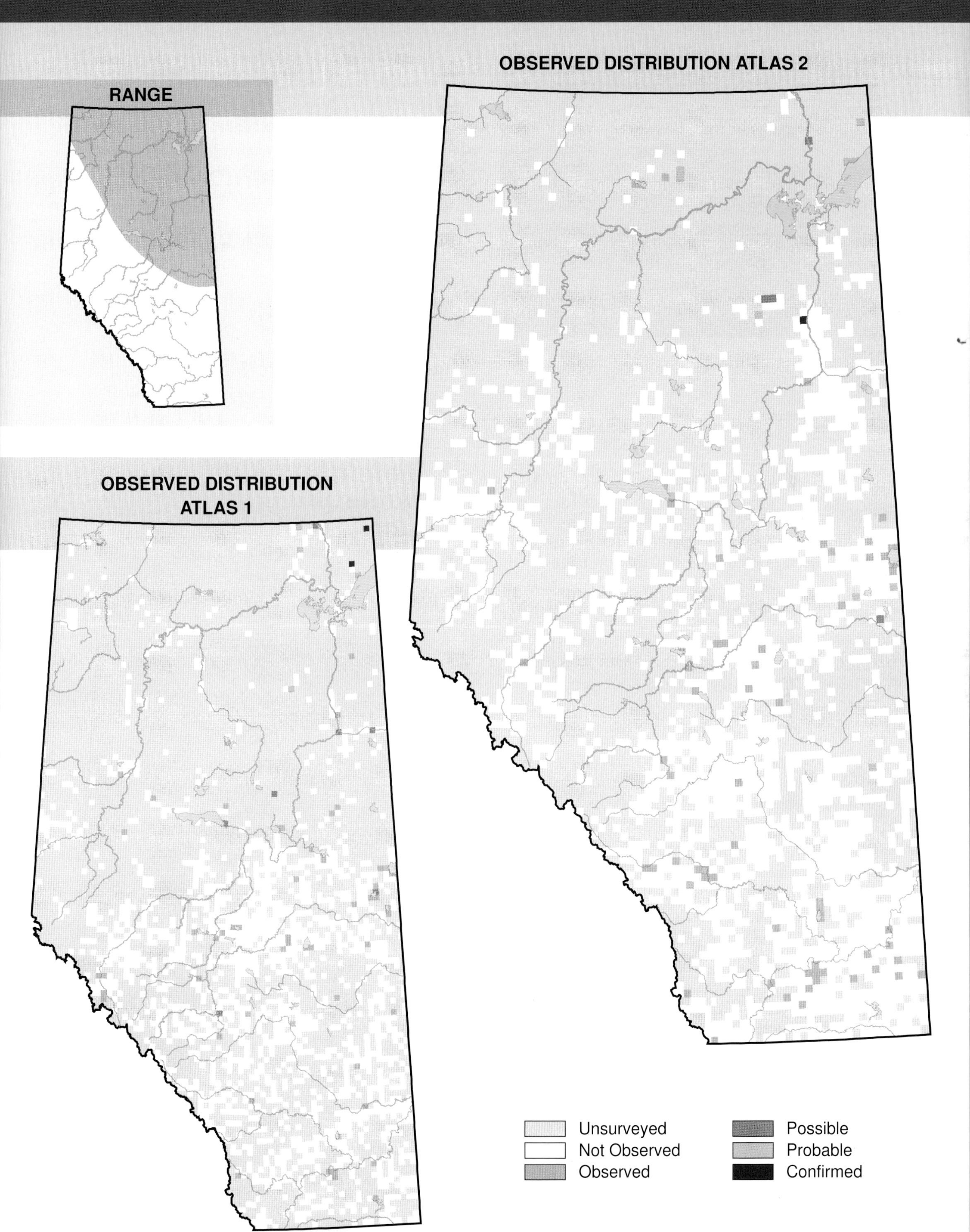

Habitat
Nest Location
Nest Type
Diet
RANGE
OBSERVED DISTRIBUTION ATLAS 2
OBSERVED DISTRIBUTION ATLAS 1
Unsurveyed
Not Observed
Observed
Possible
Probable
Confirmed

Ruddy Duck (*Oxyura jamaicensis*)

Photo: Mark Williams

NESTING

CLUTCH SIZE (EGGS):	6–10
INCUBATION (DAYS):	22–24
FLEDGING (DAYS):	56
NEST HEIGHT (METRES):	0

The Ruddy Duck breeds mainly in the Grassland, Parkland, and southern Boreal Forest Natural Regions. It is becoming more numerous in the Foothills NR with sporadic breeding records in the Rocky Mountain NR. There was no change in the observed distribution of this species from Atlas 1 to Atlas 2.

This duck prefers permanent marshes with emergent vegetation and stable water levels, in addition to open water sufficiently large for landing and taking flight. Ruddy Ducks use large marsh systems, stock ponds, reservoirs, and deep natural basins during the breeding season (Brua, 2001).

No changes in relative abundance were detected in the Parkland and Rocky Mountain NRs. Increases in relative abundance were detected in the Foothills and Grassland NRs. This may be attributed to the increase in precipitation levels in the southern part of the province during the latter part of Atlas 2. Breeding numbers correlate positively with total number of May ponds within the prairie region (Brua, 2001). Decreases in relative abundance were detected in the Boreal Forest NR. In previously drier years across the prairies, this region may have been utilized to a greater extent than in years when favourable conditions prevailed in this ducks' preferred prairie-parkland habitat. This may not reflect a decrease in duck numbers but, rather, a regional shift because the Breeding Bird Survey (1985–2005) reported no change in abundance for Alberta or across Canada.

As with other waterfowl and aquatic birds breeding in the prairie pothole region, wetland drainage and degradation continue to have negative impacts on these birds' breeding habitat (Brua, 2001). Since this bird is highly dependent on wetlands within the prairie pothole region, it is critical to protect and restore wetlands, especially within Canada. The Ruddy Duck is rated as Secure in Alberta.

TEMPORAL REPORTING RATE

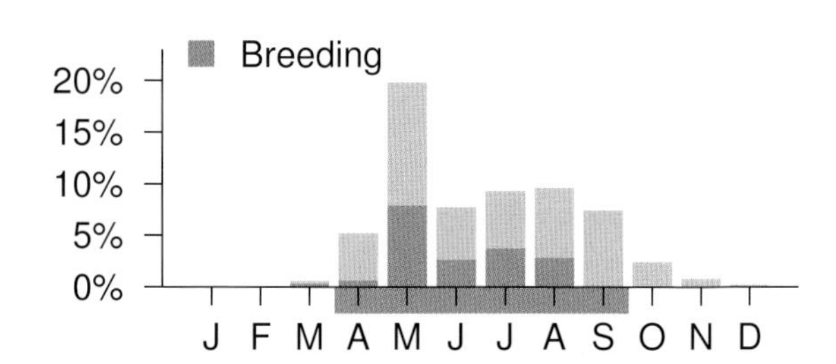

NATURAL REGIONS REPORTING RATE

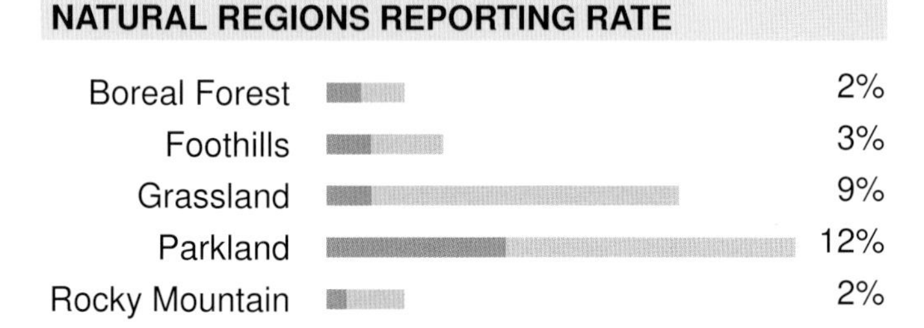

SPATIAL REPORTING RATE

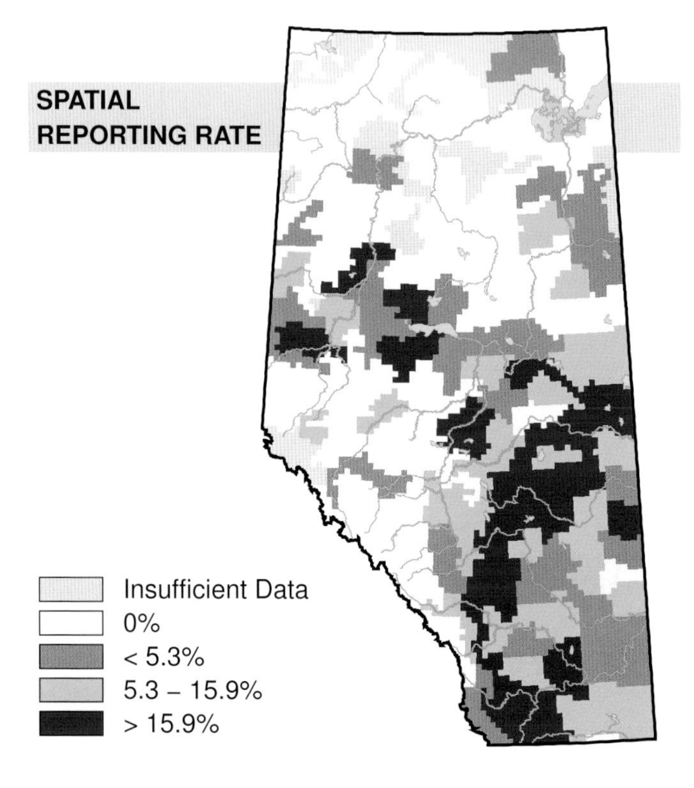

STATUS

GSWA 2000	GSWA 2005	COSEWIC Historic Rankings	COSEWIC 2007	Alberta Wildlife Act
Secure	Secure	N/A	N/A	N/A

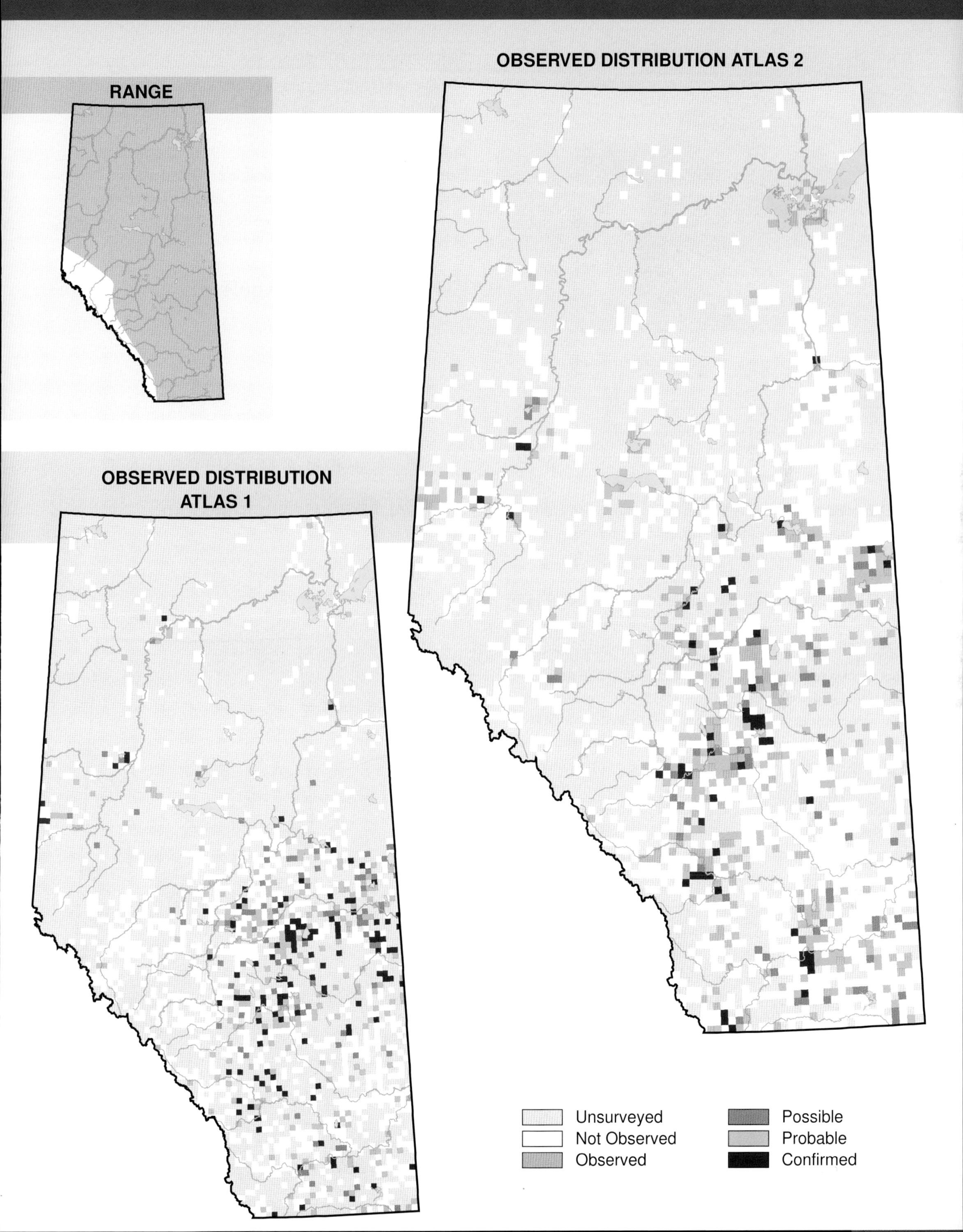
OBSERVED DISTRIBUTION ATLAS 2
RANGE
OBSERVED DISTRIBUTION
ATLAS 1
Unsurveyed
Not Observed
Observed
Possible
Probable
Confirmed

Gray Partridge (*Perdix perdix*)

Photo: Mark Williams

NESTING

CLUTCH SIZE (EGGS):	10–22
INCUBATION (DAYS):	23–25
FLEDGING (DAYS):	16
NEST HEIGHT (METRES):	0

The Gray Partridge, locally known as the Hungarian Partridge, is found mainly in the Grassland, Parkland, and southern Boreal Forest Natural Regions. Atlas 1 found that the population of this species in the Peace River Parkland Subregion of the Parkland NR was disjunct from the the main area of distribution, and this was confirmed in Atlas 2. There were no changes in distribution from Atlas 1 to Atlas 2.

This species resides in open areas of grassland and agricultural land with nearby wooded areas. It is seen frequently in and around the border between scrub and cultivated land. Gray Partridge benefit most from traditional agricultural practices, where hedgerows are maintained and the use of pesticides is limited; but, populations are declining in areas with intensive agriculture (Carroll, 1993). Carroll (1993) indicates that the maturation of shelter belts and trees in farmsteads might also contribute to increased mortality by providing more habitat for partridge predators or by shading herbaceous nesting cover.

Decreases in relative abundance were detected in the Boreal Forest and Parkland NRs. No changes in relative abundance were detected in the Grassland and Rocky Mountain NRs. The Breeding Bird Surveys (1985–2005) found no change in abundance in Alberta or across Canada. Although the populations of Gray Partridge appear to be stable in the province, these levels are susceptible to sharp drops during severe winters. This is somewhat mitigated by the species' high reproductive potential.

Management of this species in North America consists mainly of hunting regulations and habitat management programs aimed at other species, or at wildlife in general, with very limited stocking programs (Carroll, 1993). The creation of long-term set-asides, such as prairie parks and protected areas, would probably create more nesting habitat. The Gray Partridge is rated as an Exotic/Alien in the 2005 Government of Alberta Status document.

TEMPORAL REPORTING RATE

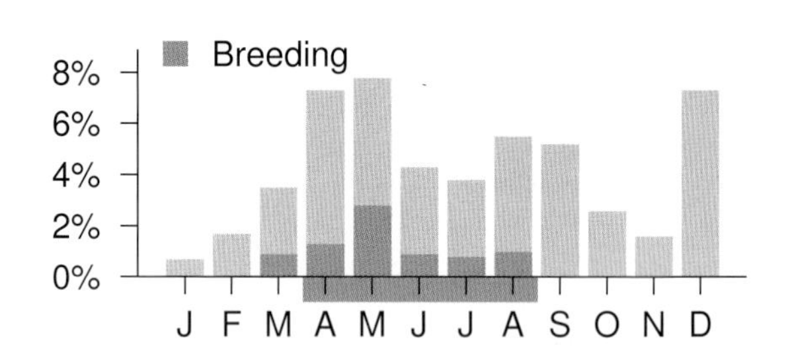

NATURAL REGIONS REPORTING RATE

Region	Rate
Boreal Forest	0.1%
Grassland	9.6%
Parkland	2.3%

SPATIAL REPORTING RATE

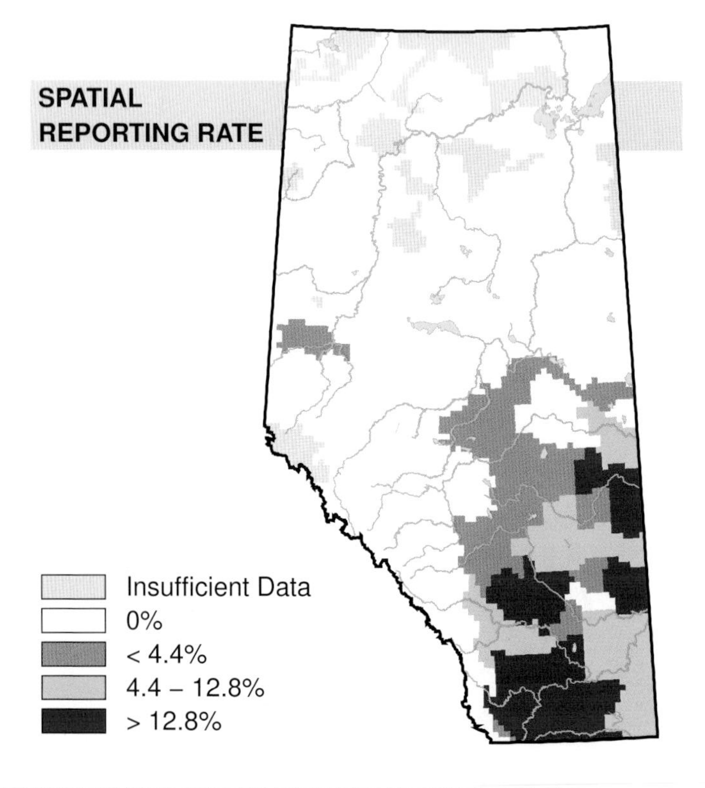

STATUS

GSWA 2000	GSWA 2005	COSEWIC Historic Rankings	COSEWIC 2007	Alberta Wildlife Act
Exotic/Alien	Exotic/Alien	N/A	N/A	N/A

RANGE

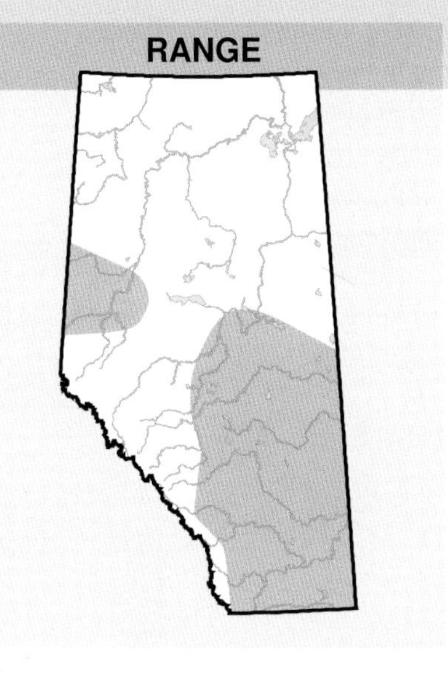

OBSERVED DISTRIBUTION ATLAS 2

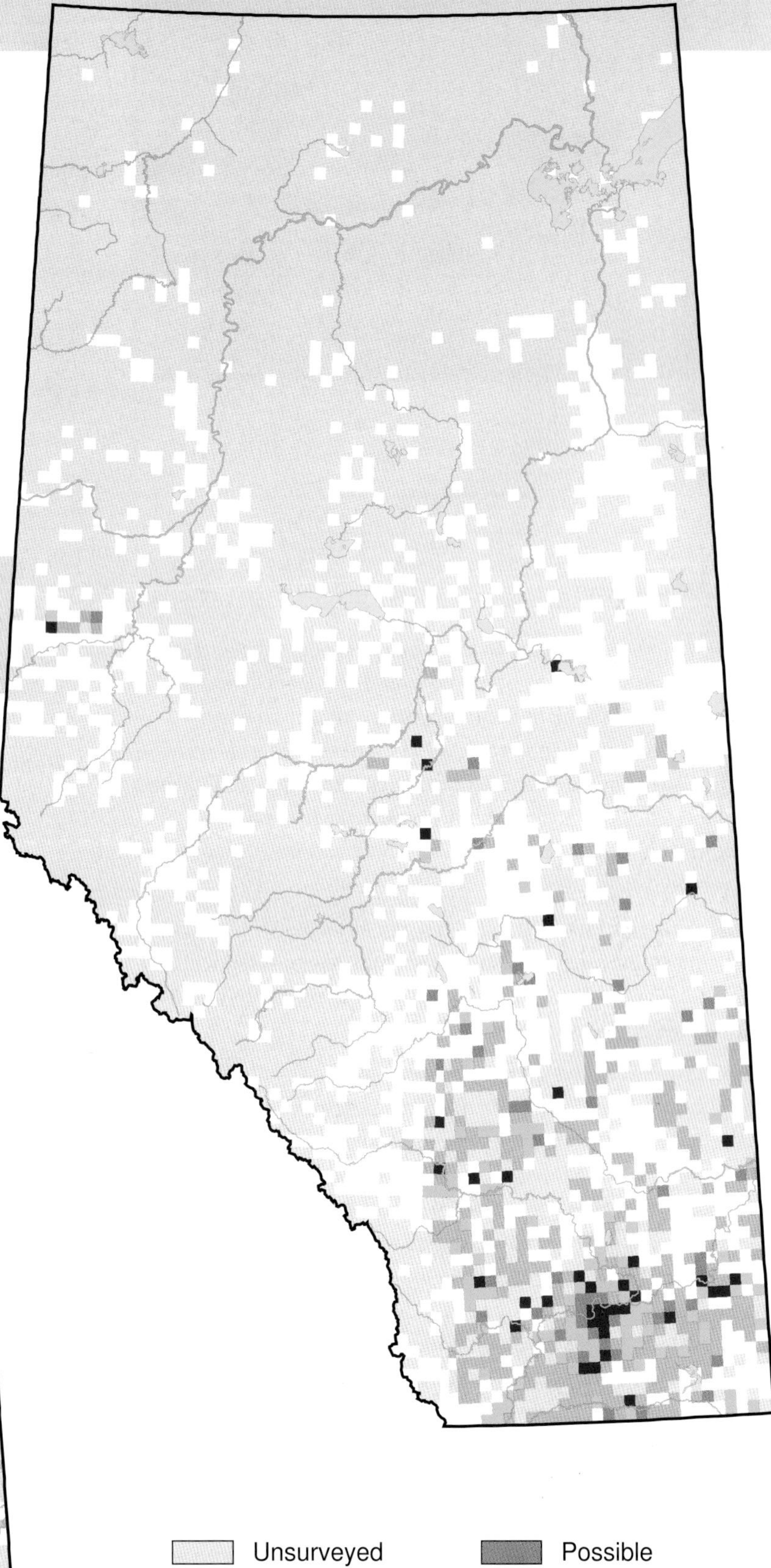

OBSERVED DISTRIBUTION ATLAS 1

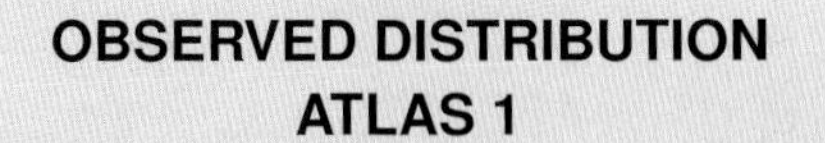

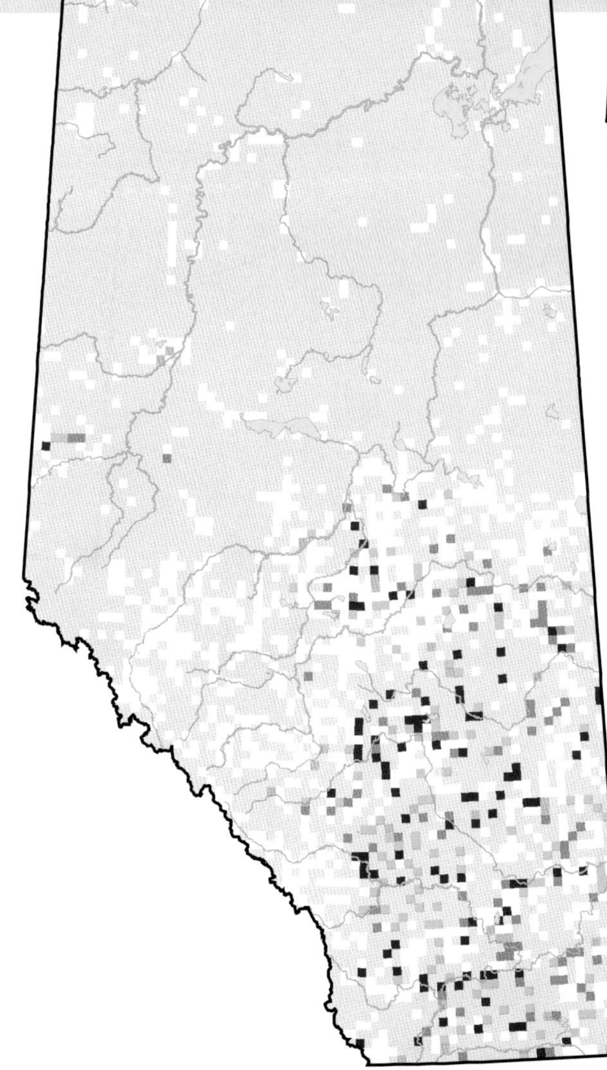

Ring-necked Pheasant (*Phasianus colchicus*)

Photo: Mark Williams

NESTING

CLUTCH SIZE (EGGS):	6–15
INCUBATION (DAYS):	22–25
FLEDGING (DAYS):	12–14
NEST HEIGHT (METRES):	0

Since being introduced into the province from Eurasia in 1908, the Ring-necked Pheasant has become a resident of the Grassland and Parkland Natural Regions of central and southern Alberta (Semenchuk, 1992). These pheasants were also released in the Peace River district where they still form a disjunct population, although there were far fewer breeding records in this area during Atlas 2. There was no change in the observed distribution between the two Atlases.

This species is found on grassland and farmland with adjacent suitable cover, especially cultivated lands interspersed with grass ditches, hedges, marshes, woodland borders, and brushy groves.

Decreases in relative abundance were detected in the Boreal Forest, Grassland, and Parkland Natural Regions. No changes in relative abundance were detected in the Foothills and Rocky Mountain NRs where it is considered a rare visitor. The Breeding Bird Surveys (1985–2005) detected no change in abundance in Alberta but did detect a decline across Canada. Despite an increase in range due to further land-clearing, local population levels are decreasing as a result of habitat degradation, mainly the reduction of cover habitats and small water bodies (Giudice and Ratti, 2001). This includes: transition from small diversified farms to large farms with fewer crops; loss of field-edge habitat; increasing use of farm pesticides; transition to "clean farming practices" (burning, weed-spraying, roadside mowing); advancement of hay-mowing dates; and overgrazing. Several studies indicate that, although populations often fluctuate with changes in weather, long-term population trends are more commonly related to habitat changes. Historically, pheasant management has included supplemental stocking and predator control but both of these programs are controversial, expensive, and have a limited capacity to increase populations (Giudice and Ratti, 2001). This species is listed as Exotic/Alien in the 2005 provincial status report.

TEMPORAL REPORTING RATE

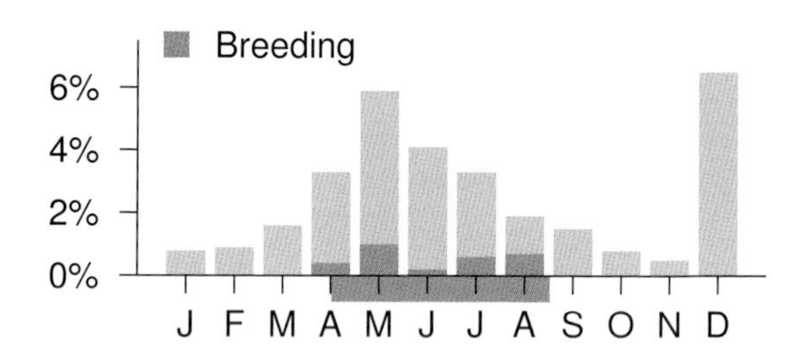

NATURAL REGIONS REPORTING RATE

SPATIAL REPORTING RATE

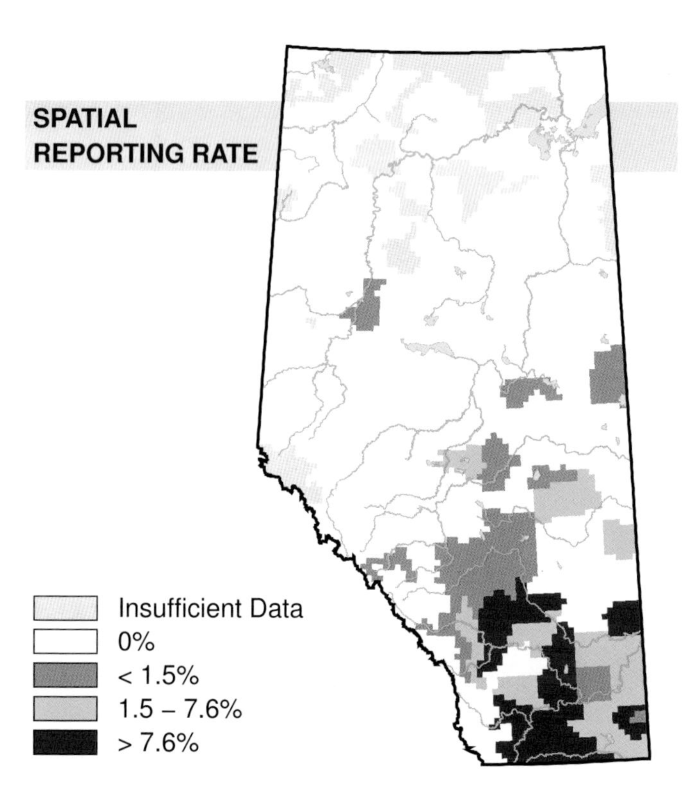

STATUS

GSWA 2000	GSWA 2005	COSEWIC Historic Rankings	COSEWIC 2007	Alberta Wildlife Act
Exotic/Alien	Exotic/Alien	N/A	N/A	N/A

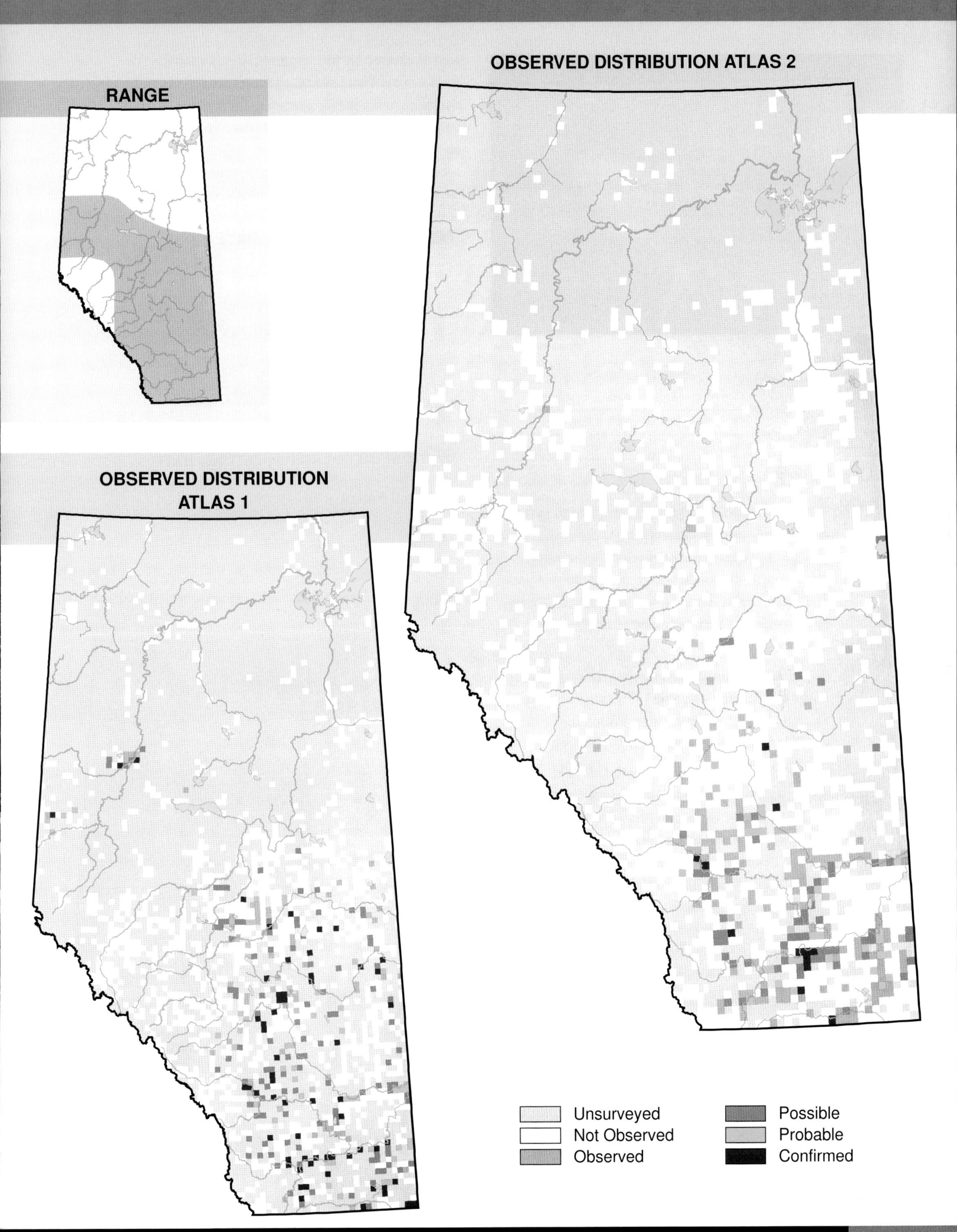

Habitat
Nest Location
Nest Type
Diet
RANGE
OBSERVED DISTRIBUTION ATLAS 2
OBSERVED DISTRIBUTION ATLAS 1
Unsurveyed
Not Observed
Observed
Possible
Probable
Confirmed

Ruffed Grouse (*Bonasa umbellus*)

Photo: Robert Gehlert

NESTING

CLUTCH SIZE (EGGS):	8–14
INCUBATION (DAYS):	24
FLEDGING (DAYS):	10–12
NEST HEIGHT (METRES):	0

The Ruffed Grouse is found mainly in central Alberta in the Parkland, Foothills and southern Boreal Forest Natural Regions, with a more widespread distribution in the northern parts. It is very scarce in the southeastern part of the province except for the Cypress Hills, where it was introduced. There are no changes in observed distribution between the two Atlases.

In Alberta, this grouse is most common in aspen-dominated and mixedwood forests, small openings. The presence of a prominent "drumming log" in shrub cover within the home range of the promiscuous male is important.

Decreases in relative abundance were detected in the Boreal Forest, Foothills, and Parkland NRs. No changes were detected in the Grassland and Rocky Mountain NRs. The Breeding Bird Surveys (1985–2005) detected no change in abundance in Alberta or across Canada, but a decline was detected across Canada for 1995–2005. Dunn (2005) states, in the National action needs for Canadian Landbird Conservation report, that the precision of the BBS is low partly due to non-linear population change of this species, and because survey techniques may not be comparable among jurisdictions. This grouse is also famous for the 10-year population cycles linked to fluctuations in snowshoe hare populations. Predators like Northern Goshawk, Great Horned Owl, and lynx prey on Ruffed Grouse when hare populations are down.

Ruffed Grouse occupy mainly early-successional deciduous forests created by fire, logging, or other large-scale disturbance. Fire control, opposition to clear-cut logging practices, and conifer management have resulted in maturation and conversion of deciduous forests and the general degradation of habitats for Ruffed Grouse. Concurrent reduction in numbers of Ruffed Grouse is thus assumed (Rusch et al., 2000). In Alberta this grouse is rated as Secure.

TEMPORAL REPORTING RATE

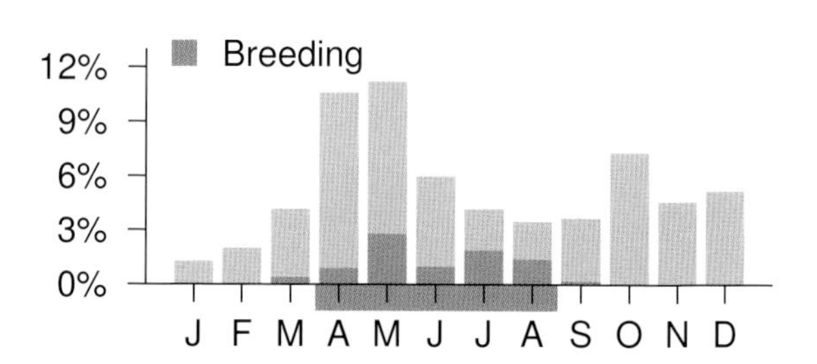

NATURAL REGIONS REPORTING RATE

Region	Rate
Boreal Forest	6%
Foothills	15%
Grassland	1%
Parkland	3%
Rocky Mountain	18%

SPATIAL REPORTING RATE

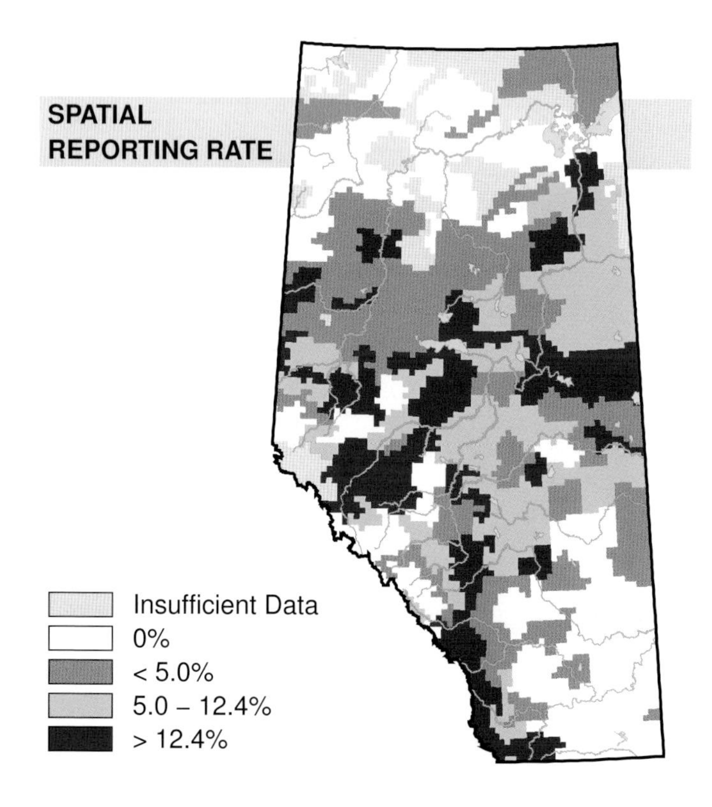

STATUS

GSWA 2000	GSWA 2005	COSEWIC Historic Rankings	COSEWIC 2007	Alberta Wildlife Act
Secure	Secure	N/A	N/A	N/A

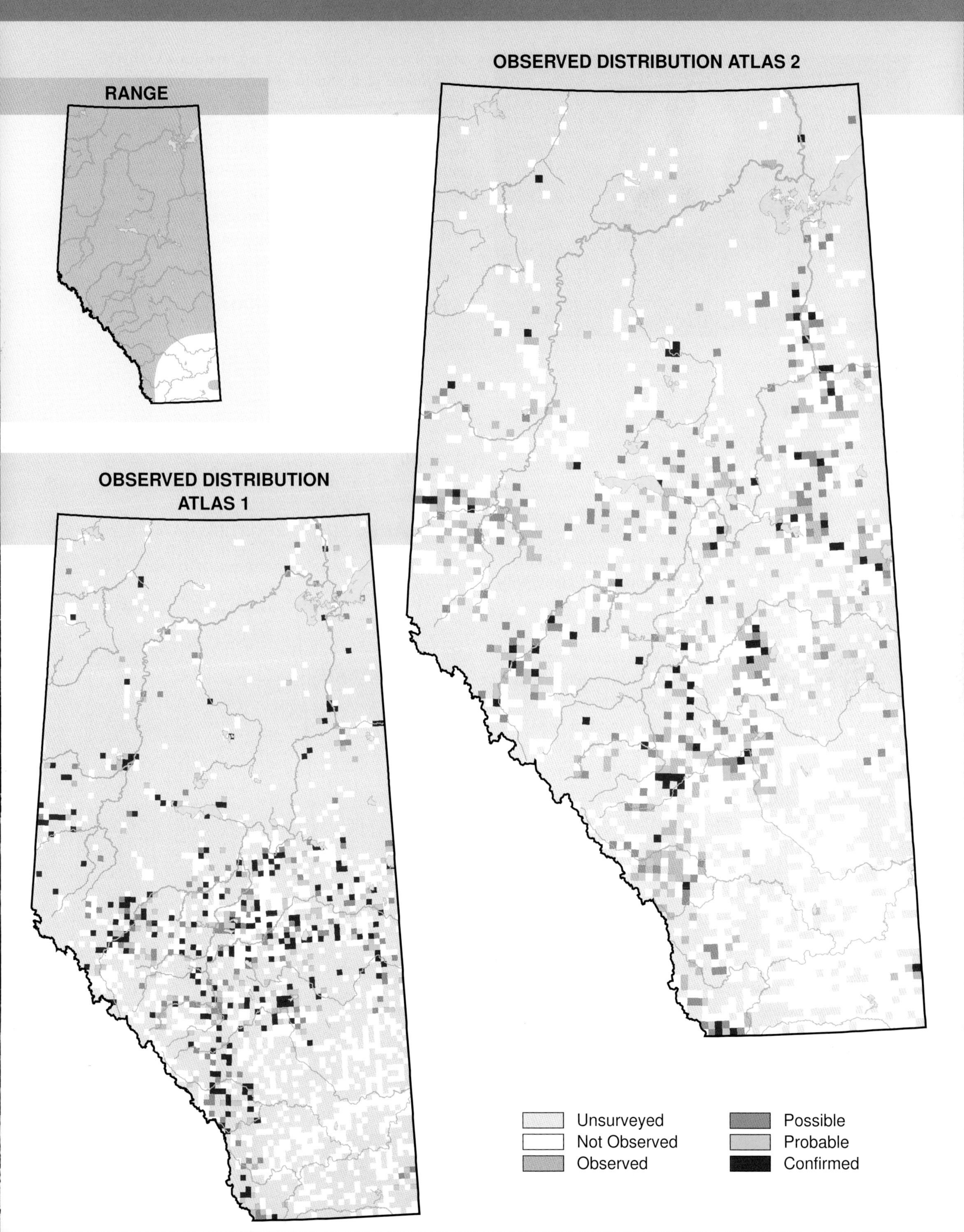
RANGE
OBSERVED DISTRIBUTION ATLAS 2
OBSERVED DISTRIBUTION ATLAS 1
Unsurveyed
Not Observed
Observed
Possible
Probable
Confirmed

Greater Sage-Grouse (*Centrocercus urophasianus*)

Photo: Gordon Court

NESTING

CLUTCH SIZE (EGGS):	6–9
INCUBATION (DAYS):	25–27
FLEDGING (DAYS):	7–10
NEST HEIGHT (METRES):	0

The extremely limited provincial distribution of the Greater Sage-Grouse was delineated in Atlas 1 and confirmed in Atlas 2. The species breeds only in the extreme southeastern corner of Alberta, east of the Milk river and south of the Cypress Hills (Semenchuk, 1992).

This grouse is closely associated with sagebrush ecosystems where the sagebrush types show tremendous natural variation in vegetative composition, habitat fragmentation, topography, substrate, weather, and frequency of fire (Schroeder et al., 1999). Many researchers report that habitat loss and degradation are the primary causes of range-wide reduction in distribution and populations of the Greater Sage-Grouse.

No changes in relative abundance were detected in the Grassland Natural Region. There are no Breeding Bird Survey (BBS) trend data available for this species. The BBS is ineffective in tracking populations of this species due to low precision (Dunn, 2005).

In the past 30 years, the Alberta/Saskatchewan population declined 66-92%, and is now one of the lowest-density populations known (Aldridge and Brigham, 2001). The Canadian decline is possibly related to limited availability of mesic habitats with high forb cover (Aldridge and Brigham, 2001). Dunn (2005) indicates that it is essential to identify an economically sustainable management strategy that creates diversity of cover, to ensure habitat availability for the full suite of native sagebrush/grassland species. COSEWIC (Committee on the Status of Endangered Wildlife in Canada) lists this grouse as Endangered. In Alberta the Greater Sage-Grouse is listed as At Risk, as the provincial population is estimated at only 350-400 individuals. There has been a drastic reduction in population and distribution, and recovery depends on the availability of sagebrush/grassland habitat. This species is designated as Endangered under the Wildlife Act and the province, in conjunction with other agencies, has an active Recovery Team in place.

TEMPORAL REPORTING RATE

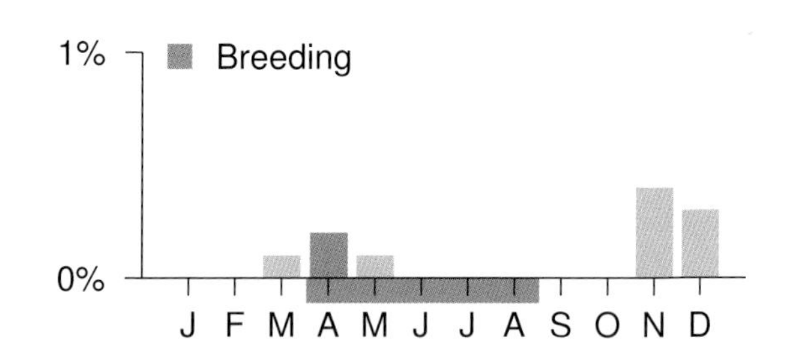

NATURAL REGIONS REPORTING RATE

SPATIAL REPORTING RATE

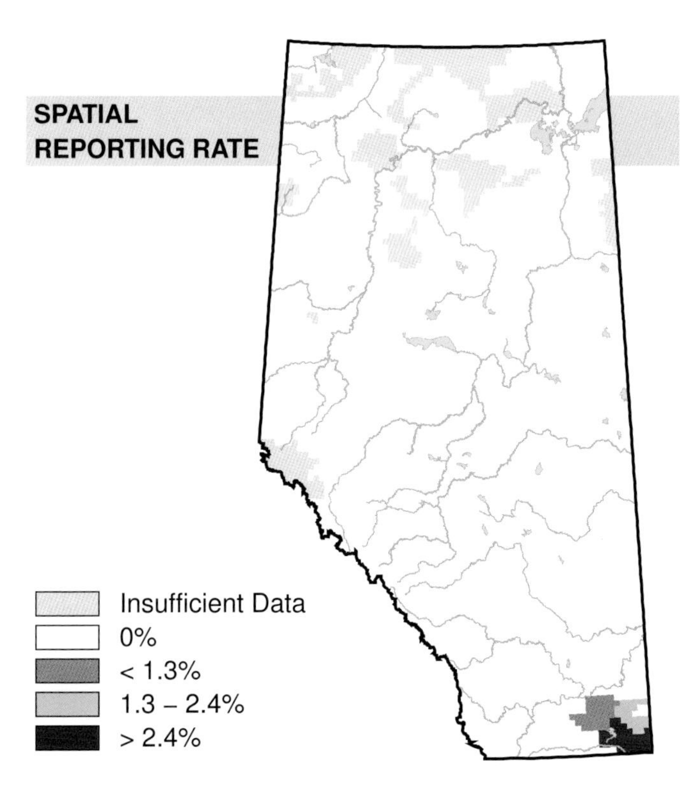

STATUS

GSWA 2000	GSWA 2005	COSEWIC Historic Rankings	COSEWIC 2007	Alberta Wildlife Act
At Risk	At Risk	Endangered	Endangered	Endangered

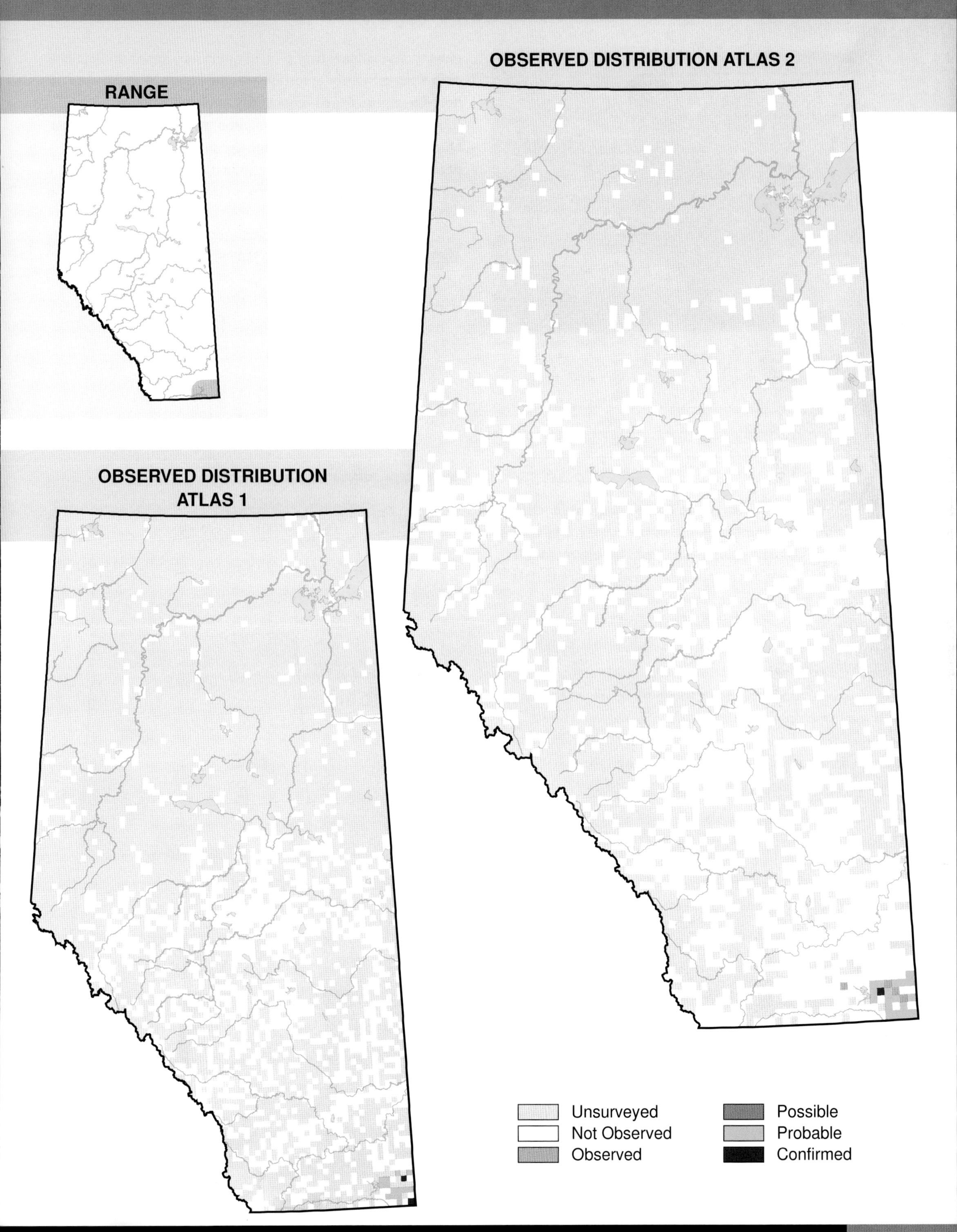
OBSERVED DISTRIBUTION ATLAS 2
RANGE
OBSERVED DISTRIBUTION
ATLAS 1
Unsurveyed
Not Observed
Observed
Possible
Probable
Confirmed

Spruce Grouse (*Falcipennis canadensis*)

Photo: Gerald Romanchuk

NESTING

CLUTCH SIZE (EGGS):	7–8
INCUBATION (DAYS):	21–24
FLEDGING (DAYS):	10
NEST HEIGHT (METRES):	0

The Spruce Grouse is found in the Rocky Mountain, Foothills, and Boreal Forest Natural Regions. Historically, it was found in all of central Alberta south to Red Deer, but the removal of the coniferous forests has forced it to retreat northward. Two subspecies of this grouse are present in Alberta: *D. c. canadensis* is found in central and northern Alberta, except in the Rocky Mountain NR, where *D. c. franklinii* is found (Semenchuk, 1992).

Spruce Grouse prefer relatively young successional stands, with small openings, in coniferous and mixedwood forests. Despite differences in species composition of conifers, these birds seem to use similarly structured forests—that is, stands that are relatively dense (Boag and Schroeder, 1992).

Decreases in relative abundance were detected in the Boreal Forest, Foothills, and Rocky Mountain NRs. These areas have seen an increase in land area affected, with the removal of tree cover by the forest and petroleum industries since Atlas 1. No change in relative abundance was detected in the Parkland NR. The Breeding Bird Survey (1985–2005) yielded insufficient data to estimate trends in Alberta. Dunn (2005) reports that the population size of this grouse is difficult to monitor due to: shifting densities as habitat changes; population fluctuation on top of habitat changes; and relatively inaccessible habitat (not sampled by the BBS). This bird is probably best tracked by broad-scale monitoring of habitat distribution and quality, combined with periodic stratified sampling of population density in standard study plots. Populations appear to fluctuate over time, primarily in response to the degree of maturation of post-fire regrowth and secondarily to predation pressure. Modern industrial forest exploitation, with its creation of open clear-cuts and subsequent single-species plantations, reduces populations locally and often eliminates them entirely (Boag and Schroeder, 1992). The Spruce Grouse is listed as Secure in Alberta.

TEMPORAL REPORTING RATE

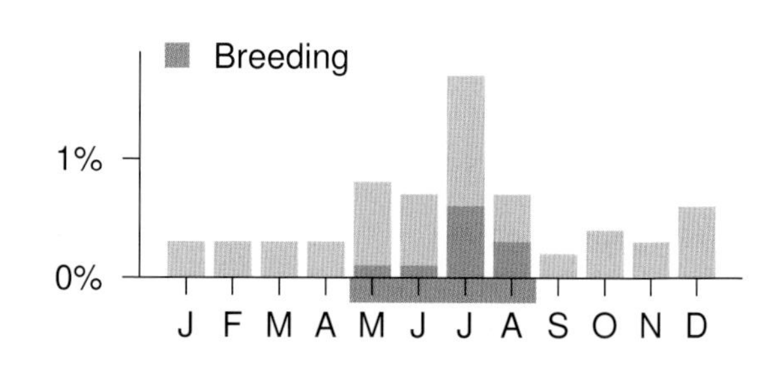

NATURAL REGIONS REPORTING RATE

SPATIAL REPORTING RATE

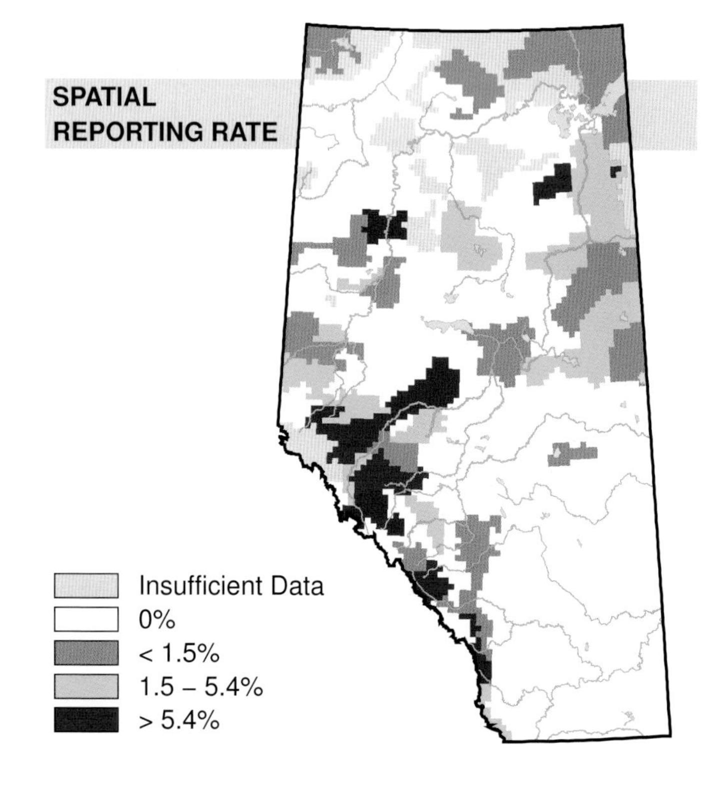

STATUS

GSWA 2000	GSWA 2005	COSEWIC Historic Rankings	COSEWIC 2007	Alberta Wildlife Act
N/A	N/A	N/A	N/A	N/A

RANGE

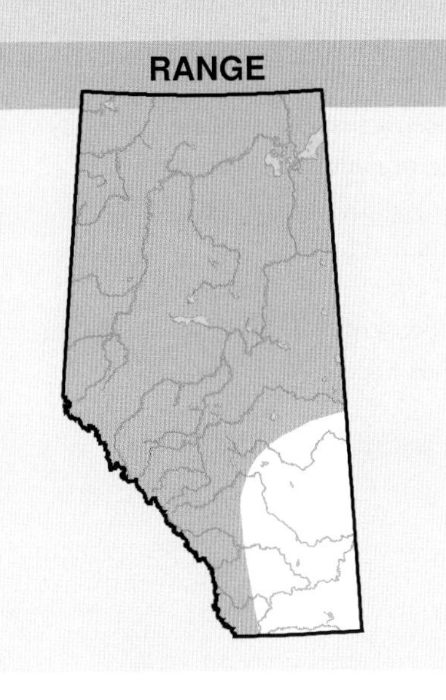

OBSERVED DISTRIBUTION ATLAS 2

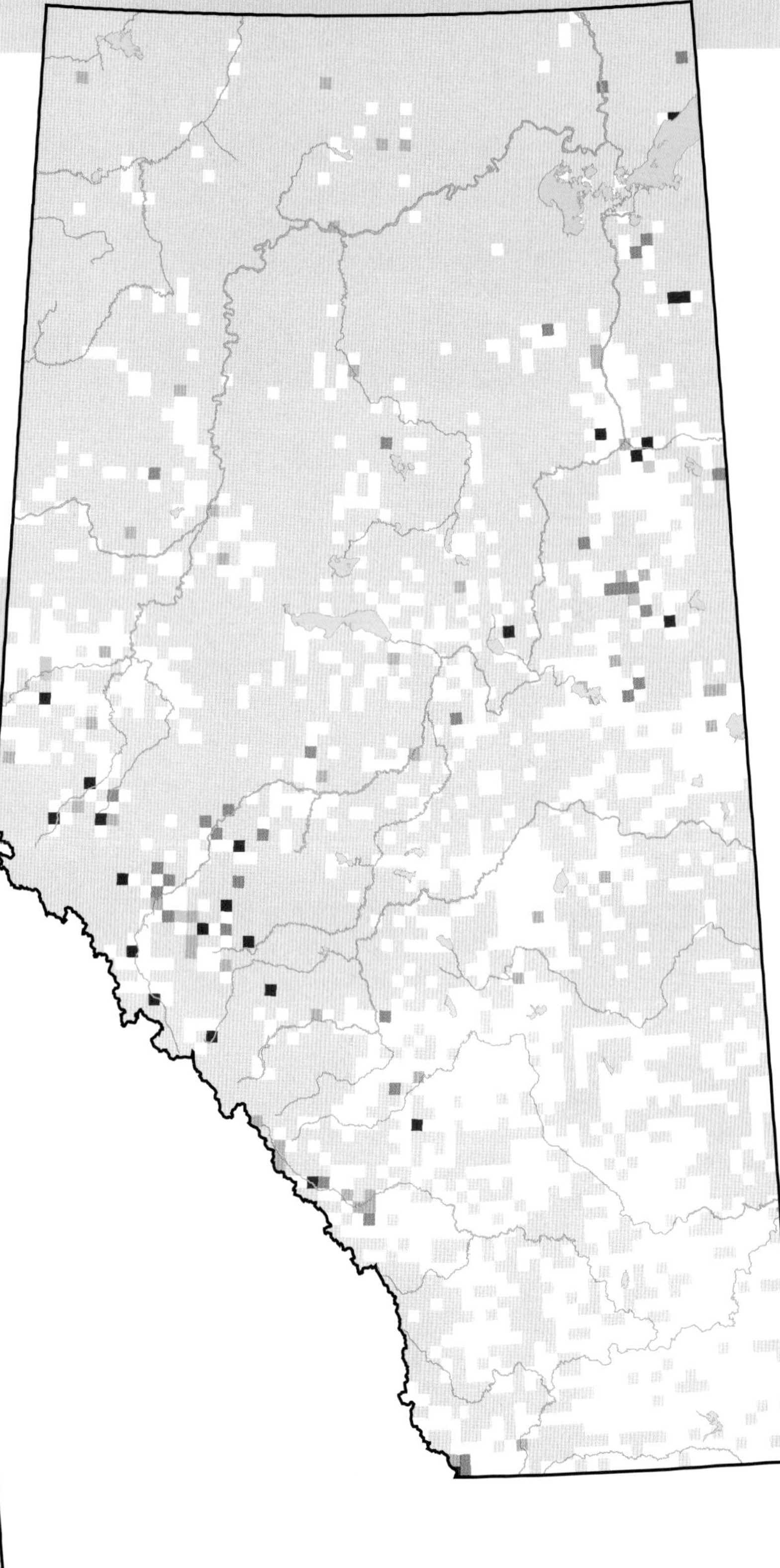

OBSERVED DISTRIBUTION ATLAS 1

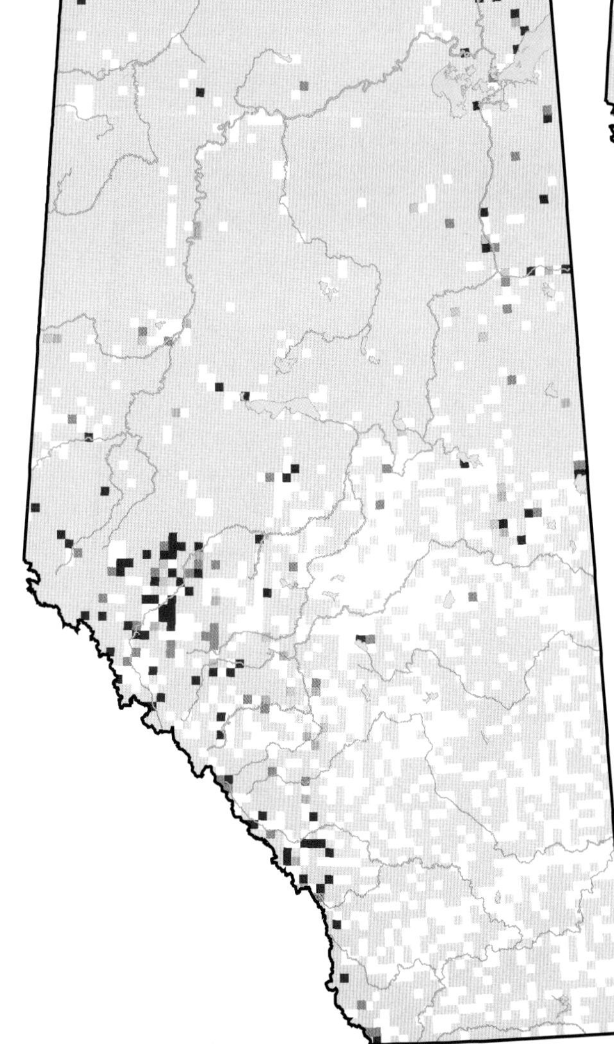

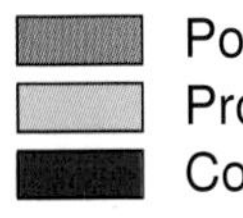

Unsurveyed
Not Observed
Observed
Possible
Probable
Confirmed

Willow Ptarmigan (*Lagopus lagopus*)

Photo: Damon Calderwood

NESTING

CLUTCH SIZE (EGGS):	7–10
INCUBATION (DAYS):	22
FLEDGING (DAYS):	12–13
NEST HEIGHT (METRES):	0

The Willow Ptarmigan is a rare breeder in a very localized part of the province (Semenchuk, 1992). Historically, the breeding range in Alberta was described as the northern part of Jasper National Park as far south as the Tonquin Valley, but may extend northward to the Kakwa area southwest of Grande Prairie. In Atlas 1 there were only three confirmed breeding records and all were in the same square. In Atlas 2 there were no confirmed breeding records. While species may be less common at the southern edge of wintering range than in the past, overall it remains widespread and abundant (Hannon et al., 1998).

This ptarmigan is the most numerous of North America's three species of ptarmigan and is normally found in arctic, subarctic, and subalpine tundra. It prefers willow/dwarf birch meadows just above timberline. In breeding season, it is recorded most often in open forests and shrubby meadows of the upper subalpine zone.

Neither the Atlas nor the Breeding Bird Survey had sufficient data to detect changes in abundance for this species. Dunn (2005), in the National action needs for Canadian Landbird Conservation, reported that there is a lack of trend data for this species. It does undergo cyclic population cycles (Hannon et al., 1998), and requires a long-term view on assessment of status. Global climate change may impact its habitat.

The Canadian Wildlife Service, in a number of its on-line fact sheets, indicates that despite the relative remoteness of ptarmigan ranges, poor land-use practices have already degraded or destroyed some of their habitat. Any land use that leads to progressive erosion, destruction of vegetation, pollution of soil, water, and air, or melting of permafrost can endanger ptarmigans and other wildlife. Although all three ptarmigan species have been studied by scientists, their levels of tolerance to human-induced changes remain largely unknown, as do many other aspects of their biology. In Alberta the Willow Ptarmigan is listed as Secure.

TEMPORAL REPORTING RATE

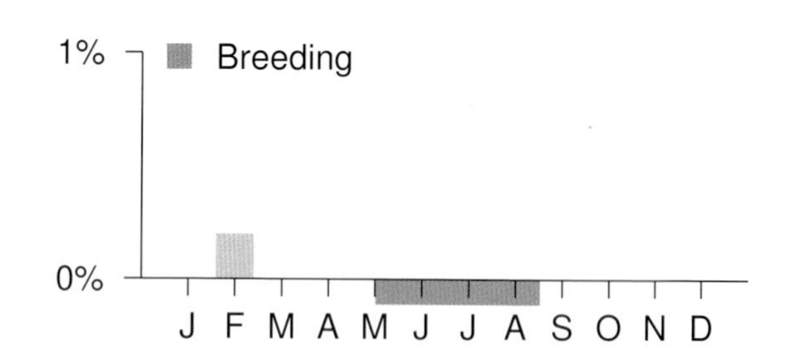

NATURAL REGIONS REPORTING RATE

Boreal Forest < 0.1%

SPATIAL REPORTING RATE

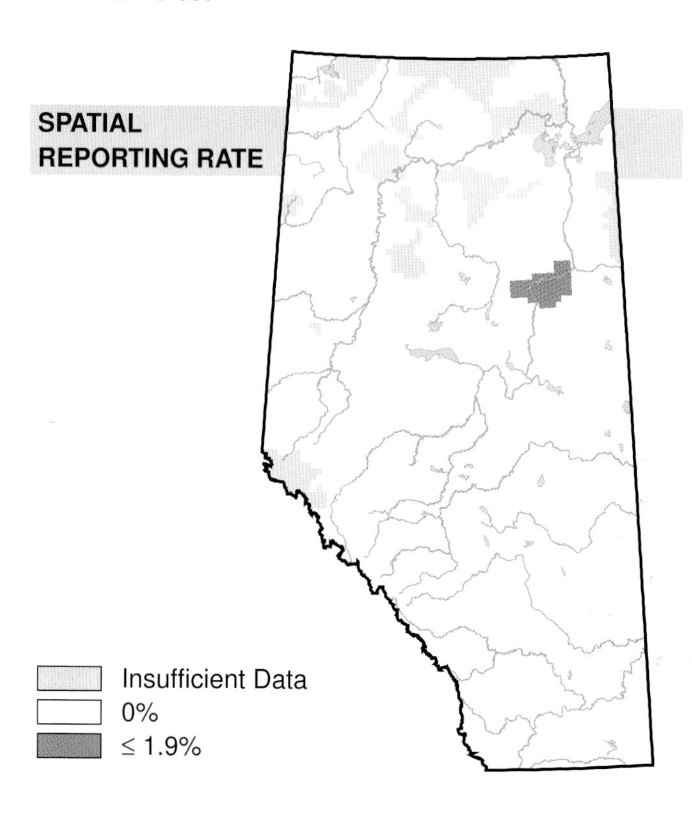

STATUS

GSWA 2000	GSWA 2005	COSEWIC Historic Rankings	COSEWIC 2007	Alberta Wildlife Act
Secure	Secure	N/A	N/A	N/A

Habitat Nest Location Nest Type Diet

RANGE

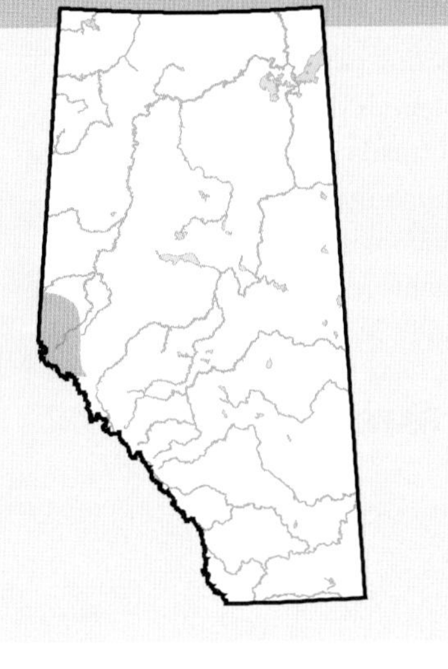
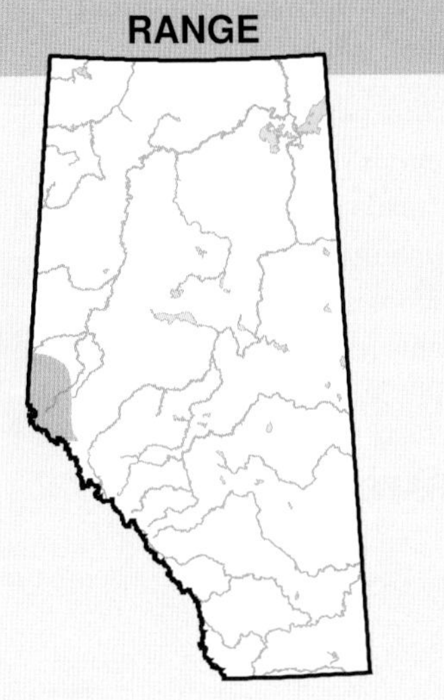

OBSERVED DISTRIBUTION ATLAS 2

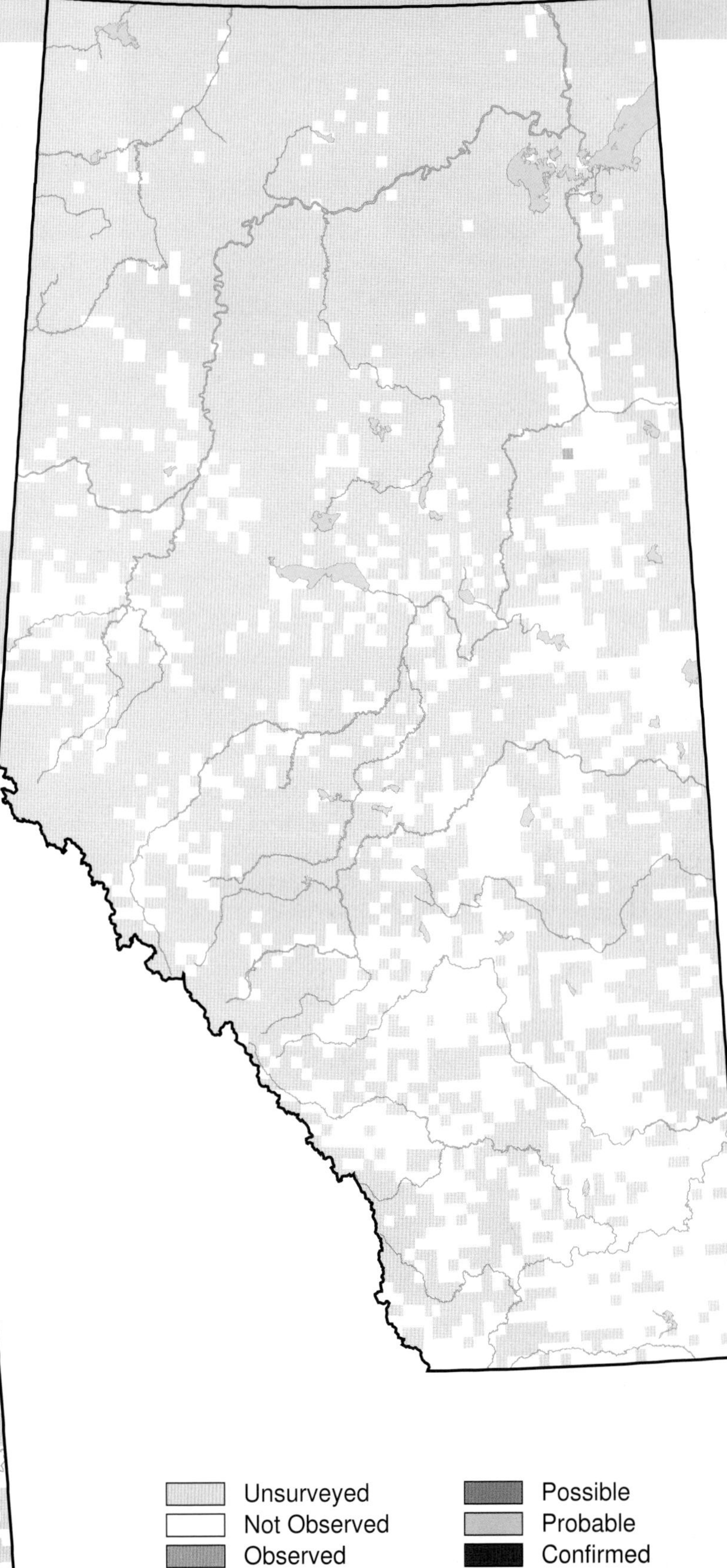

OBSERVED DISTRIBUTION ATLAS 1

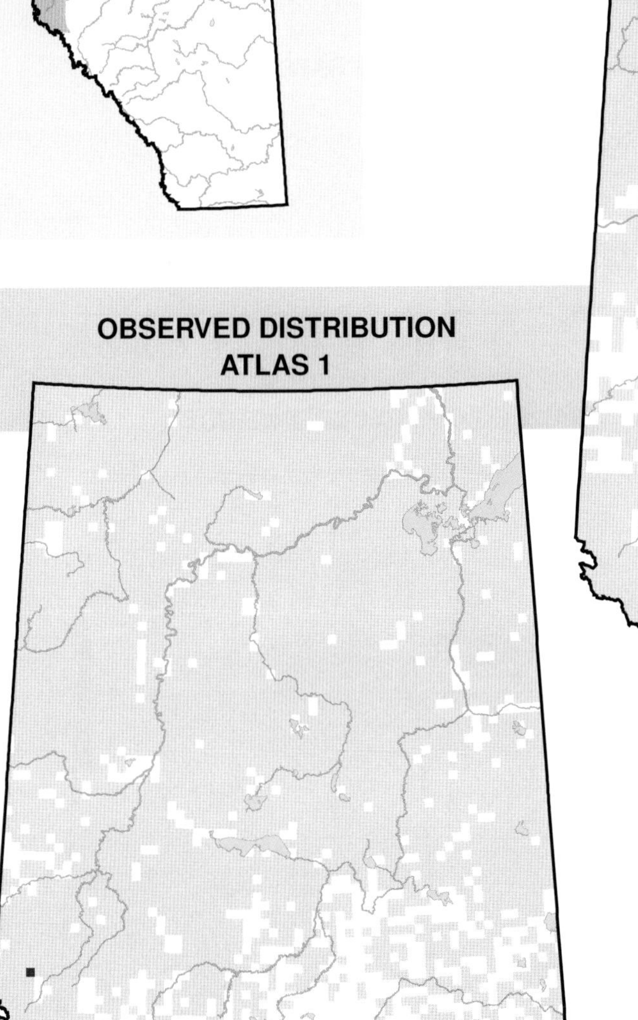

Unsurveyed
Not Observed
Observed
Possible
Probable
Confirmed

White-tailed Ptarmigan (*Lagopus leucura*)

Photo: Ian Gardiner

NESTING

CLUTCH SIZE (EGGS):	4–8
INCUBATION (DAYS):	22–23
FLEDGING (DAYS):	10
NEST HEIGHT (METRES):	0

The White-tailed Ptarmigan is a year-round resident in Alberta, considered to be widespread throughout the Rocky Mountain Natural Region from Willmore Wilderness Park south to Waterton Lakes National Park. There was no change in the distribution of the White-tailed Ptarmigan from Atlas 1.

This ptarmigan uses rocky areas and alpine meadows. Common features are open scree, rock slides, boulder fields and nearby water. In winter, a short downslope migration is made into open forests and willow meadows of the subalpine zone (Semenchuk, 1992).

No changes in relative abundance were detected in any Natural Region. The Breeding Bird Surveys (1985–2005) had no data for this species in Alberta or across Canada. Long-term studies (27 years) of hunted and unhunted populations of White-tailed Ptarmigan in Colorado indicated that populations fluctuate widely among years, with no clear evidence of population cycles (Braun et al., 1993). However, Braun et al. (1993) did indicate that, because White-tailed Ptarmigan generally occupy lands that are open to the public and managed for "multiple use", localized distribution of birds is probably affected by road construction, industrial activity, off-road vehicles, overgrazing by livestock, and ski-area developments. These factors reduce the abundance and distribution of winter food, principally willow. In addition, the opportunity for large population increases is limited because, compared to other grouse species, White-tailed Ptarmigan produce relatively few young and turnover in the breeding population is low (Hoffman, 2006).

Dunn (2005), in the National action needs for Canadian Landbird Conservation, recommends that this species has an uncertain population status in Canada and requires monitoring of harvest statistics and the availability and distribution of suitable habitat. This should be combined with a periodic assessment of the population status range-wide. In Alberta the White-tailed Ptarmigan is listed as Secure.

TEMPORAL REPORTING RATE

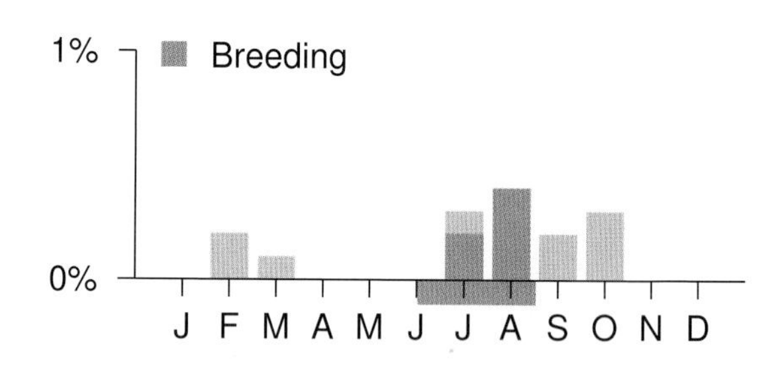

NATURAL REGIONS REPORTING RATE

Rocky Mountain 1.1%

SPATIAL REPORTING RATE

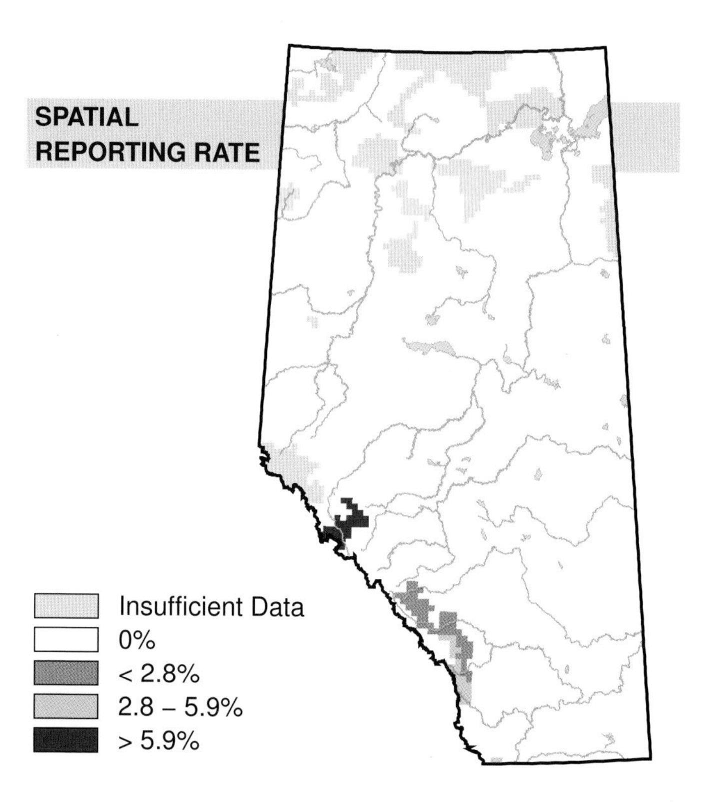

STATUS

GSWA 2000	GSWA 2005	COSEWIC Historic Rankings	COSEWIC 2007	Alberta Wildlife Act
Secure	Secure	N/A	N/A	N/A

RANGE

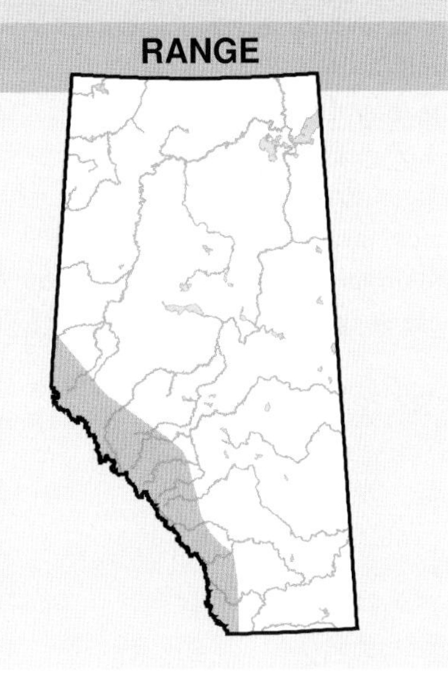

OBSERVED DISTRIBUTION ATLAS 2

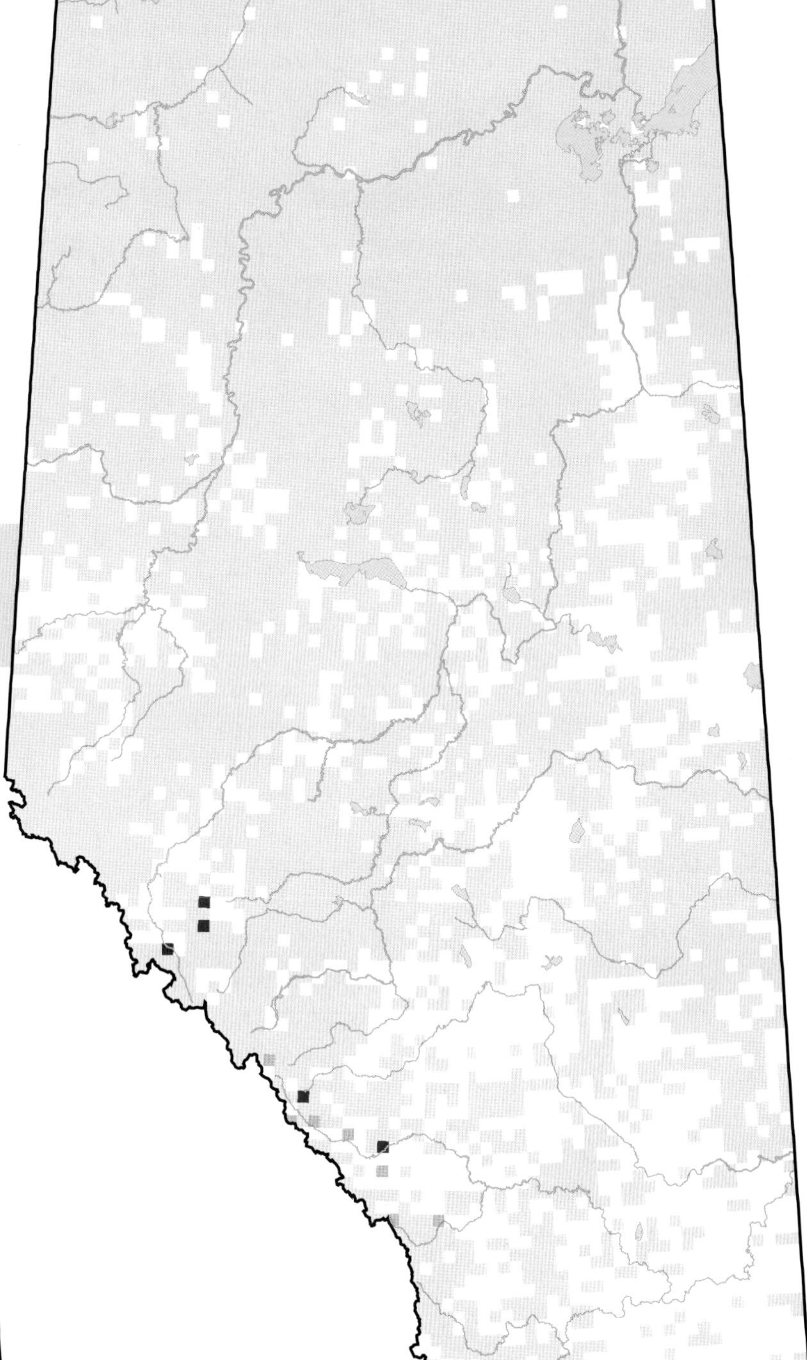

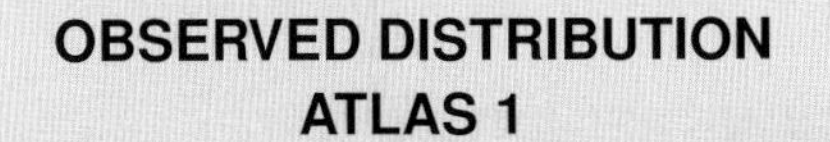

OBSERVED DISTRIBUTION ATLAS 1

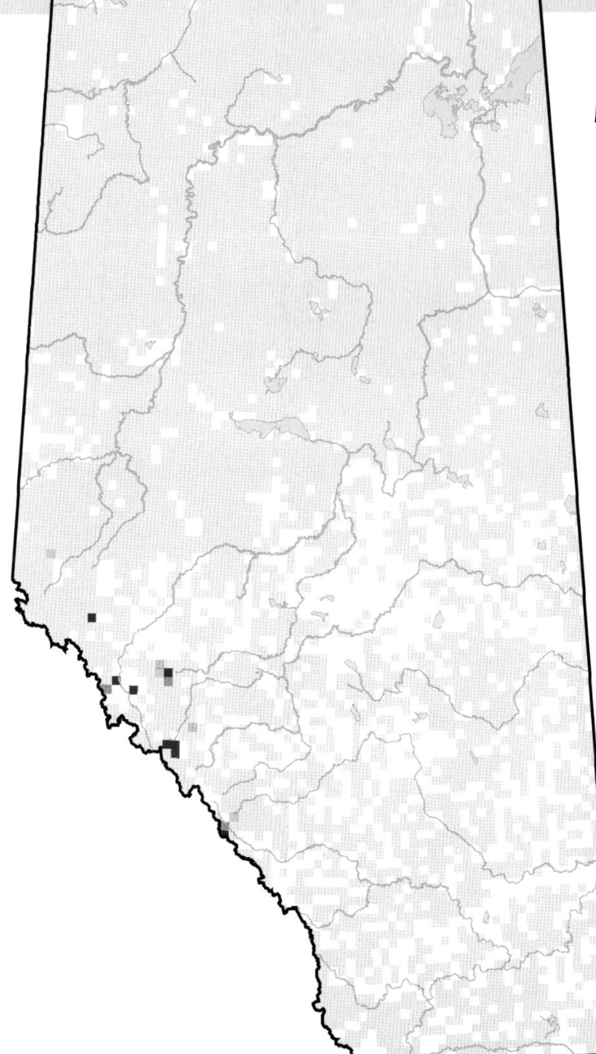

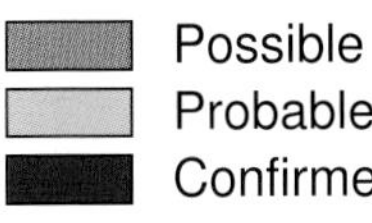

Dusky Grouse (*Dendragapus obscurus*)

Photo: Alan MacKeigan

NESTING

CLUTCH SIZE (EGGS):	5–10
INCUBATION (DAYS):	24–25
FLEDGING (DAYS):	6–7
NEST HEIGHT (METRES):	0

Based on recent mitochondrial DNA sequence data, as well as on earlier behavioural and distributional data, the 47th Supplement to the American Ornithologists' Union's Checklist of North American Birds has recognized the two groups of Blue Grouse (Dusky and Sooty) as separate species (Zwickel and Bendell, 2005). The Dusky Grouse is a permanent resident of the Rocky Mountain Natural Region and the adjoining portions of the Foothills NR. It is found from the headwaters of the Smoky River in the north to the International Boundary in the south. No changes in distribution were observed between Atlases 1 and 2.

The distribution of this grouse appears to be partly determined by the proximity of suitable breeding areas to montane forest acceptable for use in winter, where conifer needles comprise the main winter food (Zwickel and Bendell, 2005). It occupies a fairly broad vertical range, breeding at low elevations and spending fall and winter close to timberline.

No changes in relative abundance were detected in the Foothills and Rocky Mountain NRs. The Breeding Bird Surveys (1985–2005) had no data for this species in Alberta and detected no change across Canada. Blue Grouse can attain high population densities and remain distributed throughout most of their historical range. Occupation of relatively inaccessible montane forests during much of the year contributes to a generally healthy status in many areas (Zwickel and Bendell, 2005). They also note that despite intensive study of Blue Grouse during the past 60 years, our ability to predict population levels and trends remains poor. The selection of remote mountainous habitat has helped protect Blue Grouse, so that the long-term outlook for many populations is reasonably good. Zwickel and Bendell (2005) indicates that some forest practices, grazing of domestic livestock, other agrarian activities, increasing recreational inroads into montane areas, and urbanization remain threats. In Alberta the Dusky (Blue) Grouse is rated as Secure.

TEMPORAL REPORTING RATE

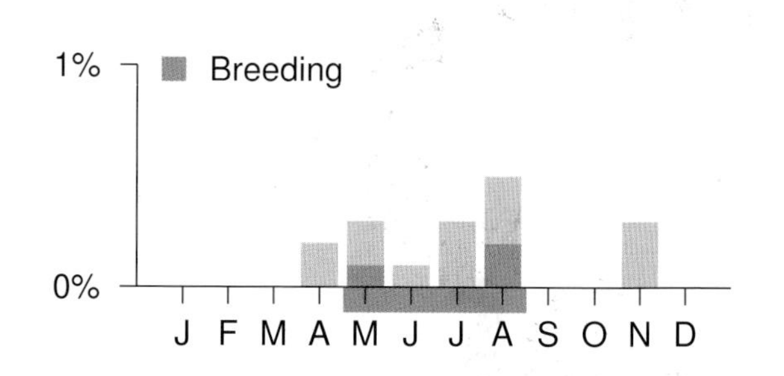

NATURAL REGIONS REPORTING RATE

SPATIAL REPORTING RATE

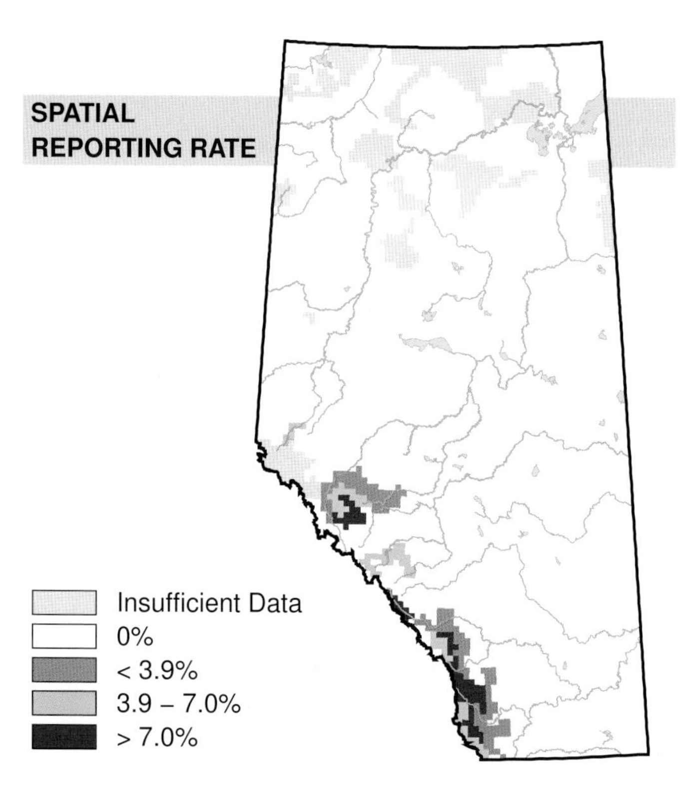

STATUS

GSWA 2000	GSWA 2005	COSEWIC Historic Rankings	COSEWIC 2007	Alberta Wildlife Act
Secure	Secure	N/A	N/A	N/A

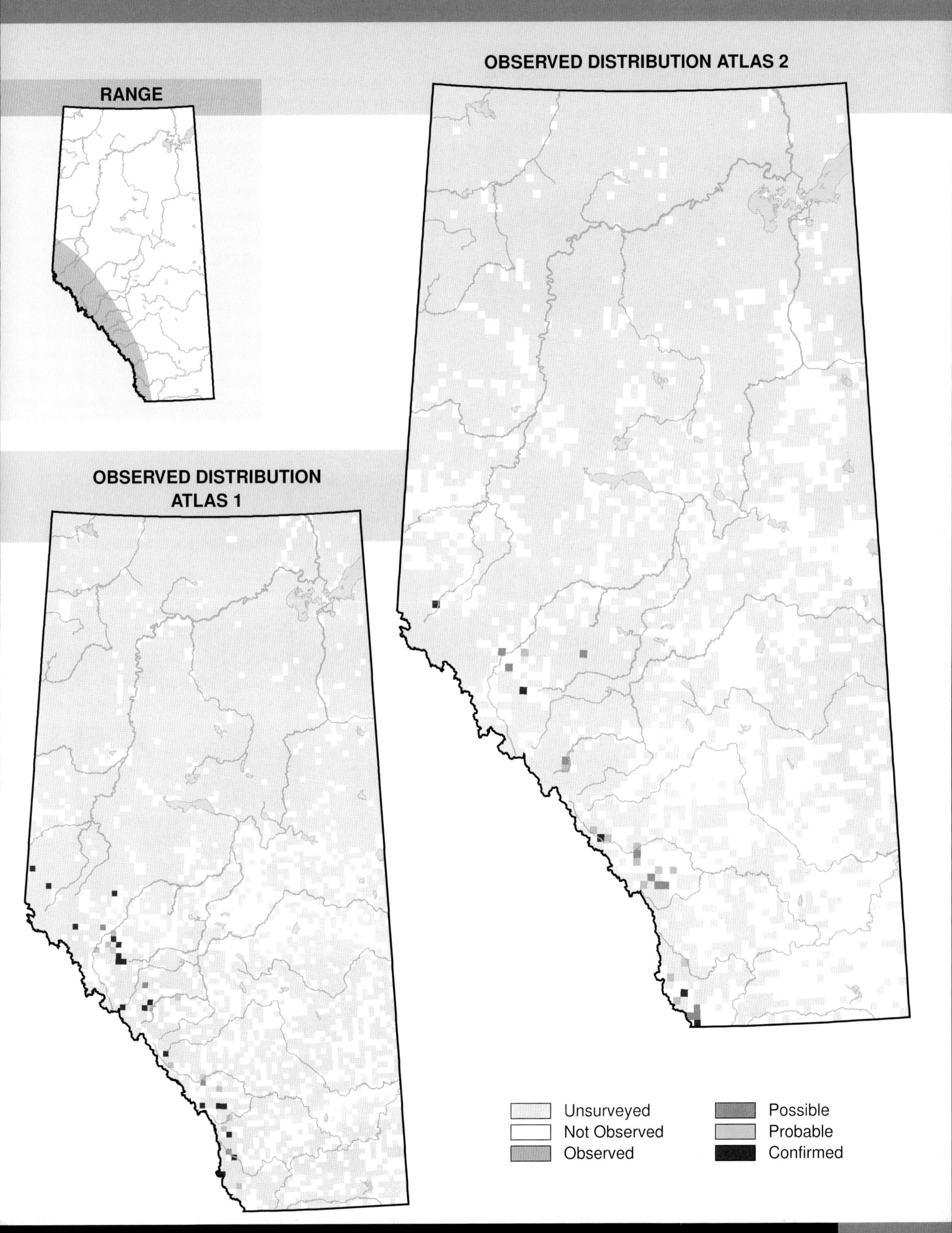
Habitat
Nest Location
Nest Type
Diet
OBSERVED DISTRIBUTION ATLAS 2
RANGE
OBSERVED DISTRIBUTION ATLAS 1
Unsurveyed
Not Observed
Observed
Possible
Probable
Confirmed

Sharp-tailed Grouse (*Tympanuchus phasianellus*)

Photo: Gordon Court

NESTING

CLUTCH SIZE (EGGS):	10–14
INCUBATION (DAYS):	23–24
FLEDGING (DAYS):	42–56
NEST HEIGHT (METRES):	0

Sharp-tailed Grouse were recorded in five of the six Natural Regions in Atlas 2, but were most commonly found in the Grassland and Parkland NRs, including the Peace River Parkland. Records for the Boreal Forest, Foothills, and Rocky Mountain NRs were localized where open, grassland-like habitats are used. The provincial distribution did not change between Atlas 1 and Atlas 2 even though the number of records shows local variation.

Reporting rates, even in the best habitat, were very low because this grouse is difficult to see or hear (except on the lek during the breeding season), and it occurs in relatively low densities. The Reporting Rate map and the Natural Regions graph provide a general guide to relative abundance but must be used cautiously. Reference to the observed distribution map demonstrates that records of this species were localized within the areas shown in the reporting rate map. Further, the inclusion of the Cypress Hills in the Rocky Mountain NR can lead to misinterpretation of the bar graph.

Increases in relative abundance were detected in the Grassland NR while decreases were detected in the Boreal Forest and Parkland NRs. Observation frequency relative to other species coincided with the detected changes in relative abundance. These changes may have been influenced by changed survey distribution and effort (e.g., there appeared to be a greater percentage of squares in poorer habitat in the Parkland for Atlas 2). No changes in relative abundance were detected in the Foothills and Rocky Mountain NRs. The Breeding Bird Survey did not detect any change but BBS researchers indicated that Sharp-tailed Grouse are poorly surveyed by this technique. The provincial status reports from 2000 and 2005 rate this bird as Sensitive and indicate that populations appear to be declining. However, supportive data are lacking.

TEMPORAL REPORTING RATE

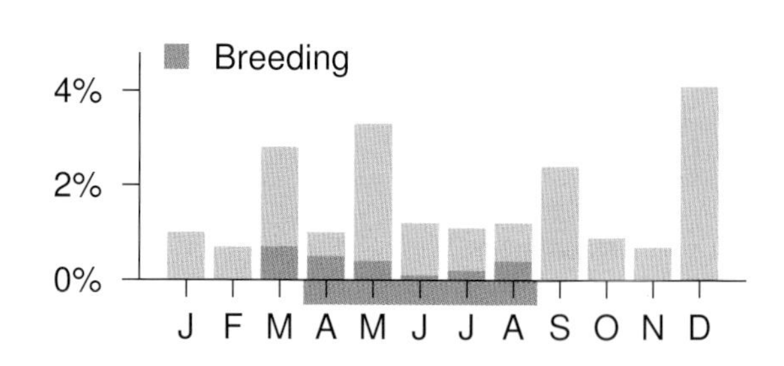

NATURAL REGIONS REPORTING RATE

SPATIAL REPORTING RATE

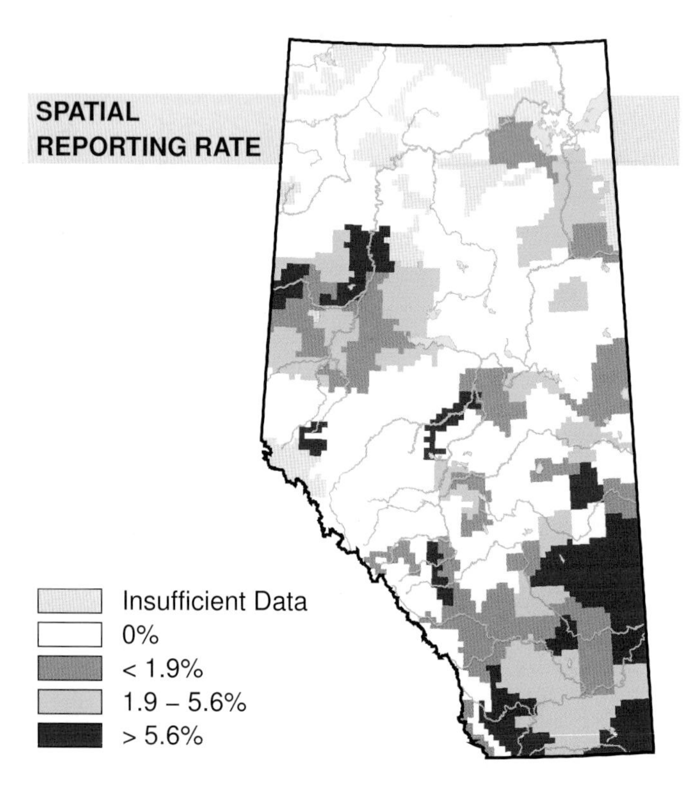

STATUS

GSWA 2000	GSWA 2005	COSEWIC Historic Rankings	COSEWIC 2007	Alberta Wildlife Act
Sensitive	Sensitive	N/A	N/A	N/A

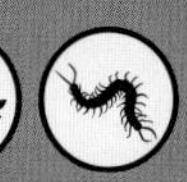

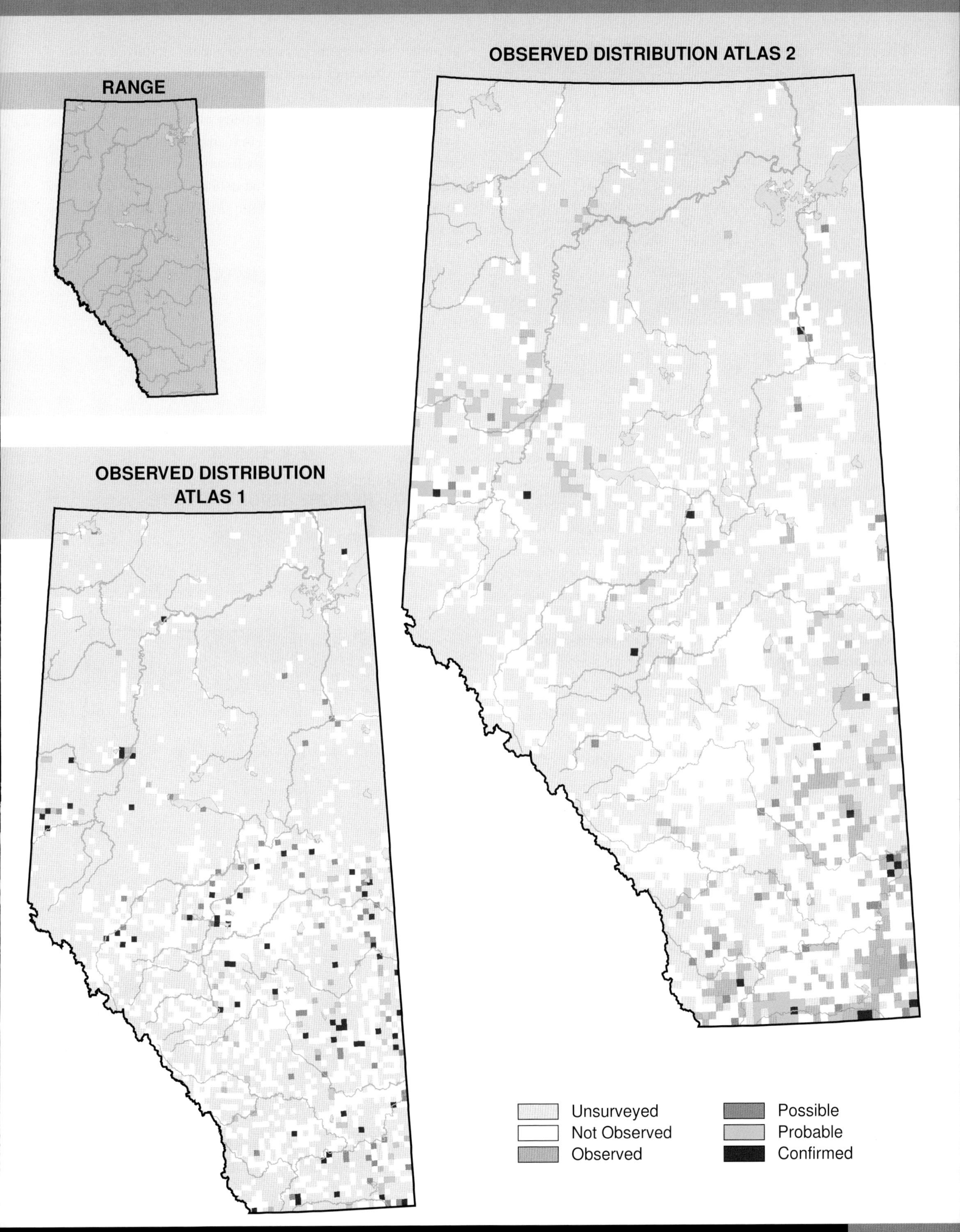
OBSERVED DISTRIBUTION ATLAS 2
RANGE
OBSERVED DISTRIBUTION
ATLAS 1
Unsurveyed
Not Observed
Observed
Possible
Probable
Confirmed

Wild Turkey (*Meleagris gallopavo*)

Photo: Glen Rowan

NESTING

CLUTCH SIZE (EGGS):	10–18
INCUBATION (DAYS):	27–28
FLEDGING (DAYS):	14
NEST HEIGHT (METRES):	0

The Wild Turkey was extirpated from its former range in the early 1900s but has been reintroduced into much of its former range in Alberta. Alberta has two main populations: one in the Porcupine Hills and one in the Cypress Hills. Turkeys were also introduced into the Lees Lake and Todd Creek areas of southern Alberta. There was no change in distribution of this species from Atlas 1, with the exception of one observation recorded west of Ponoka, a long way north of its normal range.

This one disjunct record may be a remnant of an illegal release, near Olds, of Wild Turkeys by a private individual in the fall of 1992 to "enhance local wildlife". The release was not approved by Fish and Wildlife, and subsequently the owner was charged under the Wildlife Act, convicted, and fined.

In Alberta the Wild Turkey, sometimes called Merriam's Turkey, favours open deciduous forest, forest edges and agricultural fields, and it remains on the breeding grounds over winter (Semenchuk, 1992). Roosting sites and available food and water are critical in winter; spring rainfall is critical for poult survival as precipitation determines abundance of vegetation and insects (Eaton, 1992).

No changes in relative abundance were detected in any Natural Region. There were no data from the Breeding Bird Surveys for this species in Alberta, but across Canada BBS results for 1985–2005 show an increase in abundance. Management of hunting and prevention of poaching will become more critical as human populations increase and a more urbanized human population becomes established, with reduced understanding of natural systems (Eaton, 1992). Dunn (2005) states, in the National action needs for Canadian Landbird Conservation, that this species responds well to management and re-introduction and that the population is apparently expanding, but a change in trend could quickly lead to concern for the small Canadian population. In Alberta the Wild Turkey is listed as Exotic/Alien.

TEMPORAL REPORTING RATE

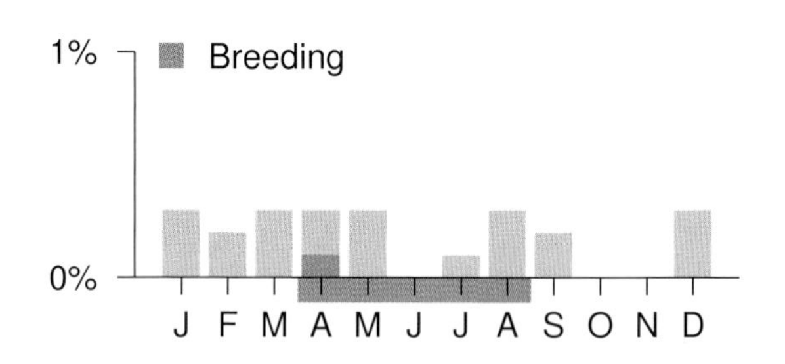

NATURAL REGIONS REPORTING RATE

SPATIAL REPORTING RATE

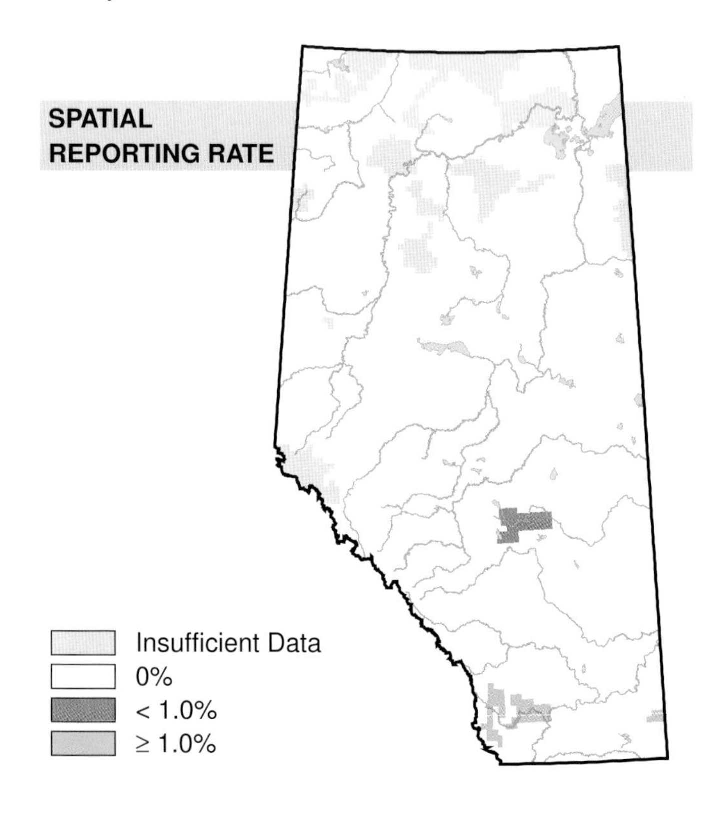

STATUS

GSWA 2000	GSWA 2005	COSEWIC Historic Rankings	COSEWIC 2007	Alberta Wildlife Act
Exotic/Alien	Exotic/Alien	N/A	N/A	N/A

OBSERVED DISTRIBUTION ATLAS 2

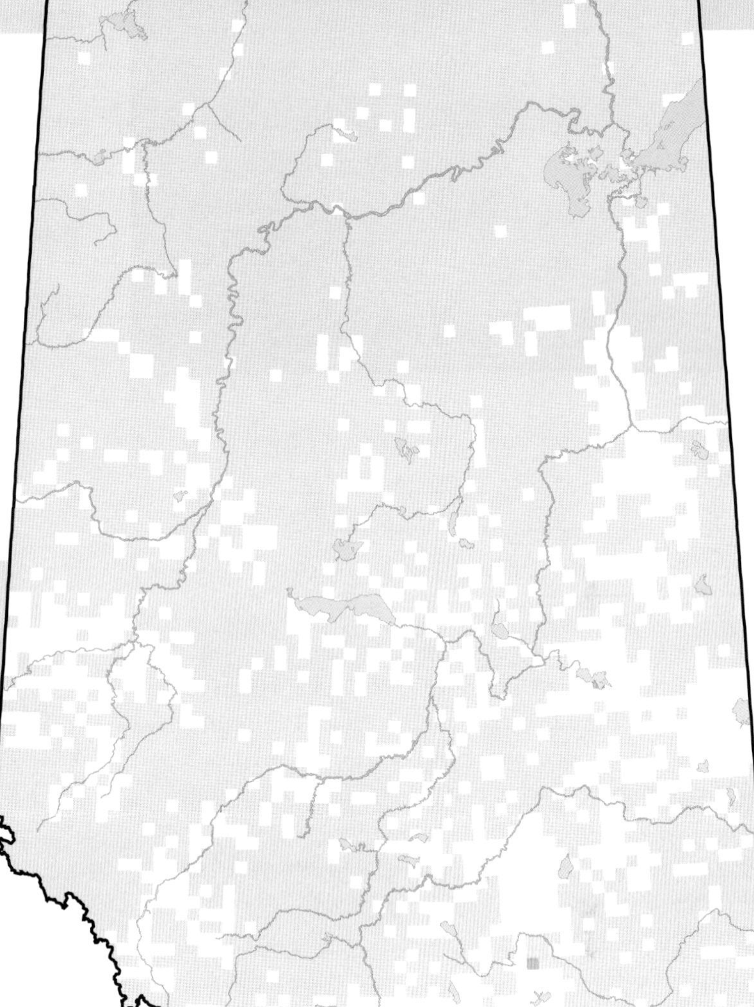

RANGE

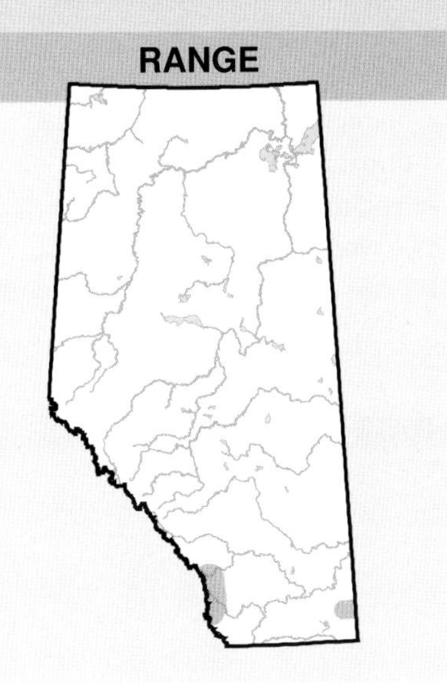

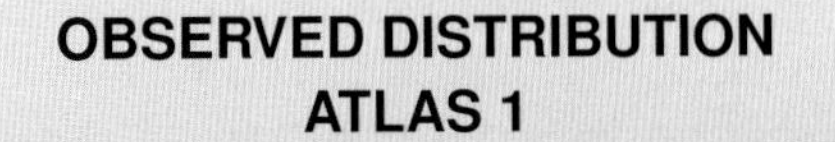

OBSERVED DISTRIBUTION ATLAS 1

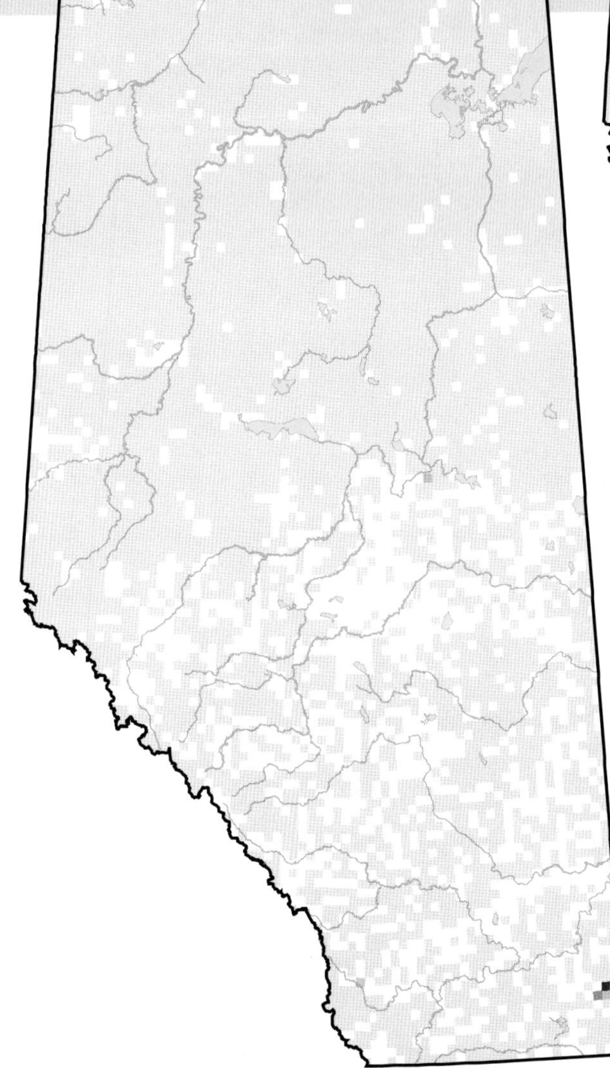

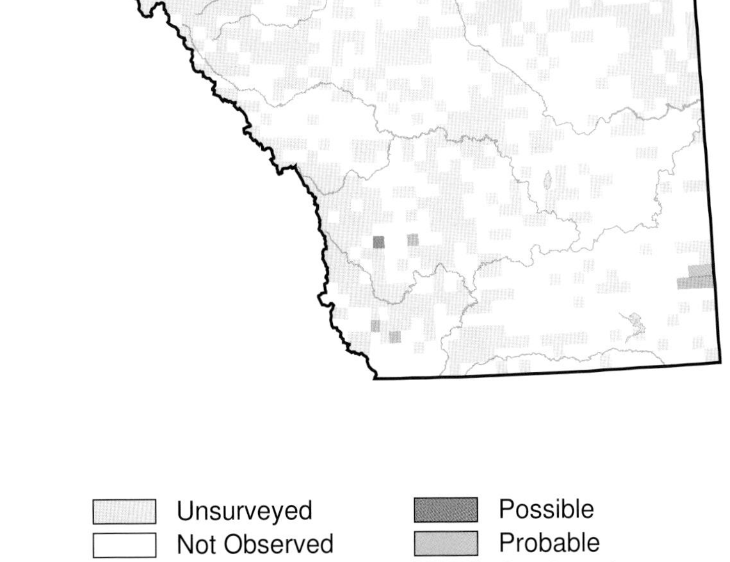

Red-throated Loon (*Gavia stellata*)

Photo: Damon Calderwood

NESTING

CLUTCH SIZE (EGGS):	1–2
INCUBATION (DAYS):	24–29
FLEDGING (DAYS):	46–50
NEST HEIGHT (METRES):	0

One of the least frequently encountered loons, the Red-throated Loon breeds only in the north-central part of the province in the Boreal Forest Natural Region. The distribution of this species did not change between Atlas 1 and Atlas 2. However, a number of more easterly breeding records were obtained during Atlas 2. This was likely the result of increased survey effort for this species rather than an actual distribution expansion.

The Red-throated Loon is a circumpolar species that breeds at high latitudes. This species tends to breed along the edges of small shallow ponds. After the young fledge, it is common for the family group to move to larger ponds or lakes nearby where fall freeze-up arrives later.

No change in relative abundance was detected although, relative to other species, the Red-throated Loon was observed more frequently in Atlas 2 than in Atlas 1. The Atlas sample size for this species was very small, so that caution must be used in the application of Atlas results. Similarly, the Breeding Bird Survey sample size for this species was too small to investigate abundance changes in Alberta and Canada. Due to the northern extent of this species' breeding distribution, targeted efforts are required to monitor this species effectively.

Numbers of this loon have declined in several parts of its range in North America, although the reason is not clear (Barr et al., 2000). Barr et al. (2000) contend that conservation measures have lagged far behind those for the Common Loon probably because, in North America, Red-throated Loons are not as prominent in the public eye, nor do they breed in populated regions. Barr and colleagues recommend that, due to declining populations in much of the Red-throated Loon's range, immediate attention to causes of and remedies for population stability is required, followed by a worldwide conservation effort.

The provincial status of this species is Secure.

TEMPORAL REPORTING RATE

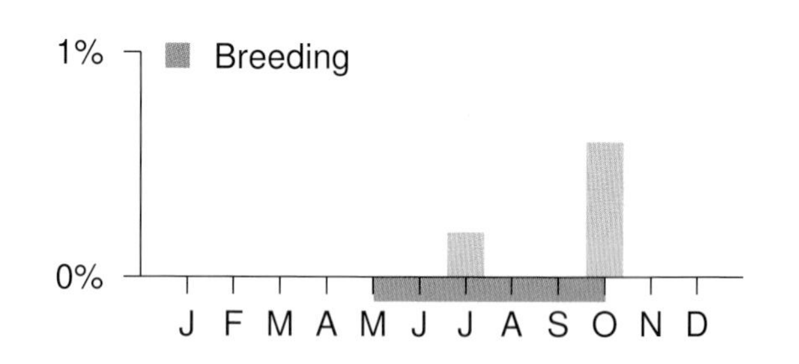

NATURAL REGIONS REPORTING RATE

SPATIAL REPORTING RATE

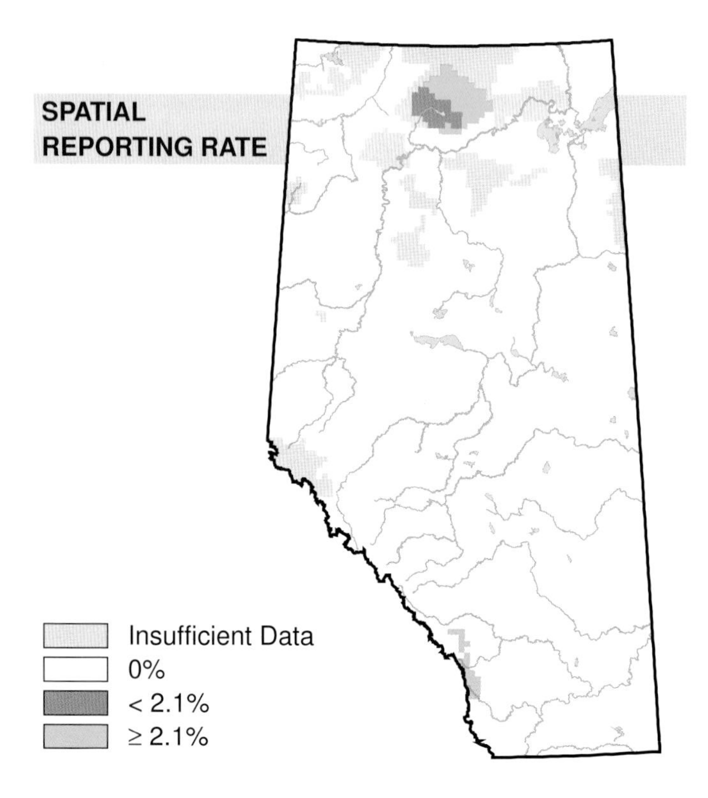

STATUS

GSWA 2000	GSWA 2005	COSEWIC Historic Rankings	COSEWIC 2007	Alberta Wildlife Act
Secure	Secure	N/A	N/A	N/A

RANGE

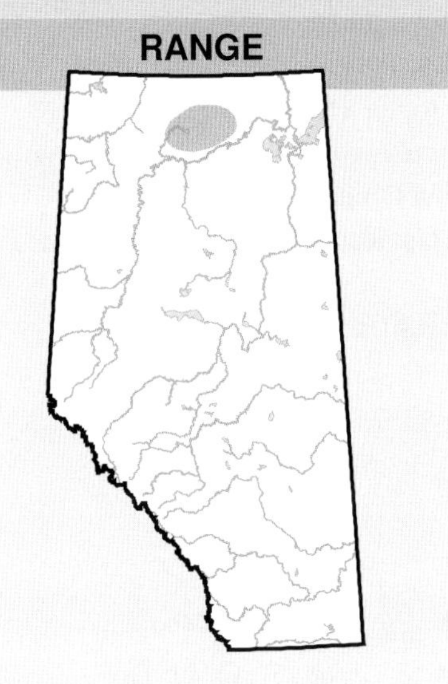

OBSERVED DISTRIBUTION ATLAS 2

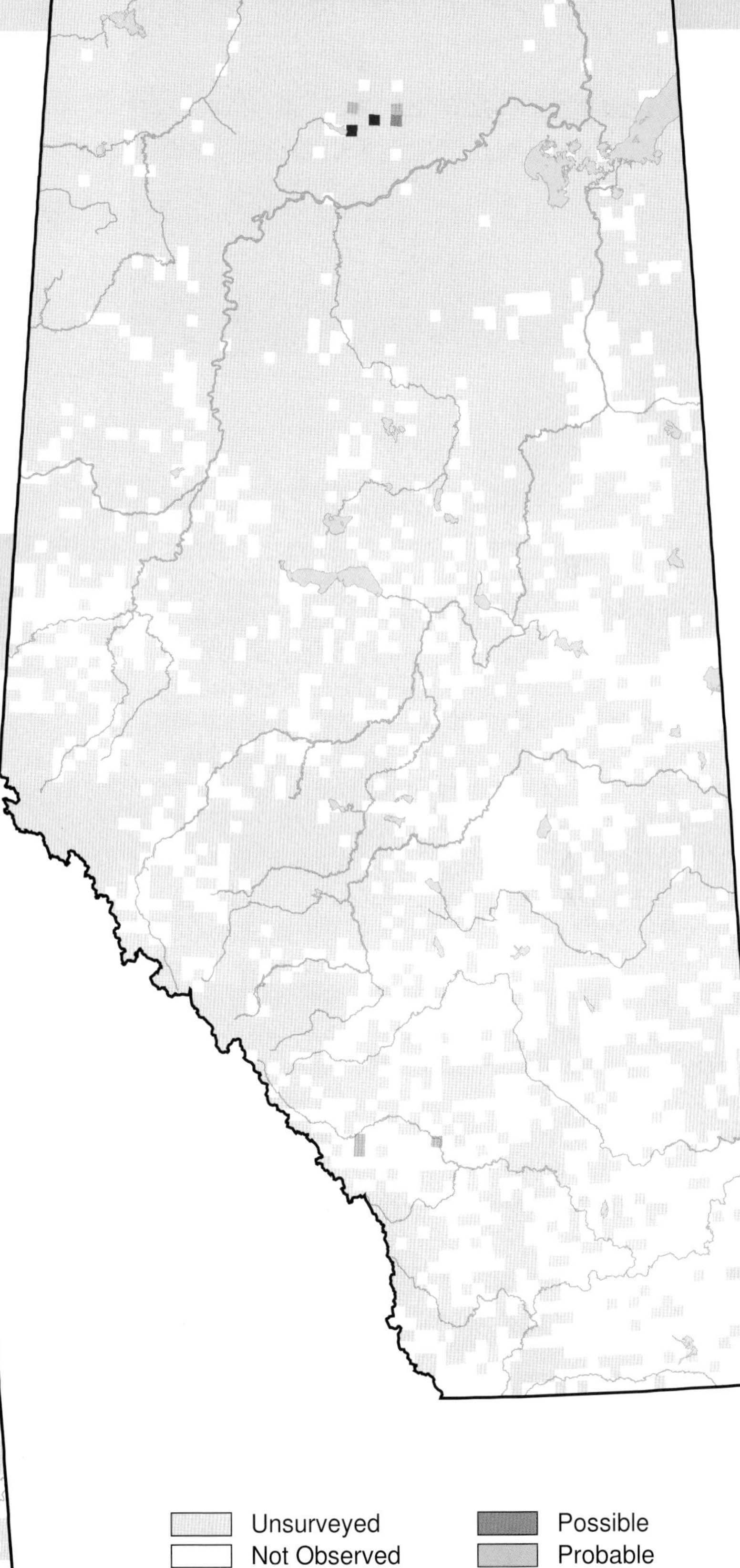

OBSERVED DISTRIBUTION ATLAS 1

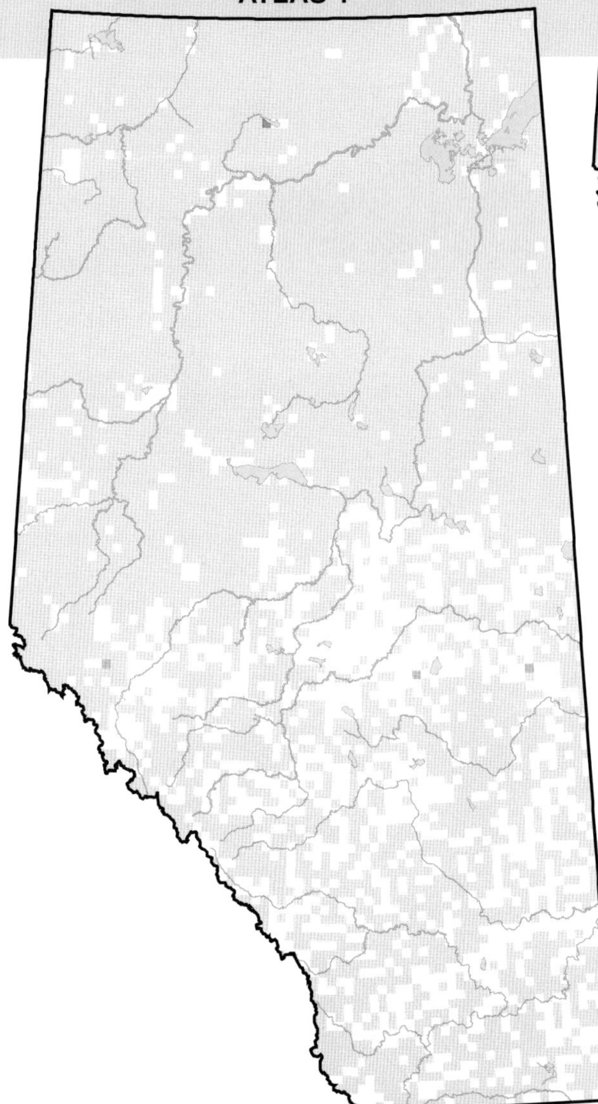

	Unsurveyed		Possible
	Not Observed		Probable
	Observed		Confirmed

Pacific Loon (*Gavia pacifica*)

Photo: Royal Alberta Museum

NESTING

CLUTCH SIZE (EGGS):	1–2
INCUBATION (DAYS):	28–30
FLEDGING (DAYS):	60
NEST HEIGHT (METRES):	0

One of the rarest birds in Alberta, the Pacific Loon breeds occasionally in the northern part of the province in the Boreal Forest Natural Region. It is unknown whether the breeding distribution of this species changed between Atlas 1 and Atlas 2 because none of the observations that were made of this species in either Atlas project has associated breeding evidence.

The Pacific Loon is well named because it spends most of its life at sea. For approximately three months every year, this bird moves inland to breed at freshwater lakes in the northern part of the continent. This species can breed at ponds and lakes that are surrounded by treeless or treed habitat. There are very few breeding records for this species in Alberta and most observations in the province occur in the interval when this bird is moving between its breeding and wintering grounds. Pacific Loons were most commonly encountered in the Rocky Mountain NR where they are sometimes encountered while they are migrating over the mountains. It was also encountered, but infrequently, in the Boreal Forest NR during Atlas 2.

There was no change in relative abundance detected in any of the NRs. The Breeding Bird Survey sample size was too small to investigate trends for the Pacific Loon in Alberta or across Canada because most survey routes are located well south of the breeding range of this species. Due to the northern range of this species' breeding distribution, the bird is not effectively monitored by surveys such as the Atlas or Breeding Bird Survey because these surveys do not adequately monitor the north. Pacific Loon-specific surveys would be needed to adequately determine population trends for this species. The provincial status of this species is Secure in Alberta.

TEMPORAL REPORTING RATE

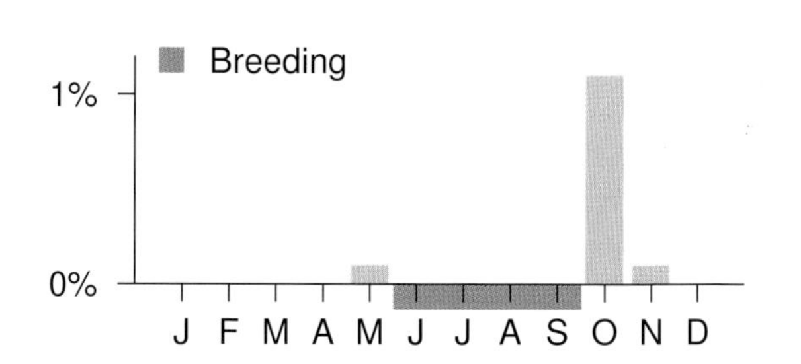

NATURAL REGIONS REPORTING RATE

SPATIAL REPORTING RATE

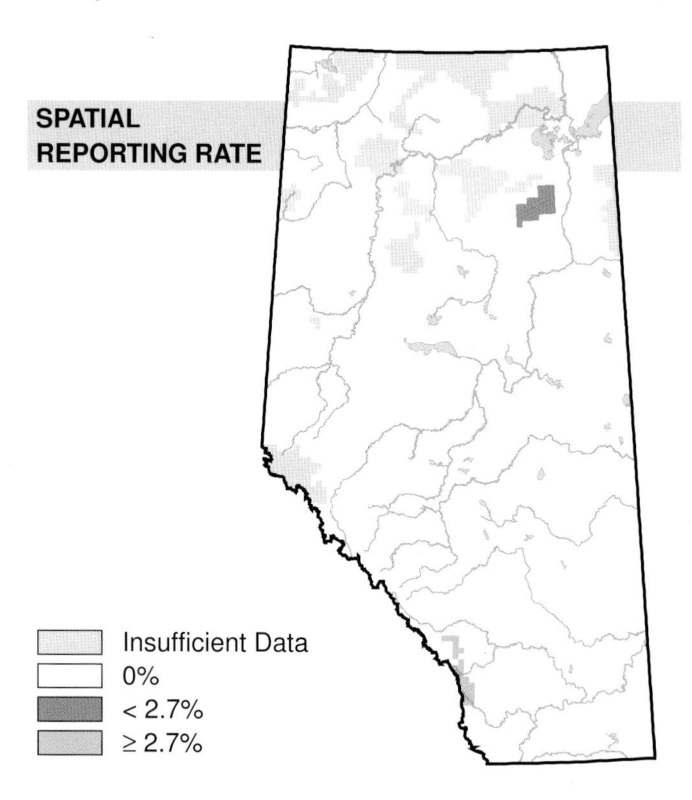

STATUS

GSWA 2000	GSWA 2005	COSEWIC Historic Rankings	COSEWIC 2007	Alberta Wildlife Act
Secure	Secure	N/A	N/A	N/A

Habitat  Nest Location Nest Type Diet 

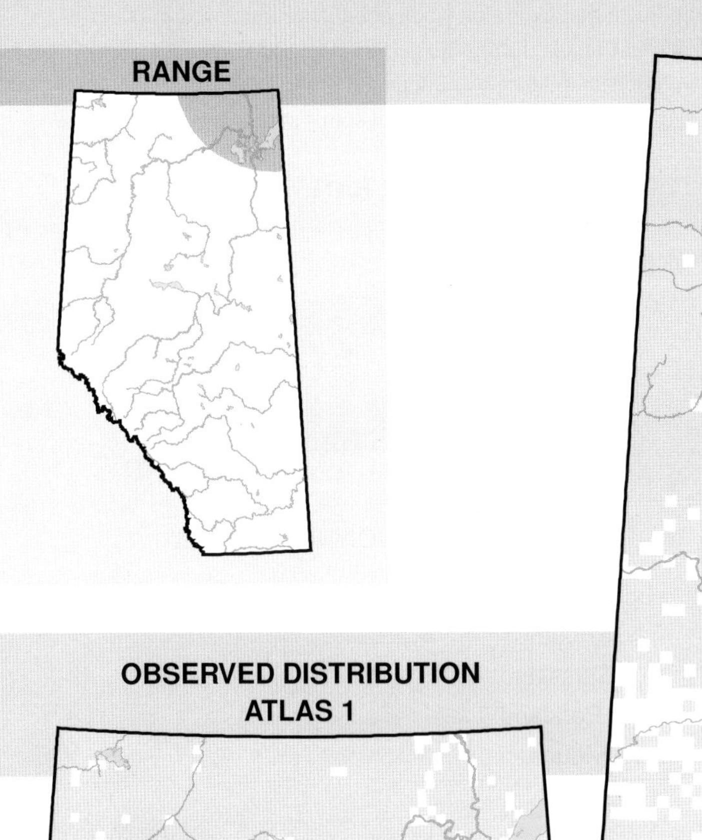

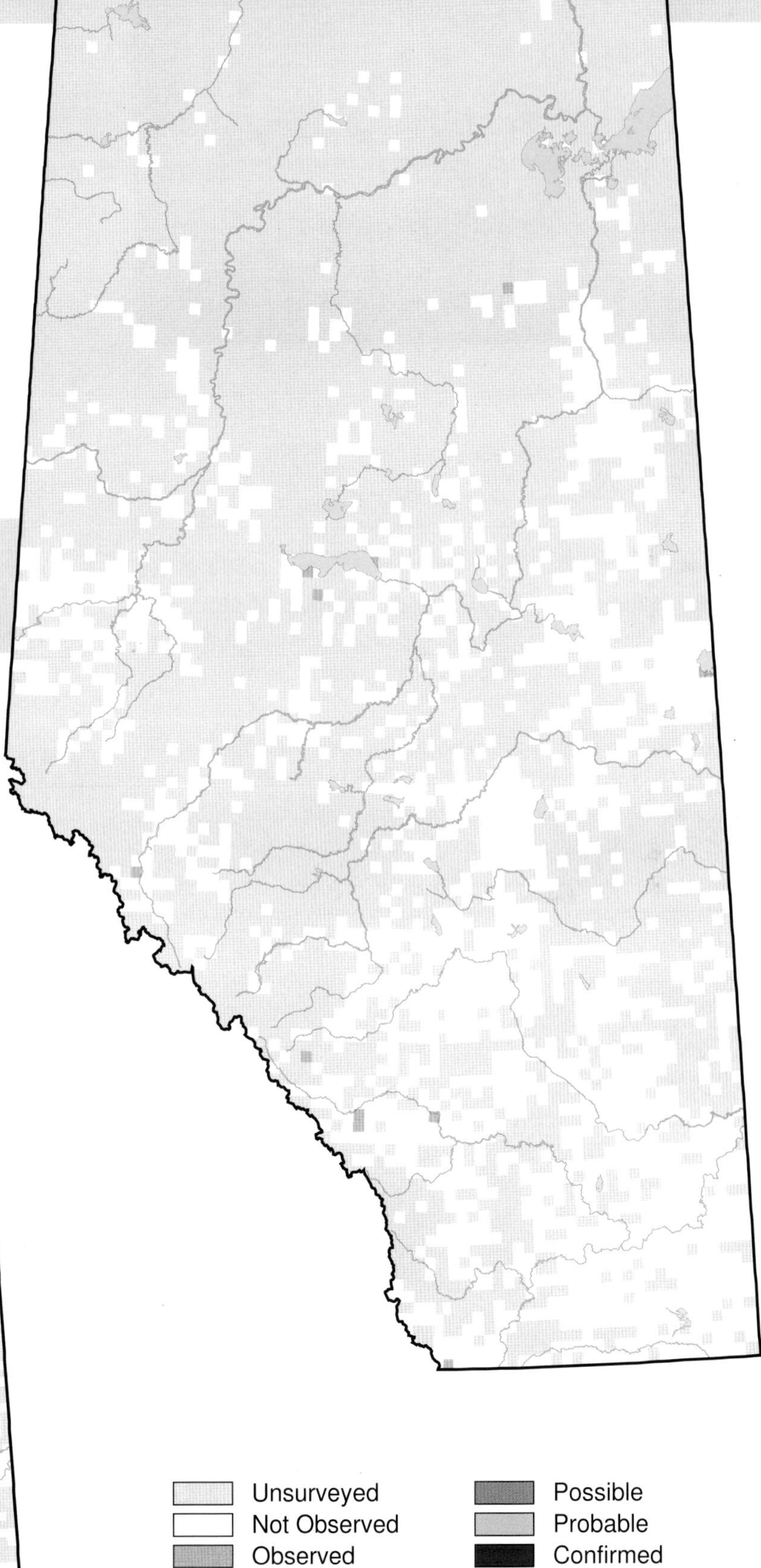

Common Loon (*Gavia immer*)

Photo: Murray Schultz

NESTING

CLUTCH SIZE (EGGS):	2
INCUBATION (DAYS):	29
FLEDGING (DAYS):	70–80
NEST HEIGHT (METRES):	0

A true representative of the Canadian wilderness, the Common Loon is found in every Natural Region in Alberta. No change in the distribution of this species was observed between Atlas 1 and Atlas 2.

The Common Loon was most often encountered in the Boreal Forest, Foothills, and Rocky Mountain NRs and was occasionally observed in the Grassland and Parkland NRs. The distribution of this species in Alberta reflects its preference to nest at clear, fish-inhabited lakes of varying sizes. Nests are usually located at the edge of lakes beside the land or within emergent vegetation. Nests are located slightly above the surface of the water; therefore, nest failures can occur when wetlands flood. Boating activity can also have a negative impact on reproductive success because the wake created by boats can wash eggs off nests.

Increases were detected in the Boreal Forest, Grassland, and Parkland NRs. In the Boreal Forest NR, relative to other species, the Common Loon was observed more frequently in Atlas 2 than in Atlas 1. In the Grassland NR, relative to other species, the Common Loon was observed at similar frequencies in Atlas 1 and Atlas 2. In the Parkland NR, however, it was observed less frequently in Atlas 2; the cause for this decline is unknown. Breeding Bird Survey data indicate that Common Loon abundance increased in Alberta and in Canada during the period 1985–2005. No changes in relative abundance were detected in the Foothills and Rocky Mountain NRs. The provincial status of this species is Secure.

TEMPORAL REPORTING RATE

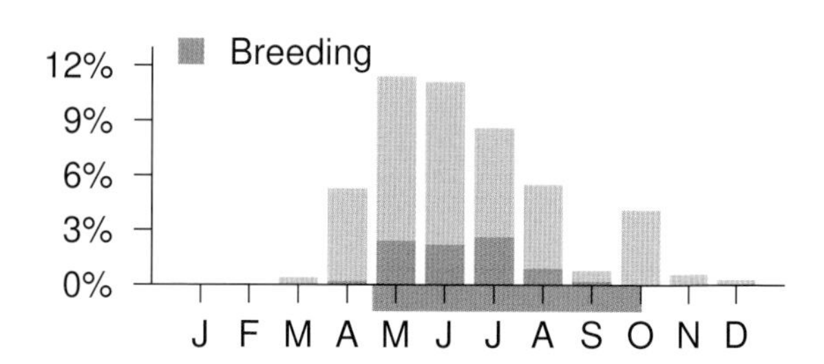

NATURAL REGIONS REPORTING RATE

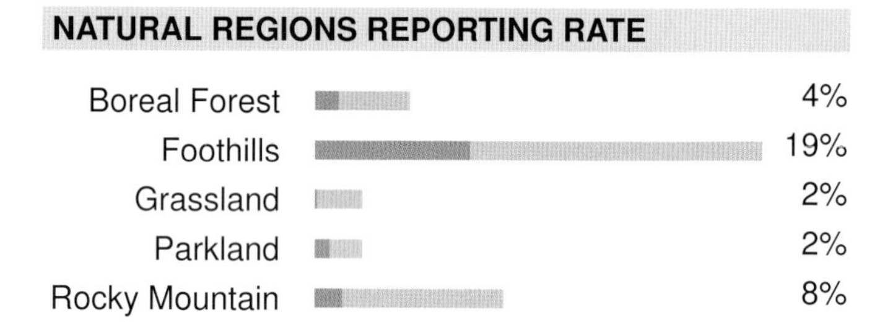

SPATIAL REPORTING RATE

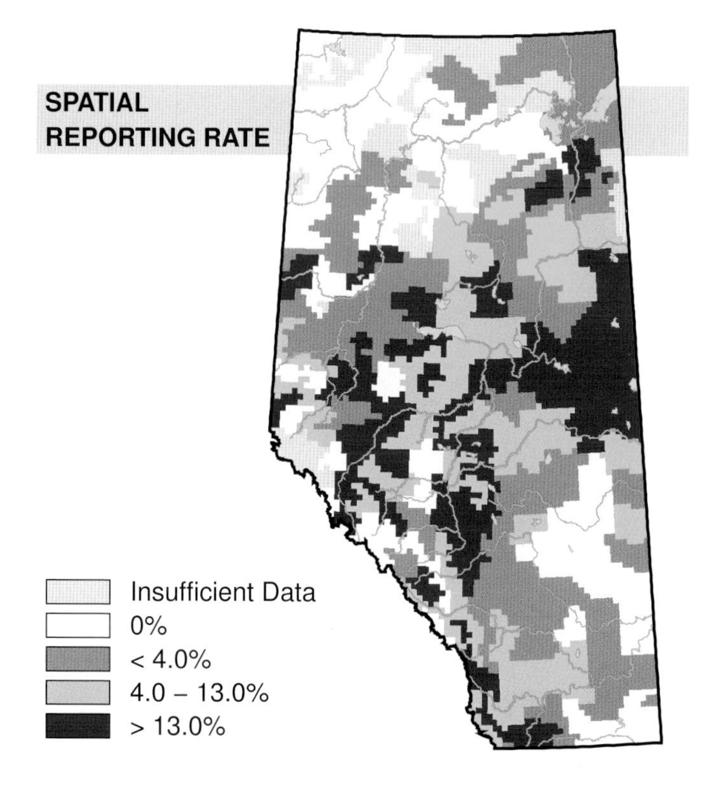

STATUS

GSWA 2000	GSWA 2005	COSEWIC Historic Rankings	COSEWIC 2007	Alberta Wildlife Act
Secure	Secure	Not At Risk	Not At Risk	N/A

RANGE

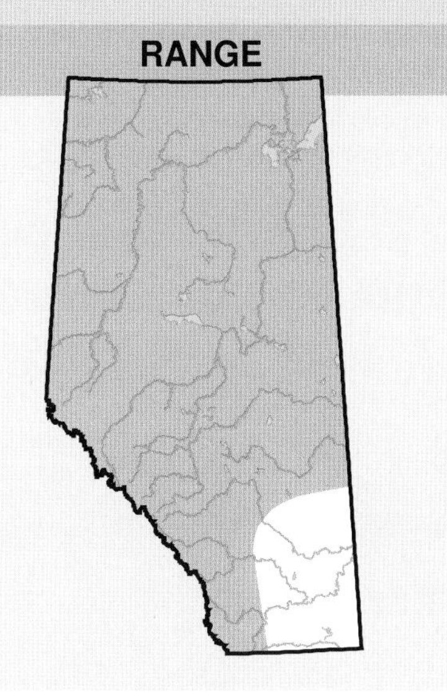

OBSERVED DISTRIBUTION ATLAS 2

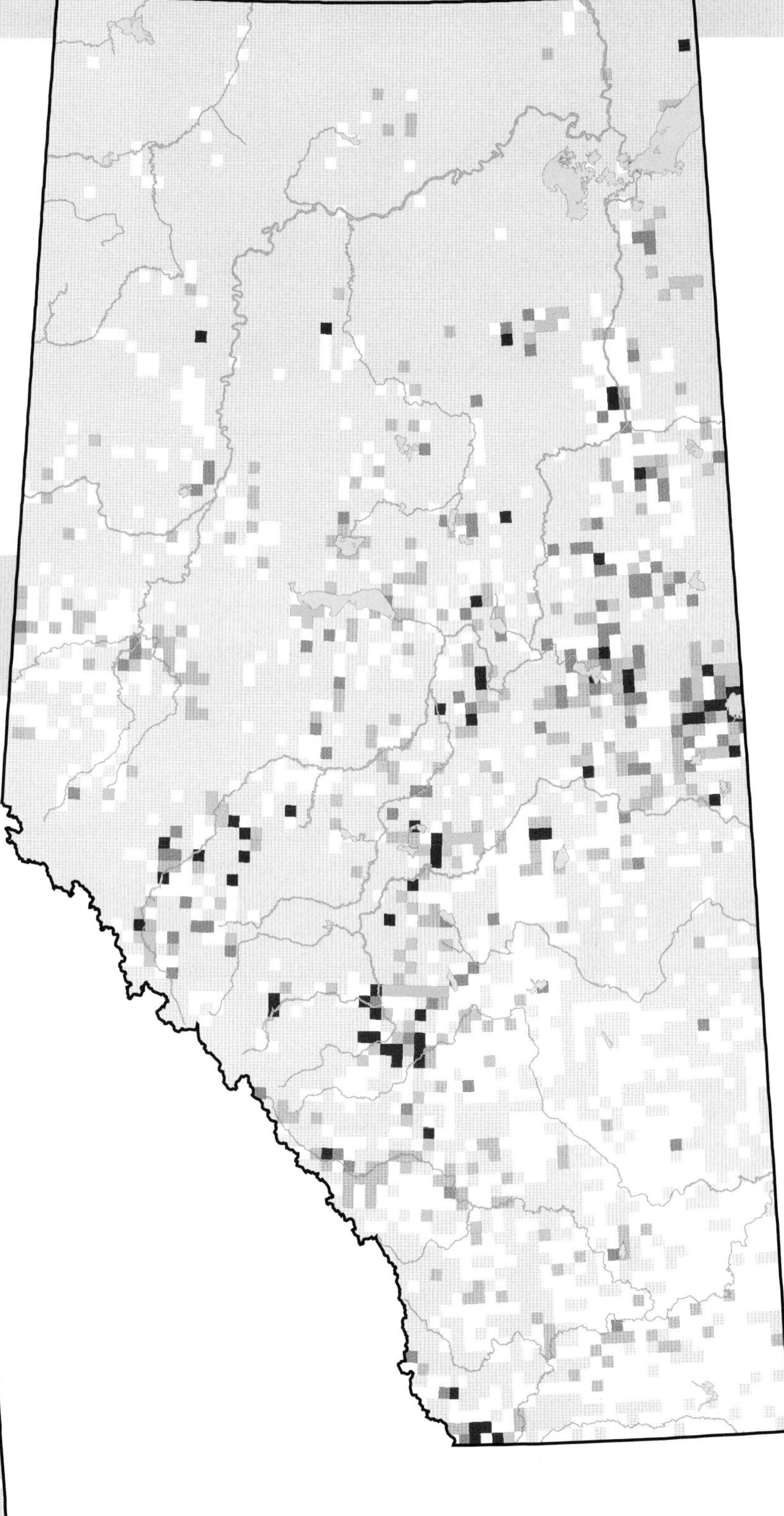

OBSERVED DISTRIBUTION ATLAS 1

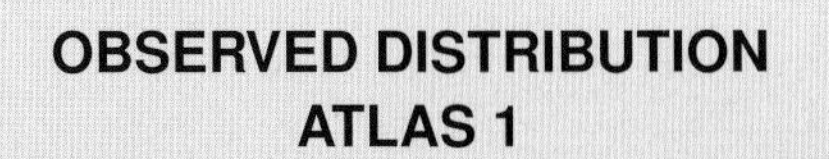

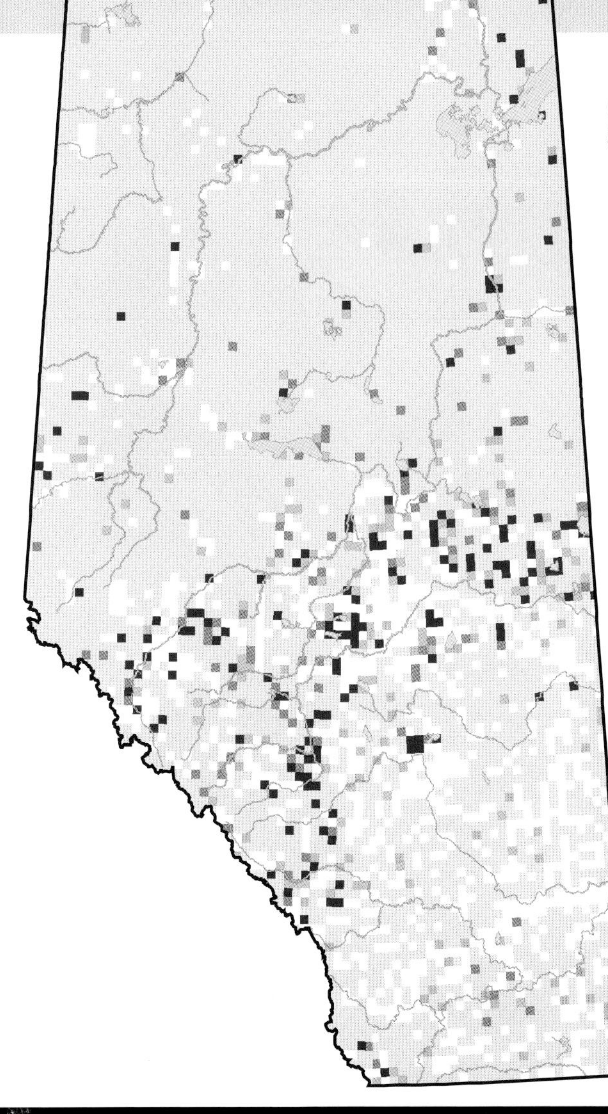

Pied-billed Grebe (*Podilymbus podiceps*)

Photo: Gerry Beyersbergen

NESTING

CLUTCH SIZE (EGGS):	6–8
INCUBATION (DAYS):	23
FLEDGING (DAYS):	25–62
NEST HEIGHT (METRES):	0

The secretive Pied-billed Grebe is found in all Natural Regions in Alberta. There was no change in the observed distribution of this species between Atlas 1 and Atlas 2.

Despite occurring in all natural regions, Pied-billed Grebes were most often detected in the Grassland and Parkland NRs and were occasionally encountered in the Boreal Forest, Foothills, and Rocky Mountain NRs. Unlike many other grebe species, Pied-billed Grebes are solitary nesters. They nest in dense stands of emergent vegetation along the periphery of marshes and ponds. The presence of vegetation is likely the main factor that influences whether or not Pied-billed Grebes will nest at a wetland.

An increase in relative abundance was detected in the Grassland NR and declines were detected in the Boreal Forest and Parkland NRs. Observation rates relative to other species mirrored these changes. No changes were detected in the Foothills and Rocky Mountain NRs. The Breeding Bird Survey detected no change in abundance in Alberta or nationally during the period 1985–2005. During Atlas 2 the Boreal Forest and Parkland NRs were drier than during Atlas 1 and the Grassland NR was wetter. Atlas-detected changes mirror those climatic differences, suggesting a spatial shift in land use rather than a change in population size. The differences between the results of the Atlas and Breeding Bird Survey analyses also make sense in light of the differences in the spatial scales used to assess change. Breeding Bird Survey data were analyzed at the provincial scale; whereas, due to sufficient sample size, it was possible to analyze Atlas data at the smaller, Natural Region scale. The provincial status of this species is Sensitive due to degradation of wetland habitat.

TEMPORAL REPORTING RATE

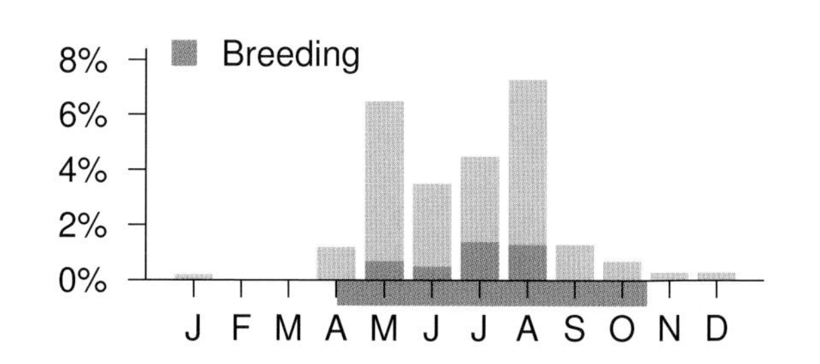

NATURAL REGIONS REPORTING RATE

SPATIAL REPORTING RATE

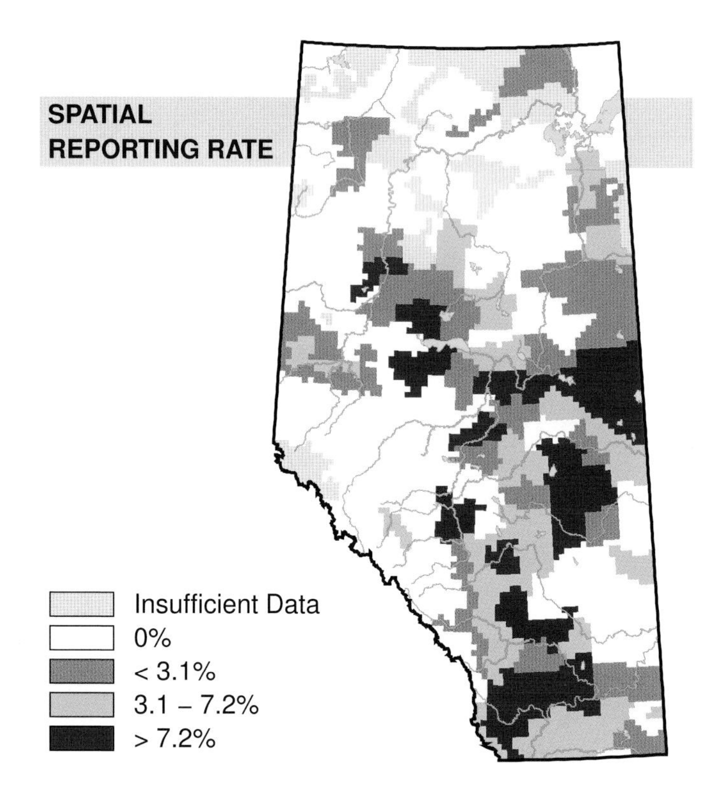

STATUS

GSWA 2000	GSWA 2005	COSEWIC Historic Rankings	COSEWIC 2007	Alberta Wildlife Act
Sensitive	Sensitive	N/A	N/A	N/A

RANGE

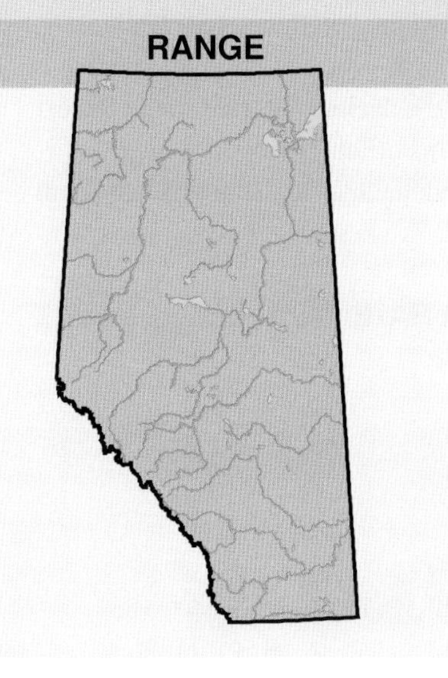

OBSERVED DISTRIBUTION ATLAS 2

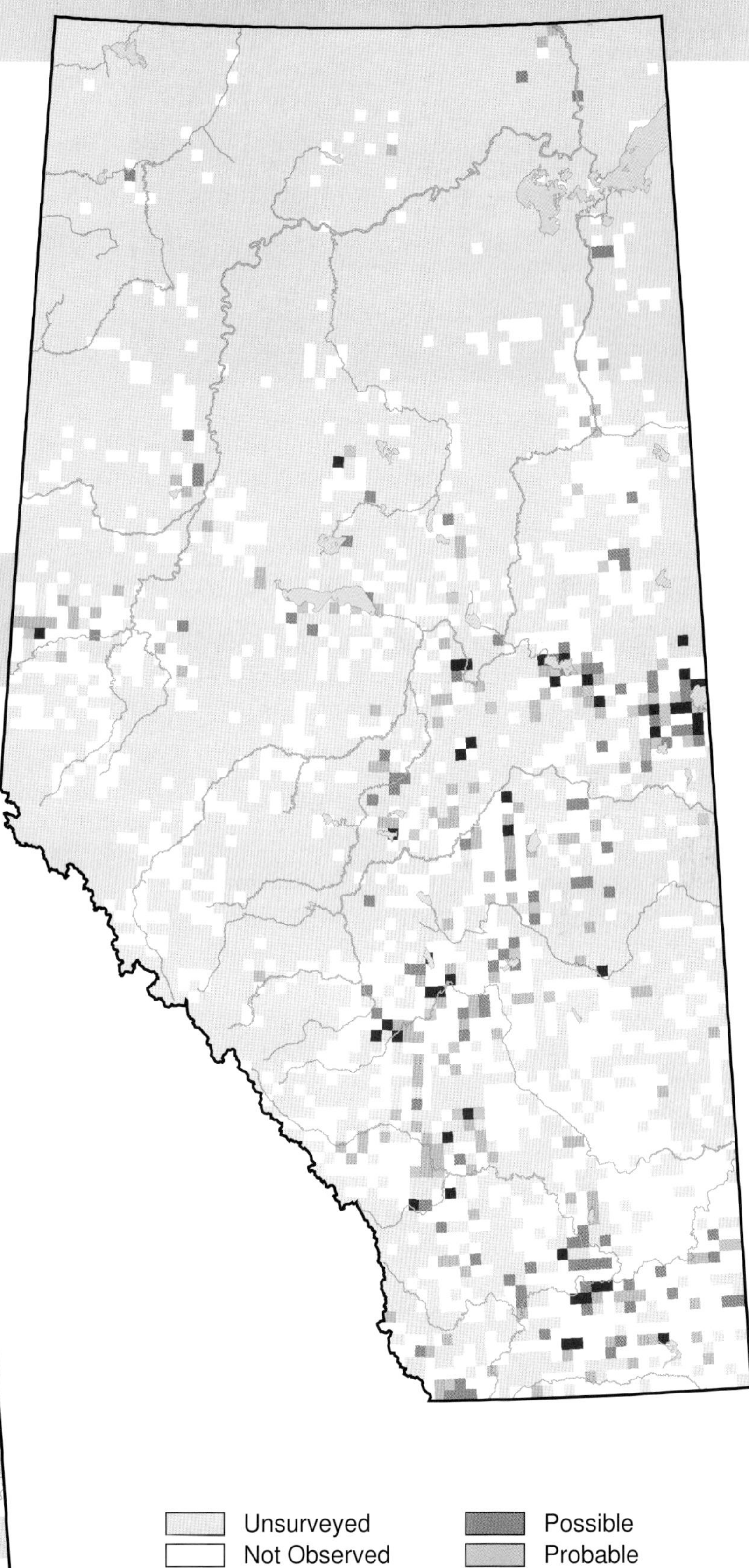

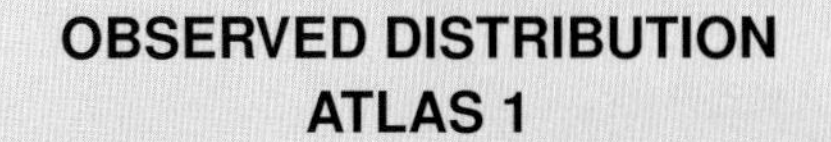

OBSERVED DISTRIBUTION ATLAS 1

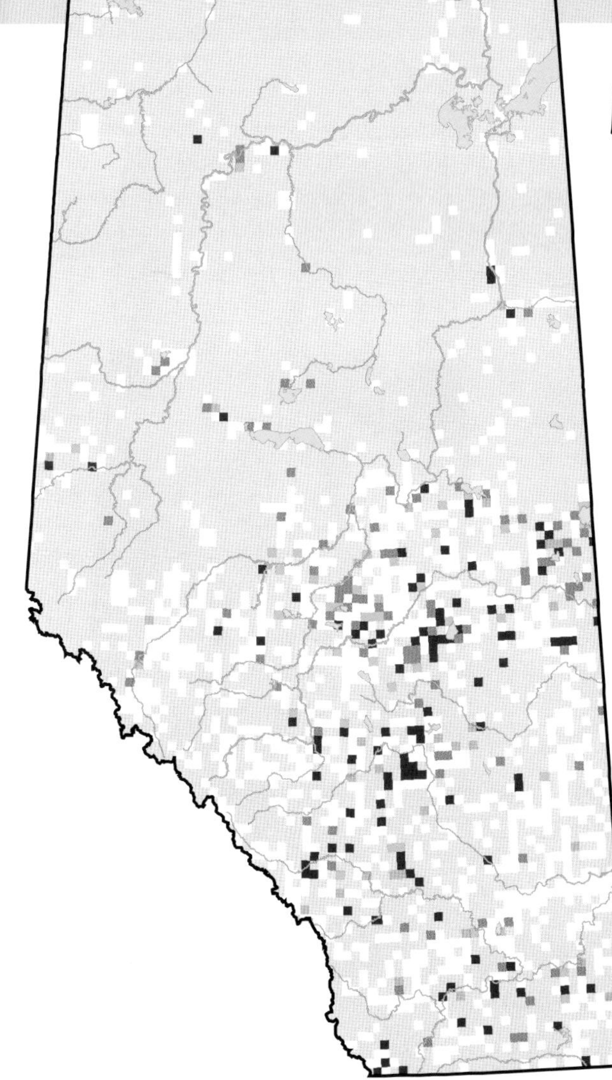

Unsurveyed
Not Observed
Observed
Possible
Probable
Confirmed

Horned Grebe (*Podiceps auritus*)

Photo: Robert Gehlert

NESTING

CLUTCH SIZE (EGGS):	3–7
INCUBATION (DAYS):	23–24
FLEDGING (DAYS):	45–60
NEST HEIGHT (METRES):	0

The Horned Grebe is found in every Natural Region in Alberta. The observed distribution of this species decreased in Atlas 2 in the northwestern part of the province around High Level and Fort Vermilion. Although there cannot be complete certainty that this species was not missed in surveys conducted in this area, the substantive loss in records in the area is suggestive of a true contraction of its Alberta distribution.

Horned Grebes often breed at shallow ponds and marshes where their nests are usually built along the edge of emergent vegetation near open water. As a result, this species was most often found in the Grassland and Parkland NRs where marshes tend to be shallow. This species was found only occasionally in the Boreal Forest, Foothills, and Rocky Mountain NRs because wetlands tend to be deeper in those NRs.

An increase in relative abundance was detected in the Grassland NR where, relative to other species, the Horned Grebe was observed more frequently in Atlas 2 than in Atlas 1. A decline was detected in the Boreal Forest NR where, relative to other species, it was observed less frequently in Atlas 2 than in Atlas 1. Changes in relative abundance were not detected in the Foothills, Parkland, or Rocky Mountain NRs. The Breeding Bird Survey found no change in abundance in Alberta, but detected an increase across Canada during the period 1995–2005. The increase in the Grassland NR is likely due to above-normal precipitation in the later half of Atlas 2, which revitalized many ephemeral and seasonal wetlands providing habitat for birds dependent on them. This heavier rainfall did not occur farther north, where conditions remained dry, and this may account for the decline detected in the Boreal Forest NR. The provincial status of this species is Sensitive due to wetland degradation and declines in the number of small ponds.

TEMPORAL REPORTING RATE

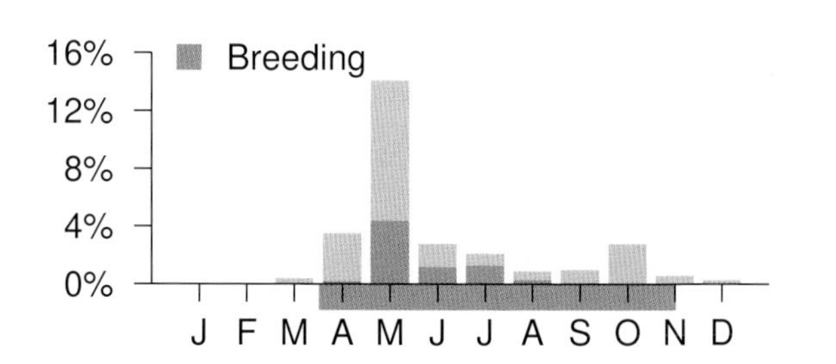

NATURAL REGIONS REPORTING RATE

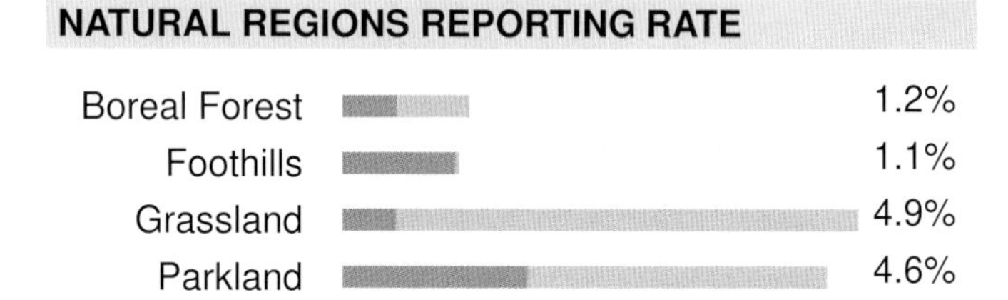

SPATIAL REPORTING RATE

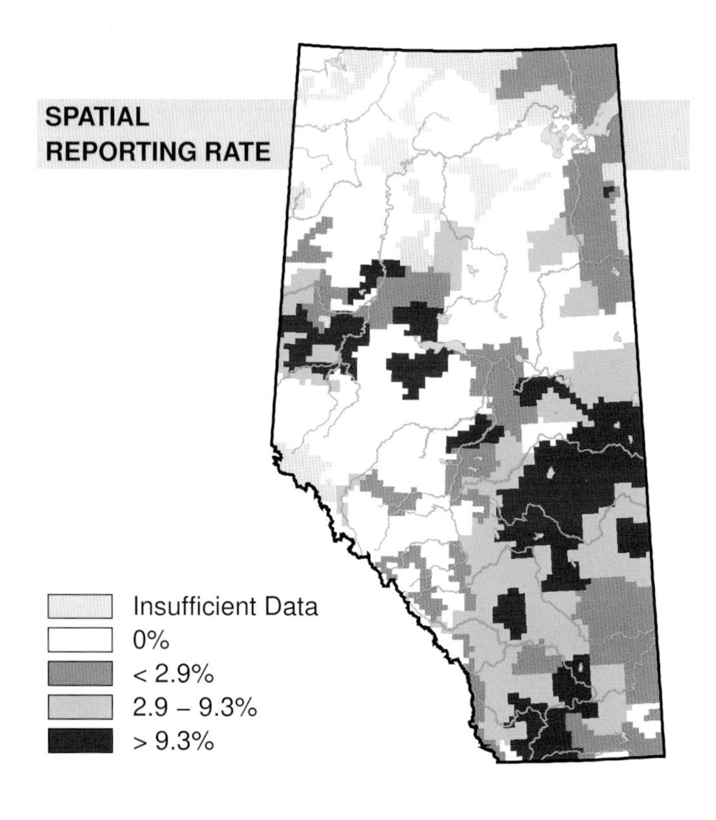

STATUS

GSWA 2000	GSWA 2005	COSEWIC Historic Rankings	COSEWIC 2007	Alberta Wildlife Act
Sensitive	Sensitive	N/A	N/A	N/A

RANGE

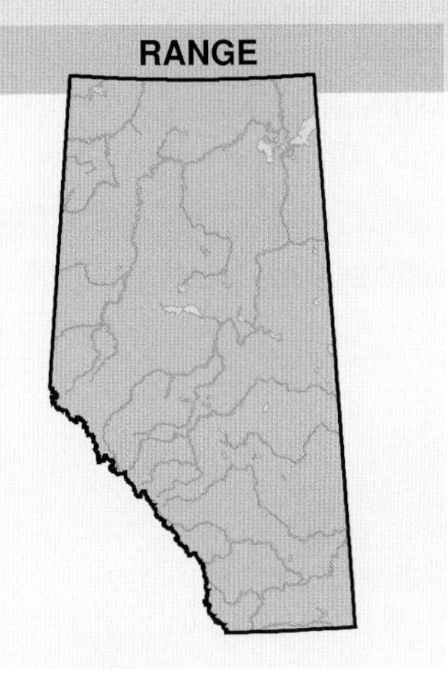

OBSERVED DISTRIBUTION ATLAS 2

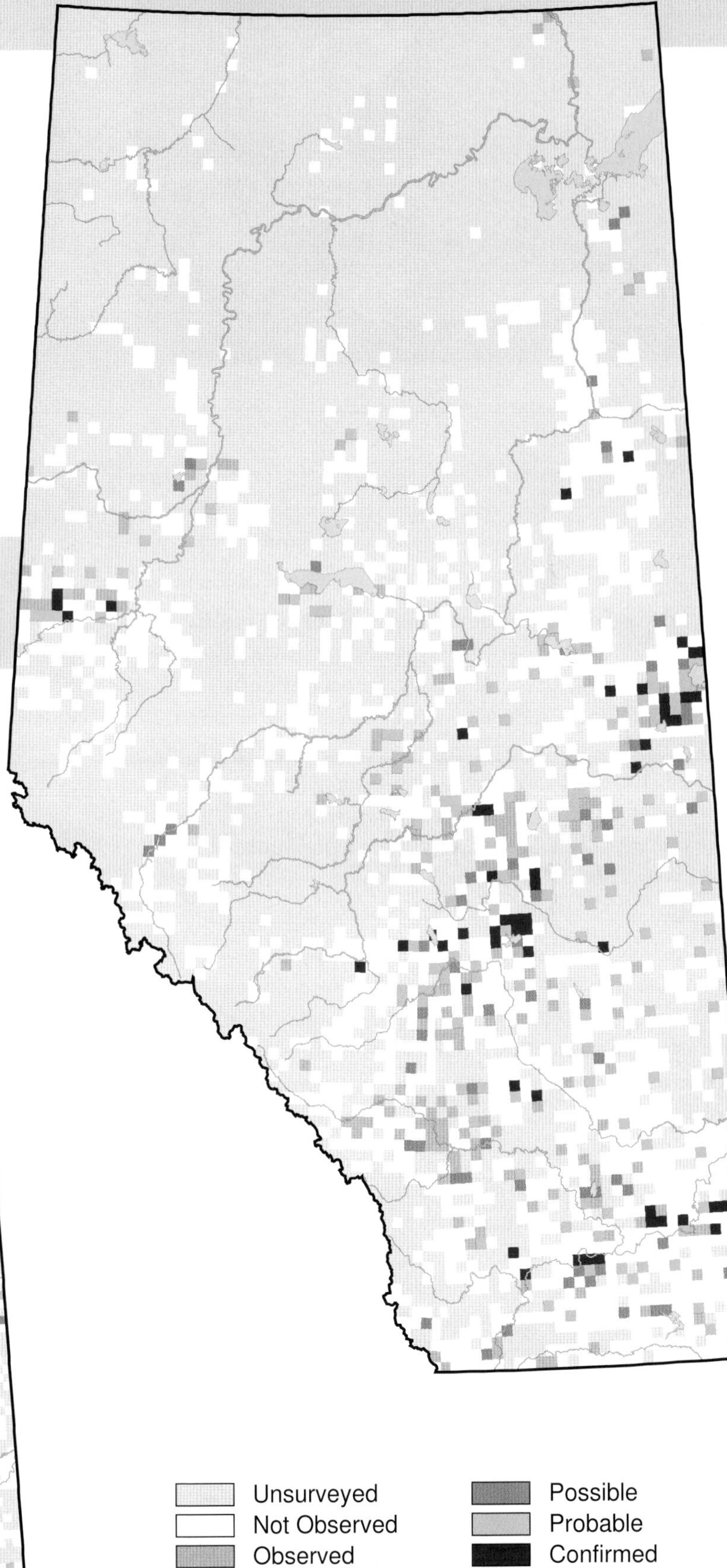

OBSERVED DISTRIBUTION ATLAS 1

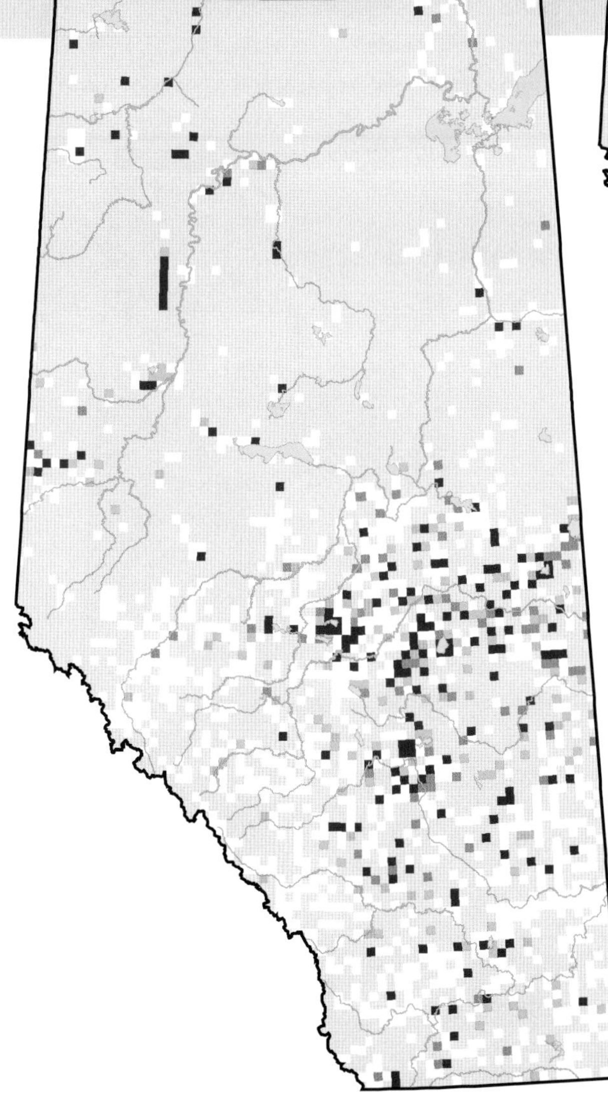

	Unsurveyed		Possible
	Not Observed		Probable
	Observed		Confirmed

Red-necked Grebe (*Podiceps grisegena*)

Photo: Alan MacKeigan

NESTING

CLUTCH SIZE (EGGS):	2–9
INCUBATION (DAYS):	22–25
FLEDGING (DAYS):	56–70
NEST HEIGHT (METRES):	0

The Red-necked Grebe is found in every Natural Region in Alberta. While the observed distribution of this species did not change between Atlas 1 and Atlas 2, the species was less common in central Alberta between Edmonton, Whitecourt, and Athabasca. It was more common in south-central Alberta between Brooks, Medicine Hat, and Taber during Atlas 2. That fewer observations were made in the central part of the province could be related to a reduction in the availability of suitable wetlands, or to increased boating activity in that area, or both.

The Red-necked Grebe can be observed to use various-sized ponds and lakes that usually have emergent vegetation present. As a result, this species was common and evenly distributed across the province. This species builds floating nests and the nature of these nesting structures makes them vulnerable to boating activities; the wake generated by watercraft can sometimes wash eggs off nests.

An increase in relative abundance was detected in the Grassland NR. This increase may be due to greater precipitation in this NR during Atlas 2 than during Atlas 1. A decline was detected in the Parkland NR which could be related to drought and drainage of wetlands or to increased boating activity. Changes in relative abundance were not detected in the Foothills and Rocky Mountain NRs. The increase in the Grassland NR and the decrease in the Parkland present the possibility that this species effected a spatial shift in its pattern of use in response to climate changes, with no reduction in population size. The Breeding Bird Survey did not find an abundance change for this species in Alberta, or nationally, during the period 1985–2005. The provincial status of this species is Secure.

TEMPORAL REPORTING RATE

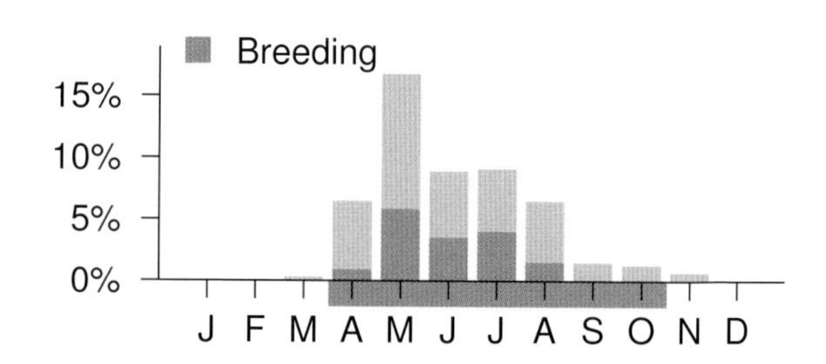

NATURAL REGIONS REPORTING RATE

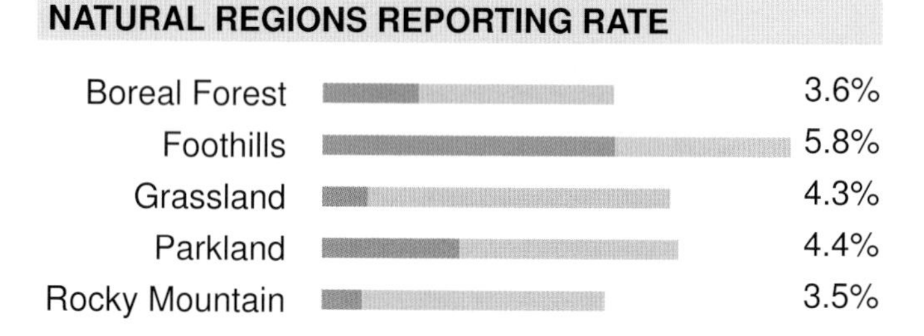

SPATIAL REPORTING RATE

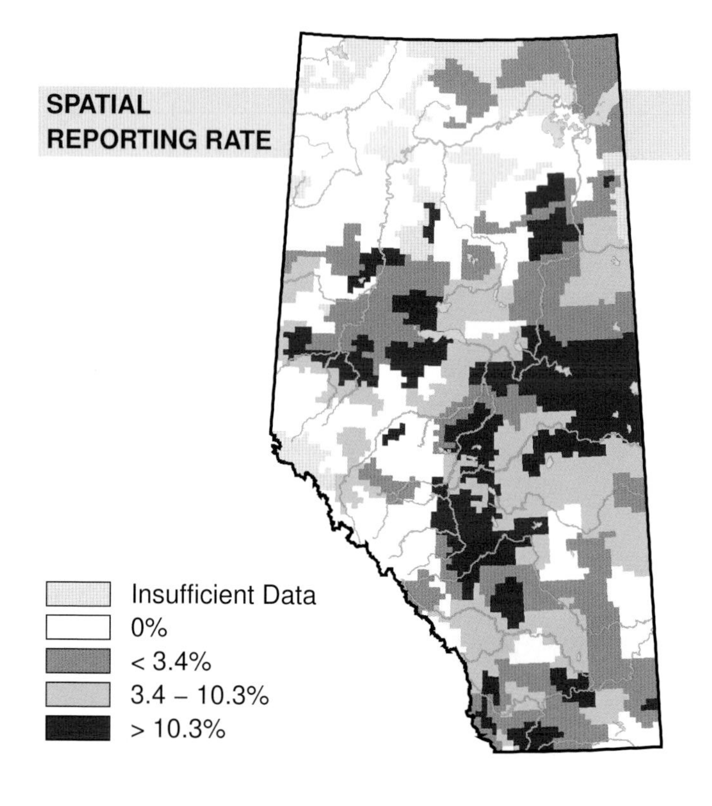

STATUS

GSWA 2000	GSWA 2005	COSEWIC Historic Rankings	COSEWIC 2007	Alberta Wildlife Act
Secure	Secure	Not At Risk	Not At Risk	N/A

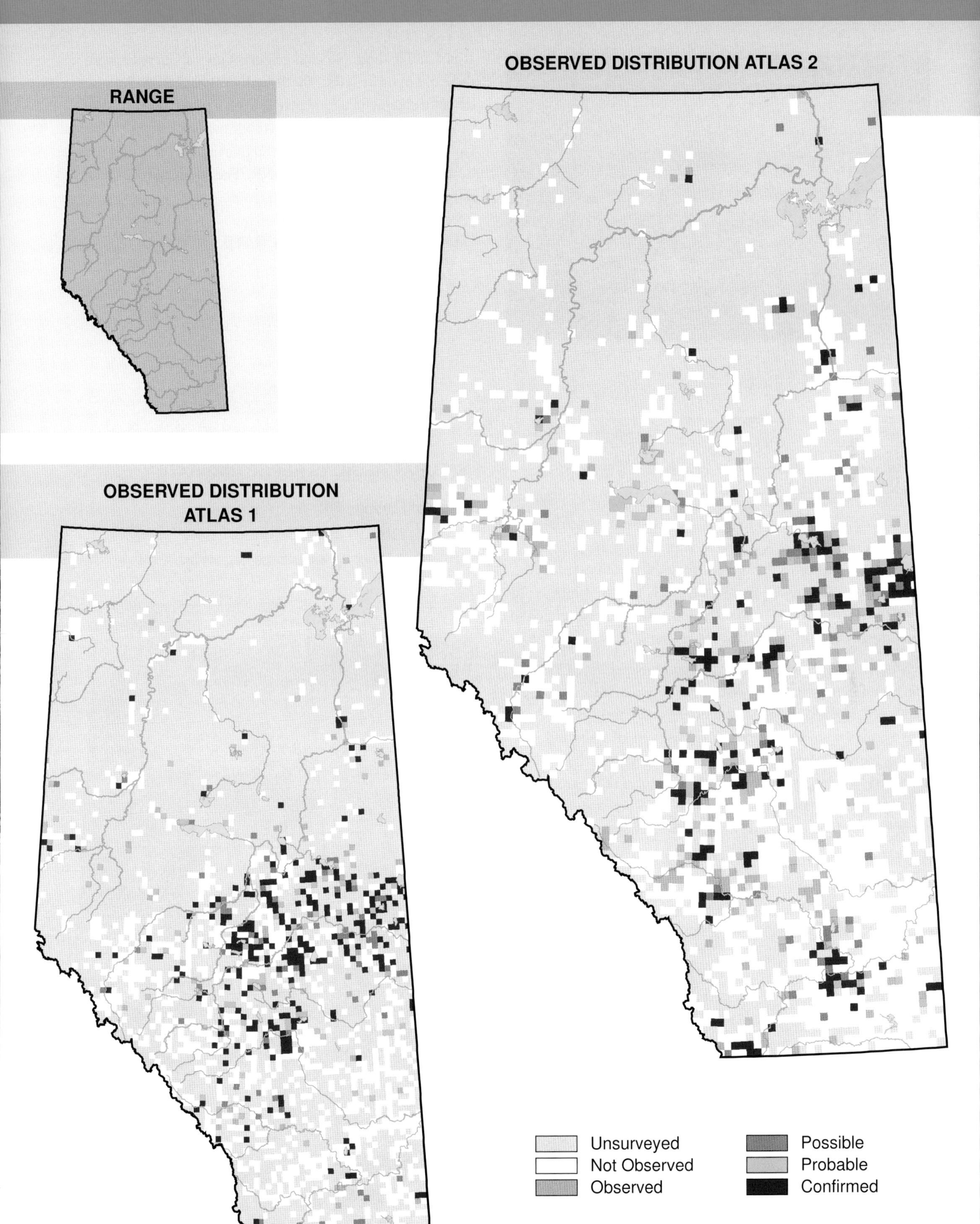
Habitat
Nest Location
Nest Type
Diet
OBSERVED DISTRIBUTION ATLAS 2
RANGE
OBSERVED DISTRIBUTION ATLAS 1
Unsurveyed
Not Observed
Observed
Possible
Probable
Confirmed

Eared Grebe (*Podiceps nigricollis*)

Photo: Royal Alberta Museum

NESTING

CLUTCH SIZE (EGGS):	1–6
INCUBATION (DAYS):	20–22
FLEDGING (DAYS):	21
NEST HEIGHT (METRES):	0

The Eared Grebe is found in every Natural Region in Alberta. The distribution of this species has expanded in the northeastern part of its Alberta range. New breeding records were documented east of Fort MacKay and south of Fitzgerald along the Slave River. The distribution shift could be related to changing conditions in northern areas due to climate change.

Despite occurring in every Natural Region, the Eared Grebe was found mainly in the Grassland and Parkland NRs and was found occasionally in the Boreal Forest and Foothills NRs. This species nests colonially in shallow parts of lakes and ponds. Eared Grebes build floating nests that are usually anchored to emergent vegetation. Unlike most other grebes, Eared Grebes do not generally occupy fish-inhabited wetlands because they feed mainly on aquatic invertebrates.

An increase in the relative abundance of Eared Grebes was detected in the Grassland NR where, relative to other species, it was observed more frequently in Atlas 2 than in Atlas 1. The above-normal precipitation in the later half of Atlas 2, which revitalized many seasonal and ephemeral wetlands in the Grassland NR, is a probable contributor to this observed increase. A decrease was detected in the Boreal Forest NR where, relative to other species, this grebe was observed less frequently in Atlas 2 than in Atlas 1. The cause for this decline is not understood. No changes in relative abundance were detected in other NRs. The Breeding Bird Survey did not detect a change in abundance for the Eared Grebe in Alberta, or across Canada. Without corroboration or known biological cause, further research is needed to evaluate the Atlas-detected decrease in the Boreal Forest NR. The status of this species in Alberta is Secure.

TEMPORAL REPORTING RATE

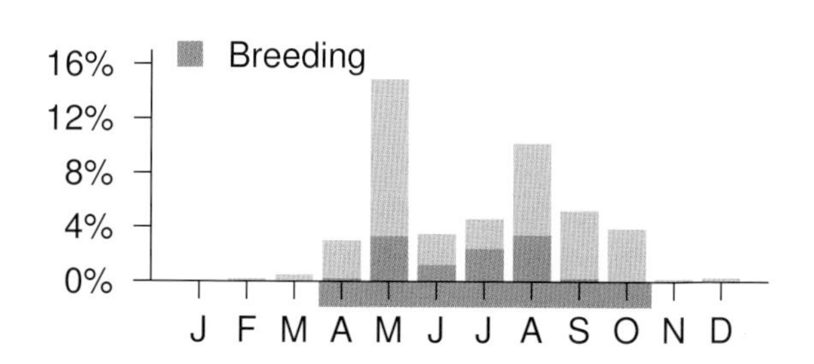

NATURAL REGIONS REPORTING RATE

SPATIAL REPORTING RATE

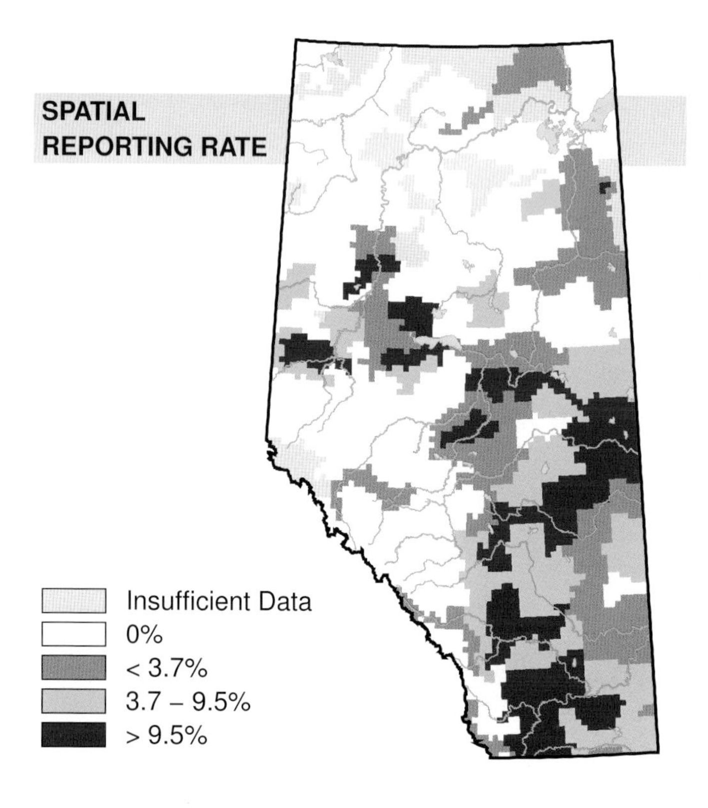

STATUS

GSWA 2000	GSWA 2005	COSEWIC Historic Rankings	COSEWIC 2007	Alberta Wildlife Act
Secure	Secure	N/A	N/A	N/A

OBSERVED DISTRIBUTION ATLAS 2

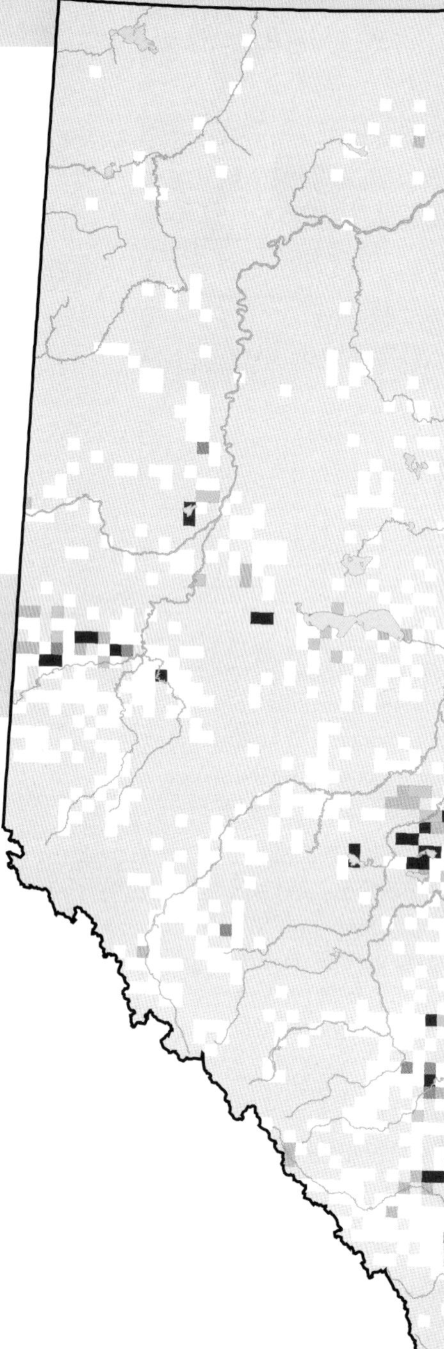

RANGE

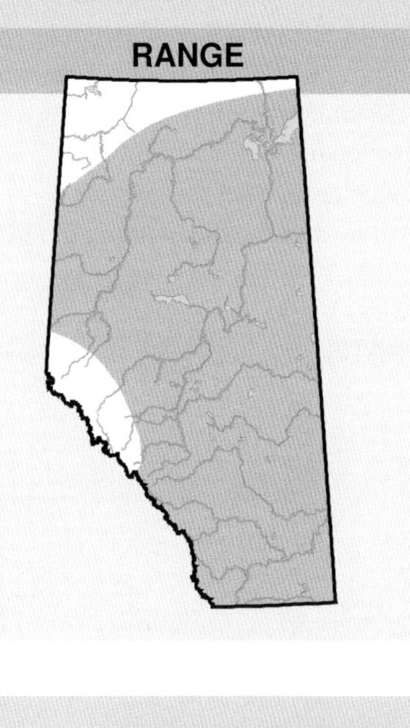

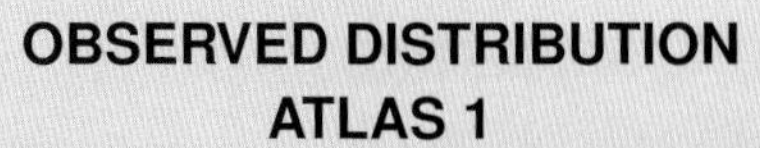

OBSERVED DISTRIBUTION ATLAS 1

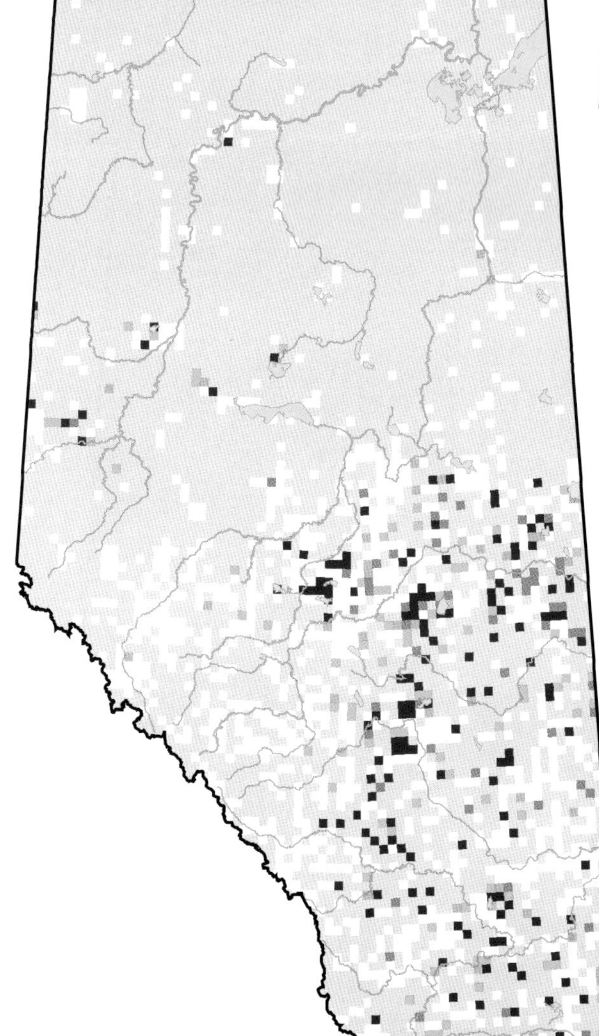

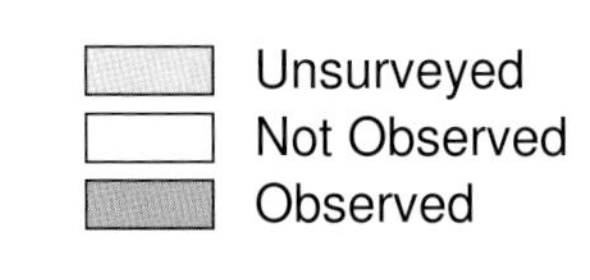

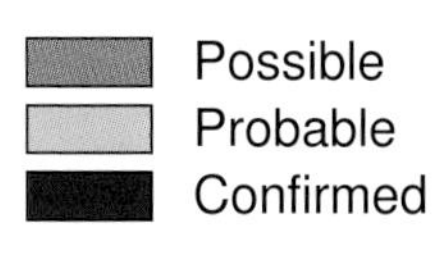

Western Grebe (*Aechmophorus occidentalis*)

Photo: Allen Cruickshank / Cornell Laboratory of Ornithology

NESTING

CLUTCH SIZE (EGGS):	3–5
INCUBATION (DAYS):	23
FLEDGING (DAYS):	49–51
NEST HEIGHT (METRES):	0

With its contrasting white and black markings and slender form, the Western Grebe is one of the most striking grebes in Alberta, and it is found in all of the province's Natural Regions except the Canadian Shield NR. The observed distribution of this species did not change between Atlas 1 and Atlas 2.

Despite occurring in almost all of the natural regions, the Western Grebe was found mainly in the Grassland NR, occasionally in the Parkland and Rocky Mountain NRs, infrequently in the Boreal Forest, and rarely in the Foothills NRs. The distribution of this species in Alberta reflects its preference to nest at lakes with large areas of open water which are surrounded by emergent vegetation.

An increase in relative abundance was detected in the Grassland NR and declines were detected in the Boreal Forest and Parkland NRs. Occurrence rates relative to other species mirrored these changes in abundance. These changes track the differences in precipitation between Atlas 1 and 2 with the Boreal Forest and Parkland NRs drier in Atlas 2 and the Grassland NR wetter. The Breeding Bird Survey sample size of Western Grebe records was too small to investigate temporal abundance changes in either Alberta or Canada. Differences between the results of the Atlas and the Breeding Bird Survey analyses could be attributed to differences in survey designs. The Breeding Bird Survey is conducted at fixed points along roads and most species are detected by sound. This results in lower detection rates for groups such as waterbirds. In contrast, some Atlas surveys were done along the perimeter of wetlands without the constraint of using fixed points. The result was a larger sample size for wetland-dependent species. Changes in relative abundance were not detected in the Foothills and Rocky Mountain NRs. The provincial status of this species is Sensitive due to its gregarious nature and the vulnerability of its nests to disturbance.

TEMPORAL REPORTING RATE

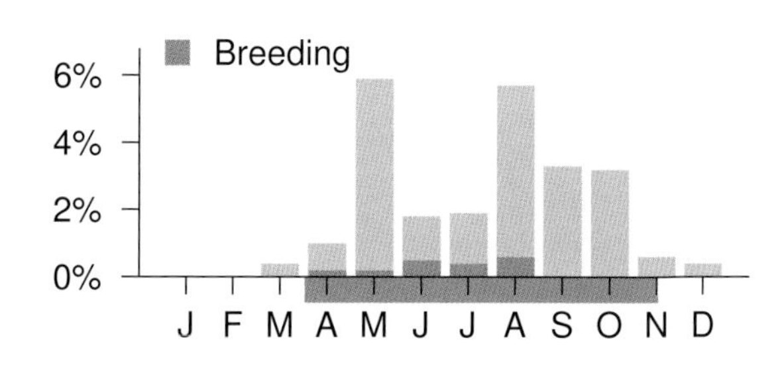

NATURAL REGIONS REPORTING RATE

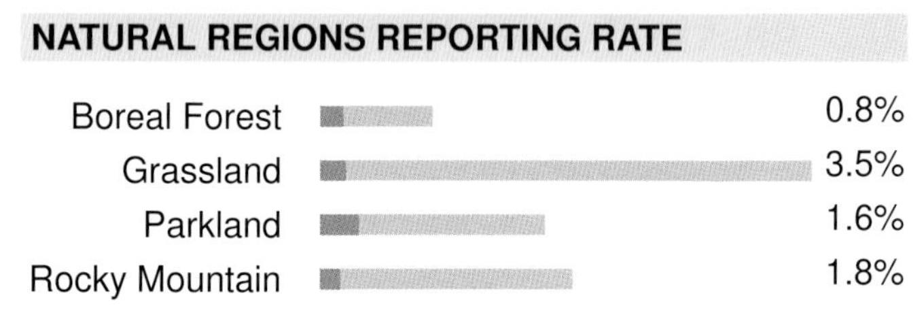

SPATIAL REPORTING RATE

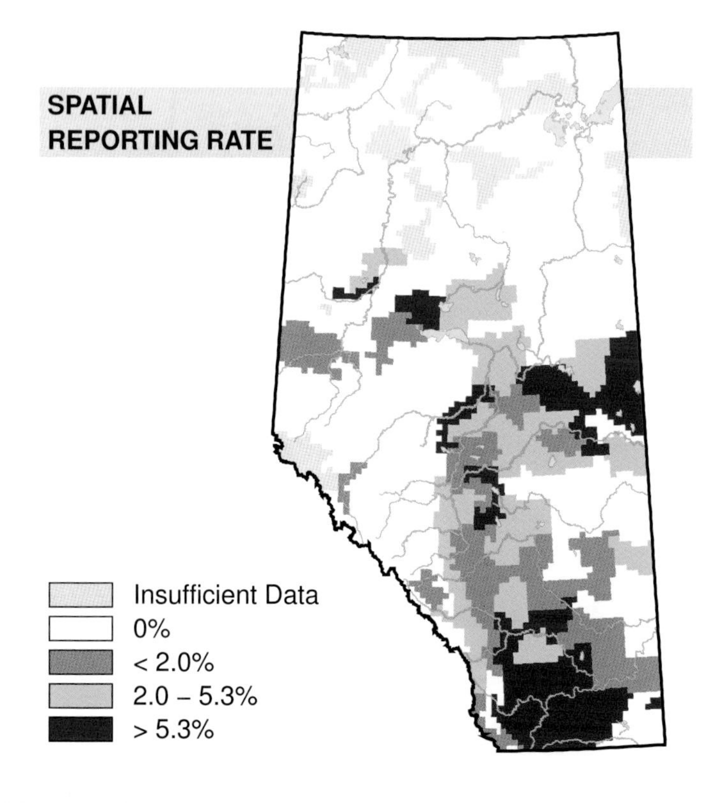

STATUS

GSWA 2000	GSWA 2005	COSEWIC Historic Rankings	COSEWIC 2007	Alberta Wildlife Act
Sensitive	Sensitive	N/A	N/A	N/A

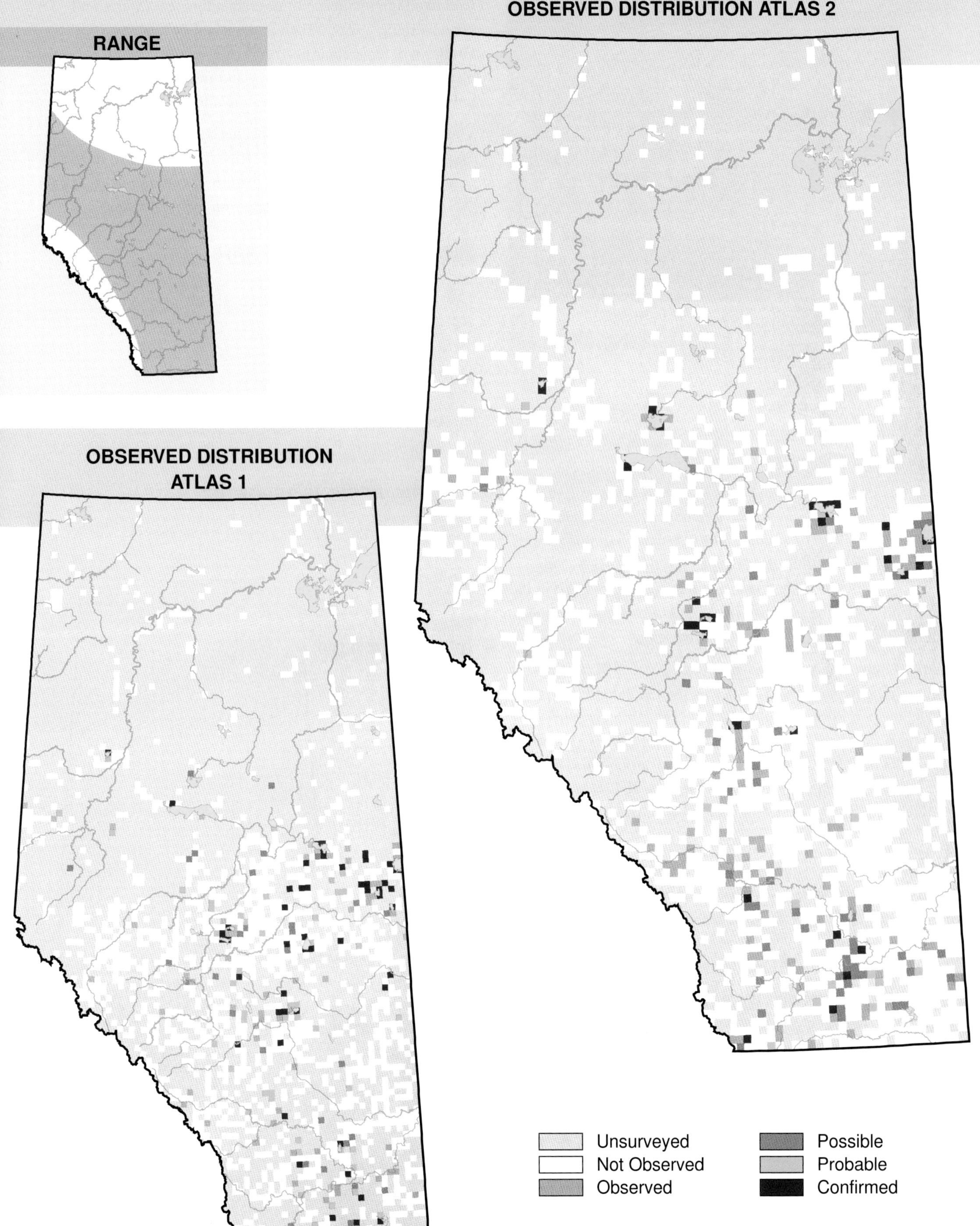

OBSERVED DISTRIBUTION ATLAS 2
RANGE
OBSERVED DISTRIBUTION ATLAS 1
Unsurveyed
Not Observed
Observed
Possible
Probable
Confirmed

Clark's Grebe *(Aechmophorus clarkii)*

Photo: Nancy Strand / Cornell Laboratory of Ornithology

NESTING

CLUTCH SIZE (EGGS):	3–5
INCUBATION (DAYS):	23
FLEDGING (DAYS):	63–77
NEST HEIGHT (METRES):	0

The rarest grebe in Alberta, Clark's Grebe is found only in the Grassland Natural Region in the province which is the northern periphery of this grebe's range. The distribution of this species extended farther north in Atlas 2 than it did in Atlas 1. This expansion could be related to above-normal precipitation in the Grassland NR in the later half of Atlas 2, providing breeding habitat that is more suitable for this species.

Clark's Grebe breeds at lakes with large areas of open water that are used by courting pairs during their elaborate courtship dance; the dance involves running side-by-side across the surface of the water. As a colonial nester, this species usually locates its nest among emergent vegetation around the edges of lakes. Clark's Grebes are considered fish specialists and, therefore, the lakes where Clark's Grebes breed must support adequate fish stocks.

No change in relative abundance was detected in Alberta. Yet, relative to other species, it was observed more frequently in Atlas 2 than in Atlas 1. Records of Clark's Grebe from the Breeding Bird Survey were too few to investigate abundance changes in Alberta, or in Canada.

As these grebes build a floating nest made from plant material anchored to emergent vegetation, the nests are vulnerable to the wake created from water craft. Repeated disturbances by curious boaters, especially in early stages of colony formation, lead to a high incidence of nest abandonment (Storer and Nuechterlein, 1992). When approached by humans, Clark's Grebes will leave their nest, exposing eggs to predation and the elements. Thus, areas with easy and frequent access are unlikely to provide suitable nesting sites.

The provincial status of this species was changed from Sensitive in 2000 to May Be at Risk in 2005, due to its small population, its intolerance for disturbance at nest sites, and its susceptibility to oil-spill mortality on the winter range.

TEMPORAL REPORTING RATE

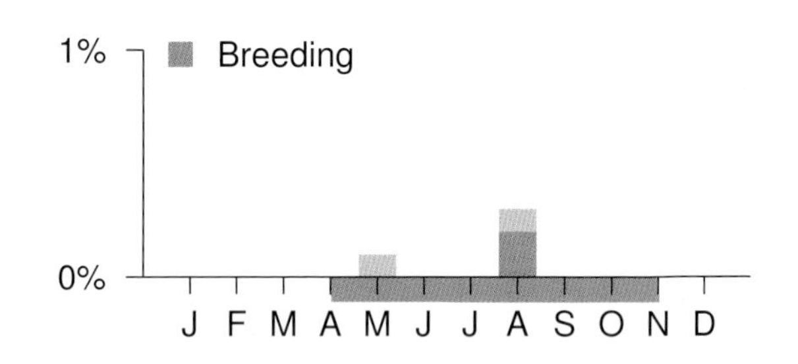

NATURAL REGIONS REPORTING RATE

Grassland 0.2%

SPATIAL REPORTING RATE

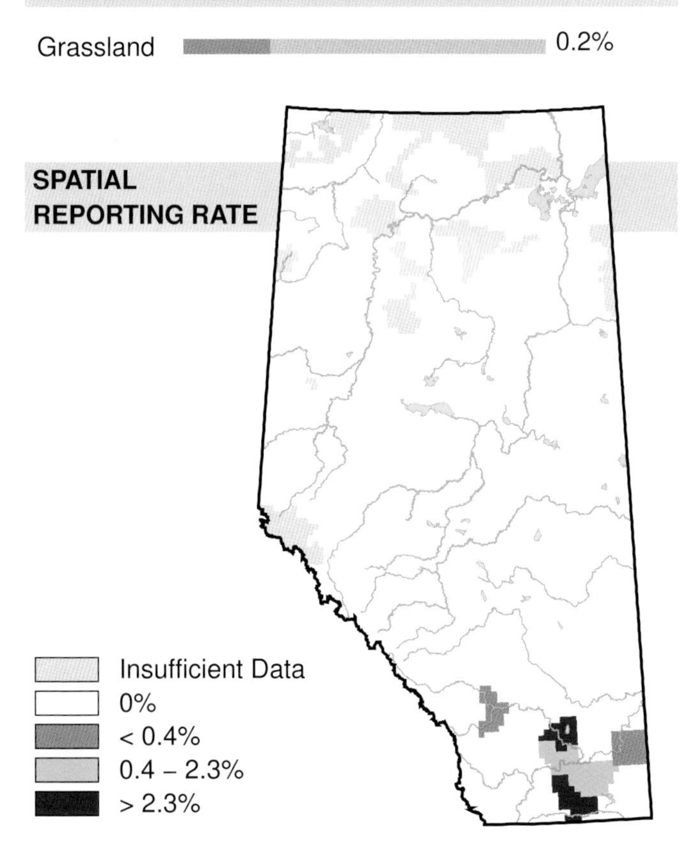

STATUS

GSWA 2000	GSWA 2005	COSEWIC Historic Rankings	COSEWIC 2007	Alberta Wildlife Act
Sensitive	May Be at Risk	N/A	N/A	N/A

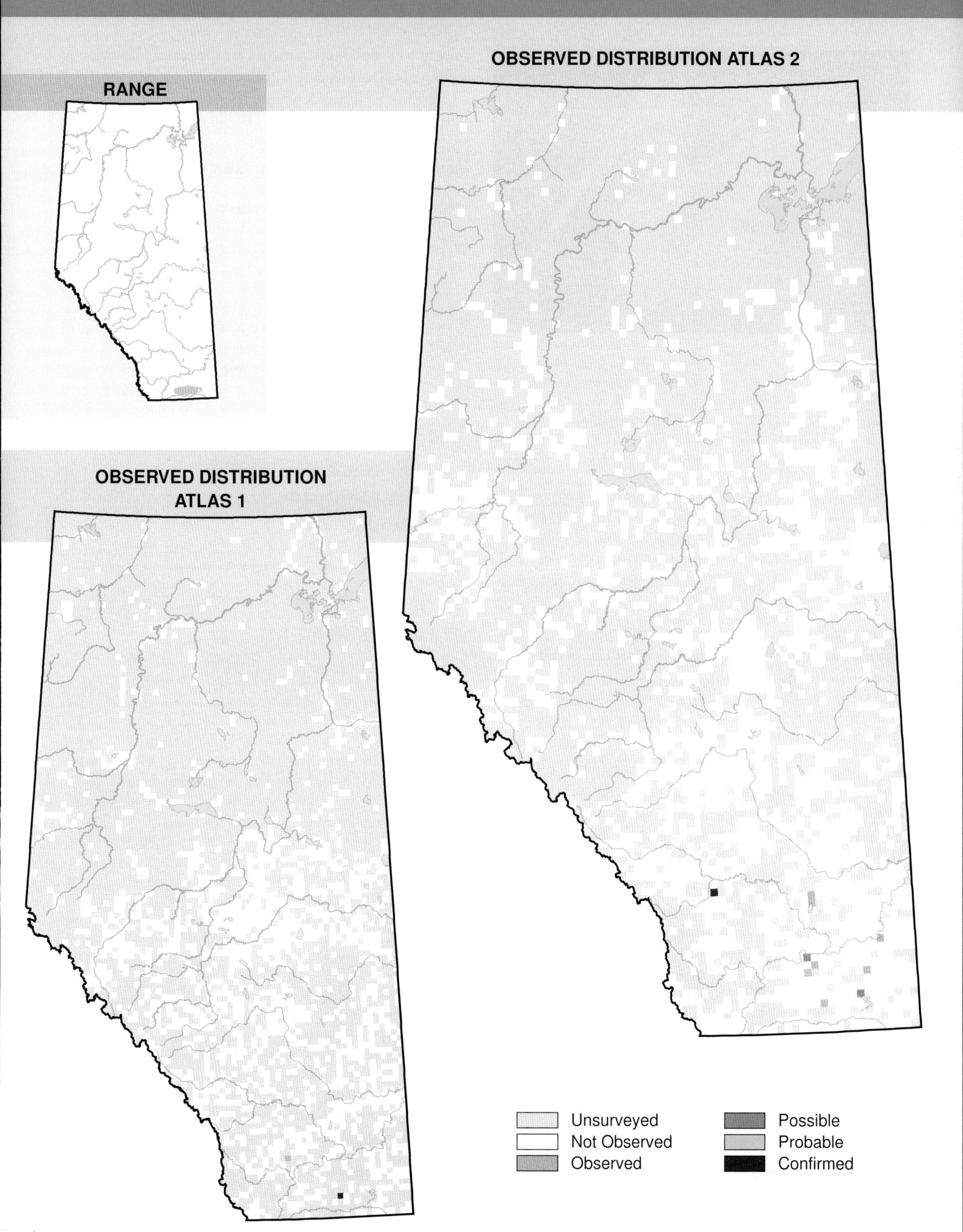
RANGE
OBSERVED DISTRIBUTION ATLAS 2
OBSERVED DISTRIBUTION ATLAS 1
Unsurveyed
Not Observed
Observed
Possible
Probable
Confirmed

American White Pelican (*Pelecanus erythrorhynchos*)

Photo: Anne Elliott

NESTING

CLUTCH SIZE (EGGS):	1–4
INCUBATION (DAYS):	29
FLEDGING (DAYS):	70
NEST HEIGHT (METRES):	0

The American White Pelican is found in every Natural Region in Alberta. No change in distribution was observed between Atlas 1 and Atlas 2 for this species.

American White Pelicans nest on the ground on islands that are surrounded by freshwater, unlike its coastal cousin, the Brown Pelican, which often nests in trees near saltwater. Drought can negatively affect American White Pelicans if land bridges form and connect islands to the mainland after water levels drop. Land bridges facilitate predator access to nests, which reduces the number of available nesting areas. This species tends to hunt in cooperative groups in shallow areas at marshes, lakes, and rivers. Fish are the main prey items but amphibians are also occasionally consumed. This species was most common in the Grassland NR and was occasionally encountered in the Boreal Forest, Foothills, Parkland, and Rocky Mountain NRs where it can be locally abundant.

Increases in relative abundance were detected in the Boreal Forest, Grassland, and Parkland NRs. Relative to other species it was observed at similar frequencies in Atlas 1 and Atlas 2 in the Boreal Forest and Parkland NRs, but in the Grassland NR it was observed less frequently in Atlas 2. With regard to the localized distribution of this colonial nester, frequency of detection relative to other species is not likely a reliable measure unless the comparisons are made using matching areas only. The Breeding Bird Survey did not detect any change in abundance in Alberta. However, as a colonial nester, this species is not well covered by the BBS. American White Pelicans are considered Sensitive in Alberta, based on a decline in the number of known colonies.

TEMPORAL REPORTING RATE

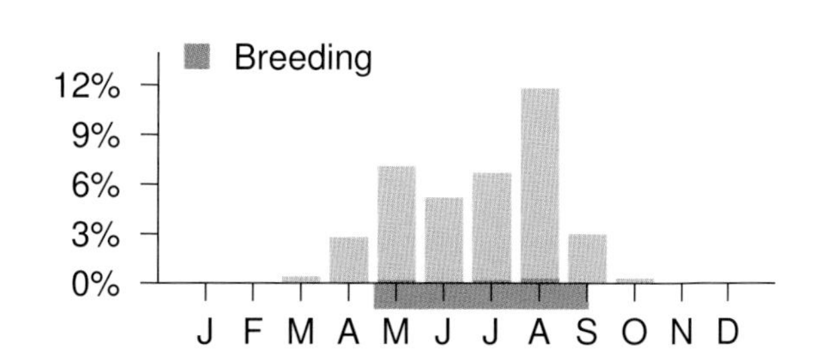

NATURAL REGIONS REPORTING RATE

Boreal Forest	1.3%
Grassland	8.7%
Parkland	1.8%

SPATIAL REPORTING RATE

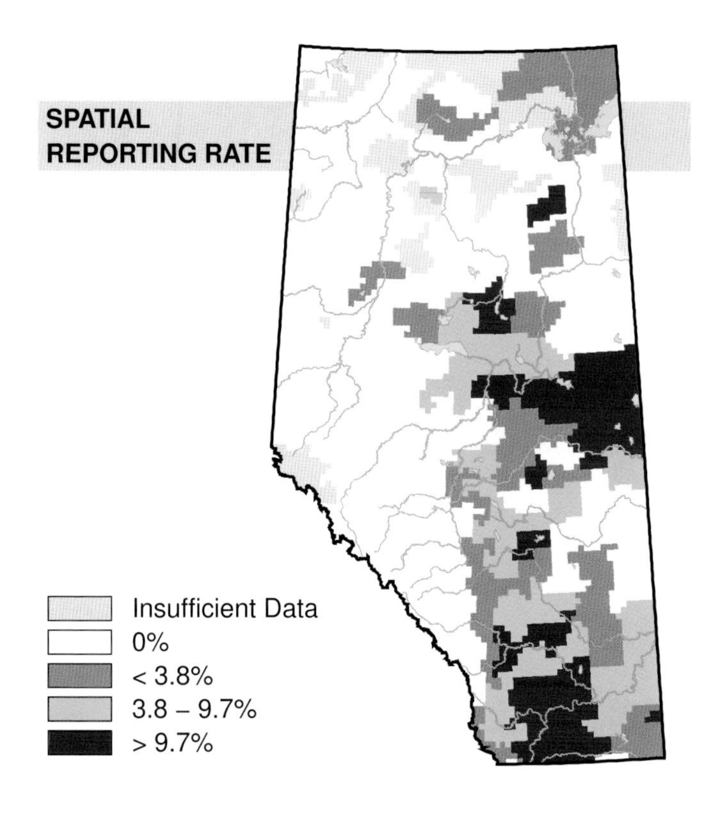

STATUS

GSWA 2000	GSWA 2005	COSEWIC Historic Rankings	COSEWIC 2007	Alberta Wildlife Act
Sensitive	Sensitive	Not At Risk	Not At Risk	N/A

RANGE

OBSERVED DISTRIBUTION ATLAS 2

OBSERVED DISTRIBUTION ATLAS 1

Unsurveyed
Not Observed
Observed
Possible
Probable
Confirmed

Double-crested Cormorant (*Phalacrocorax auritus*)

Photo: Gerald Romanchuk

NESTING

CLUTCH SIZE (EGGS):	3–4
INCUBATION (DAYS):	28
FLEDGING (DAYS):	35–42
NEST HEIGHT (METRES):	0

The Double-crested Cormorant was reported in every Natural Region in Alberta except the Canadian Shield. No change in the distribution of this species was observed in Atlas 2.

Double-crested Cormorants, feeding primarily on fish, are closely associated with aquatic habitats and often nest on islands, either on the ground or in trees. Due to their close association with water, Double-crested Cormorants have a fairly patchy distribution across Alberta. Despite their broad distribution, they were found mainly in the Grassland NR and were encountered occasionally in the Boreal Forest, Foothills, and Parkland NRs where they were locally abundant.

Increases in relative abundance were detected in the Grassland and Parkland NRs where surprisingly, relative to other species, this species was observed less frequently in Atlas 2 than in Atlas 1. Above-normal precipitation in the latter half of Atlas 2 may have flooded land bridges formed during previous periods of drought and improved nest success. Since the early 1970s, these cormorants have increased widely and it is likely that continental numbers have reached an all-time high (Reed et al., 2003). The Breeding Bird Survey did not detect any change in abundance in Alberta or Canada. No changes were observed in other NRs.

Cormorants have been persecuted throughout history; recent great increases in numbers have spurred renewed controversy, although compelling evidence that cormorants seriously damage fisheries is rare (Hatch and Weseloh, 1999). The conclusion that Double-crested Cormorants normally take an insignificant number of game fish is supported by a number of studies (Reed et al., 2003). However, in Alberta, intensive egg-oiling and adult-culling campaigns have been carried out on Lac La Biche in recent years in response to a population crash of walleye and an increase of cormorants at the lake.

The provincial status of this species is Secure.

TEMPORAL REPORTING RATE

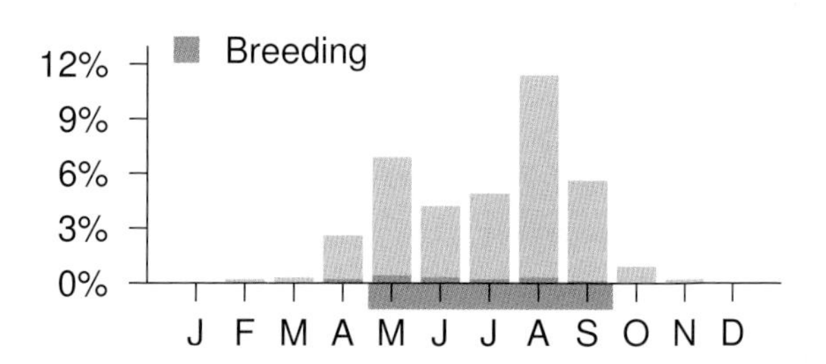

NATURAL REGIONS REPORTING RATE

SPATIAL REPORTING RATE

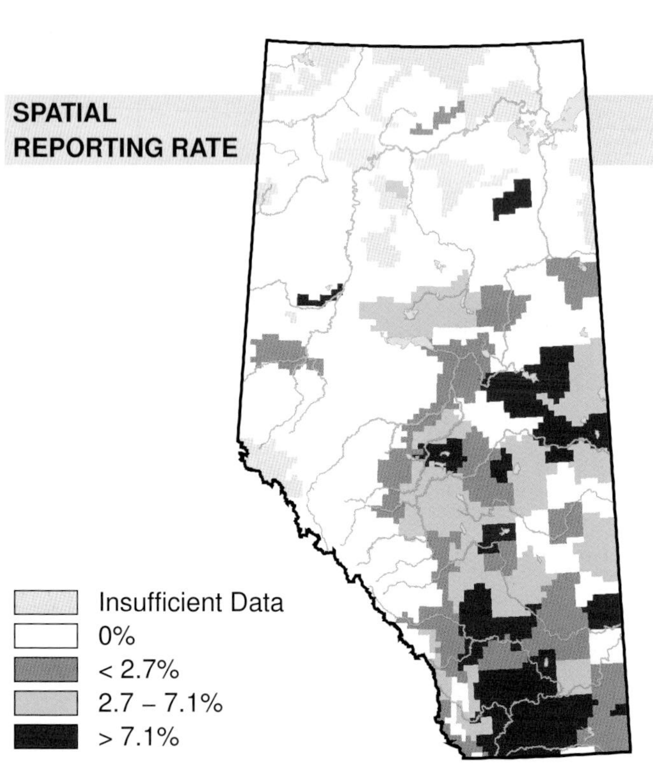

STATUS

GSWA 2000	GSWA 2005	COSEWIC Historic Rankings	COSEWIC 2007	Alberta Wildlife Act
Secure	Secure	Not At Risk	Not At Risk	N/A

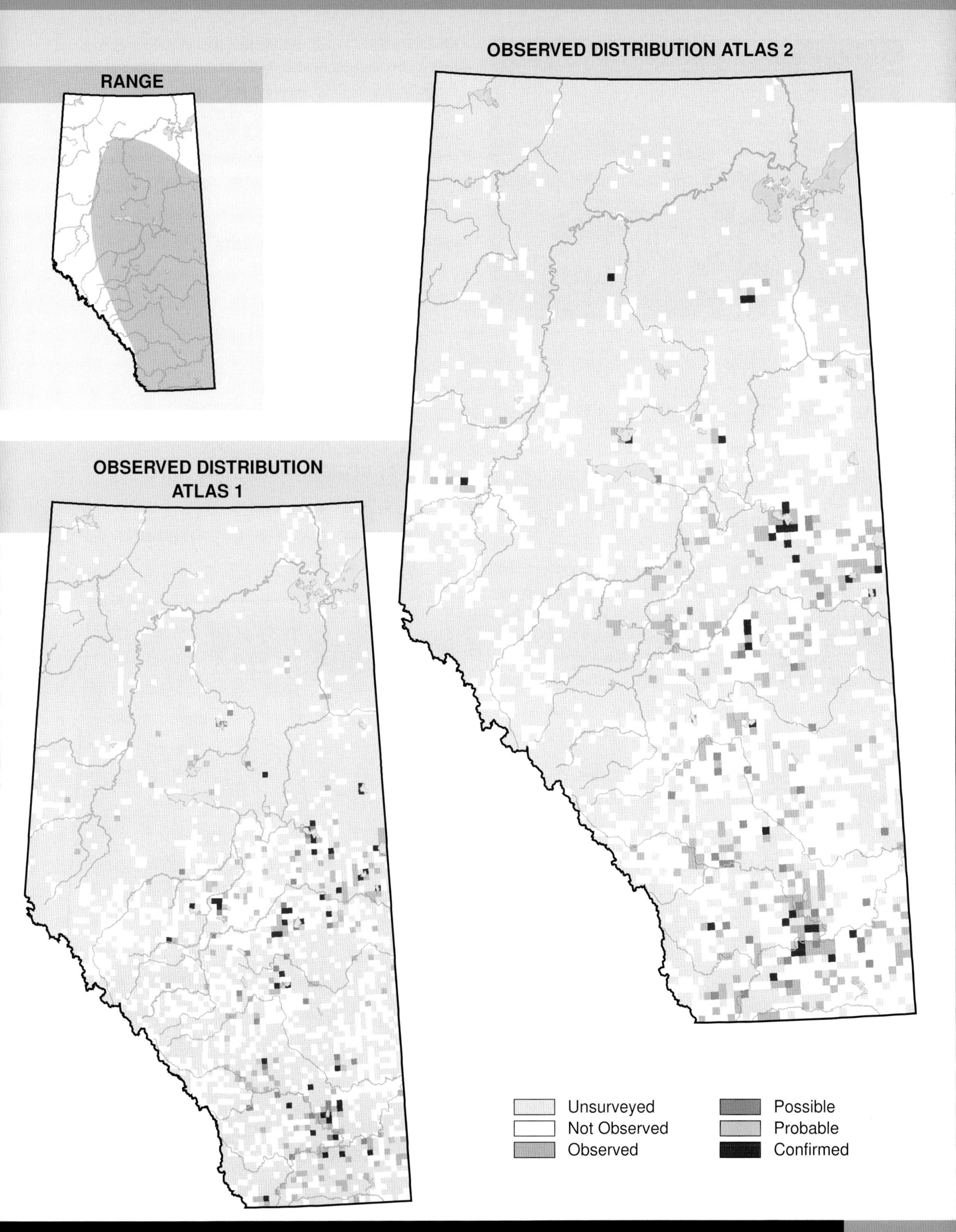
Habitat
Nest Location
Nest Type
Diet
OBSERVED DISTRIBUTION ATLAS 2
RANGE
OBSERVED DISTRIBUTION ATLAS 1
Unsurveyed
Not Observed
Observed
Possible
Probable
Confirmed

American Bittern (*Botaurus lentiginosus*)

Photo: Gordon Court

NESTING

CLUTCH SIZE (EGGS):	4–5
INCUBATION (DAYS):	24–28
FLEDGING (DAYS):	14
NEST HEIGHT (METRES):	0

The American Bittern is found in every Natural Region in Alberta. The observed distribution of this species was similar in Atlas 1 and Atlas 2, although there were fewer records in the north, especially in the northwestern part of the province. The northern coverage in Atlas 2 was more extensive than in Atlas 1, but in both efforts, northern coverage was severely limited by access. Due to this species' secretive nature, it is possible that it was actually present but not detected in northern squares surveyed in Atlas 2; therefore, the observed difference does not necessarily represent a true change in distribution.

The American Bittern requires shallow wetlands that contain large areas with tall, emergent vegetation which is used for nesting, hunting, and concealment from predators. This species was most common in the Foothills, Grassland, and Parkland NRs but was not encountered often in the Boreal Forest and Rocky Mountain NRs. Wetlands in the Grassland and Parkland NRs generally satisfy American Bittern habitat requirements because those wetlands tend to be shallow and many contain tall, emergent vegetation. In comparison, wetlands in the Boreal Forest NR are usually deep, and wetlands in the Rocky Mountain NR tend not to have abundant tall vegetation.

Declines in relative abundance were detected in the Boreal Forest and Parkland NRs where, relative to other species, it was observed less frequently in Atlas 2 than in Atlas 1. No change was detected in the Foothills, Grassland, and Rocky Mountain NRs. Data from the Breeding Bird Survey indicate that American Bittern abundance declined in Alberta during the period 1985–2005, as it did in Canada during the period 1995–2005. Declines are likely related to declines in the availability of wetlands due to drought and drainage. This species is considered Sensitive in the province because it is sensitive to wetland drainage; also, populations are difficult to monitor.

TEMPORAL REPORTING RATE

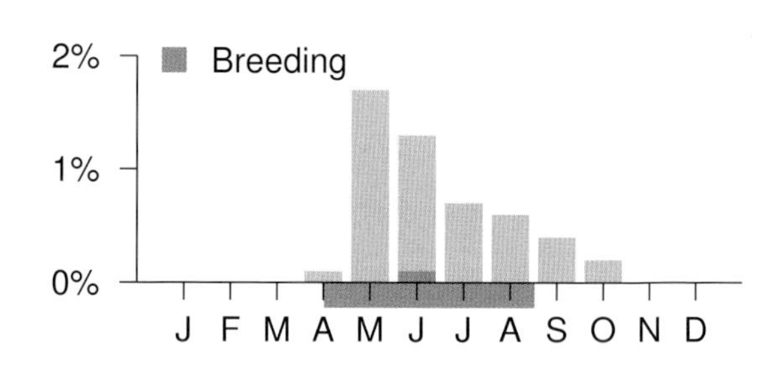

NATURAL REGIONS REPORTING RATE

SPATIAL REPORTING RATE

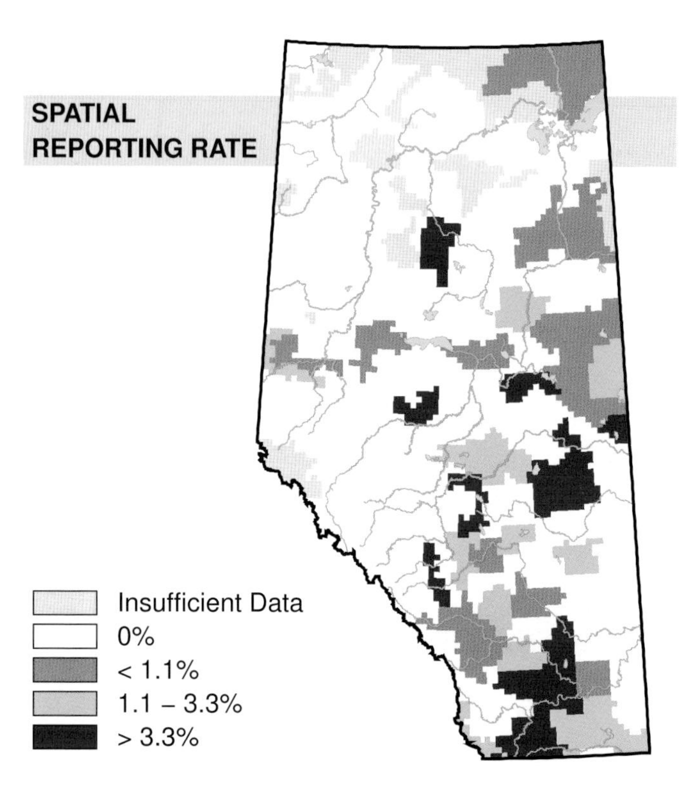

STATUS

GSWA 2000	GSWA 2005	COSEWIC Historic Rankings	COSEWIC 2007	Alberta Wildlife Act
Sensitive	Sensitive	N/A	N/A	N/A

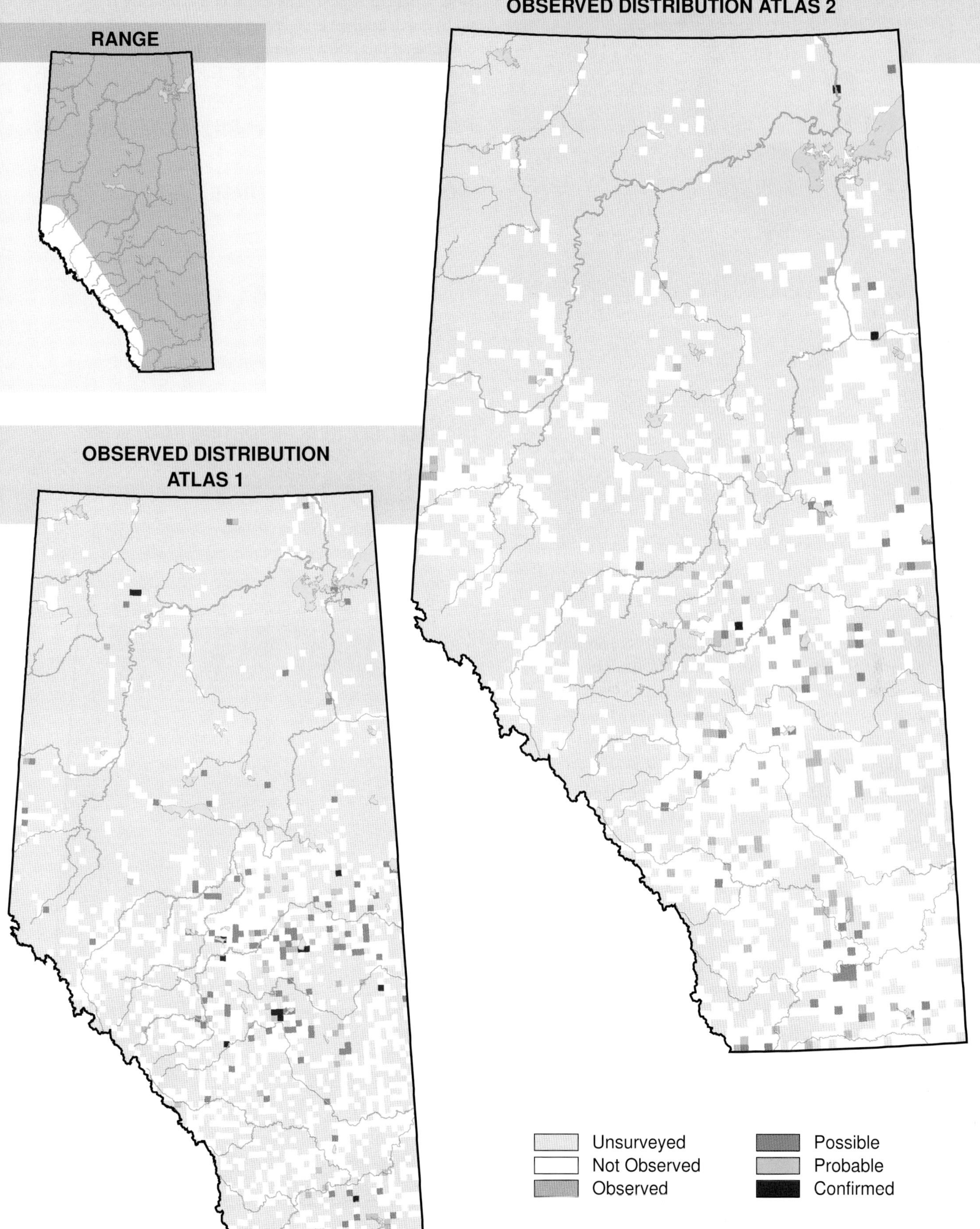
OBSERVED DISTRIBUTION ATLAS 2
RANGE
OBSERVED DISTRIBUTION
ATLAS 1
Unsurveyed
Not Observed
Observed
Possible
Probable
Confirmed

Great Blue Heron (*Ardea herodias*)

Photo: Gerald Romanchuk

NESTING

CLUTCH SIZE (EGGS):	3–5
INCUBATION (DAYS):	26–27
FLEDGING (DAYS):	53
NEST HEIGHT (METRES):	3–39.6

The Great Blue Heron is found in every Natural Region in Alberta but it is not common in the northern part of the province. No change in distribution was observed between Atlas 1 and Atlas 2.

Great Blue Herons can nest in single pairs but most often nest colonially. Nests are usually built in trees and tend to be close to wetland foraging areas. This species will occasionally nest on the ground on predator-free islands. The main food consumed by Great Blue Herons is fish, but amphibians, invertebrates, reptiles, mammals, and other birds can also be used. Because this species tends to be gregarious, its distribution across the province is fairly patchy. Reporting rates were high in the Foothills, Grassland, Parkland, and Rocky Mountain NRs and lower in the Boreal Forest NR.

Declines in relative abundance were detected for this species in the Boreal Forest, Foothills, and Grassland NRs where, relative to other species, Great Blue Herons were observed less frequently in Atlas 2 than in Atlas 1. The Breeding Bird Survey did not detect any change in abundance for this species in Alberta. However, a decline in Canada was detected for the period 1985–2005. Declines could be related to declines in the availability of wetlands. Above-normal precipitation in the later half of Atlas 2 would have had little effect on this fish-preferring species. The Atlas detected an increase in relative abundance in the Rocky Mountain NR where, relative to other species, this species was observed more frequently in Atlas 2 than in Atlas 1. The cause of this increase is not understood, as there is no apparent biological cause, and coverage of this NR was similar in both Atlases. No changes were detected in the Parkland NR. This species is considered Sensitive in the province because there are fewer than 100 known nesting colonies and this species is susceptible to disturbance at the nest site.

TEMPORAL REPORTING RATE

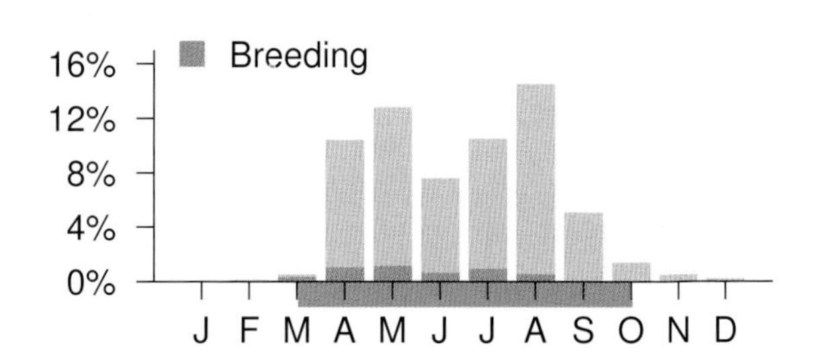

NATURAL REGIONS REPORTING RATE

Boreal Forest	2.1%
Foothills	6.3%
Grassland	6.9%
Parkland	6.6%
Rocky Mountain	7.4%

SPATIAL REPORTING RATE

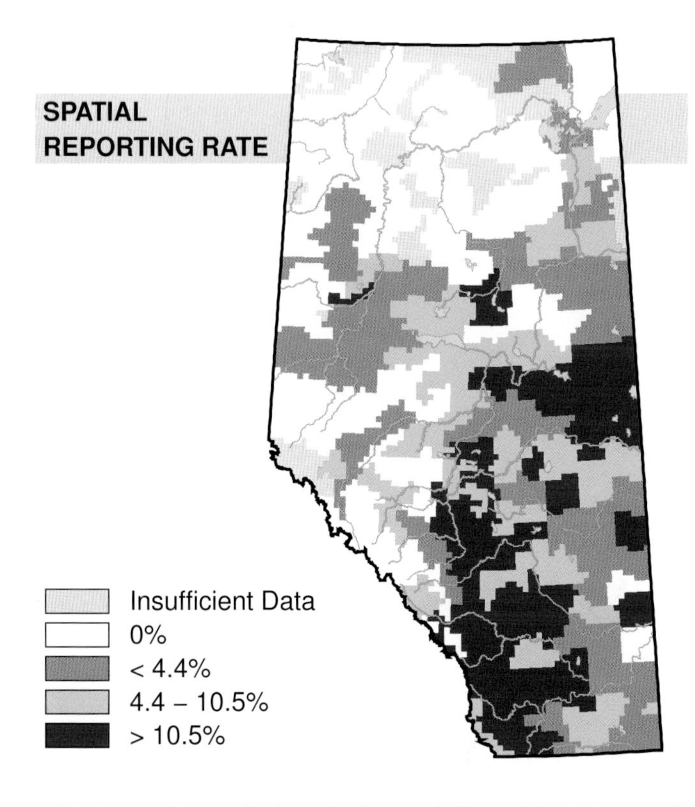

STATUS

GSWA 2000	GSWA 2005	COSEWIC Historic Rankings	COSEWIC 2007	Alberta Wildlife Act
Sensitive	Sensitive	N/A	N/A	N/A

Habitat
Nest Location
Nest Type
Diet

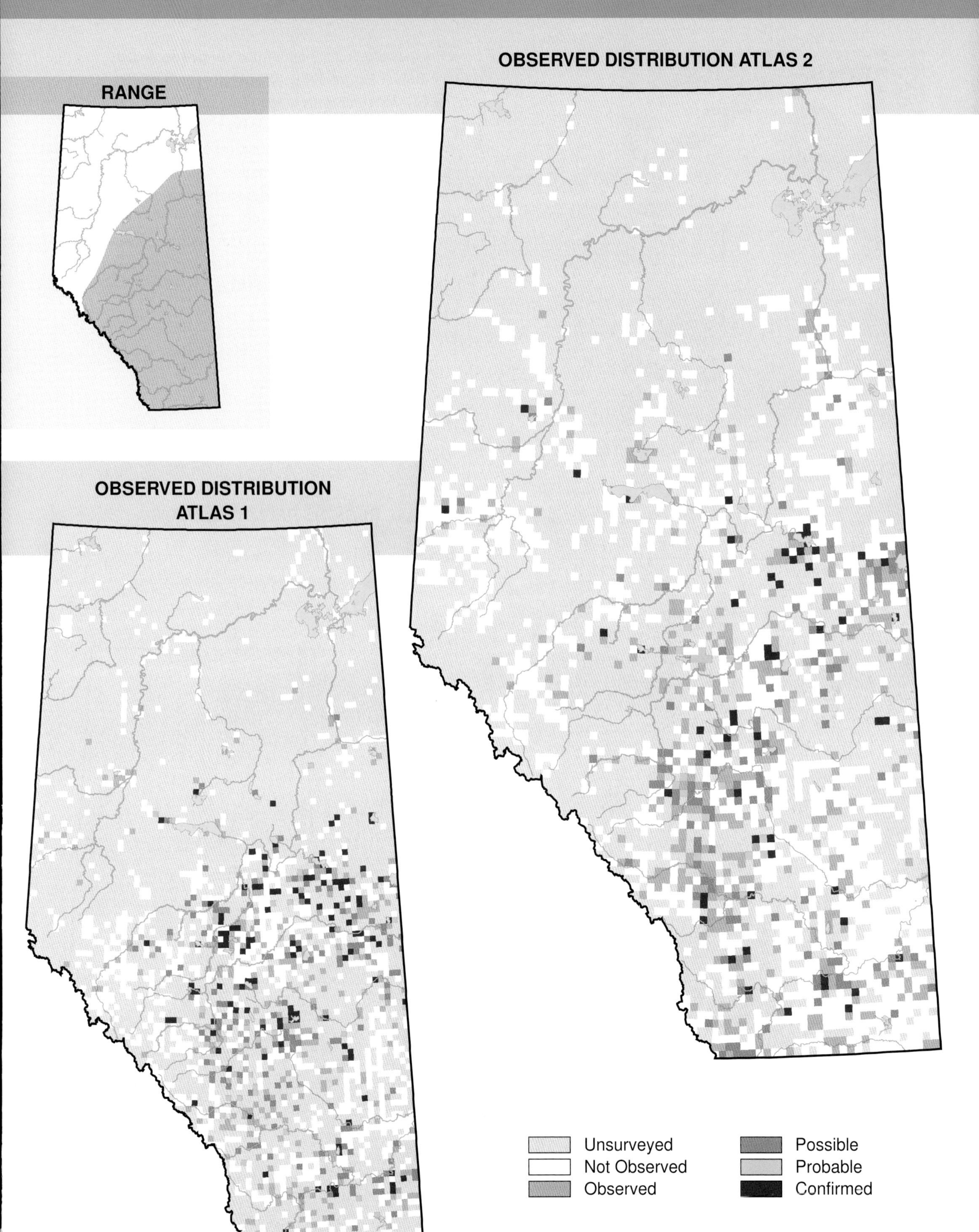
RANGE
OBSERVED DISTRIBUTION ATLAS 2
OBSERVED DISTRIBUTION ATLAS 1
Unsurveyed
Not Observed
Observed
Possible
Probable
Confirmed

Black-crowned Night-Heron (*Nycticorax nycticorax*)

Photo: Gerald Romanchuk

NESTING

CLUTCH SIZE (EGGS):	1–6
INCUBATION (DAYS):	21–28
FLEDGING (DAYS):	42
NEST HEIGHT (METRES):	2–3

One of the most striking birds in Alberta, the Black-crowned Night-Heron is found in the southeastern part of the province. No change in distribution was observed between Atlas 1 and Atlas 2 for this species, except for a single new record found near Fort McMurray. A single record does not necessarily indicate that this species is expanding north; however, birders should watch for Black-crowned Night-Herons north of the typical range, to see if more of these birds will be recorded there.

Black-crowned Night-Herons are colonial nesters. Nests are usually built in trees in isolated areas such as on islands; in addition, this species has also been known to nest on the ground or in cattails. Because of the gregarious nature of these birds, Black-crowned Night-Heron distribution is patchy across their range. Reporting rates were low everywhere, but highest in the Grasslands NR.

Decreases in relative abundance were detected in the Boreal Forest, Grassland, and Parkland NRs where, relative to other species, it was observed less frequently in Atlas 2 than in Atlas 1. The Breeding Bird Survey did not detect a change in abundance in Alberta, or in Canada. However, the Breeding Bird Survey is a road-based survey that cannot be expected to survey the breeding habitat of this species adequately. Declines in the Boreal Forest and Parkland NRs could be due in part to relatively dry conditions in these regions during Atlas 2. The cause of the decline in the Grassland NR, however, is more difficult to understand as there was above-average precipitation in this region during the later half of Atlas 2. This species is considered Sensitive in the province because it is uncommon, locally distributed, and negatively impacted by wetland conversion and human disturbance at nesting colonies.

TEMPORAL REPORTING RATE

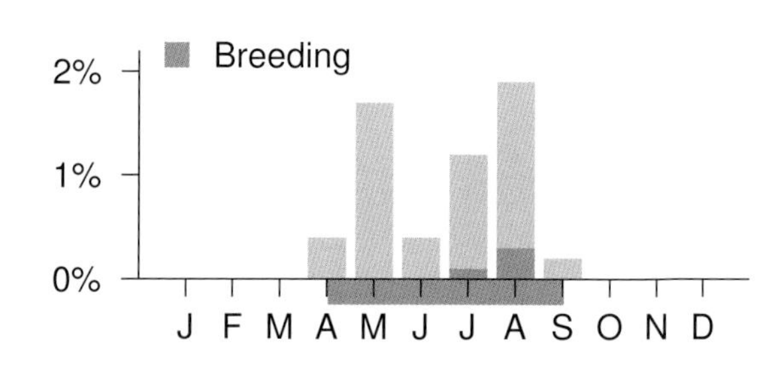

NATURAL REGIONS REPORTING RATE

SPATIAL REPORTING RATE

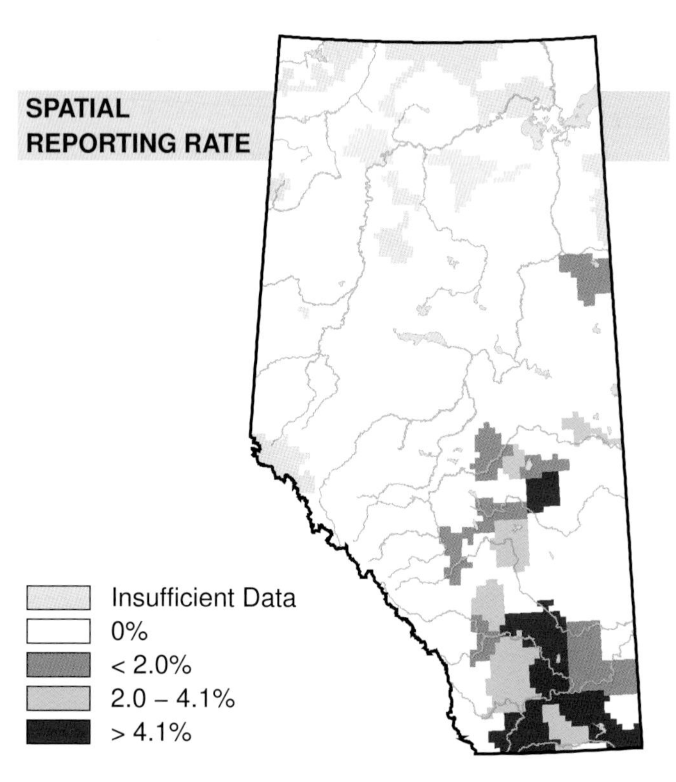

STATUS

GSWA 2000	GSWA 2005	COSEWIC Historic Rankings	COSEWIC 2007	Alberta Wildlife Act
Sensitive	Sensitive	N/A	N/A	N/A

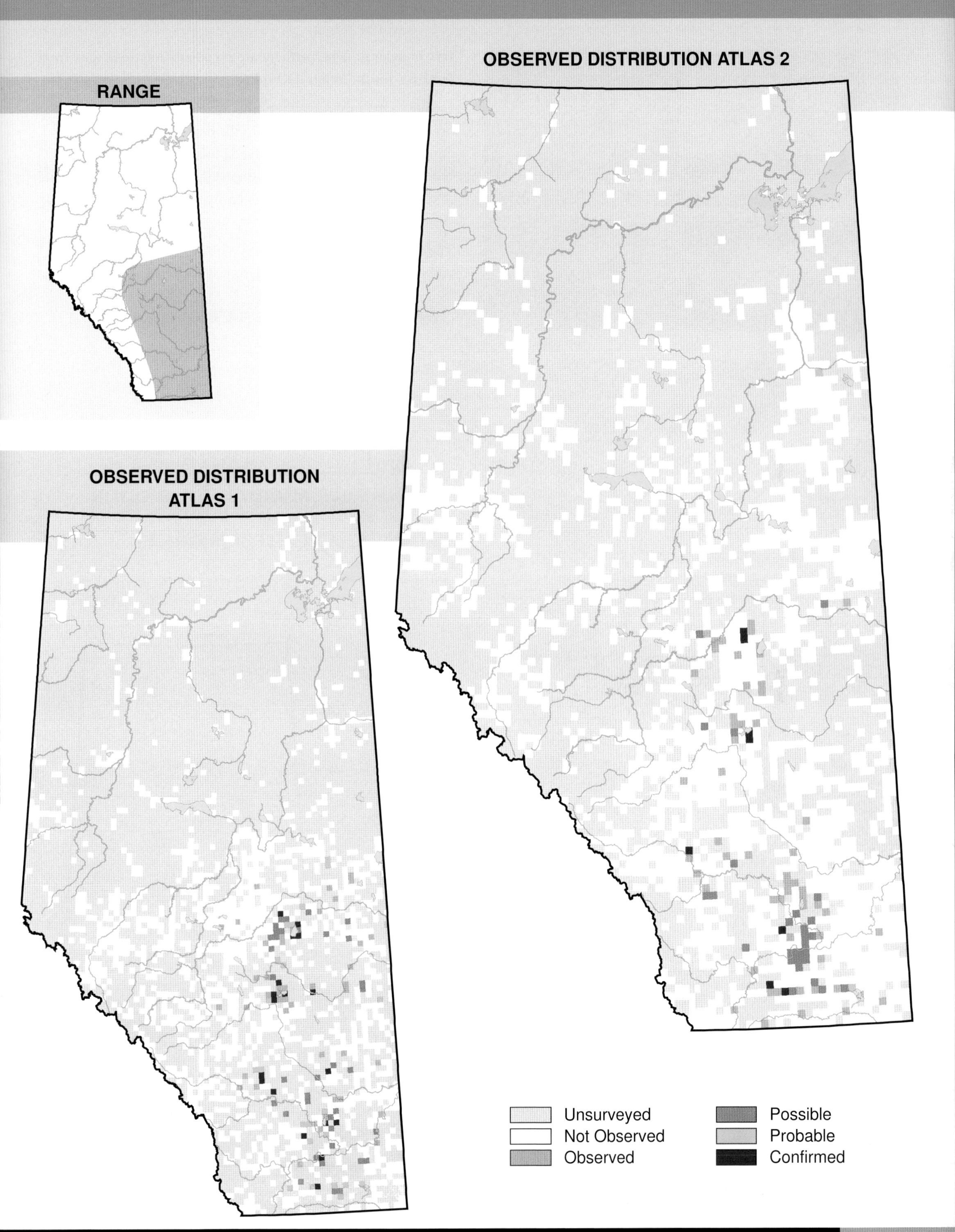
OBSERVED DISTRIBUTION ATLAS 2
RANGE
OBSERVED DISTRIBUTION ATLAS 1
Unsurveyed
Not Observed
Observed
Possible
Probable
Confirmed

White-faced Ibis (*Plegadis chihi*)

Photo: Gerald Romanchuk

NESTING

CLUTCH SIZE (EGGS):	2–4
INCUBATION (DAYS):	21–22
FLEDGING (DAYS):	28
NEST HEIGHT (METRES):	1

The observed distribution of the White-faced Ibis changed between Atlas 1 and Atlas 2. Many new records were obtained in the Grassland NR, thus expanding this species' observed distribution northward. Breeding has been confirmed in the Calgary area and there have been reports of this species being observed at Beaverhill Lake.

A species that can be nomadic in response to drought or rains, the White-faced Ibis can frequently be found north of its normal range in freshwater wetlands, especially those dominated by cattails and bulrushes. This species can also be found feeding in flooded fields, as was observed near Beaverhill Lake in 2005. Reporting rates were very low due to the bird's rarity; however, they were most common in the Grassland Natural Region. This is a new species for the Parkland NR.

Temporary loss of nesting habitat owing to drought or flooding may sharply curtail or prevent reproduction across large areas of the breeding range in some years, although the birds' apparent nomadism and ability to postpone nesting until habitat conditions improve ensure at least some breeding each year (Ryder and Manry, 1994).

Although the number of records has increased in the Grassland NR and the birds are new to the Parkland NR, no change in relative abundance was detected by the Atlas. Since Breeding Bird Surveys are limited to roads, and this species is a very rare breeder in Alberta, there are no population trend data from the BBS for the White-faced Ibis.

Owing to the birds' high mobility and tendency to shift breeding areas from one year to the next, monitoring the conservation status of White-faced Ibis populations would require coordinated surveys using standardized techniques, and repeat surveys across the bird's entire range at regular intervals (Ryder and Manry, 1994).

The status of this bird is Sensitive in Alberta, due to its rarity and its disjunction from southern populations.

TEMPORAL REPORTING RATE

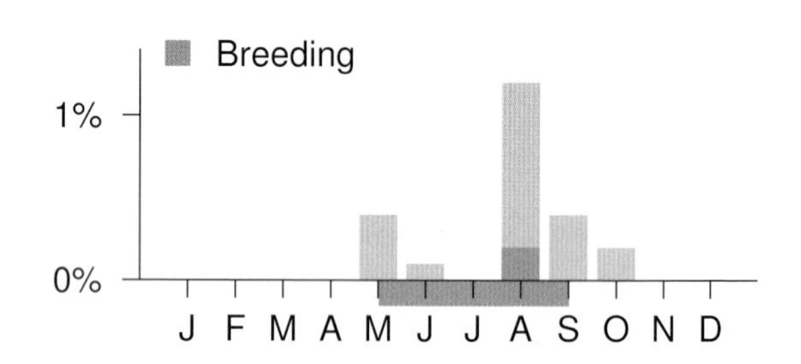

NATURAL REGIONS REPORTING RATE

Grassland 1.0%
Parkland 0.1%

SPATIAL REPORTING RATE

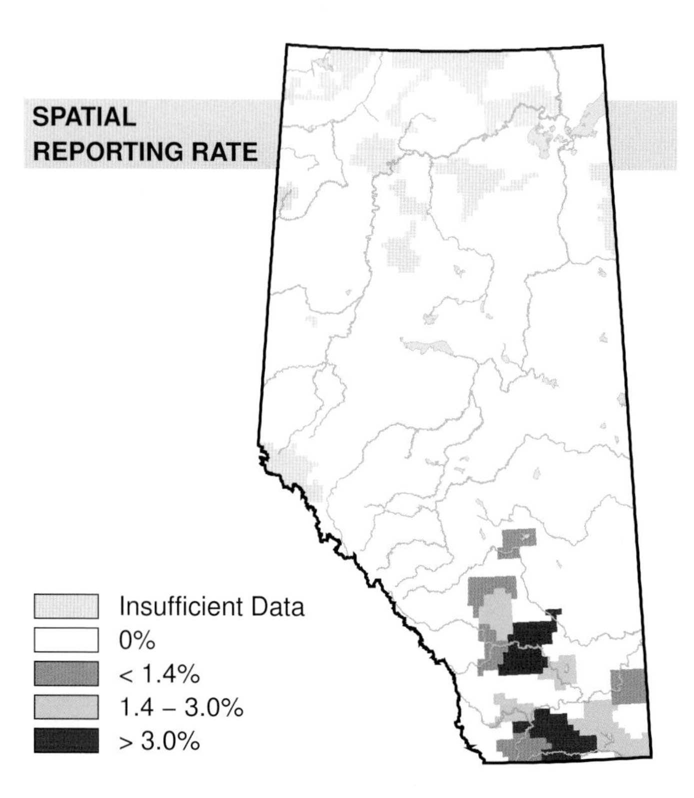

STATUS

GSWA 2000	GSWA 2005	COSEWIC Historic Rankings	COSEWIC 2007	Alberta Wildlife Act
Sensitive	Sensitive	N/A	N/A	N/A

RANGE

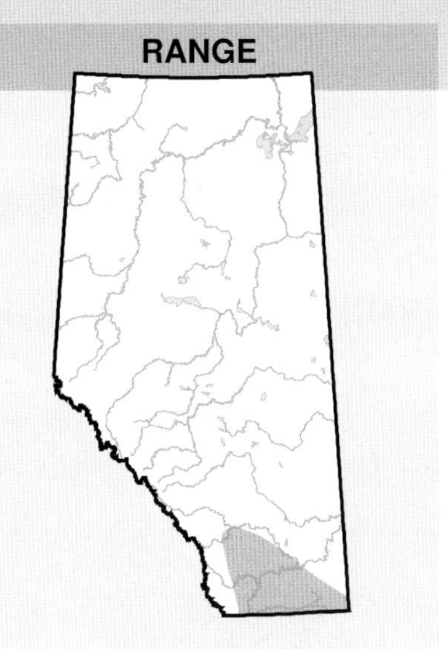

OBSERVED DISTRIBUTION ATLAS 2

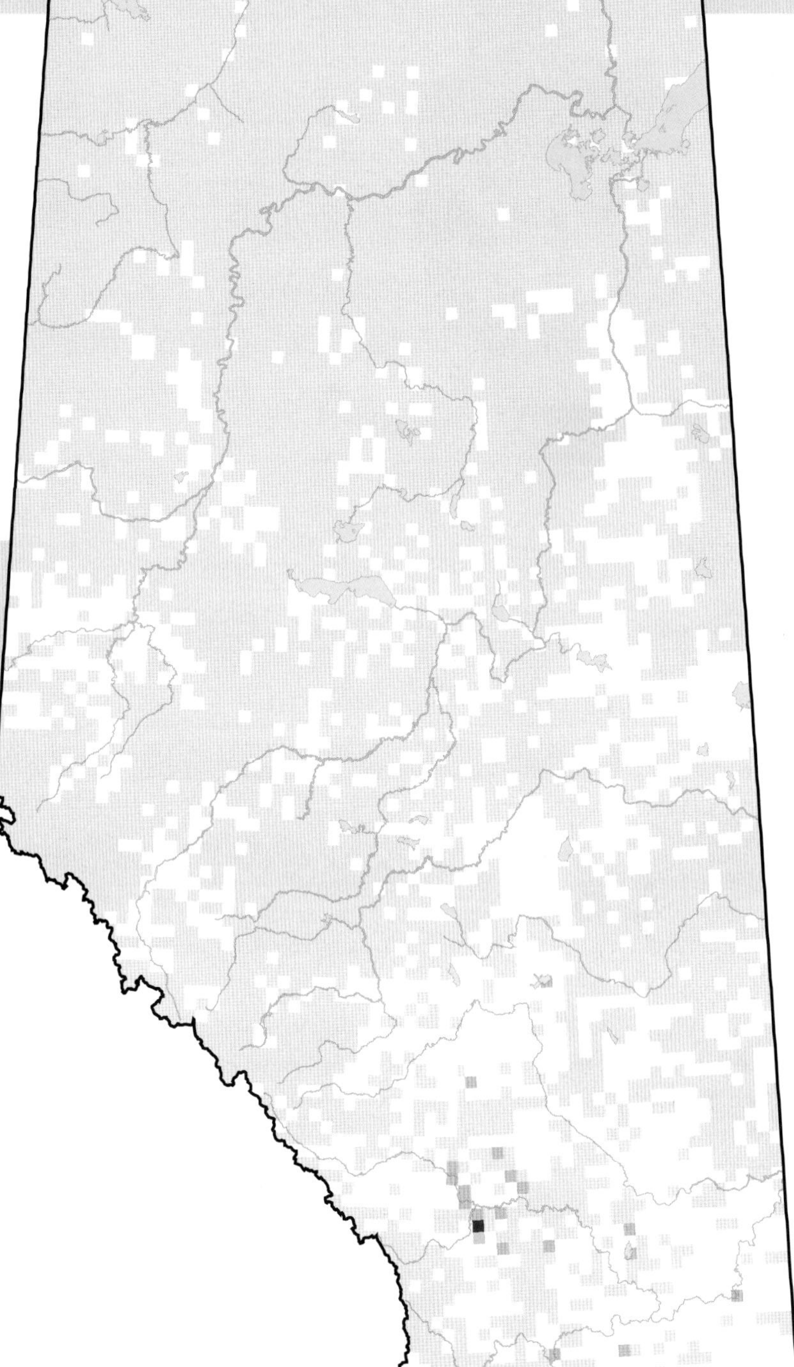

OBSERVED DISTRIBUTION ATLAS 1

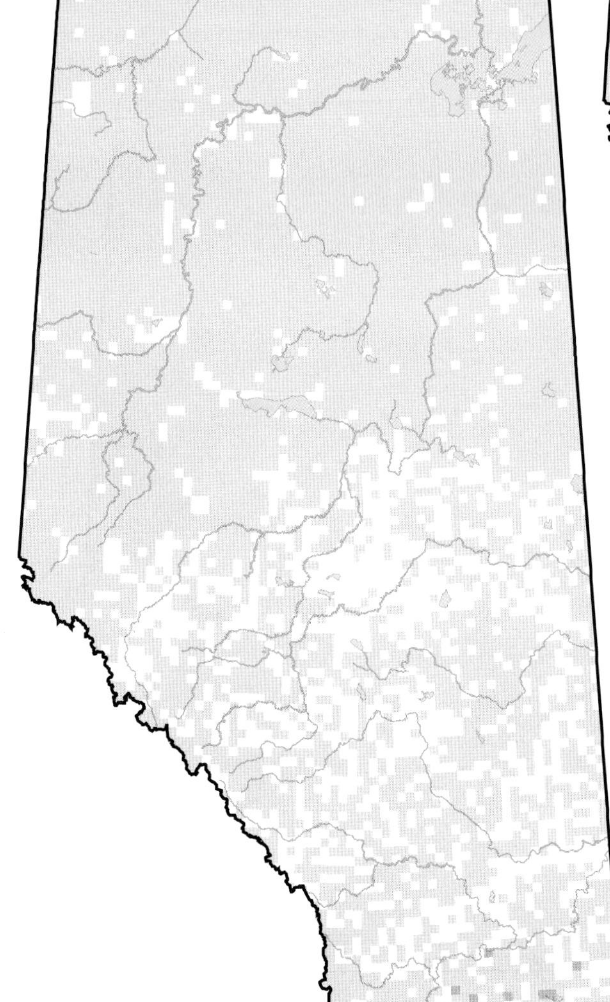

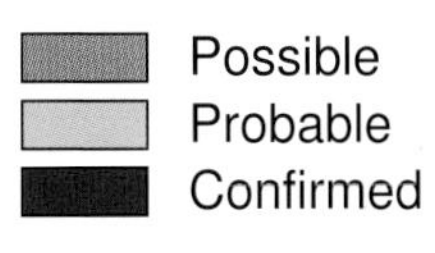

Turkey Vulture (*Cathartes aura*)

Photo: Randy Jensen

NESTING

CLUTCH SIZE (EGGS):	2
INCUBATION (DAYS):	38–41
FLEDGING (DAYS):	63
NEST HEIGHT (METRES):	0

Alberta's only vulture, the Turkey Vulture is generally found in the southeastern part of Alberta in the Grassland, Parkland, and Rocky Mountain (Cypress Hills) Natural Regions. Range expansion has been documented in other parts of its range outside Alberta, and expansion was also observed within Alberta between Atlas 1 and Atlas 2. Turkey Vultures were observed farther north and west during Atlas 2.

Turkey Vultures feed on carcasses which they locate using their highly developed olfactory organ while soaring over open areas and fragmented forests. This species roosts and nests in trees, on cliffs, and in abandoned buildings. For this reason, the Turkey Vulture was common in the Grassland and Rocky Mountain (Cypress Hills) NRs. Turkey Vultures were detected less often in the Boreal Forest, Foothills, and Parkland NRs. Occurrence in these last three natural regions could increase in the future if the range expansion of this species should continue.

An increase in relative abundance was detected in the Boreal Forest NR where, relative to other species, the Turkey Vulture was observed more frequently in Atlas 2 than in Atlas 1. The Breeding Bird Survey sample size of Turkey Vulture records was too small to investigate abundance changes in Alberta, although population increases were detected for Canada during the period 1985–2005. Abundance increases are probably related to the northern range expansion of this species. The status of this species was amended from Secure, as assigned in 2000, to Sensitive in 2005, due to its small population and the few documented breeding records in Alberta.

TEMPORAL REPORTING RATE

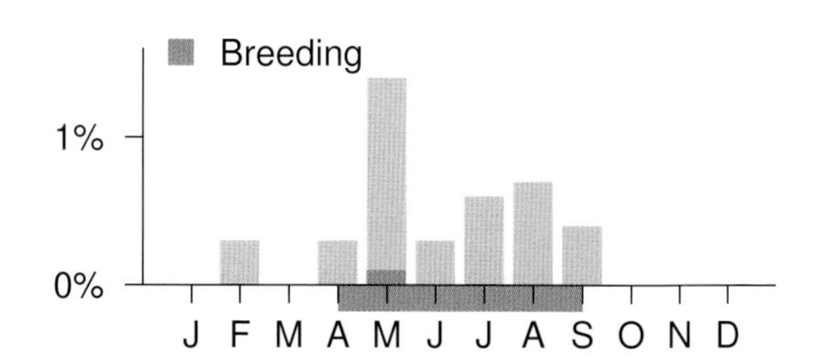

NATURAL REGIONS REPORTING RATE

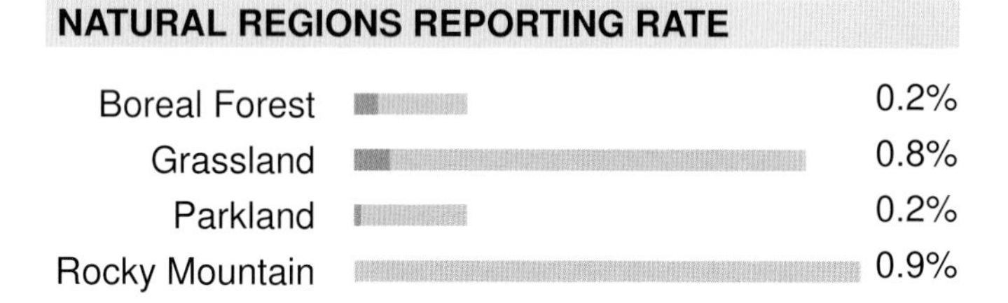

SPATIAL REPORTING RATE

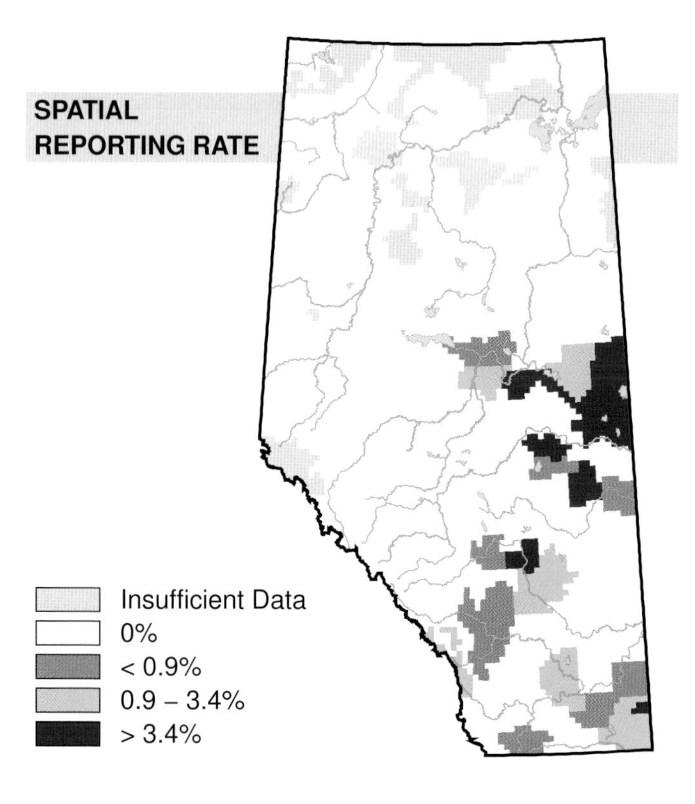

STATUS

GSWA 2000	GSWA 2005	COSEWIC Historic Rankings	COSEWIC 2007	Alberta Wildlife Act
Secure	Sensitive	N/A	N/A	N/A

Habitat 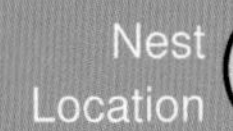 Nest Location   Nest Type  Diet  

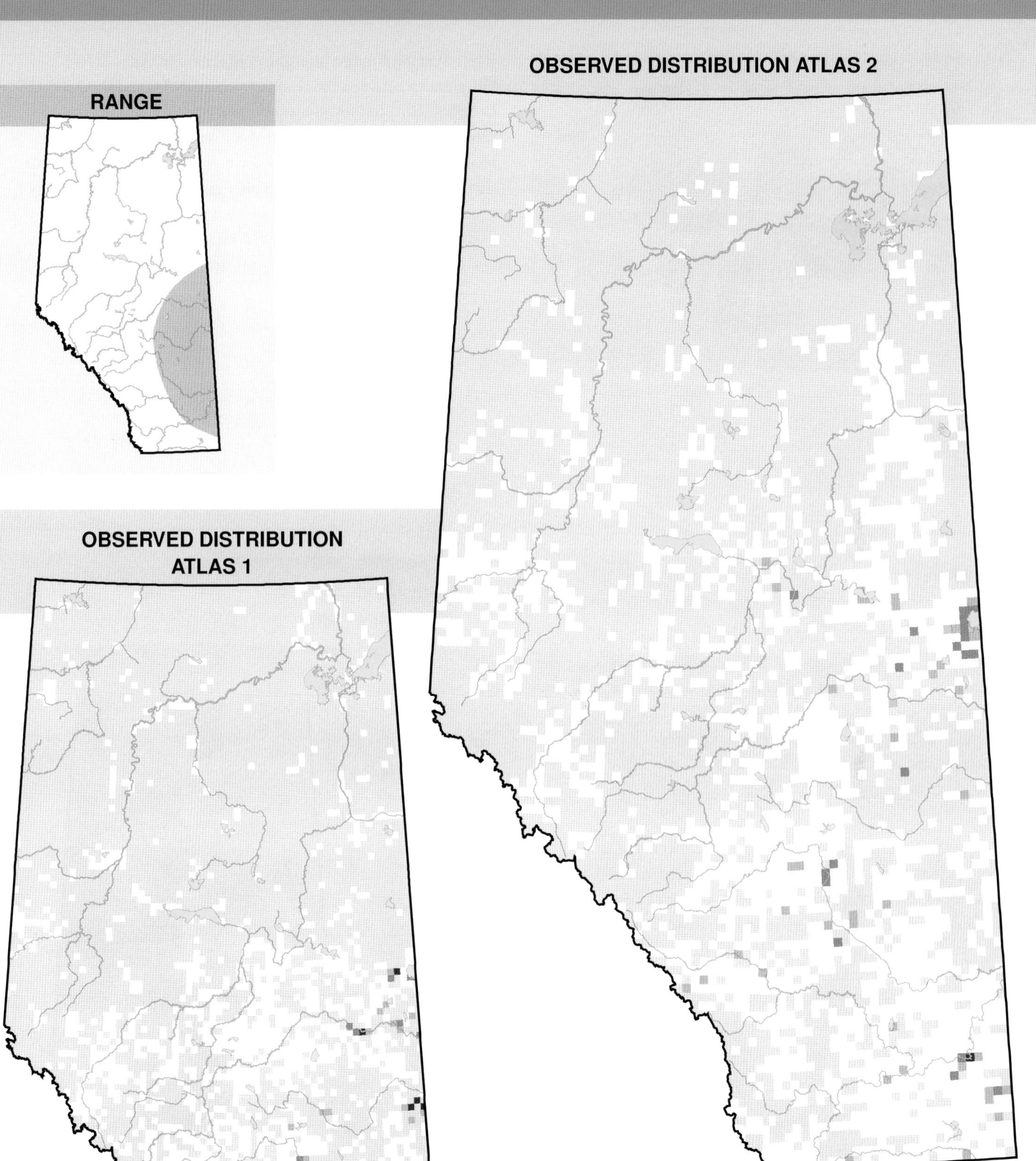

Osprey (*Pandion haliaetus*)

Photo: Mark Williams

NESTING

CLUTCH SIZE (EGGS):	2–4
INCUBATION (DAYS):	32–33
FLEDGING (DAYS):	51–59
NEST HEIGHT (METRES):	0–61

The Osprey is found in every Natural Region across the province. There was no change observed in the distribution of this species between Atlas 1 and Atlas 2.

Osprey are fish specialists and consequently their breeding areas are always associated with water bodies such as lakes and rivers. The distribution of this species reflects its need for living near water. The highest reported population densities occurred around Cold Lake, Lac La Biche, and Slave Lake, and along the Athabasca, Red Deer, and North Saskatchewan rivers.

Increases in relative abundance were detected in the Boreal Forest, Grassland, Parkland, and Rocky Mountain NRs where, relative to other species, Ospreys were observed more frequently in Atlas 2 than in Atlas 1. Changes in relative abundance were not detected in the Foothills NR. The Breeding Bird Survey did not find an abundance change in Alberta. However, an abundance increase was detected across Canada during the interval 1968–2005. Increases in Osprey abundance could be related to a population expansion of this species after the banning of DDT use in North America. The Atlas-detected increase in the Rocky Mountain NR may be due in part to the more intensive coverage of raptors in Atlas 2 in this NR; this was made possible by the inclusion of data from the Rocky Mountain Eagle Research Foundation, which initiated the Eagle Watch program after the completion of Atlas 1. Differences between the results from the Atlas and the Alberta Breeding Bird Survey analyses could be attributed to small sample size in the Breeding Bird Survey for this species. The provincial status of the Osprey is Sensitive in Alberta due to its small population and threats to habitat, especially nesting sites.

TEMPORAL REPORTING RATE

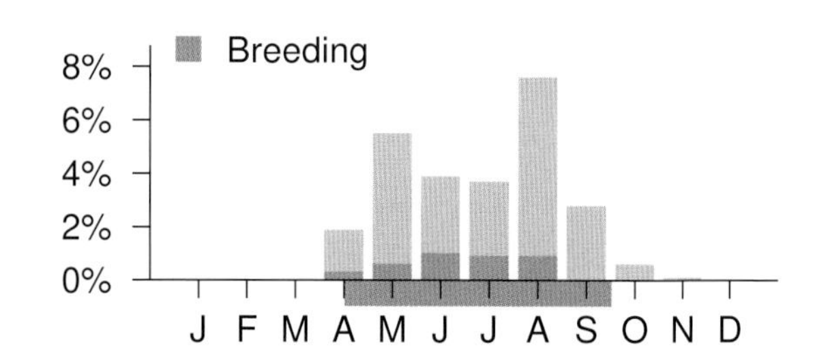

NATURAL REGIONS REPORTING RATE

Boreal Forest	1.6%
Foothills	3.6%
Grassland	2.2%
Parkland	1.4%
Rocky Mountain	8.5%

SPATIAL REPORTING RATE

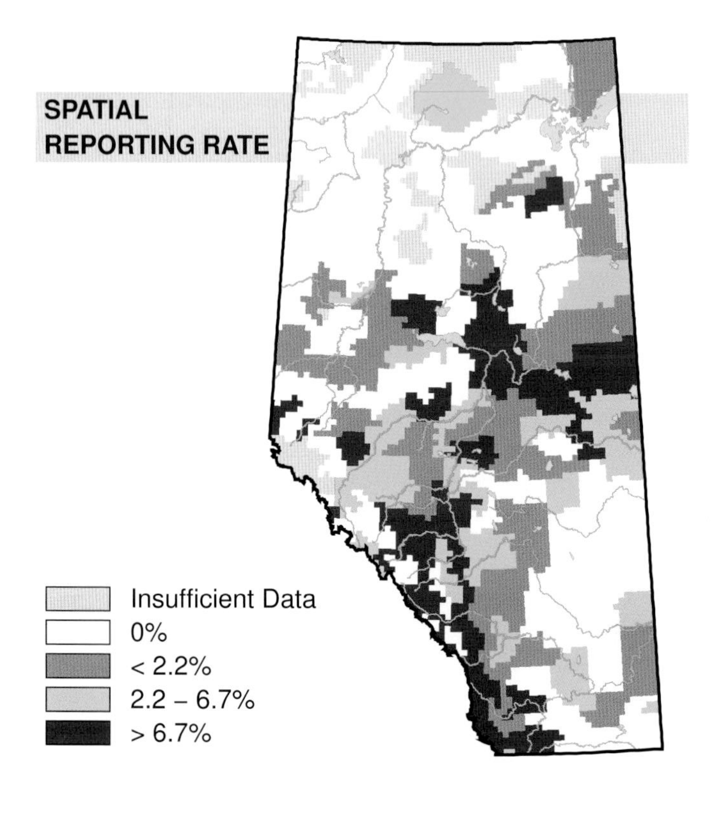

STATUS

GSWA 2000	GSWA 2005	COSEWIC Historic Rankings	COSEWIC 2007	Alberta Wildlife Act
Sensitive	Sensitive	N/A	N/A	N/A

RANGE

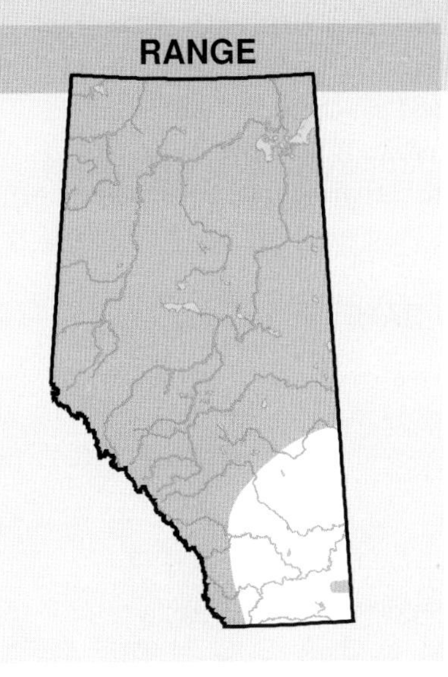

OBSERVED DISTRIBUTION ATLAS 2

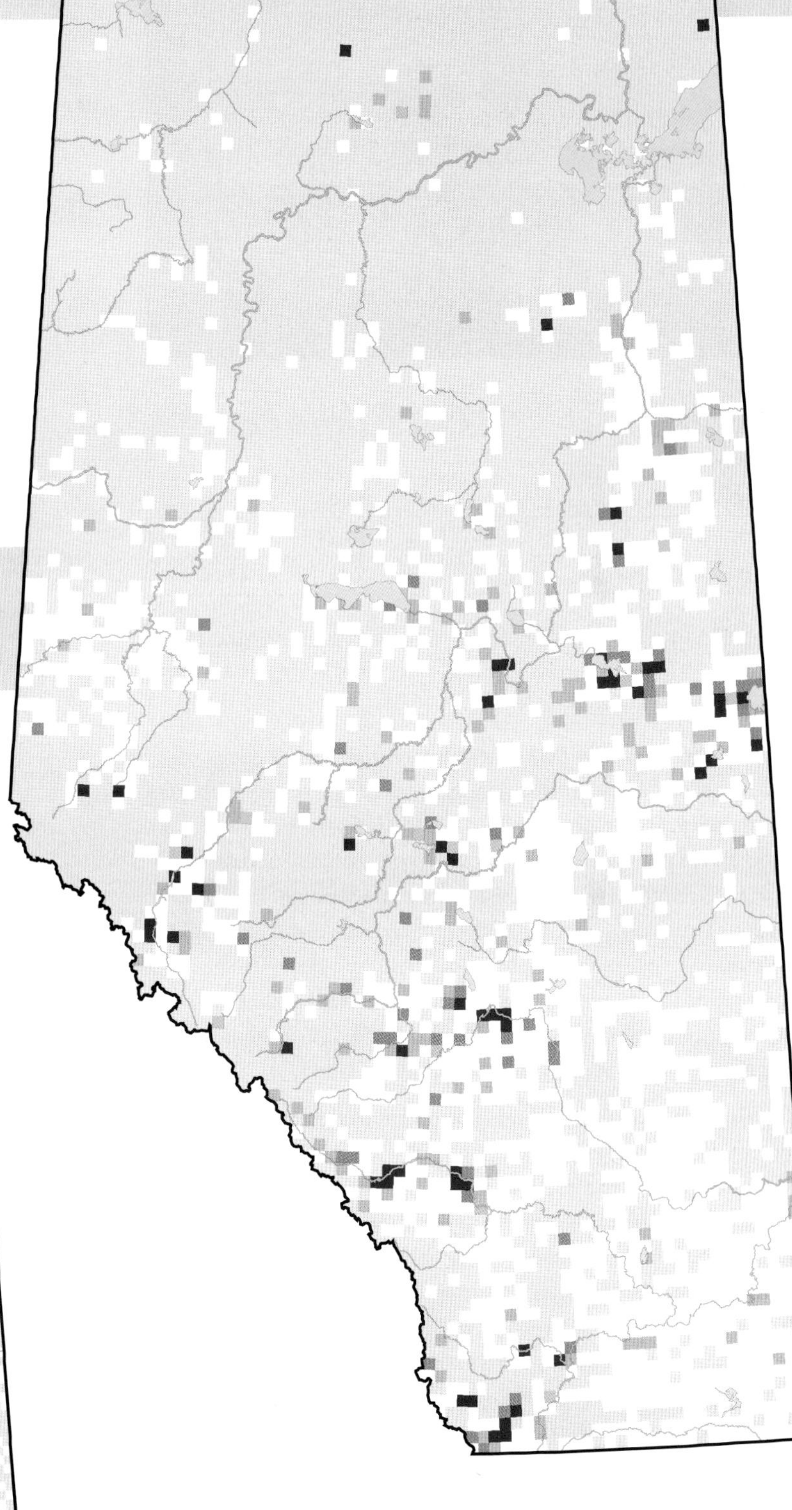

OBSERVED DISTRIBUTION ATLAS 1

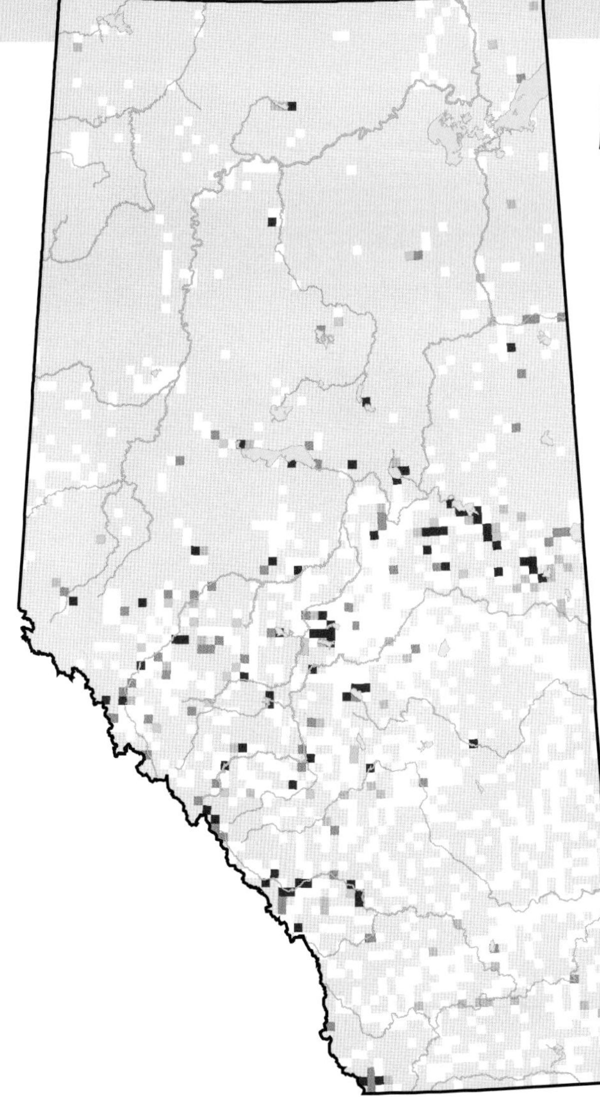

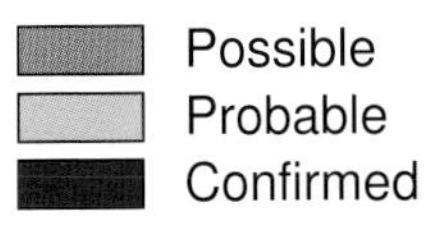

Bald Eagle (*Haliaeetus leucocephalus*)

Photo: Gordon Court

NESTING

CLUTCH SIZE (EGGS):	1–3
INCUBATION (DAYS):	43–35
FLEDGING (DAYS):	70–77
NEST HEIGHT (METRES):	3–54.8

The Bald Eagle is found in every Natural Region in Alberta. No change in distribution was observed between Atlas 1 and Atlas 2 for this species.

Bald Eagles typically nest in mature trees along the edges of forests, usually associated with large bodies of water, such as lakes, but nests are occasionally found along rivers as well. Bald Eagle distribution was not evenly distributed across the province. Highest record densities occurred around large lakes such as Cold Lake, Lac La Biche, Slave Lake and Lake Claire and along rivers such as Athabasca River and Peace River. The high reporting rate in the Rocky Mountain NR is in part a function of including considerable survey data contributed by the Rocky Mountain Eagle Research Foundation.

Increases in relative abundance were detected in the Boreal Forest, Grassland, Parkland, and Rocky Mountain NRs. Declines were detected in the Foothills NR. Differences in observation frequency relative to other species between Atlas 1 and 2 mirrored these changes. The number of Bald Eagle records in the Breeding Bird Survey was too small to detect changes in abundance in Alberta. Data from the BBS indicate that Bald Eagle abundance increased in Canada during the periods 1995–2005 and 1968–2005. Similar to Peregrine Falcon populations, Bald Eagle populations were negatively impacted by pesticide use. Increases in Bald Eagle populations in the Boreal Forest, Grassland, Parkland, and Rocky Mountain NRs could be related to a population expansion of this species after the banning of DDT use in North America. As this species is sensitive to human disturbance (Buehler, 2000), the most likely causes of the decline detected in the Foothills NR are intensive forestry and oil and gas activities since Atlas 1. The provincial status of this species is Sensitive due to its having been at risk in past and, currently, to its vulnerability to disturbance at nest sites.

TEMPORAL REPORTING RATE

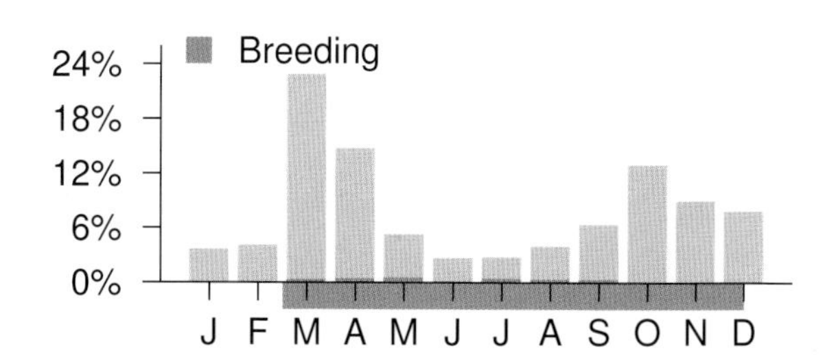

NATURAL REGIONS REPORTING RATE

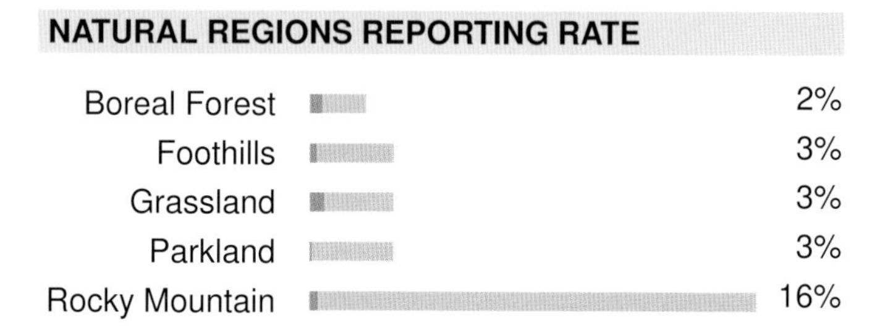

SPATIAL REPORTING RATE

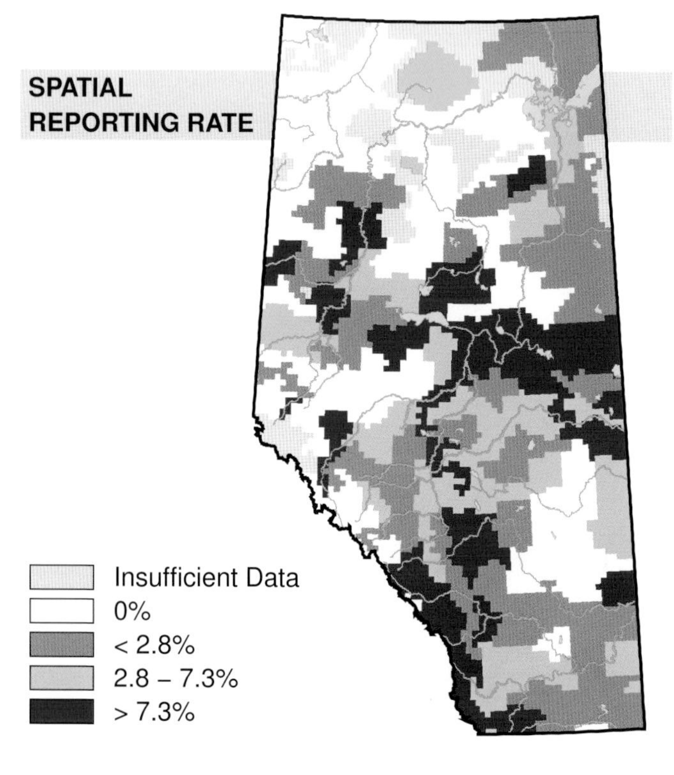

STATUS

GSWA 2000	GSWA 2005	COSEWIC Historic Rankings	COSEWIC 2007	Alberta Wildlife Act
Sensitive	Sensitive	Not At Risk	Not At Risk	N/A

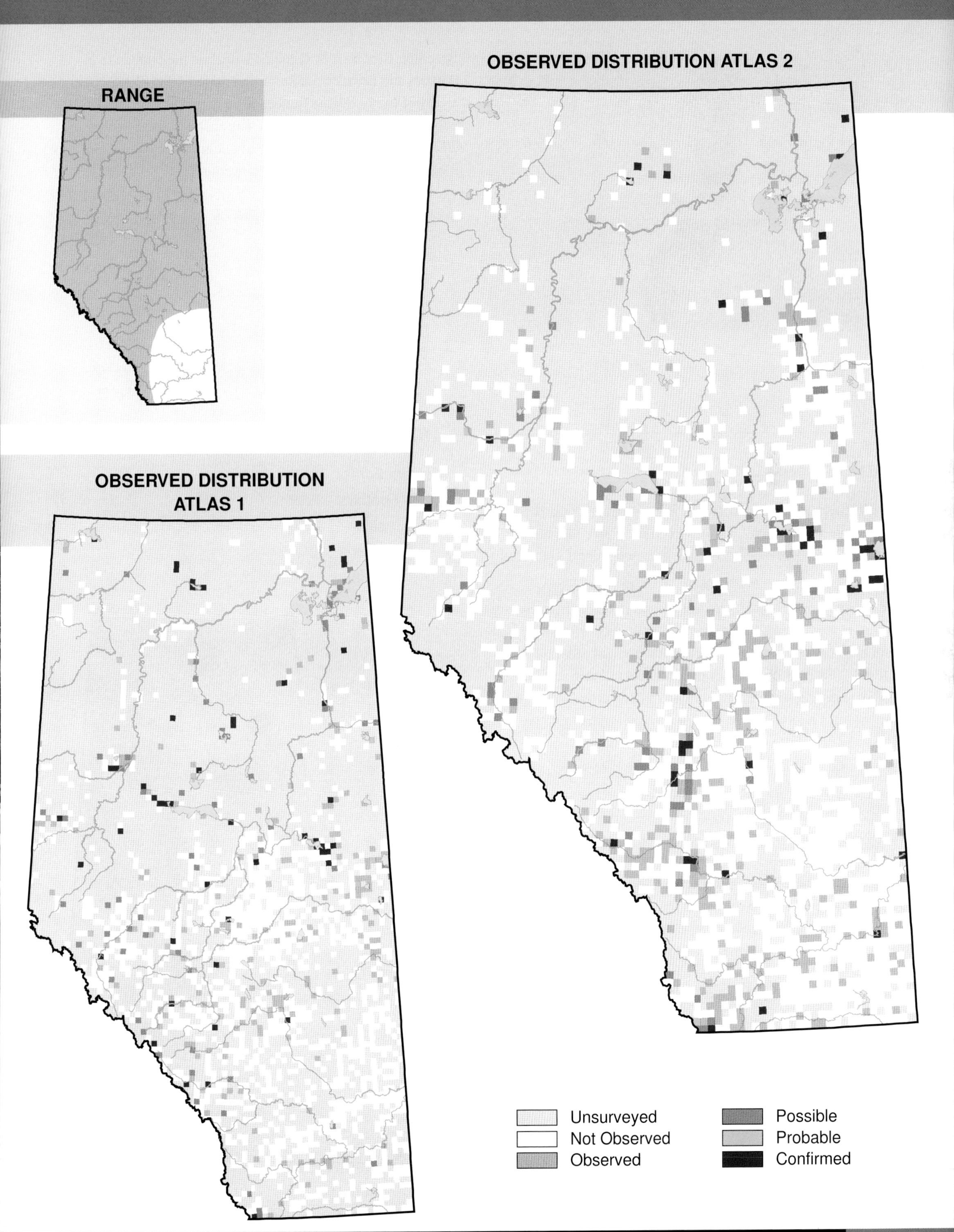
Habitat
Nest Location
Nest Type
Diet
OBSERVED DISTRIBUTION ATLAS 2
RANGE
OBSERVED DISTRIBUTION ATLAS 1
Unsurveyed
Not Observed
Observed
Possible
Probable
Confirmed

Northern Harrier (*Circus cyaneus*)

Photo: Gerald Romanchuk

NESTING

CLUTCH SIZE (EGGS):	4–6
INCUBATION (DAYS):	31–41
FLEDGING (DAYS):	30–35
NEST HEIGHT (METRES):	0

The Northern Harrier, also known as the Marsh Hawk, is found throughout Alberta in all Natural Regions. There was no change in the observed distribution between Atlas 1 and Atlas 2.

The Northern Harrier nests on the ground in open areas that are usually near wetlands or marshy meadows. Reporting rates were high in the Grassland NR, and lower in all other natural regions. This could be due, in part, to easier detection of this species when these birds are hunting in open areas. Also, there is more open habitat and shallow wetlands available in the Grassland NR for this species to use for nesting.

Increases in relative abundance were detected in the Grassland and Rocky Mountain NRs where, relative to other species, the Northern Harrier was observed more frequently in Atlas 2 than in Atlas 1. Decreases in relative abundance were detected in the Boreal Forest, Parkland, and Foothills NRs. With drier conditions in the north and wetter conditions in the south in Atlas 2, these changes may reflect a population that is shifting spatially in response to changing climatic conditions, rather than an actual change in the size of the population. It is possible, however, that the more extensive coverage of the Grassland NR in Atlas 2 and the inclusion of raptor-specific data from the Rocky Mountain Eagle Research Foundation in Atlas 2 are partly responsible for the Atlas results. The Breeding Bird Survey detected a decline in abundance in Alberta and across Canada for the period 1985–2005. Declines are most likely related to decreased availability of wetland habitats. The provincial status was changed from Secure in 2000 to Sensitive in 2005, due to declines across much of this species' range and threats to the wetlands.

TEMPORAL REPORTING RATE

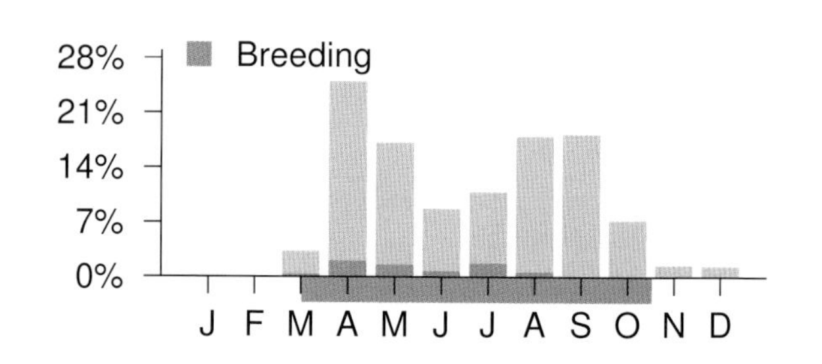

NATURAL REGIONS REPORTING RATE

SPATIAL REPORTING RATE

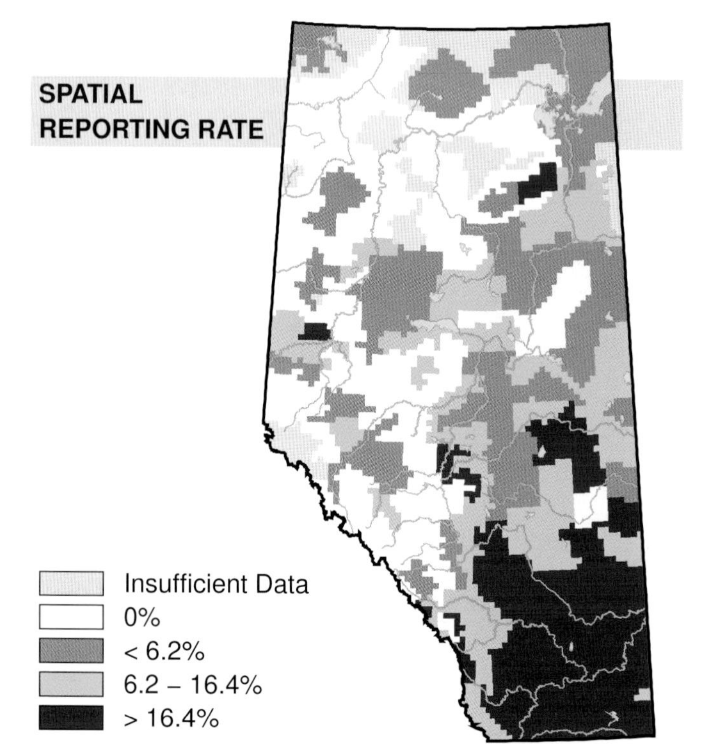

STATUS

GSWA 2000	GSWA 2005	COSEWIC Historic Rankings	COSEWIC 2007	Alberta Wildlife Act
Secure	Sensitive	Not At Risk	Not At Risk	N/A

RANGE

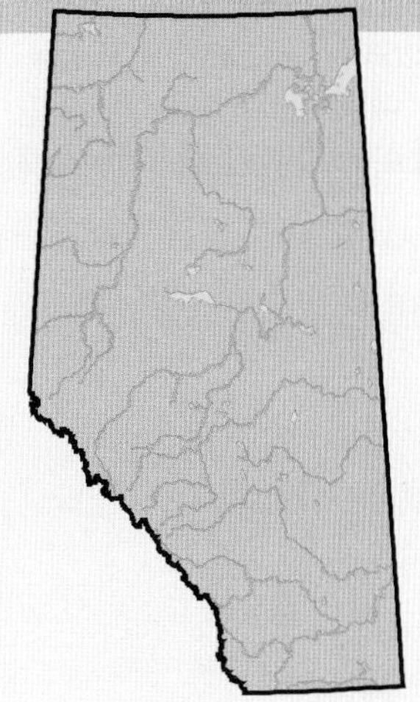

OBSERVED DISTRIBUTION ATLAS 2

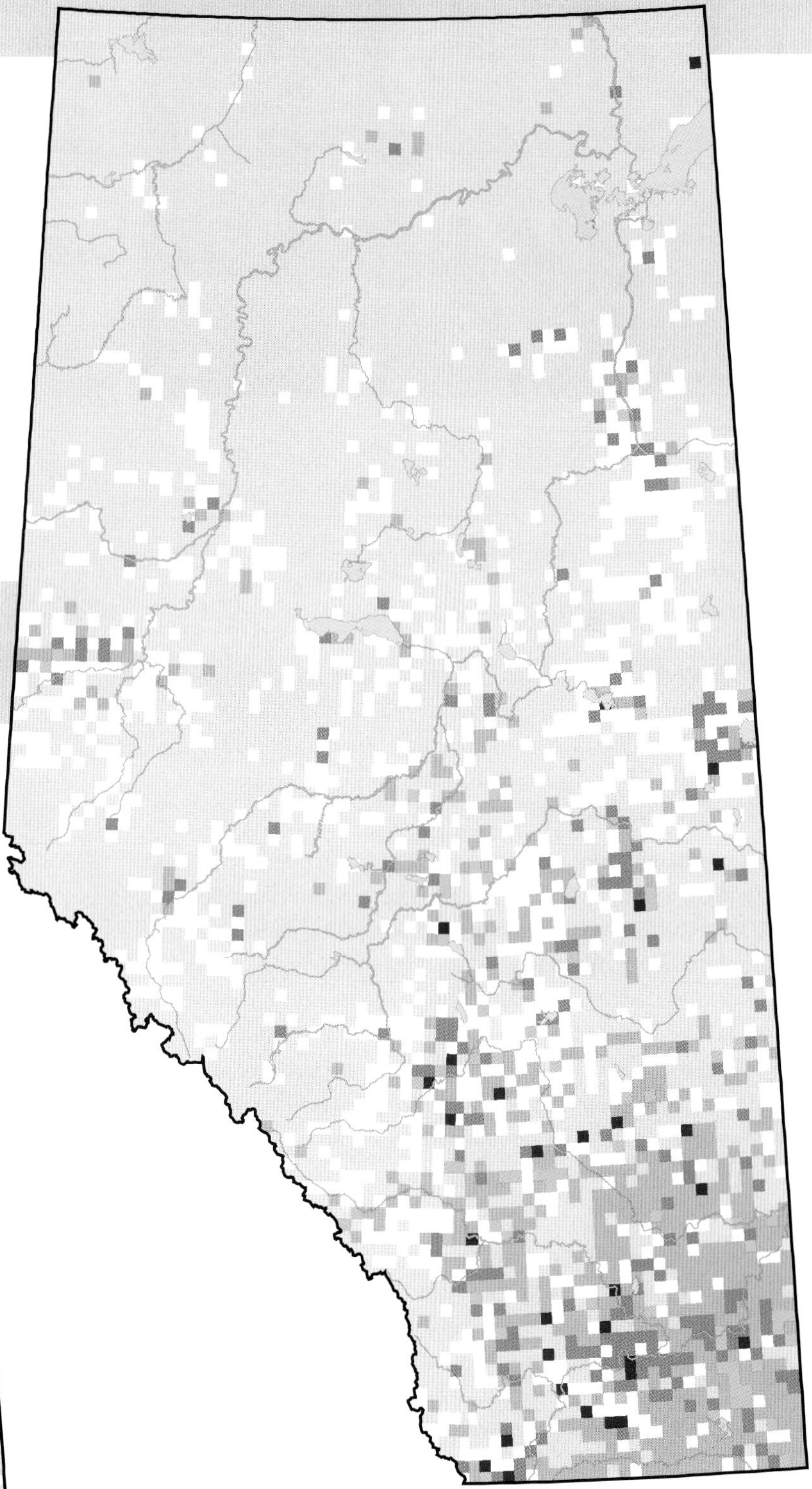

OBSERVED DISTRIBUTION ATLAS 1

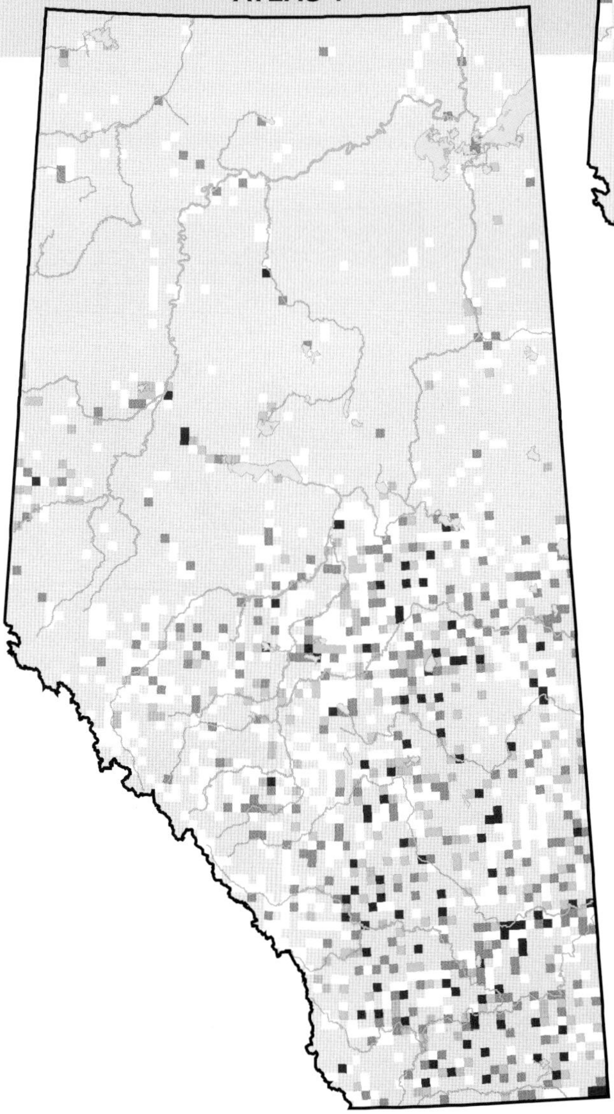

Unsurveyed
Not Observed
Observed
Possible
Probable
Confirmed

Sharp-shinned Hawk (*Accipiter striatus*)

Photo: Royal Alberta Museum

NESTING

CLUTCH SIZE (EGGS):	4–5
INCUBATION (DAYS):	34–35
FLEDGING (DAYS):	23–25
NEST HEIGHT (METRES):	3–21

The Sharp-shinned Hawk is Alberta's smallest accipiter, and it is found throughout most of the province in all Natural Regions. It is absent from southeastern Alberta except for the Cypress Hills. The distribution of this species was similar between Atlas 1 and Atlas 2.

This woodland hawk prefers dense rather than open forests and is rarely detected during the breeding season; it is considered the most difficult accipiter to census. Reporting rates were fairly uniform in the Boreal Forest, Grassland, Foothills, and Parkland NRs. High reporting rates in the Rocky Mountain NR were likely the result of the incorporation of migrational data from the Rocky Mountain Eagle Research Foundation, a group that started their raptor-targeted monitoring program in the Kananaskis area after the completion of Atlas 1. Grassland records were primarily in riparian areas where suitable habitat exists.

Declines in relative abundance were detected in the Boreal Forest and Foothills NRs. These declines may be a result either of changes in prey populations related to drier conditions in Atlas 2, or of reduced availability of suitable nesting sites resulting from resource extraction and development. Increases in relative abundance were detected in the Grassland and Parkland NRs. There is no clear biological cause for these increases. The majority of records in the Grassland and Parkland NR were observations of non-breeders, suggesting these increases could be a function of migratory variation and not an indication of abundance change. The Breeding Bird Survey data detected no change in abundance in Alberta or across Canada for the period 1985–2005, but sample size was small. Further research is needed to assess the Atlas-detected changes in abundance for this species. The provincial status of the Sharp-shinned Hawk is Secure.

TEMPORAL REPORTING RATE

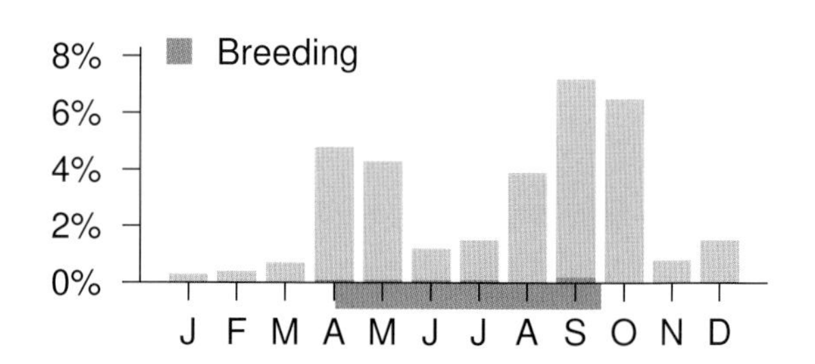

NATURAL REGIONS REPORTING RATE

Region	Rate
Boreal Forest	1.6%
Foothills	1.4%
Grassland	2.2%
Parkland	1.4%
Rocky Mountain	4.4%

SPATIAL REPORTING RATE

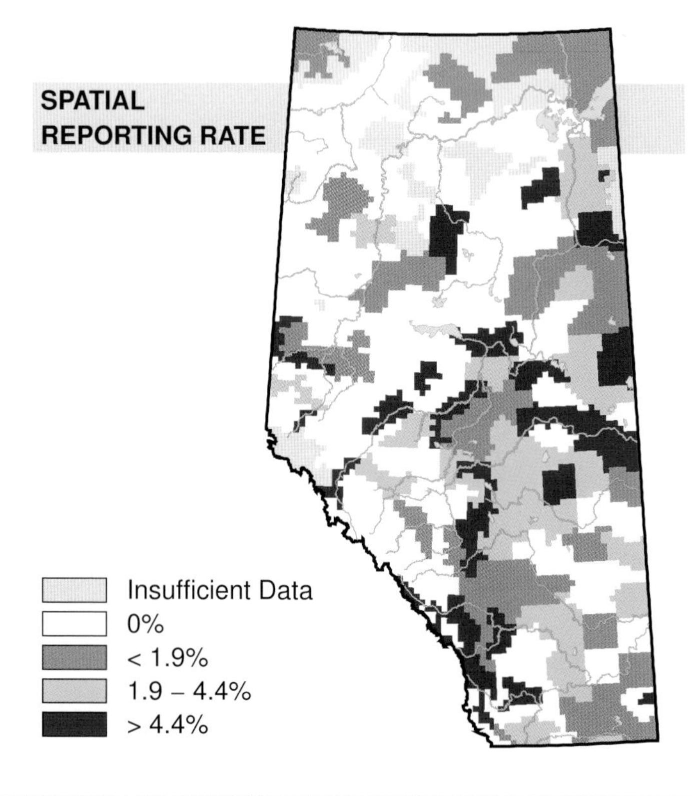

STATUS

GSWA 2000	GSWA 2005	COSEWIC Historic Rankings	COSEWIC 2007	Alberta Wildlife Act
Secure	Secure	Not At Risk	Not At Risk	N/A

RANGE

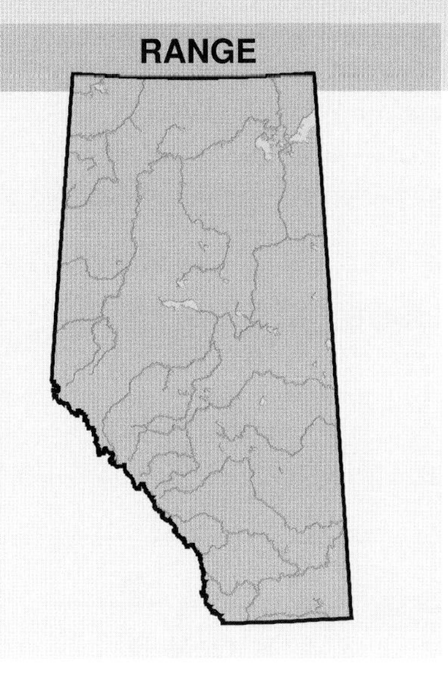

OBSERVED DISTRIBUTION ATLAS 2

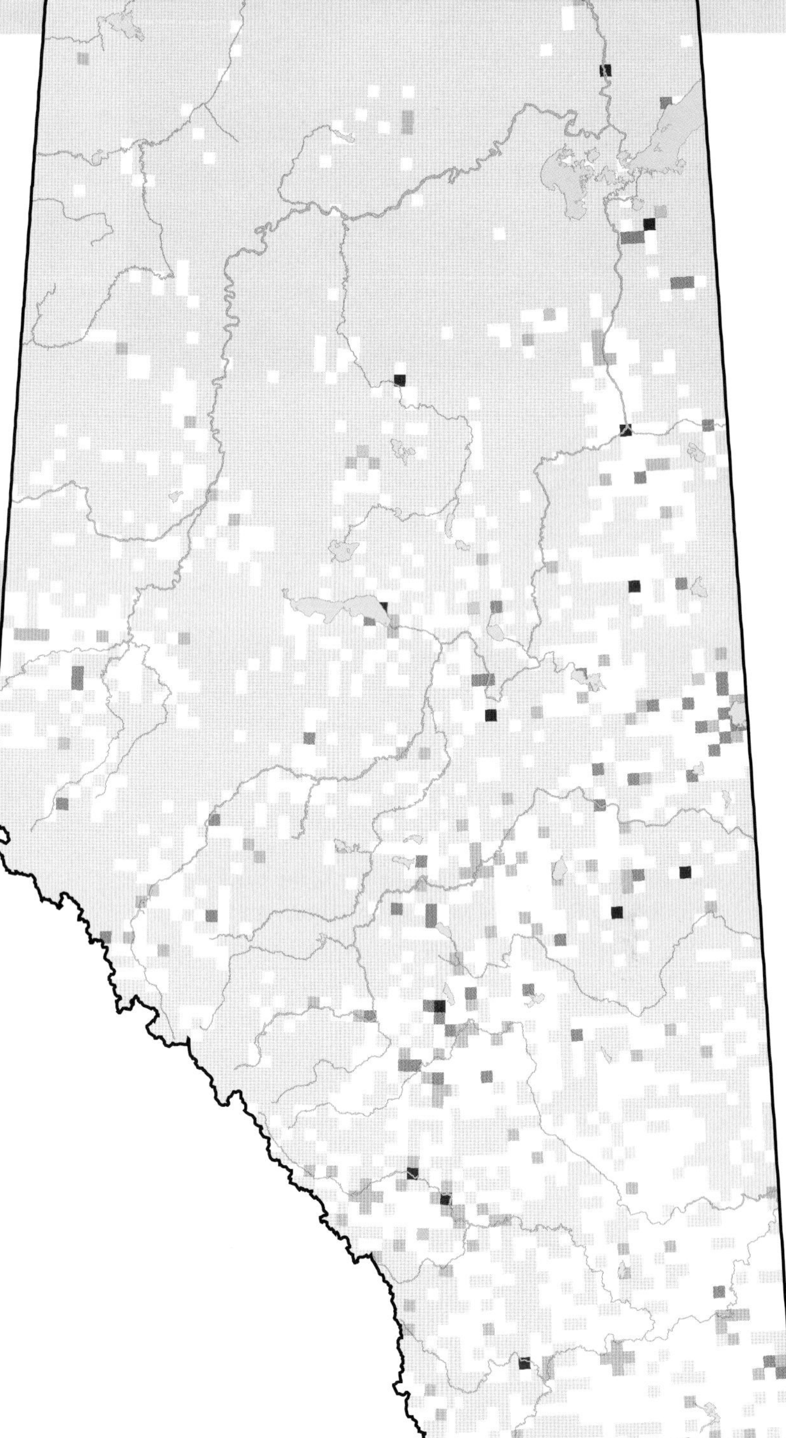

OBSERVED DISTRIBUTION ATLAS 1

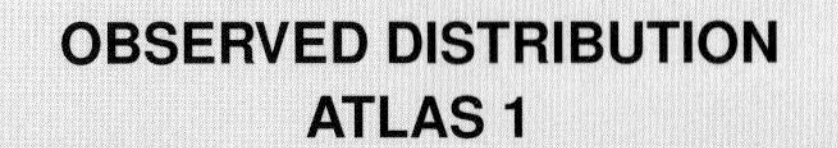

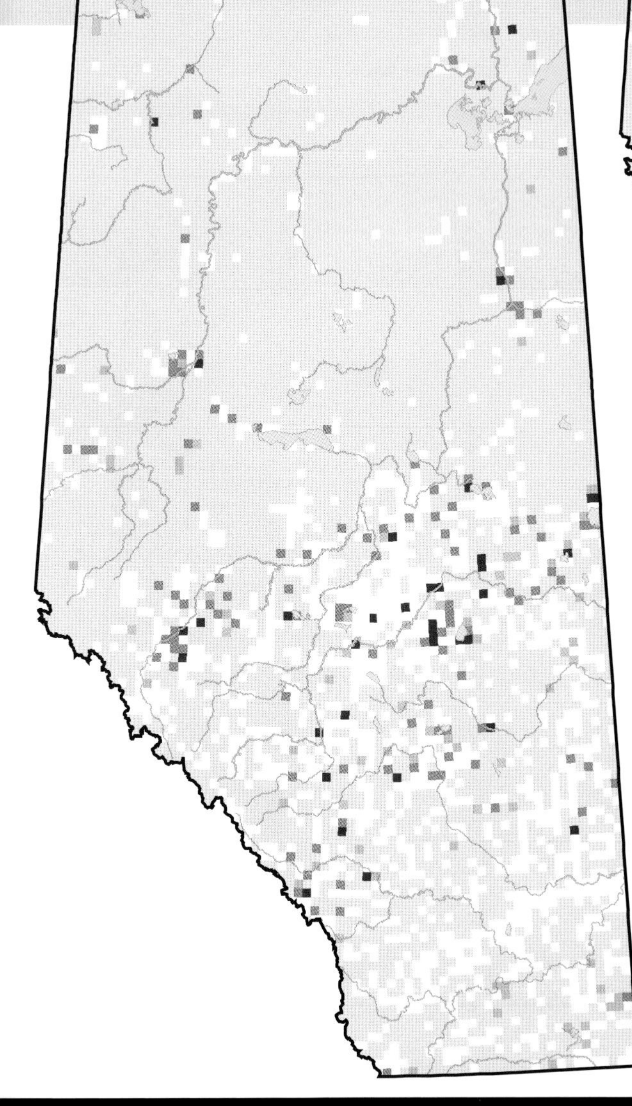

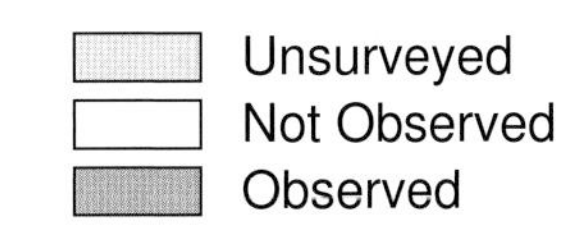

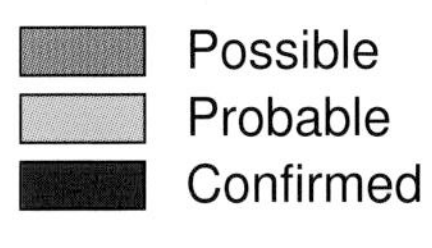

Cooper's Hawk (*Accipiter cooperii*)

Photo: Gordon Court

NESTING

CLUTCH SIZE (EGGS):	3–5
INCUBATION (DAYS):	34–36
FLEDGING (DAYS):	30–34
NEST HEIGHT (METRES):	6–20

Cooper's Hawk is distributed across central Alberta and south to the United States border. It ranges as far north as Slave Lake and Fort McMurray. No changes in distribution were observed between Atlas 1 and Atlas 2, except for two records in Atlas 1 from Wood Buffalo National Park with no matching records in Atlas 2.

This species was observed most frequently in the Rocky Mountain Natural Region. It was also observed—and recorded as breeding—in the Boreal Forest, Foothills, Grassland, and Parkland NRs. Large numbers were reported around urban centres like Edmonton and Calgary. This reflects the Cooper's Hawk's adaptability in using human-altered habitats, probably because of the abundant prey available. The low reporting rate in the Boreal Forest NR can be attributed to this hawk's distribution which does not range far north of 54 degrees latitude across Canada.

A decline in relative abundance was detected for this species in the Boreal Forest NR where, relative to other species, it was observed less frequently in Atlas 2 than in Atlas 1. This change is surprising as there is no apparent biological cause for this decrease. The Breeding Bird Survey did not detect any change in abundance of Cooper's Hawk in Alberta, or across Canada. Without corroboration or known biological cause, further research is needed to evaluate the Atlas-detected decline in the Boreal Forest NR. No change was detected in the other NRs where, relative to other species, the bird was observed less frequently in Atlas 2 than in Atlas 1 in all but the Rocky Mountain NR. The increased observation rate in the Rocky Mountain NR is likely due to data that were incorporated from the Rocky Mountain Eagle Research Foundation which monitors migrants in the Kananaskis area. The Foundation started its monitoring program after the completion of Atlas 1. This species is considered Secure in the province.

TEMPORAL REPORTING RATE

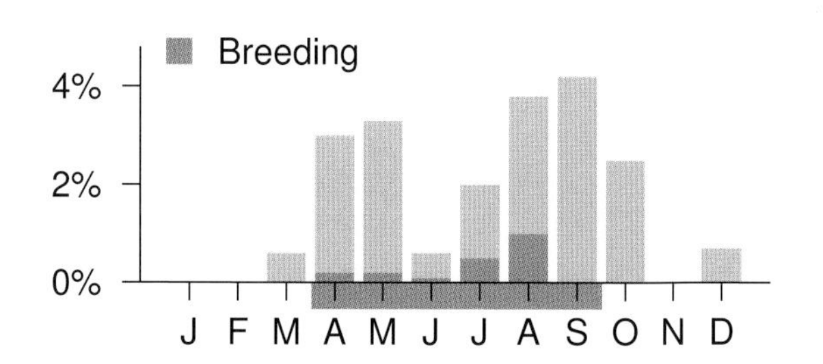

NATURAL REGIONS REPORTING RATE

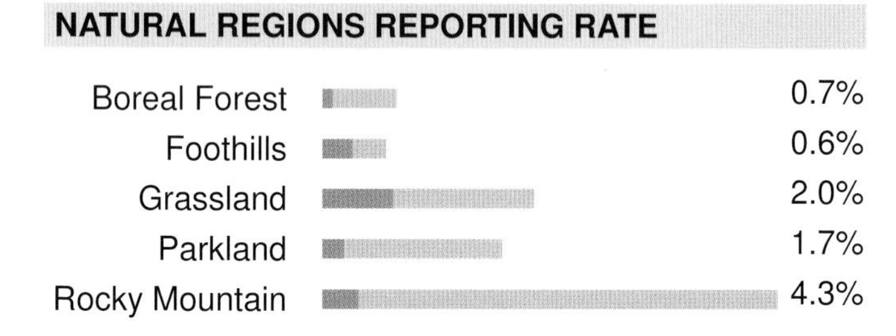

SPATIAL REPORTING RATE

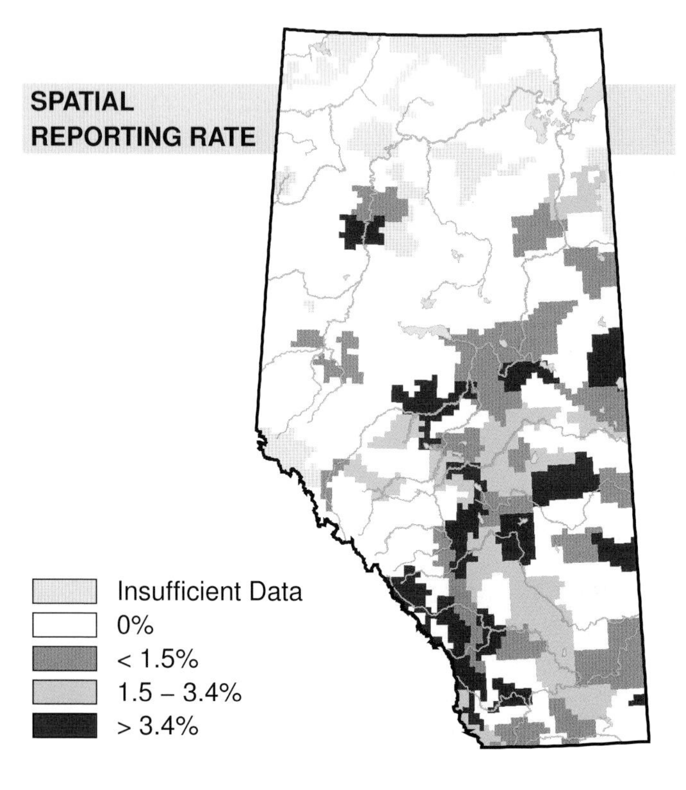

STATUS

GSWA 2000	GSWA 2005	COSEWIC Historic Rankings	COSEWIC 2007	Alberta Wildlife Act
Secure	Secure	Not At Risk	Not At Risk	N/A

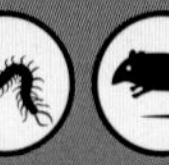

RANGE

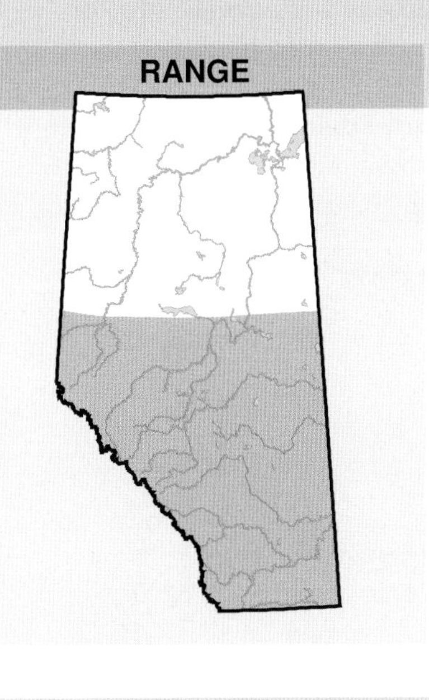

OBSERVED DISTRIBUTION ATLAS 2

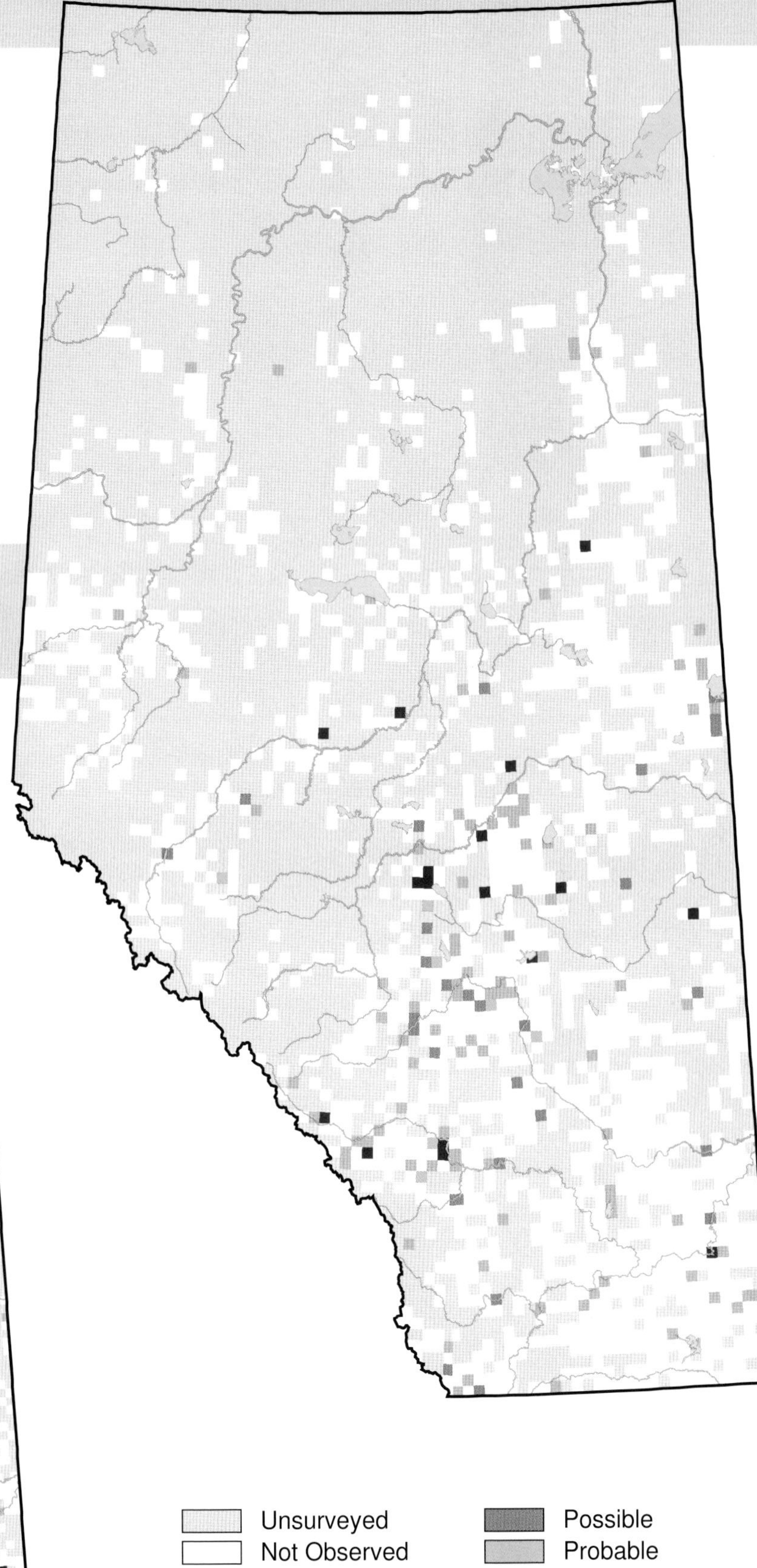

OBSERVED DISTRIBUTION ATLAS 1

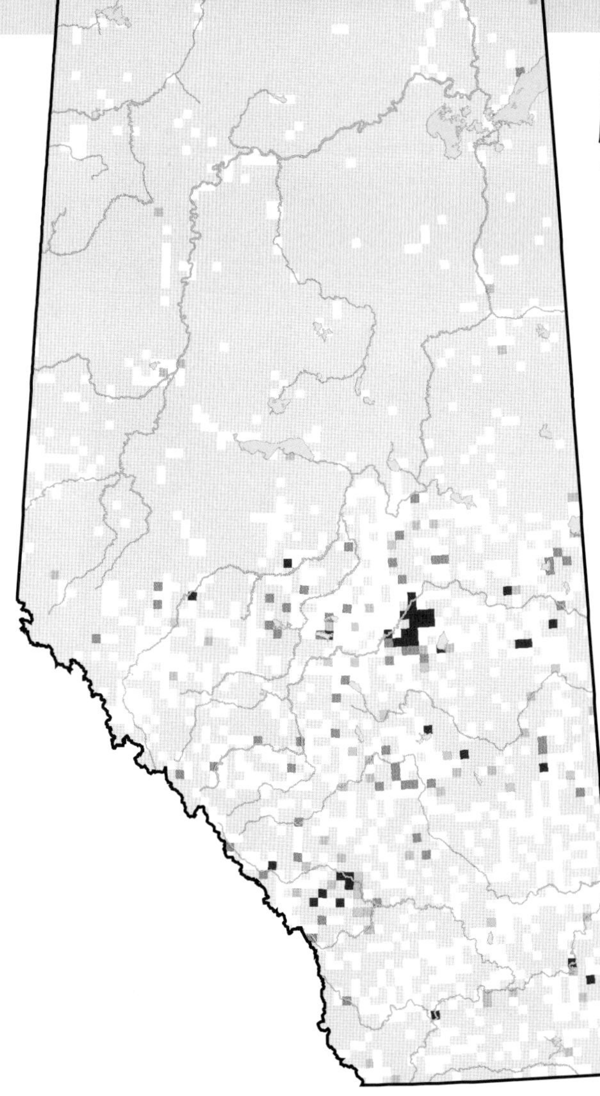

Unsurveyed
Not Observed
Observed
Possible
Probable
Confirmed

Northern Goshawk (*Accipiter gentilis*)

Photo: Gerald Romanchuk

NESTING

CLUTCH SIZE (EGGS):	2–4
INCUBATION (DAYS):	36–41
FLEDGING (DAYS):	40
NEST HEIGHT (METRES):	6–23

Alberta's largest accipiter, the Northern Goshawk is found throughout northern and central Alberta and south through the foothills. Although there were fewer records in the northeast, no change in distribution was observed between Atlas 1 and Atlas 2.

This species' secretive nature is reflected in low reporting rates in all Natural Regions except the Rocky Mountain NR. The Rocky Mountain Eagle Research Foundation shared their data with the Atlas 2 project, and the inclusion of those data is reflected in the noticeably higher reporting rate from that NR. The tendency of this species to nest in mature mixedwood forests with high canopy closure renders them almost absent from the Grassland and Parkland NRs.

Decreases in relative abundance were detected in the Boreal Forest and Foothills NRs where, relative to other species, this species was observed less frequently in Atlas 2 than in Atlas 1. The intense levels of industrial development in these regions since the time of Atlas 1 are most likely the cause of this decline; such activity reduces the amount of suitable breeding habitat for this species. No changes in relative abundance were detected in other NRs. The Breeding Bird Survey, being restricted to roadways, seldom detects this species; thus, the BBS is an inappropriate survey strategy for monitoring this species. No BBS trends were calculated for Alberta and no change in abundance was detected nationally. The Northern Goshawk is considered Sensitive in Alberta due its dependence on older-aged trees for nesting.

TEMPORAL REPORTING RATE

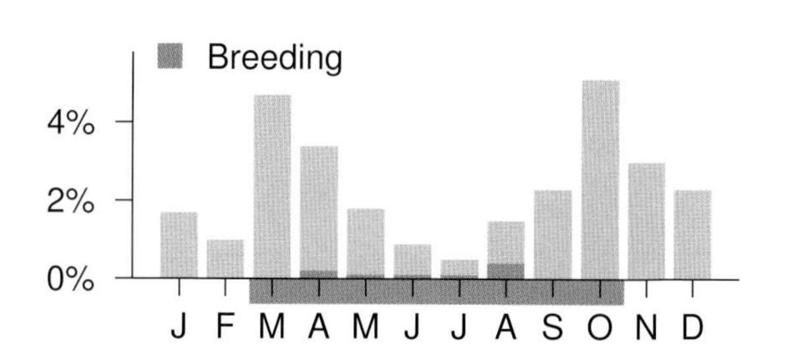

NATURAL REGIONS REPORTING RATE

Boreal Forest	0.8%
Foothills	1.8%
Grassland	0.6%
Parkland	0.4%
Rocky Mountain	6.7%

SPATIAL REPORTING RATE

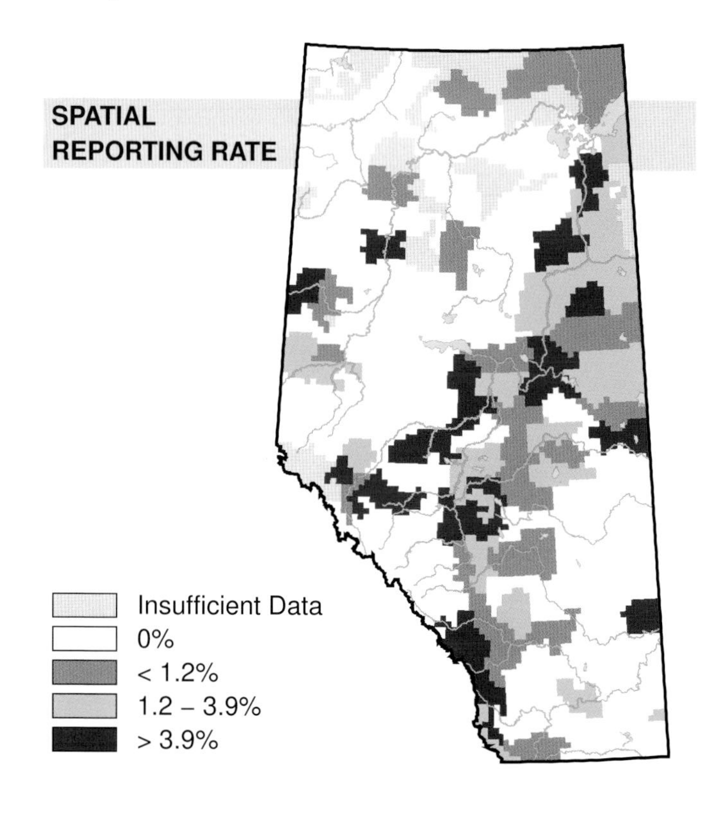

STATUS

GSWA 2000	GSWA 2005	COSEWIC Historic Rankings	COSEWIC 2007	Alberta Wildlife Act
Sensitive	Sensitive	Not At Risk	Not At Risk	N/A

OBSERVED DISTRIBUTION ATLAS 2

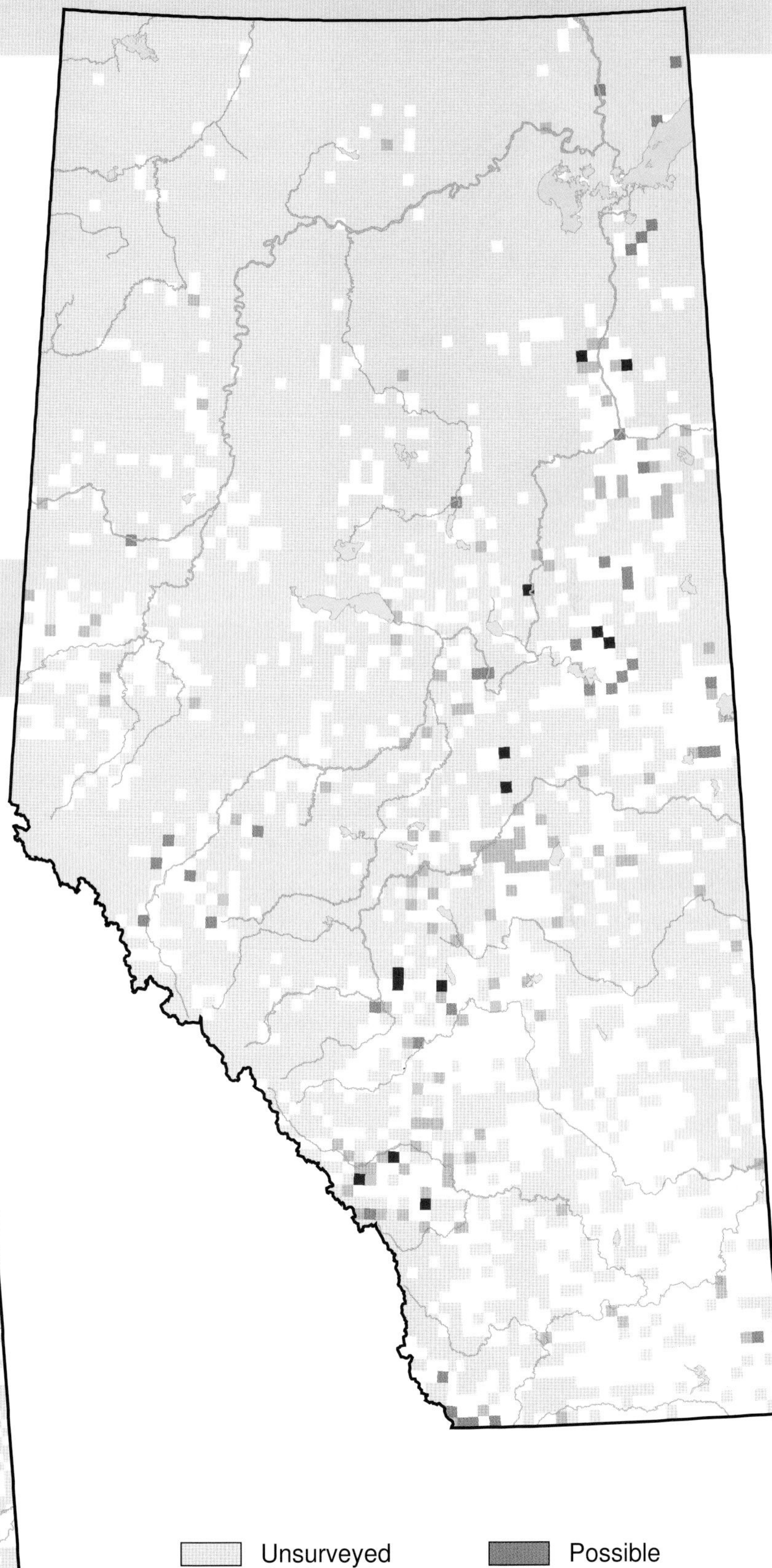

RANGE

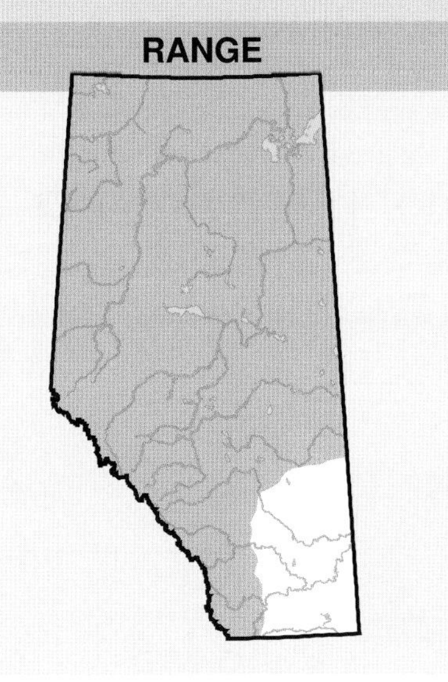

OBSERVED DISTRIBUTION ATLAS 1

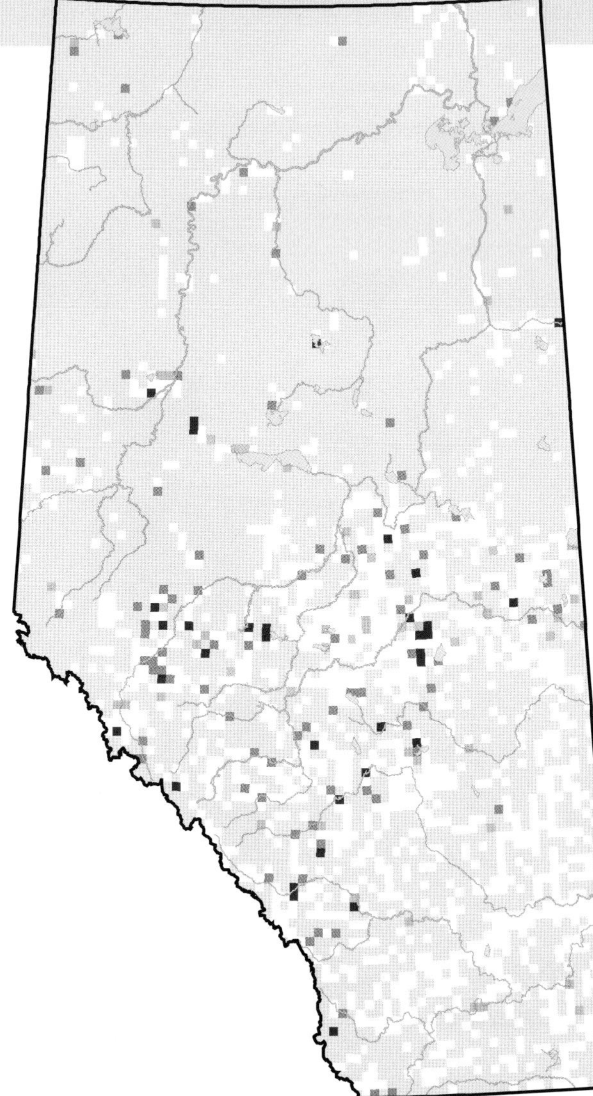

- Unsurveyed
- Not Observed
- Observed
- Possible
- Probable
- Confirmed

Broad-winged Hawk (*Buteo platypterus*)

Photo: Gerald Romanchuk

NESTING

CLUTCH SIZE (EGGS):	2–4
INCUBATION (DAYS):	23–38
FLEDGING (DAYS):	24–30
NEST HEIGHT (METRES):	6–12

Alberta's smallest buteo can be found in many parts of Alberta but it is generally considered rare in the province. There was no change observed in the distribution of the Broad-winged Hawk between Atlas 1 and Atlas 2.

Despite being found in every Natural Region in Alberta, Broad-winged Hawks were not evenly distributed across the province. Reporting rates were highest in the Rocky Mountain, Boreal Forest, and Foothills NRs, but lower in the Grassland and Parkland NRs. There were no probable or confirmed breeding records for this species in the Rocky Mountain NRs in Atlas 2, even though the reporting rate was the highest in that region. The patchy distribution of the Broad-winged Hawk is caused by the patchy distribution of mature and old-growth forests, the preferred breeding habitat of the species.

An increased relative abundance was detected in the Boreal Forest NR. This increase is surprising as this species is dependent on mature and old-growth forest which are in decline in Alberta. It is likely that this increase was a result of more complete northern survey coverage and not an indication of true population change. A decline was detected in the Parkland NR where, relative to other species, it was observed less frequently in Atlas 2 than in Atlas 1. No changes in relative abundance were detected in the other NRs. Breeding Bird Survey sample size was too small to investigate abundance change for the Broad-winged Hawk in Alberta, and no change was detected for this species in Canada. The provincial status of the Broad-winged Hawk is Sensitive due to the species' prefered breeding habitat of mature to old-growth forests experiencing reduced availability in the province.

TEMPORAL REPORTING RATE

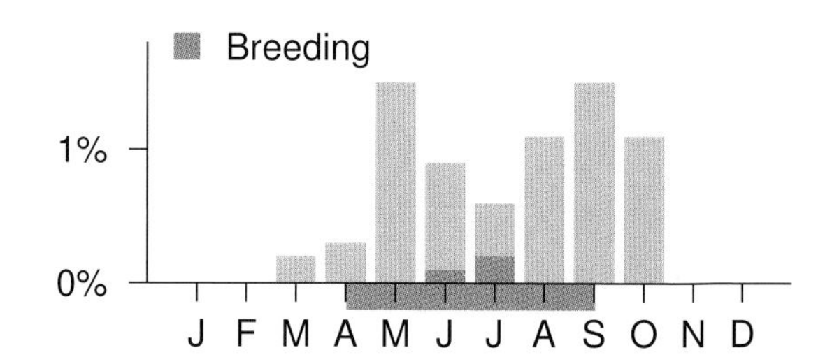

NATURAL REGIONS REPORTING RATE

Boreal Forest	0.9%
Foothills	0.7%
Grassland	0.4%
Parkland	0.3%
Rocky Mountain	1.1%

SPATIAL REPORTING RATE

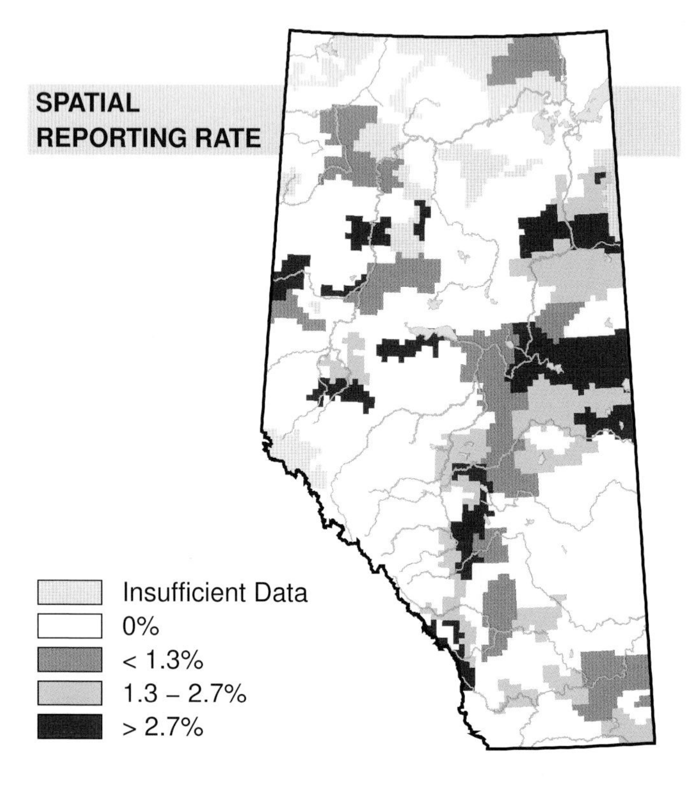

STATUS

GSWA 2000	GSWA 2005	COSEWIC Historic Rankings	COSEWIC 2007	Alberta Wildlife Act
Sensitive	Sensitive	N/A	N/A	N/A

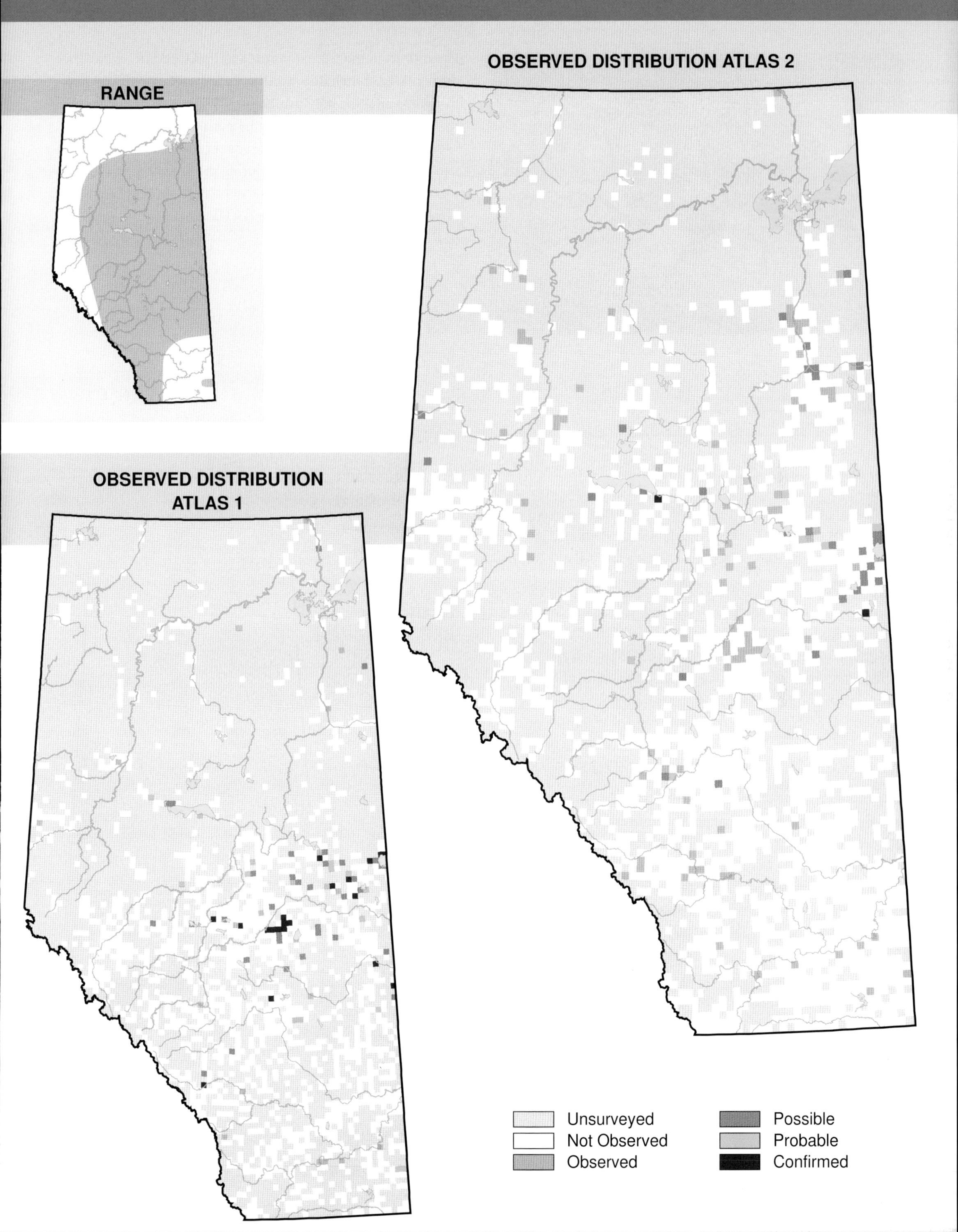
OBSERVED DISTRIBUTION ATLAS 2
RANGE
OBSERVED DISTRIBUTION ATLAS 1
Unsurveyed
Not Observed
Observed
Possible
Probable
Confirmed

Swainson's Hawk (*Buteo swainsoni*)

Photo: Jim Jacobson

NESTING

CLUTCH SIZE (EGGS):	2–3
INCUBATION (DAYS):	28
FLEDGING (DAYS):	42
NEST HEIGHT (METRES):	1–30

Swainson's Hawk is found in every Natural Region in Alberta with the exception of the Canadian Shield. No change in distribution was observed between Atlas 1 and Atlas 2 for this species.

Despite being found in almost every natural region, Swainson's Hawks were found mainly in the Grassland NR. This species was found occasionally in the Parkland NR but only infrequently in the Foothills and Rocky Mountain NRs, and rarely in the southern portion of the Boreal Forest NR. The distribution of this species in Alberta reflects its preference to nest in low trees and shrubs and also its use of open areas for hunting mammals such as Richardson's Ground Squirrels.

Declines in relative abundance were detected in the Parkland and Boreal Forest NRs. Changes in relative abundance were not detected in the Foothills, Grassland, or Rocky Mountain NRs. The Breeding Bird Survey found a decline in abundance in Alberta during the period 1985–2005. The declines detected by the Atlas and the Alberta Breeding Bird Survey are mirrored in the decline detected by the Canada-wide Breeding Bird Survey for this species during the period 1985–2005. Large-scale poisonings of Swainson's Hawks have been reported on their wintering grounds and could in part explain the population declines that have been detected in Alberta.

Conditions in the Boreal Forest and Parkland NRs were drier during Atlas 2 than during Atlas 1, and such conditions may also have affected prey availability. The status of this species in Alberta is Sensitive, due to the poisonings on the winter range and to the bird's dependency on adequate ground squirrel populations for food during the breeding period.

TEMPORAL REPORTING RATE

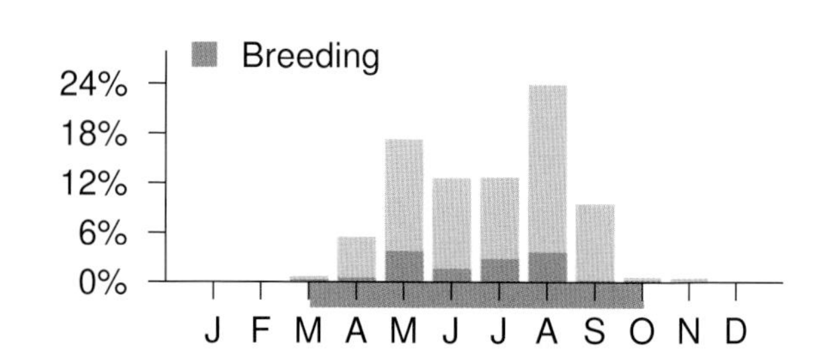

NATURAL REGIONS REPORTING RATE

SPATIAL REPORTING RATE

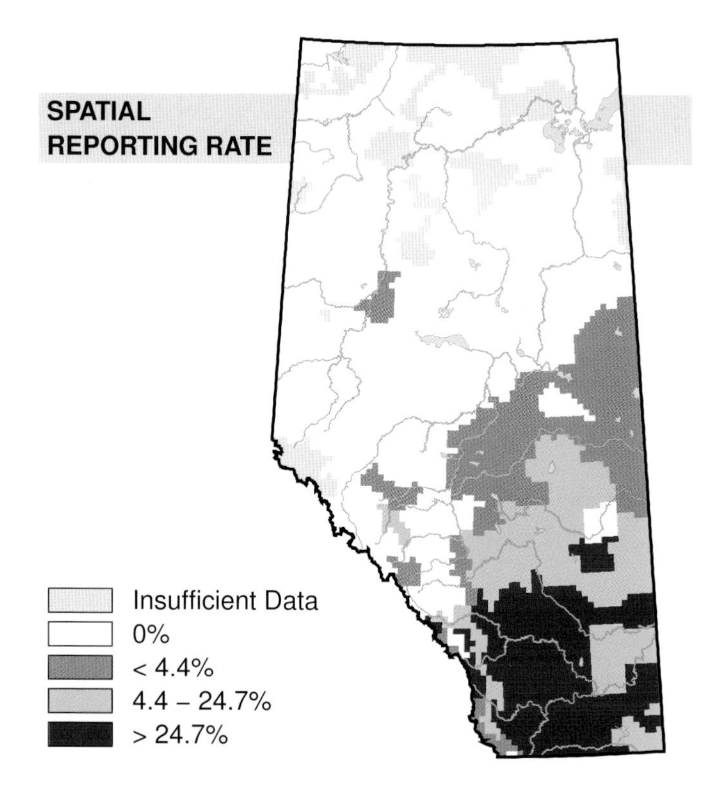

STATUS

GSWA 2000	GSWA 2005	COSEWIC Historic Rankings	COSEWIC 2007	Alberta Wildlife Act
Sensitive	Sensitive	N/A	N/A	N/A

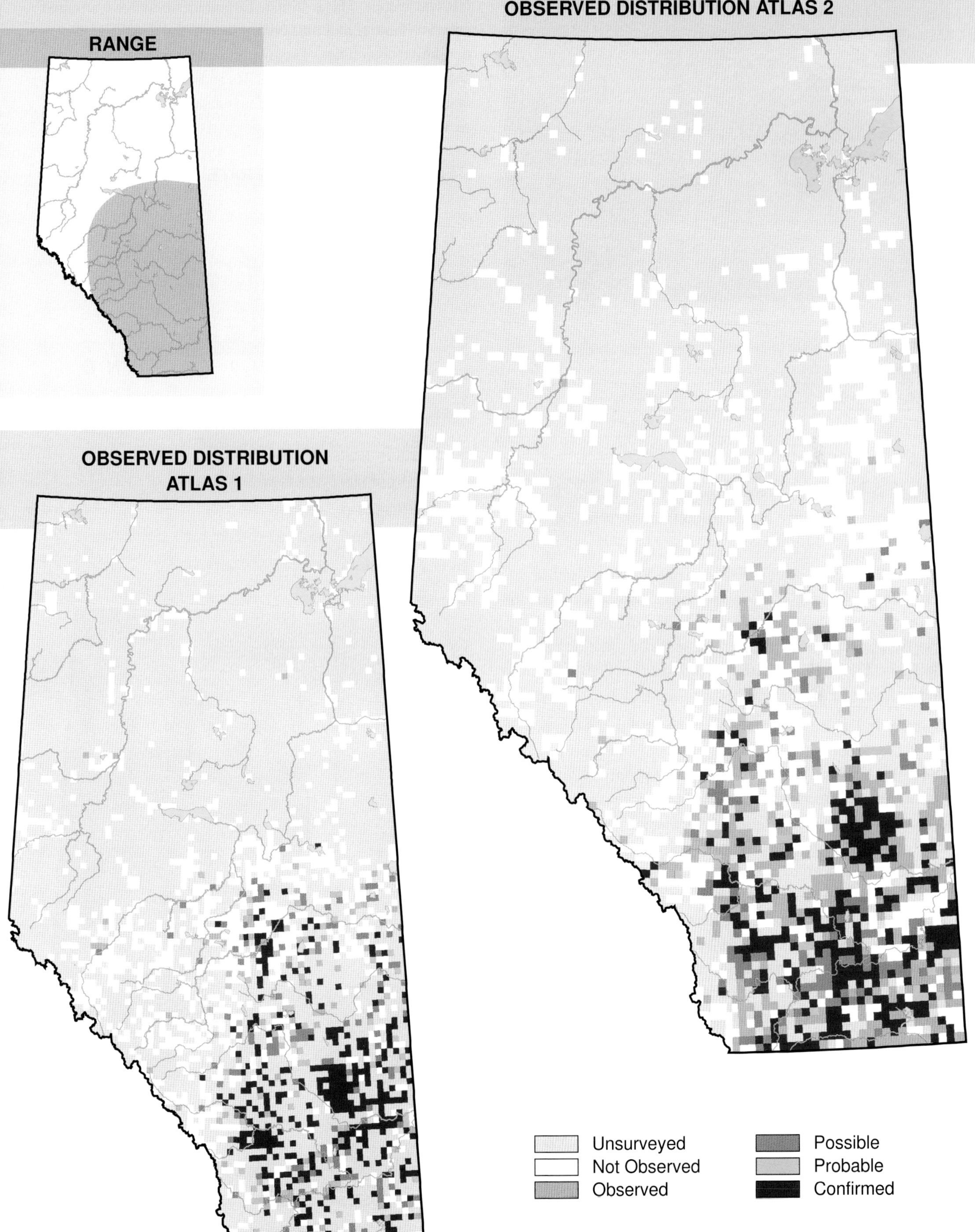
RANGE
OBSERVED DISTRIBUTION ATLAS 2
OBSERVED DISTRIBUTION ATLAS 1
Unsurveyed
Not Observed
Observed
Possible
Probable
Confirmed

Red-tailed Hawk (*Buteo jamaicensis*)

Photo: Randy Jensen

NESTING

CLUTCH SIZE (EGGS):	1–4
INCUBATION (DAYS):	28–32
FLEDGING (DAYS):	41–46
NEST HEIGHT (METRES):	0–36.6

The Red-tailed Hawk is found in every Natural Region across the province. No change in distribution was observed between Atlas 1 and Atlas 2 for this species.

Red-tailed Hawks use stick nests, usually built in the tallest tree within forest patches of varying sizes. This species can do well in forests with high fragmentation because the birds usually build their nests near forest edges. Distribution of the species is fairly uniform across the province but they were most abundant in the Parkland NR between Edmonton and Calgary.

Increases in relative abundance were detected in the Grassland and Rocky Mountain NRs where, relative to other species, Red-tailed Hawks were observed more frequently in Atlas 2 than in Atlas 1. Declines in relative abundance were detected in the Boreal Forest, Foothills, and Parkland NRs where, relative to other species, this bird was observed less frequently in Atlas 2 than in Atlas 1. The Breeding Bird Survey found no change in abundance in Alberta or Canada during the period 1985–2005. Populations could be increasing in the Grasslands because more trees are being planted there and prescribed burns in the Rocky Mountain NR may be restoring the suitability of that NR for this species. The biological cause for the Atlas-detected declines in the Boreal Forest, Foothills, and Parkland NRs is unclear. During Atlas 2, the Boreal Forest and Parkland NRs were drier than during Atlas 1 and the Grassland NR was wetter. These climate differences may have affected prey availability and thus contributed to the detected changes in relative abundance. The overall pattern of relative abundance increases in some NRs and decreases in others matches with BBS findings of no change. The status of this species is Secure in Alberta.

TEMPORAL REPORTING RATE

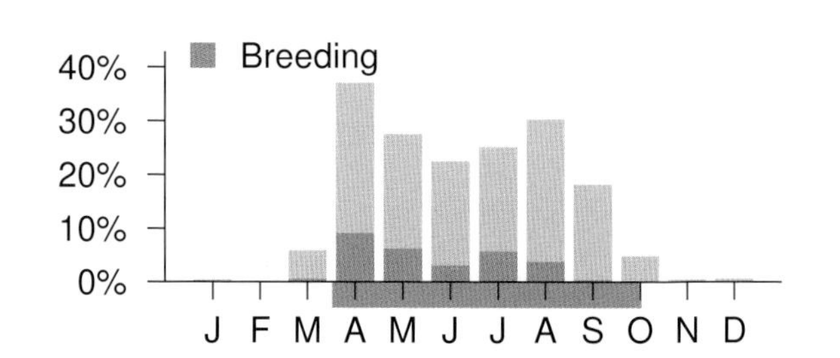

NATURAL REGIONS REPORTING RATE

SPATIAL REPORTING RATE

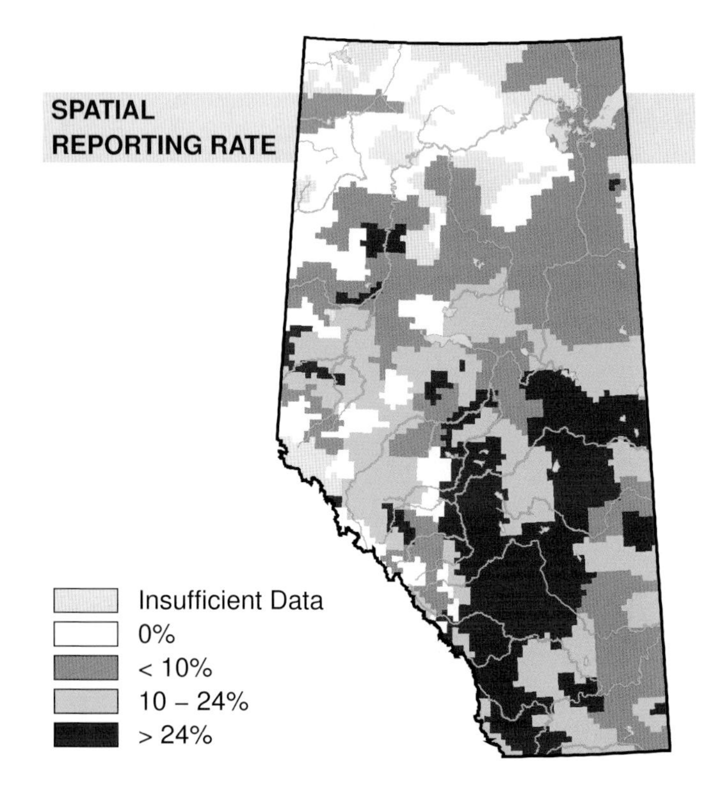

STATUS

GSWA 2000	GSWA 2005	COSEWIC Historic Rankings	COSEWIC 2007	Alberta Wildlife Act
Secure	Secure	Not At Risk	Not At Risk	N/A

Habitat
Nest Location
Nest Type
Diet

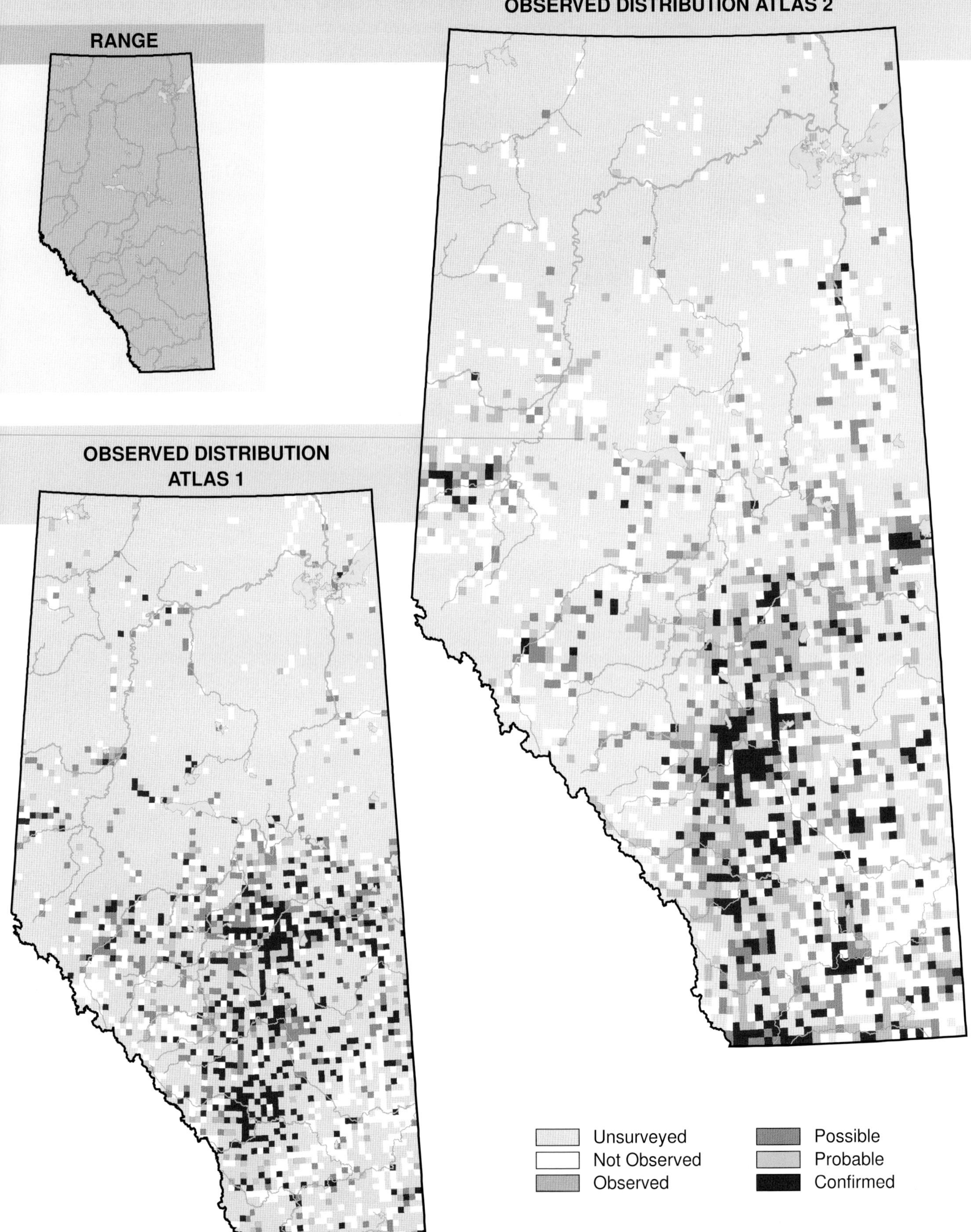
RANGE
OBSERVED DISTRIBUTION ATLAS 2
OBSERVED DISTRIBUTION ATLAS 1
Unsurveyed
Not Observed
Observed
Possible
Probable
Confirmed

Ferruginous Hawk (*Buteo regalis*)

Photo: Gerald Romanchuk

NESTING

CLUTCH SIZE (EGGS):	3–5
INCUBATION (DAYS):	36
FLEDGING (DAYS):	46
NEST HEIGHT (METRES):	1.8–16.8

Alberta's largest buteo, the Ferruginous Hawk is found in southern Alberta. There was no change observed in its distribution between Atlas 1 and Atlas 2. However, the species was encountered more often in the western part of its range. This change may be the result of increased habitat management for this species. Nesting structures have been erected across the range of the Ferruginous Hawk in an effort to increase the bird's reproductive output and to expand its range into formerly occupied areas.

The Ferruginous Hawk is an open-country species and, as such, was most often detected in the Grassland Natural Region. In addition, this species was also encountered occasionally in the Rocky Mountain NR and very rarely in the Parkland NR. Rocky Mountain NR observations were made near Crowsnest Pass and the Cypress Hills. Although not contiguous with the Rocky Mountains, the Cypress Hills are considered to be part of the Rocky Mountain NR.

Increases in relative abundance were detected in the Grassland NR where, relative to other species, the Ferruginous Hawk was observed more frequently in Atlas 2 than in Atlas 1. Breeding Bird Survey data did not reveal any change in abundance for this species in Alberta, or Canada. However, a positive trend was detected in the Prairie Potholes in the period from 1968–2005. Increases are likely the result of improved habitat management for this species in recent years. The provincial status of the Ferruginous Hawk is At Risk due to the species' vulnerability to human disturbance and habitat alterations.

TEMPORAL REPORTING RATE

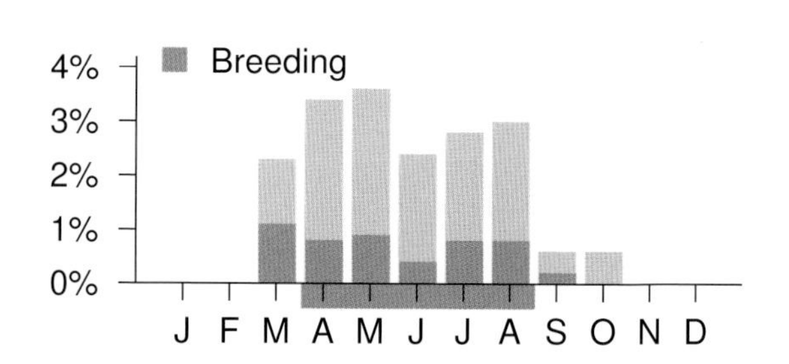

NATURAL REGIONS REPORTING RATE

Grassland  4.1%

SPATIAL REPORTING RATE

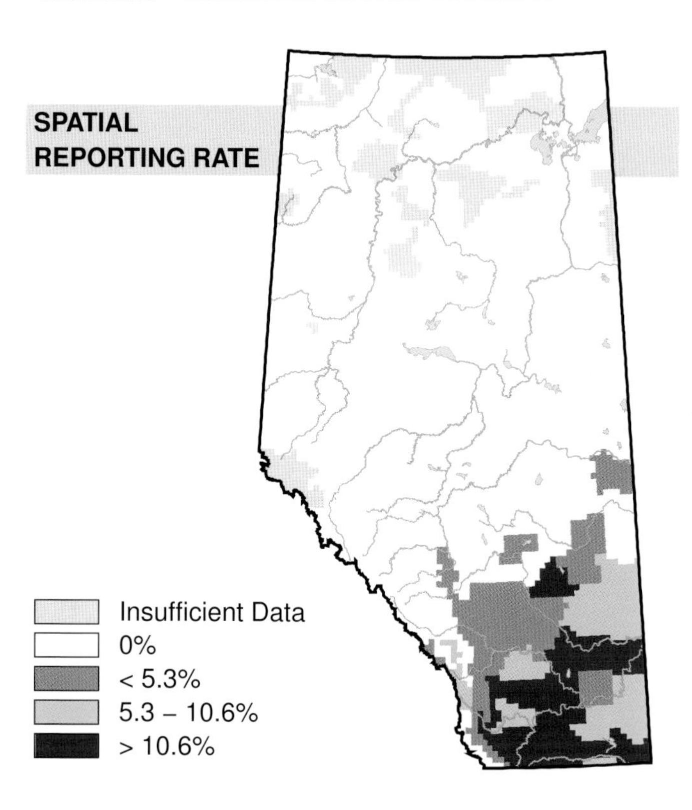

STATUS

GSWA 2000	GSWA 2005	COSEWIC Historic Rankings	COSEWIC 2007	Alberta Wildlife Act
At Risk	At Risk	Special Concern	Special Concern	Threatened

RANGE

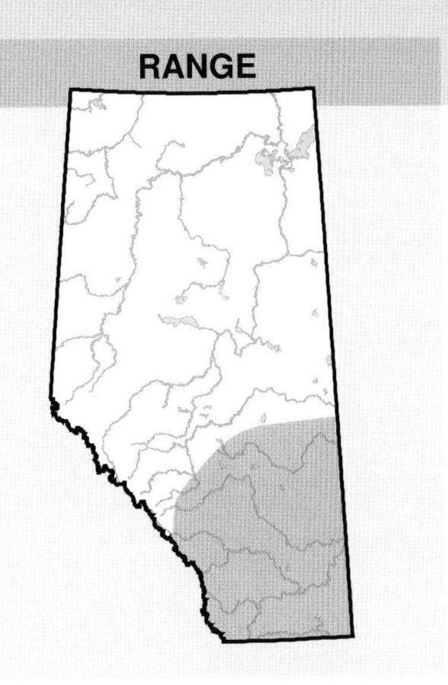

OBSERVED DISTRIBUTION ATLAS 2

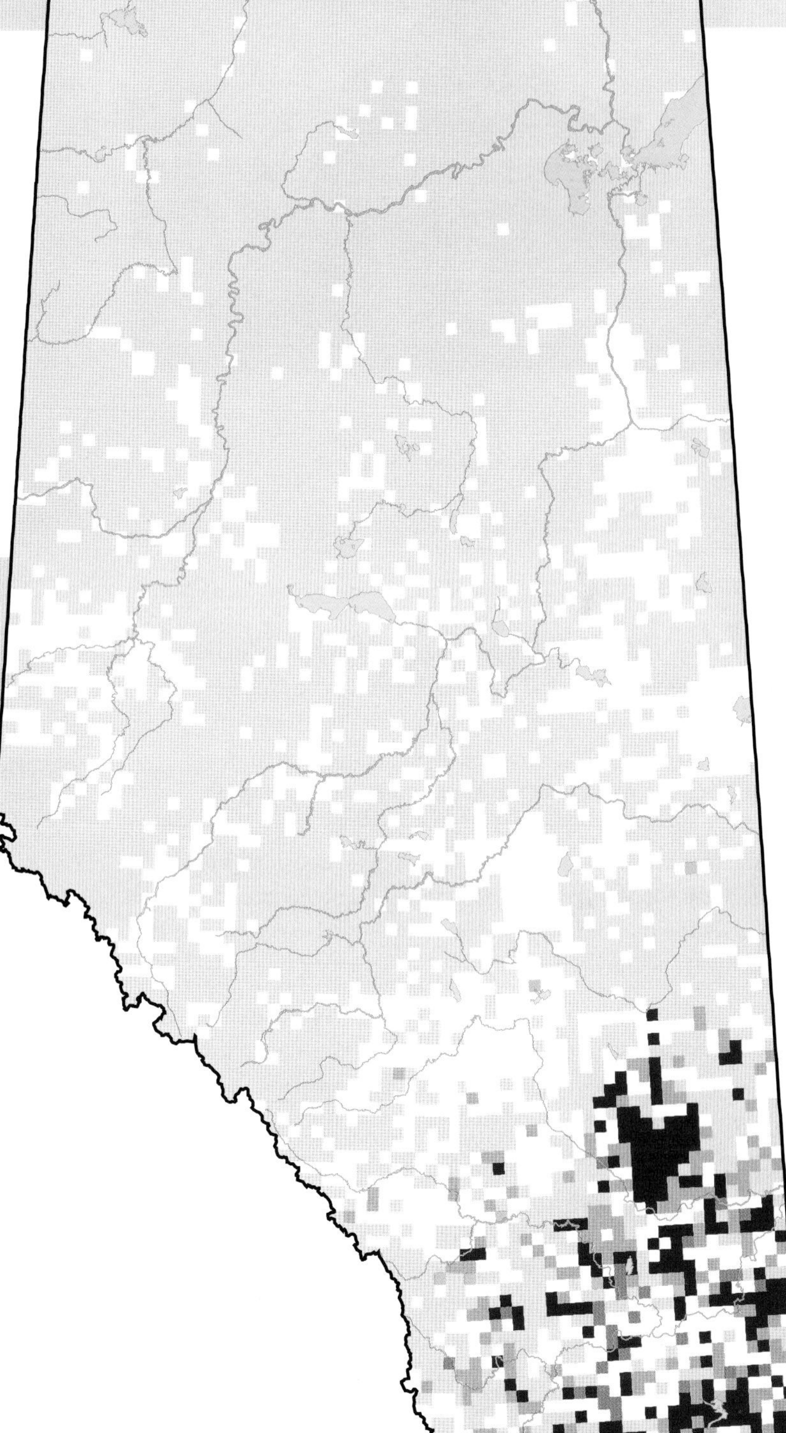

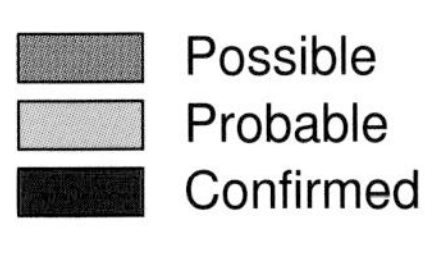

OBSERVED DISTRIBUTION ATLAS 1

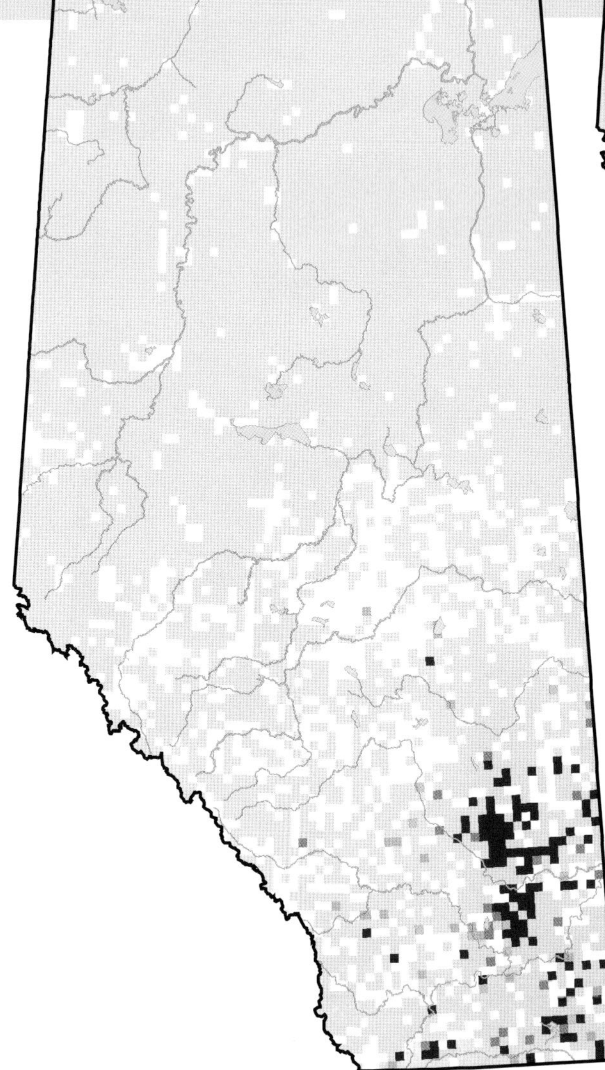

Golden Eagle (*Aquila chrysaetos*)

Photo: Gordon Court

NESTING

CLUTCH SIZE (EGGS):	1–3
INCUBATION (DAYS):	35–45
FLEDGING (DAYS):	65–75
NEST HEIGHT (METRES):	3–30.5

In Alberta, the Golden Eagle is encountered mainly in the Rocky Mountain Natural Region. There was no change observed in the distribution of this species between Atlas 1 and Atlas 2, but the species was encountered more often in the southwestern part of the province during Atlas 2. Once considered rare in Alberta, Golden Eagles have been sighted more frequently in the province in recent years, thanks to the effort afforded by the Rocky Mountain Eagle Research Foundation in monitoring the migration of this species over the Rocky Mountains.

Golden Eagles can nest on a variety of structures but tend to prefer cliff edges. They were, therefore, most often detected in the Rocky Mountain NR where these structures are widely available. This species was rarely encountered in the Foothills, Grassland, and Parkland NRs.

An increase in relative abundance was detected in the Grassland NR where, relative to other species, Golden Eagles were observed more frequently in Atlas 2 than in Atlas 1. No changes in relative abundance were detected in the other NRs. Increases may be a result of more complete survey coverage in Atlas 2 and the broader survey window of Atlas 2 than that of Atlas 1. Breeding Bird Survey sample size was too small to investigate abundance change in Alberta; however, no change was detected on a national scale for this species. The provincial status of the Golden Eagle is Sensitive due to its small population, the tendency to disperse over large areas, and its susceptibility to human disturbance at nest sites.

TEMPORAL REPORTING RATE

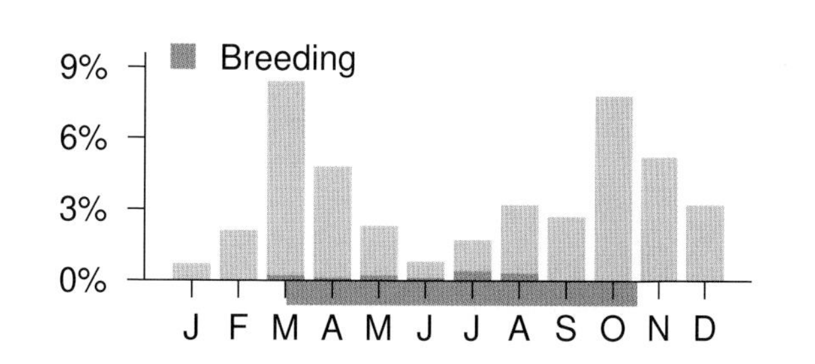

NATURAL REGIONS REPORTING RATE

SPATIAL REPORTING RATE

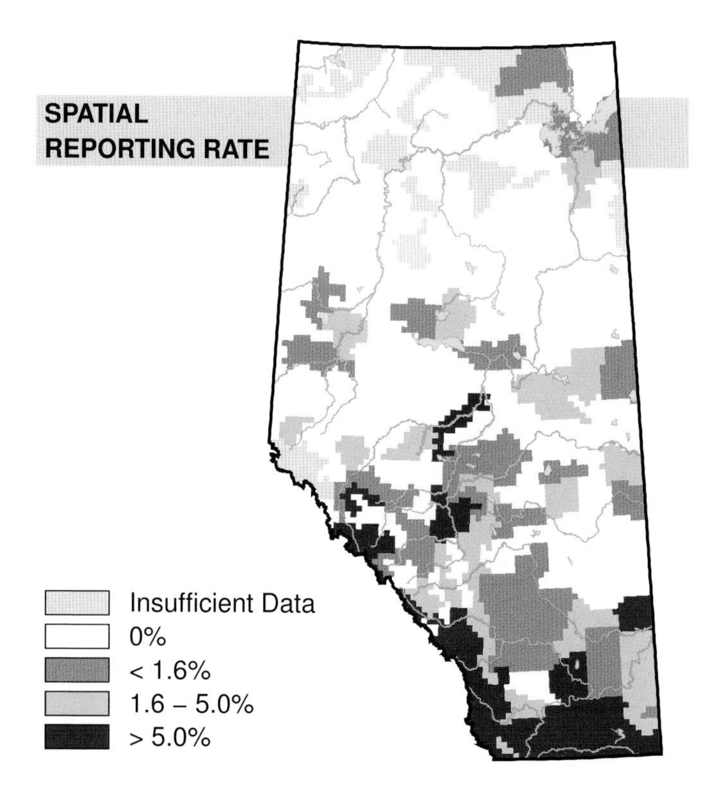

STATUS

GSWA 2000	GSWA 2005	COSEWIC Historic Rankings	COSEWIC 2007	Alberta Wildlife Act
Sensitive	Sensitive	Not At Risk	Not At Risk	N/A

RANGE

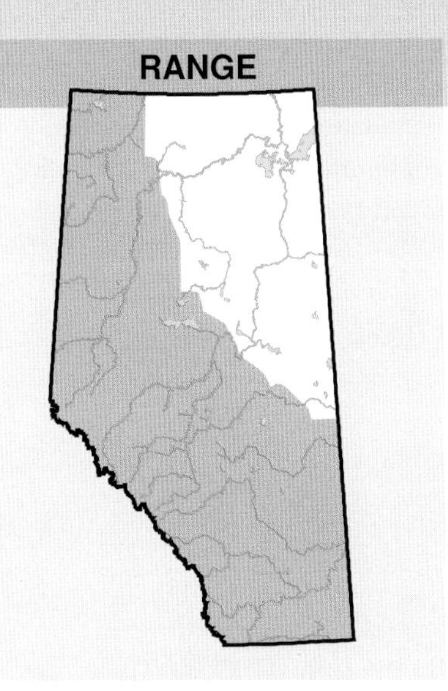

OBSERVED DISTRIBUTION ATLAS 2

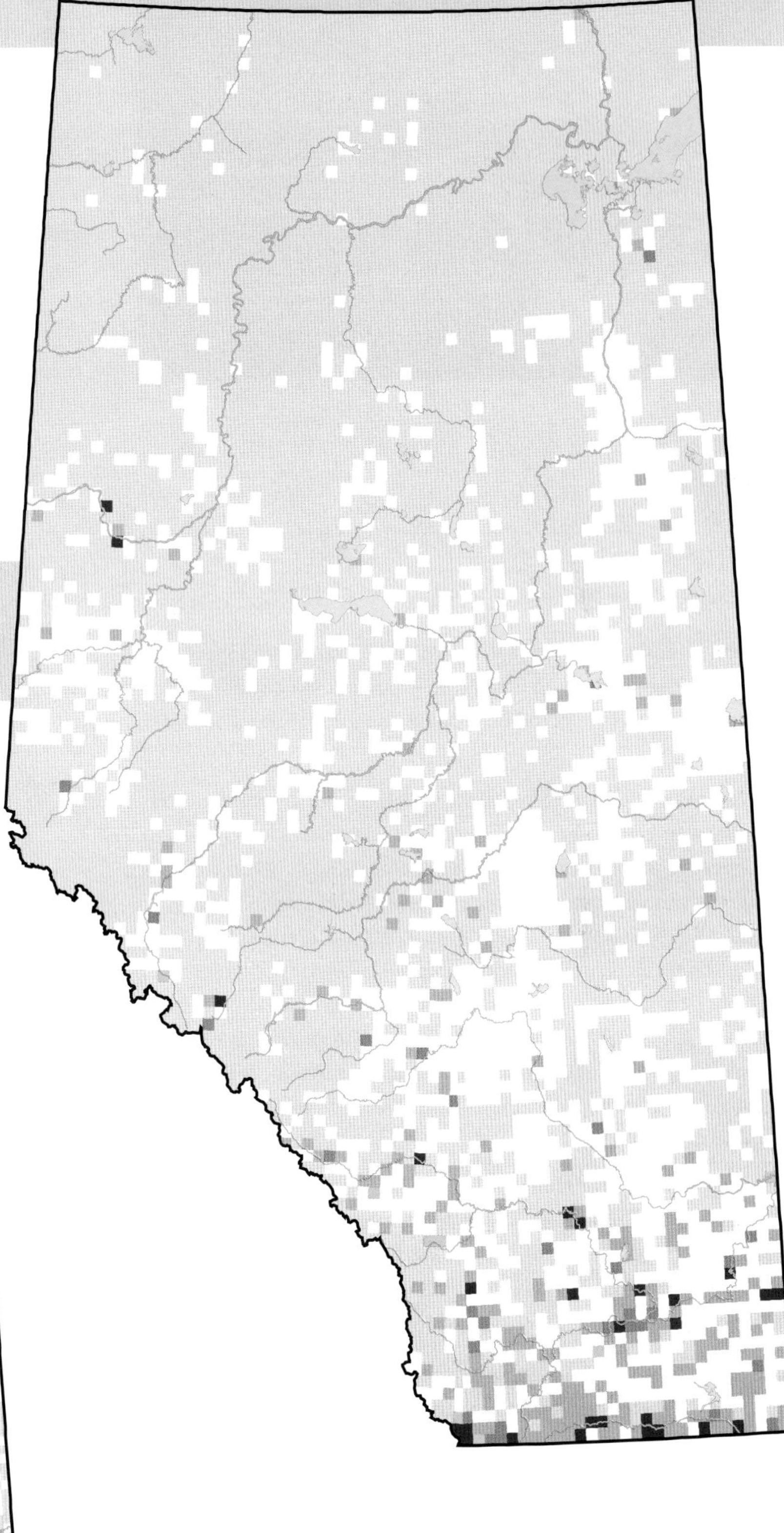

OBSERVED DISTRIBUTION ATLAS 1

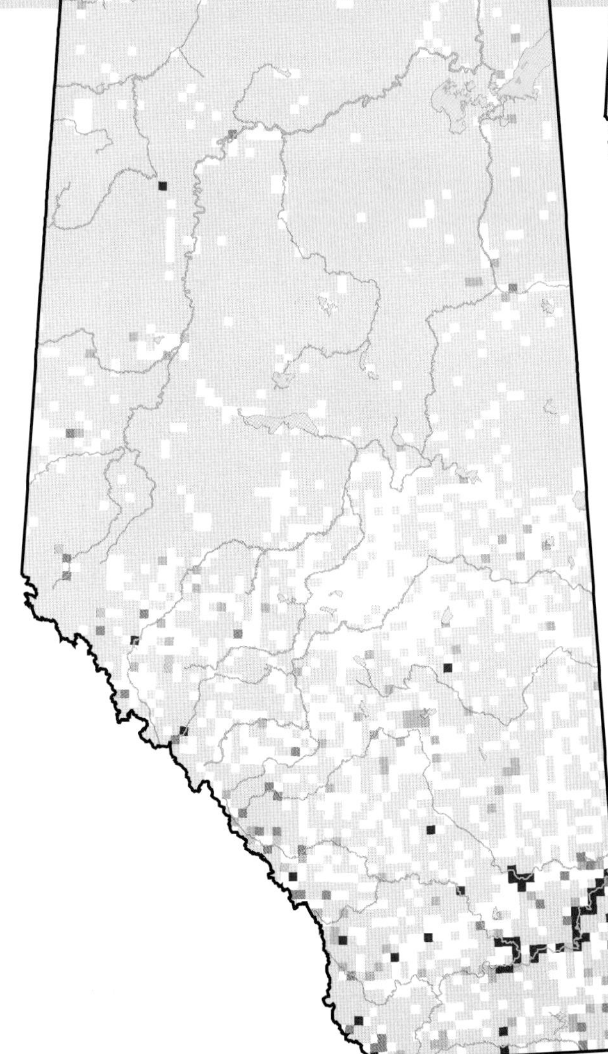

American Kestrel (*Falco sparverius*)

Photo: Gordon Court

NESTING

CLUTCH SIZE (EGGS):	4–6
INCUBATION (DAYS):	29–30
FLEDGING (DAYS):	29–31
NEST HEIGHT (METRES):	3.7–24.4

Alberta's most common falcon, the American Kestrel is found in every Natural Region across the province. No change in distribution was observed between Atlas 1 and Atlas 2 for this species.

For hunting, the American Kestrel requires open habitats such as grasslands, areas where forests have burned. It is attracted to human-modified habitats, pastures and parkland, and near areas of human activity, including heavily developed urban areas (Smallwood and Bird, 2002). It is an obligate cavity nester that relies on the cavities created by woodpeckers and fungal decay. For this reason, this species was most abundant in the Grasslands NR where it finds suitable nesting cavities along hedgerows and water courses. It is also quite common and evenly distributed in the Boreal Forest, Foothills, Parkland, and Rocky Mountain NRs.

Declines in relative abundance were detected in the Boreal Forest, Foothills, Grasslands, Parkland, and Rocky Mountain NRs where, relative to other species, it was observed less frequently in Atlas 2. Data from the Breeding Bird Survey (1995–2005) indicate that American Kestrel abundance declined in Alberta and in Canada (1985–2005). Declines in American Kestrel populations across the province could be related to declines in the availability of mature trees suitable for the construction of nest cavities. In addition to nesting cavities, limiting factors are identified as suitable prey, perches, and low, open vegetation for hunting (Smallwood and Bird, 2002). Despite the wide distribution and common status, there are indications of a population decline, at least on a regional basis (Bird et al., 2004). They identified potential causes of observed declines as increased predation by Cooper's Hawks, West Nile virus, changes in agricultural land practices, climatic changes, and/or decreases in insect availability. The provincial status of this species is Secure.

TEMPORAL REPORTING RATE

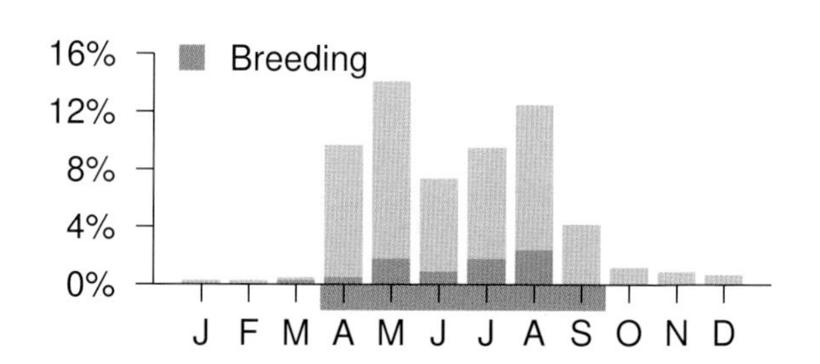

NATURAL REGIONS REPORTING RATE

Region	Rate
Boreal Forest	2.6%
Foothills	3.5%
Grassland	8.8%
Parkland	4.9%
Rocky Mountain	3.8%

SPATIAL REPORTING RATE

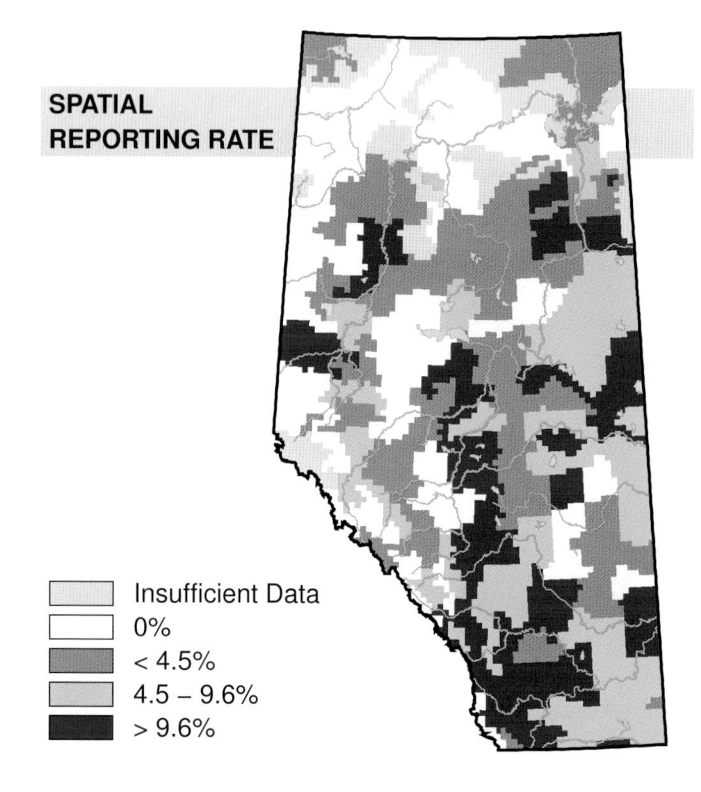

STATUS

GSWA 2000	GSWA 2005	COSEWIC Historic Rankings	COSEWIC 2007	Alberta Wildlife Act
Secure	Secure	N/A	N/A	N/A

RANGE

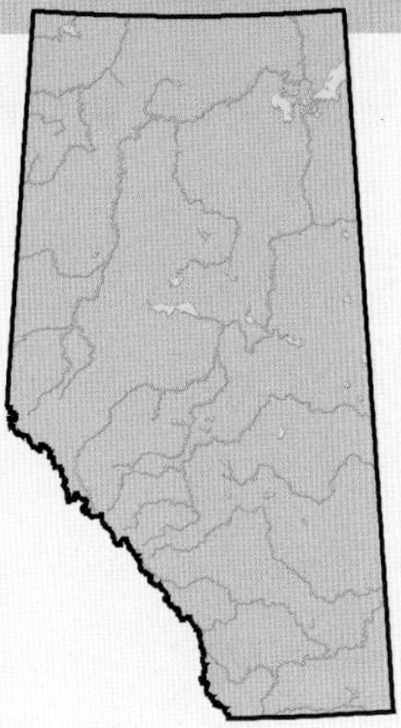

OBSERVED DISTRIBUTION ATLAS 2

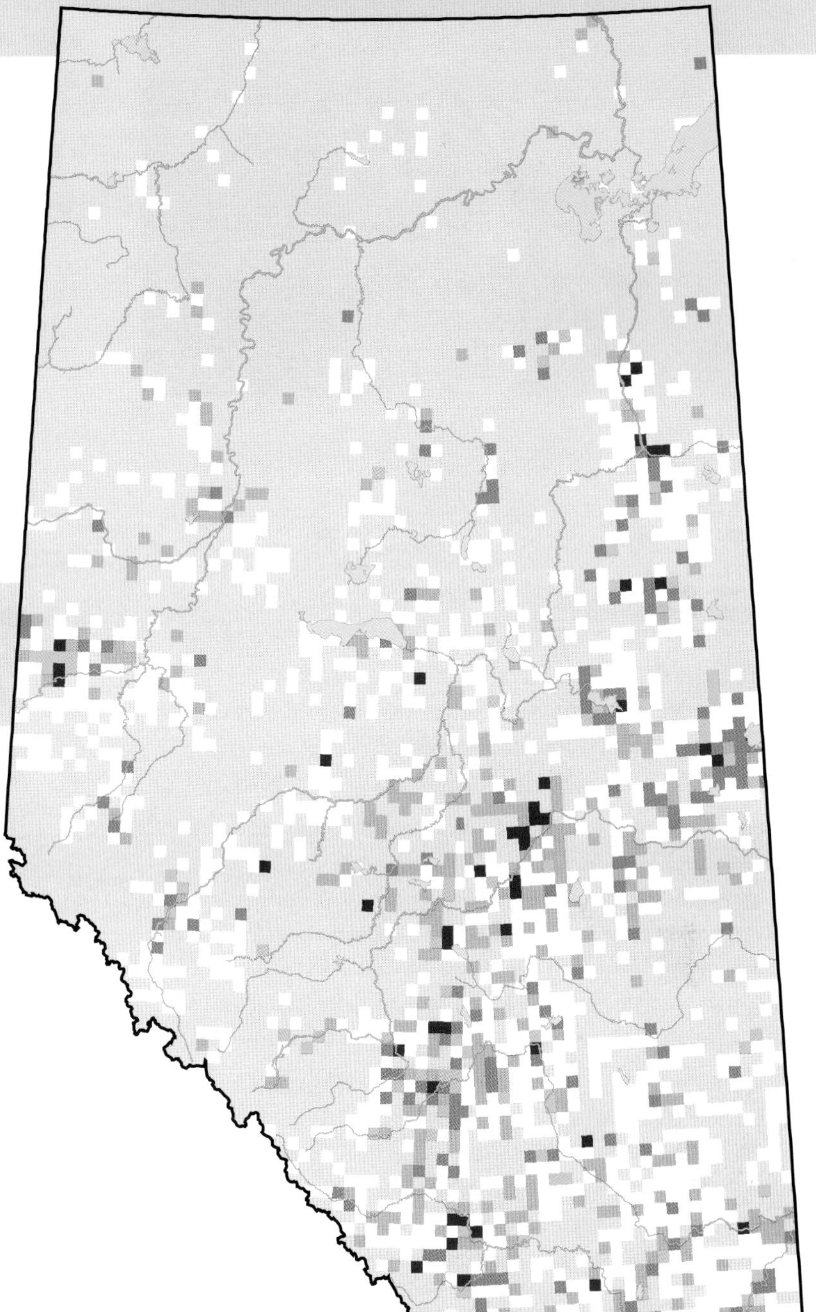

OBSERVED DISTRIBUTION ATLAS 1

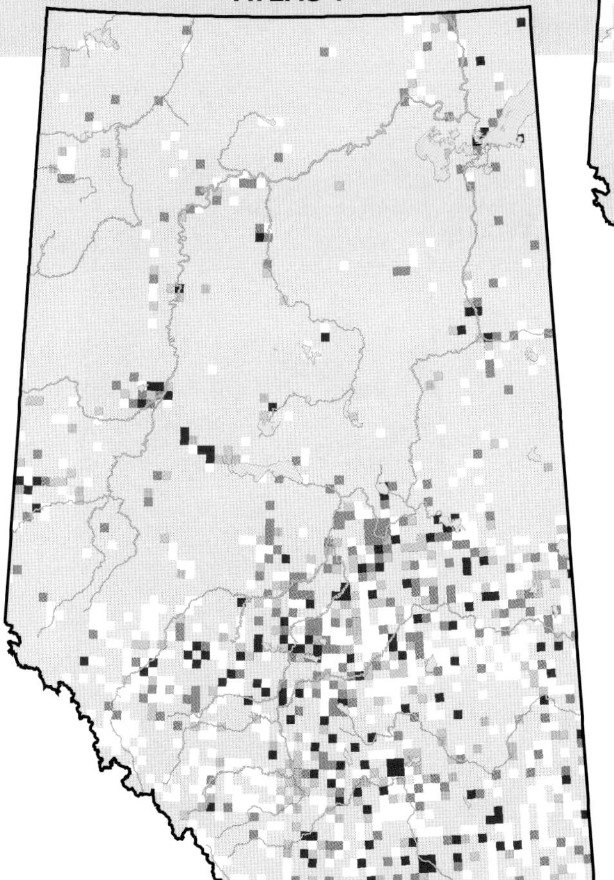

Unsurveyed
Not Observed
Observed
Possible
Probable
Confirmed

Merlin (*Falco columbarius*)

Photo: Mark Williams

NESTING

CLUTCH SIZE (EGGS):	4–6
INCUBATION (DAYS):	28–34
FLEDGING (DAYS):	25–30
NEST HEIGHT (METRES):	2.4–18.3

The Merlin is found in every Natural Region across the province. No change in distribution was observed between Atlas 1 and Atlas 2 for this species.

Merlins occupy a variety of habitats, and the bird is familiar to most people because of its ability to nest near human settlements. Merlins can be found commonly in older urban neighbourhoods or around rural housing where spruce trees have become established. Many of the observations from the Grassland and Parkland NRs were derived from those kinds of localities. Merlins are also found in the Rocky Mountain NR and, to a lesser extent, the Boreal Forest and Foothills NRs.

Increases in relative abundance were detected in the Grassland NR where, relative to other species, Merlins were observed at similar frequencies in Atlas 1 and Atlas 2. An increase was also detected in the Rocky Mountain NR where, relative to other species, this species was observed more frequently in Atlas 2 than in Atlas 1. In the Boreal Forest NR a decline was detected in relative abundance as well as in observation frequency relative to other species. No changes were detected in other NRs. The Breeding Bird Survey found no change in abundance in Alberta, but increases in abundance were detected across Canada during the period 1985–2005. Increased abundance across Canada is probably the result of the recovery of Merlin populations after DDT use was banned.

The inability of the BBS to detect a change in Alberta is likely a function of small sample size. There is no apparent biological cause for the Atlas-detected decline in the Boreal Forest NR. Lacking cause and corroboration, assessment of this finding requires further research. The provincial status of this species is Secure.

TEMPORAL REPORTING RATE

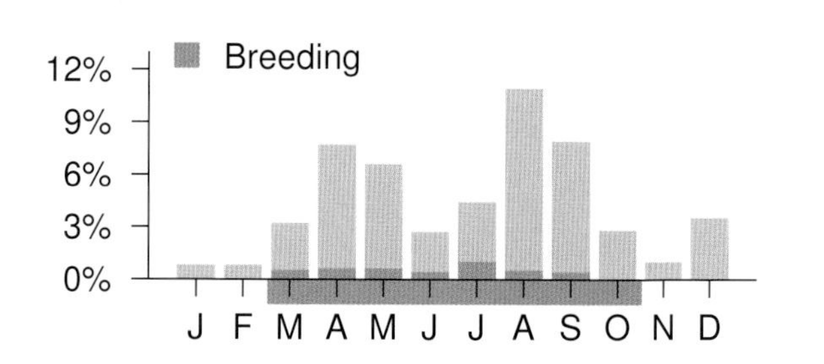

NATURAL REGIONS REPORTING RATE

SPATIAL REPORTING RATE

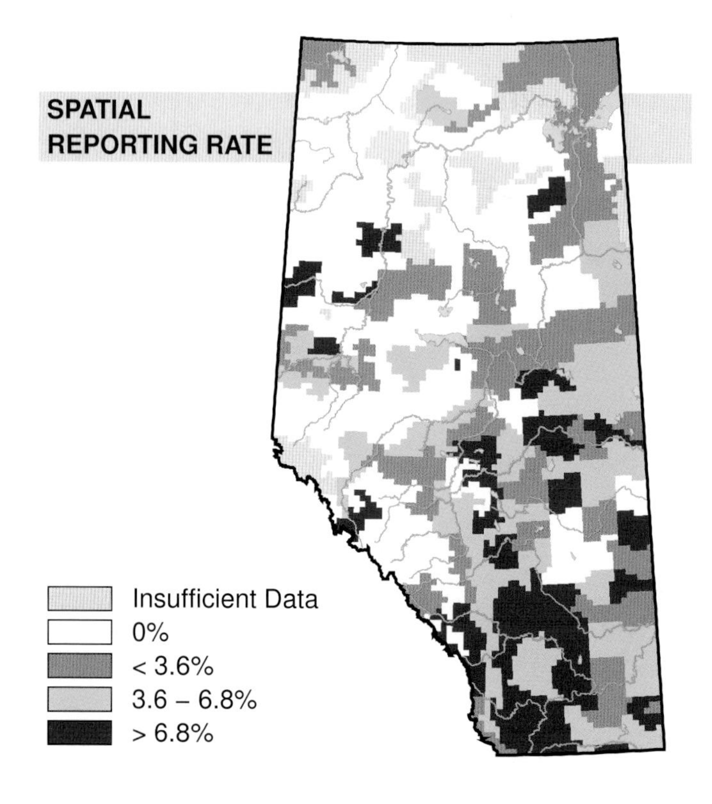

STATUS

GSWA 2000	GSWA 2005	COSEWIC Historic Rankings	COSEWIC 2007	Alberta Wildlife Act
Secure	Secure	Not At Risk	Not At Risk	N/A

OBSERVED DISTRIBUTION ATLAS 2

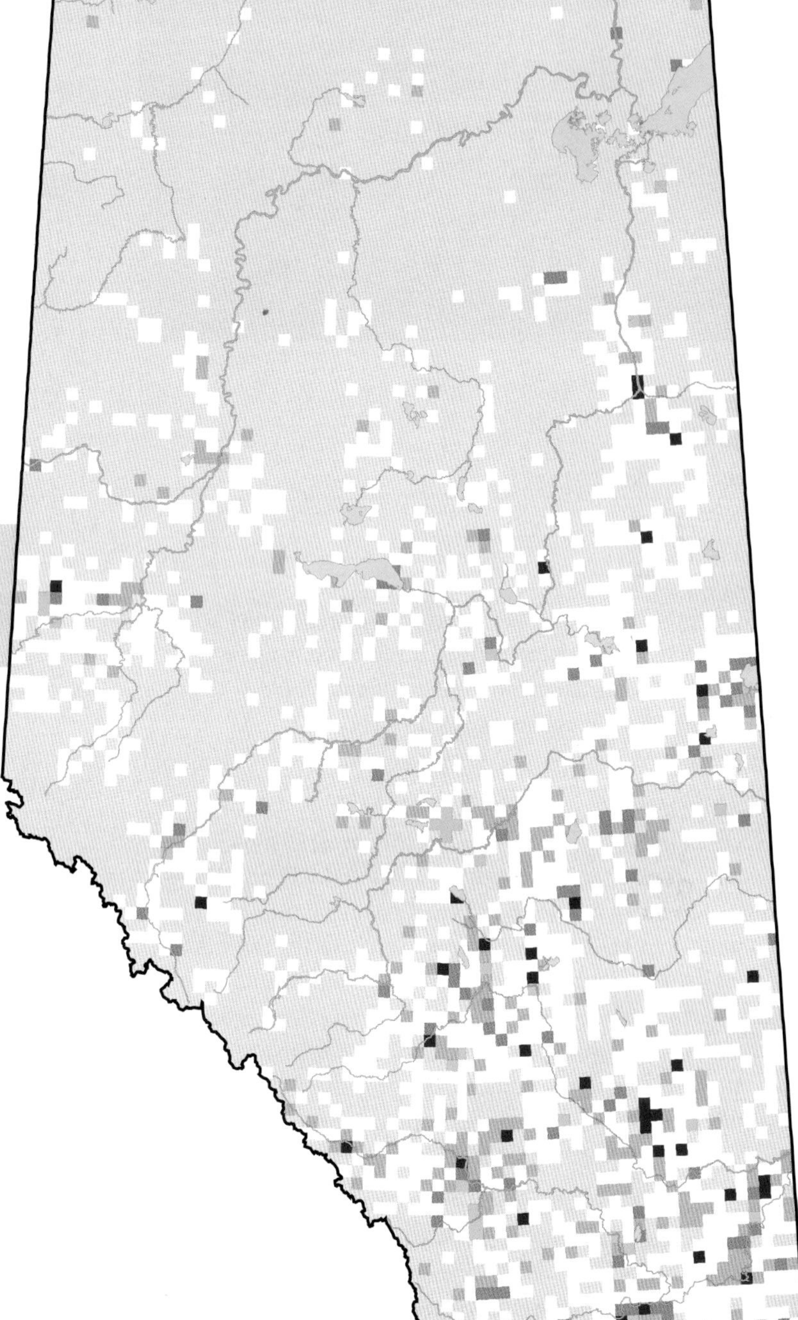

RANGE

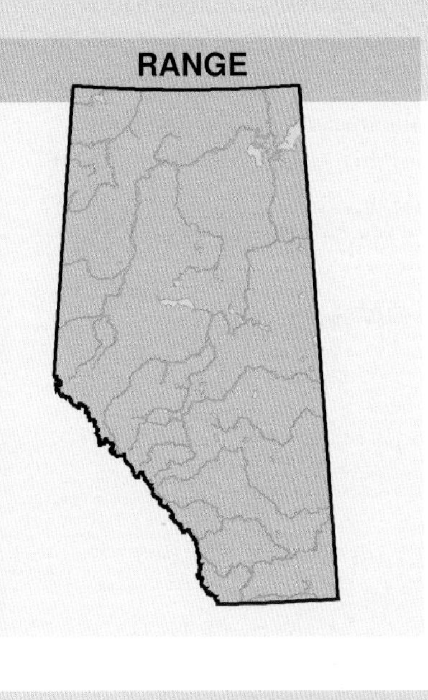

OBSERVED DISTRIBUTION ATLAS 1

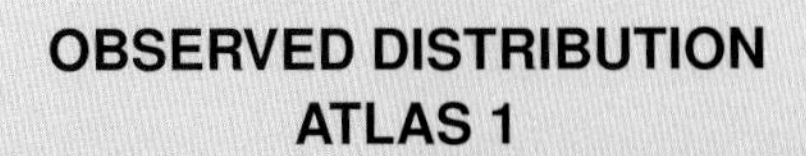

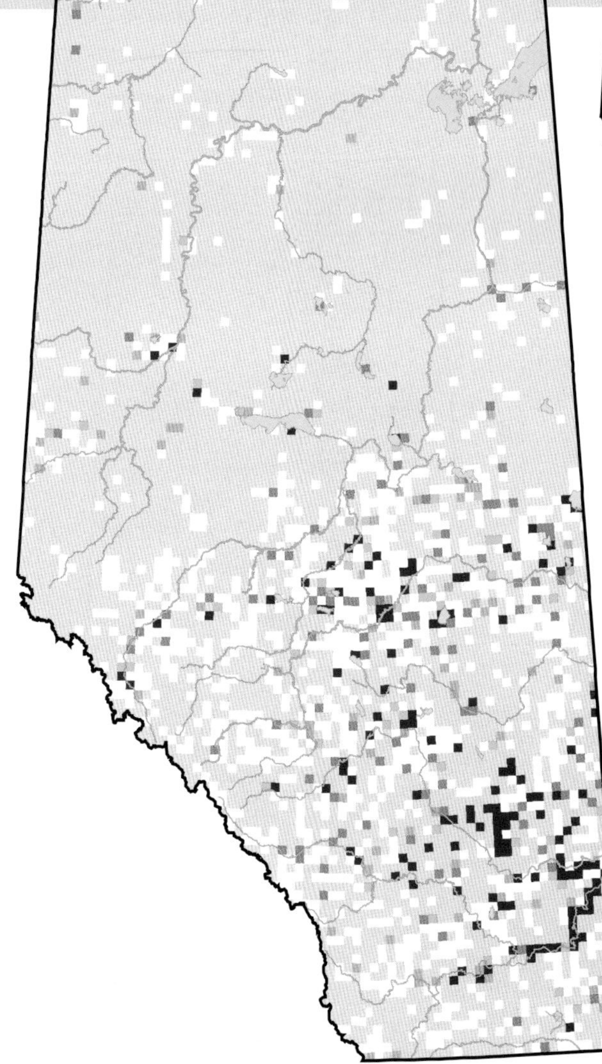

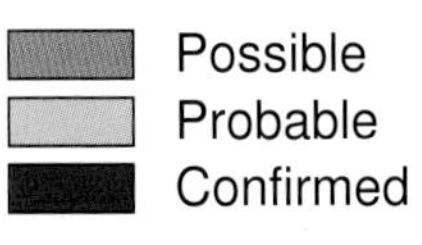

Peregrine Falcon (*Falco peregrinus*)

Photo: Gordon Court

NESTING

CLUTCH SIZE (EGGS):	3–4
INCUBATION (DAYS):	33–35
FLEDGING (DAYS):	35–40
NEST HEIGHT (METRES):	15.2–61

The Peregrine Falcon is found in every Natural Region across the province. More comprehensive distribution maps were compiled for Atlas 2 through the incorporation of species-specific data into the Atlas 2 project. However, no change in the distribution of this species was observed between Atlas 1 and Atlas 2.

Peregrine Falcons typically nest on cliff edges; so they tend to be found along rivers in Alberta. This species will also use artificial nest structures that are installed on the sides of tall buildings. Part of the population recovery strategy for this species involves installing nest platform boxes on tall, man-made structures. This strategy has been effective because sightings of Peregrine Falcons have become relatively common recently in large urban centres. The patchy breeding distribution of this species across Alberta reflects the bird's requirements for cliff-type nesting structures for breeding.

A decline in relative abundance was detected in the Boreal Forest NR. No changes were detected in other NRs. The Breeding Bird Survey sample size was too small to investigate abundance changes for this species. Given the small population and patchy breeding distribution of this species, it is clear that Peregrine Falcons are not effectively monitored by large, wide-spread, multi-species surveys such as the Atlas or Breeding Bird Survey. It seems likely that the Atlas-detected decline is, at least in part, a result of differences in survey coverage and is not indicative of change in population size.

The banning of DDT is widely recognized as having had a positive impact on this species over much of its range. Species-specific monitoring and recovery programs indicate that this species' population is increasing in Alberta (Team, 2005). The COSEWIC status of this species was downgraded from Endangered to Threatened in 2000 based on its recovery. The provincial status of this species is At Risk.

TEMPORAL REPORTING RATE

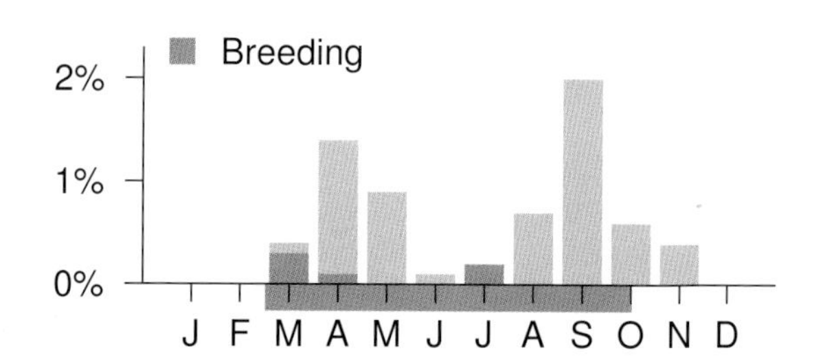

NATURAL REGIONS REPORTING RATE

Region	Rate
Boreal Forest	0.2%
Foothills	0.2%
Grassland	1.2%
Parkland	0.6%
Rocky Mountain	1.3%

SPATIAL REPORTING RATE

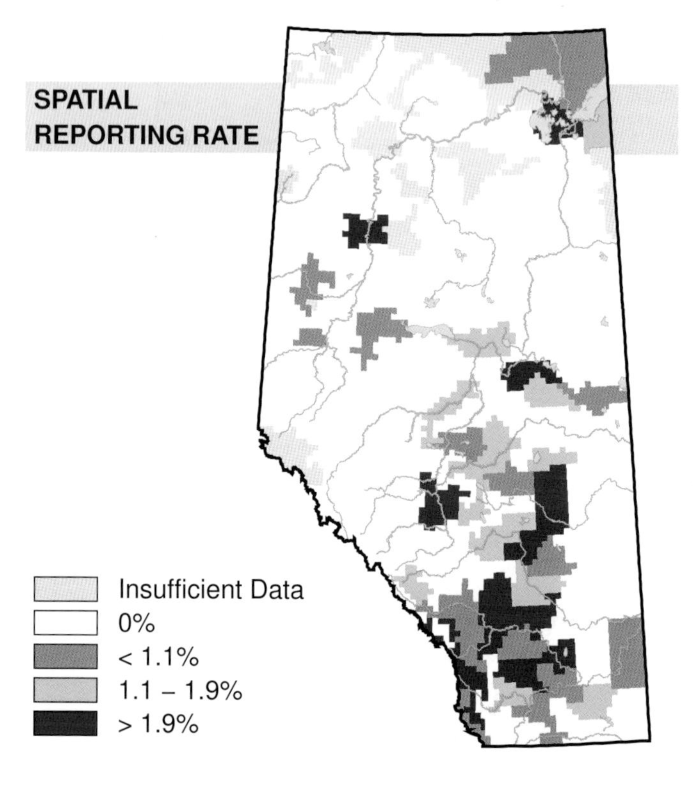

STATUS

GSWA 2000	GSWA 2005	COSEWIC Historic Rankings	COSEWIC 2007	Alberta Wildlife Act
At Risk	At Risk	Threatened	Special Concern	Threatened

Habitat Nest Location Nest Type Diet

RANGE

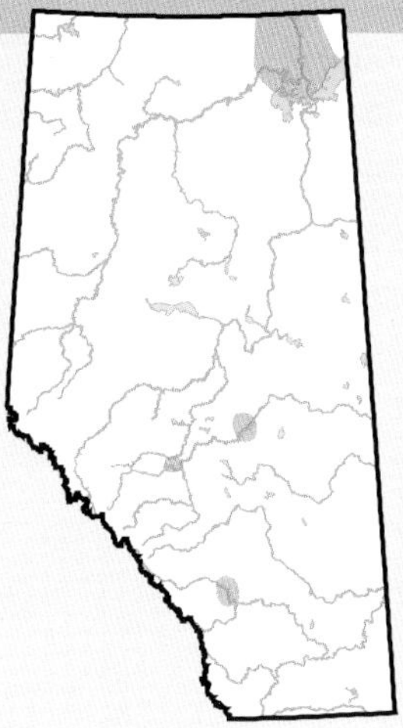

OBSERVED DISTRIBUTION ATLAS 1

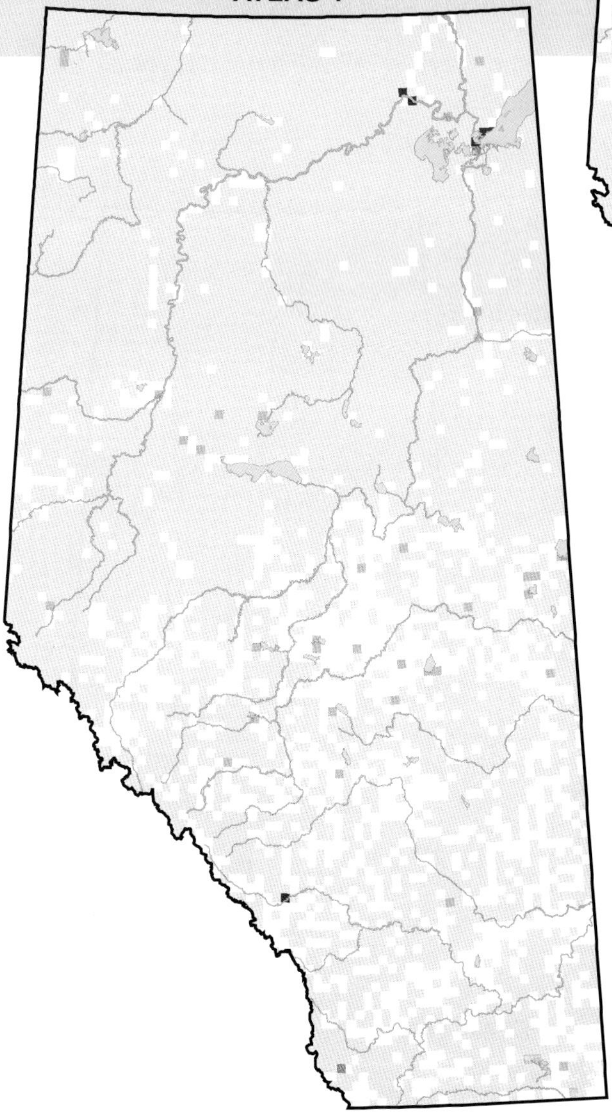

OBSERVED DISTRIBUTION ATLAS 2

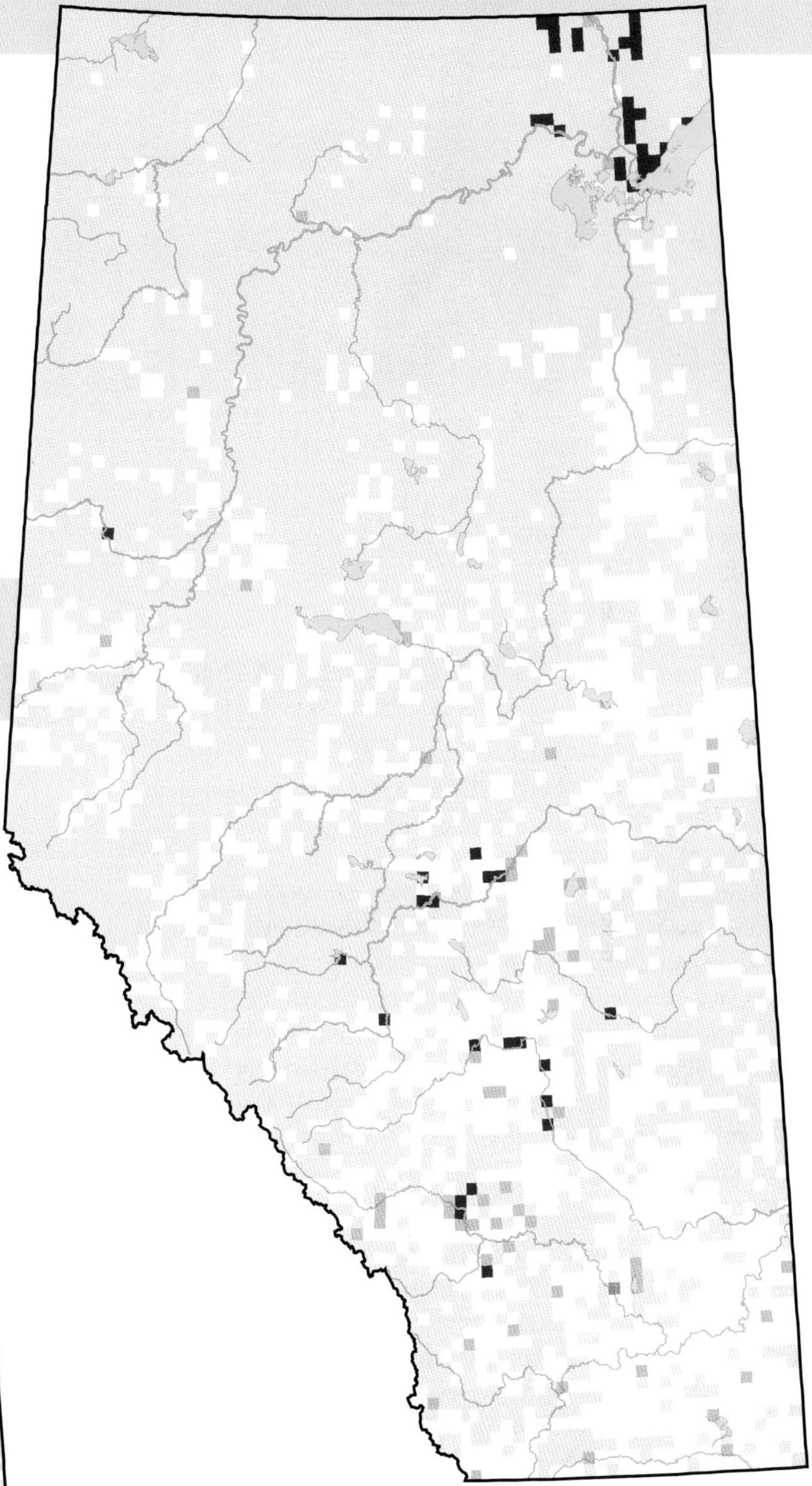

Unsurveyed
Not Observed
Observed
Possible
Probable
Confirmed

Prairie Falcon (*Falco mexicanus*)

Photo: Mark Williams

NESTING

CLUTCH SIZE (EGGS):	3–6
INCUBATION (DAYS):	33–35
FLEDGING (DAYS):	36–41
NEST HEIGHT (METRES):	6.1–122

The Prairie Falcon is found in the southern part of Alberta. No change in distribution was observed in Atlas 2 for this species.

Prairie Falcons were found most commonly in the Grassland and Rocky Mountain NRs, and were found to be uncommon in the Parkland NR. Similar to Peregrine Falcons, Prairie Falcons use cliffs for nesting. However, unlike Peregrines, Prairie Falcons consume a relatively large proportion of mammals such as Richardson's Ground Squirrels. Prairie Falcon reporting rates were highest in the southern part of the province where cliffs are located along rivers and where Richardson's Ground Squirrel populations are most abundant. Hunt (1993) reported that breeding individuals in Alberta selected native range habitat and that their core-use areas had lower proportions of irrigated cropland than expected on the basis of availability.

An increase in relative abundance was detected in the Grassland NR where surprisingly, relative to other species, the Prairie Falcon was observed at similar rates in Atlas 1 and Atlas 2. Increases in Prairie Falcon populations in this Natural Region could be related to a population expansion of this species after the banning of the use of DDT in North America. Changes in relative abundance were not detected in the Parkland and Rocky Mountain NRs. The Breeding Bird Survey sample size was too small to investigate abundance changes for this species in Alberta and Canada.

Dunn (2005), in the National action needs for Canadian Landbird Conservation, reports that Prairie Falcons are uncommon in Canada but apparently doing well. However, periodic monitoring is required to assess their status in Canada and improve our ability to track populations routinely as the status of the small population could change rapidly.

The province's status for this species is Sensitive, due to its dependence on the availability of secure nest sites and adequate ground squirrel prey.

TEMPORAL REPORTING RATE

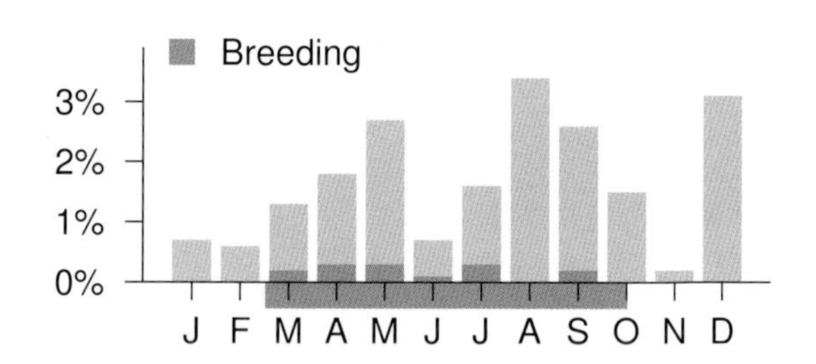

NATURAL REGIONS REPORTING RATE

SPATIAL REPORTING RATE

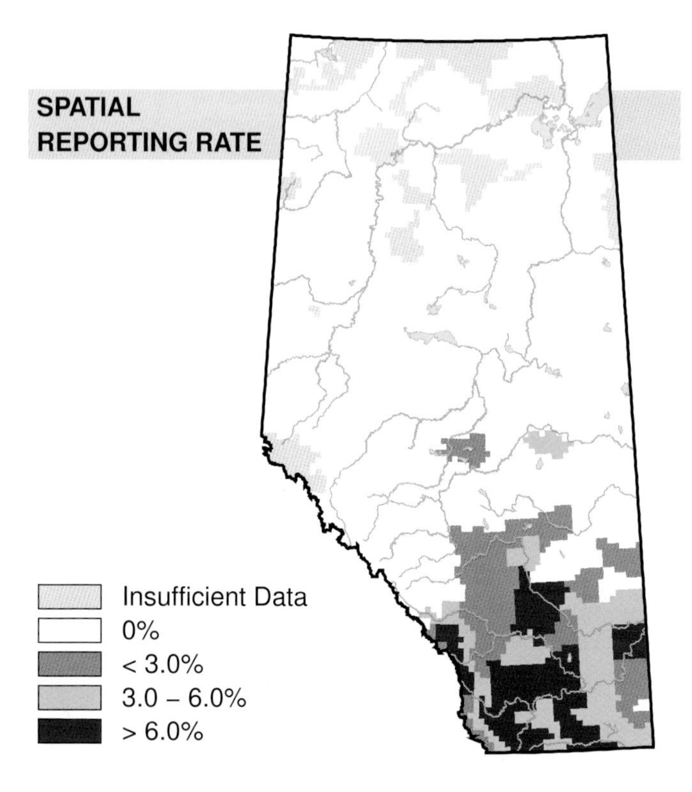

STATUS

GSWA 2000	GSWA 2005	COSEWIC Historic Rankings	COSEWIC 2007	Alberta Wildlife Act
Sensitive	Sensitive	Not At Risk	Not At Risk	N/A

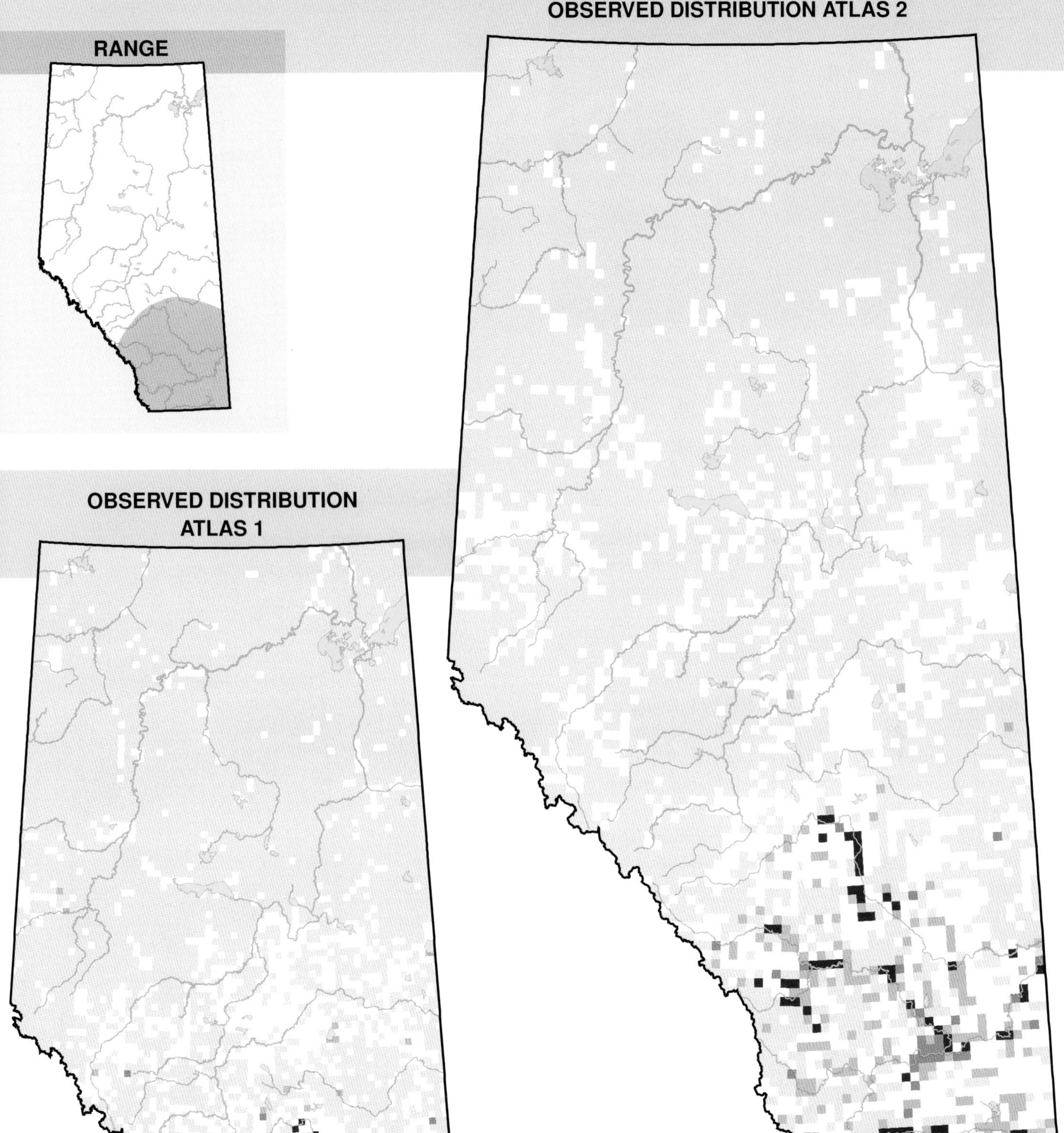
Habitat
Nest Location
Nest Type
Diet
RANGE
OBSERVED DISTRIBUTION ATLAS 2
OBSERVED DISTRIBUTION ATLAS 1
Unsurveyed
Not Observed
Observed
Possible
Probable
Confirmed

Yellow Rail (*Coturnicops noveboracensis*)

Photo: Brian E. Small / VIREO

NESTING

CLUTCH SIZE (EGGS):	6–10
INCUBATION (DAYS):	18
FLEDGING (DAYS):	35
NEST HEIGHT (METRES):	0

The Yellow Rail is sparsely distributed in wet sedge meadows of the Boreal Forest, Foothills, and Parkland Natural Regions, and is absent from the Grassland and Rocky Mountain NRs. A change in distribution was observed in Atlas 2 with new records in northwestern and southwestern Alberta.

Nests of the Yellow Rail are built in wet meadows that are dominated by sedge. The nests are rarely built among cattails or in wetlands that contain woody vegetation such as willow. Due to this bird's secretive nature and specific habitat requirements, the reporting rates recorded for the species were very low. Yellow Rails were most often detected by their sound. It is extremely difficult to observe because, like other rails, it prefers to run or hide instead of flying and commonly moves beneath procumbent vegetation (Bookhout, 1995). In addition, these birds are considered irruptive because they may be present at suitable sites one year but not the next.

No changes in relative abundance were detected, but this may be a reflection of the small number of records that were obtained. There were not enough Yellow Rail records to detect population changes using Breeding Bird Survey data in Alberta. Due to the secretive nature of this species, it is not effectively monitored by large, wide-spread, multi-species surveys such as the Atlas or Breeding Bird Survey. Yellow Rail-specific surveys, that involve the use of call-playback broadcasts, would be needed to generate enough data to evaluate abundance changes effectively for this species.

Loss of wetlands to human activity is probably the most serious factor affecting Yellow Rail populations. Drainage of marshes may explain why the southern boundary of the breeding area has moved northward in the 20th century (Bookhout, 1995).

The status of the Yellow Rail in Alberta is Undetermined, because there are insufficient data to evaluate its status reliably.

TEMPORAL REPORTING RATE

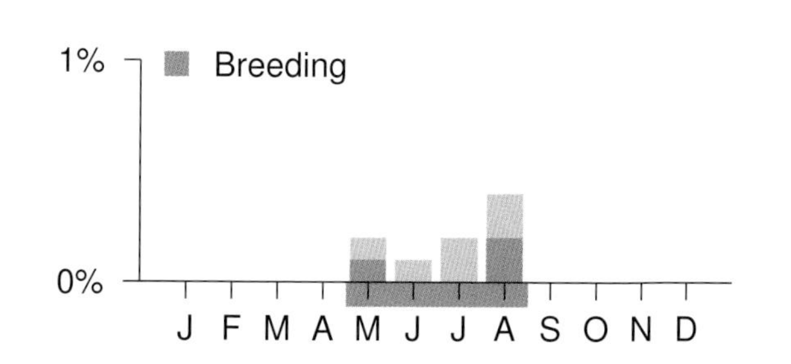

NATURAL REGIONS REPORTING RATE

SPATIAL REPORTING RATE

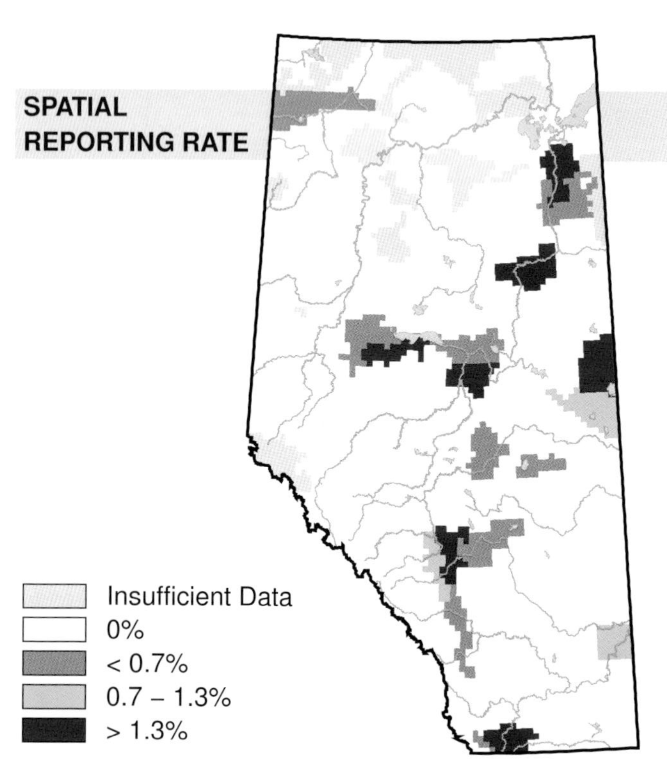

STATUS

GSWA 2000	GSWA 2005	COSEWIC Historic Rankings	COSEWIC 2007	Alberta Wildlife Act
Undetermined	Undetermined	Special Concern	Special Concern	N/A

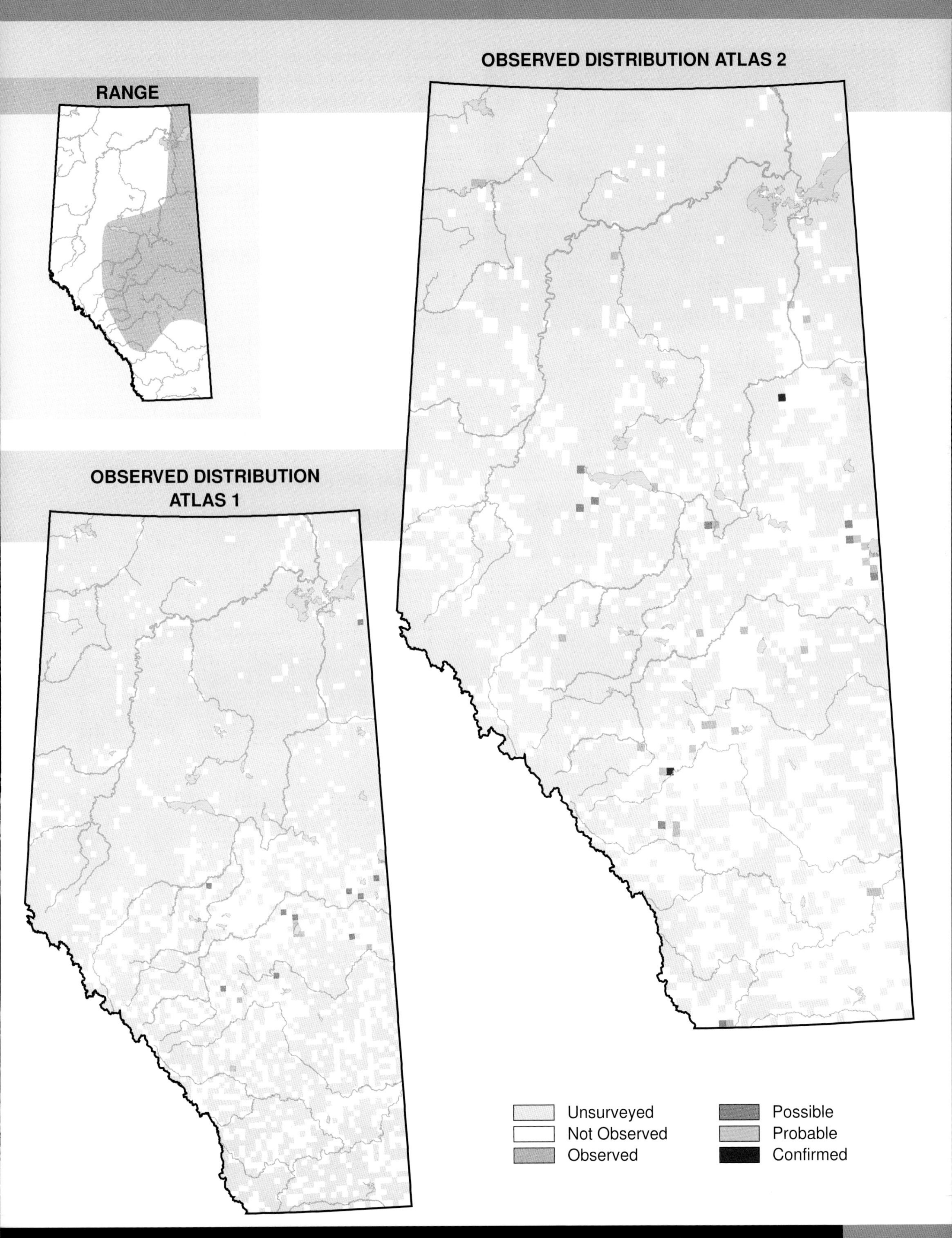
RANGE
OBSERVED DISTRIBUTION ATLAS 2
OBSERVED DISTRIBUTION ATLAS 1
Unsurveyed
Not Observed
Observed
Possible
Probable
Confirmed

Virginia Rail (*Rallus limicola*)

Photo: Alan MacKeigan

NESTING

CLUTCH SIZE (EGGS):	7–12
INCUBATION (DAYS):	19–20
FLEDGING (DAYS):	25
NEST HEIGHT (METRES):	0

Virginia Rails occur in very low numbers in the Natural Regions where they were recorded. This species is restricted to isolated wetlands, as seen from the distribution maps, but it can be locally abundant where conditions are favourable. It was observed farther north in Atlas 2 than in Atlas 1, with one observation north of Fort McMurray. This observation was within the known range of this species; so, no change in the distribution of Virginia Rails was observed in Atlas 2.

Reporting rates in the Boreal Forest, Parkland, and Rocky Mountain Natural Regions were very low. This scarcity is, without doubt, due to the Virginia Rail's secretive nature; it walks and hops through reeds and cattails. Its infrequent calling and quiet vocalizations contribute further to the difficulty in detecting the presence of Virginia Rails.

A decline in relative abundance was detected in the Boreal Forest NR where, relative to other species, the Virginia Rail was observed less frequently in Atlas 2 than in Atlas 1. During Atlas 2, conditions drier than those encountered during Atlas 1 in this NR would have reduced wetland availability, and this seems a likely cause for the detected decline. No change was detected in other NRs, but the rarity of this species placed it below the statistical threshold of our analytical methods in all NRs except the Boreal Forest. The Breeding Bird Survey sample size was also too small to investigate abundance changes in Alberta and no change was detected on the national scale. Considering the low detectability of this species, it is clear that neither the Atlas nor the Breeding Bird Survey is adequate for monitoring this species, and targeted efforts will be required to determine its status. The Alberta Provincial status report lists the Virginia Rail as Undetermined, due to insufficient data. However, this species' dependence on wetlands, coupled with ongoing loss of wetlands, raises concern for the long-term welfare of this species in Alberta.

TEMPORAL REPORTING RATE

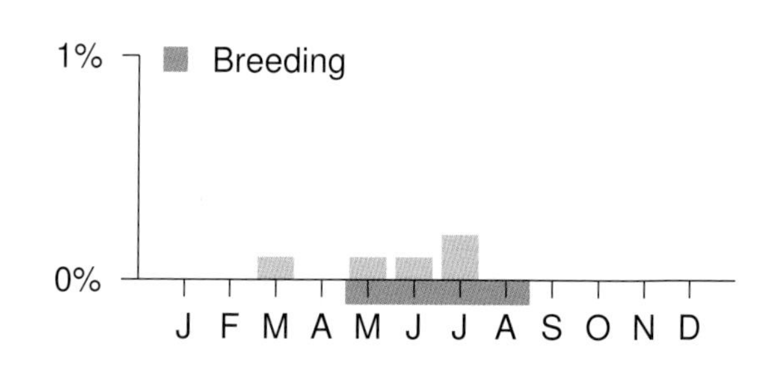

NATURAL REGIONS REPORTING RATE

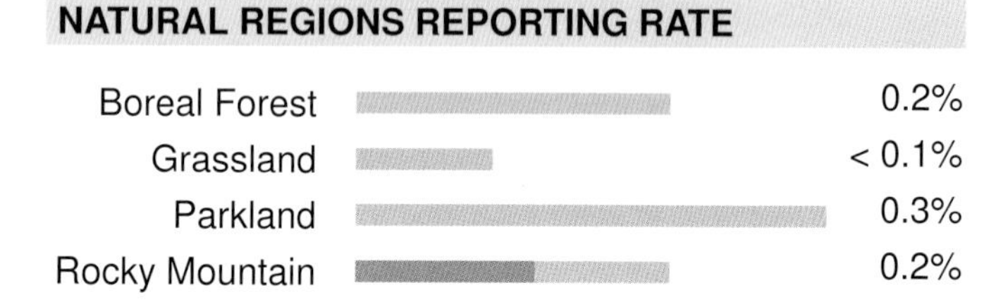

SPATIAL REPORTING RATE

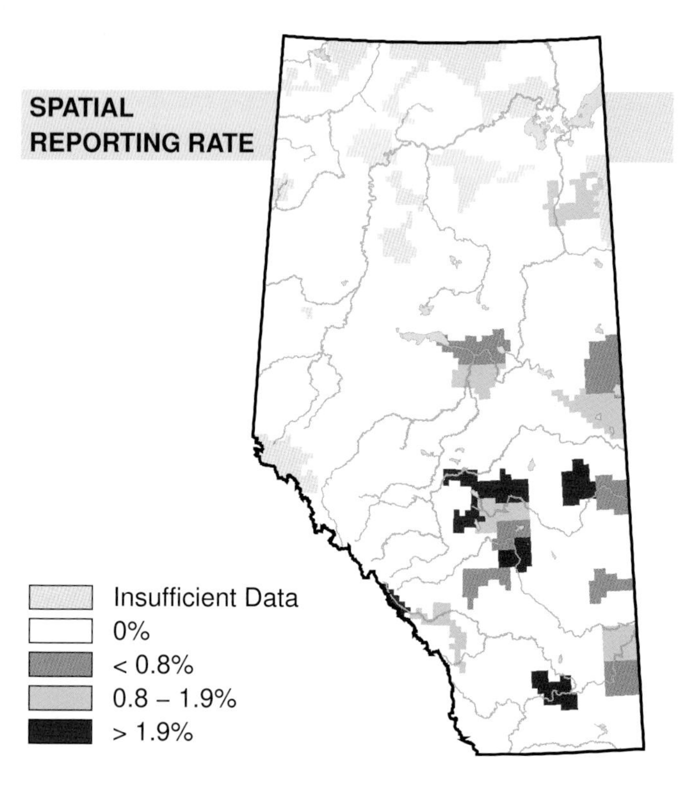

STATUS

GSWA 2000	GSWA 2005	COSEWIC Historic Rankings	COSEWIC 2007	Alberta Wildlife Act
Undetermined	Undetermined	N/A	N/A	N/A

Habitat 

Nest Location 
Nest Type 
Diet 

RANGE

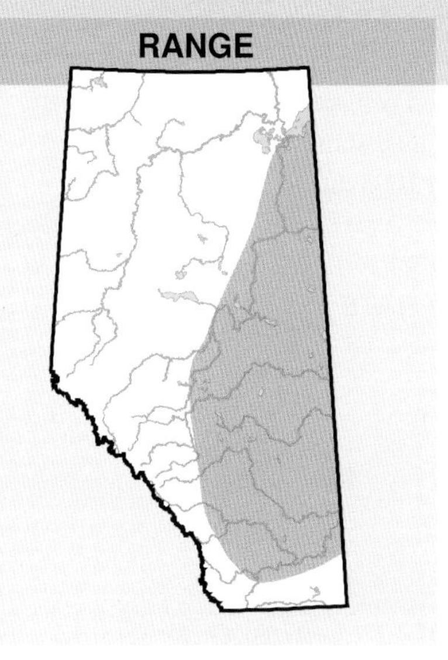

OBSERVED DISTRIBUTION ATLAS 2

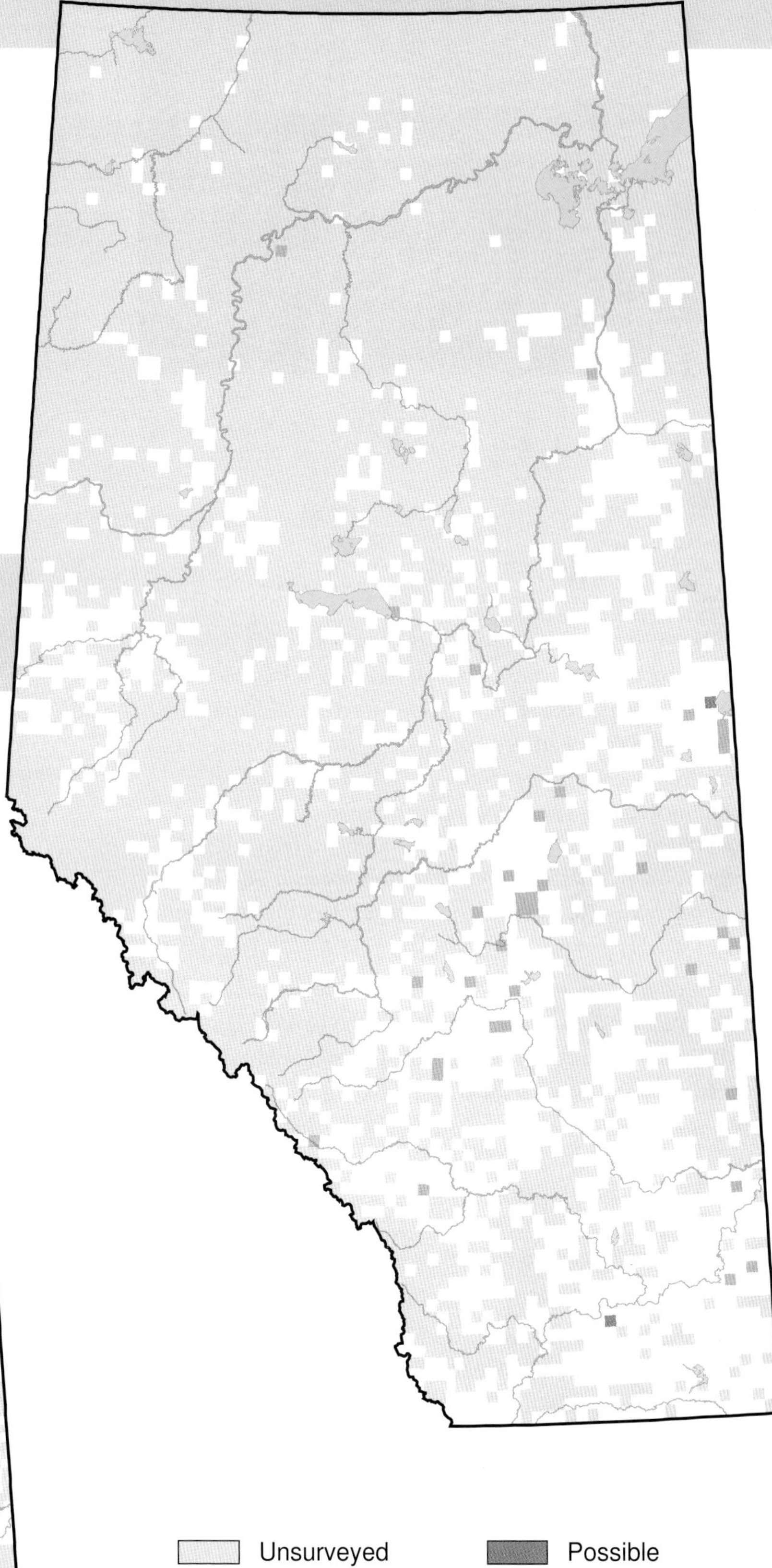

OBSERVED DISTRIBUTION ATLAS 1

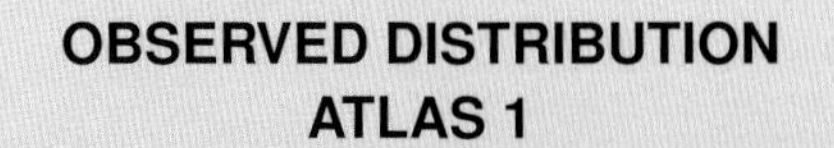

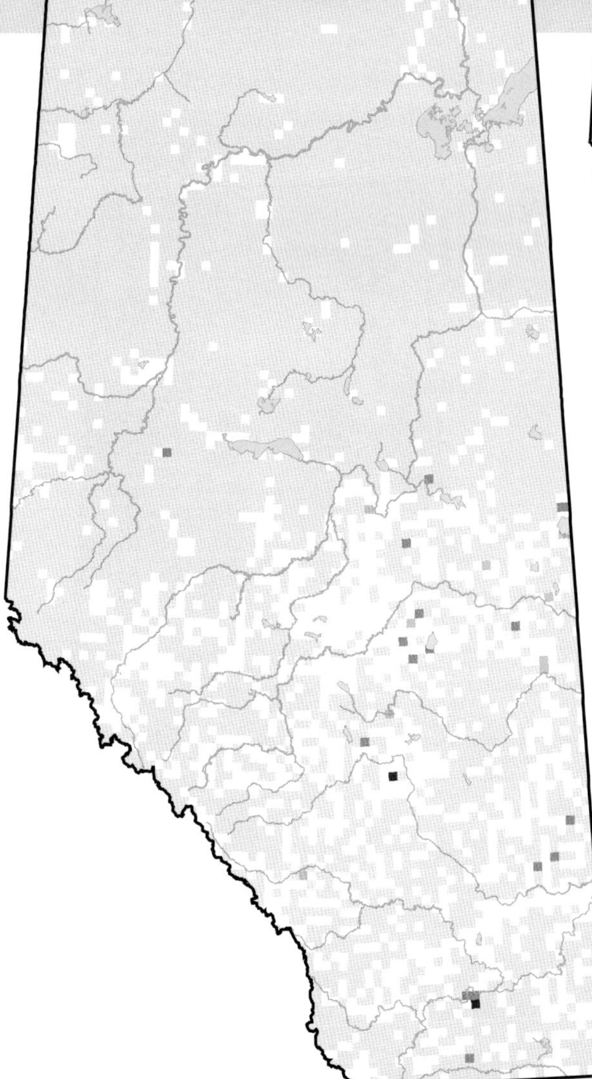

	Unsurveyed		Possible
	Not Observed		Probable
	Observed		Confirmed

Sora (*Porzana carolina*)

Photo: Gerald Romanchuk

NESTING

CLUTCH SIZE (EGGS):	7–13
INCUBATION (DAYS):	16–20
FLEDGING (DAYS):	21–25
NEST HEIGHT (METRES):	0.15–0.30

The Sora is the best known of the three Alberta rails. A species heard more often than seen (75% of records), it was detected in all Natural Regions in Alberta. The observed distribution did not change between Atlas 1 and Atlas 2. However, there was a notable absence of Sora records in the La Crete and Fort Vermilion region in north-central Alberta because there was a gap in survey coverage there. Data are not available for much of the northern Boreal Forest NR for this species. This is likely the result of insufficient survey coverage rather than an absence of this species from the north.

Reporting rates for the Sora were noticeably lower in the Boreal Forest and Foothills NRs than in the Grassland, Parkland, and Rocky Mountain NRs. The preferred habitat of this species is a mix of shallow and moderately deep water with emergent vegetation present. Grassland and Parkland NR wetlands, as well as valley wetlands found in the Rocky Mountain NR, would satisfy Sora requirements because these wetlands tend to be shallow. However, many Boreal Forest and Foothills NR wetlands would not be suitable because there the wetlands are usually deep.

Declines in relative abundance were detected in all natural regions except the Rocky Mountain NR. These declines were matched by results of the Breeding Bird Survey which also detected an abundance decline in Alberta during the period 1985–2005. Wetland loss is likely the main contributor to the decline of this species. An increase in relative abundance was detected in the Rocky Mountain NR but has no known cause, and it appears a questionable trend in view of the general declines in other NRs. Targeted research is required to evaluate this finding. The Sora's provincial status was changed from Secure in 2000 to Sensitive in 2005 due to large declines in their populations.

TEMPORAL REPORTING RATE

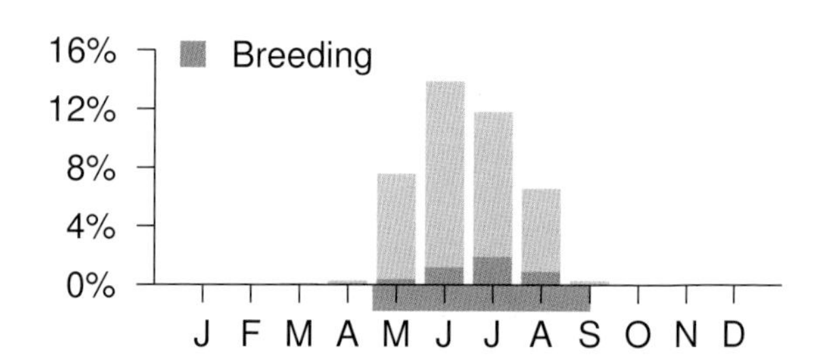

NATURAL REGIONS REPORTING RATE

Region	Rate
Boreal Forest	4.3%
Foothills	3.9%
Grassland	6.1%
Parkland	9.1%
Rocky Mountain	6.5%

SPATIAL REPORTING RATE

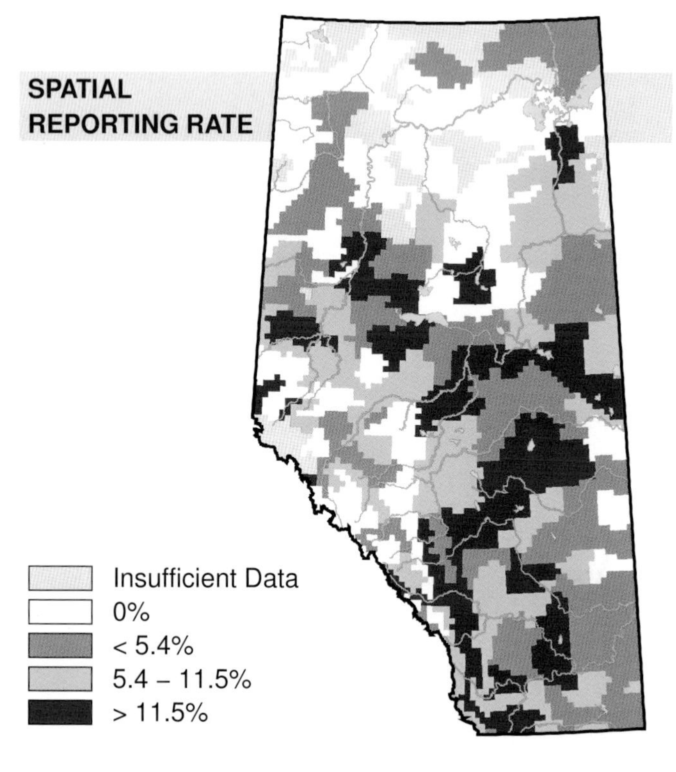

STATUS

GSWA 2000	GSWA 2005	COSEWIC Historic Rankings	COSEWIC 2007	Alberta Wildlife Act
Secure	Sensitive	N/A	N/A	N/A

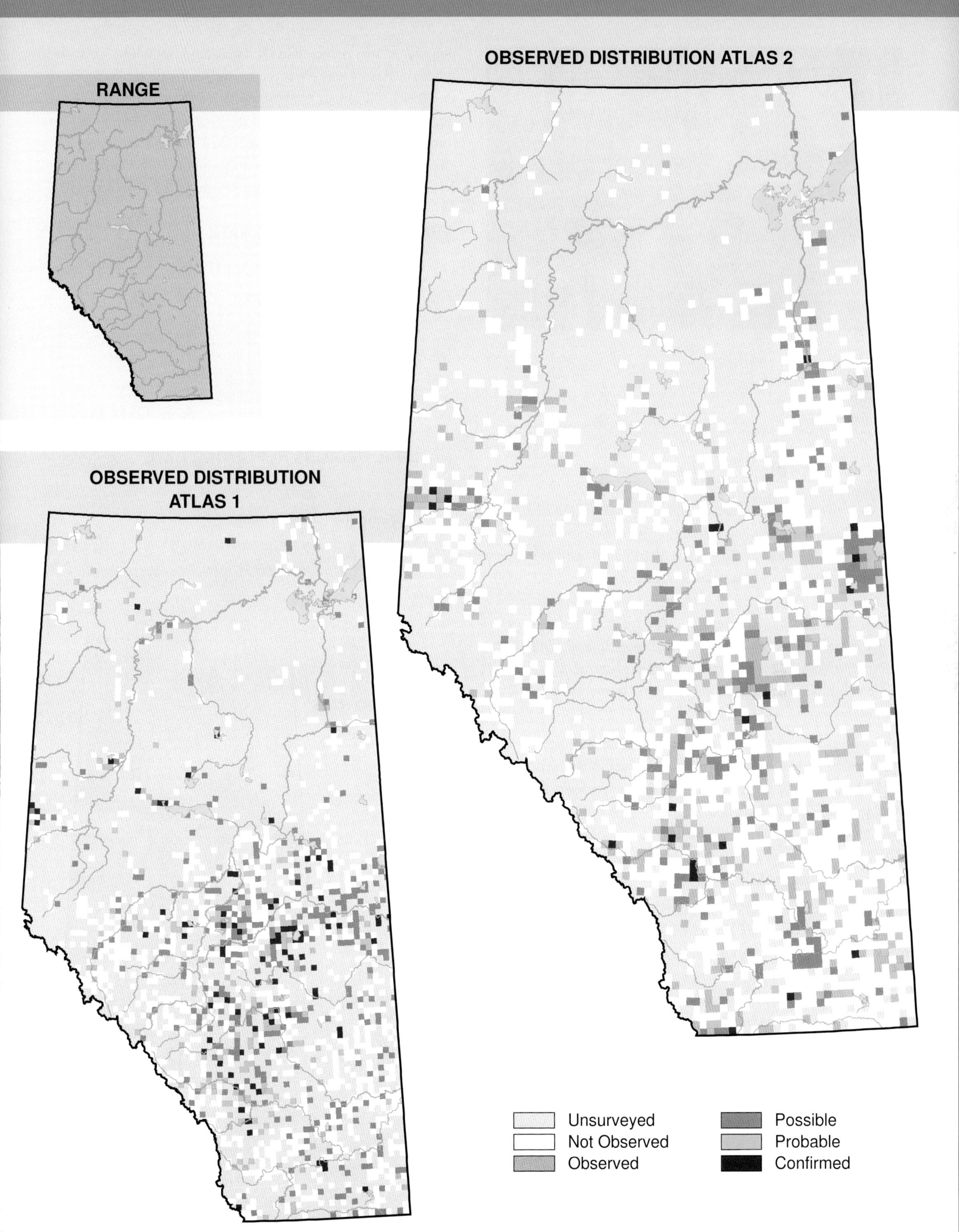
OBSERVED DISTRIBUTION ATLAS 2
RANGE
OBSERVED DISTRIBUTION
ATLAS 1
Unsurveyed
Not Observed
Observed
Possible
Probable
Confirmed

American Coot (*Fulica americana*)

Photo: Mark Williams

NESTING

CLUTCH SIZE (EGGS):	8–12
INCUBATION (DAYS):	23–24
FLEDGING (DAYS):	49–56
NEST HEIGHT (METRES):	0.1–0.3

The American Coot is found in every Natural Region in Alberta. It is distributed across central Alberta and south to the United States border. The observed distribution for this species was similar in Atlas 1 and Atlas 2. However, American Coots were detected less often in the northern part of the province during Atlas 2. Atlas 2 had more extensive coverage than Atlas 1, but in both efforts northern coverage was limited. Thus, it is difficult to know if the observed difference represents a true range contraction or is an artifact of the differences in survey coverage.

Although found in every natural region in Alberta, the American Coot was most often found in the Grassland and Parkland NRs. The preferred habitat of this species is shallow marshes and lakes. Water bodies in the Grassland and Parkland NRs tend to be shallow due to the topography in those Regions. This species was less common in the Boreal Forest, Foothills, and Rocky Mountain NRs where deeper water bodies occur.

Declines in relative abundance were detected in the Boreal Forest and Parkland NRs where, relative to other species, it was observed less frequently in Atlas 2 than in Atlas 1. The Breeding Bird Survey data do not indicate any change in the abundance of American Coots in Alberta; however, declines were detected across Canada during the period 1995–2005. Population declines could be related to habitat loss from drought and drainage of wetlands. An increase in relative abundance was detected in the Grassland NR where, relative to other species, it was observed more frequently in Atlas 2 than in Atlas 1. Above-normal precipitation in the later half of Atlas 2 revitalized many wetlands in the Grasslands NR and appears to be the probable cause for the increase in this area. No change was detected in the Foothills and Rocky Mountain NRs. This species is considered Secure in the province.

TEMPORAL REPORTING RATE

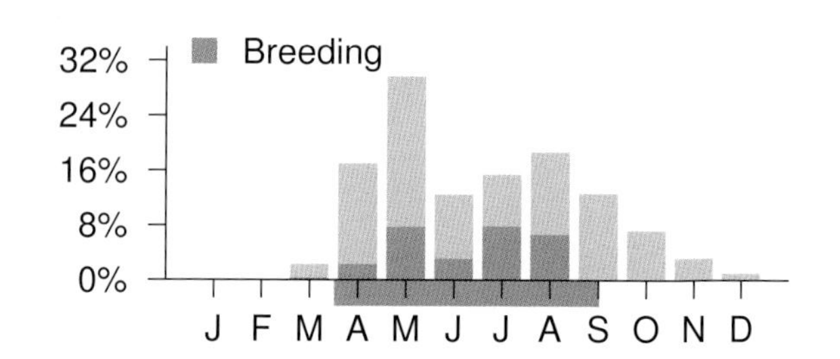

NATURAL REGIONS REPORTING RATE

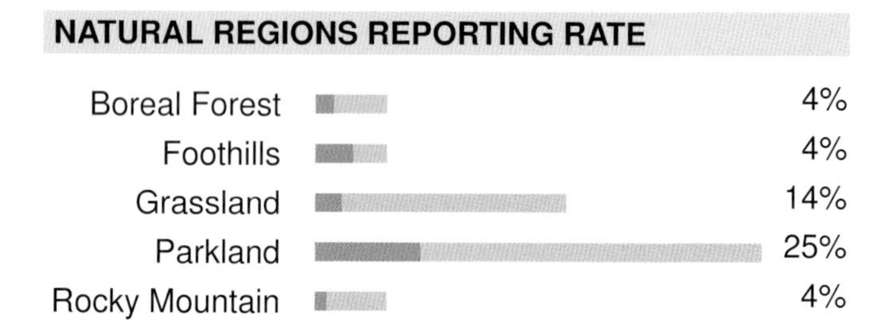

SPATIAL REPORTING RATE

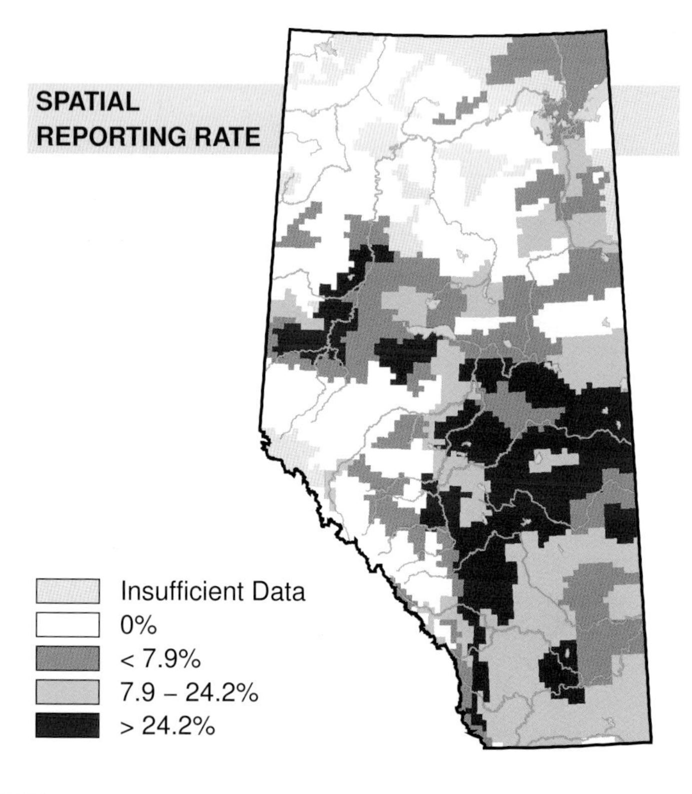

STATUS

GSWA 2000	GSWA 2005	COSEWIC Historic Rankings	COSEWIC 2007	Alberta Wildlife Act
Secure	Secure	Not At Risk	Not At Risk	N/A

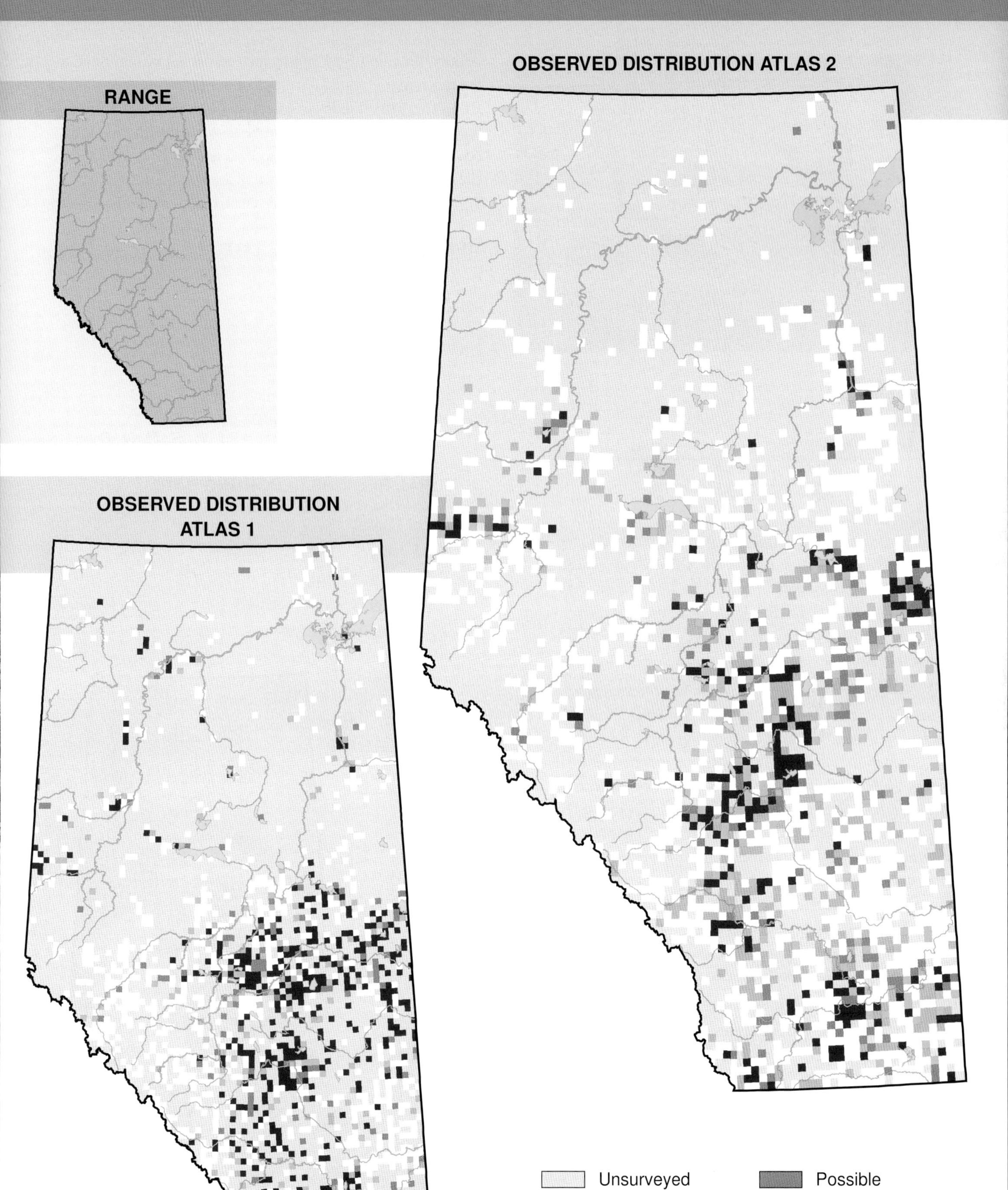
RANGE
OBSERVED DISTRIBUTION ATLAS 2
OBSERVED DISTRIBUTION ATLAS 1
Unsurveyed
Not Observed
Observed
Possible
Probable
Confirmed

Sandhill Crane (*Grus canadensis*)

Photo: Gordon Court

NESTING

CLUTCH SIZE (EGGS):	2
INCUBATION (DAYS):	29–32
FLEDGING (DAYS):	70
NEST HEIGHT (METRES):	0

Sandhill Cranes breed in the Boreal Forest, Foothills, and Rocky Mountain Natural Regions in Alberta. No change in distribution was observed in Atlas 2 for this species. Few records were obtained in the northwestern part of the province, likely the result of less extensive survey coverage in this area rather than an actual range contraction.

Sandhill Cranes nest on the ground in wet forested areas usually near small ponds or marshes. For this reason, this species was observed most often during the breeding season in the Boreal Forest and Foothills NRs and less frequently in the Rocky Mountain NR. Observations in the Grassland and Parkland NRs were mostly non-breeders, except for a few breeding records obtained east of Waterton Lakes National Park and in the transition zone from the Foothills NR, near Rocky Mountain House. Unlike Alberta's other crane species, the Whooping Crane, Sandhill Crane pairs are usually solitary nesters. For this reason, Sandhill Cranes were widely distributed across their range. In contrast, Whooping Crane breeding distribution was limited to a small concentrated area of northern Alberta.

An increase in relative abundance was detected in the Boreal Forest NR. No changes were detected in the Foothills and Rocky Mountain NRs. The Breeding Bird Survey did not detect an abundance change in Alberta, but an abundance increase was detected across Canada during the period 1995–2005. Differences between the results of the Atlas and Alberta Breeding Bird Survey analyses could be attributed to the much more limited coverage of northern Alberta by the BBS. The Atlas did detect a decline in the Parkland NR, but as records in that NR were primarily associated with migration, this change must be cautiously considered. The provincial status of this species is Sensitive in Alberta because it is vulnerable to wetland loss and sensitive to human disturbance.

TEMPORAL REPORTING RATE

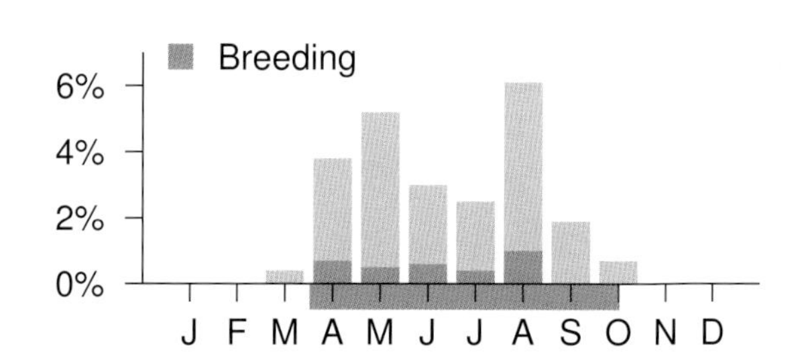

NATURAL REGIONS REPORTING RATE

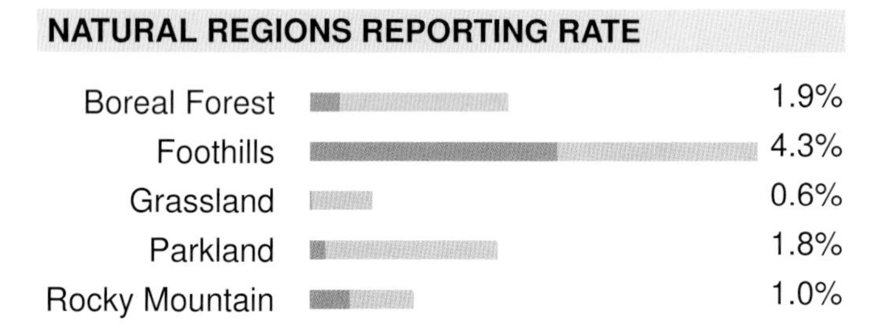

SPATIAL REPORTING RATE

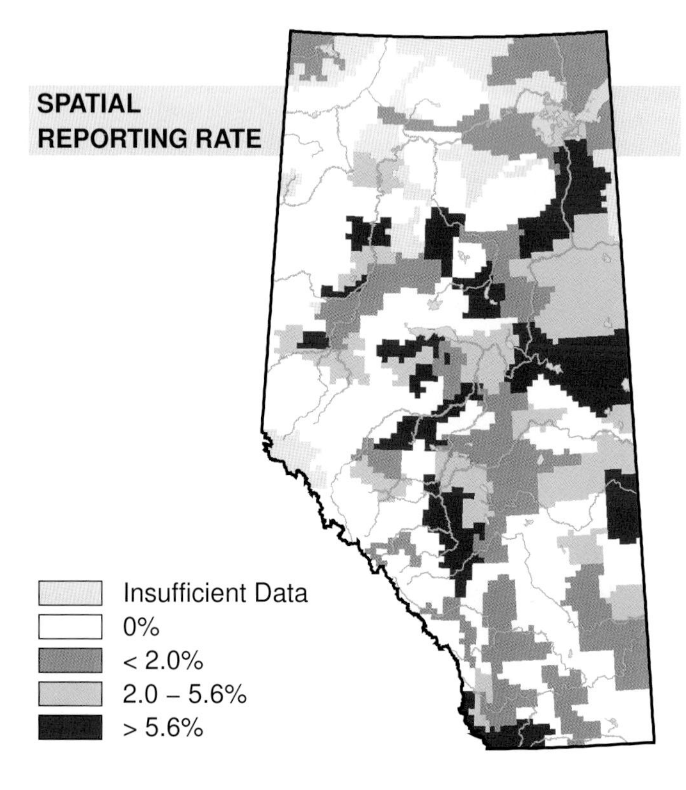

STATUS

GSWA 2000	GSWA 2005	COSEWIC Historic Rankings	COSEWIC 2007	Alberta Wildlife Act
Sensitive	Sensitive	N/A	N/A	N/A

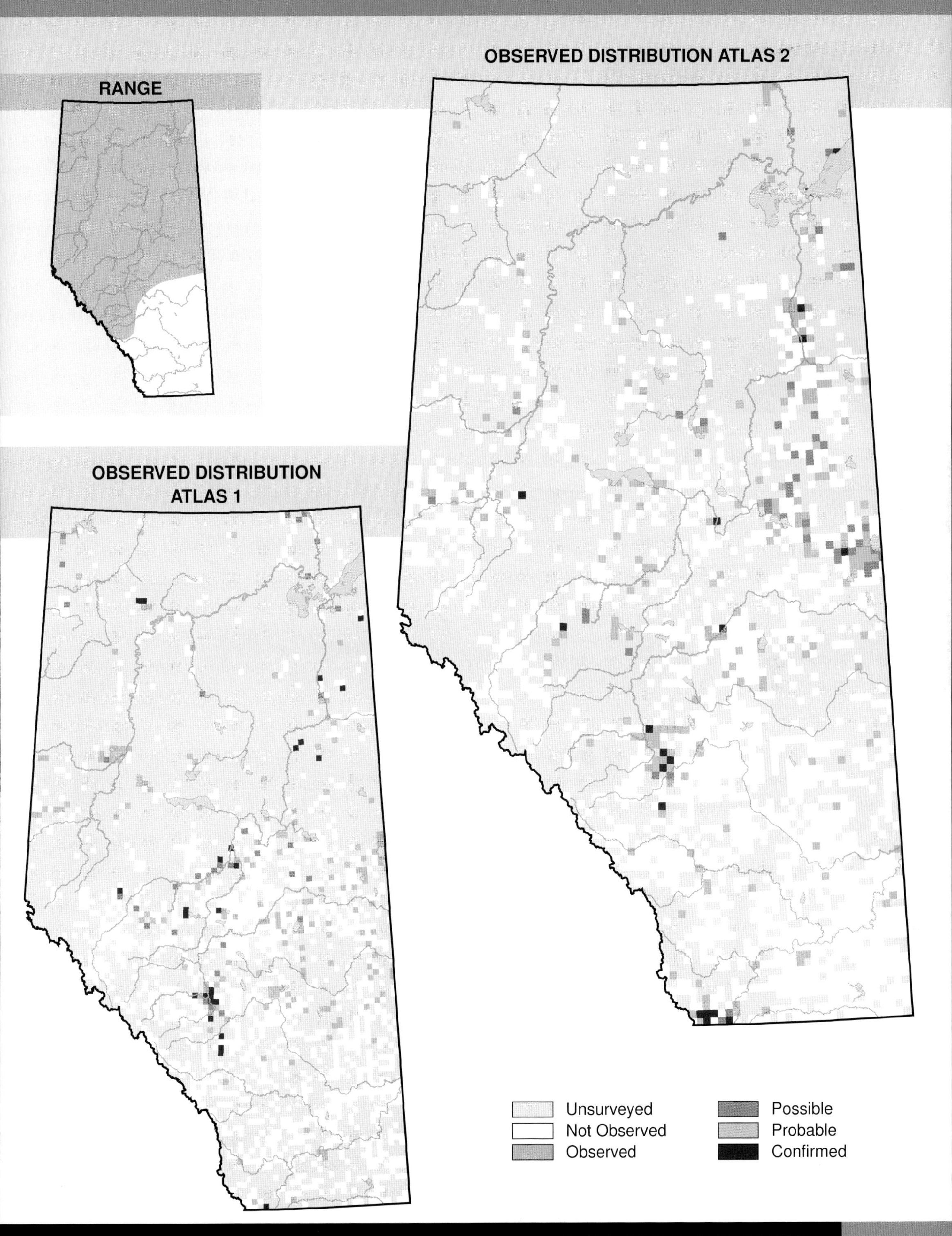
Habitat
Nest Location
Nest Type
Diet
RANGE
OBSERVED DISTRIBUTION ATLAS 2
OBSERVED DISTRIBUTION ATLAS 1
Unsurveyed
Not Observed
Observed
Possible
Probable
Confirmed

Whooping Crane (*Grus americana*)

Photo: Gordon Court

NESTING

CLUTCH SIZE (EGGS):	2
INCUBATION (DAYS):	29–31
FLEDGING (DAYS):	14
NEST HEIGHT (METRES):	0

A true symbol of conservation, the Whooping Crane is one of the rarest birds in North America. Historically, Whooping Crane populations declined due to over-harvest, degradation of habitat, and human disturbance. In Alberta, this species breeds only in the northern part of the Boreal Forest Natural Region in Wood Buffalo National Park. The observed distribution of this species did not change between Atlas 1 and Atlas 2.

Whooping Cranes nest on the ground in areas surrounded by shallow wetlands of varying sizes. The breeding area is typically a Boreal Forest wetland matrix and is interspersed with stands of White Spruce, Black Spruce, Tamarack, and willows. Availability of this habitat does not limit population growth of this species because the prevalence of this habitat-type is abundant around the area that is currently used for nesting.

No changes in Whooping Crane relative abundance were detected. The Breeding Bird Survey sample size of Whooping Crane records was too small to investigate temporal abundance changes. Due to its small population and the limited breeding distribution of this species, it is not effectively monitored by large, wide-spread, multi-species surveys such as the Atlas or Breeding Bird Survey. Whooping Crane-specific surveys indicate that populations are fairly stable. Population increases would happen slowly: this species is fairly long-lived, is not able to reproduce until after it is 4 years old, and it has a low yearly reproductive output when it does breed. The provincial status of the Whooping Crane is At Risk because of its small population and the potential threats to wetlands.

TEMPORAL REPORTING RATE

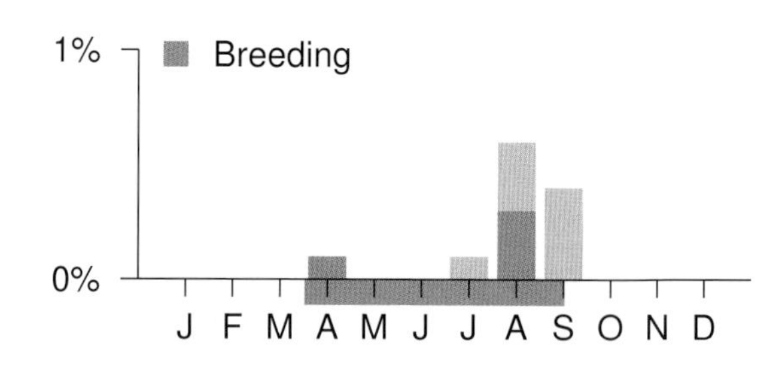

NATURAL REGIONS REPORTING RATE

SPATIAL REPORTING RATE

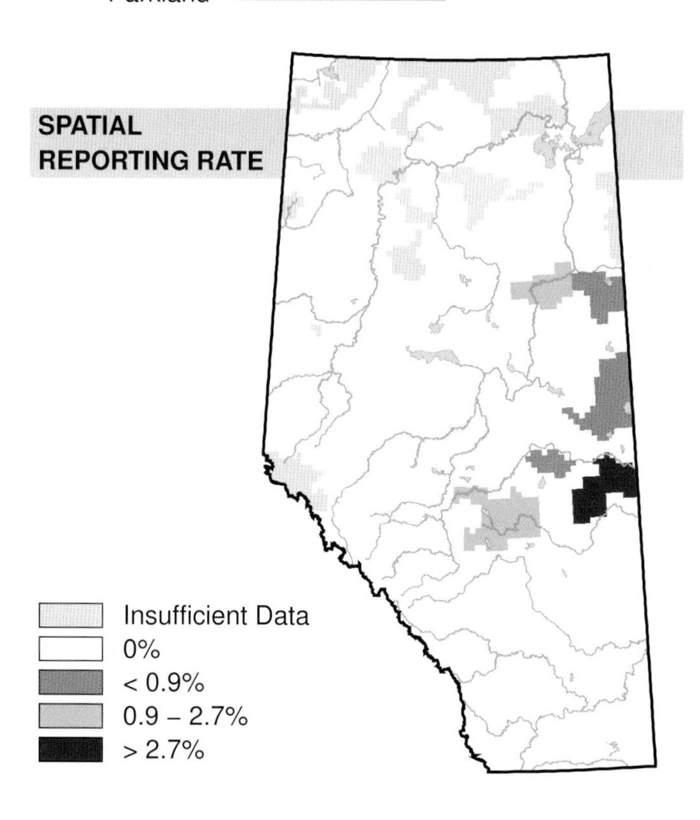

STATUS

GSWA 2000	GSWA 2005	COSEWIC Historic Rankings	COSEWIC 2007	Alberta Wildlife Act
At Risk	At Risk	Endangered	Endangered	Endangered

Habitat

Nest
Location

Nest
Type

Diet

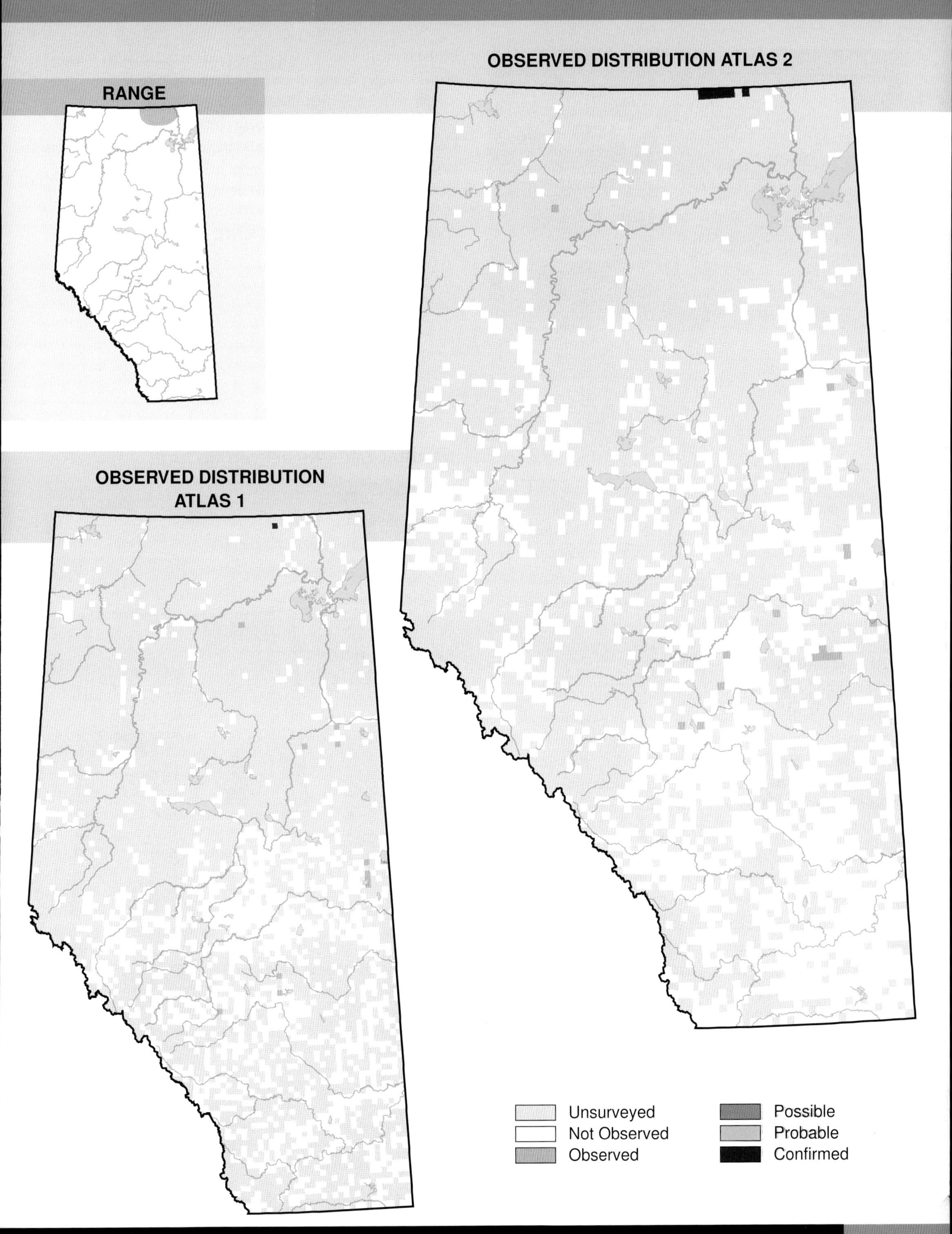
OBSERVED DISTRIBUTION ATLAS 2
RANGE
OBSERVED DISTRIBUTION
ATLAS 1
Unsurveyed
Not Observed
Observed
Possible
Probable
Confirmed

Semipalmated Plover (*Charadrius semipalmatus*)

Photo: Gerry Beyersbergen

NESTING

CLUTCH SIZE (EGGS):	4
INCUBATION (DAYS):	23
FLEDGING (DAYS):	22–31
NEST HEIGHT (METRES):	0

The Semipalmated Plover has been known to breed only in the northeastern part of Alberta, in the Boreal Forest Natural Region. No breeding records for this species were obtained in Atlas 1 or Atlas 2. No change in distribution was observed for this species.

In Alberta, Semipalmated Plovers were most often encountered during the spring and fall migrations when they were moving between their coastal winter range and their sub-Arctic and Arctic breeding range. These migrants were most often encountered in the Grassland NR but were also occasionally observed in the Boreal Forest, Foothills, Parkland, and Rocky Mountain NRs. The southern extent of the Semipalmated Plover's breeding range dips into the northeast corner of Alberta. Even though there are a few breeding records from this area, it is not known if Semipalmated Plovers are regular breeders in the province, given the remote nature of their potential breeding areas and the difficulties of access for monitoring them.

Increases in relative abundance were detected in the Grassland NR but there were decreases in the Boreal Forest and Parkland NRs. With no breeding records and the majority of records being recorded during migration, detected changes must be interpreted with caution. The broader survey window of Atlas 2, as well as the inherent variability of migration records, could be the primary causes for the observed changes, rather than actual changes in abundance. The Breeding Bird Survey sample size for the Semipalmated Plover was too small to investigate abundance changes, in either Alberta or Canada. Due to the northern range of this species' breeding distribution, targeted efforts are needed to monitor this species effectively. The status of this species is Secure in Alberta.

TEMPORAL REPORTING RATE

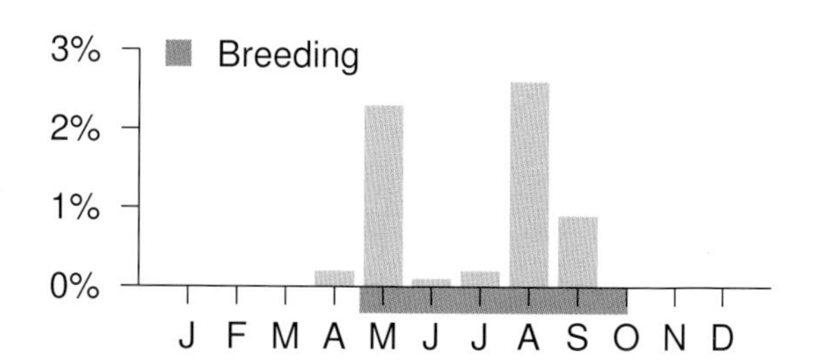

NATURAL REGIONS REPORTING RATE

Region	Rate
Boreal Forest	0.3%
Grassland	1.5%
Parkland	0.7%
Rocky Mountain	0.8%

SPATIAL REPORTING RATE

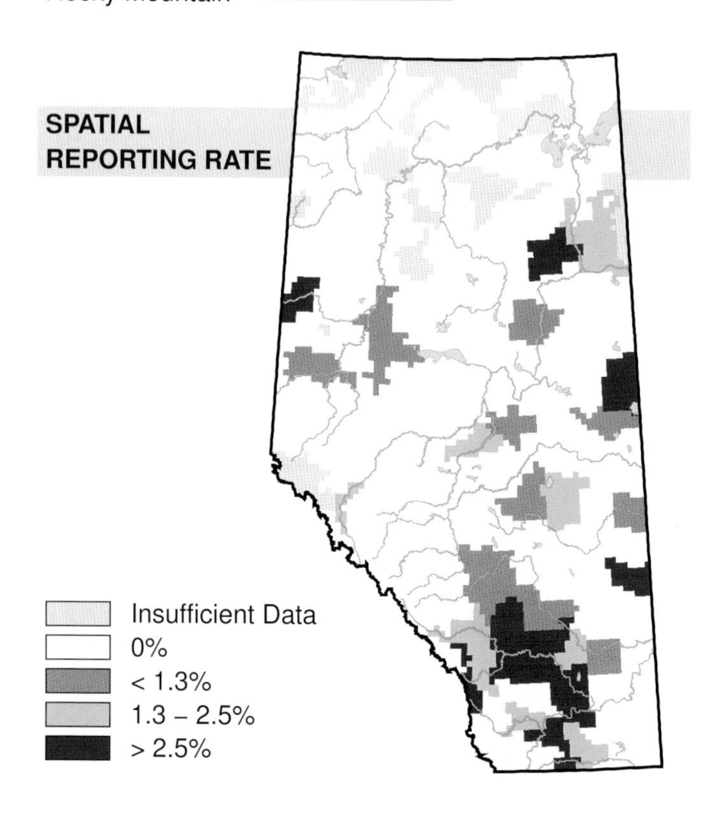

STATUS

GSWA 2000	GSWA 2005	COSEWIC Historic Rankings	COSEWIC 2007	Alberta Wildlife Act
Secure	Secure	N/A	N/A	N/A

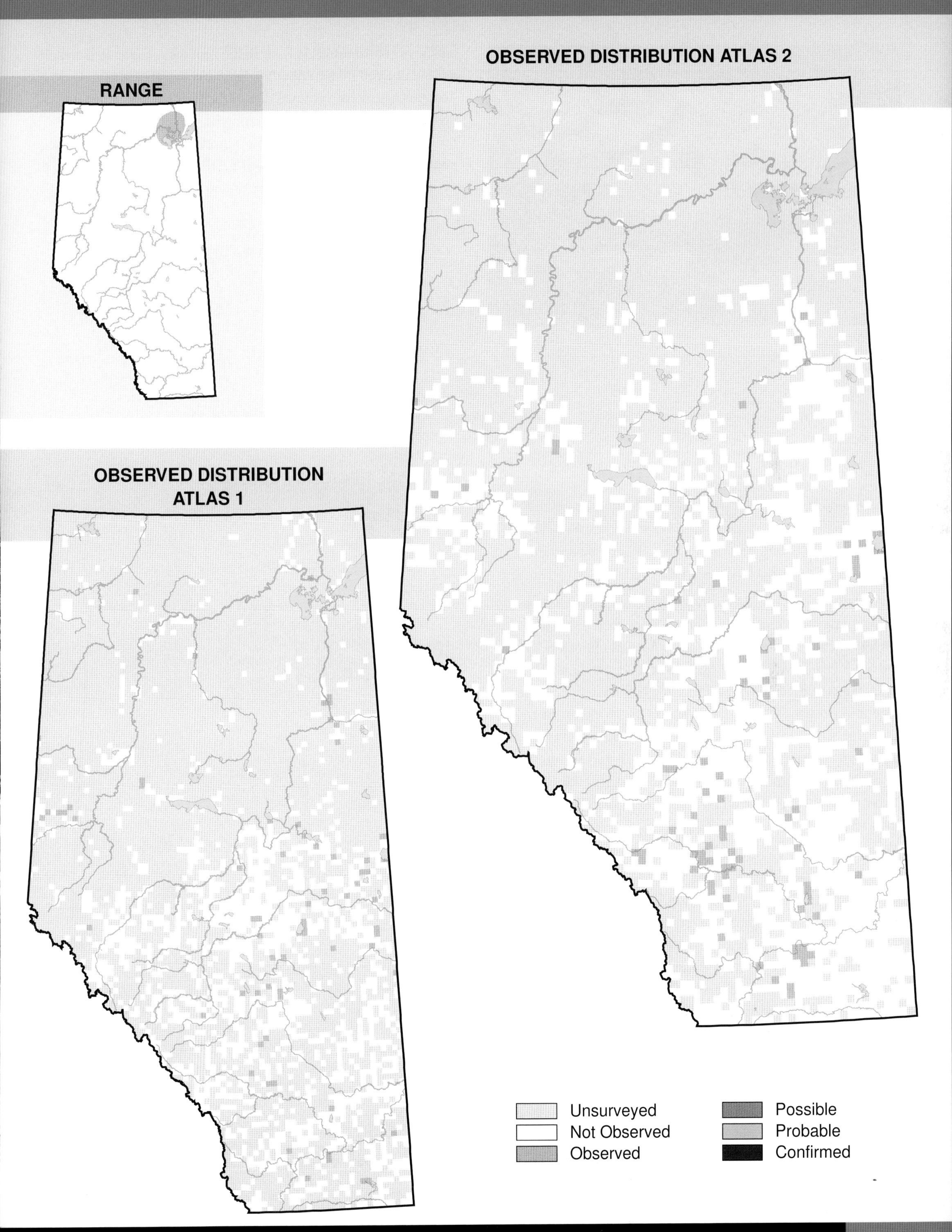
Habitat
Nest Location
Nest Type
Diet
RANGE
OBSERVED DISTRIBUTION ATLAS 2
OBSERVED DISTRIBUTION ATLAS 1
Unsurveyed
Not Observed
Observed
Possible
Probable
Confirmed

Piping Plover (*Charadrius melodus*)

Photo: Gordon Court

NESTING

CLUTCH SIZE (EGGS):	3–5
INCUBATION (DAYS):	27–31
FLEDGING (DAYS):	28–35
NEST HEIGHT (METRES):	0

The Piping Plover is found in the Parkland and Grassland Natural Regions of Alberta and at one site—Muriel Lake—in the Boreal Forest NR. The distribution of this species did not change between Atlas 1 and Atlas 2.

In Alberta, the Piping Plover inhabits shorelines of alkaline lakes. Nests are typically built on wide, exposed, sand or gravel beaches sparsely vegetated with small clumps of grass, and are located away from water. Nesting depends on the maintenance of lake levels sufficiently high to keep the vegetation down. These birds forage within 5 metres of waterline where they glean invertebrates from rocks and vegetation. Because of this special habitat requirement, they are found most commonly in the Parkland NR and less commonly in the Grassland NR. Only one breeding site is known to exist in the Boreal Forest NR.

No changes in relative abundance were detected by the Atlas. The Breeding Bird Survey failed to provide trend data for Alberta or Canada, for any time periods, due to low sample sizes. Only 200 individuals are estimated to breed in Alberta. Due to the low abundance of this species and its secretive nature, it is not effectively monitored by large, wide-spread, multi-species surveys such as the Atlas or Breeding Bird Survey. Piping Plover-specific surveys would be needed to determine population trends adequately for this species. This species is considered At Risk in the province, and is designated as Endangered under Alberta's Wildlife Act. These designations were assigned due to the bird's dependence on specific habitat types that are being negatively affected by human disturbance, and to their clumped distribution. Declines are thought to be caused by poor nesting success and low adult survivorship which, in turn, are due to predators, drought, and water management.

TEMPORAL REPORTING RATE

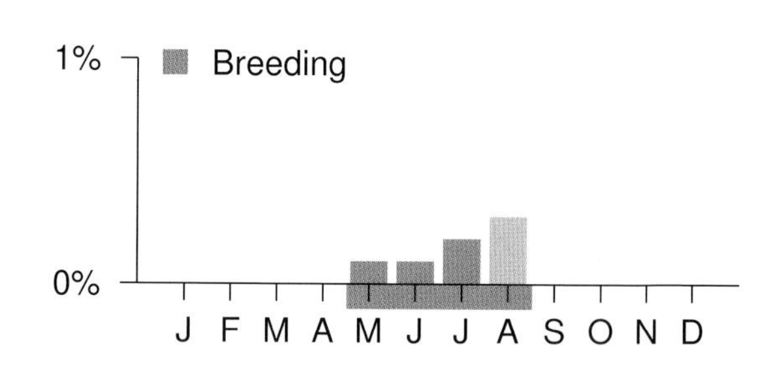

NATURAL REGIONS REPORTING RATE

SPATIAL REPORTING RATE

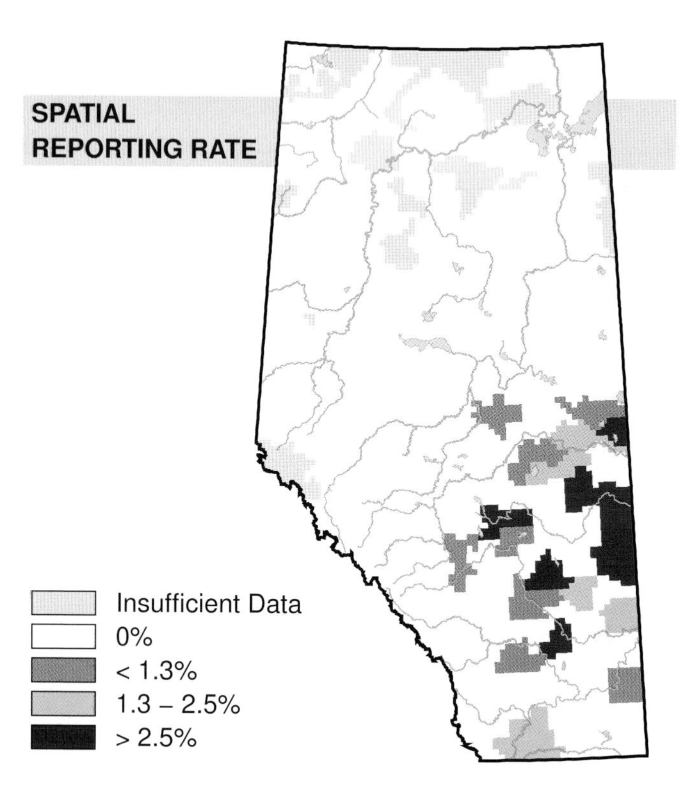

STATUS

GSWA 2000	GSWA 2005	COSEWIC Historic Rankings	COSEWIC 2007	Alberta Wildlife Act
At Risk	At Risk	Endangered	Endangered	Endangered

OBSERVED DISTRIBUTION ATLAS 2

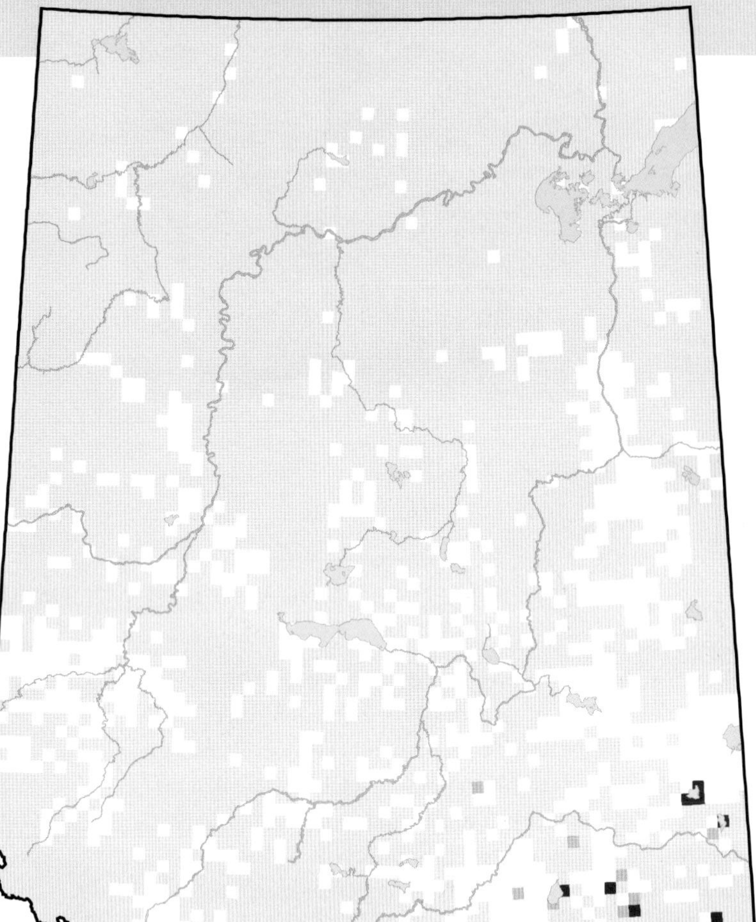

RANGE

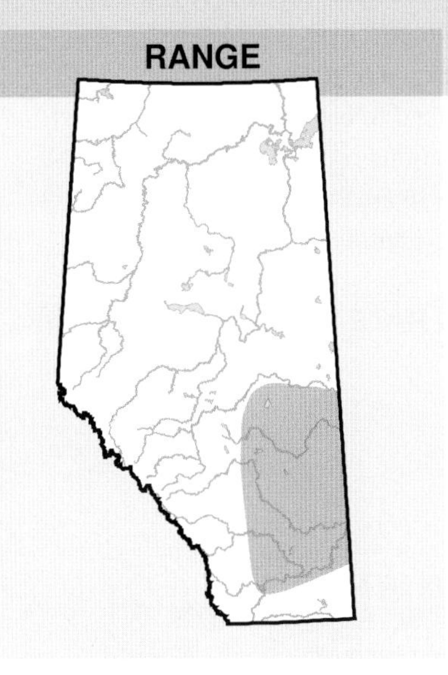

OBSERVED DISTRIBUTION ATLAS 1

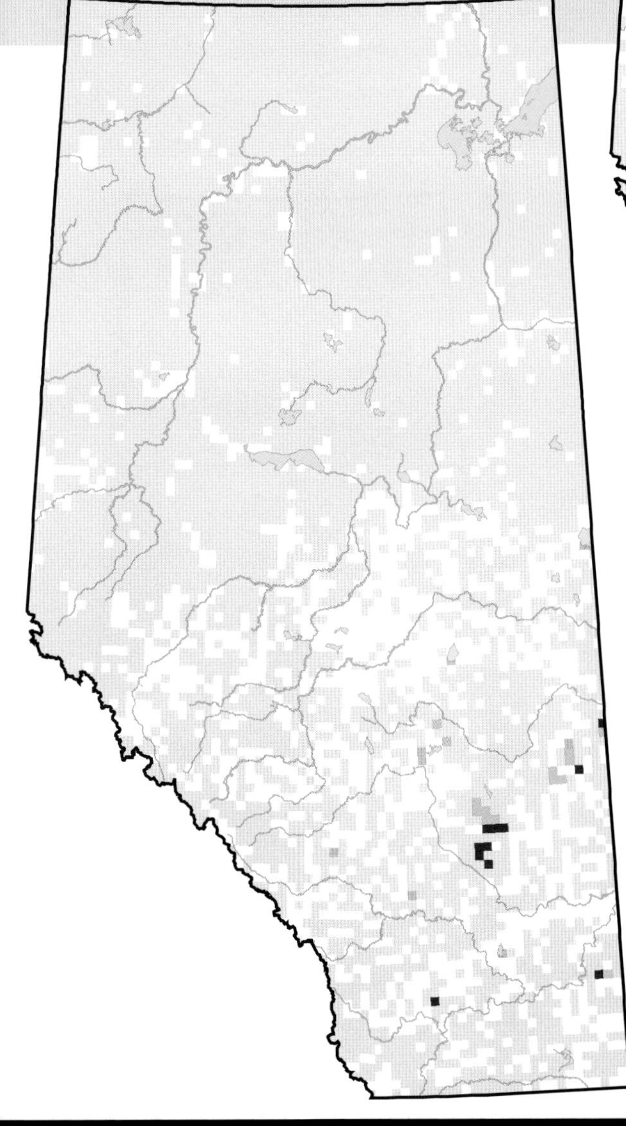

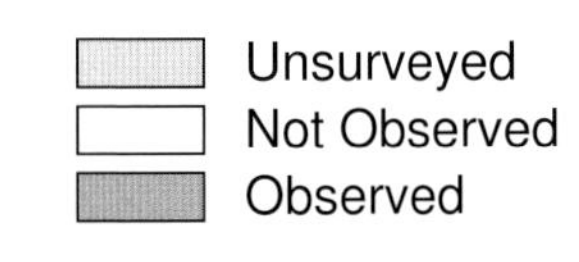

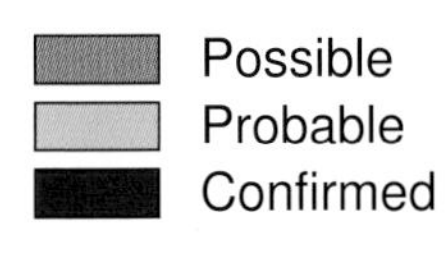

Killdeer (*Charadrius vociferus*)

Photo: Gary Kurtz

NESTING

CLUTCH SIZE (EGGS):	4
INCUBATION (DAYS):	24–26
FLEDGING (DAYS):	30
NEST HEIGHT (METRES):	0

The Killdeer is found in every Natural Region in Alberta. The observed distribution of this species decreased in Atlas 2 in the northwestern part of the province around High Level and Fort Vermilion. It is not possible to be completely certain that the presence of the Killdeer was not missed in surveys conducted in the area, but the substantive loss in records in this area suggests that this may be a true contraction in its Alberta distribution.

The Killdeer nests on the ground in open areas such as sandbars, mudflats and fields with short vegetation. So, this species tends to be found in many human-modified environments, such as on gravel roads and parking lots, and in sports fields, cultivated fields, and pastureland. In contrast to the situation with most shorebirds, Killdeer breeding sites are not required to be near wetlands. Even though Killdeer occur in every natural region in the province, they were found most commonly in the Grassland and Parkland NRs, which likely reflects the bird's preference to nest in open areas.

Declines in relative abundance were detected in the Boreal Forest, Foothills, and Parkland NRs where, relative to other species, Killdeer were observed less frequently in Atlas 2 than in Atlas 1. No changes were detected in the Grassland and Rocky Mountain NRs. The Breeding Bird Survey found a decline in Killdeer abundance in Alberta during the period 1985–2005. The decline found by the Atlas and the Alberta Breeding Bird Survey for this species is mirrored in the decline identified in the Canada-wide Breeding Bird Survey. It is unclear why Killdeer population declines have been detected. However, it is probably not related to reduced habitat availability because Killdeer tend to breed in human-modified environments. Killdeer populations could be declining because the birds feed on invertebrates in areas where humans use pesticides. This species is considered Secure in Alberta.

TEMPORAL REPORTING RATE

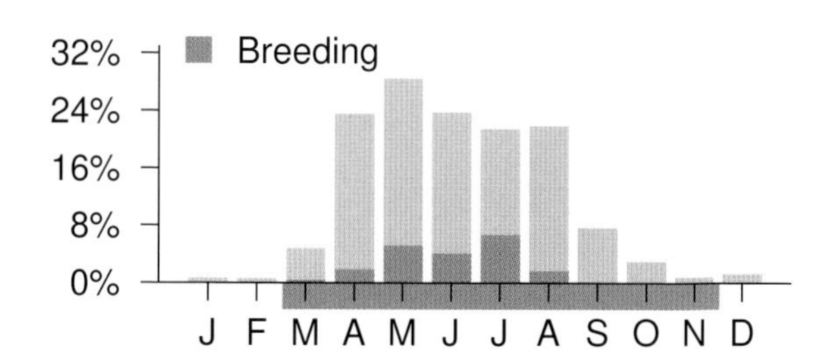

NATURAL REGIONS REPORTING RATE

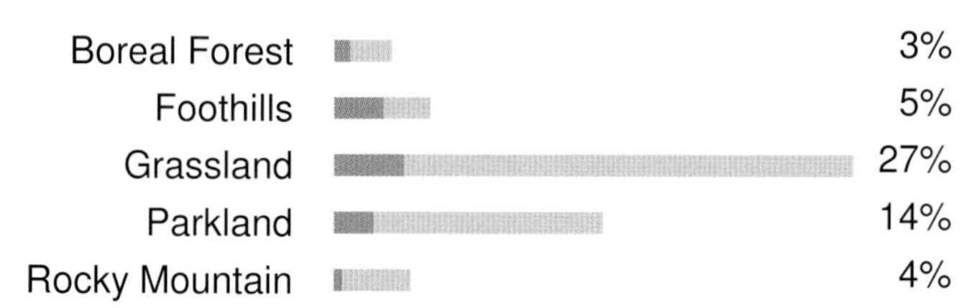

SPATIAL REPORTING RATE

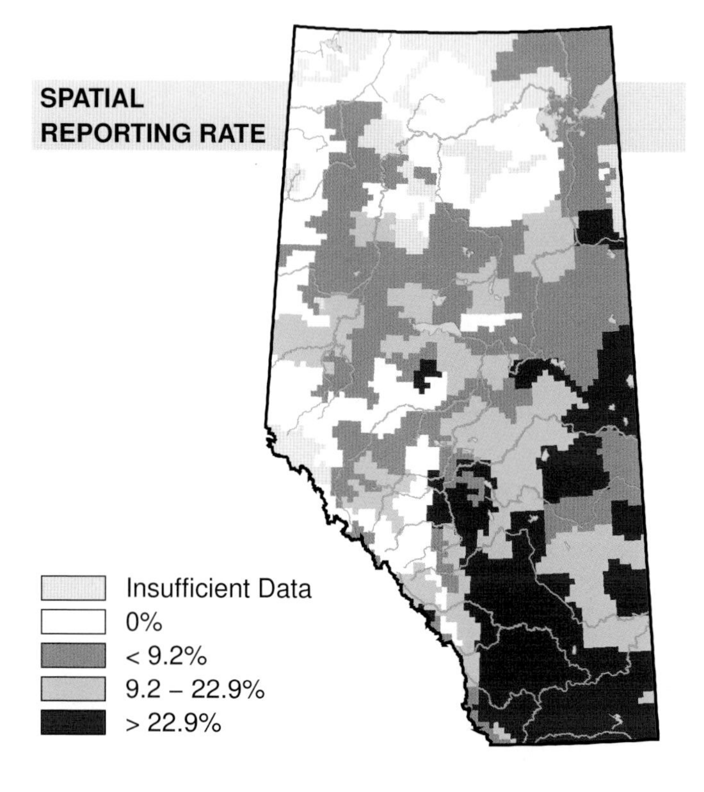

STATUS

GSWA 2000	GSWA 2005	COSEWIC Historic Rankings	COSEWIC 2007	Alberta Wildlife Act
Secure	Secure	N/A	N/A	N/A

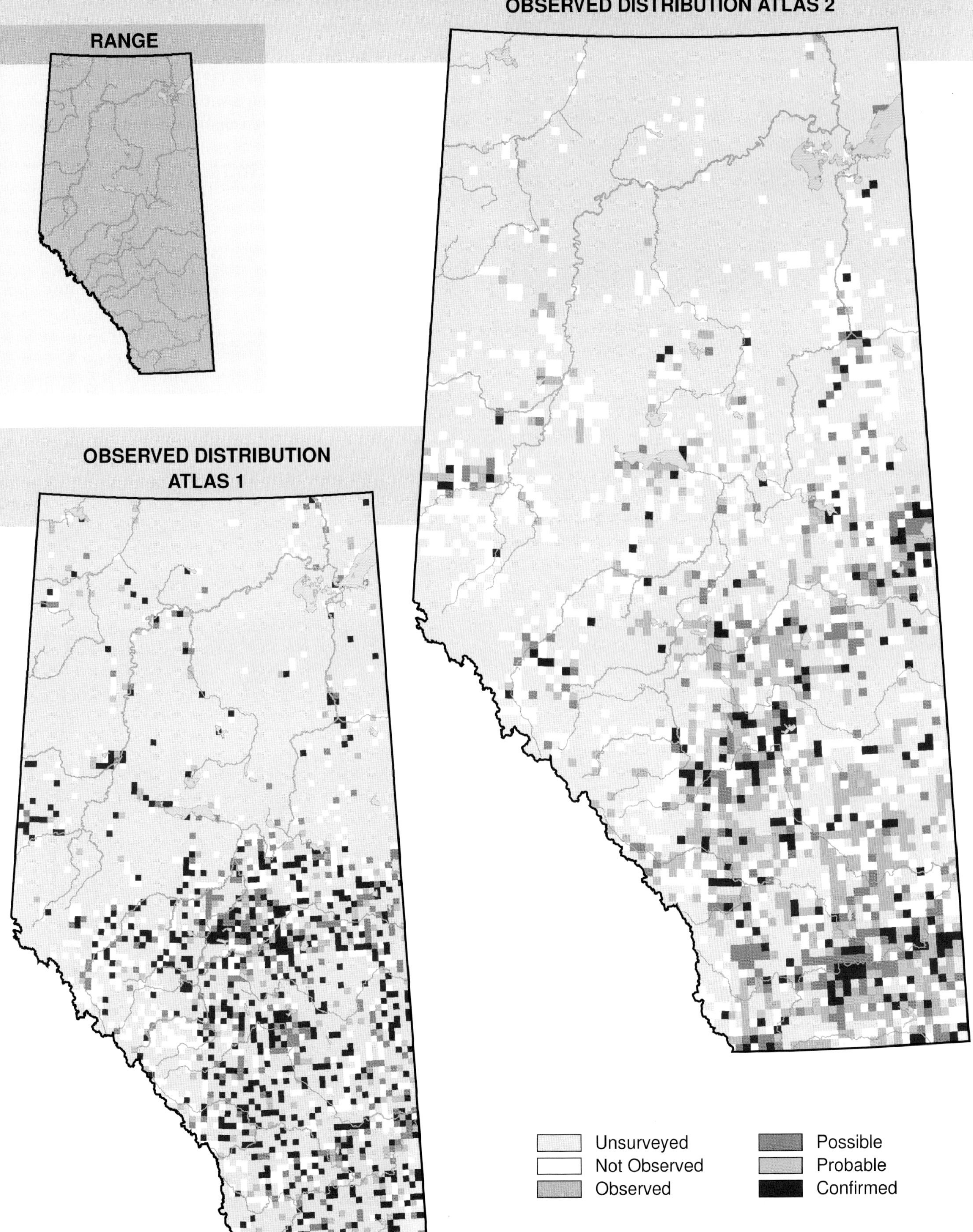
OBSERVED DISTRIBUTION ATLAS 2
RANGE
OBSERVED DISTRIBUTION
ATLAS 1
Unsurveyed
Not Observed
Observed
Possible
Probable
Confirmed

Mountain Plover (*Charadrius montanus*)

Photo: Gordon Court

NESTING

CLUTCH SIZE (EGGS):	1–4
INCUBATION (DAYS):	29
FLEDGING (DAYS):	33–34
NEST HEIGHT (METRES):	0

One of Alberta's rarest shorebirds, the Mountain Plover breeds in the southeastern part of the province's Grassland Natural Region. No change in distribution was observed between Atlas 1 and Atlas 2 for this species.

Unlike many other shorebirds that are associated with wet areas, the Mountain Plover breeds in dry prairie habitats with sparse, low vegetation. This species is able to nest in human-modified environments such as cultivated areas or pasture land, provided that the area is flat and vegetation is either not present or low-growing. As a result, this species was found only in the Grassland NR.

No change in relative abundance was detected in Alberta and observation rates for this species in both Atlas 1 and Atlas 2 were very low. The Breeding Bird Survey sample size of Mountain Plover records was too small in Alberta and Canada to investigate abundance changes. Survey-wide analyses of BBS data detected no change in abundance for the period 1987–2005. Due to the small population and limited breeding distribution of this species, it is not effectively monitored by large, wide-spread, multi-species surveys such as the Atlas or the Breeding Bird Survey. Mountain Plover-specific surveys would be needed to determine whether its population is changing. The provincial status was changed from Sensitive in 2000 to At Risk in 2005, and nationally it is ranked as Endangered by COSEWIC. These rankings are so assessed because the species' population is small and it has a very strong requirement for open nest sites in native grasslands.

TEMPORAL REPORTING RATE

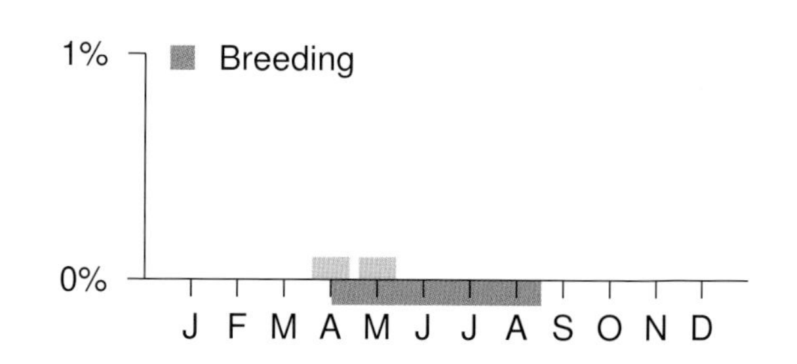

NATURAL REGIONS REPORTING RATE

Grassland  < 0.1%

SPATIAL REPORTING RATE

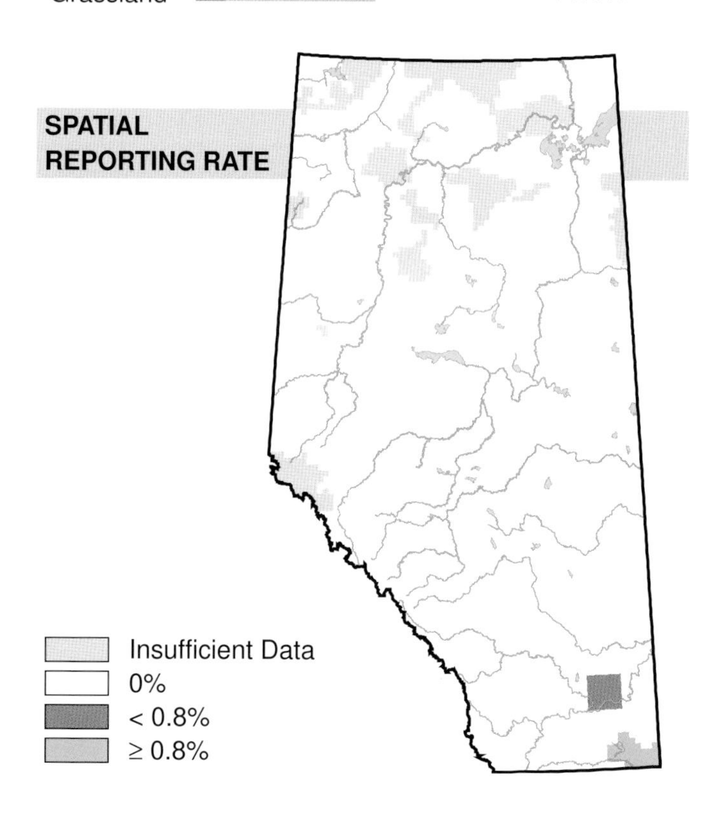

STATUS

GSWA 2000	GSWA 2005	COSEWIC Historic Rankings	COSEWIC 2007	Alberta Wildlife Act
Sensitive	At Risk	Endangered	Endangered	Endangered

Habitat

Nest
Location

Nest
Type

Diet

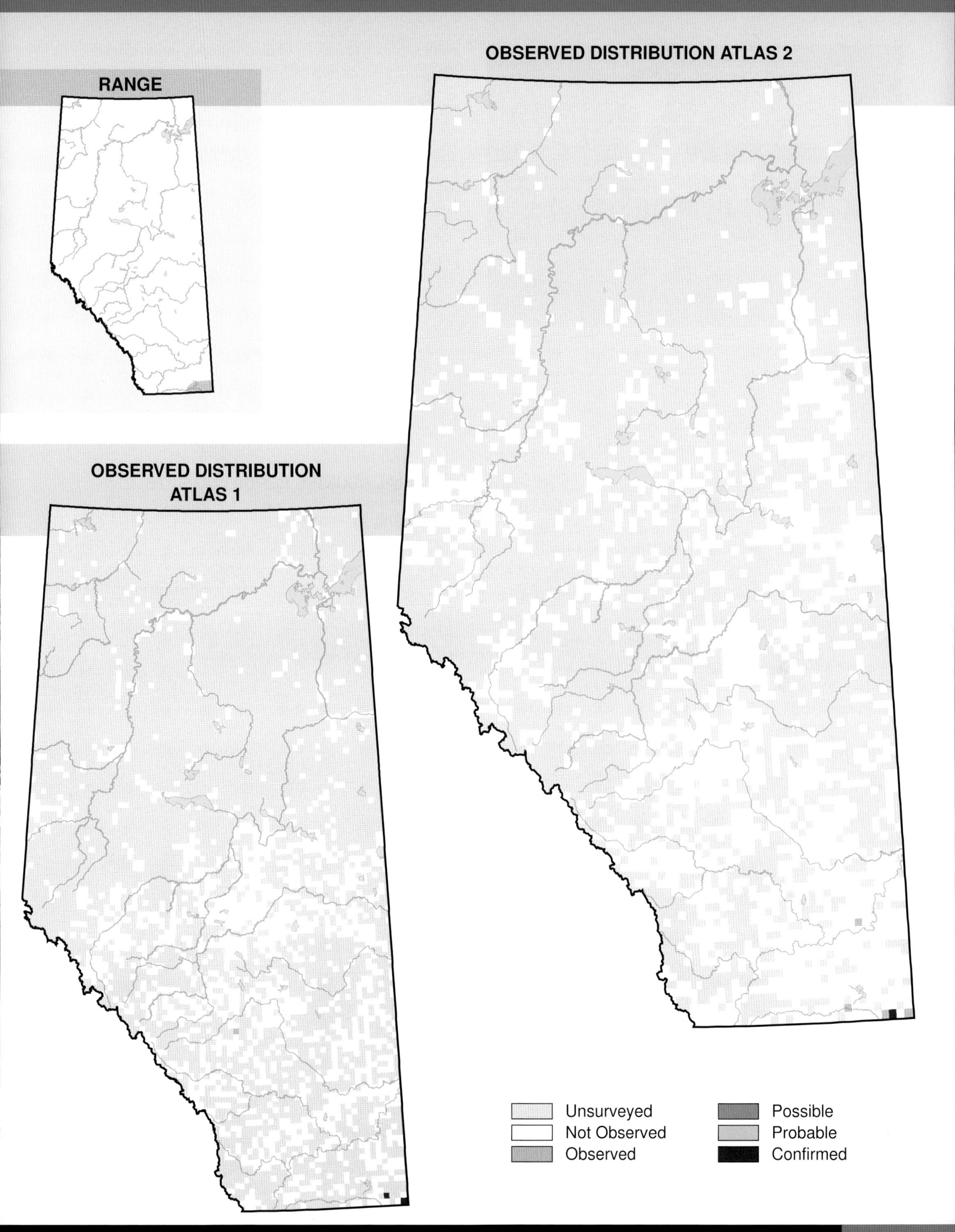
RANGE
OBSERVED DISTRIBUTION ATLAS 2
OBSERVED DISTRIBUTION
ATLAS 1
Unsurveyed
Not Observed
Observed
Possible
Probable
Confirmed

Black-necked Stilt (*Himantopus mexicanus*)

Photo: Gerald Romanchuk

NESTING

CLUTCH SIZE (EGGS):	4
INCUBATION (DAYS):	25–26
FLEDGING (DAYS):	28–32
NEST HEIGHT (METRES):	0

The Black-necked Stilt is a rather new addition to Alberta's avifauna. The first confirmed report in the province was of a bird seen near Langdon on 13 May 1970 (Weseloh 1972). There were nine more records in the province in 1972, 1977, 1979, and 1980, including breeding records at Beaverhill Lake in 1977 and New Dayton in 1980 (Pinel et al., 1991). Over the next twenty years, an ongoing expansion of the breeding range in the western United States was noted (Robinson et al., 1999), which corresponded to increased observations in Alberta (Chapman et al., 1985) and Saskatchewan (Salisbury and Salisbury, 1989; Smith, 1996; Wedgwood and Taylor, 1988) in the 1980s. During Atlas 1 there were four confirmed breeding records in the Grassland Natural Region and one in the Parkland Natural Region, confirming the expansion of the breeding range in Alberta. In Atlas 2, many more breeding observations were recorded within the expected distribution of this species.

In Alberta, the species is most often found on shallow wetlands, particularly on those with some emergent vegetation such as cattails and bulrushes. Black-necked Stilts may be semi-colonial and several pairs may be observed on the same wetland. A dozen or more pairs may be present on larger cattail marshes such as Frank Lake and the Kininvie Marsh east of Brooks. The species may show a preference for man-made wetlands (Robinson et al., 1999), which appears to be the case on Ducks Unlimited wetlands in the irrigation districts of southern Alberta.

No changes in relative abundance were detected for this species. Having been so rare, this species was below the change detection threshold of our statistical methods. Despite its expansion, Black-necked Stilts are rated as Sensitive in the province because their small and localized population makes them vulnerable to wetland disturbance or loss.

TEMPORAL REPORTING RATE

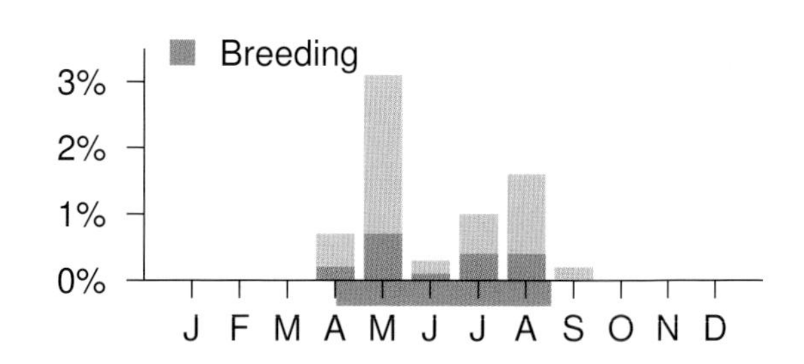

NATURAL REGIONS REPORTING RATE

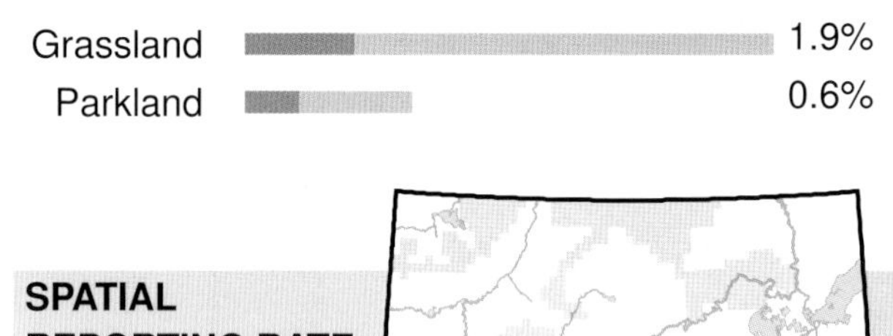

SPATIAL REPORTING RATE

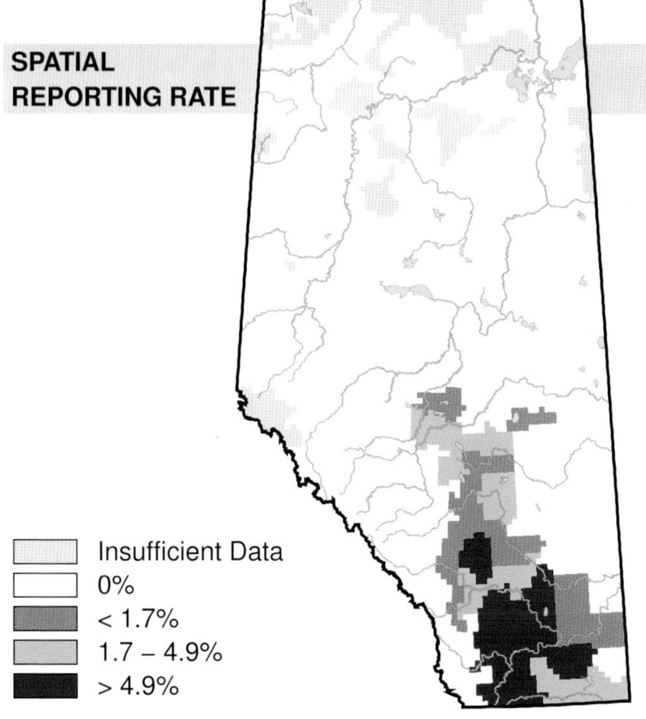

STATUS

GSWA 2000	GSWA 2005	COSEWIC Historic Rankings	COSEWIC 2007	Alberta Wildlife Act
Sensitive	Sensitive	N/A	N/A	N/A

RANGE

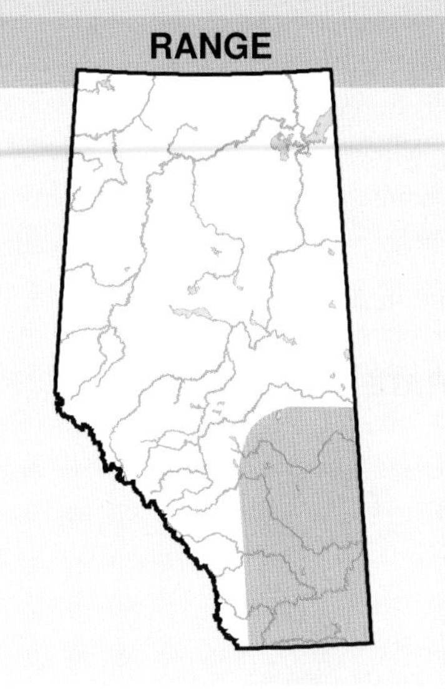

OBSERVED DISTRIBUTION ATLAS 2

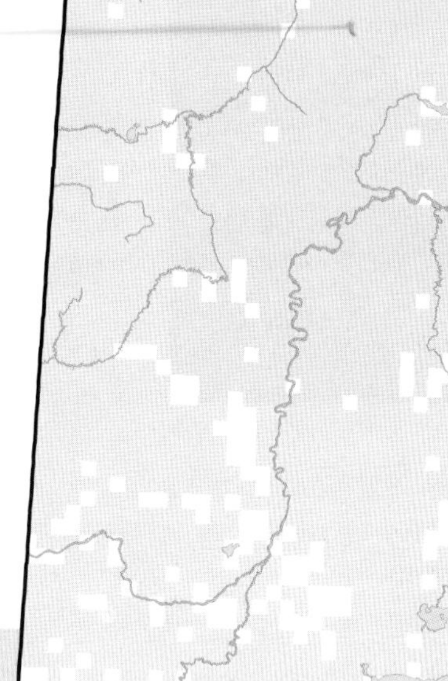

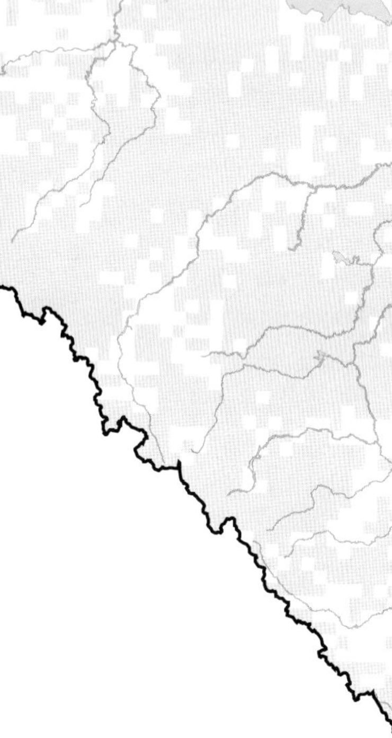

OBSERVED DISTRIBUTION ATLAS 1

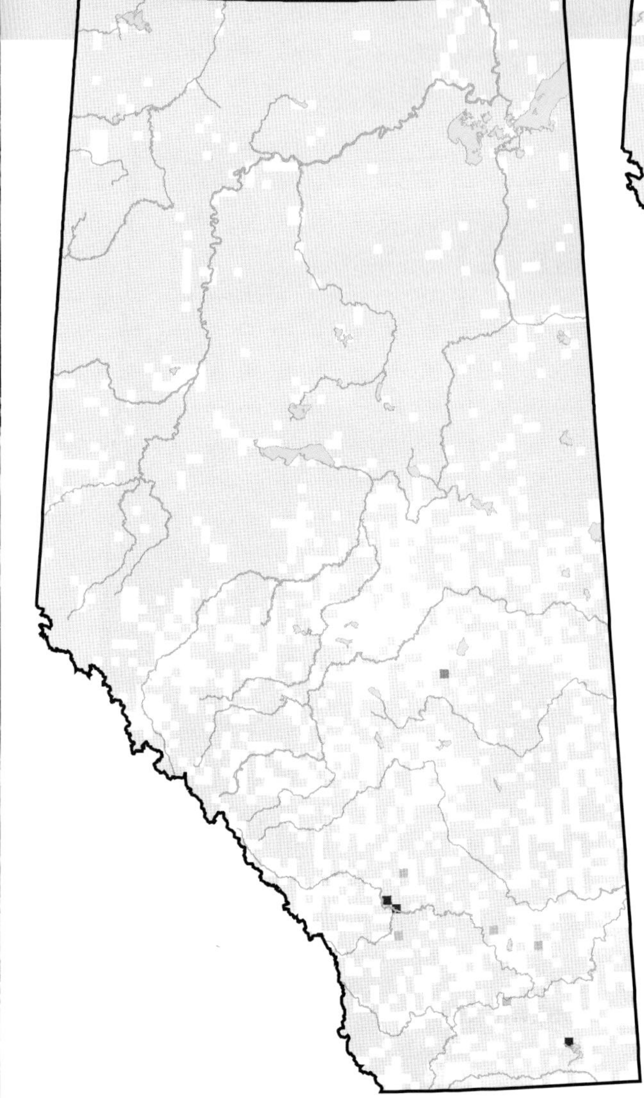

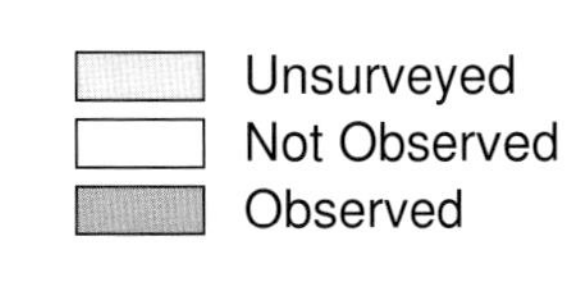

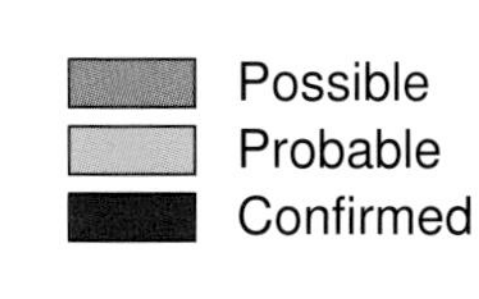

American Avocet (*Recurvirostra americana*)

Photo: Gerald Romanchuk

NESTING

CLUTCH SIZE (EGGS):	3–5
INCUBATION (DAYS):	22–24
FLEDGING (DAYS):	28–35
NEST HEIGHT (METRES):	0

One of the most striking shorebirds in North America due to its red head and neck in the breeding season, the American Avocet is found in all Natural Regions of Alberta except the Rocky Mountain NR. There was no change in distribution observed between Atlas 1 and Atlas 2. One new possible breeding record was found in northeastern Alberta and one probable breeding record was found in the Hinton area. However, these individual records do not necessarily indicate a true range expansion.

These birds were most common in the Grassland NR because they prefer shallow, alkaline wetlands. This species also uses mud flats around lakes; such habitat is more common in the Parkland NR than in the Boreal Forest or Foothills NRs. The distribution of the American Avocet reflects this, being more common in the Parkland NR than in the Boreal Forest or Foothills NRs. To attract American Avocets, wetlands must provide cattails, bulrushes, sedges, and open shallow areas for foraging, a rare combination in the Rocky Mountain NR, which may account for this species' absence in that Natural Region.

There was no relative abundance change detected in any of the Natural Regions. Relative to other species, it was observed with similar frequencies in Atlas 1 and Atlas 2 except in the Foothills NR where no records were obtained in Atlas 1. No change was detected using Breeding Bird Survey data for this species in Alberta, or Canada.

Wetland losses and conversions have affected some traditional breeding areas across the American Avocet's range. Robinson et al. (1997) contend that it can readily respond to shifts in habitat availability, but populations will eventually decline when habitat is permanently lost. Improvement in water quality or availability of nesting islands in this bird's breeding range would make up for some habitat loss and likely increase populations (Robinson et al., 1997).

This species is considered Secure in Alberta.

TEMPORAL REPORTING RATE

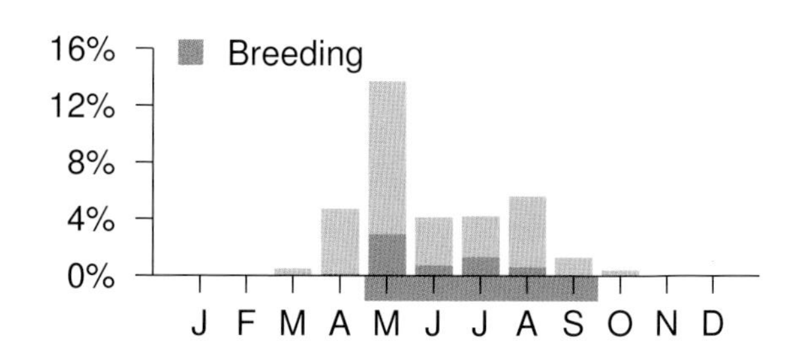

NATURAL REGIONS REPORTING RATE

SPATIAL REPORTING RATE

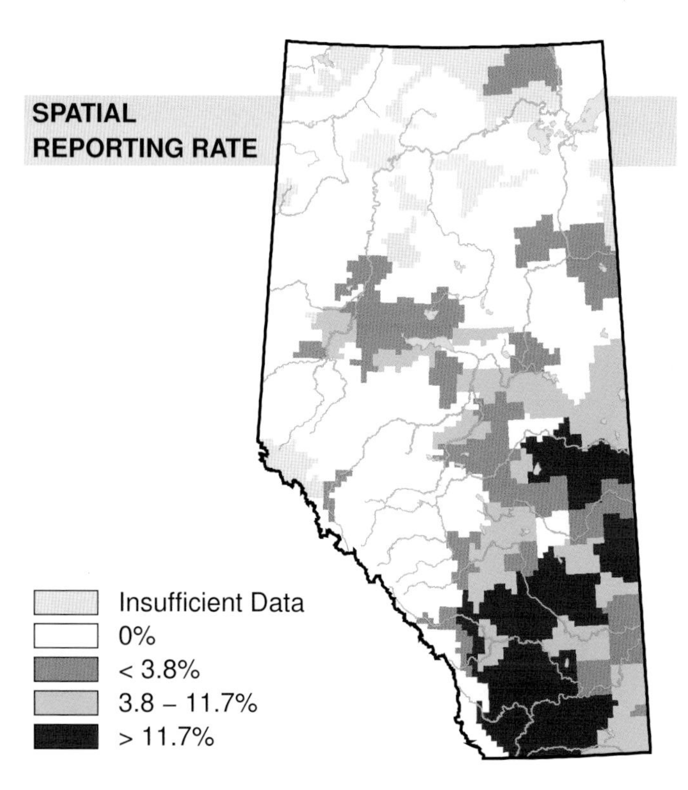

STATUS

GSWA 2000	GSWA 2005	COSEWIC Historic Rankings	COSEWIC 2007	Alberta Wildlife Act
Secure	Secure	N/A	N/A	N/A

Habitat  Nest Location Nest Type Diet 

OBSERVED DISTRIBUTION ATLAS 2

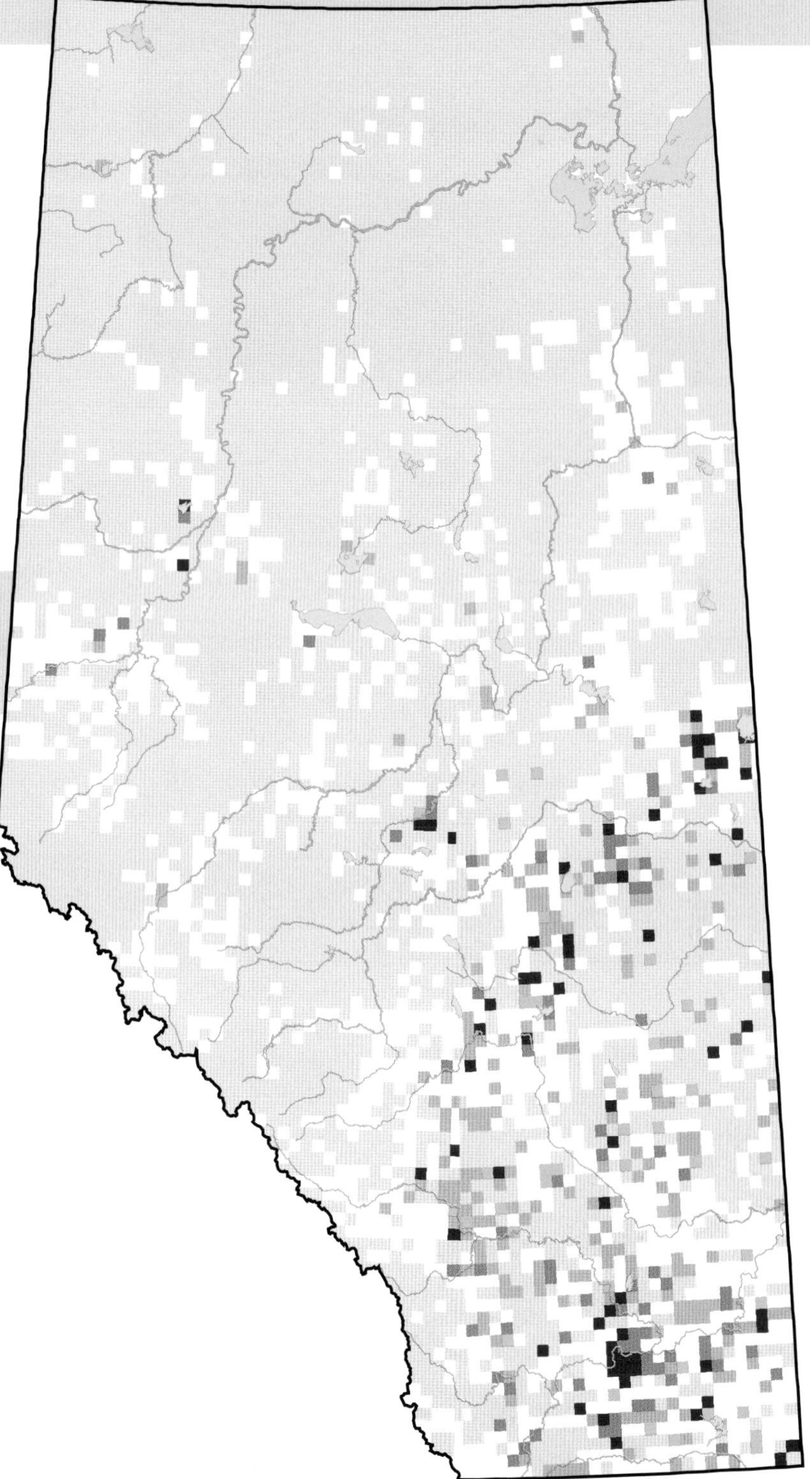

RANGE

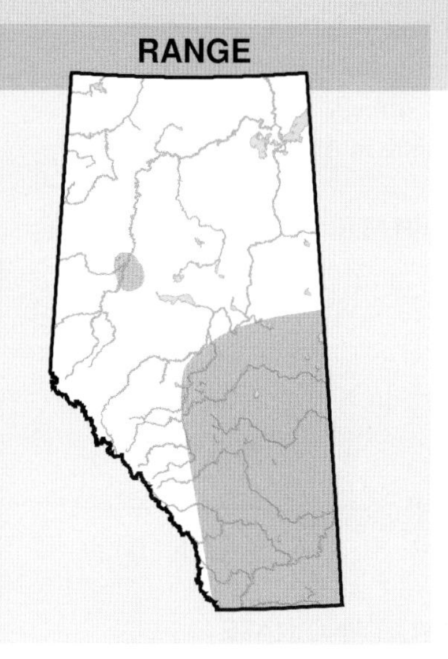

OBSERVED DISTRIBUTION ATLAS 1

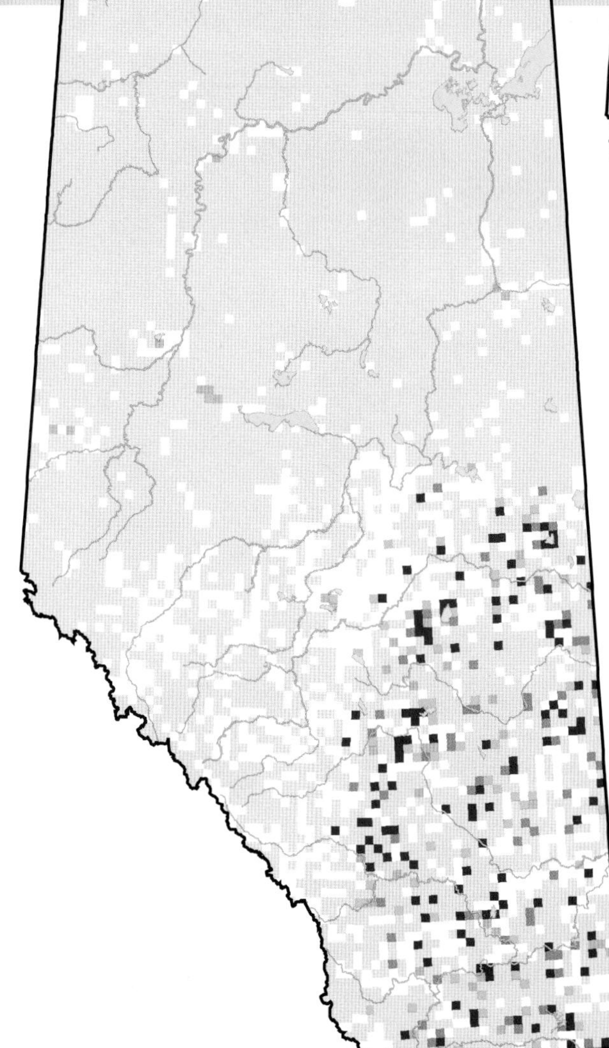

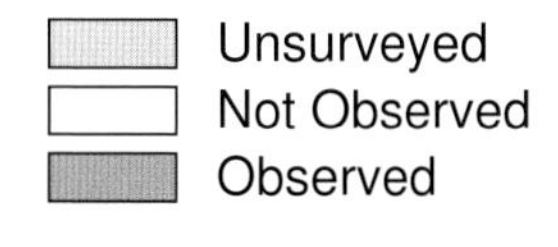

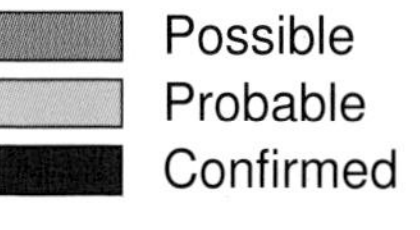

Spotted Sandpiper *(Actitis macularius)*

Photo: Raymond Toal

NESTING

CLUTCH SIZE (EGGS):	4
INCUBATION (DAYS):	20–24
FLEDGING (DAYS):	16–18
NEST HEIGHT (METRES):	0

The Spotted Sandpiper is found in every Natural Region in Alberta. No change in distribution was observed between Atlas 1 and Atlas 2 for this species.

Compared with other shorebirds, the Spotted Sandpiper can breed in a wide variety of habitat types such as sagebrush, grasslands, and forests. Ideally, the breeding sites of this species contain a shoreline, semi-open habitat, and patches of dense vegetation. Predator avoidance is one of the main defensive strategies that influence where Spotted Sandpipers build their nests. In areas where predation rates are high, Spotted Sandpiper nests are built within denser vegetation. This species has a fairly even distribution across the province; however, it was most frequently encountered in the Foothills and Rocky Mountain NRs.

Declines in relative abundance were detected in the Boreal Forest, Foothills, and Rocky Mountain NRs where, relative to other species, the Spotted Sandpiper was observed less frequently in Atlas 2 than in Atlas 1. No change was detected in the Grassland and Parkland NRs. The Breeding Bird Survey did not find an abundance change for this species in Alberta, but the declines detected in Alberta by the Atlas were mirrored in the abundance decline that was detected across Canada by the Breeding Bird Survey during the period 1985–2005. Differences between what was found by the Atlas and the Alberta Breeding Bird Survey could be explained given that the Atlas had more extensive coverage, especially in the north. Abundance declines could be related to declines in the availability of wetlands due to drought and wetland drainage. This species is considered Secure in Alberta.

TEMPORAL REPORTING RATE

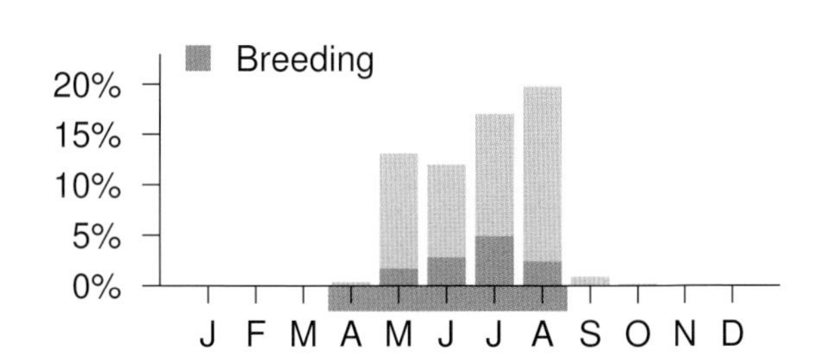

NATURAL REGIONS REPORTING RATE

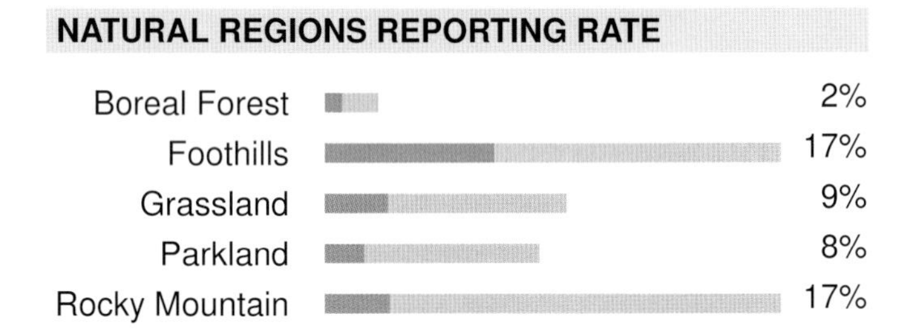

SPATIAL REPORTING RATE

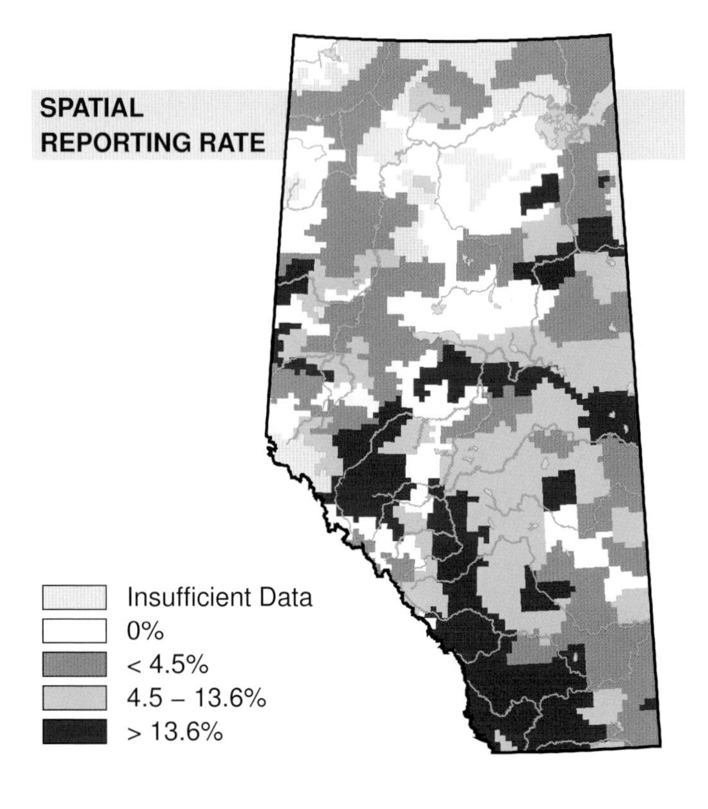

STATUS

GSWA 2000	GSWA 2005	COSEWIC Historic Rankings	COSEWIC 2007	Alberta Wildlife Act
Secure	Secure	N/A	N/A	N/A

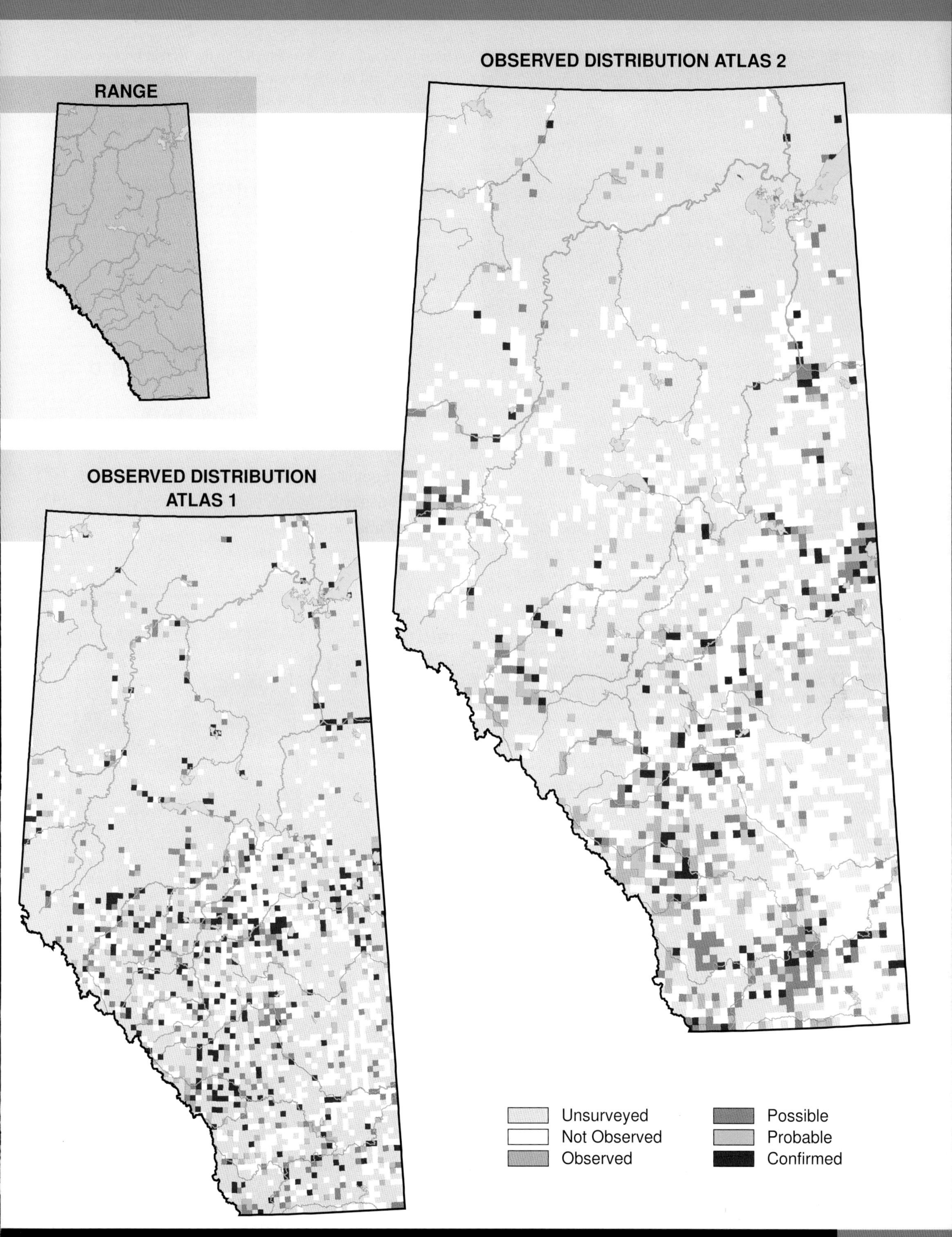

Habitat
Nest Location
Nest Type
Diet
RANGE
OBSERVED DISTRIBUTION ATLAS 2
OBSERVED DISTRIBUTION ATLAS 1
Unsurveyed
Not Observed
Observed
Possible
Probable
Confirmed

Solitary Sandpiper (*Tringa solitaria*)

Photo: Gerald Romanchuk

NESTING

CLUTCH SIZE (EGGS):	4
INCUBATION (DAYS):	23–24
FLEDGING (DAYS):	unknown
NEST HEIGHT (METRES):	1–12.2

The Solitary Sandpiper breeds in the Boreal Forest, Foothills, Parkland, and Rocky Mountain Natural Regions. Although it is seen less frequently in the northwest, no change in distribution was observed between Atlas 1 and Atlas 2 for this species.

Unlike most shorebirds, the Solitary Sandpiper lays its eggs above the ground in tree nests that were originally built by songbirds such as the American Robin. Solitary Sandpipers breed near wetlands that are surrounded by coniferous trees. Despite occurring in four natural regions, the Solitary Sandpiper was found mainly in the Foothills NR in areas where there was an abundance of wetland habitat. Records from the Grassland NR were collected mostly during migration when this species was moving between its South American winter range and its breeding range. A single Grassland NR breeding record was obtained in the transition zone between the Grassland NR and the Parkland NR.

An increase in relative abundance was detected in the Grassland and Parkland NRs. As Grassland NR records were primarily those of migrating birds, this change is most likely an artifact of a broader survey window in Atlas 2, or migratory variability. The biological cause for the increase detected in the Parkland NR is unknown and internal evidence, such as occurrence relative to other species, does not support this finding. No change was detected in the Boreal Forest, Foothills, and Rocky Mountain NRs. The Breeding Bird Survey found no change in abundance for this species in Alberta or Canada. Solitary Sandpipers are considered Secure in the province.

TEMPORAL REPORTING RATE

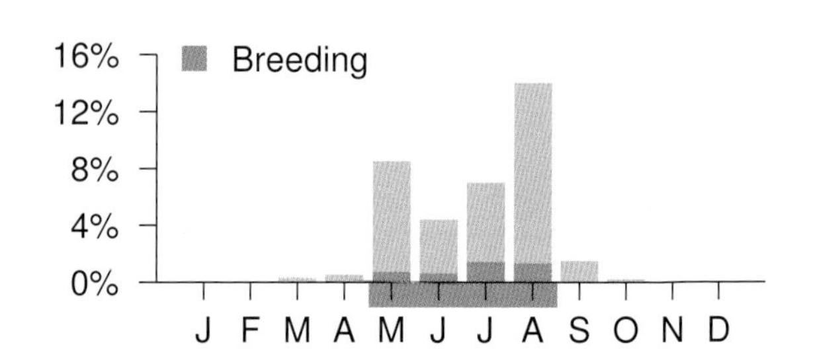

NATURAL REGIONS REPORTING RATE

Boreal Forest	2%
Foothills	12%
Grassland	4%
Parkland	3%
Rocky Mountain	3%

SPATIAL REPORTING RATE

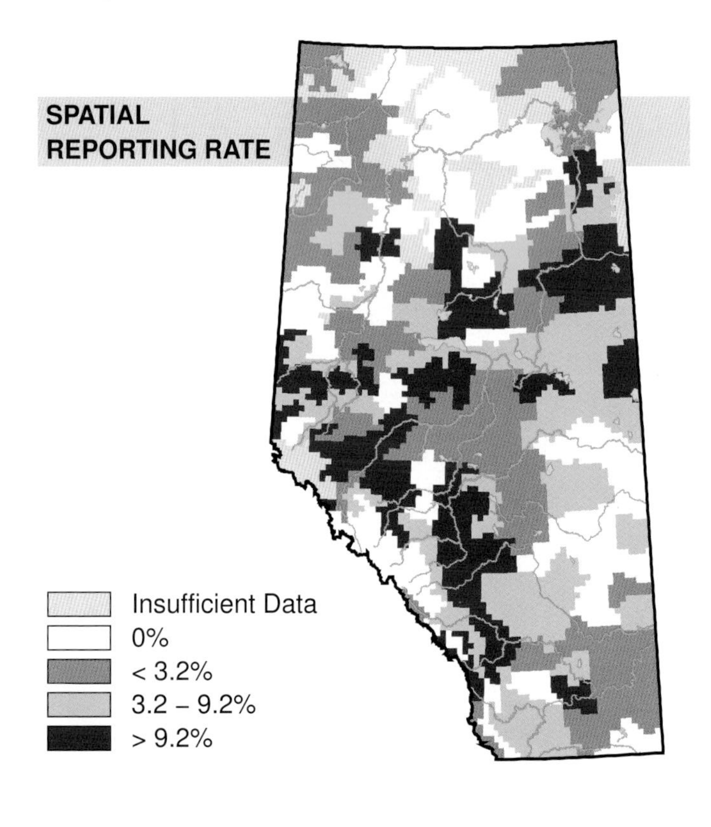

STATUS

GSWA 2000	GSWA 2005	COSEWIC Historic Rankings	COSEWIC 2007	Alberta Wildlife Act
Secure	Secure	N/A	N/A	N/A

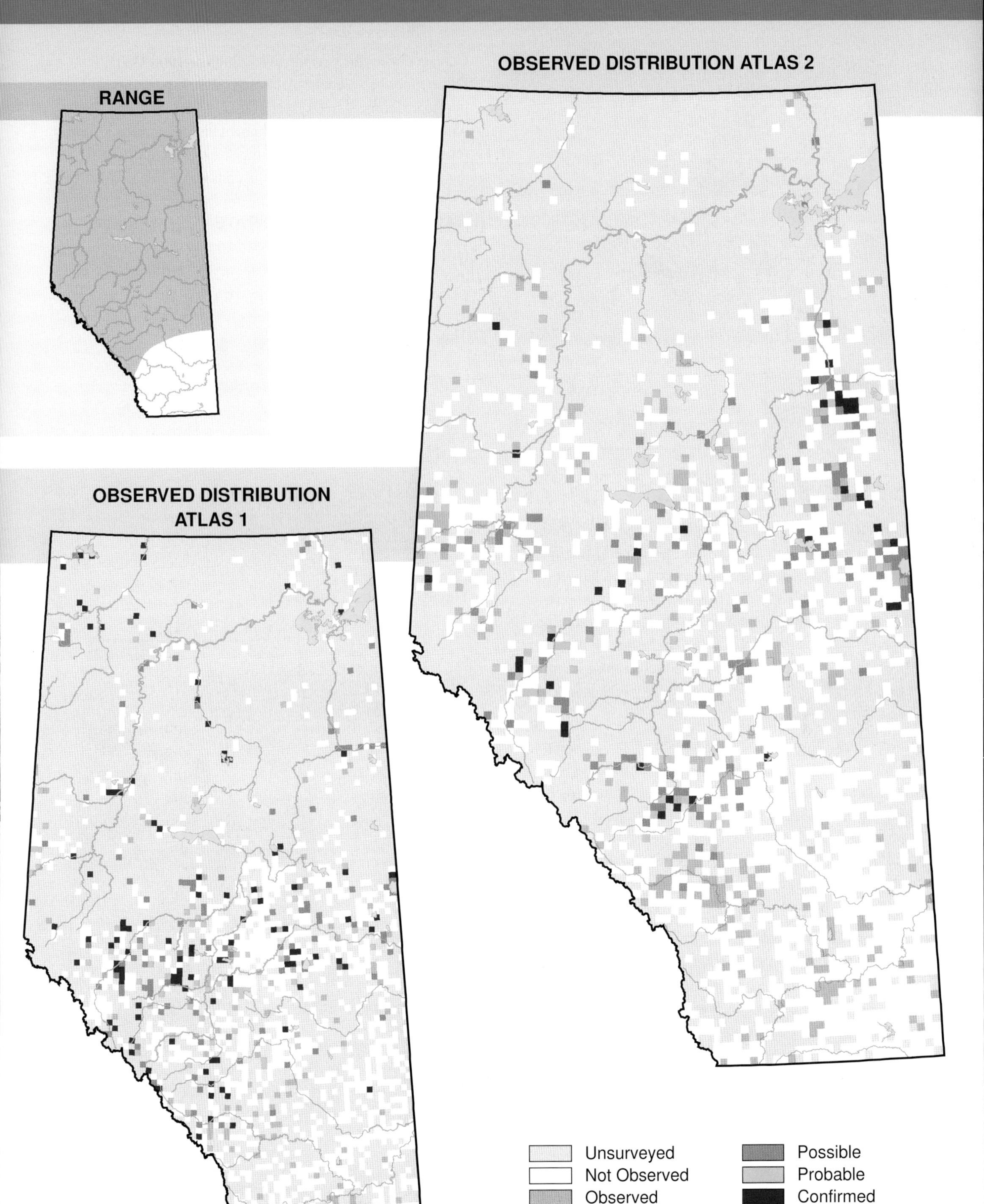

RANGE
OBSERVED DISTRIBUTION ATLAS 2
OBSERVED DISTRIBUTION ATLAS 1
Unsurveyed
Not Observed
Observed
Possible
Probable
Confirmed

Greater Yellowlegs (*Tringa melanoleuca*)

Photo: Duane Boone

NESTING

CLUTCH SIZE (EGGS):	4
INCUBATION (DAYS):	23
FLEDGING (DAYS):	18–20
NEST HEIGHT (METRES):	0

The Greater Yellowlegs is found in the Boreal Forest, Foothills, Parkland, and Rocky Mountain Natural Regions of Alberta. No change in distribution was observed between Atlas 1 and Atlas 2 for this species. However, these birds were detected more often in the northeastern part of the province between Fort McMurray and Lac La Biche during Atlas 2. This was likely a localized population increase that occurred because a large fire that had burned recently in that area created suitable habitat for this species.

The Greater Yellowlegs breeds in bogs with a combination of open areas, small conifer patches, and small ponds. Fires can create a patchy forest structure that is suitable breeding habitat for this species. As a result, this species was most common–and fairly evenly distributed–in the Boreal Forest, Foothills, and Parkland NRs, and was only infrequently encountered in the Rocky Mountain NR. Records from the Grassland NR were migratory.

An increase in relative abundance was detected in the Parkland NR where, relative to other species, the Greater Yellowlegs was observed more frequently in Atlas 2 than in Atlas 1. The cause for this increase is unknown. No change was detected in the Boreal Forest, Foothills, and Rocky Mountain NRs. Data from the Breeding Bird Survey indicate that the relative abundance of Greater Yellowlegs has not changed in Alberta or Canada during the periods 1985–2005 or 1968–2005. Differences between the Atlas and BBS results may be due in part to differences in the spatial scale of the analyses. However, without corroboration or any known biological cause, further research is needed to evaluate the Atlas-detected increase in the Parkland NR. This species is considered Secure in Alberta.

TEMPORAL REPORTING RATE

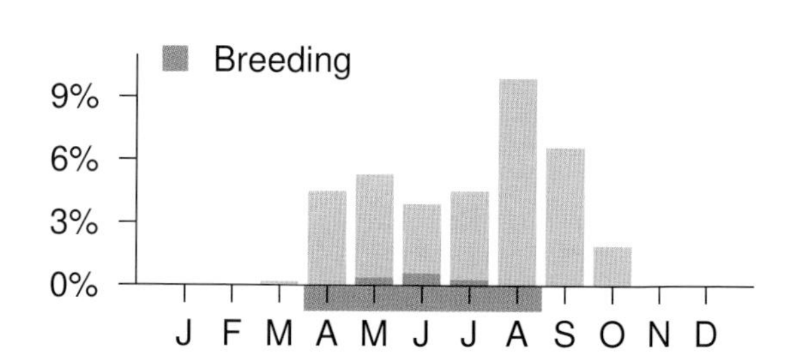

NATURAL REGIONS REPORTING RATE

Region	Rate
Boreal Forest	1.8%
Foothills	4.5%
Grassland	3.1%
Parkland	2.4%
Rocky Mountain	1.4%

SPATIAL REPORTING RATE

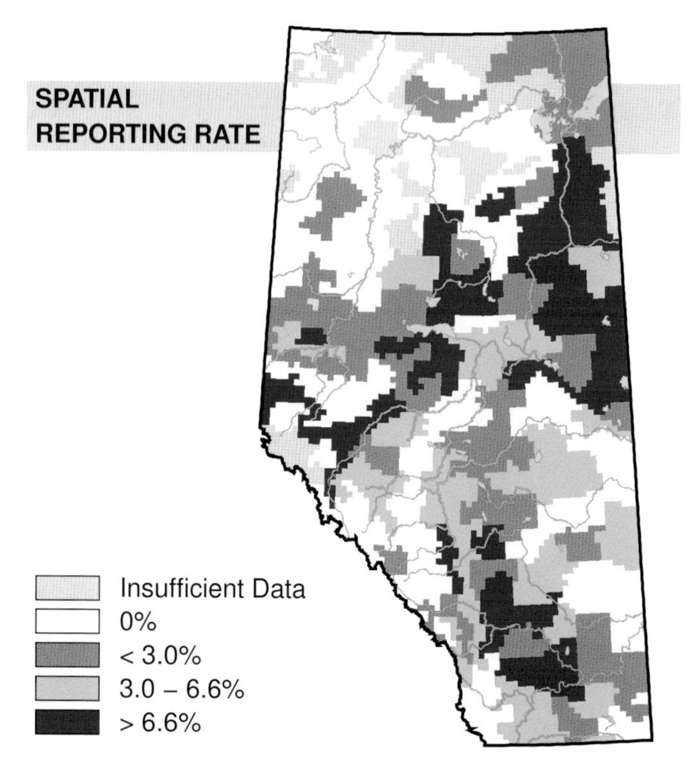

STATUS

GSWA 2000	GSWA 2005	COSEWIC Historic Rankings	COSEWIC 2007	Alberta Wildlife Act
Secure	Secure	N/A	N/A	N/A

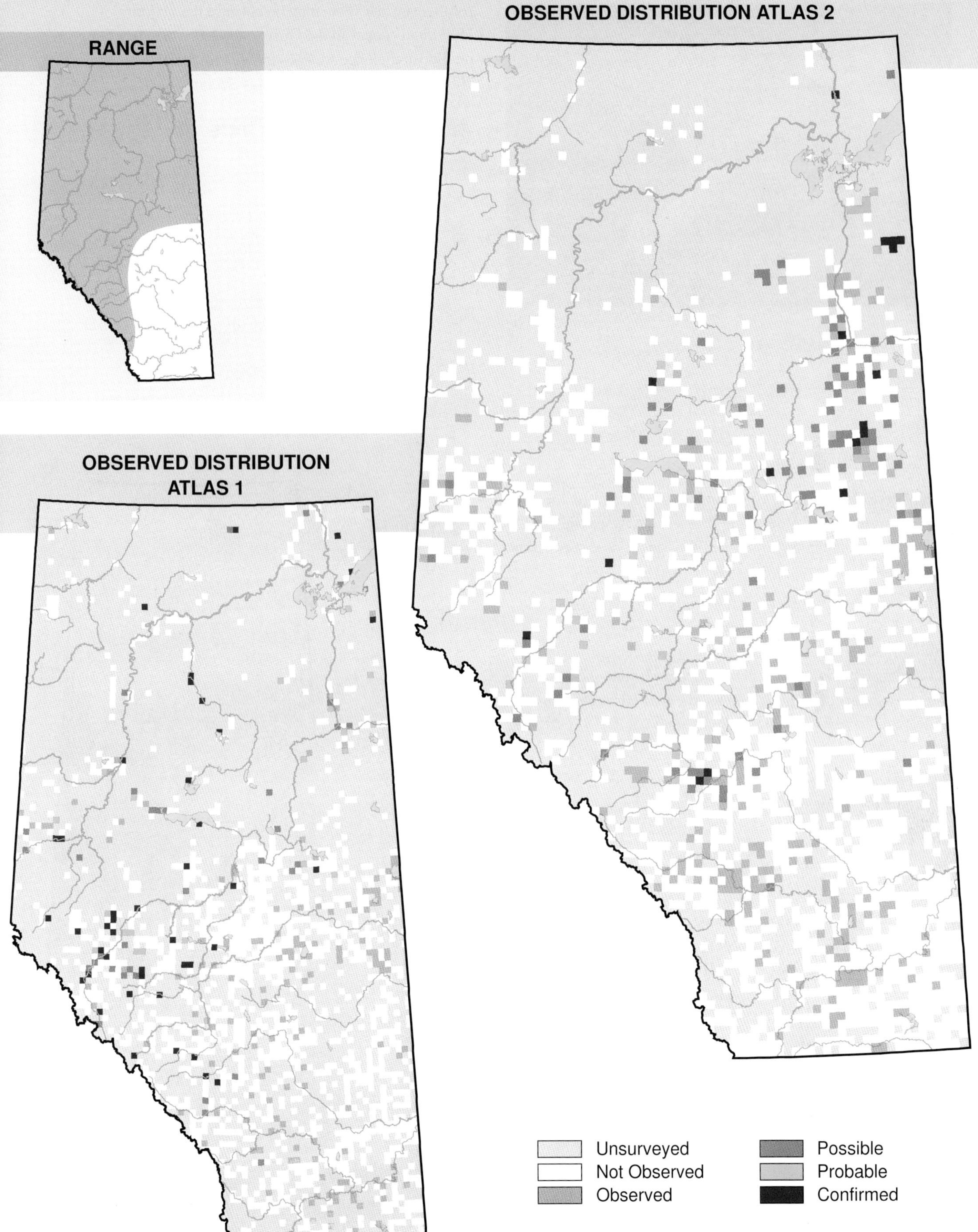
RANGE
OBSERVED DISTRIBUTION ATLAS 2
OBSERVED DISTRIBUTION
ATLAS 1
Unsurveyed
Not Observed
Observed
Possible
Probable
Confirmed

Willet (*Tringa semipalmata*)

Photo: Gerald Romanchuk

NESTING

CLUTCH SIZE (EGGS):	4
INCUBATION (DAYS):	21–29
FLEDGING (DAYS):	30–35
NEST HEIGHT (METRES):	0

The Willet is found in the Grassland, Parkland, and southern Boreal Forest Natural Regions in Alberta. There was no observable change in distribution of this species between Atlas 1 and Atlas 2.

Willets were reported almost exclusively in the Grassland and Parkland NRs. The highest reporting rates were from the centre of the Grassland NR where the occurrence of wetlands and reasonable amounts of remaining grassland coincide. Wetlands are needed for feeding but the western subspecies of Willet usually nests in short grass cover and, to a lesser degree, crops (Lowther et al., 2001).

Decreases in relative abundance were detected in the Parkland NR where conditions were drier in Atlas 2 than in Atlas 1. In addition to a decline in wetland area of about 7.5%, grassland decreased about 20% between 1985 and 1999 in the Parkland NR (Watmough and Schmoll, 2007); so, both habitat elements required for breeding were reduced in that NR. No changes in relative abundance were detected in other NRs. The Breeding Bird Survey did not detect a trend in Alberta or across Canada for this species. This may be because the majority of individuals occur in the Grasslands NR where no change in relative abundance was observed. Provincial status reports in 2000 and 2005 rate the Willet as Secure.

TEMPORAL REPORTING RATE

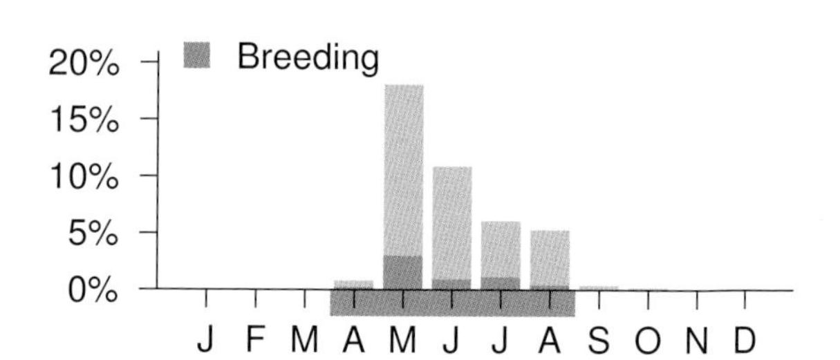

NATURAL REGIONS REPORTING RATE

SPATIAL REPORTING RATE

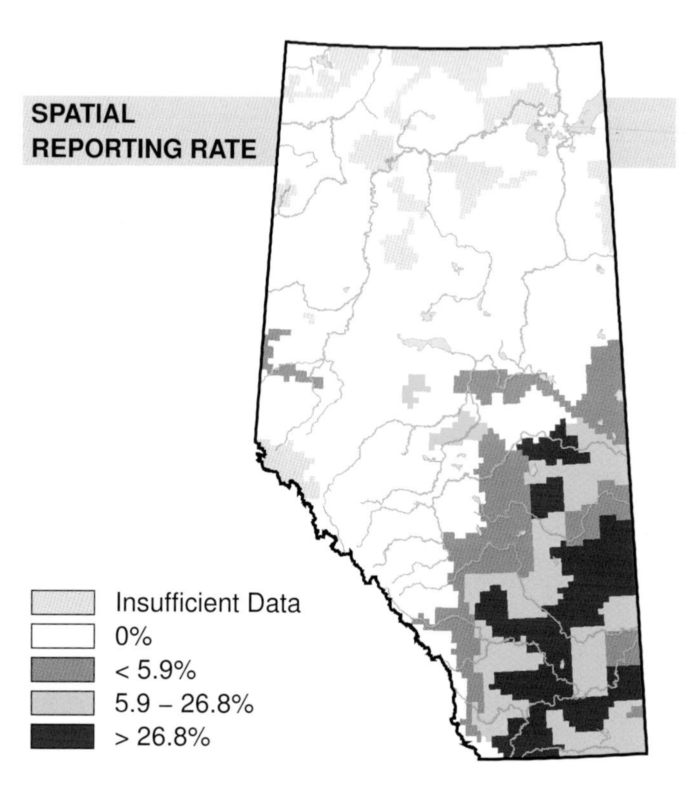

STATUS

GSWA 2000	GSWA 2005	COSEWIC Historic Rankings	COSEWIC 2007	Alberta Wildlife Act
Secure	Secure	N/A	N/A	N/A

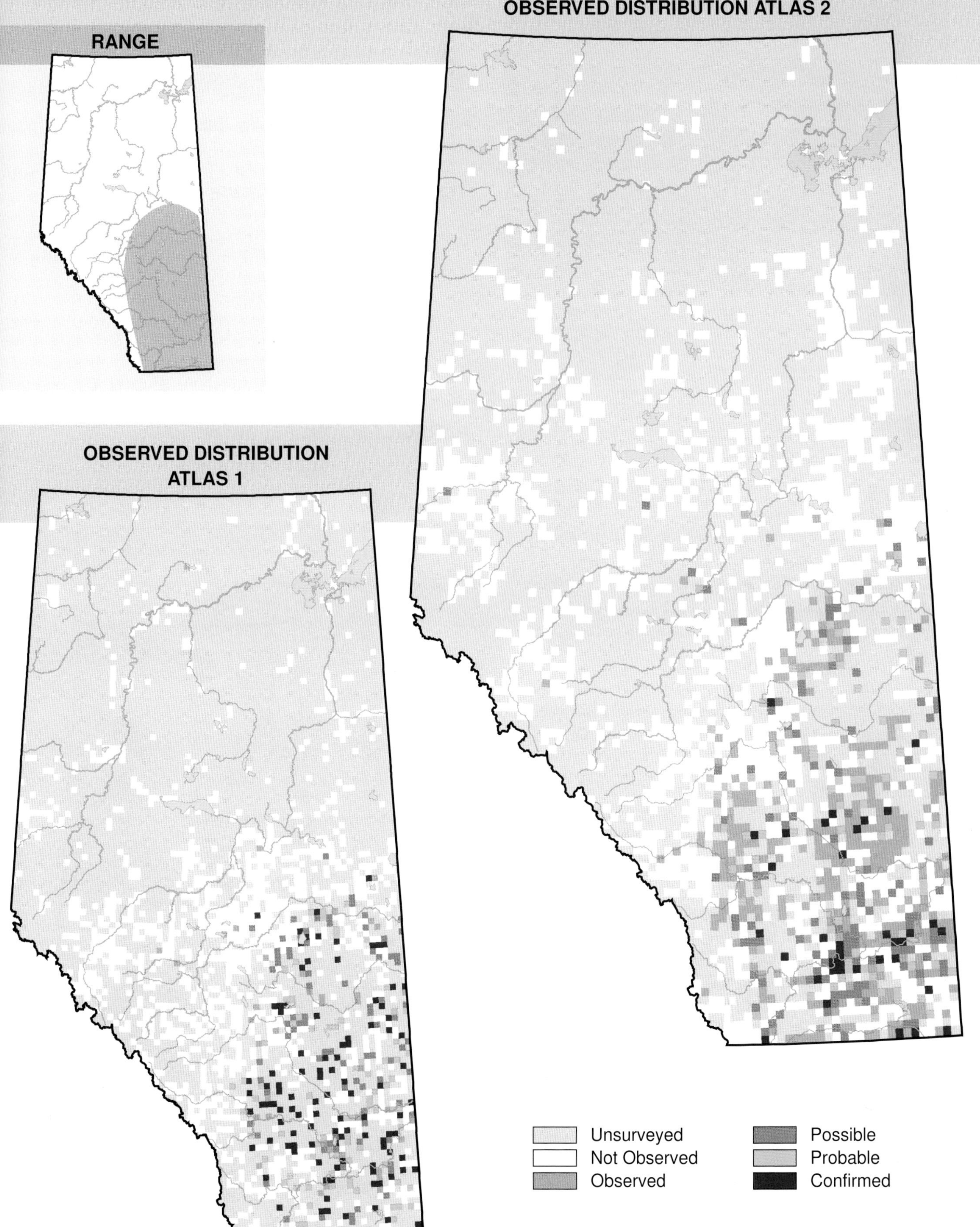
RANGE
OBSERVED DISTRIBUTION ATLAS 2
OBSERVED DISTRIBUTION ATLAS 1
Unsurveyed
Not Observed
Observed
Possible
Probable
Confirmed

Lesser Yellowlegs (*Tringa flavipes*)

Photo: Gerald Romanchuk

NESTING

CLUTCH SIZE (EGGS):	4
INCUBATION (DAYS):	22–23
FLEDGING (DAYS):	23
NEST HEIGHT (METRES):	0

The Lesser Yellowlegs breeds in the Boreal Forest, Foothills, Parkland, and Rocky Mountain Natural Regions in Alberta. No change in distribution was observed between Atlas 1 and Atlas 2 for this species. However, they were detected more often in the northeastern part of the province between Fort McMurray and Lac La Biche during Atlas 2. This is likely a localized population increase that occurred because a large fire recently burned there.

Similar to the Greater Yellowlegs, the Lesser Yellowlegs typically breeds in areas that provide open spaces and patches of trees, and where wetlands are present. However, breeding areas of the Lesser Yellowlegs are generally drier and contain more vegetation than those of the Greater Yellowlegs. This species can do well in second-growth forests after disturbances such as fire. As a result, this species was commonly observed in the Boreal Forest, Foothills, and Parkland NRs but was encountered less frequently in the Rocky Mountain NR.

Declines in relative abundance were detected in the Boreal Forest and Parkland NRs, where relative to other species, the Lesser Yellowlegs was observed less frequently in Atlas 2 than in Atlas 1. Drier conditions in these NRs in Atlas 2 may have contributed to this decline. No changes were detected in the Foothills and Rocky Mountain NRs. The Breeding Bird Survey found no change in abundance in Alberta. However, a decline was detected in Canada during the period 1985–2005. Differences between the results of Atlas and Breeding Bird Survey analyses are likely due to the use of different scales for analysis and more extensive coverage of northern areas in the Atlas than in the BBS. Breeding Bird Survey data were analyzed at the provincial scale; whereas, provided with a sufficient sample size, it was possible to analyze Atlas data at the smaller Natural Region scale. This species is considered Secure in Alberta.

TEMPORAL REPORTING RATE

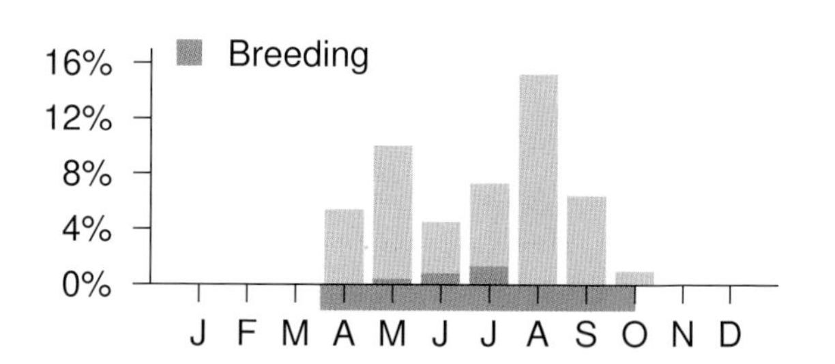

NATURAL REGIONS REPORTING RATE

Region	Rate
Boreal Forest	2.4%
Foothills	5.0%
Grassland	5.1%
Parkland	4.2%
Rocky Mountain	1.6%

SPATIAL REPORTING RATE

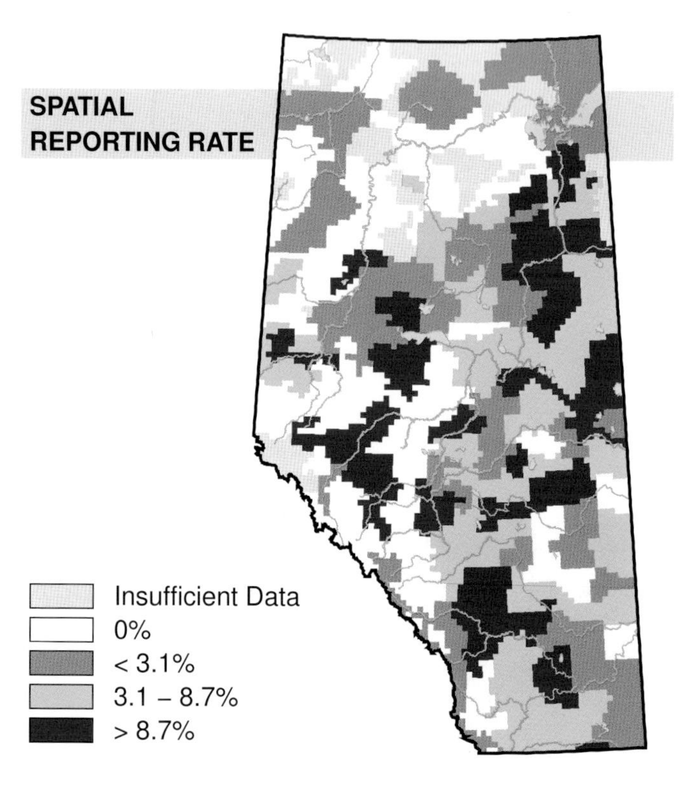

STATUS

GSWA 2000	GSWA 2005	COSEWIC Historic Rankings	COSEWIC 2007	Alberta Wildlife Act
Secure	Secure	N/A	N/A	N/A

Habitat Nest Location Nest Type Diet 

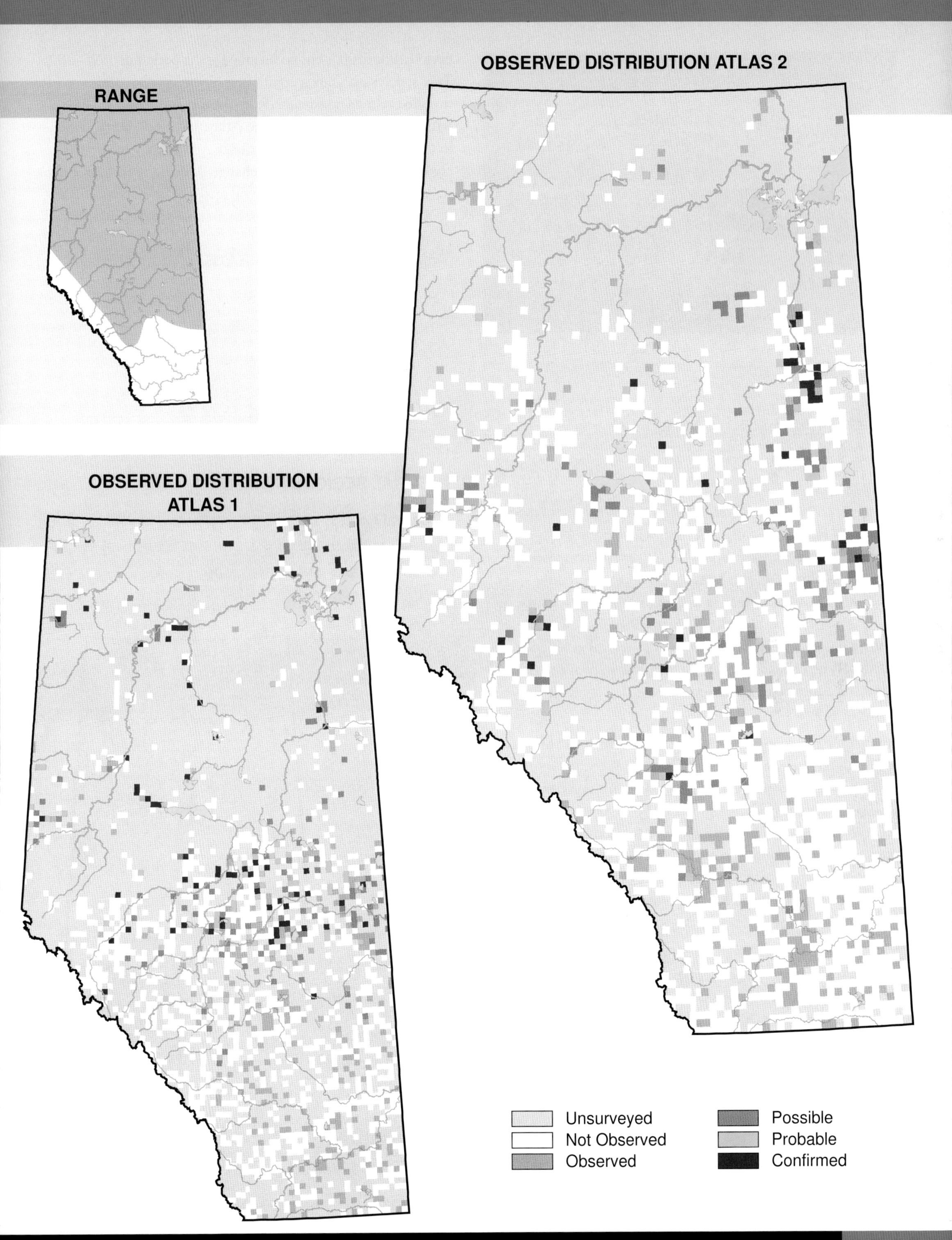

Upland Sandpiper (*Bartramia longicauda*)

Photo: Gerald Romanchuk

NESTING

CLUTCH SIZE (EGGS):	4
INCUBATION (DAYS):	24
FLEDGING (DAYS):	32–34
NEST HEIGHT (METRES):	0

Although Upland Sandpiper distribution lies mainly in the Grassland Natural Region, this species also occurs in grassy habitats throughout Alberta. Of the four most common large upland shorebirds, this bird has the broadest distribution, reflecting its use of a broad array of grassy habitats across North America; the other three (Long-billed Curlew, Marbled Godwit, and Willet) occur mainly in the central grasslands. There was no change in observed distribution between Atlas 1 and Atlas 2, although fewer squares revealed breeding evidence in Atlas 2.

Although it is widely distributed in the province, the highest reporting rates are in the northern two-thirds of the Grassland NR. This is consistent with the bird's need for grassland cover of various heights (to accommodate display, nesting, and brood-rearing) with few or no shrubs (Houston and Bowen, 2001). The southernmost portion of the Grassland NR has plenty of intact grassland but will not always have tall enough cover, while the Parkland NR will often have too many shrubs and trees to suit this species.

Decreases in relative abundance were detected in the Parkland NR where, relative to other species, this species was observed less frequently in Atlas 2 than in Atlas 1. Habitat monitoring shows a loss of about 20% of the native grassland in this NR between 1985 and 1999 (Watmough and Schmoll, 2007). No changes in relative abundance were detected in other NRs. The Breeding Bird Survey detected a decline in Alberta for the period 1985–2005, but no change for the period of 1995–2005. This species presents a conflicting picture, with a decline detected in the Grassland NR by the standardized BBS but with no change detected by the Atlas. Increased precipitation in the Grassland NR in Atlas 2 over Atlas 1 may have contributed to there being no change detected in this NR by the Atlas or by the shorter BBS trend analysis. In Alberta it is rated Sensitive but has never been assessed federally.

TEMPORAL REPORTING RATE

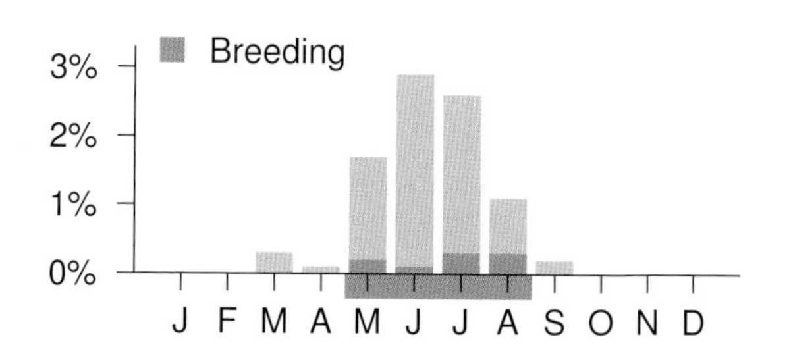

NATURAL REGIONS REPORTING RATE

Boreal Forest	< 0.1%
Grassland	4.3%
Parkland	0.3%
Rocky Mountain	0.7%

SPATIAL REPORTING RATE

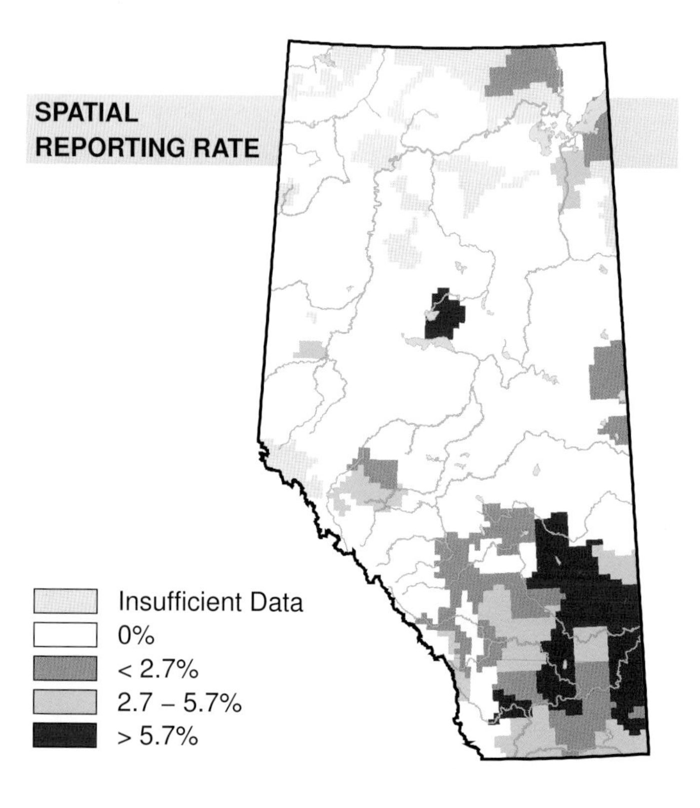

STATUS

GSWA 2000	GSWA 2005	COSEWIC Historic Rankings	COSEWIC 2007	Alberta Wildlife Act
Sensitive	Sensitive	N/A	N/A	N/A

Habitat

Nest
Location

Nest
Type

Diet

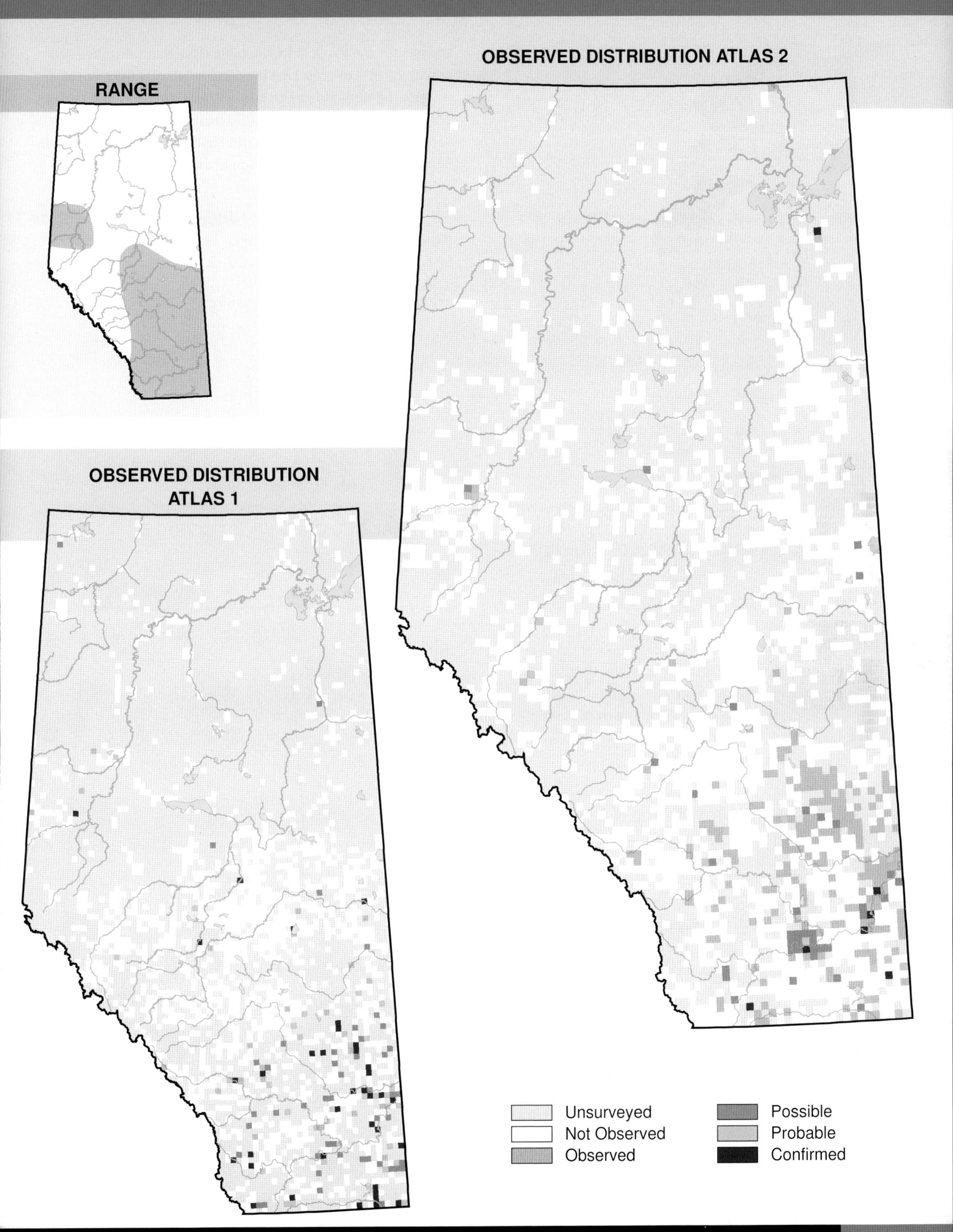

RANGE
OBSERVED DISTRIBUTION ATLAS 2
OBSERVED DISTRIBUTION
ATLAS 1
Unsurveyed
Not Observed
Observed
Possible
Probable
Confirmed

Long-billed Curlew (*Numenius americanus*)

Photo: Gerald Romanchuk

NESTING

CLUTCH SIZE (EGGS):	4
INCUBATION (DAYS):	27–30
FLEDGING (DAYS):	41–45
NEST HEIGHT (METRES):	0

The distribution of the Long-billed Curlew is largely confined to the Grassland Natural Region. There is no appreciable change in the distribution between Atlas 1 and Atlas 2.

Within the Grassland NR this curlew was observed most commonly in areas with predominantly sandy soil and rolling topography. This fits well with the bird's habitat preference for rolling shortgrass or mixedgrass prairie with low cover and little or no shrubby vegetation; their breeding distribution is bounded by regions of high winter precipitation in the west and of low winter temperatures along the northern edge (Dugger and Dugger, 2002).

Increases in relative abundance were detected in the Grassland NR where, relative to other species, Long-billed Curlews were observed more frequently in Atlas 2 than in Atlas 1. This increase is clearly visible in the comparison of the Atlas 1 and Atlas 2 maps. Many squares in the northern part of their range that were checked in Atlas 1 but had no record now show at least one observation in Atlas 2. Breeding records, concentrated in the south and east, remain about the same. No changes in relative abundance were detected in the other NRs. The Breeding Bird Survey did not detect a trend, either in Alberta or nationally, but coverage of this and other grassland species is poor and thus statistical power is low.

There is no accurate estimate of the current population size, but the species is considered vulnerable throughout its range and continued loss of grassland breeding habitats is thought to be the greatest threat to population stability (Dugger and Dugger, 2002). Excessive vehicular traffic (particularly off-road vehicles) and recreational use of breeding habitats can result in nest abandonment and disruption of critical parental behaviours.

The provincial status of this species improved from May Be at Risk in 2000, to Sensitive in 2005. Federally, it is listed as a Species of Special Concern.

TEMPORAL REPORTING RATE

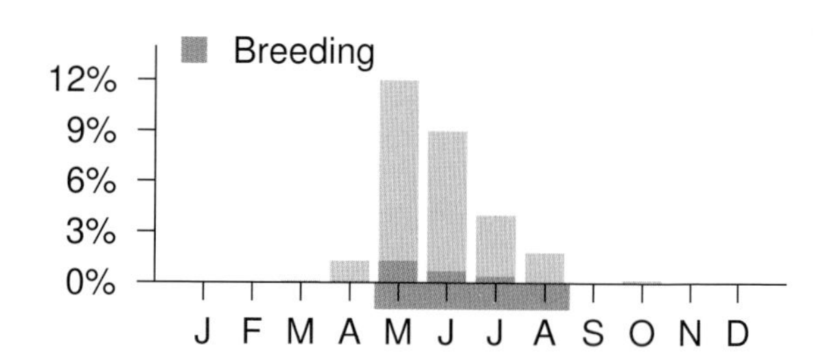

NATURAL REGIONS REPORTING RATE

Grassland 16%

SPATIAL REPORTING RATE

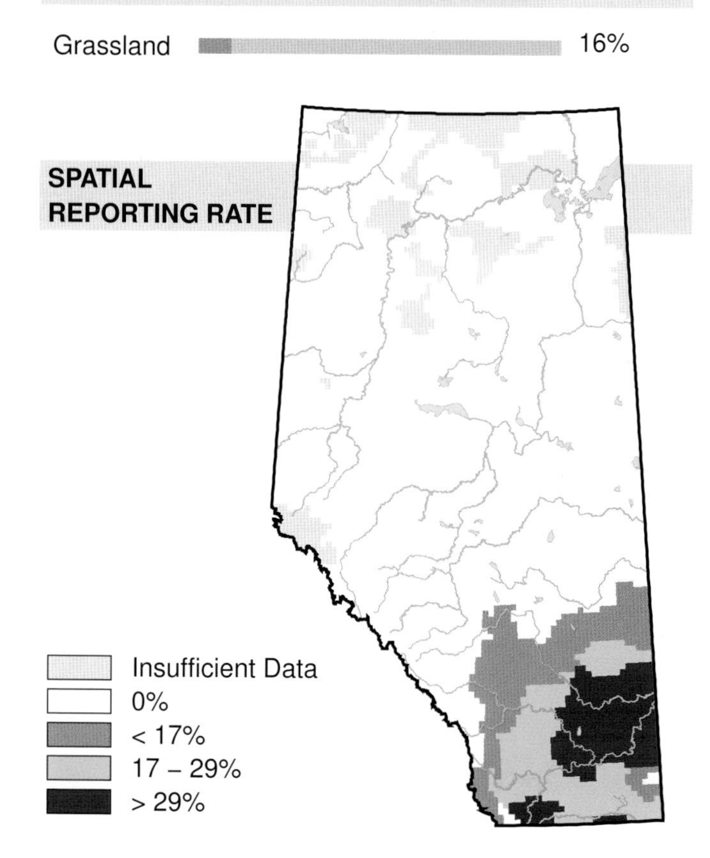

STATUS

GSWA 2000	GSWA 2005	COSEWIC Historic Rankings	COSEWIC 2007	Alberta Wildlife Act
May Be at Risk	Sensitive	Special Concern	Special Concern	N/A

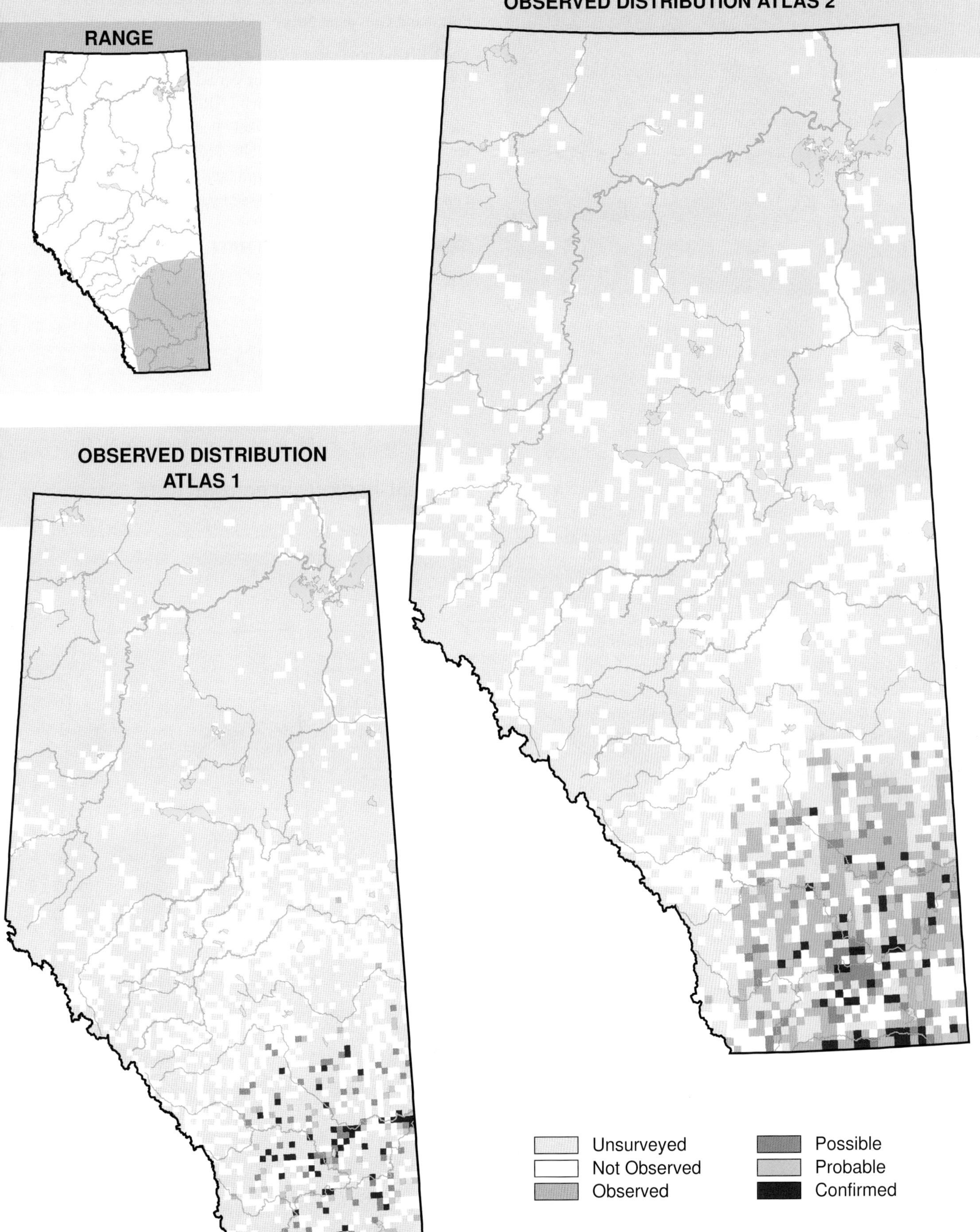
RANGE
OBSERVED DISTRIBUTION ATLAS 2
OBSERVED DISTRIBUTION ATLAS 1
Unsurveyed
Not Observed
Observed
Possible
Probable
Confirmed

Marbled Godwit (*Limosa fedoa*)

Photo: Gerald Romanchuk

NESTING

CLUTCH SIZE (EGGS):	4
INCUBATION (DAYS):	23
FLEDGING (DAYS):	21
NEST HEIGHT (METRES):	0

The Marbled Godwit occurs mainly in the Grassland Natural Region, with some occurrences in Parkland and Boreal Forest NRs. There was no observable change in distribution between Atlas 1 and Atlas 2.

The areas with the highest reporting rates are in the central and northern portions of the Grassland NR and the extreme southern part of the Parkland NR. These areas offer a mix of grass and wetland habitats. Marbled Godwit occur less often in the southern grasslands because wetlands are less common, while in Parkland and Boreal Forest NRs it is probably limited by shrubs and trees which they avoid (Gratto-Trevor, 2000).

Decreases in relative abundance were detected in the Parkland NR where, relative to other species, this species was observed less frequently in Atlas 2 than in Atlas 1. The change in this region may be associated with drier conditions there and in northern portions of the province during Atlas 2. Habitat monitoring shows about 7.5% of wetlands in this natural area (and about 6% province-wide) have been lost since 1985 (Watmough and Schmoll, 2007). Marbled Godwits require landscapes with at least 5% of the area in wetlands (Ryan et al., 1984) and need seasonal wetlands or better in periods of drought (Gratto-Trevor, 2000). Their use of the Parkland NR is already limited by their aversion to shrub-, tree-, and tall dense grass-cover (Gratto-Trevor, 2000); reduced wetland availability would render more of this natural region unacceptable. No changes in relative abundance were detected in the Boreal Forest and Grassland NRs.

The BBS did not detect a 20-year decline, but did detect a 5.4% abundance decline for the period 1995–2005. If Marbled Godwit numbers in the Grassland NR have declined but the species is still common, it is likely that the Atlas is less sensitive than the BBS in this NR. Provincial status reports from both 2000 and 2005 accord the species a status of Secure. This bird has never been assessed by COSEWIC.

TEMPORAL REPORTING RATE

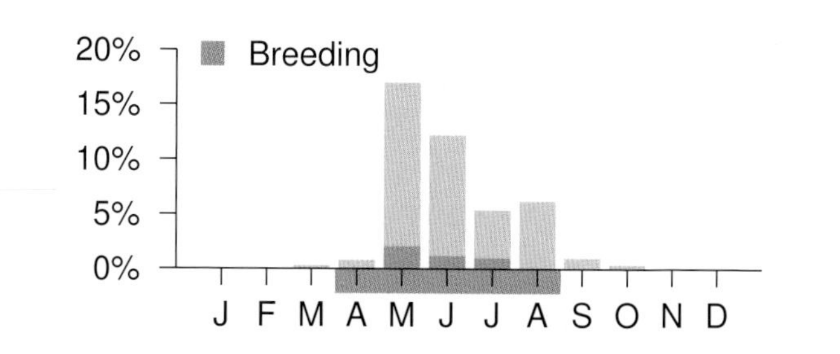

NATURAL REGIONS REPORTING RATE

SPATIAL REPORTING RATE

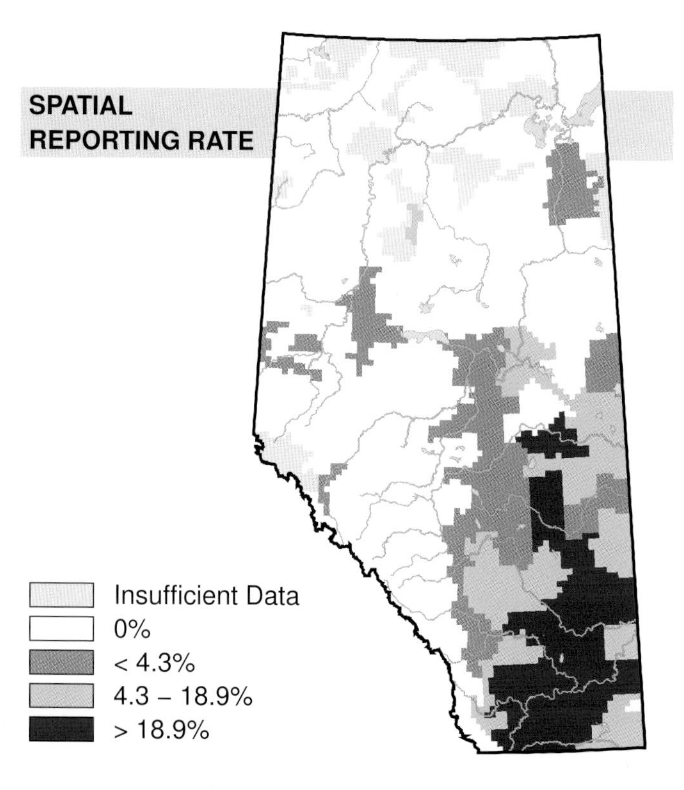

STATUS

GSWA 2000	GSWA 2005	COSEWIC Historic Rankings	COSEWIC 2007	Alberta Wildlife Act
Secure	Secure	N/A	N/A	N/A

RANGE

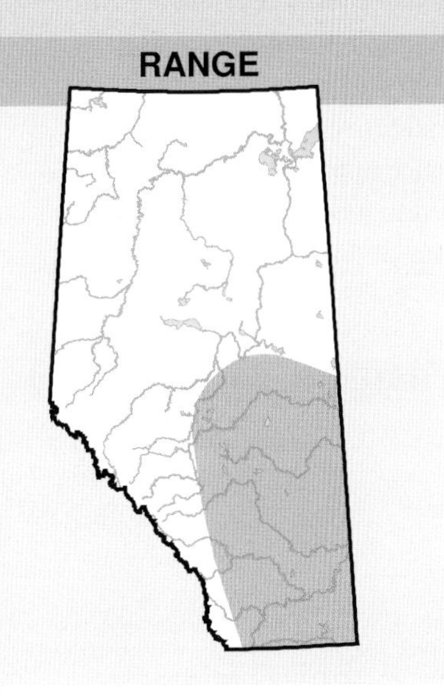

OBSERVED DISTRIBUTION ATLAS 1

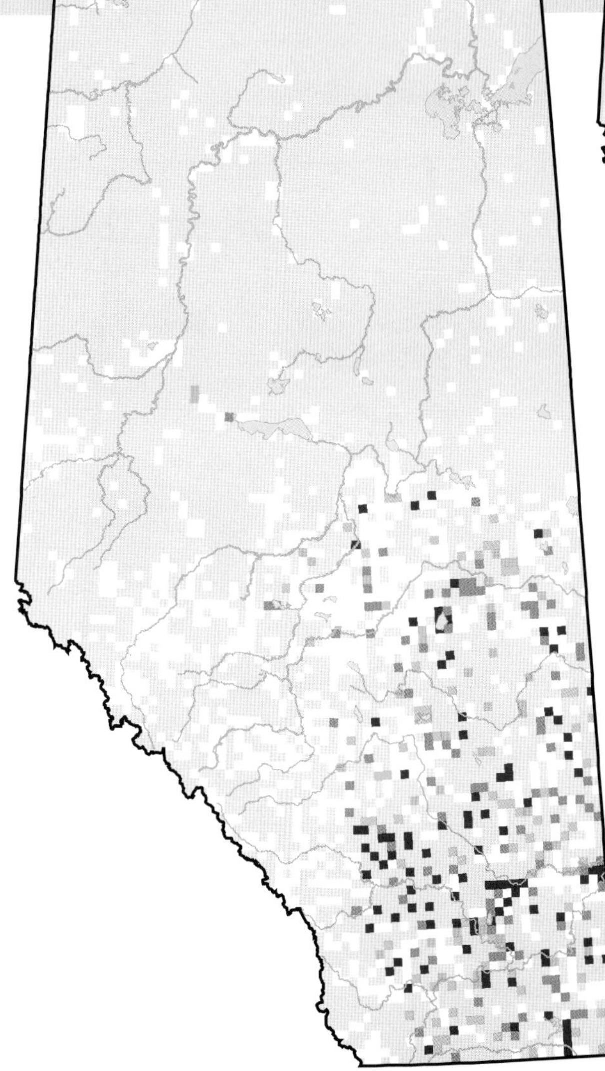

OBSERVED DISTRIBUTION ATLAS 2

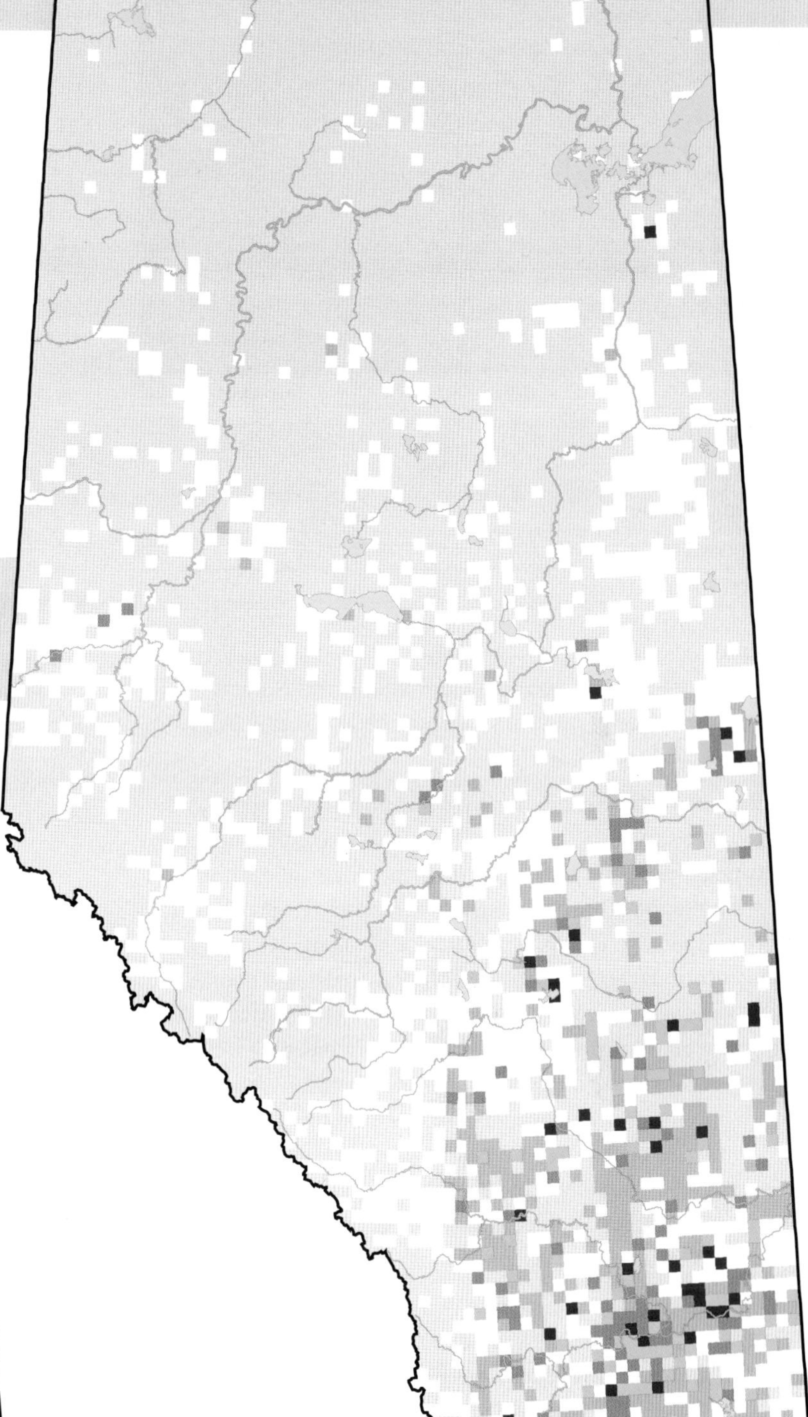

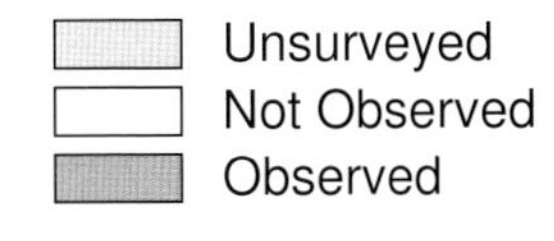

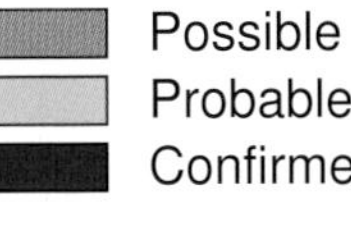

Least Sandpiper (*Calidris minutilla*)

Photo: Gerald Romanchuk

NESTING

CLUTCH SIZE (EGGS):	4
INCUBATION (DAYS):	20–23
FLEDGING (DAYS):	18
NEST HEIGHT (METRES):	0

The Least Sandpiper has been known to breed only in the northeastern part of Alberta in the Boreal Forest Natural Region. No change in distribution was observed between Atlas 1 and Atlas 2 for this species.

In Alberta, Least Sandpipers were most often encountered during the spring and fall migrations when they were moving between their winter range and their sub-Arctic and Arctic breeding range. These migrants were most often encountered in the Grassland, Parkland, and Rocky Mountain NRs but were also occasionally observed in the Boreal Forest and Foothills NRs. No breeding records were confirmed during Atlas 2. The southern periphery of the Least Sandpiper's breeding range barely dips into the northeast corner of Alberta. Even though there have been a few breeding records from this area, it is not known if this species is a regular breeder in the province because their potential breeding area is so remote and difficult to monitor. Least Sandpipers nest in bogs, muskeg or sedge meadows, usually near open water.

A decline in relative abundance was detected in the Boreal Forest NR, and increases were detected in the Parkland and Grassland NRs. This pattern was mirrored in the species' rate of occurrence relative to other species. Detected changes could be a function of the broader survey window of Atlas 2; or of possible changes in migratory behaviour due to climatic change; or of the inherent variability of migration records. As a result of these factors, changes detected for this species by the Atlas must be cautiously considered. The Breeding Bird Survey sample size for Least Sandpiper was too small to investigate abundance changes in Alberta and Canada. Although the Atlas had more extensive coverage of northern Alberta than the BBS, the absence of breeding records suggests that targeted efforts are required to monitor this species effectively. The provincial status of this species is Secure in Alberta.

TEMPORAL REPORTING RATE

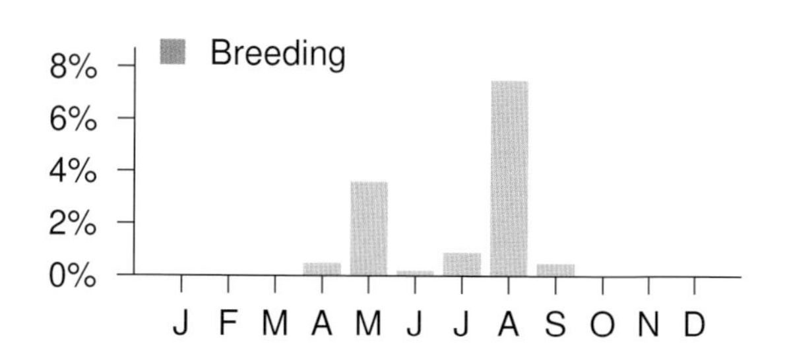

NATURAL REGIONS REPORTING RATE

Boreal Forest	0.4%
Grassland	1.9%
Parkland	1.2%
Rocky Mountain	0.6%

SPATIAL REPORTING RATE

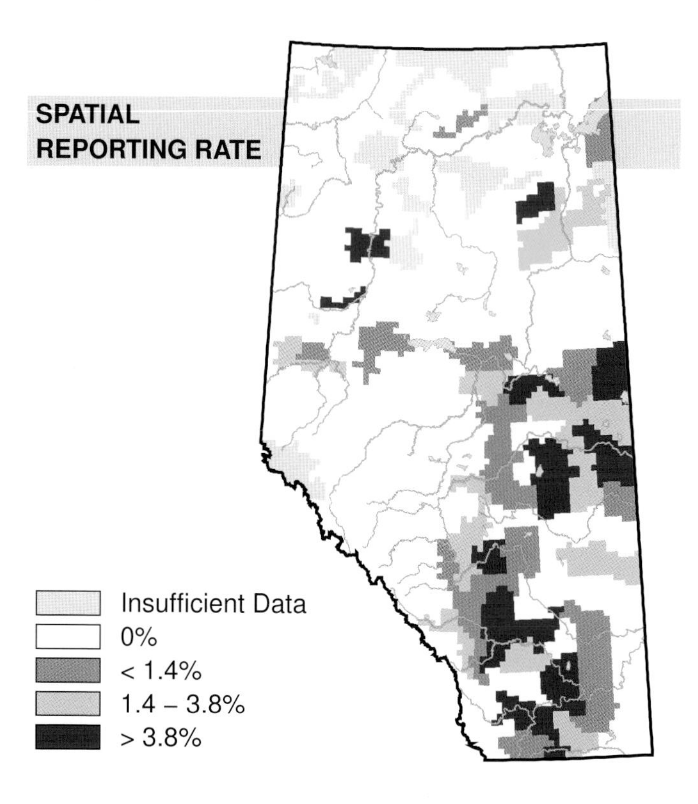

STATUS

GSWA 2000	GSWA 2005	COSEWIC Historic Rankings	COSEWIC 2007	Alberta Wildlife Act
Secure	Secure	N/A	N/A	N/A

RANGE

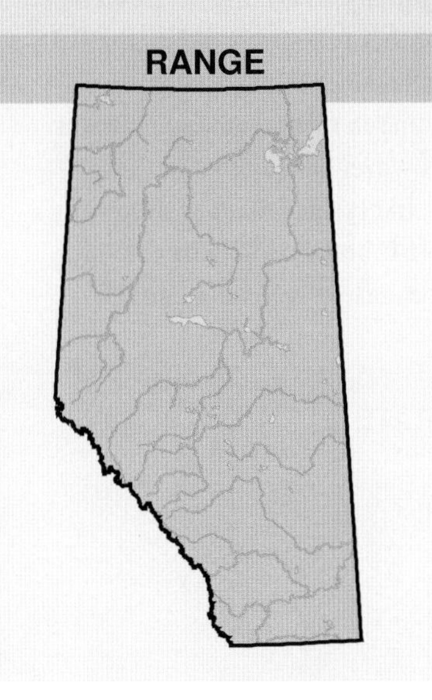

OBSERVED DISTRIBUTION ATLAS 2

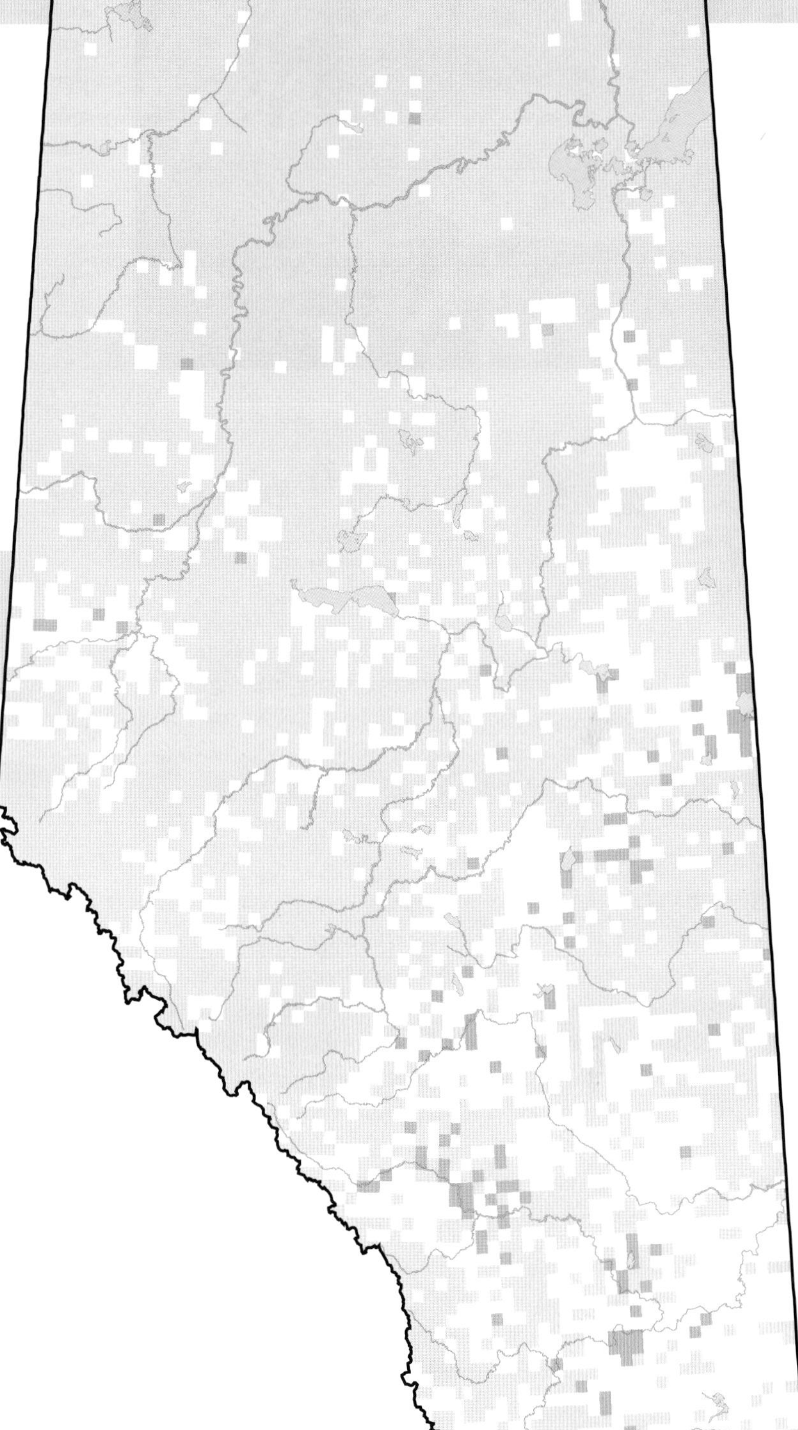

OBSERVED DISTRIBUTION ATLAS 1

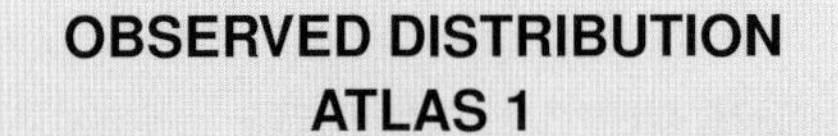

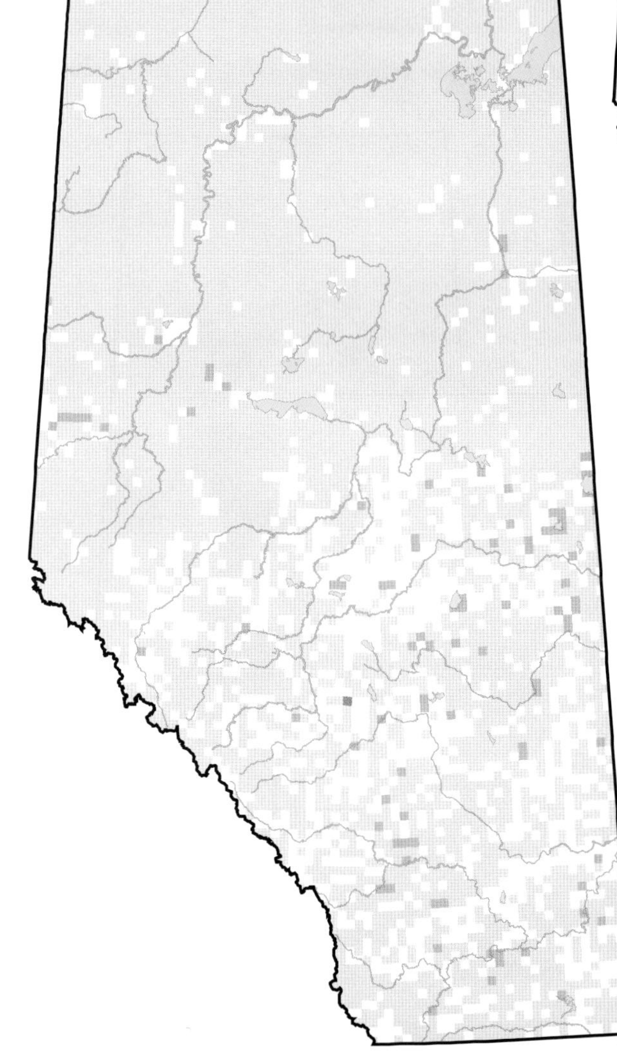

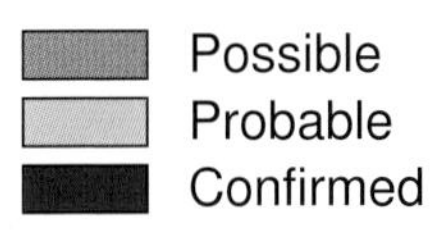

Short-billed Dowitcher (*Limnodromus griseus*)

Photo: Royal Alberta Museum

NESTING

CLUTCH SIZE (EGGS):	4–5
INCUBATION (DAYS):	21
FLEDGING (DAYS):	12–17
NEST HEIGHT (METRES):	0

The Short-billed Dowitcher is known to breed in the Boreal Forest and Foothills Natural Regions in Alberta, although no records of this bird were obtained in the Foothills NR in Atlas 2. Despite the lack of records from the Foothills NR, the area of observed distribution for this species did not change substantially from Atlas 1 to Atlas 2, as the Foothills NR is at the periphery of this species' range.

Short-billed Dowitchers nest in black-spruce bogs where other shorebirds, such as Lesser Yellowlegs, Greater Yellowlegs, and Solitary Sandpipers, also breed. Nests are built near the edge of wetlands on small mounds that are usually surrounded with, or adjacent to, woody vegetation. During Atlas 2, no probable or confirmed breeding evidence was obtained in the Boreal Forest NR for this species. A nest-visiting record and a territorial nesting behaviour record for this species were obtained in the Parkland NR. Observations in the Grassland NR were non-breeders.

A decline in relative abundance was detected in the Boreal Forest NR where, relative to other species, this dowitcher species was observed less frequently in Atlas 2 than in Atlas 1. Population declines could be related to declines in the availability of wetlands due to drought and drainage. The Breeding Bird Survey sample size of Short-billed Dowitcher records was too small to investigate abundance changes, in either Alberta or Canada. Differences between the results of Atlas and Breeding Bird Survey analyses are likely due in part to the Atlas's more extensive northern coverage. Also, the BBS is a road-based survey, which introduces a bias against water-dependent nesters. A decline in relative abundance was detected in the Parkland NR. However, since most of these were non-breeding records, it is likely an artifact of migratory variation. The status of the Short-billed Dowitcher is Undetermined in the province because there is not enough information about this species.

TEMPORAL REPORTING RATE

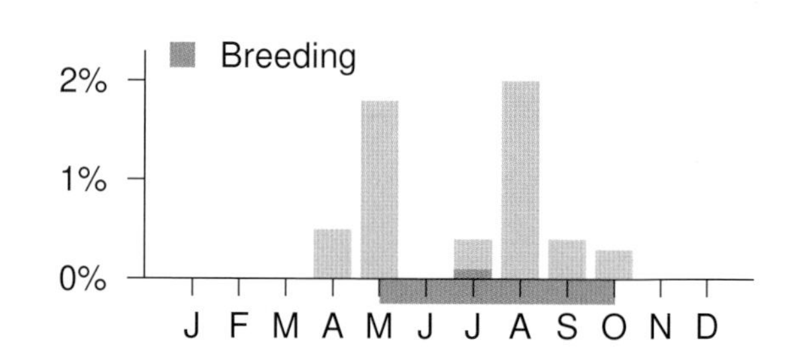

NATURAL REGIONS REPORTING RATE

SPATIAL REPORTING RATE

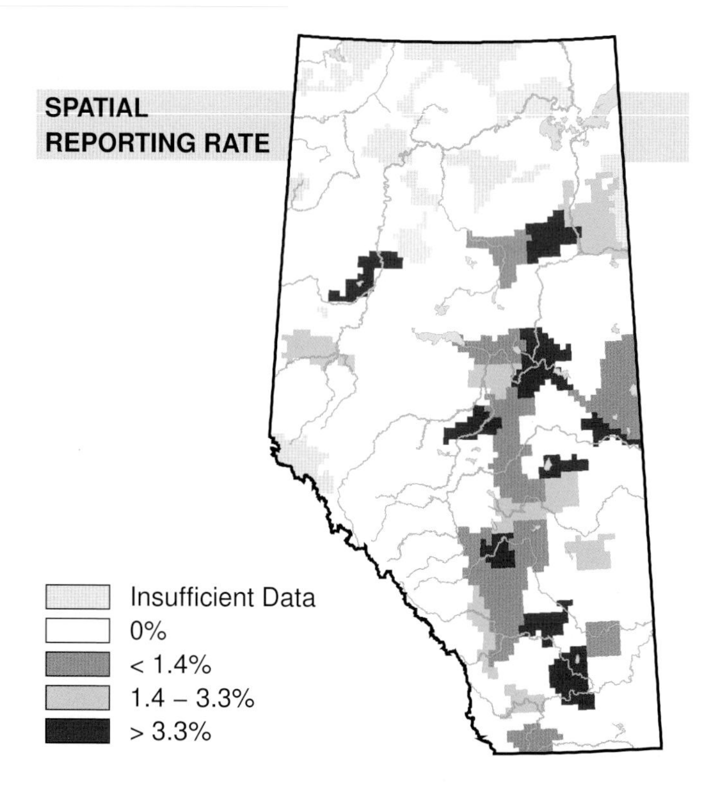

STATUS

GSWA 2000	GSWA 2005	COSEWIC Historic Rankings	COSEWIC 2007	Alberta Wildlife Act
Undetermined	Undetermined	N/A	N/A	N/A

OBSERVED DISTRIBUTION ATLAS 2

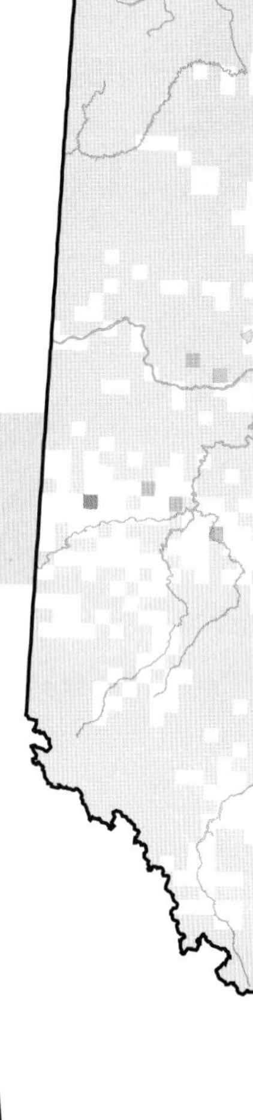

RANGE

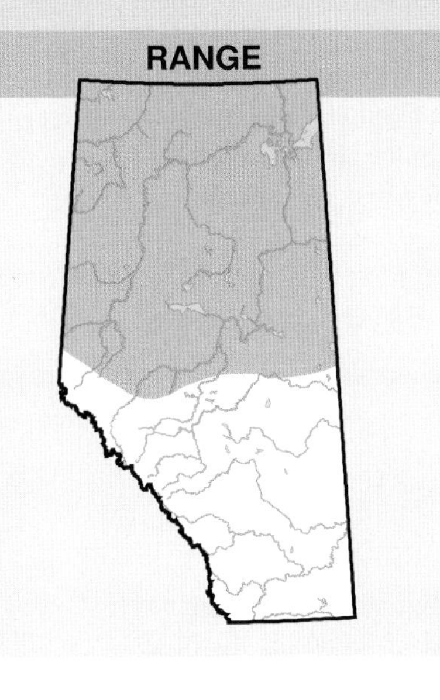

OBSERVED DISTRIBUTION ATLAS 1

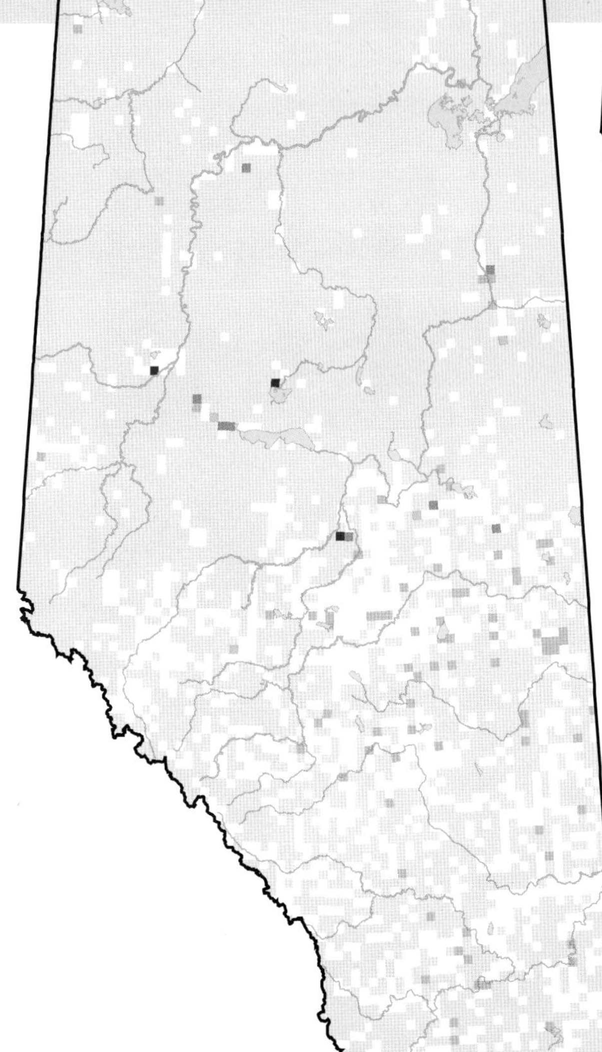

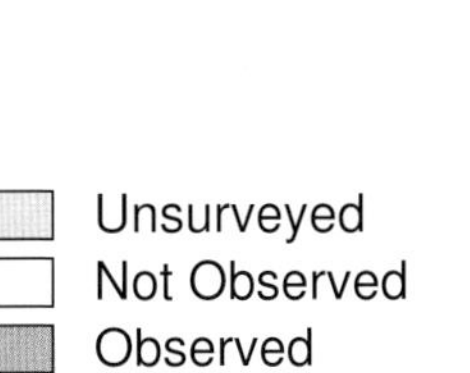

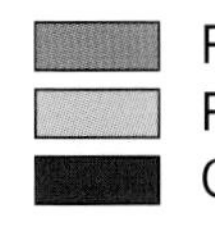

Wilson's Snipe (*Gallinago delicata*)

Photo: Gerald Romanchuk

NESTING

CLUTCH SIZE (EGGS):	4
INCUBATION (DAYS):	18–20
FLEDGING (DAYS):	19–20
NEST HEIGHT (METRES):	0

One of the most common shorebirds in Alberta, Wilson's Snipe is found in every Natural Region in the province. No change in distribution was observed between Atlas 1 and Atlas 2 for this species.

Wilson's Snipe is a ground-nester that breeds in wet areas with clumped vegetation. This species tends to breed at sites with short vegetation so that predators can be detected when they approach the nest. Wilson's Snipe uses its long bill to feed on invertebrates, its main prey, which it usually finds just below the surface of wet organic soils. Even though this bird was fairly evenly distributed across the province, it was found most often in the Foothills NR and least often in the Boreal Forest and Grassland NRs. This observation was in part expected because the Foothills NR tends to be a fairly wet region while the Grassland NR is the driest natural region in Alberta. Low observation rates in the Boreal Forest NR were unexpected, but could be a function of less extensive coverage than was afforded the southern parts of the province.

Declines in relative abundance were detected in the Boreal Forest and Parkland NRs where, relative to other species, this bird was observed less frequently in Atlas 2 than in Atlas 1. As these NRs were drier in Atlas 2 than in Atlas 1, the observed declines could be related to reductions in habitat. Relative abundance increases were detected in the Grassland and Rocky Mountain NRs. During Atlas 2, the Grassland NR was wetter than during Atlas 1, which may have provided more suitable habitat for this species. The increase in the Rocky Mountain NR, however, may be related to differences in coverage between Atlas 1 and 2, as there is no apparent biological cause. No change was detected in the Foothills NR. The Breeding Bird Survey found no change in the abundance of this species in Alberta, or Canada in recent years. This species is considered Secure in Alberta.

TEMPORAL REPORTING RATE

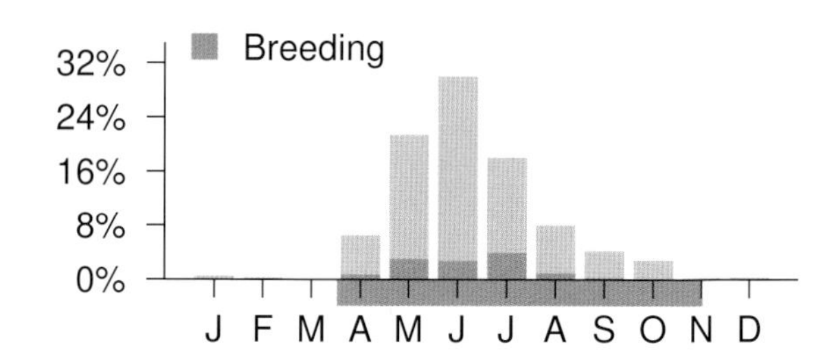

NATURAL REGIONS REPORTING RATE

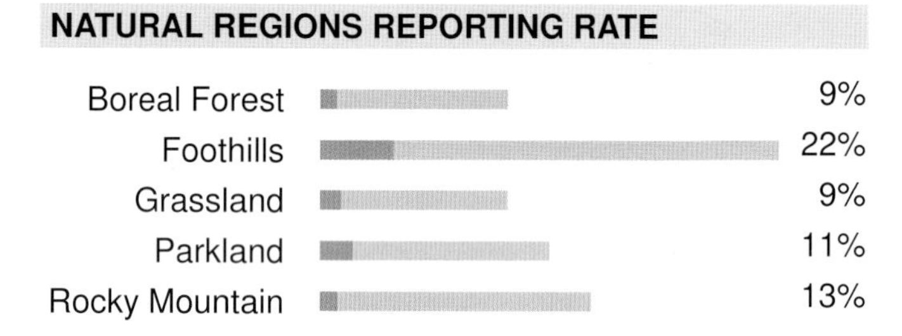

SPATIAL REPORTING RATE

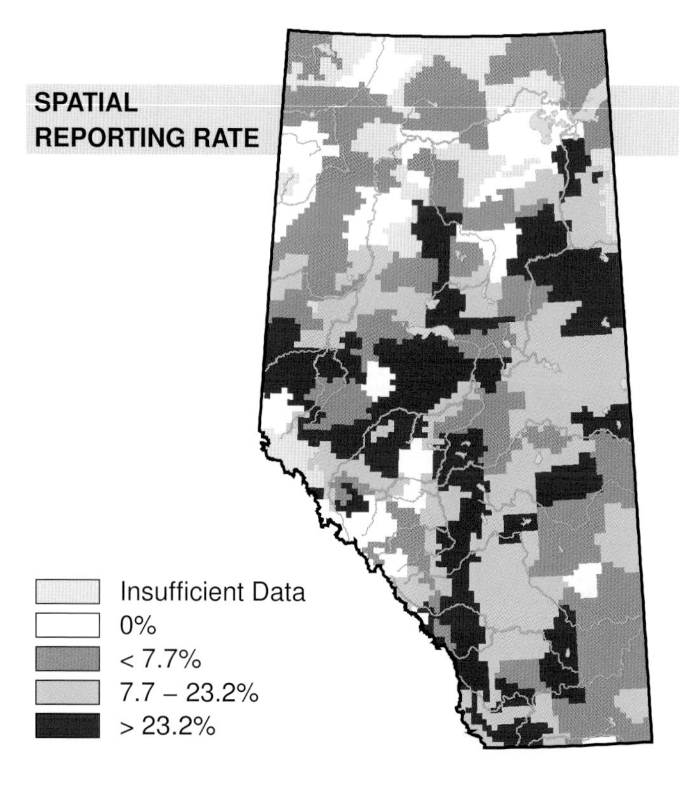

STATUS

GSWA 2000	GSWA 2005	COSEWIC Historic Rankings	COSEWIC 2007	Alberta Wildlife Act
N/A	N/A	N/A	N/A	N/A

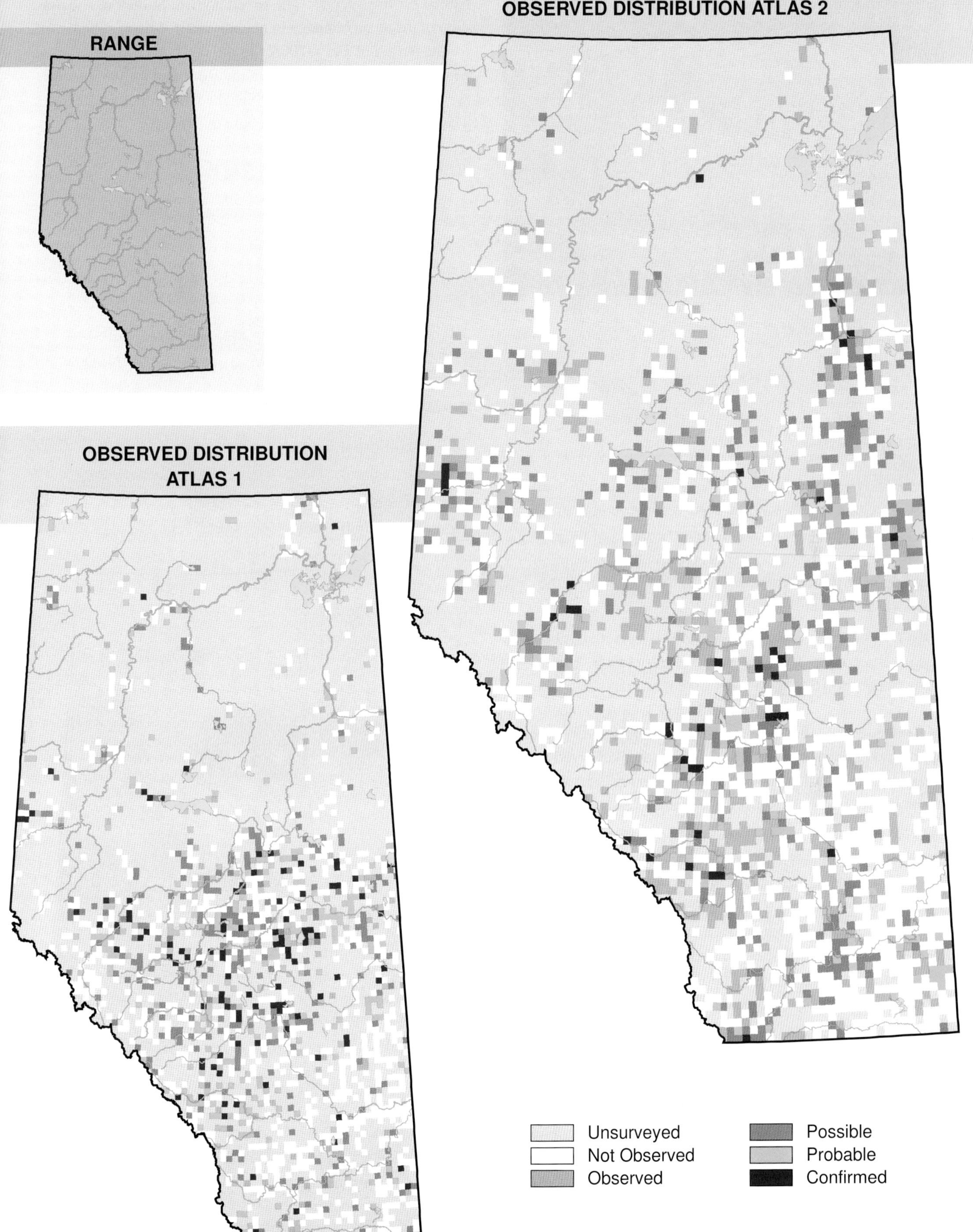
OBSERVED DISTRIBUTION ATLAS 2
RANGE
OBSERVED DISTRIBUTION ATLAS 1
Unsurveyed
Not Observed
Observed
Possible
Probable
Confirmed

Wilson's Phalarope (*Phalaropus tricolor*)

Photo: Gerald Romanchuk

NESTING

CLUTCH SIZE (EGGS):	4
INCUBATION (DAYS):	16–21
FLEDGING (DAYS):	unknown
NEST HEIGHT (METRES):	0

The largest phalarope, Wilson's Phalarope is found in every Natural Region in Alberta. The observed distribution of this species decreased in Atlas 2 in the northern part of the province. It is unclear whether this represents a range contraction or if the presence of Wilson's Phalarope in the north was missed during Atlas 2.

Wilson's Phalarope is closely associated with wetlands because this species consumes mainly aquatic invertebrates. Unlike other phalaropes, this species does not nest immediately beside wetlands. Rather, these birds usually nest within 100 metres of wetland edges in tall, dense vegetation. Despite occurring in all natural regions, this species was most often observed in the Grassland and Parkland NRs, but were found only occasionally in the Boreal Forest and Rocky Mountain NRs.

Declines in relative abundance were detected in the Boreal Forest and Parkland NRs where, relative to other species, Wilson's Phalarope was observed less frequently in Atlas 2 than in Atlas 1. These NRs were drier in Atlas 2 than in Atlas 1, creating conditions that would have reduced the amount of suitable habitat for this species. An increase in relative abundance was detected in the Grassland NR where, relative to other species, this bird was observed more frequently in Atlas 2 than in Atlas 1. This NR was wetter during Atlas 2 than during Atlas 1, which would have provided more suitable habitat for this species. The Breeding Bird Survey in Alberta did not detect any change in abundance of Wilson's Phalarope for the period of 1985–2005, but did detect an increase for the period 1995–2005. As many of the BBS data for this species are gathered in the Grassland NR, the BBS trend is expected. The status of this species is Secure in Alberta.

TEMPORAL REPORTING RATE

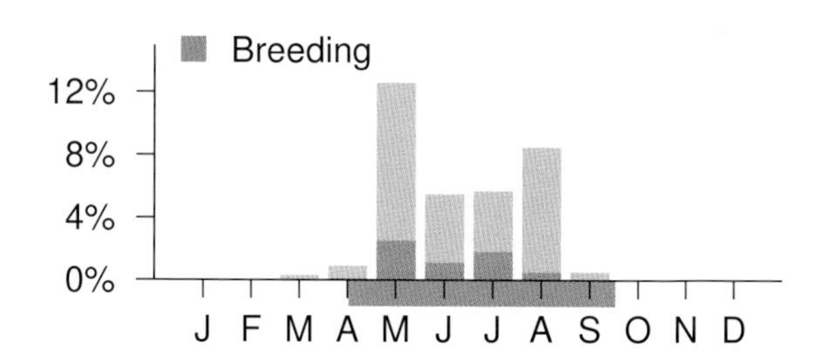

NATURAL REGIONS REPORTING RATE

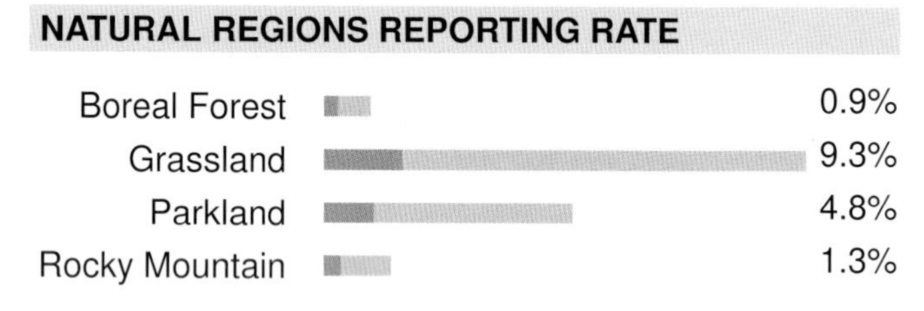

SPATIAL REPORTING RATE

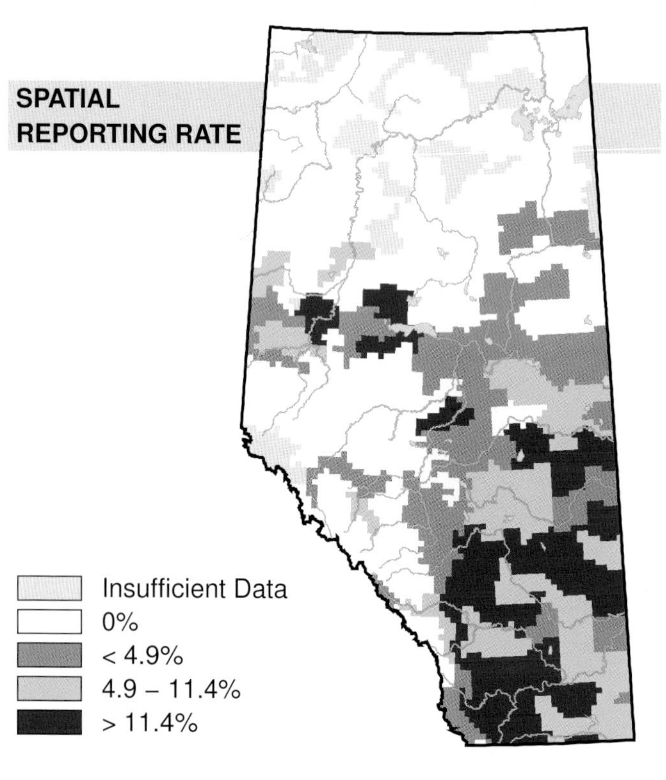

STATUS

GSWA 2000	GSWA 2005	COSEWIC Historic Rankings	COSEWIC 2007	Alberta Wildlife Act
Secure	Secure	N/A	N/A	N/A

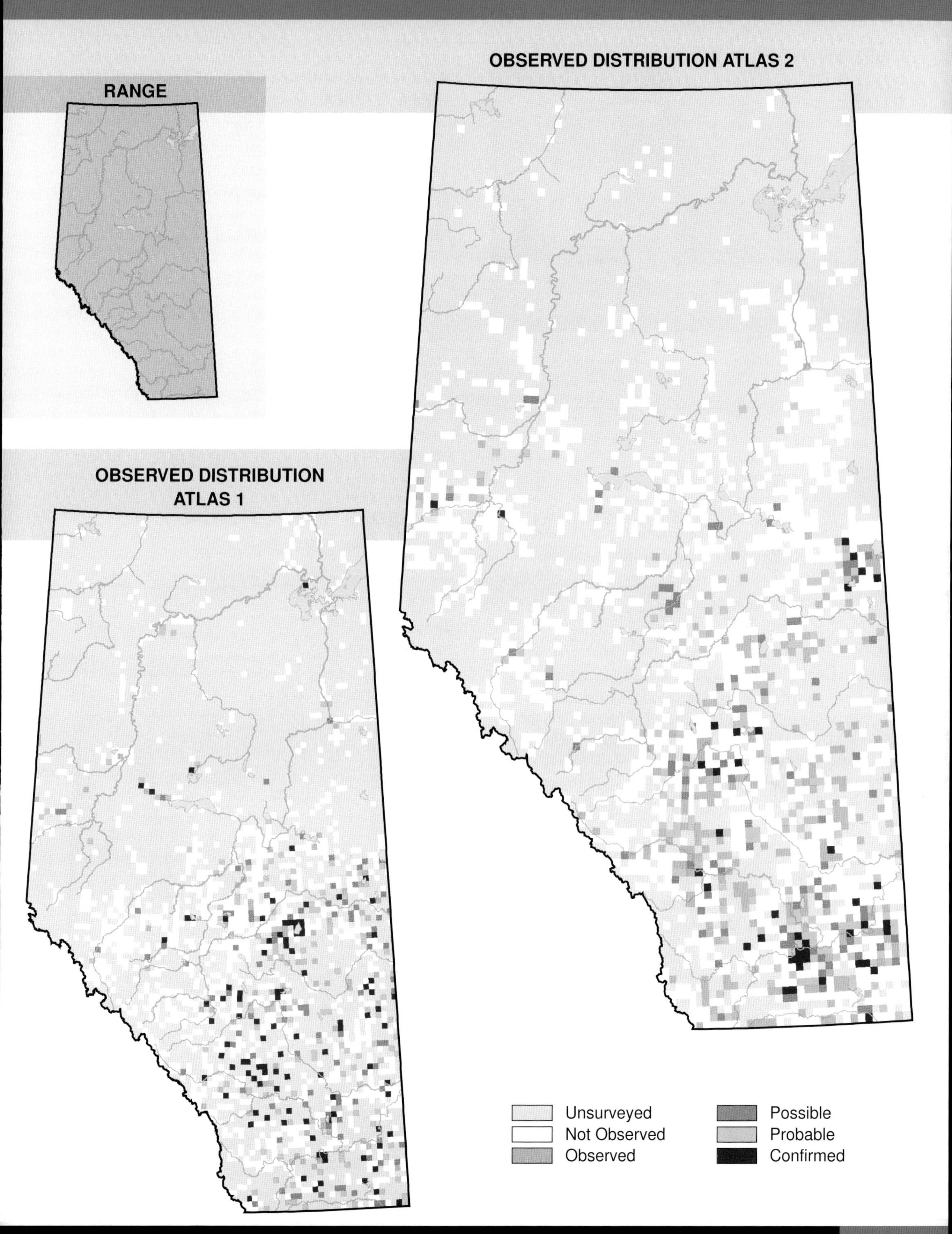
Habitat
Nest Location
Nest Type
Diet
RANGE
OBSERVED DISTRIBUTION ATLAS 2
OBSERVED DISTRIBUTION ATLAS 1
Unsurveyed
Not Observed
Observed
Possible
Probable
Confirmed

Red-necked Phalarope (*Phalaropus lobatus*)

Photo: Gerald Romanchuk

NESTING

CLUTCH SIZE (EGGS):	4
INCUBATION (DAYS):	17–23
FLEDGING (DAYS):	20
NEST HEIGHT (METRES):	0

One of Alberta's rarest breeding birds, the Red-necked Phalarope is known to breed in the northern part of the Boreal Forest Natural Region. The observed distribution of this species did not change between Atlas 1 and Atlas 2.

In Alberta, Red-necked Phalaropes are usually encountered during spring and fall migrations when they are moving between their South American wintering grounds and their Arctic breeding grounds. During Atlas 2, migrant Red-necked Phalaropes were most often observed in the Grassland NR. The southern periphery of their breeding range extends into the northern part of Alberta, but there are only a few documented cases of Red-necked Phalaropes breeding in the province. This species builds its nest along the edges of wetlands on a raised mound that is usually surrounded with clumps of grasses, sedges, or forbs. Compared to Wilson's Phalarope, which is more common in Alberta, the Red-necked Phalarope generally nests closer to water and within vegetation that is shorter and less dense.

A decline in relative abundance was detected in the Boreal Forest NR and an increase was detected in the Grassland NR. However, with no breeding records and the recognition that most sightings were recorded in May, August, and September, it appears that all the Atlas records of this species were secured during migration. In such a case, the detected changes could just as likely be an artifact of the broader survey window in Atlas 2, or the result of the inherent variability of migration records, as they could be indicative of actual changes in abundance. The Breeding Bird Survey, lacking extensive northern coverage, provided a sample size that was too small to investigate abundance changes in Alberta or Canada. The status of this species is Secure in Alberta.

TEMPORAL REPORTING RATE

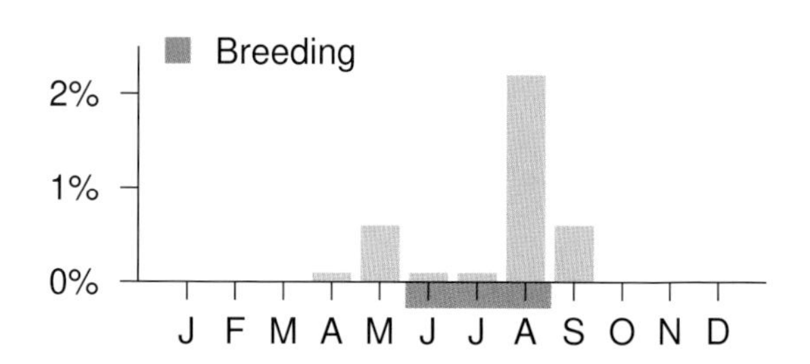

NATURAL REGIONS REPORTING RATE

Boreal Forest	< 0.1%
Grassland	0.7%
Parkland	0.2%

SPATIAL REPORTING RATE

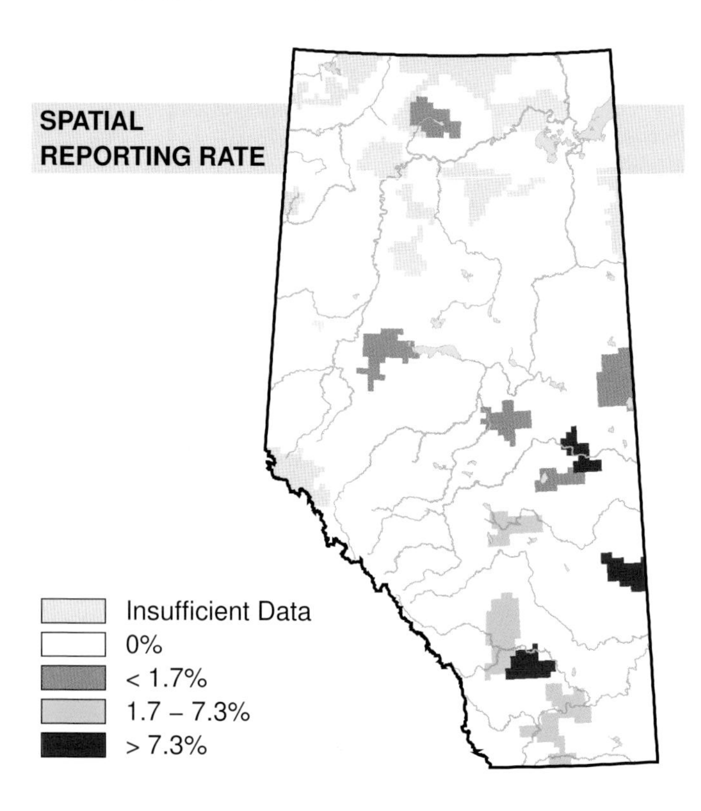

STATUS

GSWA 2000	GSWA 2005	COSEWIC Historic Rankings	COSEWIC 2007	Alberta Wildlife Act
Secure	Secure	N/A	N/A	N/A

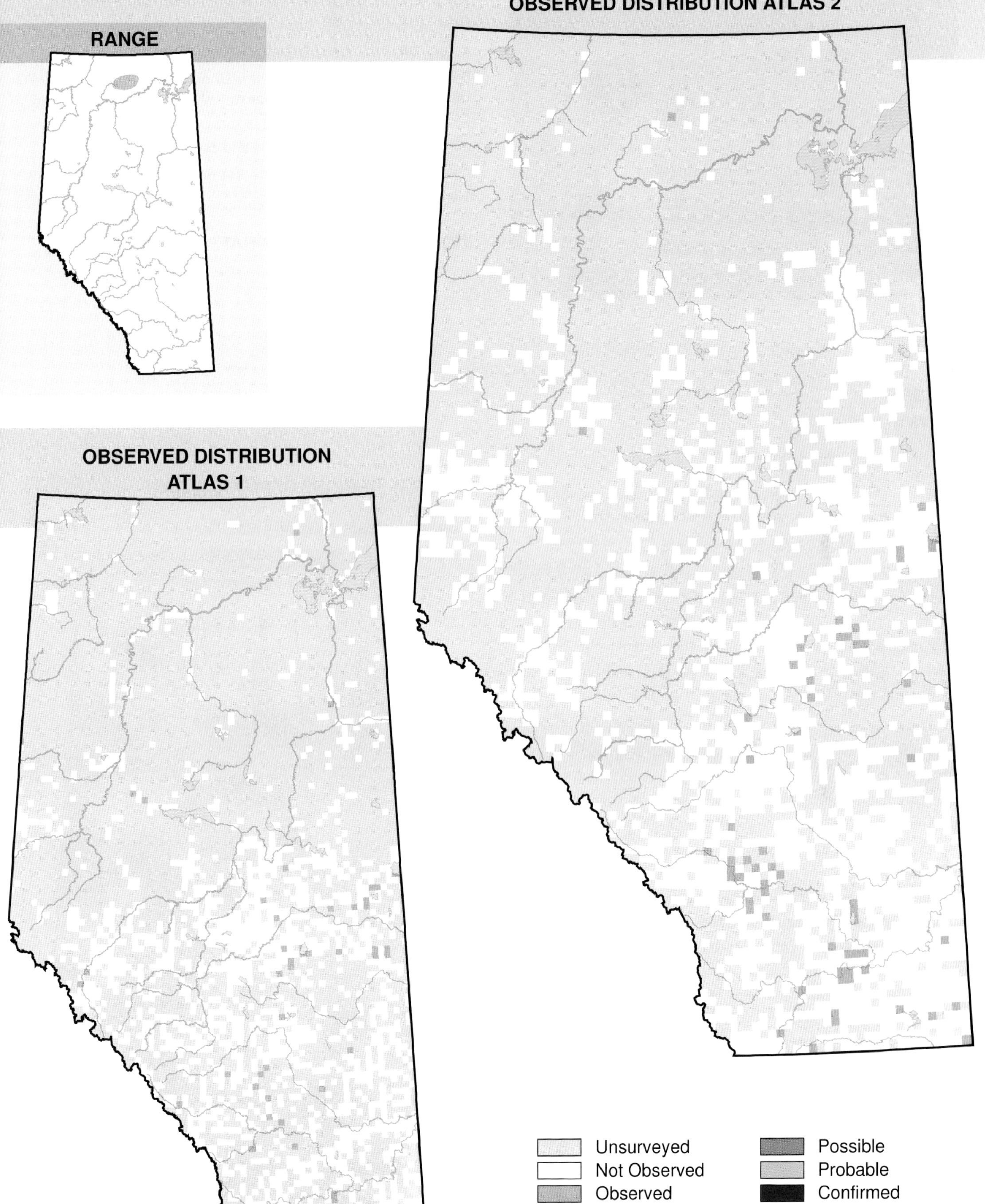
RANGE
OBSERVED DISTRIBUTION ATLAS 2
OBSERVED DISTRIBUTION ATLAS 1
Unsurveyed
Not Observed
Observed
Possible
Probable
Confirmed

Franklin's Gull (*Larus pipixcan*)

Photo: Gerald Romanchuk

NESTING

CLUTCH SIZE (EGGS):	1–3
INCUBATION (DAYS):	21–28
FLEDGING (DAYS):	28–30
NEST HEIGHT (METRES):	0

Franklin's Gull breeds in every Natural Region across Alberta. No change in distribution was detected between Atlas 1 and Atlas 2.

Despite being found in all the natural regions, this species was found most frequently in the Grassland NR. It was found only occasionally in the Rocky Mountain NR. The distribution is explained by this species' dependency on extensive prairie marshes for nesting. In these areas this species builds its nests on floating mats built among emergent vegetation.

Declines in relative abundance were detected in the Boreal and Parkland NRs where, relative to other species, this gull was observed less frequently in Atlas 2 than in Atlas 1. Drought or duck habitat management could be the main reason for these declines. Franklin's Gulls appear to respond to total expanse of water, depth of water, density and dispersion of vegetation, and size and dispersion of open-water areas. Any changes in these patterns result in colony desertion. An increase in relative abundance was detected in the Grassland NR where, relative to other species, the bird was observed more frequently in Atlas 2 than in Atlas 1. Above-normal precipitation in the later half of Atlas 2 is the likely cause of this increase. The Breeding Bird Survey did not find abundance changes for this species in Alberta, or across Canada. There has been considerable controversy surrounding these surveys because Franklin's Gulls nest in remote areas that Breeding Bird Survey routes may not cover. Therefore, many of the observations made during Breeding Bird Surveys are probably non-breeding individuals, and any change that would be detected would indicate change to only the non-breeding population. The Atlas was able to detect change because, unlike the strategy used in the Breeding Bird Survey, Atlas surveyors were not restricted to surveying at specified points and could, therefore, access remote breeding colonies. This species is considered Secure in Alberta.

TEMPORAL REPORTING RATE

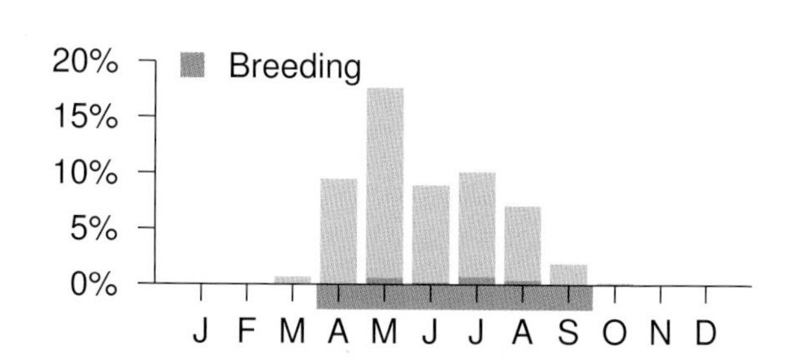

NATURAL REGIONS REPORTING RATE

Boreal Forest	2%
Grassland	11%
Parkland	6%

SPATIAL REPORTING RATE

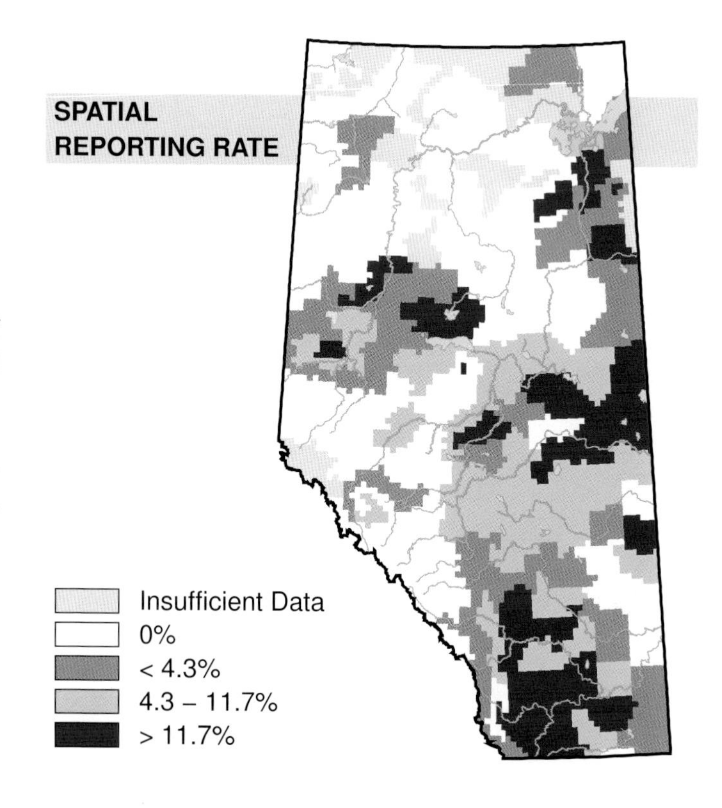

STATUS

GSWA 2000	GSWA 2005	COSEWIC Historic Rankings	COSEWIC 2007	Alberta Wildlife Act
Secure	Secure	N/A	N/A	N/A

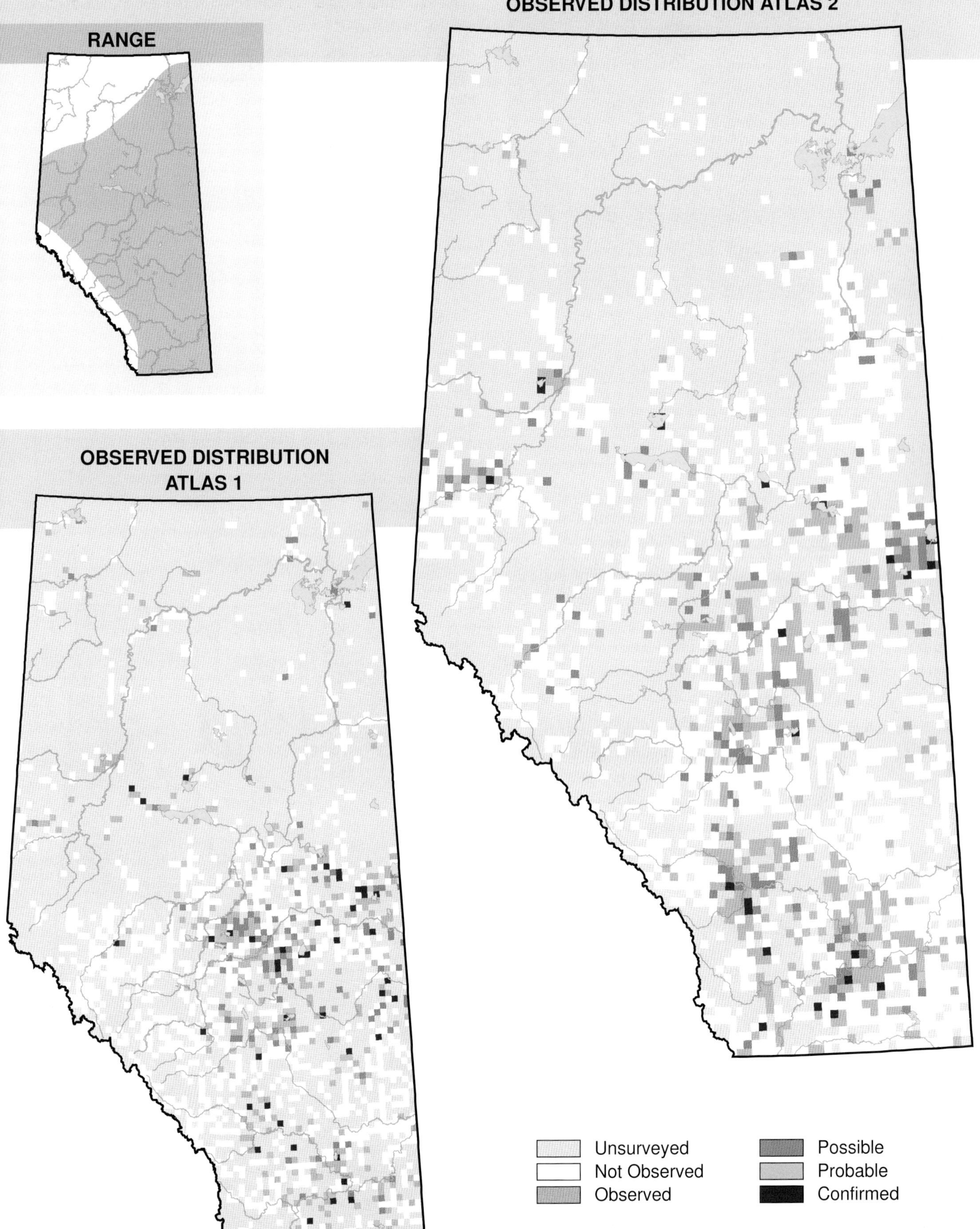
OBSERVED DISTRIBUTION ATLAS 2
RANGE
OBSERVED DISTRIBUTION ATLAS 1
Unsurveyed
Not Observed
Observed
Possible
Probable
Confirmed

Bonaparte's Gull *(Larus philadelphia)*

Photo: Gerald Romanchuk

NESTING

CLUTCH SIZE (EGGS):	2–4
INCUBATION (DAYS):	24
FLEDGING (DAYS):	unknown
NEST HEIGHT (METRES):	2–7

Bonaparte's Gull breeds in the Boreal Forest and Foothills Natural Regions in Alberta. The observed distribution of this species did not change between Atlas 1 and Atlas 2.

Bonaparte's Gull is unique among most other breeding gulls in Alberta because it nests off the ground in trees. Nests are usually located around the perimeter of lakes and marshes in conifer trees. For this reason, this species was found in the Boreal Forest and Foothills NRs where this type of habitat can be found. Observations made in other NRs were non-breeders.

Declines in relative abundance were detected in the Boreal Forest and Foothills NRs where, relative to other species, Bonaparte's Gull was observed less frequently in Atlas 2 than in Atlas 1. These declines could be related to drought or wetland drainage. The Breeding Bird Survey did not find abundance changes in Alberta or Canada. Differences between the results of Atlas and Breeding Bird Survey analyses could be attributed to differences in survey designs. The Breeding Bird Survey is conducted at fixed points along roads and most species are identified by sound. This results in lower detection rates for groups such as waterbirds. In contrast, some Atlas surveys were conducted along the perimeter of wetlands without the constraint of using fixed points. The result was a larger sample size for birds that are associated with wetlands. In the Grassland NR, an increase in relative abundance was detected, while a decrease was detected in the Parkland NR. As records in these NRs were primarily those of non-breeders and likely migrants, these changes are likely a function of migratory variation and not an indication of true changes in abundance. No change in relative abundance was detected in the Rocky Mountain NR. This species is considered Secure in Alberta.

TEMPORAL REPORTING RATE

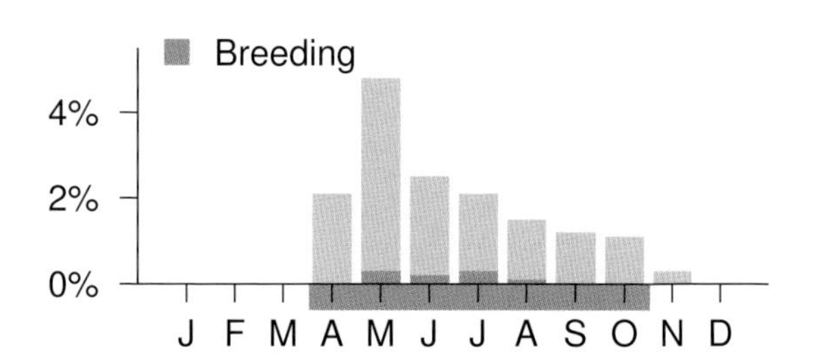

NATURAL REGIONS REPORTING RATE

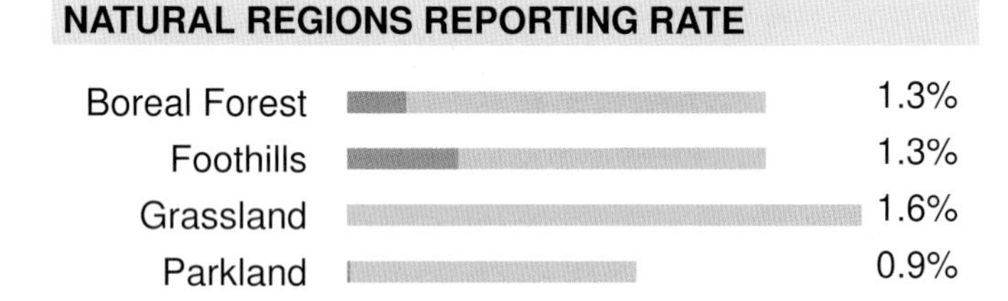

SPATIAL REPORTING RATE

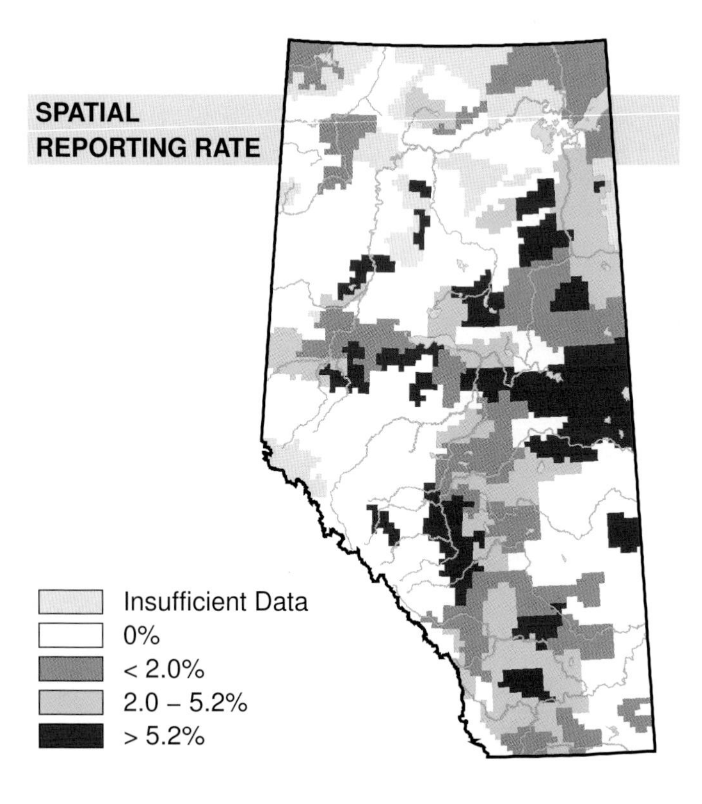

STATUS

GSWA 2000	GSWA 2005	COSEWIC Historic Rankings	COSEWIC 2007	Alberta Wildlife Act
Secure	Secure	N/A	N/A	N/A

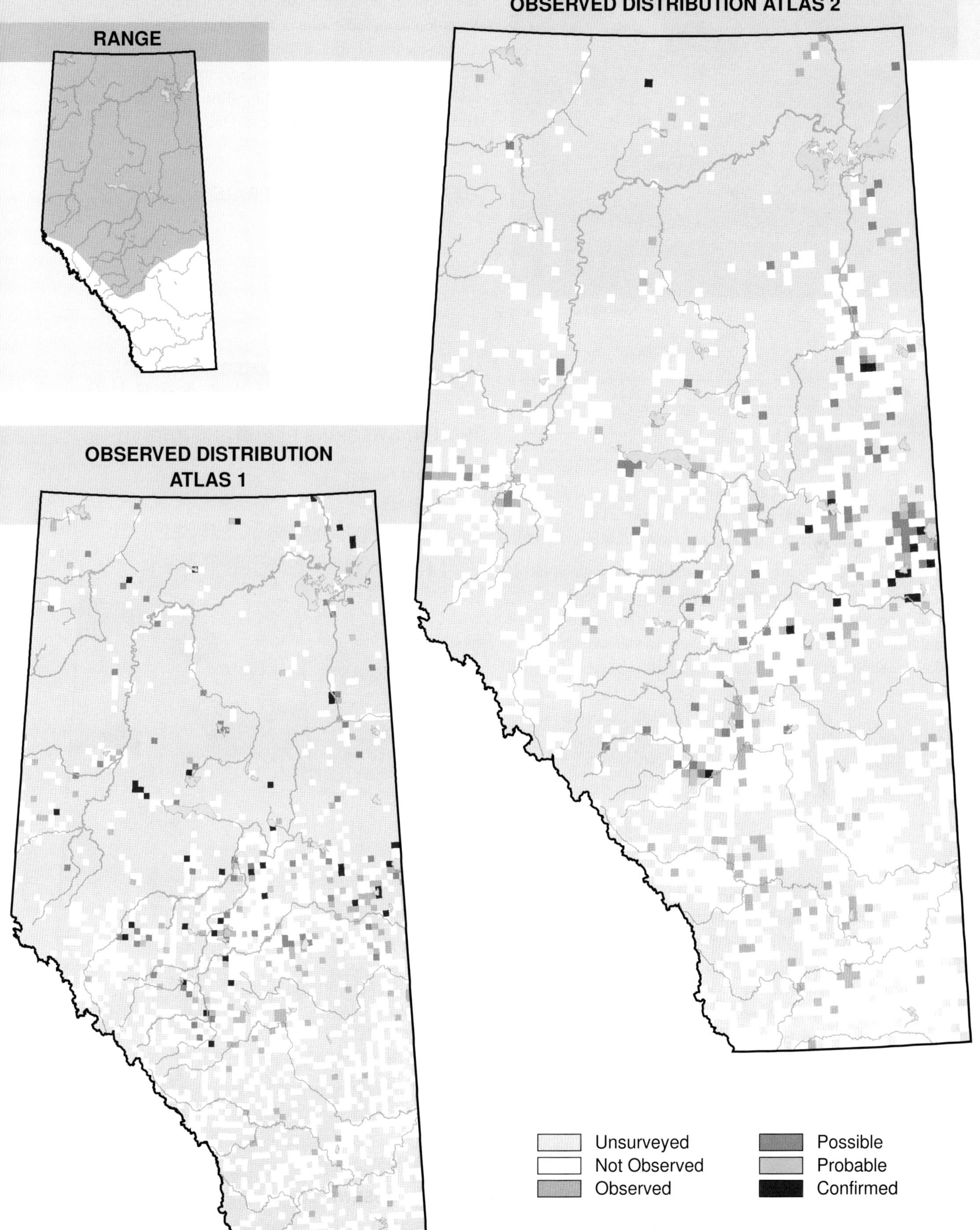
Habitat
Nest Location
Nest Type
Diet
RANGE
OBSERVED DISTRIBUTION ATLAS 2
OBSERVED DISTRIBUTION ATLAS 1
Unsurveyed
Not Observed
Observed
Possible
Probable
Confirmed

Mew Gull (*Larus canus*)

Photo: Royal Alberta Museum

NESTING

CLUTCH SIZE (EGGS):	2–5
INCUBATION (DAYS):	23–26
FLEDGING (DAYS):	28–35
NEST HEIGHT (METRES):	0

One of the rarest breeding gulls in Alberta, the Mew Gull breeds only in the Boreal Forest Natural Region. The distribution of this species did not change in Atlas 2.

In Alberta, Mew Gulls are encountered during spring and fall migration when they are moving between their wintering grounds along the Pacific Coast and their northern breeding grounds, mainly in Alaska and northwestern Canada. During Atlas 2, migrating Mew Gulls were most often observed in the Grassland NR; however, migrants and non-breeders were also encountered infrequently in the Boreal Forest, Parkland, and Rocky Mountain NRs. The southern periphery of their breeding range occurs in the northern part of the Boreal Forest NR in Alberta, with a few breeding records from there during Atlas 2. In non-coastal areas this species usually nests on the ground on islands. However, they can also build nests off the ground in trees provided these are near water.

A relative abundance decline was detected in the Boreal Forest NR where, relative to other species, the Mew Gull was observed less frequently in Atlas 2 than in Atlas 1. No change was detected in other NRs. The Breeding Bird Survey sample size was too small to investigate Mew Gull abundance change in Alberta. However, no abundance decline was detected across Canada. The Atlas sample size of breeding records for this species was very small in both Atlases; so, it is difficult to determine if the detected decline is a function of migratory variation or an indication of a real change in the population. Due to the distribution of this species' breeding range in northern Alberta, it is not effectively monitored by surveys such as the Atlas or the Breeding Bird Survey because these surveys do not adequately monitor the far north. This species is considered Secure in Alberta.

TEMPORAL REPORTING RATE

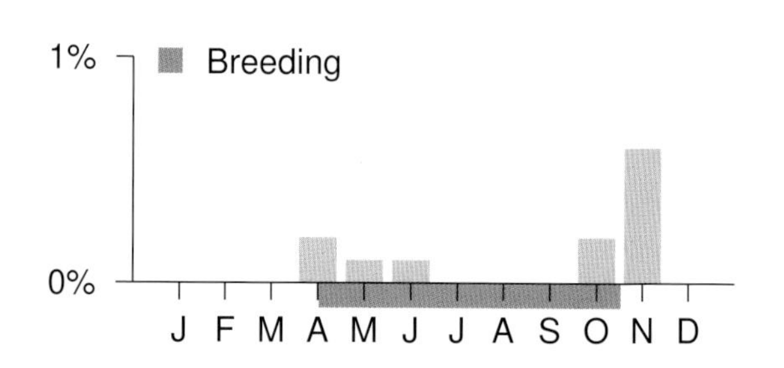

NATURAL REGIONS REPORTING RATE

SPATIAL REPORTING RATE

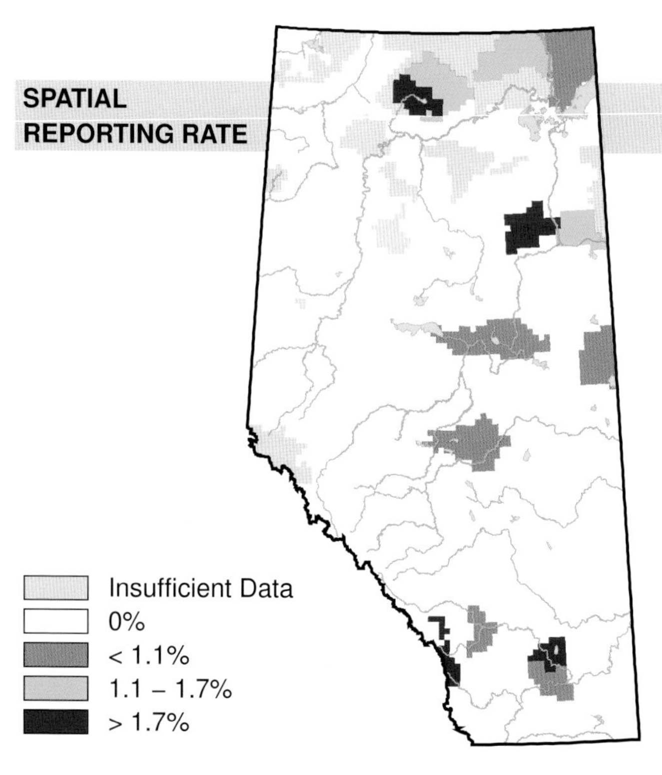

STATUS

GSWA 2000	GSWA 2005	COSEWIC Historic Rankings	COSEWIC 2007	Alberta Wildlife Act
Secure	Secure	N/A	N/A	N/A

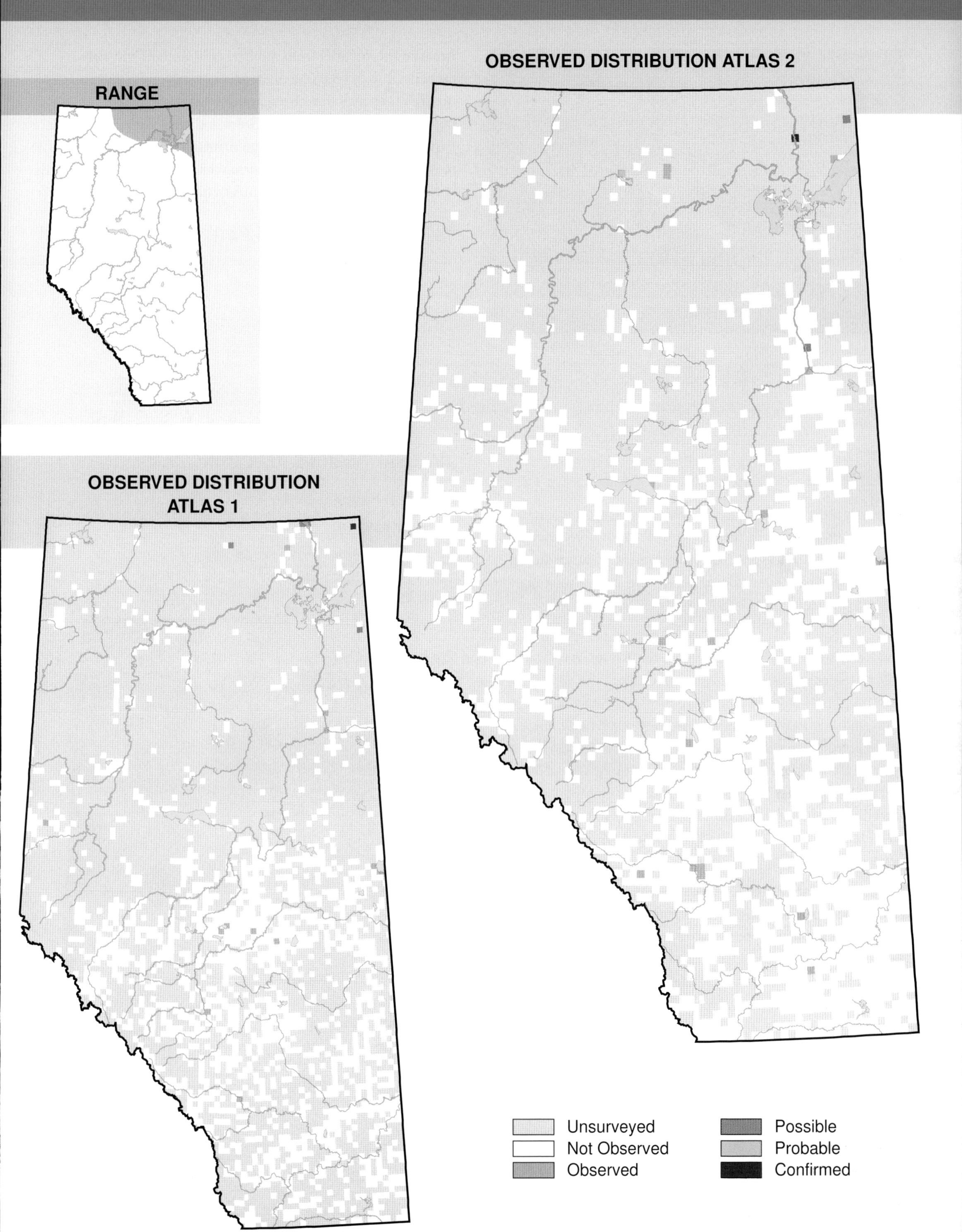

OBSERVED DISTRIBUTION ATLAS 2
RANGE
OBSERVED DISTRIBUTION ATLAS 1
Unsurveyed
Not Observed
Observed
Possible
Probable
Confirmed

Ring-billed Gull (*Larus delawarensis*)

Photo: Duane Boone

NESTING

CLUTCH SIZE (EGGS):	2–4
INCUBATION (DAYS):	25
FLEDGING (DAYS):	37
NEST HEIGHT (METRES):	0

Alberta's most common gull, the Ring-billed Gull is found in every Natural Region in the province. It is possible that the breeding range of this species has expanded because, during Atlas 2, new breeding evidence was found farther west than during Atlas 1, around Grande Prairie, west of Peace River, and west of High Level.

The Ring-billed Gull is a colonial nester that tends to nest near human settlements such as towns and cities. Nests are usually built on islands with varying degrees of cover that can range from sparse woody vegetation to no vegetation at all. Ground substrate can also vary to include soil, sand, or rock. This species was most commonly encountered in the Grassland NR, and it was common in the Parkland and Rocky Mountain NRs but only infrequent in the Boreal Forest NR. In the Foothills NR, few records were obtained in total and, of those, none was a breeding record.

Declines in relative abundance were detected in the Boreal Forest and Parkland NRs. It is unclear why these declines were detected, because this species does well in human-modified environments and human development has expanded in those regions. It is possible, however, that reduced precipitation in these NRs during Atlas 2—compared to precipitation levels recorded during Atlas 1—affected this species negatively. An increase in relative abundance was detected in the Rocky Mountain NR. This increase should be interpreted with caution because few records from the Rocky Mountain NR were associated with breeding evidence; consequently, these observations may have been non-breeders. No change was detected in the Foothills and Grassland NRs. The Breeding Bird Survey reported no change in abundance for this species in Alberta or across Canada during the period 1985–2005. This species is considered Secure in Alberta.

TEMPORAL REPORTING RATE

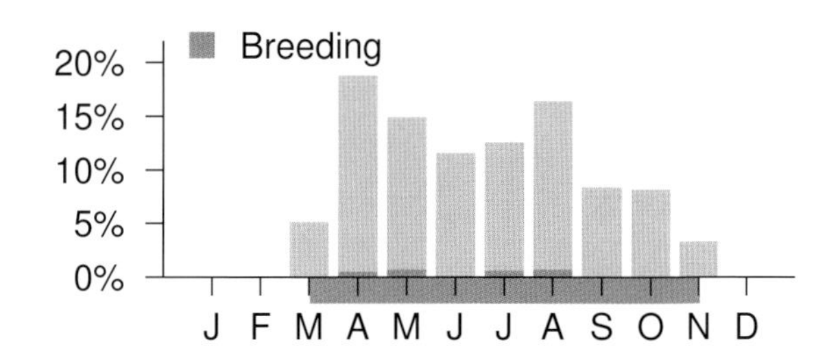

NATURAL REGIONS REPORTING RATE

Region	Rate
Boreal Forest	2%
Grassland	15%
Parkland	8%
Rocky Mountain	3%

SPATIAL REPORTING RATE

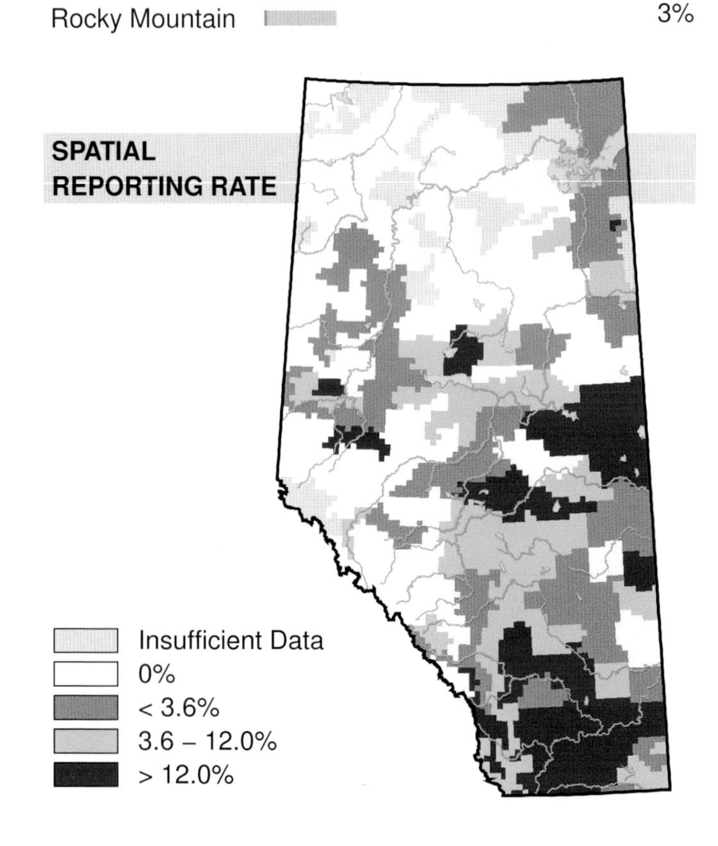

STATUS

GSWA 2000	GSWA 2005	COSEWIC Historic Rankings	COSEWIC 2007	Alberta Wildlife Act
Secure	Secure	N/A	N/A	N/A

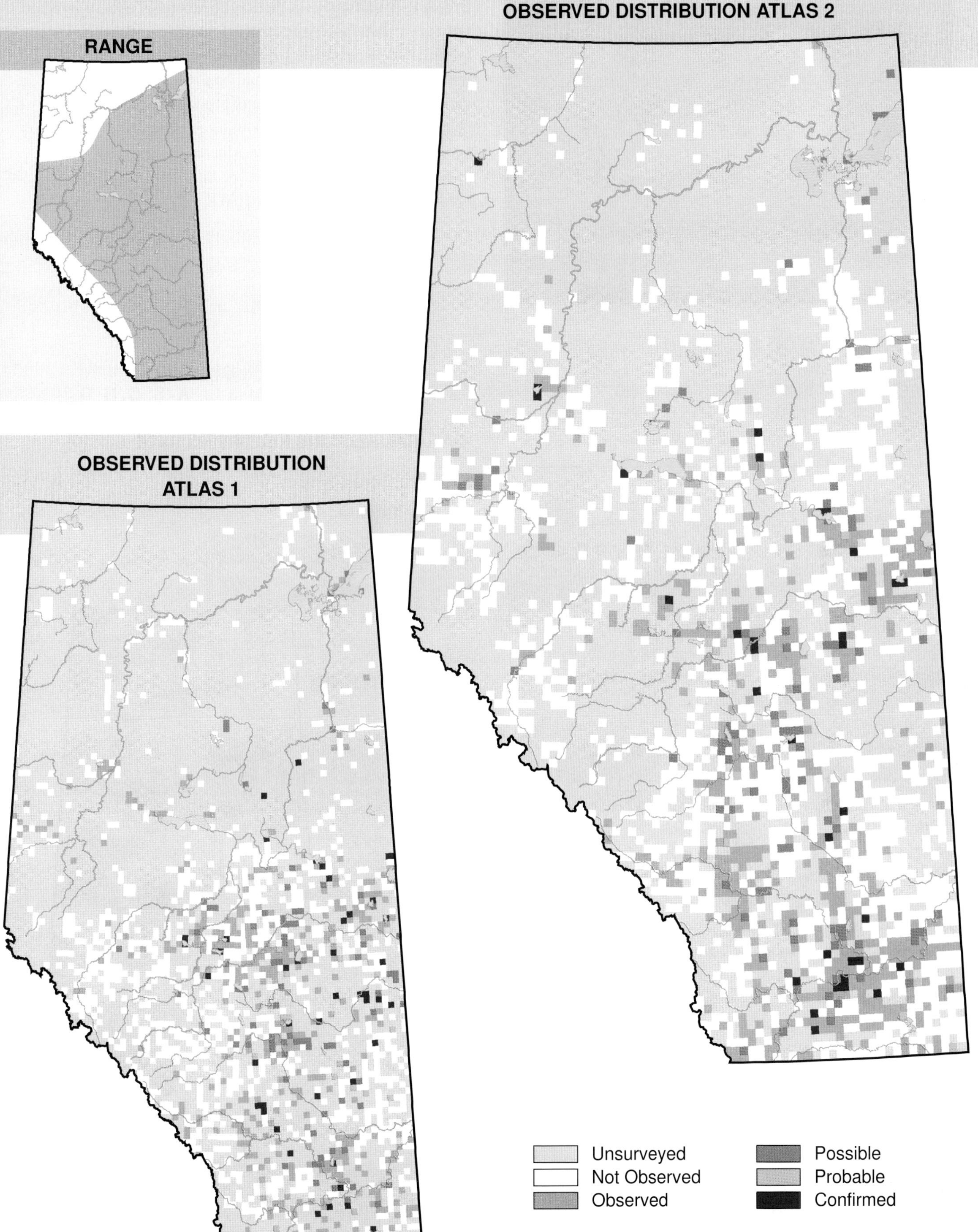

RANGE
OBSERVED DISTRIBUTION ATLAS 2
OBSERVED DISTRIBUTION ATLAS 1
Unsurveyed
Not Observed
Observed
Possible
Probable
Confirmed

California Gull (*Larus californicus*)

Photo: Randy Jensen

NESTING

CLUTCH SIZE (EGGS):	2–3
INCUBATION (DAYS):	25–27
FLEDGING (DAYS):	40
NEST HEIGHT (METRES):	0

The California Gull breeds in the Boreal Forest, Foothills, Grassland, and Parkland Natural Regions in Alberta. A few breeding records were found farther west during Atlas 2, near Grande Prairie and Peace River, than were reported during Atlas 1.

The California Gull is a ground nester that usually nests on treeless islands which may be located on lakes or in rivers. For this reason, this species was found most often in the Grassland NR, but it was also found in the Boreal Forest, Foothills, and Parkland NRs. Observations from the Rocky Mountain NR were non-breeders.

Declines in relative abundance were detected in the Boreal Forest and Parkland NRs where, relative to other species, California Gulls were observed less frequently in Atlas 2 than in Atlas 1. Declines could be related to drought. When drought occurs, the land bridges that can form between the mainland and breeding islands often make these sites unsuitable for nesting due to increased predator access. In contrast, the Breeding Bird Survey found an abundance increase in Alberta during the period 1985–2005. It is unclear why there are differences between what was found by the Atlas and by the BBS. These differences could be partially explained by differences in the spatial distribution of surveys. The Atlas had more widespread survey coverage in the north, than the Breeding Bird Survey. In the Rocky Mountain NR, an increase in relative abundance was detected. Despite agreement with BBS results, these records were non-breeders and mostly migrants. So, this increase could be a result of migratory variation and not an indication of abundance change. No change was detected in the Foothills and Grassland NRs. This species is considered Secure in Alberta.

TEMPORAL REPORTING RATE

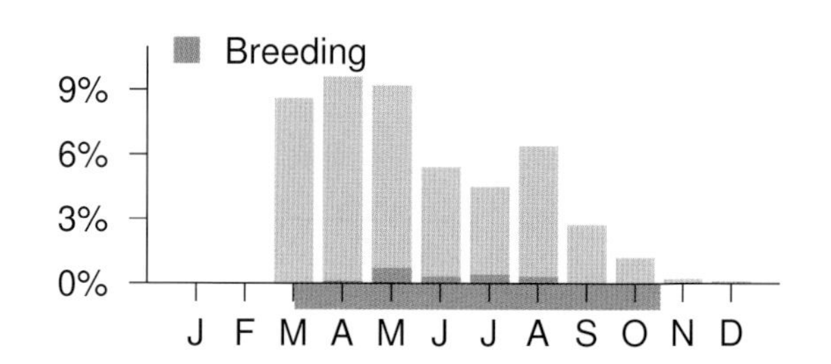

NATURAL REGIONS REPORTING RATE

Region	Rate
Boreal Forest	0.7%
Grassland	8.1%
Parkland	1.9%
Rocky Mountain	2.2%

SPATIAL REPORTING RATE

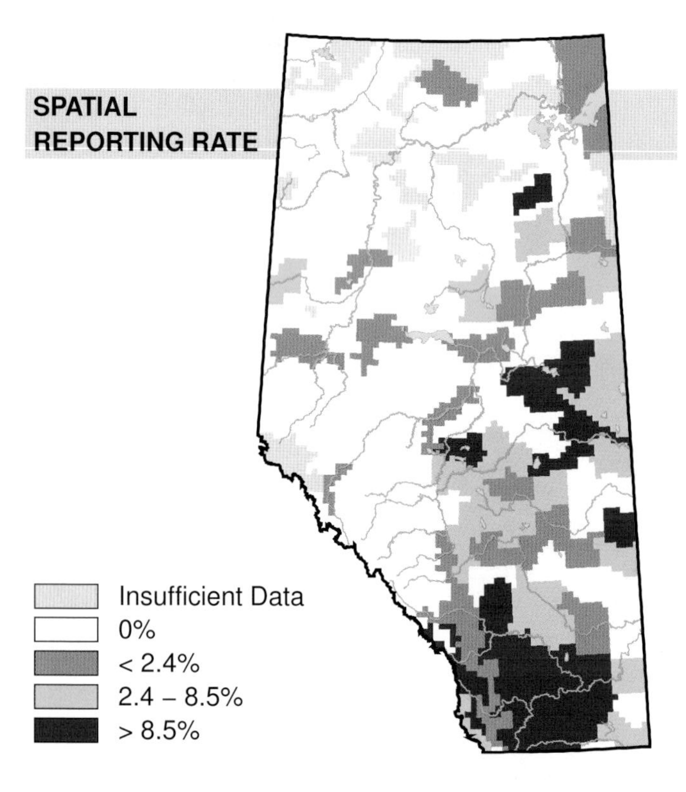

STATUS

GSWA 2000	GSWA 2005	COSEWIC Historic Rankings	COSEWIC 2007	Alberta Wildlife Act
Secure	Secure	N/A	N/A	N/A

OBSERVED DISTRIBUTION ATLAS 2

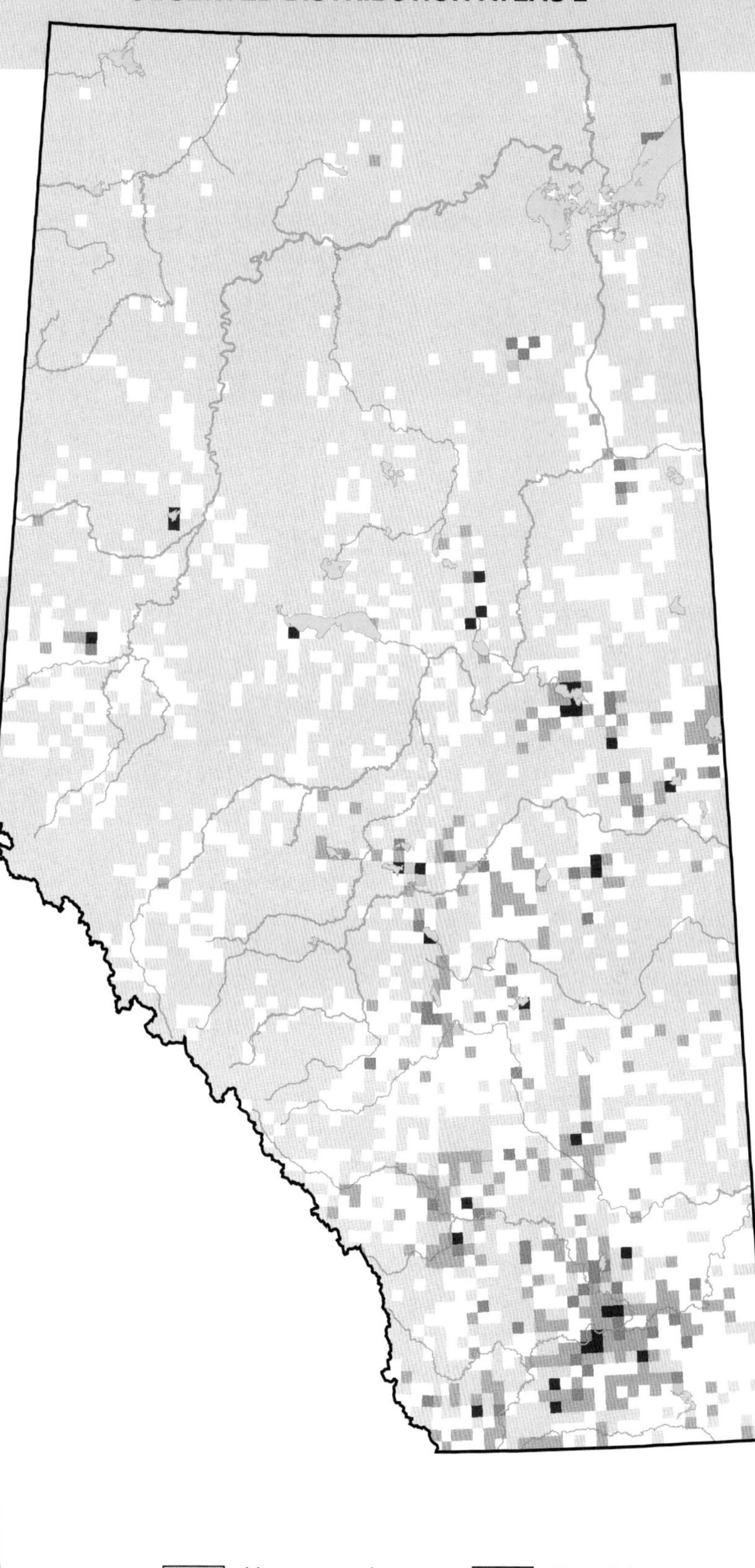

RANGE

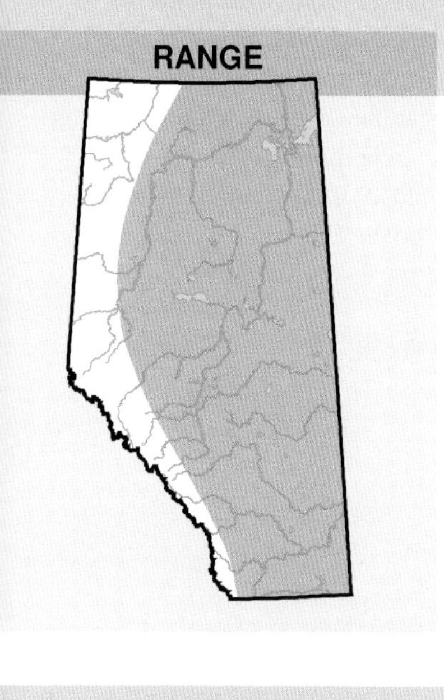

OBSERVED DISTRIBUTION ATLAS 1

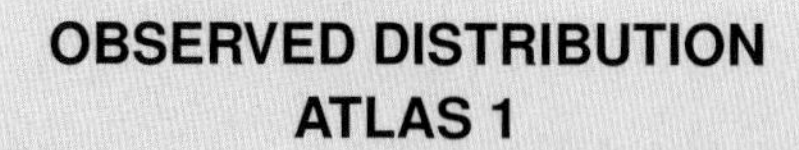

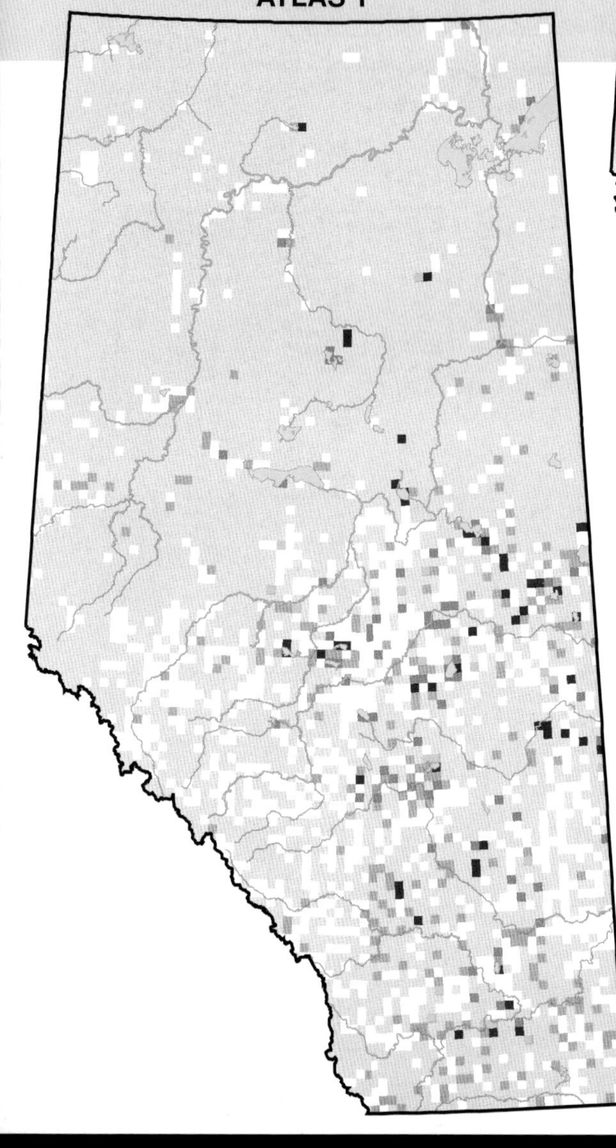

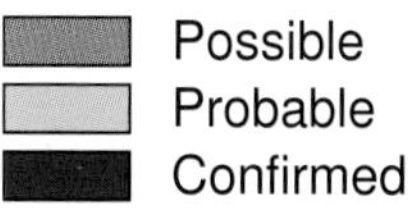

Herring Gull (*Larus argentatus*)

Photo: David Fairless

NESTING

CLUTCH SIZE (EGGS):	2–3
INCUBATION (DAYS):	25–28
FLEDGING (DAYS):	35
NEST HEIGHT (METRES):	0

Alberta's largest Gull, the Herring Gull is found in the Boreal Forest, Foothills, and Parkland Natural Regions. The new breeding evidence obtained near Whitecourt could indicate that the distribution of this species has expanded in that area.

The Herring Gull is a ground nester that usually uses sparsely vegetated islands for nesting. Being a fairly uncommon breeder in Alberta, this species had a patchy distribution across the Boreal Forest, Foothills, and Parkland NRs. Observations made in the Grassland and Rocky Mountain NRs were those of non-breeders.

A decline in relative abundance was detected in the Boreal Forest NR where, relative to other species, Herring Gulls were observed less frequently in Atlas 2 than in Atlas 1. No change was detected in the Foothills, Grassland, Parkland, and Rocky Mountain NRs. The Breeding Bird Survey sample size was too small to investigate Herring Gull abundance change in Alberta. However, a decline was found across Canada during the period 1985–2005. Declines in the Boreal Forest NR could be related to drought, as it was drier there during Atlas 2 than during Atlas 1. This species is sensitive to nest predation because it nests in the open on the ground; therefore, drought has a negative impact on reproductive success because land bridges, that facilitate predator movements, form between the mainland and breeding islands.

By the 1960s, the populations may even have exceeded historical numbers, possibly the result of plentiful food derived from human refuse, but the numbers levelled off in the mid-1970s and 1980s as dumps closed and changed, and as overfishing destroyed fish stocks (Pierotti and Good, 1994). Herring Gulls are very adaptable; they will eat almost anything and will nest almost anywhere, in both natural areas and the human landscape. As a consequence, it is expected that the Herring Gull will continue to thrive. This species is considered Secure in Alberta.

TEMPORAL REPORTING RATE

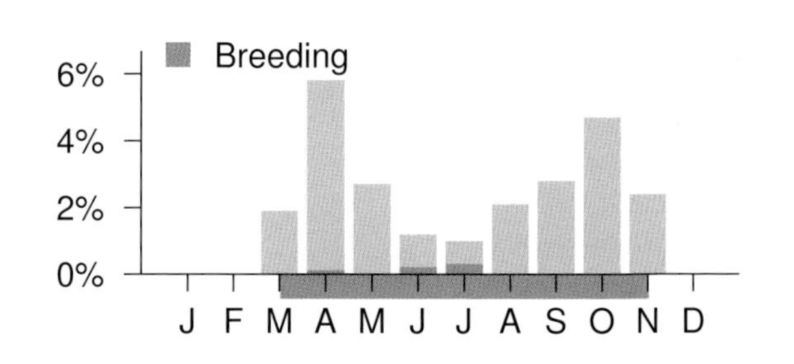

NATURAL REGIONS REPORTING RATE

Boreal Forest	0.6%
Grassland	2.1%
Parkland	0.8%
Rocky Mountain	1.3%

SPATIAL REPORTING RATE

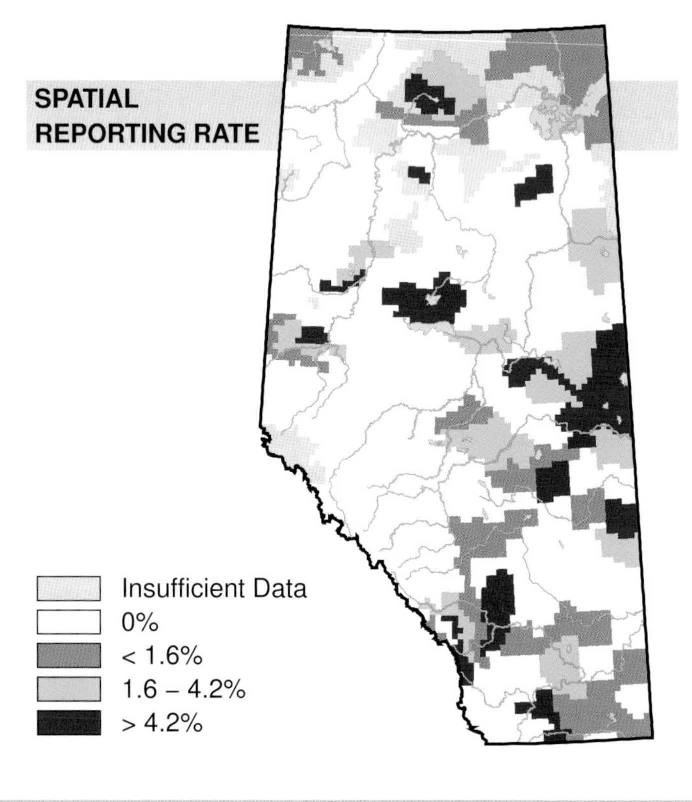

STATUS

GSWA 2000	GSWA 2005	COSEWIC Historic Rankings	COSEWIC 2007	Alberta Wildlife Act
Secure	Secure	N/A	N/A	N/A

Habitat Nest Location Nest Type Diet

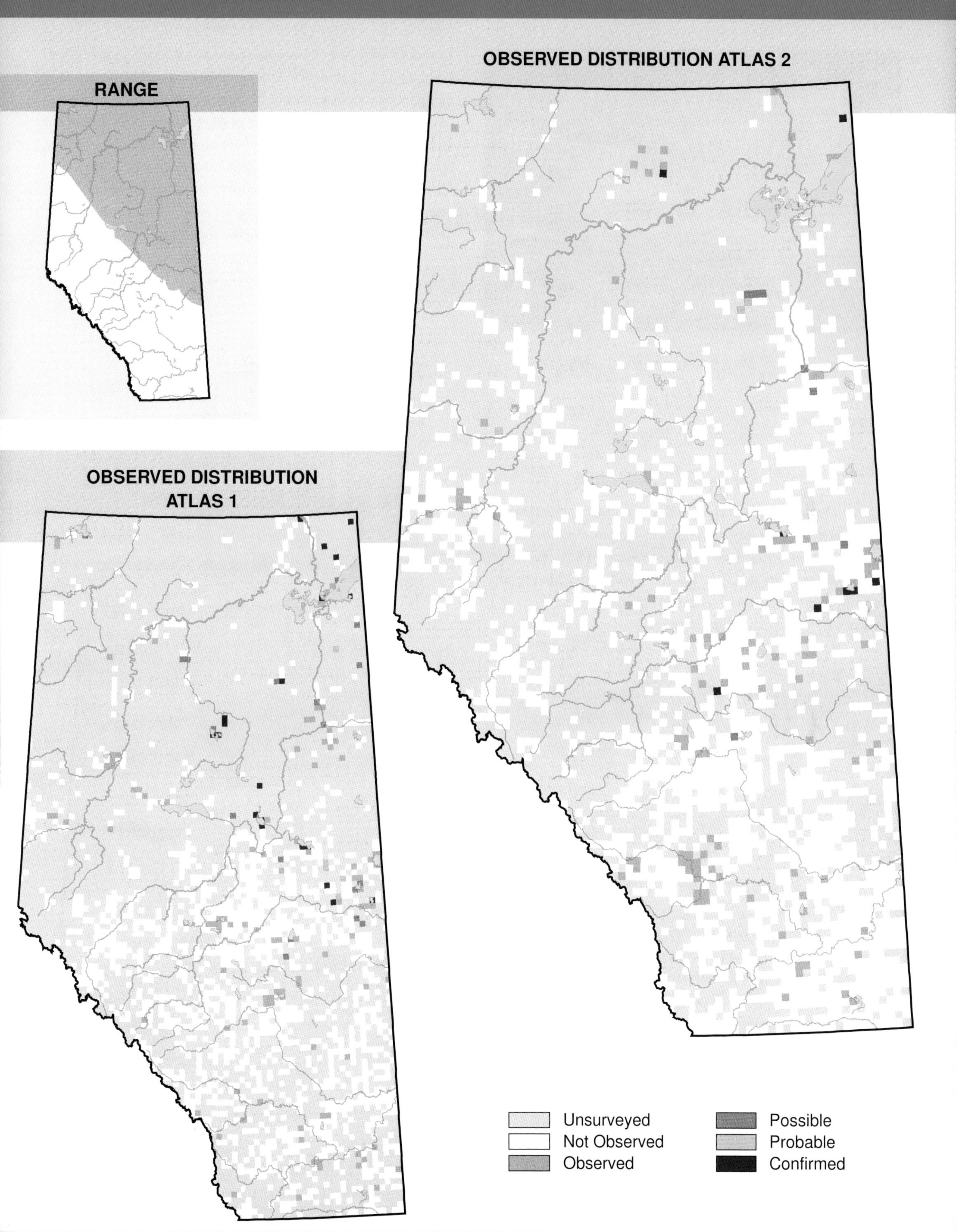

Caspian Tern (*Hydroprogne caspia*)

Photo: Royal Alberta Museum

NESTING

CLUTCH SIZE (EGGS):	2–3
INCUBATION (DAYS):	26
FLEDGING (DAYS):	37
NEST HEIGHT (METRES):	0

Across its range the Caspian Tern has a patchy distribution and the pattern holds true in Alberta. This species breeds in the Boreal Forest and Grassland Natural Regions. The new evidence of breeding near Lac La Biche could indicate that the distribution of this species has expanded in that area, considering that only non-breeders were recorded there during Atlas 1. No evidence of breeding was found in the northeastern part of the province during Atlas 2. It is unclear if this was because the range contracted in that area or whether this species was present but missed there during Atlas 2.

The Caspian Tern is a ground nester that usually nests on sparsely vegetated islands. This species can be found nesting near other birds such as Herring Gulls, California Gulls, and Ring-billed Gulls. This species was found in the Boreal Forest and Grassland NRs where treeless islands occur. The patchy distribution of this species can be explained partially by the strong fidelity that breeding adults have to previous breeding locations.

An increase in relative abundance was detected in the Grassland NR where, relative to other species, the Caspian Tern was observed more frequently in Atlas 2 than in Atlas 1. Above-average precipitation in this NR in the later half of Atlas 2 may have provided more habitat for this species. However, noting its strong breeding-location fidelity, it is unknown how soon new areas would be used if available. The Breeding Bird Survey sample size was too small to investigate Caspian Tern abundance change in Alberta and no change was detected on a national scale. With no known biological cause or corroboration, and given the more intensive coverage of the Grassland NR in Atlas 2, this increase must be interpreted with caution. No change was detected in the Boreal Forest NR. This species is considered Sensitive in Alberta due to the small number of colonies and the vulnerability of these colonies to disturbance and water level fluctuations.

TEMPORAL REPORTING RATE

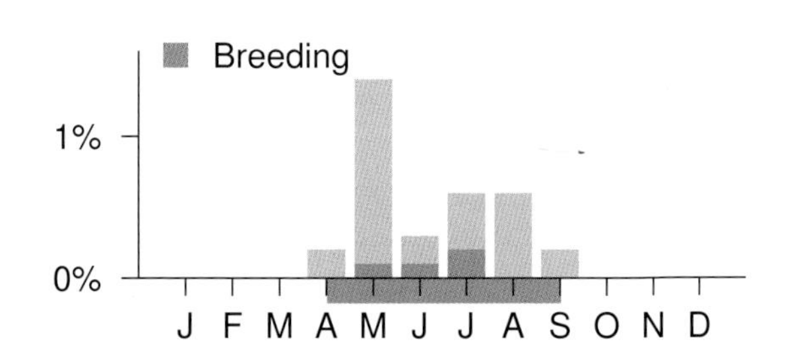

NATURAL REGIONS REPORTING RATE

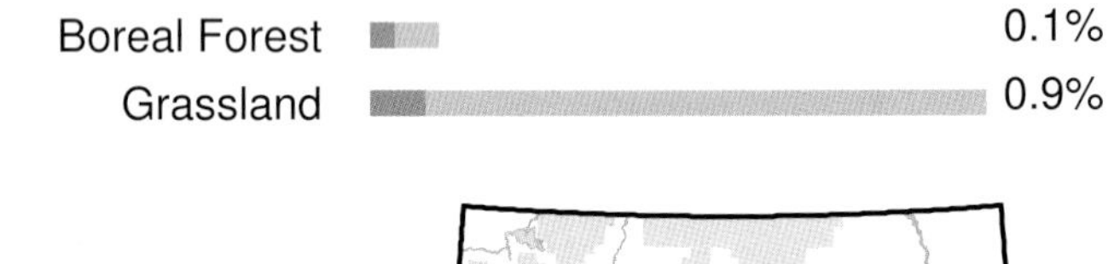

SPATIAL REPORTING RATE

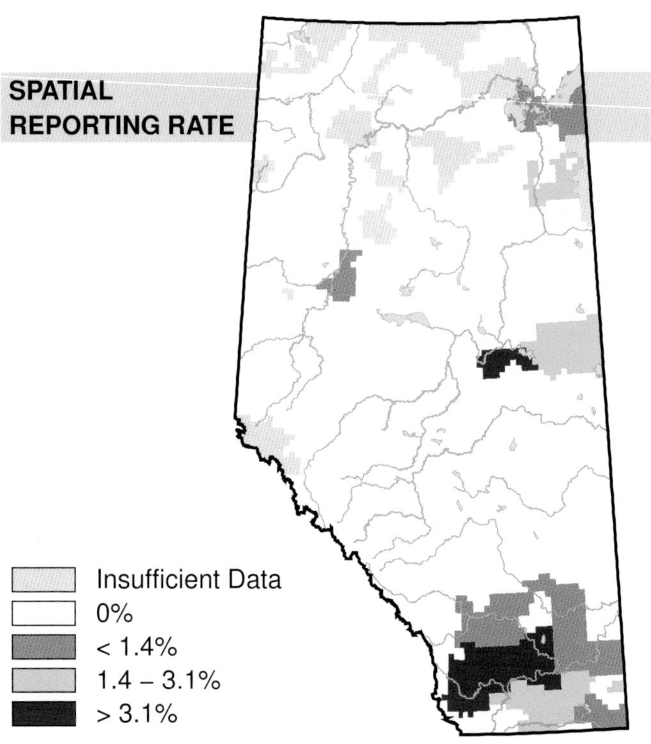

STATUS

GSWA 2000	GSWA 2005	COSEWIC Historic Rankings	COSEWIC 2007	Alberta Wildlife Act
Sensitive	Sensitive	Not At Risk	Not At Risk	N/A

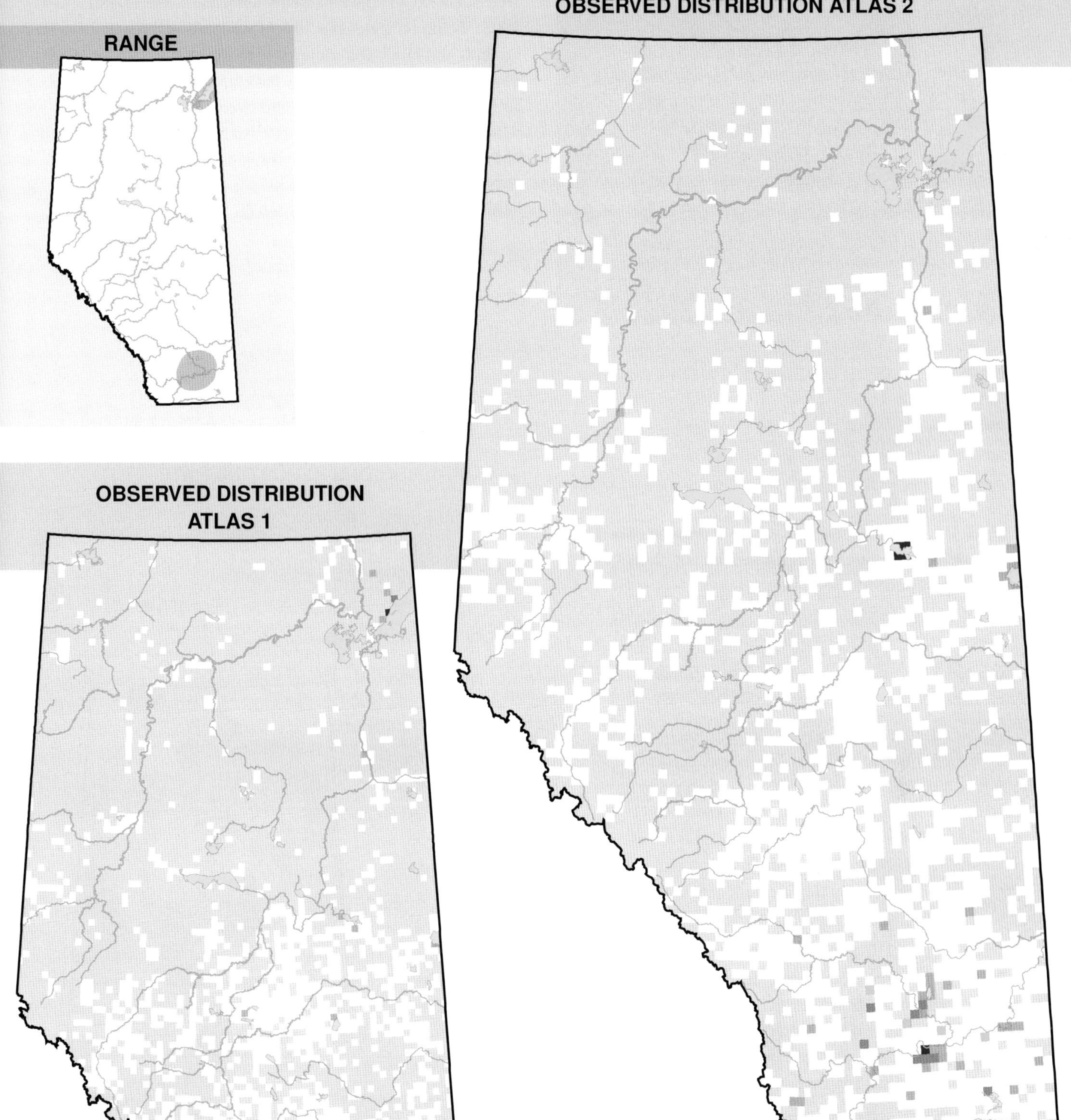
Habitat
Nest Location
Nest Type
Diet
RANGE
OBSERVED DISTRIBUTION ATLAS 2
OBSERVED DISTRIBUTION ATLAS 1
Unsurveyed
Not Observed
Observed
Possible
Probable
Confirmed

Black Tern (*Chlidonias niger*)

Photo: Gerald Romanchuk

NESTING

CLUTCH SIZE (EGGS):	2–4
INCUBATION (DAYS):	21–22
FLEDGING (DAYS):	18–21
NEST HEIGHT (METRES):	0

The Black Tern is found in every Natural Region in Alberta. The observed distribution of this species did not change between Atlas 1 and Atlas 2.

The Black Tern breeds in shallow wetlands with emergent vegetation. Nests are flimsy, often floating, and are easily destroyed by wind or changing water levels. Reproductive success is highly variable (Dunn, 2005).

Despite occurring in every Natural Region, this species was found most often in the Foothills, Grassland, and Parkland NRs where shallow wetlands can be locally abundant. Reporting rates were low in the Boreal Forest NR likely because many of the wetlands are surrounded by trees and shrubs rather than emergent vegetation such as cattails. Reporting rates were low in the Rocky Mountain NR because, due to topography, wetlands tend to be deep and often do not contain a large component of emergent vegetation.

Declines in relative abundance were detected in the Boreal Forest and Parkland NRs and relative to other species, Black Terns were observed less frequently in Atlas 2 than in Atlas 1. No changes were detected in the Foothills, Grassland, or Rocky Mountain NRs. The Breeding Bird Survey reported a decline in abundance for this species in Alberta during the period 1968–2005. Declines could have occurred because of drought and wetland drainage because this species is closely associated with wetlands. Populations of this tern in North America have declined markedly since the 1960s (Dunn, 2005). Dunn indicates that loss of wetlands on breeding grounds and migration routes is probably a major cause of decline, but food supplies may have been reduced through agricultural control of insects and overfishing in the marine winter range. The Black Tern is rated Sensitive in Alberta as it breeds in wetland habitats.

TEMPORAL REPORTING RATE

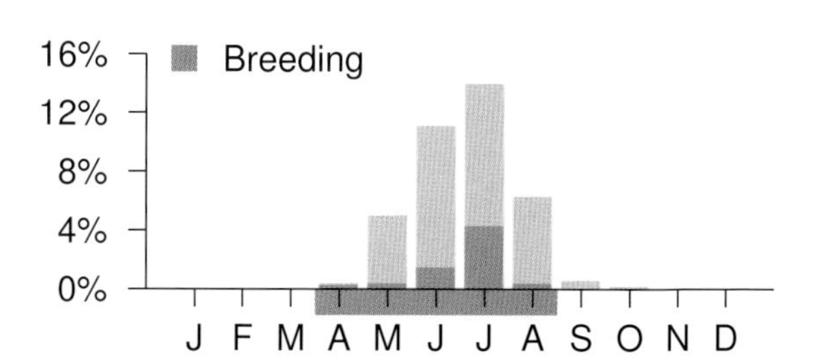

NATURAL REGIONS REPORTING RATE

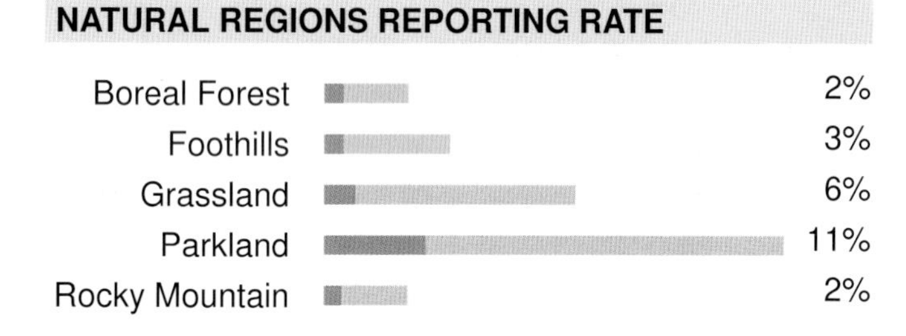

SPATIAL REPORTING RATE

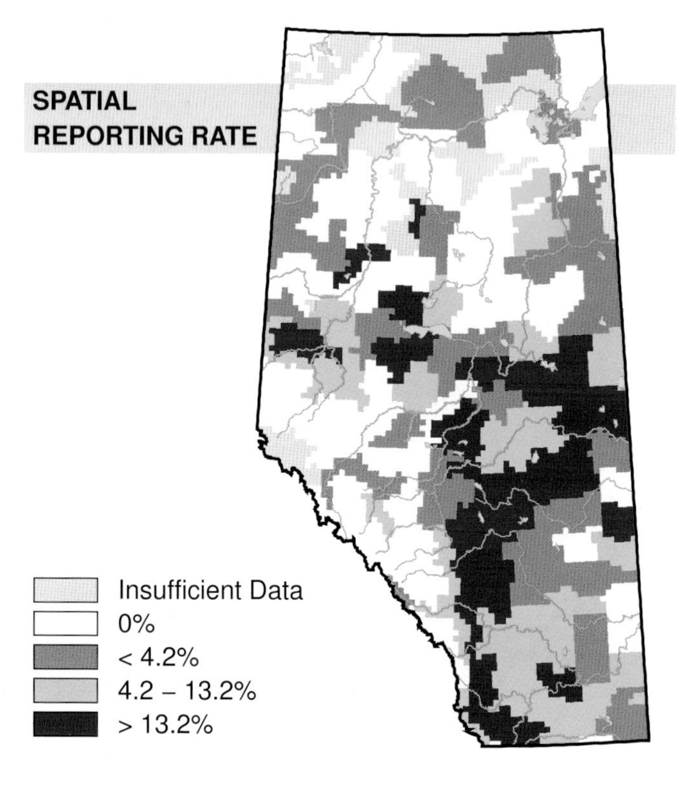

STATUS

GSWA 2000	GSWA 2005	COSEWIC Historic Rankings	COSEWIC 2007	Alberta Wildlife Act
Sensitive	Sensitive	Not At Risk	Not At Risk	N/A

Habitat 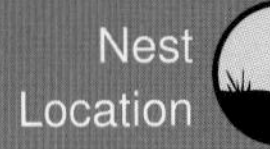 

Nest Location 

Nest Type 

Diet 

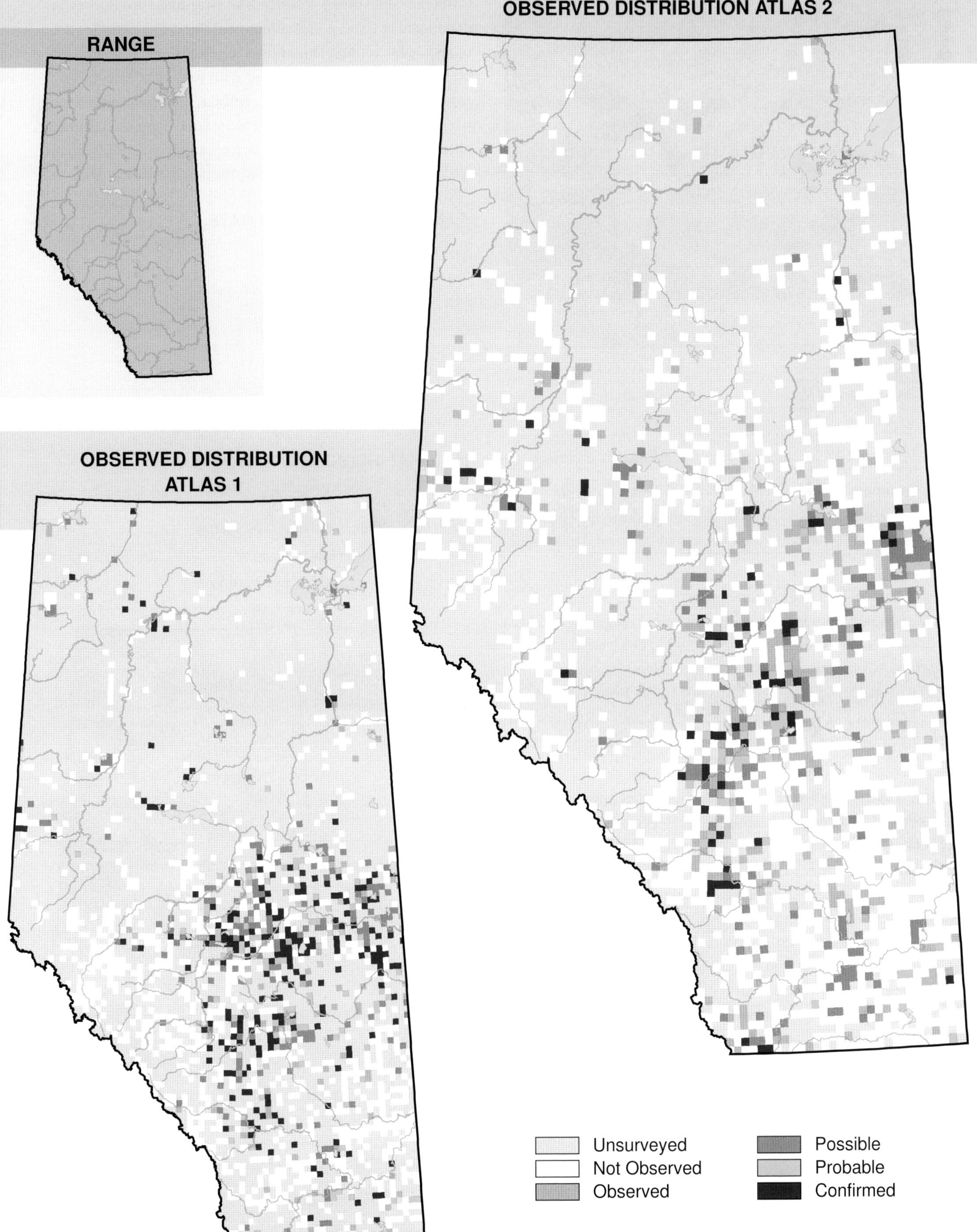

Common Tern (*Sterna hirundo*)

Photo: Uve Hublitz / Cornell Laboratory of Ornithology

NESTING

CLUTCH SIZE (EGGS):	2–4
INCUBATION (DAYS):	21–30
FLEDGING (DAYS):	28
NEST HEIGHT (METRES):	0

The Common Tern breeds in the Boreal Forest, Foothills, Grassland, and Parkland Natural Regions in Alberta. The distribution of this species did not change between Atlas 1 and Atlas 2.

The Common Tern is a colonial ground nester that usually nests on sandy islands supporting only sparse vegetation. This species was most common in the Foothills, Grassland, and Parkland NRs where sparsely vegetated islands occur. In comparison, the reporting rate was low in the Boreal Forest NR. This was likely because many of the islands in that natural region are treed and not, therefore, suitable breeding habitat for this species.

An increase in relative abundance was detected in the Grassland NR, although relative to other species, this tern was observed less frequently in Atlas 2 than in Atlas 1. This increase could be related to unusually high precipitation in the Grassland NR in the later half of the Atlas field program. A decrease in relative abundance was detected in the Parkland NR where, relative to other species, Common Terns were observed less frequently in Atlas 2 than in Atlas 1. No change was detected in the Boreal Forest and Foothills NRs. The Breeding Bird Survey did not find an abundance change in Alberta; however, a Canada-wide decline was detected during the period 1985–2005. Differences between the results of Atlas and Breeding Bird Survey analyses could be attributed to differences in survey designs. The Breeding Bird Survey is conducted at fixed points along roads, and most species are observed by sound. This results in lower detection rates for birds that are closely associated with wetlands. In contrast, some Atlas surveys were done without the constraint of using fixed points, some of which were located along the perimeter of wetlands. The result was a larger sample size for wetland birds. This species is considered Secure in Alberta.

TEMPORAL REPORTING RATE

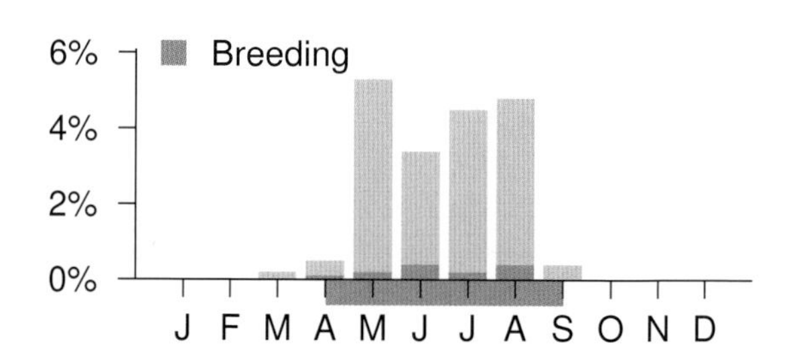

NATURAL REGIONS REPORTING RATE

Region	Rate
Boreal Forest	1.1%
Foothills	2.2%
Grassland	4.0%
Parkland	2.1%

SPATIAL REPORTING RATE

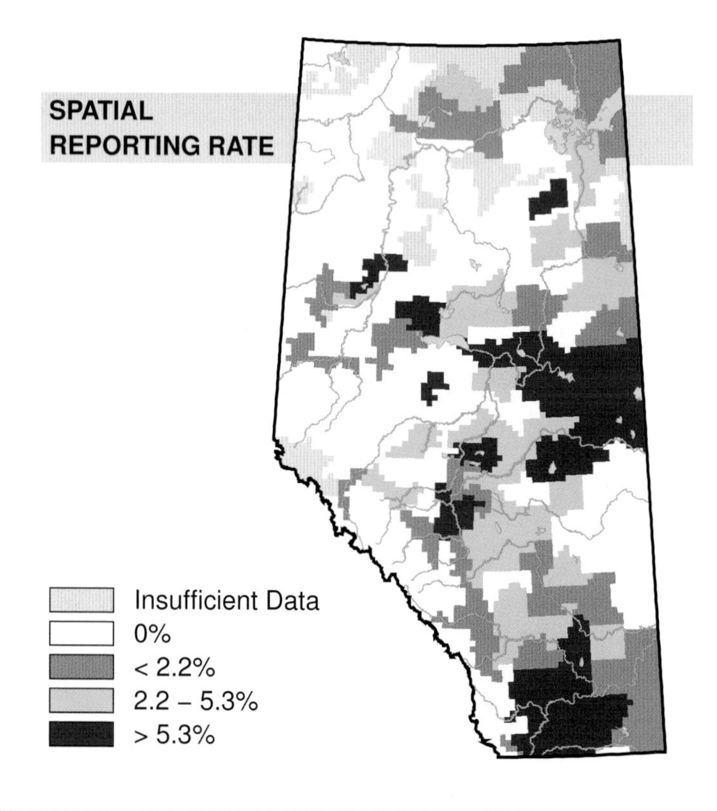

STATUS

GSWA 2000	GSWA 2005	COSEWIC Historic Rankings	COSEWIC 2007	Alberta Wildlife Act
Secure	Secure	Not At Risk	Not At Risk	N/A

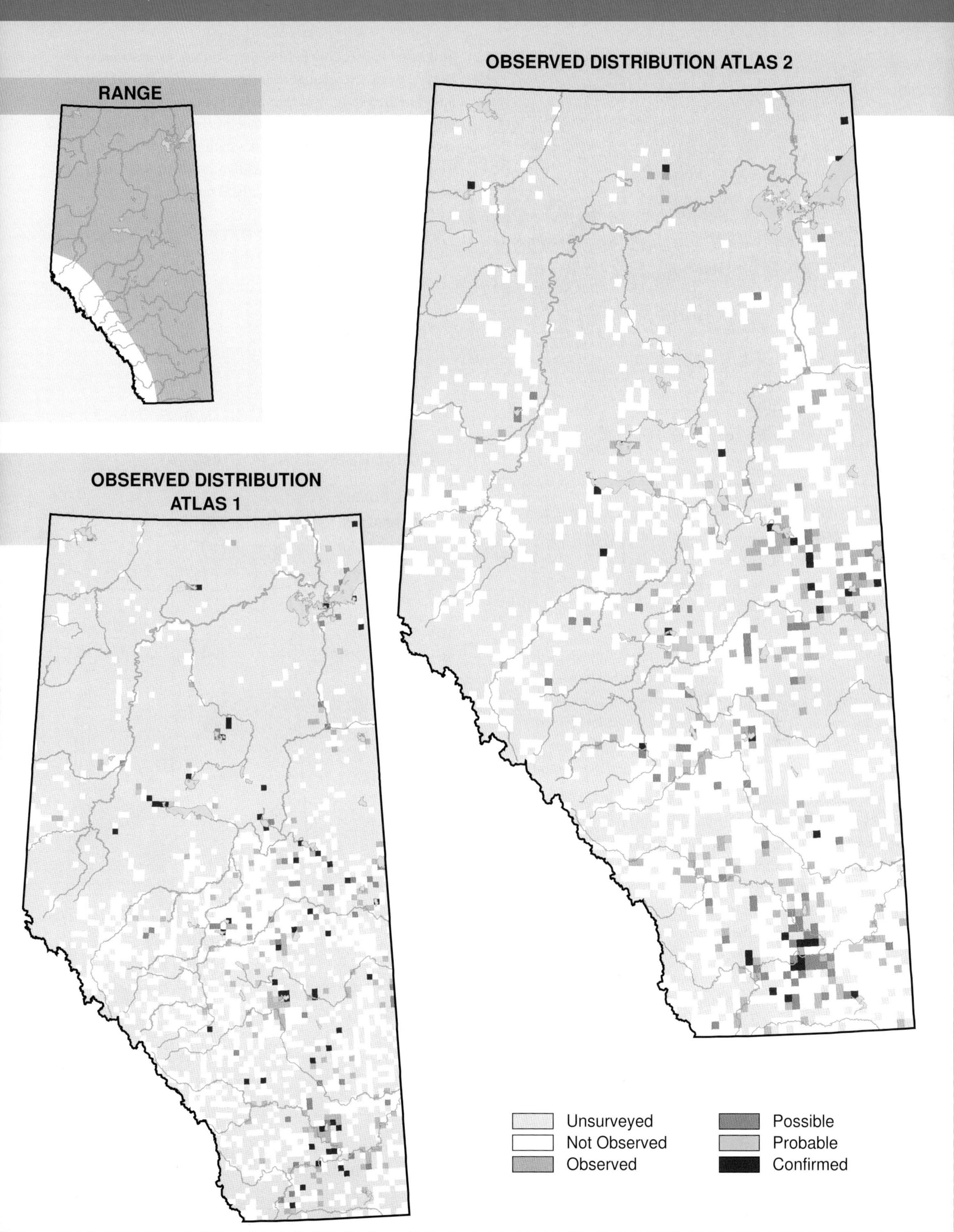
Habitat
Nest Location
Nest Type
Diet
OBSERVED DISTRIBUTION ATLAS 2
RANGE
OBSERVED DISTRIBUTION ATLAS 1
Unsurveyed
Not Observed
Observed
Possible
Probable
Confirmed

Forster's Tern (*Sterna forsteri*)

Photo: Gerald Romanchuk

NESTING

CLUTCH SIZE (EGGS):	2–4
INCUBATION (DAYS):	23–25
FLEDGING (DAYS):	unknown
NEST HEIGHT (METRES):	0

Forster's Tern breeds in the Boreal Forest, Grassland, and Parkland Natural Regions. The distribution of this species did not change between Atlas 1 and Atlas 2. However, these birds were observed less frequently in east-central Alberta near Beaverhill Lake. This was likely the result of drought in that area.

Unlike many other terns that build their nests on sparsely vegetated islands, Forster's Tern is a ground nester that usually builds its nest among emergent vegetation in wetlands. This species was most common in the southern Boreal Forest, Grassland, and Parkland NRs where shallow wetlands occur. This species was infrequently encountered in the Foothills NR.

An increase in relative abundance was detected in the Grassland NR where, relative to other species, Forster's Tern was observed more frequently in Atlas 2 than in Atlas 1. Above-average precipitation in the later half of Atlas 2 efforts is a probable cause for this increase. A decline in relative abundance was detected in the Parkland NR, where relative to other species, it was observed less frequently in Atlas 2 than in Atlas 1. Declines could be related to drought or wetland drainage. Considering these changes together suggests the possibility that the population did not actually change, but rather that birds that had previously migrated north over the drought-stricken wetlands of the Grassland NR were able to find suitable breeding habitat there during the later half of Atlas 2 work. No change was detected in the Boreal Forest NR. The Breeding Bird Survey sample size was too small to investigate changes in abundance of Forster's Tern in Alberta, and no change was detected on a national scale in Canada. This species is considered Sensitive in Alberta due to degradation of, and loss of, wetland habitats.

TEMPORAL REPORTING RATE

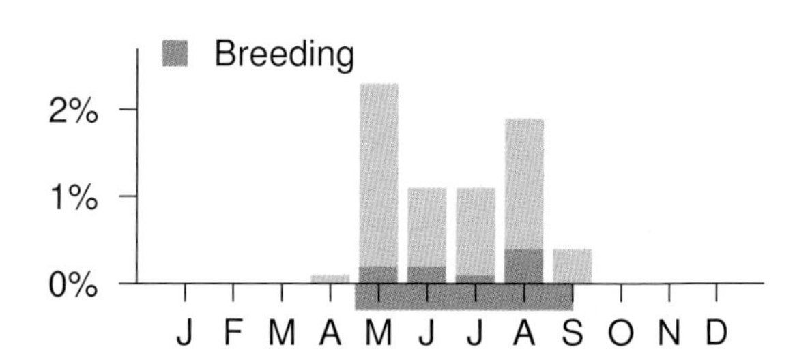

NATURAL REGIONS REPORTING RATE

SPATIAL REPORTING RATE

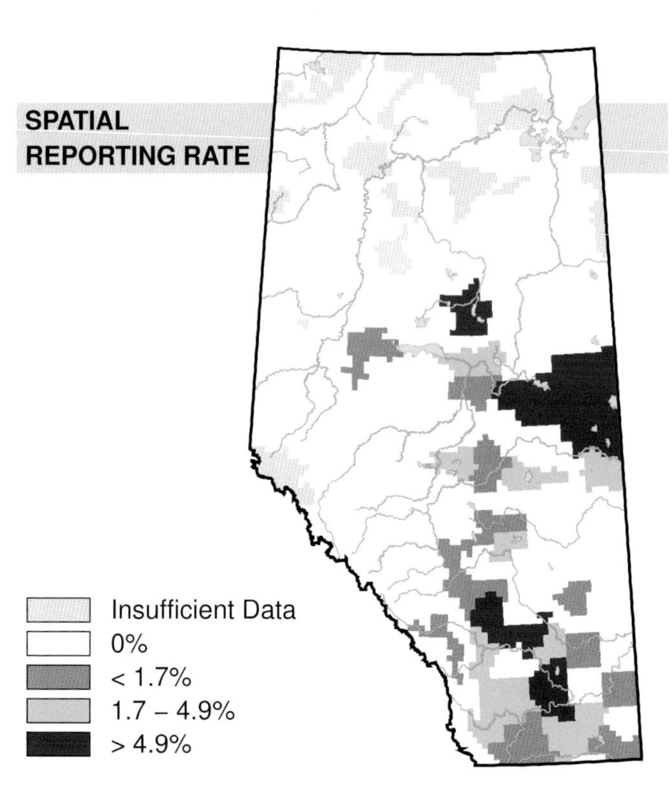

STATUS

GSWA 2000	GSWA 2005	COSEWIC Historic Rankings	COSEWIC 2007	Alberta Wildlife Act
Sensitive	Sensitive	Data Deficient	Data Deficient	N/A

OBSERVED DISTRIBUTION ATLAS 2

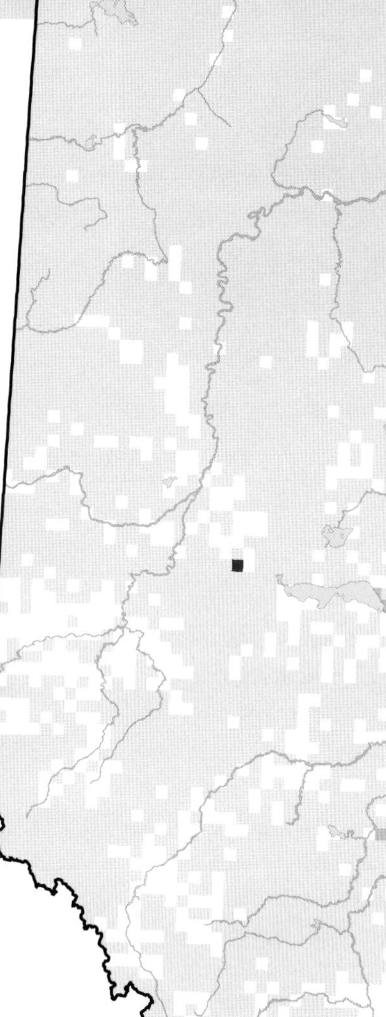

RANGE

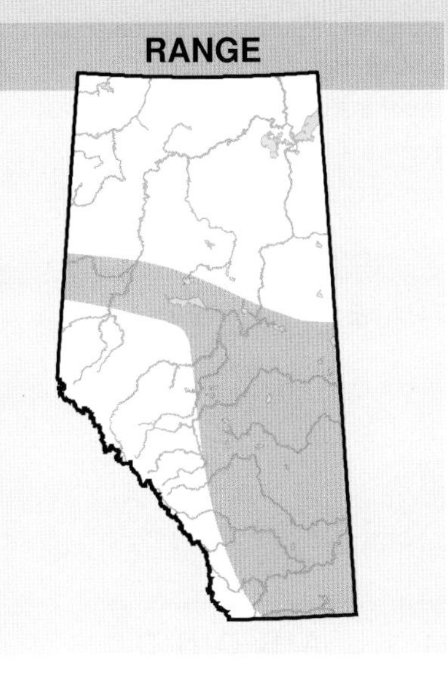

OBSERVED DISTRIBUTION ATLAS 1

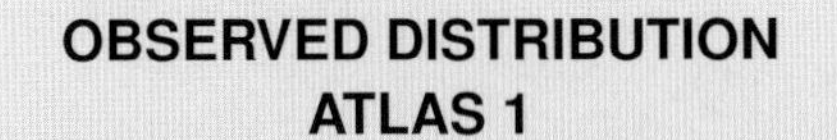

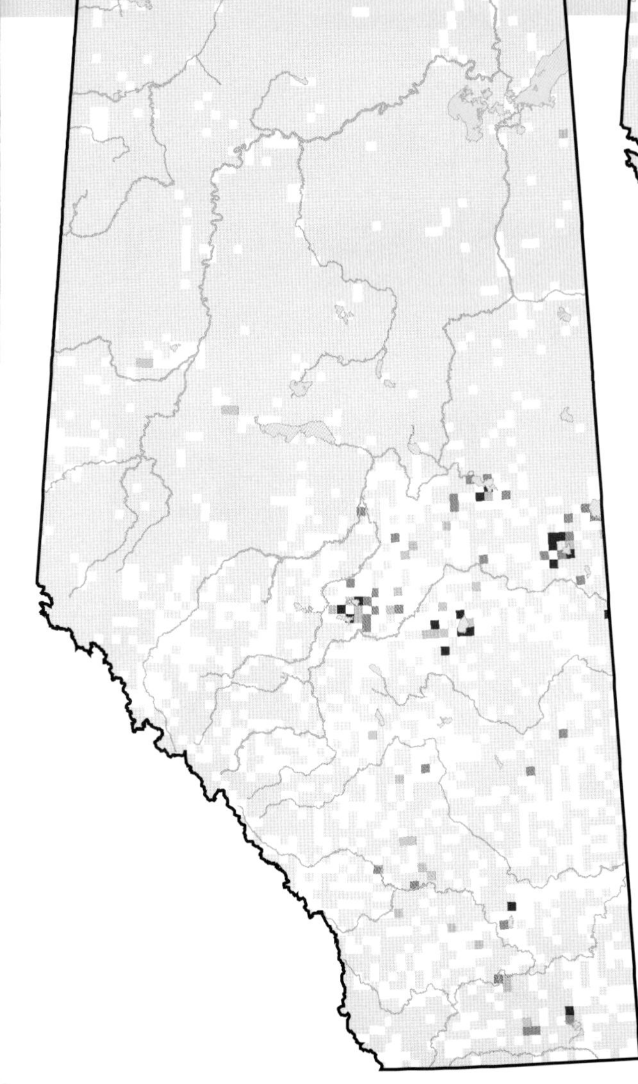

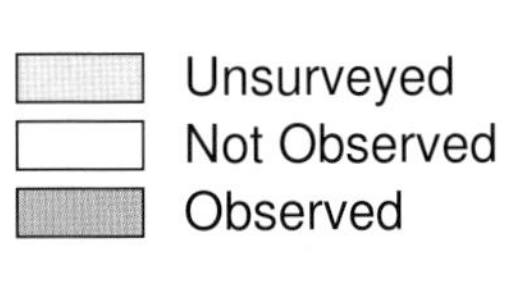

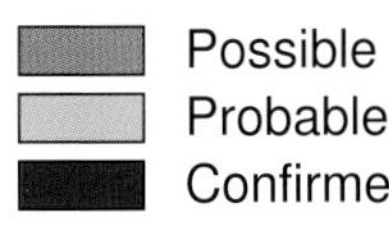

Rock Pigeon (*Columba livia*)

Photo: Jean-Guy Dallaire

NESTING

CLUTCH SIZE (EGGS):	2
INCUBATION (DAYS):	17–19
FLEDGING (DAYS):	25–26
NEST HEIGHT (METRES):	0–30

A common sight in urban areas throughout the world, the Rock Pigeon (formerly named Rock Dove) is found in all the Natural Regions of Alberta except the Canadian Shield. The observed distribution of this species did not change between Atlas 1 and Atlas 2.

Introduced into North America in the early 1600s, this species has been very successful in spreading across the continent. They are most often associated with cities, towns, and farms, using ledges in farm buildings and on bridges, sky scrapers, and other tall buildings for nesting. The only requirement for nesting is a flat surface, usually under cover. They may forage near breeding sites, or may travel a few kilometres to feed on seeds, fruits, and occasionally on insects. On that account, it was most commonly found in the Grassland NR, and was less commonly found in the Parkland NR. It was least common in the Boreal Forest, Foothills, and Rocky Mountain NRs. Large populations are found in and around major cities or near large farms in the province.

Declines in relative abundance were detected in the Boreal Forest and Parkland NR and an increase in relative abundance was found in the Rocky Mountain NR. Observation rates relative to other species mirrored these changes. The Breeding Bird Survey found no change in abundance in Alberta or Canada-wide during the period 1985–2005. There is no apparent biological cause which would explain the Atlas-detected declines in this species. The Rock Pigeon tends to nest in groups; so, it is possible that the changes in abundance that were detected by the Atlas are a result of variation in the specific breeding sites that were visited between Atlases. The Rock Pigeon is considered Exotic/Alien in the province.

TEMPORAL REPORTING RATE

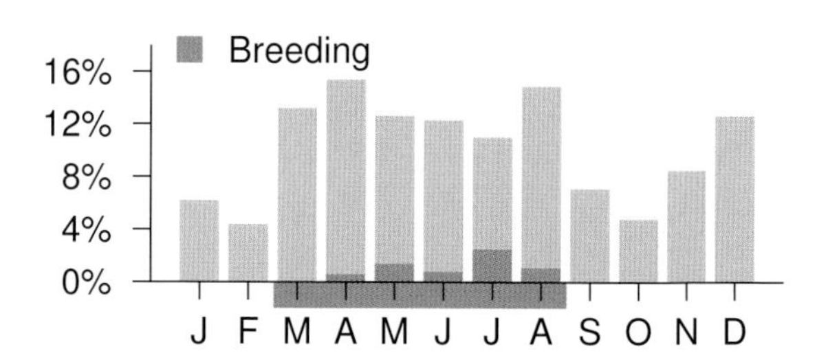

NATURAL REGIONS REPORTING RATE

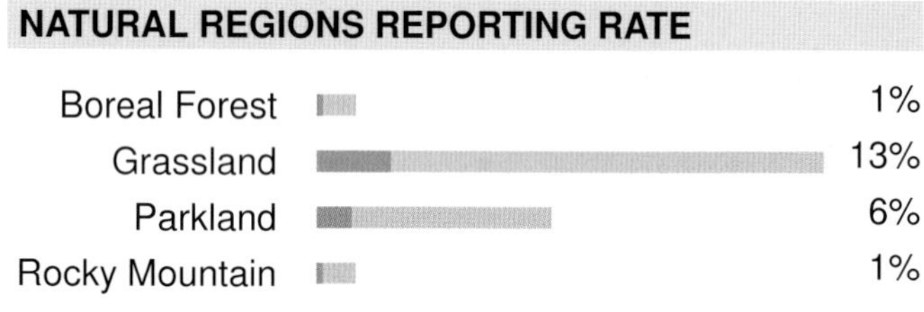

SPATIAL REPORTING RATE

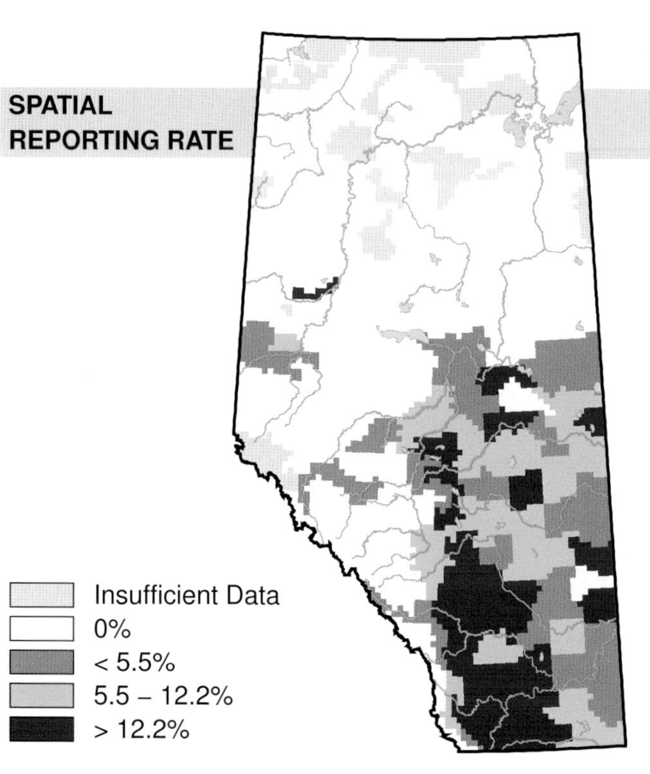

STATUS

GSWA 2000	GSWA 2005	COSEWIC Historic Rankings	COSEWIC 2007	Alberta Wildlife Act
Exotic/Alien	Exotic/Alien	N/A	N/A	N/A

Habitat  Nest Location  Nest Type  Diet  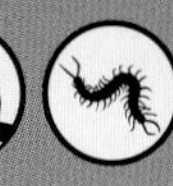 

RANGE

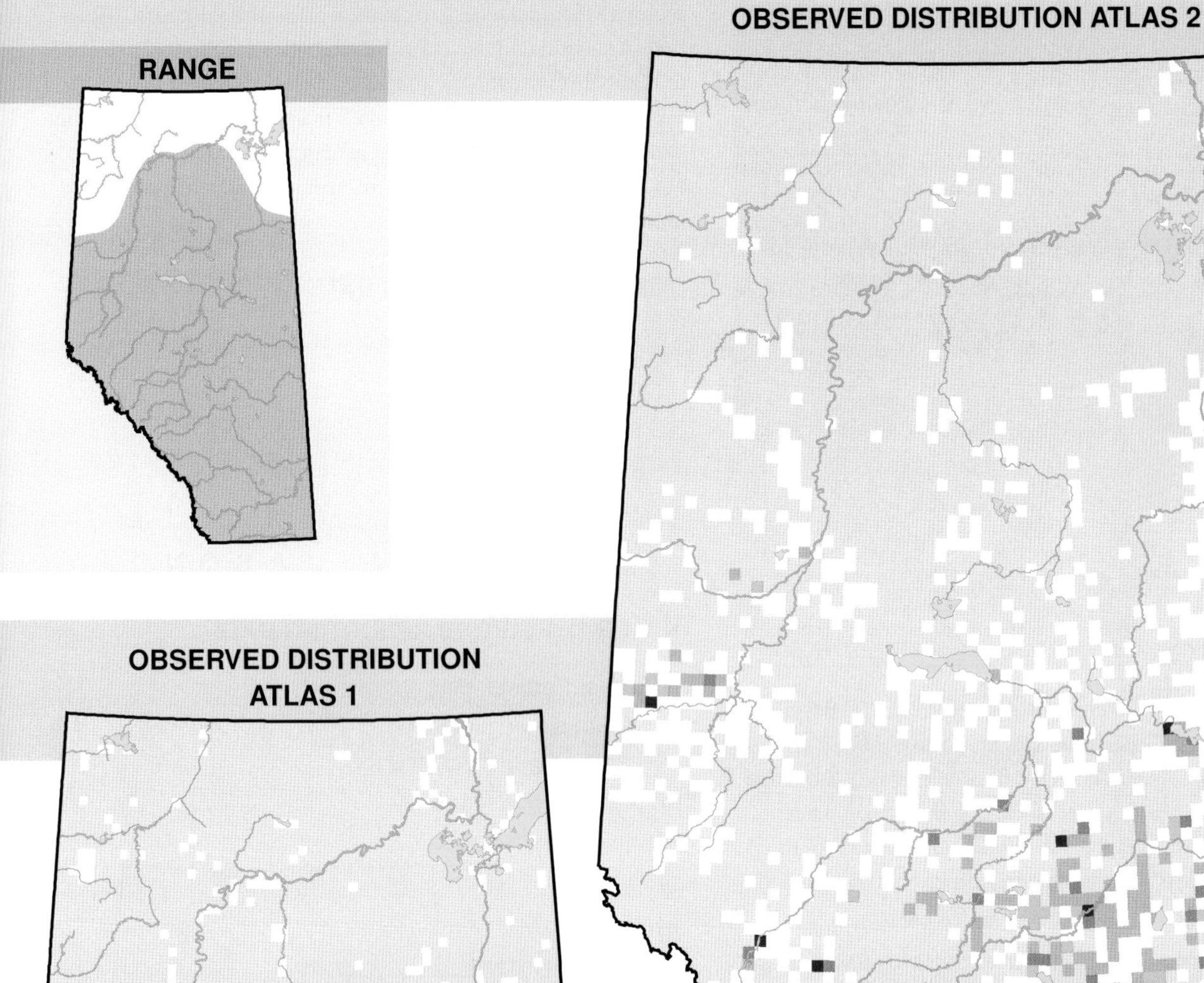

OBSERVED DISTRIBUTION ATLAS 2

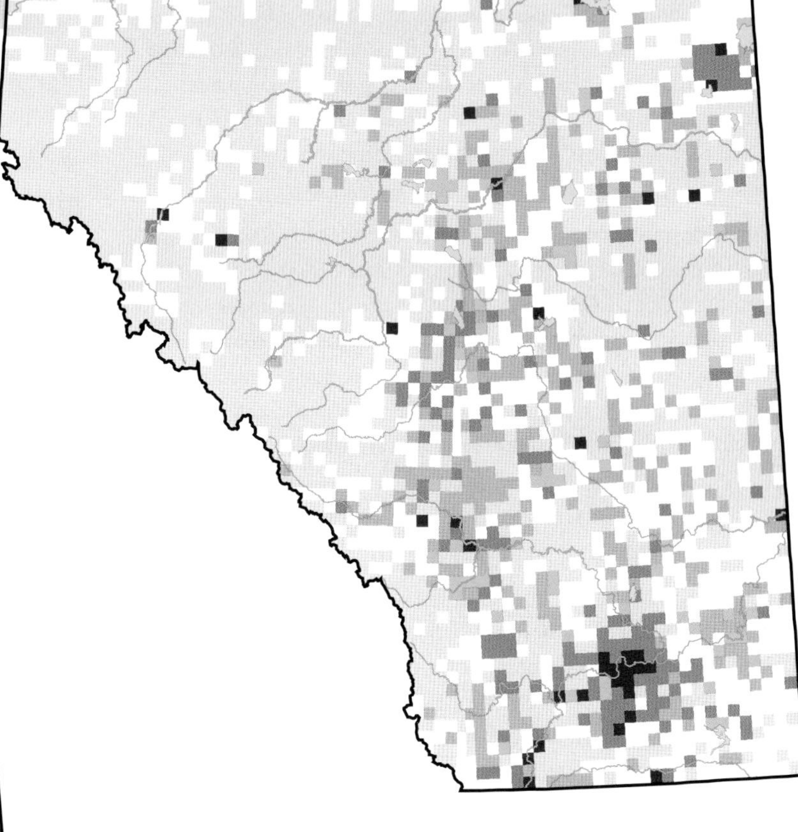

OBSERVED DISTRIBUTION ATLAS 1

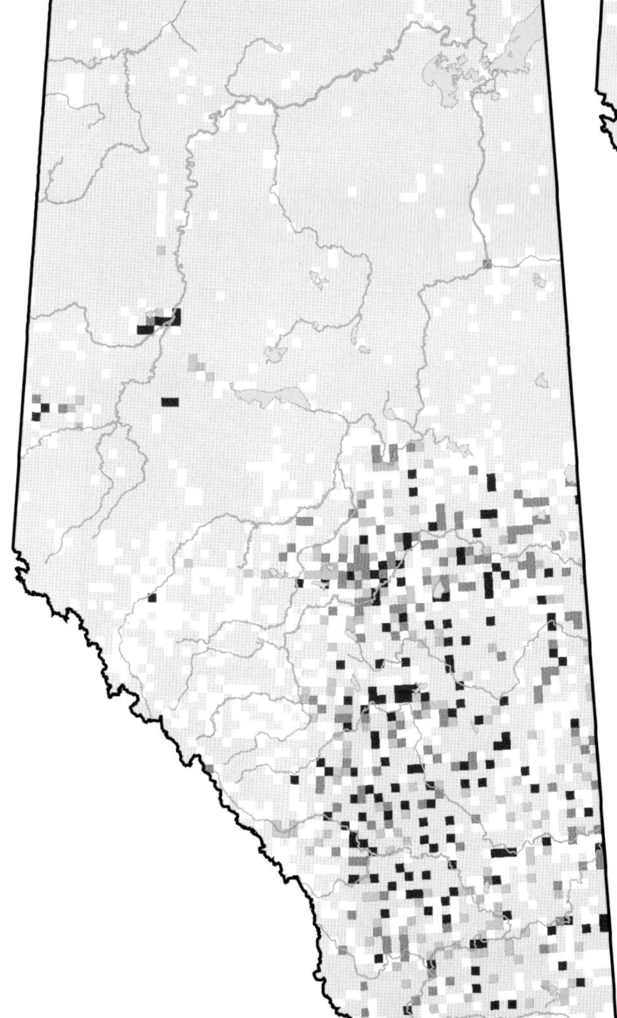

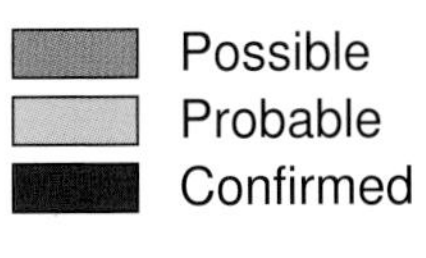

Mourning Dove (*Zenaida macroura*)

Photo: Gerald Romanchuk

NESTING

CLUTCH SIZE (EGGS):	2
INCUBATION (DAYS):	13–16
FLEDGING (DAYS):	12–15
NEST HEIGHT (METRES):	0–12.2

The Mourning Dove is found in the Boreal Forest, Grassland, Parkland, and Rocky Mountain NRs in Alberta. The distribution of this species changed between Atlas 1 and Atlas 2, particularly in the northwest of the province where there were fewer probable and confirmed breeding records. In the northeast, new records of possible breeding were obtained south of Lake Athabasca and near Mariana Lakes.

This widespread and abundant dove inhabits woodland edges and open woods, but is not found in closed canopy forests. The Mourning Dove has no aversion to nesting near humans, and feeds almost entirely on the ground on herbaceous plants in early successional stages. It was, therefore, most common in the Grassland NR, less common in the Parkland and Rocky Mountain NRs, and least common in the Boreal Forest NR.

Declines in relative abundance of Mourning Doves were detected in Boreal Forest and Parkland NRs, and an increase was detected in the Rocky Mountain NR. Reporting rates relative to other species mirrored these changes. The Breeding Bird Survey did not detect an abundance change during the period 1985–2005 in Alberta or Canada; however, an increase was detected in Alberta between 1995 and 2005. It is unclear why the Atlas found declines given that the Mourning Dove is generally thought to benefit from anthropogenic modifications of original vegetation, such as the planting of trees and shrubs in cities and towns or the creation of openings in forested regions. The sparseness of records in the Boreal Forest NR in both Atlases may have contributed to the detection of a decline in this NR. However, the marked difference in distribution of this species between Atlas 1 and 2 makes it difficult to discount these declines entirely. Further research is needed to evaluate the Atlas-detected declines. This species is considered Secure in the province.

TEMPORAL REPORTING RATE

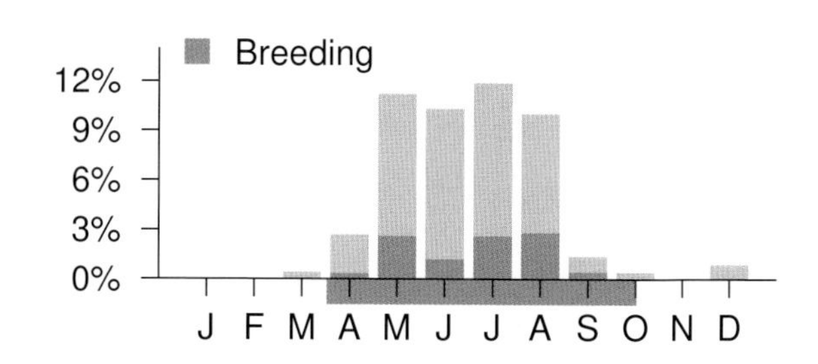

NATURAL REGIONS REPORTING RATE

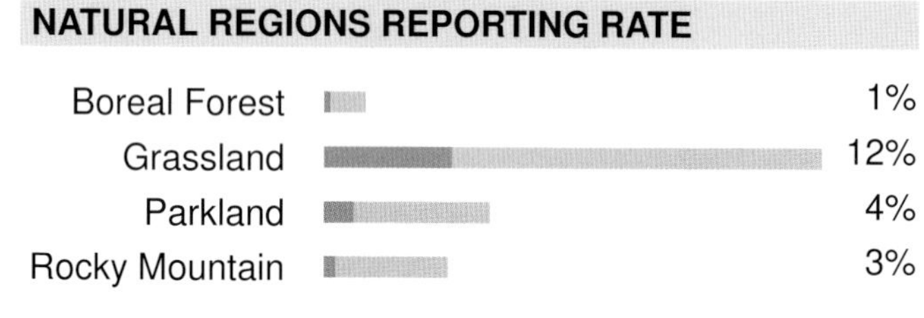

SPATIAL REPORTING RATE

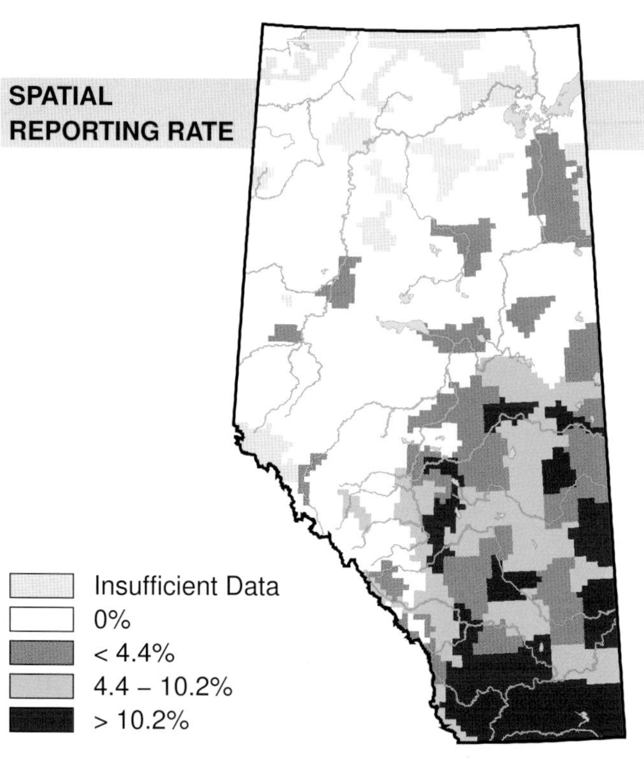

STATUS

GSWA 2000	GSWA 2005	COSEWIC Historic Rankings	COSEWIC 2007	Alberta Wildlife Act
Secure	Secure	N/A	N/A	N/A

RANGE

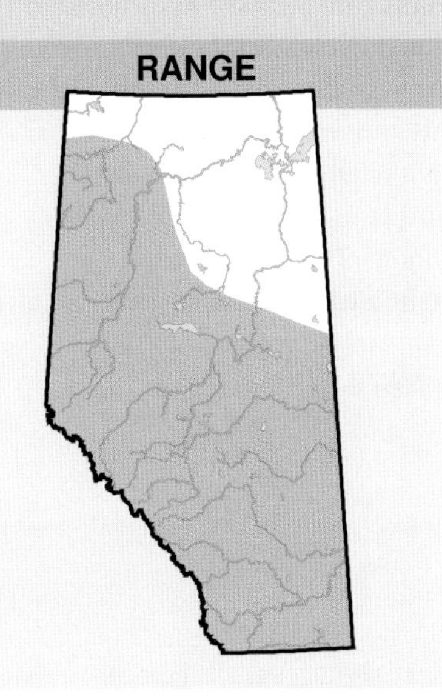

OBSERVED DISTRIBUTION ATLAS 1

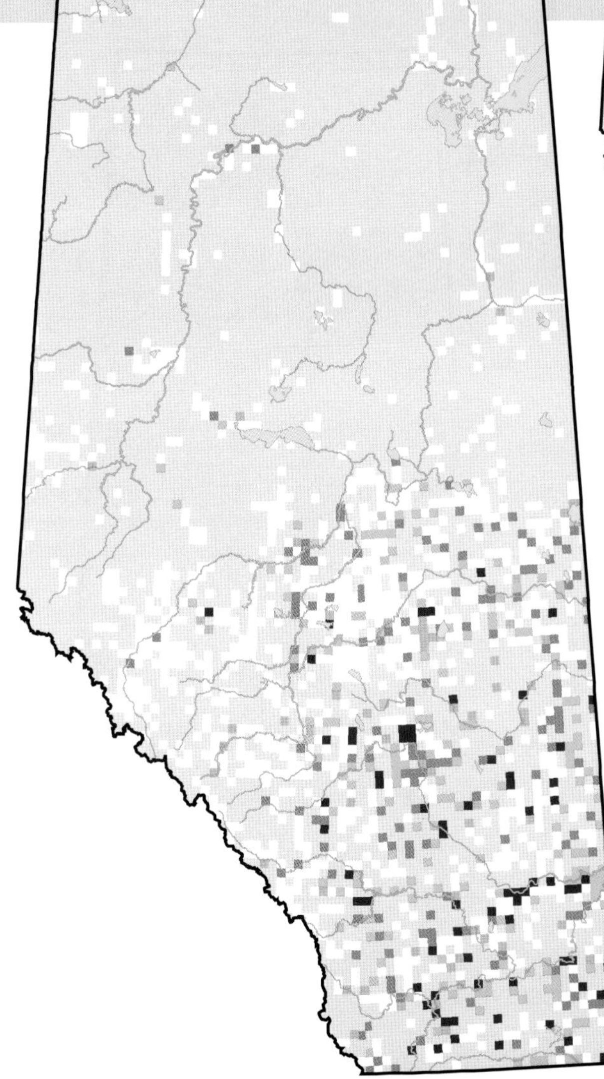

OBSERVED DISTRIBUTION ATLAS 2

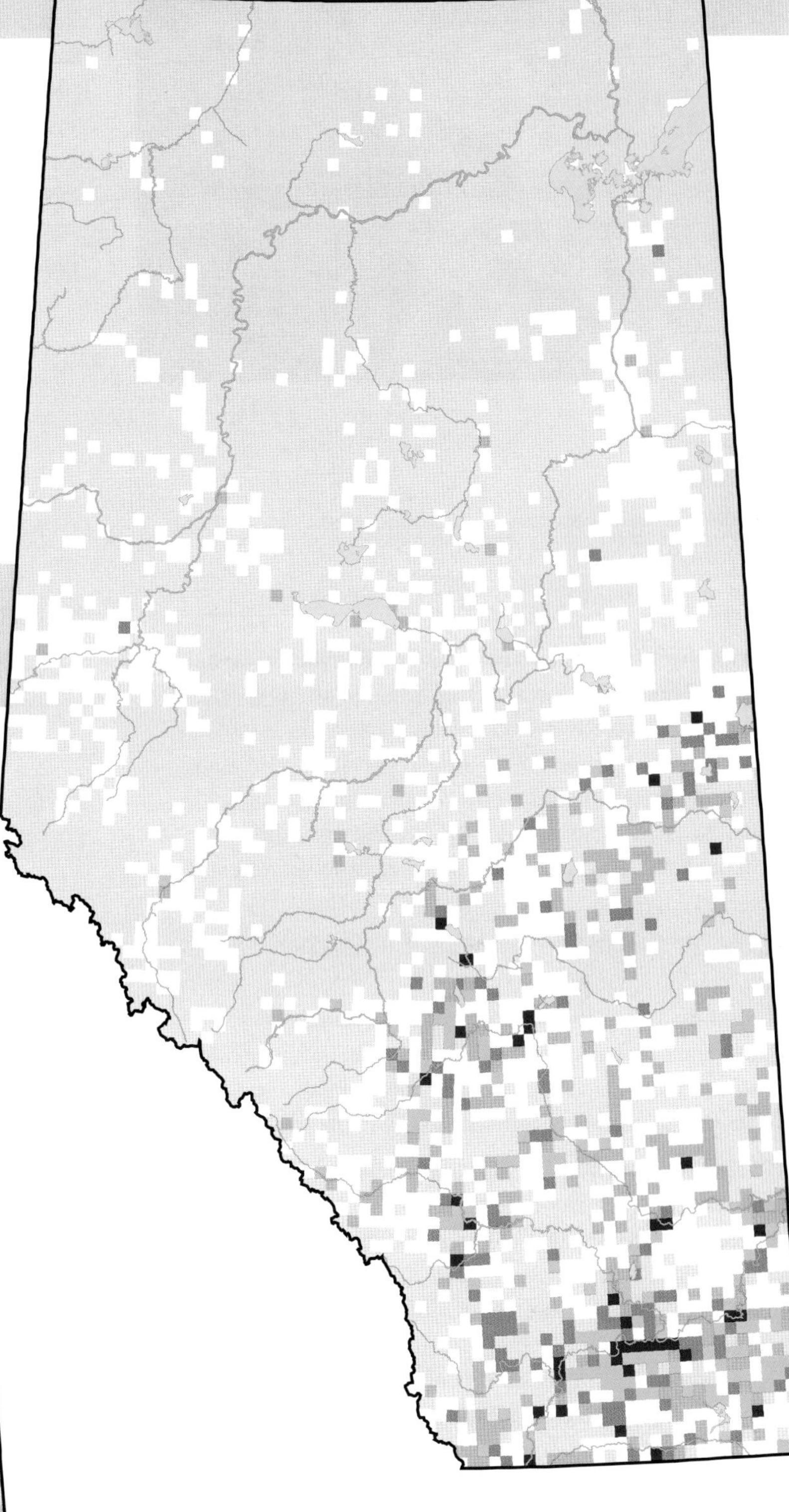

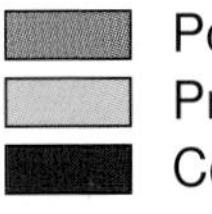

Black-billed Cuckoo (*Coccyzus erythropthalmus*)

Photo: Royal Alberta Museum

NESTING

CLUTCH SIZE (EGGS):	2–4
INCUBATION (DAYS):	10–13
FLEDGING (DAYS):	14–17
NEST HEIGHT (METRES):	0.1–6.1

The elusive Black-billed Cuckoo is found in the Grassland Natural Region in Alberta. The distribution of this species changed dramatically between Atlas 1 and Atlas 2. In Atlas 1 this species was present in the Boreal Forest and Parkland NRs, ranging from Red Deer to Barrhead and east through Athabasca to St. Paul and Bonnyville, and throughout the Grassland NR. During Atlas 2, breeding observations were obtained only in the Medicine Hat and Bindloss areas. A few other non-breeding observations were made north of these sites.

Black-billed Cuckoos prefer dense thickets in coulees, along roads, and near streams. They also breed in open woodlands with willow and alder, and occasionally in garden hedges with thick vegetation. They were found only in the Grassland NR.

No change in relative abundance was detected in Atlas 2, but altogether the only records obtained were one incidental observation from the Boreal Forest and four from the Grassland NR in Atlas 2. With so few records, this species was below the change detection threshold of our statistical methods. Relative to other species, the Black-billed Cuckoo was observed much less frequently during Atlas 2 than during Atlas 1. Declines in abundance were detected in Saskatchewan and Canada-wide by the Breeding Bird Survey during the period 1985–2005. No change in abundance was detected for Alberta; however, similar to the Atlas, sample size was also small. The reason for these declines is unknown. This species may be sensitive to habitat fragmentation, and might require a minimum-sized grove of trees for breeding. Further research is needed to assess why this species is declining. The Black-billed Cuckoo's status is rated Undetermined in the province.

TEMPORAL REPORTING RATE

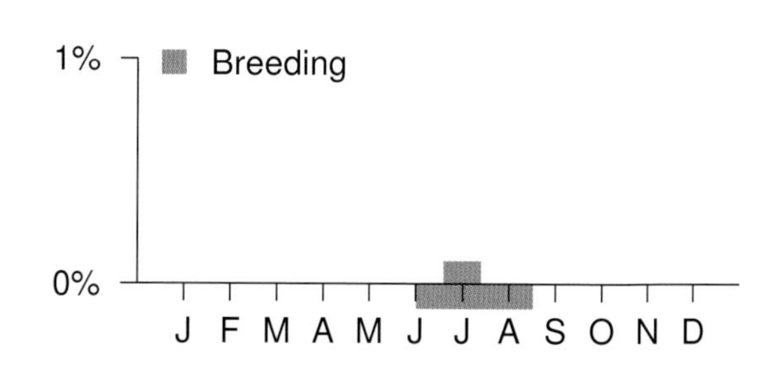

NATURAL REGIONS REPORTING RATE

SPATIAL REPORTING RATE

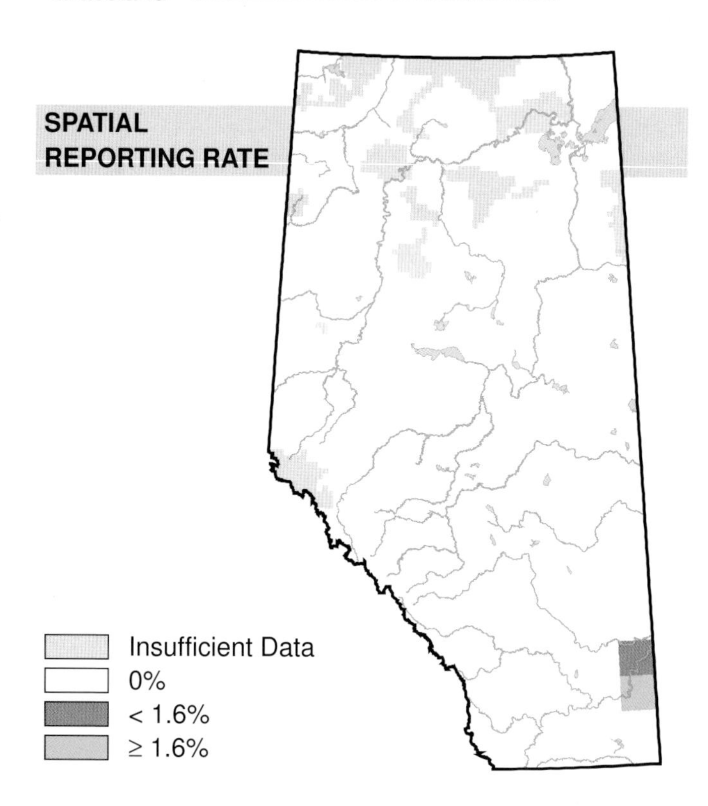

STATUS

GSWA 2000	GSWA 2005	COSEWIC Historic Rankings	COSEWIC 2007	Alberta Wildlife Act
Undetermined	Undetermined	N/A	N/A	N/A

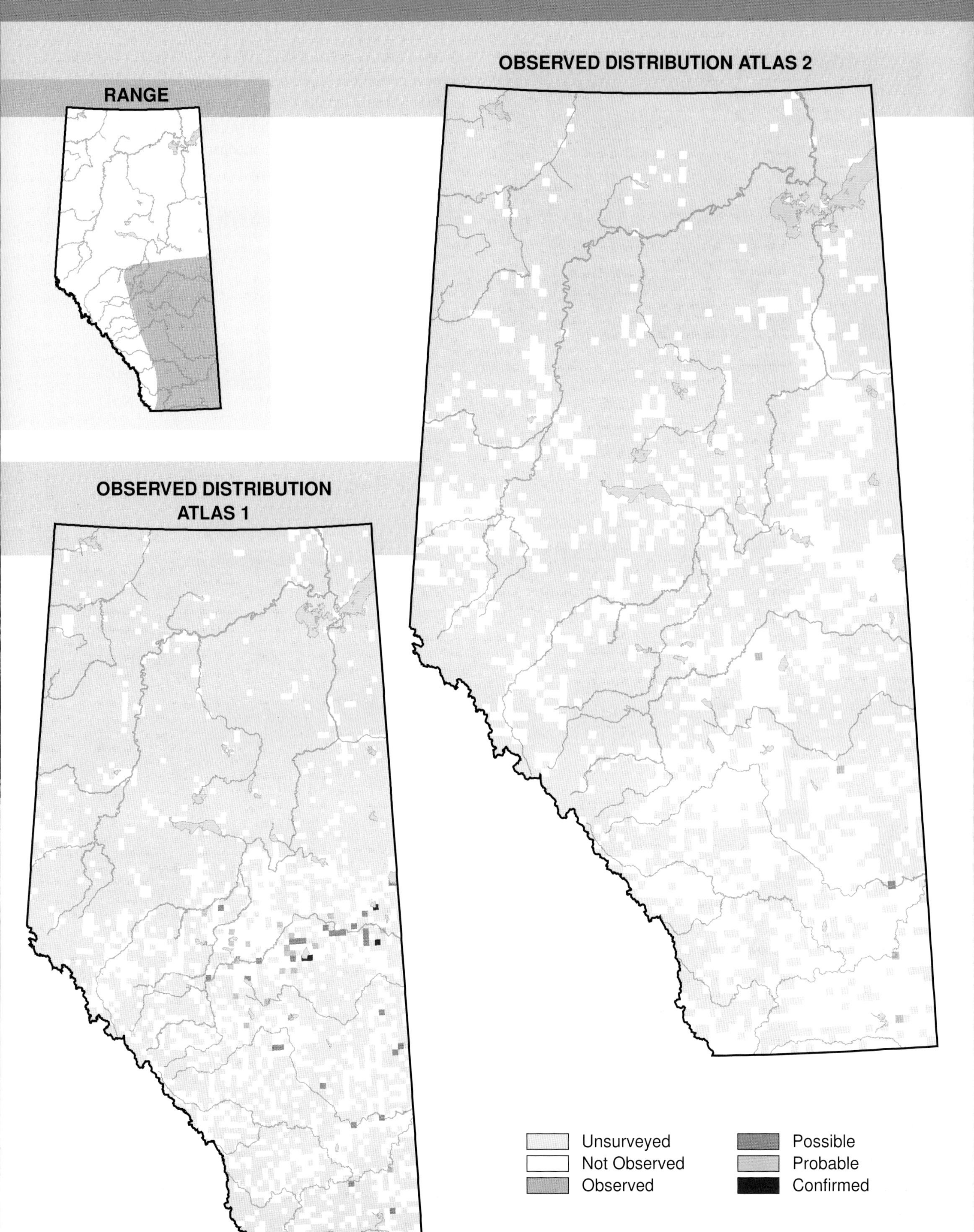
Habitat
Nest
Location
Nest
Type
Diet
RANGE
OBSERVED DISTRIBUTION ATLAS 2
OBSERVED DISTRIBUTION
ATLAS 1
Unsurveyed
Not Observed
Observed
Possible
Probable
Confirmed

Great Horned Owl (*Bubo virginianus*)

Photo: Murray Schultz

NESTING

CLUTCH SIZE (EGGS):	2–3
INCUBATION (DAYS):	30–35
FLEDGING (DAYS):	63–70
NEST HEIGHT (METRES):	4.6–21.3

Alberta's official provincial bird, the Great Horned Owl, is found in every Natural Region across the province. It is more prevalent in the southern Boreal Forest, Parkland, and Grassland NRs than in the other regions. There was no change observed in the distribution of this species from Atlas 1.

Despite being able to occupy a wide variety of habitats, Great Horned Owls prefer open and fragmented areas such as second-growth forests and agricultural lands. In these areas, a wide variety of structures can be used for nesting, including stick nests constructed by other species, tree and cliff cavities, rock outcrops on cliffs, and abandoned buildings.

The Great Horned Owl was observed less frequently in all natural regions in Atlas 2 than in Atlas 1. This difference is likely due to the lack of distinction made in Atlas 1 between data gathered in species-specific surveys, such as owl prowls, and the general surveys. In Atlas 2, species-specific surveys, such as the Alberta Nocturnal Owl Survey, were treated in reporting rate calculations as incidental records. As a result, the lower daytime detectability of owls was reflected in Atlas 2.

In many parts of the Great Horned Owl's range, populations are stable due to the bird's generalist nature and preference for human-modified habitats. In the Boreal Forest NR, a decline in relative abundance was detected, but in other natural regions this owl appears to be maintaining stable populations. The observed decline is likely due to fewer owl surveys from the northern part of the province, especially around High Level and Fort Vermilion, rather than to an actual population change. Data from the Breeding Bird Survey did not indicate any change in Great Horned Owl abundance in Alberta. This bird is considered Secure in Alberta.

TEMPORAL REPORTING RATE

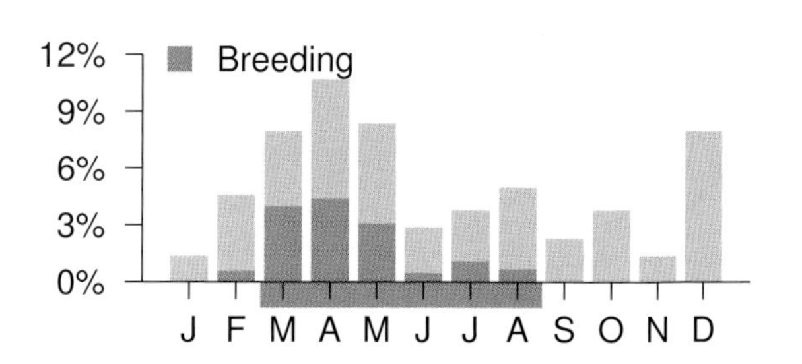

NATURAL REGIONS REPORTING RATE

Region	Rate
Boreal Forest	1.0%
Foothills	1.9%
Grassland	6.8%
Parkland	4.8%
Rocky Mountain	3.9%

SPATIAL REPORTING RATE

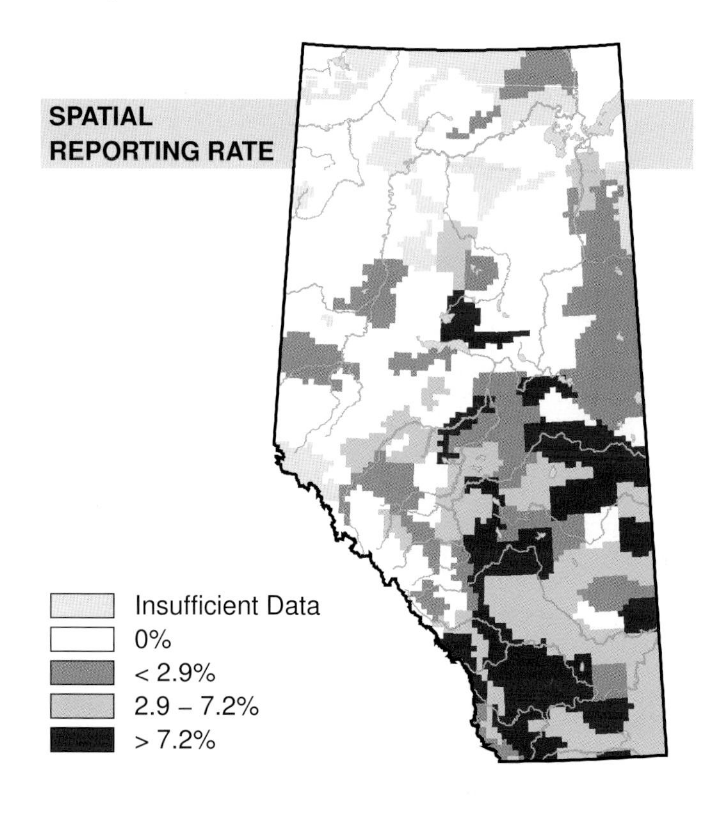

STATUS

GSWA 2000	GSWA 2005	COSEWIC Historic Rankings	COSEWIC 2007	Alberta Wildlife Act
Secure	Secure	N/A	N/A	N/A

OBSERVED DISTRIBUTION ATLAS 2

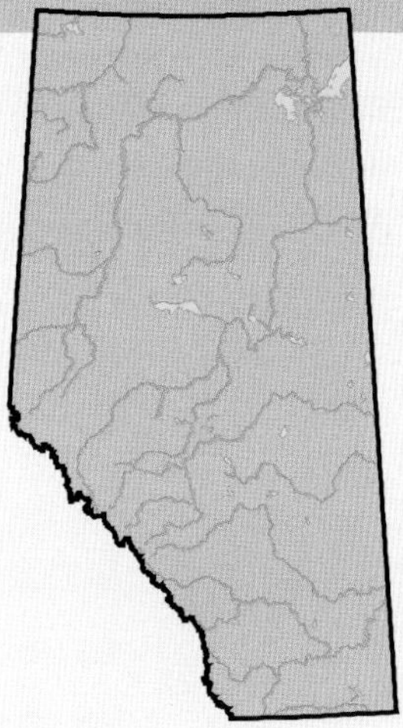

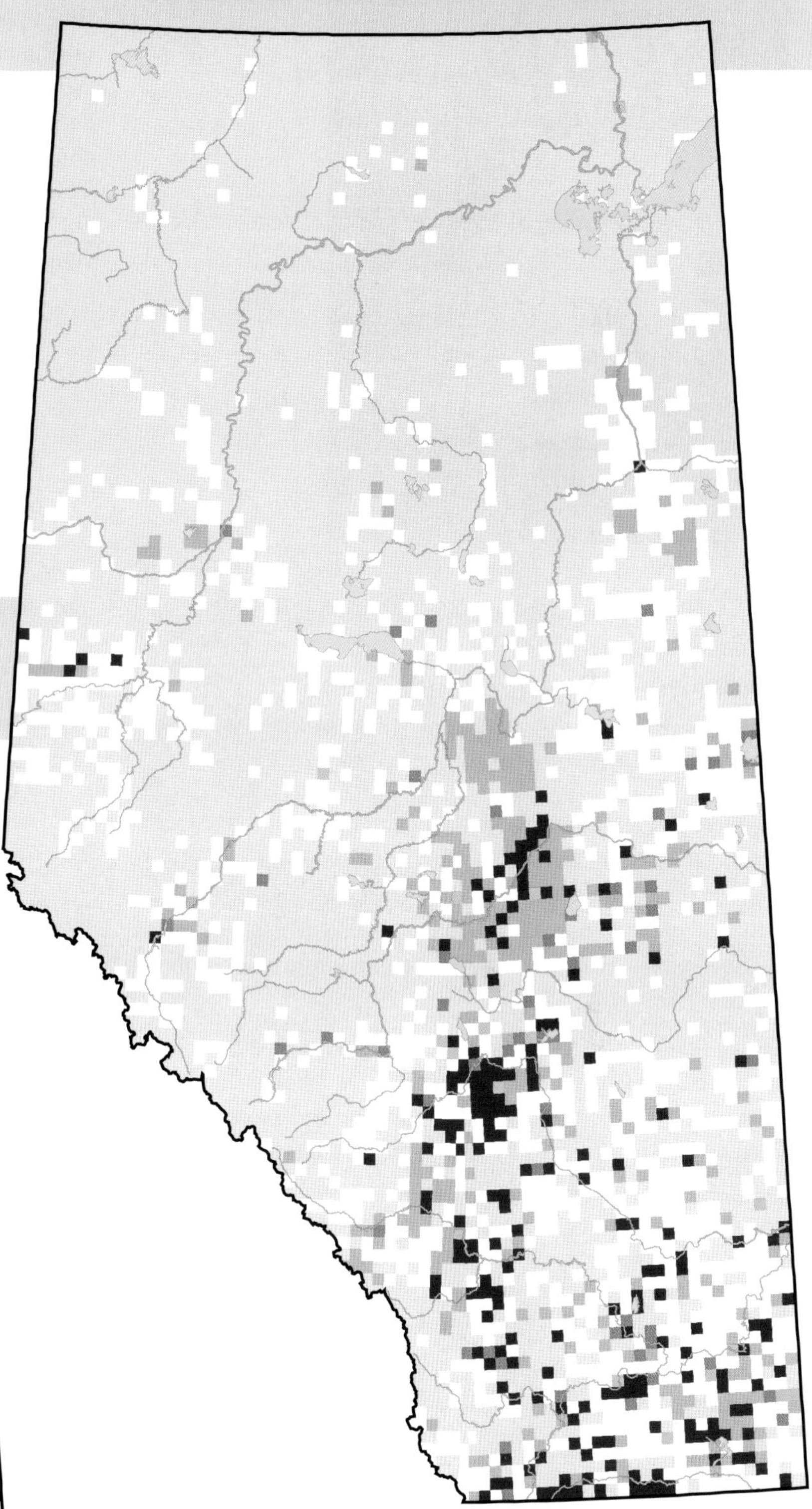

RANGE

OBSERVED DISTRIBUTION ATLAS 1

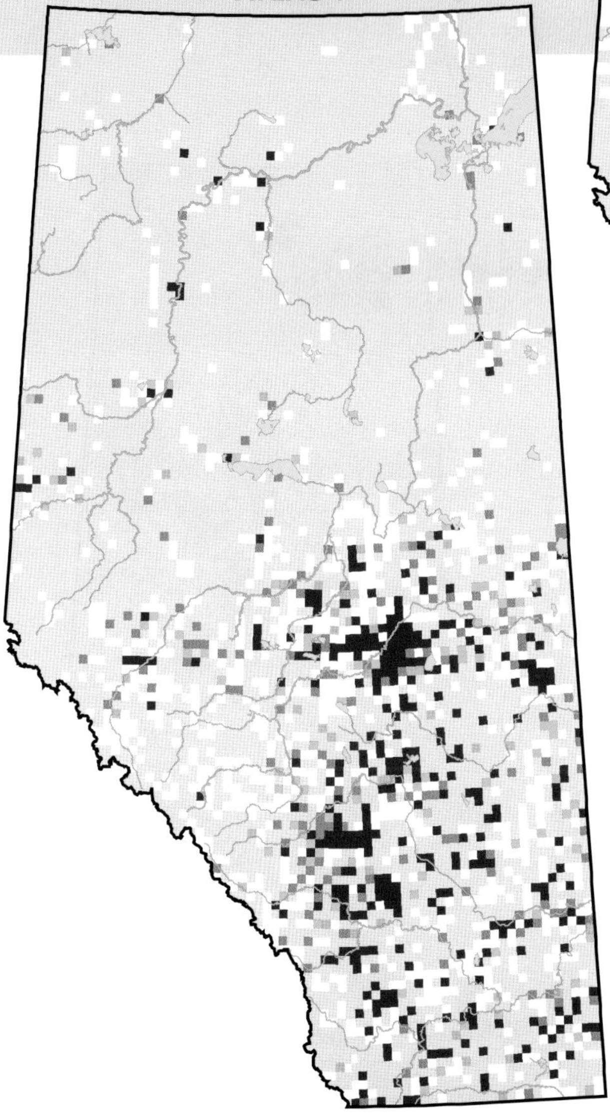

Unsurveyed
Not Observed
Observed
Possible
Probable
Confirmed

Northern Hawk Owl (*Surnia ulula*)

Photo: Gordon Court

NESTING

CLUTCH SIZE (EGGS):	3–9
INCUBATION (DAYS):	25–30
FLEDGING (DAYS):	25–35
NEST HEIGHT (METRES):	3–12.2

The Northern Hawk Owl is found mainly in the Boreal Forest Natural Region. In addition to this NR, they are occasionally found in the Foothills, Parkland, and Rocky Mountain NRs but are found only infrequently in the Grassland NR. No change in distribution was observed between Atlas 1 and Atlas 2 for this species.

During the fall, Hawk Owls often move away from their breeding areas to congregate in the winter in areas where prey density is high. Increased attention has been given to this movement in recent years by local raptor banders and photographers. Large numbers of birders head north in the winter to observe influxes of Northern Hawk Owls. This was the likely cause of the higher reporting rates at the southern extent of its range during Atlas 2, especially around Fort Saskatchewan, Westlock, and Athabasca.

A decline in the Northern Hawk Owl's relative abundance was detected in the Boreal Forest NR where, relative to other species, the bird was observed less frequently in Atlas 2 than in Atlas 1. No changes were detected in other NRs. This decline is likely a result of habitat loss in this NR ascribed to human activities such as forestry and oil and gas extraction. The Breeding Bird Survey records of Hawk Owl were too few to investigate temporal abundance changes. In 2005, the status of the Northern Hawk Owl in Alberta was changed to Sensitive due to its low population numbers, difficulties in detecting population trends, their requirement of mature forests, and their vulnerability to certain forestry practices.

TEMPORAL REPORTING RATE

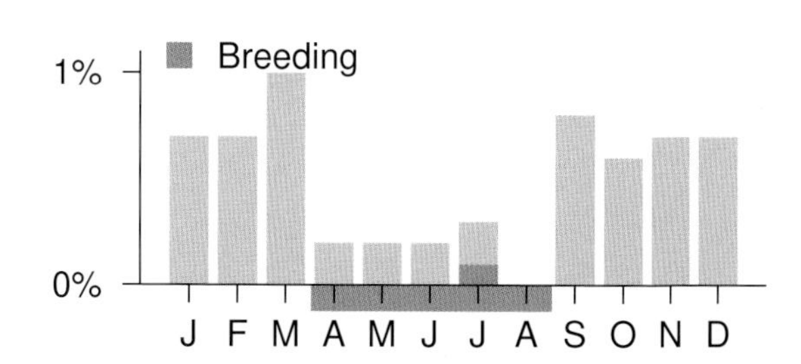

NATURAL REGIONS REPORTING RATE

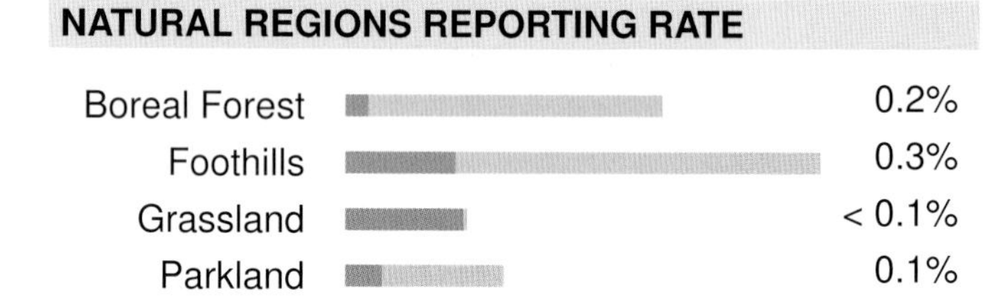

SPATIAL REPORTING RATE

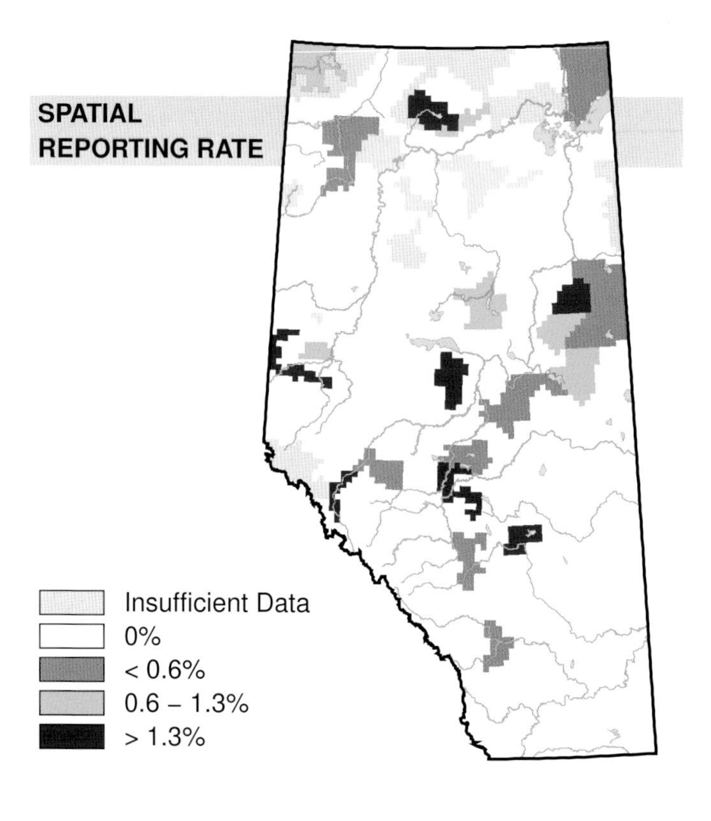

STATUS

GSWA 2000	GSWA 2005	COSEWIC Historic Rankings	COSEWIC 2007	Alberta Wildlife Act
Secure	Sensitive	Not At Risk	Not At Risk	N/A

Habitat

Nest Location

Nest Type

Diet

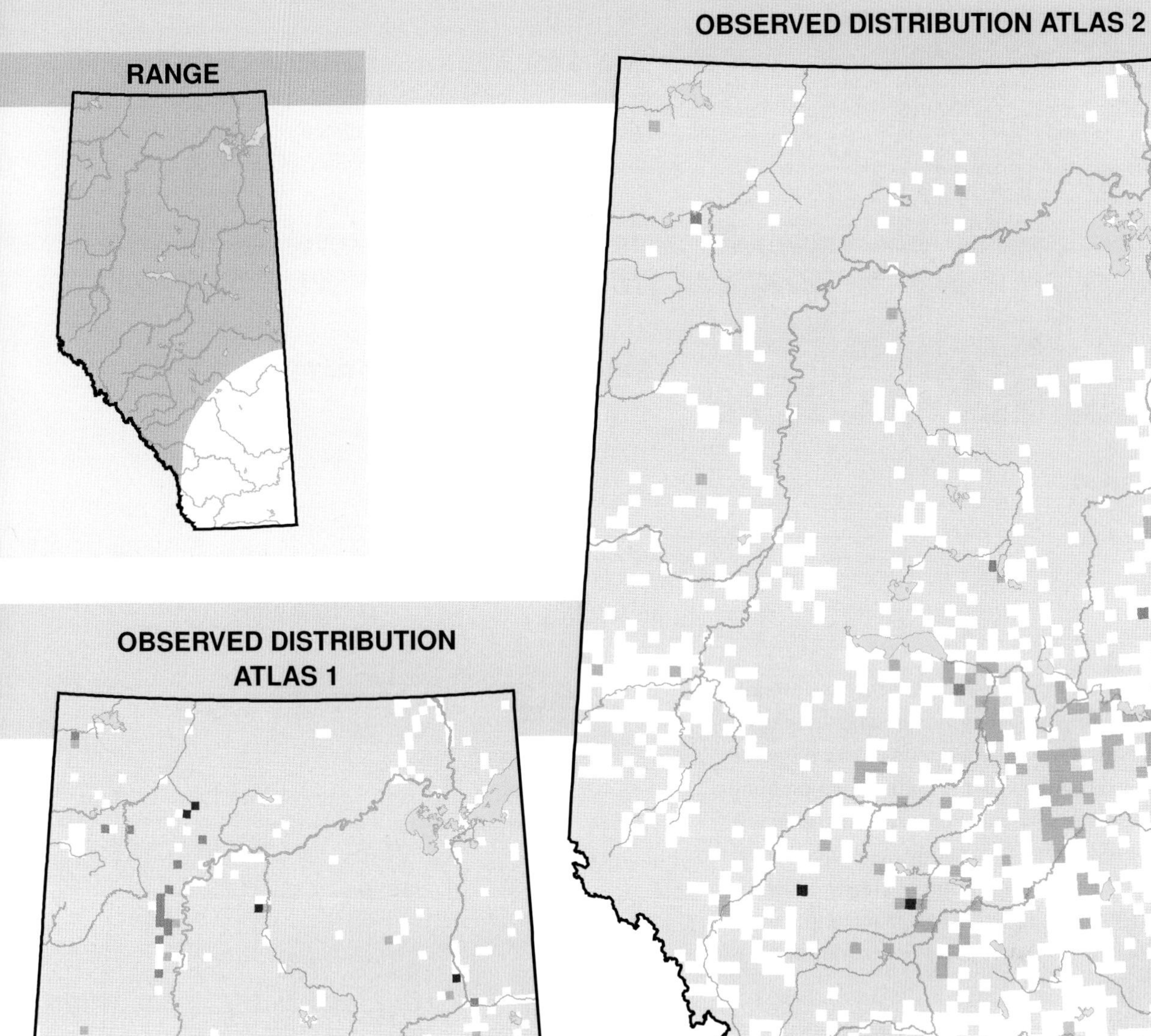

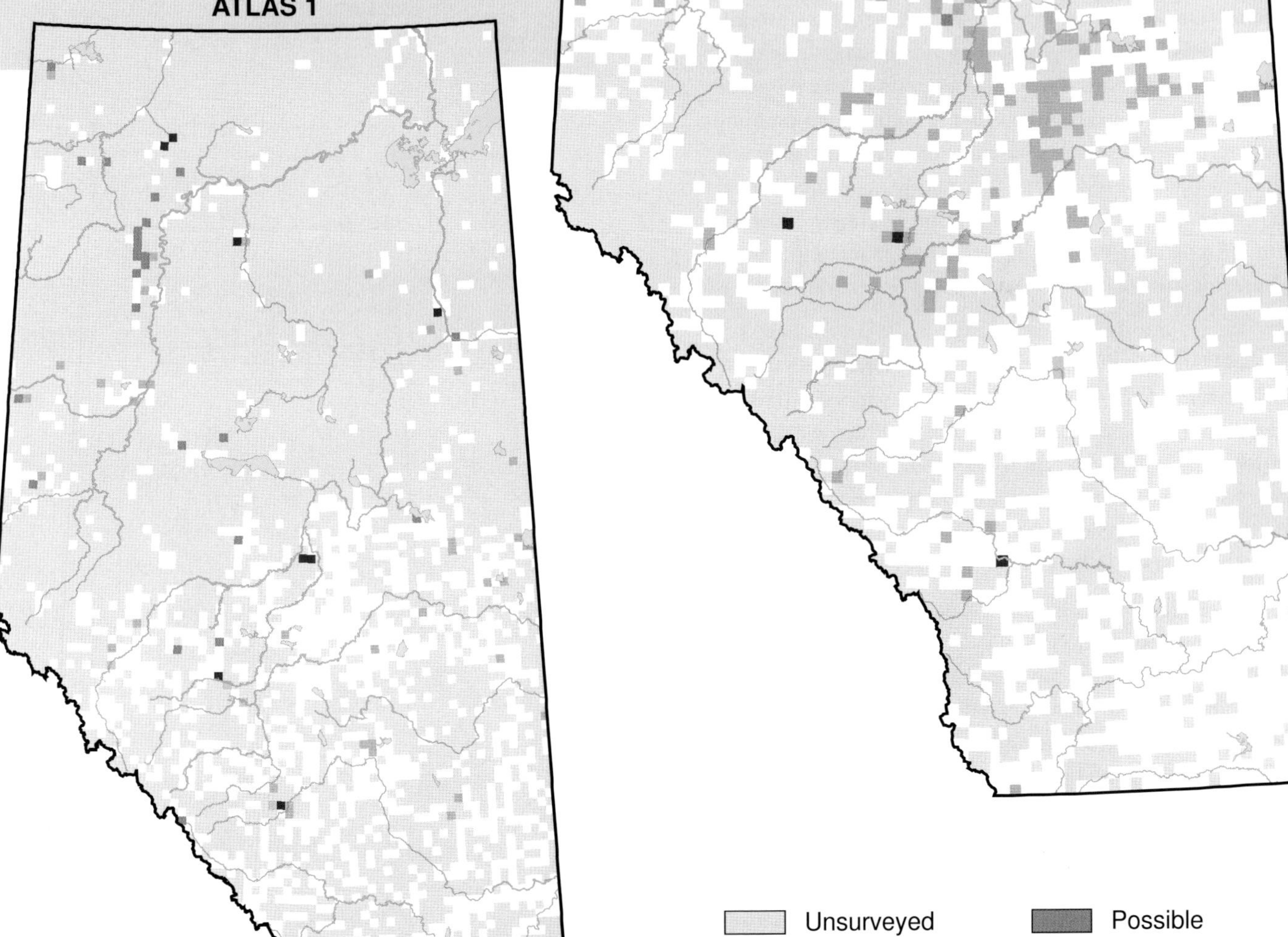

Northern Pygmy-Owl (*Glaucidium gnoma*)

Photo: Gordon Court

NESTING

CLUTCH SIZE (EGGS):	3–6
INCUBATION (DAYS):	28
FLEDGING (DAYS):	30
NEST HEIGHT (METRES):	2.4–6.1

Alberta's smallest owl is found mainly in the Rocky Mountain and Foothills Natural Regions, but it is occasionally found in the Boreal Forest NR. The Northern Pygmy-Owl's observed distribution increased in Atlas 2 in the northeastern portion of its range. Nevertheless, it is unlikely that this expansion is an actual population expansion into formerly unoccupied areas. Rather, the expansion is likely the result of greater atlassing effort afforded the Northern Pygmy-Owls in areas that were not previously thought to be within the species' range.

This species was detected very rarely in the Boreal Forest and Parkland NRs and more frequently, but still rarely, in the Rocky Mountain NR. Most records for this species were non-breeding records from outside the breeding season. It is rarely seen during the breeding season, but is more commonly observed during the non-breeding season as individuals move into towns and hunt birds and small mammals during the day (Holt and Petersen, 2000).

No change in relative abundance was detected for this species in any of the NRs where it occurs. Even so, relative to other species, it was observed more frequently in Atlas 2 than in Atlas 1 in the Rocky Mountain NR. This was likely a result of the fact that more birders have become aware that Northern Pygmy-Owls can often be observed around Kananaskis and Canmore. The Breeding Bird Survey recorded too few Northern Pygmy-Owls to permit investigation of temporal abundance changes in Alberta or across Canada for the period 1985–2005.

Dunn (2005), in the National action needs for Canadian Landbird Conservation, indicates that the rarity of this species and the potential for negative effects of forestry justify better knowledge of population status in Canada.

The status of Northern Pygmy-Owls in Alberta is Sensitive due to its small population, to difficulties in detecting population trends, and to the bird's requirement of mature forests for nesting.

TEMPORAL REPORTING RATE

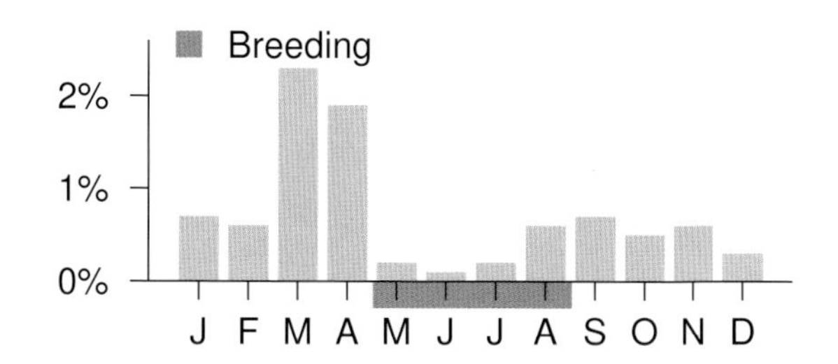

NATURAL REGIONS REPORTING RATE

Boreal Forest	< 0.1%
Parkland	< 0.1%
Rocky Mountain	1.7%

SPATIAL REPORTING RATE

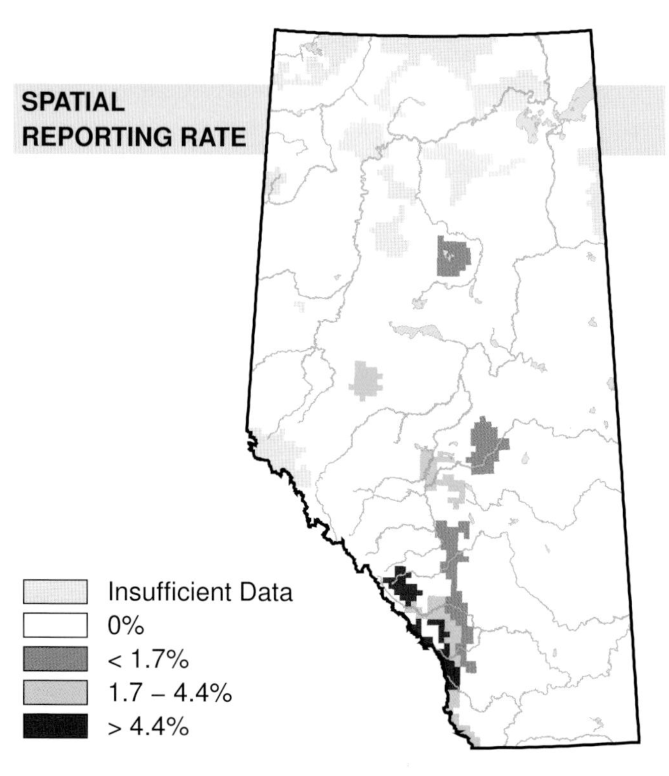

STATUS

GSWA 2000	GSWA 2005	COSEWIC Historic Rankings	COSEWIC 2007	Alberta Wildlife Act
Sensitive	Sensitive	N/A	N/A	N/A

Habitat 
Nest Location 
Nest Type 
Diet  

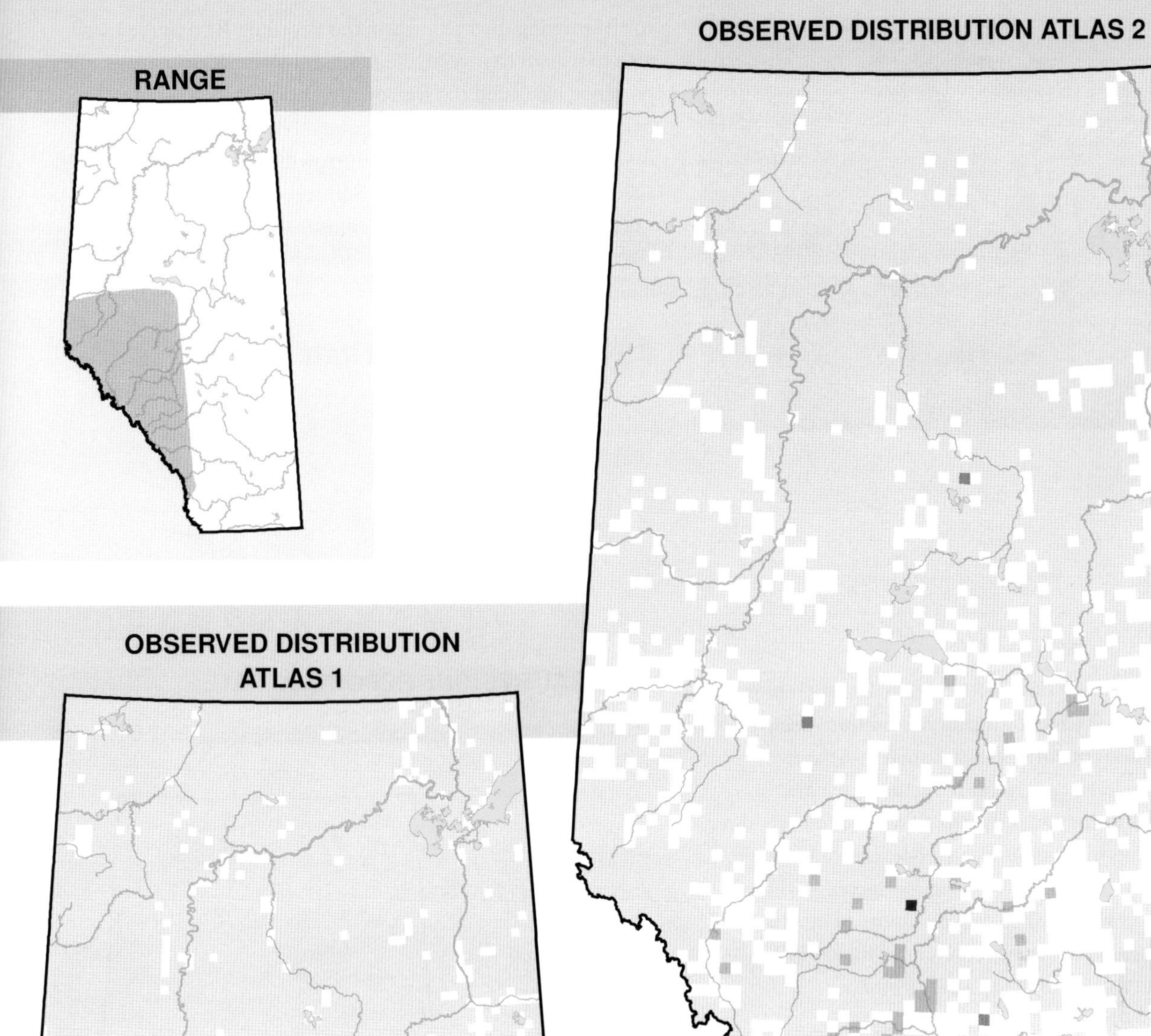

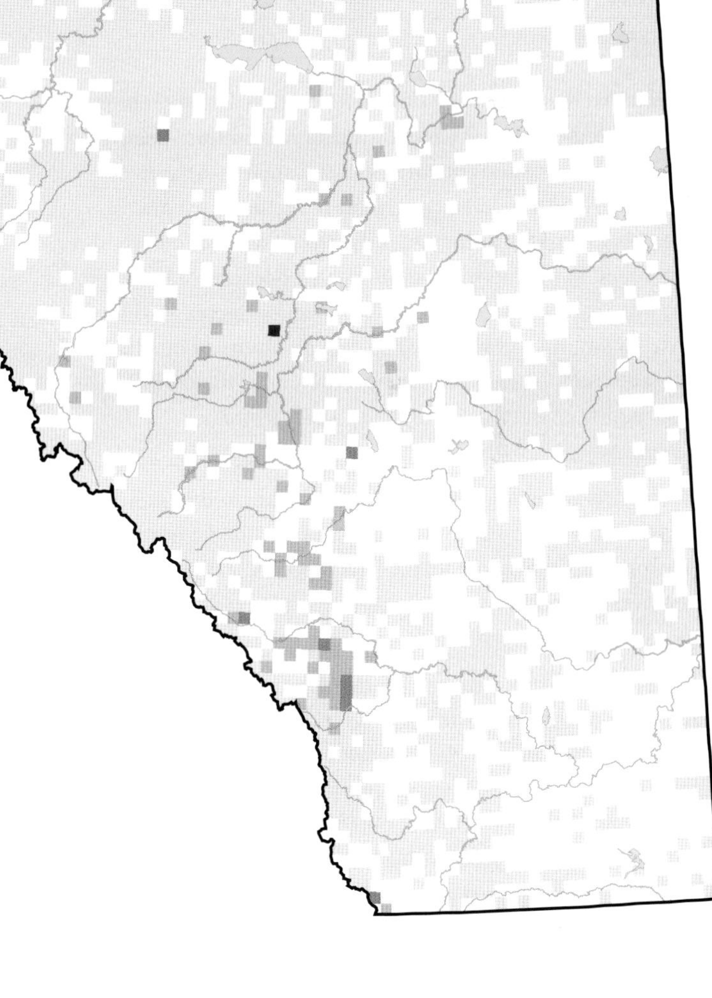

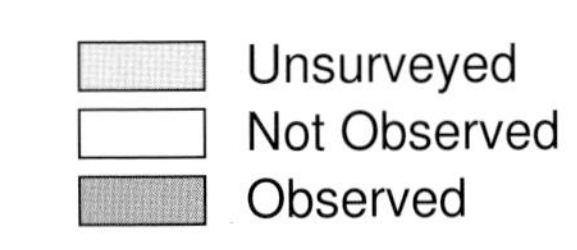

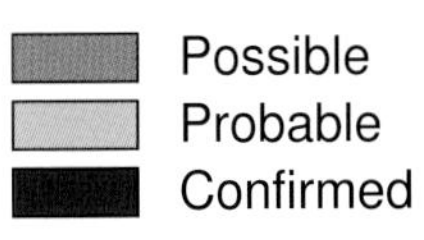

Burrowing Owl (*Athene cunicularia*)

Photo: Gordon Court

NESTING

CLUTCH SIZE (EGGS):	5–7
INCUBATION (DAYS):	28–29
FLEDGING (DAYS):	28
NEST HEIGHT (METRES):	0

In Alberta, the Burrowing Owl is found only in the Grassland Natural Region. The observed distribution for this species decreased in the western portion of its range in Atlas 2 compared to Atlas 1. The character of the habitat at the periphery of a species' range is usually of a lower quality than that found in the more centrally located portions of its range. For this reason, when a species' abundance decreases, its distribution often contracts. The change observed in Alberta was consistent with contractions reported for the peripheral distribution of the Burrowing Owl's range in other jurisdictions.

Burrowing Owls were not evenly distributed within the Grassland NR. Most records were observed in the northeastern portion of its range in Alberta. The reporting rate from the southeastern portion of its range was relatively low.

Despite the observed range contraction, a change in relative abundance was not detected for the Burrowing Owl. It was, however, observed less frequently, relative to other species, in Atlas 2 than in Atlas 1. It is known that the Burrowing Owl population in Alberta is small. Further, Burrowing Owls are difficult to detect due to their small size, nocturnal habits and shy nature. These facts resulted in a relatively small number of records for this species which, in turn, placed it below the sensitivity threshold of our change analysis. The sample size for Burrowing Owl records in the Breeding Bird Survey was also too small to investigate temporal abundance changes. The province has listed the Burrowing Owl as At Risk due to habitat degradation in the Grassland NR and also to loss of fossorial species. In Alberta, Burrowing Owls rely on animals such as Richardson's Ground Squirrel and American Badger to excavate the burrows. For this reason, Burrowing Owl populations are affected by the presence, or absence, of these mammals.

TEMPORAL REPORTING RATE

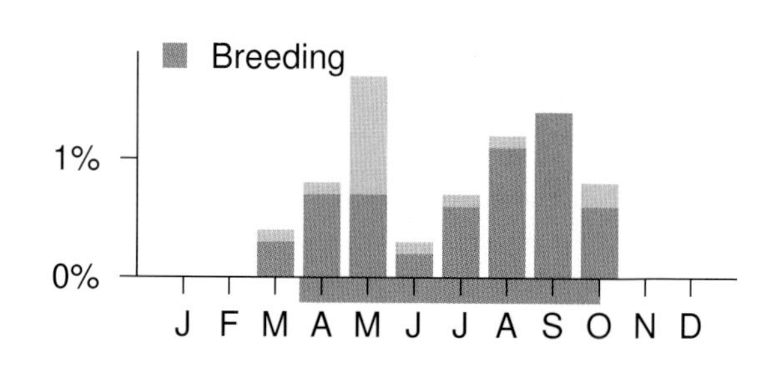

NATURAL REGIONS REPORTING RATE

Grassland 1.6%

SPATIAL REPORTING RATE

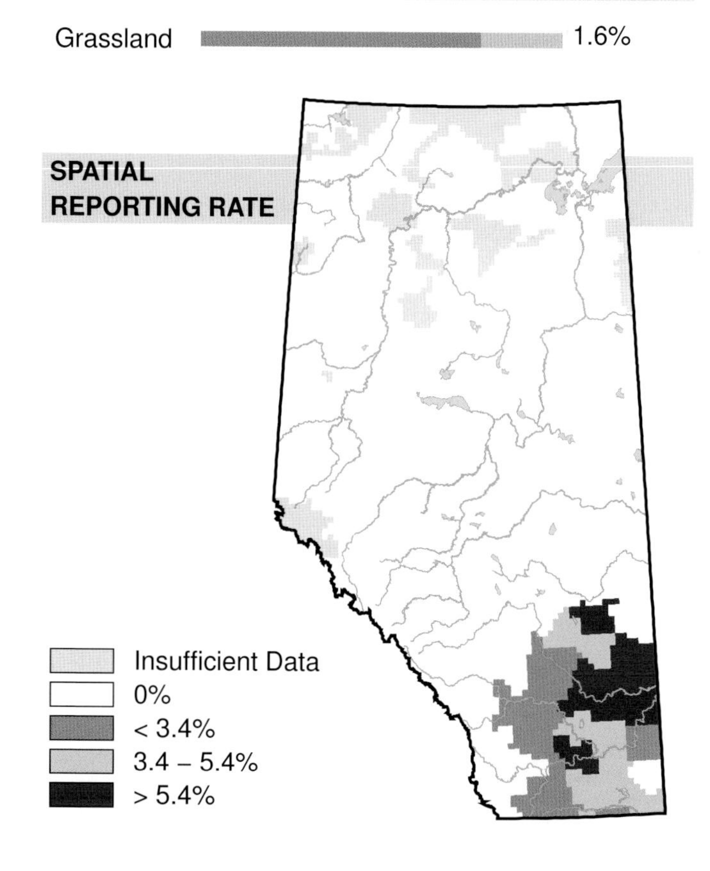

STATUS

GSWA 2000	GSWA 2005	COSEWIC Historic Rankings	COSEWIC 2007	Alberta Wildlife Act
At Risk	At Risk	Endangered	Endangered	Threatened

Habitat  Nest Location  Nest Type  Diet  

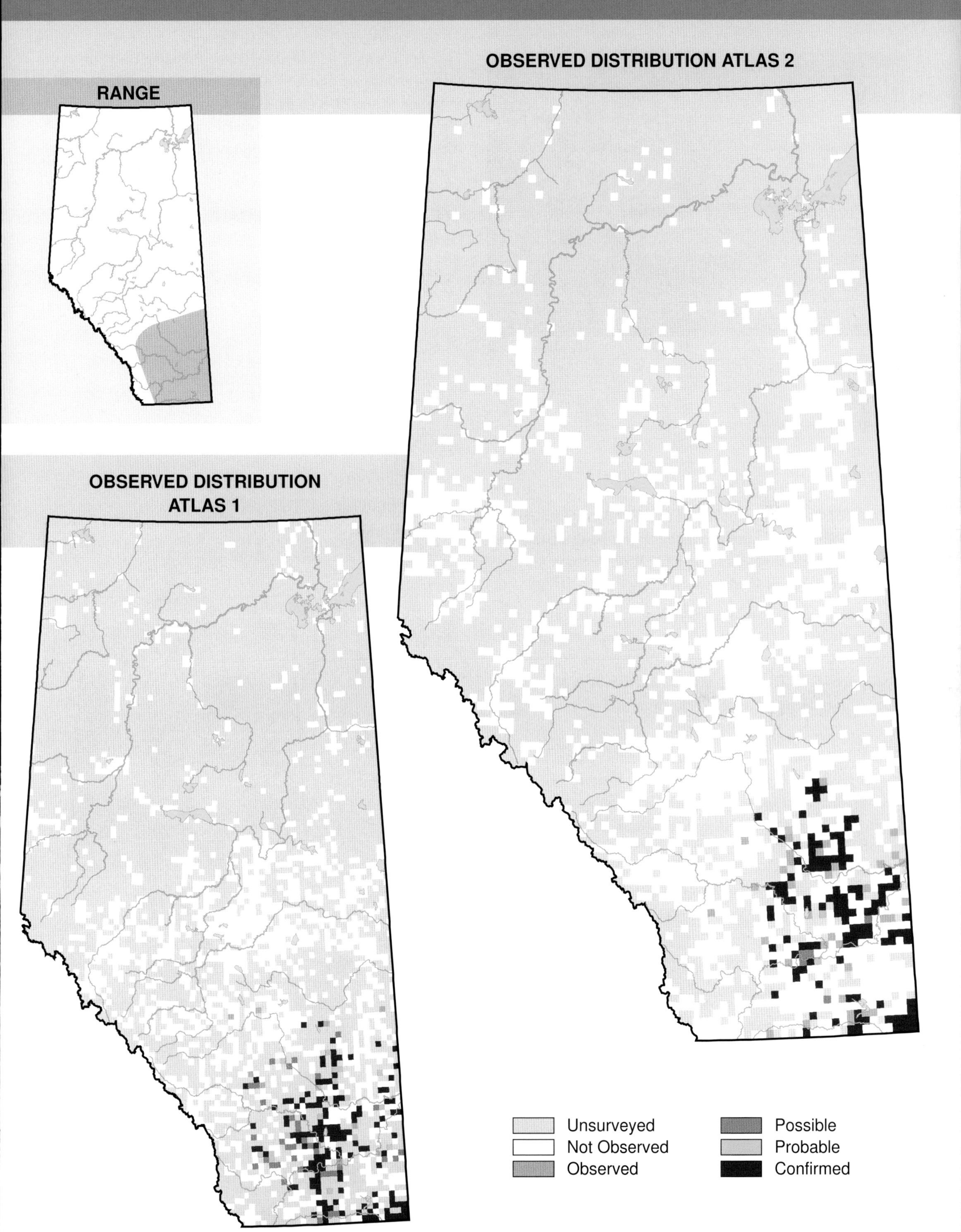

Barred Owl (*Strix varia*)

Photo: Gordon Court

NESTING

CLUTCH SIZE (EGGS):	2–3
INCUBATION (DAYS):	28–33
FLEDGING (DAYS):	42
NEST HEIGHT (METRES):	<10

The Barred Owl is distributed over much of Alberta's forested regions, with most occurrences in the Boreal Forest, Foothills, and Rocky Mountain Natural Regions. No changes in distribution were observed between Atlas 1 and Atlas 2.

Large balsam poplar cavities are their preferred nesting sites. As a result, Barred Owls tend to show a clumped breeding distribution, particularly in riparian habitat where this habitat type is prevalent. The distribution maps from both Atlases reflect this propensity to breed near rivers and lakes.

Reporting rates are high in some areas of the northern Boreal Forest and Foothills NRs, particularly around Hinton, Fort McMurray, Lac La Biche, and Athabasca. Nocturnal owls have received increased attention from various groups interested in acquiring baseline data on their populations. The volunteer-based Alberta Nocturnal Owl Survey was initiated in 2002 after Atlas 1 was completed, and has produced many new records of Barred Owls for the province. Increased access to remote areas and isolated woodlands also contributed to higher numbers.

The Barred Owl appears to be fairly stable throughout its range, although an increase in relative abundance was detected in the Boreal Forest NR where, relative to other species, it was observed more frequently in Atlas 2 than in Atlas 1. The increase is likely due to increased survey effort in that region.

Too few Barred Owls are detected on Breeding Bird Surveys to assess abundance trends in Alberta and no change has been detected by the BBS on the Canada-wide scale. The Barred Owl is considered a Sensitive species in Alberta, due to its dependence on mature and old-growth mixedwood forests, a habitat-type that is in decline.

TEMPORAL REPORTING RATE

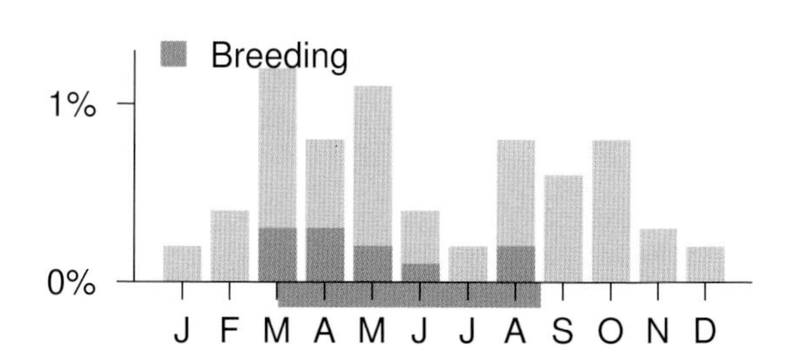

NATURAL REGIONS REPORTING RATE

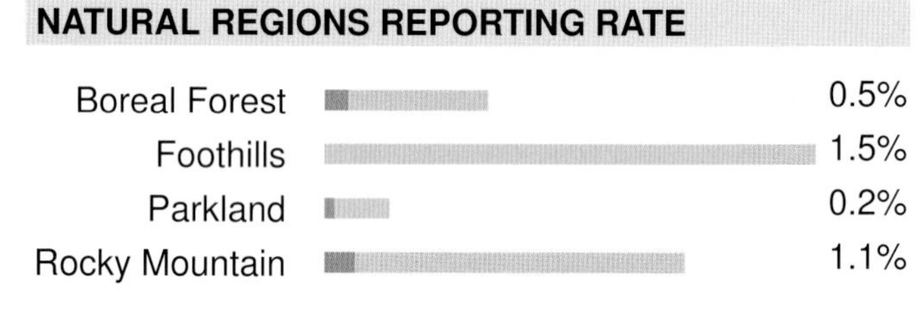

SPATIAL REPORTING RATE

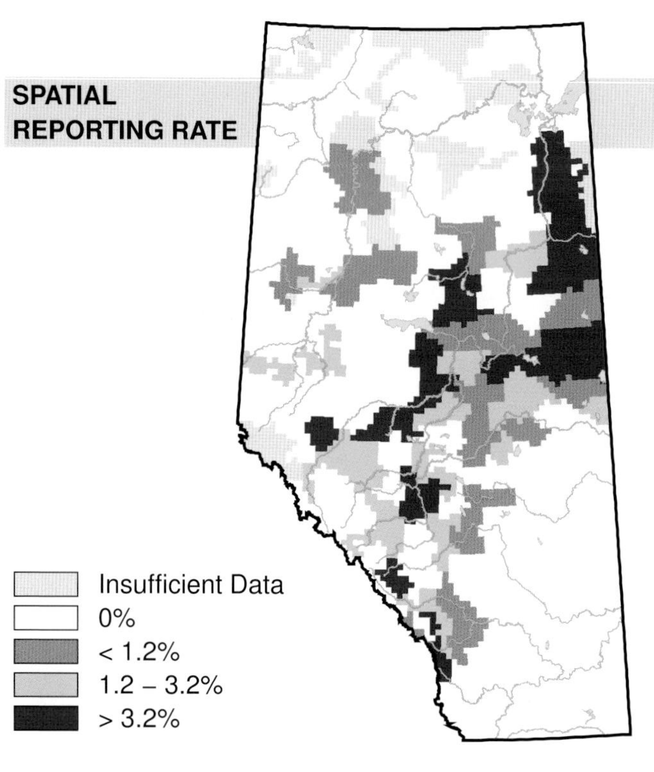

STATUS

GSWA 2000	GSWA 2005	COSEWIC Historic Rankings	COSEWIC 2007	Alberta Wildlife Act
Sensitive	Sensitive	N/A	N/A	N/A

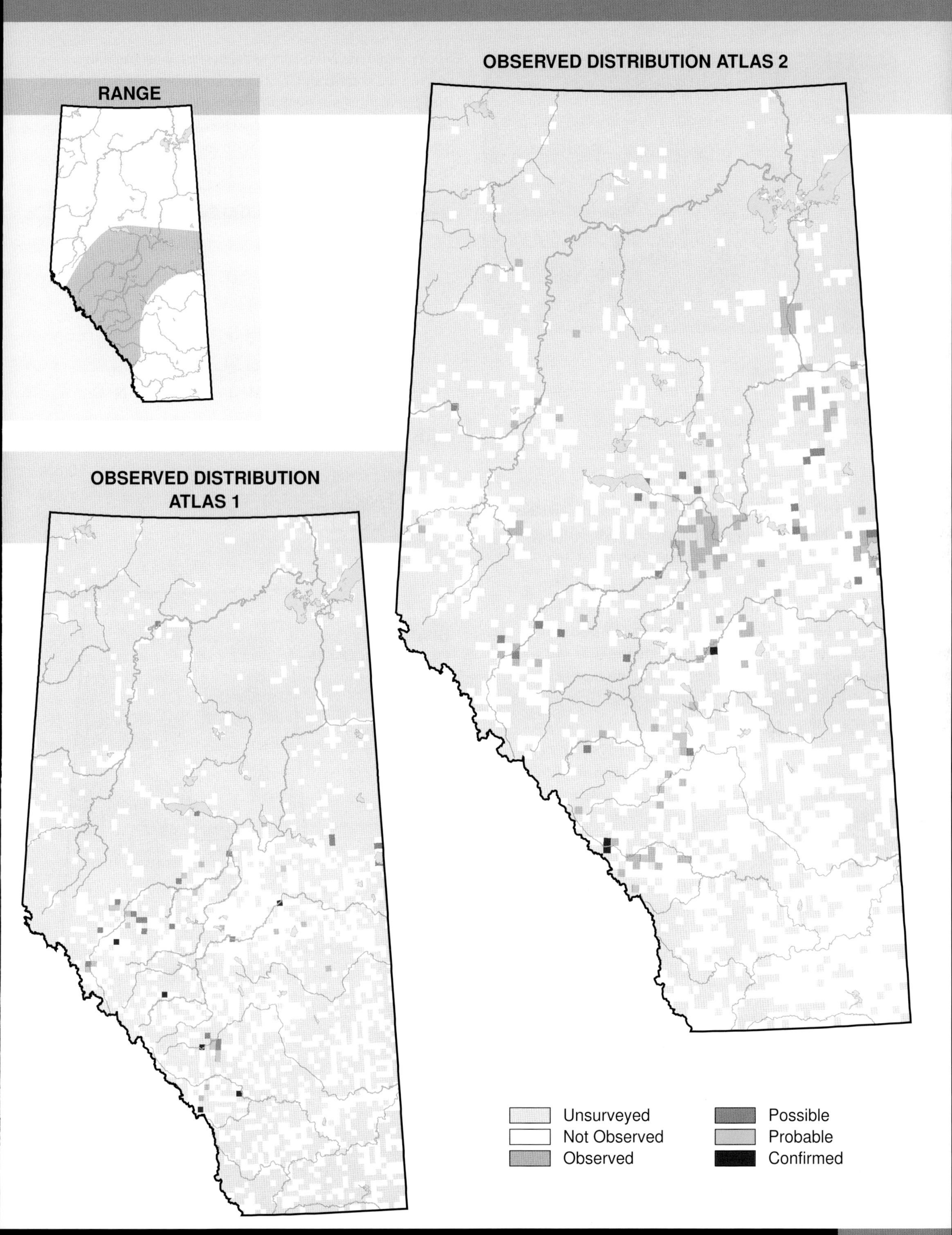
RANGE
OBSERVED DISTRIBUTION ATLAS 2
OBSERVED DISTRIBUTION ATLAS 1
Unsurveyed
Not Observed
Observed
Possible
Probable
Confirmed

Great Gray Owl (*Strix nebulosa*)

Photo: Jim Slobodian

NESTING

CLUTCH SIZE (EGGS):	2–5
INCUBATION (DAYS):	30
FLEDGING (DAYS):	21–28
NEST HEIGHT (METRES):	2–15

The Great Gray Owl's breeding range is holarctic, and these birds are closely associated with boreal forests during the nesting period. There was no change observed in the distribution of this species between Atlas 1 and Atlas 2.

Gray Gray Owls normally use stick nests that were built by hawks and ravens. However, this species will occasionally use tops of broken trees or man-made platforms. This owl was most common in the Boreal Forest, Foothills, and Rocky Mountain Natural Regions. It was rare in the Grassland and Parkland NRs.

Increases in relative abundance of Great Gray Owls were detected in the Boreal Forest NR where, surprisingly, relative to other species, it was observed less frequently in Atlas 2 than in Atlas 1. The increase in the Boreal Forest NR is likely a reflection of increased observer detection, rather than actual population change; road access in the north has increased in recent years, and Atlas 2 had more extensive coverage in northern Alberta than Atlas 1. A decline in relative abundance was detected in the Foothills NR where, relative to other species, the bird was observed less frequently in Atlas 2 than in Atlas 1. Populations could be declining in the Foothills NR due to reduced availability of breeding habitat in this NR. No changes in relative abundance were detected in the Parkland or Rocky Mountain NRs. Due to the nocturnal nature of this species, the number of Great Gray Owl records in the Breeding Bird Survey was too small to investigate temporal abundance changes. The status of the Great Gray Owl in Alberta is Sensitive due to its natural scarcity and its requirement for mature forests during the nesting period.

TEMPORAL REPORTING RATE

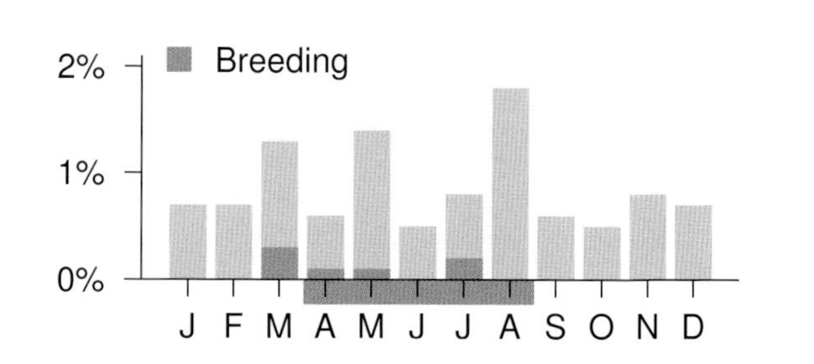

NATURAL REGIONS REPORTING RATE

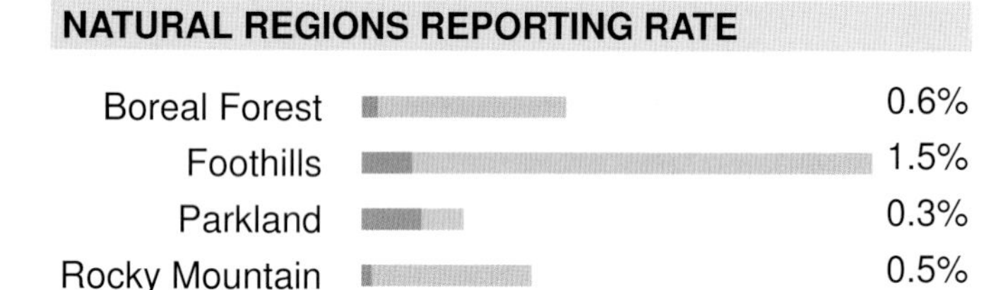

SPATIAL REPORTING RATE

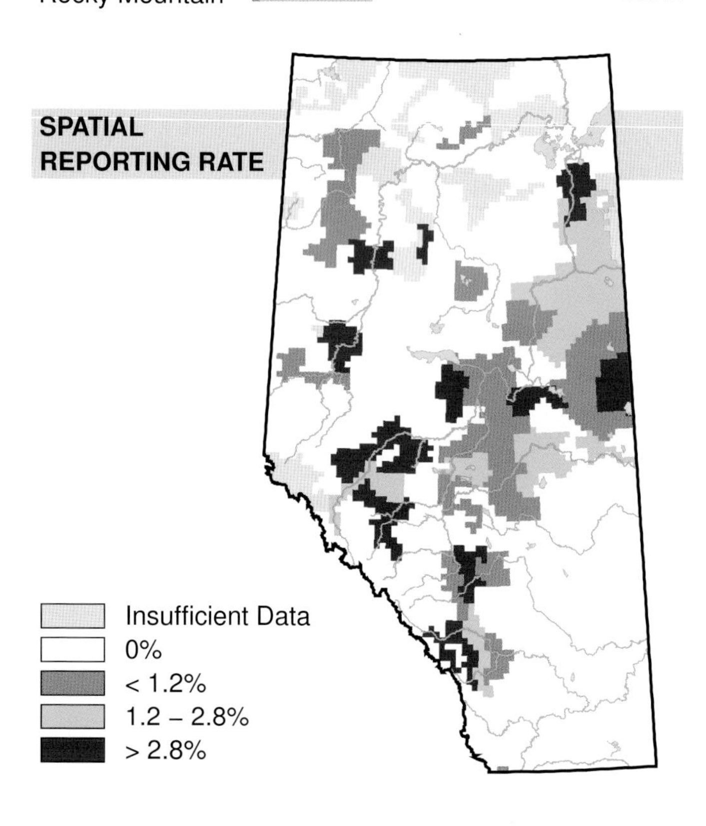

STATUS

GSWA 2000	GSWA 2005	COSEWIC Historic Rankings	COSEWIC 2007	Alberta Wildlife Act
Sensitive	Sensitive	Not At Risk	Not At Risk	N/A

RANGE

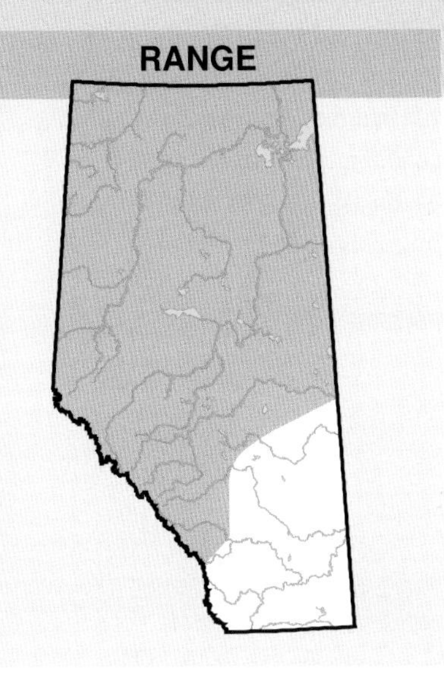

OBSERVED DISTRIBUTION ATLAS 2

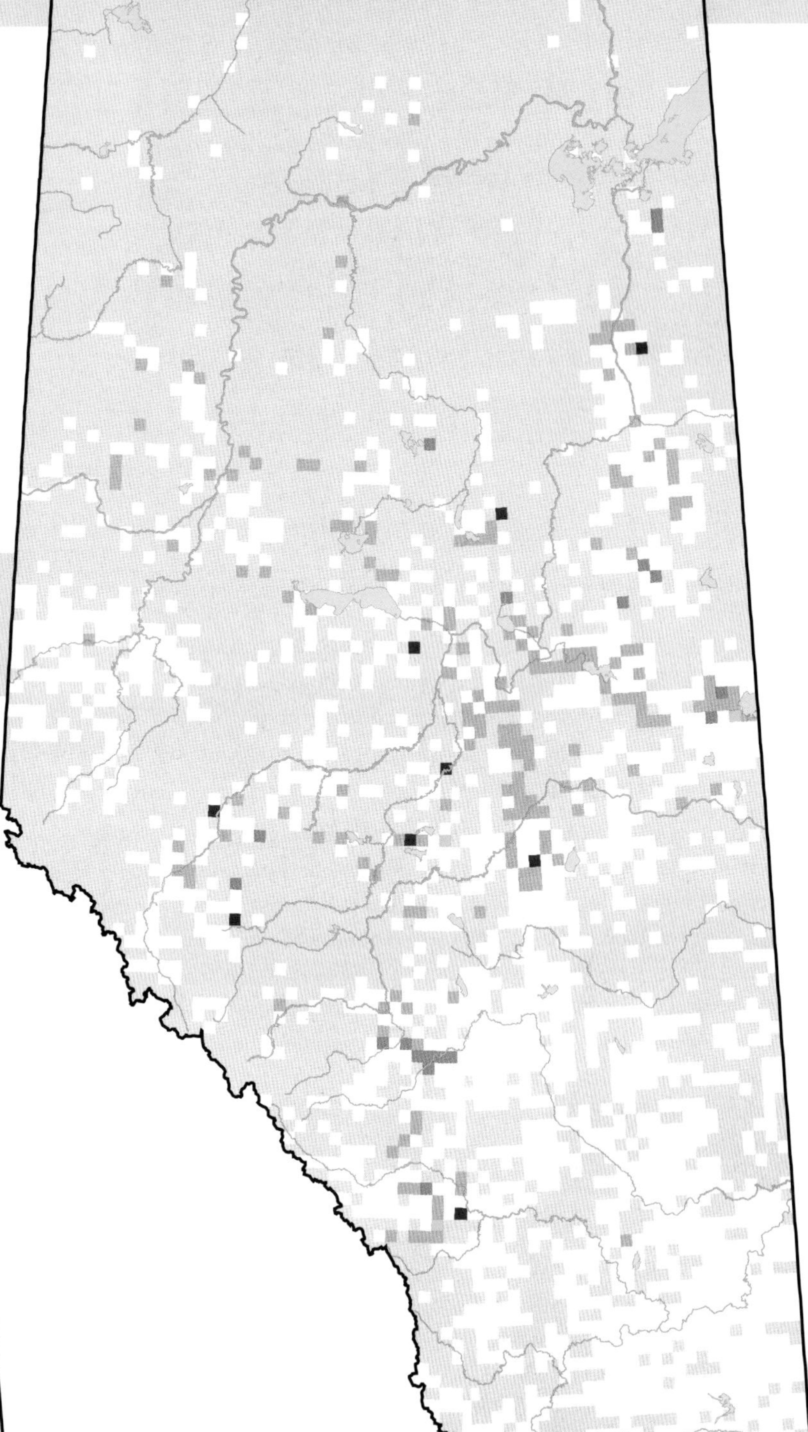

OBSERVED DISTRIBUTION ATLAS 1

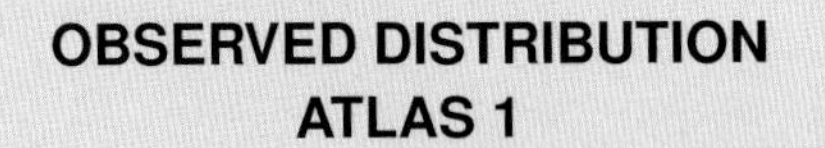

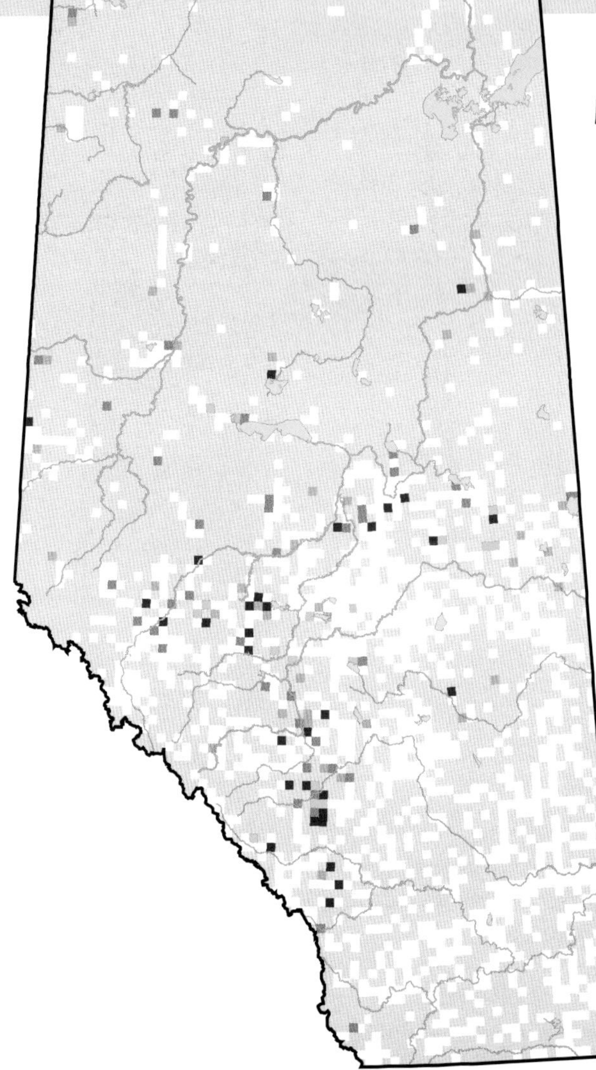

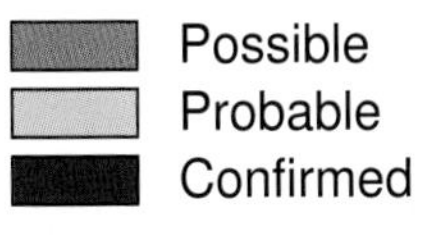

Long-eared Owl (*Asio otus*)

Photo: Gerald Romanchuk

NESTING

CLUTCH SIZE (EGGS):	3–6
INCUBATION (DAYS):	28
FLEDGING (DAYS):	35
NEST HEIGHT (METRES):	>6

The Long-eared Owl is found in the Boreal Forest, Foothills, Grassland, Parkland, and Rocky Mountain Natural Regions. The distribution of this species increased in the northeastern part of its range in Atlas 2. This expansion could be the result of increased forest fragmentation in that area.

Long-eared Owls generally prefer fragmented habitats containing dense woodlands for roosting and nesting, and open areas for hunting at night. For this reason, when Long-eared Owls are detected, they are often found in agricultural areas where shelter belts or forest patches have been retained. Long-eared Owls occur in low density and tend to have a fairly even distribution in Alberta. However, the extent of their distribution in the Boreal Forest NR is poorly understood due to their secretive nature. Being elusive, they were detected infrequently, and this fact is reflected in the very low values shown in the reporting rate map and graphs for this species.

A decline in the relative abundance of Long-eared Owl was detected in the Parkland NR where, relative to other species, this species was observed less frequently in Atlas 2 than in Atlas 1. In Atlas 1, species-specific surveys were not classified separately from regular surveys. In Atlas 2 this discrimination was made; so, this decline could be an artifact. However, increased conversion of forests for agricultural use or housing developments in that NR may be reducing the amount of suitable habitat for this species. No changes in relative abundance were detected in the Boreal Forest, Foothills, or Grassland NRs. Due to the Long-eared Owl's nocturnal nature, the sample size in the Breeding Bird Survey was too small to investigate temporal abundance changes. Further research is needed to properly assess the status of this species in Alberta. The provincial status of the Long-eared Owl is Secure.

TEMPORAL REPORTING RATE

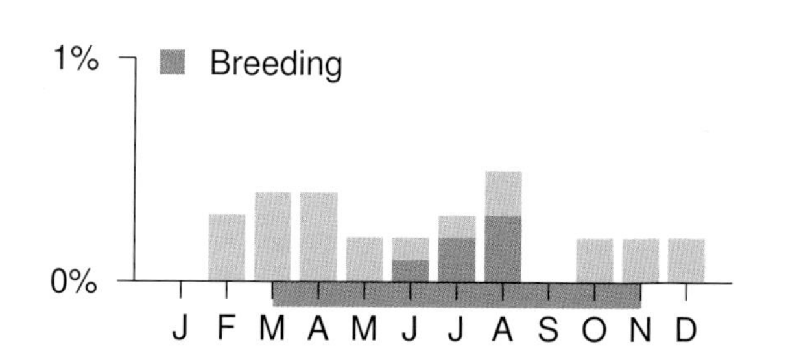

NATURAL REGIONS REPORTING RATE

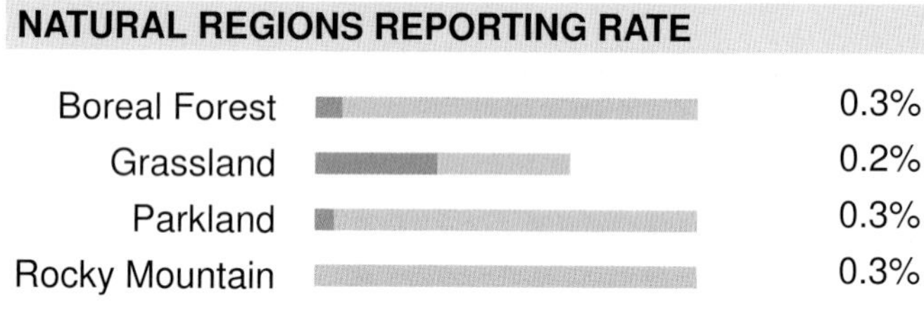

SPATIAL REPORTING RATE

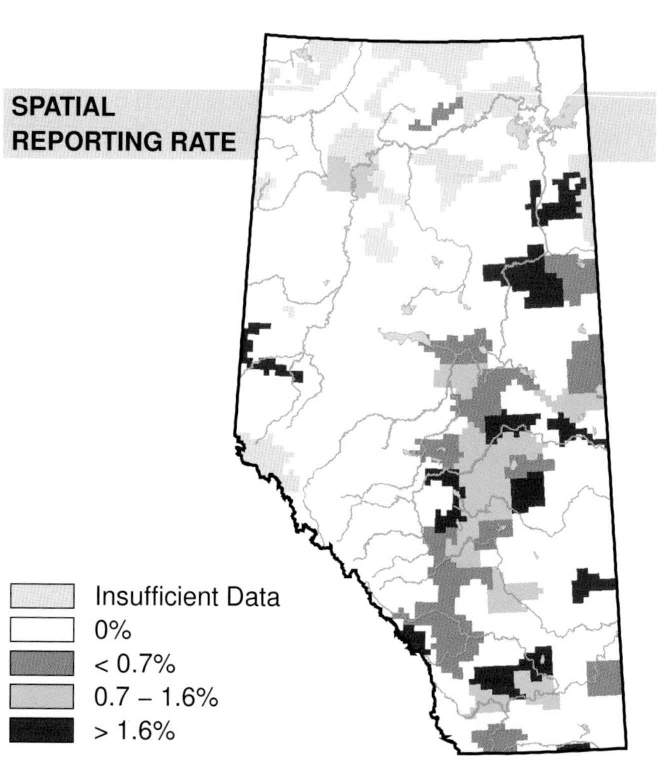

STATUS

GSWA 2000	GSWA 2005	COSEWIC Historic Rankings	COSEWIC 2007	Alberta Wildlife Act
Secure	Secure	N/A	N/A	N/A

RANGE

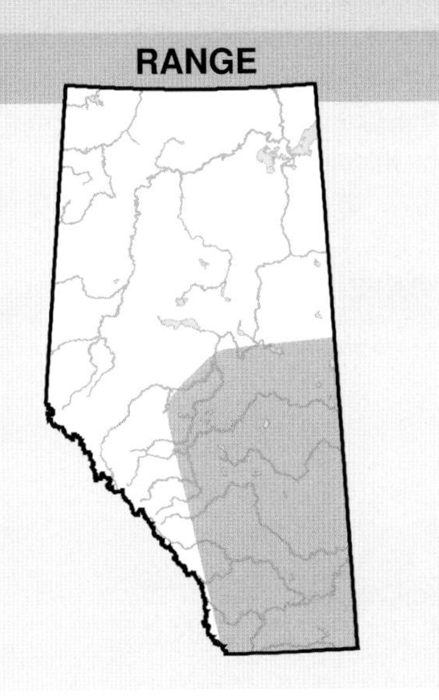

OBSERVED DISTRIBUTION ATLAS 2

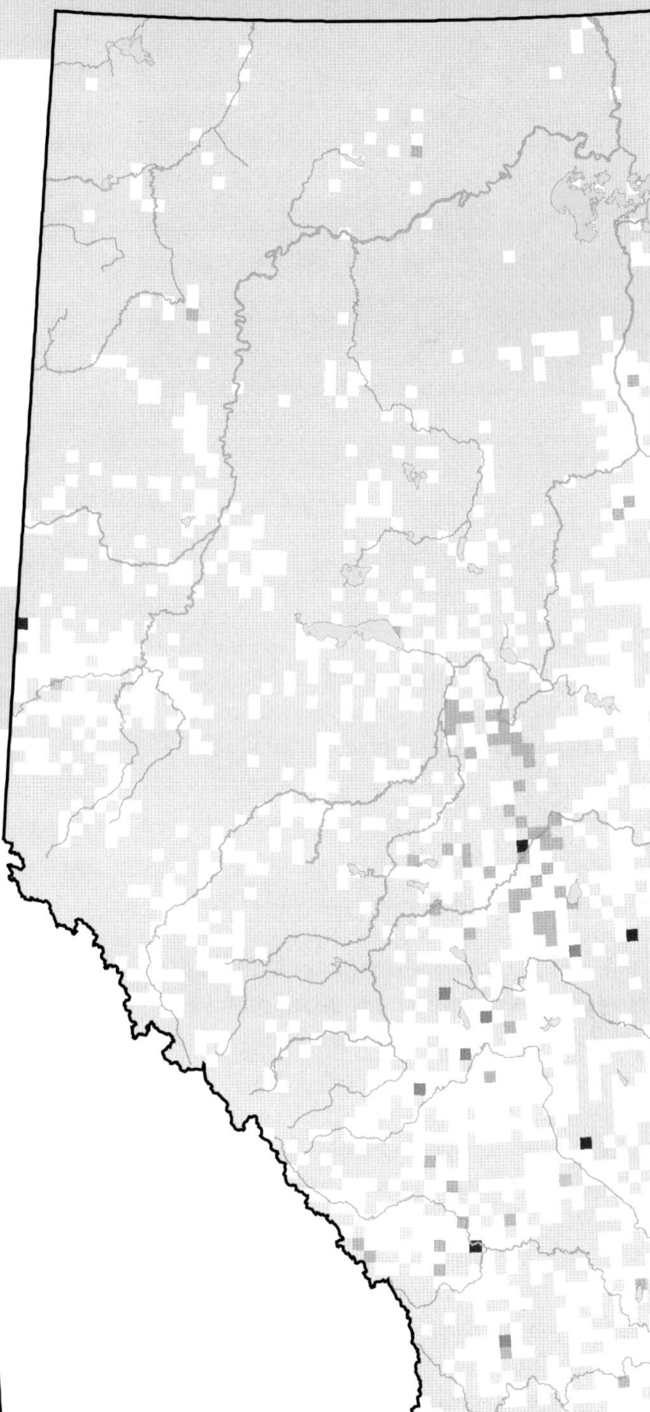

OBSERVED DISTRIBUTION ATLAS 1

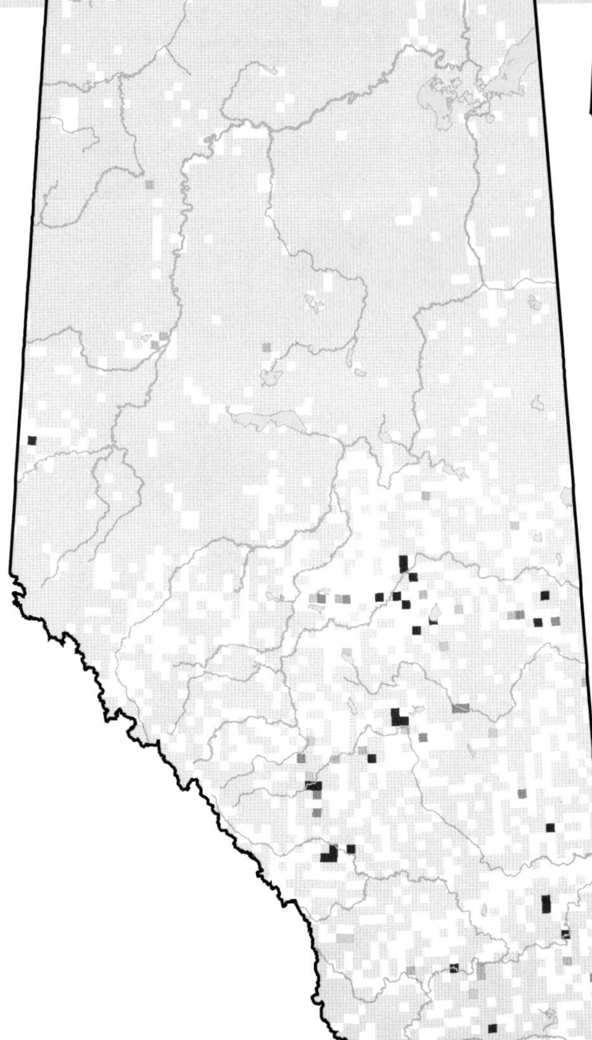

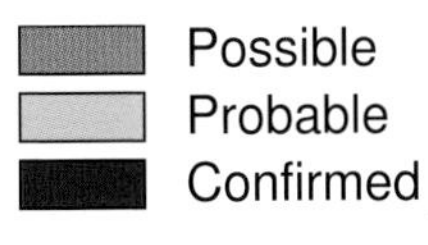

Short-eared Owl (*Asio flammeus*)

Photo: Gordon Court

NESTING

CLUTCH SIZE (EGGS):	5–7
INCUBATION (DAYS):	24–28
FLEDGING (DAYS):	24–27
NEST HEIGHT (METRES):	0

The most widely distributed owl in the world, the Short-eared Owl occurs mainly in the Grassland Natural Region and to a lesser extent in the Parkland, Boreal Forest, Foothills, and Rocky Mountain NRs in Alberta. There was no change observed in the distribution of this species between Atlas 1 and Atlas 2.

Throughout its range, the Short-eared Owl is found in open areas where it nests on the ground and hunts during the day or night. As was expected, the reporting rate for the Grassland NR was highest of all the NRs; in the other natural regions where this species occurs, lower rates were recorded.

Decreases in Short-eared Owl relative abundance were detected in the Boreal Forest and Parkland NRs. Short-eared Owls are vulnerable to the conversion and fragmentation of open habitats. Increasing cultivation in these NRs may be having a negative impact on this species. No changes in relative abundance were detected in the Foothills, Grassland, or Rocky Mountain NRs. The Breeding Bird Survey did not detect any change in abundance in Alberta or across Canada during the period 1985–2005. The Short-eared Owl in Alberta is listed as May Be at Risk because its population is thought to be in decline, although the cause is unknown. COSEWIC last assessed this species in 1994 and assigned it the status of Special Concern because of past declines related to losses of its preferred habitat. Short-eared Owl abundance trends are difficult to determine because these birds are nomadic and also prone to large annual abundance fluctuations related to prey availability.

TEMPORAL REPORTING RATE

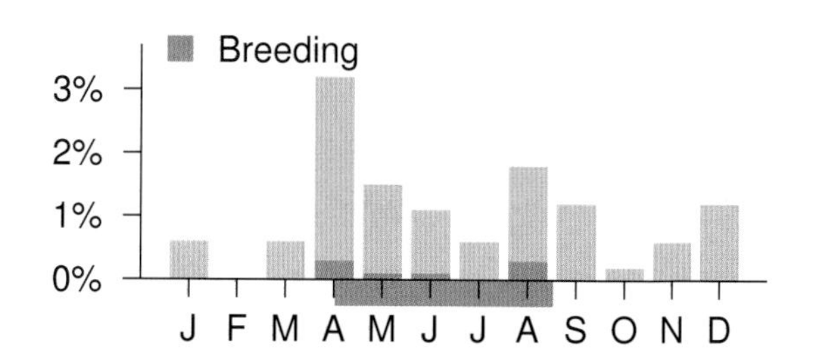

NATURAL REGIONS REPORTING RATE

SPATIAL REPORTING RATE

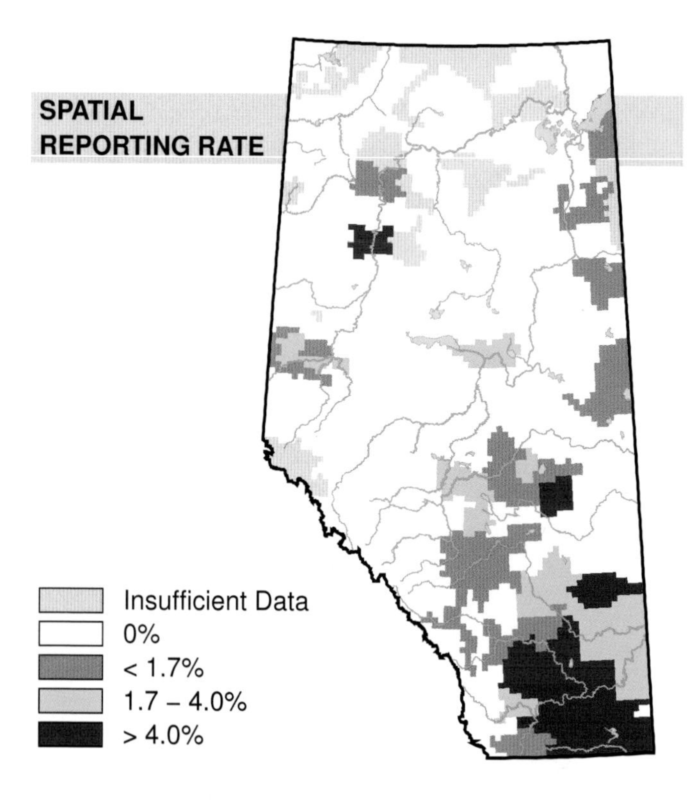

STATUS

GSWA 2000	GSWA 2005	COSEWIC Historic Rankings	COSEWIC 2007	Alberta Wildlife Act
May Be at Risk	May Be at Risk	Special Concern	Special Concern	N/A

RANGE

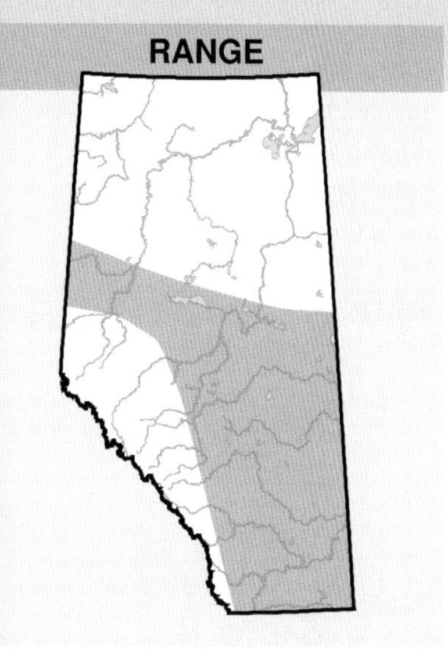

OBSERVED DISTRIBUTION ATLAS 2

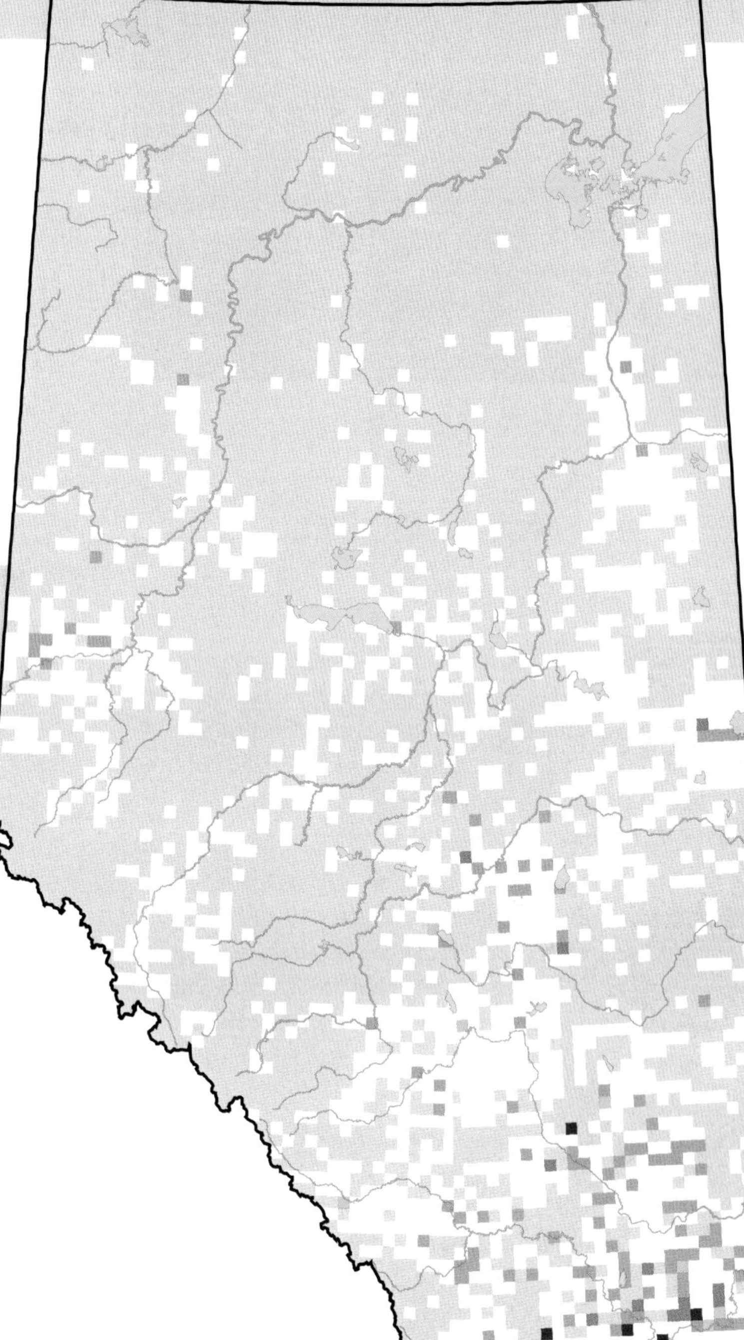

OBSERVED DISTRIBUTION ATLAS 1

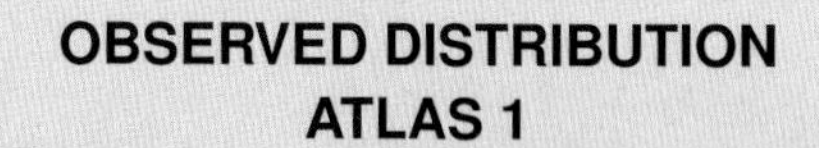

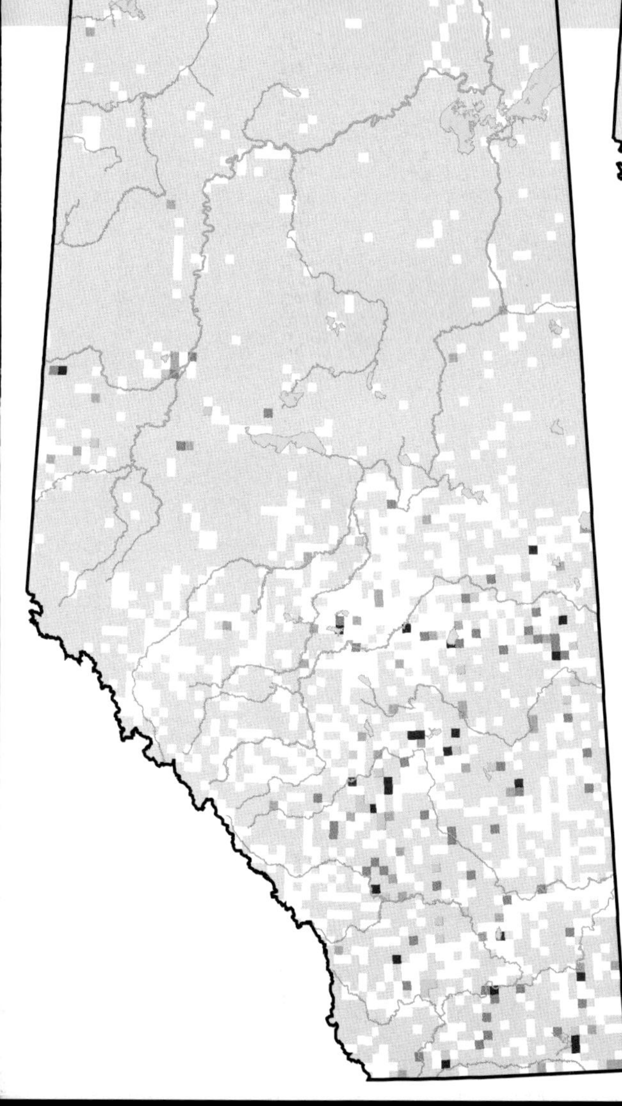

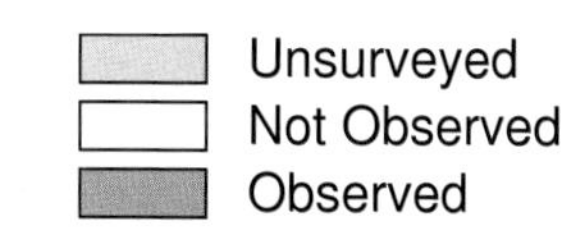

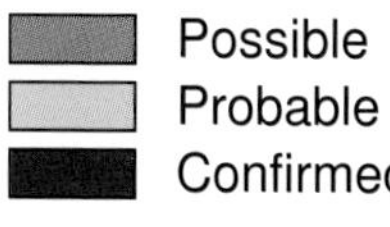

Boreal Owl (*Aegolius funereus*)

Photo: Alan MacKeigan

NESTING

CLUTCH SIZE (EGGS):	4–6
INCUBATION (DAYS):	26–28
FLEDGING (DAYS):	30–35
NEST HEIGHT (METRES):	3–8

As the name suggests, the Boreal Owl is found mainly in the boreal forest in Alberta and around the world. Found mainly in the Boreal Forest Natural Region, this species can also be found in the Rocky Mountain, Foothills, and Parkland NRs. There was no change observed in the distribution of this species from Atlas 1.

Boreal Owl density is not uniform across its range. Highest densities were observed in the northern part of the Rocky Mountain NR and in the northern part of its range. In these areas, Boreal Owls nest in old trees in mature stands. Similar to other small Alberta owls, the Boreal Owl is an obligate cavity nester that relies on the presence of cavities created by large woodpeckers and fungal decay.

Increases in relative abundance were detected in the Boreal Forest NR where, relative to other species, Boreal Owls were observed more frequently in Atlas 2 than in Atlas 1. This increase was unexpected, because owl-specific surveys were appropriately separated in Atlas 2 for reporting rate calculations, but not in Atlas 1, creating the expectation of artifactual declines. Although speculative, it is possible that the increase that was detected in the Boreal Forest NR is related to the increase in relative abundance of the Pileated Woodpecker that was detected by the Atlas in that NR. Pileated Woodpeckers create nest cavities that are ideal for use by Boreal Owls that nest in such secondary cavities. However, it is likely that the detected increase is in part due to more complete northern survey coverage in Atlas 2. The Breeding Bird Survey sample size was too small to examine any trends for this species, due largely to the fact that specialized surveys are required to properly assess this species, and owls in general. No changes in relative abundance were detected in other NRs. The status of the Boreal Owl in Alberta is Secure.

TEMPORAL REPORTING RATE

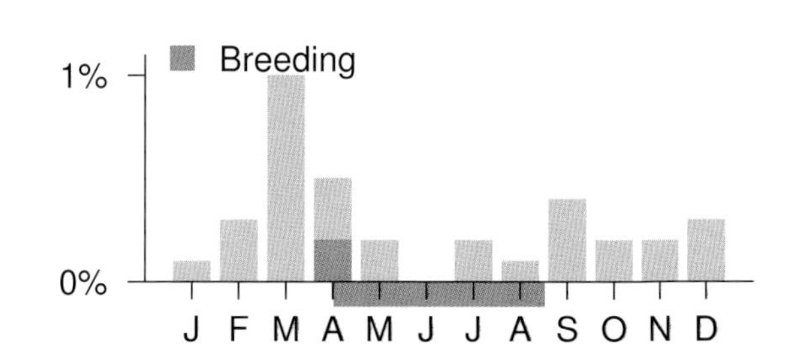

NATURAL REGIONS REPORTING RATE

Boreal Forest	0.2%
Foothills	0.5%
Parkland	0.2%
Rocky Mountain	0.5%

SPATIAL REPORTING RATE

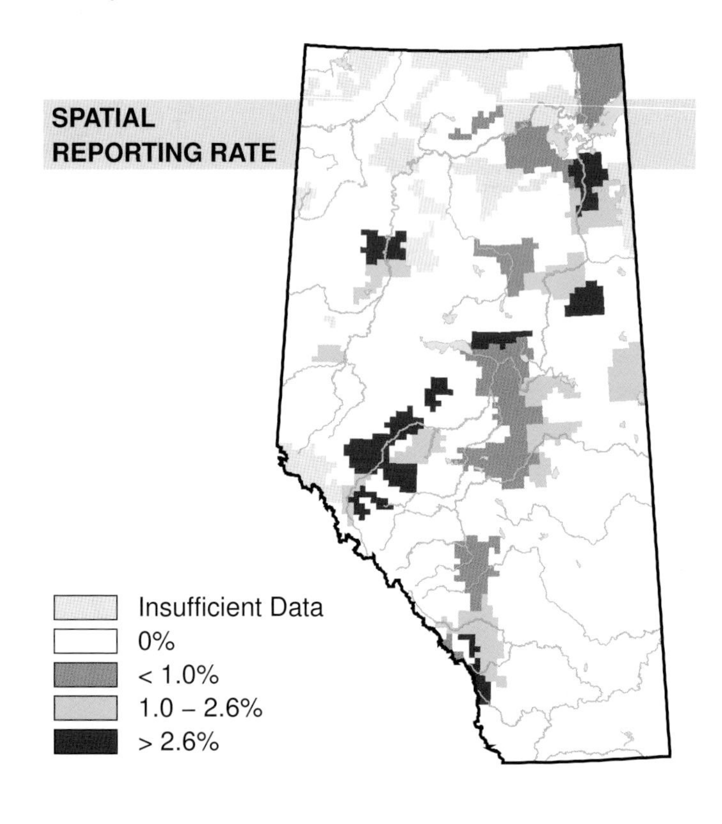

STATUS

GSWA 2000	GSWA 2005	COSEWIC Historic Rankings	COSEWIC 2007	Alberta Wildlife Act
Secure	Secure	Not At Risk	Not At Risk	N/A

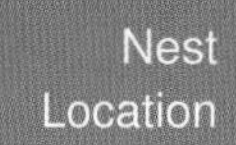

Habitat
Nest Location
Nest Type
Diet

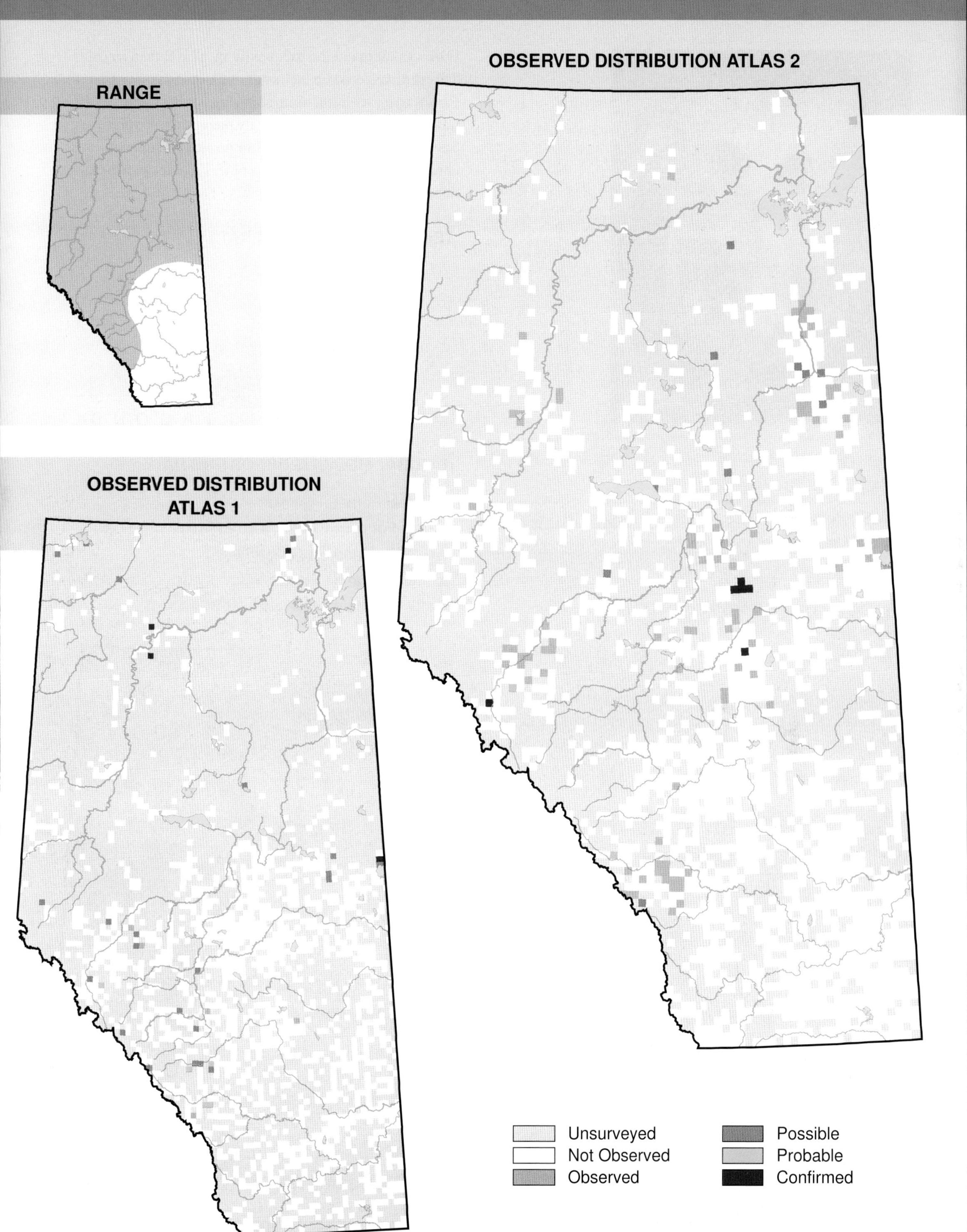

RANGE
OBSERVED DISTRIBUTION ATLAS 2
OBSERVED DISTRIBUTION ATLAS 1
Unsurveyed
Not Observed
Observed
Possible
Probable
Confirmed

Northern Saw-whet Owl (*Aegolius acadicus*)

Photo: Gordon Court

NESTING

CLUTCH SIZE (EGGS):	5–6
INCUBATION (DAYS):	26–28
FLEDGING (DAYS):	34
NEST HEIGHT (METRES):	4.3–18.3

With the exception of the northern part of the Boreal Forest Natural Region, the Northern Saw-whet Owl can be found in almost any forested area in Alberta where this bird can locate the mature trees preferred for nesting. More of these diminutive owls were observed in the northwestern part of their range, especially around Peace River and Grande Prairie, during Atlas 2 than during Atlas 1. This is mostly likely due to the incorporation of data from the Alberta Nocturnal Owl Survey which was initiated after the completion of Atlas 1. Despite this, the observed distribution did not change between Atlas 1 and Atlas 2.

Throughout their range, these owls are evenly distributed in the areas where suitable habitat occurs. The Foothills and Rocky Mountain NRs provide fairly uniform habitat for these small owls, and so the reporting rates from these Regions were similar. However, reporting rates in the Boreal Forest, Parkland, and Grassland NRs were lower because these Regions contain habitat that is less suitable for Northern Saw-whet Owls. In the Grassland NR, they are usually present only along creeks and rivers where there are large trees.

A decrease in relative abundance was detected in the Parkland NR where, relative to other species, the Northern Saw-whet Owl was observed less frequently in Atlas 2 than in Atlas 1. This decline could be the result of on-going conversion of forests to agricultural use or housing developments. The number of Breeding Bird Survey records of Northern Saw-whet Owls was too small to investigate temporal abundance changes. The status of this species in Alberta is Secure.

TEMPORAL REPORTING RATE

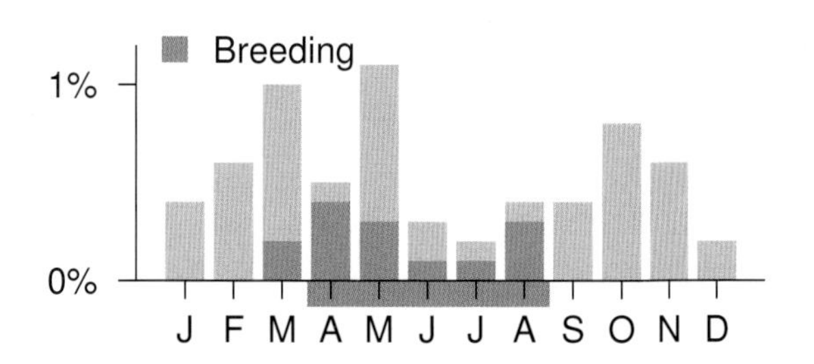

NATURAL REGIONS REPORTING RATE

Region	Rate
Boreal Forest	0.7%
Foothills	2.9%
Grassland	0.5%
Parkland	1.0%
Rocky Mountain	2.1%

SPATIAL REPORTING RATE

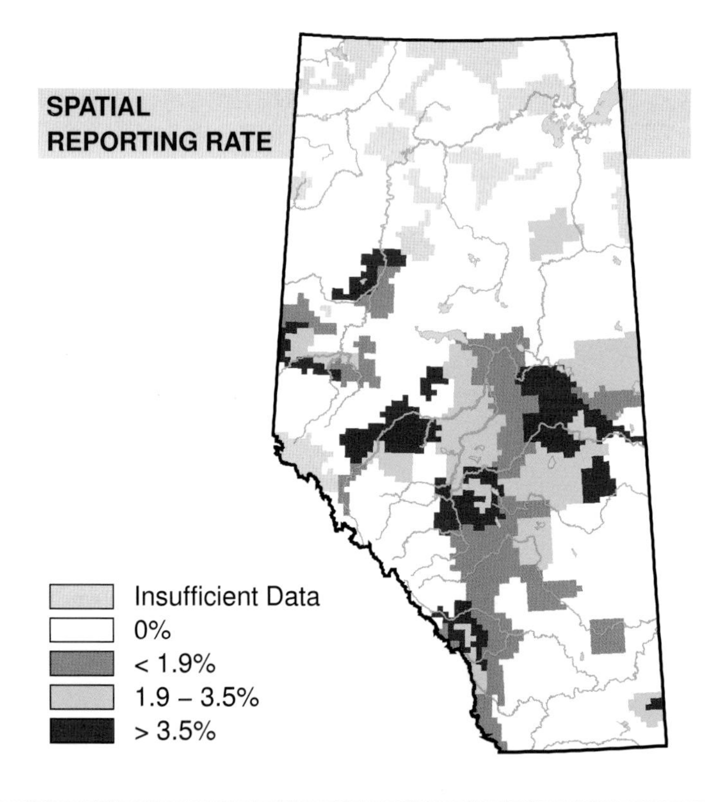

STATUS

GSWA 2000	GSWA 2005	COSEWIC Historic Rankings	COSEWIC 2007	Alberta Wildlife Act
Secure	Secure	N/A	N/A	N/A

OBSERVED DISTRIBUTION ATLAS 2

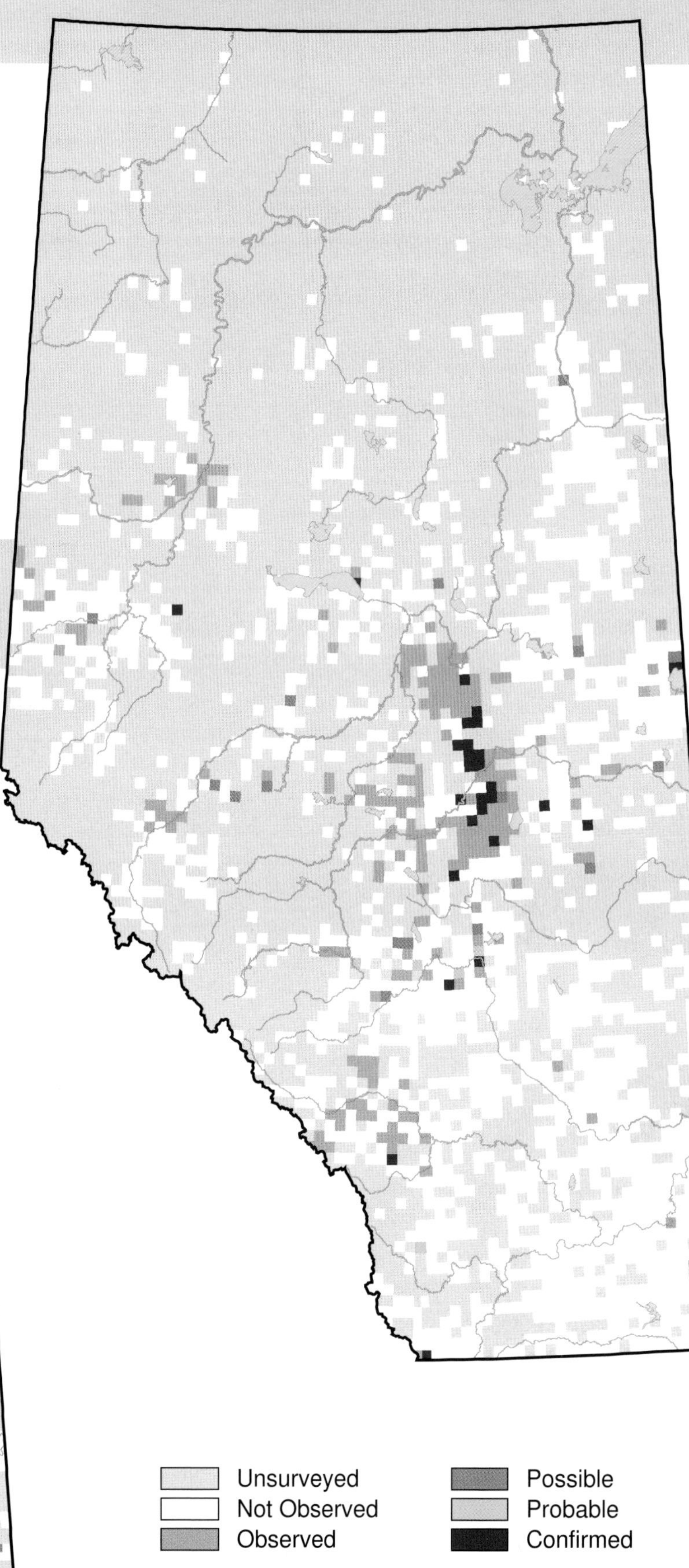

RANGE

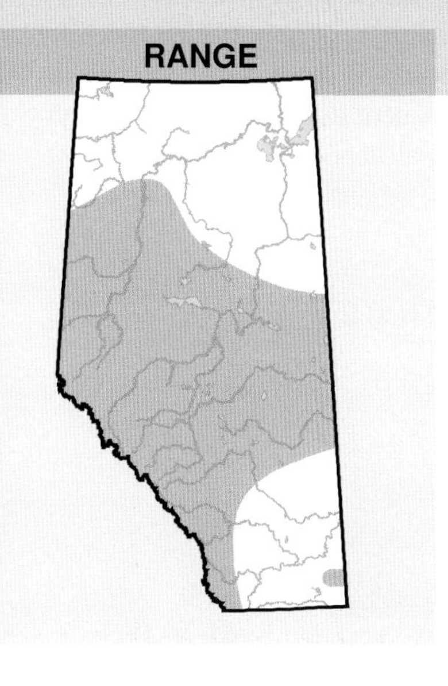

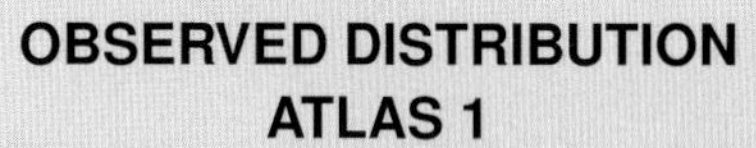

OBSERVED DISTRIBUTION ATLAS 1

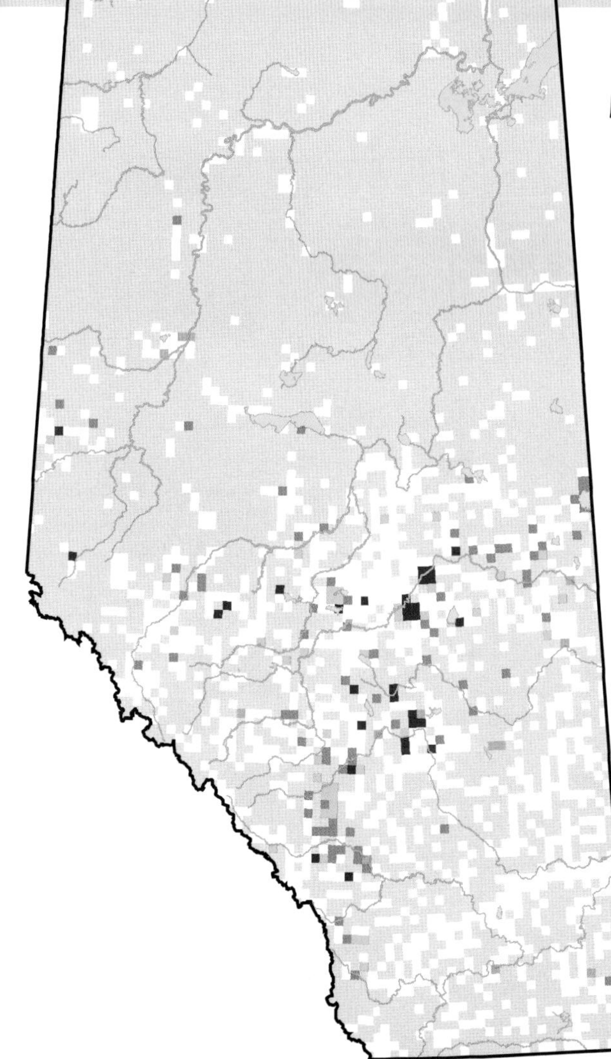

Unsurveyed	Possible
Not Observed	Probable
Observed	Confirmed

Common Nighthawk (*Chordeiles minor*)

Photo: Gerald Romanchuk

NESTING

CLUTCH SIZE (EGGS):	2
INCUBATION (DAYS):	19
FLEDGING (DAYS):	23
NEST HEIGHT (METRES):	0

The Common Nighthawk is found in every Natural Region in Alberta. The distribution of this species did not change from Atlas 1.

This common nightjar uses logged or slashburned areas of forest sites, woodland clearings, prairies and plains, open forests, rock outcrops, and flat gravel rooftops of city buildings during the breeding season. Nests are usually in the open near logs, boulders, grassy clumps, and shrubs. Therefore, this species is found most often in the Grassland NR, less commonly in the Foothills and Rocky Mountain NRs, and least commonly in the Boreal Forest and Parkland NRs where treed habitats are more closed. Open habitats are also needed for foraging in flight during dawn and dusk.

Declines in relative abundance were detected in the Boreal Forest, Foothills, Grassland, and Parkland NRs and, relative to other species, the Common Nighthawk was observed less frequently in Atlas 2 in these NRs and in the Rocky Mountain NR. Declines in abundance were also found across Canada by the Breeding Bird Survey from 1985–2005.

As the Common Nighthawk is subject to strong fluctuation in local abundance, and the BBS has a low detection rate, the true population status is poorly known (Poulin et al., 1996). Dunn (2005) indicates that available evidence shows significant, persistent, range-wide decline. Little is known of the population status and ecology in the northern part of its range, so further investigation is required, to track status and developing population targets for use in conservation action decisions. This species is considered Sensitive in the province due to declines across North America since 1966. Speculation as to the causes of these declines implicates increased predation, habitat loss, and use of pesticides on their food source.

TEMPORAL REPORTING RATE

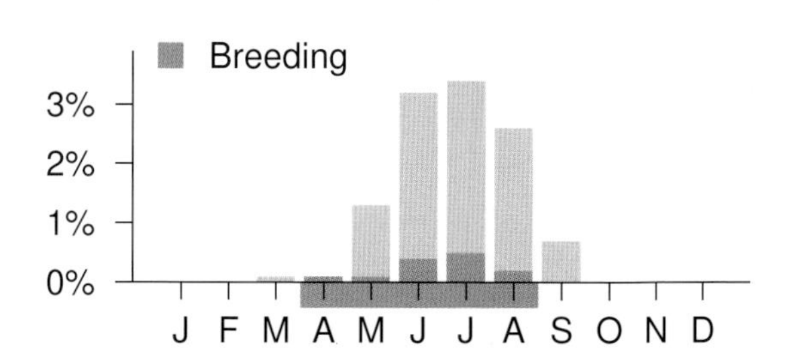

NATURAL REGIONS REPORTING RATE

Boreal Forest	0.8%
Foothills	1.8%
Grassland	2.8%
Parkland	0.7%
Rocky Mountain	1.1%

SPATIAL REPORTING RATE

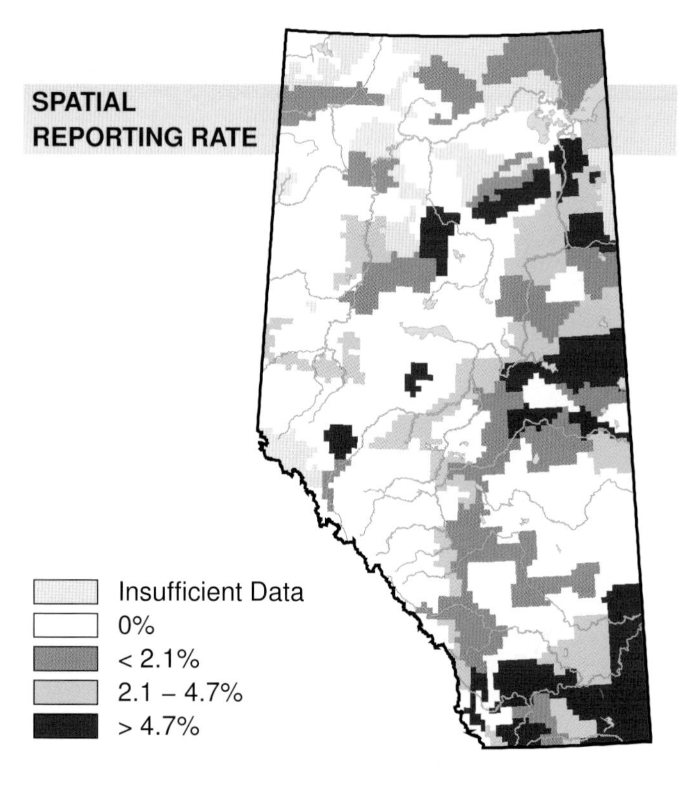

STATUS

GSWA 2000	GSWA 2005	COSEWIC Historic Rankings	COSEWIC 2007	Alberta Wildlife Act
Sensitive	Sensitive	N/A	Threatened	N/A

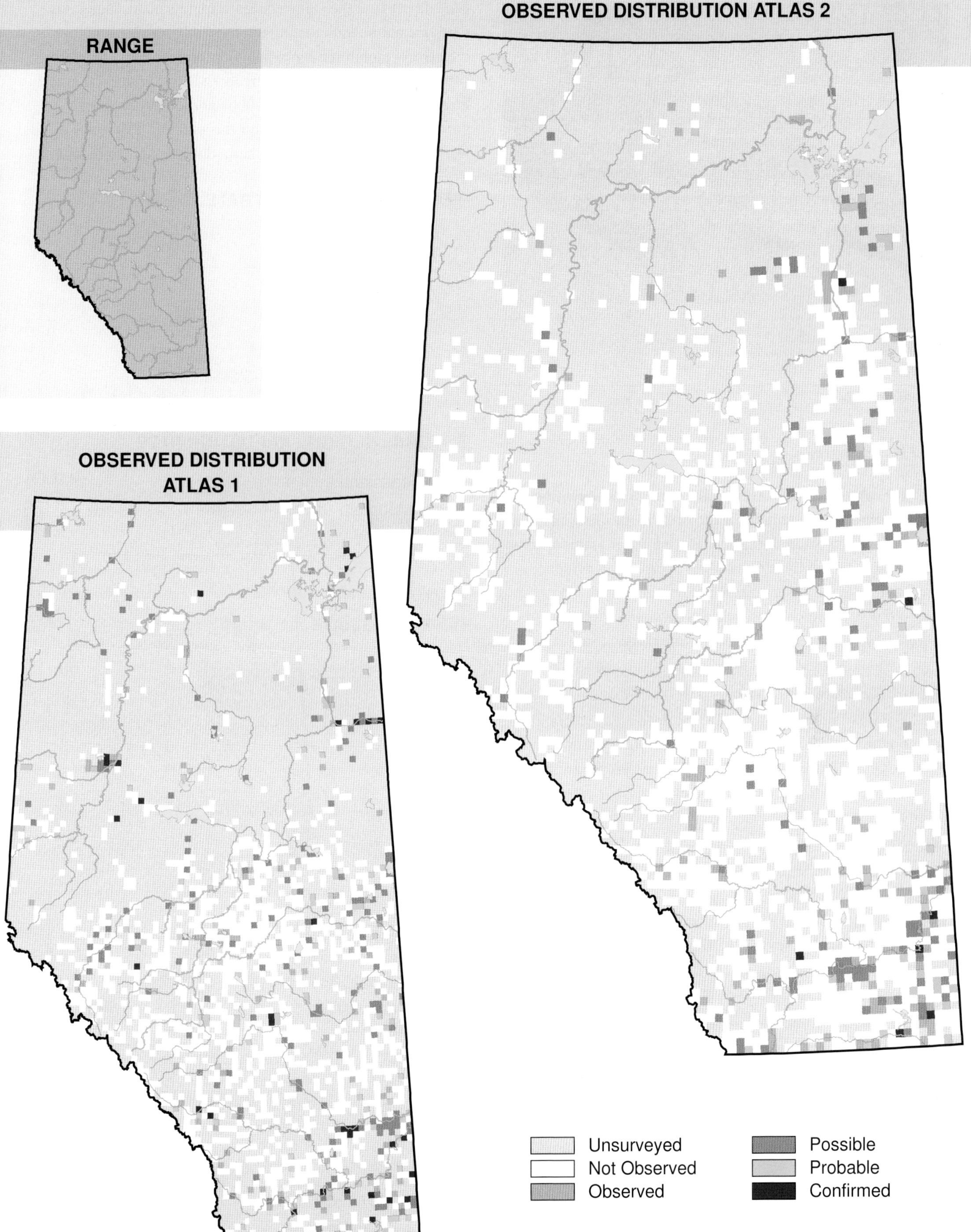
OBSERVED DISTRIBUTION ATLAS 2
RANGE
OBSERVED DISTRIBUTION
ATLAS 1
Unsurveyed
Not Observed
Observed
Possible
Probable
Confirmed

Common Poorwill (*Phalaenoptilus nuttallii*)

Photo: Royal Alberta Museum

NESTING

CLUTCH SIZE (EGGS):	2
INCUBATION (DAYS):	unknown
FLEDGING (DAYS):	unknown
NEST HEIGHT (METRES):	0

One of the rarest birds in Alberta, the Common Poorwill is found in the southern part of the province in the Grassland and Rocky Mountain Natural Regions. During Atlas 1, this species was found infrequently and there was no breeding evidence associated with the observations that were reported. The observed distribution of this species was similar in Atlases 1 and 2. However, more observations were made during Atlas 2, among which were reported possible, probable, and confirmed breeding evidence.

The Common Poorwill is found in dry, open areas where it nests on the ground, usually at the base of a tree or shrub, and it forages at night for invertebrates. This species benefits from cattle grazing and logging because these activities create additional nesting habitat when vegetation is cleared or kept low. Due to its preference for open areas, this species was found in the Grassland NR and in the Rocky Mountain NR around Cypress Hills.

There was no change in relative abundance detected in any of the NRs. However, being so rare, this species was below the statistical threshold of our data. Relative to other species, the Common Poorwill was observed more often in the Grassland NR in Atlas 2 than in Atlas 1. The Breeding Bird Survey was not able to investigate abundance change in Alberta or across Canada because the sample sizes were too small. Due to its secretive nature and nocturnal habits, this species is not effectively monitored by large, wide-spread, multi-species surveys such as the Atlas or Breeding Bird Survey. Common Poorwill-specific surveys would be needed to determine population trends adequately for this species. The status of this species is Undetermined in Alberta due to insufficient data.

TEMPORAL REPORTING RATE

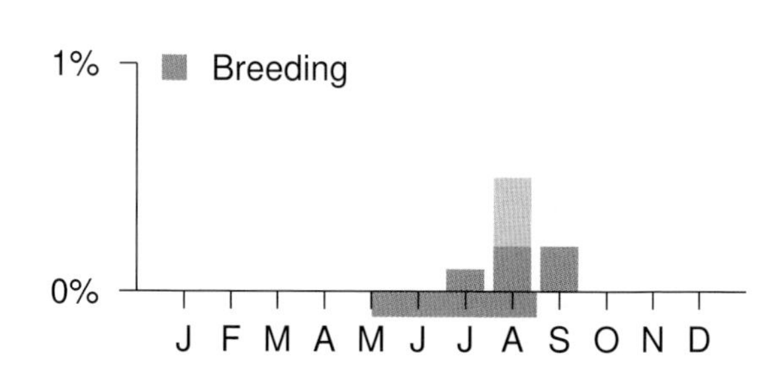

NATURAL REGIONS REPORTING RATE

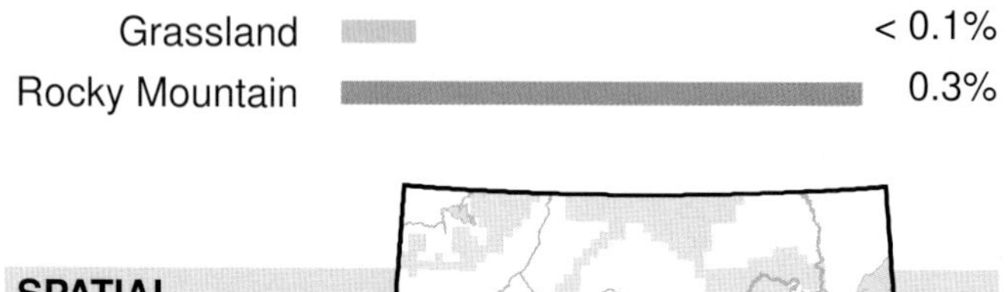

SPATIAL REPORTING RATE

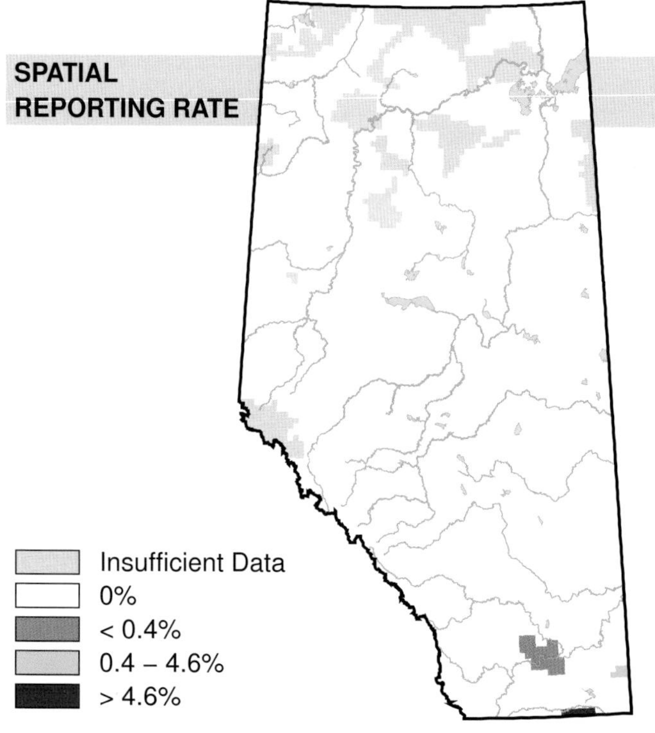

STATUS

GSWA 2000	GSWA 2005	COSEWIC Historic Rankings	COSEWIC 2007	Alberta Wildlife Act
Undetermined	Undetermined	Data Deficient	Data Deficient	N/A

RANGE

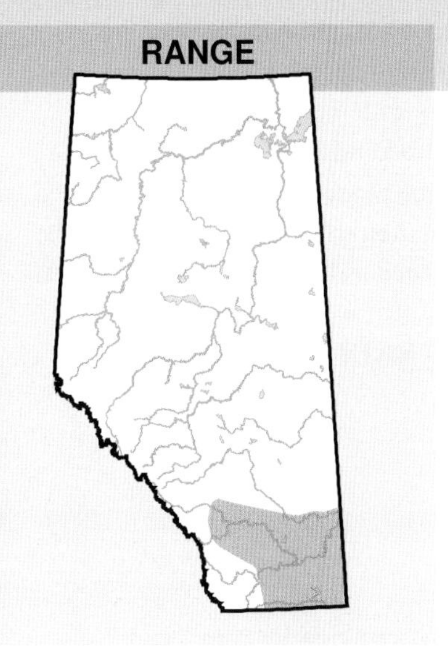

OBSERVED DISTRIBUTION ATLAS 2

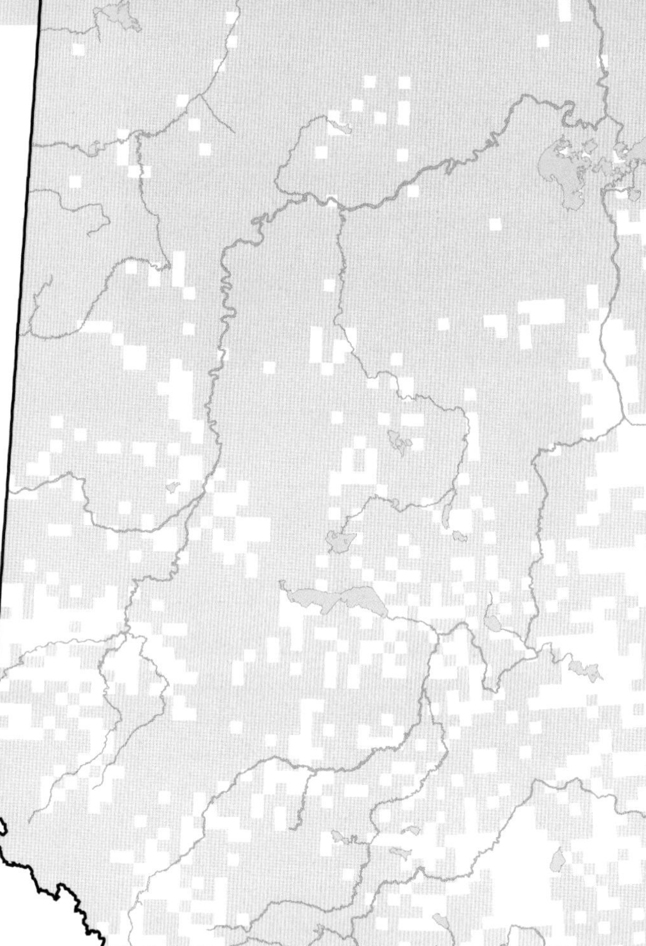

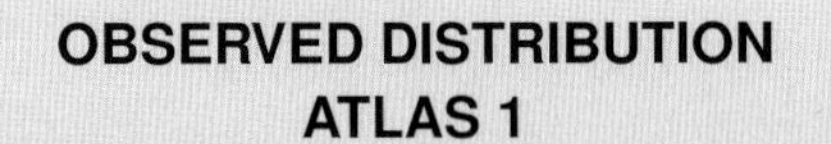

OBSERVED DISTRIBUTION ATLAS 1

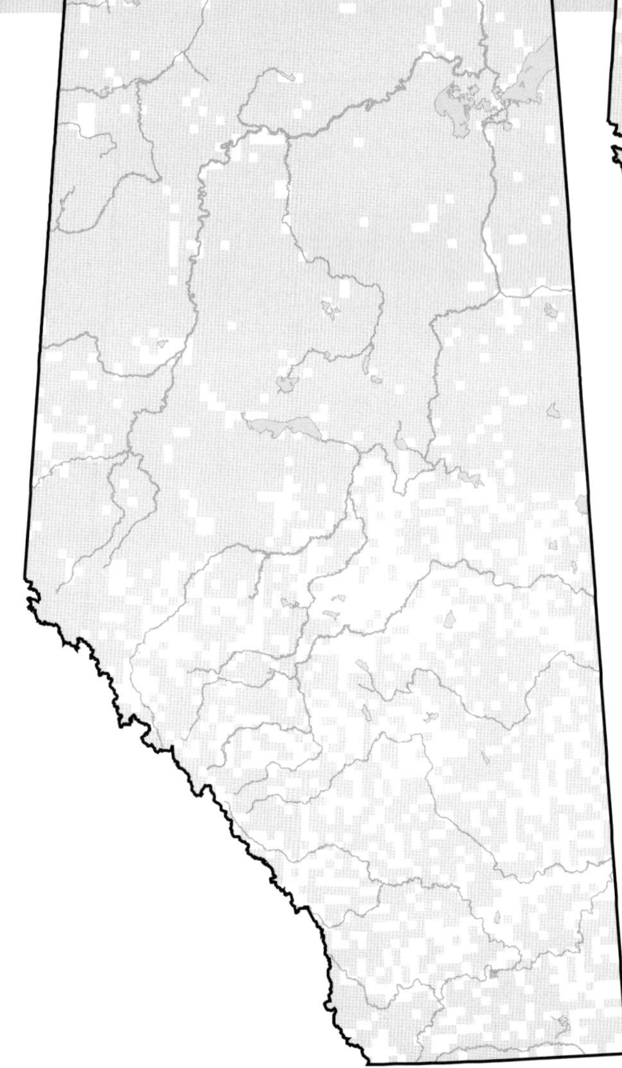

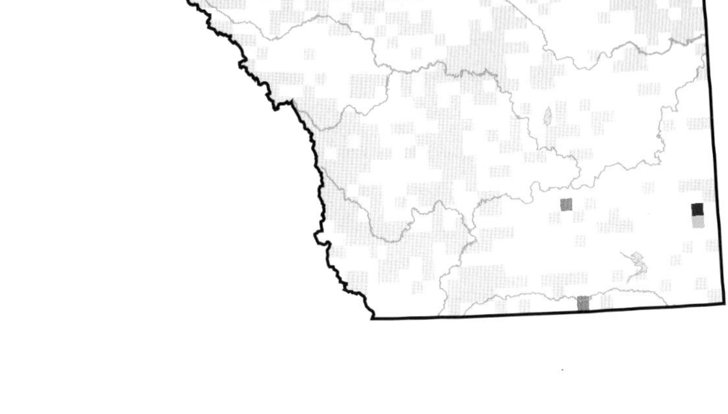

Black Swift (*Cypseloides niger*)

Photo: G Holroyd

NESTING

CLUTCH SIZE (EGGS):	1
INCUBATION (DAYS):	24–27
FLEDGING (DAYS):	31
NEST HEIGHT (METRES):	unknown

The largest swift breeding in North America, the Black Swift is found in the Rocky Mountain Natural Region in Alberta. The observed distribution of this species changed slightly between Atlas 1 and Atlas 2. No observations were recorded in Atlas 2 around Saskatchewan River Crossing where they had been sighted previously in Atlas 1. However, this species is rare in the province, and the remoteness of their breeding sites makes them largely inaccessible. Therefore, the presence of Black Swifts could have been missed there during Atlas 2.

Found in montane habitats, this species nests on ledges or in shallow caves on steep rock faces and canyons, usually near waterfalls. Due to their habit of flying high and nesting in remote areas, this species has not been extensively studied or monitored. As a result, little information exists on their breeding and behaviour. This species was most commonly encountered in the Rocky Mountain NR. The single record from near Hanna, in the Grassland NR, is not considered to have been a breeder.

No change in relative abundance was detected in the Rocky Mountain NR and, relative to other species, observation frequencies in both Atlas 1 and Atlas 2 were similar. The Breeding Bird Survey found no abundance change in Canada, and it was not able to assess population trends for this species in Alberta because the sample size was too small. Due to the very specific habitat requirements of this species' and the remote distribution of its preferred habitat, the bird is not effectively monitored by large, wide-ranging, multi-species surveys such as the Atlas or Breeding Bird Survey. Surveys designed specifically to monitor Black Swifts would be needed to adequately determine population trends for this species. The status of this species is designated as Undetermined in Alberta.

TEMPORAL REPORTING RATE

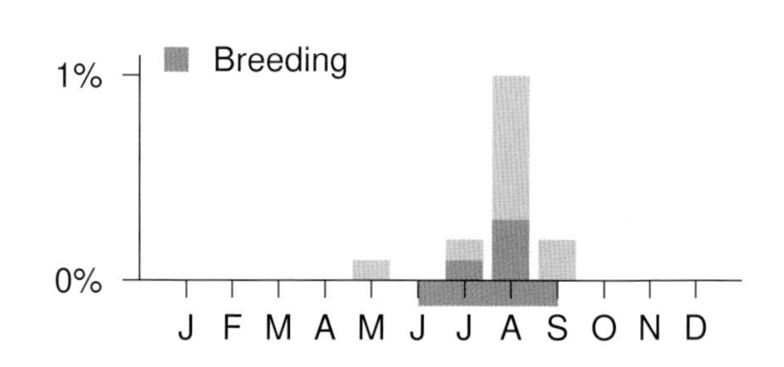

NATURAL REGIONS REPORTING RATE

SPATIAL REPORTING RATE

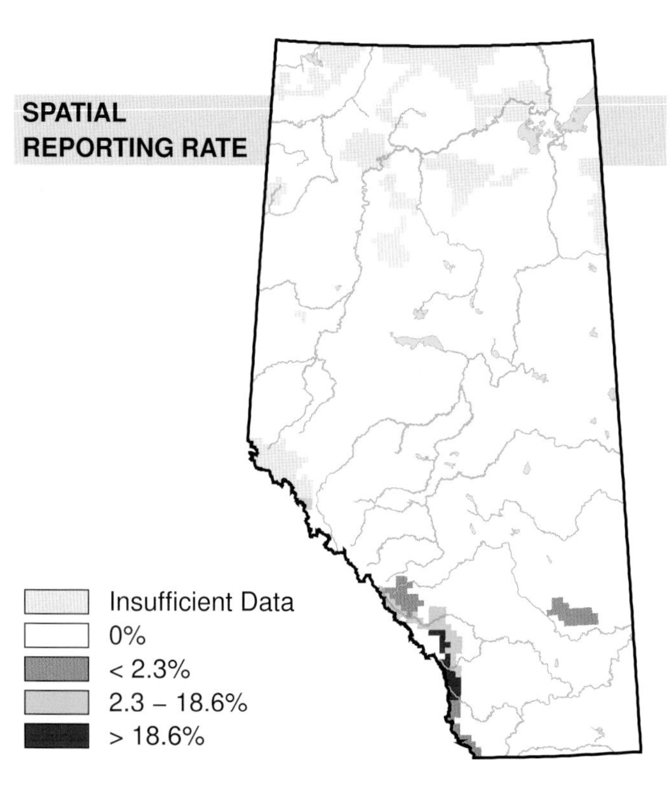

STATUS

GSWA 2000	GSWA 2005	COSEWIC Historic Rankings	COSEWIC 2007	Alberta Wildlife Act
Undetermined	Undetermined	N/A	N/A	N/A

Habitat Nest Location Nest Type Diet

RANGE

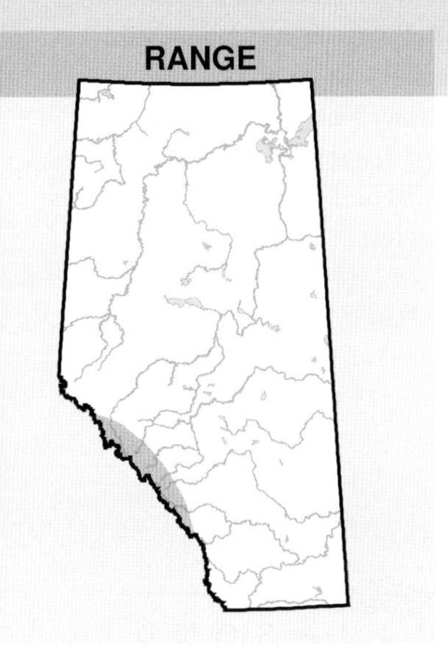

OBSERVED DISTRIBUTION ATLAS 2

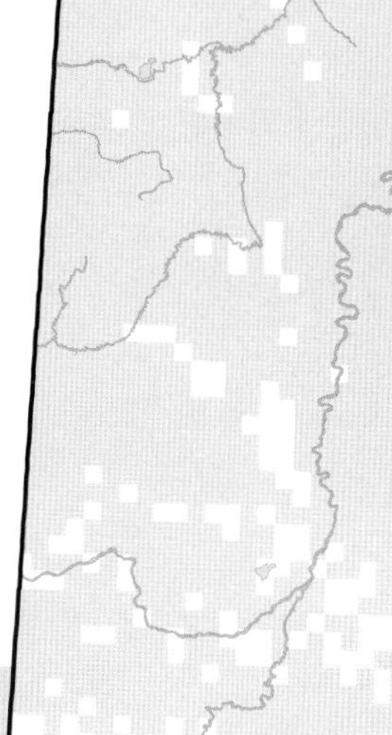

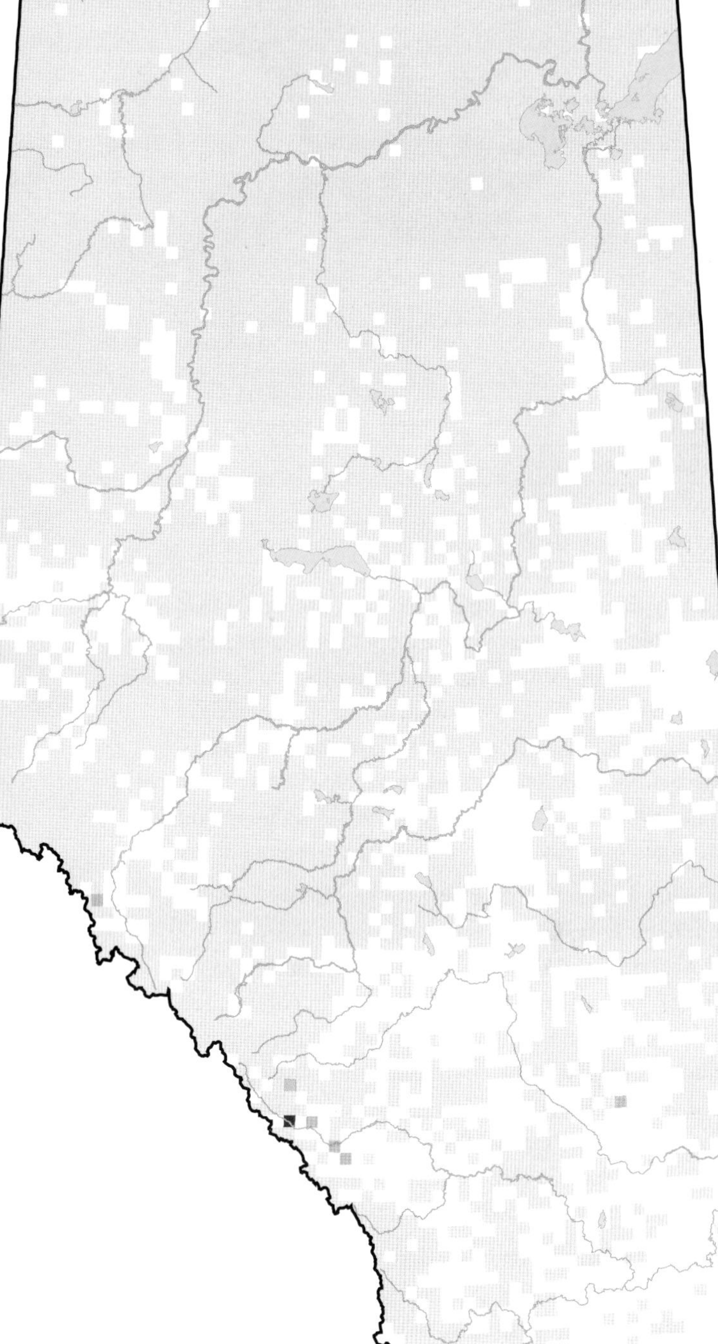

OBSERVED DISTRIBUTION ATLAS 1

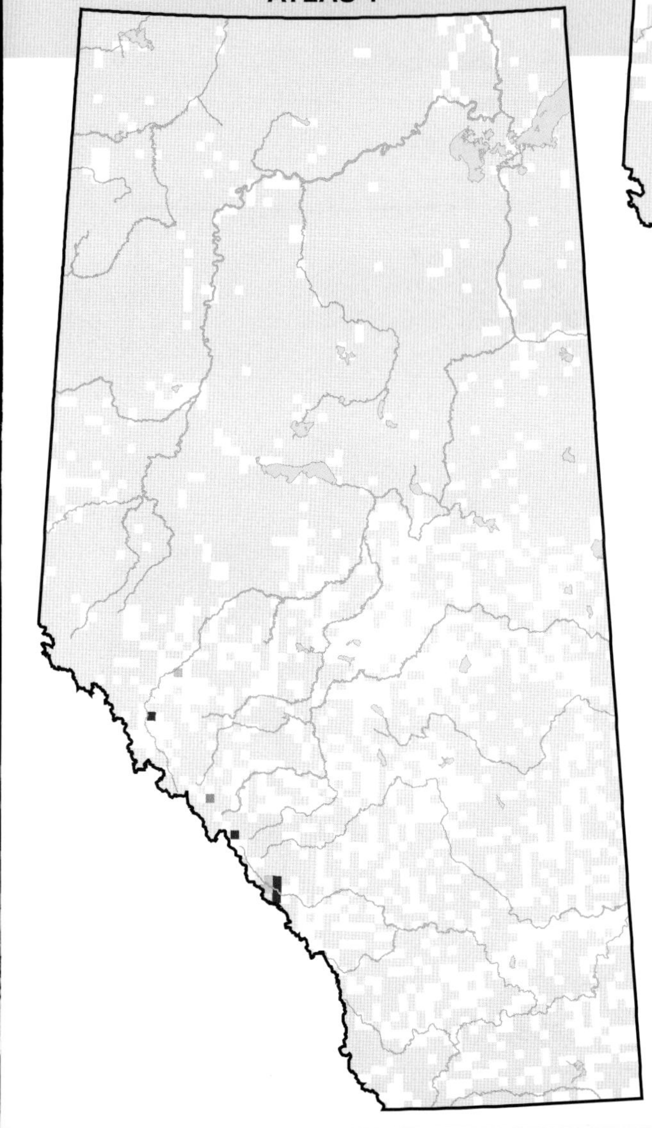

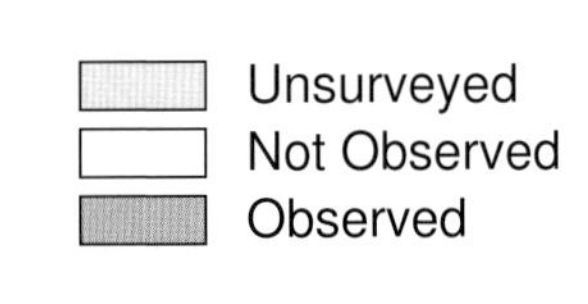

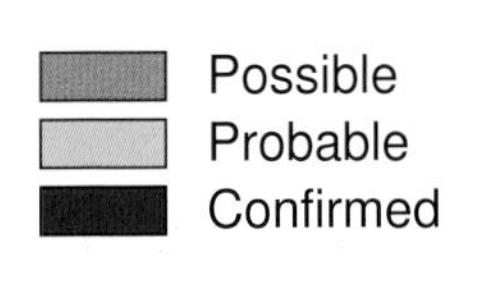

Ruby-throated Hummingbird (*Archilochus colubris*)

Photo: Gerald Romanchuk

NESTING

CLUTCH SIZE (EGGS):	2
INCUBATION (DAYS):	16
FLEDGING (DAYS):	19
NEST HEIGHT (METRES):	3–7

The Ruby-throated Hummingbird is found breeding in every Natural Region in the province except the Canadian Shield. This bird does not range much farther north than 56°latitude in Alberta. The observed distribution of this species did not change between Atlas 1 and Atlas 2. Records from the Grassland NR north and west of Medicine Hat have changed from possible breeding to only observed. This was also the case in the Peace River and Grande Prairie regions.

Having the largest breeding range of any North American hummingbird, the Ruby-throated species is found in mixedwood forests and is associated with forest edges and clearings, gardens, and orchards. Nesting in poplar, pine, birch, or oak, this hummingbird also needs to be close to floral nectar, particularly nectar produced by tubular red flowers. Small insects and sap from wells created by sapsuckers can also be consumed. For this reason, this species was observed most often in the Boreal Forest, Foothills, and Parkland NRs. It was observed less often in the Rocky Mountain and Grassland NRs.

Declines in relative abundance were detected in the Boreal Forest and Parkland NRs. No change was detected in the other NRs. Relative to other species, this hummingbird was observed less frequently in the Boreal Forest and Parkland NRs. The Breeding Bird Survey did not detect any changes in its abundance in Canada during the period 1985–2005. Breeding Bird Survey trends could not be determined for this species in Alberta because sample sizes were too small.

Plantings of flowers in domestic gardens has increased breeding habitat throughout Ruby-throated Hummingbird range; so it is unclear why declines are being observed in the Atlas. However, little research has been done on potential causes of population change for this species. More research is needed to assess the status of this bird in Alberta. This species is considered Secure in the province.

TEMPORAL REPORTING RATE

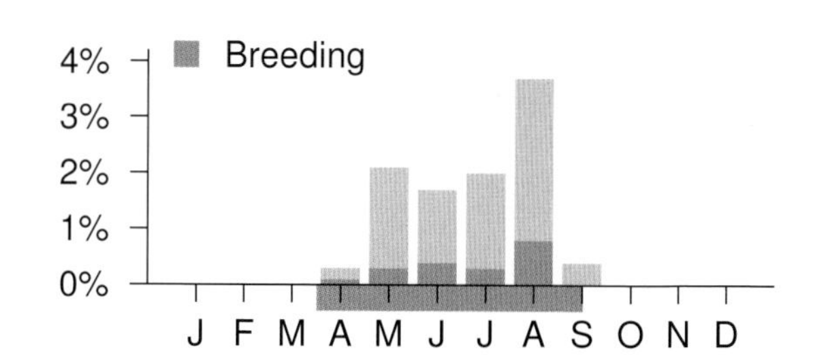

NATURAL REGIONS REPORTING RATE

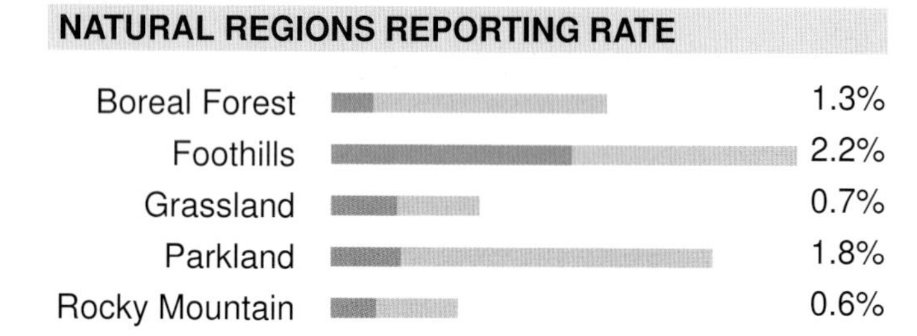

SPATIAL REPORTING RATE

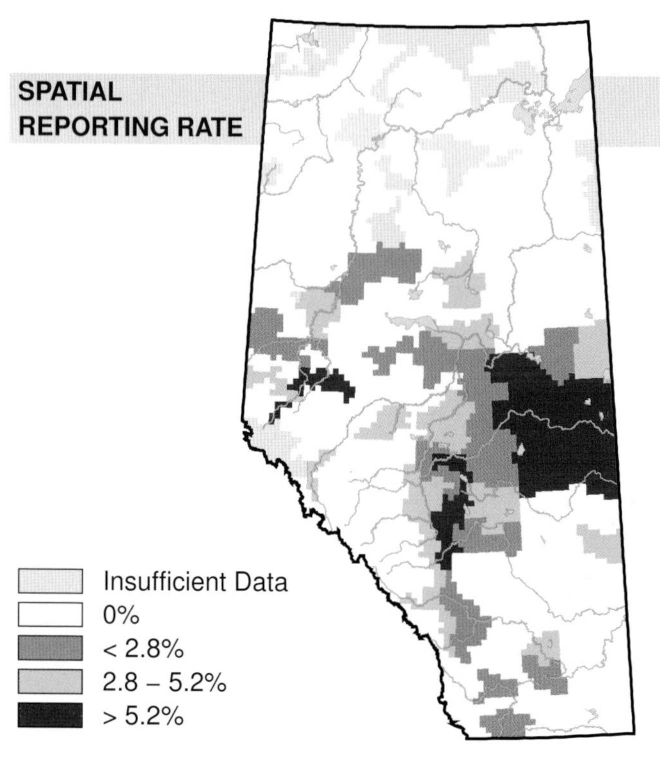

STATUS

GSWA 2000	GSWA 2005	COSEWIC Historic Rankings	COSEWIC 2007	Alberta Wildlife Act
Secure	Secure	N/A	N/A	N/A

Habitat

Nest
Location

Nest
Type

Diet

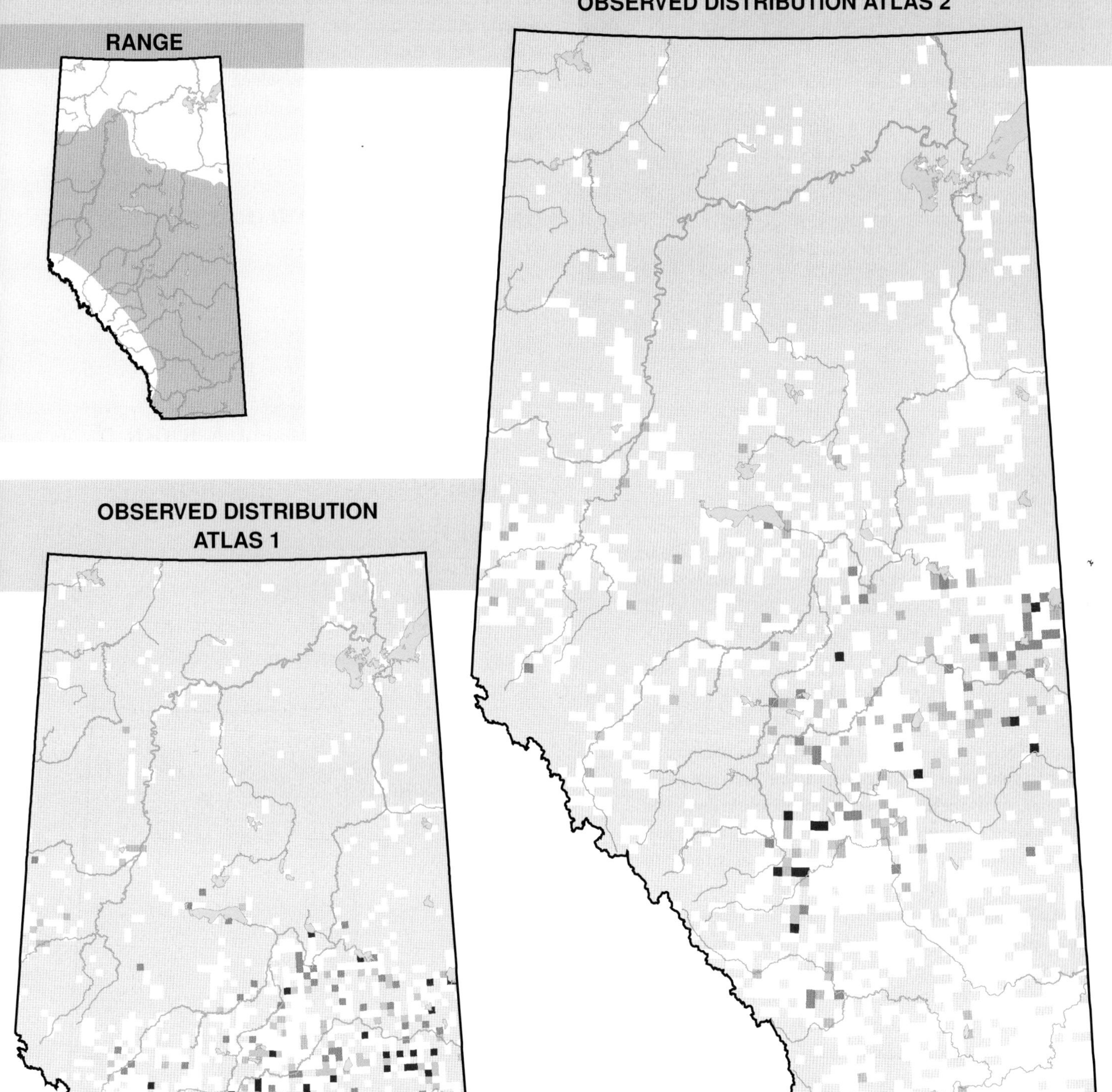
OBSERVED DISTRIBUTION ATLAS 2
RANGE
OBSERVED DISTRIBUTION
ATLAS 1

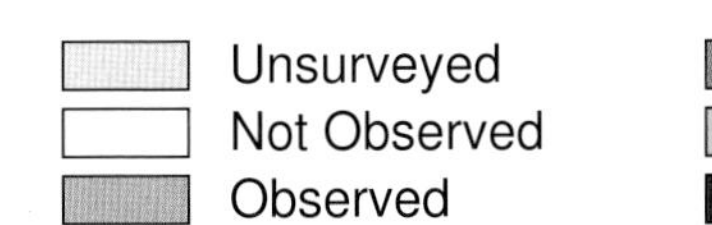
Unsurveyed
Not Observed
Observed

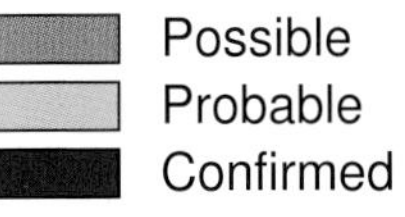
Possible
Probable
Confirmed

Calliope Hummingbird (*Stellula calliope*)

Photo: Gerald Romanchuk

NESTING

CLUTCH SIZE (EGGS):	2
INCUBATION (DAYS):	15–16
FLEDGING (DAYS):	18–22
NEST HEIGHT (METRES):	0.5–21.3

The smallest bird to breed north of Mexico, the Calliope Hummingbird is found in the Rocky Mountain, Foothills, Grassland, and Parkland Natural Regions. No change in distribution was observed between Atlas 1 and Atlas 2.

This northwestern montane species nests in Lodgepole and Ponderosa pines, Douglas Fir, Engelmann Spruce, and alder in Alberta, and is found usually above 1200-metres elevation. It tends to be associated with forest edges and openings. The nest is often constructed at the base of an old pinecone, thus to camouflage the nest and make it appear like a cone, and difficult to find. They were most often encountered in the Rocky Mountain NR and were less commonly found in the Foothills NR. A few records were obtained from the Peace River Parkland where this species' range extends eastward from British Columbia. Breeding records from the Boreal Forest and Grassland NRs were obtained along the periphery of these NRs. The increased use of hummingbird feeders might have helped to expand the non-breeding range of this species and/or the increased use of feeders could have helped increase detection rates in new areas.

No change in relative abundance was found in any of the NRs but, relative to other species, the Calliope Hummingbird was observed more frequently in the Rocky Mountain NR. No trend data are available for Alberta or Canada due to low Breeding Bird Survey detection rates. Calliope Hummingbirds have never been found on Breeding Bird Survey routes in Alberta and they have been found on only 37 routes from 1966–2005 in British Columbia. Since this species responds positively to short-term logging which opens up the canopy, it is doubtful that this species is declining. Increased use of hummingbird feeders may increase observations of this species. This species is considered Secure in the province.

TEMPORAL REPORTING RATE

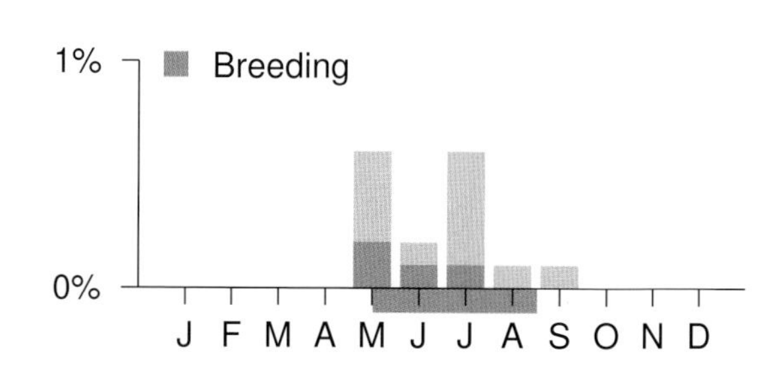

NATURAL REGIONS REPORTING RATE

Natural Region	Rate
Boreal Forest	< 0.1%
Foothills	0.2%
Grassland	0.8%
Parkland	0.3%
Rocky Mountain	4.0%

SPATIAL REPORTING RATE

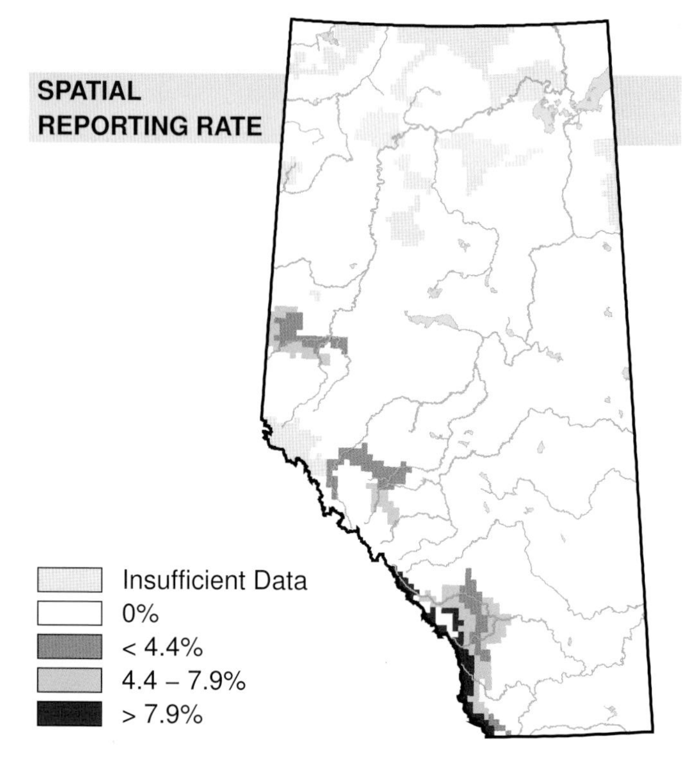

STATUS

GSWA 2000	GSWA 2005	COSEWIC Historic Rankings	COSEWIC 2007	Alberta Wildlife Act
Secure	Secure	N/A	N/A	N/A

OBSERVED DISTRIBUTION ATLAS 2

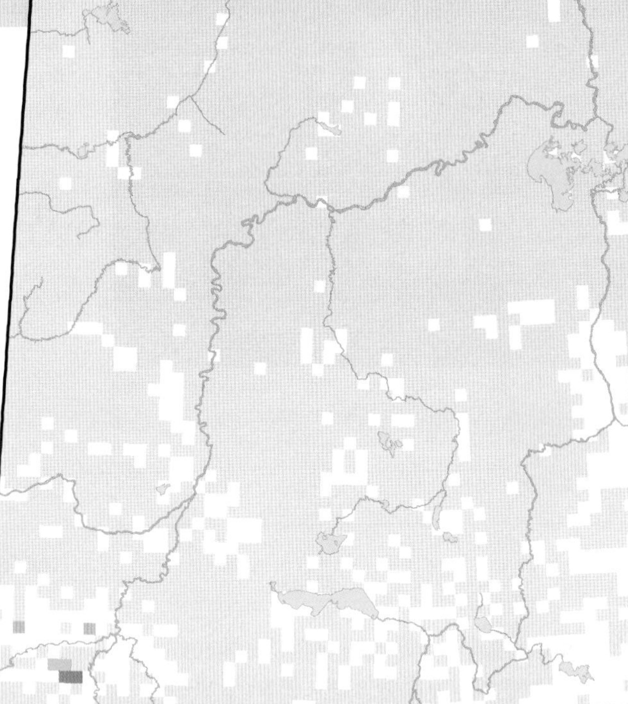

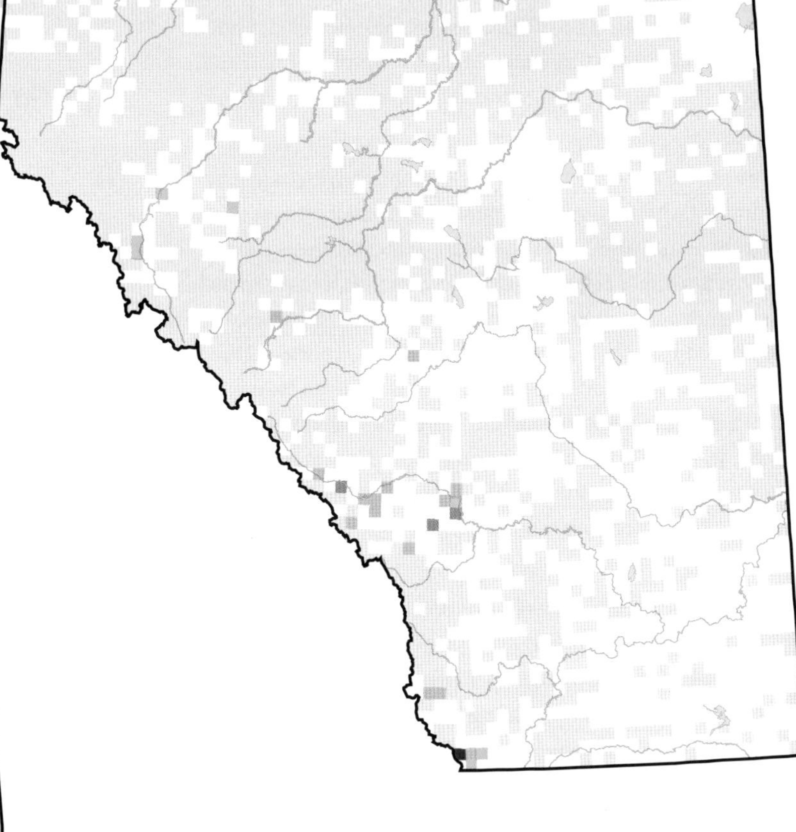

RANGE

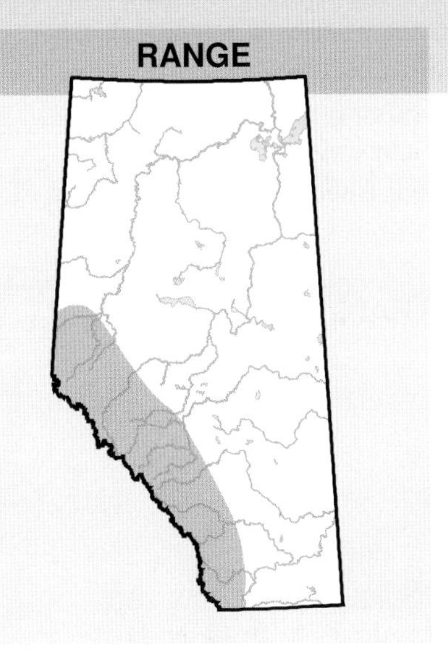

OBSERVED DISTRIBUTION ATLAS 1

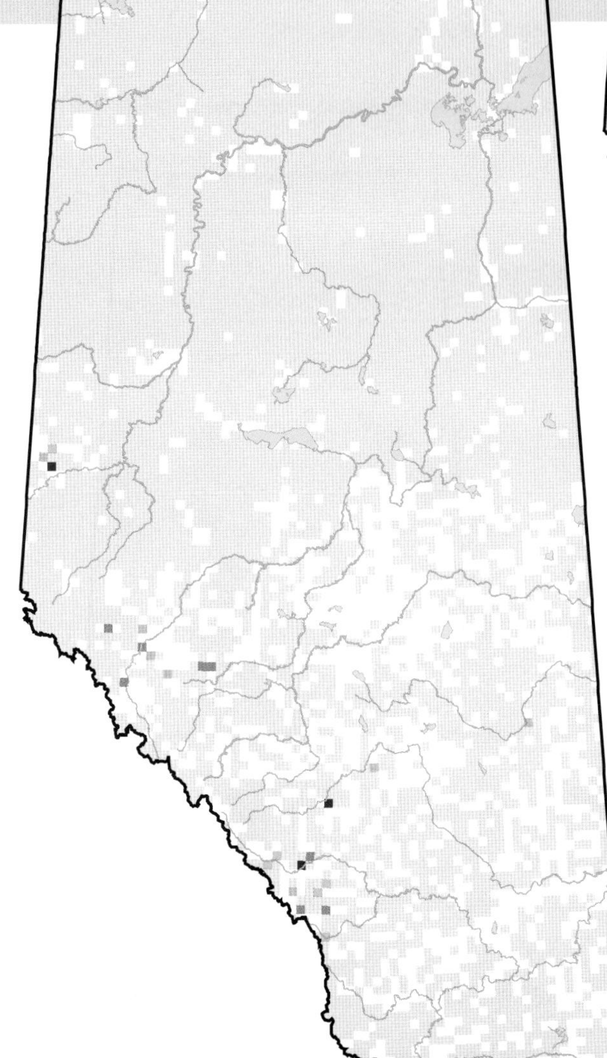

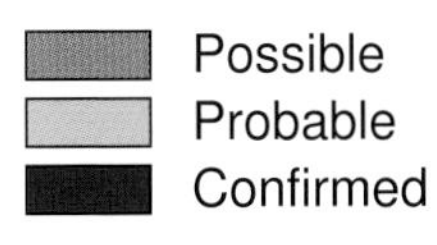

Rufous Hummingbird (*Selasphorus rufus*)

Photo: Gerald Romanchuk

NESTING

CLUTCH SIZE (EGGS):	2
INCUBATION (DAYS):	12
FLEDGING (DAYS):	21
NEST HEIGHT (METRES):	0.3–15

The Rufous Hummingbird, which is aggressive by nature, is found in the Rocky Mountain, Foothills, and Boreal Forest Natural Regions. The distribution of this species did not change between Atlas 1 and Atlas 2. One probable breeding record northeast of Edmonton is not considered a range extension.

This species of hummingbird is found in a wide range of habitats. Although mostly found in secondary-succession forests, it can also breed in mature and old forests greater than 120 years old. This montane species builds its nest in a shrub or on a lower branch of an oak or pine tree. Colonies of nests separated by only a few metres have been reported. Due to their preference for mountainous areas, Rufous Hummingbirds were most commonly found in the Rocky Mountain NR, and were less common in the Foothills NR. They were rare in the Parkland and Boreal Forest NRs. Breeding in the Grassland NR was reported only on the periphery of the region.

An increase in relative abundance was detected in the Parkland NR. However, relative to other species, the Rufous Hummingbird was found less frequently in this NR during Atlas 2 than during Atlas 1. Change in this NR should be interpreted with caution, though, because this species does not regularly breed there and most of the records from that region are not associated with breeding evidence. No change was detected in any of the other NRs. The Breeding Bird Survey did not detect a change in abundance on a Canada-wide scale during the period 1985–2005. There were not enough Breeding Bird Survey records to determine whether or not the abundance of this species had changed in Alberta. The use of feeders is thought to help supplement food early in the season when flowers are not blooming and may elevate local populations above natural levels. The Rufous Hummingbird is considered Secure in Alberta.

TEMPORAL REPORTING RATE

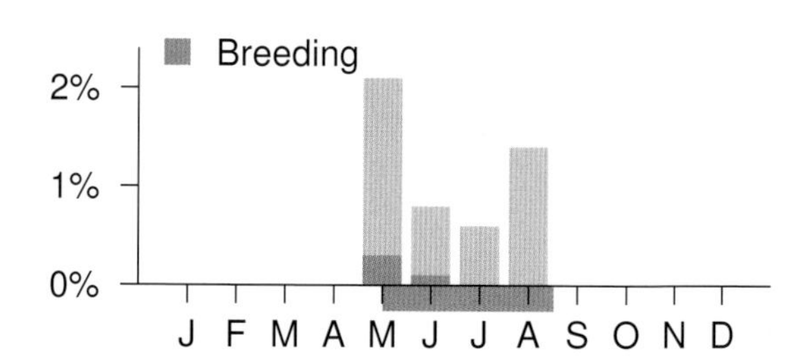

NATURAL REGIONS REPORTING RATE

Region	Rate
Boreal Forest	0.1%
Foothills	0.8%
Grassland	1.2%
Parkland	1.1%
Rocky Mountain	7.8%

SPATIAL REPORTING RATE

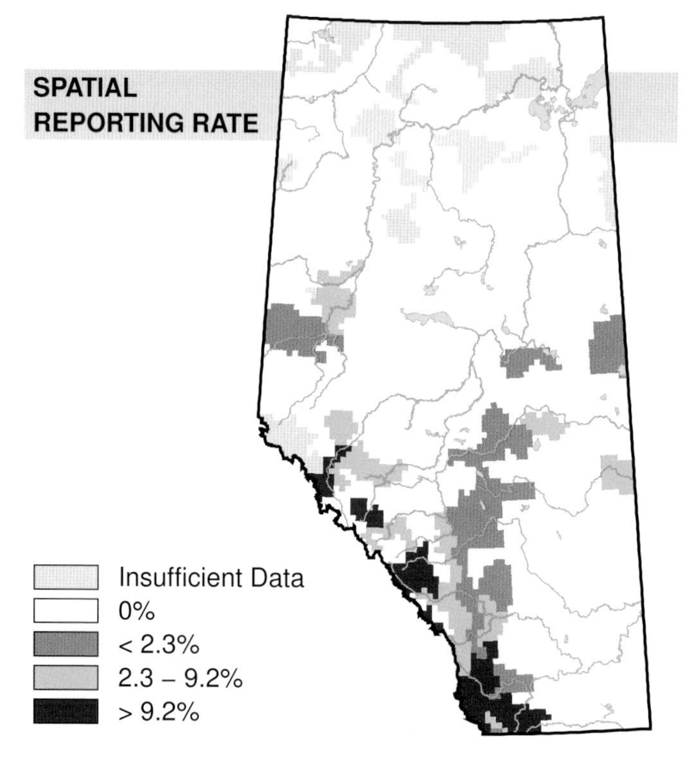

STATUS

GSWA 2000	GSWA 2005	COSEWIC Historic Rankings	COSEWIC 2007	Alberta Wildlife Act
Secure	Secure	N/A	N/A	N/A

RANGE

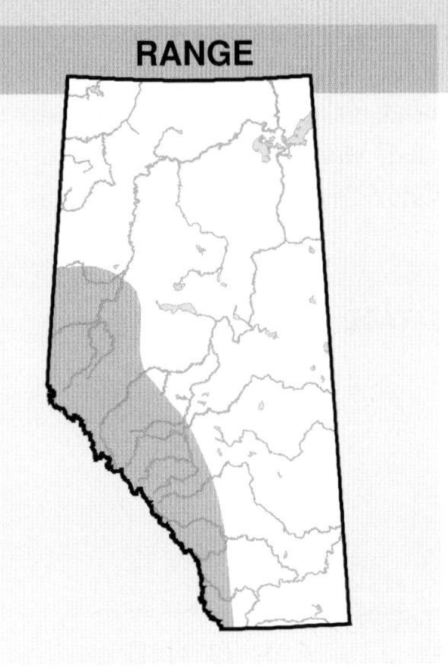

OBSERVED DISTRIBUTION ATLAS 2

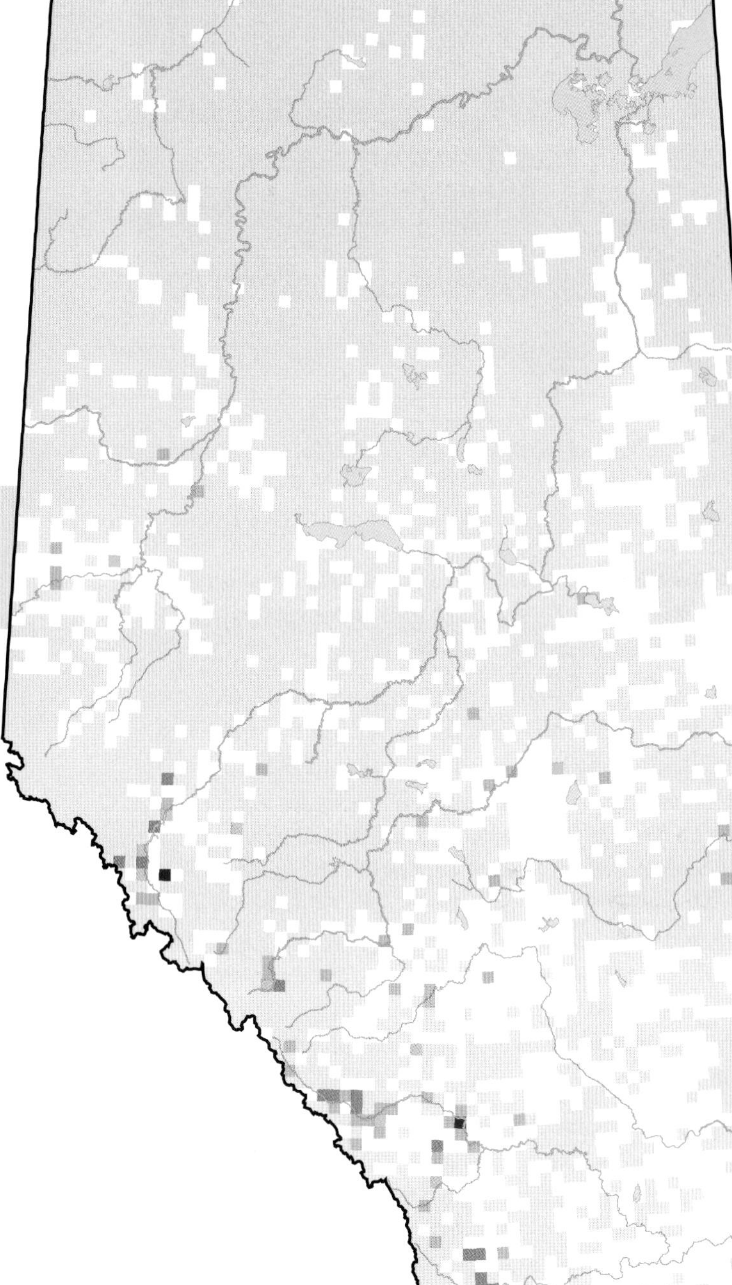

OBSERVED DISTRIBUTION ATLAS 1

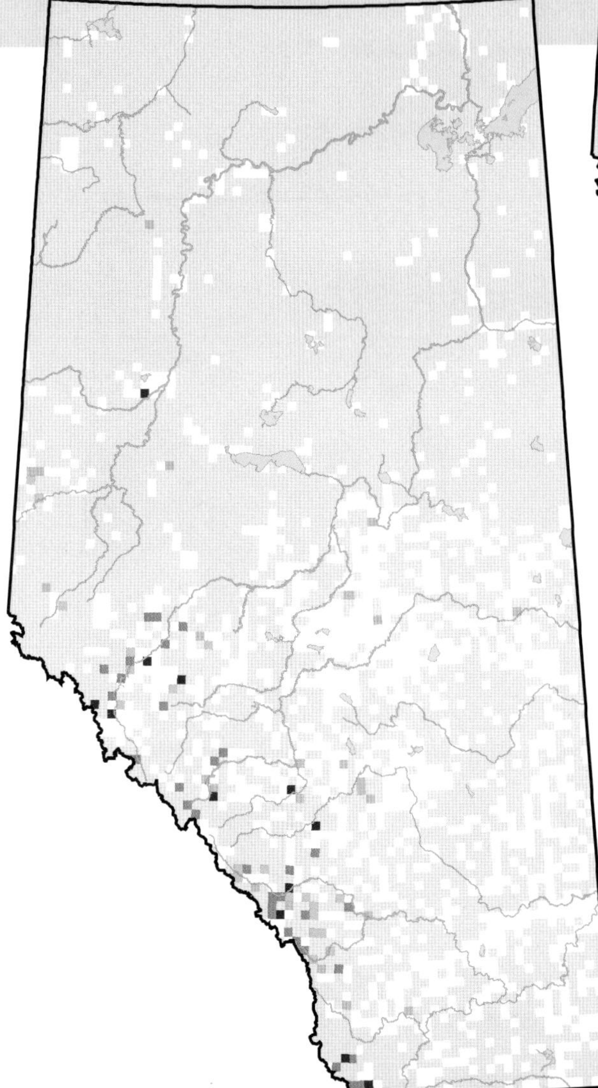

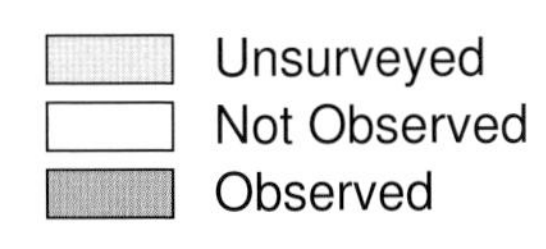

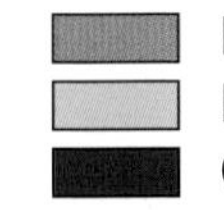

Belted Kingfisher (*Ceryle alcyon*)

Photo: Gerald Romanchuk

NESTING

CLUTCH SIZE (EGGS):	6–8
INCUBATION (DAYS):	23–24
FLEDGING (DAYS):	30–35
NEST HEIGHT (METRES):	0

The Belted Kingfisher breeds in all the Natural Regions of Alberta. There was no observed change in this species' distribution between Atlas 1 and Atlas 2.

The Belted Kingfisher inhabits streams, rivers, ponds, and lakes. These water bodies must be calm so that this species can locate fish, its main prey. They prefer water that is not obscured or overgrown with vegetation. Nests are in earthen banks with no vegetation, and are generally near water such as in a river bank. Therefore, they were most common in the Rocky Mountain NR where the topography facilitates nesting. They were less common in the Foothills and Grassland NRs, and least common in the Boreal Forest and Parkland NRs. Water in the Boreal Forest and Parkland NRs tends to be overgrown with plants and is murky with tannins, making visibility poor for locating prey.

Declines in the relative abundance of this species were detected in the Boreal Forest, Foothills, and Parkland NRs. Relative to other species, the Belted Kingfisher was observed less often in the Boreal Forest, Foothills, Grassland, and Parkland NRs. The Breeding Bird Survey found a decline in abundance Canada-wide during the period 1985–2005; however, no change in abundance was detected for Alberta. It is unknown why declines are occurring, since the Belted Kingfisher is not as susceptible to contaminants as other piscivorous birds. Water quality, cover, and availability of suitable nest sites are considered the most important factors affecting the Belted Kingfisher. They are also sensitive to disturbance during the breeding season and may abandon the nest if disturbed frequently, but this is not likely the main cause of the declines that were detected. More research is needed to properly assess this species' status. The province has rated this species' status as Secure.

TEMPORAL REPORTING RATE

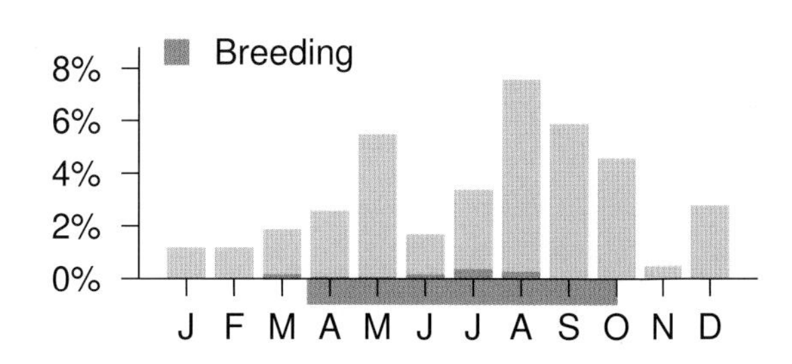

NATURAL REGIONS REPORTING RATE

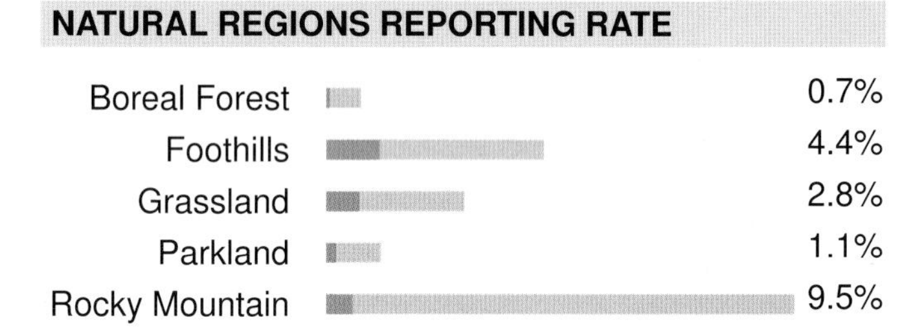

SPATIAL REPORTING RATE

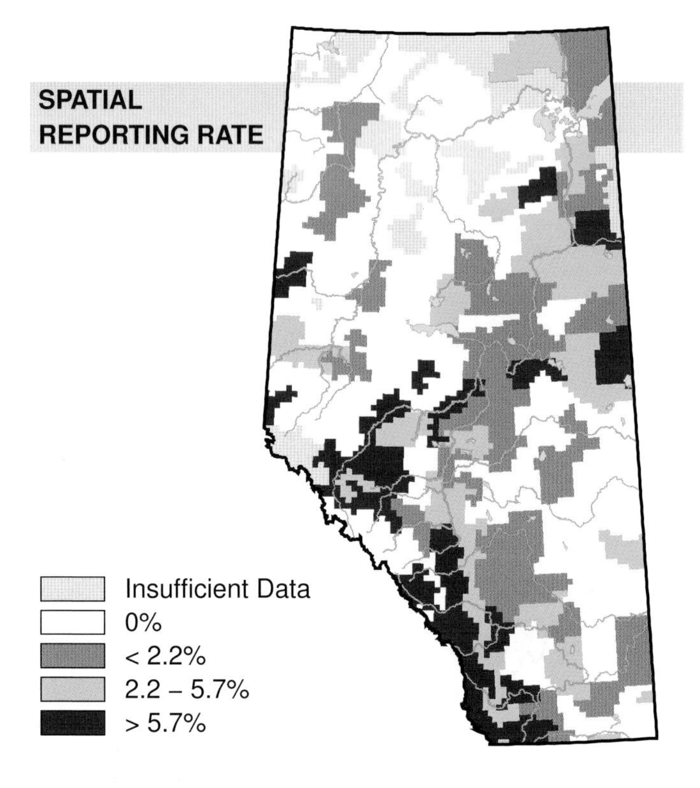

STATUS

GSWA 2000	GSWA 2005	COSEWIC Historic Rankings	COSEWIC 2007	Alberta Wildlife Act
Secure	Secure	N/A	N/A	N/A

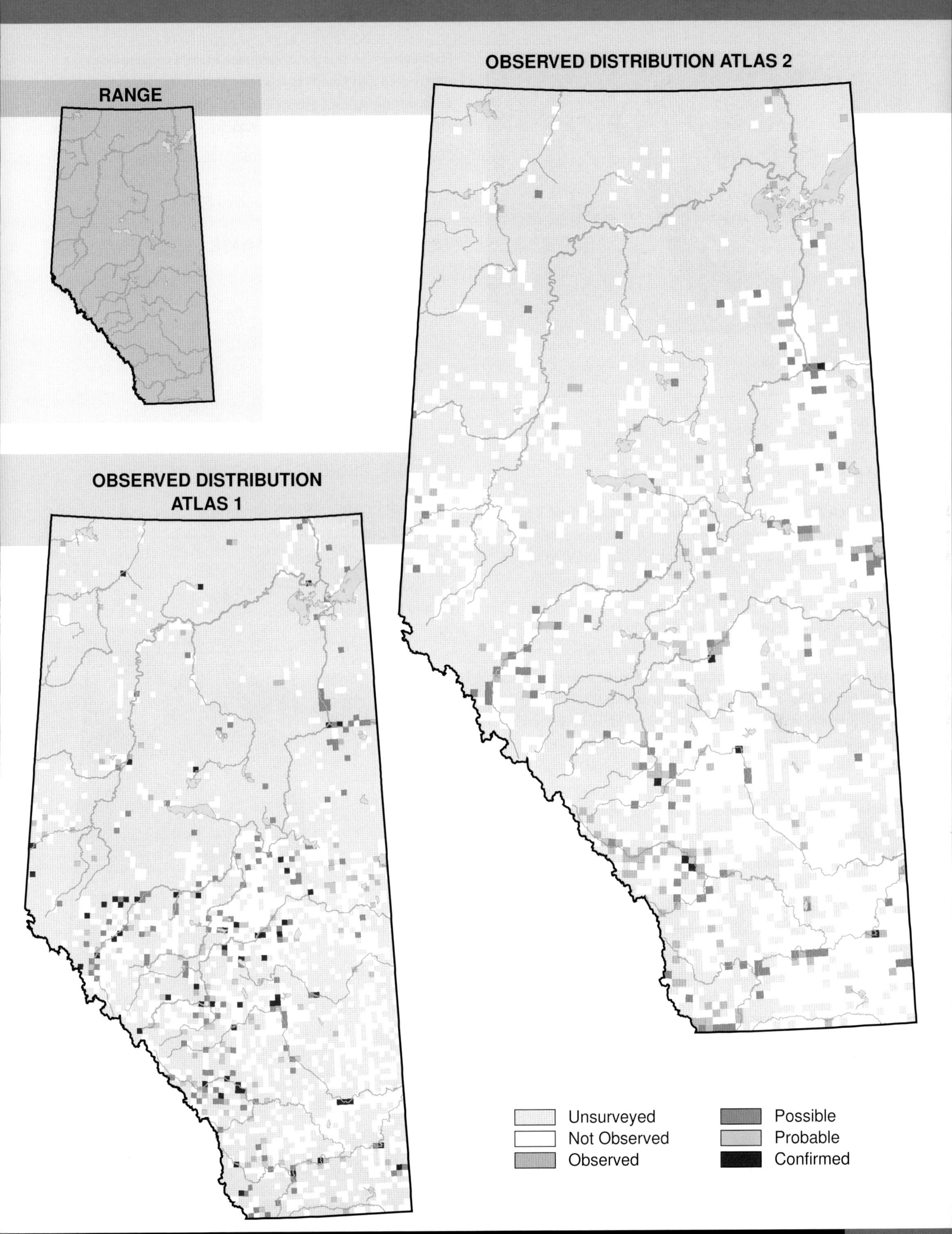
Habitat
Nest Location
Nest Type
Diet
RANGE
OBSERVED DISTRIBUTION ATLAS 2
OBSERVED DISTRIBUTION ATLAS 1
Unsurveyed
Not Observed
Observed
Possible
Probable
Confirmed

Yellow-bellied Sapsucker (*Sphyrapicus varius*)

Photo: Robert Gehlert

NESTING

CLUTCH SIZE (EGGS):	4–7
INCUBATION (DAYS):	12–13
FLEDGING (DAYS):	14
NEST HEIGHT (METRES):	2.4–12.2

The Yellow-bellied Sapsucker occurs in every Natural Region of the province, except most of the Rocky Mountain NR and the extreme southwest, where it is replaced by the Red-naped Sapsucker. There was no change in the distribution of this species from Atlas 1, although during Atlas 2 sapsuckers were observed in more squares in the Grassland NR, where the species is largely transient, and in northern Alberta.

A denizen of deciduous or mixedwood forests, where it feeds on the sap of a variety of deciduous trees and shrubs that it neatly incises for that purpose, the Yellow-bellied Sapsucker was observed most frequently in the Foothills and Boreal Forest NRs. The highest concentrations of sapsuckers were in the Boreal Forest NR south of Grande Prairie, and in the Boreal Forest and Foothills NRs around Lesser Slave Lake, and west of Edmonton and Red Deer.

An increase in relative abundance was detected in the Foothills NR, but a decrease was detected in the Parkland NR. Reporting rates relative to other species mirrored these changes. No changes in relative abundance were detected in the Boreal Forest, Grassland, and Rocky Mountain NRs. Specialized surveys carried out in the Foothills NR to document reproductive interactions between this species and the Red-naped species (the Royal Alberta Museum's Project Sapsucker) may account for some of the observed increase in that NR. The species may also benefit from moderate timber-harvesting activity, in that part of the province.

The decrease in relative abundance in the Parkland NR is unexpected and harder to explain. Perhaps a succession of dry years during Atlas 2 affected the production of sap at sapwells, resulting in an exodus of sapsuckers.

The Breeding Bird Survey did not detect a change in Yellow-bellied Sapsucker abundance in Alberta between 1985 and 2005. The species is considered Secure in Alberta.

TEMPORAL REPORTING RATE

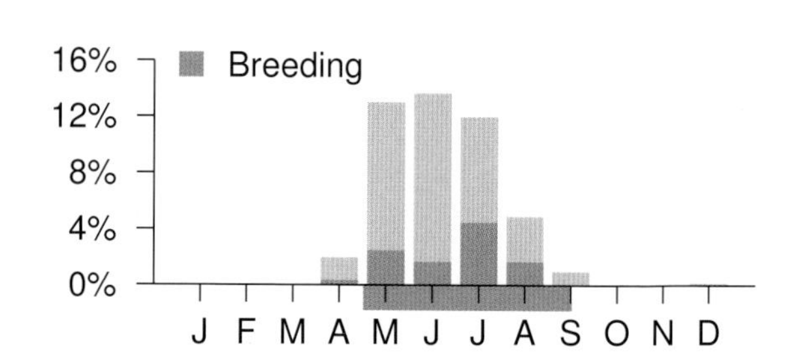

NATURAL REGIONS REPORTING RATE

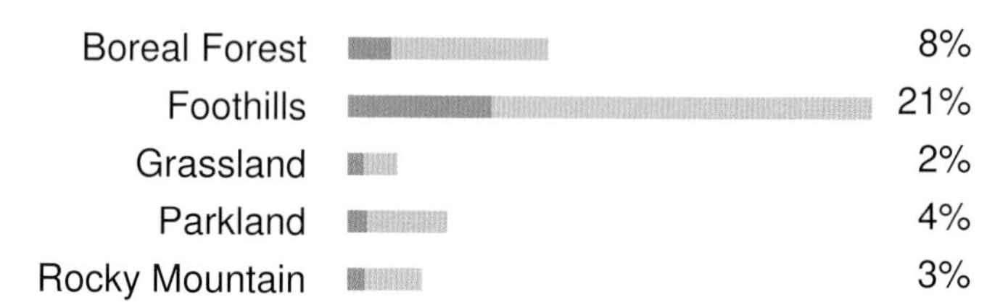

SPATIAL REPORTING RATE

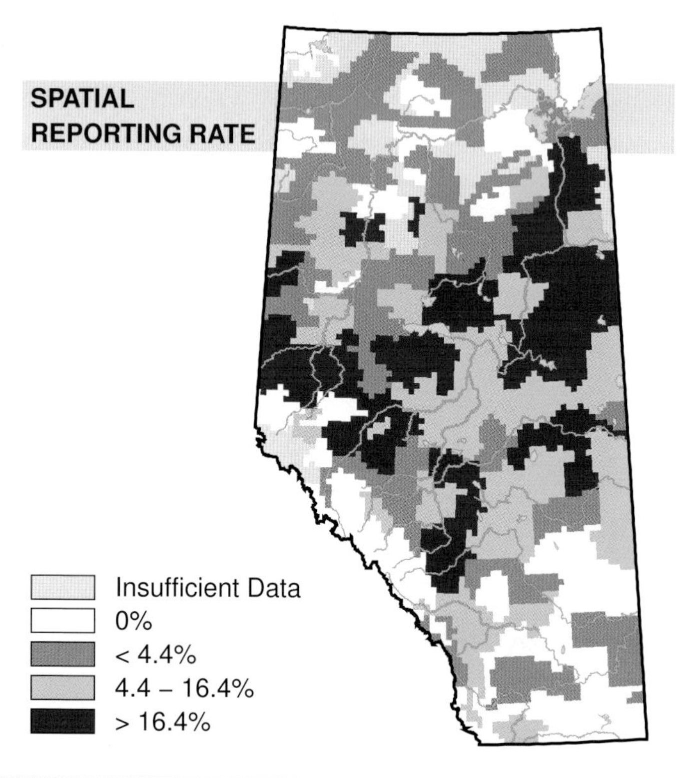

STATUS

GSWA 2000	GSWA 2005	COSEWIC Historic Rankings	COSEWIC 2007	Alberta Wildlife Act
Secure	Secure	N/A	N/A	N/A

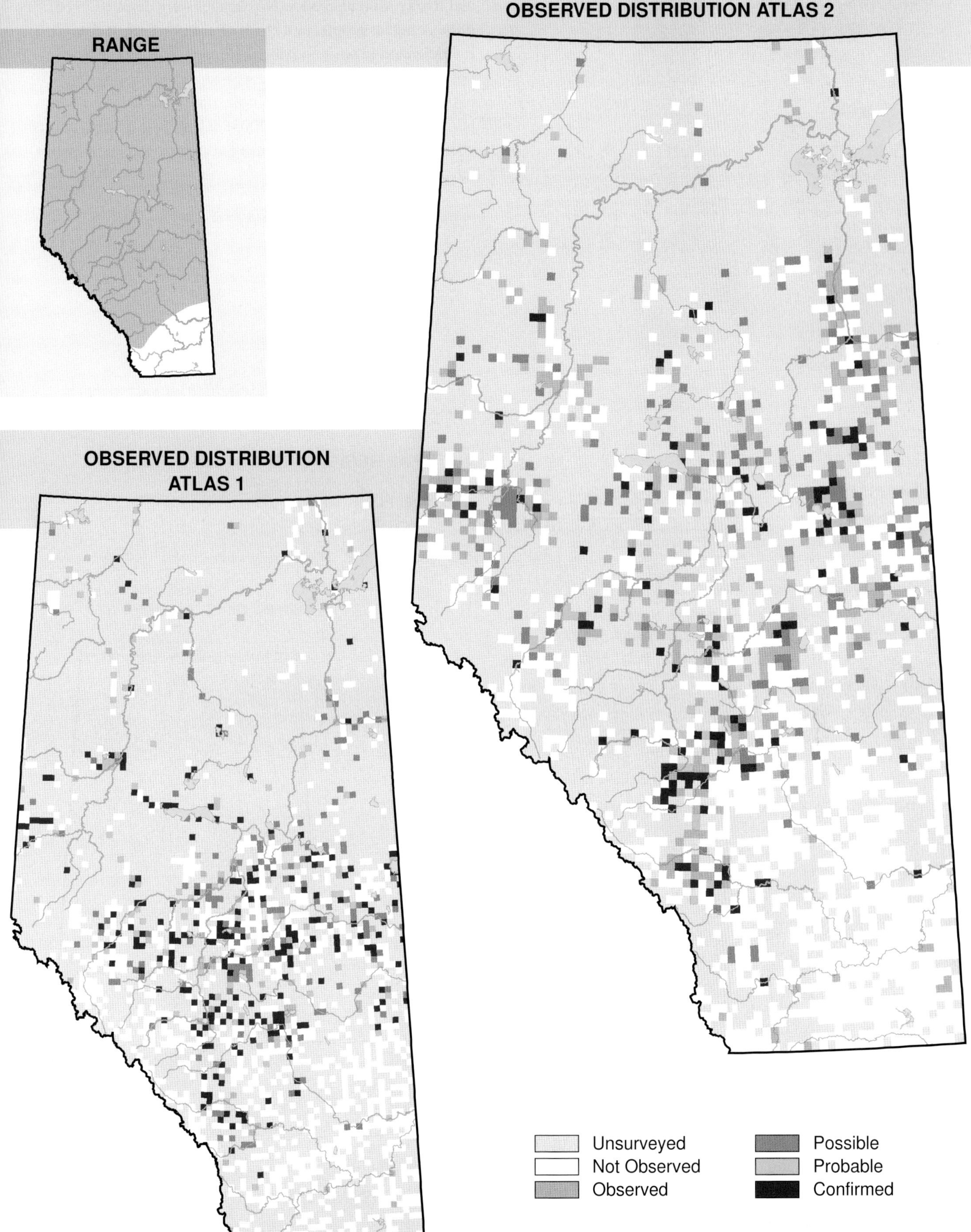
OBSERVED DISTRIBUTION ATLAS 2
RANGE
OBSERVED DISTRIBUTION ATLAS 1
Unsurveyed
Not Observed
Observed
Possible
Probable
Confirmed

Red-naped Sapsucker (*Sphyrapicus nuchalis*)

Photo: Randy Jensen

NESTING

CLUTCH SIZE (EGGS):	4–5
INCUBATION (DAYS):	12–13
FLEDGING (DAYS):	25–29
NEST HEIGHT (METRES):	0.9–10.7

The Red-naped Sapsucker is a close cousin of the Yellow-bellied Sapsucker, which it replaces in the Rocky Mountain Natural Region (including the Cypress Hills) and the southwestern corner of Alberta. Where the two species meet, loosely in the Forest Reserve, from about the Clearwater River, west of Caroline, to the Porcupine Hills, interbreeding occurs that results in a narrow hybrid zone. The presence of hybrids, sometimes incorrectly identified as one or the other type of sapsucker, limits our ability to determine the exact limits of distribution of these forms. In Atlas 2, the Red-naped Sapsucker bred along the Red Deer River in the Grasslands NR, representing a possible extension from Atlas 1. However, the Yellow-bellied Sapsucker also appeared there. It is probable that the two species are broadly equivalent in the area and may even interbreed.

An inhabitant of mixedwood forests of western Alberta and the Cypress Hills, this species was noted most frequently in the Rocky Mountain NR, with appearances in southern sections of the Foothills, Grassland, and Parkland NRs. Highest concentrations were adjacent to the Rocky Mountain NR in extreme southwestern Alberta.

Increases in relative abundance were detected in the Parkland and Rocky Mountain NRs. No changes were detected in other NRs. A higher profile, as a result of a study of sapsucker hybridization in western Alberta (Project Sapsucker), or possibly of benefits associated with moderate logging activity in the foothills, may account for the observed change. The Breeding Bird Survey (1985-2005) did not detect a change. Its status is Undetermined, until more accurate population figures have been secured.

TEMPORAL REPORTING RATE

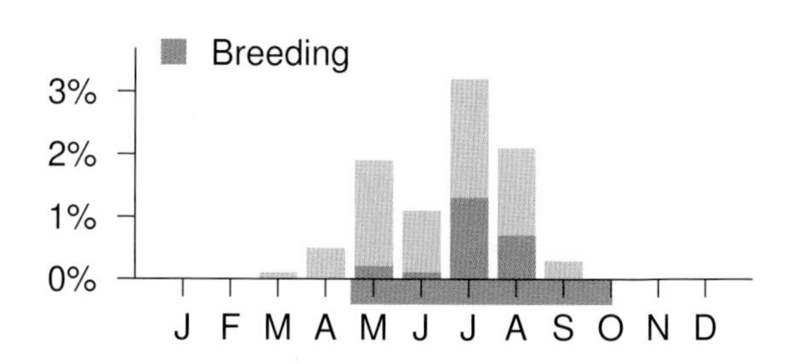

NATURAL REGIONS REPORTING RATE

Region	Rate
Boreal Forest	< 0.1%
Foothills	1.1%
Grassland	1.3%
Parkland	1.1%
Rocky Mountain	9.0%

SPATIAL REPORTING RATE

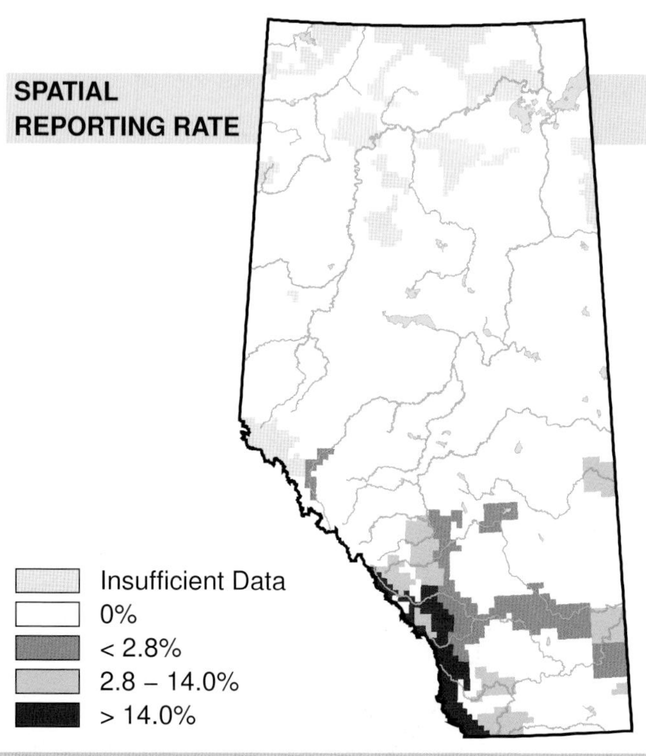

STATUS

GSWA 2000	GSWA 2005	COSEWIC Historic Rankings	COSEWIC 2007	Alberta Wildlife Act
Undetermined	Undetermined	N/A	N/A	N/A

RANGE

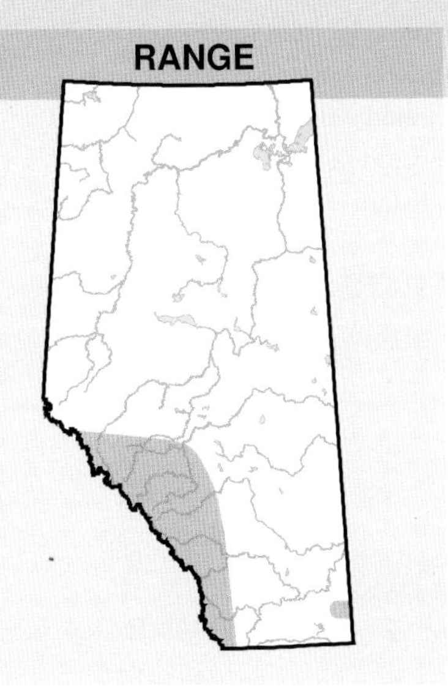

OBSERVED DISTRIBUTION ATLAS 2

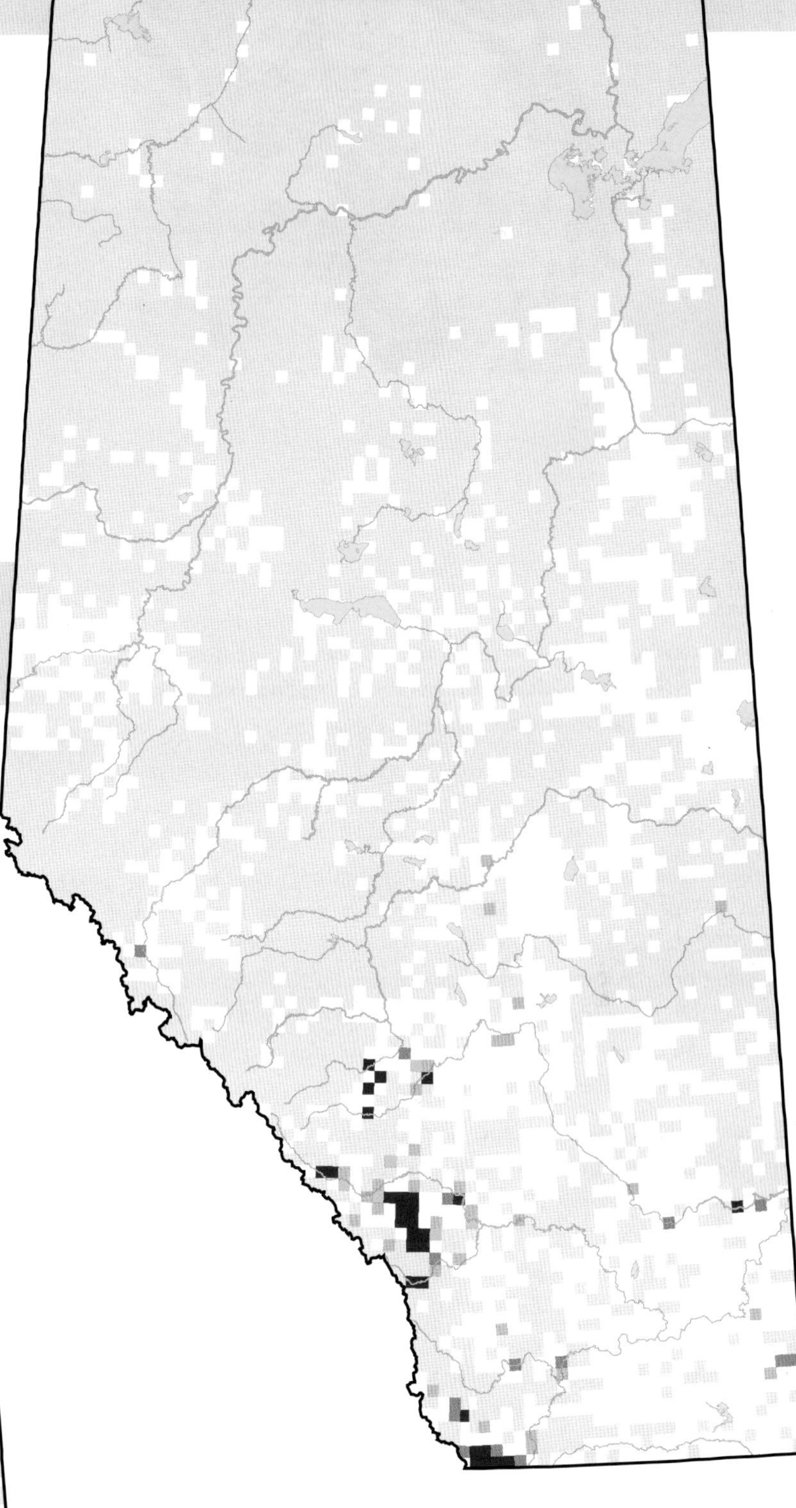

OBSERVED DISTRIBUTION ATLAS 1

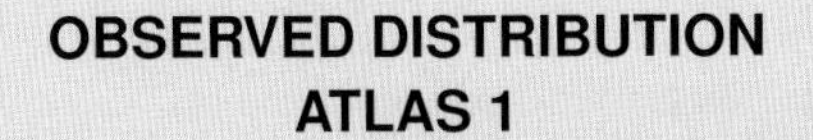

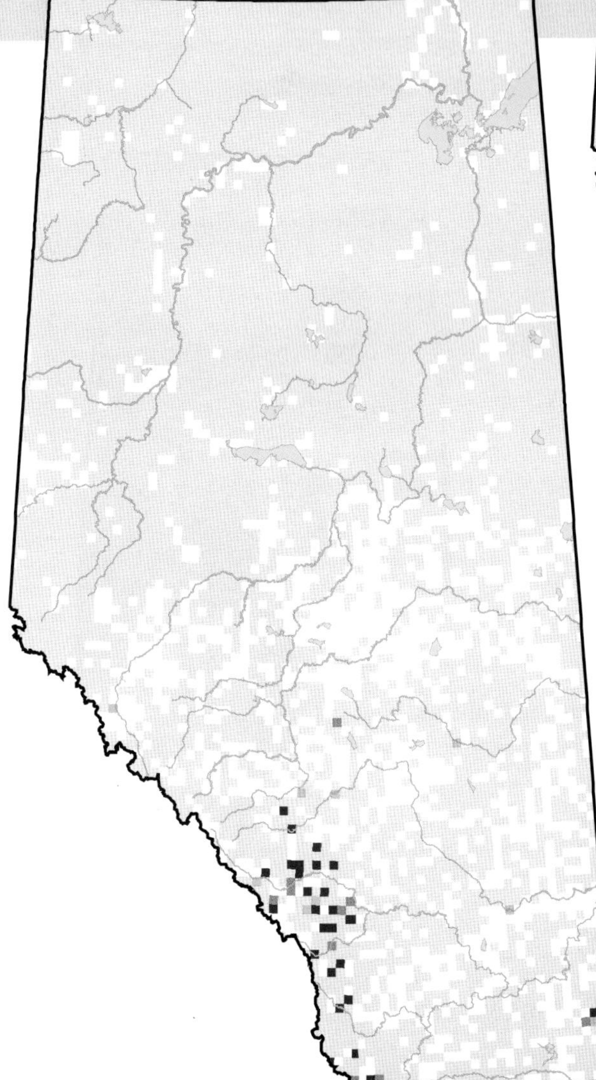

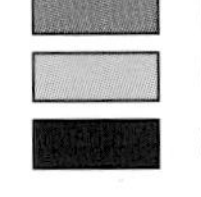

Downy Woodpecker (*Picoides pubescens*)

Photo: Randy Jensen

NESTING

CLUTCH SIZE (EGGS):	4–5
INCUBATION (DAYS):	12
FLEDGING (DAYS):	21–25
NEST HEIGHT (METRES):	0.9–15

The Downy Woodpecker is found in every Natural Region in Alberta. No change in distribution was observed between Atlas 1 and Atlas 2 for this species.

The Downy Woodpecker is associated with deciduous forests that have varying stand ages. Unlike other woodpeckers, this species is less abundant in forests with high proportions of conifer trees. Despite occurring in all natural regions, the Downy Woodpecker was observed most often in the Parkland NR. The distribution of this species was fairly uniform in the Boreal Forest, Foothills, and Rocky Mountain NRs where it was relatively common. Distribution in the Grassland NR was patchy and tended to be associated with wet areas, such as along the South Saskatchewan River.

Increases in relative abundance were detected in the Parkland and Rocky Mountain NRs. Relative to other species, the Downy Woodpecker was observed more frequently in Atlas 2 in the Rocky Mountain NR, but at similar frequencies in the Parkland NR. The cause for these increases is unknown. A decline was detected in the Boreal Forest NR where, relative to other species, this bird was observed less frequently in Atlas 2 than in Atlas 1. The Breeding Bird Survey did not detect any change in abundance for this species in Alberta, but Canada-wide declines were detected in the periods from 1968–2005 and from 1985–2005. The cause for the decline in the Boreal Forest NR is not understood, but the Canada-wide decline is in agreement. The Atlas-detected increases in the Parkland and Rocky Mountain NRs have no corroborative evidence thus should be interpreted with caution. No changes in relative abundance were detected in the Foothills and Grassland NRs. The provincial status of this species is Secure.

TEMPORAL REPORTING RATE

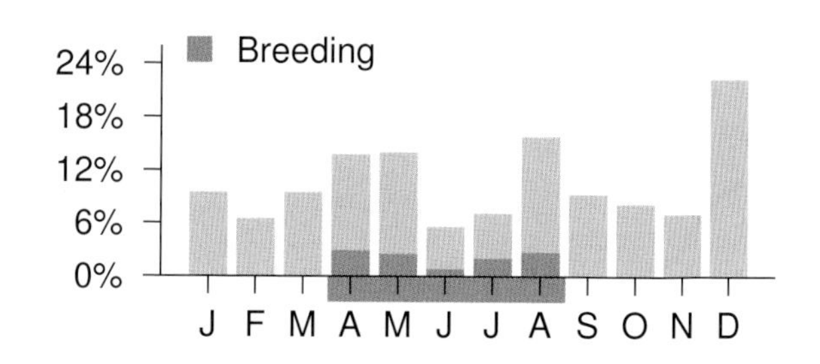

NATURAL REGIONS REPORTING RATE

SPATIAL REPORTING RATE

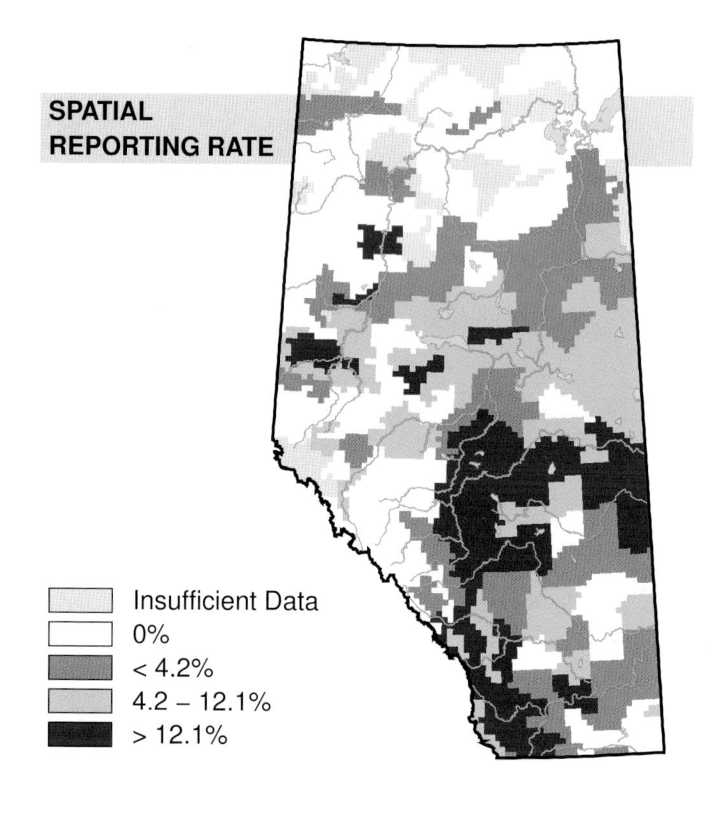

STATUS

GSWA 2000	GSWA 2005	COSEWIC Historic Rankings	COSEWIC 2007	Alberta Wildlife Act
Secure	Secure	N/A	N/A	N/A

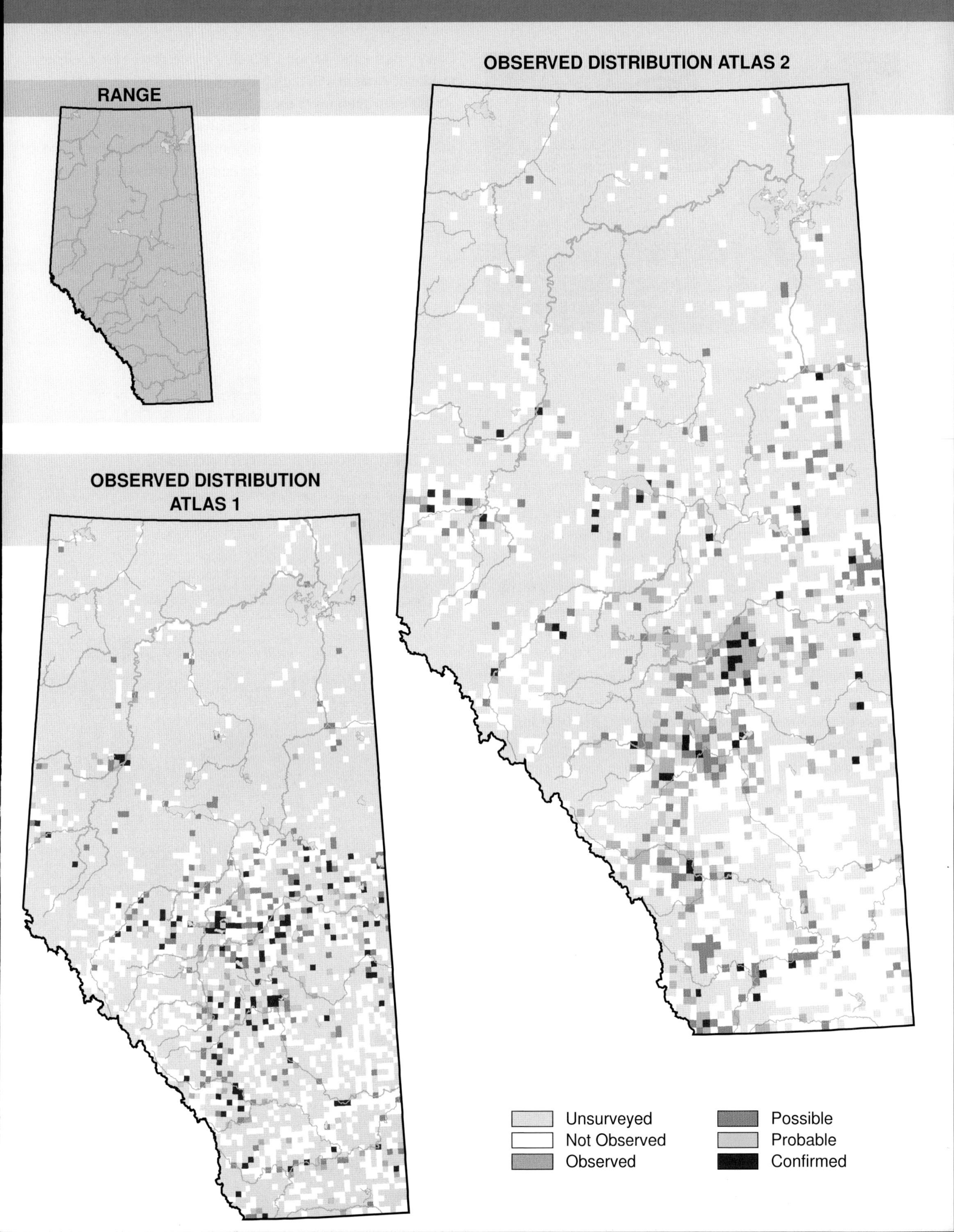
RANGE
OBSERVED DISTRIBUTION ATLAS 2
OBSERVED DISTRIBUTION ATLAS 1
Unsurveyed
Not Observed
Observed
Possible
Probable
Confirmed

Hairy Woodpecker (*Picoides villosus*)

Photo: James Potter

NESTING

CLUTCH SIZE (EGGS):	3–6
INCUBATION (DAYS):	11–15
FLEDGING (DAYS):	24–30
NEST HEIGHT (METRES):	2–6

The Hairy Woodpecker is found in every Natural Region in Alberta. No change in the distribution of this species was observed between Atlas 1 and Atlas 2.

Unlike the Downy Woodpecker, which will occupy forests of varying ages, the Hairy Woodpecker tends to be associated with mature forests. Hairy Woodpeckers tend to excavate cavities in mature live trees with heart rot. Their nests are usually built near open areas or in a part of the forest with low tree density. As a result, this species was widely distributed across Boreal Forest, Foothills, Parkland, and Rocky Mountain NRs. Grassland NR distribution was patchy and tended to be associated with rivers where mature trees are present, such as along the Red Deer River.

Increases in relative abundance were detected in the Boreal Forest and Rocky Mountain NRs where, relative to other species, the Hairy Woodpecker was observed more frequently in Atlas 2 than in Atlas 1. These changes could be related to increased forest fragmentation because this species tends to nest near open areas. A decline was detected in the Parkland NR where, relative to other species, this bird was observed less frequently in Atlas 2 than in Atlas 1. This could be the result of reduced availability of mature trees. Changes in relative abundance were not detected in the Foothills and Grassland NRs. The Breeding Bird Survey found no change in abundance in Alberta, but the abundance of this species increased across Canada during the period 1995–2005. Differences between the results of Atlas and Breeding Bird Survey analyses could be attributed to analyses of the data at different spatial scales. Breeding Bird Survey data were analyzed at the provincial scale; whereas, due to sufficient sample size, it was possible to analyze Atlas data at the smaller Natural Region scale. The provincial status of this species is Secure.

TEMPORAL REPORTING RATE

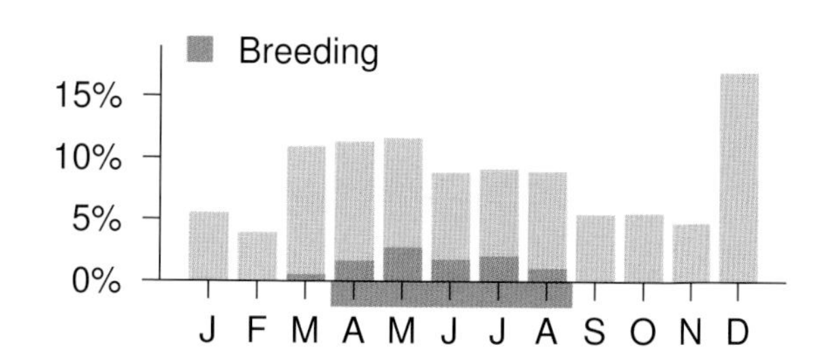

NATURAL REGIONS REPORTING RATE

Boreal Forest	7.3%
Foothills	8.7%
Grassland	2.5%
Parkland	5.8%
Rocky Mountain	9.7%

SPATIAL REPORTING RATE

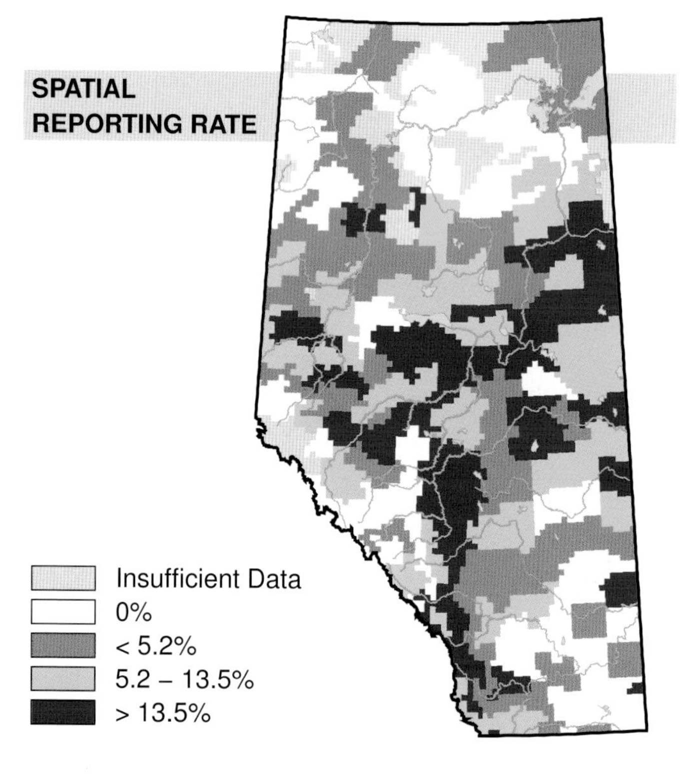

STATUS

GSWA 2000	GSWA 2005	COSEWIC Historic Rankings	COSEWIC 2007	Alberta Wildlife Act
Secure	Secure	N/A	N/A	N/A

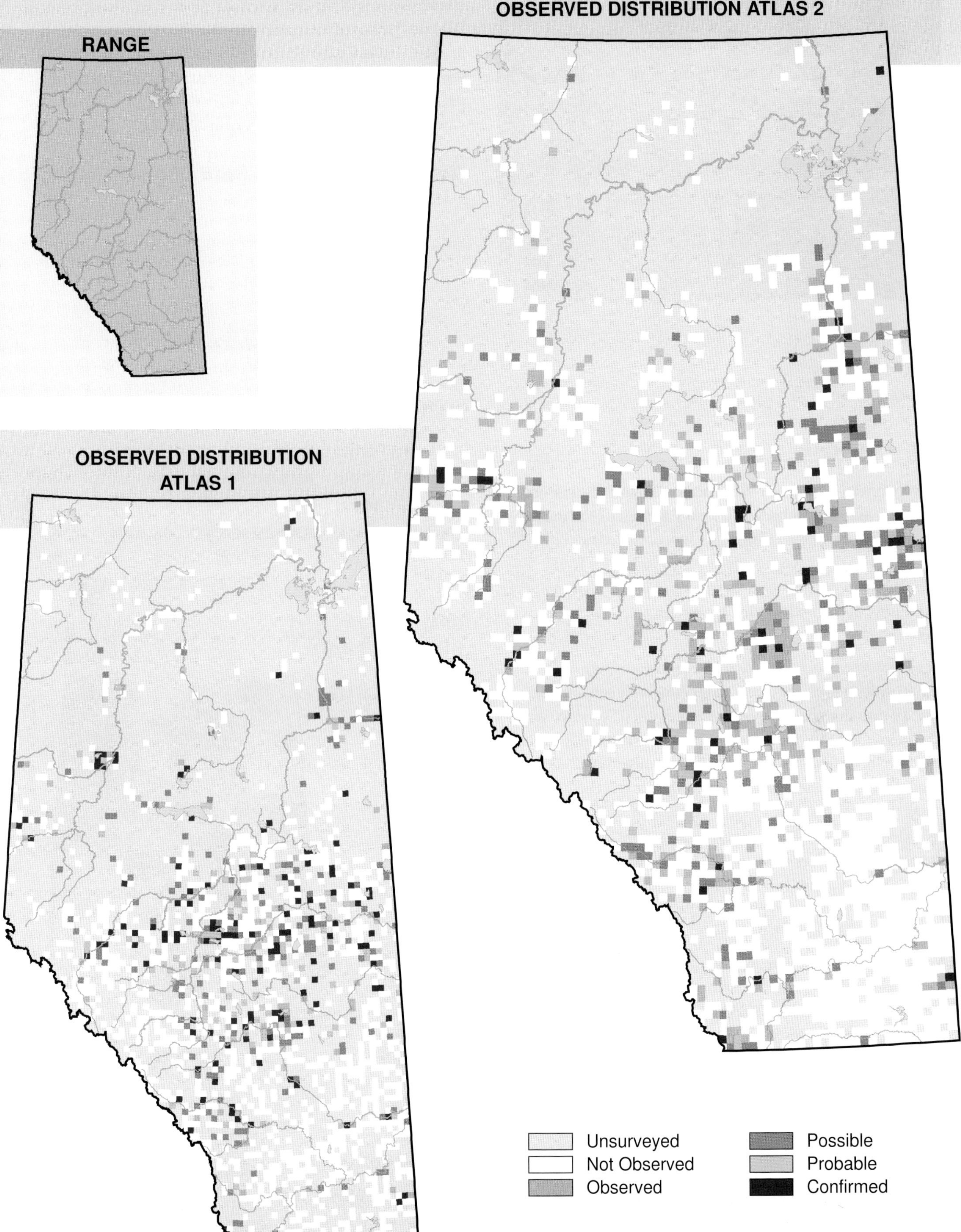
RANGE
OBSERVED DISTRIBUTION ATLAS 2
OBSERVED DISTRIBUTION ATLAS 1
Unsurveyed
Not Observed
Observed
Possible
Probable
Confirmed

American Three-toed Woodpecker (*Picoides dorsalis*)

Photo: Robert Gehlert

NESTING

CLUTCH SIZE (EGGS):	4–5
INCUBATION (DAYS):	14
FLEDGING (DAYS):	18–23
NEST HEIGHT (METRES):	0.3–13.7

The American Three-toed Woodpecker is found in the Boreal Forest, Foothills, Parkland, and Rocky Mountain Natural Regions. No change in distribution was observed between Atlas 1 and Atlas 2 for this species. However, they were detected more often in the area between Lac La Biche and Fort McMurray during Atlas 2, likely because a fire had burned there recently.

The American Three-toed Woodpecker uses mature or old-growth coniferous forests and prefers a higher stand density than does its cousin, the Black-backed Woodpecker. The three-toed species is also known to use areas with habitats that have been disturbed by fire, flooding, disease, insects and wind damage. In these areas, they feed on insects in dead or dying trees and excavate cavities in mature trees for nesting. Despite occurring in four natural regions, this species was mainly found in the Boreal Forest, Foothills, and Rocky Mountain NRs. Habitat in the Parkland NR is not well suited to this species. Therefore, it was found there only infrequently.

An increase in relative abundance was detected in the Boreal Forest NR where, relative to other species, it was observed more frequently in Atlas 2 than in Atlas 1. No change was detected in the Foothills, Parkland, and Rocky Mountain NRs. Breeding Bird Survey sample size was too small to investigate abundance trends for the American Three-toed Woodpecker in Alberta. However, an increase in abundance was detected across Canada for the period 1995–2005. Abundance of this species is expected to increase, at least in the short-term, with the range expansion of the Mountain Pine Beetle. This woodpecker is considered Secure in Alberta.

TEMPORAL REPORTING RATE

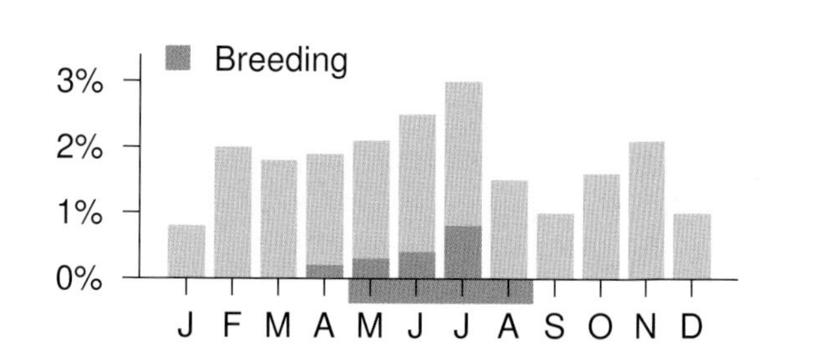

NATURAL REGIONS REPORTING RATE

Region	Rate
Boreal Forest	1.1%
Foothills	3.9%
Parkland	0.1%
Rocky Mountain	5.7%

SPATIAL REPORTING RATE

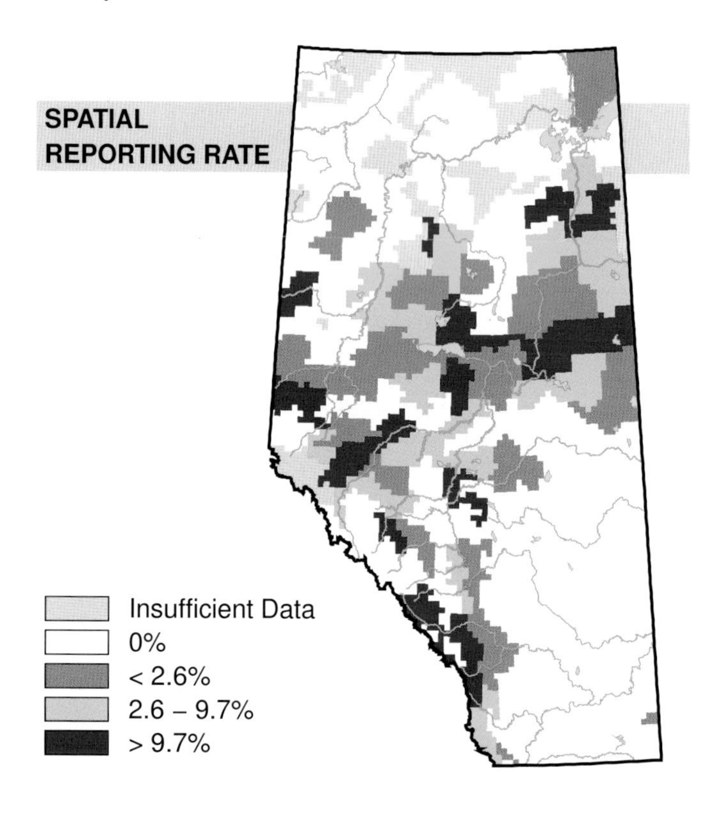

STATUS

GSWA 2000	GSWA 2005	COSEWIC Historic Rankings	COSEWIC 2007	Alberta Wildlife Act
N/A	N/A	N/A	N/A	N/A

Habitat Nest Location Nest Type Diet

RANGE

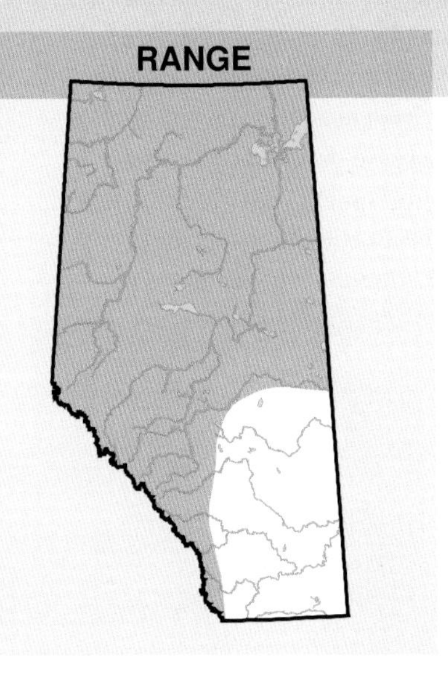

OBSERVED DISTRIBUTION ATLAS 1

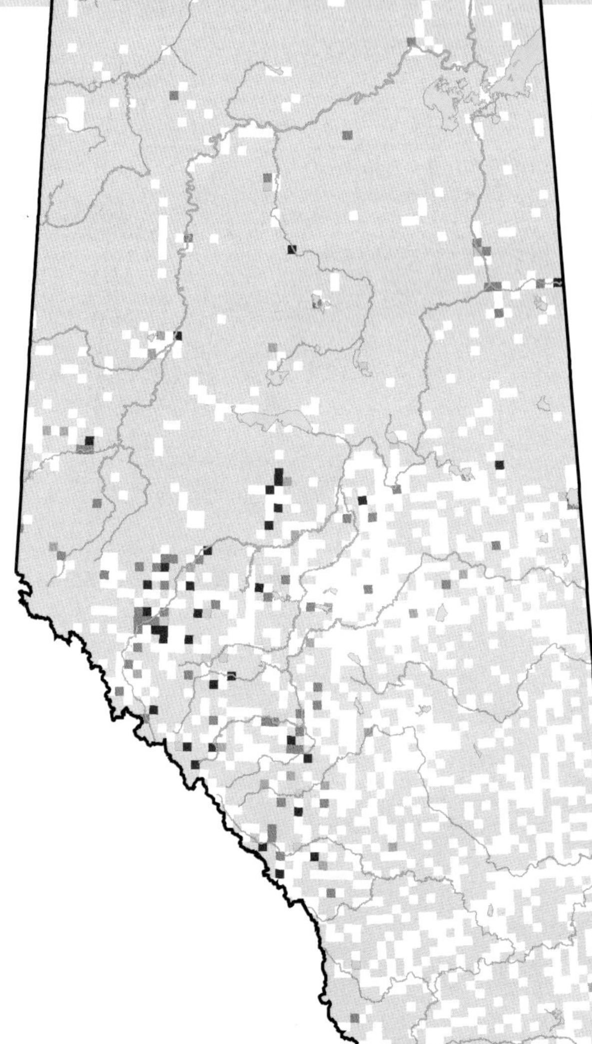

OBSERVED DISTRIBUTION ATLAS 2

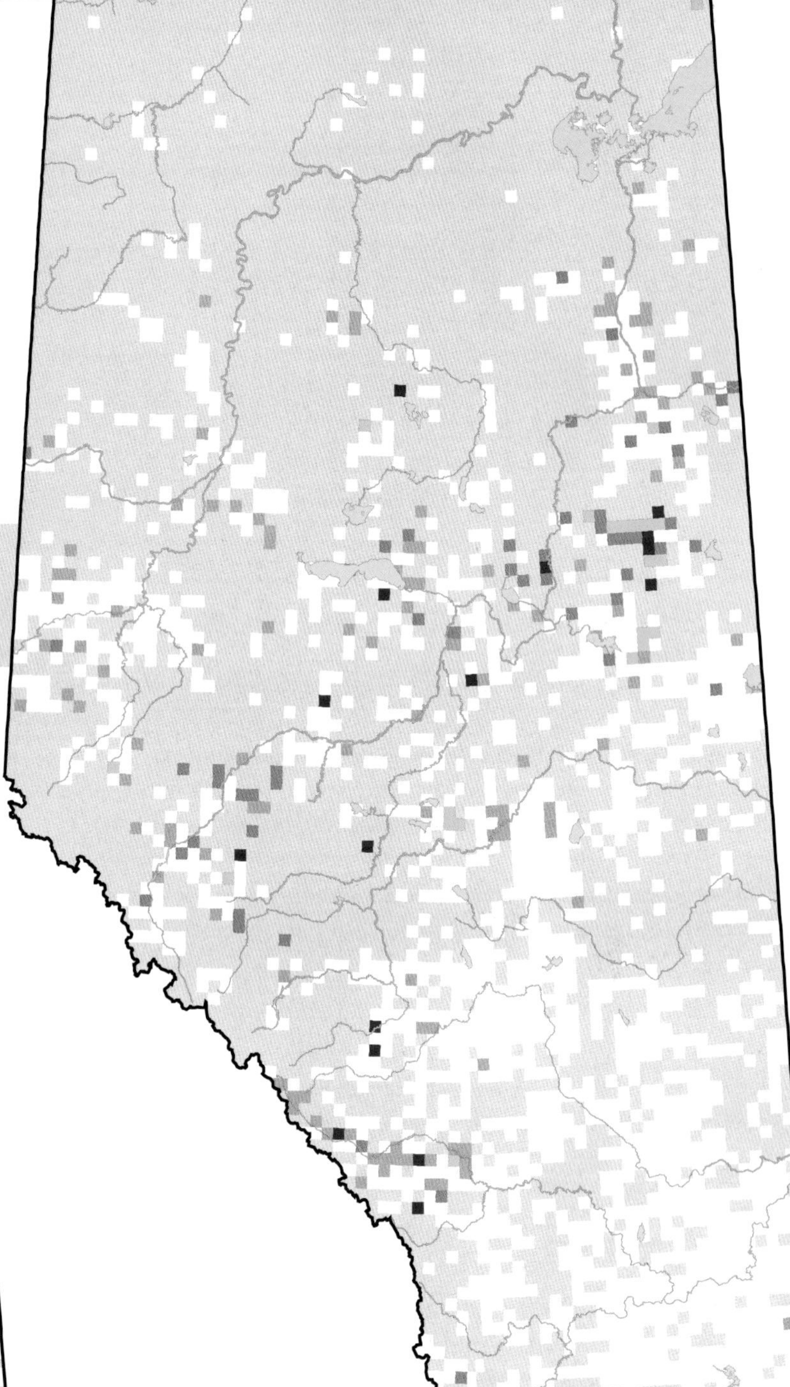

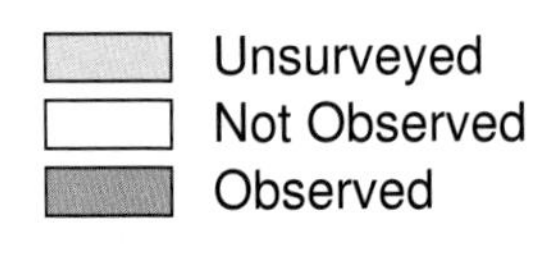

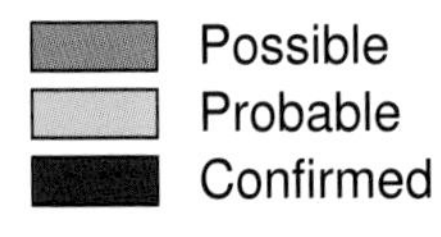

Black-backed Woodpecker (*Picoides arcticus*)

Photo: Gerald Romanchuk

NESTING

CLUTCH SIZE (EGGS):	3–4
INCUBATION (DAYS):	2–6
FLEDGING (DAYS):	25
NEST HEIGHT (METRES):	<5

The Black-backed Woodpecker is found in the Boreal Forest, Foothills, Parkland, and Rocky Mountain Natural Regions. No change in distribution was observed between Atlas 1 and Atlas 2 for this species. However, they were detected more often in the area between Lac La Biche and Fort McMurray during Atlas 2 efforts, likely because a large fire had recently burned there.

The Black-backed Woodpecker generally favours coniferous forests that have been disturbed by fire. However, this species will also use forests that have been disturbed by insects, disease and wind damage. Beetle larvae that hatch and develop within trees are the main food consumed by the Black-backed Woodpecker. Nest cavities are excavated in mature trees that are within, or adjacent to, disturbed forest habitat. Despite occurring in four NRs, this species was found mainly in the Boreal Forest NR, found occasionally in the Foothills and Parkland NRs, and found only rarely in the Rocky Mountain NR.

An increase in relative abundance was detected in the Boreal Forest NR where, relative to other species, it was observed more frequently in Atlas 2 than in Atlas 1. No change was detected in the Foothills, Parkland, or Rocky Mountain NRs. The Breeding Bird Survey sample size was too small to investigate changes in Black-backed Woodpecker abundance in Alberta. However, a decline was detected across Canada for the period 1985–2005. The increase detected by the Atlas in the Boreal Forest NR is not surprising because of the range expansion of the Mountain Pine Beetle. This species is considered Sensitive in Alberta due to its dependence on mature coniferous forests.

TEMPORAL REPORTING RATE

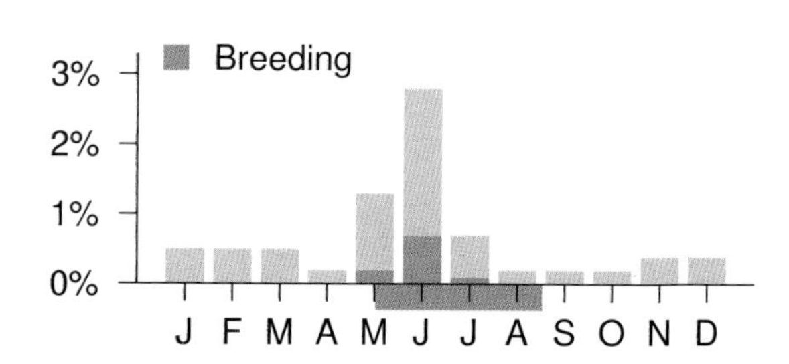

NATURAL REGIONS REPORTING RATE

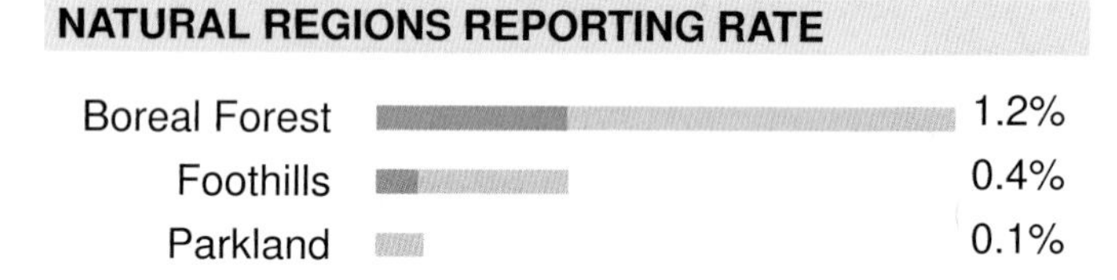

SPATIAL REPORTING RATE

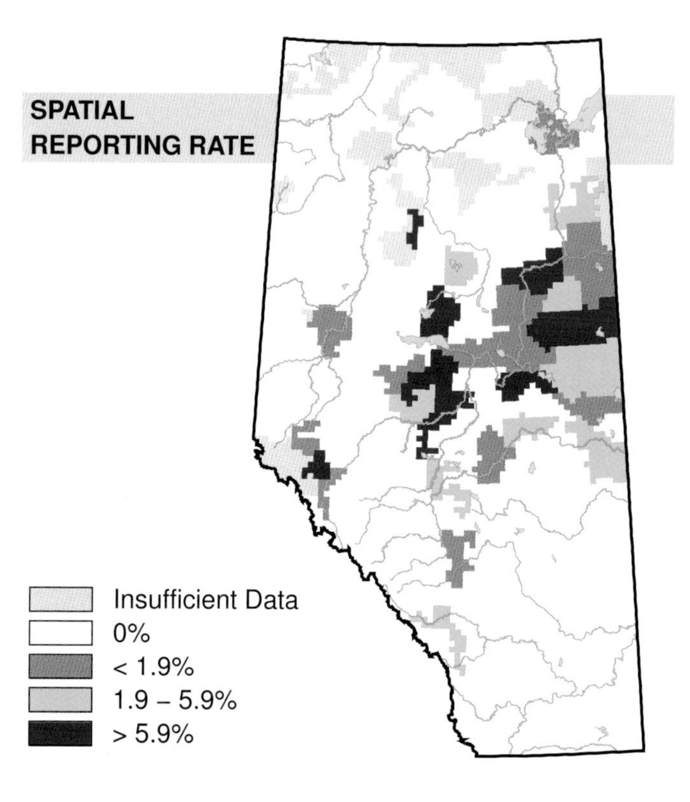

STATUS

GSWA 2000	GSWA 2005	COSEWIC Historic Rankings	COSEWIC 2007	Alberta Wildlife Act
Sensitive	Sensitive	N/A	N/A	N/A

OBSERVED DISTRIBUTION ATLAS 2

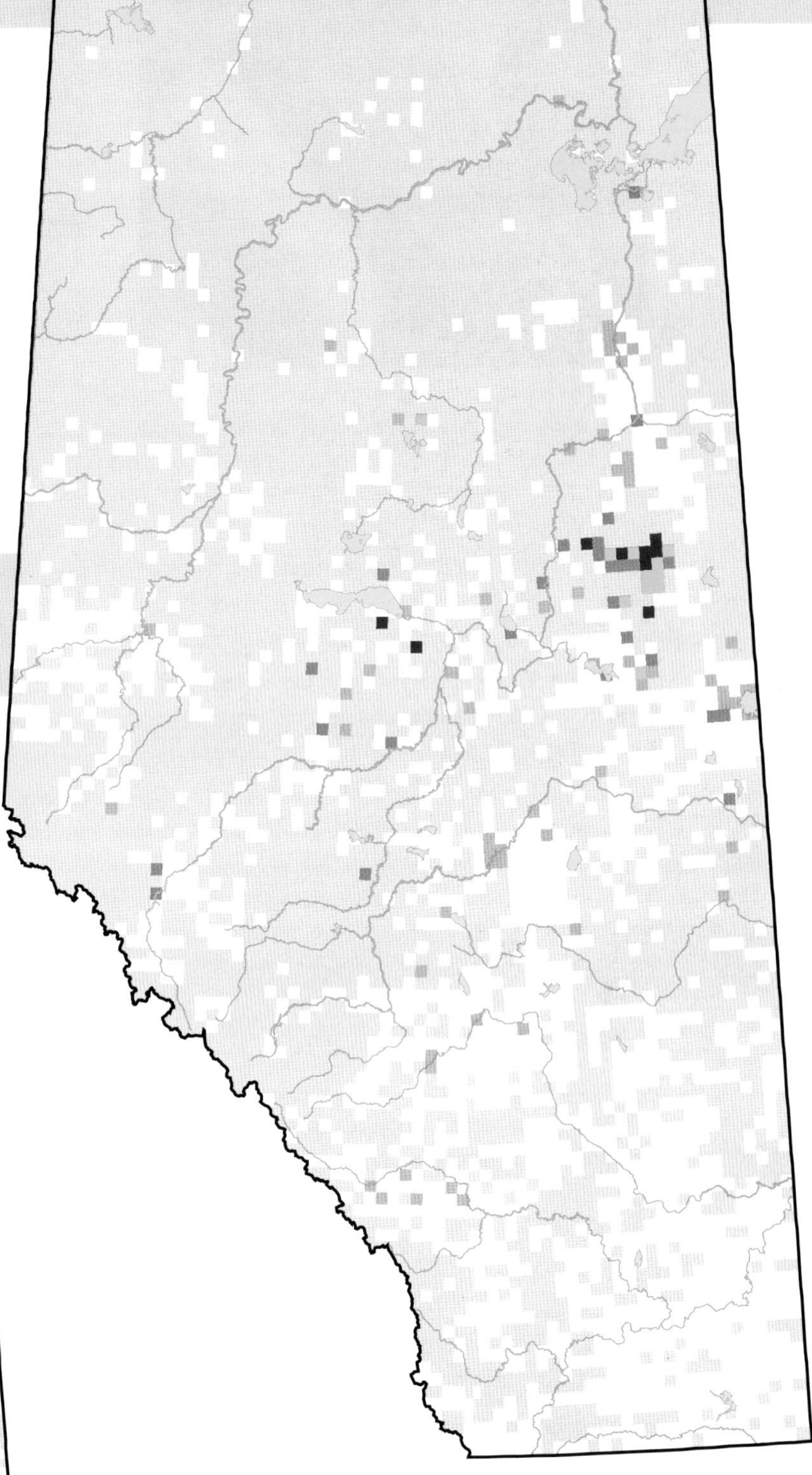

RANGE

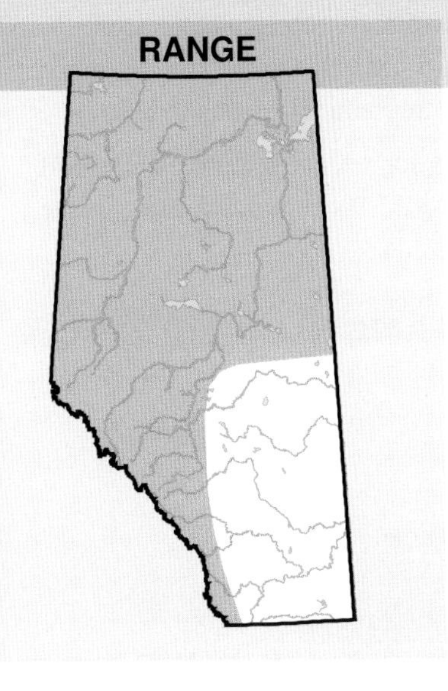

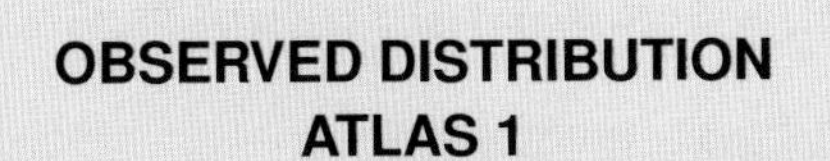

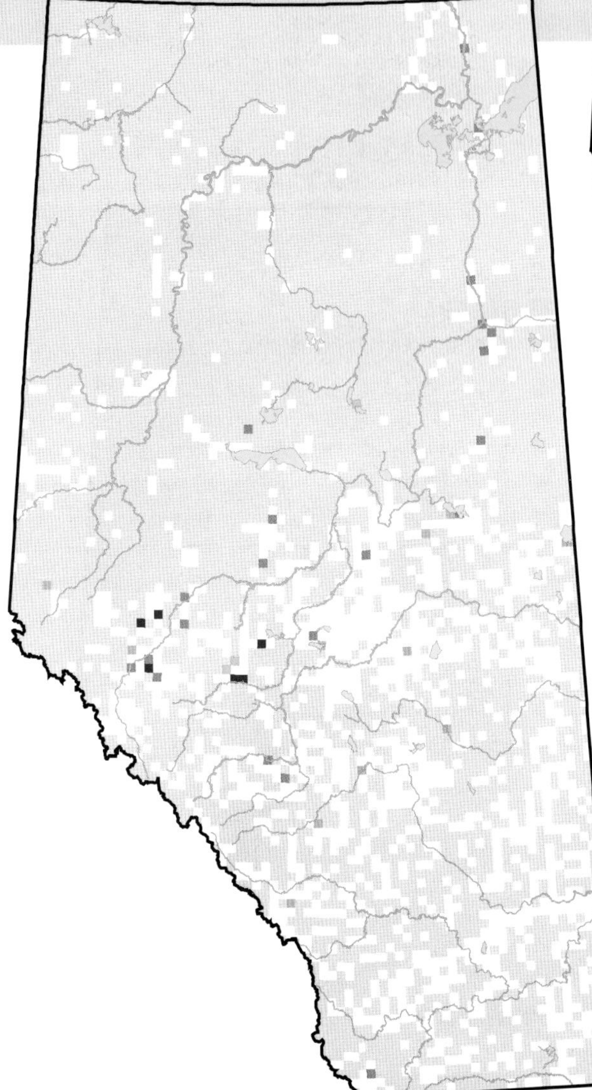

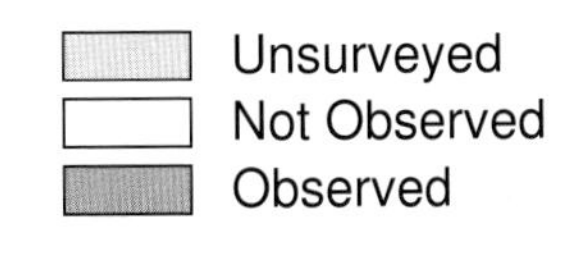

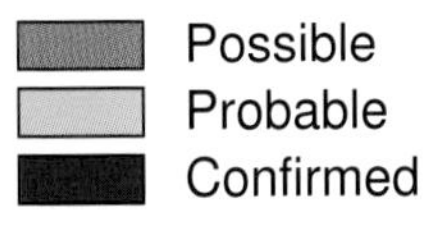

Northern Flicker (*Colaptes auratus*)

Photo: Gerald Romanchuk

NESTING

CLUTCH SIZE (EGGS):	3–12
INCUBATION (DAYS):	11–13
FLEDGING (DAYS):	25–28
NEST HEIGHT (METRES):	3

The Northern Flicker is at home in every Natural Region of the province. Yellow-shafted individuals found in the eastern part of the province give way to red-shafted ones to the west, but not before the two intermix in a broad hybrid zone in western Alberta.

Formerly a species of clement months, flickers in growing numbers are opting to stay for Alberta's winters, assisted by food provided at bird feeders. There was no change in the observed distribution of this species from Atlas 1, though the records were more evenly distributed in Atlas 2.

An obligate cavity-nester, the Northern Flicker nonetheless is not dependent on large tracts of forests for breeding, even thriving in open spaces with few trees. This is perhaps not surprising given the species' predilection for ants and their larvae, which it finds on the ground.

The Northern Flicker was observed more frequently in the Rocky Mountain, Boreal Forest, and Foothills Natural Regions, with highest concentrations in the southern Foothills NR and the Rocky Mountain NR. High concentrations of flickers were also found in the southern Boreal Forest NR south and west of Calgary, including Kananaskis Country.

Decreases in relative abundance were detected in the Boreal Forest, Foothills, and Parkland NRs. No change in relative abundance was detected in the Grassland and Rocky Mountain NRs. Observation rates relative to other species mirrored these changes. There are no biological causes apparent for these declines. At least in the Boreal Forest NR, improved access to pristine patches of habitats, less attractive to the species, could have produced an apparent decrease in relative abundance.

The Breeding Bird Survey did not detect a temporal change for Alberta for the period 1985–2005. The provincial status of the Northern Flicker is Secure.

TEMPORAL REPORTING RATE

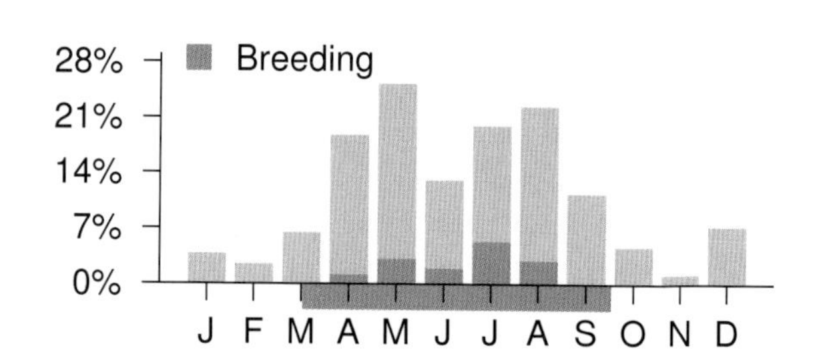

NATURAL REGIONS REPORTING RATE

Region	Rate
Boreal Forest	16%
Foothills	20%
Grassland	9%
Parkland	13%
Rocky Mountain	33%

SPATIAL REPORTING RATE

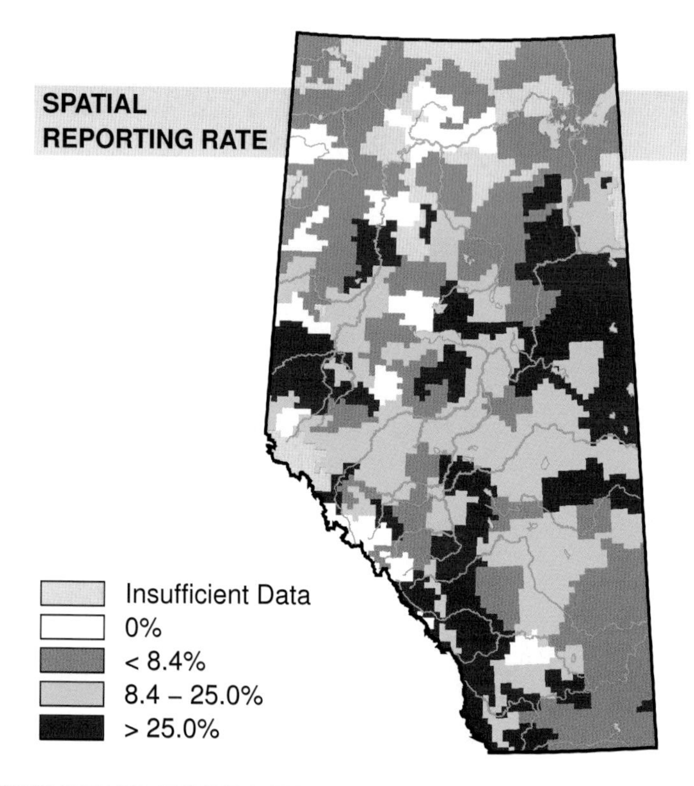

STATUS

GSWA 2000	GSWA 2005	COSEWIC Historic Rankings	COSEWIC 2007	Alberta Wildlife Act
Secure	Secure	N/A	N/A	N/A

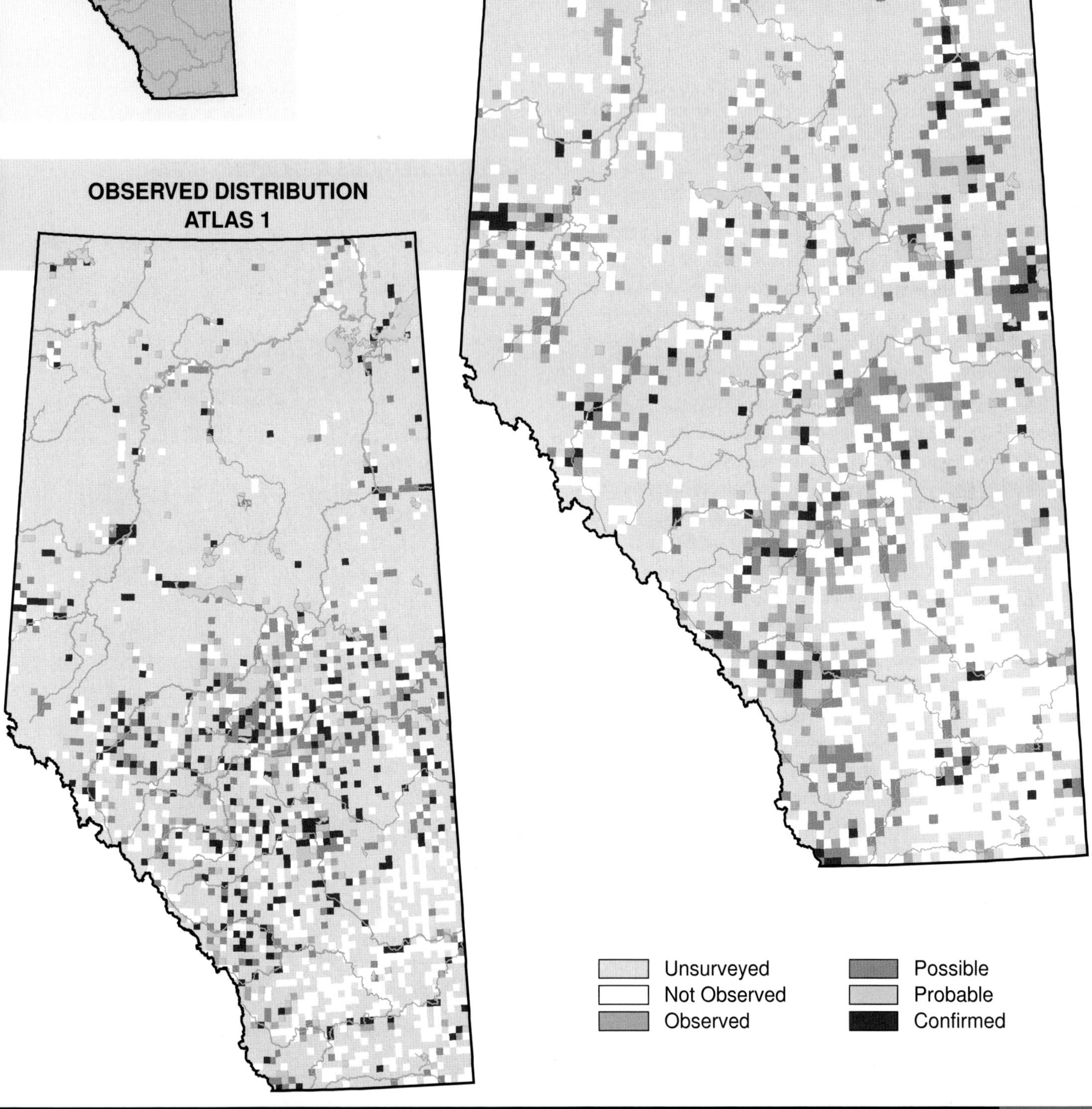
RANGE
OBSERVED DISTRIBUTION ATLAS 2
OBSERVED DISTRIBUTION ATLAS 1
Unsurveyed
Not Observed
Observed
Possible
Probable
Confirmed

Pileated Woodpecker (*Dryocopus pileatus*)

Photo: Glen Rowan

NESTING

CLUTCH SIZE (EGGS):	3–4
INCUBATION (DAYS):	18
FLEDGING (DAYS):	22–26
NEST HEIGHT (METRES):	4.6–24.4

The largest woodpecker in Alberta, the Pileated Woodpecker is found in every Natural Region in the province. No change in the distribution of this species was observed between Atlas 1 and Atlas 2, but this species was observed more often in the northern part of its range during Atlas 2.

Despite occurring in all Natural Regions, the Pileated Woodpecker was found mainly in the Boreal Forest, Foothills, Parkland, and Rocky Mountain NRs, and was only infrequently encountered in the Grassland NR. The distribution of this species in Alberta reflects its preference to nest in mature mixedwood forests. Nests are excavated in large trees and, therefore, the presence of large trees is an essential component of this bird's breeding habitat. This species can breed in young forests provided some large, typically older, trees are interspersed among the young stands.

Declines in relative abundance were detected in the Foothills and Parkland NRs where, relative to other species, the Pileated Woodpecker was observed less frequently in Atlas 2 than in Atlas 1. Declines may be the result of reduced availability of mature trees for nesting in those NRs. Changes in relative abundance were not detected in the Grassland and Rocky Mountain NRs, but an increase was detected in the Boreal Forest NR. The increase in the Boreal Forest NR has no apparent biological cause and appears to be a function of the more extensive coverage of the north in Atlas 2. The Breeding Bird Survey detected an increase in abundance in Alberta during the period 1985–2005, but no change was detected for the period 1995–2005. Given the scale of habitat loss in Alberta, it is puzzling why this species appears to be increasing in some areas. Further research is needed to assess the discrepancies between the Atlas and BBS results. The provincial status of this species is Sensitive due to its requirement for mature trees for nesting.

TEMPORAL REPORTING RATE

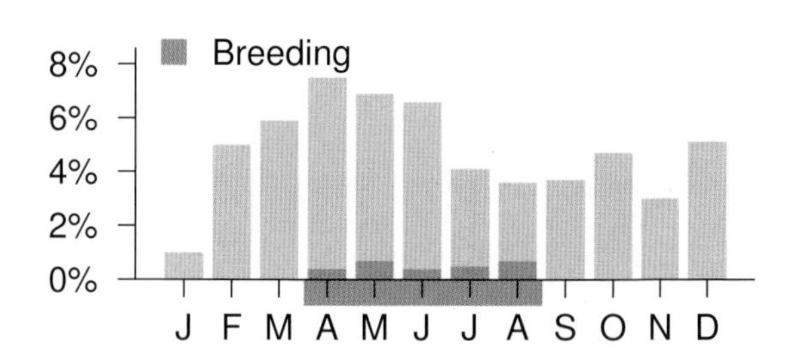

NATURAL REGIONS REPORTING RATE

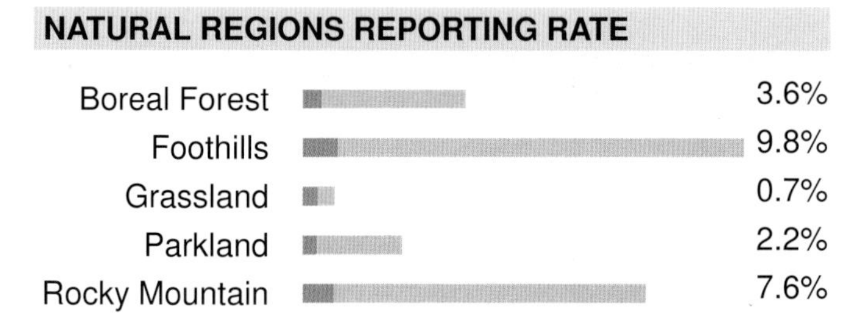

SPATIAL REPORTING RATE

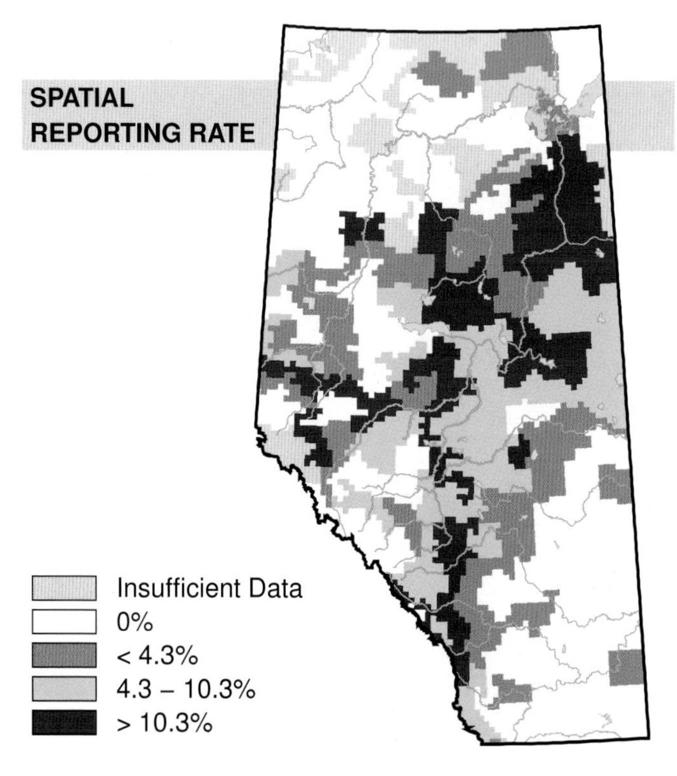

STATUS

GSWA 2000	GSWA 2005	COSEWIC Historic Rankings	COSEWIC 2007	Alberta Wildlife Act
Sensitive	Sensitive	N/A	N/A	N/A

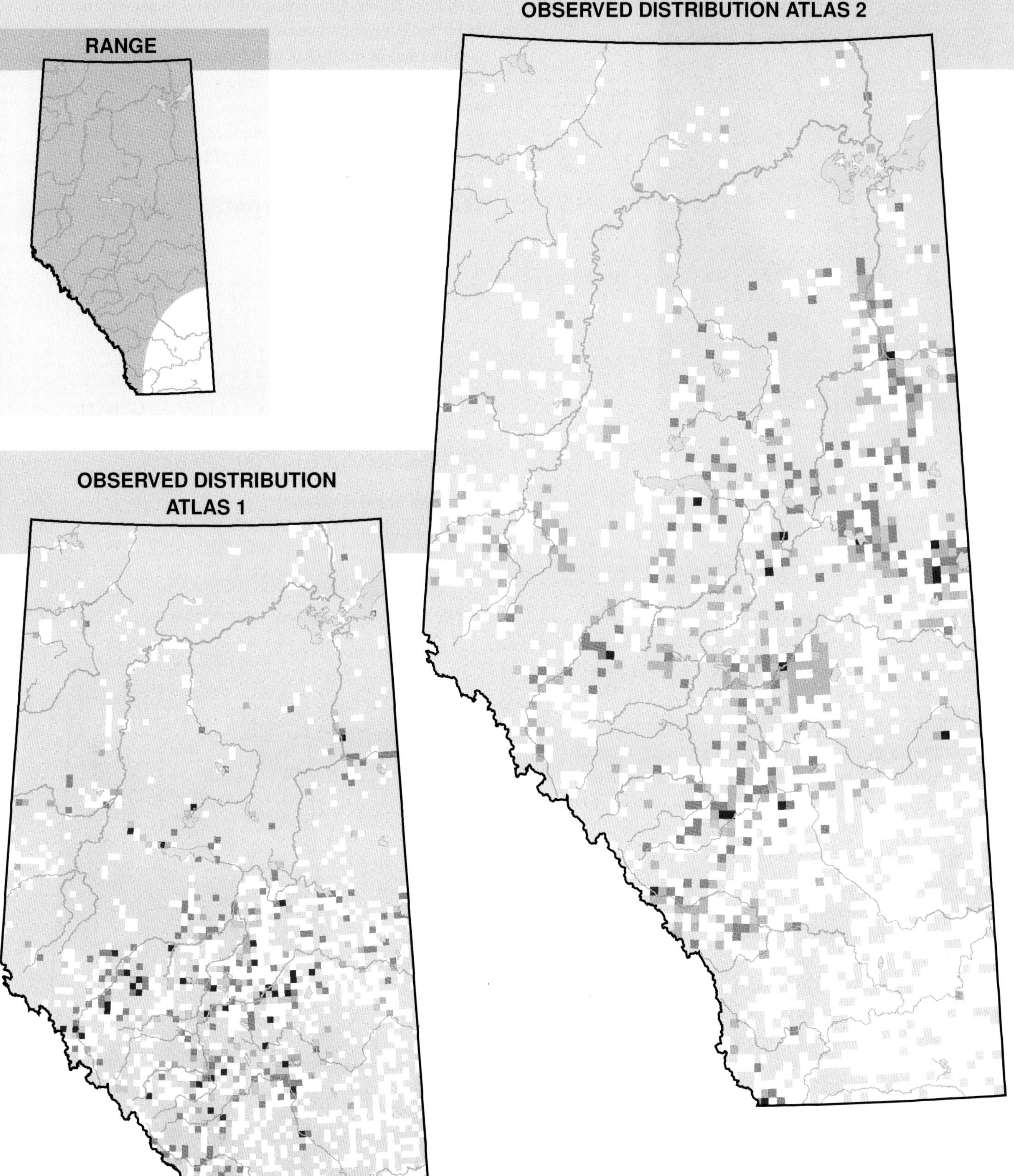
RANGE
OBSERVED DISTRIBUTION ATLAS 2
OBSERVED DISTRIBUTION ATLAS 1
Unsurveyed
Not Observed
Observed
Possible
Probable
Confirmed

Olive-sided Flycatcher (*Contopus cooperi*)

Photo: Royal Alberta Museum

NESTING

CLUTCH SIZE (EGGS):	3–4
INCUBATION (DAYS):	16–17
FLEDGING (DAYS):	21–23
NEST HEIGHT (METRES):	>2

The Olive-sided Flycatcher is found in the northern and central parts of the province south to the North Saskatchewan River in the east, and in the west south through the Rocky Mountain and Foothills Natural Regions to Waterton Lakes. There was no change in distribution in Atlas 2.

It breeds in semi-open coniferous and mixedwood forests along edges and openings, often near water. Tall, prominent trees and snags, which serve as singing and foraging perches, and unobstructed air space for foraging are common features of all nesting habitats (Altman and Sallabanks, 2000).

No changes in relative abundance were detected in the Rocky Mountain NR, increases were detected in the Grassland NR, and decreases were detected in the Boreal Forest, Foothills, and Parkland NRs. The Grassland NR is a marginal part of its overall range in Alberta with mostly migratory observations. The Breeding Bird Survey (1985–2005) indicated a decline in this flycatcher in Alberta and across Canada. As this flycatcher responds positively to certain types of harvested forest, a contradiction appears to exist between the increase in this type of habitat in the breeding range and widespread declines in populations (Altman and Sallabanks, 2000). One hypothesis suggests that populations are affected mostly by loss or alteration of habitat on wintering grounds. Altman and Sallabanks (2000) also indicate that this species may have evolved to depend on natural disturbances that create forest openings and naturally patchy habitat with abundant edge. Fire-suppression policies may reduce suitable habitat, especially for breeding. Preliminary assessment of the importance of post-fire habitat to nest success indicates the need to avoid or minimize salvage of burned trees.

TEMPORAL REPORTING RATE

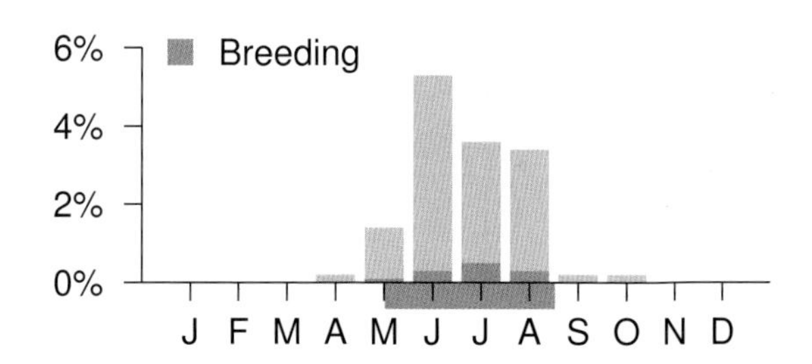

NATURAL REGIONS REPORTING RATE

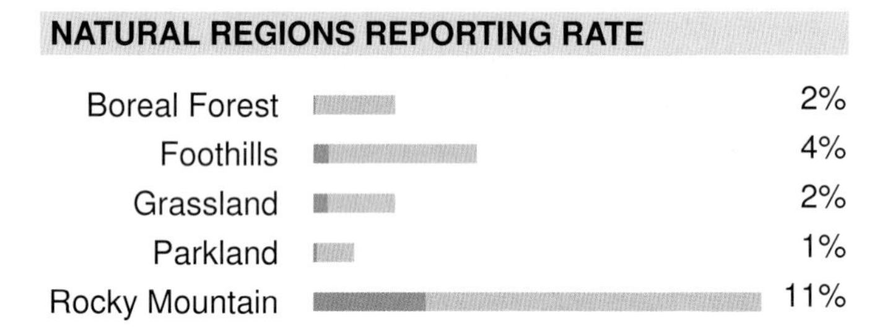

SPATIAL REPORTING RATE

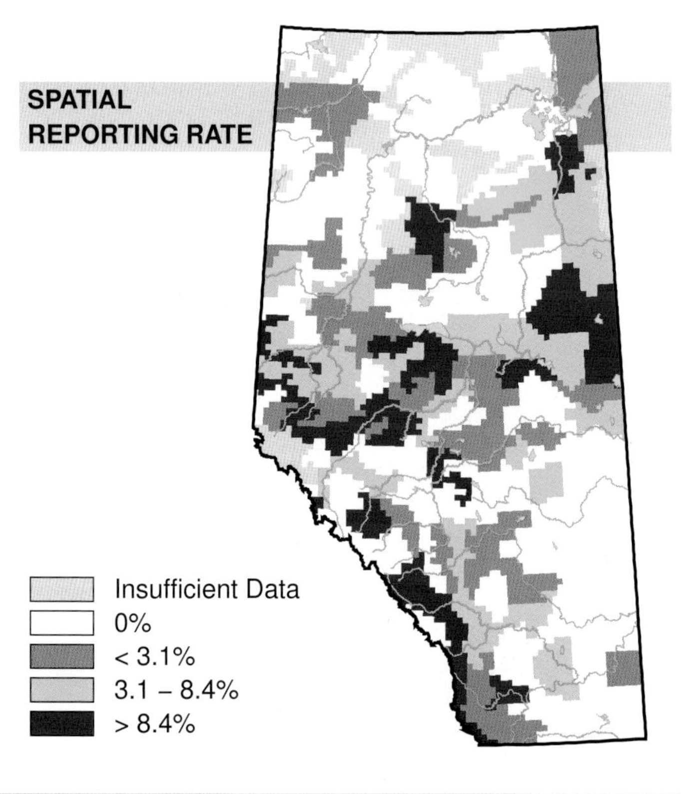

STATUS

GSWA 2000	GSWA 2005	COSEWIC Historic Rankings	COSEWIC 2007	Alberta Wildlife Act
N/A	N/A	N/A	N/A	N/A

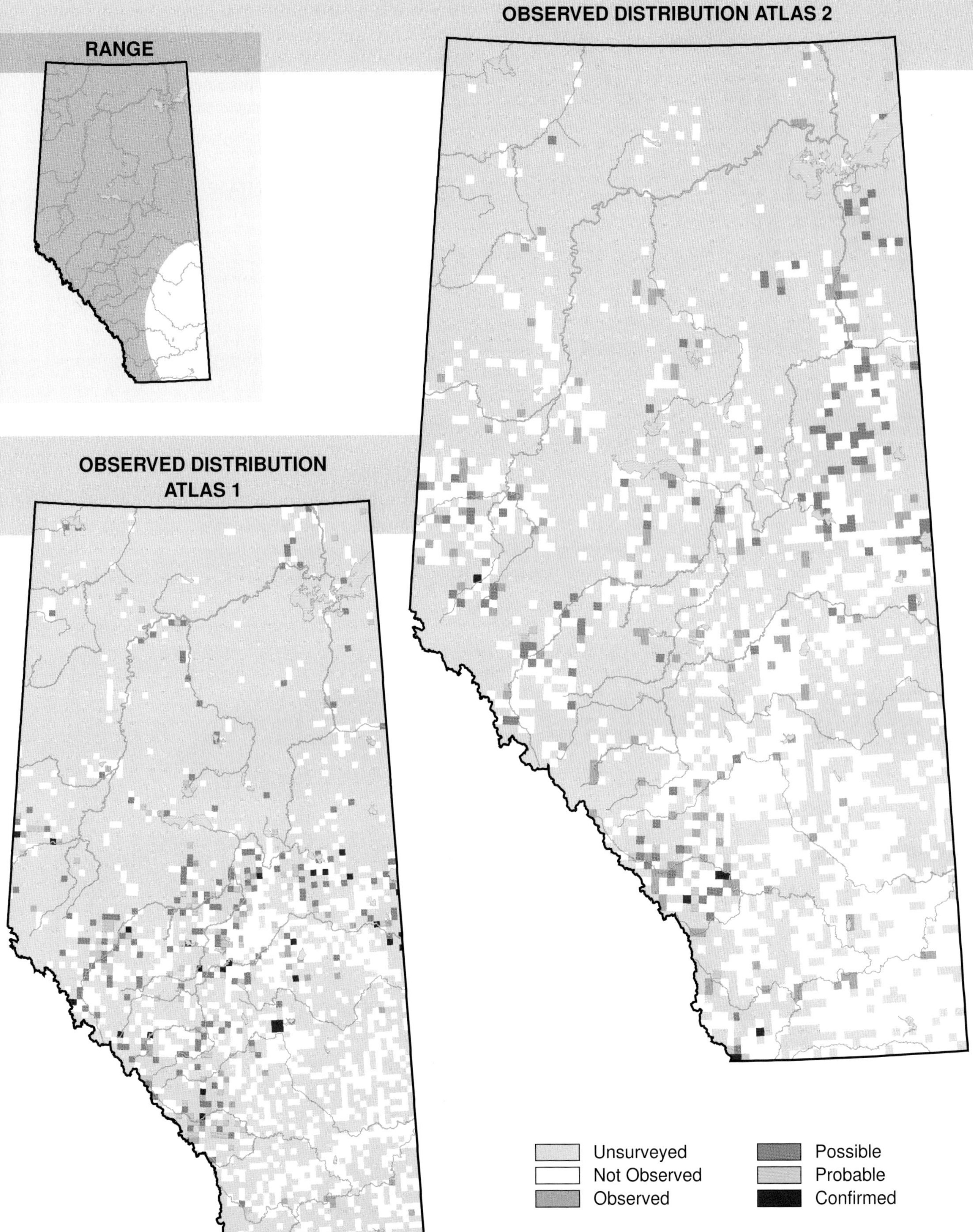
RANGE
OBSERVED DISTRIBUTION ATLAS 1
OBSERVED DISTRIBUTION ATLAS 2
Unsurveyed
Not Observed
Observed
Possible
Probable
Confirmed

Western Wood-Pewee (*Contopus sordidulus*)

Photo: Gerald Romanchuk

NESTING

CLUTCH SIZE (EGGS):	2–4
INCUBATION (DAYS):	12–13
FLEDGING (DAYS):	14–18
NEST HEIGHT (METRES):	3–15

The Western Wood-Pewee is widely distributed in the province. It is found most commonly in the Foothills, Rocky Mountain, and Grassland Natural Regions. There was no change in the observed distribution.

This flycatcher is a habitat generalist, widespread in woodlands and forests, especially in forest edges and riparian zones, but is absent from dense forests (Bemis and Rising, 1999). In the prairie areas of Alberta it nests in well-wooded coulees, cottonwoods along rivers, or farm windbreaks.

Decreases in relative abundance were detected in the Boreal Forest, Foothills, and Parkland NRs. No changes in relative abundance were detected in the Grassland and Rocky Mountain NRs. The Breeding Bird Survey (1985–2005) indicated a decline in relative abundance of the Western Wood-Pewee in Alberta and across Canada.

This species was reviewed under the National action needs for Canadian Landbird Conservation because of the magnitude and persistence of the population decline Dunn (2005).

Several studies have documented the effects of loss of breeding habitat on the Western Wood-Pewee. This loss includes: the elimination of preferred riparian habitats by agriculture, including the continued practice of grazing in these zones; riparian loss to urbanization; and the deleterious effects of clear-cut logging on riparian zones. Another suspected cause identified by researchers is, quite naturally, the rapid destruction of tropical forests where many migrants overwinter. The non-breeding distribution of the Western Wood-Pewee is not that well known; consequently, the reseach priority should be to gather information on this bird's winter distribution and wintering habitats (Bemis and Rising, 1999).

The Western Wood-Pewee is listed as Secure in Alberta.

TEMPORAL REPORTING RATE

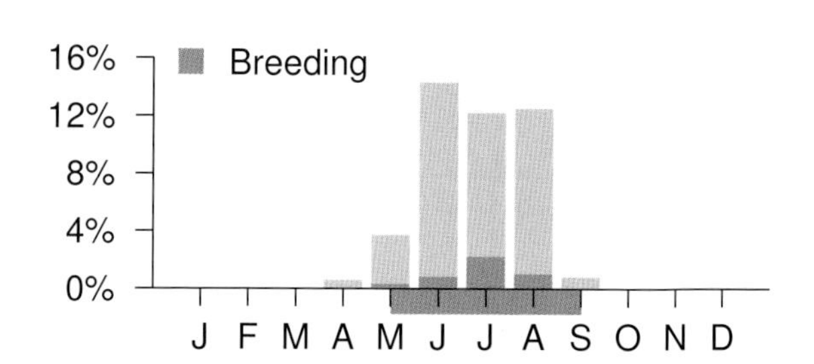

NATURAL REGIONS REPORTING RATE

SPATIAL REPORTING RATE

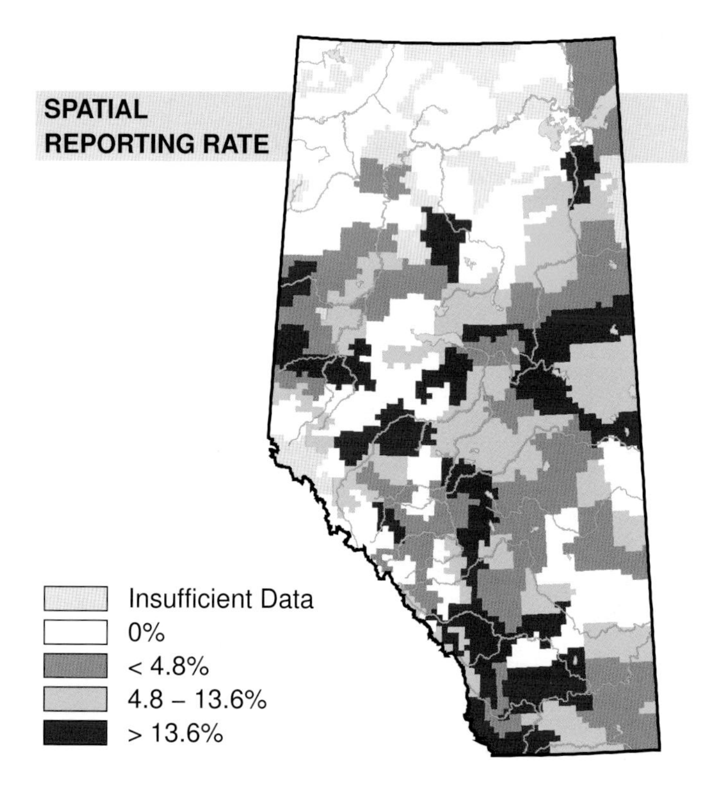

STATUS

GSWA 2000	GSWA 2005	COSEWIC Historic Rankings	COSEWIC 2007	Alberta Wildlife Act
Secure	Secure	N/A	N/A	N/A

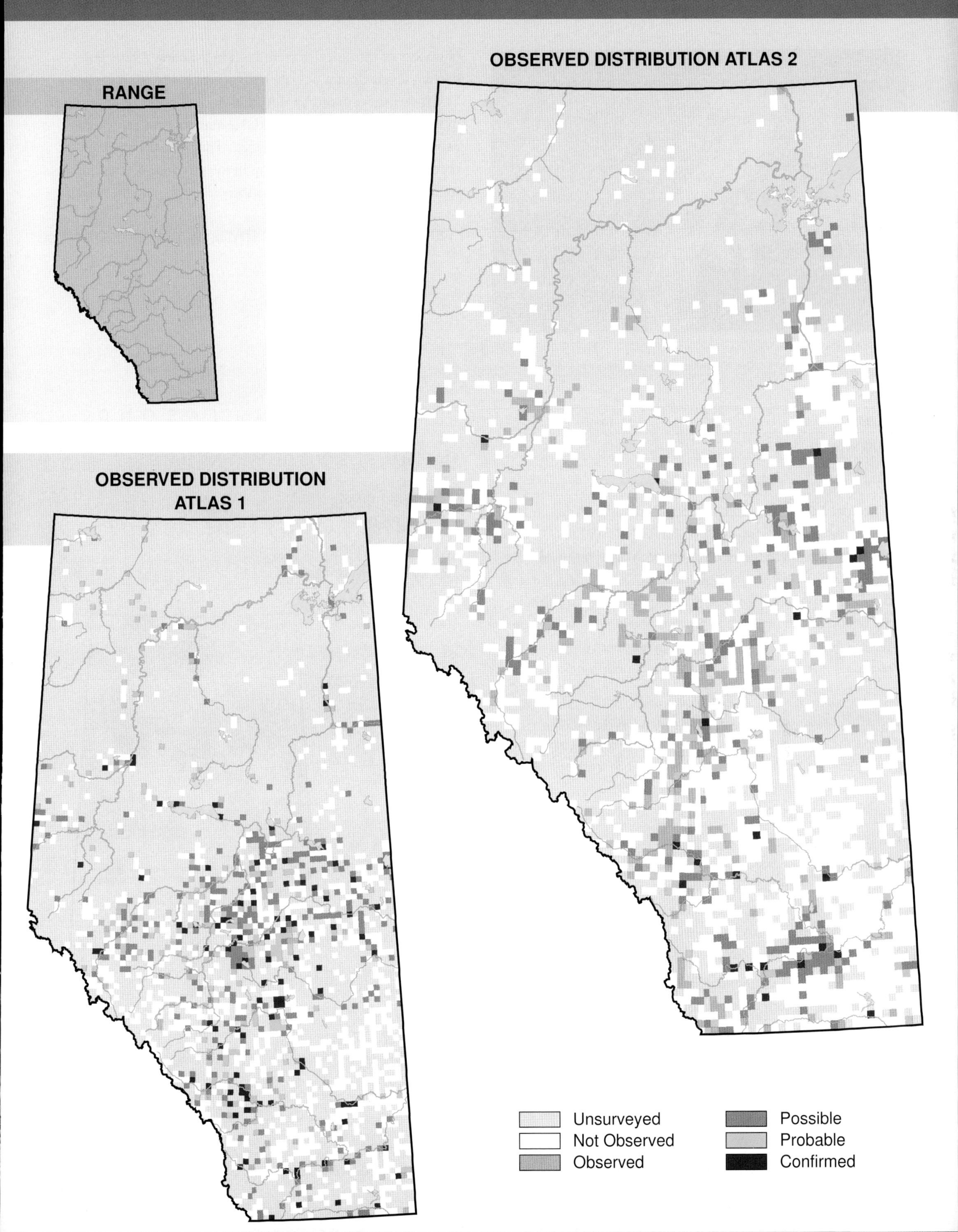
Habitat
Nest
Location
Nest
Type
Diet
RANGE
OBSERVED DISTRIBUTION ATLAS 2
OBSERVED DISTRIBUTION
ATLAS 1
Unsurveyed
Not Observed
Observed
Possible
Probable
Confirmed

Yellow-bellied Flycatcher (*Empidonax flaviventris*)

Photo: Royal Alberta Museum

NESTING

CLUTCH SIZE (EGGS):	3–5
INCUBATION (DAYS):	12–14
FLEDGING (DAYS):	13–14
NEST HEIGHT (METRES):	0

The Yellow-bellied Flycatcher is not common in Alberta. It breeds in the Boreal Forest and Foothills NRs and is occasionally found in the Rocky Mountain and Parkland NRs. Its distribution in the Prairies and British Columbia is not well defined. Its preference for boggy, mosquito-infested habitats and its habit of arriving late in the spring probably contribute to its low detectability. The population numbers of this neotropical migrant could be affected by conditions experienced on its wintering grounds.

With all Empidonax flycatchers there is concern about misidentification. If this uncertainty does introduce a bias, it would likely be to increase the number of observations of the more common Least Flycatcher and reduce the numbers of Yellow-bellied Flycatchers, given their somewhat similar songs.

This species shows three pockets of distribution in Alberta: Cypress Hills, south of Fort McMurray, and the area around Grande Cache. It is not common in any habitat but was observed more frequently in Atlas 2 in the Boreal Forest, Foothills, and Rocky Mountain NRs than in Atlas 1.

An increase in relative abundance was detected in the Boreal Forest NR. No pattern of increase or decline of Yellow-bellied Flycatchers was detected by the Breeding Bird Survey and there are insufficient data for an Alberta status summary.

Numbers of this species were high in the early 1980s but current levels do not deviate much from long-term averages. The increase recorded by the Atlas could be a function of the more extensive coverage in Atlas 2 and may not represent actual increases in population size. This is a species that continues to require monitoring in the province to determine its specific habitat requirements, its distribution, and its status.

TEMPORAL REPORTING RATE

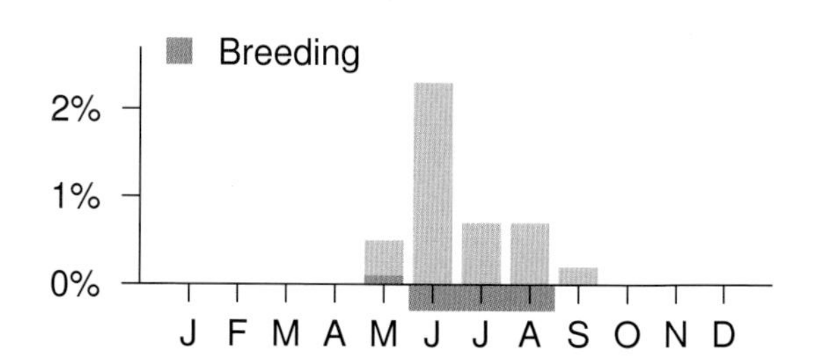

NATURAL REGIONS REPORTING RATE

SPATIAL REPORTING RATE

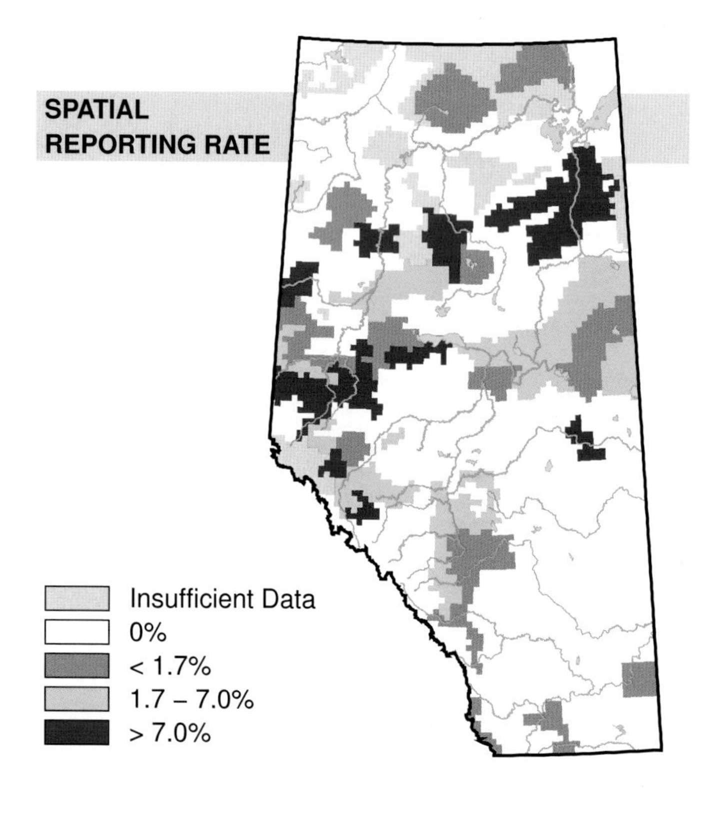

STATUS

GSWA 2000	GSWA 2005	COSEWIC Historic Rankings	COSEWIC 2007	Alberta Wildlife Act
Undetermined	Undetermined	N/A	N/A	N/A

Habitat Nest Location Nest Type Diet 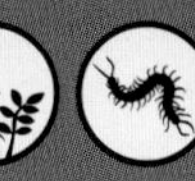

RANGE

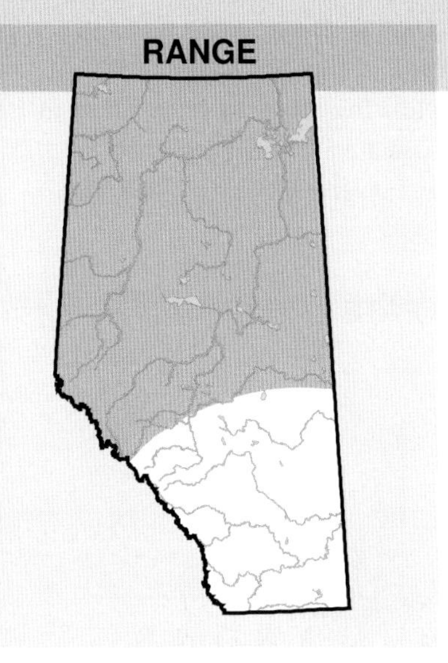

OBSERVED DISTRIBUTION ATLAS 2

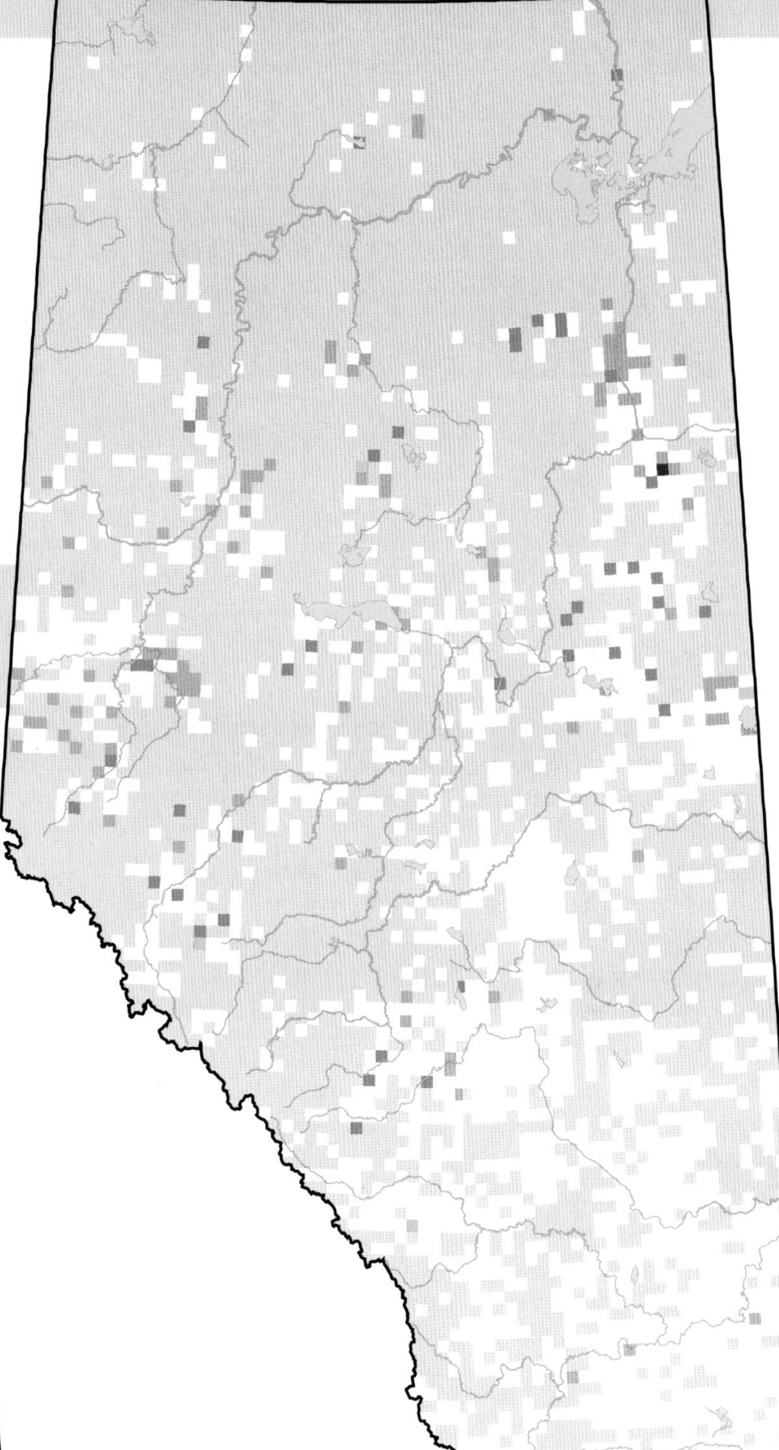

OBSERVED DISTRIBUTION ATLAS 1

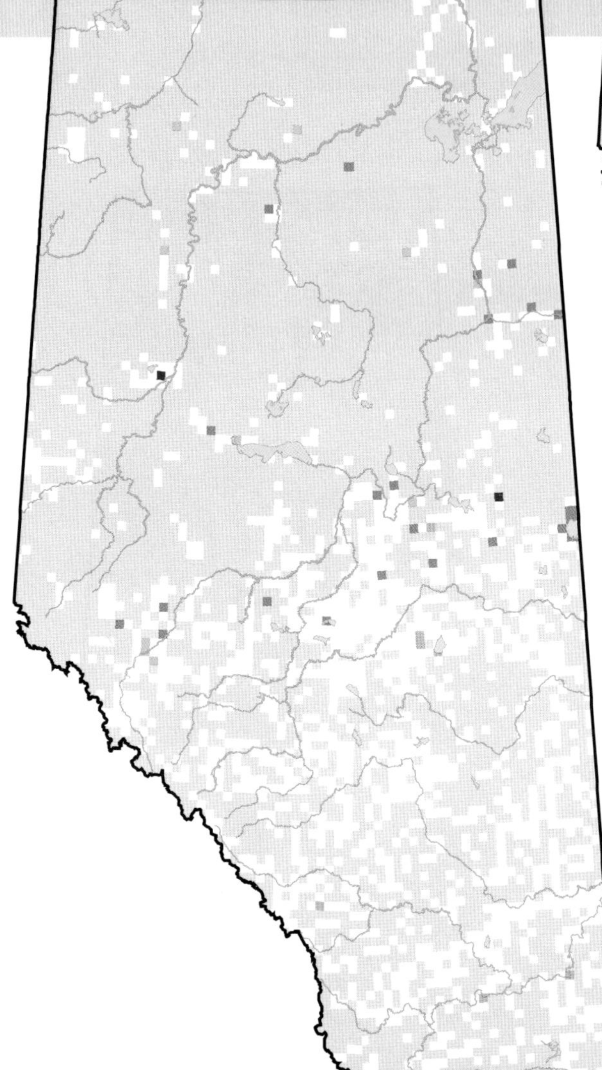

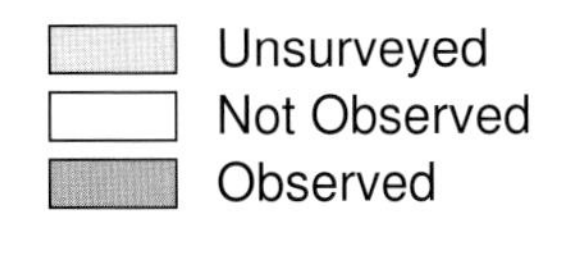

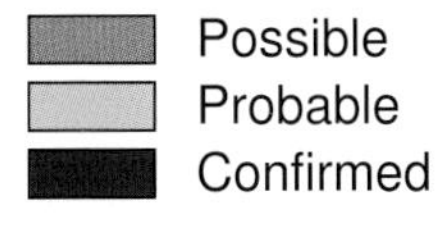

Alder Flycatcher (*Empidonax alnorum*)

Photo: Gerald Romanchuk

NESTING

CLUTCH SIZE (EGGS):	3–4
INCUBATION (DAYS):	12–13
FLEDGING (DAYS):	13–14
NEST HEIGHT (METRES):	0.3–9.1

The Alder Flycatcher breeds in all regions of Alberta except the dry grasslands of the southeast. It is most likely encountered in the Boreal Forest and Foothills Natural Regions but is also a regular breeder in the Parkland NR. This species cannot be distinguished from the Willow Flycatcher in the field except by its call. The three-syllable "fee-bee-o" of the Alder Flycatcher is diagnostic, but its two-syllable "zwee-oo" call can be confused with the two-syllable "fitz-bew" call of the Willow Flycatcher. Observations of either species, where they occur sympatrically, have to be accepted with caution. The observed distribution of the Alder Flycatcher has not changed substantively since Atlas 1.

Alder Flycatchers can be found throughout the range of the Willow Flycatcher in Alberta. In an ideal world the Alder Flycatcher would be found in association with alder bushes and the Willow Flycatcher with willow bushes, but unfortunately this is not so. Any wet, boggy environment with thick, low, woody bushes for nesting and tall bushes or trees for song perches suits this species. The area of highest concentrations of this species in Atlas 2 were in areas around, and to the north of, Cold Lake-Lac la Biche and the region around Grande Cache.

A decrease in relative abundance was detected in the Grassland NR. This reduction is consistent with reductions in abundance observed through Breeding Bird Surveys in Alberta. The downward trend observed in Alberta is also seen in Saskatchewan but not consistently elsewhere in Canada.

Drainage of wetlands, loss of habitat due to development, and loss of wintering habitat are all possible causes for a decline. An increase was detected in the Parkland NR, but relative to other species, it was observed less frequently in Atlas 2 than in Atlas 1. Reasons for this increase are not understood and without corroboration, must be viewed cautiously. The status of this species is Secure.

TEMPORAL REPORTING RATE

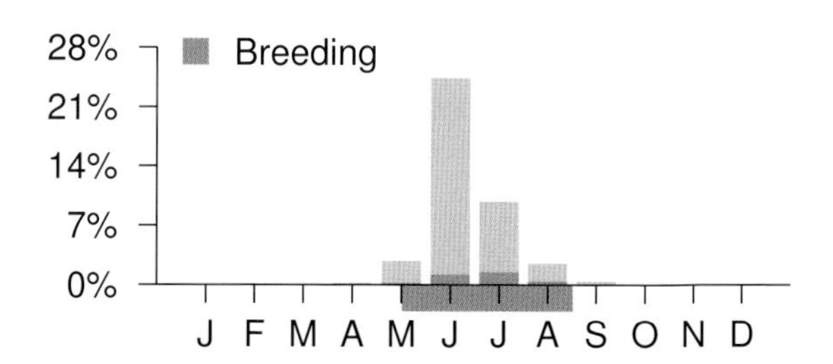

NATURAL REGIONS REPORTING RATE

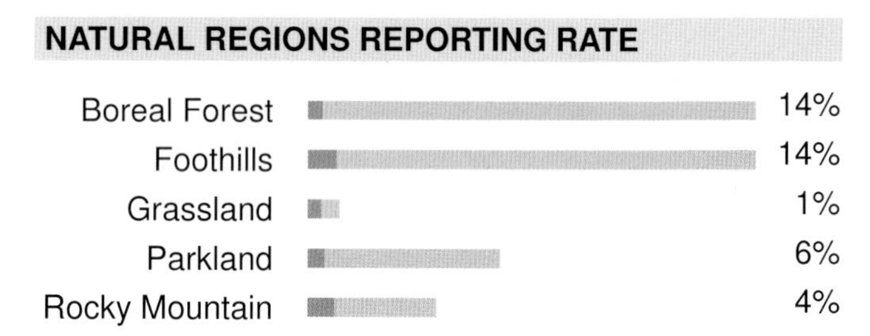

SPATIAL REPORTING RATE

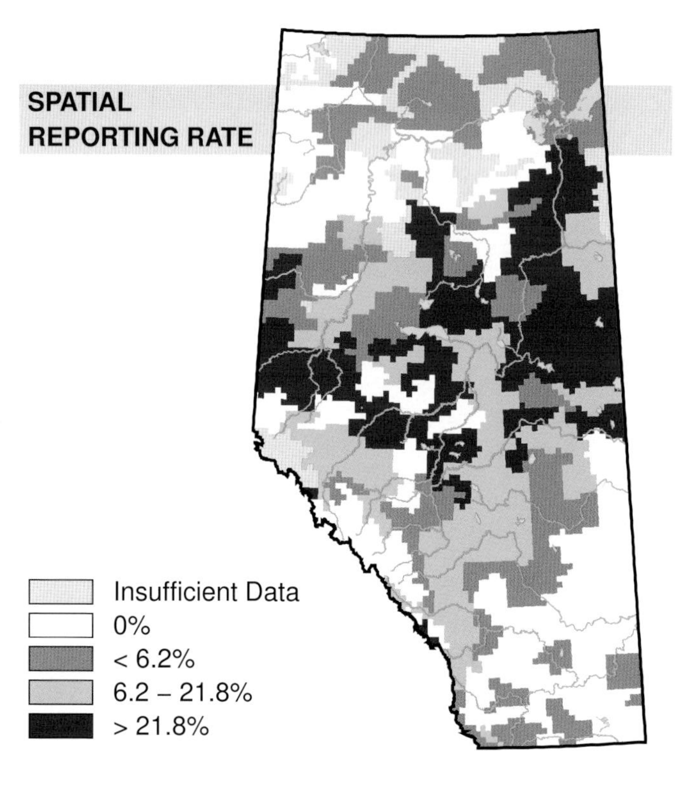

STATUS

GSWA 2000	GSWA 2005	COSEWIC Historic Rankings	COSEWIC 2007	Alberta Wildlife Act
Secure	Secure	N/A	N/A	N/A

RANGE

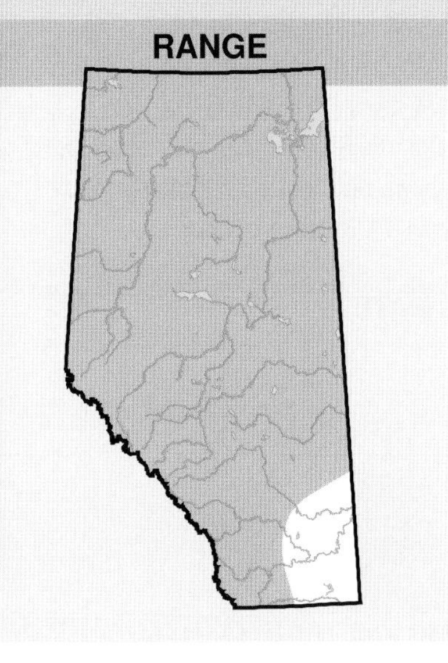

OBSERVED DISTRIBUTION ATLAS 2

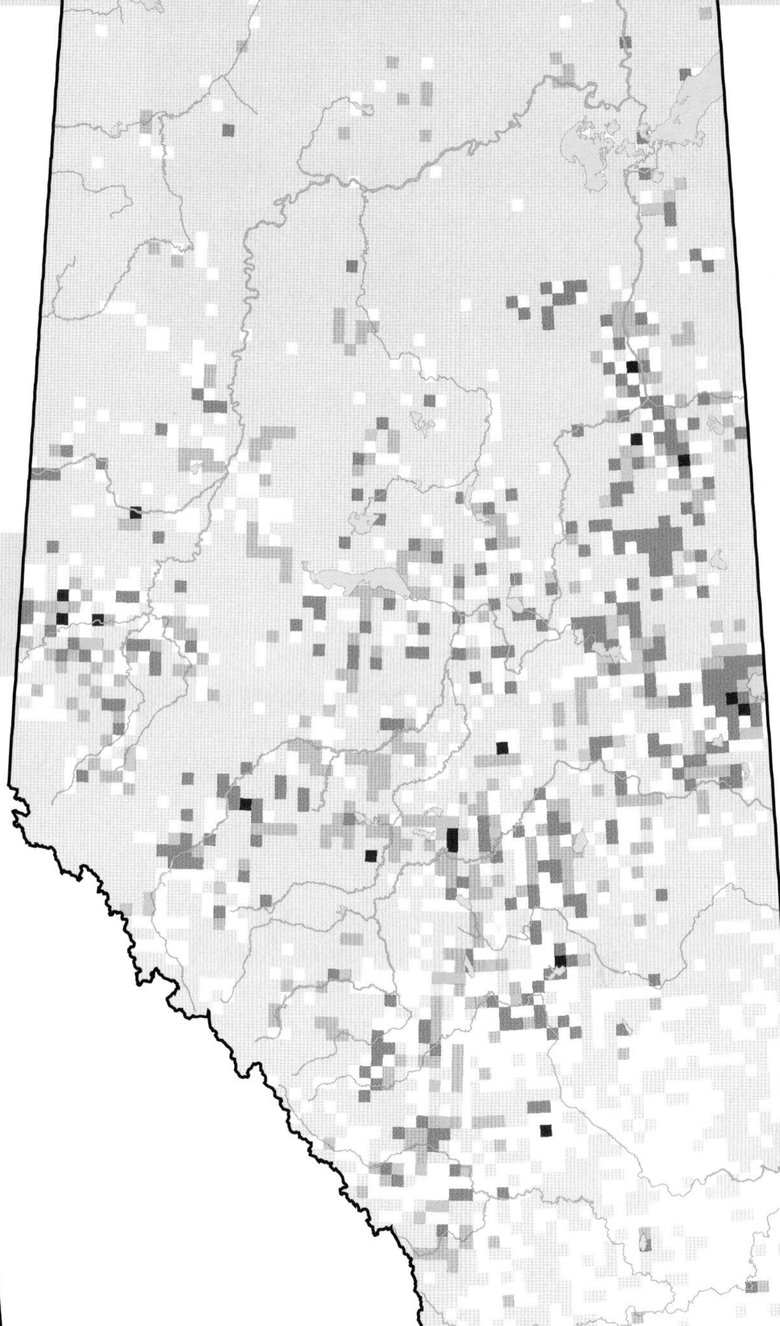

OBSERVED DISTRIBUTION ATLAS 1

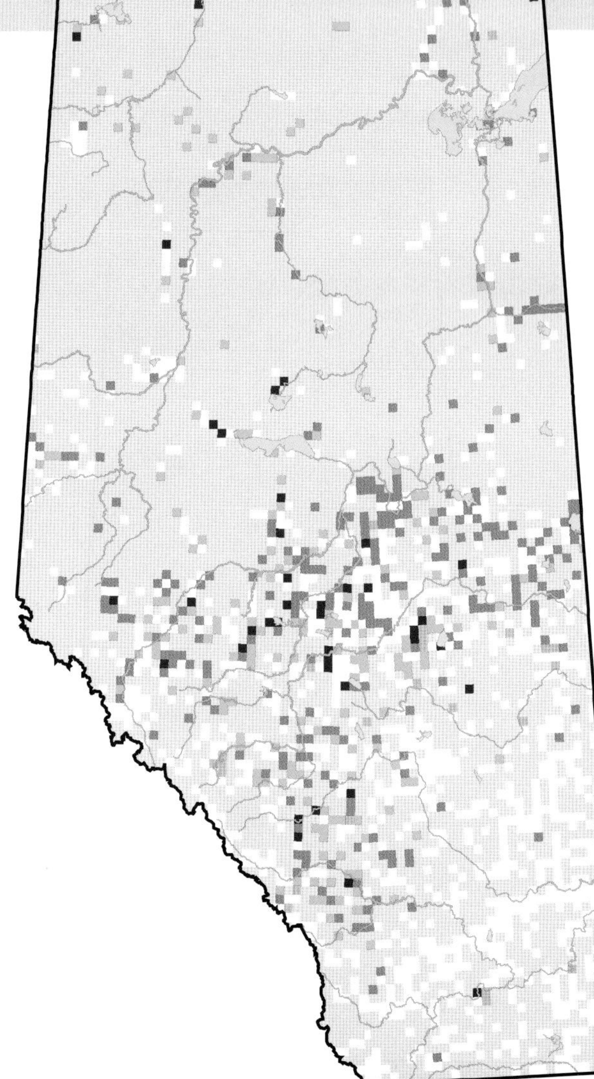

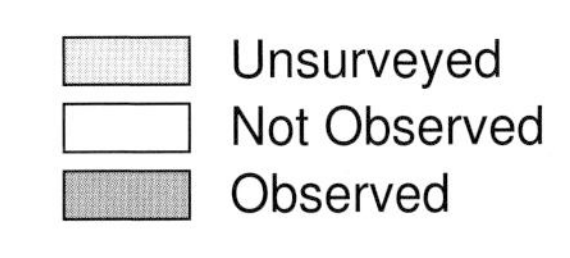

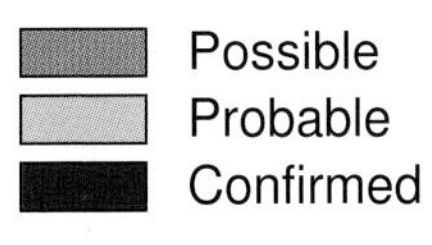

Willow Flycatcher (*Empidonax traillii*)

Photo: Damon Calderwood

NESTING

CLUTCH SIZE (EGGS):	3–4
INCUBATION (DAYS):	12–13
FLEDGING (DAYS):	12–14
NEST HEIGHT (METRES):	2

The Willow Flycatcher breeds in a limited area of foothills (Bow Valley) and mountain (Banff to Jasper) habitats. This species cannot be distinguished from the Alder Flycatcher in the field except by its call. The Alder's three-syllable "fee-bee-o" is diagnostic, but its two-syllable "zwee-oo" call can be confused with the Willow's two-syllable "fitz-bew" call. Where these species occur sympatrically, records must be accepted with caution.

Observations of Willow Flycatchers recorded outside their core range during Atlas 2 were probably mis-identified Alder Flycatchers. The two species overlap extensively in habitat use, even to the point of usurping song perches in consecutive years. Other studies elsewhere have found that Willow Flycatchers prefer drier habitats, but this is clearly not true in Alberta. Wet, boggy environments with thick, low, woody bushes for nesting and tall bushes or trees for song perches suit this species throughout its limited Alberta distribution.

The area of highest concentration of this species in Atlas 2 was in the vicinity of Bow Valley Park and northern Kananaskis country. Scattered observations were made farther north to the latitude of Jasper. Seemingly similar habitats and many that support Alder Flycatchers elsewhere in Alberta are not used by this species. Currently there is no good explanation for the distribution of Willow Flycatchers in Alberta.

No change in relative abundance was detected in any NR. The Breeding Bird Survey did not detect any change provincially or nationally for the period 1985–2005. The very limited distribution of this species in Alberta increases its vulnerability to habitat loss or change. The Willow Flycatcher, like the Alder Flycatcher, faces obstacles due to drainage of wetlands, loss of habitat due to development, and loss of wintering habitat. The status of this species is Secure in Alberta.

TEMPORAL REPORTING RATE

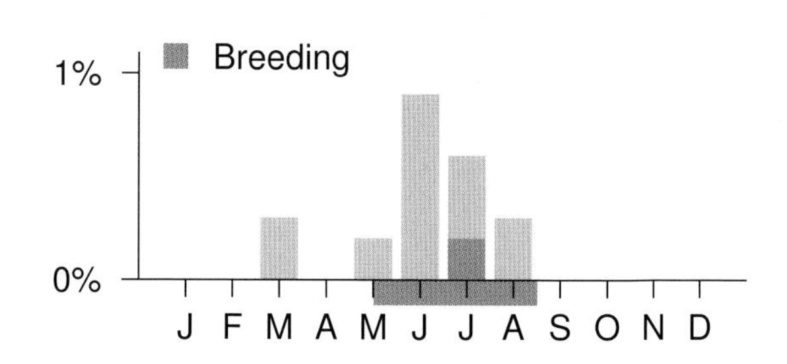

NATURAL REGIONS REPORTING RATE

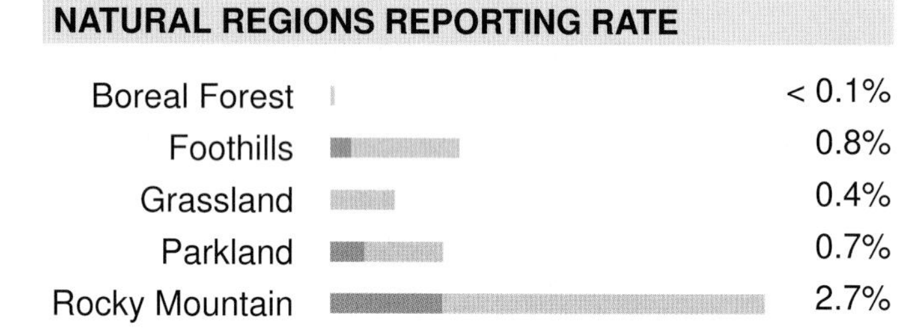

SPATIAL REPORTING RATE

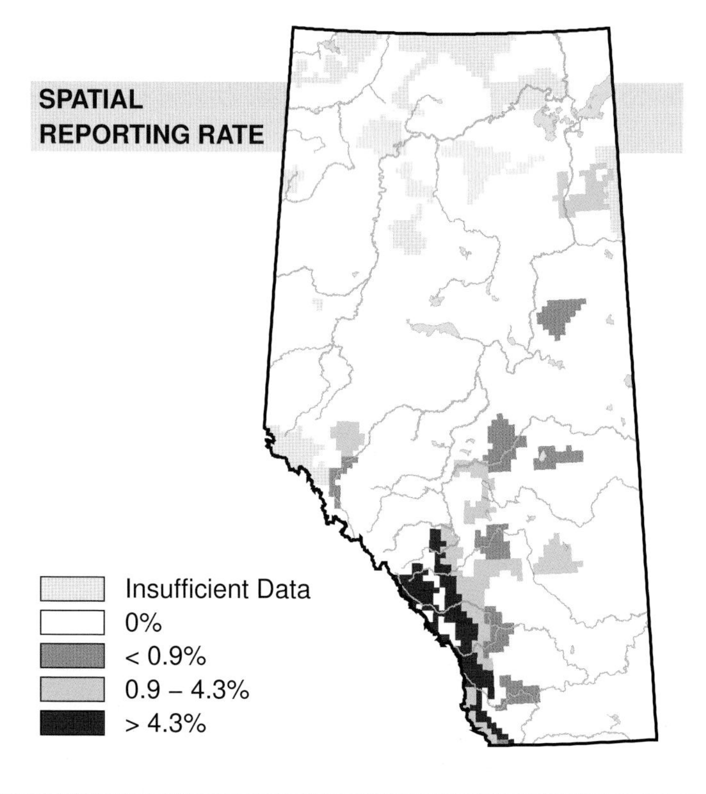

STATUS

GSWA 2000	GSWA 2005	COSEWIC Historic Rankings	COSEWIC 2007	Alberta Wildlife Act
Secure	Secure	N/A	N/A	N/A

RANGE

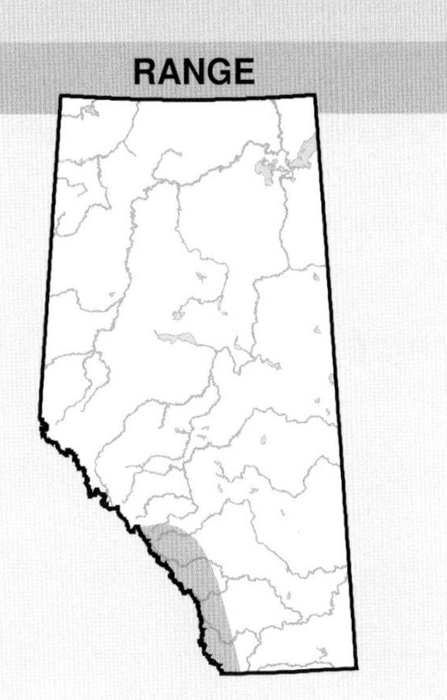

OBSERVED DISTRIBUTION ATLAS 2

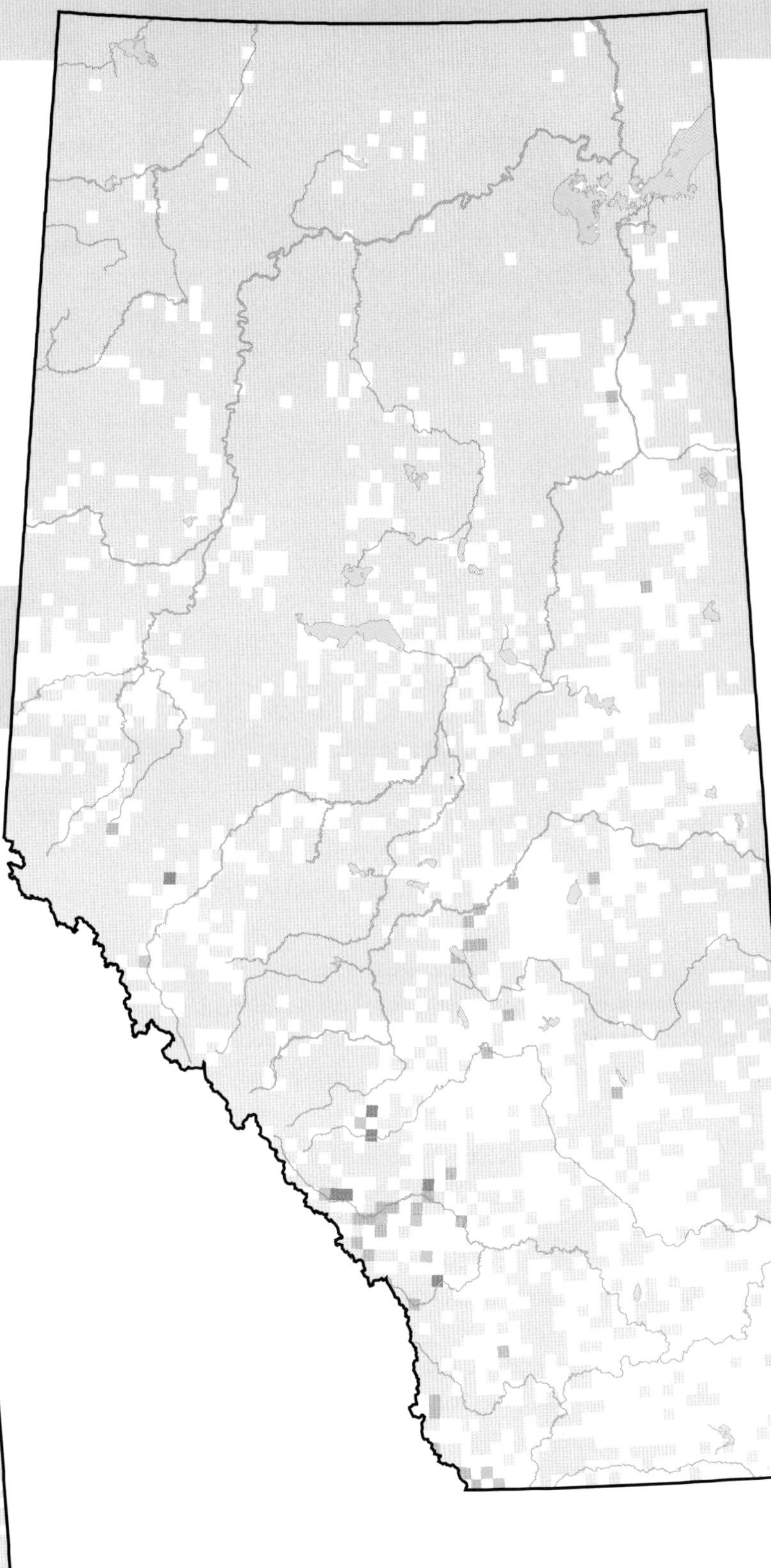

OBSERVED DISTRIBUTION ATLAS 1

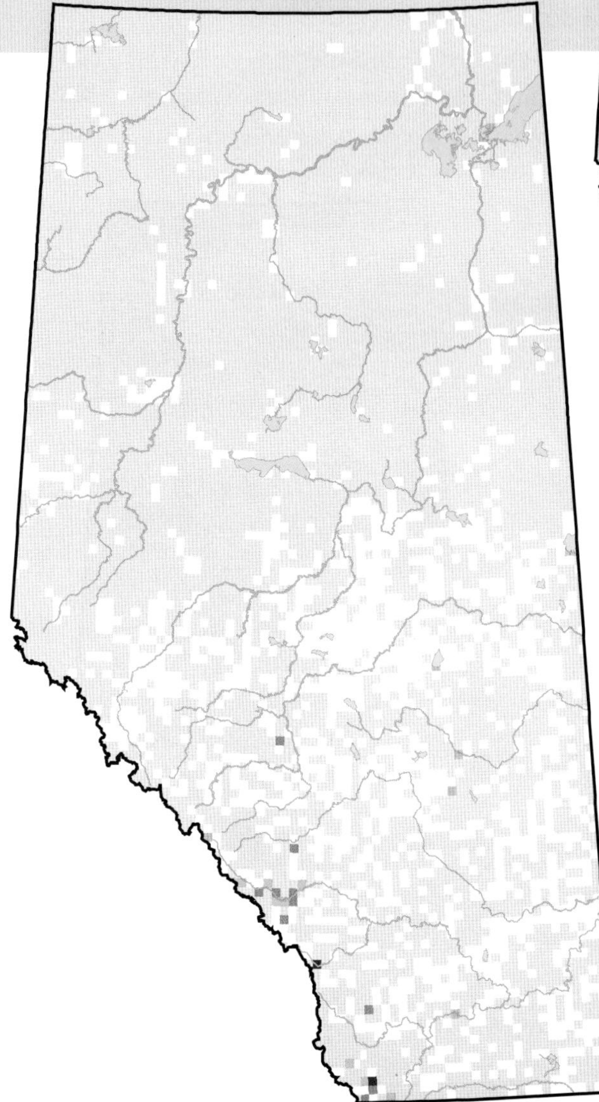

Unsurveyed
Not Observed
Observed
Possible
Probable
Confirmed

Least Flycatcher (*Empidonax minimus*)

Photo: Gerald Romanchuk

NESTING

CLUTCH SIZE (EGGS):	3–6
INCUBATION (DAYS):	12–16
FLEDGING (DAYS):	13–16
NEST HEIGHT (METRES):	2–8

The Least Flycatcher is the most common forest flycatcher in Alberta. In both Atlas projects, it was found to breed in every Natural Region. The Least Flycatcher winters in the Neotropics, arrives in Alberta in May, and is gone in September. Sightings outside this period, which corresponds to the availability of flying insects, are either misidentifications or individuals that have failed to migrate and will likely perish.

At a general level there is little difference between the results of the two Atlas projects. The Least Flycatcher was commonly seen in all forested habitats, with the Parkland and Boreal Forest NRs being the regions of highest concentration. Least Flycatchers are common urban birds, too, because they find parks, golf courses, and even mature backyards to their liking.

A decline in relative abundance was detected in the Boreal Forest, Grassland, and Parkland NRs where, relative to other species, this flycatcher was observed less frequently in Atlas 2 than in Atlas 1. This change is also reflected in Breeding Bird Survey Data and provincial status reports. In contrast, an increase in relative abundance was detected in the Rocky Mountain NR where, relative to other species, it was observed more frequently in Atlas 2 than in Atlas 1. Breeding Bird Survey data show evidence of a decline in the numbers of this species from Ontario westward beginning in the late 1960s.

This species is a good example of one whose decline has been masked by the fact that is still seen regularly. Although identifying the cause of the decline in the numbers of neotropical migrants is difficult, the ability of this species to tolerate humans and disturbance and to breed in secondary forest growth, suggest that the cause of its decline may well be outside Alberta. Increases in the Rocky Mountain NR may be related to opening of the predominantly coniferous forests through fire, logging, and other disturbance, creating more suitable habitat for this flycatcher.

TEMPORAL REPORTING RATE

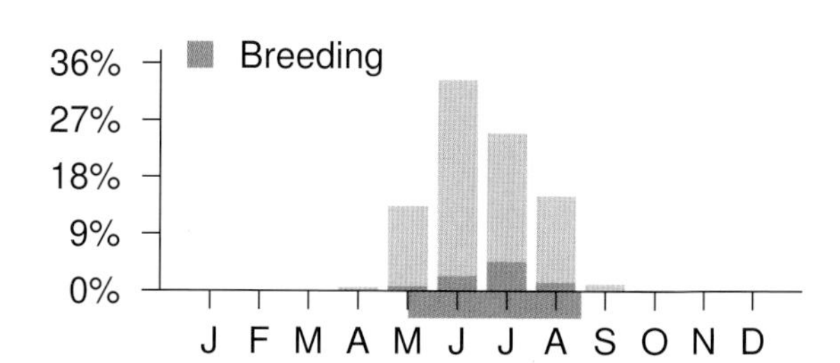

NATURAL REGIONS REPORTING RATE

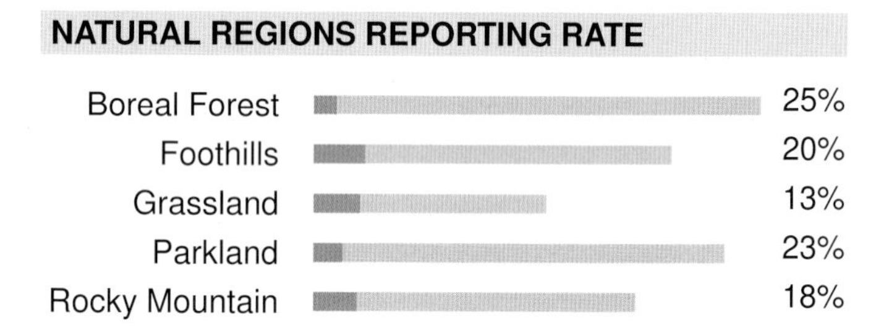

SPATIAL REPORTING RATE

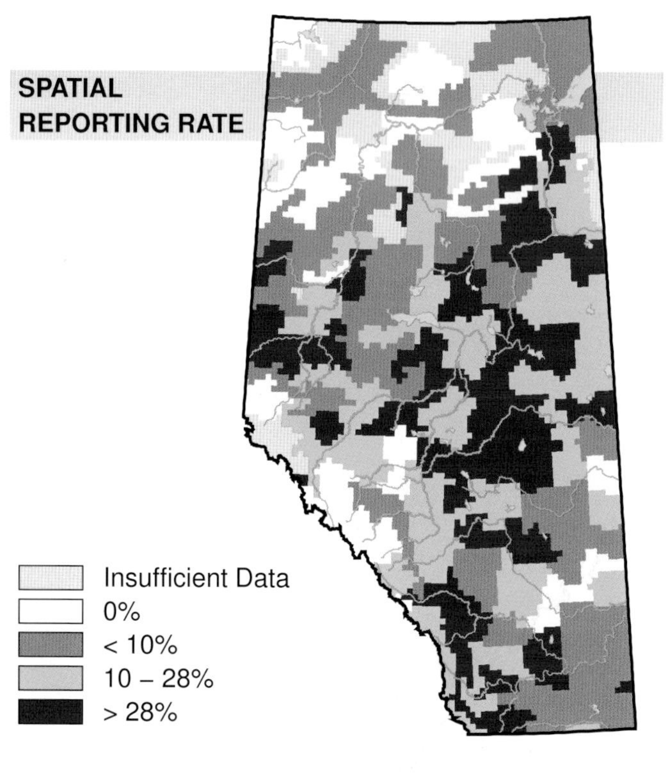

STATUS

GSWA 2000	GSWA 2005	COSEWIC Historic Rankings	COSEWIC 2007	Alberta Wildlife Act
Secure	Sensitive	N/A	N/A	N/A

Habitat

Nest
Location

Nest
Type

Diet

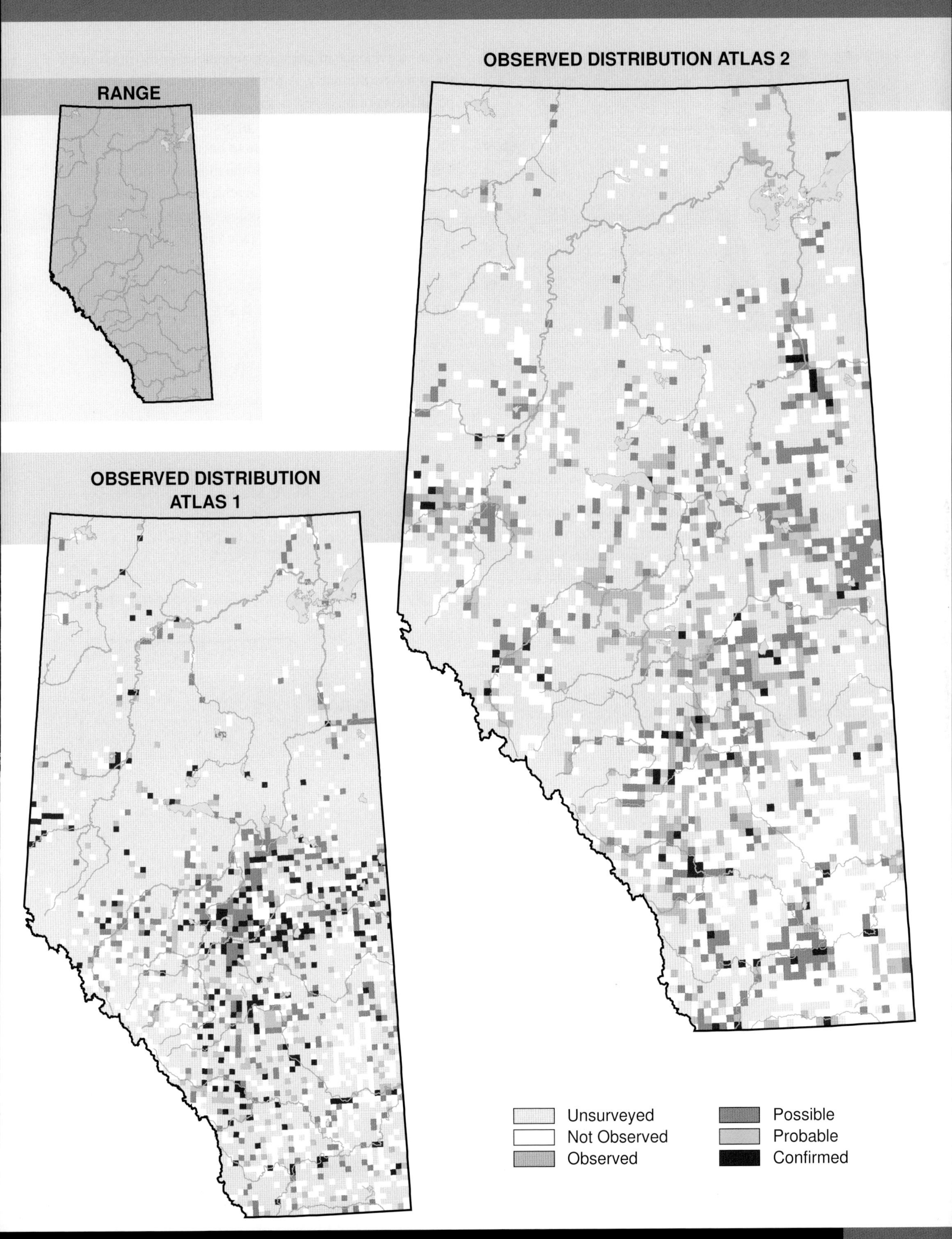
OBSERVED DISTRIBUTION ATLAS 2
RANGE
OBSERVED DISTRIBUTION
ATLAS 1
Unsurveyed
Not Observed
Observed
Possible
Probable
Confirmed

Hammond's Flycatcher (*Empidonax hammondii*)

Photo: Damon Calderwood

NESTING

CLUTCH SIZE (EGGS):	3–4
INCUBATION (DAYS):	12–15
FLEDGING (DAYS):	17–18
NEST HEIGHT (METRES):	8–18

Hammond's Flycatcher has a breeding distribution similar to that of the Dusky Flycatcher throughout its western North American range. In Canada it breeds throughout much of British Columbia, but in Alberta it is found only in the Rocky Mountain Natural Region. Sightings outside this region must be considered accidentals or misidentifications. It was not detected in the Cypress Hills in either Atlas. It is easily confused with the Dusky Flycatcher unless the observer is familiar with the distinctive aspects of their songs. Like the rest of the Empidonax-group, Hammond's Flycatcher is a neotropical migrant and subject to loss of wintering habitat. It prefers cool, forested regions in central Mexico and Central America for its winter range. It migrates northward earlier in the spring than the other members of its genus. It is, perhaps, better able to tolerate inclement weather than its relative, the Dusky Flycatcher.

Hammond's Flycatcher prefers coniferous forests and is quite common just west of the Alberta/BC border, in the Columbia River valley where Douglas Fir, Western Larch, and Ponderosa Pine predominate. There were few observations of this species in either Atlas and no substantive difference noted between them. Observers recorded this species in Waterton Lakes, Banff, and Jasper national parks.

No change in relative abundance was detected by the Atlas. Breeding Bird Survey data primarily from British Columbia show no evidence of changing abundance or distribution for this species. Similarly, Alberta's status report indicates that populations of this species are Secure. However, having been recorded in such low numbers in both Atlases, Hammond's Flycatcher is clearly vulnerable to habitat change, such as Mountain Pine Beetle damage–the moreso given the bird's restricted range and association with habitats dominated by Lodgepole Pine in Alberta. Ironically, these same changes may enhance the suitability of montane habitats for the Dusky Flycatcher.

TEMPORAL REPORTING RATE

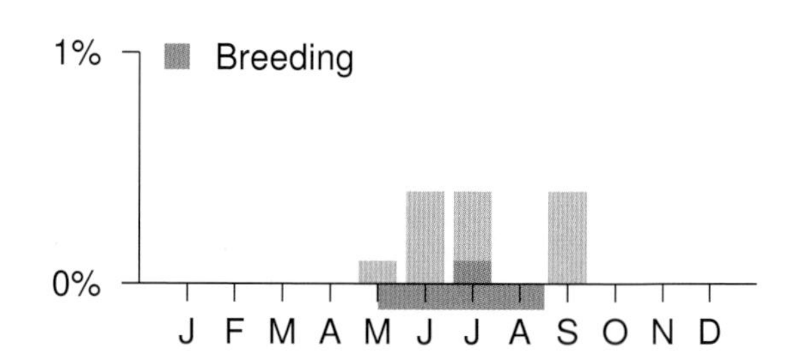

NATURAL REGIONS REPORTING RATE

SPATIAL REPORTING RATE

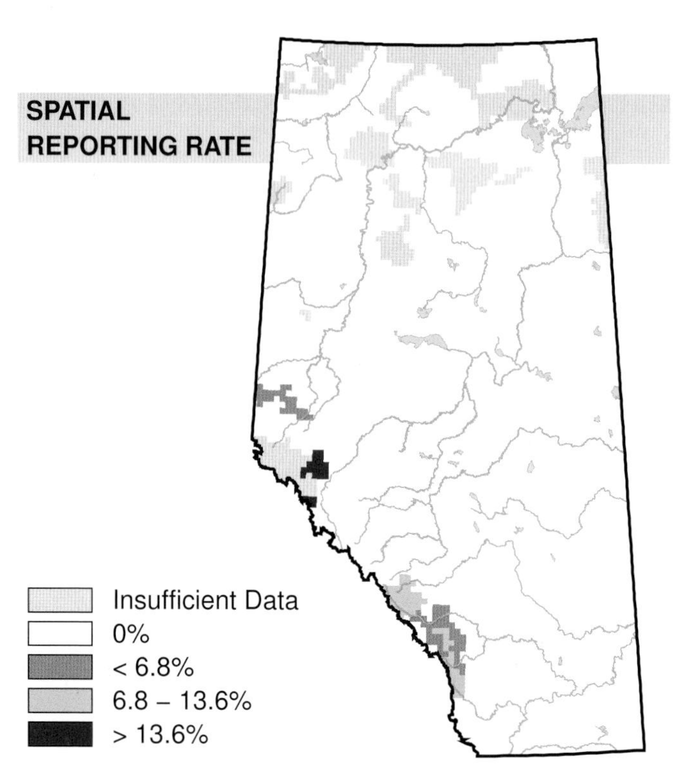

STATUS

GSWA 2000	GSWA 2005	COSEWIC Historic Rankings	COSEWIC 2007	Alberta Wildlife Act
Secure	Secure	N/A	N/A	N/A

Nest
Type

Diet

RANGE

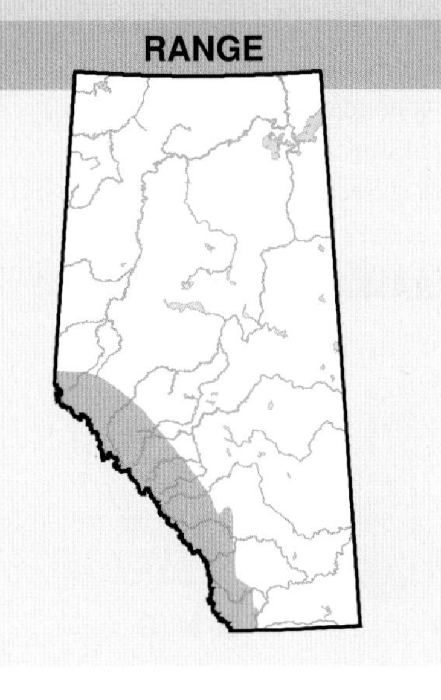

OBSERVED DISTRIBUTION ATLAS 2

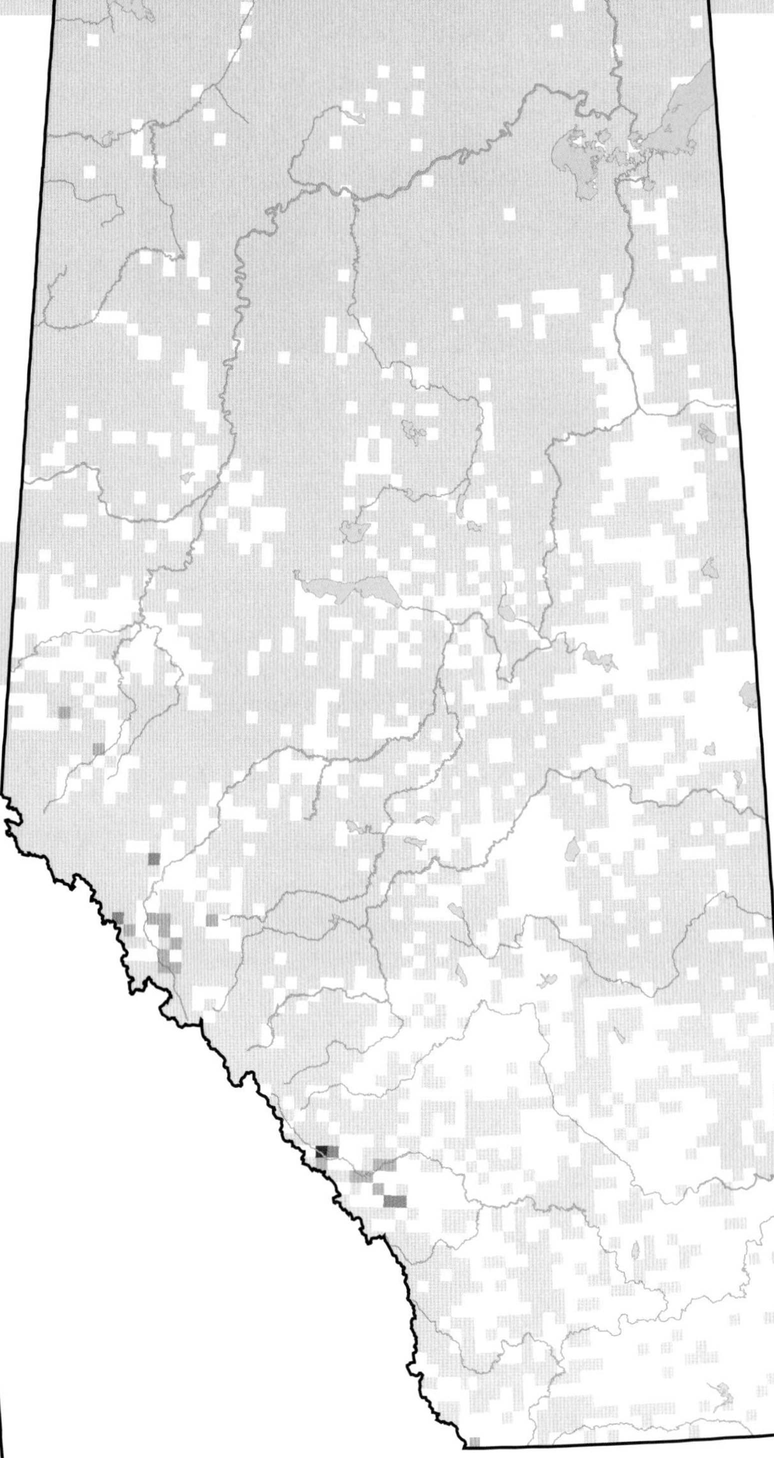

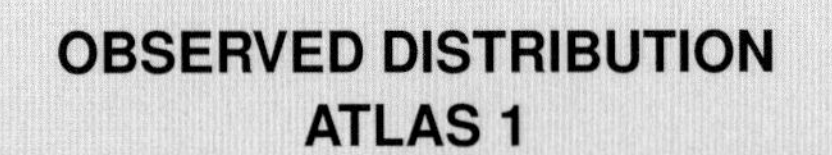

OBSERVED DISTRIBUTION ATLAS 1

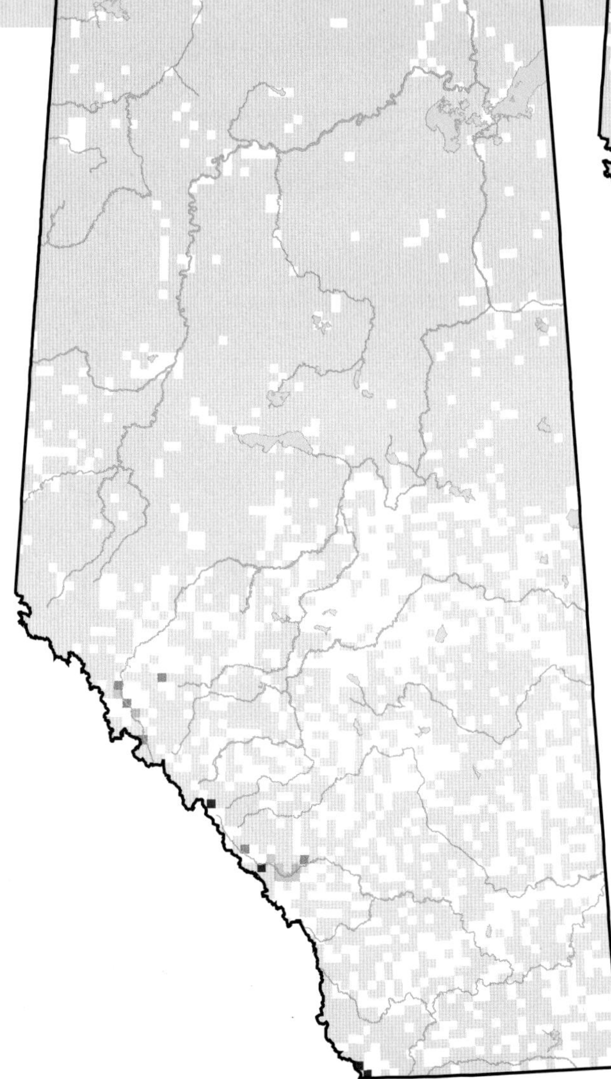

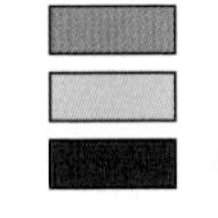

Unsurveyed
Not Observed
Observed
Possible
Probable
Confirmed

Dusky Flycatcher (*Empidonax oberholseri*)

Photo: Damon Calderwood

NESTING

CLUTCH SIZE (EGGS):	3–4
INCUBATION (DAYS):	12–15
FLEDGING (DAYS):	18
NEST HEIGHT (METRES):	1–2

The Dusky Flycatcher breeds in western North America. In Canada, it is most common in British Columbia, but it occurs in Alberta in the Rocky Mountain and Foothills Natural Regions. This species breeds in the Cypress Hills, as well, opening up the possibility of sightings of birds between these hills and the Rocky Mountain foothills to the west. Observations east of the foothills are potentially either migrants, incorrect identifications, or accidentals.

There is little difference in the distributions of this species based on observations in Atlases 1 and 2. There are more observations of this species in central Alberta in Atlas 2 which may suggest a movement eastward, but additional confirmed records would be needed to rule out misidentifications. The Dusky Flycatcher is found in aspen groves, riparian willows, and other deciduous forest patches at lower elevations in the mountains. It nests low to the ground in deciduous bushes. The call of the Dusky Flycatcher is similar to that of Hammond's Flycatcher, making positive identifications challenging for inexperienced birders. Due to its range and its dependence on flying insects for food, the Dusky species is vulnerable to late spring storms and cold weather.

An increase in the relative abundance of this species was noted in the Rocky Mountain NR with a particular concentration west of Calgary. Relative to other species, the Dusky Flycatcher was observed more frequently in Atlas 2 than in Atlas 1. The Breeding Bird Survey did not detect any trend in abundance. This flycatcher is relatively common in the right habitat throughout its range, but it remains a species that is not well studied in Alberta. There is no indication that its habitat is threatened by human activities in the mountains. The provincial status for this species is Secure.

TEMPORAL REPORTING RATE

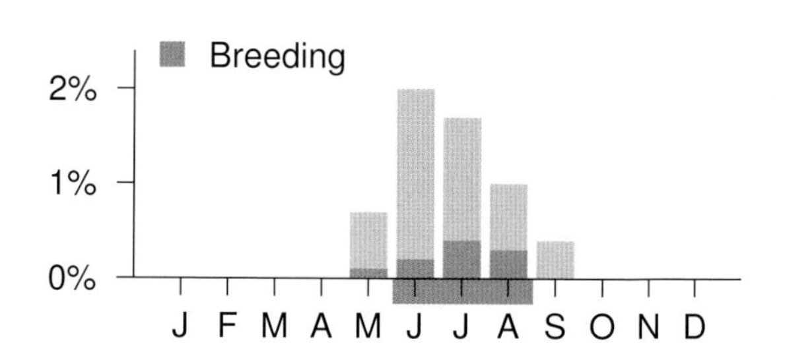

NATURAL REGIONS REPORTING RATE

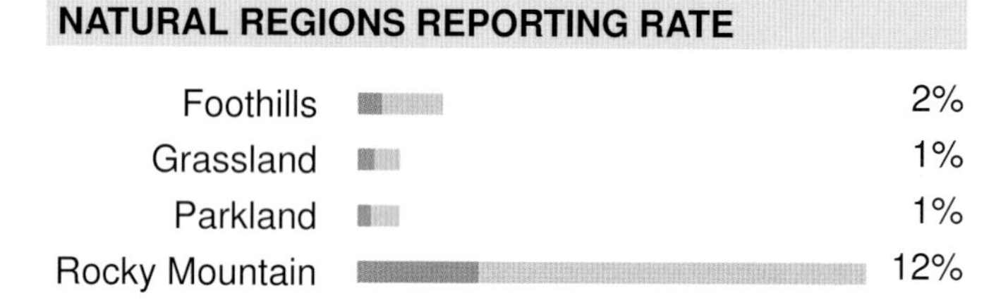

SPATIAL REPORTING RATE

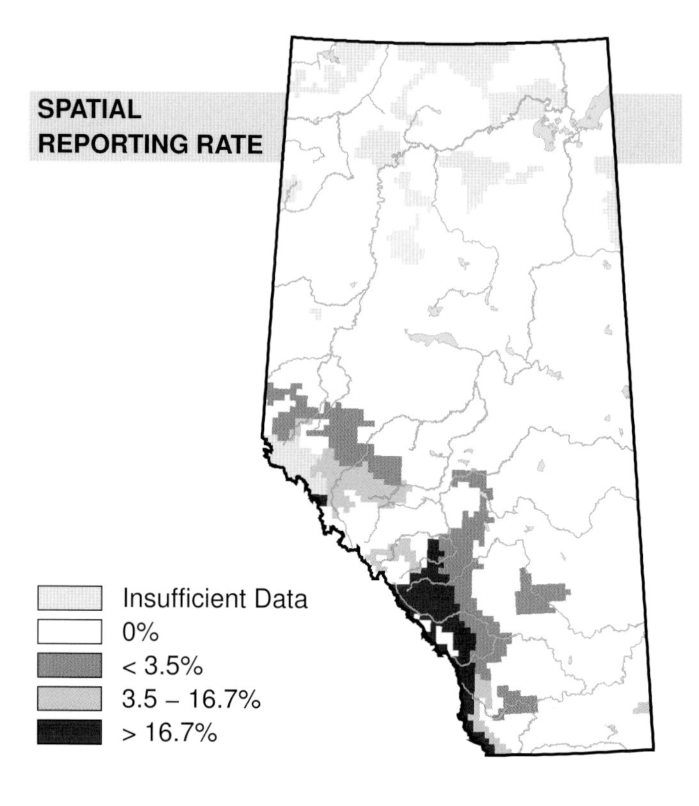

STATUS

GSWA 2000	GSWA 2005	COSEWIC Historic Rankings	COSEWIC 2007	Alberta Wildlife Act
Secure	Secure	N/A	N/A	N/A

RANGE

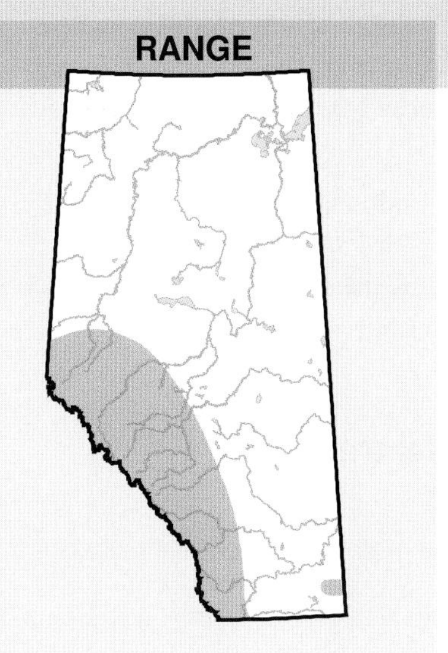

OBSERVED DISTRIBUTION ATLAS 1

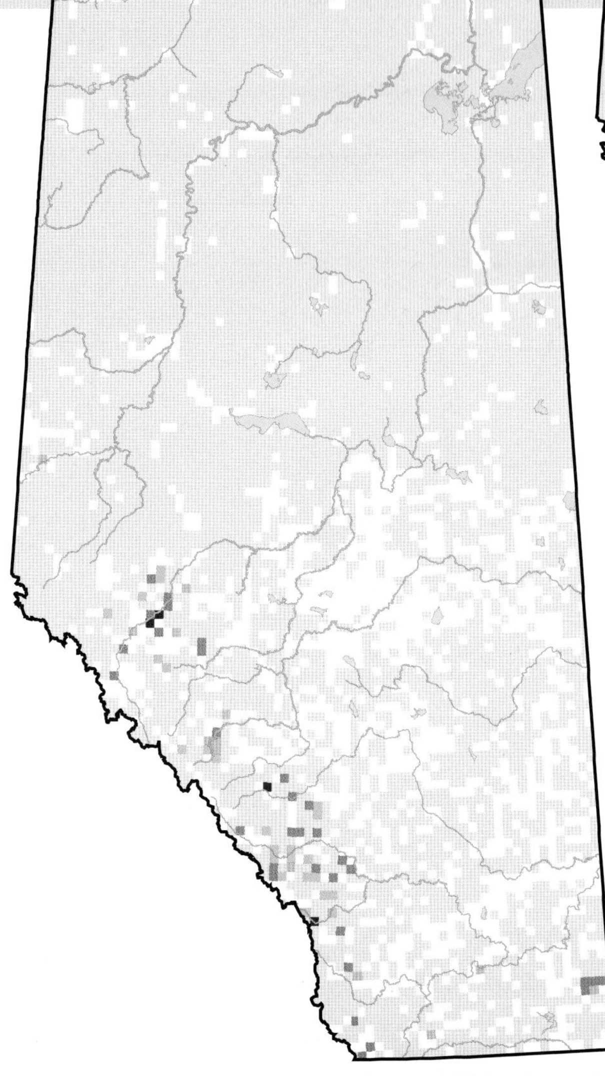

OBSERVED DISTRIBUTION ATLAS 2

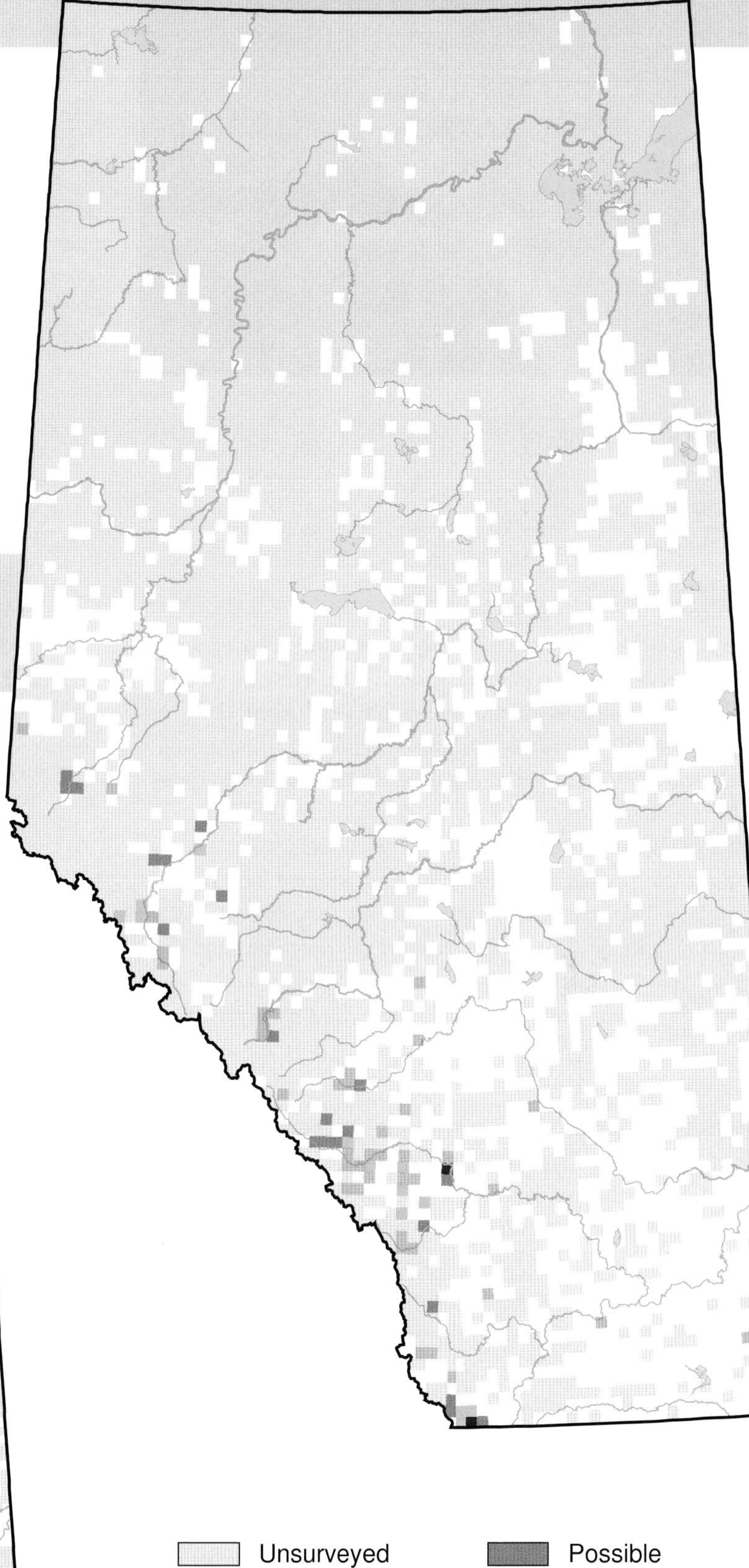

Cordilleran Flycatcher (*Empidonax occidentalis*)

Photo: Royal Alberta Museum

NESTING

CLUTCH SIZE (EGGS):	3–4
INCUBATION (DAYS):	14–15
FLEDGING (DAYS):	14–18
NEST HEIGHT (METRES):	0–9.1

The Cordilleran Flycatcher and the Pacific-slope Flycatcher are two forms of what was previously named the Western Flycatcher. Alberta is identified as an area of possible sympatry of the two forms, and recent studies have suggested that Pacific-slope Flycatchers may be the more common type in the province. The two forms are distinguished by subtle differences in their calls which, even when recorded and analyzed, are debated by experts. For purposes of this Atlas, records will be presumed to represent only the complex, not a particular species. These two forms are distributed widely in western North America. In Alberta, they are restricted to mountain regions usually near a fast-moving stream. Their vocalizations are distinctive but often masked by the sounds of moving water.

Observations made east of the foothills in Alberta are likely misidentifications. The "Western Flycatcher" can be confused with a Yellow-bellied Flycatcher, as both show yellowish tinges to their plumage. Both forms have been observed near Jasper. The Cordilleran/Pacific-slope complex is not commonly seen in any part of Alberta but observations were made throughout the mountain parks with concentrations north of Jasper.

A decline in relative abundance was detected in the Rocky Mountain NR where, relative to other species, this complex was observed less frequently in Atlas 2 than in Atlas 1. As with all the Empidonax-group species, this decline could be attributable to reductions in wintering habitat. The Pacific-slope form winters in coastal regions of Mexico where heavy resort development is occurring. The Cordilleran form is thought to winter in the interior of Mexico.

There is evidence of a decline in numbers of this species-complex in the Northern Rockies over the last 40 years as measured through Breeding Bird Survey data. This observation must be tempered by the uncertainty of historical identifications. The provincial status is Undetermined due to the paucity of records.

TEMPORAL REPORTING RATE

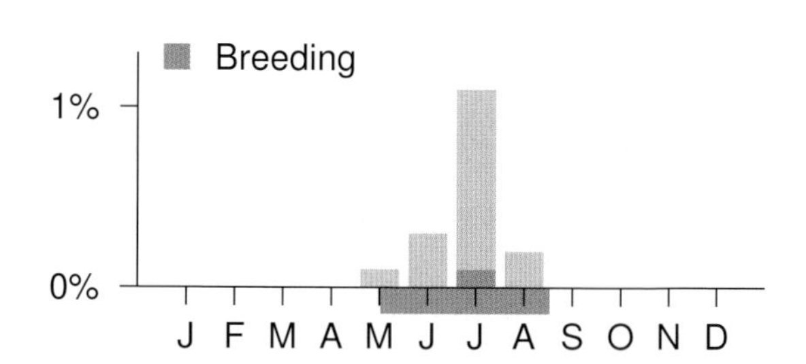

NATURAL REGIONS REPORTING RATE

SPATIAL REPORTING RATE

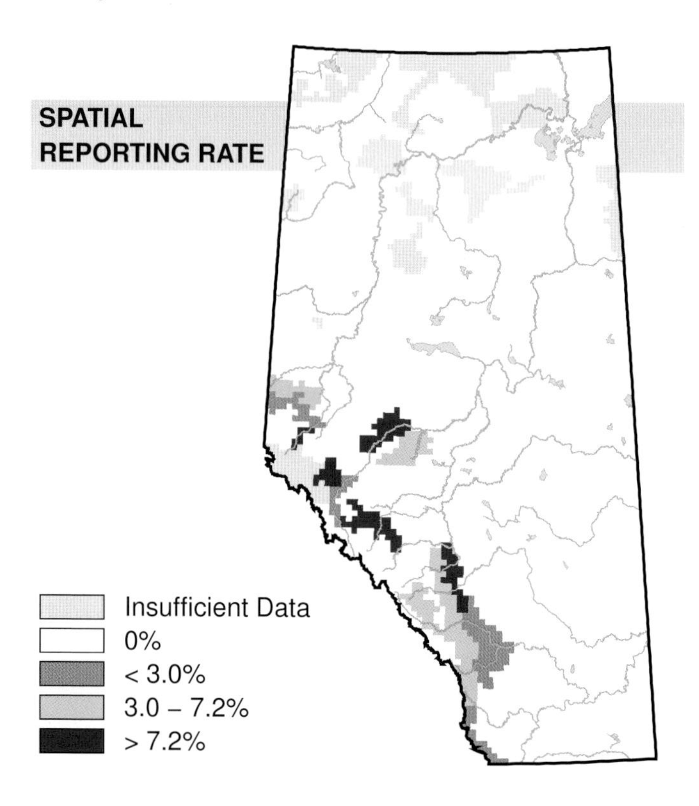

STATUS

GSWA 2000	GSWA 2005	COSEWIC Historic Rankings	COSEWIC 2007	Alberta Wildlife Act
Undetermined	Undetermined	N/A	N/A	N/A

RANGE

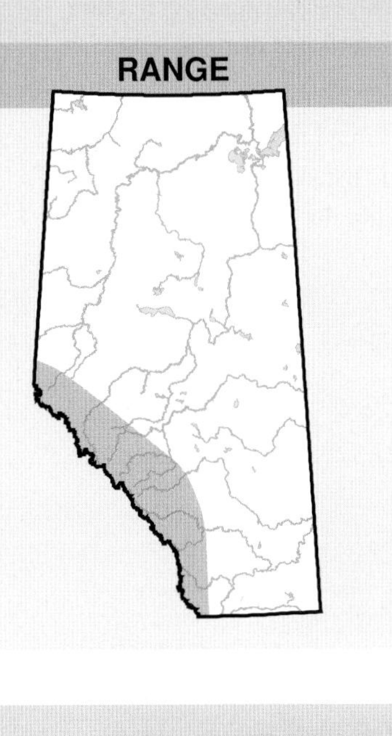

OBSERVED DISTRIBUTION ATLAS 2

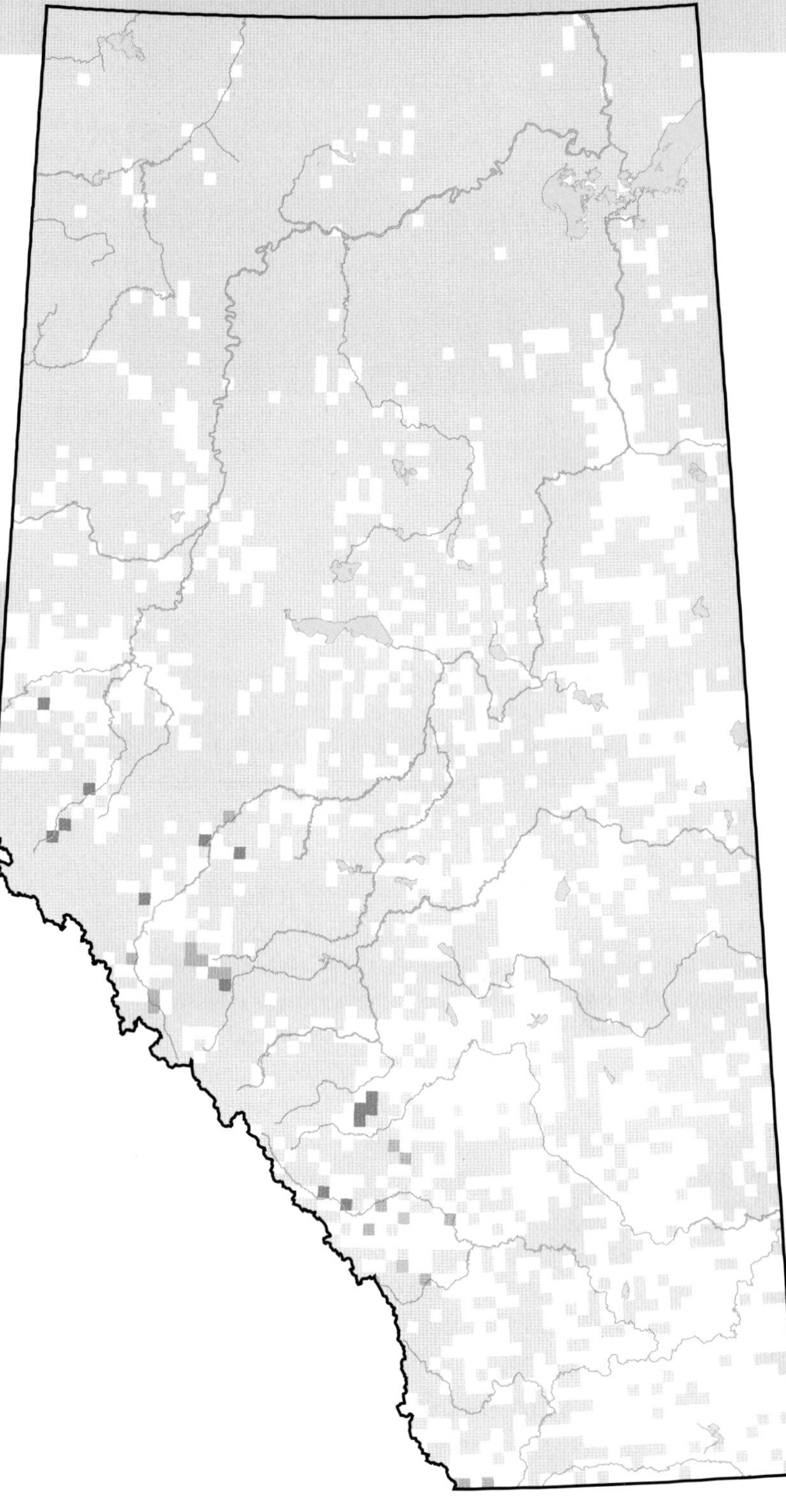

OBSERVED DISTRIBUTION ATLAS 1

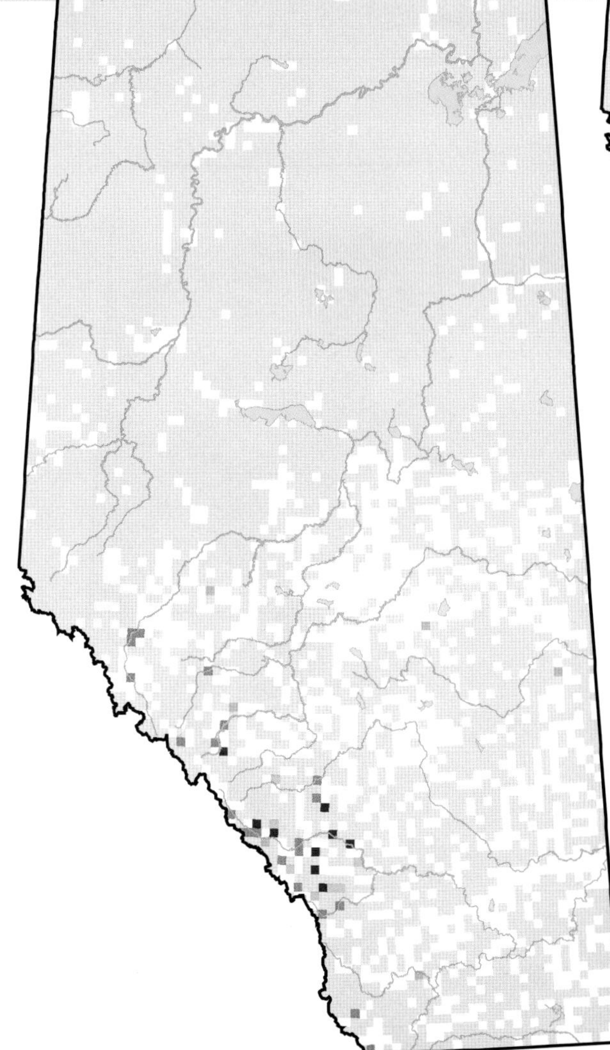

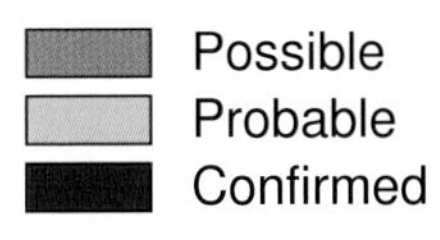

Eastern Phoebe (*Sayornis phoebe*)

Photo: Alan MacKeigan

NESTING

CLUTCH SIZE (EGGS):	5
INCUBATION (DAYS):	13–20
FLEDGING (DAYS):	15–17
NEST HEIGHT (METRES):	0–6.1

The Eastern Phoebe is found in most parts of Alberta but only locally in the Grassland and Rocky Mountain Natural Regions. It is most commonly found in the Foothills, Parkland, and Boreal Forest NRs. The observed distribution of this species did not change in Atlas 2.

Eastern Phoebes breed in open wooded areas, often in the vicinity of lakes and streams and most usually at the forest edge or near open areas (Semenchuk, 1992). Natural sites chosen for nesting include cliff ledges, caves, and earth-bank overhangs. Within a short distance of the nest site extensive woody vegetation also seems to be required. Buildings, bridges, and culverts provided nest sites and the accompanying modification of the landscape provided trees and shrubs that allowed Eastern Phoebes to move into areas of the Prairie Provinces not previously occupied (Weeks, 1994).

Decreases in relative abundance were detected in the Boreal Forest, Foothills, and Parkland NRs. No changes in relative abundance were detected in the Grassland and Rocky Mountain NRs. The Breeding Bird Survey (1995–2005) indicates a decline in the Eastern Phoebe in Alberta and across Canada.

Large areas within this phoebe's range may be unoccupied because suitable nesting substrate is lacking. The adoption of man-made structures for nesting also brings dependence. Changing of bridge and culvert design can significantly reduce available nest sites and thus suitable territories in many regions. This is especially true when new designs involve replacement of small wooden or concrete bridges and square concrete culverts with unusable circular, corrugated-metal pipes (Weeks, 1994). The clearing of woody plants in the vicinity of these structures, also eliminates these sites for nesting.

In Alberta the Eastern Phoebe is listed as Sensitive, as populations are declining possibly due to loss of habitat on wintering range.

TEMPORAL REPORTING RATE

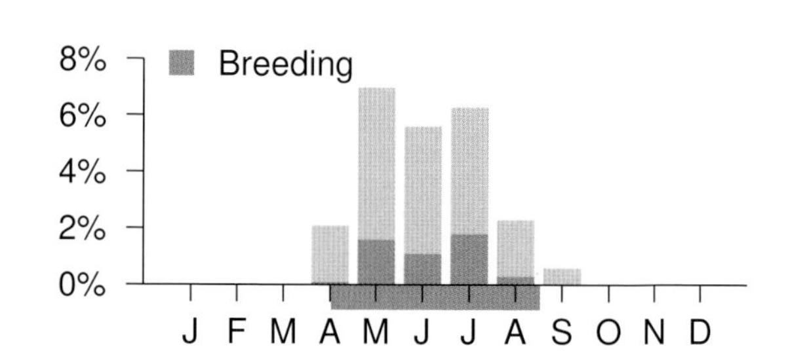

NATURAL REGIONS REPORTING RATE

Boreal Forest	2.9%
Foothills	4.6%
Grassland	1.1%
Parkland	3.3%
Rocky Mountain	1.5%

SPATIAL REPORTING RATE

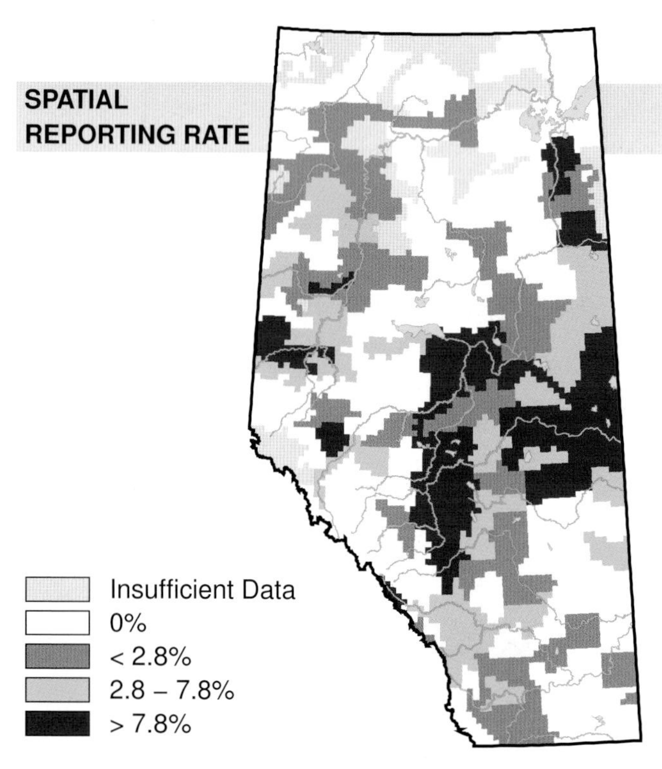

STATUS

GSWA 2000	GSWA 2005	COSEWIC Historic Rankings	COSEWIC 2007	Alberta Wildlife Act
Secure	Sensitive	N/A	N/A	N/A

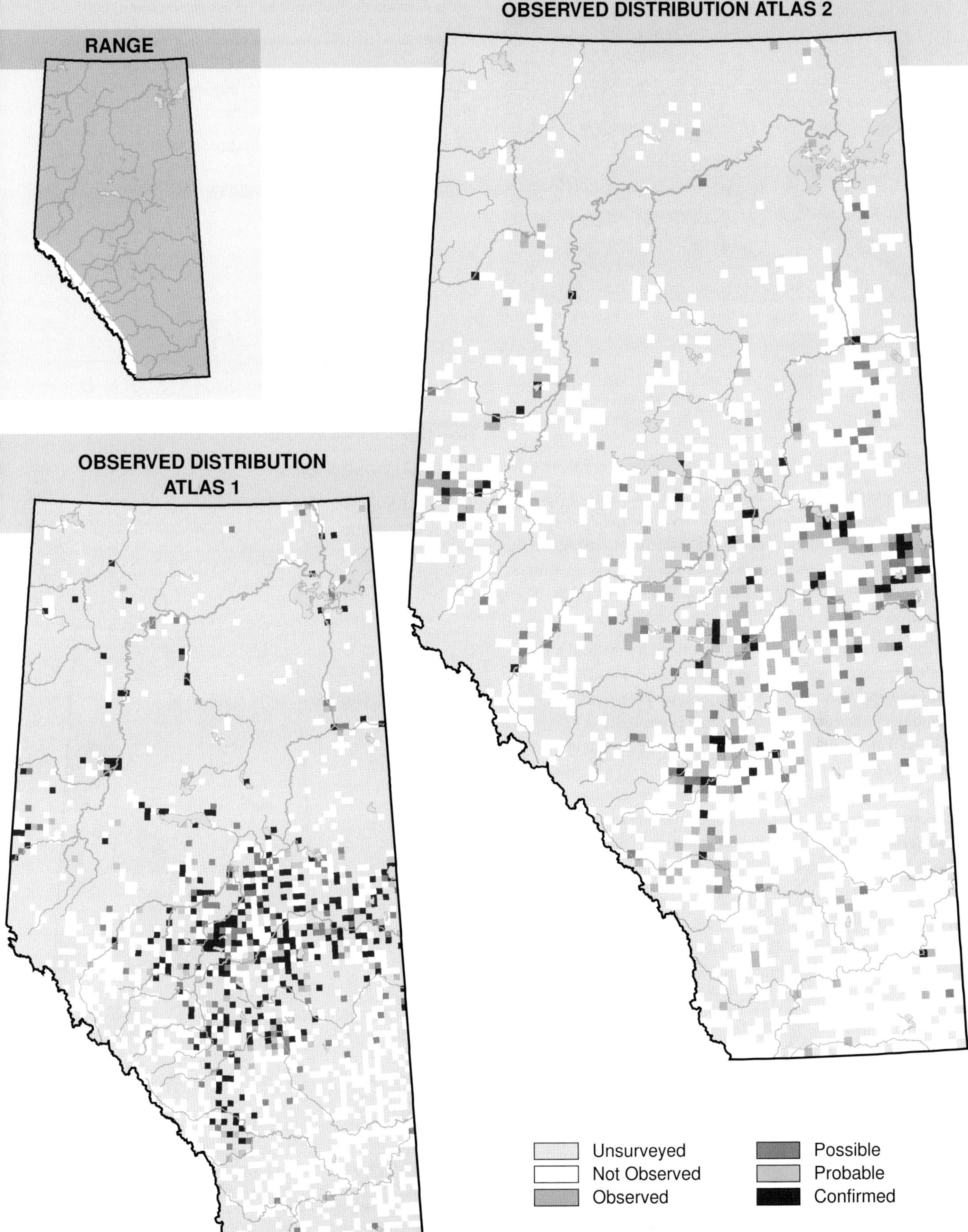
RANGE
OBSERVED DISTRIBUTION ATLAS 2
OBSERVED DISTRIBUTION ATLAS 1
Unsurveyed
Not Observed
Observed
Possible
Probable
Confirmed

Say's Phoebe (*Sayornis saya*)

Photo: Gerald Romanchuk

NESTING

CLUTCH SIZE (EGGS):	3–7
INCUBATION (DAYS):	12–14
FLEDGING (DAYS):	14–16
NEST HEIGHT (METRES):	0

Say's Phoebe breeds most commonly in the Grassland Natural Region. Extralimital spring migration records in late May/early June suggest a minor migration route, east of the Rockies, of birds heading to breeding areas in northern British Columbia, Yukon, Northwest Territories and Alaska. No changes were observed in this species' distribution in Atlas 2.

This is a bird of dry open country, particularly the badlands, prairie coulees, river banks, and farms where it is often seen perching on a hoodoo, a low bush, a fence post, wire, or a building (Semenchuk, 1992). Nesting is generally in a sheltered area with some overhang including natural ledges, holes in earth-banks, natural cavities in trees, in a building, or under a bridge. Like other phoebes, Say's Phoebe has undoubtedly benefited from use of human-made structures that provide suitable nest sites, as it is apparently constrained only by lack of suitable nest sites (Schukman and Wolf, 1998).

An increase in relative abundance was detected in the Parkland NR. No changes were detected in the other NRs. Although the Breeding Bird Surveys (1985–2005) yielded insufficient data for this species in Alberta to permit a determination for the province, no change was detected nationally. Dunn (2005) states there there is currently an uncertain population trend for this species. The BBS is considered poor for this species as it often stops singing before the BBS takes place. This difficulty would apply also to many Atlas surveys casting doubt on the Atlas-detected increase in the Parkland NR. A possible cross-Canada decline must be investigated (Dunn, 2005).

Despite this species' extensive breeding range, most aspects of its biology remain poorly known (Schukman and Wolf, 1998). Future research should focus on population dynamics, migration and habitat requirements.

The Say's Phoebe is listed as Secure in Alberta.

TEMPORAL REPORTING RATE

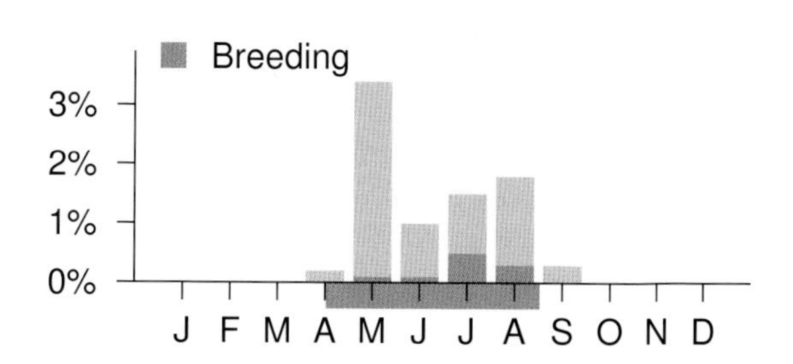

NATURAL REGIONS REPORTING RATE

Region	Rate
Boreal Forest	0.5%
Foothills	0.7%
Grassland	2.8%
Parkland	0.9%
Rocky Mountain	1.1%

SPATIAL REPORTING RATE

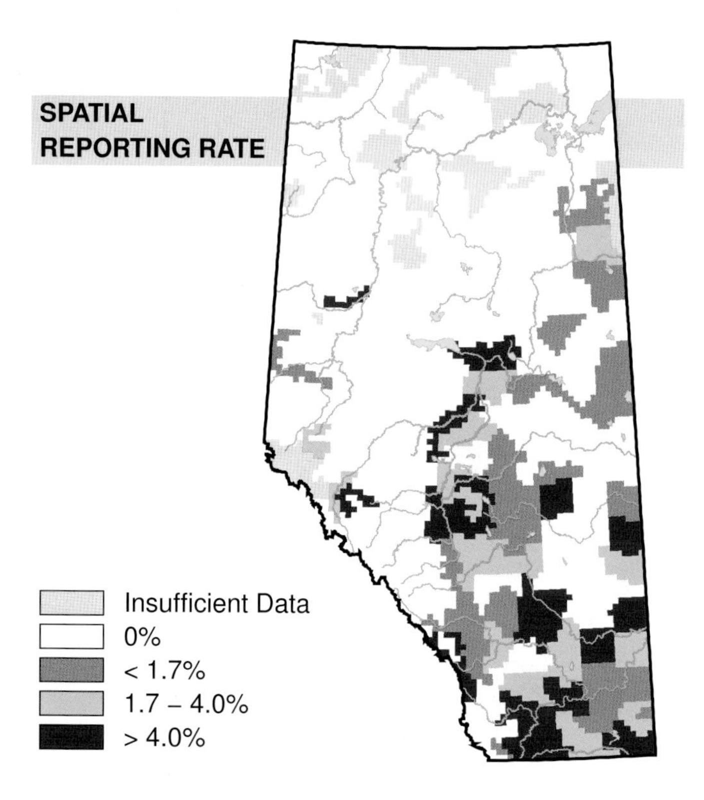

STATUS

GSWA 2000	GSWA 2005	COSEWIC Historic Rankings	COSEWIC 2007	Alberta Wildlife Act
Secure	Secure	N/A	N/A	N/A

Habitat  Nest Location   Nest Type  Diet 

RANGE

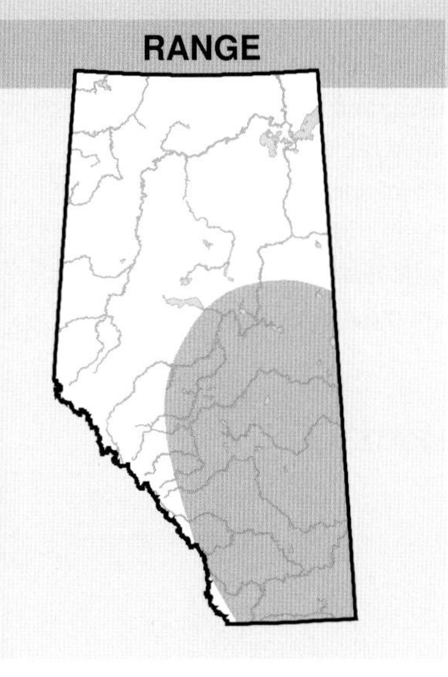

OBSERVED DISTRIBUTION ATLAS 2

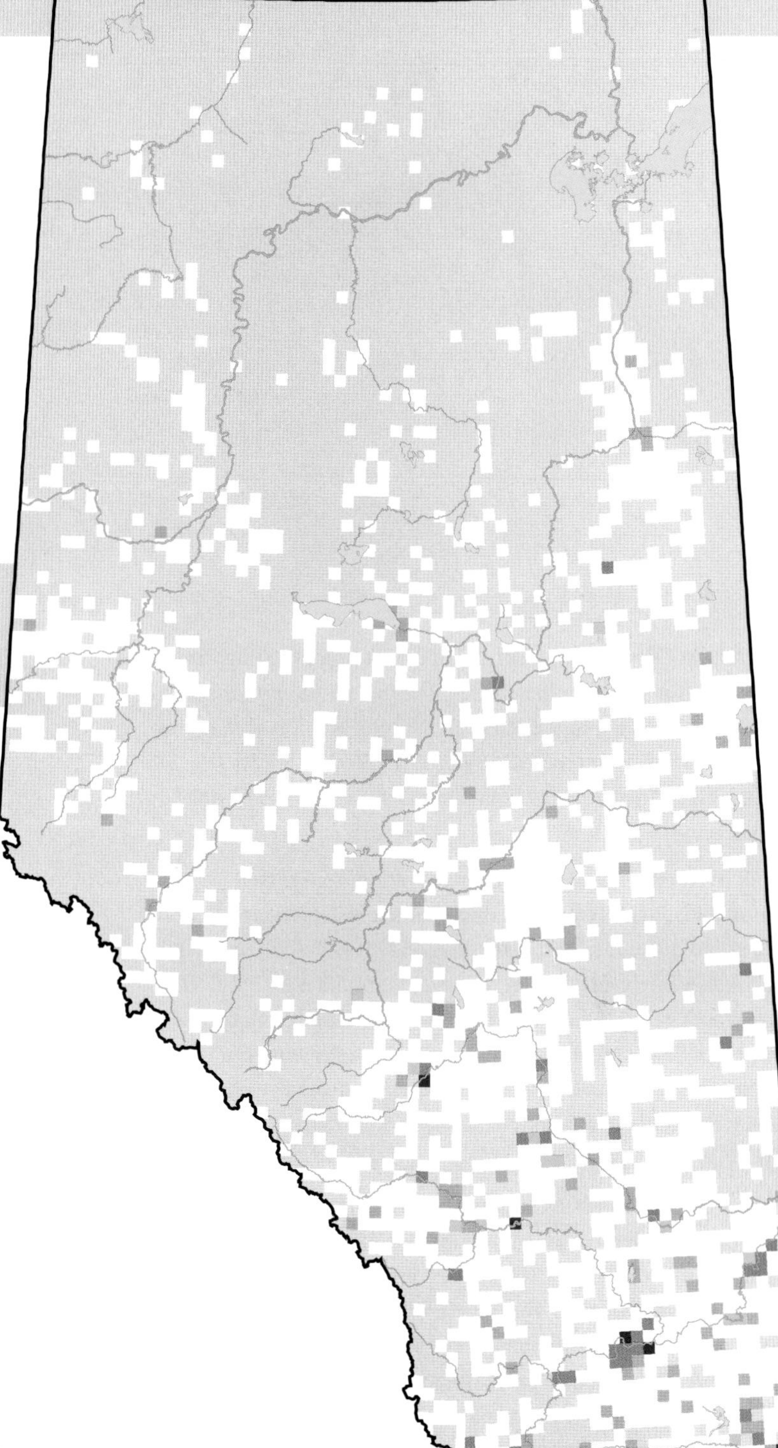

OBSERVED DISTRIBUTION ATLAS 1

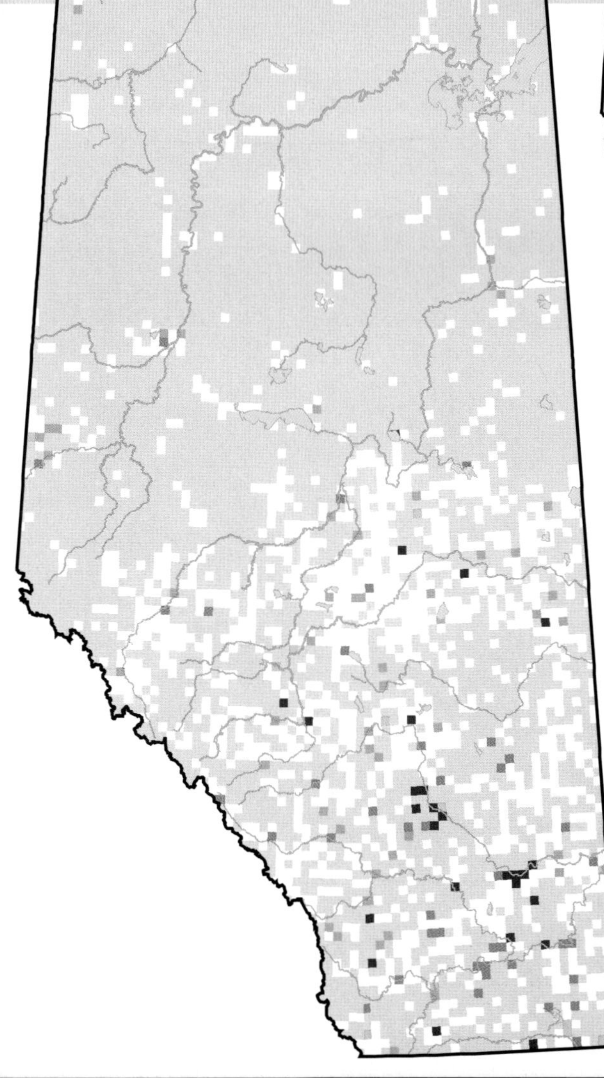

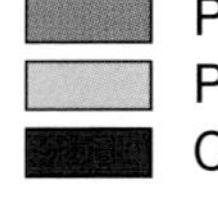

Great Crested Flycatcher (*Myiarchus crinitus*)

Photo: Gerald Romanchuk

NESTING

CLUTCH SIZE (EGGS):	4–5
INCUBATION (DAYS):	12–15
FLEDGING (DAYS):	12–21
NEST HEIGHT (METRES):	3–7

The Great Crested Flyatcher is not a common breeding bird in Alberta. It is found in the east-central part of the province, almost exclusively in the Parkland Natural Region with some records in the southeastern corner of the Boreal Forest NR. The observed distribution has not changed between Atlas 1 and Atlas 2.

This flycatcher breeds in mature deciduous and mixed woodlands, usually in more open parts, close to clearings or the woodland edge (Semenchuk, 1992). It is an obligatory but secondary cavity-nester, preferring a natural cavity but it will use woodpecker holes. This bird is seldom found in deep forest, but has benefited instead from the fragmentation of deciduous forest and subsequent increase in small woodlots and woodland edges (Lanyon, 1997).

No changes in relative abundance were detected in the Boreal Forest or Parkland NRs, although relative to other species, it was observed less frequently in Atlas 2 than in Atlas 1. The sample size was very small, however, making change detection difficult. The Breeding Bird Survey (1985–2005) has no data for the Great Crested Flycatcher in Alberta but it does indicate a decline for this species across Canada. The BBS data indicate no clear overall trend, with slight declines in some areas and slight increases in others (Lanyon, 1997).

In 2000, the Great Crested Flycatcher was listed as Secure in Alberta, but the 2005 status update now lists this species as Sensitive. The justification for this change recognizes that: there are probably fewer than 300 breeding pairs in the province; the species may be vulnerable to habitat loss or deterioration by various land uses; and it is unclear what effect habitat fragmentation has on this species. Lanyon (1997), however, indicates that populations may be increasing where habitat is "improved" by fragmentation of woodlands and increased edge effect, and may be decreasing locally because of destruction of dead snags and other "clean" forestry practices.

TEMPORAL REPORTING RATE

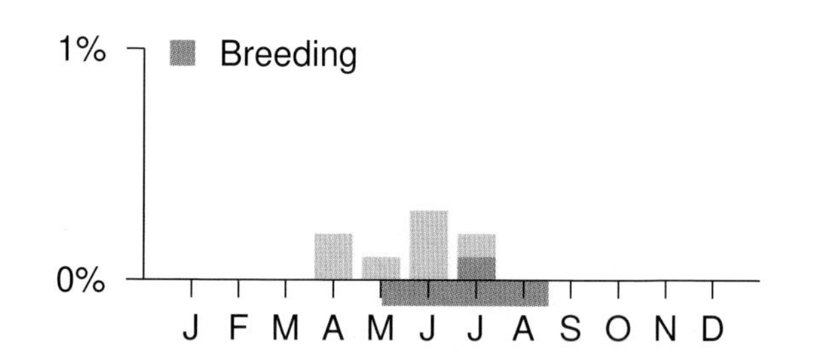

NATURAL REGIONS REPORTING RATE

SPATIAL REPORTING RATE

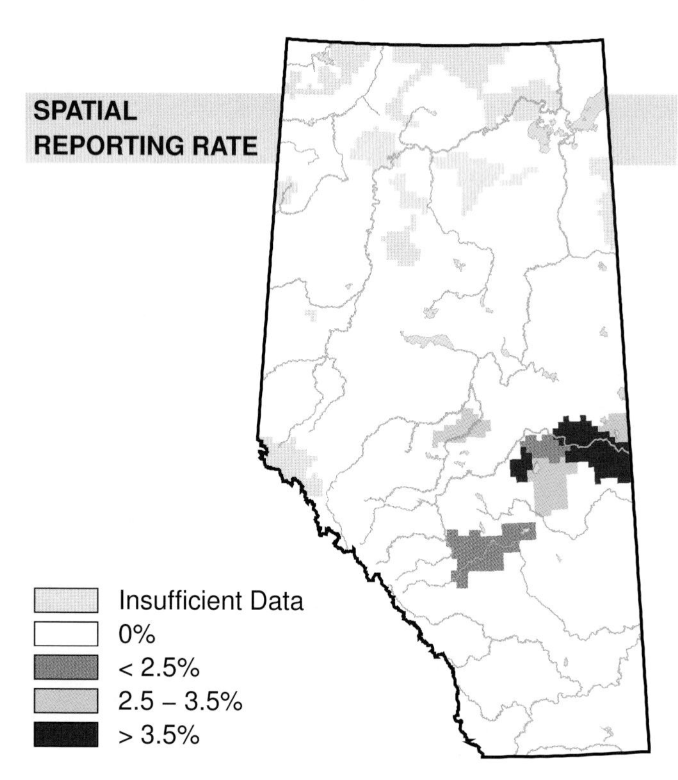

STATUS

GSWA 2000	GSWA 2005	COSEWIC Historic Rankings	COSEWIC 2007	Alberta Wildlife Act
Secure	Sensitive	N/A	N/A	N/A

Habitat  Nest Location  Nest Type  Diet 

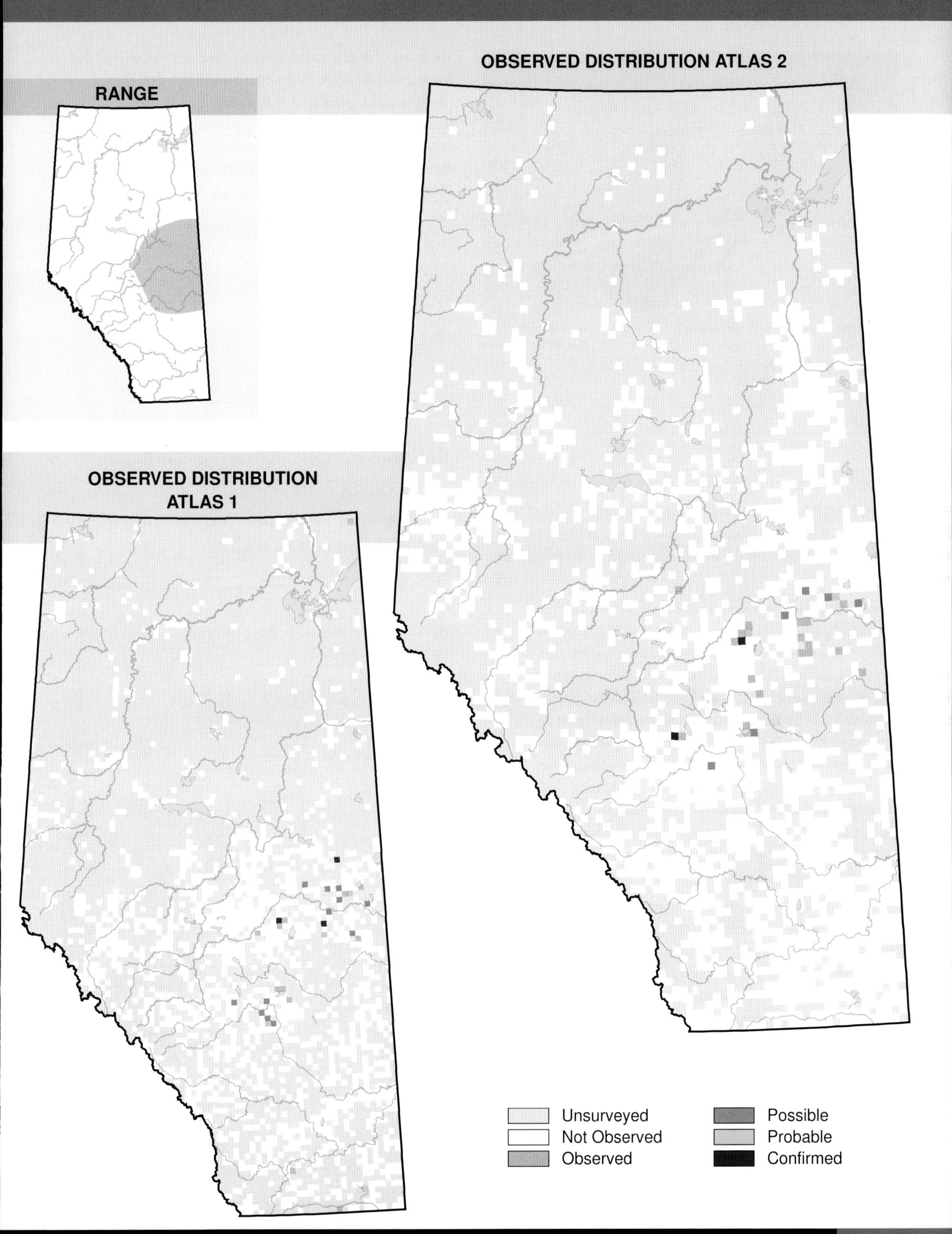

Western Kingbird (*Tyrannus verticalis*)

Photo: Gerald Romanchuk

NESTING

CLUTCH SIZE (EGGS):	3–5
INCUBATION (DAYS):	12–14
FLEDGING (DAYS):	16–17
NEST HEIGHT (METRES):	1.5–12.2

The Western Kingbird breeds in the southern part of the province in the Grassland and Parkland Natural Regions. The distribution of this species did not change between Atlas 1 and Atlas 2.

The Western Kingbird is found in open areas such as grasslands, cultivated fields, urban areas and grazed areas. Breeding areas are usually interspersed with vegetation and structures such as homes, barns, power poles, and fences. For this reason, human habitation in the prairies has likely caused an increase in the abundance of this species because people tend to plant trees and erect buildings on the land. Such structures can be used for nest placement and they often provide perching locations between aerial feeding bouts. Due to its preference for open habitats, this species was found mainly in the Grassland NR but it was also occasionally found in the Parkland NR, and only infrequently observed in the southern part of the Boreal Forest NR.

Declines in relative abundance were detected in the Grassland and Parkland NRs where, relative to other species, the Western Kingbird was found less frequently during Atlas 2 than Atlas 1. The Breeding Bird Survey detected no change in abundance in Alberta or across Canada during the period 1985–2005, although an increase was detected in Alberta for the period of 1995–2005. It is unclear why the Atlas and Breeding Bird Survey trends differ. From a biological point of view, the positive trend found by the Breeding Bird Survey is reasonable. The Western Kingbird tends to be associated with human-modified environments and the availability of those habitat types is increasing. This species is considered Secure in Alberta.

TEMPORAL REPORTING RATE

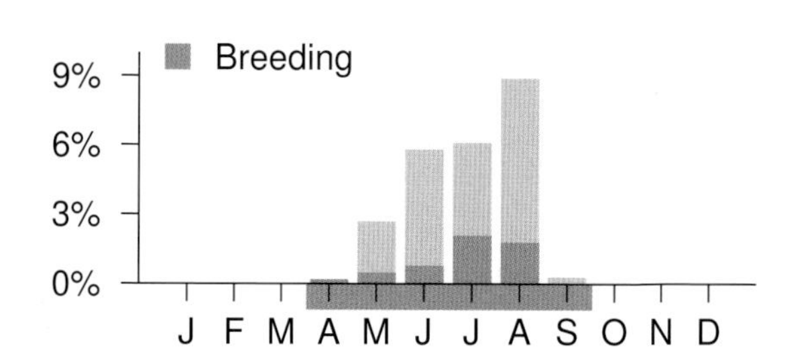

NATURAL REGIONS REPORTING RATE

Boreal Forest	< 0.1%
Grassland	4.3%
Parkland	0.8%

SPATIAL REPORTING RATE

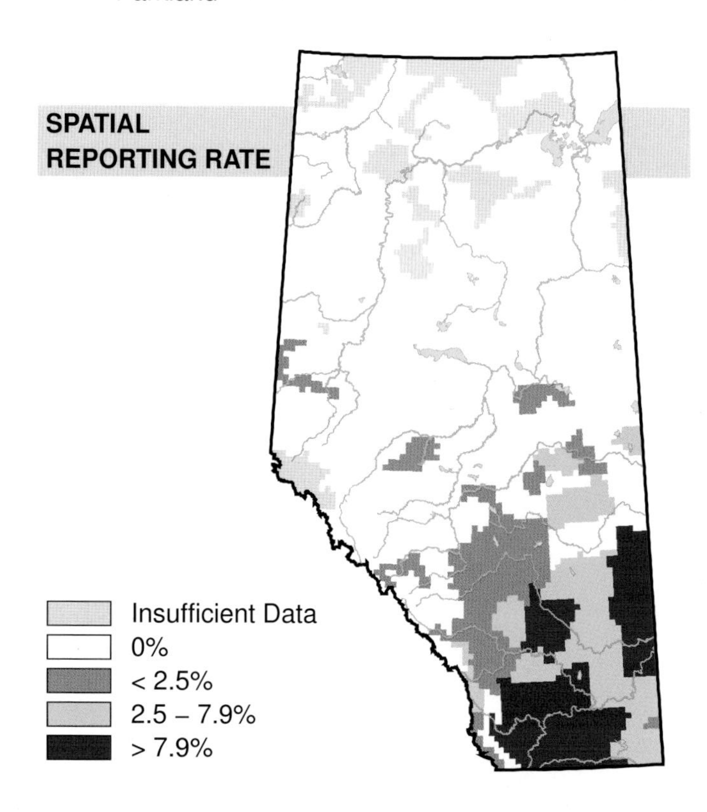

STATUS

GSWA 2000	GSWA 2005	COSEWIC Historic Rankings	COSEWIC 2007	Alberta Wildlife Act
Secure	Secure	N/A	N/A	N/A

RANGE

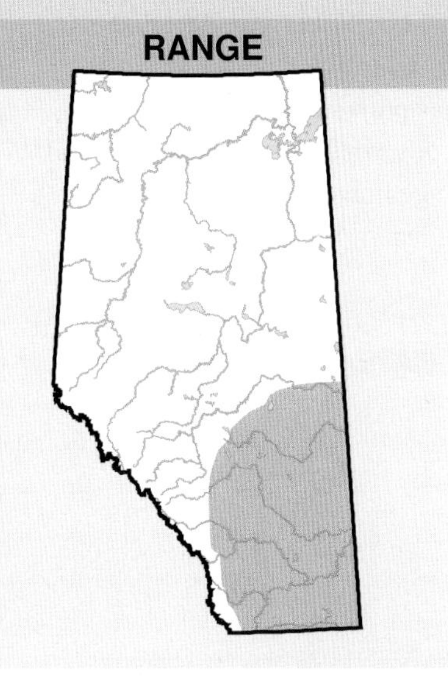

OBSERVED DISTRIBUTION ATLAS 2

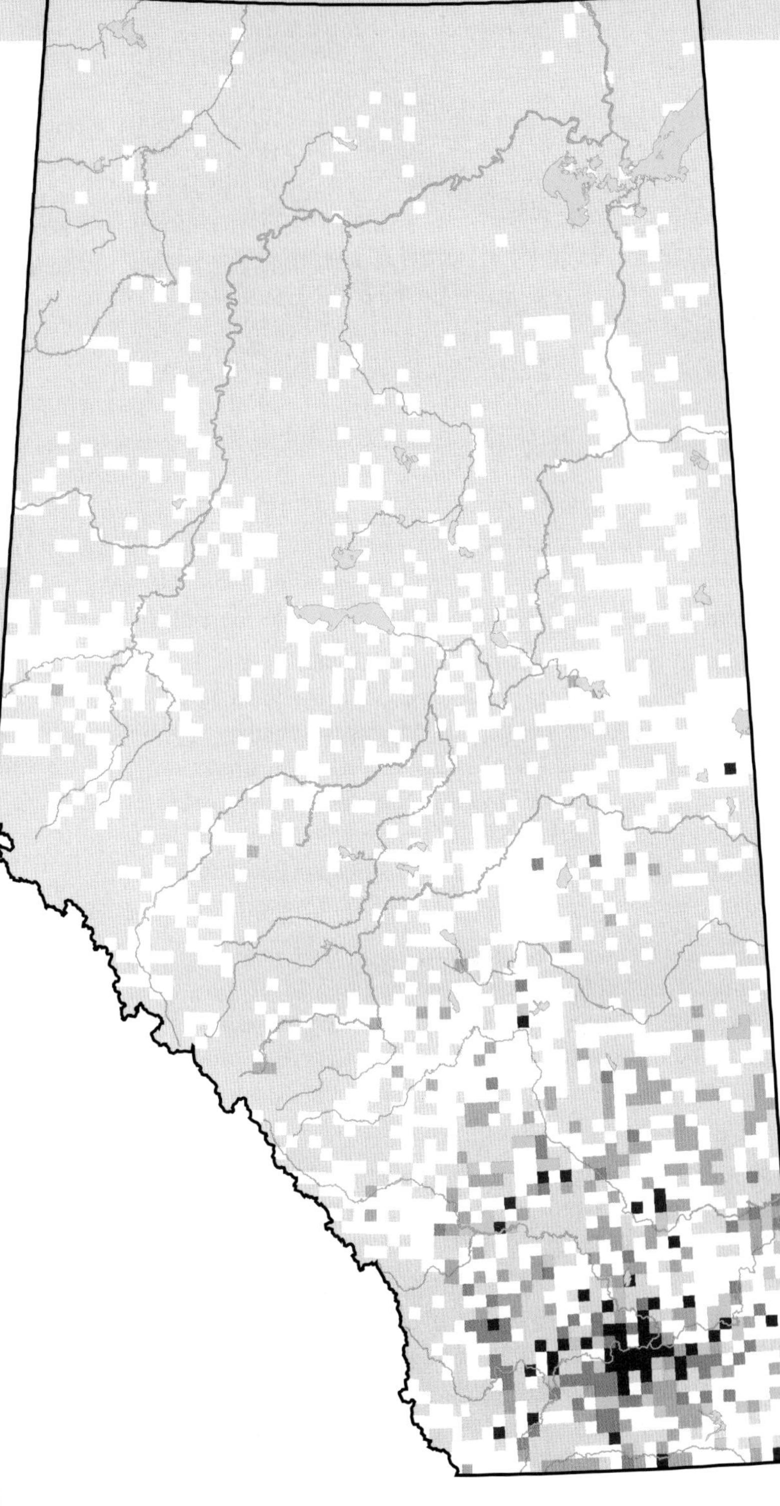

OBSERVED DISTRIBUTION ATLAS 1

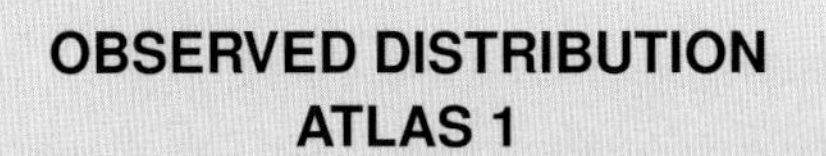

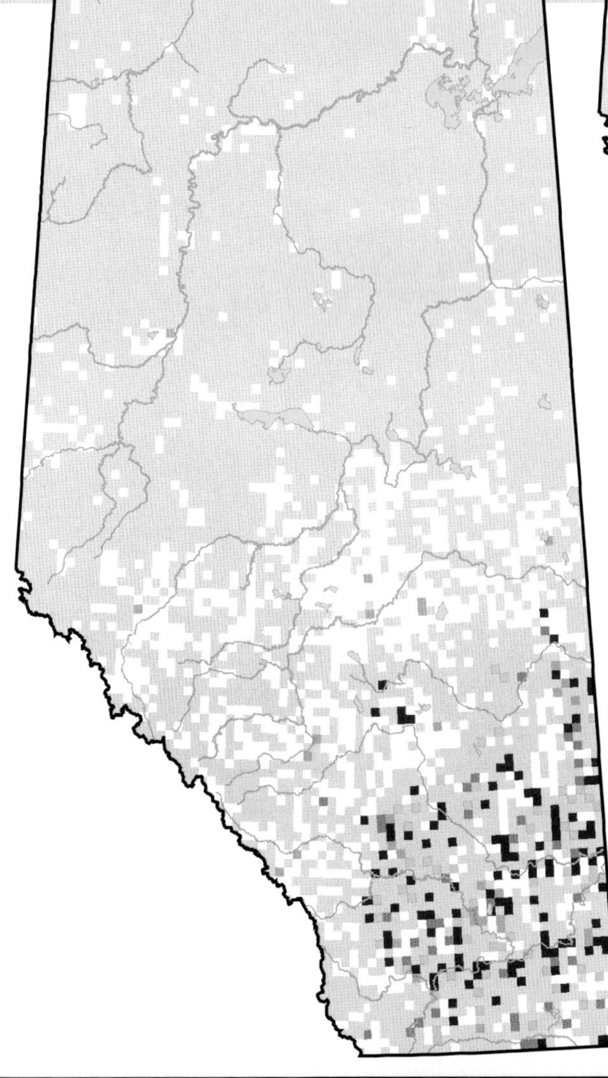

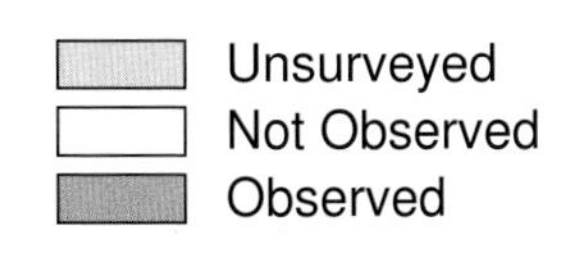

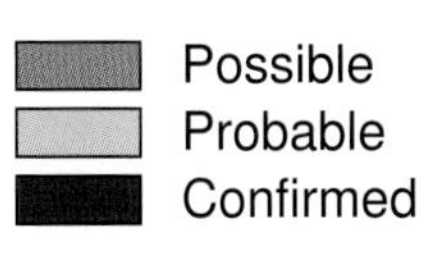

Eastern Kingbird (*Tyrannus tyrannus*)

Photo: Raymond Toal

NESTING

CLUTCH SIZE (EGGS):	3–4
INCUBATION (DAYS):	12–13
FLEDGING (DAYS):	16–18
NEST HEIGHT (METRES):	<6

Despite its common name, the Eastern Kingbird breeds abundantly in the western part of the continent and its range extends to the Pacific Ocean. This flycatcher is found throughout the province although it is scarce in the northern Boreal Forest and Rocky Mountain Natural Regions. It is most common in the Grassland NR and less common in the Parkland NR. There is no change in observed distribution between Atlas 1 and Atlas 2.

It prefers open country with scattered trees, as well as woodland edges, shrubby fields, and pastures. In the forested areas it utilizes clearings such as roadsides, burns, beaver ponds, or other wetlands with standing dead trees. It often nests densely in trees that overhang water or in dead, standing snags surrounded by water (Murphy, 1996).

Decreases in relative abundance were detected in the Boreal Forest, Foothills, Grassland, Parkland, and Rocky Mountain NRs. The Breeding Bird Surveys (1968–2005) indicated a decline for the Eastern Kingbird in Alberta and across Canada. For the period 1995–2005, the BBS showed no change in the province, but a decline nationally. Dunn (2005) indicates that this species requires attention as it is experiencing possible population declines. Changes in agricultural practices have resulted in a loss of habitat. There is also some concern that pesticides may affect this species.

Murphy (1996) indicates that factors affecting adult survival during migration and/or on wintering grounds are probably more important for determining changes in population size. Another limiting factor identified by Murphy (1996) is the decline in the number of small farms over the last several decades, a phenomenon that has probably affected kingbirds negatively because of the loss of open space. Shelter belts and riparian environments in its range must be protected to maintain its populations. The Eastern Kingbird is listed as Secure in Alberta.

TEMPORAL REPORTING RATE

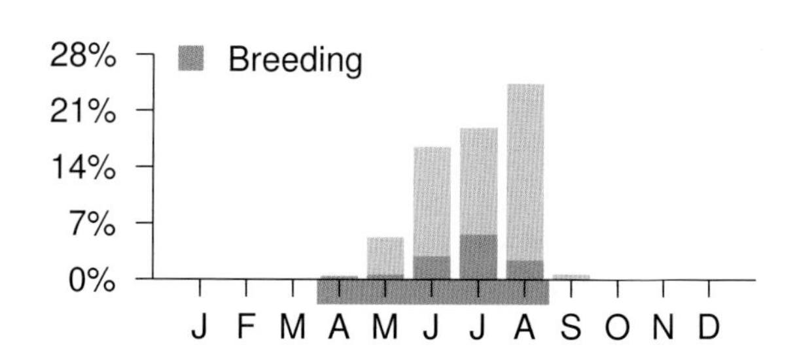

NATURAL REGIONS REPORTING RATE

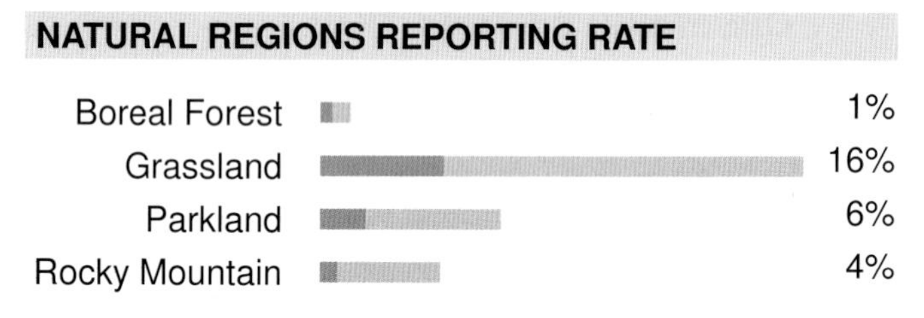

SPATIAL REPORTING RATE

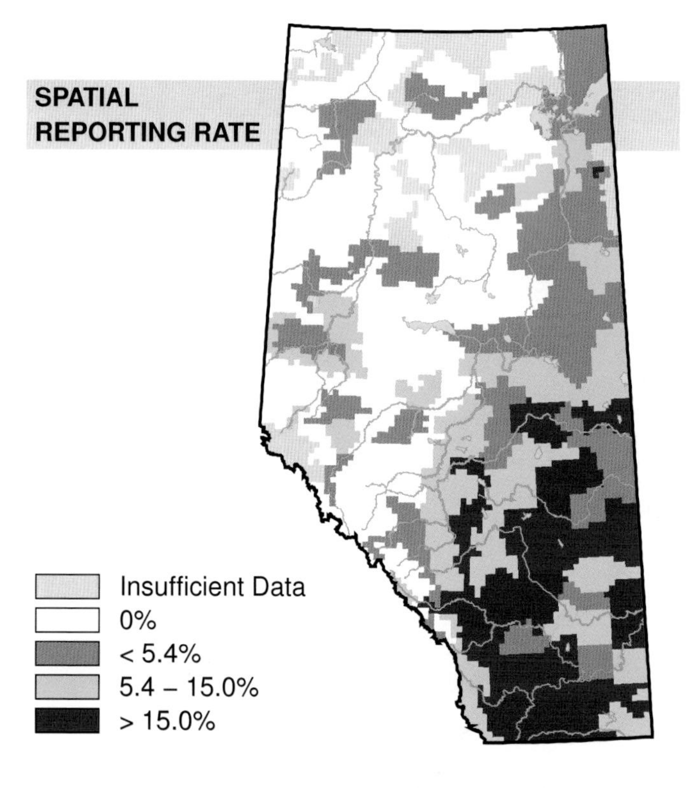

STATUS

GSWA 2000	GSWA 2005	COSEWIC Historic Rankings	COSEWIC 2007	Alberta Wildlife Act
Secure	Secure	N/A	N/A	N/A

RANGE

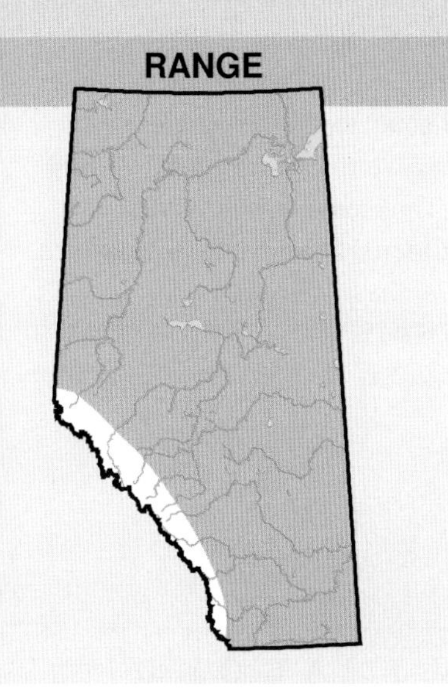

OBSERVED DISTRIBUTION ATLAS 1

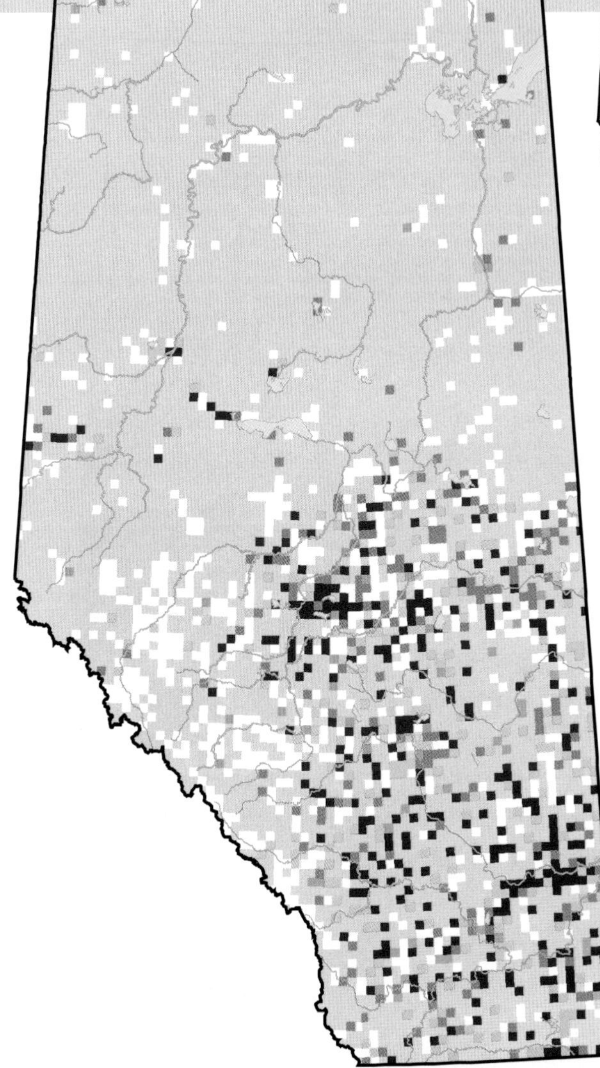

OBSERVED DISTRIBUTION ATLAS 2

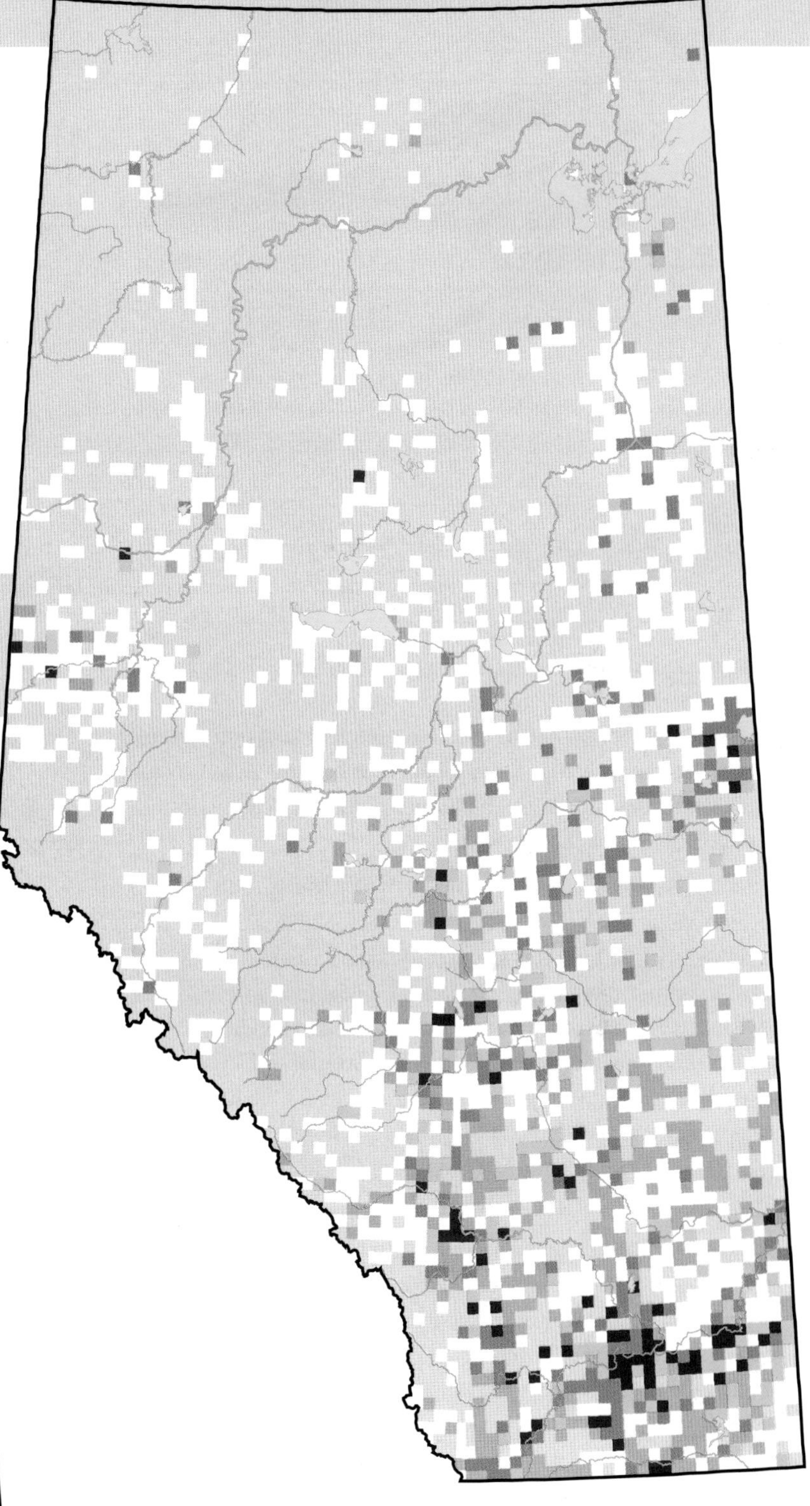

Unsurveyed
Not Observed
Observed
Possible
Probable
Confirmed

Loggerhead Shrike (*Lanius ludovicianus*)

Photo: Gerald Romanchuk

NESTING

CLUTCH SIZE (EGGS):	4–7
INCUBATION (DAYS):	14–16
FLEDGING (DAYS):	17–21
NEST HEIGHT (METRES):	1–4

Endemic to North America, the Loggerhead Shrike is found in the Grassland and Parkland Natural Regions in Alberta. The observed distribution of this species in the Parkland and Boreal Forest NRs decreased in Atlas 2. The range contracted to south of the Battle River, whereas its former range lay slightly north of the North Saskatchewan River.

The Loggerhead Shrike is at home in open country with short vegetation, pastures with hedgerows, riparian areas, and open woodlands. Breeding sites are usually located near isolated trees or large shrubs. Foraging sites are open, interspersed with low shrubs and trees for hunting arthropods, amphibians, reptiles, small mammals, and birds. This species will use utility lines and poles, and fencelines; so, it is frequently observed along roadsides. This species' preferred habitat is most prevalent in the prairies, and it was therefore most often observed in the Grassland NR.

Declines in relative abundance were detected in the Grassland and Parkland NRs where, relative to other species, the Loggerhead Shrike was observed less frequently in Atlas 2 than in Atlas 1. In both of these NRs there was substantial loss of native grasslands between Atlas 1 and Atlas 2 (Watmough and Schmoll, 2007). No change in abundance was detected by the Breeding Bird Survey in Alberta or Canada-wide during the period 1985–2005, although sample size was small. Reduction of suitable habitat, pesticide residues, and increased human disturbance are all factors that could cause this species to decline. The Committee on the Status of Endangered Wildlife in Canada lists this species as Threatened in western Canada. The Loggerhead Shrike is considered Sensitive in the province.

TEMPORAL REPORTING RATE

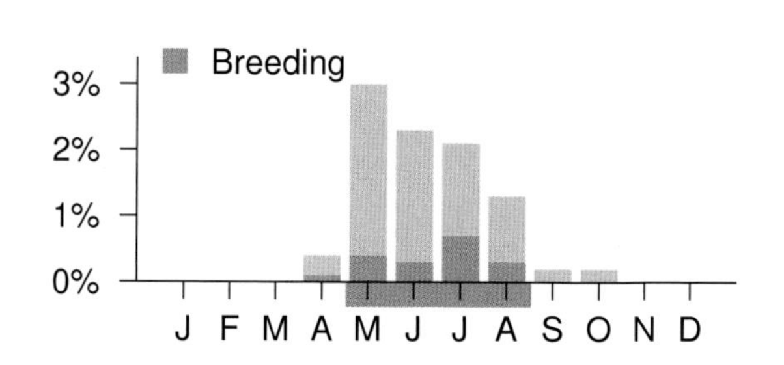

NATURAL REGIONS REPORTING RATE

SPATIAL REPORTING RATE

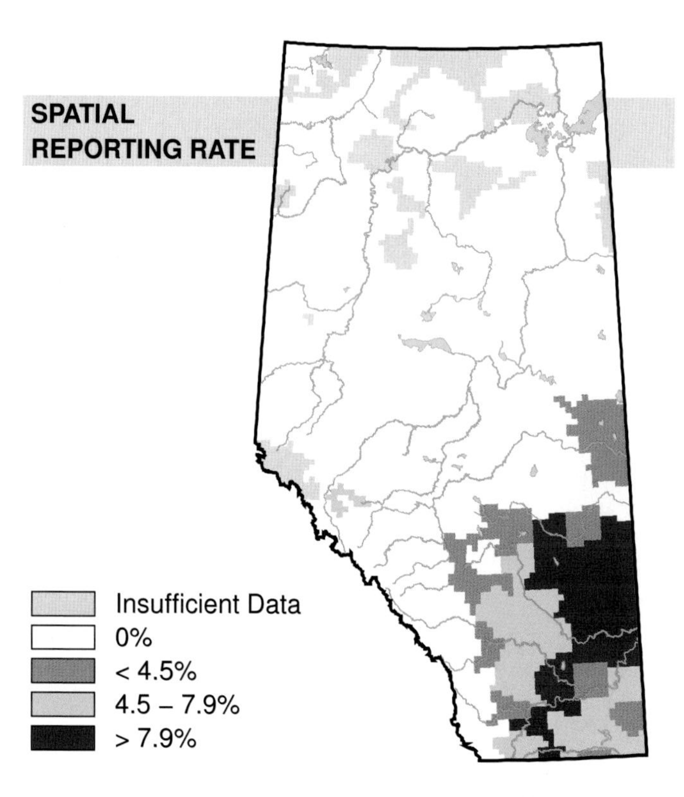

STATUS

GSWA 2000	GSWA 2005	COSEWIC Historic Rankings	COSEWIC 2007	Alberta Wildlife Act
Sensitive	Sensitive	Threatened	Threatened	N/A

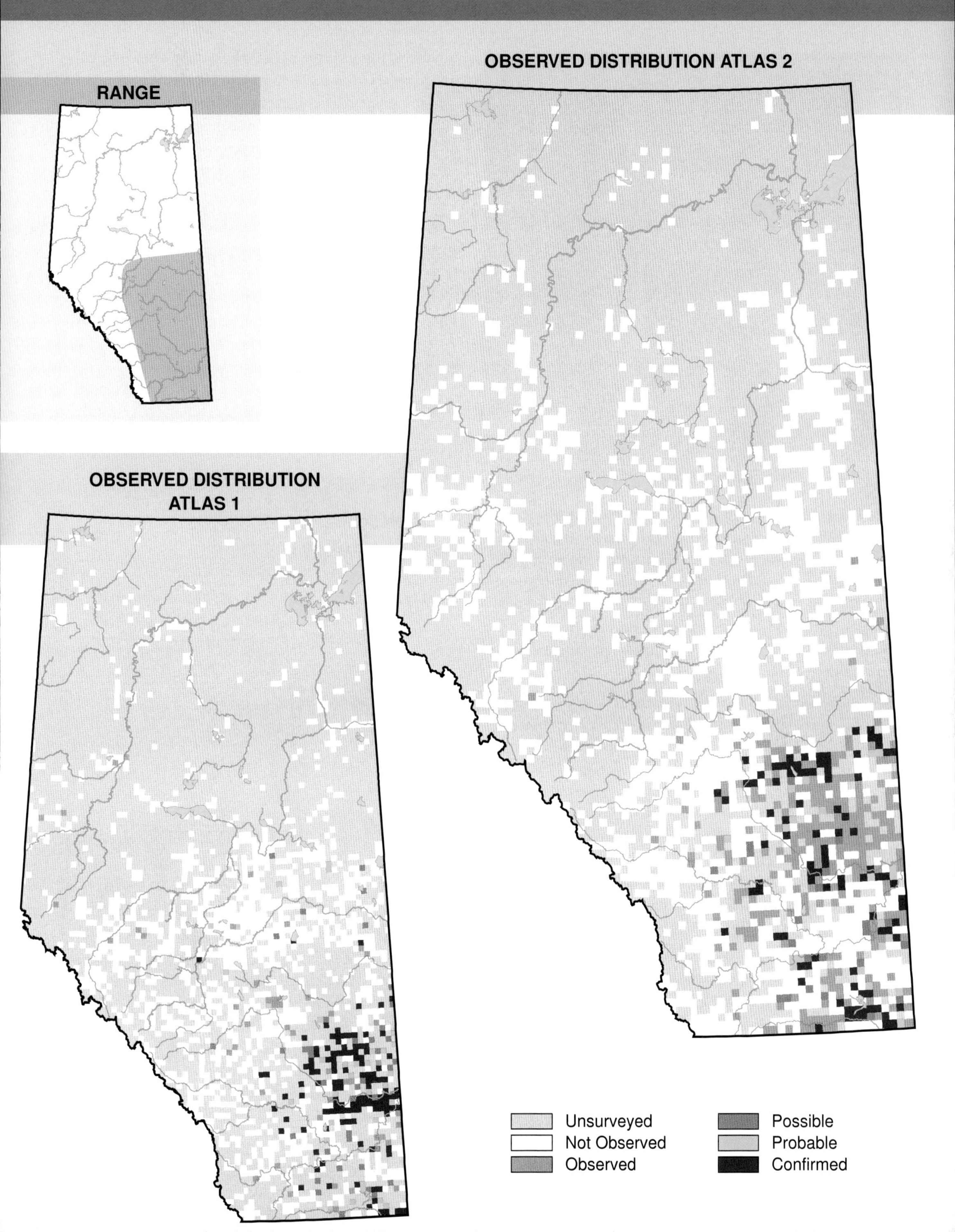
OBSERVED DISTRIBUTION ATLAS 2
RANGE
OBSERVED DISTRIBUTION ATLAS 1
Unsurveyed
Not Observed
Observed
Possible
Probable
Confirmed

Northern Shrike (*Lanius excubitor*)

Photo: Royal Alberta Museum

NESTING

CLUTCH SIZE (EGGS):	5–7
INCUBATION (DAYS):	15–16
FLEDGING (DAYS):	19–20
NEST HEIGHT (METRES):	2–4

The Northern Shrike breeds in the northern part of Alberta in the Boreal Forest Natural Region. It is unknown whether the breeding distribution of this species changed between Atlas 1 and Atlas 2 because no breeding evidence was obtained during Atlas 2 and the breeding observations reported in Atlas 1 were probably misidentifications. Non-breeding, mainly winter, observations were made across the southern part of the province during both Atlas projects.

As its name implies, the Northern Shrike nests in the north. The main breeding range of this species occurs in the tundra zone, beyond the northern edge of the spruce treeline. Due to the remoteness of its breeding range, this species is encountered mainly when it is wintering in central and southern Alberta. During the winter, Northern Shrikes can be fairly common in rural areas and they are seen occasionally feeding on birds around bird feeders. This species was found mainly in the Grassland, Parkland, and Rocky Mountain NRs and was found rarely in the Boreal Forest NR.

A decline in relative abundance was detected in the Boreal Forest NR and increases were detected in the Grassland and Parkland NRs. Observation rates relative to other species mirrored these changes. The increases are, at least in part, a function of the broader survey window of Atlas 2 than that of Atlas 1. The decline in the Boreal Forest NR must be considered cautiously because the sample size is relatively small and the coverage within this species' known breeding range is very sparse. The Breeding Bird Survey sample size was too small to investigate trends for the Northern Shrike in Alberta or across Canada because most survey routes lie south of the breeding range of this species. This species is not effectively monitored by surveys such as the Atlas or Breeding Bird Survey because these surveys do not adequately monitor the north. This species is considered Secure in Alberta.

TEMPORAL REPORTING RATE

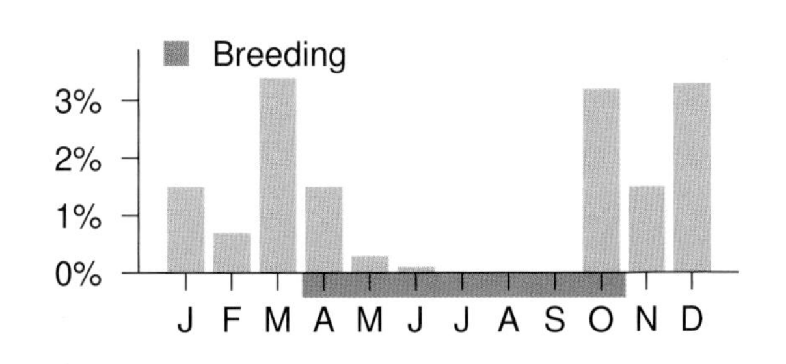

NATURAL REGIONS REPORTING RATE

SPATIAL REPORTING RATE

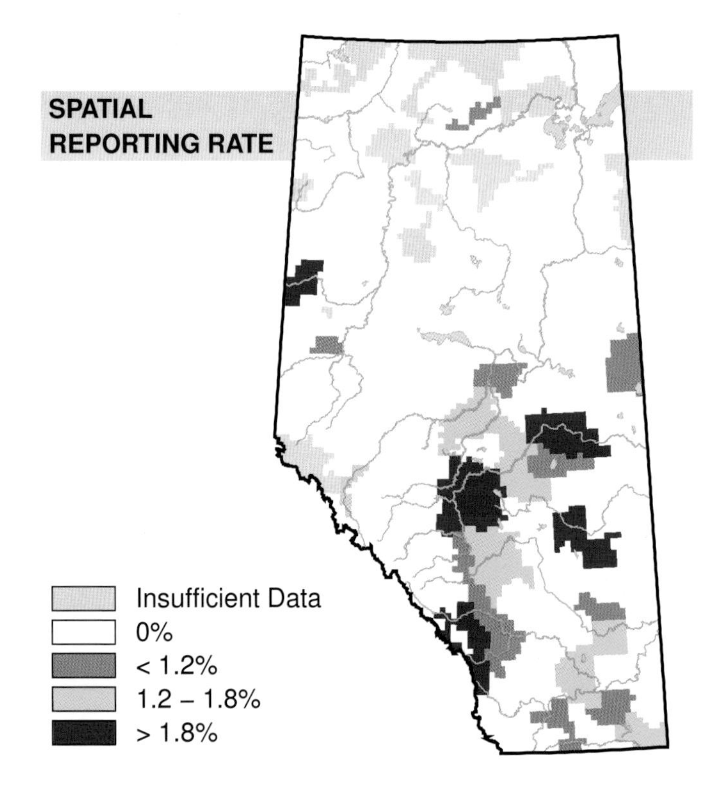

STATUS

GSWA 2000	GSWA 2005	COSEWIC Historic Rankings	COSEWIC 2007	Alberta Wildlife Act
Secure	Secure	N/A	N/A	N/A

Habitat Nest Location Nest Type Diet 

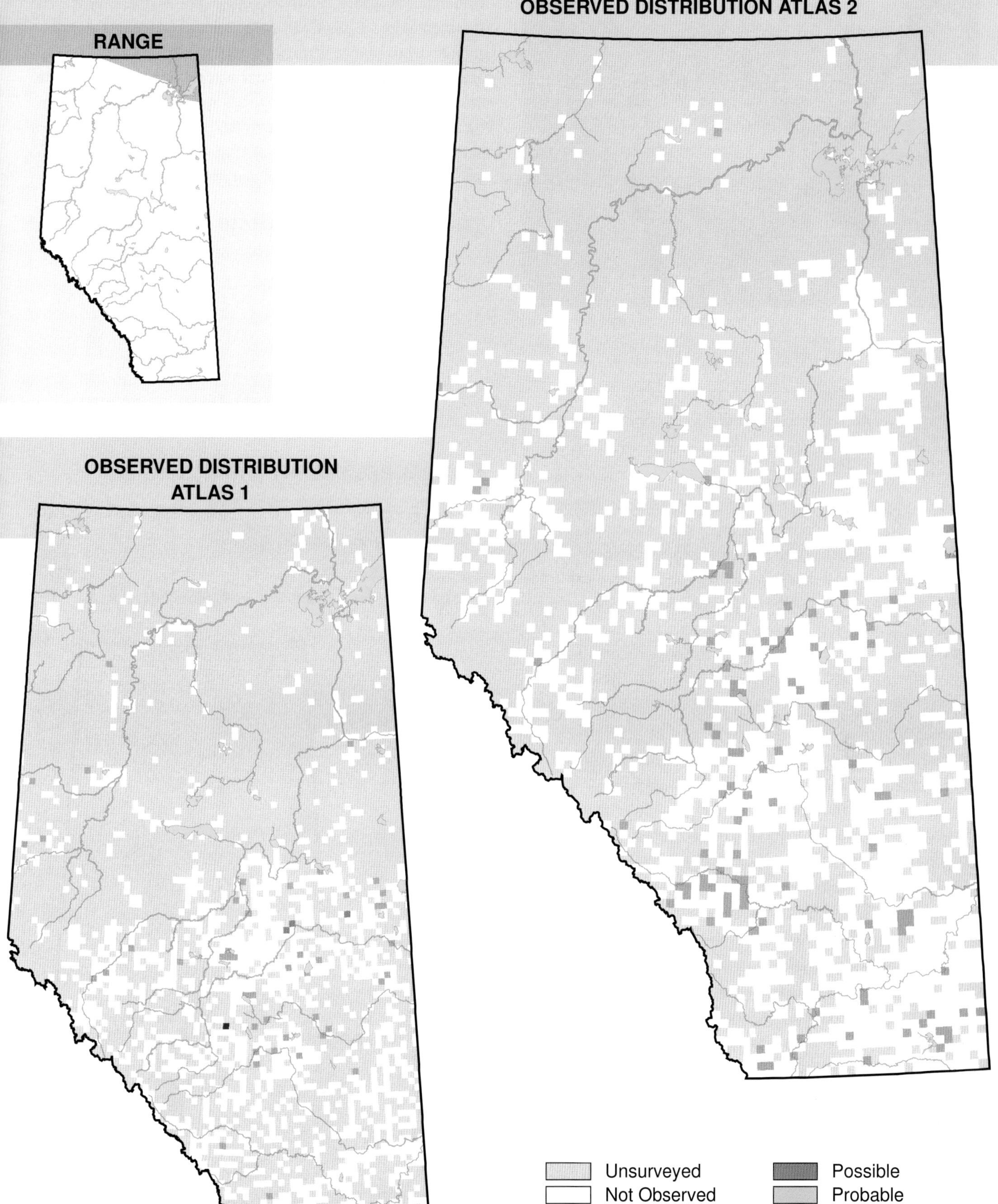

Cassin's Vireo (*Vireo cassinii*)

Photo: Cindy Creighton

NESTING

CLUTCH SIZE (EGGS):	3–5
INCUBATION (DAYS):	14
FLEDGING (DAYS):	14
NEST HEIGHT (METRES):	<4.5

Cassin's Vireo is found in the Foothills, Parkland, and Rocky Mountain Natural Regions in Alberta. After Atlas 1 was completed, the Solitary Vireo taxon was split into three species. Two of these species, the Blue-headed Vireo and Cassin's Vireo, occur in Alberta. Cassin's Vireo occurs in the southwestern part of the province; whereas, the Blue-headed Vireo is present farther north along the Rocky Mountains, east into the Foothills and Parkland NRs, and up into the Boreal Forest Natural Region. The two species occur sympatrically in the Grassland, Parkland, Foothills, and Rocky Mountain NRs.

Despite occurring in three natural regions, Cassin's Vireo was found mainly in the Rocky Mountain NR, and only infrequently encountered in the southern Foothills and Parkland NRs. The distribution of this species in Alberta reflects its preference to nest at higher elevations in mixedwood forests. Compared to its cousin, the Blue-headed Vireo, Cassin's Vireo tends to be found in forests with a larger proportion of deciduous trees.

Frequency of detection and change analysis calculations for this species were made using the combined Cassin's and Blue-headed vireo records from Atlas 2 and Solitary Vireo records from Atlas 1. Data were analyzed in this way because the Cassin's and Blue-headed vireos had not been recognized as distinct species during Atlas 1.

An increase in relative abundance was detected in the Rocky Mountain NR where, relative to other species, these birds were observed more frequently in Atlas 2 than in Atlas 1. Changes in relative abundance were not detected in the other NRs. Breeding Bird Survey sample size was too small to investigate changes in abundance for Cassin's Vireo in Alberta. However, the BBS did detect an increase in the abundance of this species across Canada during the period 1968–2005. The status of this species in Alberta is Undetermined due to insufficient data.

TEMPORAL REPORTING RATE

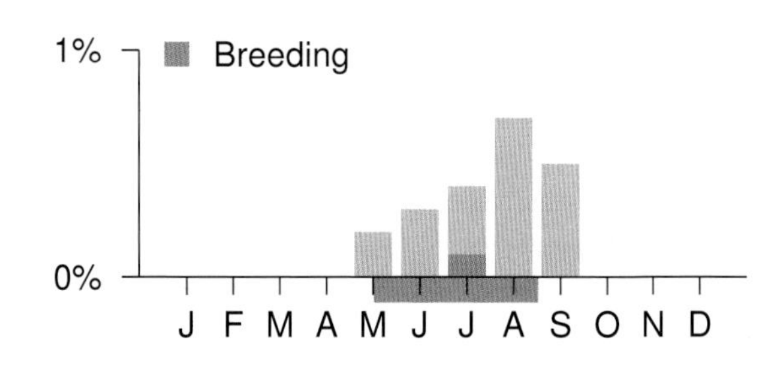

NATURAL REGIONS REPORTING RATE

Foothills	0.2%
Grassland	0.6%
Parkland	0.2%
Rocky Mountain	4.9%

SPATIAL REPORTING RATE

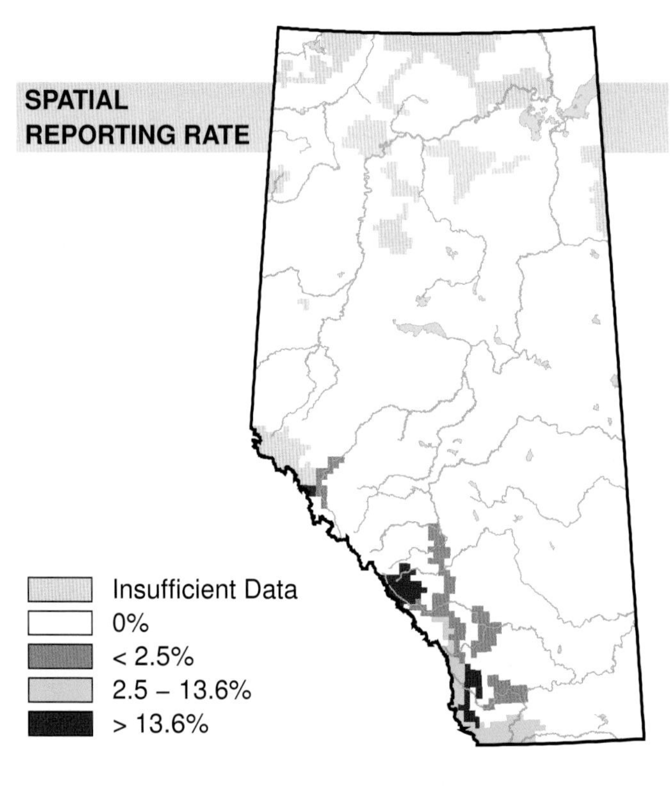

STATUS

GSWA 2000	GSWA 2005	COSEWIC Historic Rankings	COSEWIC 2007	Alberta Wildlife Act
Undetermined	Undetermined	N/A	N/A	N/A

RANGE

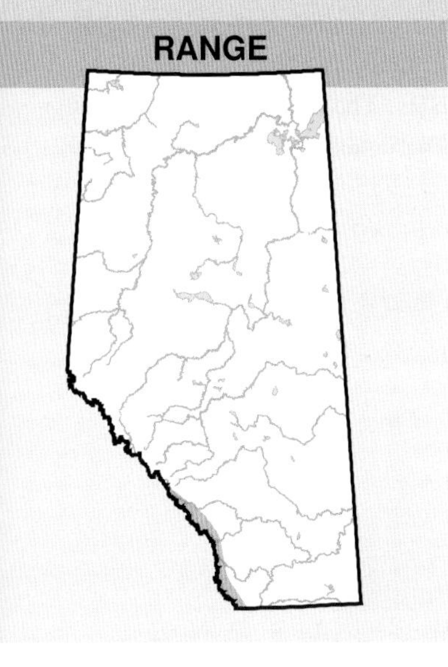

OBSERVED DISTRIBUTION ATLAS 2

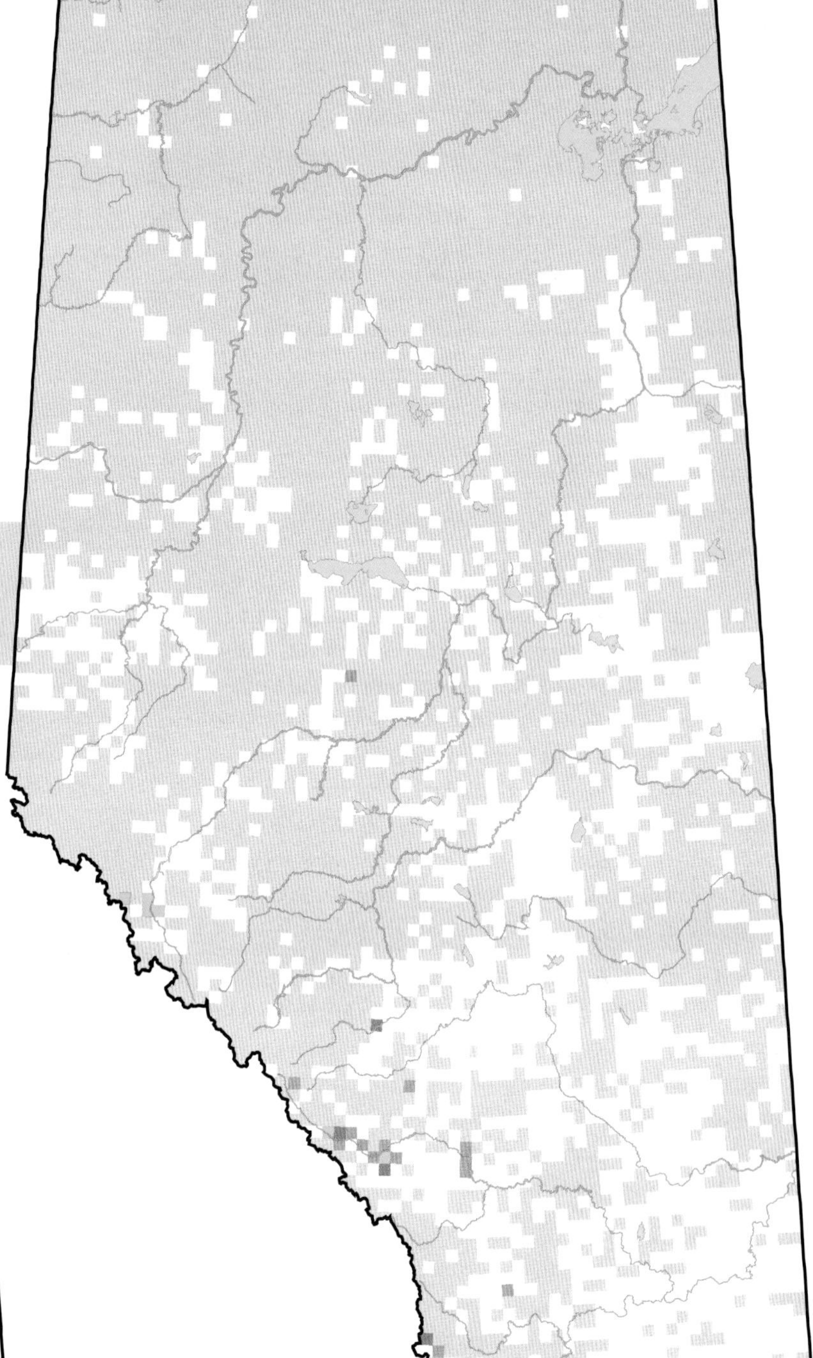

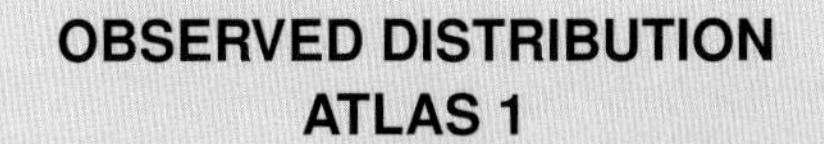

OBSERVED DISTRIBUTION ATLAS 1

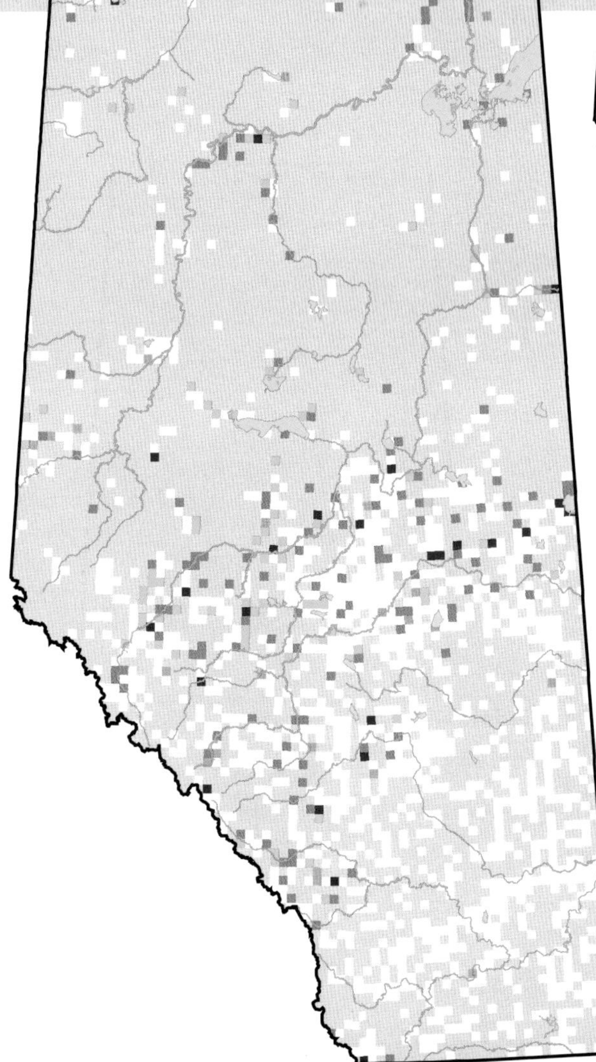

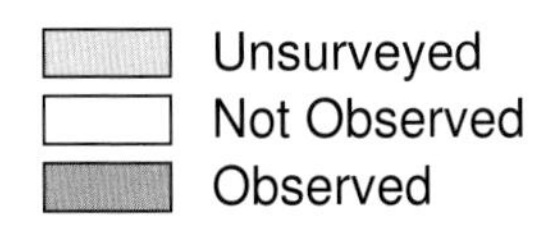

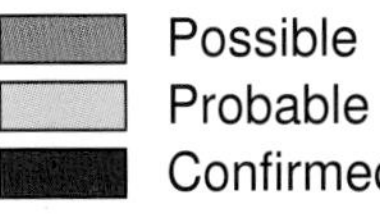

Blue-headed Vireo (*Vireo solitarius*)

Photo: Royal Alberta Museum

NESTING

CLUTCH SIZE (EGGS):	3–5
INCUBATION (DAYS):	14
FLEDGING (DAYS):	14
NEST HEIGHT (METRES):	<4.5

The Blue-headed Vireo is found in every Natural Region across the province. No change in distribution was observed between Atlas 1 and Atlas 2. Subsequent to Atlas 1, the Solitary Vireo taxon was split into three species. Two of these species, the Blue-headed Vireo and Cassin's Vireo occur in Alberta. The Blue-headed Vireo does not occur in the southwestern part of the province where the Cassin's Vireo is present and the Cassin's Vireo does not occur in the Boreal Forest NR. They occur sympatrically in the Grassland, Parkland, Foothills, and Rocky Mountain NRs.

Despite being found in every NR, Blue-headed Vireos were found mainly in the Boreal Forest and Foothills NRs, but were found only infrequently in the Grassland, Parkland, and Rocky Mountain NRs. This distribution reflects its preference to nest in mature conifer and mixedwood forests. Blue-headed Vireos were widely distributed across their range and were not usually found nesting in areas that are occupied by people.

Frequency of detection and change analysis calculations for this species were made using the combined records of Blue-headed and Cassin's vireos from Atlas 2 and Solitary Vireo records from Atlas 1. Data were analyzed in this way because the Blue-headed and Cassin's vireos had not been recognized as two distinct species during Atlas 1.

Increases in relative abundance were detected in the Boreal Forest and Rocky Mountain NRs where, relative to other species, this species was observed more frequently in Atlas 2. As the Cassin's Vireo does not occur in the Boreal Forest NR, this increase relates specifically to the Blue-headed Vireo. No change was detected in other NRs. The Breeding Bird Survey did not indicate a change in Alberta abundance, but did detect an increase across Canada from 1985–2005. The provincial status of this species is Undetermined due to insufficient data.

TEMPORAL REPORTING RATE

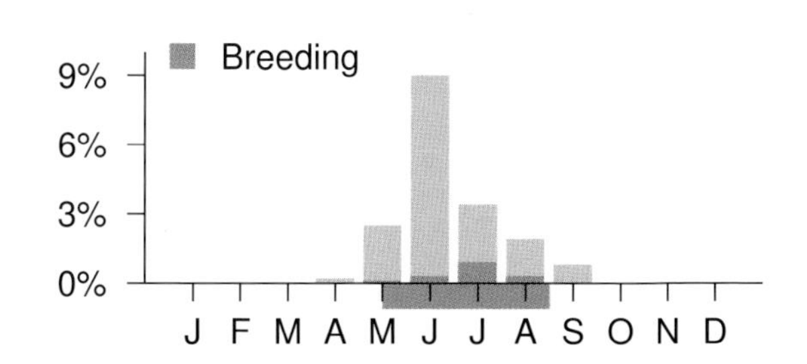

NATURAL REGIONS REPORTING RATE

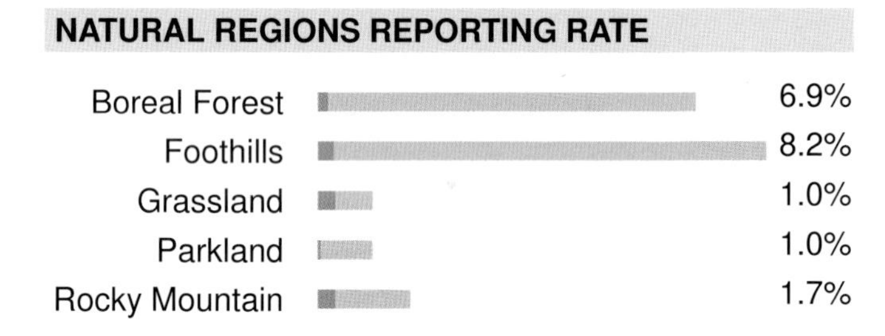

SPATIAL REPORTING RATE

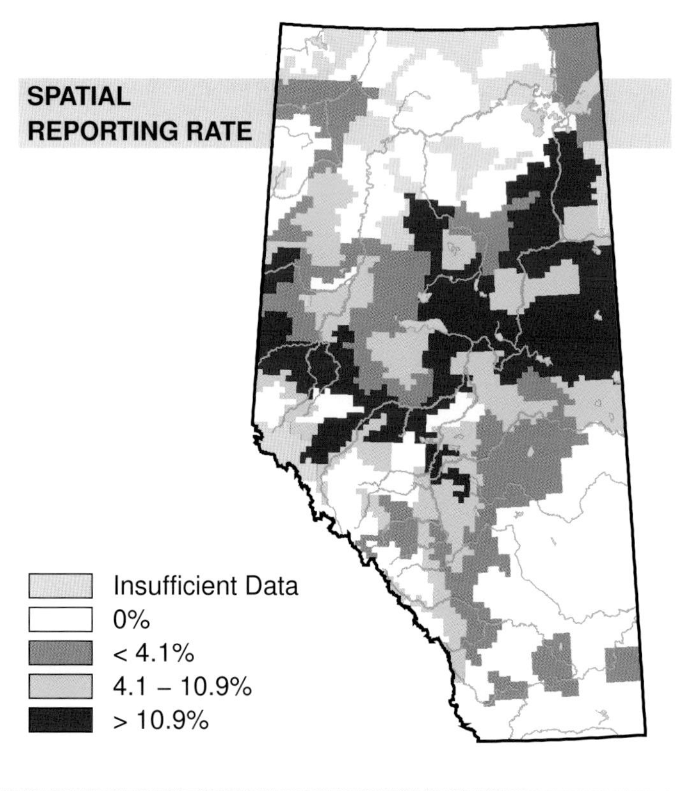

STATUS

GSWA 2000	GSWA 2005	COSEWIC Historic Rankings	COSEWIC 2007	Alberta Wildlife Act
Secure	Secure	N/A	N/A	N/A

Habitat  Nest Location  Nest Type  Diet  

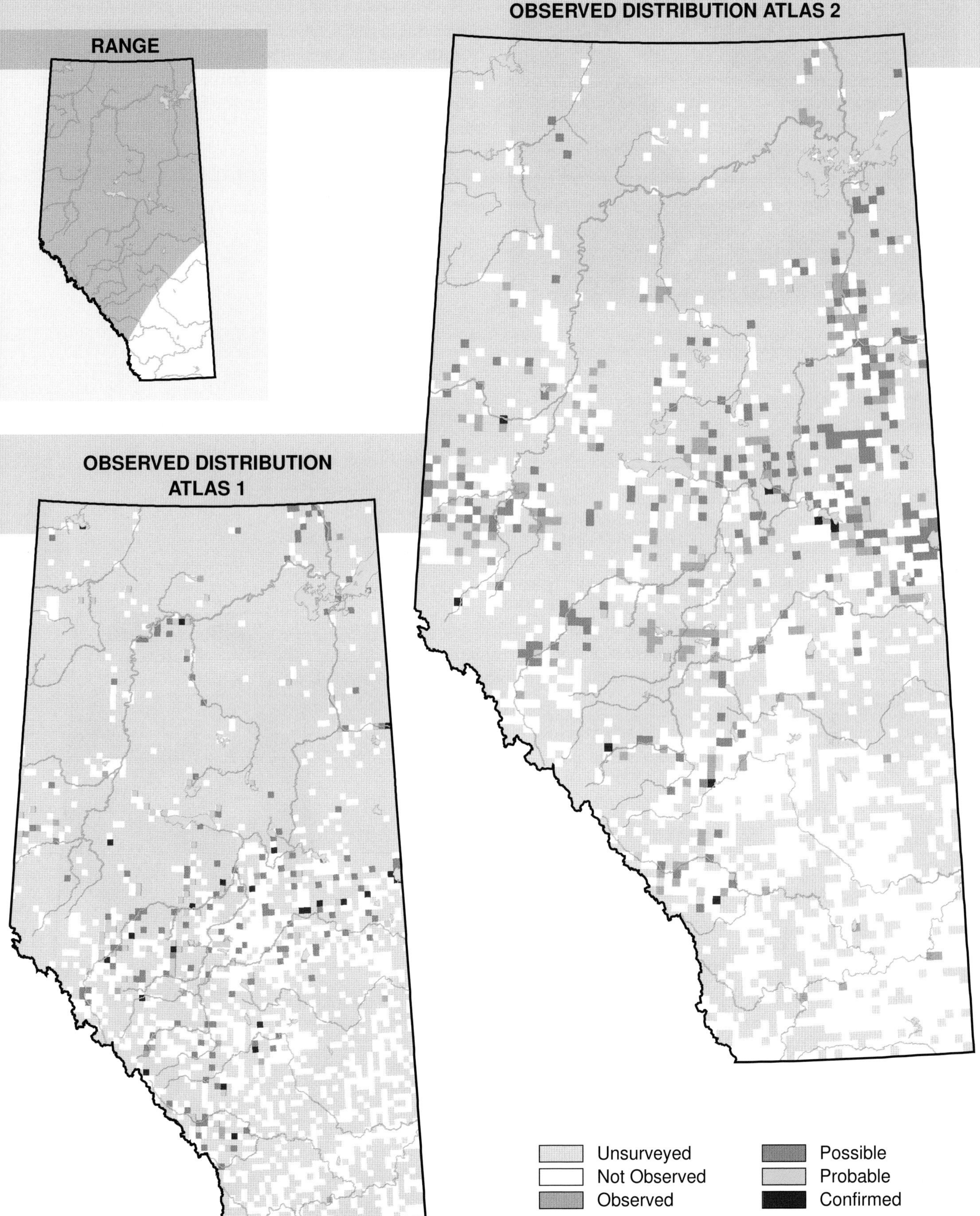

Warbling Vireo (*Vireo gilvus*)

Photo: Gary Kurtz

NESTING

CLUTCH SIZE (EGGS):	3–5
INCUBATION (DAYS):	12–14
FLEDGING (DAYS):	16
NEST HEIGHT (METRES):	0

The Warbling Vireo is found in every Natural Region across the province. No change in distribution was observed between Atlas 1 and Atlas 2 for this species.

Despite being found in every natural region, Warbling Vireos were found mainly in the Foothills and Rocky Mountain NRs. This species was occasionally found in the Boreal Forest, Grassland, and Parkland NRs. The patchy distribution of Warbling Vireos reflects their preference to nest in mature deciduous forests near water.

Declines in relative abundance were detected in the Boreal Forest, Grassland, and Parkland NRs. Changes in relative abundance were not detected in the Foothills NR but an increase was detected in the Rocky Mountain NR. Differences in observation frequency relative to other species between Atlas 1 and Atlas 2 mirrored these changes. The Breeding Bird Survey detected no change in Alberta or nationally in the period 1985–2005. However, an increase was detected in British Columbia and a decrease in Saskatchewan for the period 1985–2005. The biological causes for the Atlas-detected declines in the Boreal Forest, Grassland, and Parkland NRs are unclear, but reductions in mature deciduous forests may be occurring with expanding agriculture and/or resource extraction in these NRs. The cause for the increase in the Rocky Mountain NR is also unclear as amounts of suitable habitat for this species have not changed substantially in this NR. If this species is indeed increasing in parts of Alberta and decreasing in others, this may explain the discrepancy between the Atlas and the Alberta BBS results. Further research is needed to properly assess the status of this species in Alberta. Currently, the status of Warbling Vireos in Alberta is rated Secure.

TEMPORAL REPORTING RATE

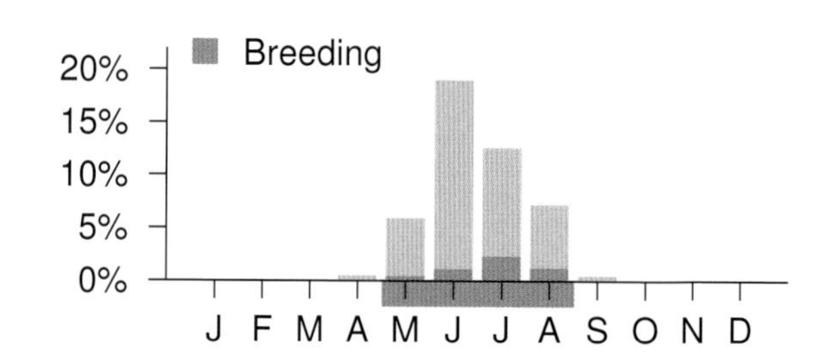

NATURAL REGIONS REPORTING RATE

Region	Rate
Boreal Forest	2%
Foothills	23%
Grassland	5%
Parkland	6%
Rocky Mountain	33%

SPATIAL REPORTING RATE

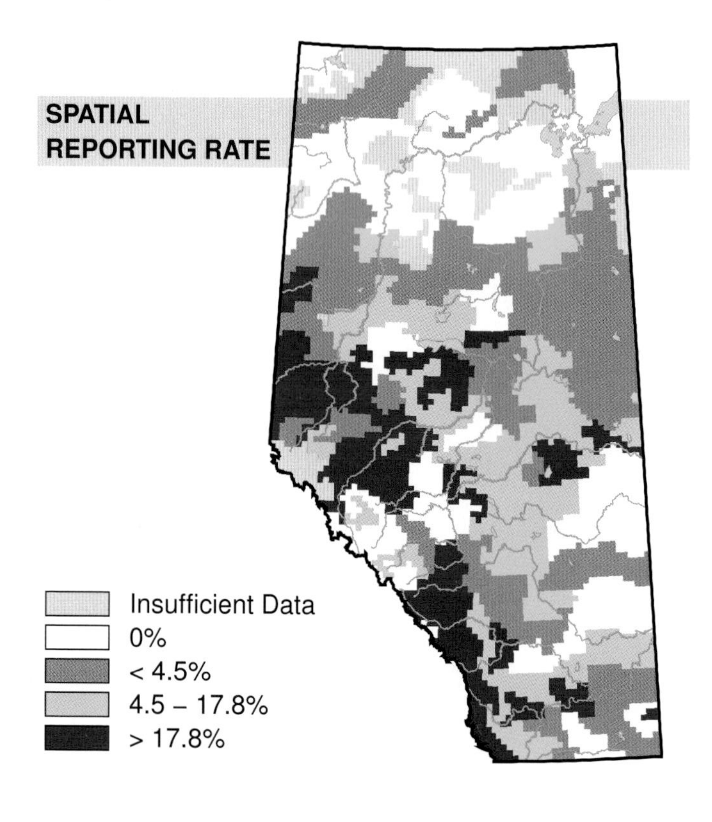

STATUS

GSWA 2000	GSWA 2005	COSEWIC Historic Rankings	COSEWIC 2007	Alberta Wildlife Act
Secure	Secure	N/A	N/A	N/A

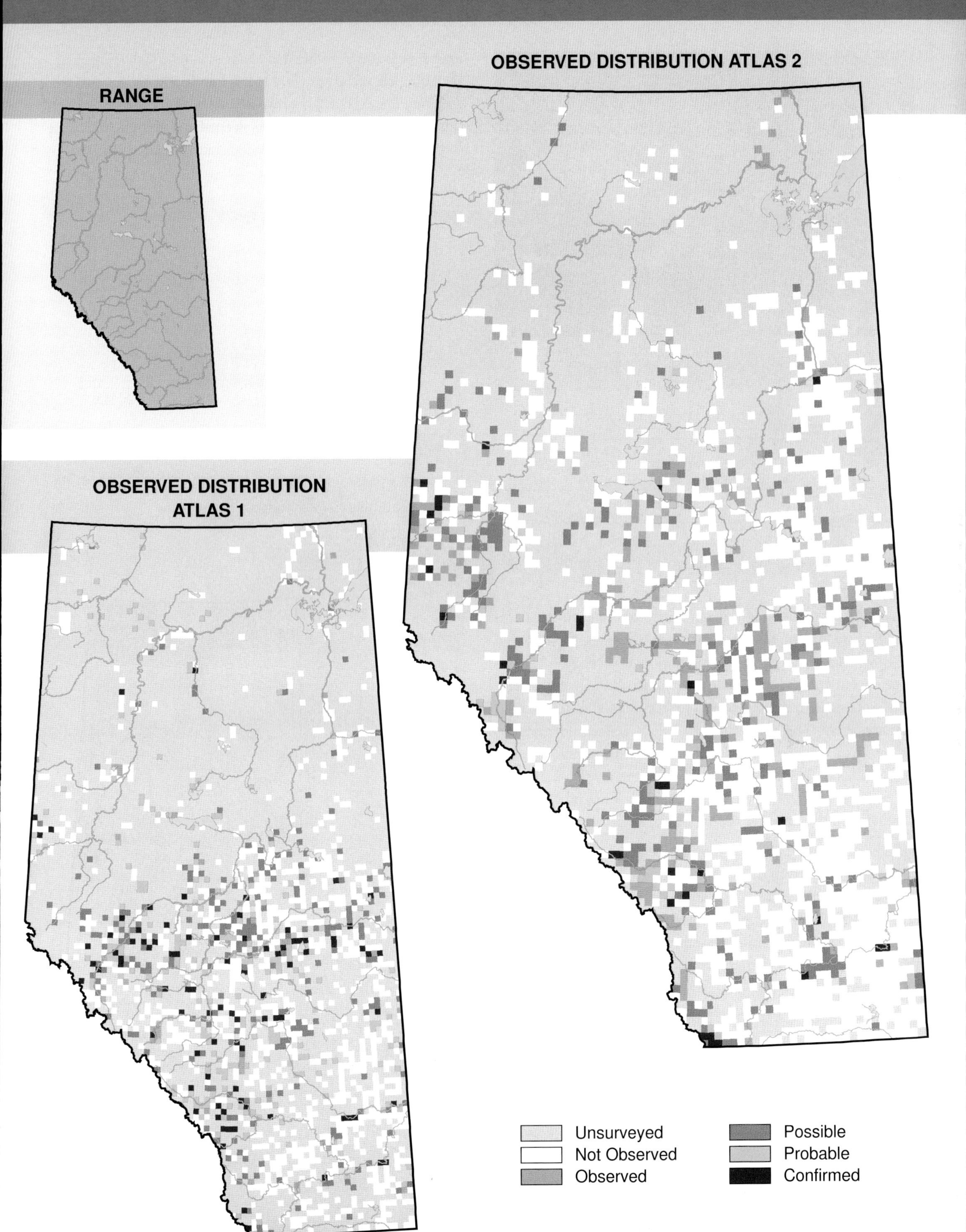
RANGE
OBSERVED DISTRIBUTION ATLAS 2
OBSERVED DISTRIBUTION ATLAS 1
Unsurveyed
Not Observed
Observed
Possible
Probable
Confirmed

Philadelphia Vireo (*Vireo philadelphicus*)

Photo: Debbie Godkin

NESTING

CLUTCH SIZE (EGGS):	3–5
INCUBATION (DAYS):	11–14
FLEDGING (DAYS):	12–14
NEST HEIGHT (METRES):	<14

The Philadelphia Vireo is found in the Boreal Forest, Foothills, Parkland, and Rocky Mountain Natural Regions. No change in distribution was observed between Atlas 1 and Atlas 2 for this species.

Despite being found in four natural regions, Philadelphia Vireos were found mainly in the Boreal Forest and Foothills NRs. This species was infrequently found in other NRs. The species' range reflects its tendency to associate with early- and mid-succession deciduous forests that develop after forestry activity or fire. This species also uses forest edges and gaps in forest cover during the breeding season. For these reasons, populations of this species tend not to be negatively affected by the extraction of renewable resources, provided habitat regeneration is allowed to occur.

Relative abundance changes were not detected in any NR. The Breeding Bird Survey found an increase in abundance for this species in Alberta and across Canada during the period 1985–2005. The BBS-detected increase is expected, given that this species generally prefers fragmented and post-disturbance habitats such as those found after forest harvesting. The differences between the Atlas and BBS results are confusing because Atlas 2 had broader coverage than the BBS over much of this species' range; yet, it failed to detect any change in abundance. It is possible that changes in regrowth strategies, including chemical spraying to control deciduous growth, are reducing the expected positive impact of forest harvesting for this species. Further research is needed to determine the cause of this discrepancy. The provincial status of this species is Secure.

TEMPORAL REPORTING RATE

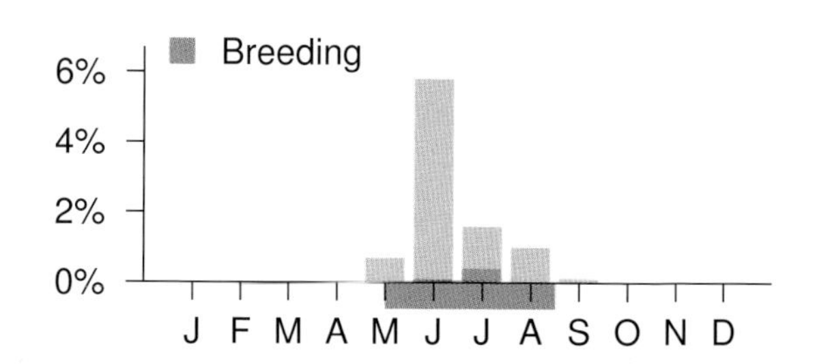

NATURAL REGIONS REPORTING RATE

SPATIAL REPORTING RATE

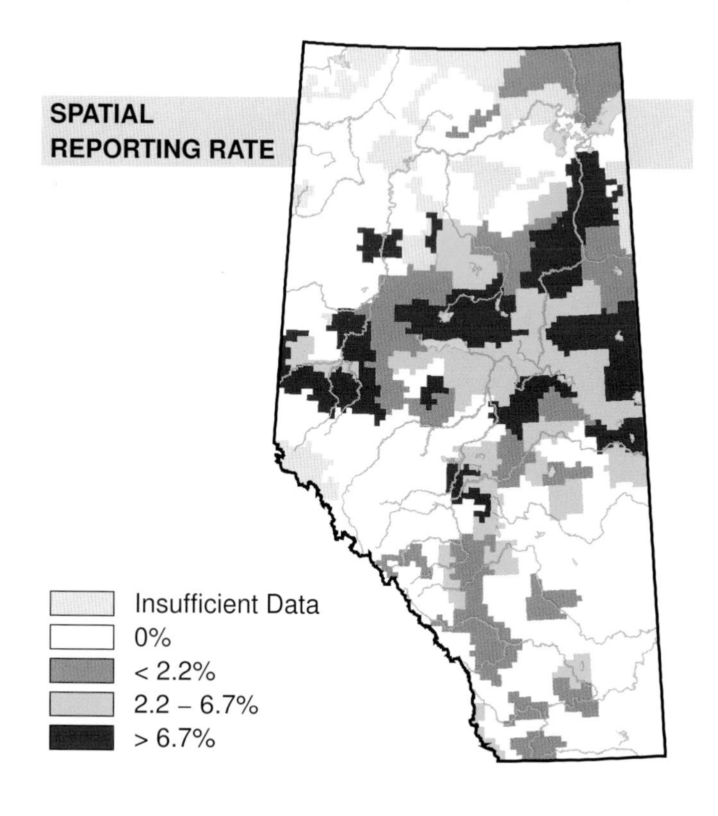

STATUS

GSWA 2000	GSWA 2005	COSEWIC Historic Rankings	COSEWIC 2007	Alberta Wildlife Act
Secure	Secure	N/A	N/A	N/A

Habitat

Nest
Location
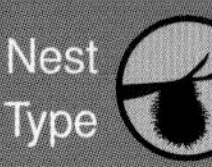
Nest
Type

Diet

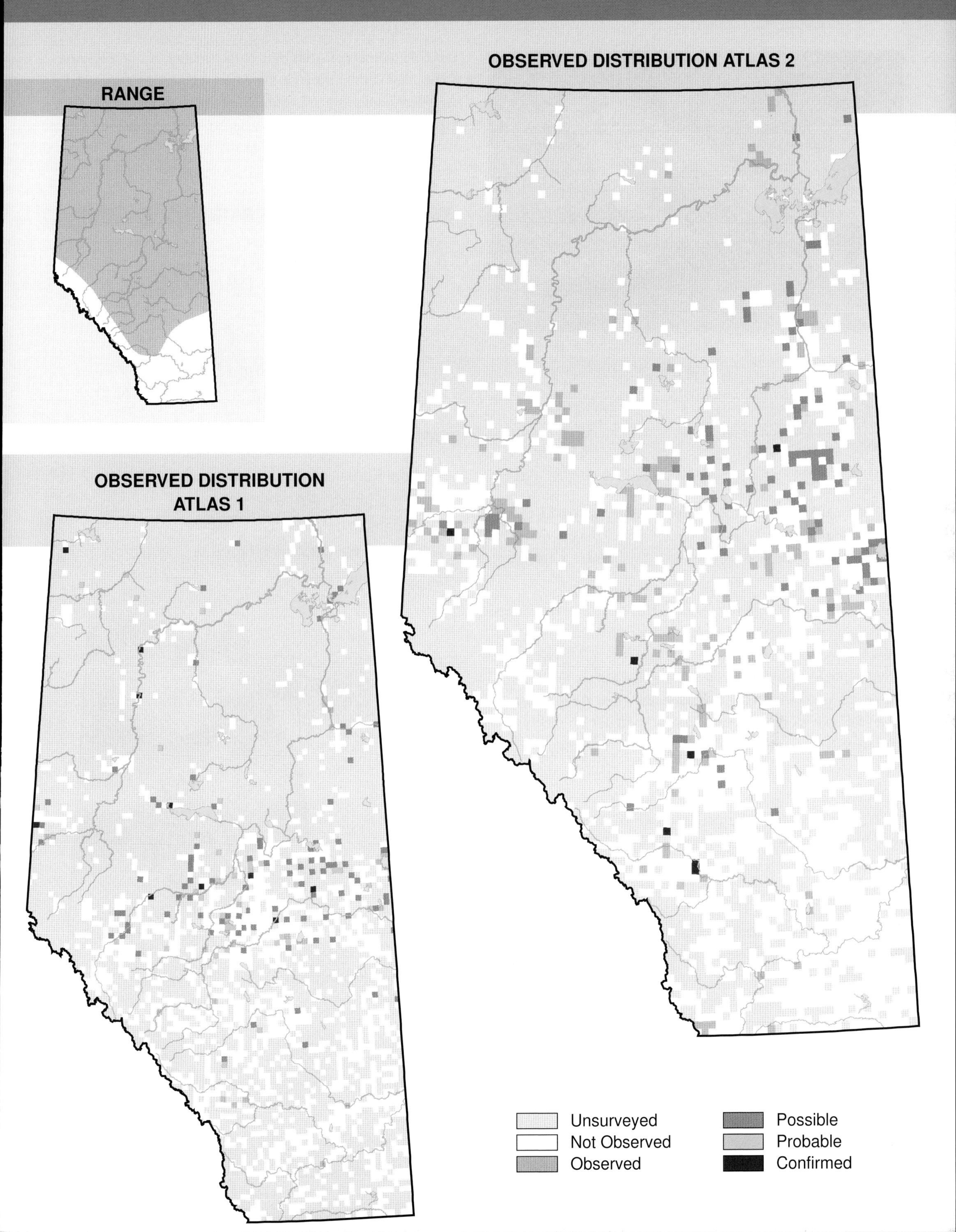
RANGE
OBSERVED DISTRIBUTION ATLAS 2
OBSERVED DISTRIBUTION
ATLAS 1
Unsurveyed
Not Observed
Observed
Possible
Probable
Confirmed

Red-eyed Vireo (*Vireo olivaceus*)

Photo: Gerald Romanchuk

NESTING

CLUTCH SIZE (EGGS):	3–4
INCUBATION (DAYS):	12–14
FLEDGING (DAYS):	12
NEST HEIGHT (METRES):	1.5–3

One of the most abundant woodland songbirds, the Red-eyed Vireo is found in every Natural Region across the province. No change in distribution was observed between Atlas 1 and Atlas 2 for this species.

Despite being found in every Natural Region, Red-eyed Vireos were found mainly in the Boreal Forest and Foothills NRs and only occasionally in the Grassland, Parkland, and Rocky Mountain NRs. Unlike its cousin the Philadelphia Vireo, the Red-eyed Vireo tends to be observed less often along forest edges and more often in interior forest habitat.

Declines in relative abundance were detected in the Boreal Forest and Parkland NRs where, relative to other species, the Red-eyed Vireo was observed less frequently in Atlas 2 than in Atlas 1. Declines in those natural regions could be associated with increased forest fragmentation because this species tends to be associated with interior forest habitat. Changes in relative abundance were not detected in the Foothills, Grassland, and Rocky Mountain NRs. The Breeding Bird Survey did not find any recent change in abundance for this species in Alberta or Canada. Differences between the results of Atlas and Breeding Bird Survey analyses could be attributed to the different spatial scales on which the data were analyzed, and to differences in field methods. Breeding Bird Survey data were analyzed at the provincial scale; whereas, due to sufficient sample size, it was possible to analyze Atlas data at the smaller, Natural Region scale. In addition, because this vireo is a species associated with forest interiors, it is likely that the Atlas had better coverage for this species, because Atlas surveys are not limited to roadsides, as the BBS is. The status of this species is Secure in Alberta.

TEMPORAL REPORTING RATE

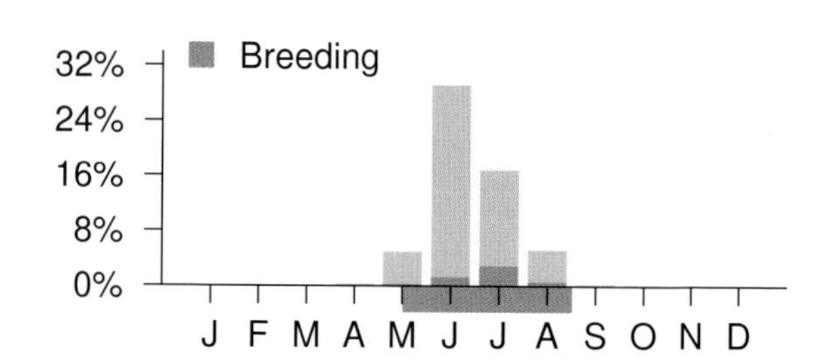

NATURAL REGIONS REPORTING RATE

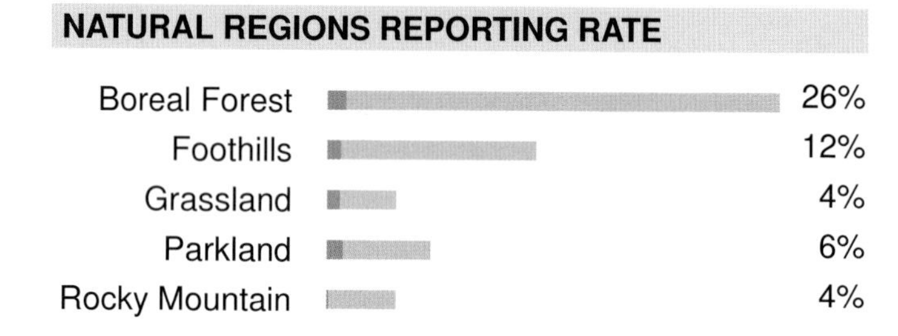

SPATIAL REPORTING RATE

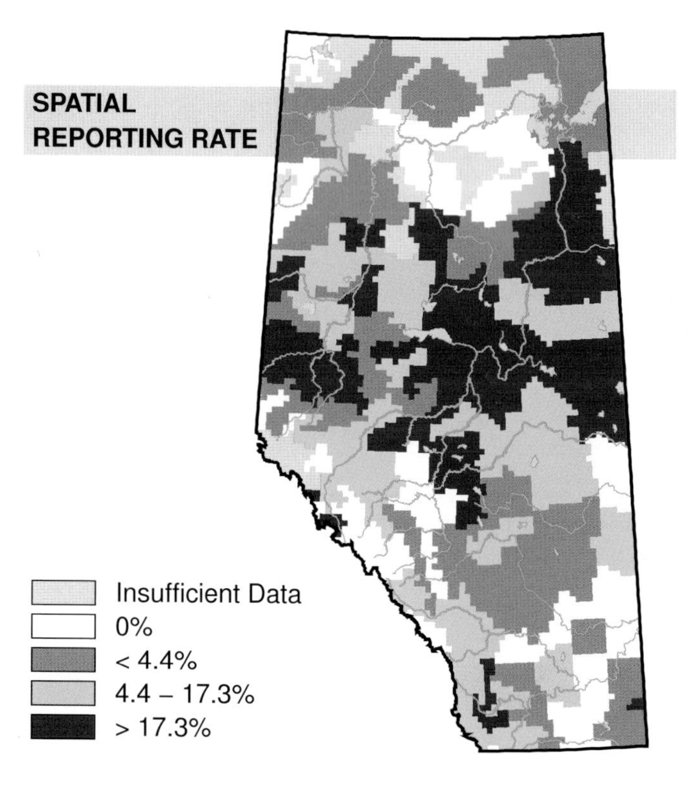

STATUS

GSWA 2000	GSWA 2005	COSEWIC Historic Rankings	COSEWIC 2007	Alberta Wildlife Act
Secure	Secure	N/A	N/A	N/A

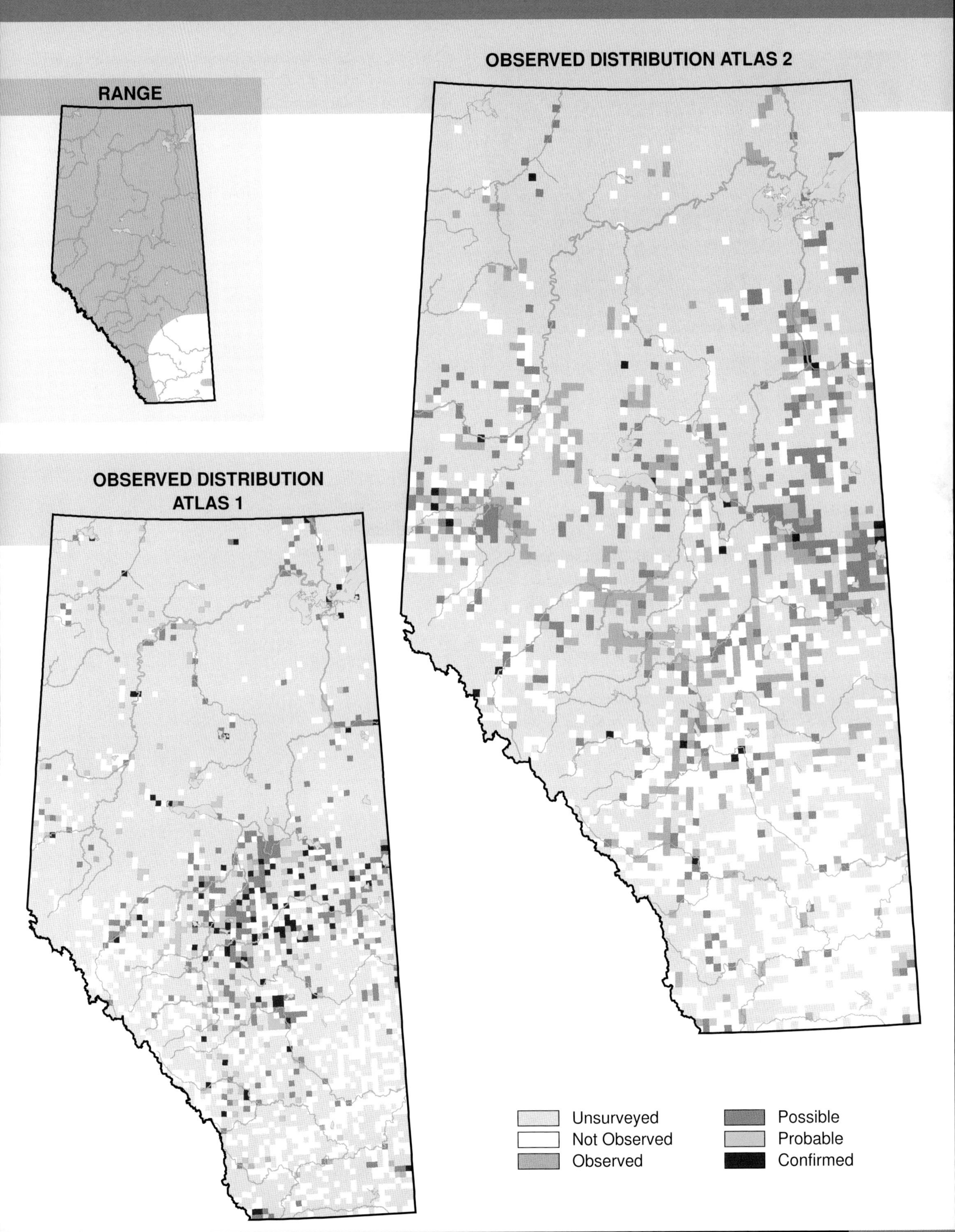

OBSERVED DISTRIBUTION ATLAS 2
RANGE
OBSERVED DISTRIBUTION
ATLAS 1
Unsurveyed
Not Observed
Observed
Possible
Probable
Confirmed

Gray Jay (*Perisoreus canadensis*)

Photo: Robert Gehlert

NESTING

CLUTCH SIZE (EGGS):	2–6
INCUBATION (DAYS):	16–18
FLEDGING (DAYS):	17–20
NEST HEIGHT (METRES):	1.5–4.5

The inquisitive Gray Jay is a familiar sight at picnic areas and campgrounds in Alberta's Mountain Parks and northern areas. No change was observed in the distribution of this species between Atlas 1 and Atlas 2. However, there were more records in the northeastern part of the province in Atlas 2.

Coniferous and mixedwood forests with a spruce component are the preferred habitat of the Gray Jay. This species is a year-round resident and is one of the earliest nesters in Alberta. Gray Jays often build their nests along the southern edge of forest stands, perhaps taking advantage of the sunshine through increased southern exposure of the nests. The Gray Jay's variable diet, which is not thought to limit its distribution, can include invertebrates, nestling birds, eggs, fungi, berries, and carrion. Gray Jays were most commonly found in the Boreal Forest, Foothills, and Rocky Mountain Natural Regions where spruce can be common. This species was rare in the Grassland and Parkland NRs.

An increase in relative abundance was detected in the Parkland NR and a decline was detected in the Rocky Mountain NR. Differences in observation frequency relative to other species between Atlas 1 and 2 mirrored these changes. Differences in landscape composition between Atlas 1 and 2 provide little clue as to the cause for these changes. No changes were detected in the other NRs. The Breeding Bird Survey did not detect any change in abundance for this species in Alberta, but a decline in Canada was detected for the period from 1985–2005. Further research is needed to evaluate the Atlas-detected changes. This species is considered Secure in Alberta.

TEMPORAL REPORTING RATE

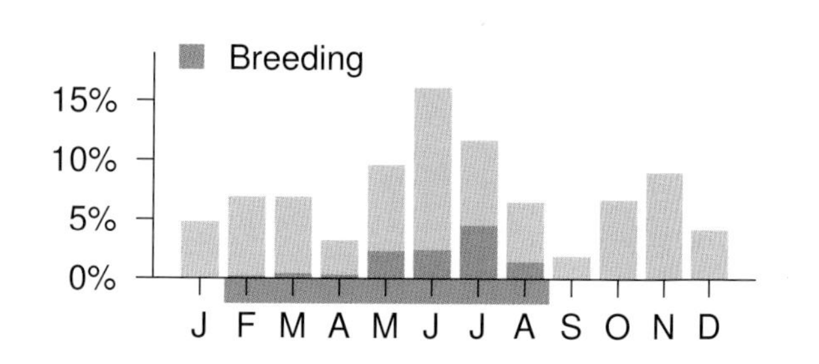

NATURAL REGIONS REPORTING RATE

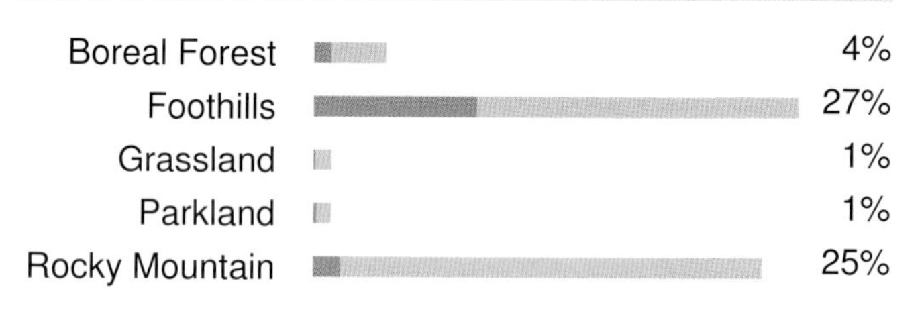

SPATIAL REPORTING RATE

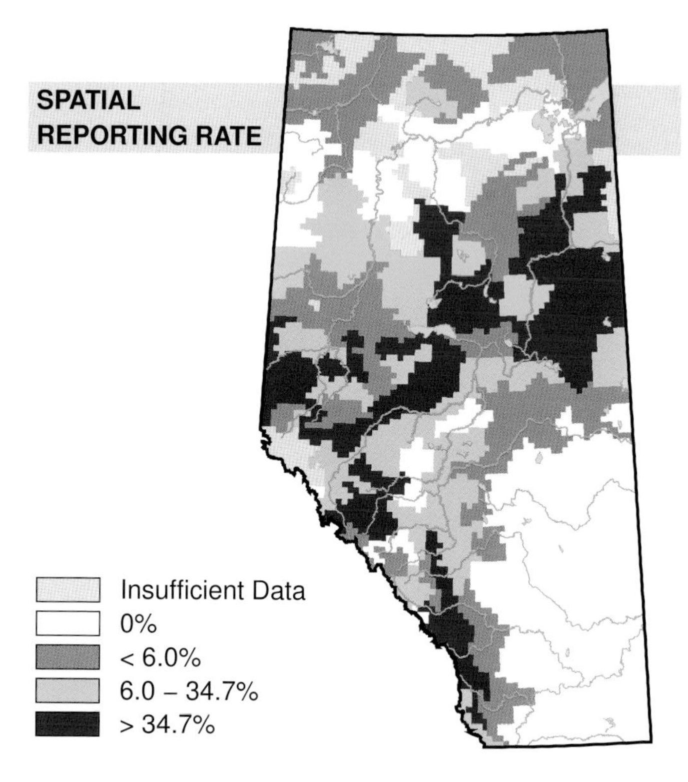

STATUS

GSWA 2000	GSWA 2005	COSEWIC Historic Rankings	COSEWIC 2007	Alberta Wildlife Act
Secure	Secure	N/A	N/A	N/A

RANGE

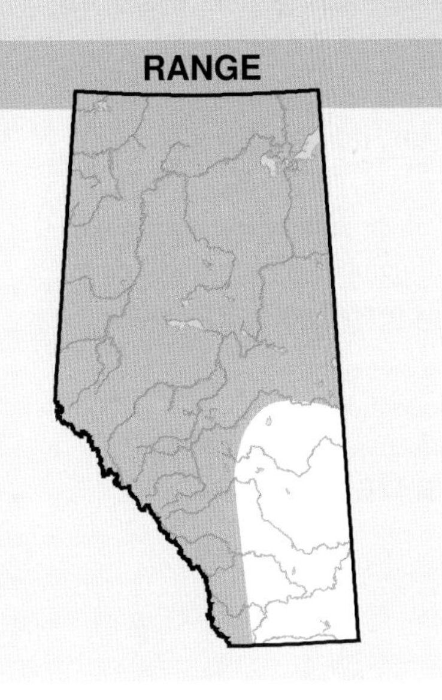

OBSERVED DISTRIBUTION ATLAS 2

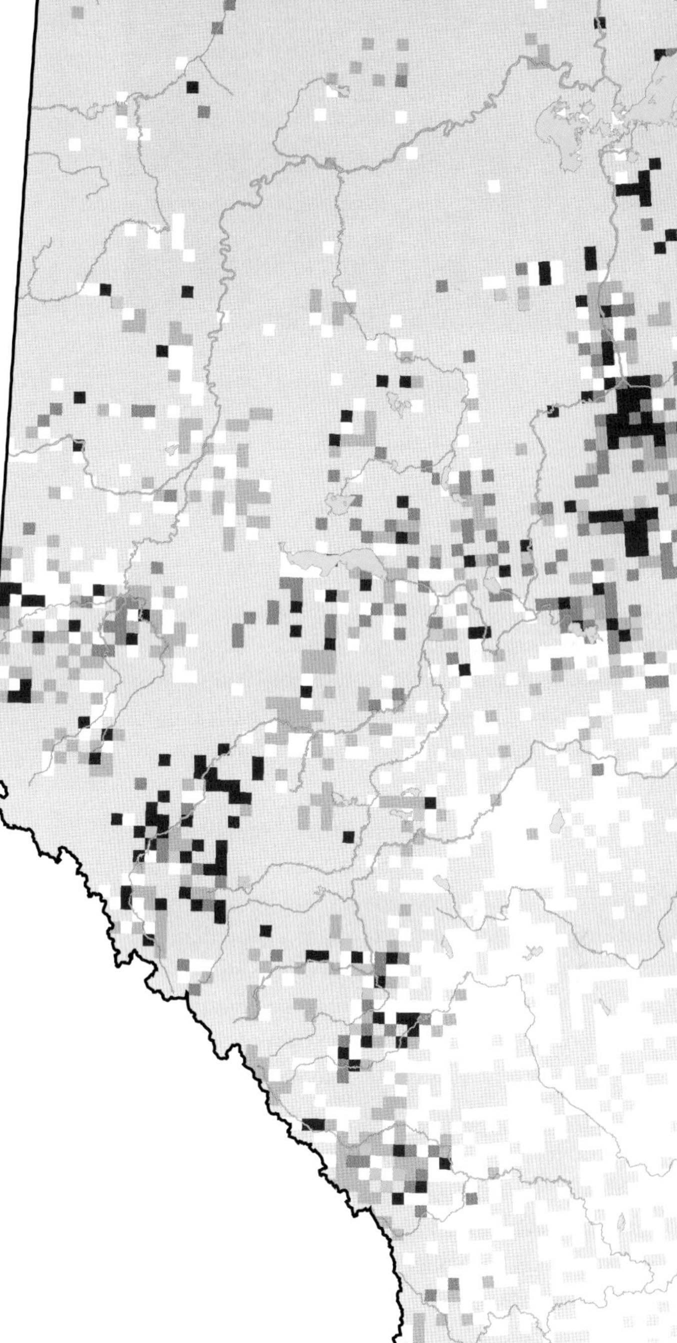

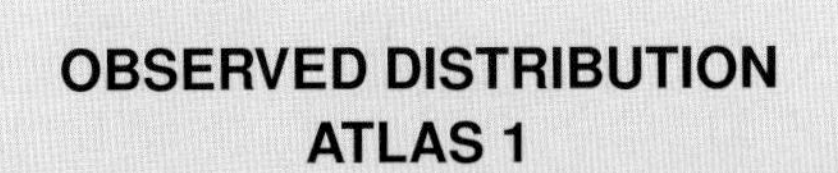

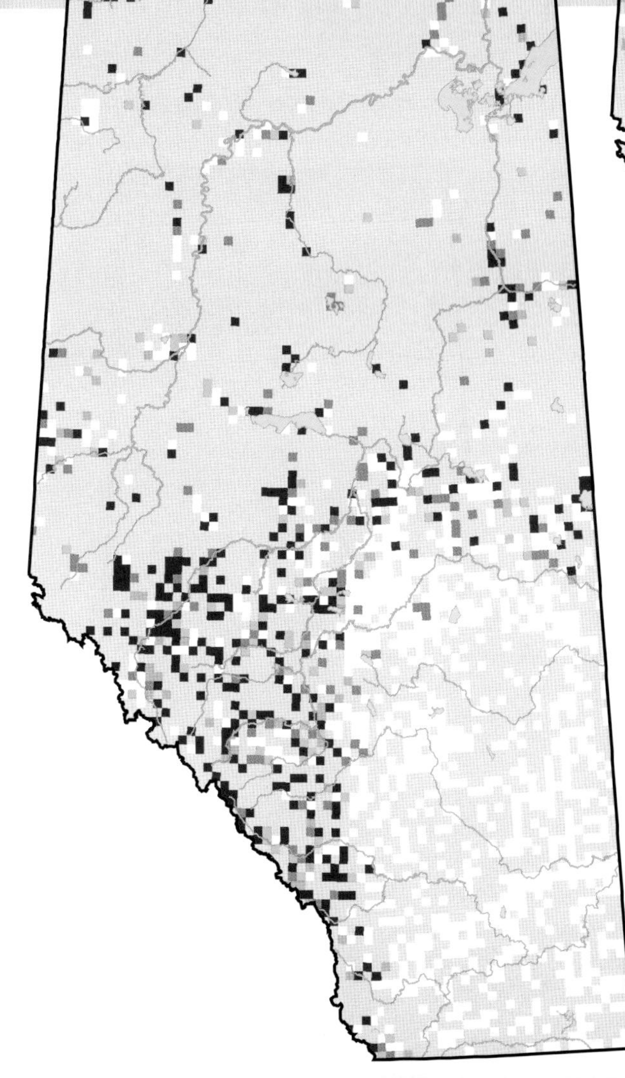

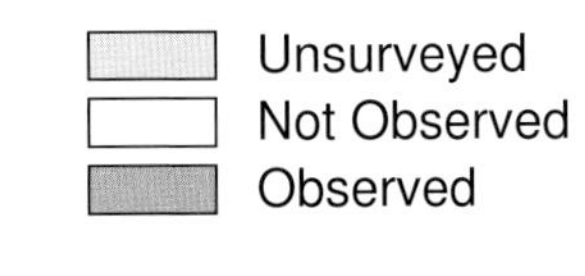

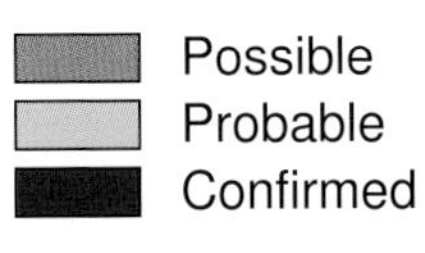

Steller's Jay (*Cyanocitta stelleri*)

Photo: Stan Gosche

NESTING

CLUTCH SIZE (EGGS):	3–5
INCUBATION (DAYS):	16–17
FLEDGING (DAYS):	22–30
NEST HEIGHT (METRES):	2.5–8

Found mostly on the west side of the Rocky Mountains where they are common, Steller's Jays are becoming more abundant in Alberta in the Rocky Mountain Natural Region. Although there were few records in the northern part of the species' range in Atlas 2, the distribution of this species did not change substantially.

Steller's Jay is commonly found across western, central, and southern British Columbia where it is usually associated with coniferous and mixedwood forests. This jay was detected more often in Atlas 2 in the southern part of the Rocky Mountain NR where its range is expanding into Alberta. Low-elevation corridors across the southern Rockies facilitate this expansion. Observations were obtained east of Waterton Lakes National Park in the Parkland and Grassland NRs. Further expansion of this species along the eastern slopes of the Rocky Mountains is expected to occur in future.

Periodic irruptions of large flocks (mainly young birds) bring this jay into areas and habitats not normally occupied; these may follow years of good breeding success and/or poor food supply (Greene et al., 1998).

An increase in relative abundance was detected in the Rocky Mountain NR where, relative to other species, this bird was observed more frequently in Atlas 2 than in Atlas 1. This change is likely the result of a range expansion in that area. The Breeding Bird Survey sample size of Steller's Jay records was too small to investigate abundance changes in Alberta. However, population increases were detected for Canada during the period 1968–2005.

Greene et al. (1998) indicate that BBS patterns should be interpreted with caution, especially for estimates based on a small number of routes. Greene et al. (1998) also state that even estimates of trends based on large sample sizes can be strongly influenced by the small number of routes, and thus may not accurately represent the overall population trend in a census region.

The status of this species is Secure in Alberta.

TEMPORAL REPORTING RATE

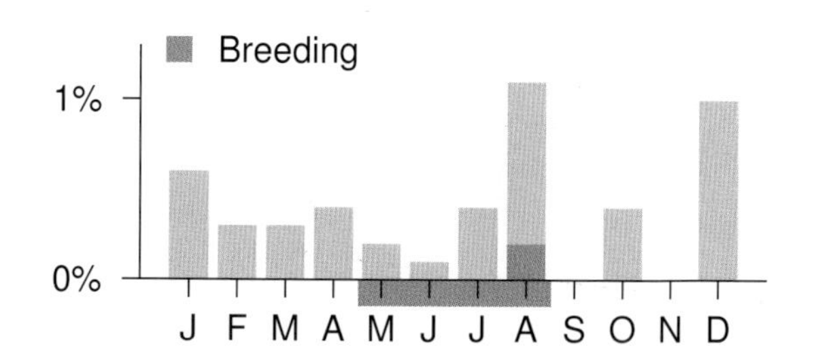

NATURAL REGIONS REPORTING RATE

SPATIAL REPORTING RATE

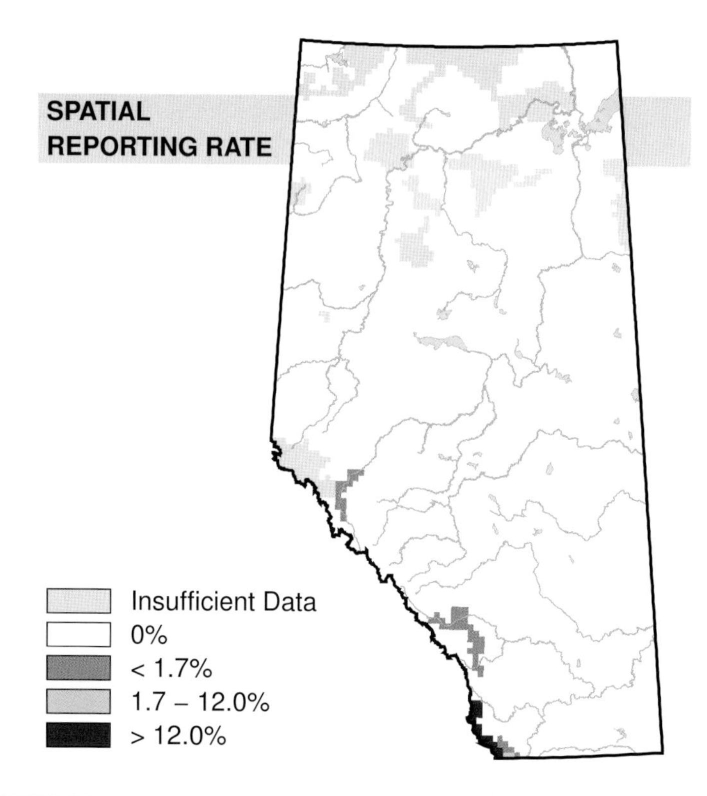

STATUS

GSWA 2000	GSWA 2005	COSEWIC Historic Rankings	COSEWIC 2007	Alberta Wildlife Act
Secure	Secure	N/A	N/A	N/A

RANGE

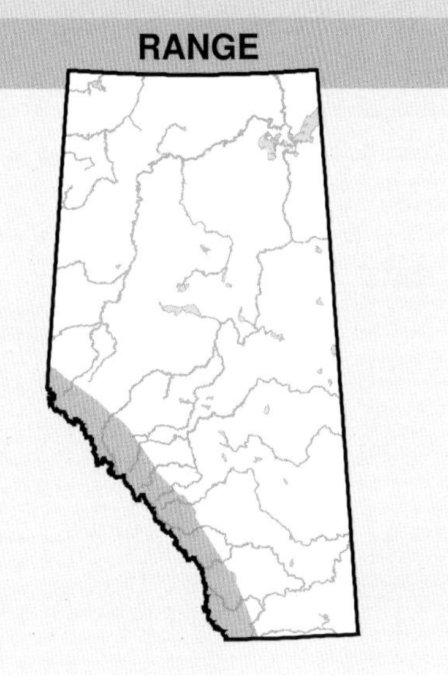

OBSERVED DISTRIBUTION ATLAS 1

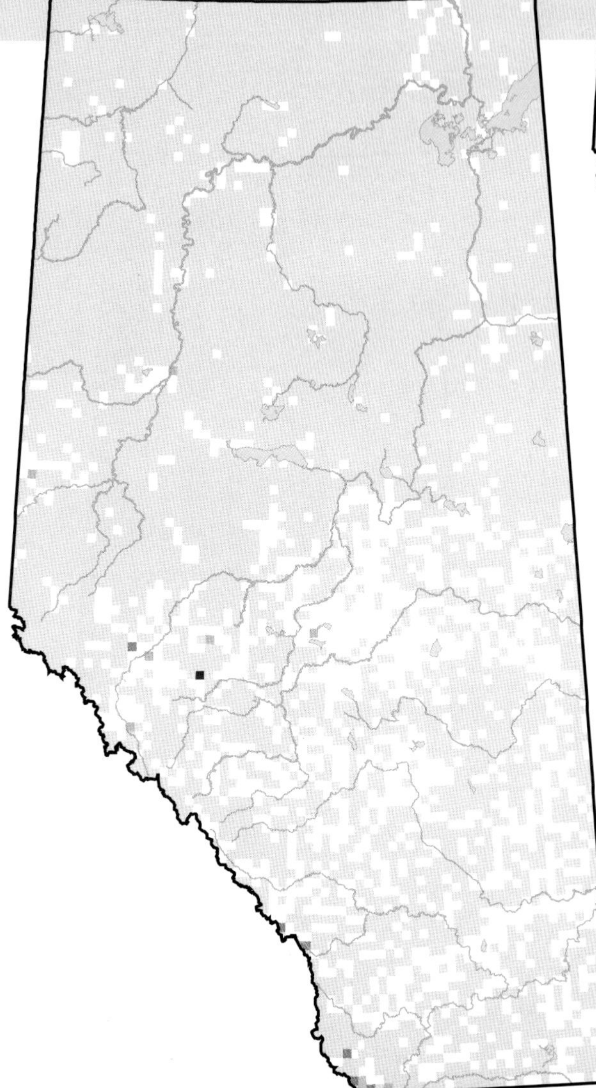

OBSERVED DISTRIBUTION ATLAS 2

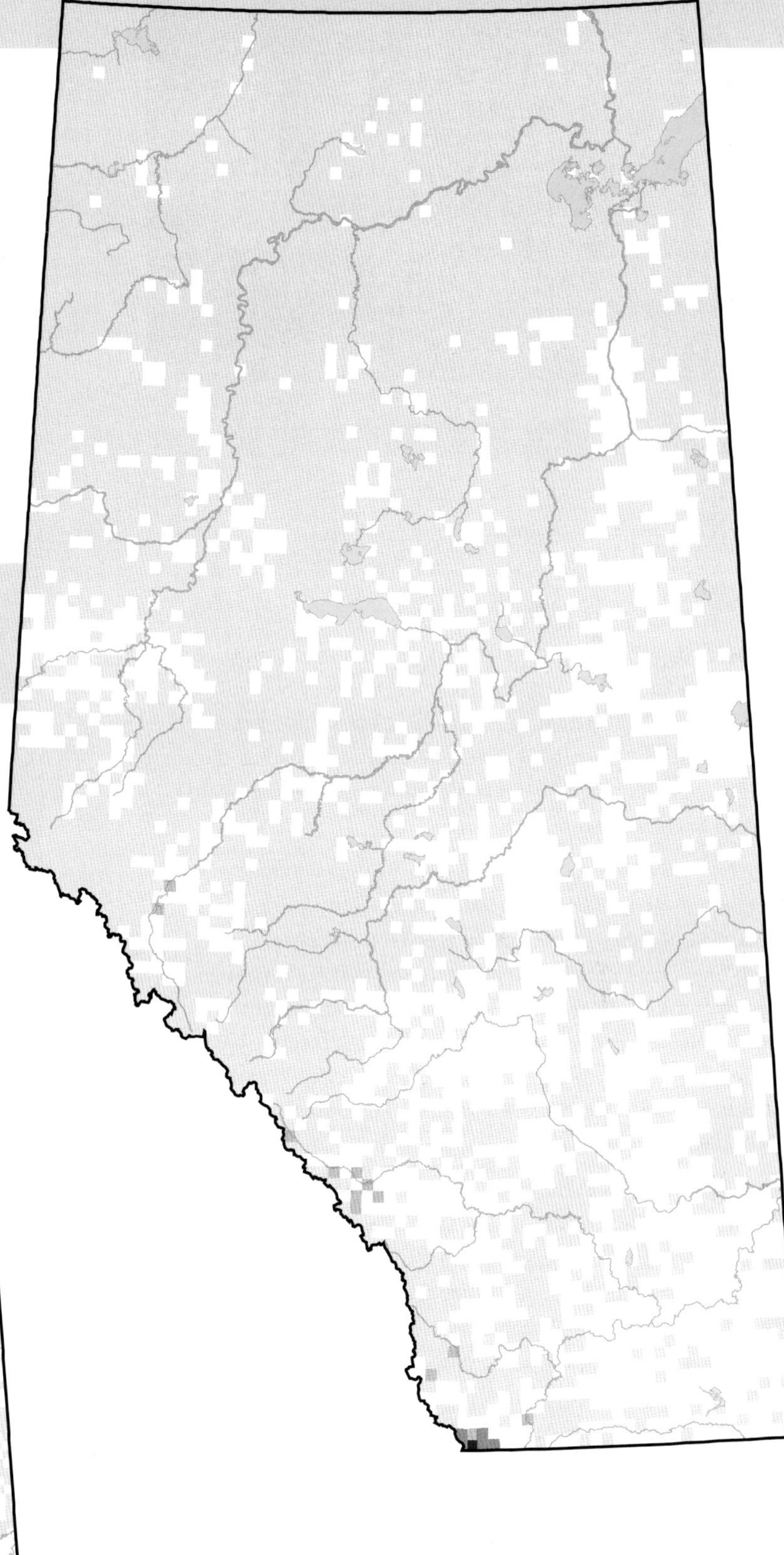

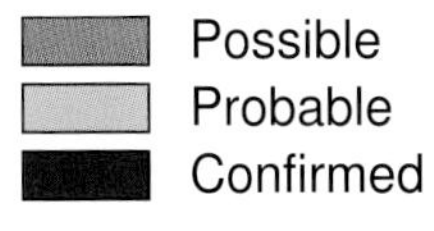

Blue Jay (*Cyanocitta cristata*)

Photo: Gerald Romanchuk

NESTING

CLUTCH SIZE (EGGS):	4–6
INCUBATION (DAYS):	17–18
FLEDGING (DAYS):	17–21
NEST HEIGHT (METRES):	2.4–7.5

The raucous sounds of the Blue Jay are familiar to most people in Alberta because this species is found in every Natural Region across the province. This species was observed more often in the southern and western parts of the province during Atlas 2 which is likely the result of the continued westward expansion of the Blue Jay's range.

The Blue Jay can be common in towns and residential neighbourhoods where they use established trees for nesting, find an abundance of food at bird feeders, and nest in areas with fewer nest predators than are found in natural habitats. In natural settings, Blue Jays tend to be associated with coniferous and mixedwood forests. For this reason, Blue Jays were most common in the Boreal Forest, Foothills, and Parkland NRs. This species was less common in the Grassland and Rocky Mountain NRs.

Declines in relative abundance were detected in the Boreal Forest and Parkland NRs where, relative to other species, it was observed less frequently in Atlas 2 than in Atlas 1. Reasons for these declines are unclear; however, changes to forest composition in these natural regions could be having an impact on this species. Increases were detected in the Grassland and Rocky Mountain NRs where, relative to other species, Blue Jays were observed more frequently in Atlas 2 than in Atlas 1. Increases in these NRs are likely the result of the expansion of Blue Jay distribution into the western and southern parts of the province. No change was detected in the Foothills NR. The Breeding Bird Survey did not detect an abundance change for this species in Alberta, but across Canada a decline was detected for the period 1968–1985. This species is considered Secure in Alberta.

TEMPORAL REPORTING RATE

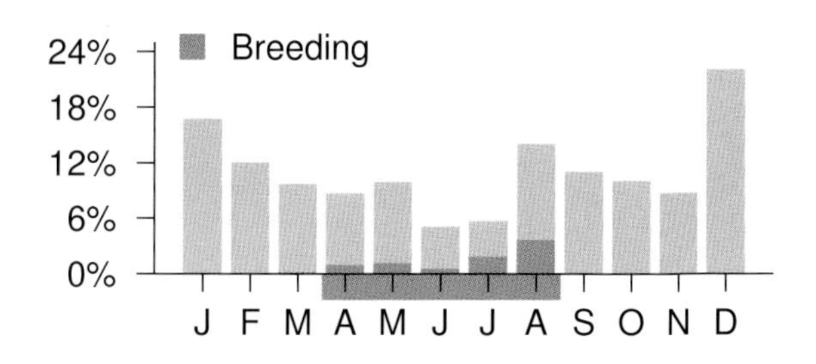

NATURAL REGIONS REPORTING RATE

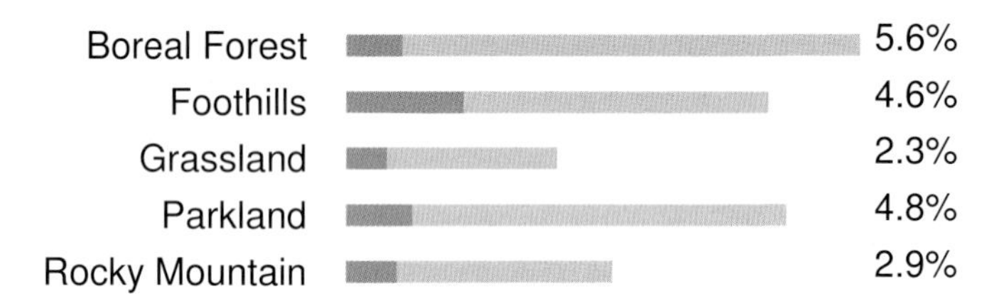

SPATIAL REPORTING RATE

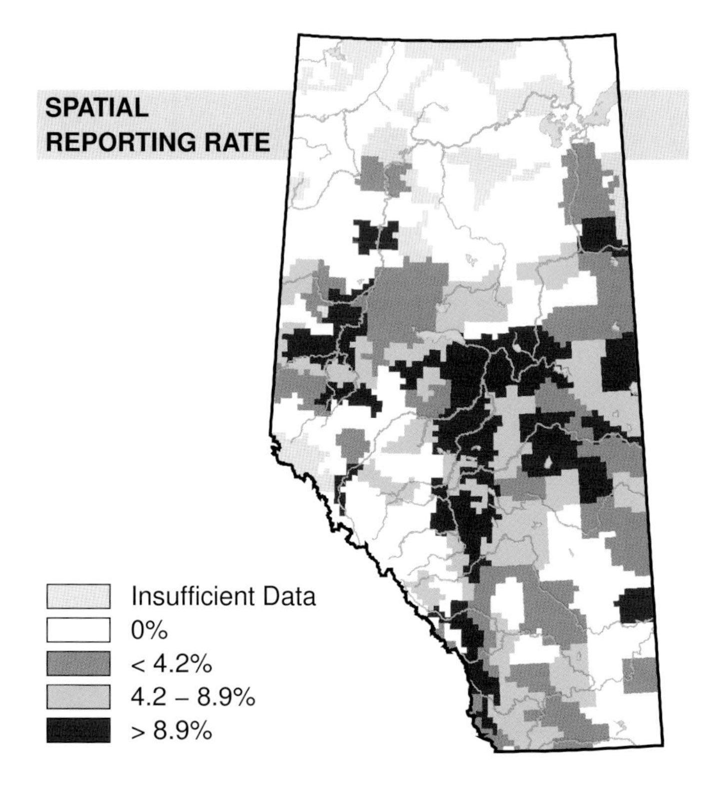

STATUS

GSWA 2000	GSWA 2005	COSEWIC Historic Rankings	COSEWIC 2007	Alberta Wildlife Act
Secure	Secure	N/A	N/A	N/A

Habitat  Nest Location  Nest Type  Diet 

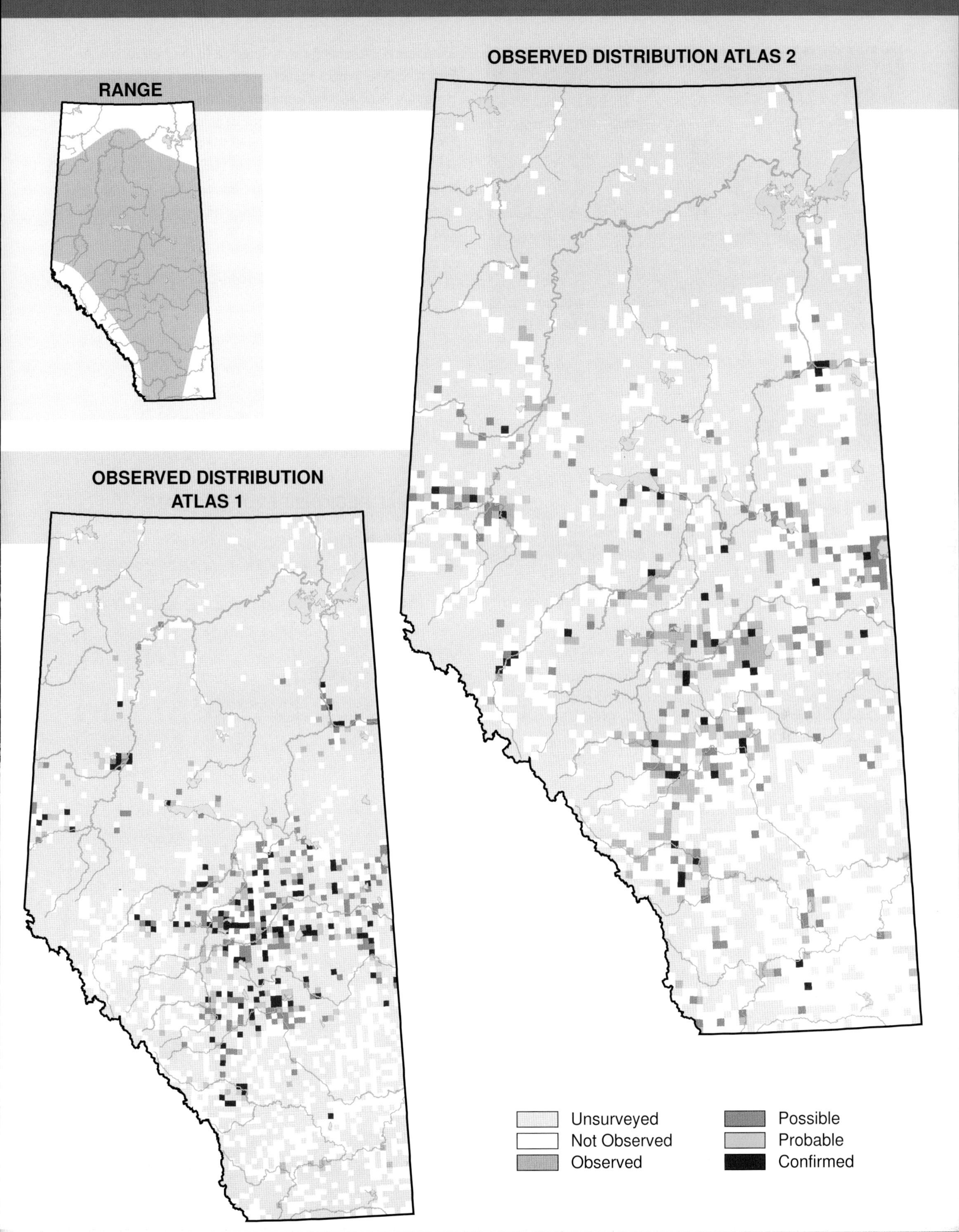

Clark's Nutcracker (*Nucifraga columbiana*)

Photo: Jim Jacobson

NESTING

CLUTCH SIZE (EGGS):	2–4
INCUBATION (DAYS):	16–17
FLEDGING (DAYS):	22
NEST HEIGHT (METRES):	3–17

A typical representative of high mountainous places, Clark's Nutcracker is found mainly in the Rocky Mountains in western Alberta. No change in distribution was observed between Atlas 1 and Atlas 2 for this species; however, the species was detected more often in the southern part of its Alberta range during Atlas 2.

Clark's Nutcracker occupies coniferous forests in mountainous areas where it specializes in collecting seeds produced by large-seeded pines. Seeds that are harvested by nutcrackers are either immediately consumed or cached for future consumption. Seeds that are not retrieved often germinate and grow, making Clark's Nutcracker one of the main seed dispersers for certain species of pines. The principal factor, therefore, that limits Clark's Nutcracker populations is the presence of large-seeded pines. These birds inhabit mainly the Rocky Mountain Natural Region in Alberta, because that is where large-seeded pines occur. Although observed in other NRs, the majority of breeding records for this species came from the Rocky Mountain NR.

An increase in relative abundance was detected in the Rocky Mountain NR where, relative to other species, this bird was observed more frequently in Atlas 2 than in Atlas 1. There is no known biological cause for this increase; so, it may relate to better coverage in Atlas 2, rather than to an actual change in abundance. No change was detected in the Foothills and Parkland NRs where reporting rates remained very low. Breeding Bird Survey sample size for Clark's Nutcracker was too small to investigate changes in abundance in Alberta. However, no change was detected for this species on a national scale. The status of this species in Alberta was amended from Secure, as assigned in 2000, to Sensitive in 2005, due to its restricted distribution, dependency on declining pine species, and its potential susceptibility to West Nile Virus.

TEMPORAL REPORTING RATE

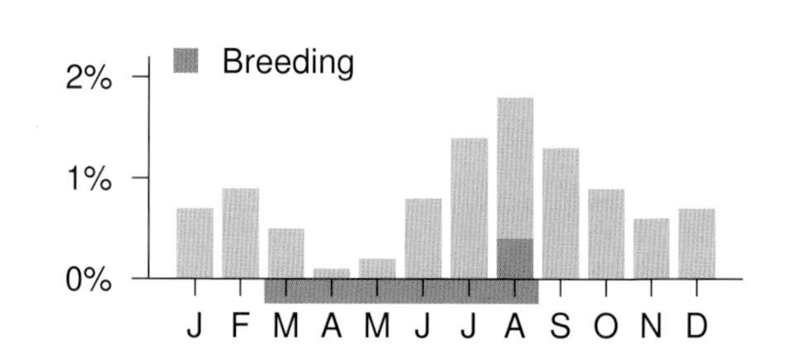

NATURAL REGIONS REPORTING RATE

SPATIAL REPORTING RATE

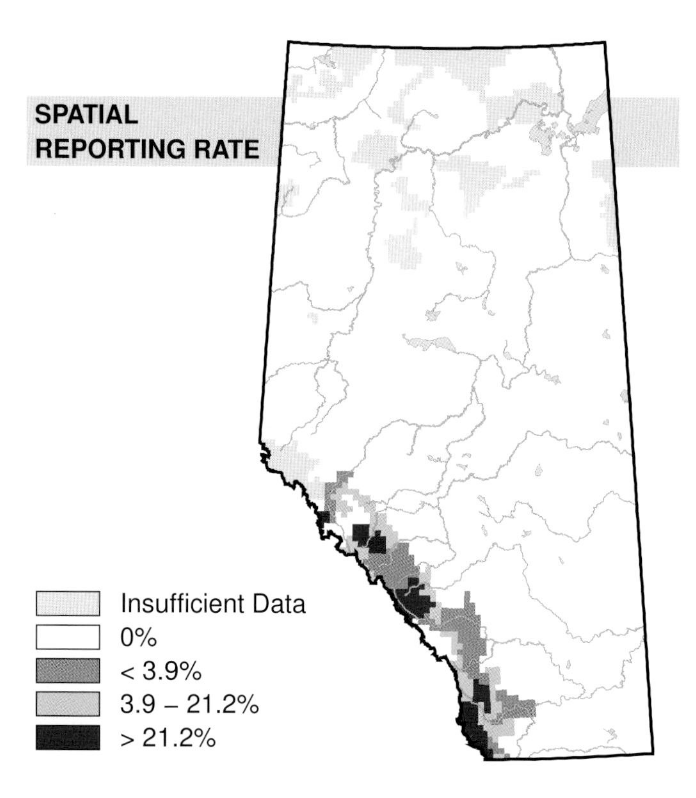

STATUS

GSWA 2000	GSWA 2005	COSEWIC Historic Rankings	COSEWIC 2007	Alberta Wildlife Act
Secure	Sensitive	N/A	N/A	N/A

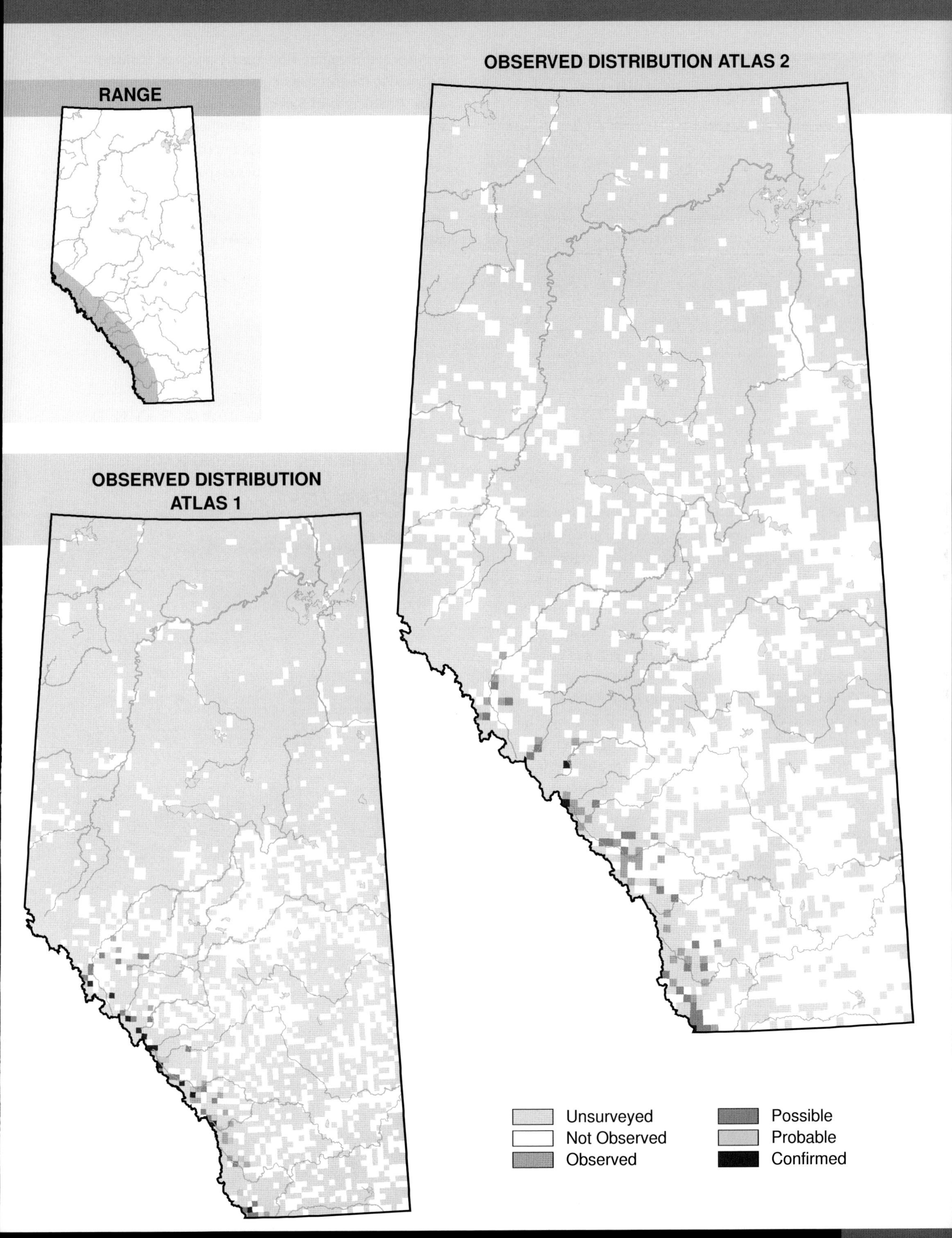

RANGE
OBSERVED DISTRIBUTION ATLAS 2
OBSERVED DISTRIBUTION ATLAS 1
Unsurveyed
Not Observed
Observed
Possible
Probable
Confirmed

Black-billed Magpie (*Pica hudsonia*)

Photo: Gerald Romanchuk

NESTING

CLUTCH SIZE (EGGS):	6–9
INCUBATION (DAYS):	16–18
FLEDGING (DAYS):	22–28
NEST HEIGHT (METRES):	<7

A good representative bird from western North America, the Black-billed Magpie is found in every Natural Region in Alberta. No change in distribution was observed for this species between Atlas 1 and Atlas 2.

The Black-billed Magpie does quite well in human-modified environments. This species is commonly encountered near human settlements in both urban and rural settings. Black-billed Magpies usually nest in fragmented areas such as along the edge of patchy forest stands near open habitat, or along hedgerows beside agricultural fields. Due to their habitat preference, Black-billed Magpies were most often encountered in the Parkland and Grassland NRs. This species was less commonly detected in the Boreal Forest, Foothills, and Rocky Mountain NRs.

Declines in relative abundance were detected in the Boreal Forest and Parkland NRs where, relative to other species, it was observed less frequently in Atlas 2 than in Atlas 1. The cause for these changes is unknown, but the expansion of Common Ravens back into these regions may be placing pressure on American Crows and Black-billed Magpies. An increase was detected in the Rocky Mountain NR where, relative to other species, it was observed more frequently in Atlas 2 than in Atlas 1. No change was detected in the Foothills and Grassland NRs. Data from the Breeding Bird Survey indicate that Black-billed Magpie abundance increased in Alberta during the period 1985–2005. Differences between the Atlas and Breeding Bird Survey analyses could be attributed to the fact that data were analyzed at different spatial scales. Breeding Bird Survey data were analyzed at the provincial level; whereas, due to sufficient sample size, it was possible to analyze Atlas data at the smaller and ecologically defined Natural Region scale. This species is considered Secure in Alberta.

TEMPORAL REPORTING RATE

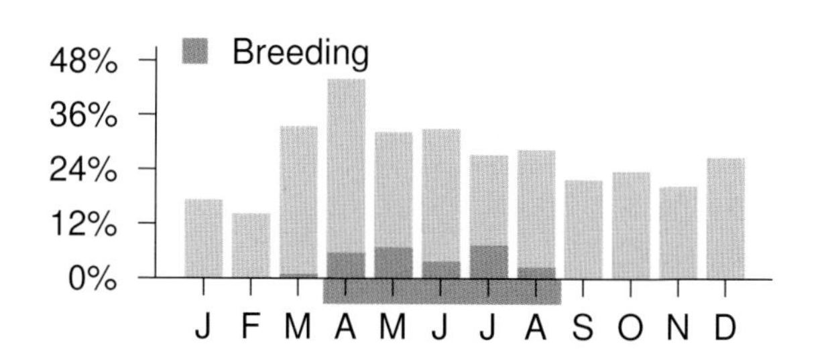

NATURAL REGIONS REPORTING RATE

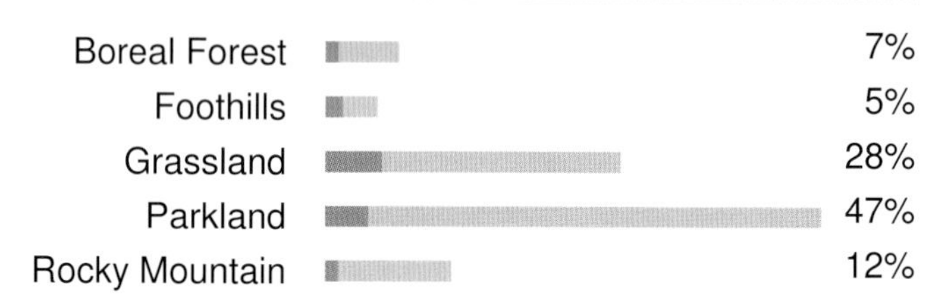

SPATIAL REPORTING RATE

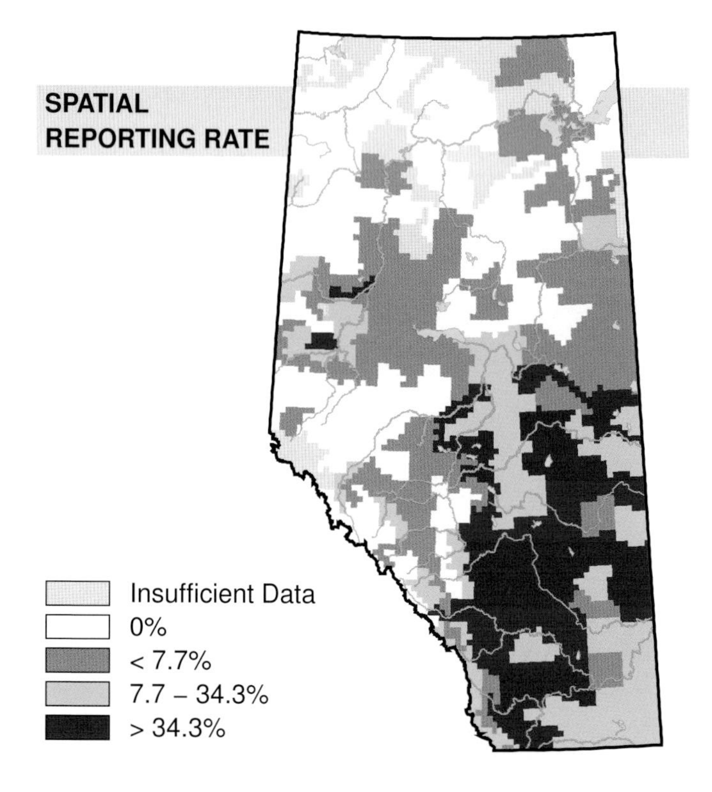

STATUS

GSWA 2000	GSWA 2005	COSEWIC Historic Rankings	COSEWIC 2007	Alberta Wildlife Act
N/A	N/A	N/A	N/A	N/A

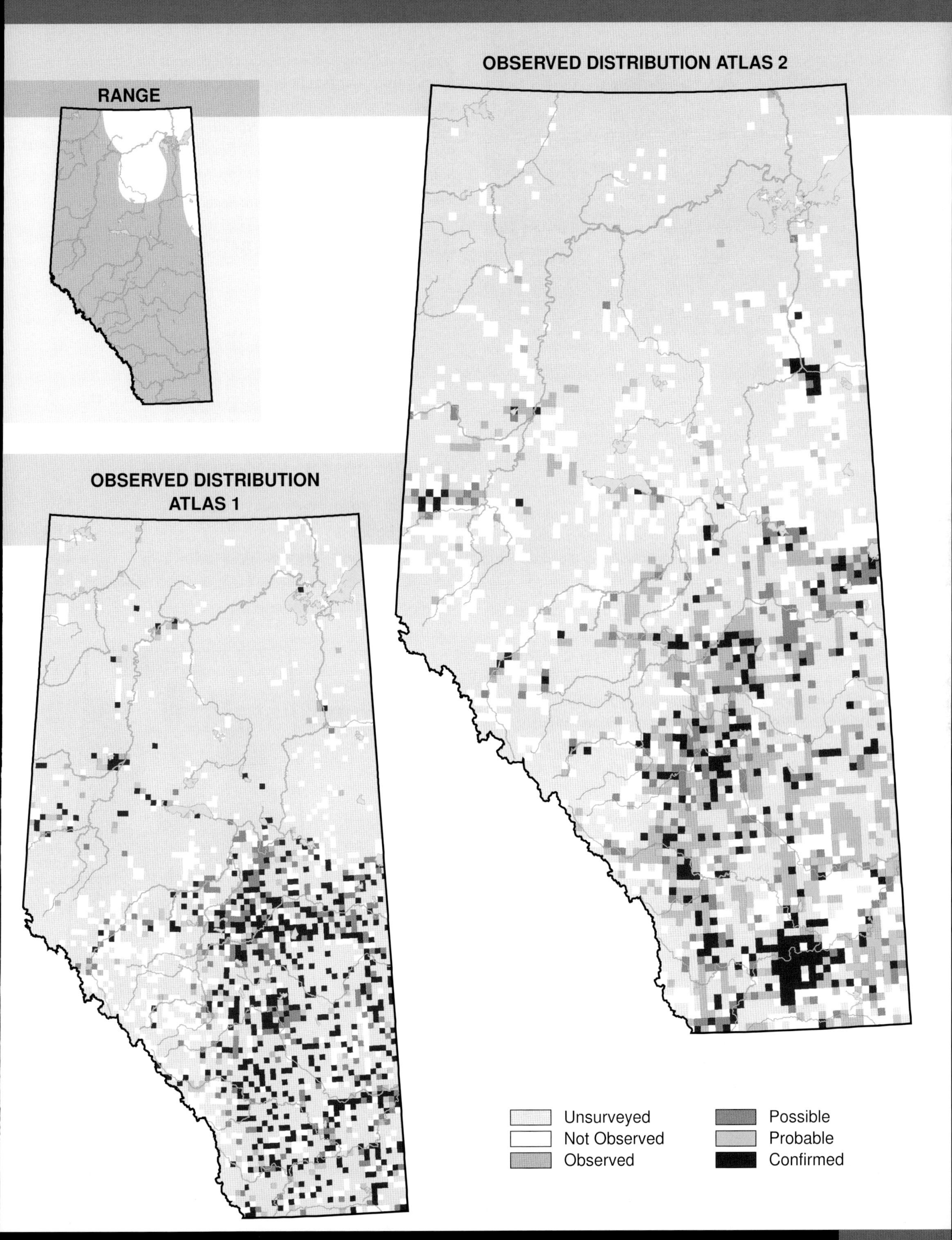
OBSERVED DISTRIBUTION ATLAS 2
RANGE
OBSERVED DISTRIBUTION ATLAS 1
Unsurveyed
Not Observed
Observed
Possible
Probable
Confirmed

American Crow (*Corvus brachyrhynchos*)

Photo: Gerald Romanchuk

NESTING

CLUTCH SIZE (EGGS):	4–7
INCUBATION (DAYS):	17–20
FLEDGING (DAYS):	35
NEST HEIGHT (METRES):	0–21.3

The American Crow is found across the province in all of the Natural Regions, and is considered one of the most widespread birds in North America. There was no change in the observed distribution between Atlas 1 and Atlas 2.

Requiring open areas for ground feeding, and scattered trees for nesting and roosting, crows are—not surprisingly—the most common birds in the Parkland Natural Region. During breeding, reporting rates were lower for the Grassland and Rocky Mountain NRs, and even lower for the Boreal Forest and Foothills NRs. This may reflect competition with its cousin, the Common Raven, whose highest reporting rate was recorded in the Rocky Mountain NR, and whose lowest reporting rate was from the Grassland NR.

Declines in relative abundance were detected in the Boreal, Grassland, and Parkland NRs where, relative to other species, this crow was observed less frequently in Atlas 2 than in Atlas 1. In Alberta, the Breeding Bird Survey did not detect any change in abundance; yet, in Saskatchewan, a decline was detected during the period 1968–2005. An increase in relative abundance was detected in the Rocky Mountain NR where, relative to other species, it was observed more frequently in Atlas 2 than in Atlas 1.

The American Crow appears to be the biggest victim of West Nile virus, a disease introduced to North America in 1999. Crows die within one week of infection, and few seem able to survive exposure. Using BBS data, researchers demonstrated significant changes in population trajectories for seven bird species from four families including the American Crow population which declined by up to 45% since West Nile Virus arrival (LaDeau et al., 2007). No other North American bird is dying at the same rate from the disease, and the loss of crows in some areas has been severe.

The provincial status of the American Crow is Secure.

TEMPORAL REPORTING RATE

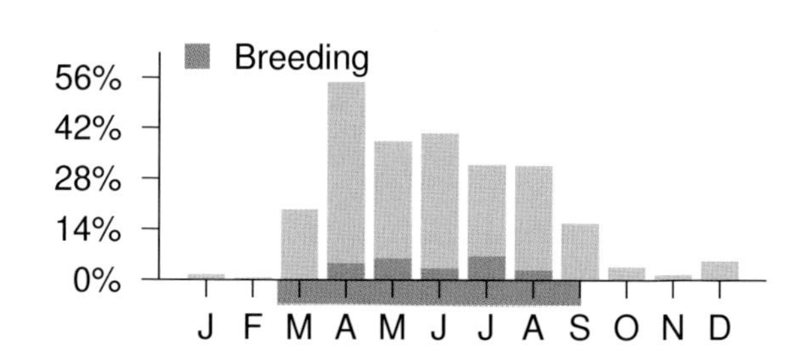

NATURAL REGIONS REPORTING RATE

SPATIAL REPORTING RATE

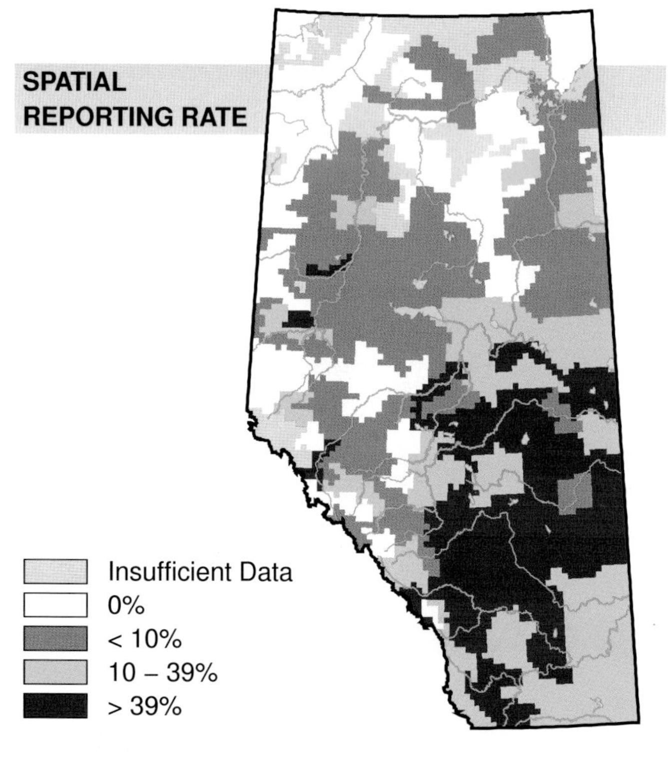

STATUS

GSWA 2000	GSWA 2005	COSEWIC Historic Rankings	COSEWIC 2007	Alberta Wildlife Act
Secure	Secure	N/A	N/A	N/A

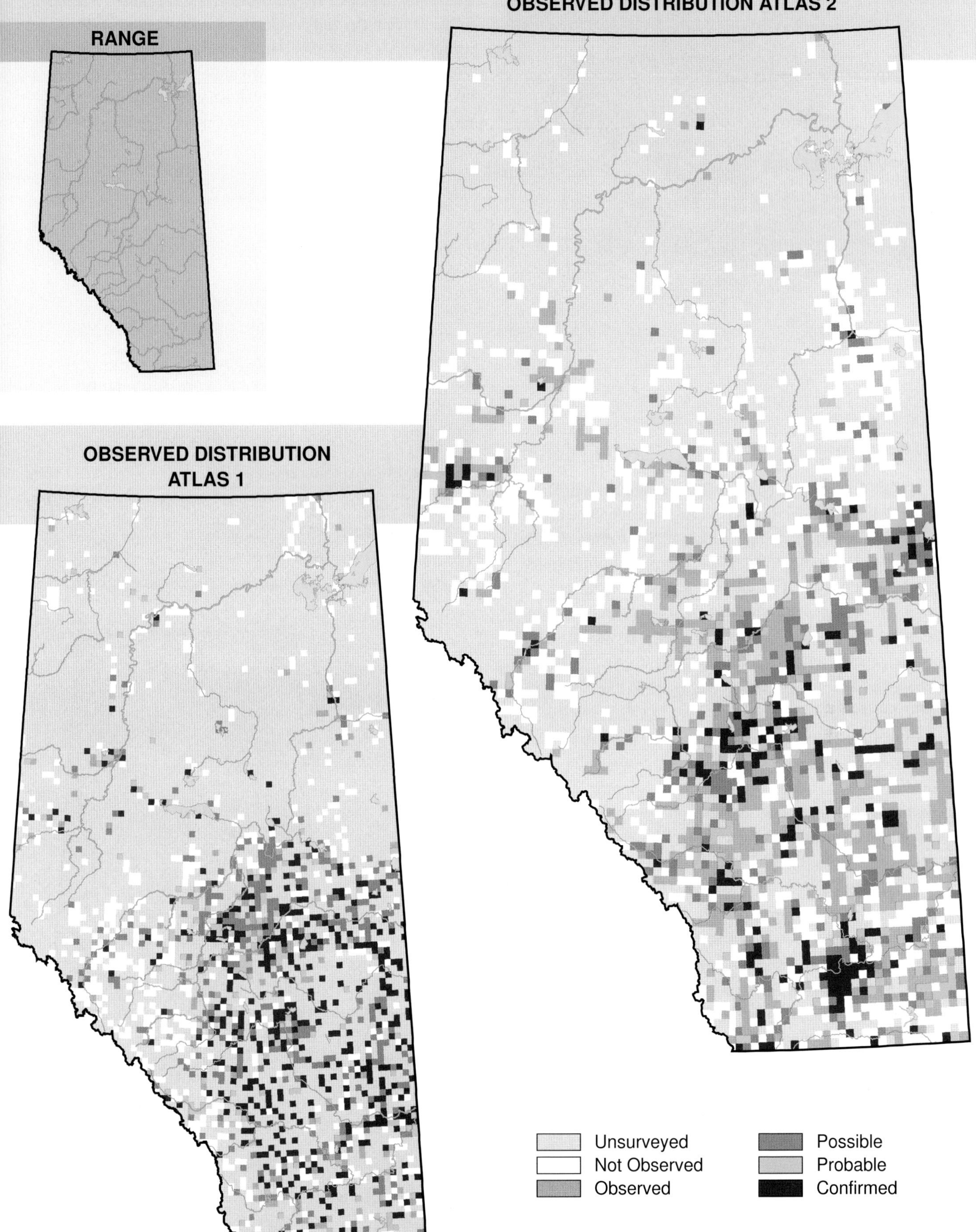
Habitat
Nest
Location
Nest
Type
Diet
RANGE
OBSERVED DISTRIBUTION ATLAS 2
OBSERVED DISTRIBUTION
ATLAS 1
Unsurveyed
Not Observed
Observed
Possible
Probable
Confirmed

Common Raven (*Corvus corax*)

Photo: Gerald Romanchuk

NESTING

CLUTCH SIZE (EGGS):	4–7
INCUBATION (DAYS):	20–21
FLEDGING (DAYS):	35–42
NEST HEIGHT (METRES):	30.5

The largest, and one of the most widespread of the songbirds in the world, the Common Raven is found throughout the entire province of Alberta. A change in distribution was observed in the Grassland Natural Region, where more records of ravens were found in Atlas 2. This may be an expansion of their distribution or it may reflect inclusion of some misidentified Common Crows which are more common in the prairies. Ravens, however, are very adaptable, and have moved into new areas in the past 50 years, particularly in areas where they were heavily persecuted.

Reporting rates were highest in the Foothills and Rocky Mountain NRs. Common Ravens were less common in the Boreal Forest and Parkland NRs, and least common in the Grassland NR. This species requires suitable trees for nesting and prefers contoured landscapes with cliffs and hills that provide thermals for their flights, a habitat type that is most available in the Foothills and Rocky Mountain NRs.

Increases in relative abundance were detected in the Boreal Forest, Grassland, and Parkland NRs where, relative to other species, this bird was observed more frequently in Atlas 2 than in Atlas 1. These increases appear to be due in part to the expanding distribution of this species. Breeding Bird Survey data indicate that Common Raven abundance increased in Alberta and across Canada during the period 1985–2005. The Atlas detected a decline in relative abundance in the Foothills NR where, relative to other species, it was observed less frequently in Atlas 2 than in Atlas 1. This decline is puzzling, as the heavy industrial development in the Foothills NR in recent years was not expected to affect this species adversely. Without corroboration or known biological cause, further research is needed to evaluate the Atlas-detected decrease in the Foothills NR. The status of the Common Raven is Secure in Alberta.

TEMPORAL REPORTING RATE

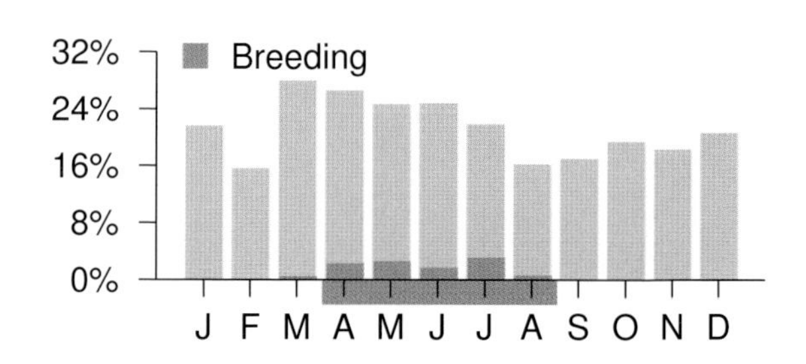

NATURAL REGIONS REPORTING RATE

SPATIAL REPORTING RATE

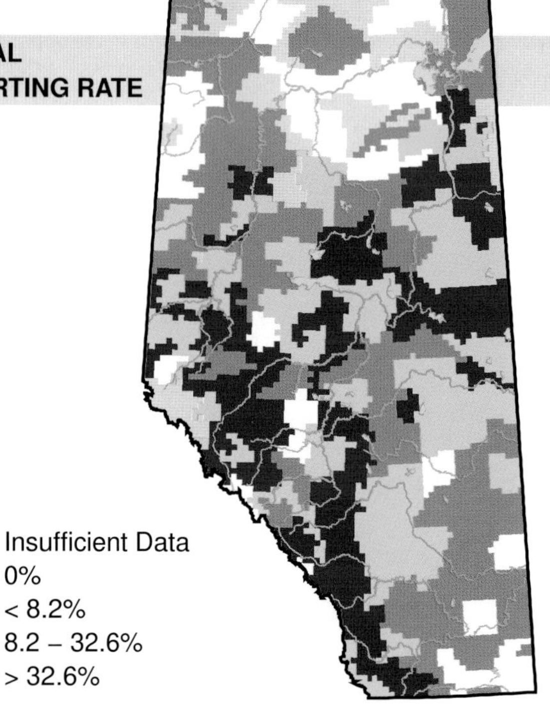

STATUS

GSWA 2000	GSWA 2005	COSEWIC Historic Rankings	COSEWIC 2007	Alberta Wildlife Act
Secure	Secure	N/A	N/A	N/A

RANGE

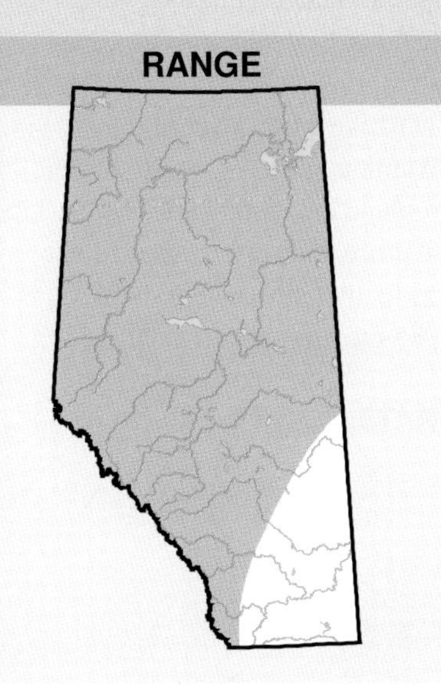

OBSERVED DISTRIBUTION ATLAS 2

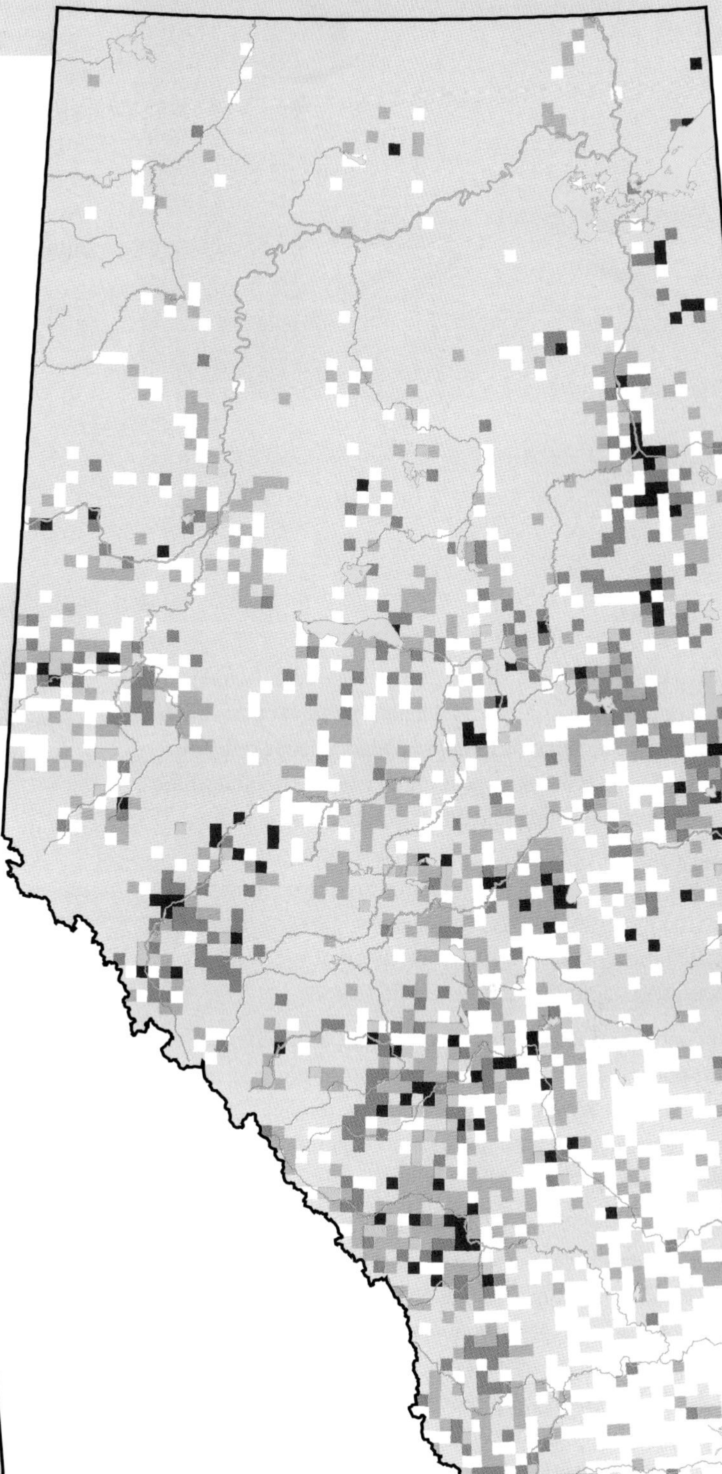

OBSERVED DISTRIBUTION ATLAS 1

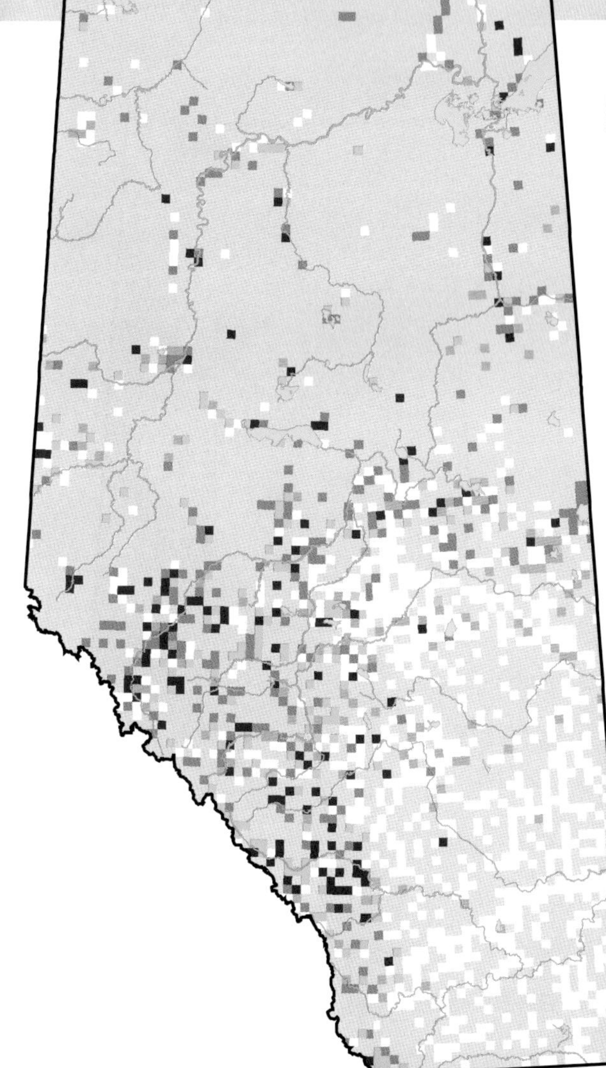

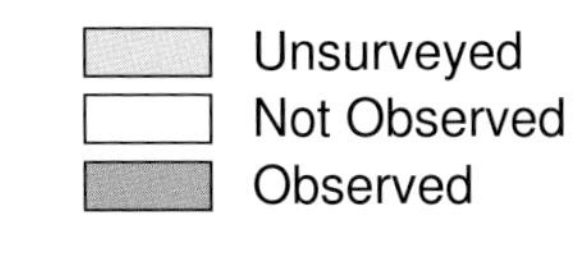

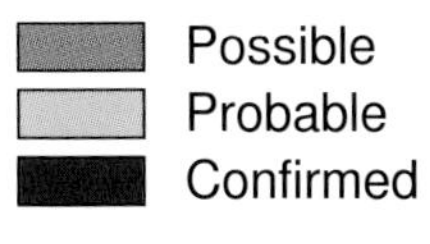

Horned Lark (*Eremophila alpestris*)

Photo: Raymond Toal

NESTING

CLUTCH SIZE (EGGS):	3–4
INCUBATION (DAYS):	10–12
FLEDGING (DAYS):	3–5
NEST HEIGHT (METRES):	0

Horned Larks are widely distributed in all open landscapes in Alberta. Some will overwinter in the southernmost parts of the province. The observed distribution appears unchanged from Atlas 1 to Atlas 2. It might appear to have contracted but many squares between the Battle and North Saskatchewan rivers, where it was observed in Atlas 1, were not surveyed in Atlas 2. Like many other grassland birds, there were fewer confirmed breeding records in Atlas 2 for this species. Fewer squares were visited multiple times; so confirmation of breeding was very dependent on finding nests or seeing young.

Highest concentrations are in the grasslands, especially the southeast portion. This lark's observed distribution and relative abundance are in keeping with its known preference for bare or sparsely vegetated sites with low ground cover and a lack of woody vegetation (Beason, 1995).

Increases in relative abundance were detected in the Grassland NR while decreases were detected in the Boreal Forest and Parkland NRs. No changes in relative abundance were detected in the Foothills and Rocky Mountain NRs. It was observed less frequently in Atlas 2, relative to other species, than in Atlas 1 in all NRs. The Breeding Bird Survey detected a decline for Horned Larks while Grassland Bird Monitoring (unpublished Canadian Wildlife Service data) found no trend. This is a species well monitored by BBS and Grassland Bird Monitoring, and the disagreement in trend findings suggests that the observed increase in the Grassland NR in Atlas 2 may not be real. Other evidence to support an overall decline includes substantial decreases in native grass cover in the Parkland and Grassland NRs (Watmough and Schmoll, 2007). This species will utilize bare cropland but that, too, is less common due to a shift to second crops and conservation tillage. Overall habitat for this species is likely reduced. Provincial status reports from 2000 and 2005 rate the species Secure.

TEMPORAL REPORTING RATE

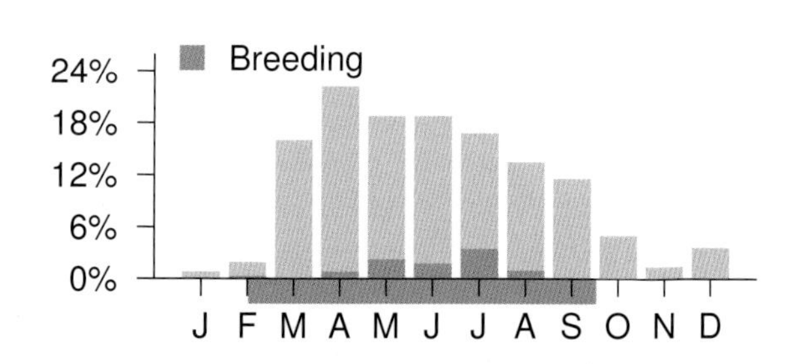

NATURAL REGIONS REPORTING RATE

SPATIAL REPORTING RATE

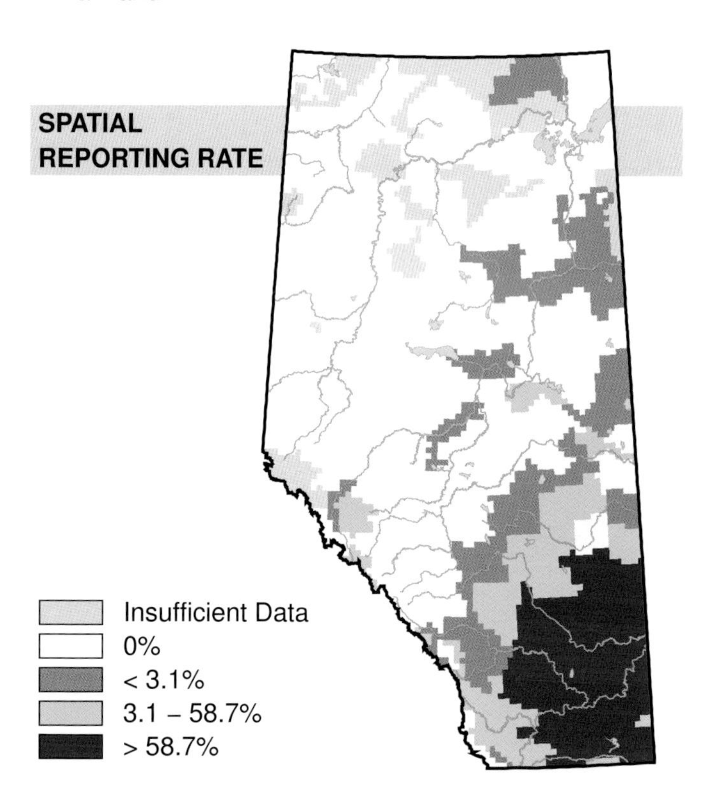

STATUS

GSWA 2000	GSWA 2005	COSEWIC Historic Rankings	COSEWIC 2007	Alberta Wildlife Act
Secure	Secure	N/A	N/A	N/A

RANGE

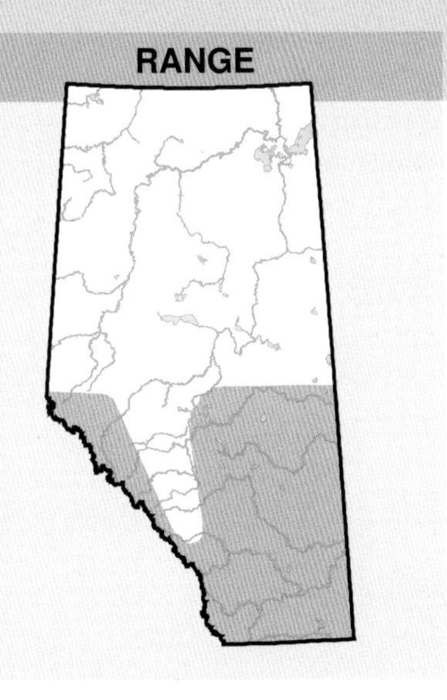

OBSERVED DISTRIBUTION ATLAS 2

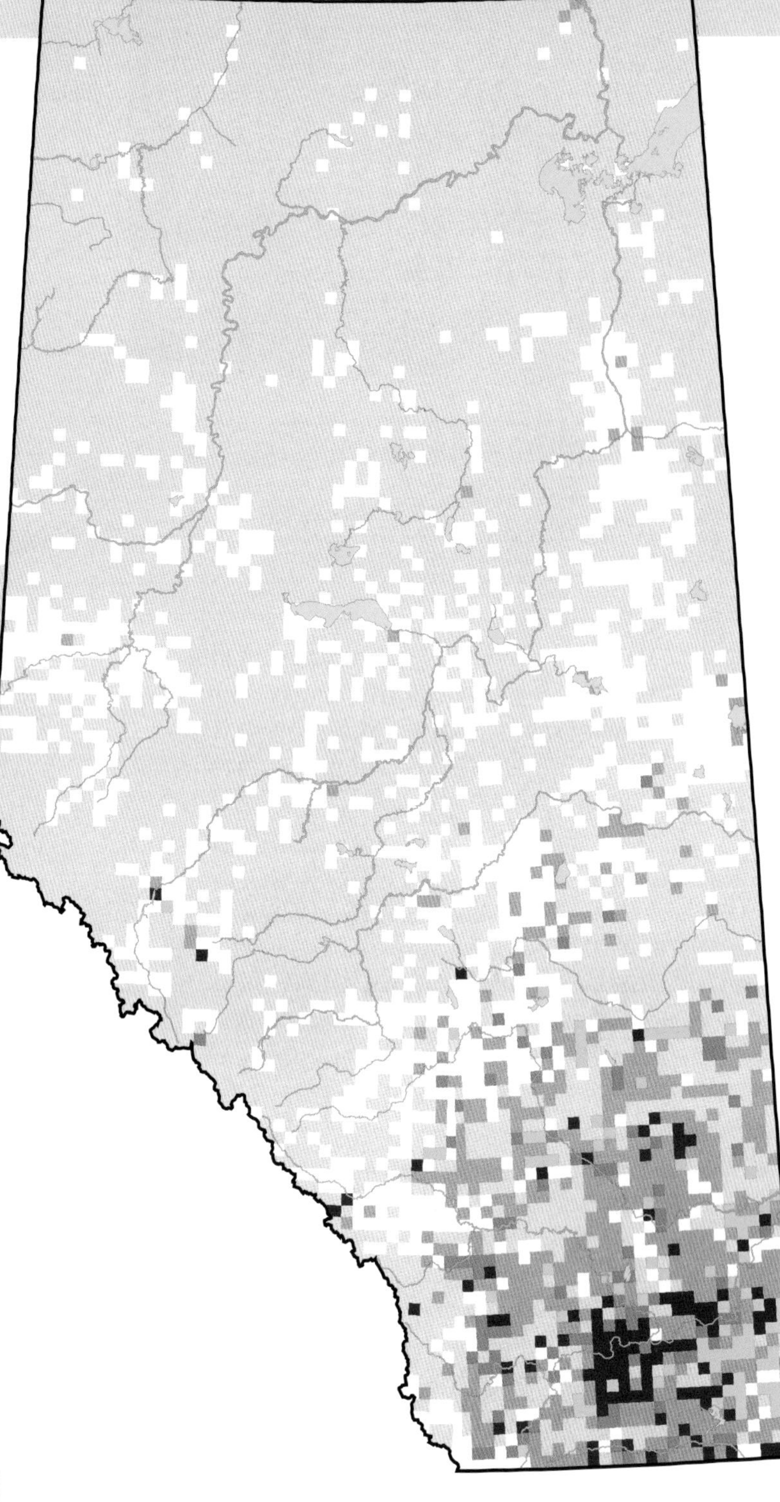

OBSERVED DISTRIBUTION ATLAS 1

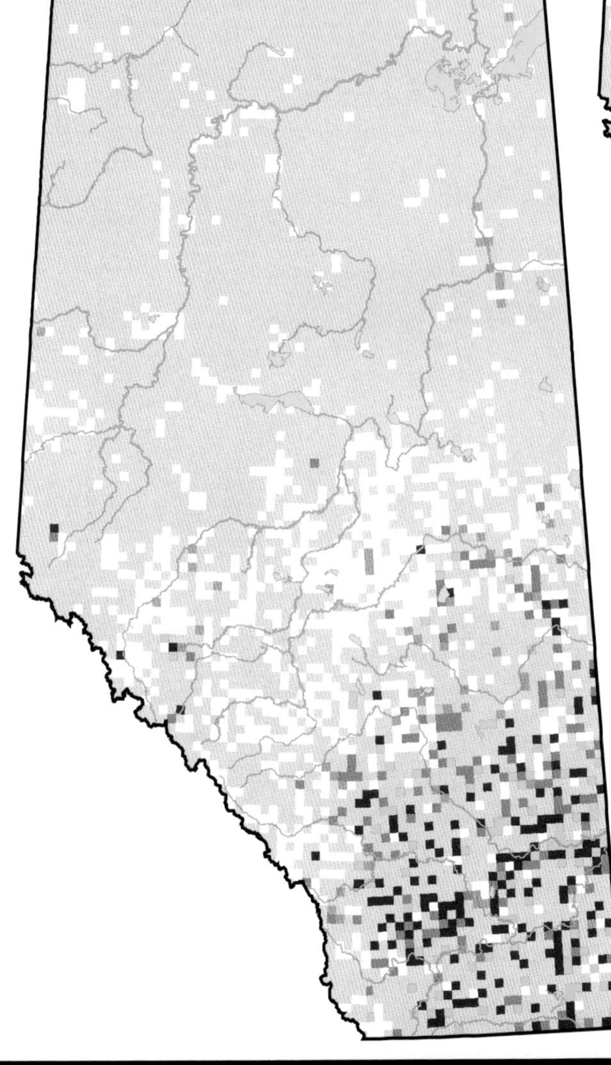

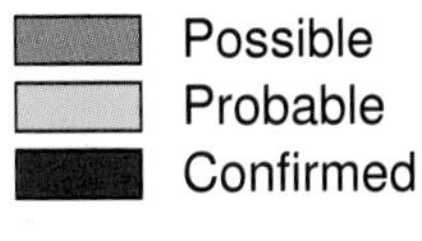

Purple Martin (*Progne subis*)

Photo: Gerald Romanchuk

NESTING

CLUTCH SIZE (EGGS):	4–6
INCUBATION (DAYS):	15–17
FLEDGING (DAYS):	24–28
NEST HEIGHT (METRES):	>1.5

The Purple Martin is found in the Parkland, Grassland, and Boreal Forest Natural Regions in Alberta. The observed distribution of this species expanded slightly between Atlas 1 and Atlas 2, as new breeding observations were obtained in the Hinton and Grande Prairie areas.

The Purple Martin uses snags for nesting; thus, burned forests, logged areas with retained snags, and muskeg with many burned snags are used. Purple Martins have also adapted well to using "Purple Martin apartment" nest boxes erected in cities, towns and around other human settlements. In fact, these birds are now the almost sole occupants of these man-made structures east of the Rocky Mountains. This species was most often observed in the Parkland NR and was less common in the Grassland and Boreal Forest NRs.

Declines in relative abundance of the Purple Martin were detected in the Boreal Forest and Parkland NRs and, relative to other species, it was observed less often in these NRs. The Breeding Bird Survey found no change in abundance in Alberta or Canada-wide during the period 1985–2005. The nest boxes that people have built to attract Purple Martins have been beneficial to this species; however, the loss of natural nests due to logging or selective removal can affect this species negatively. Also, the European Starling and House Sparrow out-compete martins for nest sites, which may be one reason for the observed declines during the course of the Atlas project. This species is considered Sensitive in the province due to loss of nest sites and competition from House Sparrows and European Starlings.

TEMPORAL REPORTING RATE

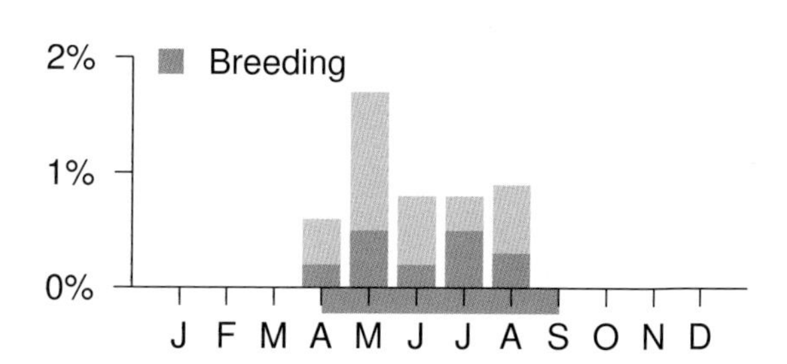

NATURAL REGIONS REPORTING RATE

SPATIAL REPORTING RATE

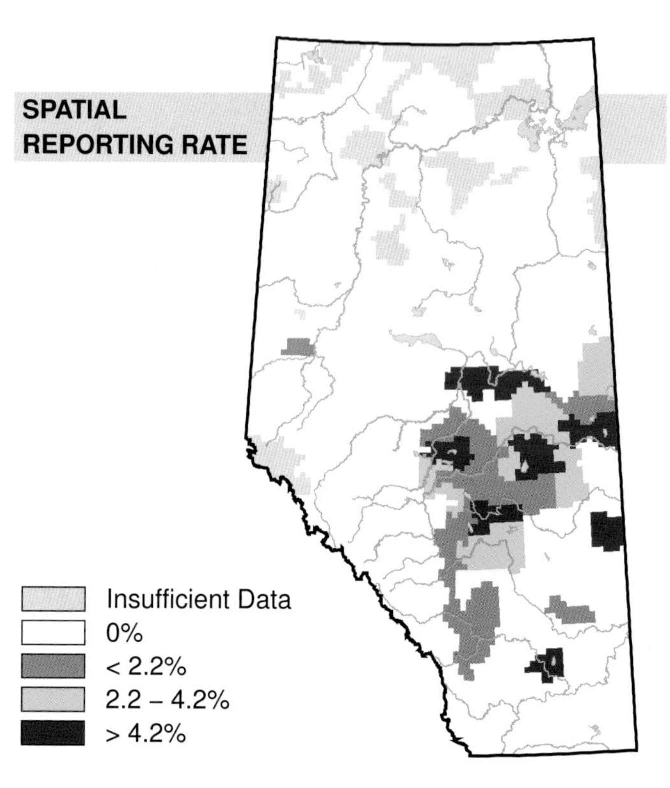

STATUS

GSWA 2000	GSWA 2005	COSEWIC Historic Rankings	COSEWIC 2007	Alberta Wildlife Act
Sensitive	Sensitive	N/A	N/A	N/A

OBSERVED DISTRIBUTION ATLAS 2

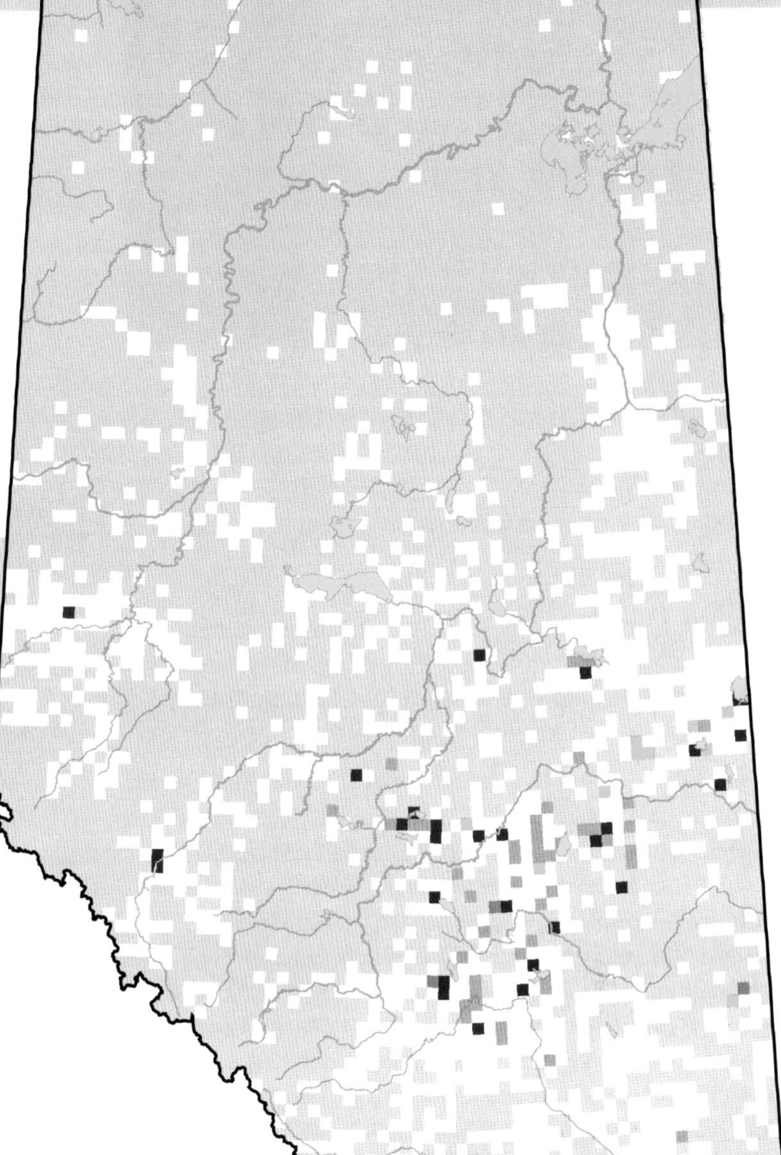

RANGE

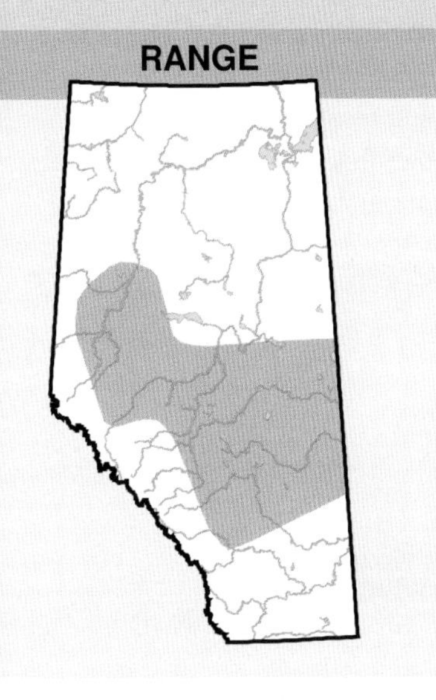

OBSERVED DISTRIBUTION ATLAS 1

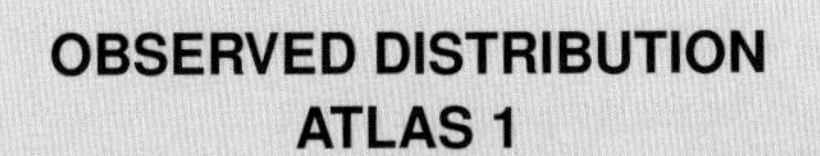

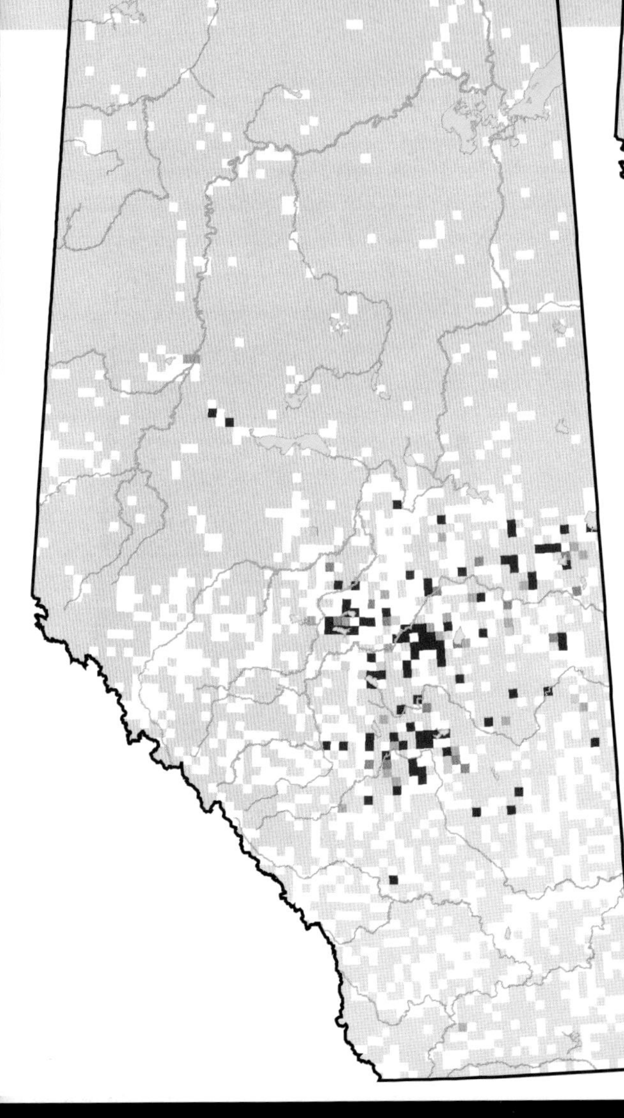

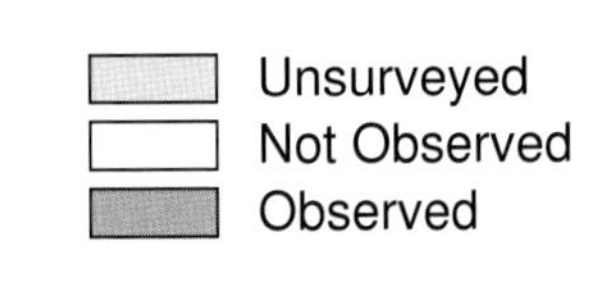

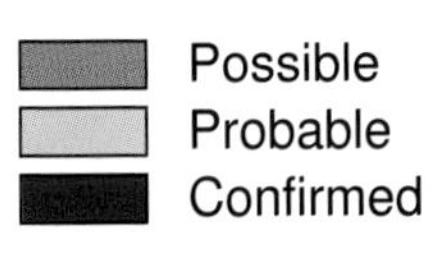

Tree Swallow (*Tachycineta bicolor*)

Photo: Raymond Toal

NESTING

CLUTCH SIZE (EGGS):	4–6
INCUBATION (DAYS):	13–16
FLEDGING (DAYS):	16–24
NEST HEIGHT (METRES):	>1.5

Although limited by the availability of suitable cavities for nesting and of insects for food, the Tree Swallow is found in every Natural Region of Alberta. The observed distribution did not change between Atlas 1 and Atlas 2.

This swallow is found in open areas that are usually near water. It prefers fields, marshes, shorelines, streams, and wooded swamps with standing dead trees that contain cavities for nesting. The Tree Swallow will also readily use nest boxes in open areas, and has probably expanded its range in that manner. These birds require substantial open areas around the nest site for foraging on flying insects. For this reason, this species was found most commonly in the Parkland, Foothills, and Rocky Mountain NRs. Tree Swallows were less common in the Grassland NR because trees are less numerous there, and they were least common in the Boreal Forest NR because the forest there is more closed.

Declines in relative abundance were detected in the Boreal Forest, Foothills, and Parkland NRs. Increases in relative abundance were detected in the Grassland and Rocky Mountain NRs. Tree Swallow observation rates relative to other species mirrored these changes. The Breeding Bird Survey found an abundance decline in Alberta during the period 1995–2005 and across Canada during the period 1985–2005. One of the main factors that is thought to be having a negative effect on Tree Swallow populations is the competition with introduced species for nest sites. The House Sparrow is the main nest competitor, and not only will this species occupy cavities that could be used by the Tree Swallow, it will also kill swallows that are nesting in cavities. This species is considered Secure in the province.

TEMPORAL REPORTING RATE

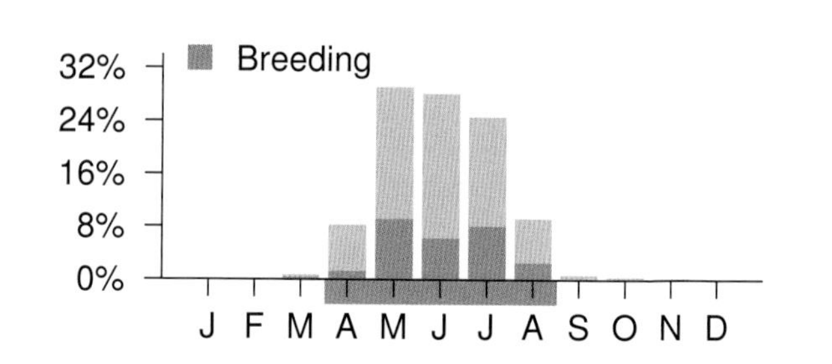

NATURAL REGIONS REPORTING RATE

SPATIAL REPORTING RATE

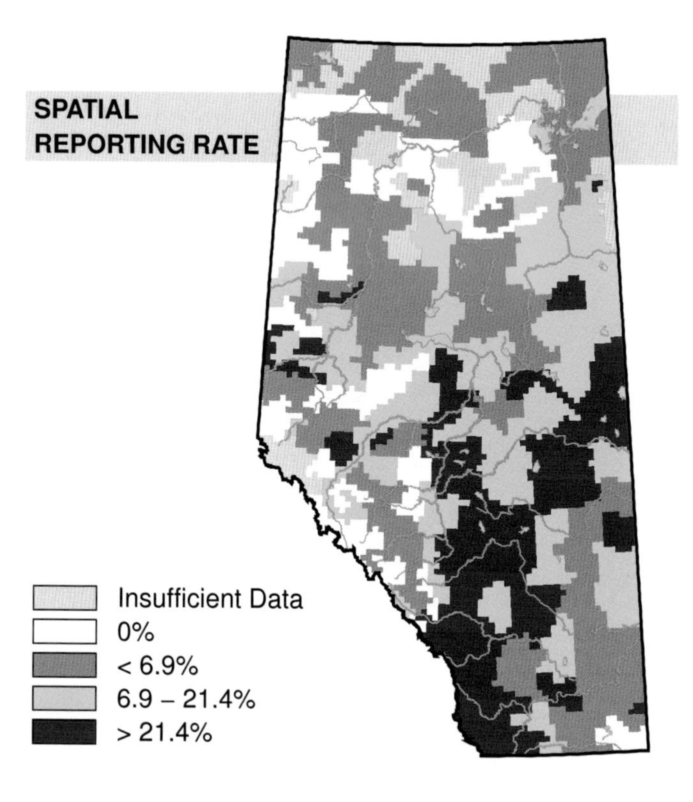

STATUS

GSWA 2000	GSWA 2005	COSEWIC Historic Rankings	COSEWIC 2007	Alberta Wildlife Act
Secure	Secure	N/A	N/A	N/A

Habitat

Nest
Location

Nest
Type

Diet

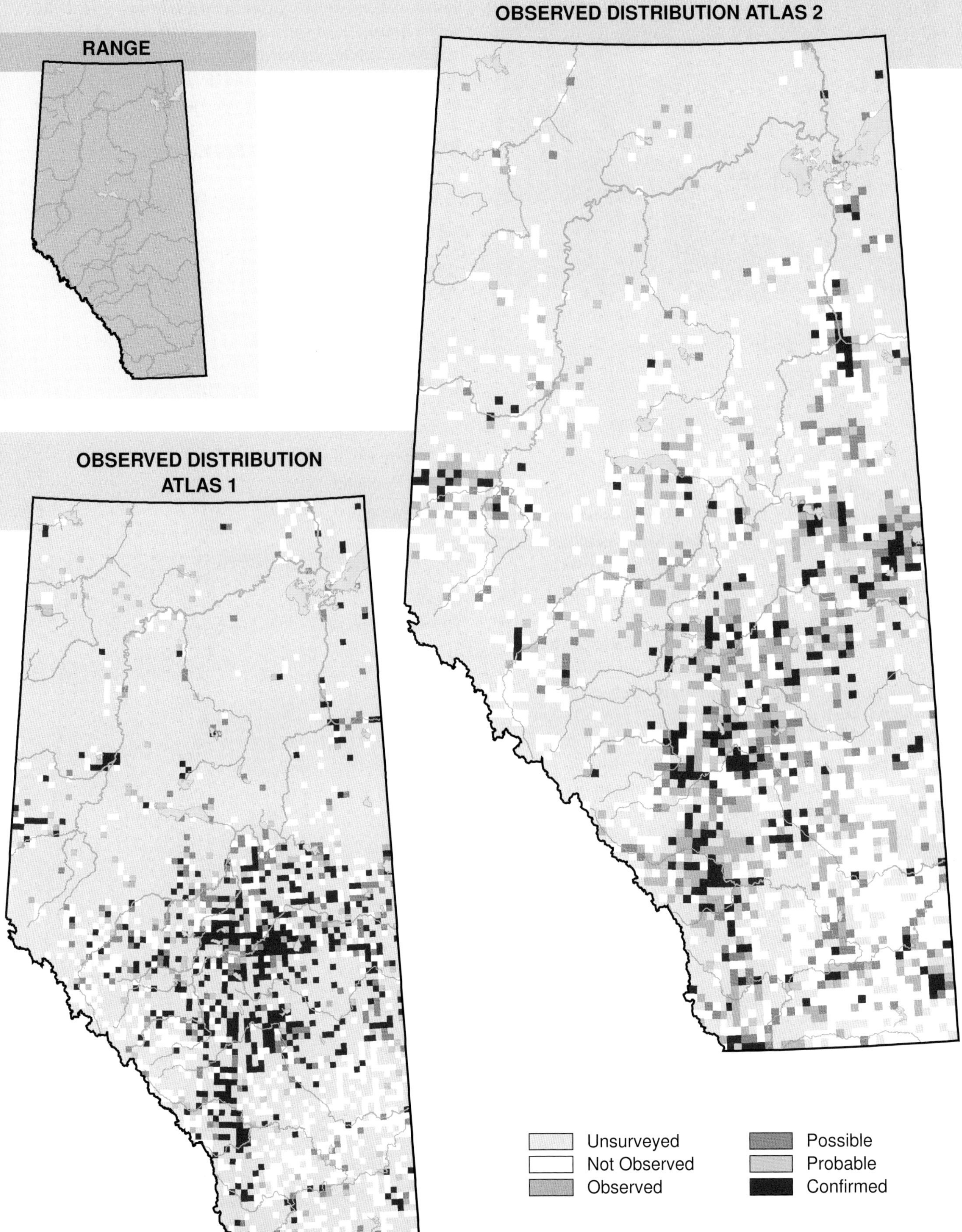
RANGE
OBSERVED DISTRIBUTION ATLAS 2
OBSERVED DISTRIBUTION
ATLAS 1
Unsurveyed
Not Observed
Observed
Possible
Probable
Confirmed

Violet-green Swallow (*Tachycineta thalassina*)

Photo: Royal Alberta Museum

NESTING

CLUTCH SIZE (EGGS):	4–5
INCUBATION (DAYS):	15
FLEDGING (DAYS):	23–25
NEST HEIGHT (METRES):	>1.5

Despite an extensive distribution in other parts of its North American range, the Violet-green Swallow is found only in the Rocky Mountain and Foothills Natural Regions and a small portion of the Grassland NR in Alberta. The observed distribution of this species did not change between Atlas 1 and Atlas 2.

This little-known swallow prefers open, montane, coniferous, deciduous, and mixedwood forests. The Violet-green Swallow nests in cavities in trees, cliffs, sandbanks, and nest boxes. This species will nest singly or in colonies of up to 25 pairs. For these reasons, it was encountered most often in the Rocky Mountain NR. It was encountered less often in the Grassland NR in southern Alberta, and there were a few records ranging into the Boreal Forest NR near Grande Prairie. Parkland NR observations are not considered to be breeding records.

An increase in relative abundance was detected in the Grassland NR where, relative to other species, the Violet-green Swallow was observed more frequently in Atlas 2 than in Atlas 1. The Breeding Bird Survey detected no change in abundance in Canada for the period 1985–2005, but the sample size is not large enough to permit determination of trends in Alberta. There is no information regarding factors that would affect the conservation status of Violet-green Swallows. The status of this species may remain secure in the future due to its ability to nest in isolated areas and near humans, but there may be negative effects near urban areas from competition for nest sites with European Starlings and House Sparrows. This species is considered Secure in Alberta.

TEMPORAL REPORTING RATE

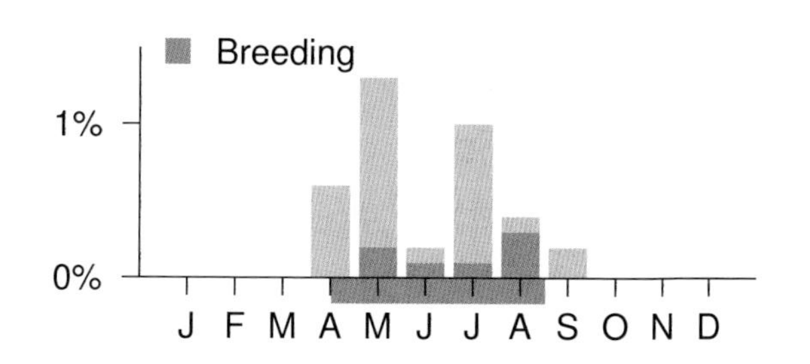

NATURAL REGIONS REPORTING RATE

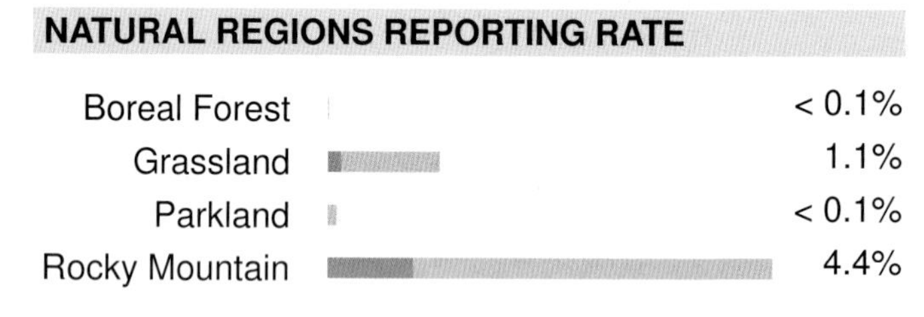

SPATIAL REPORTING RATE

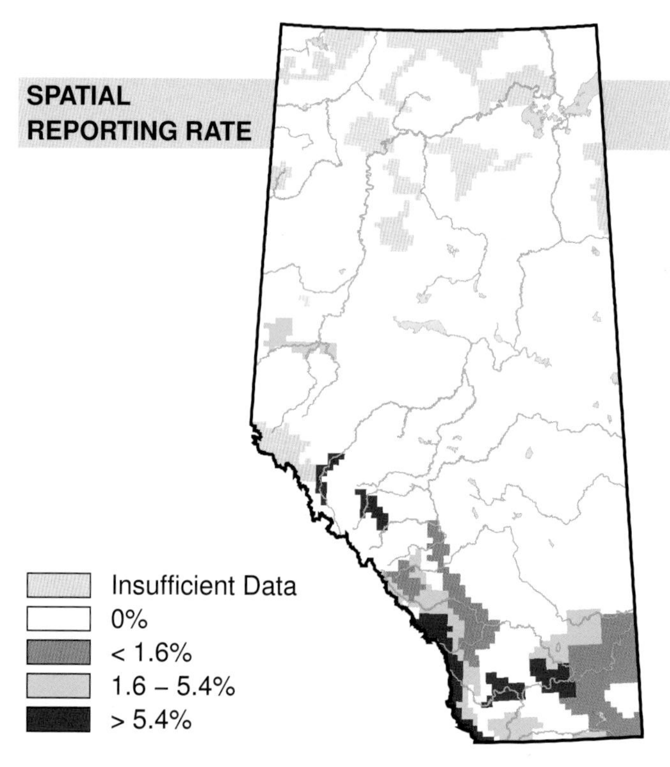

STATUS

GSWA 2000	GSWA 2005	COSEWIC Historic Rankings	COSEWIC 2007	Alberta Wildlife Act
Secure	Secure	N/A	N/A	N/A

OBSERVED DISTRIBUTION ATLAS 2

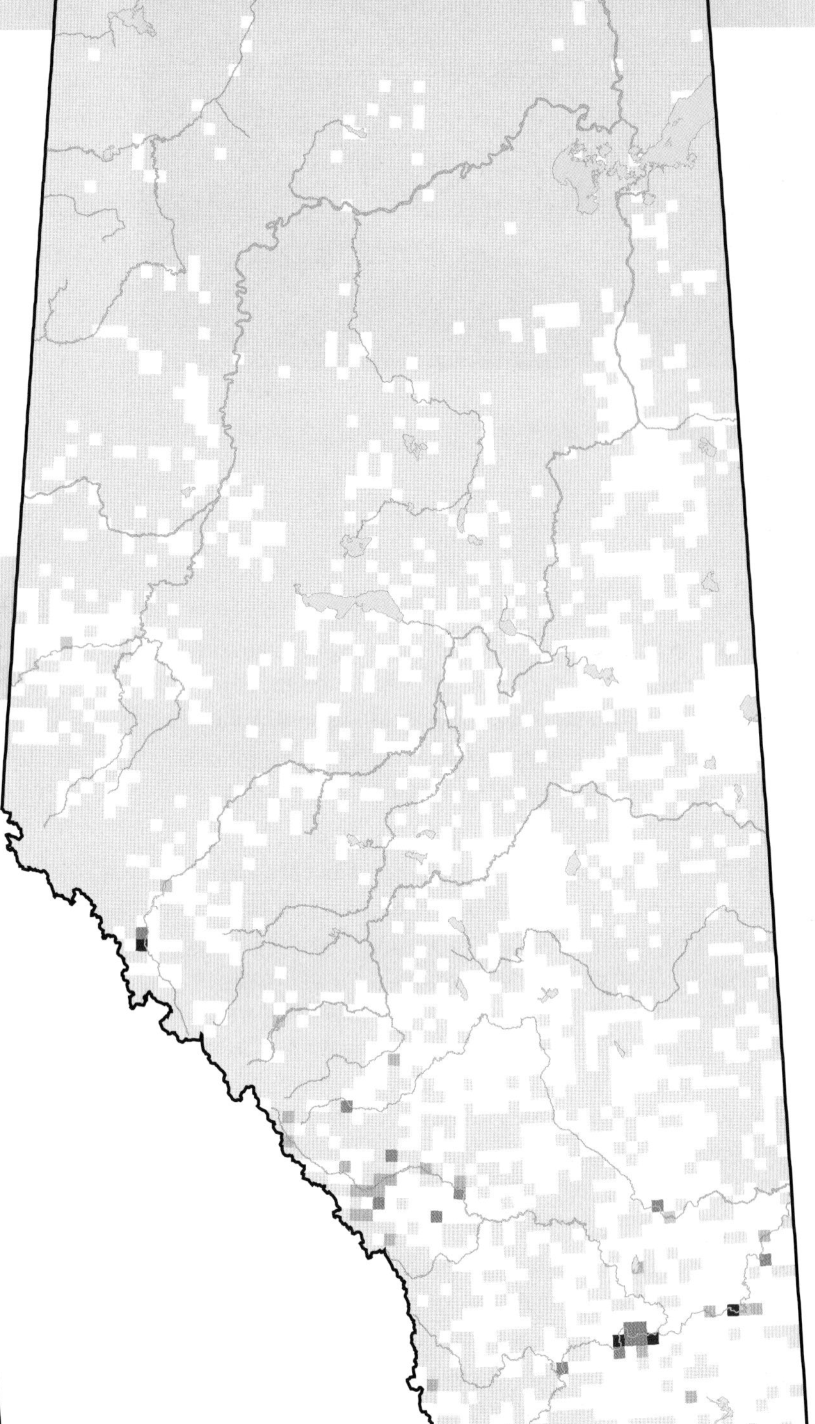

RANGE

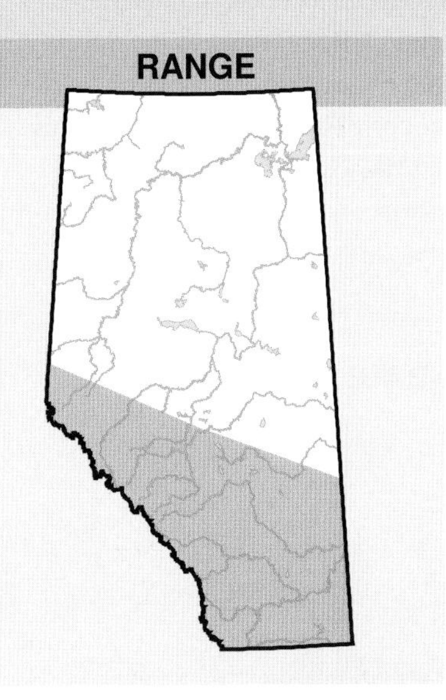

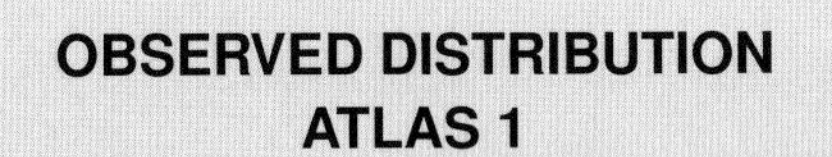

OBSERVED DISTRIBUTION ATLAS 1

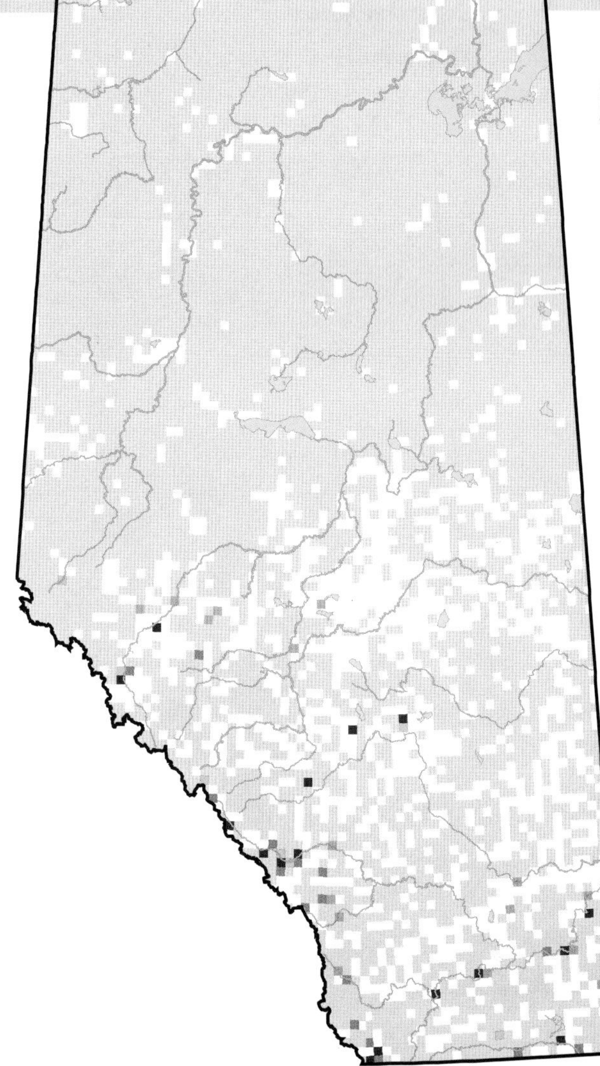

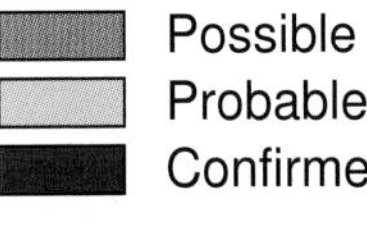

Unsurveyed
Not Observed
Observed
Possible
Probable
Confirmed

Northern Rough-winged Swallow (*Stelgidopteryx serripennis*)

Photo: Royal Alberta Museum

NESTING

CLUTCH SIZE (EGGS):	5–7
INCUBATION (DAYS):	16
FLEDGING (DAYS):	8–21
NEST HEIGHT (METRES):	>1.2

The Northern Rough-winged Swallow, known for the stiffened barbs on the leading edge of the first primary feather, is found in the Grassland, Parkland, and Rocky Mountain Natural Regions, and on the southern edge of the Boreal Forest NR. The observed distribution of this species did not change between Atlas 1 and Atlas 2.

The Northern Rough-winged Swallow prefers open areas which can include open woodlands. It nests singly or in small colonies, and uses cavities in rock gorges, shale banks, stony road cuts, railway embankments, gravel pits, and eroded margins of streams. This species is often associated with water where it spends time foraging for insects above the water, although it is also able to pick insects right off the surface of the water. This swallow was found most often in the Rocky Mountain and Grassland NRs in southern Alberta, and was encountered less commonly in the Parkland and Boreal Forest NRs where open areas are not as prevalent.

A decline in relative abundance of the Northern Rough-winged Swallow was detected in the Grassland and Parkland NRs. Relative to other species, it was observed less often in these NRs. Declines in abundance were also found Canada-wide by the Breeding Bird Survey during the period 1985–2005. No trends could be determined by the BBS in Alberta due to small sample size. Reasons for these declines are uncertain. Habitat destruction and water pollution can affect populations; however, little study has been done to document these effects for this species. The Atlas detected an increase in relative abundance in the Rocky Mountain NR, but most of the data from this NR comprise non-breeding records. Further, there is no apparent biological cause or corroborating evidence for this increase. This species, like the Bank Swallow and Cliff Swallow, can benefit from structures built by humans, as these buildings can create artificial nesting sites for them. It is rated Secure in the province.

TEMPORAL REPORTING RATE

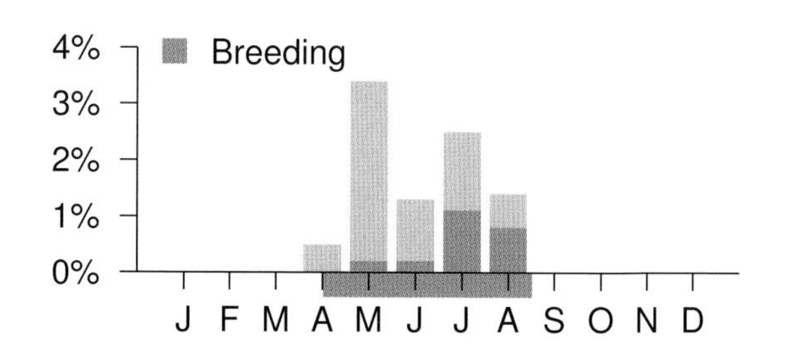

NATURAL REGIONS REPORTING RATE

SPATIAL REPORTING RATE

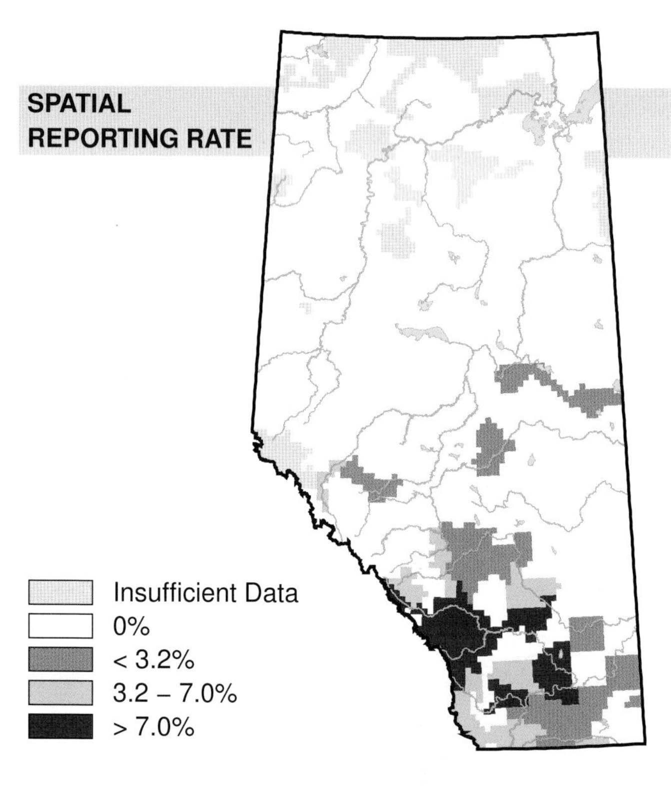

STATUS

GSWA 2000	GSWA 2005	COSEWIC Historic Rankings	COSEWIC 2007	Alberta Wildlife Act
Secure	Secure	N/A	N/A	N/A

Habitat Nest Location Nest Type Diet

RANGE

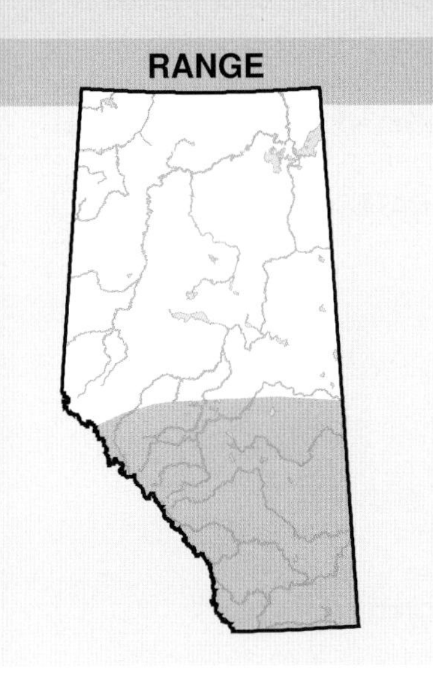

OBSERVED DISTRIBUTION ATLAS 2

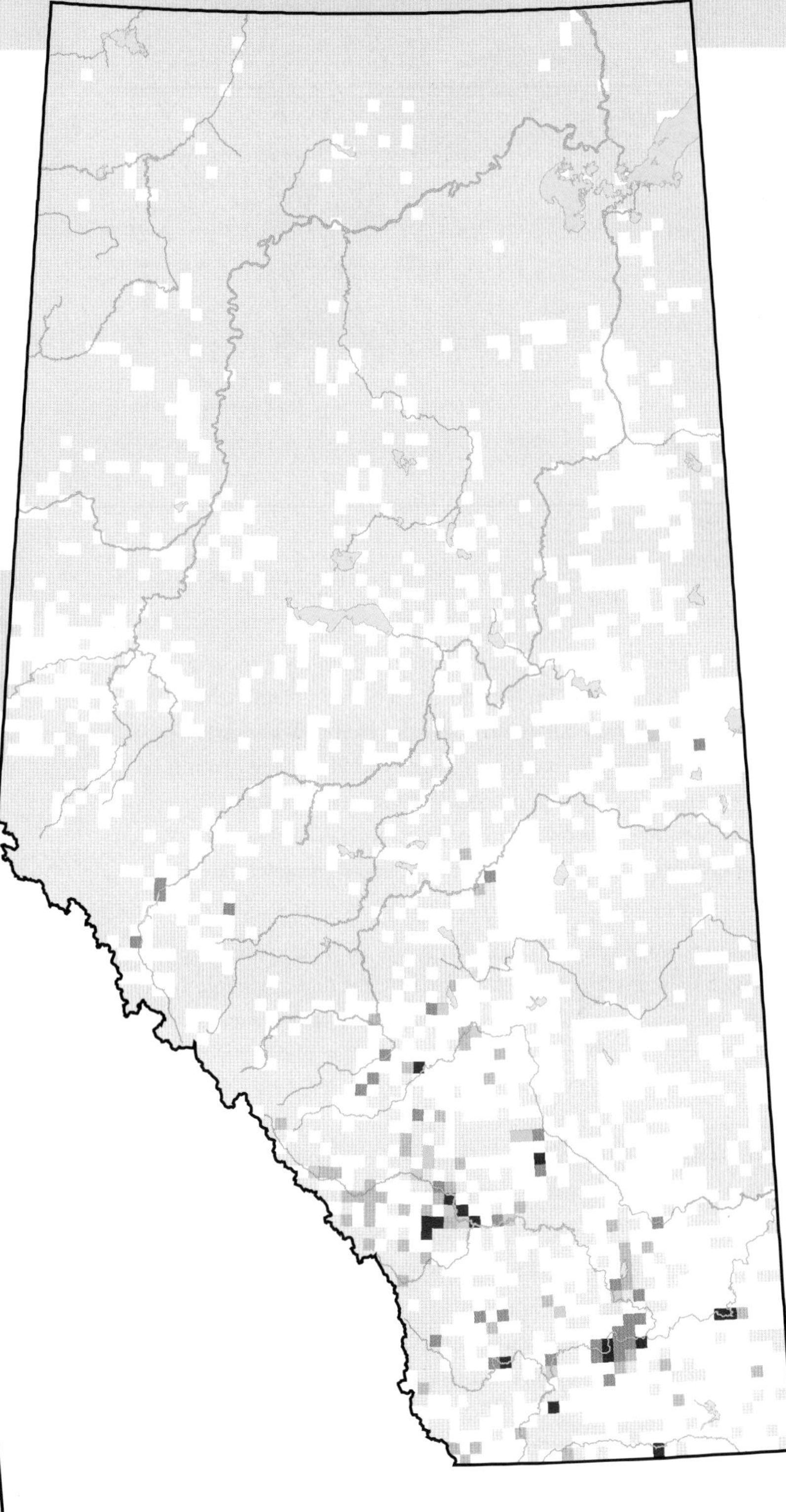

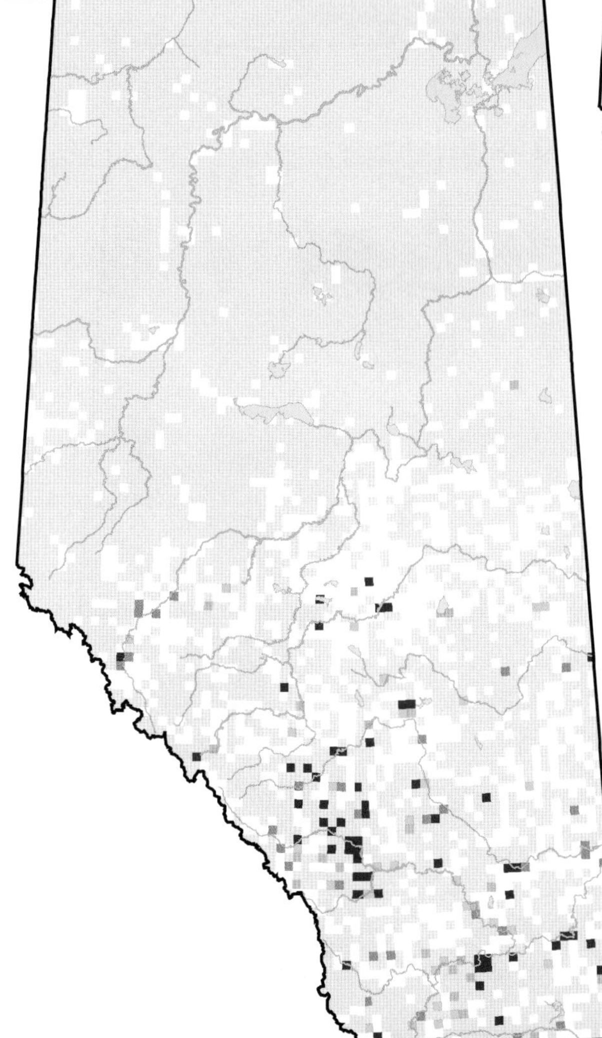

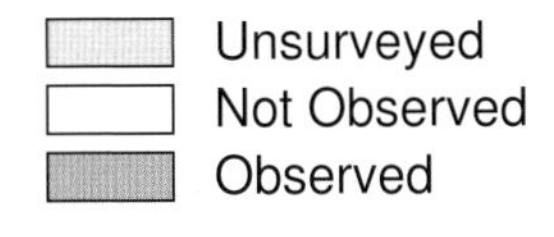

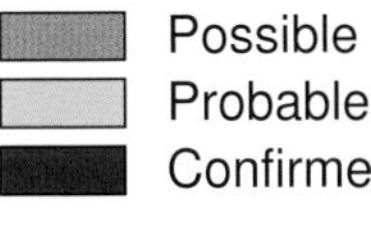

Bank Swallow (*Riparia riparia*)

Photo: Gerald Romanchuk

NESTING

CLUTCH SIZE (EGGS):	4–6
INCUBATION (DAYS):	14–16
FLEDGING (DAYS):	14
NEST HEIGHT (METRES):	>1.2

One of the most widely distributed species of swallows in the world, the Bank Swallow breeds in all Natural Regions in Alberta. The observed distribution changed between Atlas 1 and Atlas 2. No observations were recorded in Atlas 2 north of 57°where they had been detected previously. In Atlas 1, the Bank Swallow was observed in Wood Buffalo National Park, and near High Level, Fort Vermilion, and Rainbow Lake.

A lowland breeder, the Bank Swallow is found along rivers, streams, lakes, wetlands, and reservoirs near large open areas. It prefers to nest in colonies of varied sizes that can range from 10 to several thousand pairs. They excavate their own nests with their bills and feet. Nests are most often parallel to the ground and perpendicular to the bank wall. This species was most often observed in the Grassland and Rocky Mountain NRs and less often in the Foothills and Parkland NRs. The Bank Swallow was least common in the Boreal Forest NR.

Declines in relative abundance were detected in the Boreal Forest, Foothills, Grassland, and Parkland NRs. Relative to other species, Bank Swallows were observed less often in all NRs. Declines in abundance were also detected by the Breeding Bird Survey on a Canada-wide and Alberta-wide scale for the period 1985–2005. Two major causes of colony losses include road-building and water-flow regulation which can increase erosion of the riverbanks. The Bank Swallow is difficult to monitor due to the species' habit of nesting in colonies, which are regularly relocated due to changes in bank suitability for nesting. This species is rated Secure in Alberta.

TEMPORAL REPORTING RATE

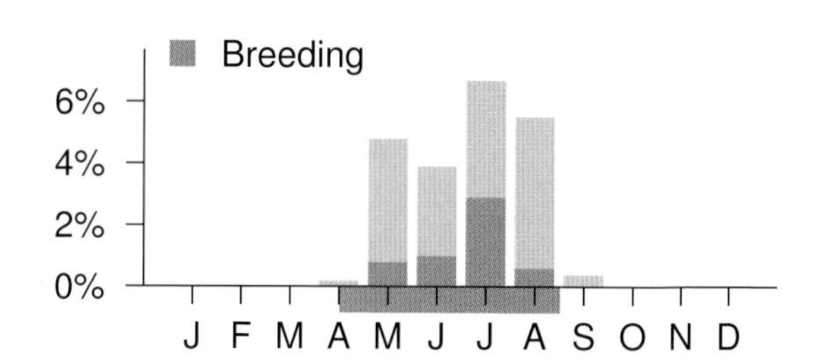

NATURAL REGIONS REPORTING RATE

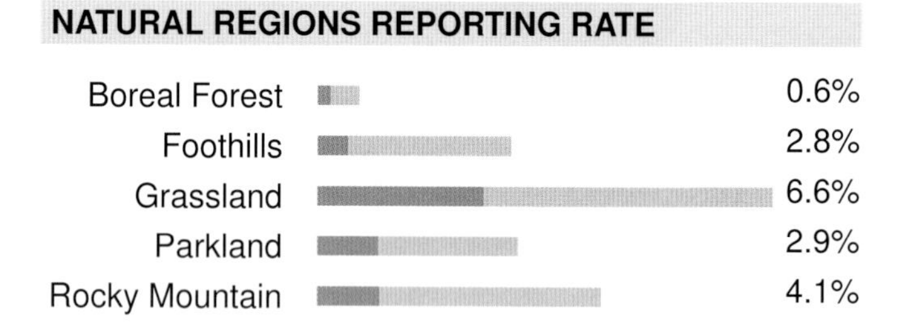

SPATIAL REPORTING RATE

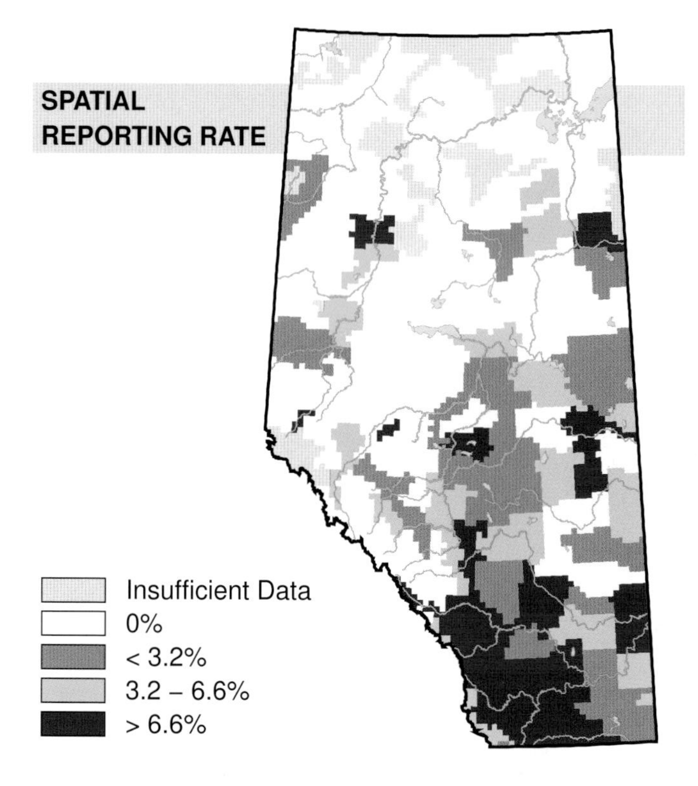

STATUS

GSWA 2000	GSWA 2005	COSEWIC Historic Rankings	COSEWIC 2007	Alberta Wildlife Act
Secure	Secure	N/A	N/A	N/A

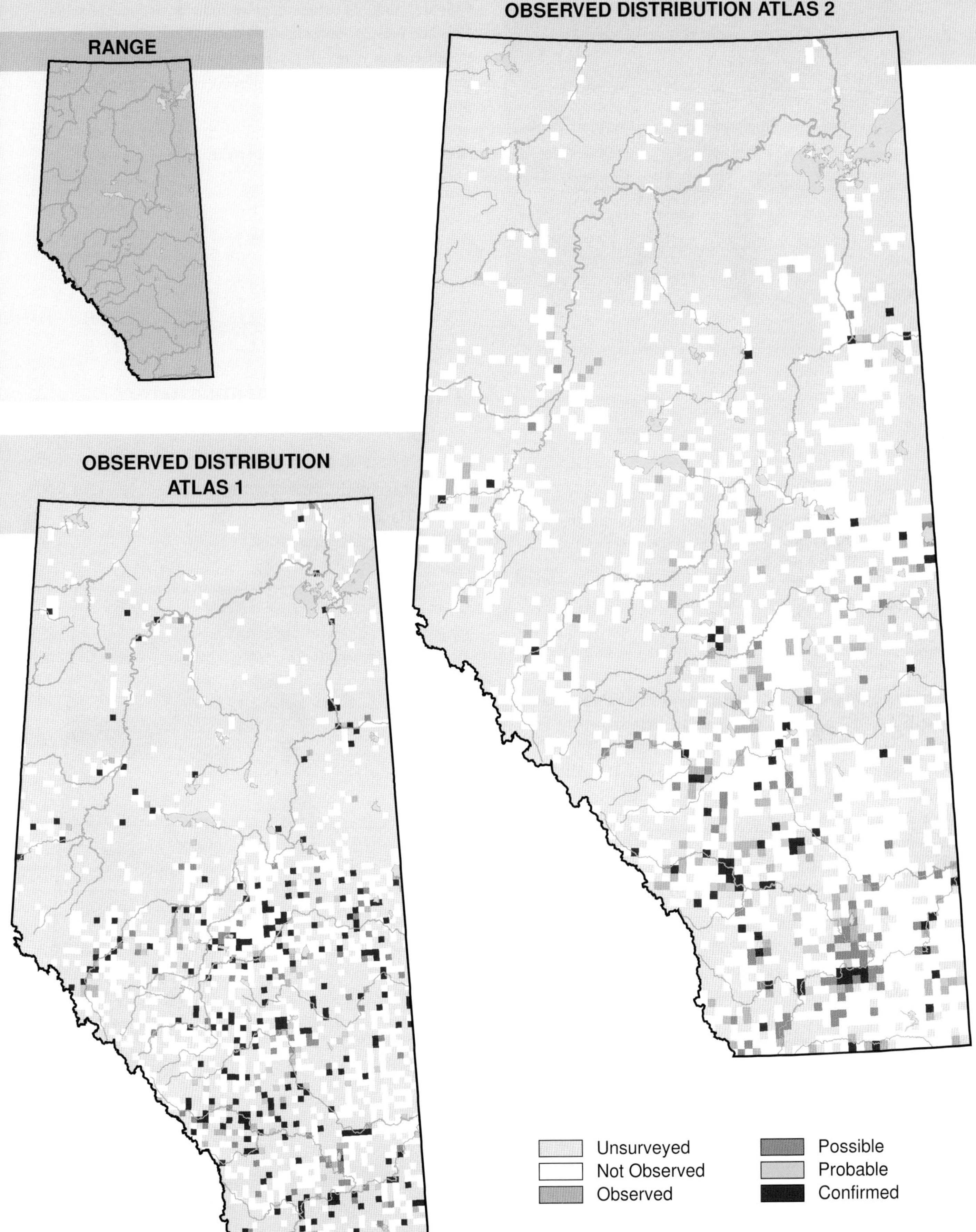
OBSERVED DISTRIBUTION ATLAS 2
RANGE
OBSERVED DISTRIBUTION
ATLAS 1
Unsurveyed
Not Observed
Observed
Possible
Probable
Confirmed

Cliff Swallow (*Petrochelidon pyrrhonota*)

Photo: Royal Alberta Museum

NESTING

CLUTCH SIZE (EGGS):	4–5
INCUBATION (DAYS):	12–14
FLEDGING (DAYS):	21–24
NEST HEIGHT (METRES):	>0.9

The Cliff Swallow is a colonial nester found in every Natural Region in Alberta. The observed distribution of this species did not change between Atlas 1 and Atlas 2.

This species uses almost any vertical cliff face with horizontal overhang for nesting. Preferred nesting locations include canyons, escarpments, foothills, river valleys, and man-made structures like bridges and buildings. Being one of the most social birds in North America, this species will nest in colonies as large as 3500 nests. The Cliff Swallow avoids heavy forest, desert, and alpine areas. It was therefore found most commonly in the Grassland NR, and less commonly in the Rocky Mountain and Parkland NRs. It was least commonly encountered in the Boreal Forest and Foothills NRs, where the forests tend to be more continuous.

Declines in relative abundance of the Cliff Swallow were detected in all NRs. Relative to other species, it was also observed less often in all NRs. Declines were found by the Breeding Bird Survey on Alberta-wide and Canada-wide scales during the period 1985–2005. Reasons for these declines are not understood. This species' nesting habitat has been enhanced by humans, due to the construction of alternate nesting structures like bridges, culverts, and buildings. It can, however, be difficult to study population changes for species such as the Cliff Swallow because they are colonial nesters. Colonies, that often contain many individuals, may abandon sites and move to new sites that do not fall within survey areas in subsequent years. If this occurs, the survey results could be misinterpreted. The Cliff Swallow is rated as Secure in the province.

TEMPORAL REPORTING RATE

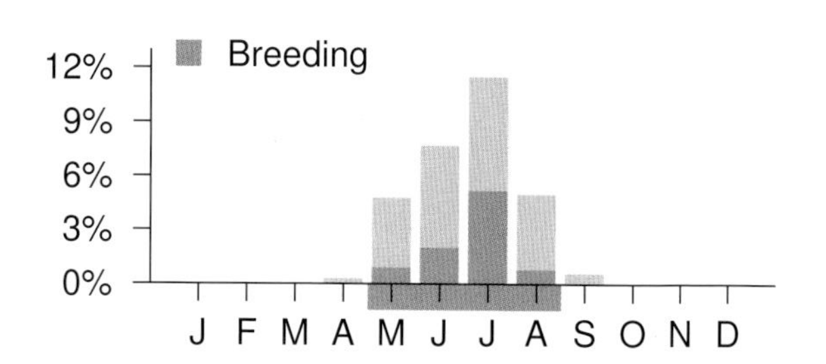

NATURAL REGIONS REPORTING RATE

SPATIAL REPORTING RATE

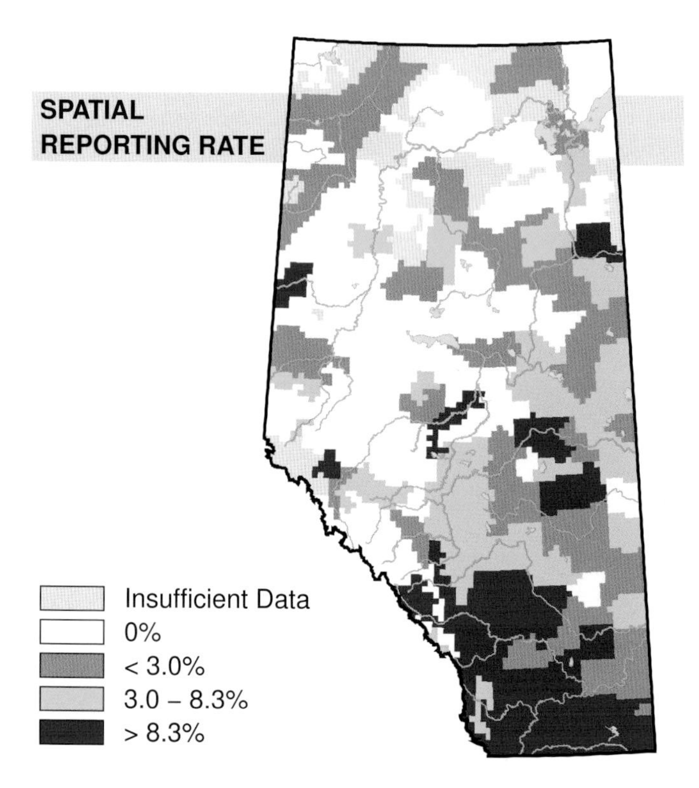

STATUS

GSWA 2000	GSWA 2005	COSEWIC Historic Rankings	COSEWIC 2007	Alberta Wildlife Act
N/A	N/A	N/A	N/A	N/A

RANGE

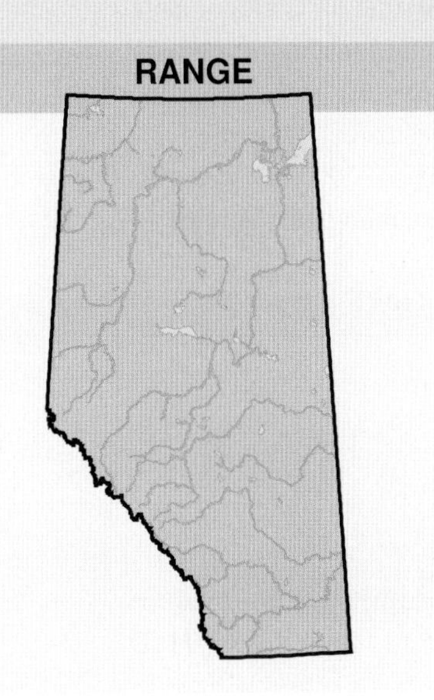

OBSERVED DISTRIBUTION ATLAS 2

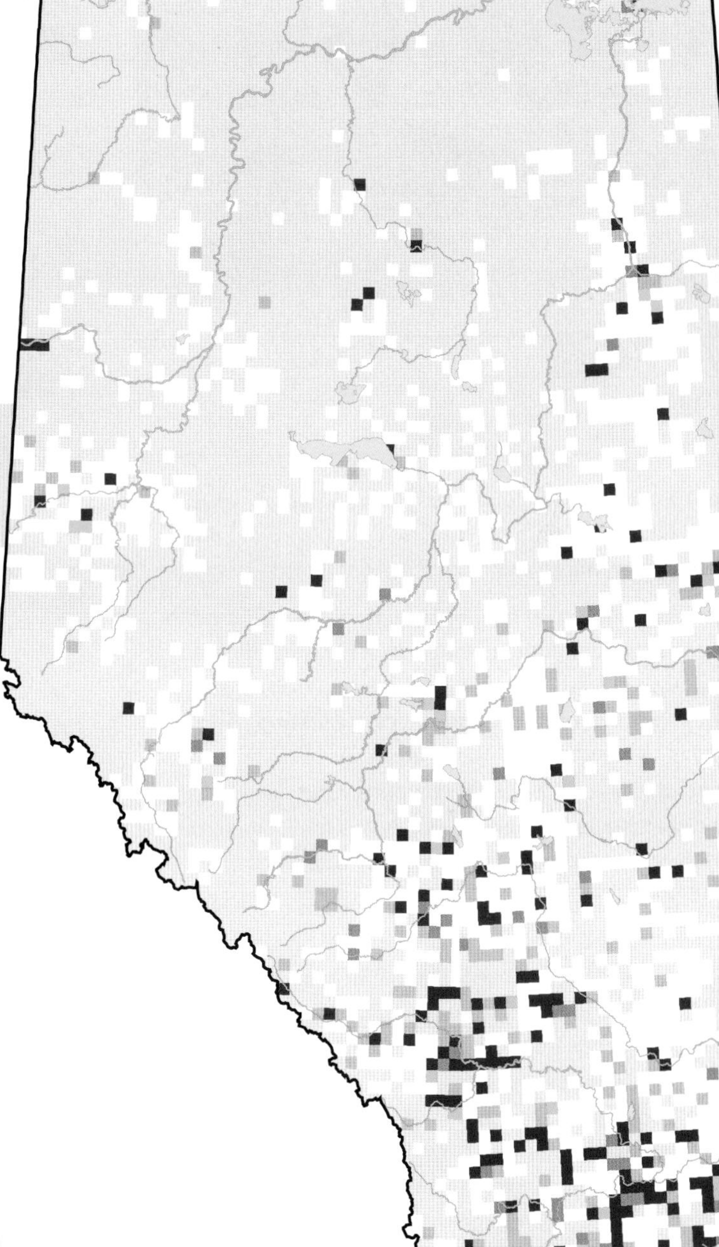

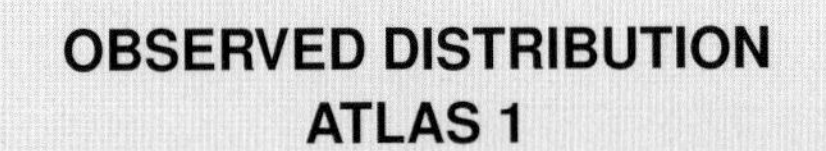

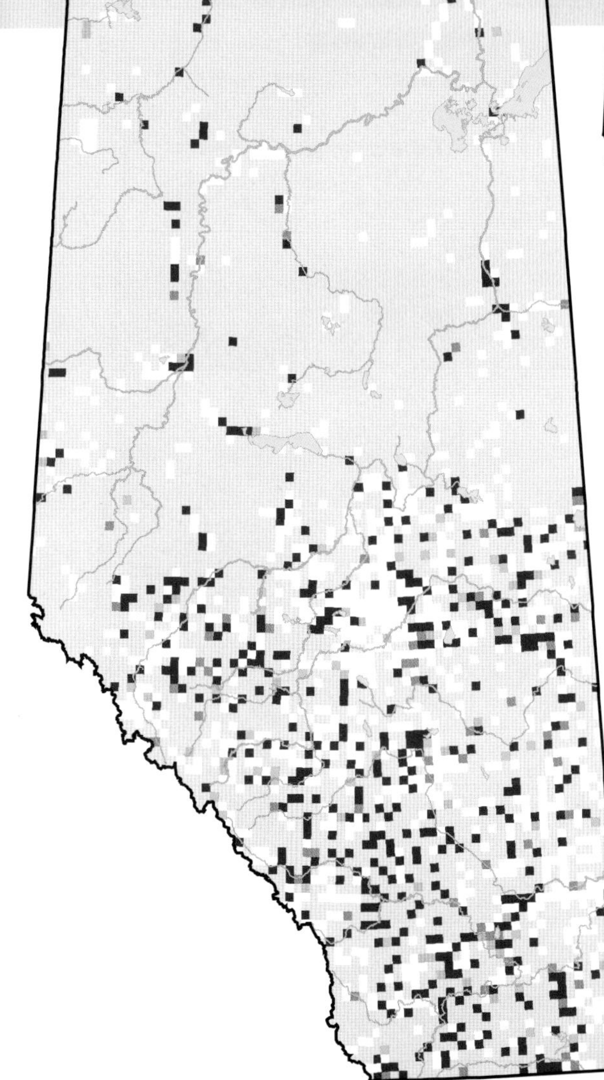

Unsurveyed
Not Observed
Observed

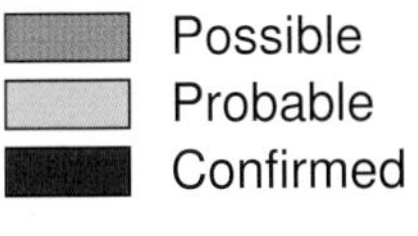

Barn Swallow (*Hirundo rustica*)

Photo: Gordon Court

NESTING

CLUTCH SIZE (EGGS):	4–6
INCUBATION (DAYS):	14–15
FLEDGING (DAYS):	17–24
NEST HEIGHT (METRES):	1.8–12.2

The Barn Swallow is found in every Natural Region across the province. The observed distribution of this species did not change between Atlas 1 and Atlas 2.

This species tends to nest near open areas, which are used for foraging, such as agricultural fields, open areas in cities, and along highways. Nests are often built beneath overhanging structures such as eaves along the edges of buildings. Because of this, this swallow is often found near sites of human activity. This species was common in the Foothills, Grassland, Parkland, and Rocky Mountain NRs. It was less common in the north because there are fewer people living there.

Declines in relative abundance were detected in every NR and, relative to other species, Barn Swallows were observed less frequently during Atlas 2 than in Atlas 1 in every NR. Similarly, the Breeding Bird Survey found a decline in abundance in Alberta and across Canada during the period 1985–2005. In addition, provincial-scale declines were found in British Columbia, Saskatchewan, Manitoba, Ontario, Quebec, New Brunswick, and Nova Scotia during the period 1985–2005. One of the factors that could be contributing to these declines is recent changes in agricultural practices. Links have been found between Barn Swallow declines and reductions in dairy farming operations. It has been shown that when dairy farming ceases, invertebrate abundance tends to decline. When this happens, less food is available for Barn Swallows; so, their numbers decline. In addition, large industrial farm operations have become more common. As a result, nesting structures such as barns and outbuildings have become rarer. The provincial status of this species changed from Secure in 2000 to Sensitive in 2005 due to the declines detected in Alberta and in surrounding jurisdictions.

TEMPORAL REPORTING RATE

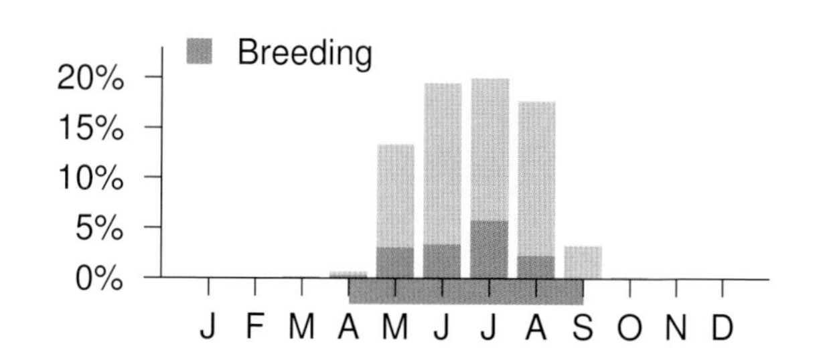

NATURAL REGIONS REPORTING RATE

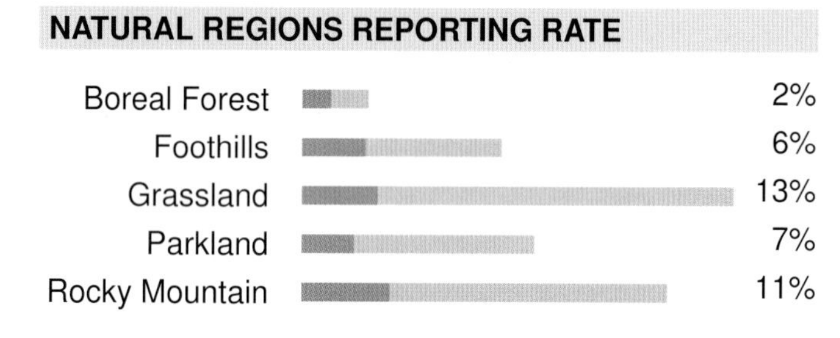

SPATIAL REPORTING RATE

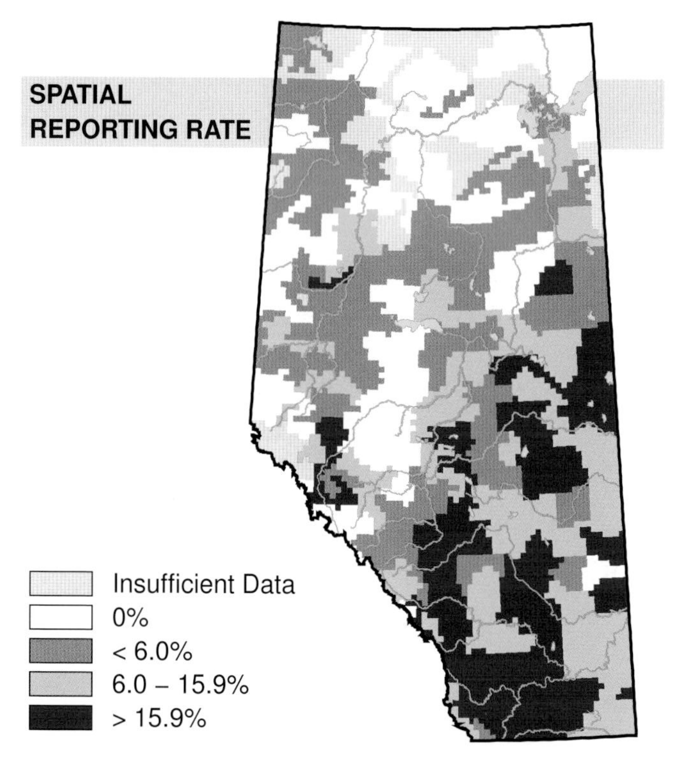

STATUS

GSWA 2000	GSWA 2005	COSEWIC Historic Rankings	COSEWIC 2007	Alberta Wildlife Act
Secure	Sensitive	N/A	N/A	N/A

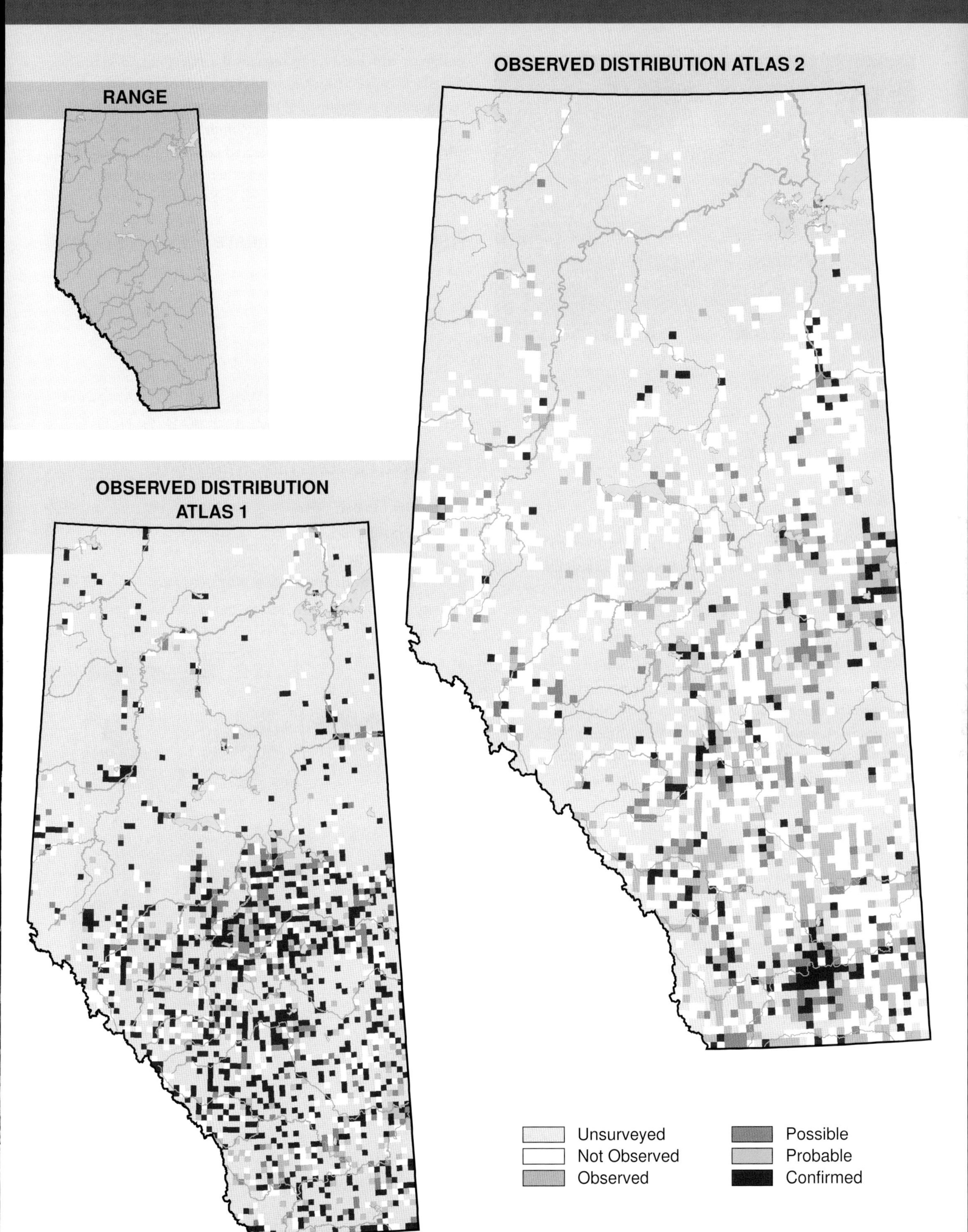
RANGE
OBSERVED DISTRIBUTION ATLAS 2
OBSERVED DISTRIBUTION ATLAS 1
Unsurveyed
Not Observed
Observed
Possible
Probable
Confirmed

Black-capped Chickadee (*Poecile atricapillus*)

Photo: Gerald Romanchuk

NESTING

CLUTCH SIZE (EGGS):	6–8
INCUBATION (DAYS):	11–13
FLEDGING (DAYS):	16
NEST HEIGHT (METRES):	1.2–7.0

The Black-capped Chickadee is found in suitable habitat in all Natural Regions of the province. It favours the Boreal Forest, Parkland, Foothills, and Rocky Mountain NRs. In southern Alberta, it is usually restricted to wooded coulees and river valleys when nesting, and is common in the Cypress Hills.

Preferred habitat comprises deciduous or mixed woods where this cavity nester usually excavates its own nest in a deciduous tree with broken bark exposing the soft inner wood. It also uses disturbed areas, such as old fields or suburban areas, where suitable nest sites are available with sufficient foliage to support adequate food for dependent offspring (Smith, 1993). In areas with cold winters, feeders often enhance Black-capped Chickadee survival, particularly in disturbed areas where food supplies are limited.

No changes in relative abundance were detected in the Boreal Forest and Foothills NRs. Increases in relative abundance were detected in the Rocky Mountain NR. This NR is dominated by parks and protected areas where there is no industrial activity, and where the removal of suitable habitat is limited to infrastructure additions. The resulting decadent forests would provide abundant cavities for populations that are limited by the availability of suitable nest sites. Decreases in relative abundance were detected in the Grassland and Parkland NRs. Clearing of aspen-dominated forests and groves for forestry, agriculture, country residential development, and urban expansion are prevalent in these regions. Removal of snags and cull trees with dead limbs reduces the availability of nest sites for Black-capped Chickadees and the lack of cottonwood regeneration is detrimental to the long-term stability of cavity-nesting bird populations (Smith, 1993). The Breeding Bird Surveys (1985–2005) detected no change in abundance in Alberta, but did detect an increase across Canada. The Black-capped Chickadee is listed as Secure in Alberta.

TEMPORAL REPORTING RATE

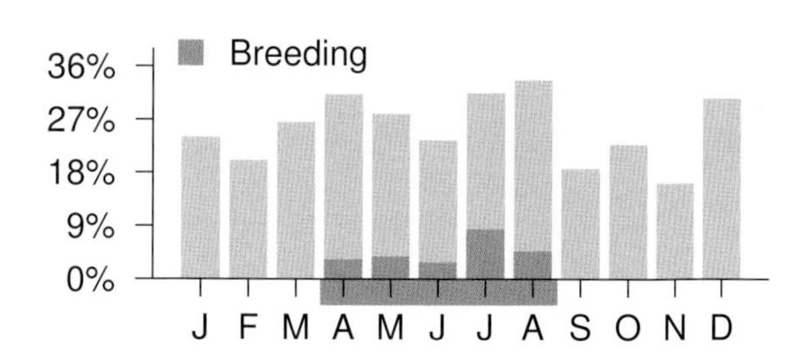

NATURAL REGIONS REPORTING RATE

Region	Rate
Boreal Forest	28%
Foothills	33%
Grassland	6%
Parkland	27%
Rocky Mountain	37%

SPATIAL REPORTING RATE

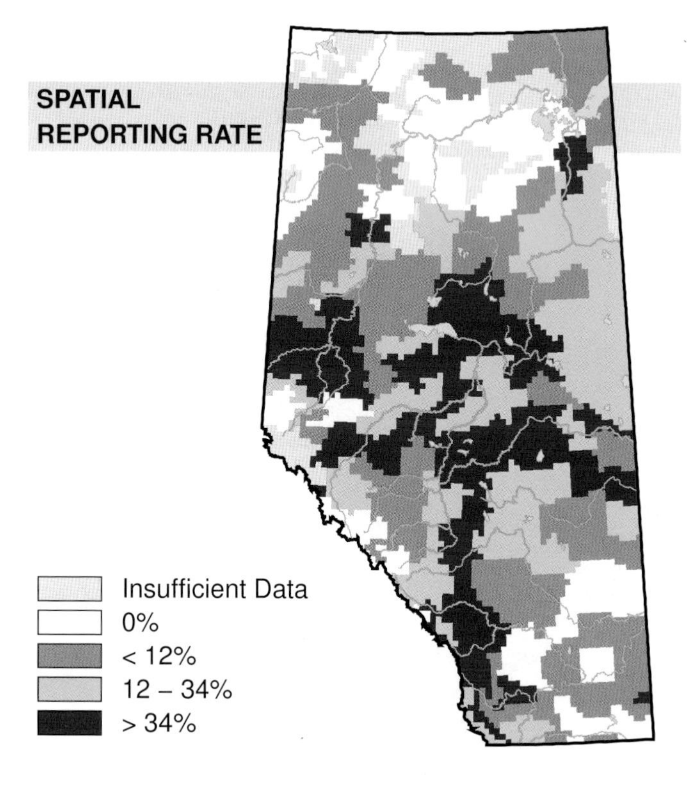

STATUS

GSWA 2000	GSWA 2005	COSEWIC Historic Rankings	COSEWIC 2007	Alberta Wildlife Act
N/A	N/A	N/A	N/A	N/A

RANGE

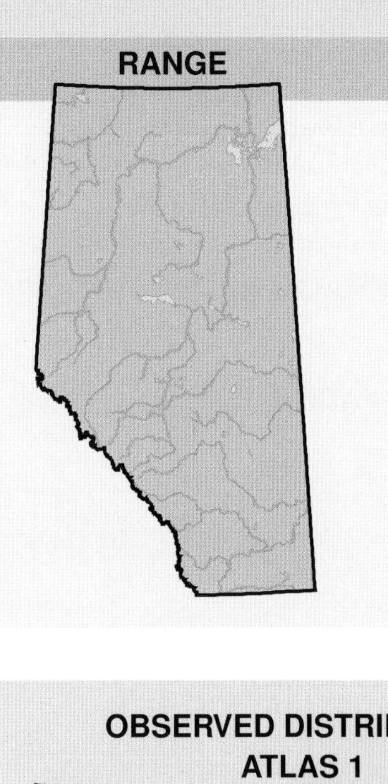

OBSERVED DISTRIBUTION ATLAS 2

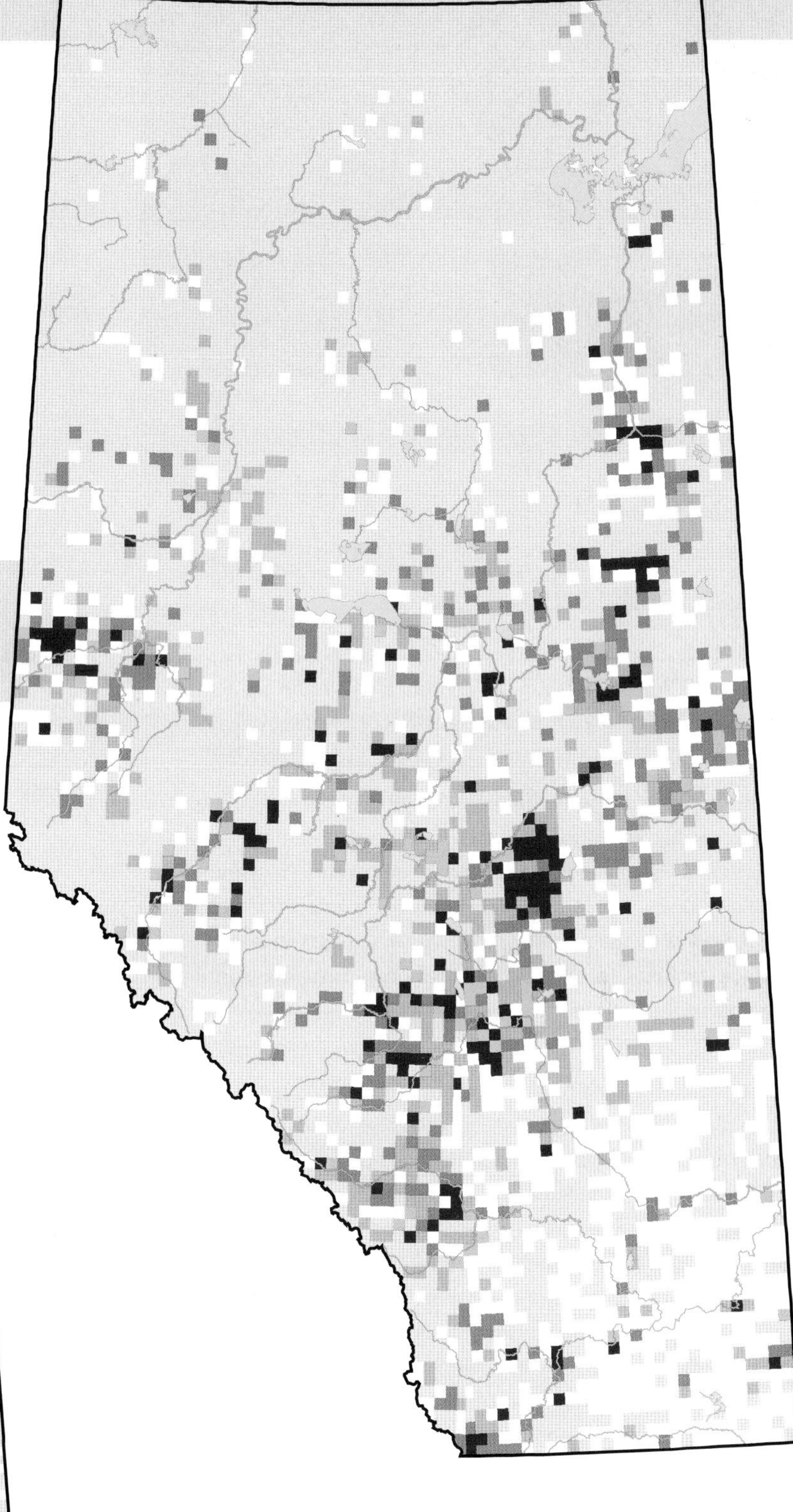

OBSERVED DISTRIBUTION ATLAS 1

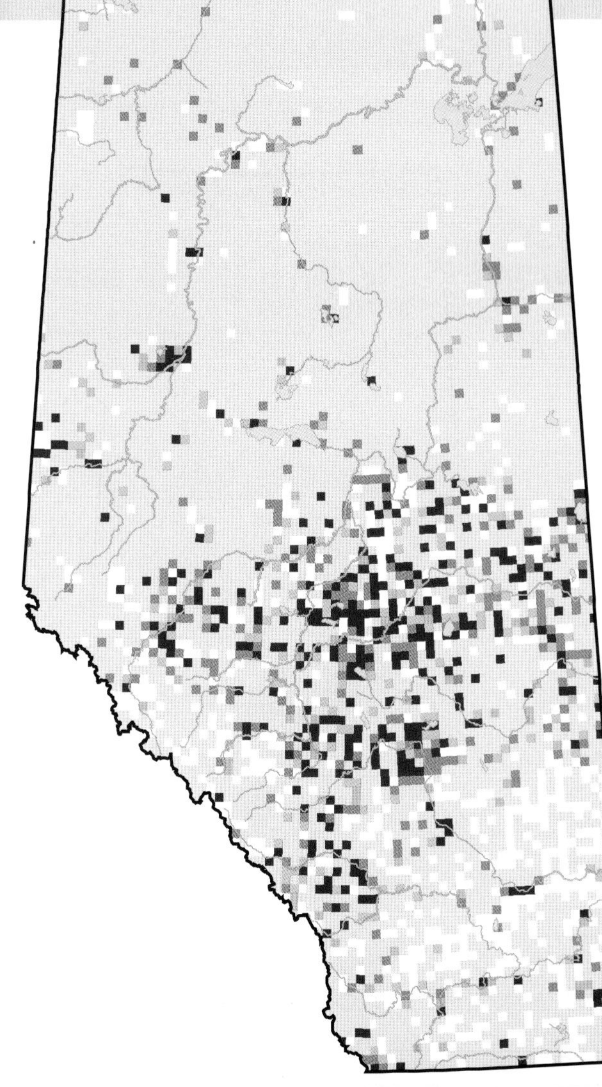

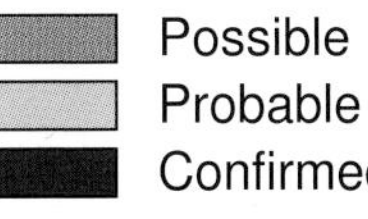

Mountain Chickadee (*Poecile gambeli*)

Photo: Raymond Toal

NESTING

CLUTCH SIZE (EGGS):	6–9
INCUBATION (DAYS):	14
FLEDGING (DAYS):	20
NEST HEIGHT (METRES):	<3

Mountain Chickadees are permanent residents of the Rocky Mountain Natural Region in Alberta. They are found also in the western portion of the Foothills NR. The observed distribution of this species did not change in Atlases 2.

This chickadee is found in mountainous habitat where it uses open coniferous and mixedwood forests for foraging. Mountain Chickadees are secondary cavity nesters using old nesting cavities of woodpeckers and Black-capped Chickadees, usually low to the ground. In winter they may join flocks of Black-capped Chickadees, and have been recorded as far east as Elk Island National Park.

Increases in relative abundance were detected in the Rocky Mountain NR. This may be a result of the older age profile of this region's forests; such decadent forests contribute to an increase in cavity abundance. No changes in relative abundance were detected in other NRs. The Breeding Bird Surveys (1995–2005) found an increase in abundance across Canada, but the BBS sample size in Alberta was too small to estimate abundance changes. Density varies widely from year to year depending on conifer seed crops (McCallum et al., 1999); so, the BBS may be biased. Local populations experience occasional food-related crashes, forcing immigration among populations with different seed-crop schedules. Human-caused changes in the landscape, such as reduced availability of dispersal corridors and of seasonally important food sources, could reduce the abundance of this species (McCallum et al., 1999).

Dunn (2005), in the National action needs for Canadian Landbird Conservation, reports that, although this species is common and widespread, it has been declining steadily in the U.S. for the last 40 years and in Canada for the last 20 years. The report recommends that the steady decline of the Mountain Chickadee population merits investigation, although confirmation of decline should be part of any investigation. The Mountain Chickadee is rated as Secure in Alberta.

TEMPORAL REPORTING RATE

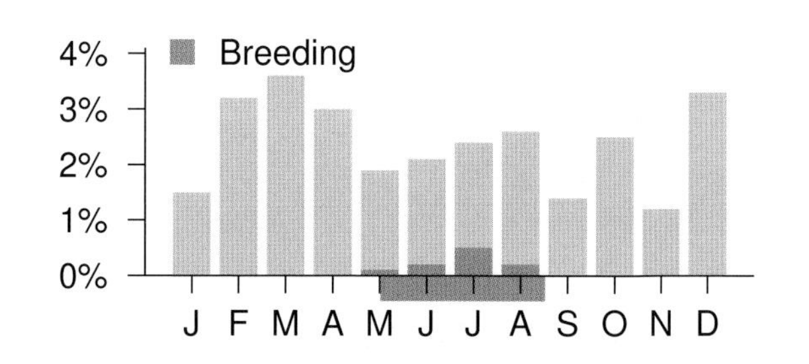

NATURAL REGIONS REPORTING RATE

SPATIAL REPORTING RATE

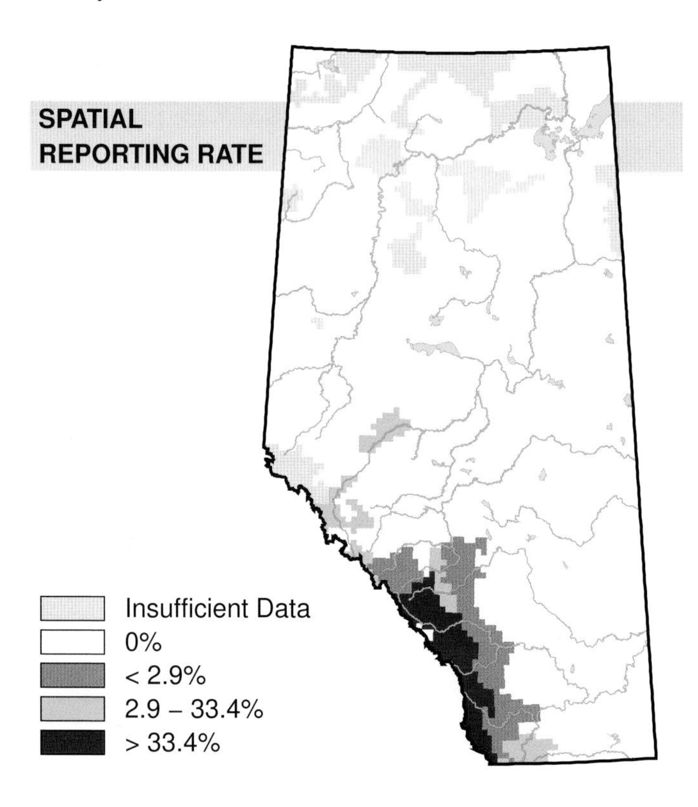

STATUS

GSWA 2000	GSWA 2005	COSEWIC Historic Rankings	COSEWIC 2007	Alberta Wildlife Act
N/A	N/A	N/A	N/A	N/A

RANGE

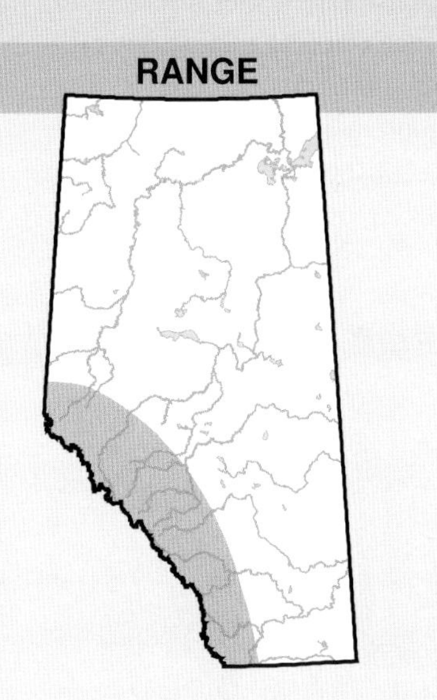

OBSERVED DISTRIBUTION ATLAS 2

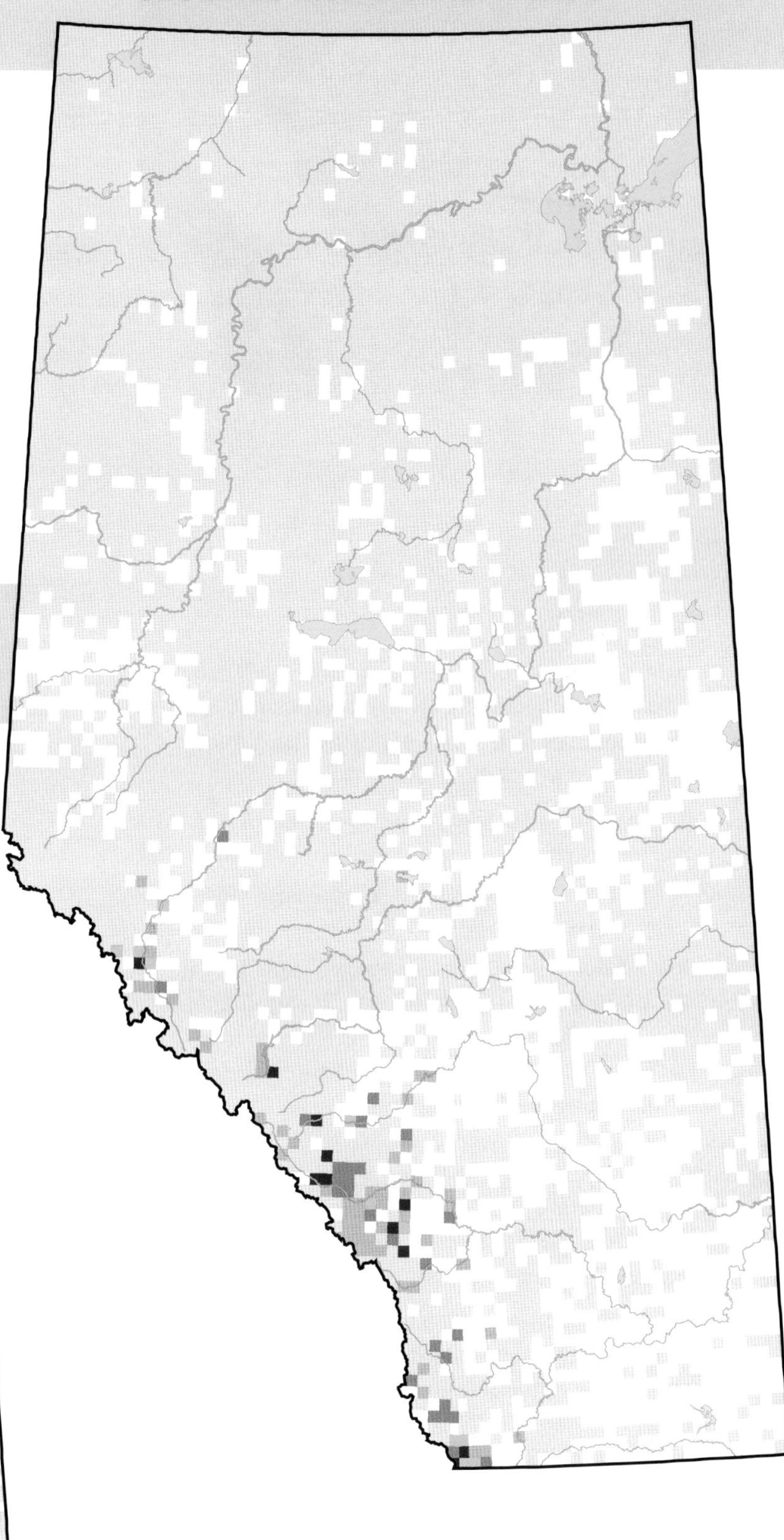

OBSERVED DISTRIBUTION ATLAS 1

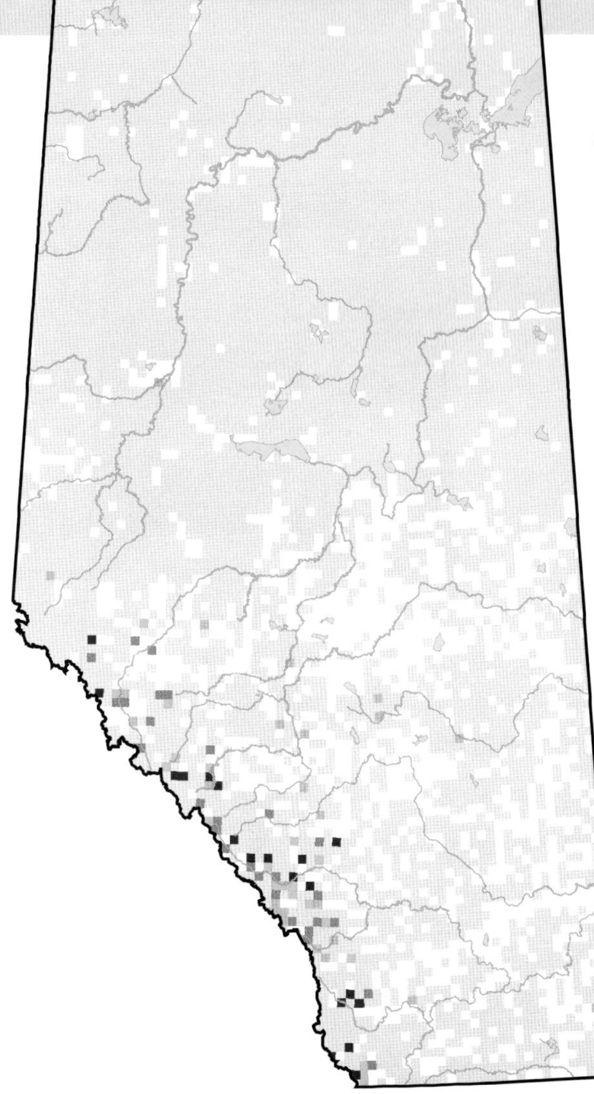

Unsurveyed
Not Observed
Observed
Possible
Probable
Confirmed

Boreal Chickadee (*Poecile hudsonica*)

Photo: Gerald Romanchuk

NESTING

CLUTCH SIZE (EGGS):	4–7
INCUBATION (DAYS):	15
FLEDGING (DAYS):	18
NEST HEIGHT (METRES):	1–4

The Boreal Chickadee is found predominantly in the southern Boreal Forest, Foothills, and Rocky Mountain Natural Regions. The observed distribution of this species did not change from Atlas 1 to Atlas 2.

This bird is a year-round resident of dense coniferous and mixedwood forests where it prefers spruce and fir habitats. The tree species present appear to be more important for habitat selection than the age of the forest.

Increases in relative abundance were detected in the Boreal Forest NR. This increase may be related to climate differences between Atlases 1 and 2 that have led to increases in food supply and cavity abundance. The Breeding Bird Surveys (1985–2005) detected no change in abundance in Alberta or across Canada, although a national decline was detected for the period of 1968–2005. However, there are few reliable population trends available for this species, and census data indicate that this species is never abundant in any part of its extensive boreal forest range (Ficken et al., 1996).

Ficken et al. (1996) indicated that little is known about the conservation status of the Boreal Chickadee; little has changed. Dunn (2005), in the National action needs for Canadian Landbird Conservation, states that this species is experiencing long-term declines, which have been particularly consistent over the past three decades. The report indicates that status information is the main need, but apparent large declines and high Canadian stewardship responsibility justify research. In particular, the research should focus on determining critical habitat and on studying the long- and short-term effects of forestry practices that affect important portions of their range. Birder observations, such as checklist program data, may prove the most practical means of tracking its status. In Alberta this species is rated as Secure.

TEMPORAL REPORTING RATE

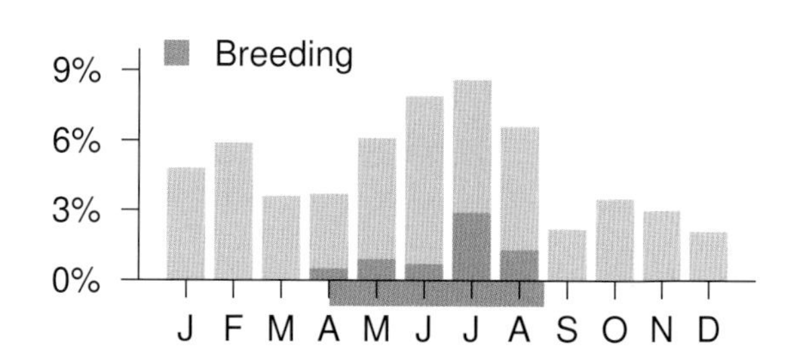

NATURAL REGIONS REPORTING RATE

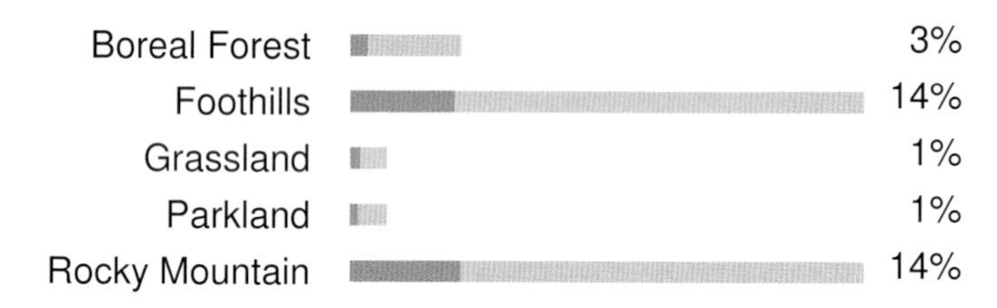

SPATIAL REPORTING RATE

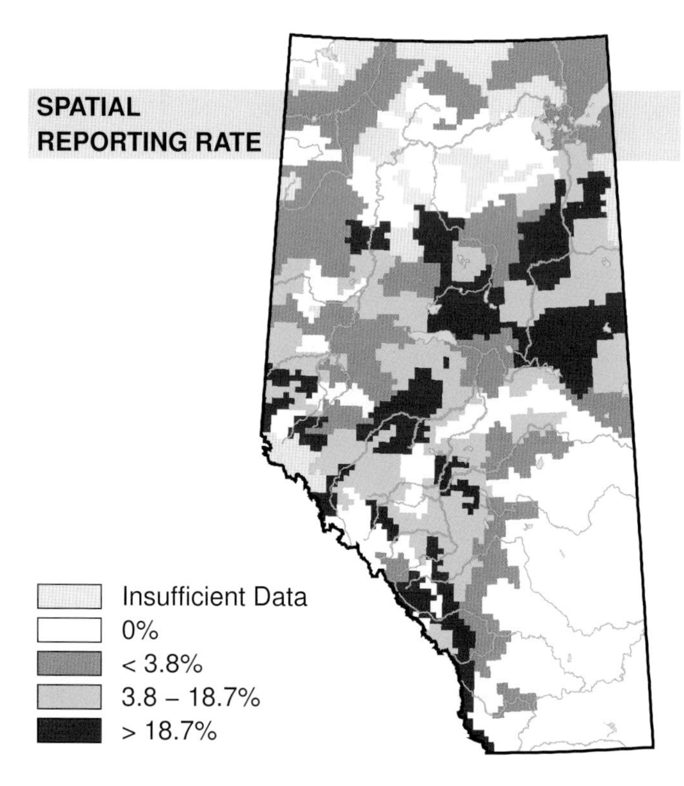

STATUS

GSWA 2000	GSWA 2005	COSEWIC Historic Rankings	COSEWIC 2007	Alberta Wildlife Act
N/A	N/A	N/A	N/A	N/A

OBSERVED DISTRIBUTION ATLAS 2

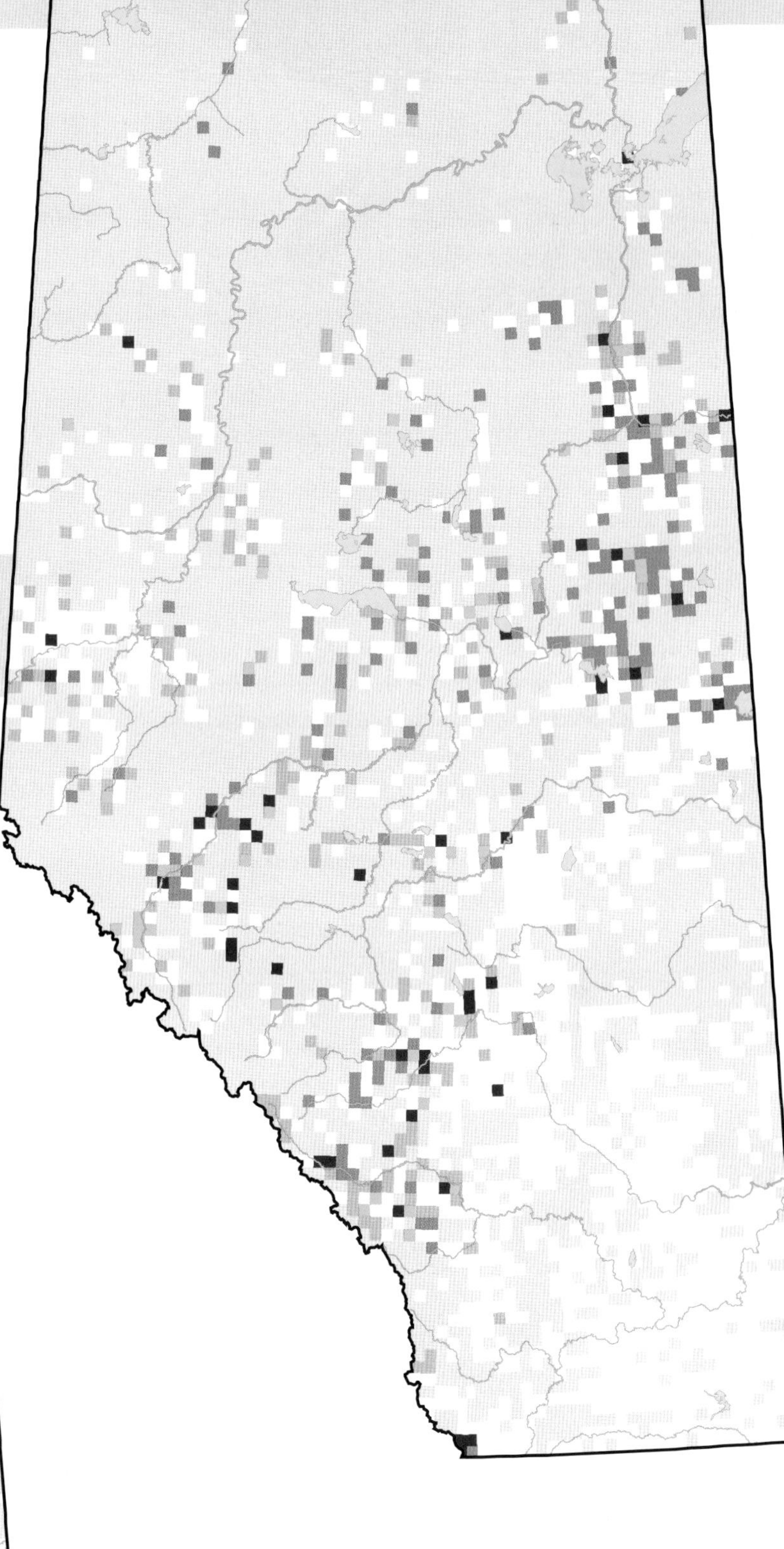

RANGE

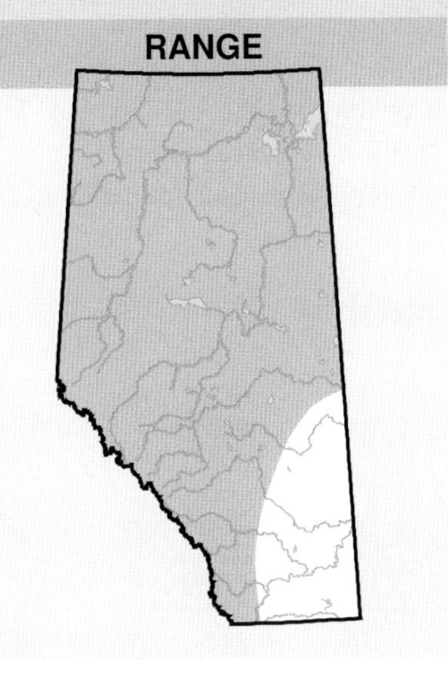

OBSERVED DISTRIBUTION ATLAS 1

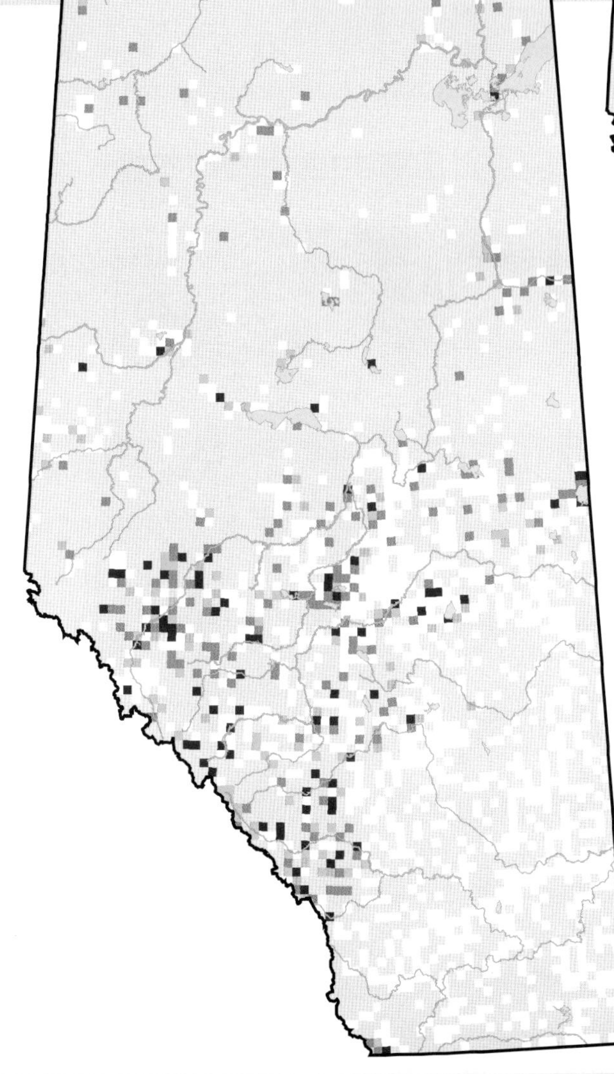

	Unsurveyed		Possible
	Not Observed		Probable
	Observed		Confirmed

Red-breasted Nuthatch (*Sitta canadensis*)

Photo: Raymond Toal

NESTING

CLUTCH SIZE (EGGS):	4–7
INCUBATION (DAYS):	12
FLEDGING (DAYS):	18–21
NEST HEIGHT (METRES):	2–12

The Red-breasted Nuthatch is found in every Natural Region in Alberta. The observed distribution of this species decreased in Atlas 2 in the northwestern part of the province, around High Level and Fort Vermilion. It is unclear whether this represents an actual range contraction or whether the presence of Red-breasted Nuthatches in the northwest was missed during Atlas 2. The latter is possible because there were not many observations recorded in the northwest during Atlas 1 and the records that were found were sparsely distributed.

The Red-breasted Nuthatch is usually found in mature coniferous forests or mixedwood forests that are conifer-dominated. This species often excavates its own cavities for nesting in trees. Unlike many woodpeckers that are strong cavity excavators, Red-breasted Nuthatches are considered weak excavators; therefore, they require dead and decaying trees for nest building. Due to its close association with coniferous forests, this species was found mainly in the Boreal Forest, Foothills, and Rocky Mountain NRs and was less frequently found in the Grassland and Parkland NRs.

Increases in relative abundance were detected in all NRs (except the Canadian Shield) where, relative to other species, the Red-breasted Nuthatch was observed more frequently in Atlas 2 than in Atlas 1. The Breeding Bird Survey did not detect an abundance change in Alberta, but an increase in abundance was detected on a Canada-wide scale. Fire suppression could be having a positive effect on this species because fires often reduce the numbers of old dead trees that this species could use for nesting. Bird-feeding could also being having an influence on the abundance of this species. Bird-feeding has increased in popularity and the Red-breasted Nuthatch is one of the species that will visit bird feeders, especially when suet or peanut butter is used. This species is considered Secure in Alberta.

TEMPORAL REPORTING RATE

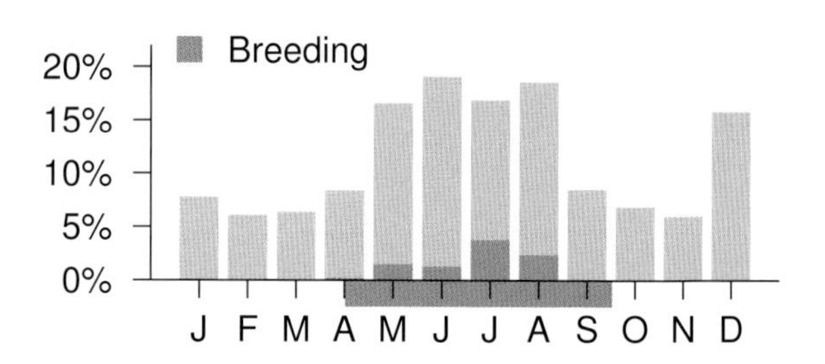

NATURAL REGIONS REPORTING RATE

Boreal Forest	13%
Foothills	42%
Grassland	4%
Parkland	5%
Rocky Mountain	55%

SPATIAL REPORTING RATE

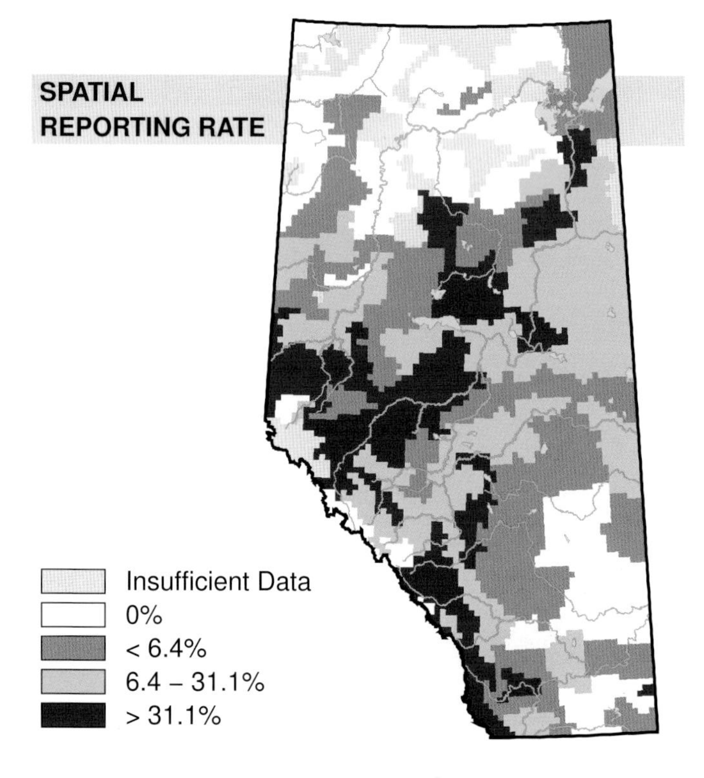

STATUS

GSWA 2000	GSWA 2005	COSEWIC Historic Rankings	COSEWIC 2007	Alberta Wildlife Act
Secure	Secure	N/A	N/A	N/A

Habitat  Nest Location  Nest Type  Diet  

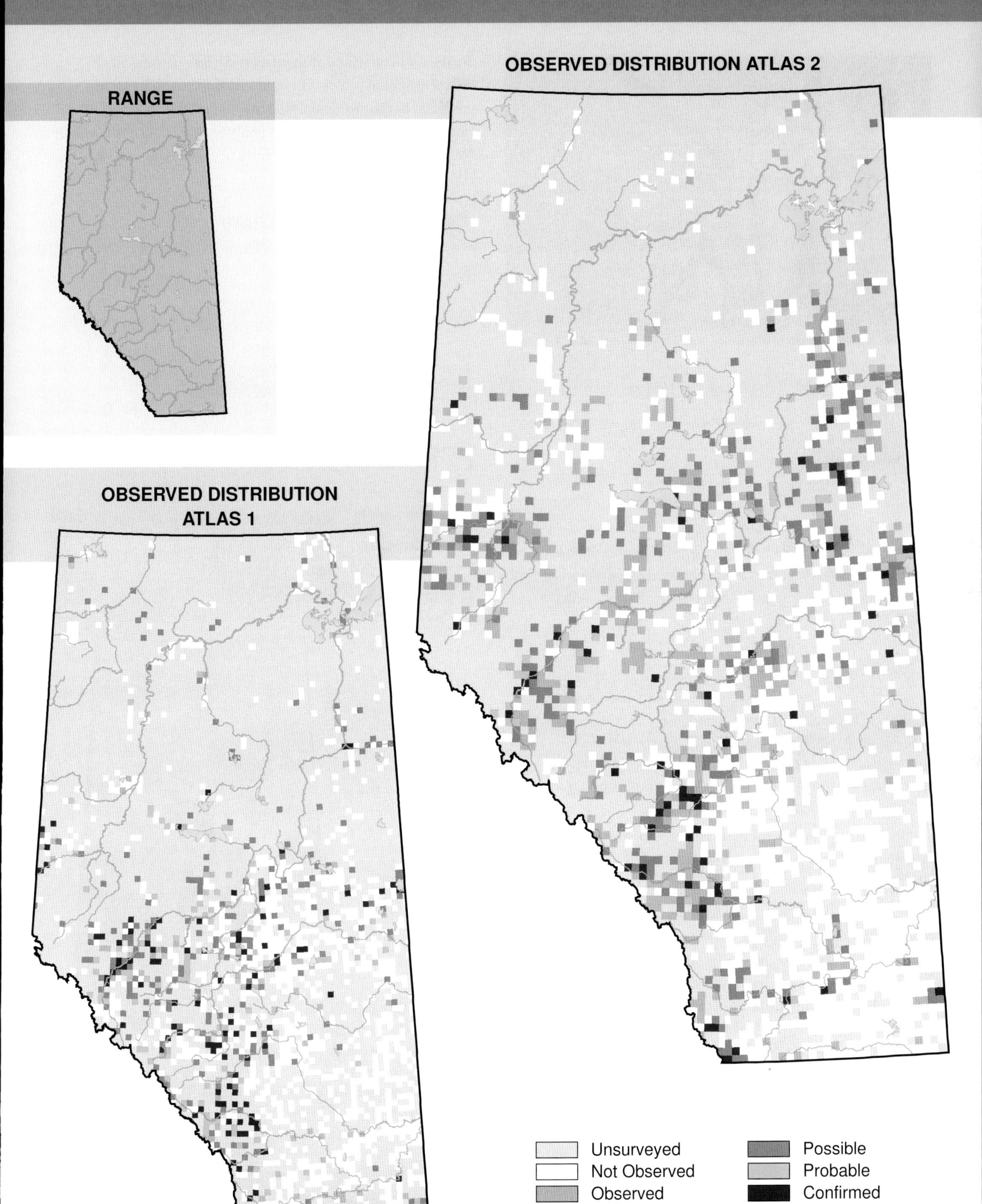

White-breasted Nuthatch (*Sitta carolinensis*)

Photo: Jim Jacobson

NESTING

CLUTCH SIZE (EGGS):	5–8
INCUBATION (DAYS):	12
FLEDGING (DAYS):	14
NEST HEIGHT (METRES):	2–10

The White-breasted Nuthatch is found across the western and central part of Alberta and occurs, at least in part, in every Natural Region except the Canadian Shield. The observed distribution of this species did not change between Atlas 1 and Atlas 2.

The White-breasted Nuthatch is often found in forests that are dominated by deciduous trees. This species generally avoids young deciduous forests and forests that are dominated by conifers. This species is often found near edges that can be either natural, such as along wetlands and meadows, or artificial, such as along roads, fields and clearings. This species was found mainly along rivers in the Foothills NR and throughout the Parkland NR. It was less common in the southern part of the Boreal Forest and along rivers in the Grassland NR and in the Rocky Mountain NR.

Increases in relative abundance were detected in the Grassland and Parkland NRs where, relative to other species, the White-breasted Nuthatch was observed more frequently in Atlas 2 than in Atlas 1. Increases could be caused by increased fragmentation of forests because this species tends to breed near edges. No change was detected in the other NRs. The Breeding Bird Survey did not detect an abundance change for this species in Alberta or Canada. Differences between the results of Atlas and Breeding Bird Survey analyses could be attributed to analyses of the data that were performed on different spatial scales. Breeding Bird Survey data were analyzed at the provincial scale; whereas, due to sufficient sample size, it was possible to analyze Atlas data at the smaller Natural Region scale. This species is considered Secure in Alberta.

TEMPORAL REPORTING RATE

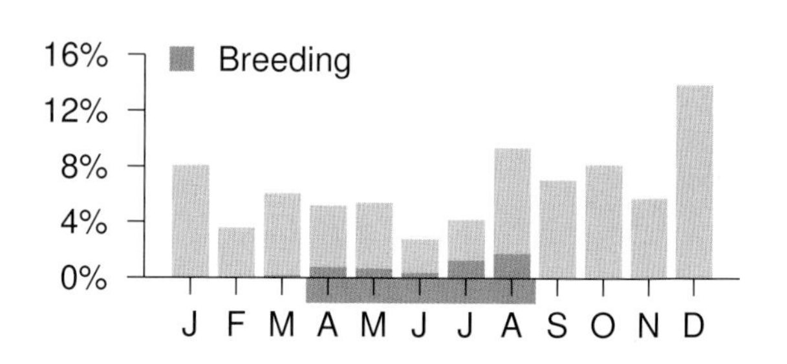

NATURAL REGIONS REPORTING RATE

Boreal Forest	2.2%
Foothills	5.8%
Grassland	1.6%
Parkland	3.0%
Rocky Mountain	1.8%

SPATIAL REPORTING RATE

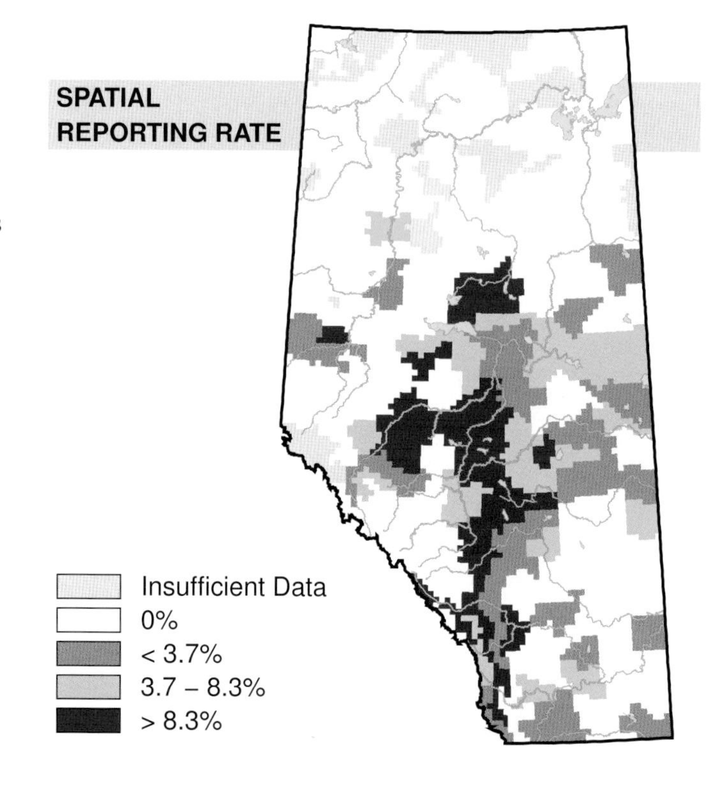

STATUS

GSWA 2000	GSWA 2005	COSEWIC Historic Rankings	COSEWIC 2007	Alberta Wildlife Act
Secure	Secure	N/A	N/A	N/A

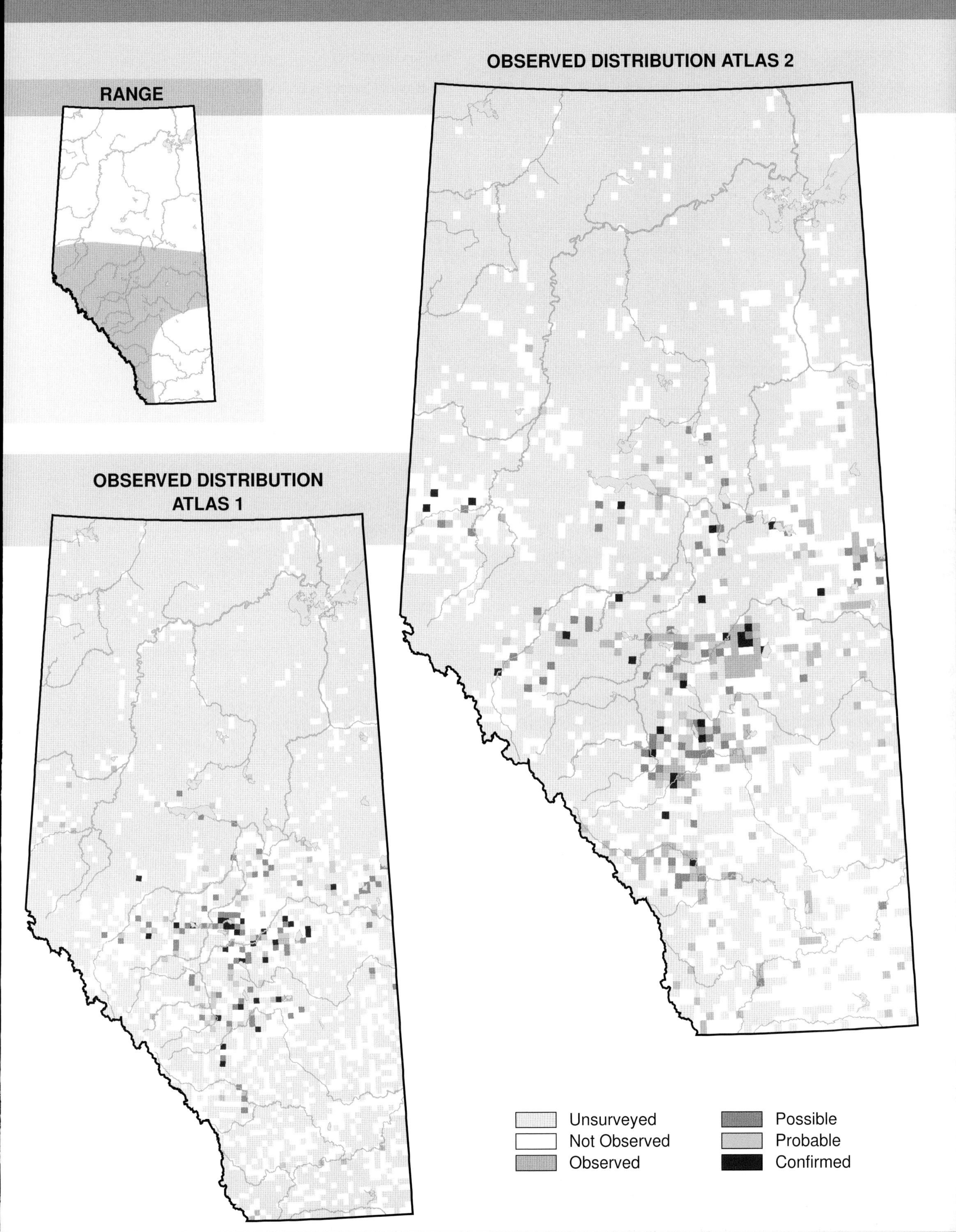
RANGE
OBSERVED DISTRIBUTION ATLAS 2
OBSERVED DISTRIBUTION ATLAS 1
Unsurveyed
Not Observed
Observed
Possible
Probable
Confirmed

Brown Creeper (*Certhia americana*)

Photo: Gerald Romanchuk

NESTING

CLUTCH SIZE (EGGS):	5–6
INCUBATION (DAYS):	14–15
FLEDGING (DAYS):	14–16
NEST HEIGHT (METRES):	1–15

The inconspicuous Brown Creeper breeds in the Boreal Forest, Foothills, and Rocky Mountain Natural Regions. No change in distribution was observed in Atlas 2; however, records from the Manning and Peace River areas changed from observed to possible breeding.

Brown Creepers favour closed-canopy, mature, and old-growth forests with a large abundance of dead and dying trees which they use for nesting, and large trees which they rely on to provide bark-dwelling invertebrates for food. This species was found most frequently in the Rocky Mountain and Foothills NRs and less frequently in the Boreal Forest NR. Brown Creepers observed in the Grassland and Parkland NRs are not considered to be breeders. When found in harvested plots in mixedwood boreal forests in Alberta, the bird tended to be in plots with many large residual trees and snags and in plots where trees and snags were left in clumps (Schieck et al., 2000).

Their relative abundance increased in the Boreal Forest, Foothills, and Rocky Mountain NRs. Like the Winter Wren, the Brown Creeper prefers late successional stages of coniferous and mixed coniferous-deciduous forests. These forest types have been increasing, particularly in some parts of the Foothills NR and throughout much of the Rocky Mountain NR, due to fire suppression. The Breeding Bird Survey (1985–2005) detected no change in abundance in Canada, although the sample size was too small to determine abundance changes in Alberta.

The Brown Creeper is considered Sensitive in Alberta due to its dependence on mature forests and its vulnerability to forest fragmentation. Retaining continuous, unfragmented areas of unlogged mature and old-growth forests will provide optimum habitat, as well as allowing forests to regrow and actively restoring "old-growth" forests (Hejl et al., 2002).

TEMPORAL REPORTING RATE

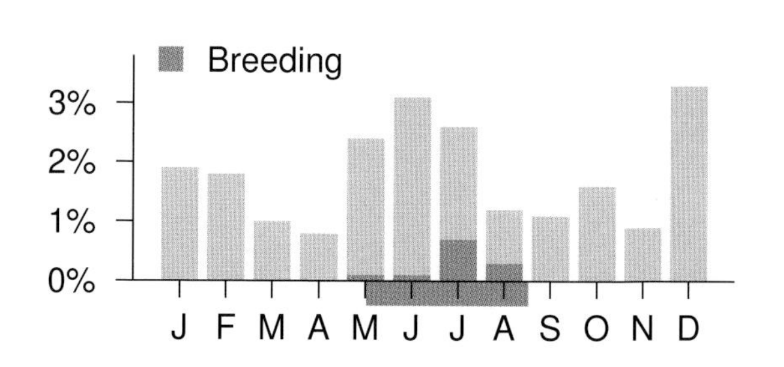

NATURAL REGIONS REPORTING RATE

Region	Rate
Boreal Forest	1.3%
Foothills	4.6%
Grassland	0.3%
Parkland	0.2%
Rocky Mountain	4.2%

SPATIAL REPORTING RATE

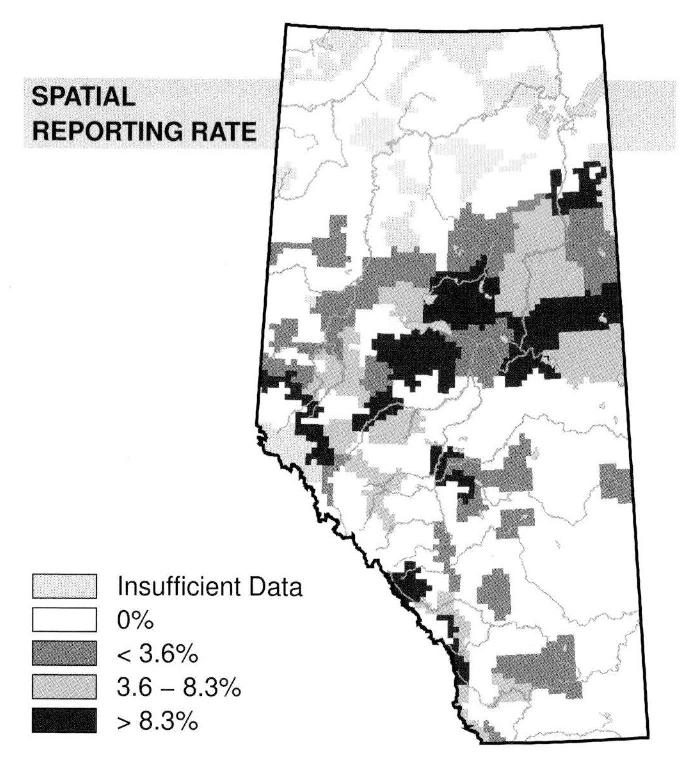

STATUS

GSWA 2000	GSWA 2005	COSEWIC Historic Rankings	COSEWIC 2007	Alberta Wildlife Act
Undetermined	Sensitive	N/A	N/A	N/A

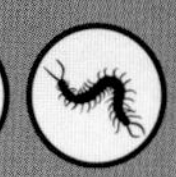

RANGE

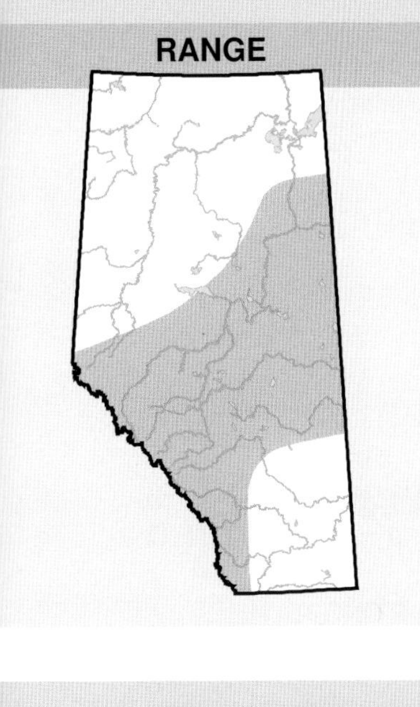

OBSERVED DISTRIBUTION ATLAS 2

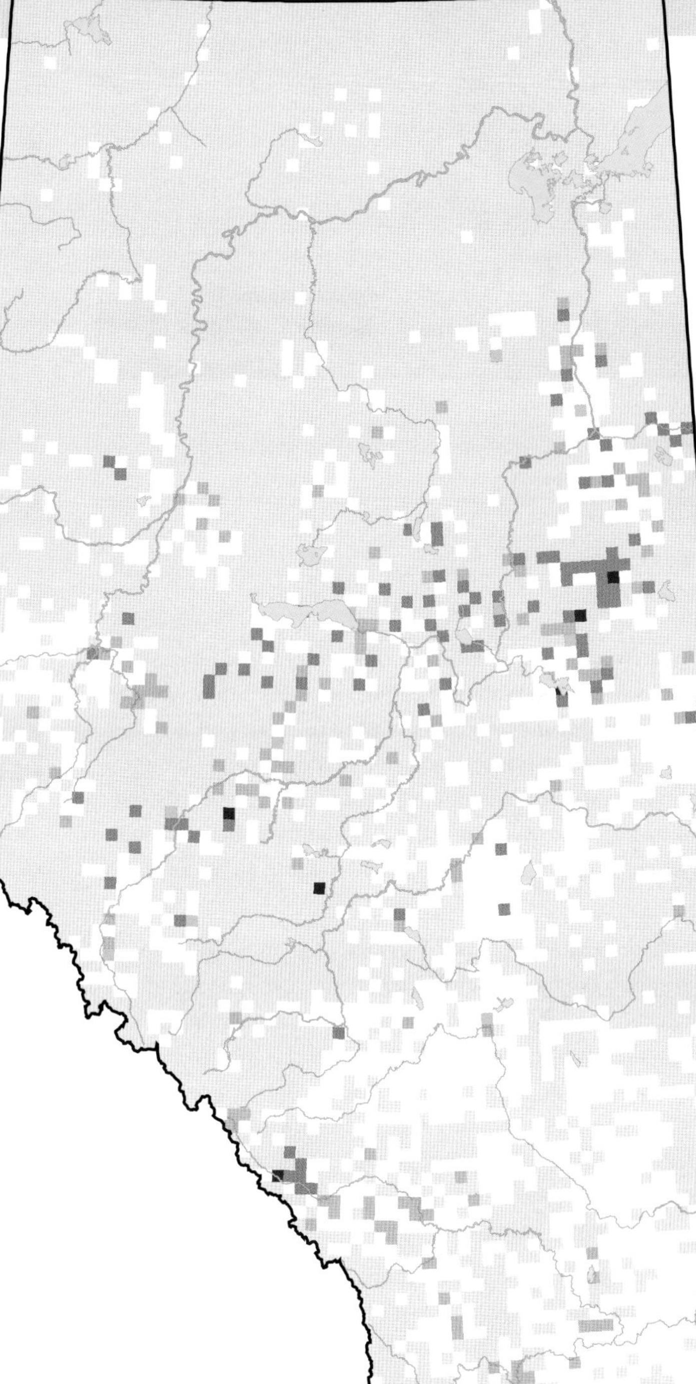
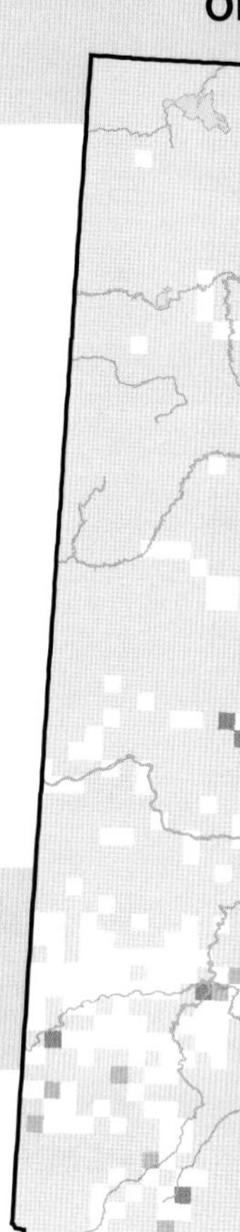

OBSERVED DISTRIBUTION ATLAS 1

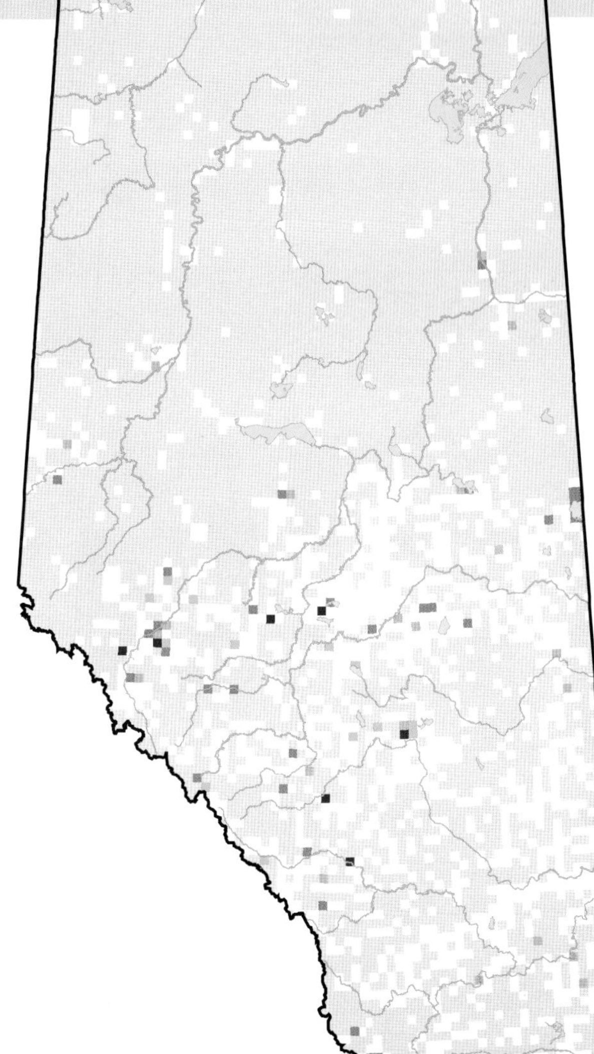

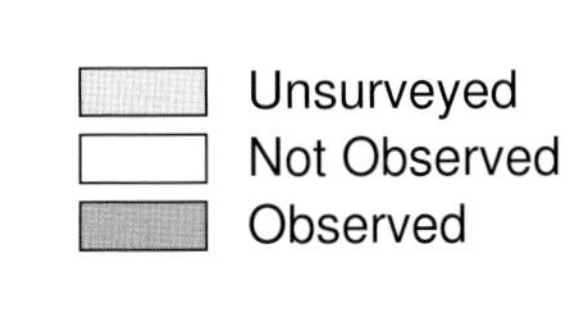

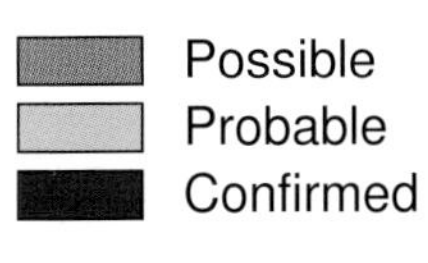

Rock Wren (*Salpinctes obsoletus*)

Photo: Randy Jensen

NESTING

CLUTCH SIZE (EGGS):	5–6
INCUBATION (DAYS):	12–16
FLEDGING (DAYS):	12–16
NEST HEIGHT (METRES):	0

The Rock Wren is found in southern Alberta in the Grassland and Rocky Mountain Natural Regions. In Atlas 2 they were not found as far north and west as they were in Atlas 1. It is not clear if this represents a true range contraction, because there were few and sparsely distributed records in the northwestern part of this species' range in Atlas 1. Therefore, the presence of this species could have been missed in those areas during Atlas 2.

As its name implies, the Rock Wren is found in rocky areas. Nesting areas tend to be quite arid and they can range from alpine habitats to dry prairie habitats. The main factor that seems to limit the distribution of this species is the presence of exposed rocks such as rock piles and scree slopes. For this reason, this species was found mainly in the Grassland NR along rivers where there is enough topographical variation to create rock piles, and also in the Rocky Mountain NR where scree slopes can be common.

No change was detected in the relative abundance of this species in any of the NRs where it occurs. The Breeding Bird Survey sample size of Rock Wren records was too small to investigate abundance change in Alberta. However, no abundance change was detected for this species in British Columbia or Canada. Rock Wren populations will not likely be affected by human activity in the foreseeable future because this species nests in areas where human disturbance is minimal. This species is considered Secure in Alberta.

TEMPORAL REPORTING RATE

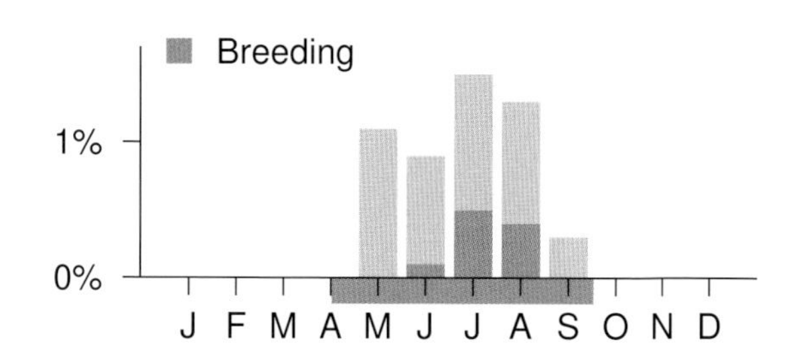

NATURAL REGIONS REPORTING RATE

SPATIAL REPORTING RATE

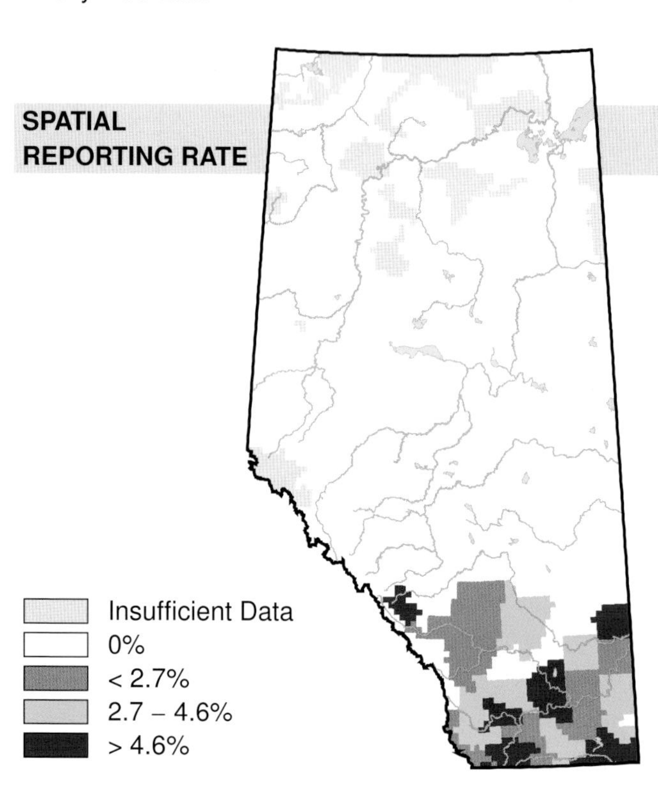

STATUS

GSWA 2000	GSWA 2005	COSEWIC Historic Rankings	COSEWIC 2007	Alberta Wildlife Act
Secure	Secure	N/A	N/A	N/A

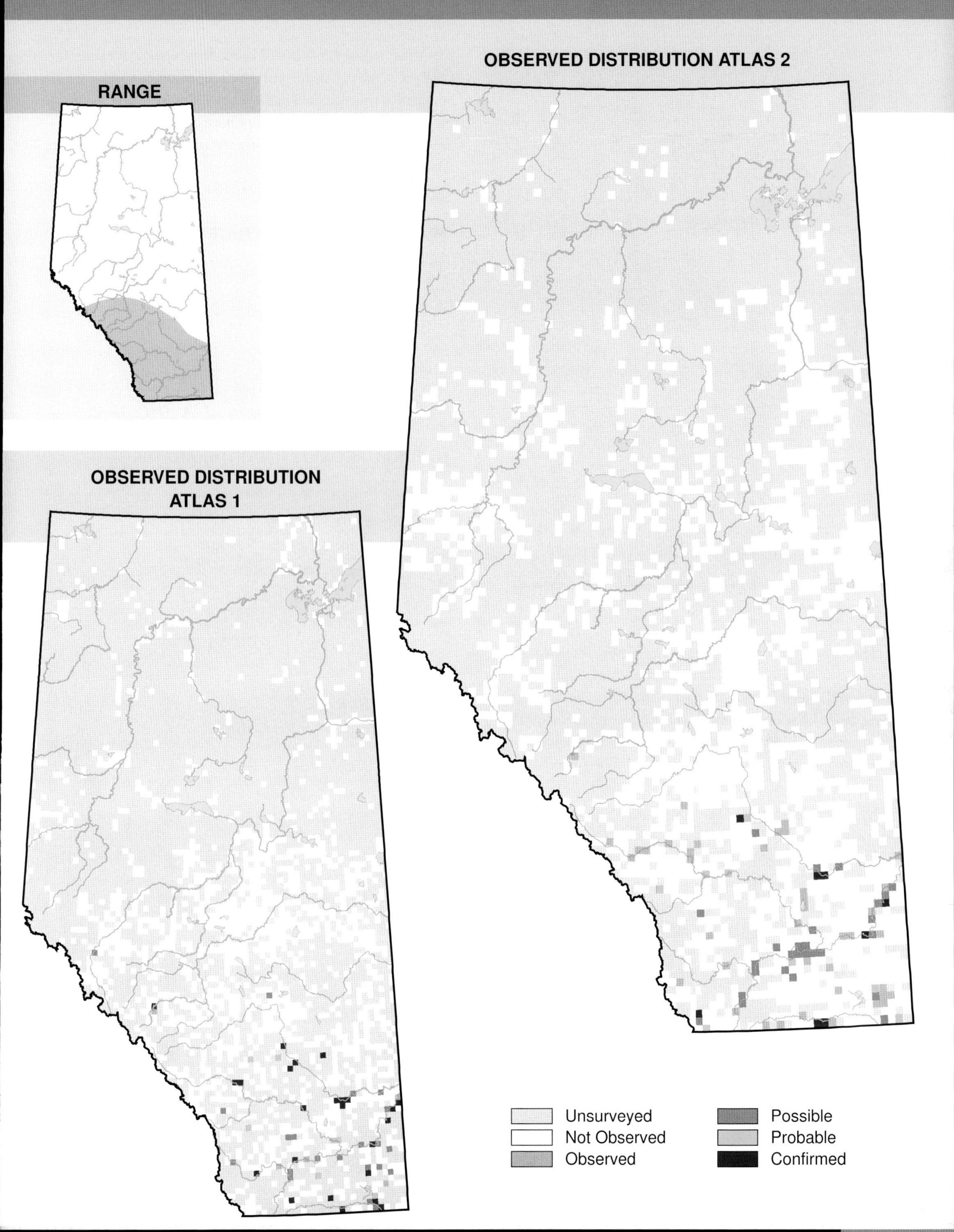
OBSERVED DISTRIBUTION ATLAS 2
RANGE
OBSERVED DISTRIBUTION
ATLAS 1
Unsurveyed
Not Observed
Observed
Possible
Probable
Confirmed

House Wren (*Troglodytes aedon*)

Photo: Gerald Romanchuk

NESTING

CLUTCH SIZE (EGGS):	6–8
INCUBATION (DAYS):	14
FLEDGING (DAYS):	12–18
NEST HEIGHT (METRES):	0–5

The House Wren is found in every Natural Region in the province. The observed distribution of this species did not change between Atlas 1 and Atlas 2.

The House Wren often breeds near the edge of fragmented deciduous forests. This species is a secondary cavity nester and, therefore, it will occupy artificial nestboxes, provided deciduous forest habitat is nearby. The bubbly sound of this species is familiar to people because these birds often nest near human settlements such as in backyards, and in nestboxes along fence lines in rural areas. Due to their preference for fragmented deciduous forests, this species was encountered most often in the Parkland NR. It was also fairly common in the southern part of the Boreal Forest NR and in the Foothills, Grassland, and Rocky Mountain NRs.

Declines in relative abundance were detected in the Boreal Forest, Foothills, Grassland, and Parkland NRs where, relative to other species, the House Wren was observed less frequently in Atlas 2 than in Atlas 1. These declines are surprising because House Wrens tend to nest in fragmented forests. The Breeding Bird Survey did not detect a change in abundance in Alberta for the period 1985–2005, but did detect a decrease nationally for the same time period. A national decline was also found for the period 1995–2005, which was matched in Saskatchewan. Although the biological cause for these declines is unknown, the BBS results do support the Atlas-detected declines. Further research is needed to assess the validity of the Atlas findings and, if validated, their biological cause. In contrast to the rest of the province, an increase in relative abundance was detected in the Rocky Mountain NR. Although prescribed burns in this NR may have contributed to this increase, the distribution maps suggest that better coverage in the southern parts of this NR was also a contributor to this finding. This species is considered Secure in Alberta.

TEMPORAL REPORTING RATE

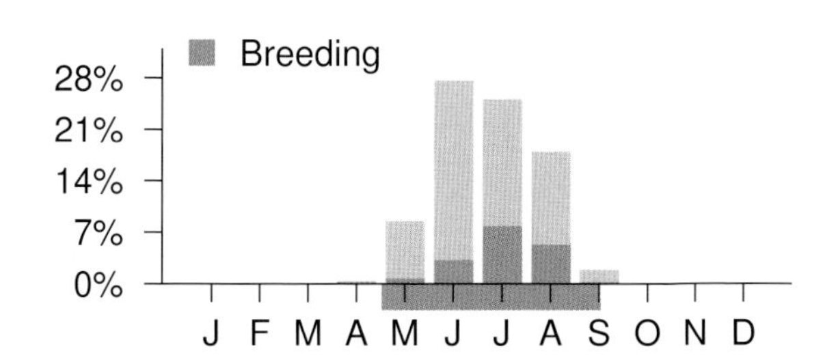

NATURAL REGIONS REPORTING RATE

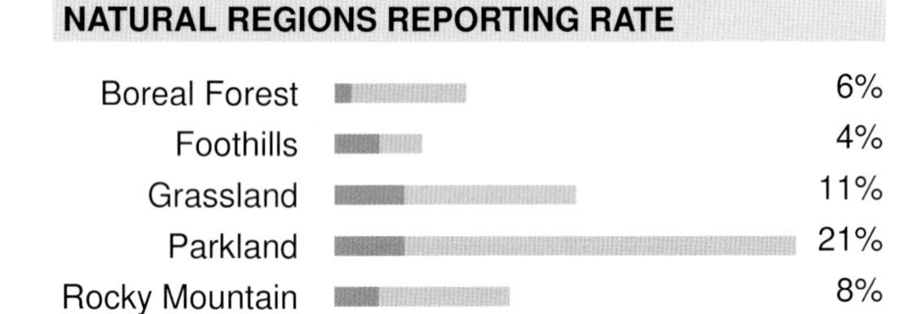

SPATIAL REPORTING RATE

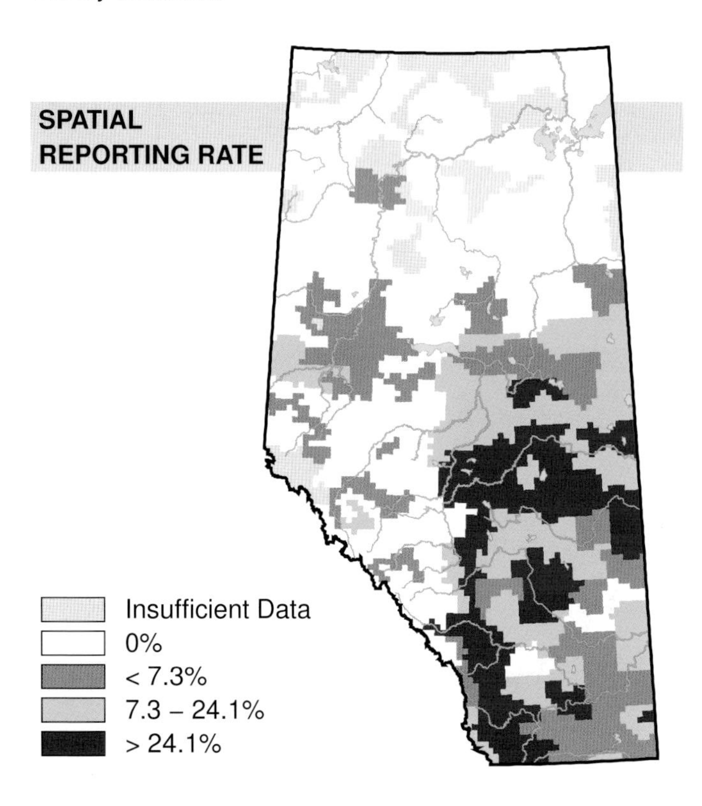

STATUS

GSWA 2000	GSWA 2005	COSEWIC Historic Rankings	COSEWIC 2007	Alberta Wildlife Act
Secure	Secure	N/A	N/A	N/A

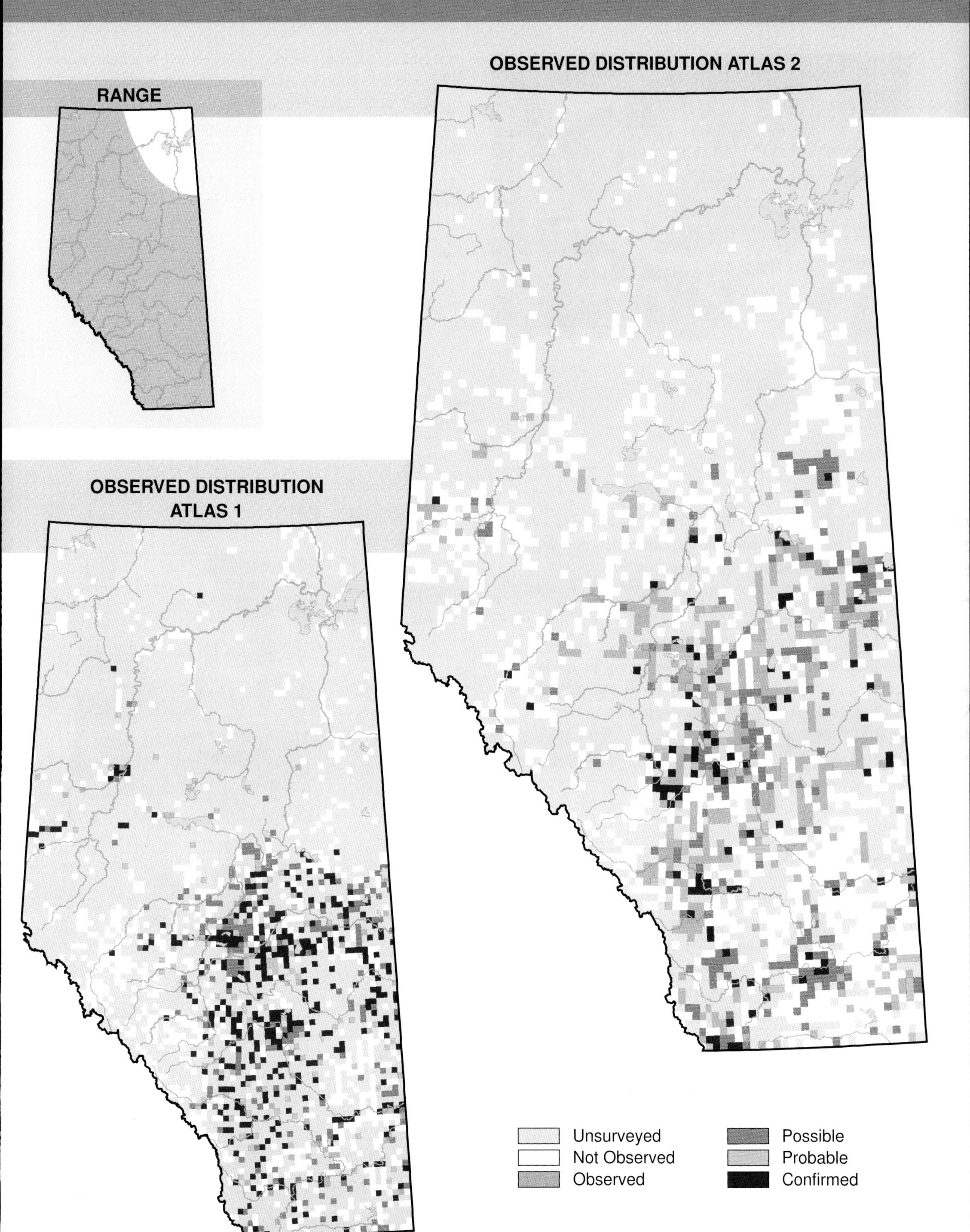

OBSERVED DISTRIBUTION ATLAS 2
RANGE
OBSERVED DISTRIBUTION ATLAS 1
Unsurveyed
Not Observed
Observed
Possible
Probable
Confirmed

Winter Wren (*Troglodytes troglodytes*)

Photo: Debbie Godkin

NESTING

CLUTCH SIZE (EGGS):	5–6
INCUBATION (DAYS):	14–16
FLEDGING (DAYS):	19
NEST HEIGHT (METRES):	0

Winter Wrens breed in the Boreal Forest, Foothills, and Rocky Mountain Natural Regions. There are some new possible breeding records west of Manning and the Peace River. However, these indicate no notable change in the observed distribution of this species between Atlas 1 and Atlas 2.

The Winter Wren is associated with dark, moist forests that are considered mature or old-growth. They are found frequently near water, particularly streams. As a result, this species was distributed fairly evenly across the Boreal Forest, Foothills, and Rocky Mountain NRs.

Relative abundance increased in the Boreal Forest, Foothills, and Rocky Mountain NRs and, relative to other species, the Winter Wren was observed more frequently in Atlas 2 than Atlas 1 in these NRs. The Breeding Bird Survey also found an increase in abundance in Canada in the interval 1985–2005. No abundance change was found by the Breeding Bird Survey for Alberta. An important component of the Winter Wren's breeding habitat includes dead wood in the form of fallen logs, standing dead trees, stumps, and slash piles. Forests have become older due to fire suppression in recent years which may, in part, account for the increase in abundance. In addition, survey coverage was more extensive in the north during Atlas 2; therefore, increases are also likely related, at least in part, to increased field effort. Forest harvesting and management, on the other hand, generally reduce the amount of older-aged forest, and so this activity may negatively affect populations in the future. This species is considered Secure in Alberta.

TEMPORAL REPORTING RATE

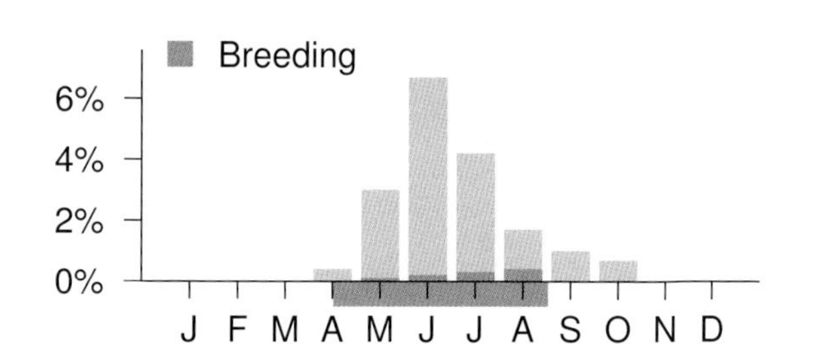

NATURAL REGIONS REPORTING RATE

SPATIAL REPORTING RATE

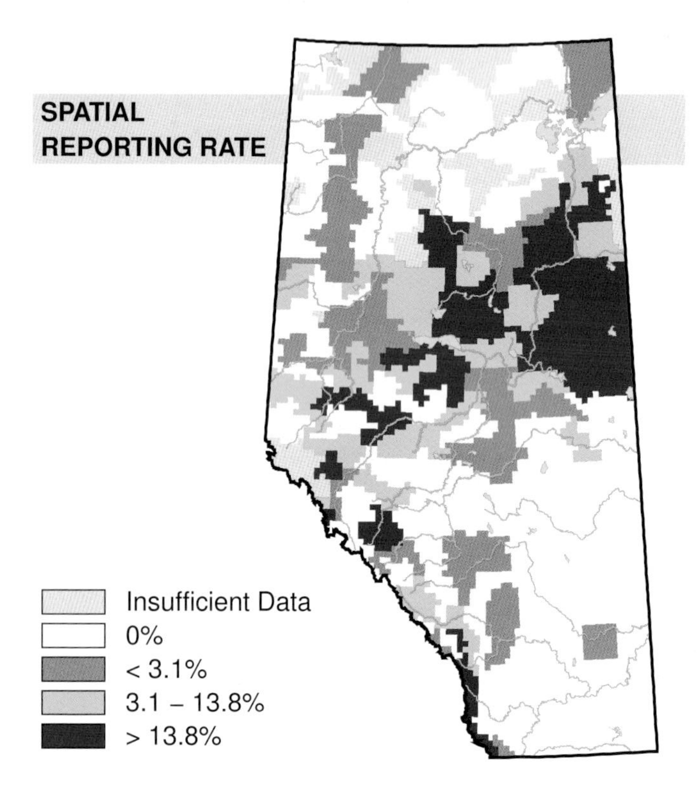

STATUS

GSWA 2000	GSWA 2005	COSEWIC Historic Rankings	COSEWIC 2007	Alberta Wildlife Act
Secure	Secure	N/A	N/A	N/A

RANGE

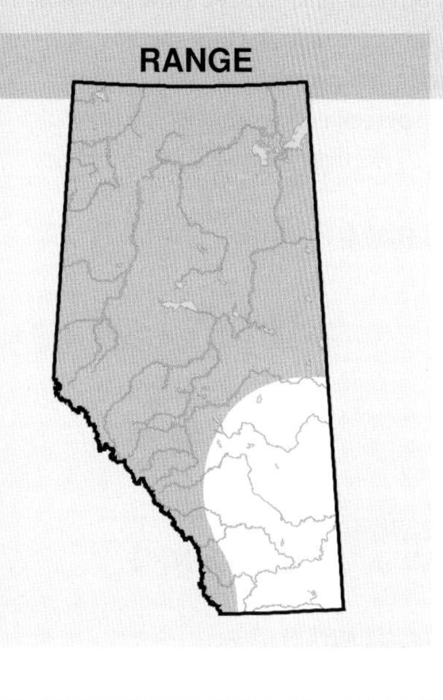

OBSERVED DISTRIBUTION ATLAS 2

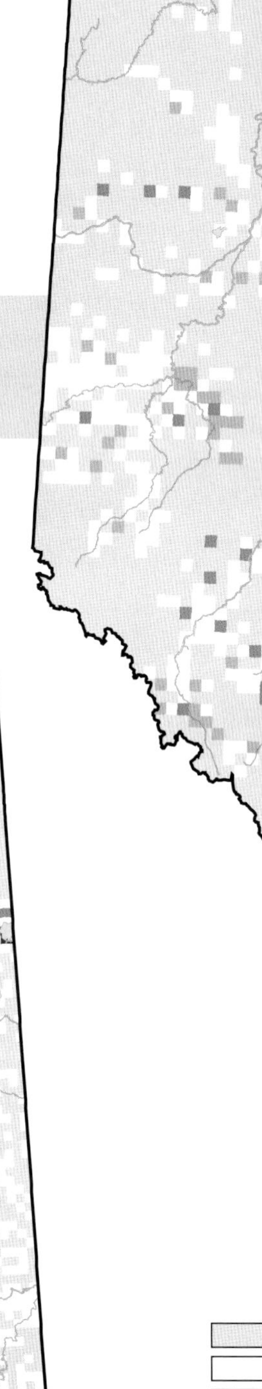

OBSERVED DISTRIBUTION ATLAS 1

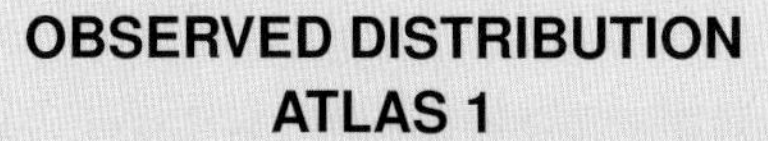

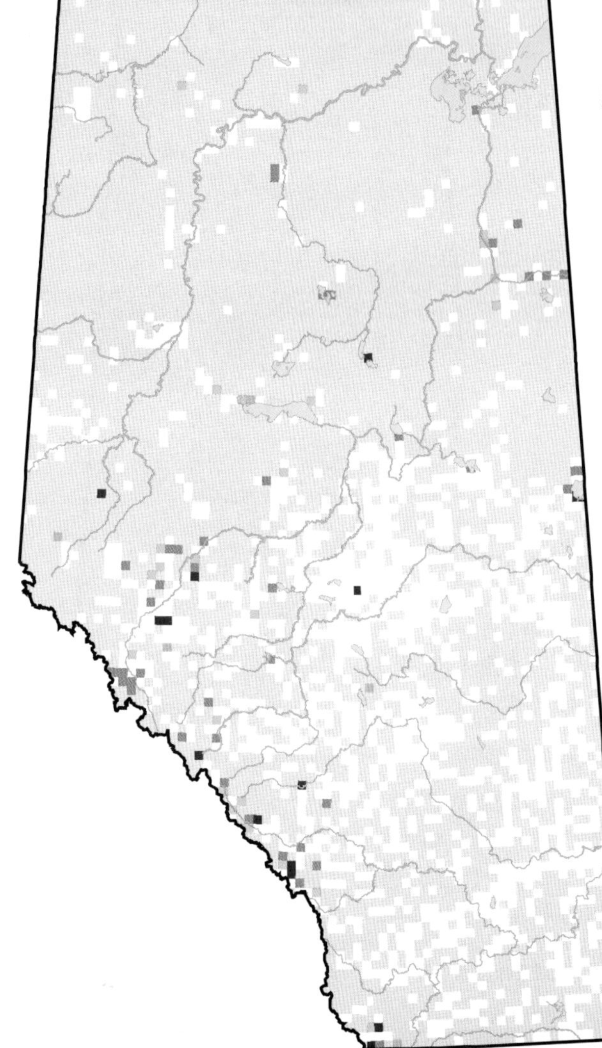

Sedge Wren (*Cistothorus platensis*)

Photo: Robert Gehlert

NESTING

CLUTCH SIZE (EGGS):	5–7
INCUBATION (DAYS):	12–14
FLEDGING (DAYS):	12–14
NEST HEIGHT (METRES):	0.2–0.5

The Sedge Wren is sparsely distributed across the east-central part of Alberta in the Boreal Forest, Foothills, and Parkland Natural Regions. The observed distribution of this species did not change between Atlas 1 and Atlas 2.

As its name implies, the Sedge Wren tends to nest on the edge of wetlands or in wet meadows in vegetation that is usually composed of sedges or tall grasses. Unlike the closely related Marsh Wren, the Sedge Wren tends to avoid wetlands that are dominated by cattails. The Sedge Wren breeds in areas that are unstable because these areas are susceptible to drying and flooding. As a result, this species is highly nomadic both between and within seasons. This species was found mainly near water in the western part of the Boreal Forest NR and in the Parkland NR. Records from the Foothills were those of non-breeders.

A decline in relative abundance was detected in the Boreal Forest NR where, relative to other species, the Sedge Wren was observed less frequently in Atlas 2 than in Atlas 1. Declines could be related to the loss and drainage of wetlands. It is easy to convert the wetland habitats that are preferred by the Sedge Wren to human use because the water in these meadows is usually shallow. No change in relative abundance was detected in the Parkland and Foothills NRs. The Breeding Bird Survey sample size was too small to detect trends for this species in Alberta. No trend was detected by the Breeding Bird Survey on a Canada-wide scale. This species is considered Sensitive in Alberta because it is sensitive to drought and wetland drainage.

TEMPORAL REPORTING RATE

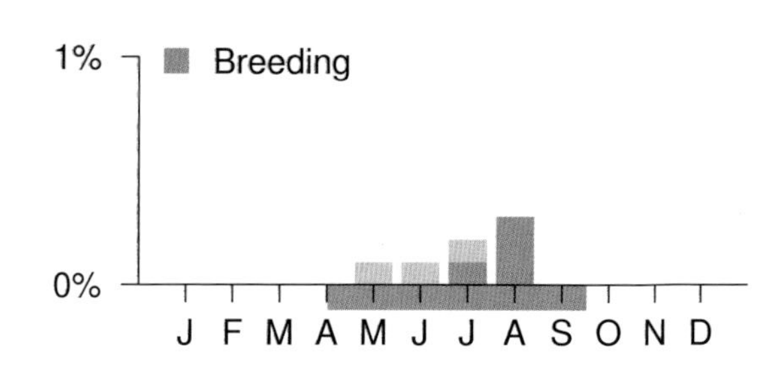

NATURAL REGIONS REPORTING RATE

SPATIAL REPORTING RATE

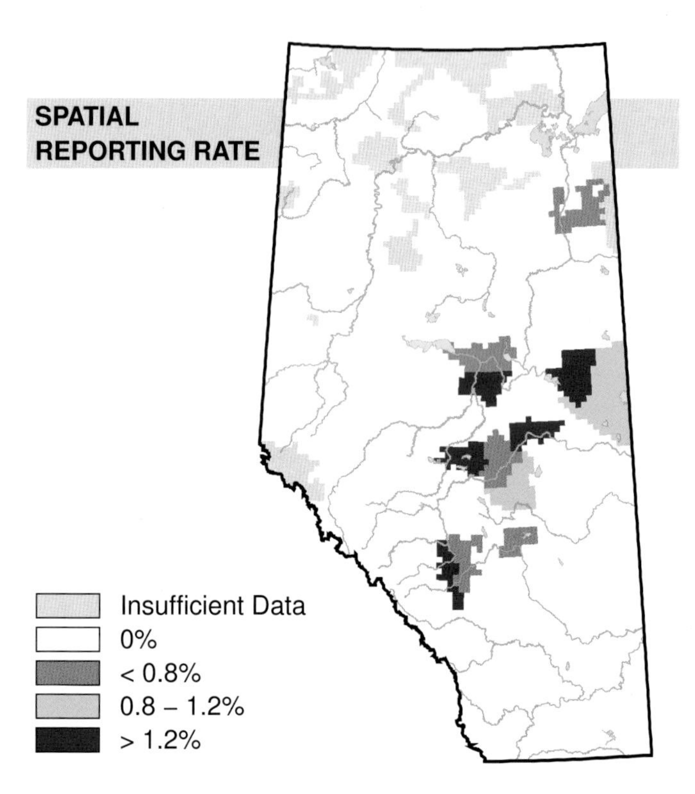

STATUS

GSWA 2000	GSWA 2005	COSEWIC Historic Rankings	COSEWIC 2007	Alberta Wildlife Act
Sensitive	Sensitive	Not At Risk	Not At Risk	N/A

Habitat

Nest
Location

Nest
Type

Diet

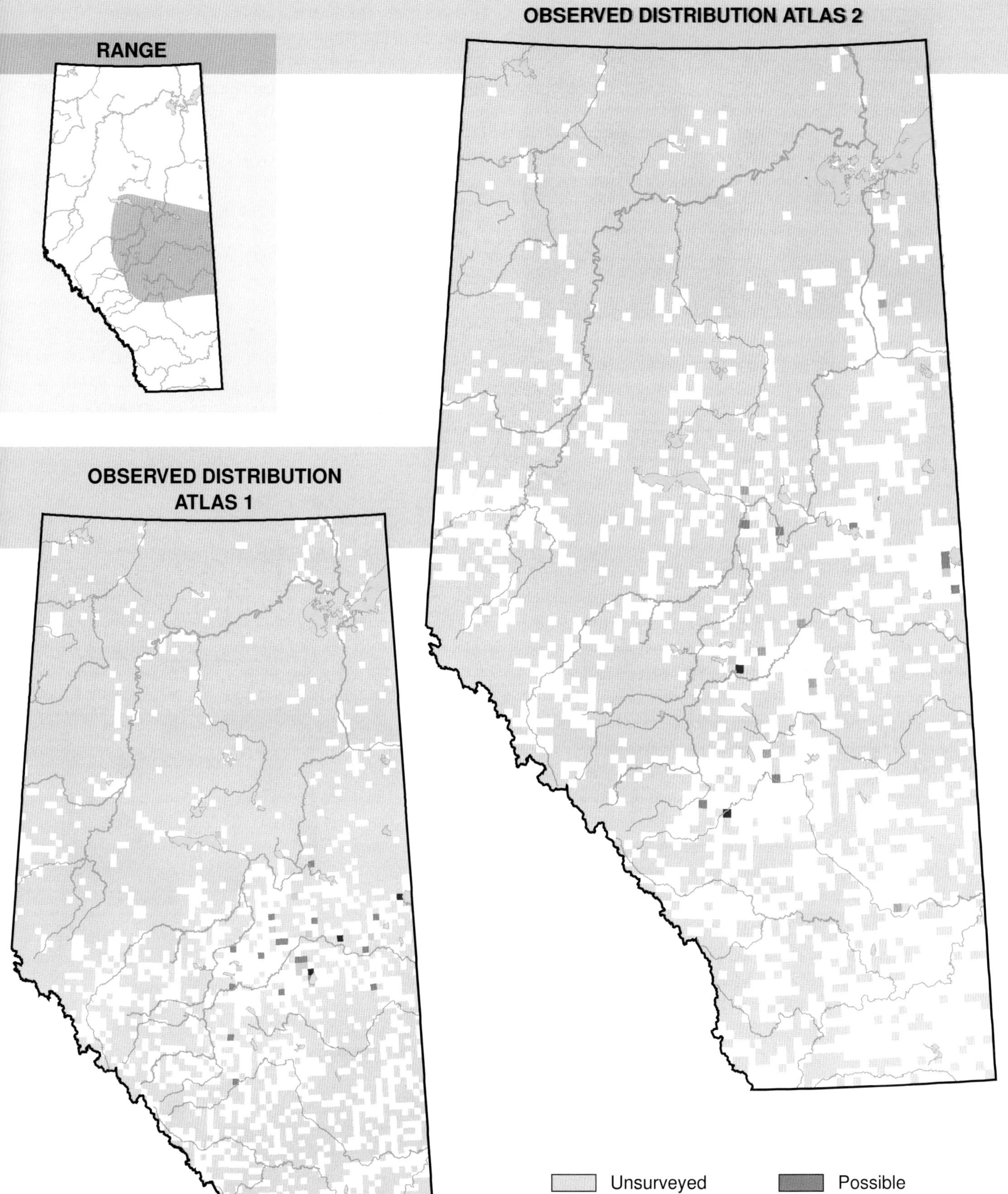
OBSERVED DISTRIBUTION ATLAS 2
RANGE
OBSERVED DISTRIBUTION
ATLAS 1
Unsurveyed
Not Observed
Observed
Possible
Probable
Confirmed

Marsh Wren (*Cistothorus palustris*)

Photo: Dawn Hall

NESTING

CLUTCH SIZE (EGGS):	4–6
INCUBATION (DAYS):	12–14
FLEDGING (DAYS):	13–15
NEST HEIGHT (METRES):	0.5–0.9

The Marsh Wren is found across the province in every Natural Region. The observed distribution of this species did not change between Atlas 1 and Atlas 2.

As its name implies, the Marsh Wren is closely associated with wetlands. Nests are built above the water in emergent vegetation that often consists of cattails or bulrushes. Marshes are the preferred habitat of this species at all times of the year including the breeding season, migration, and during the winter. This species feeds opportunistically on terrestrial and aquatic invertebrates among the emergent vegetation within marsh habitats. Due to its small size and secretive nature, this species is not often seen. During the breeding season, however, the loud sewing machine-type breeding call often gives away the presence of these small birds. Most of the breeding evidence for this species was gathered from the Parkland and Foothills NRs, but the Marsh Wren was also fairly common in the other NRs.

Declines in relative abundance were detected in the Boreal Forest and Parkland NRs where, relative to other species, Marsh Wrens were observed less often in Atlas 2 than in Atlas 1. These declines are likely the result of drier conditions in these NRs during Atlas 2. In Atlas 1, quite a few Marsh Wrens were recorded in the east-central part of the province, especially around Beaverhill Lake. However, during Atlas 2 this region was quite dry and few records of this species were obtained. An increase in relative abundance was detected in the Grassland NR. This increase may be a result of wetter conditions in this NR in Atlas 2 than in Atlas 1. The Breeding Bird Survey did not find a change in abundance in Alberta or across Canada. Together, these findings suggest the possibility that the population of Marsh Wrens is shifting spatially in response to changing climate conditions, and not actually changing size. This species is considered Secure in Alberta.

TEMPORAL REPORTING RATE

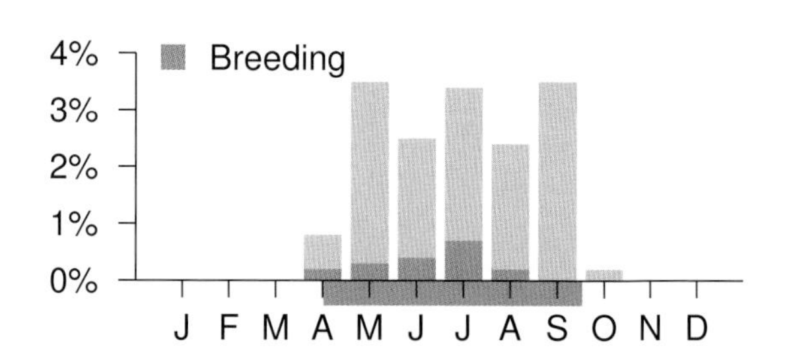

NATURAL REGIONS REPORTING RATE

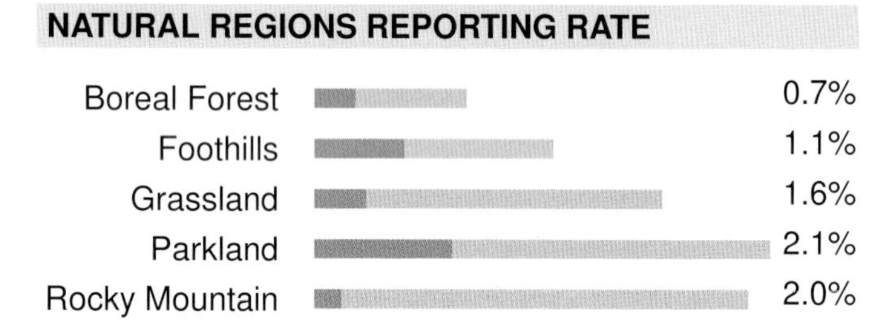

SPATIAL REPORTING RATE

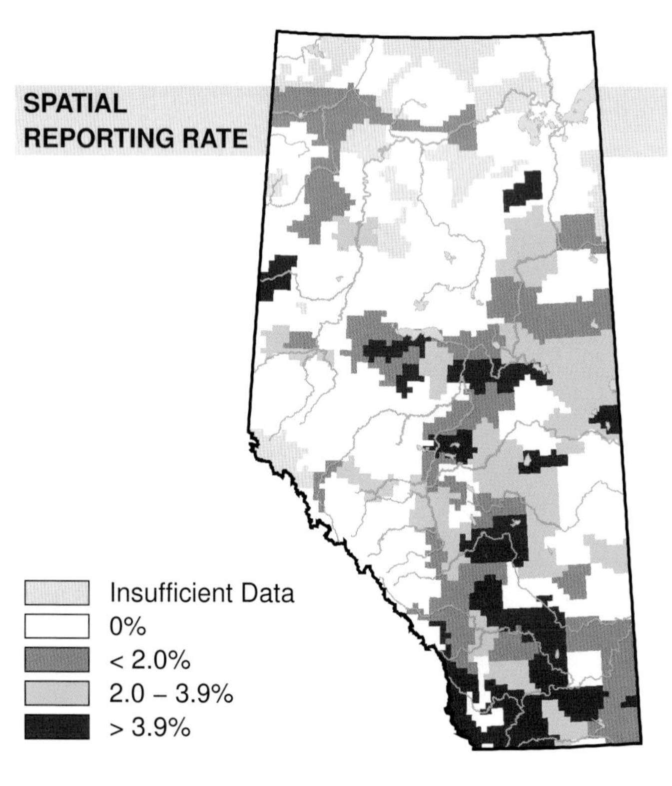

STATUS

GSWA 2000	GSWA 2005	COSEWIC Historic Rankings	COSEWIC 2007	Alberta Wildlife Act
Secure	Secure	N/A	N/A	N/A

Habitat  Nest Location  Nest Type  Diet 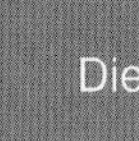 

RANGE

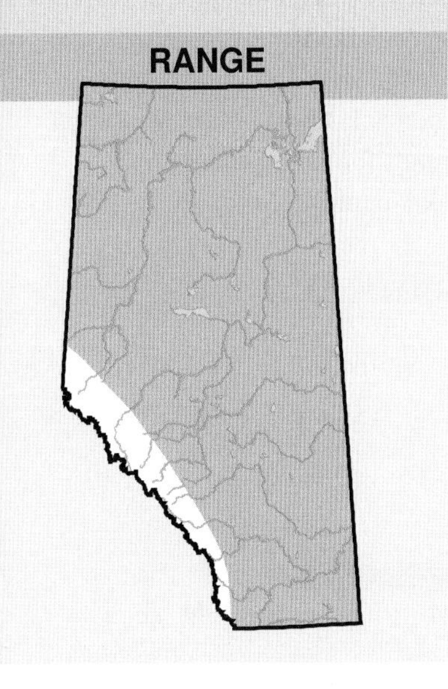

OBSERVED DISTRIBUTION ATLAS 2

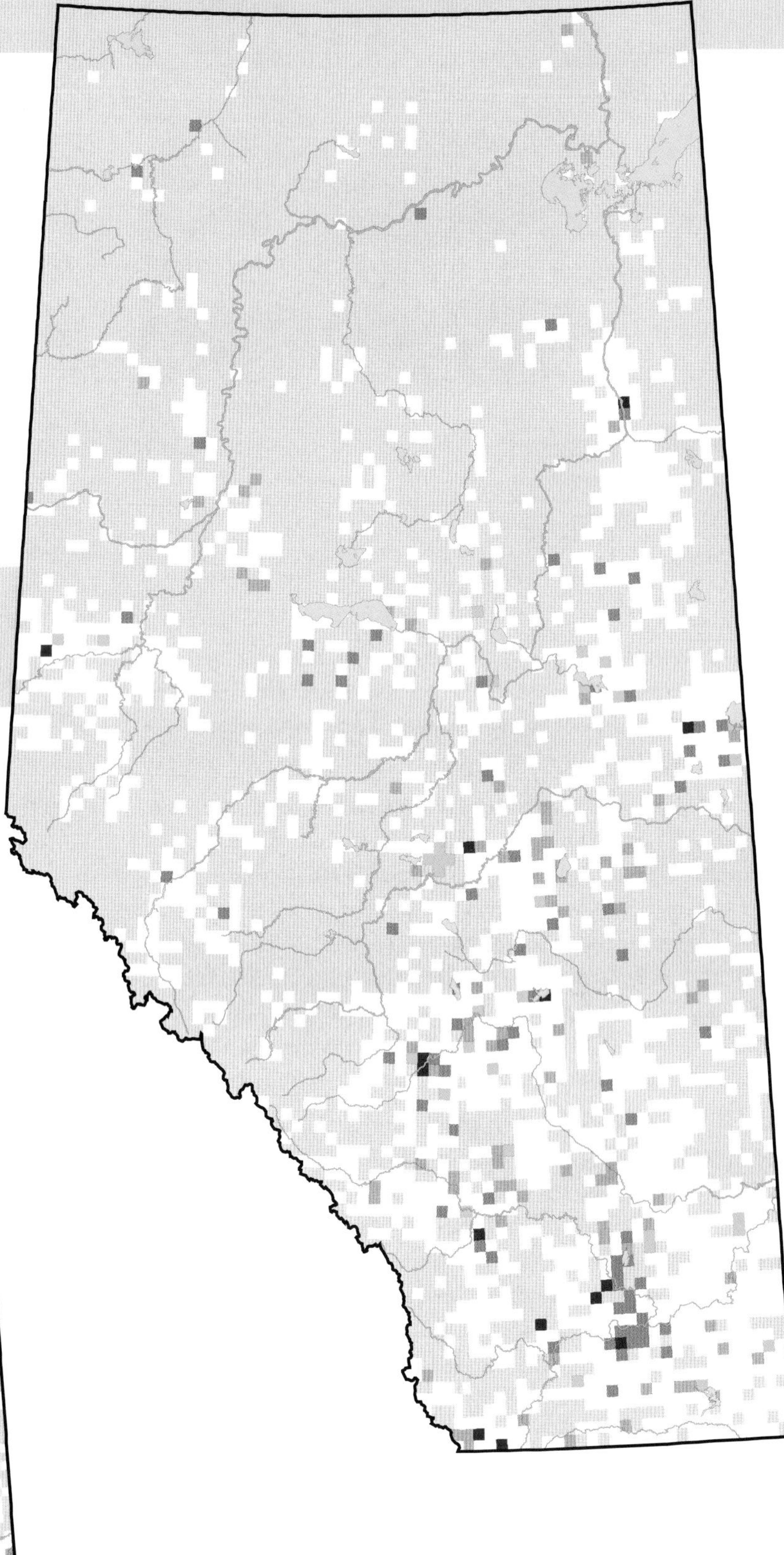

OBSERVED DISTRIBUTION ATLAS 1

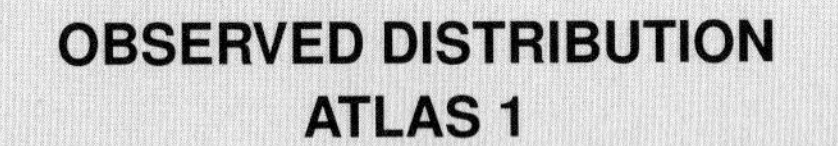

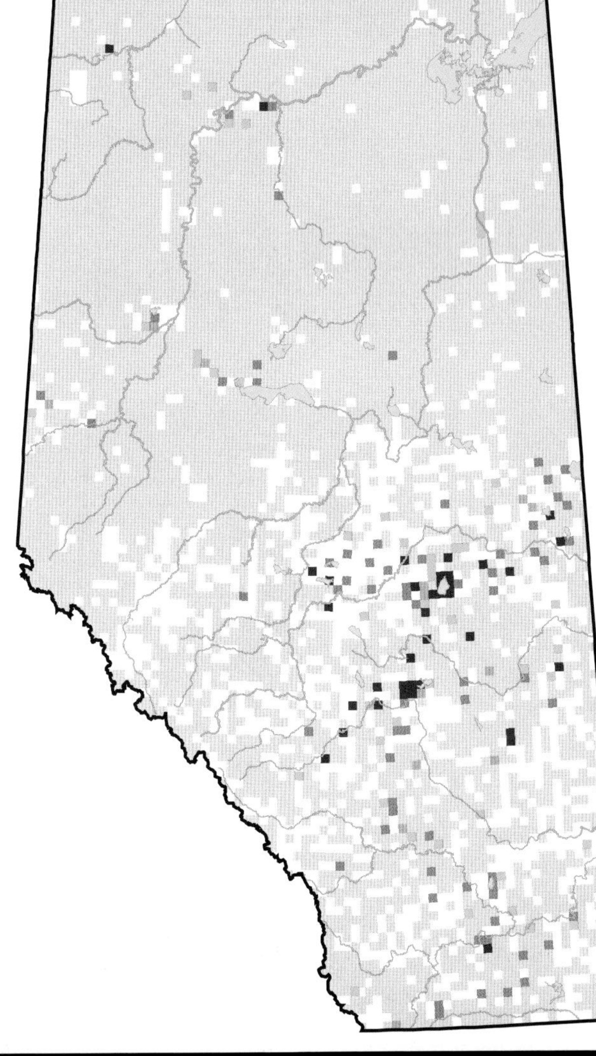

Unsurveyed	Possible
Not Observed	Probable
Observed	Confirmed

American Dipper (*Cinclus mexicanus*)

Photo: Raymond Toal

NESTING

CLUTCH SIZE (EGGS):	3–6
INCUBATION (DAYS):	18
FLEDGING (DAYS):	24–26
NEST HEIGHT (METRES):	0

The American Dipper is found in the Rocky Mountain and Foothills Natural Regions in Alberta. The observed distribution of this species did not change between Atlas 1 and Atlas 2.

The American Dipper is closely associated with fast-moving streams. Preferred streams tend to be clear, to contain rocks on the bottom, and to have banks that consist of cliffs, boulders, or overhanging logs and trees. This species nests close to these streams, usually below overhanging structures beside the water. The American Dipper feeds mainly on aquatic invertebrates which it catches in the water while walking, swimming, or diving. Because of its requirement for fast-moving streams, this species was found mainly in the Rocky Mountain NR, but was found occasionally in the Foothills and Parkland NRs along the perimeter of the Rocky Mountain NR.

Relative abundance changes were not detected in any NR. However, this result should be interpreted with caution: the sample size for American Dippers was small because, in Alberta, they tend to be found in remote areas. Relative to other species, American Dippers were observed more frequently in Atlas 2 than in Atlas 1 in the Rocky Mountain and Foothills NRs. The Breeding Bird Survey sample size was too small to investigate abundance change in Alberta; there was no change detected on a Canada-wide scale. Due to the remote distribution of this species' breeding range, it is clearly not monitored effectively by wide-spread, multi-species surveys such as the Atlas or Breeding Bird Survey. A survey that specifically targets the American Dipper would be needed to determine effectively any trends for this species. This species is considered Secure in Alberta.

TEMPORAL REPORTING RATE

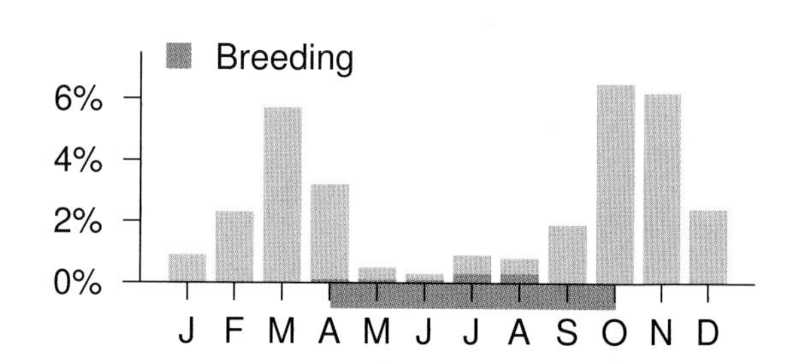

NATURAL REGIONS REPORTING RATE

SPATIAL REPORTING RATE

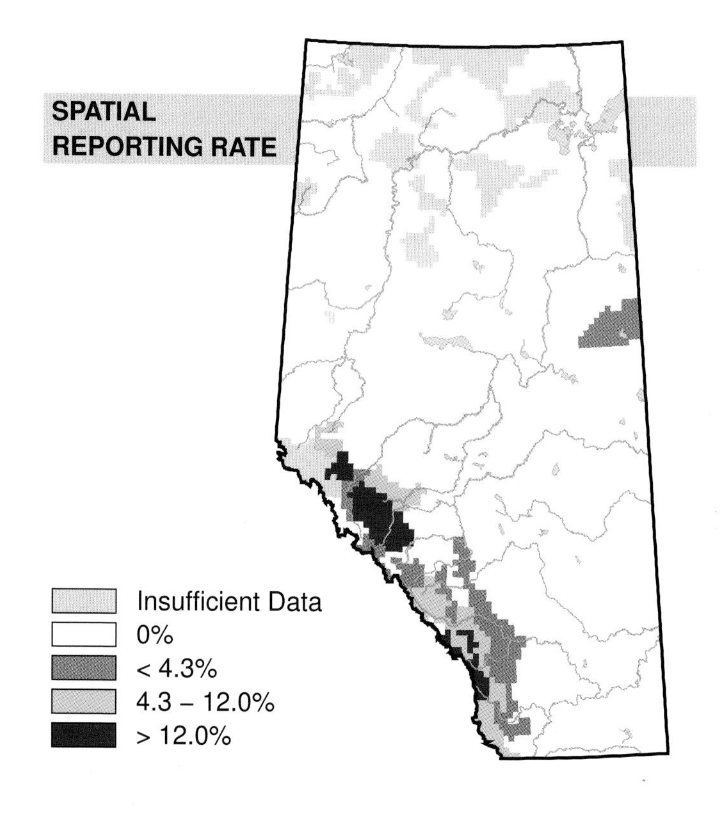

STATUS

GSWA 2000	GSWA 2005	COSEWIC Historic Rankings	COSEWIC 2007	Alberta Wildlife Act
Secure	Secure	N/A	N/A	N/A

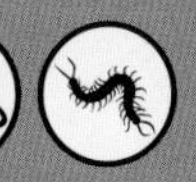

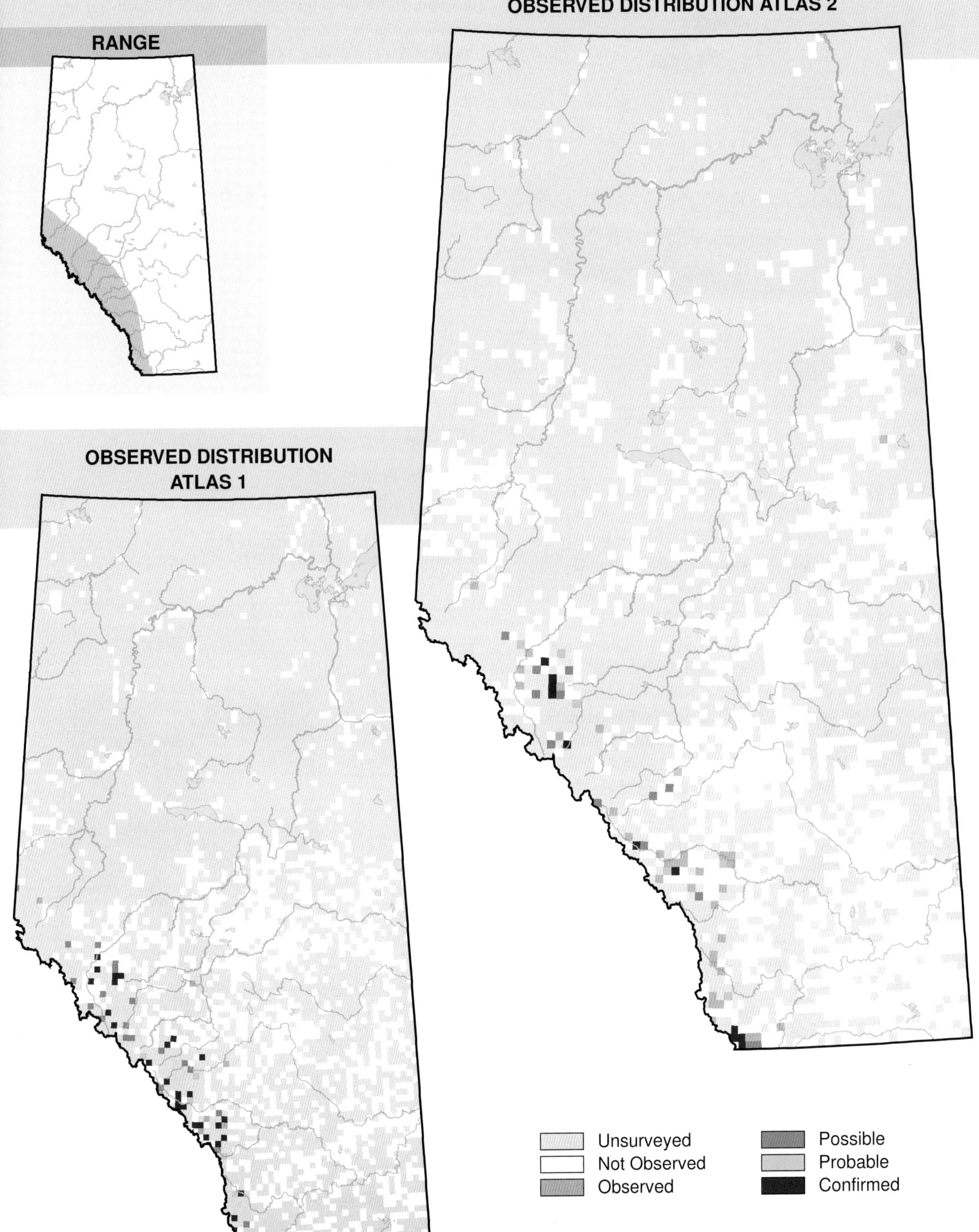
RANGE
OBSERVED DISTRIBUTION ATLAS 2
OBSERVED DISTRIBUTION ATLAS 1
Unsurveyed
Not Observed
Observed
Possible
Probable
Confirmed

Golden-crowned Kinglet (*Regulus satrapa*)

Photo: Gerald Romanchuk

NESTING

CLUTCH SIZE (EGGS):	5–10
INCUBATION (DAYS):	14–15
FLEDGING (DAYS):	14–19
NEST HEIGHT (METRES):	>1.5

The Golden-crowned Kinglet is found in the Boreal Forest, Foothills, and Rocky Mountain Natural Regions in Alberta. Many more observations of this species were recorded in the Boreal Forest NR during Atlas 2. In addition, breeding evidence was found for this species in the Cypress Hills, where breeding evidence was not found in Atlas 1.

One of the province's smallest songbirds, the Golden-crowned Kinglet nests mainly in coniferous forests. However, it will occasionally nest in mixedwood and deciduous forests. This species does not seem to be adversely affected by forest fragmentation, but population densities tend to diminish in areas that have been harvested or where forest fires have occurred. Due to its preference for coniferous trees this species was mainly found in the Foothills and Rocky Mountain NRs

Increases in relative abundance were detected in the Grassland and Boreal Forest NRs where, relative to other species, the Golden-crowned Kinglet was found more frequently in Atlas 2 than in Atlas 1. The increase in the Grassland NR is most likely a function of migratory variation as most of the records from this NR were of non-breeders. This species' habitat requirements are primarily limited to conifer availability. Therefore, in regard to the increase of this kinglet species in the Boreal Forest NR, forest regeneration activities after harvesting have likely provided additional suitable habitat for this bird. However, in part, this increase is probably related to the more complete coverage of the north in Atlas 2. No change was detected in the other NRs. The Breeding Bird Survey did not find an abundance change in Alberta. However, an abundance increase was detected on a Canada-wide scale during the period 1968–2005. This species is considered Secure in Alberta.

TEMPORAL REPORTING RATE

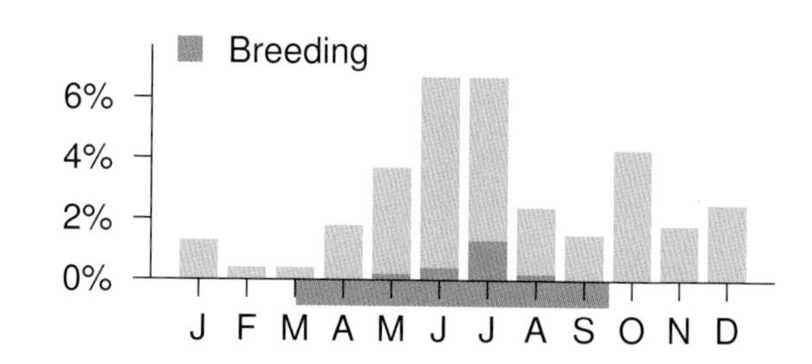

NATURAL REGIONS REPORTING RATE

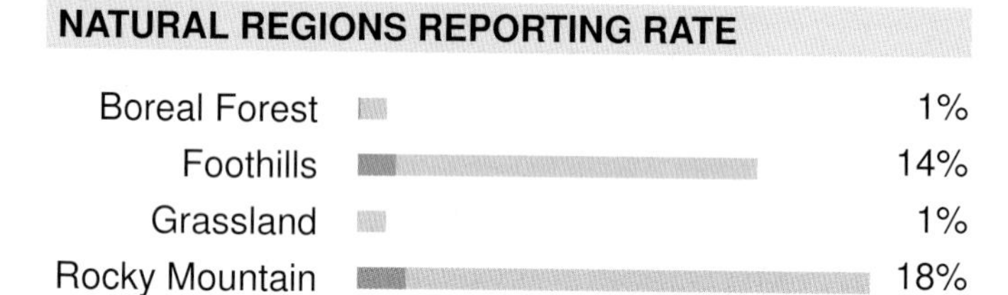

SPATIAL REPORTING RATE

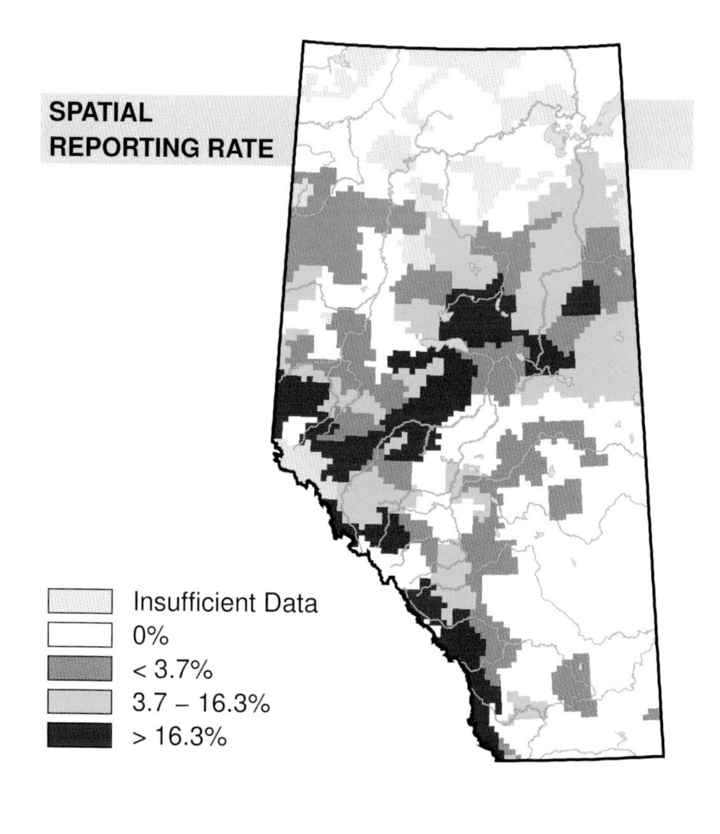

STATUS

GSWA 2000	GSWA 2005	COSEWIC Historic Rankings	COSEWIC 2007	Alberta Wildlife Act
Secure	Secure	N/A	N/A	N/A

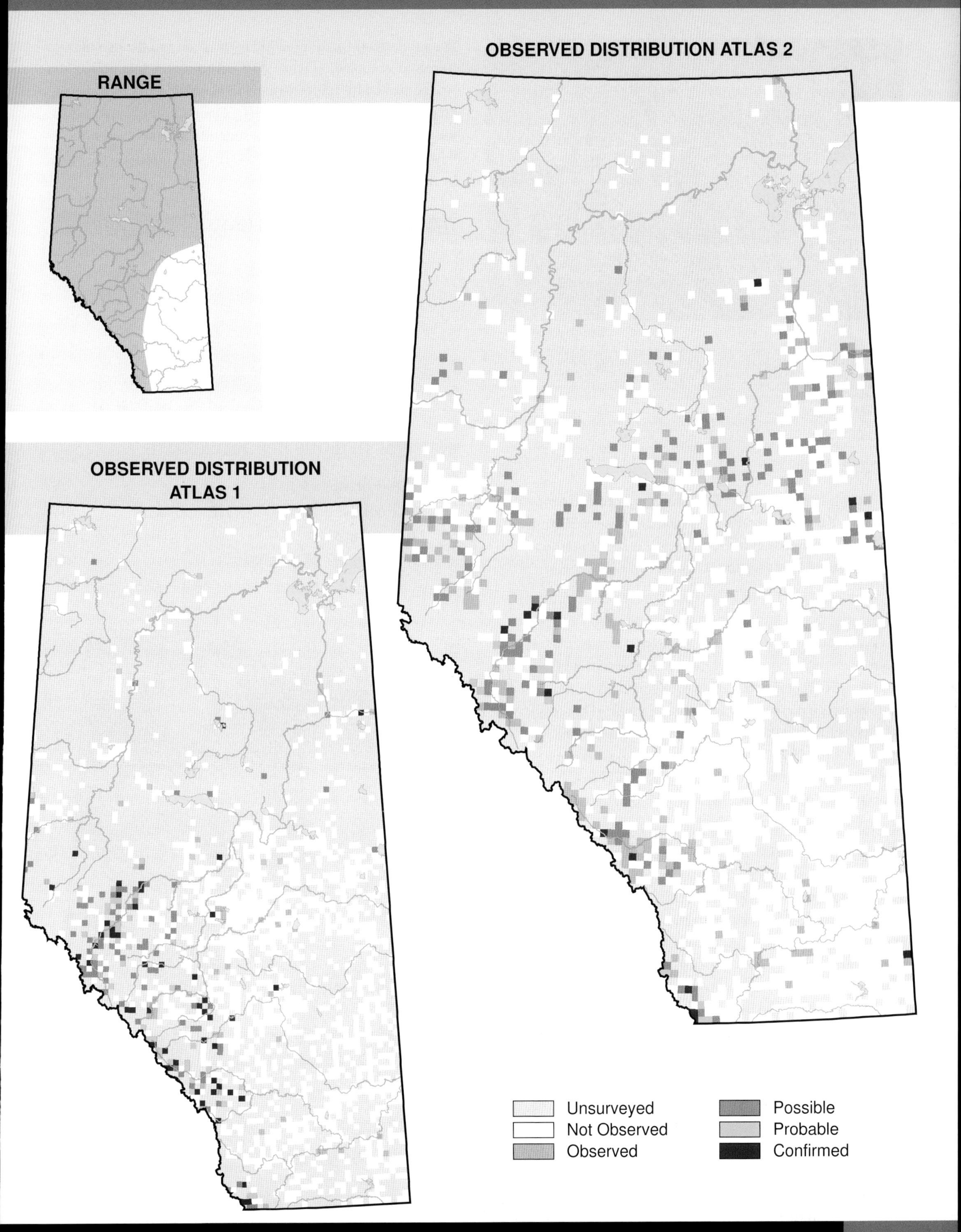
OBSERVED DISTRIBUTION ATLAS 2
RANGE
OBSERVED DISTRIBUTION ATLAS 1
Unsurveyed
Not Observed
Observed
Possible
Probable
Confirmed

Ruby-crowned Kinglet (*Regulus calendula*)

Photo: Royal Alberta Museum

NESTING

CLUTCH SIZE (EGGS):	7–8
INCUBATION (DAYS):	12–15
FLEDGING (DAYS):	12
NEST HEIGHT (METRES):	0.6–30.5

The Ruby-crowned Kinglet breeds in the Boreal Forest, Foothills, Parkland, and Rocky Mountain Natural Regions in Alberta. The observed distribution of this species did not change between Atlas 1 and Atlas 2.

The Ruby-crowned Kinglet breeds in coniferous or mixedwood forests that have a large coniferous component. This species tends to build its nest far above the ground among the top branches of conifers where it can be well concealed from predators. This also makes it difficult to survey or gather reproductive data (Ingold and Wallace, 1994). This species also tends to forage near the tops of densely distributed coniferous trees. Due to its preference for coniferous forests, this species was found mainly in the Boreal Forest, Foothills, and Rocky Mountain NRs, and it was found only infrequently in the Grassland and Parkland NRs.

An increase in relative abundance was detected in the Boreal Forest NR where, relative to other species, the Ruby-crowned Kinglet was observed more frequently in Atlas 2 than Atlas 1. This increase could be due, in part, to more extensive northern coverage during Atlas 2. Relative abundance changes were not detected in the other NRs. Similar to what was found for the Boreal Forest NR by the Atlas, the Breeding Bird Survey found an increase in Alberta during the period 1968–2005 and across Canada during the period 1985–2005.

There is an overall North American decline in the BBS trend for Ruby-crowned Kinglets that is the result of the significant declining trend for 1980–2000 in the U.S., but this is not mirrored in Canada (Dunn, 2005). The continental population was impacted by severe winters in its southern U.S. wintering areas (Ingold and Wallace, 1994). Dunn (2005) contends that the decline will not persist and suggests that there are no current conservation needs. This species is considered Secure in Alberta.

TEMPORAL REPORTING RATE

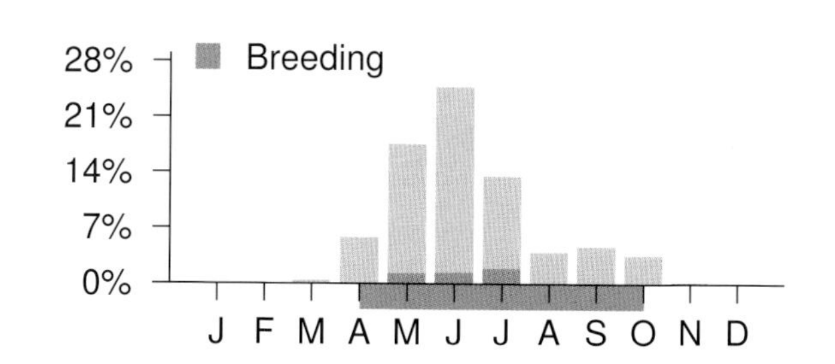

NATURAL REGIONS REPORTING RATE

Region	Rate
Boreal Forest	14%
Foothills	40%
Grassland	2%
Parkland	2%
Rocky Mountain	56%

SPATIAL REPORTING RATE

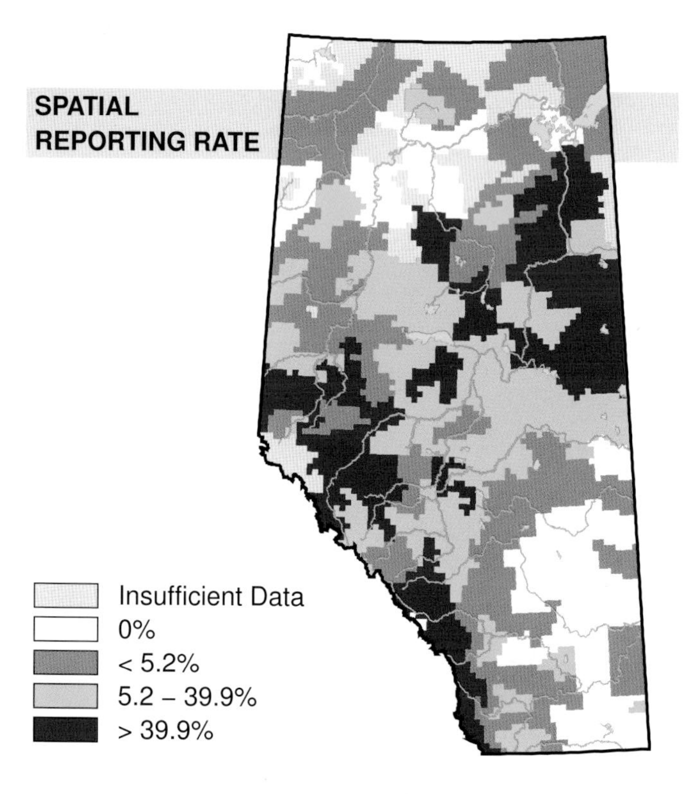

STATUS

GSWA 2000	GSWA 2005	COSEWIC Historic Rankings	COSEWIC 2007	Alberta Wildlife Act
Secure	Secure	N/A	N/A	N/A

Habitat   Nest Location  Nest Type  Diet

RANGE

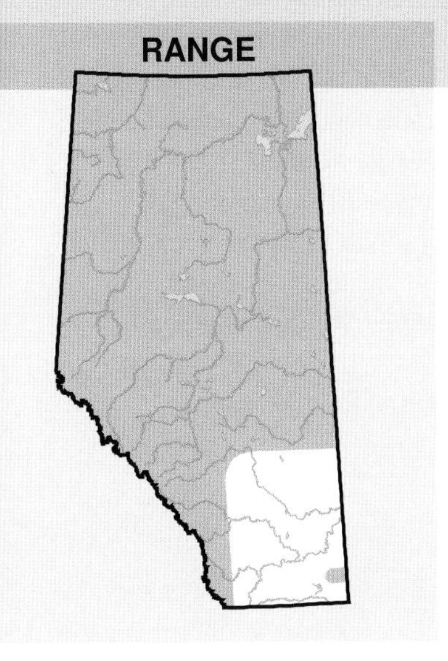

OBSERVED DISTRIBUTION ATLAS 2

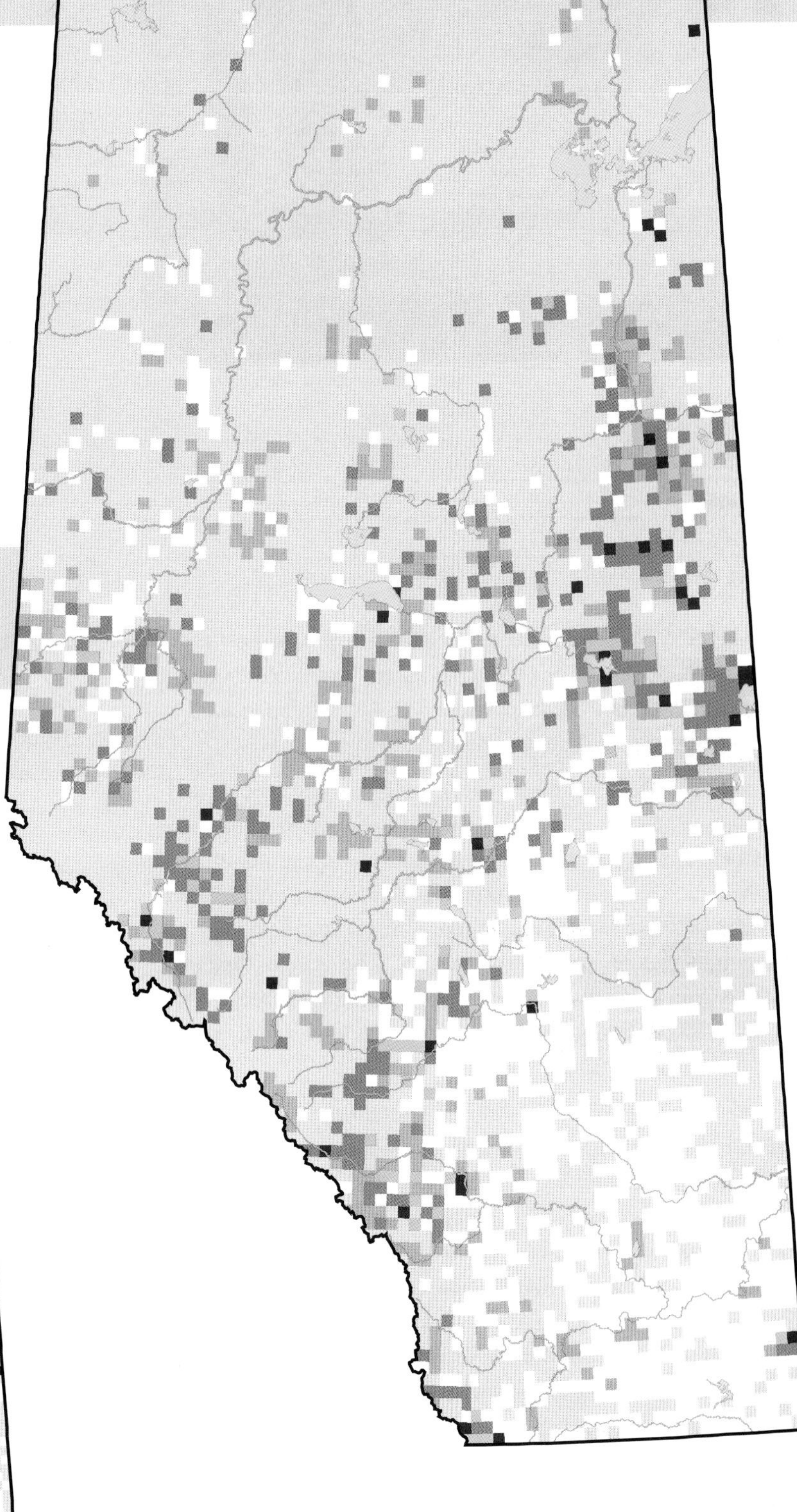

OBSERVED DISTRIBUTION ATLAS 1

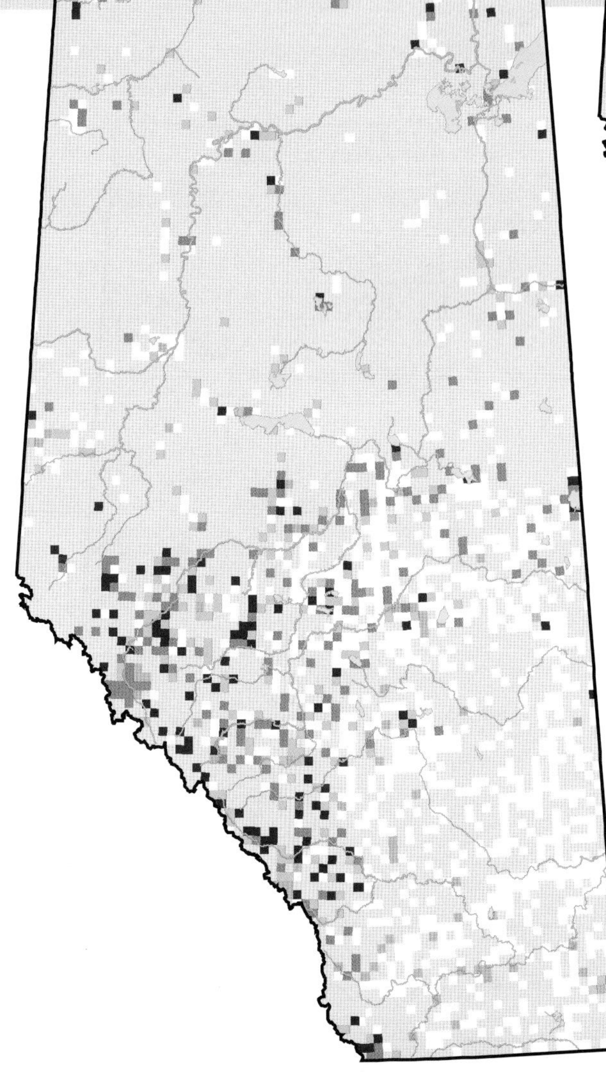

	Unsurveyed		Possible
	Not Observed		Probable
	Observed		Confirmed

Eastern Bluebird (*Sialia sialis*)

Photo: Isidor Jeklin / Cornell Laboratory of Ornithology

NESTING

CLUTCH SIZE (EGGS):	4
INCUBATION (DAYS):	12–16
FLEDGING (DAYS):	15–20
NEST HEIGHT (METRES):	0.6–15

The Eastern Bluebird is rarely encountered in Alberta. In Canada, the range of this species typically extends west only as far as south-central Saskatchewan. The distribution of this species was patchy in both Atlas 1 and Atlas 2.

Eastern Bluebirds are a secondary cavity-nesting species that often uses nestboxes. The presence of this species would be easily detected because there is extensive nestbox monitoring in Alberta. Despite the potential for both Eastern and Western bluebirds to use the boxes, the reporting rate for Eastern Bluebirds was very low during Atlas 2. This species tends to prefer open habitats; therefore, they were encountered in the Grassland and Parkland NRs. No breeding evidence was found for this species during Atlas 2.

Changes in relative abundance were not detected for this species in any natural region in Alberta. However, the rarity of this species placed it below the change detection threshold of our statistical methods. Similarly, the Breeding Bird Survey sample size was too small to investigate Eastern Bluebird abundance change in Alberta, but an increase in abundance was detected on a national scale in Canada during the period 1985–2005. This increase was likely the result of the conversion of forests into open habitat and the increasing number of nestboxes that are available. Alberta Breeding Bird Survey and Atlas sample sizes were small because this species is at the periphery of its range in Alberta.

LaDeau et al. (2007) reported that the Eastern Bluebird suffered quite significant declines after 2002, dropping down by 44%, which is tightly correlated with the human epidemic of West Nile virus in the northeastern United States. This species had not been studied previously in lab or field experiments in regard to its West Nile virus exposure; so the fact that this species was so significantly impacted was a surprise. This species is considered Secure in Alberta.

TEMPORAL REPORTING RATE

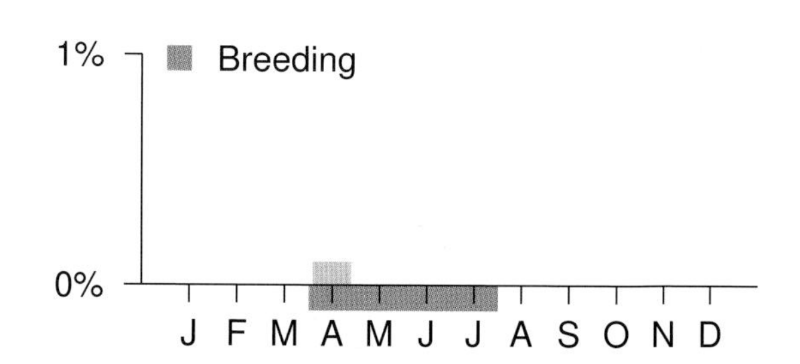

NATURAL REGIONS REPORTING RATE

SPATIAL REPORTING RATE

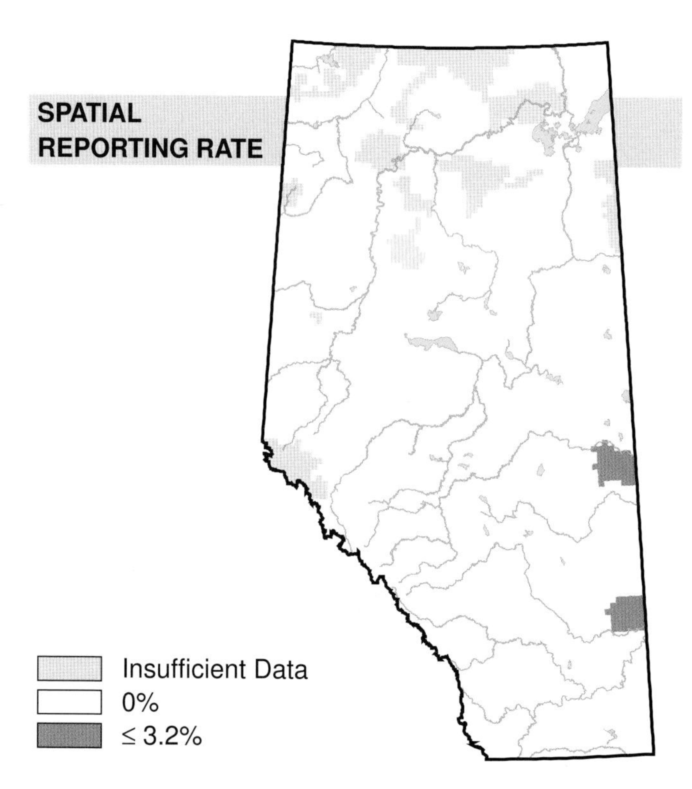

STATUS

GSWA 2000	GSWA 2005	COSEWIC Historic Rankings	COSEWIC 2007	Alberta Wildlife Act
Secure	Secure	Not At Risk	Not At Risk	N/A

Habitat
Nest Location
Nest Type
Diet

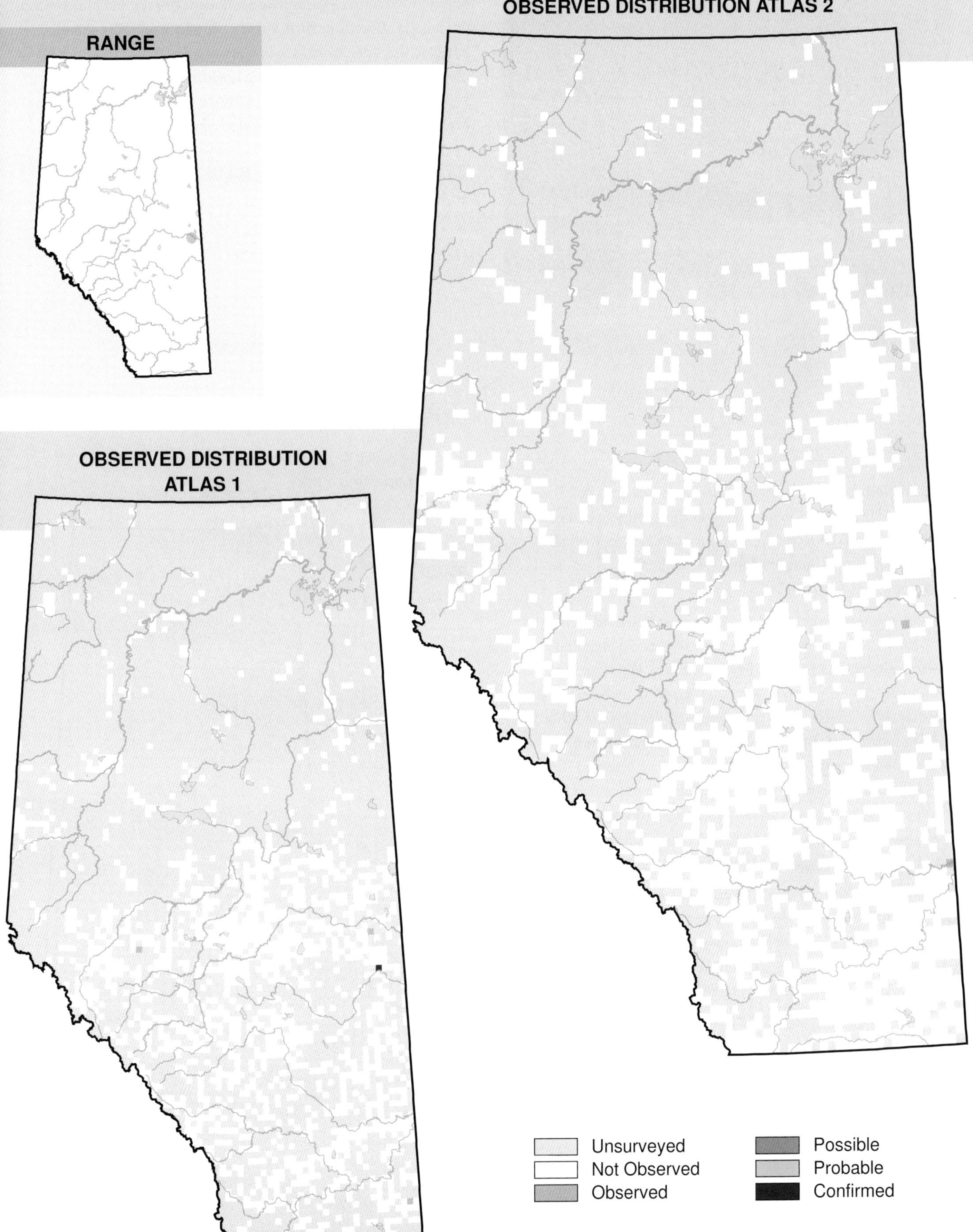

RANGE
OBSERVED DISTRIBUTION ATLAS 2
OBSERVED DISTRIBUTION ATLAS 1
Unsurveyed
Not Observed
Observed
Possible
Probable
Confirmed

Western Bluebird (*Sialia mexicana*)

Photo: Alan MacKeigan

NESTING

CLUTCH SIZE (EGGS):	4–6
INCUBATION (DAYS):	12–16
FLEDGING (DAYS):	18–25
NEST HEIGHT (METRES):	1–9

The Western Bluebird is one of the rarest thrushes in Alberta. This species is occasionally found in the southwestern part of the province in the Rocky Mountain and Parkland Natural Regions. The distribution of this species remained patchy between Atlas 1 and Atlas 2.

Western Bluebird records often have breeding evidence associated with them. This is because Bluebird breeding season is extensively monitored in Alberta. Bird banders, naturalists, and rural landowners have put up thousands of nestboxes for bluebirds in many parts of the province. For this reason, if Western Bluebird populations were expanding, the increase in their distribution would likely be detected. It does not appear that the range of this species has expanded because confirmed breeding occurred only in the Parkland NR, and there were a couple of non-breeding observations made in the southern part of the Rocky Mountain NR during Atlas 2.

Changes in relative abundance were not detected for this species in any natural region in Alberta. However, with such low reporting rates, any change would have been beyond the detection threshold of the analytical methods used in the Atlas. Neither were there enough Western Bluebird records to detect population changes using Breeding Bird Survey data in Alberta or Canada. Breeding Bird Survey sample sizes were small because this species is rare in Canada. The Western Bluebird is much more common during the breeding season in the western and southwestern United States and in northern and central Mexico. This species is considered Secure in Alberta.

TEMPORAL REPORTING RATE

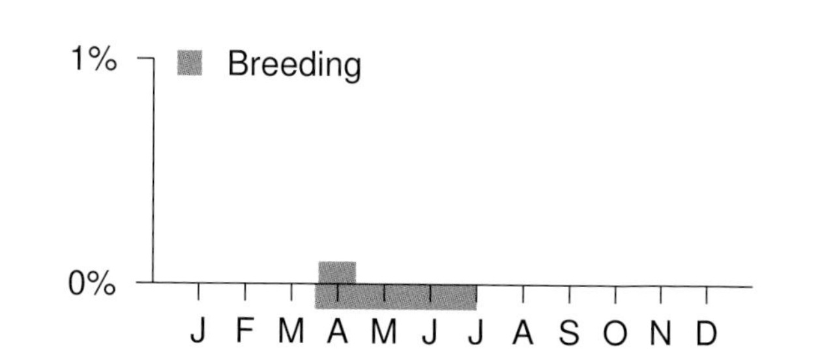

NATURAL REGIONS REPORTING RATE

SPATIAL REPORTING RATE

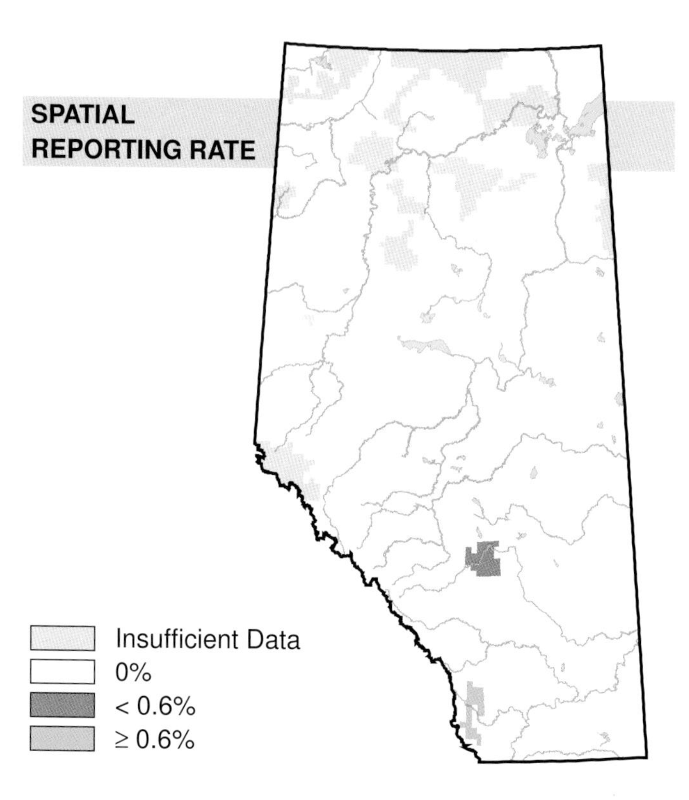

STATUS

GSWA 2000	GSWA 2005	COSEWIC Historic Rankings	COSEWIC 2007	Alberta Wildlife Act
Secure	Secure	N/A	N/A	N/A

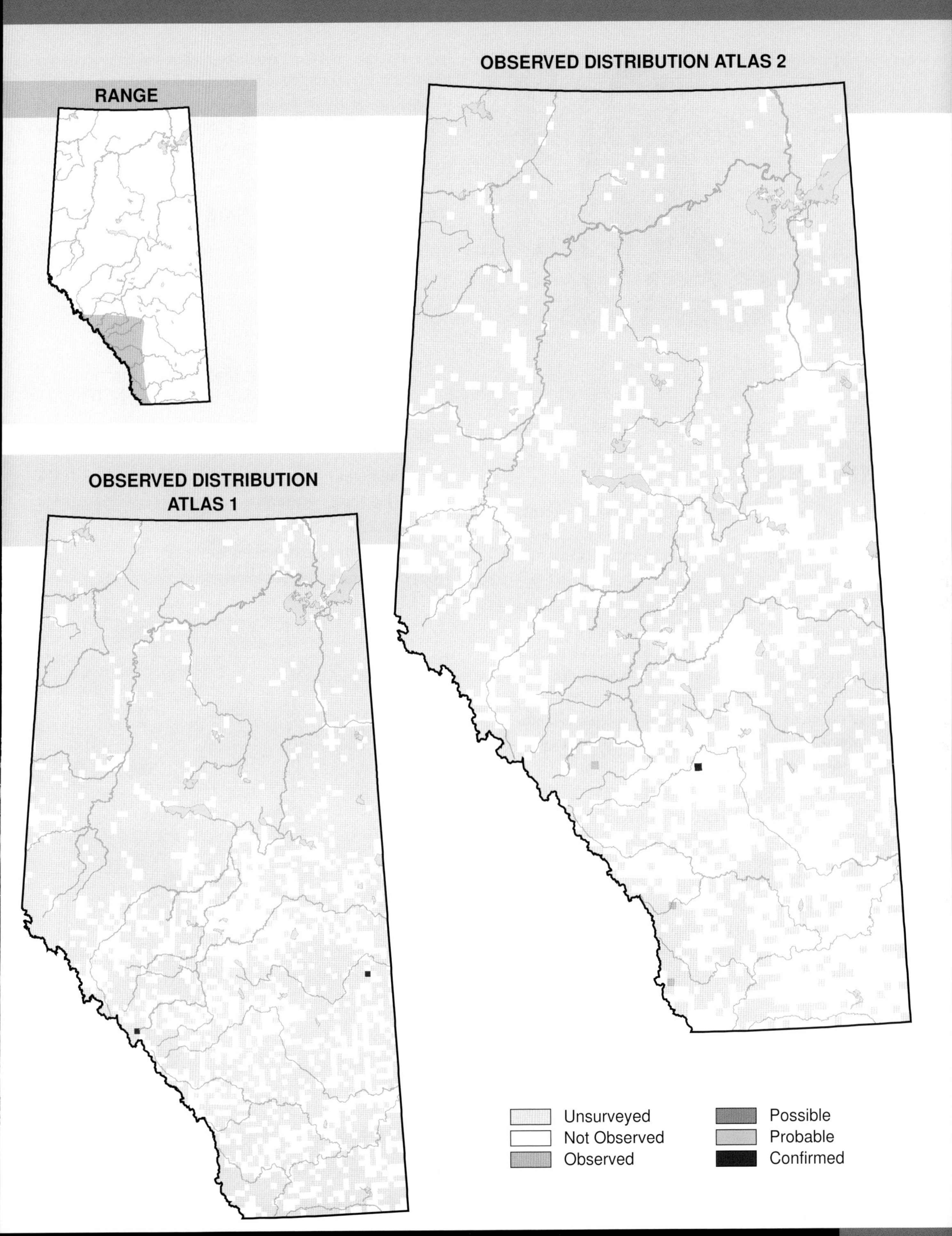

OBSERVED DISTRIBUTION ATLAS 2
RANGE
OBSERVED DISTRIBUTION ATLAS 1
Unsurveyed
Not Observed
Observed
Possible
Probable
Confirmed

Mountain Bluebird (*Sialia currucoides*)

Photo: Gerald Romanchuk

NESTING

CLUTCH SIZE (EGGS):	4–7
INCUBATION (DAYS):	13–14
FLEDGING (DAYS):	22–23
NEST HEIGHT (METRES):	0.6–15.2

Alberta's most common bluebird, the Mountain Bluebird, is found across Alberta in every Natural Region. No change in the distribution of this species was observed between Atlas 1 and Atlas 2.

The Mountain Bluebird breeds in dry open areas with clumps of trees or shrubs. Therefore, this species benefits from forestry practices and the expansion of agriculture because these activities tend to increase the amount of habitat that this species prefers. This species was fairly common in all natural regions except the Boreal Forest NR where reporting rates were low. This was likely because open habitat tends to be rare in that natural region.

Declines in relative abundance were detected in the Boreal Forest, Foothills, and Parkland NRs where, relative to other species, the Mountain Bluebird was observed less frequently in Atlas 2 than in Atlas 1. No change was detected in the Grassland and Rocky Mountain NRs. The Breeding Bird Survey detected no change in abundance in Alberta, or across Canada, during the period of 1985–2005. The Atlas-detected declines are problematic, though, because this species benefits from more recent forestry practices, the conversion of forests to agriculture, and the increased number of nestboxes that are available. Atlas 2 provided a generally more extensive coverage than did Atlas 1, and this raises concerns that the better coverage in Atlas 2 would be expected to bring a positive bias to the data, not a negative one. The distribution maps for this species clearly show broad-scale reduction throughout its core range in Alberta in Atlas 2. Further research is needed to evaluate Atlas findings regarding Mountain Bluebirds in Alberta. The provincial status of this species is Secure.

TEMPORAL REPORTING RATE

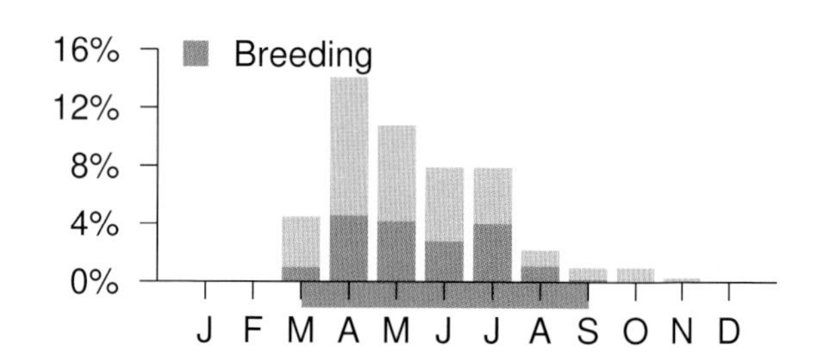

NATURAL REGIONS REPORTING RATE

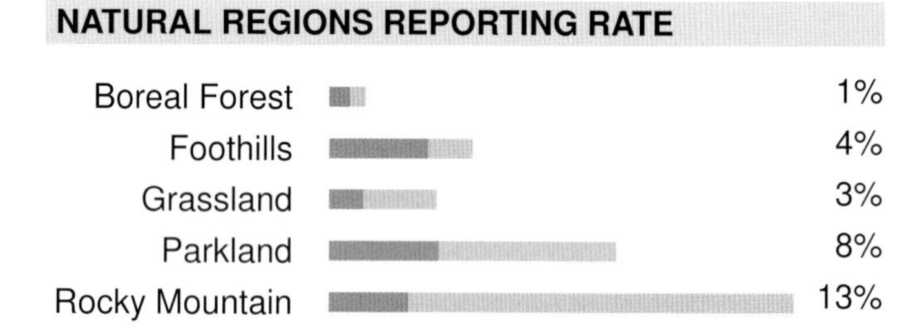

SPATIAL REPORTING RATE

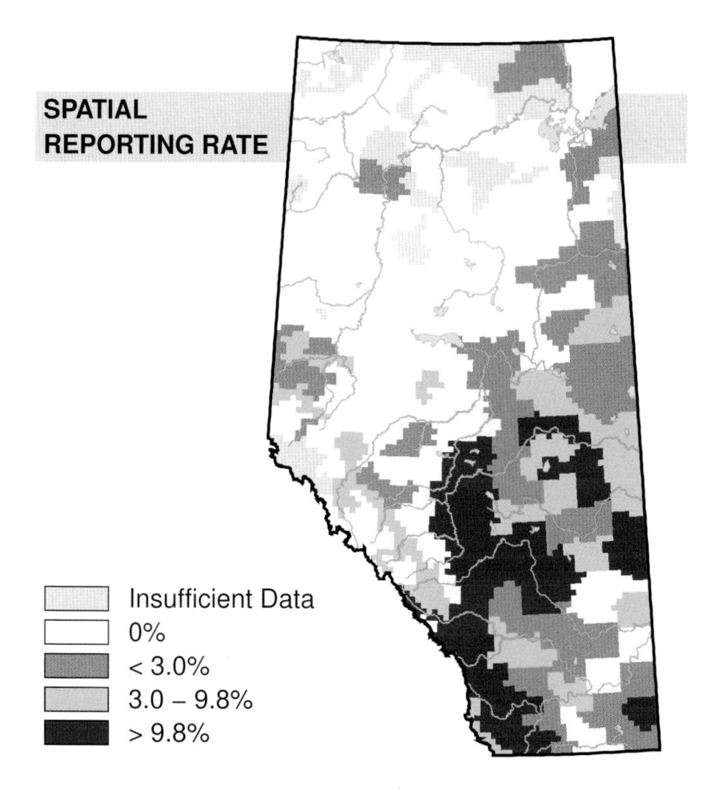

STATUS

GSWA 2000	GSWA 2005	COSEWIC Historic Rankings	COSEWIC 2007	Alberta Wildlife Act
Secure	Secure	N/A	N/A	N/A

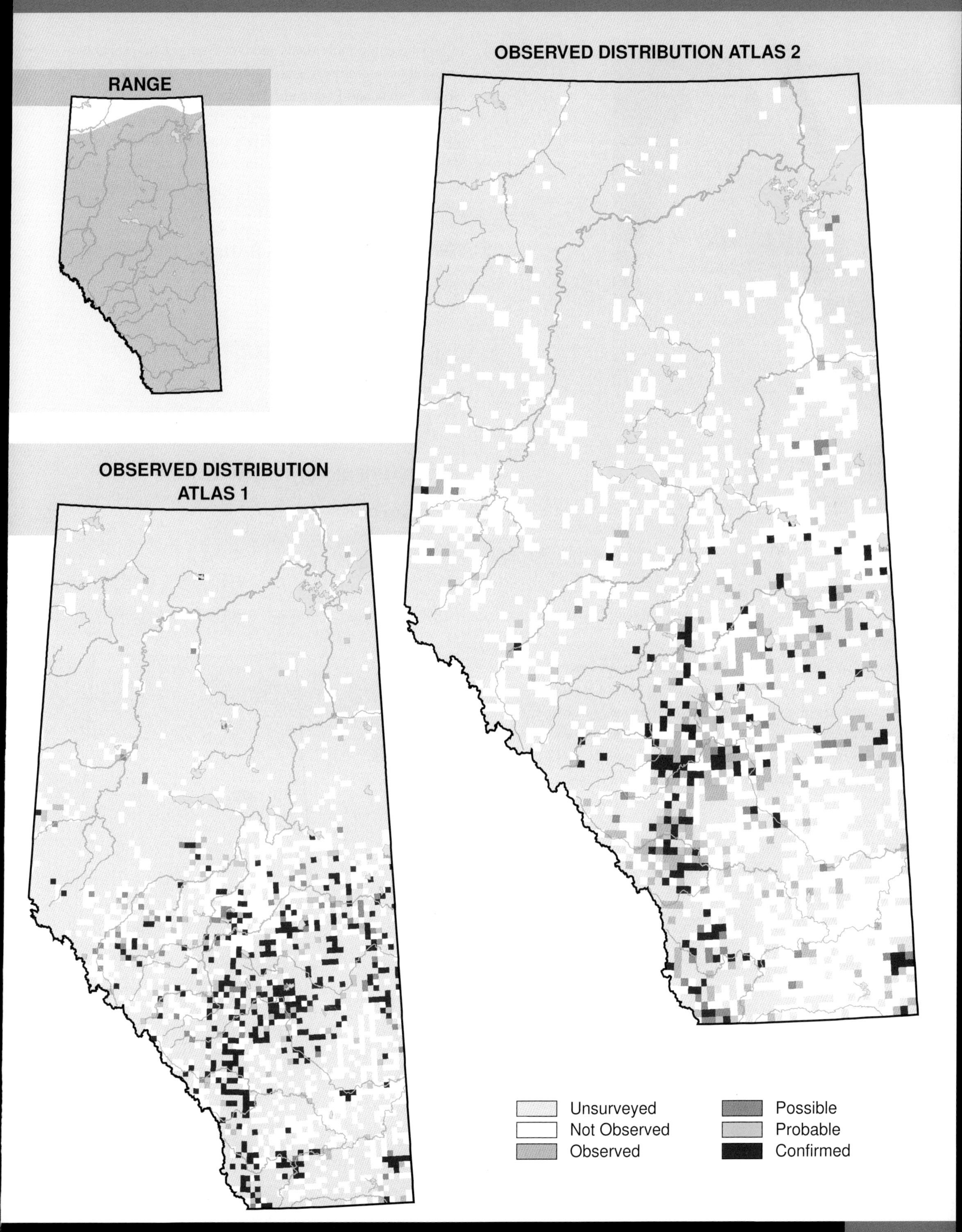

RANGE
OBSERVED DISTRIBUTION ATLAS 2
OBSERVED DISTRIBUTION ATLAS 1
Unsurveyed
Not Observed
Observed
Possible
Probable
Confirmed

Townsend's Solitaire (*Myadestes townsendi*)

Photo: Randy Jensen

NESTING

CLUTCH SIZE (EGGS):	3–5
INCUBATION (DAYS):	11–14
FLEDGING (DAYS):	10–14
NEST HEIGHT (METRES):	0

Townsend's Solitaire breeds in the Rocky Mountain and Foothills Natural Regions. The observed distribution of this species did not change in Atlas 2.

Townsend's Solitaire is a ground nester that prefers to breed in coniferous forests of varying ages. They do not avoid forest edges and are often found in the mountains near tree line. Due to the preference for mountainous and hilly areas, reporting rates for this species were highest in the Foothills and Rocky Mountain NRs, although limited access to this rugged terrain renders it difficult to assess the status of the population. Observations made in the Parkland NR were of non-breeders, except for a few located on the periphery, adjacent to the Foothills NR. Observations made in the Grassland NR were records of non-breeders.

There were declines in relative abundance in the Foothills and Rocky Mountain NRs. These declines are surprising as the precipitation was similar in these NRs in Atlas 1 and 2, and prescribed burns or forestry would not be expected to impact this species negatively. It appears that differences in coverage in these NRs between Atlas 1 and 2 are the likely cause of the detected changes. An increase in relative abundance was detected in the Grassland NR, but observations in this NR were of non-breeding individuals and thus changes are probably a function of migratory variation. No change was detected in the Boreal Forest and Parkland NRs. The Breeding Bird Survey found no change for this species in Alberta or Canada. Solitaires are detected on relatively few BBS routes in many states and provinces; so improved sampling is needed before more fine-grained analysis of population trends will be possible (Bowen, 1997).

TEMPORAL REPORTING RATE

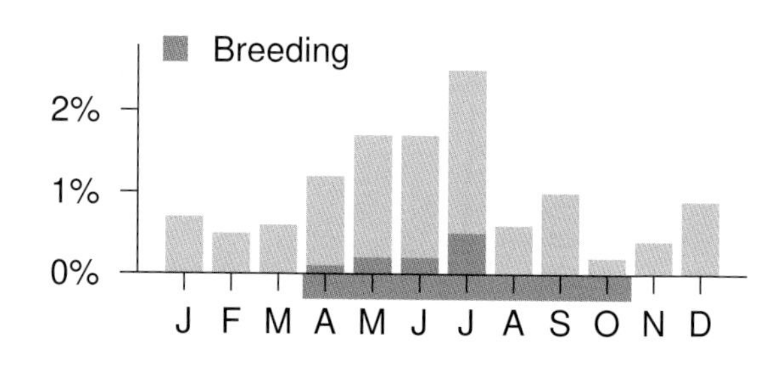

NATURAL REGIONS REPORTING RATE

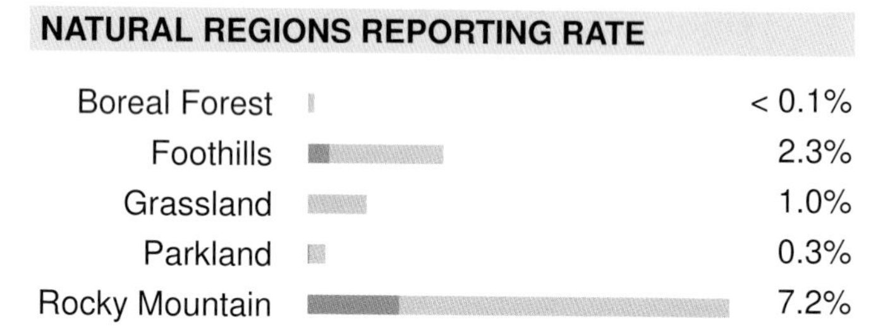

SPATIAL REPORTING RATE

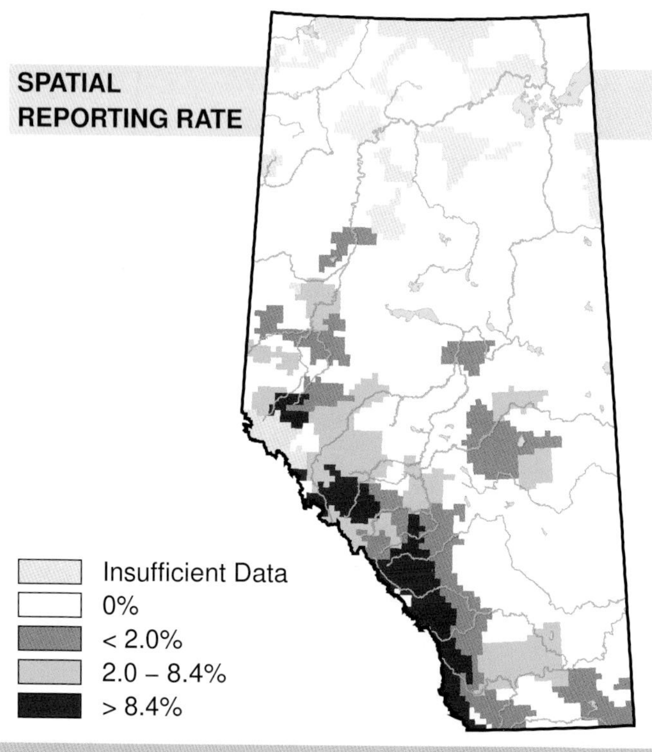

STATUS

GSWA 2000	GSWA 2005	COSEWIC Historic Rankings	COSEWIC 2007	Alberta Wildlife Act
Secure	Secure	N/A	N/A	N/A

Habitat
Nest Location
Nest Type
Diet

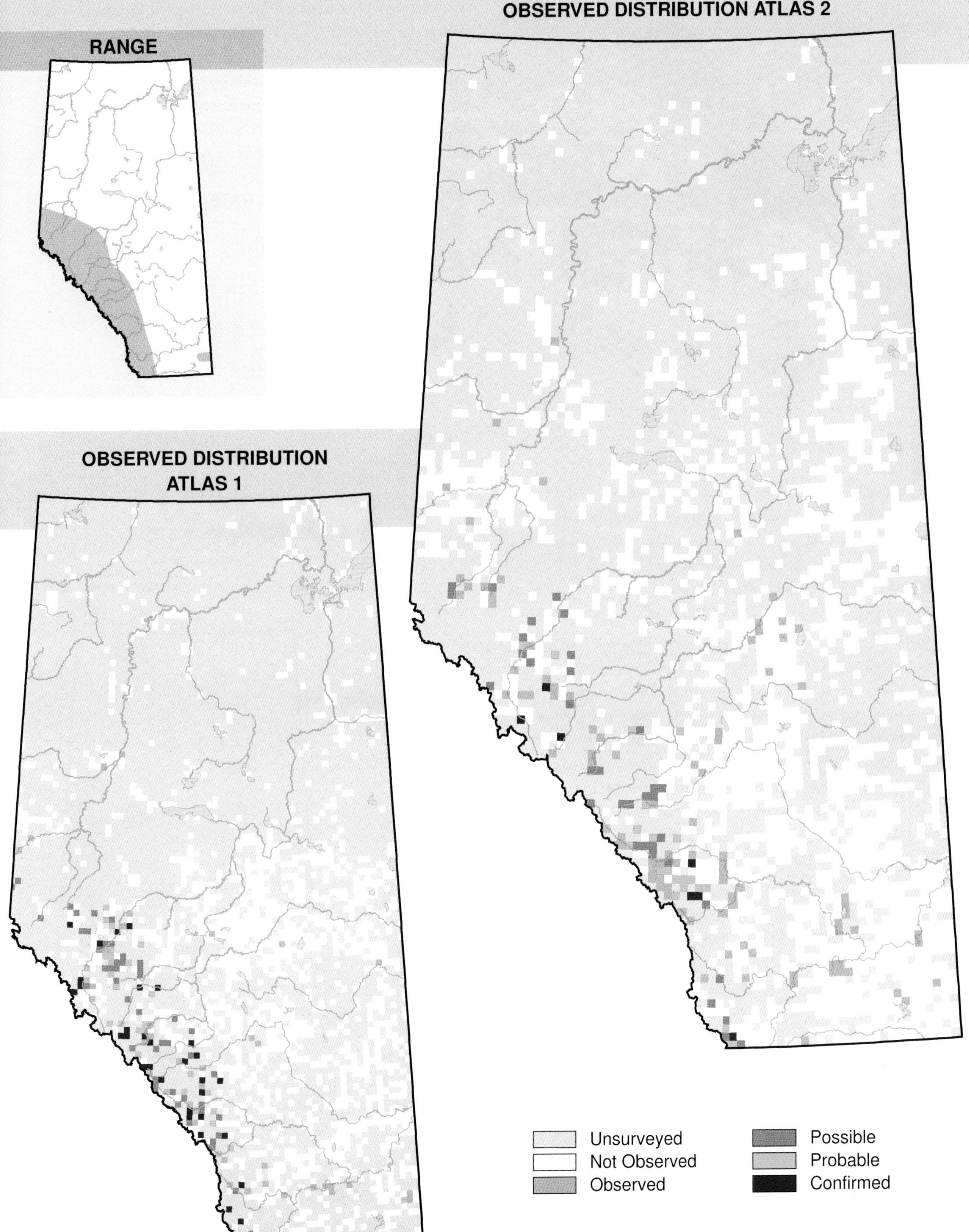
OBSERVED DISTRIBUTION ATLAS 2
RANGE
OBSERVED DISTRIBUTION ATLAS 1
Unsurveyed
Not Observed
Observed
Possible
Probable
Confirmed

Veery (*Catharus fuscescens*)

Photo: Randy Jensen

NESTING

CLUTCH SIZE (EGGS):	3–5
INCUBATION (DAYS):	10–13
FLEDGING (DAYS):	10–12
NEST HEIGHT (METRES):	0

The Veery breeds in all the Natural Regions across Alberta except the Canadian Shield NR. Extra-limital observations were recorded north of Manning and in Wood Buffalo National Park, but these records lacked breeding evidence. With only a few observations logged, it is premature to interpret these as an indication of a change in the distribution of this species.

Reporting rates were low for all the natural regions, but they were higher for the Rocky Mountain, Parkland, and Grassland NRs. Veerys inhabit deciduous forests, with a strong preference for riparian areas. They prefer areas with dense understoreys, which are more prevalent in the Parkland and Grassland NRs. The higher reporting rate from the Rocky Mountain NR appears to be a function of sample size and not necessarily an indication of habitat selection, as this species is found only locally in Waterton Lakes National Park and on the edge of Jasper and Banff national parks.

Declines in relative abundance were detected for the Boreal Forest, Grassland, and Parkland NRs, and an increase was detected for the Rocky Mountain NR. Observation rates, relative to other species, mirrored these changes. The increase in the Rocky Mountain NR may be a result of high reporting rates in the Waterton Lakes National Park area and these, in turn, are a result of greater volunteer effort; the elevated rates may not, therefore, indicate an increase in abundance. The Breeding Bird Survey found no abundance change for this species in Alberta; whereas, a Canada-wide decline was found during the period 1985–2005. The declines found provincially in the Boreal Forest, Grassland, and Parkland NRs by the Atlas, and also on a national scale by the Breeding Bird Survey, could be the result of increased forest fragmentation because this species is sensitive to this type of habitat change. This species is considered Secure in Alberta.

TEMPORAL REPORTING RATE

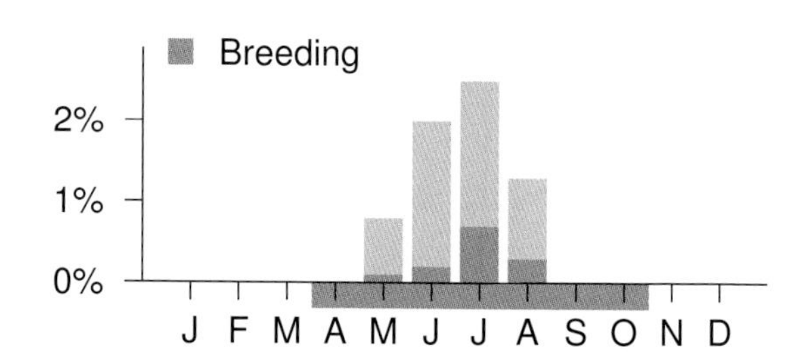

NATURAL REGIONS REPORTING RATE

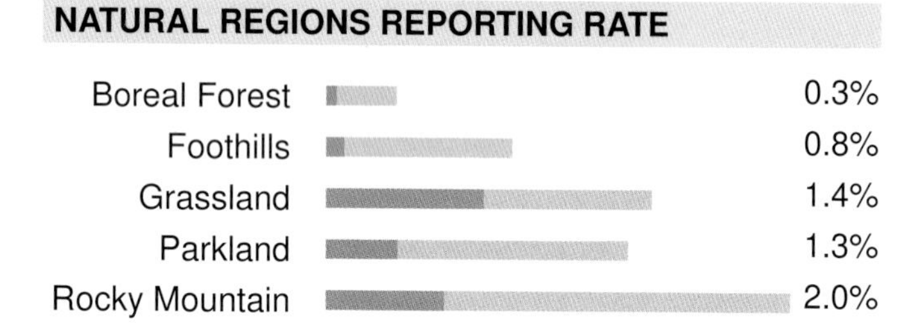

SPATIAL REPORTING RATE

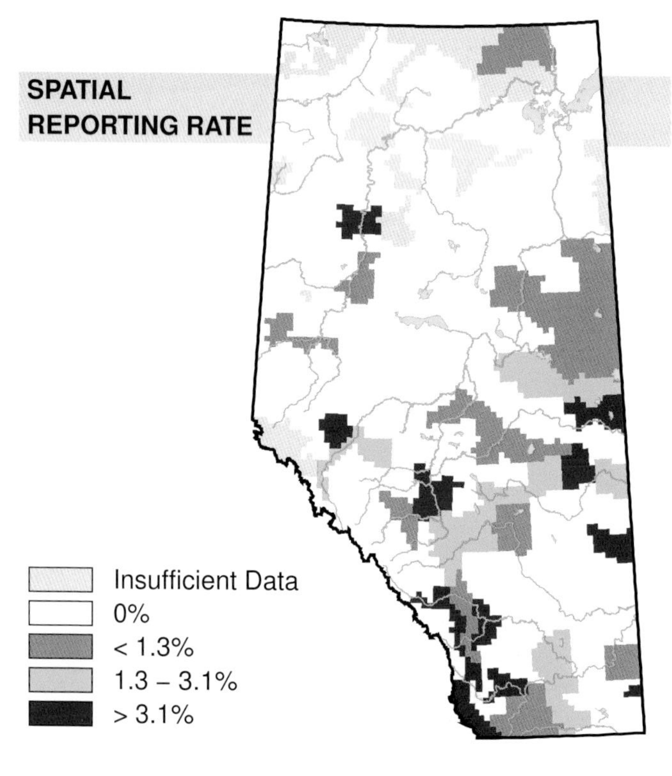

STATUS

GSWA 2000	GSWA 2005	COSEWIC Historic Rankings	COSEWIC 2007	Alberta Wildlife Act
Secure	Secure	N/A	N/A	N/A

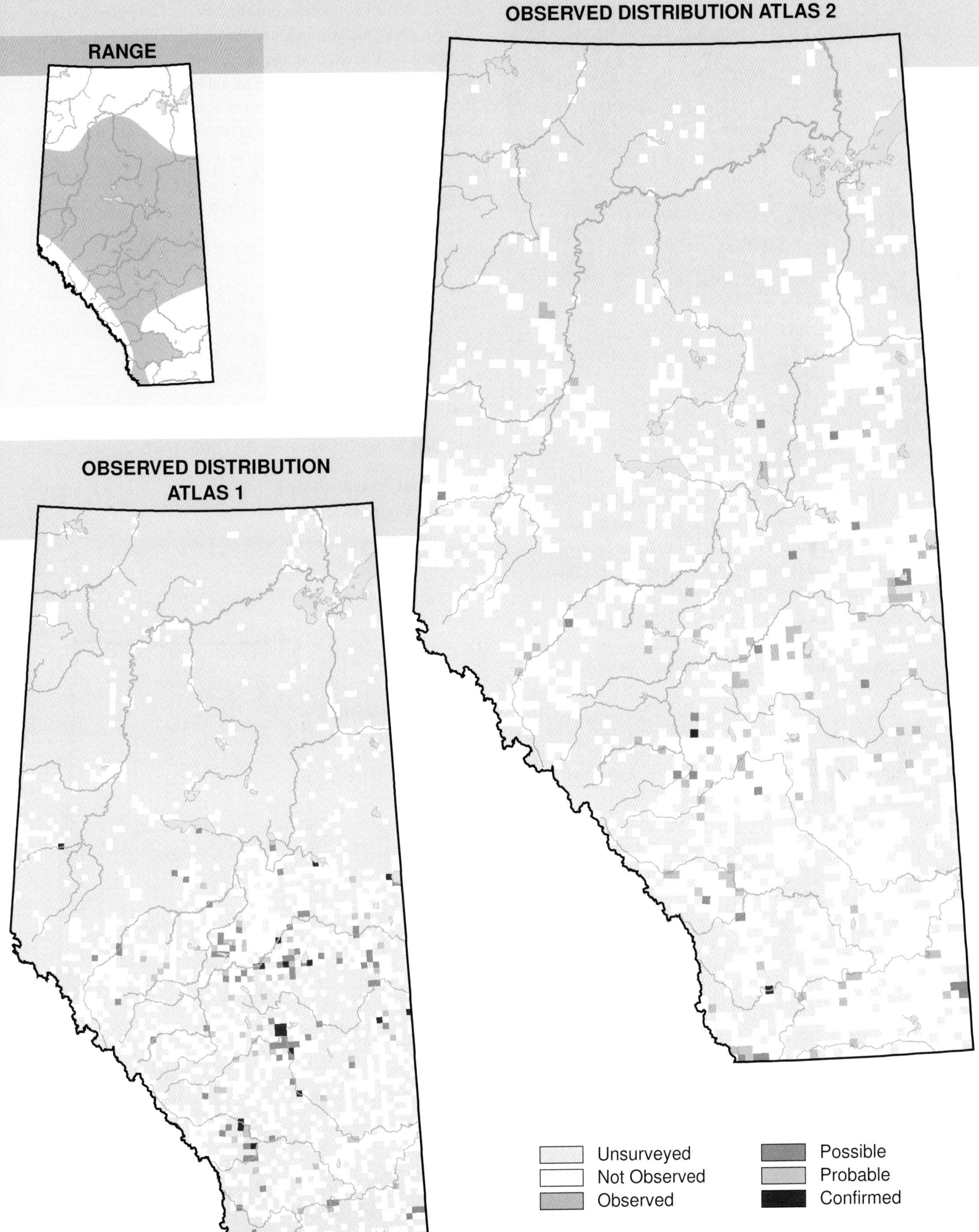

RANGE
OBSERVED DISTRIBUTION ATLAS 2
OBSERVED DISTRIBUTION ATLAS 1
Unsurveyed
Not Observed
Observed
Possible
Probable
Confirmed

Gray-cheeked Thrush (*Catharus minimus*)

Photo: Chester Olson

NESTING

CLUTCH SIZE (EGGS):	3–5
INCUBATION (DAYS):	12–14
FLEDGING (DAYS):	11–13
NEST HEIGHT (METRES):	0–3.3

The Gray-cheeked Thrush breeds only in the northern part of the Boreal Forest Natural Region of Alberta; however, no breeding records were found in this NR in Atlas 2. One observation was made in the Caribou Mountains, but this is not considered a range extension as this species is known to have bred in that area. Records from the Foothills, Grassland, and Rocky Mountain NRs, and from the Cold Lake and Lac La Biche regions, are not considered to be breeders. Reporting rates were very low in the Foothills, Grassland, Parkland, and Rocky Mountain NRs. These records were all considered to be of non-breeders.

No change in relative abundance was detected, although this thrush was observed with higher frequency in Atlas 2, relative to other species, than in Atlas 1. None of the observations for this species had breeding evidence associated with them in either Atlas project. The only breeding records obtained for this species in Alberta have come from the Caribou Mountains. This species breeds in the far north of the province, an area that was afforded few surveys during the Atlas. The Breeding Bird Survey sample size was too small to investigate Gray-cheeked Thrush abundance change in Alberta; however, an increase was found across Canada during the period 1995–2005. Lowther et al. (2001) report that although abundance data are most easily acquired on the breeding grounds, the current BBS program does not sample the subarctic distribution of this species accurately.

The shrub habitat used for breeding by the Gray-cheeked Thrush is extensive throughout the northern taiga and is under no significant threat from human activities, but work to obtain baseline data on its abundance would be useful and critical to evaluating this species' population status and to detect any status changes (Lowther et al., 2001). The status of this species is Undetermined in Alberta due to insufficient data.

TEMPORAL REPORTING RATE

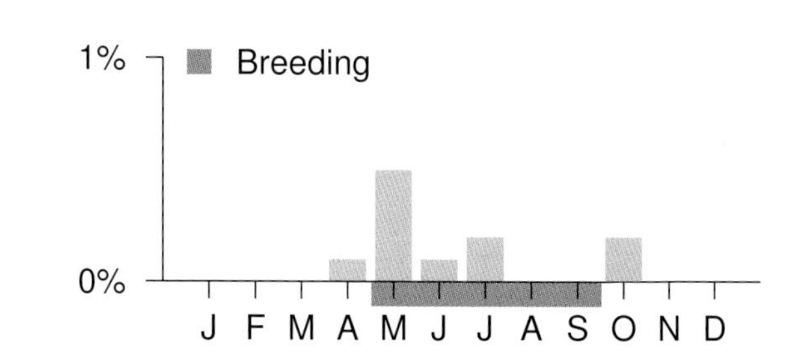

NATURAL REGIONS REPORTING RATE

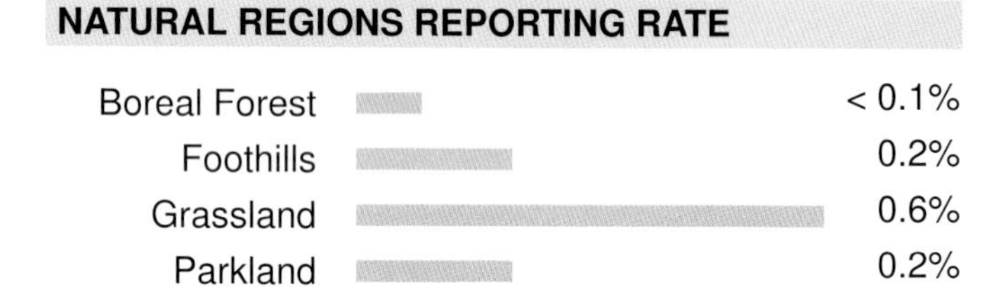

SPATIAL REPORTING RATE

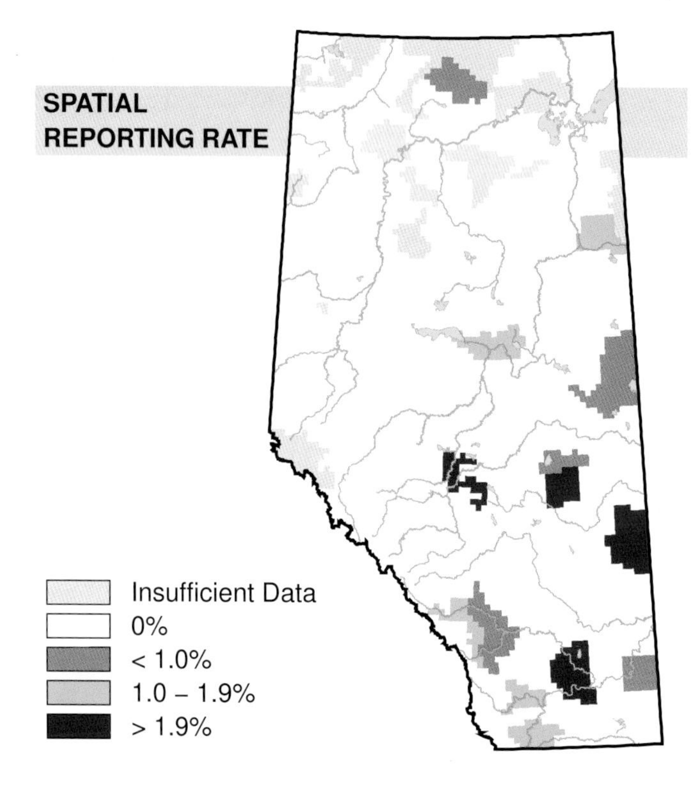

STATUS

GSWA 2000	GSWA 2005	COSEWIC Historic Rankings	COSEWIC 2007	Alberta Wildlife Act
Undetermined	Undetermined	N/A	N/A	N/A

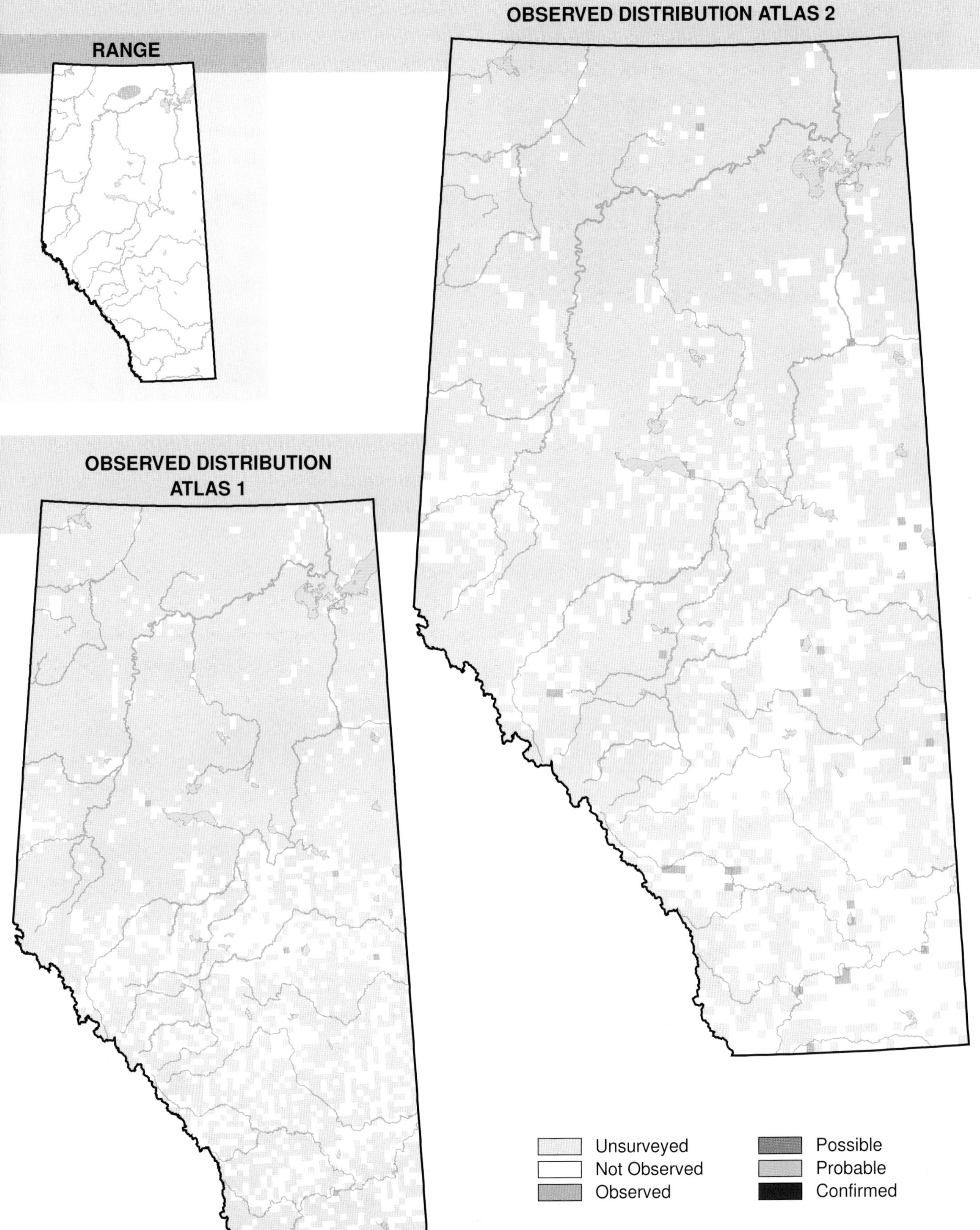
RANGE
OBSERVED DISTRIBUTION ATLAS 2
OBSERVED DISTRIBUTION
ATLAS 1
Unsurveyed
Not Observed
Observed
Possible
Probable
Confirmed

Swainson's Thrush (*Catharus ustulatus*)

Photo: Stan Gosche

NESTING

CLUTCH SIZE (EGGS):	3–5
INCUBATION (DAYS):	12–14
FLEDGING (DAYS):	10–12
NEST HEIGHT (METRES):	0–2.2

Swainson's Thrush breeds in the Boreal Forest, Foothills, Parkland, and Rocky Mountain Natural Regions in Alberta. The observed distribution of this species did not change between Atlas 1 and Atlas 2.

Swainson's Thrush breeds in coniferous and mixedwood stands that can range from old-growth to young, early-successional forests. This species tends to prefer forests that have closed canopies, dense understorey cover, high tree density, and conifer trees. As a result, this species was most abundant in the Boreal Forest, Foothills, and Rocky Mountain NRs where these conditions can be common. Reporting rates were low in the Grassland and Parkland NRs because conifer trees tend to be rare in those regions.

Increases in relative abundance were detected in the Boreal Forest and Grassland NRs where, relative to other species, Swainson's Thrush was observed more frequently in Atlas 2 than in Atlas 1. Increases in the Boreal Forest NR could be related to forestry practices. The thick undergrowth that grows post-harvest provides good nesting habitat for this species. In addition, conifers—the particular tree-type with which this species is closely associated—are often replanted post-harvest. The biological cause for changes in the Grassland NR is unclear, although wetter conditions in this NR in Atlas 2 than in Atlas 1 may have contributed to increased food availability. However, as the majority of the records from the Grassland NR were of non-breeders, the detected increase in this NR is likely a function of migratory variation. No change was detected in the Foothills, Parkland, and Rocky Mountain NRs. The Breeding Bird Survey did not find a change in abundance in Alberta, or nationally, for the period 1985–2005, but did detect an increase nationally for the period 1995–2005. This species is considered Secure in Alberta.

TEMPORAL REPORTING RATE

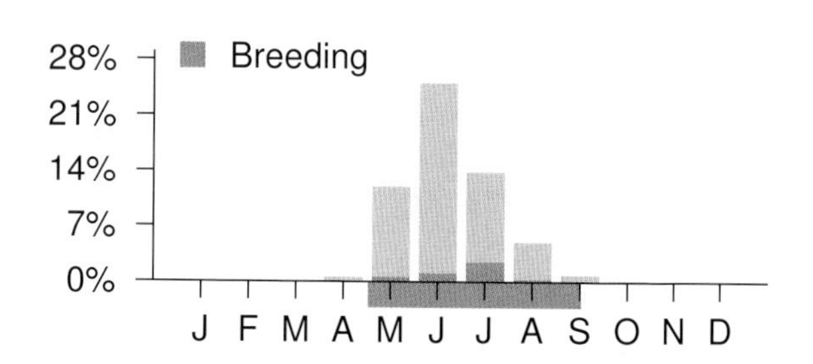

NATURAL REGIONS REPORTING RATE

Region	Rate
Boreal Forest	14%
Foothills	35%
Grassland	4%
Parkland	4%
Rocky Mountain	31%

SPATIAL REPORTING RATE

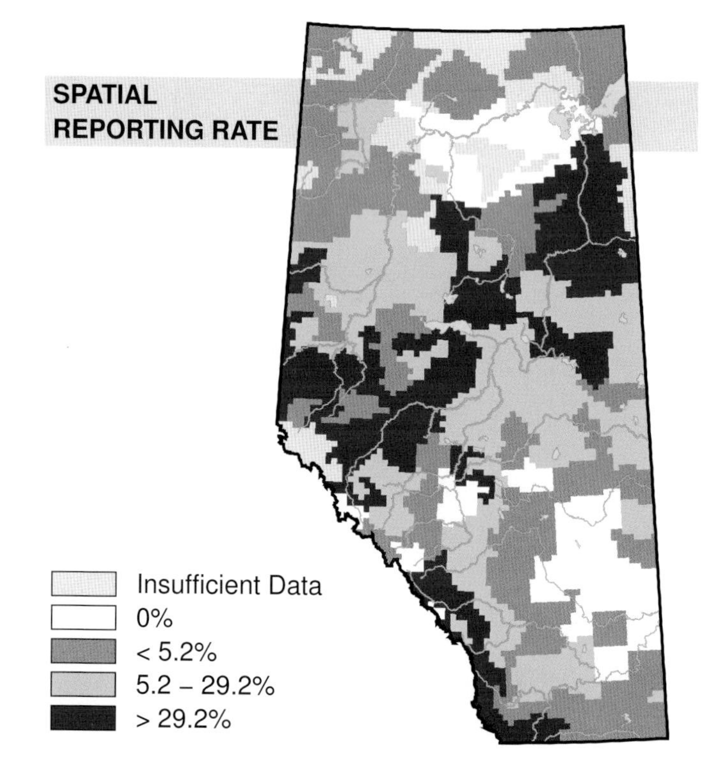

STATUS

GSWA 2000	GSWA 2005	COSEWIC Historic Rankings	COSEWIC 2007	Alberta Wildlife Act
Secure	Secure	N/A	N/A	N/A

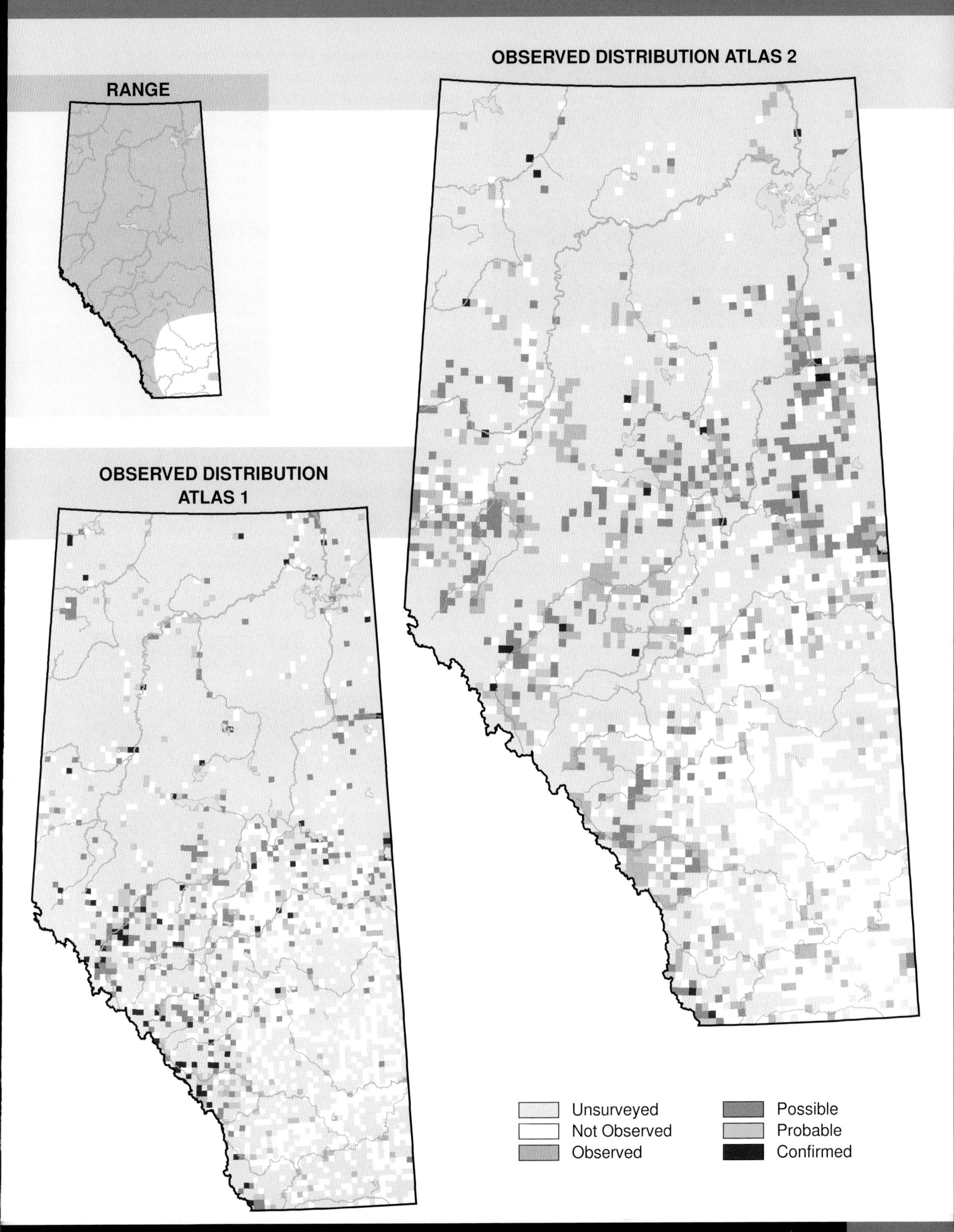
Habitat
Nest Location
Nest Type
Diet
RANGE
OBSERVED DISTRIBUTION ATLAS 2
OBSERVED DISTRIBUTION ATLAS 1
Unsurveyed
Not Observed
Observed
Possible
Probable
Confirmed

Hermit Thrush (*Catharus guttatus*)

Photo: Randy Jensen

NESTING

CLUTCH SIZE (EGGS):	3–6
INCUBATION (DAYS):	12
FLEDGING (DAYS):	10–12
NEST HEIGHT (METRES):	0

The Hermit Thrush breeds in the Boreal Forest, Foothills, Parkland, and Rocky Mountain Natural Regions in Alberta. The observed distribution of this species did not change between Atlas 1 and Atlas 2.

Across its range, the Hermit Thrush nests in diverse forest types and can nest along forest edges. However, it is usually found in dense forests that contain conifer trees. For this reason, this species was most often encountered in the Boreal Forest, Foothills, and Rocky Mountain NRs. This species was less frequently encountered in the Parkland NR and in the northern part of the Grassland NR.

An increase in relative abundance was detected in the Boreal Forest NR where, relative to other species, the Hermit Thrush was observed more frequently in Atlas 2 than in Atlas 1. There is no apparent biological cause for this increase; so, it is likely a function of more extensive survey coverage of the north in Atlas 2 than in Atlas 1. Declines were detected in the Parkland and Rocky Mountain NRs where, relative to other species, it was observed less frequently in Atlas 2. The breeding density of this species tends to decrease post-fire or when forests are thinned. Declines in the Rocky Mountain NR could be related to change in the structure of forests. In the Parkland NR, declines could have been caused by the continued conversion of forests to open country for agriculture or acreage development. No change was detected in the Foothills and Grassland NRs. The Breeding Bird Survey did not find an abundance change in Alberta, but a Canada-wide abundance increase was detected during the period 1968–2005. Although the Atlas-detected increase in the Boreal Forest NR is weakly corroborated by the Canada-wide BBS trend, it is clear that, lacking a probable biological cause, further research is needed to evaluate this finding. This species is considered Secure in Alberta.

TEMPORAL REPORTING RATE

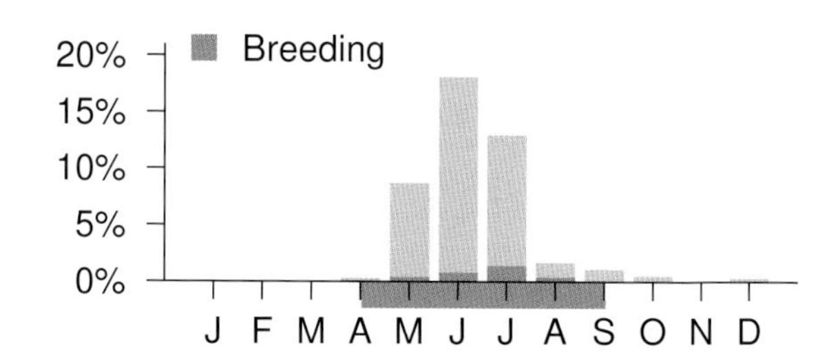

NATURAL REGIONS REPORTING RATE

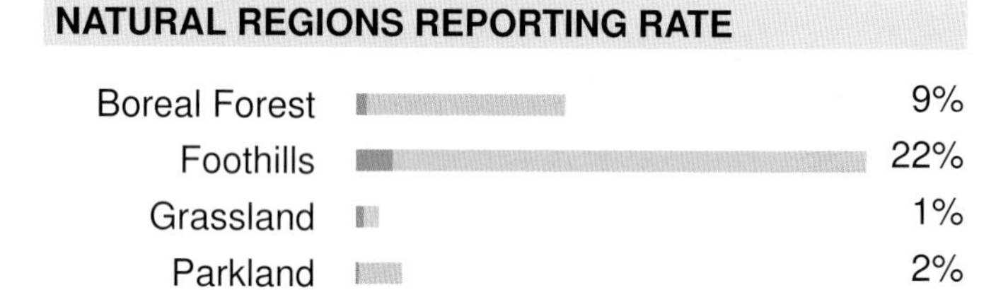

SPATIAL REPORTING RATE

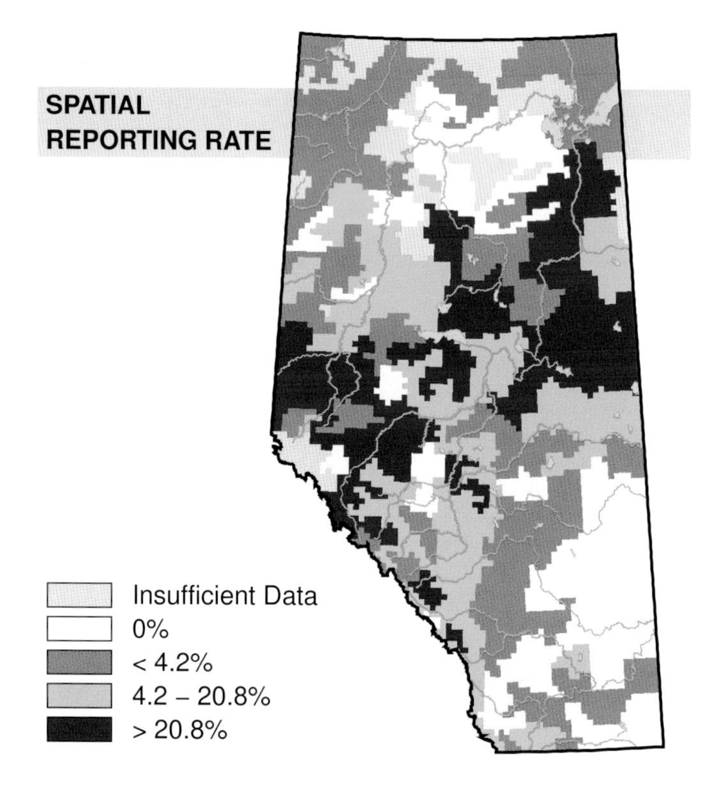

STATUS

GSWA 2000	GSWA 2005	COSEWIC Historic Rankings	COSEWIC 2007	Alberta Wildlife Act
Secure	Secure	N/A	N/A	N/A

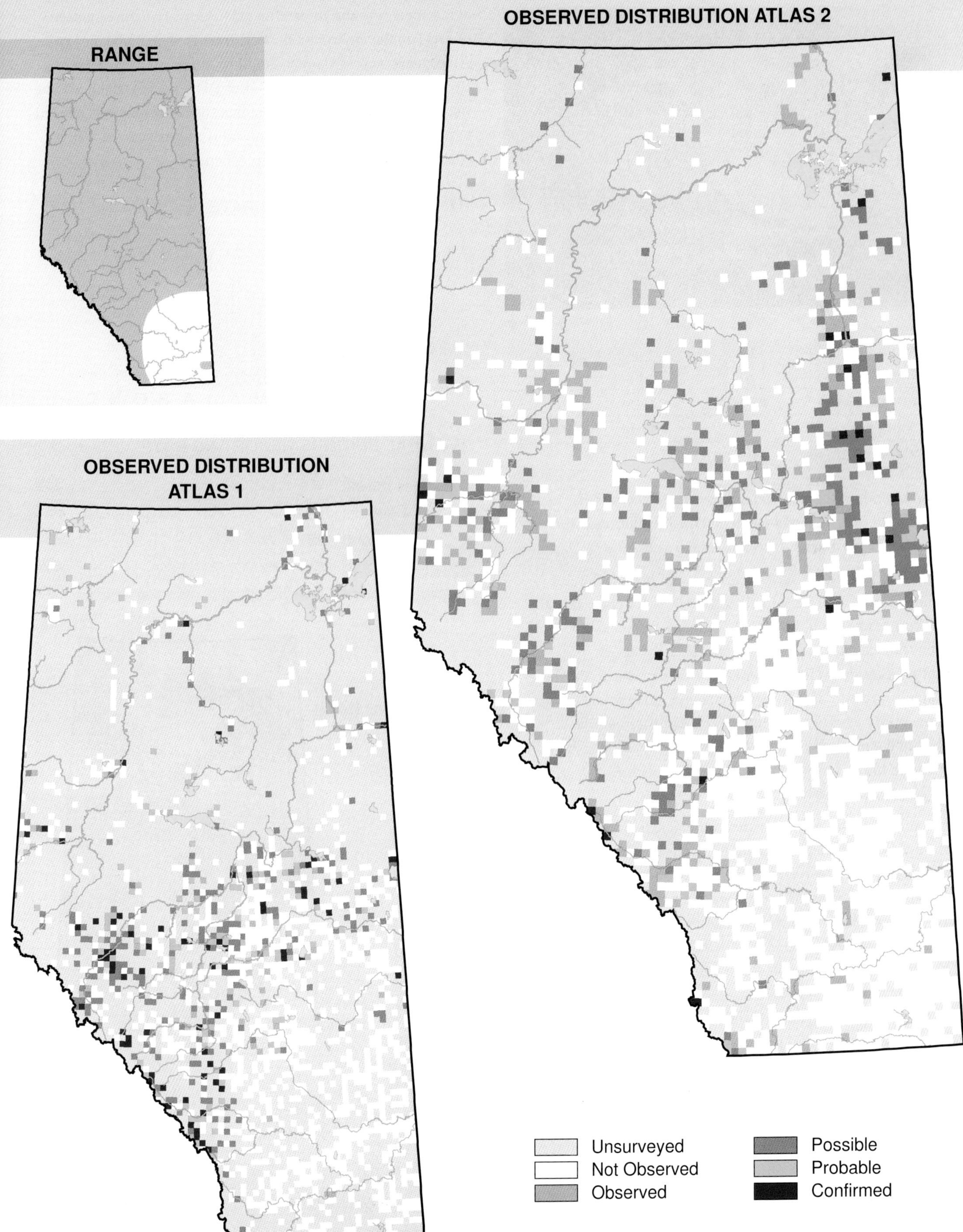
Habitat
Nest Location
Nest Type
Diet
RANGE
OBSERVED DISTRIBUTION ATLAS 2
OBSERVED DISTRIBUTION ATLAS 1
Unsurveyed
Not Observed
Observed
Possible
Probable
Confirmed

American Robin (*Turdus migratorius*)

Photo: Robert Gehlert

NESTING

CLUTCH SIZE (EGGS):	3–5
INCUBATION (DAYS):	11–14
FLEDGING (DAYS):	14–16
NEST HEIGHT (METRES):	1–5

The American Robin is widely distributed across Alberta and is found in every Natural Region. The distribution of this species did not change between Atlas 1 and Atlas 2.

The American Robin nests in a variety of habitats but it is usually found in areas with a mixture of short-grass and pockets of shrubs and trees. This species frequents residential areas but can also be found in riparian areas and in early-successional forests post-disturbance. Due to the generalist nature of this species' habitat requirements, it was found fairly evenly across the Boreal Forest, Foothills, Parkland, and Rocky Mountain NRs. The reporting rate was lower in the Grassland NR because this species needs trees within its breeding habitat.

Decreases in relative abundance were detected in the Boreal Forest, Foothills, and Parkland NRs where, relative to other species, it was observed less frequently in Atlas 2 than in Atlas 1. The Breeding Bird Survey detected a Canada-wide increase from 1968–1985, but no change from 1985–2005, and then a decline from 1995–2005. In Alberta an increase was detected during the period of 1968–2005, but no change was found in other time periods. Despite the apparent trend shifting from increase to no change to possible decline, and lacking both corroboration and identified biological cause, the detected declines of this species in Alberta must be regarded with caution. Historically, the American Robin was among the first avian species found to be affected negatively by the presence of DDT in the environment. Past increases in abundance in Alberta and across Canada could be the result of American Robin populations recovering after DDT use was banned. The findings from the Atlas suggest that further research is needed to properly evaluate the current status of this species. This species is considered Secure in Alberta.

TEMPORAL REPORTING RATE

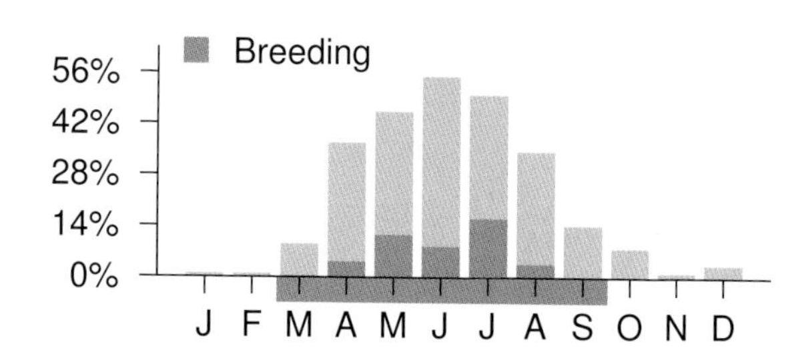

NATURAL REGIONS REPORTING RATE

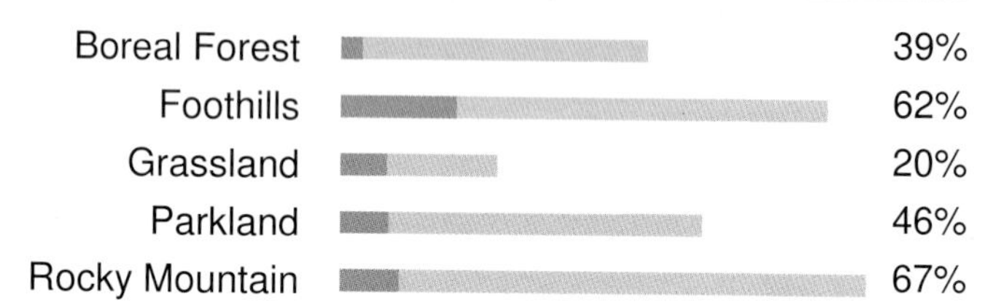

SPATIAL REPORTING RATE

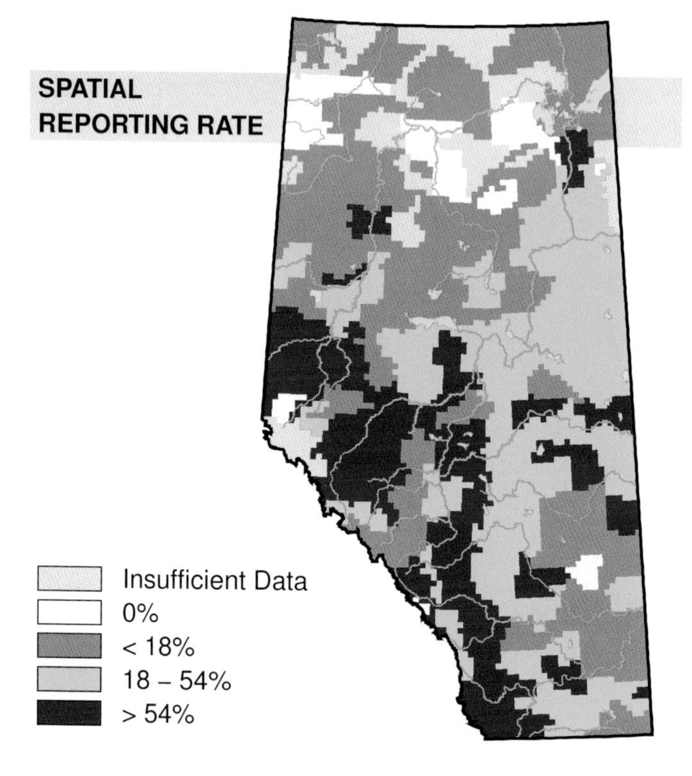

STATUS

GSWA 2000	GSWA 2005	COSEWIC Historic Rankings	COSEWIC 2007	Alberta Wildlife Act
Secure	Secure	N/A	N/A	N/A

Habitat  Nest Location  Nest Type  Diet  

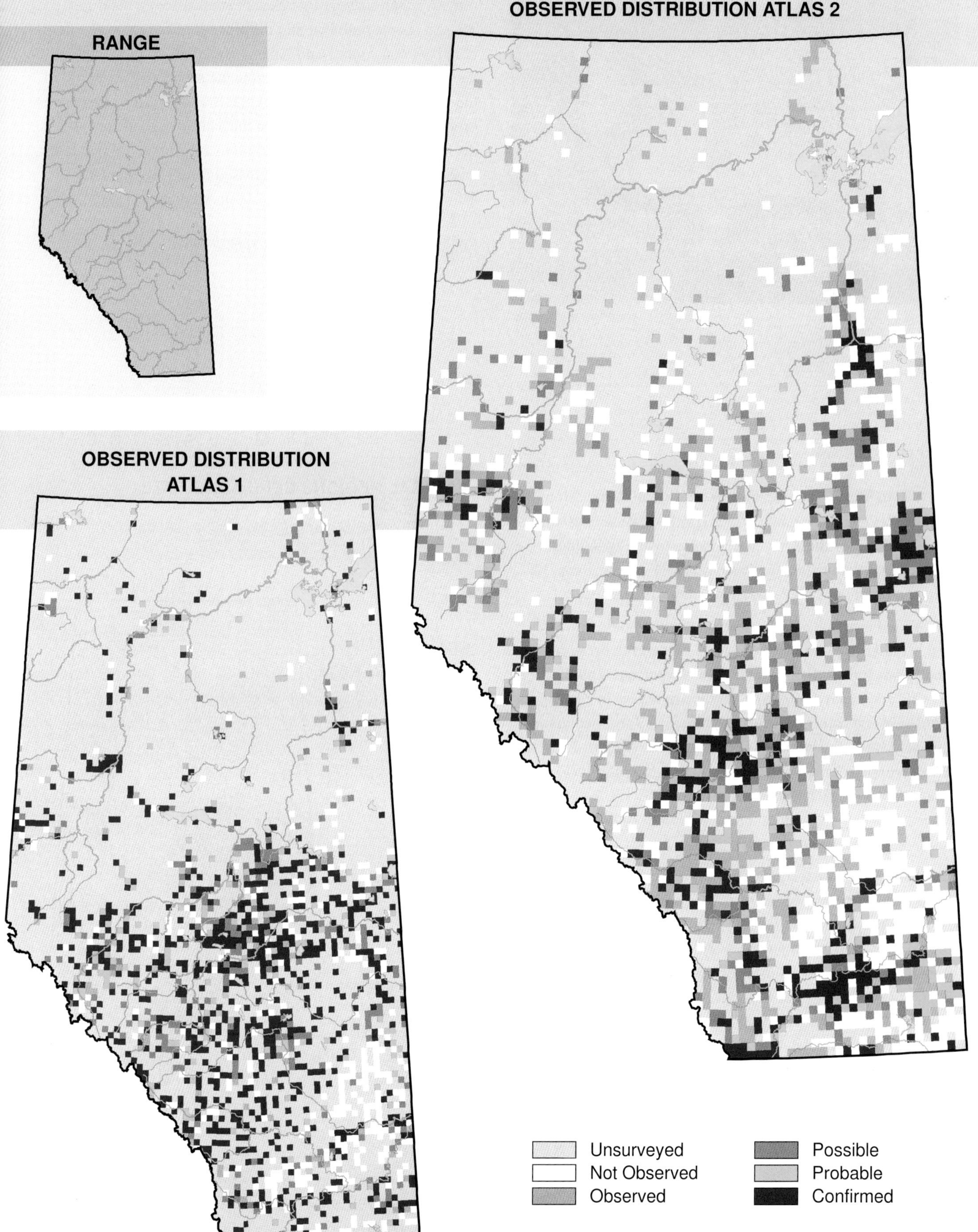

Varied Thrush (*Ixoreus naevius*)

Photo: Royal Alberta Museum

NESTING

CLUTCH SIZE (EGGS):	3–4
INCUBATION (DAYS):	13–14
FLEDGING (DAYS):	13–15
NEST HEIGHT (METRES):	1–5

The Varied Thrush breeds in the Foothills and Rocky Mountain Natural Regions. The range of this species, which is usually thought of as a western species, appears to be expanding eastward. New breeding evidence was found in the Swan Hills area, thus farther northeastward in Atlas 2 than in Atlas 1.

The Varied Thrush tends to prefer mature and old-growth stands rather than younger forests; they prefer to breed in more contiguous forests than in small forested patches; and they tend to avoid fragmented forests. This species breeds in coniferous or mixedwood forests. Due to its preference for mountainous and hilly areas, this species is portrayed as having its highest reporting rates in the Foothills and Rocky Mountain NRs. Records from the Grassland and Parkland NRs lacked breeding evidence with the exception of a few records of birds singing in suitable habitat.

No change in relative abundance was detected in any of the NRs. However, observation rates relative to other species in Atlas 2, compared with Atlas 1, were lower in both the Foothills and Rocky Mountain NRs. The Breeding Bird Survey found a decline in abundance for this species in Alberta during the period 1995–2005. It is unclear why there is a discrepancy between the Breeding Bird Survey that detected a change and the Atlas that did not detect a change. By employing a frequency-based measure, the Atlas is statistically less sensitive than surveys that use abundance-based measures. Thus, it seems probable that differences in coverage of these NRs between Atlas 1 and Atlas 2, together with the scale of the declines, placed the change below the sensitivity threshold of the Atlas. Declines detected by the Breeding Bird Survey are probably related to a decline in the average age of forests; declines in the number of contiguous forest patches that are available for breeding; and increased forest fragmentation. This species is considered Secure in Alberta.

TEMPORAL REPORTING RATE

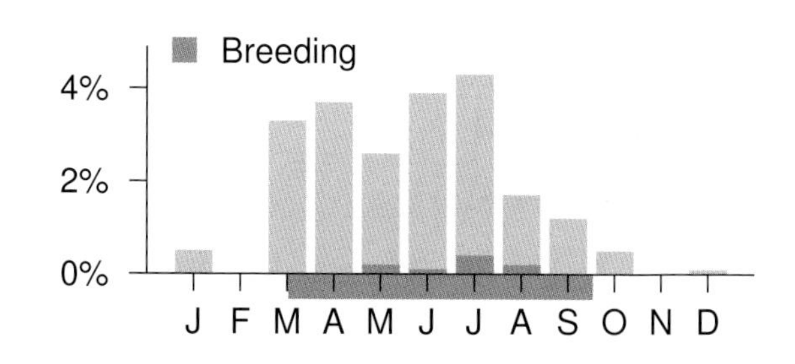

NATURAL REGIONS REPORTING RATE

SPATIAL REPORTING RATE

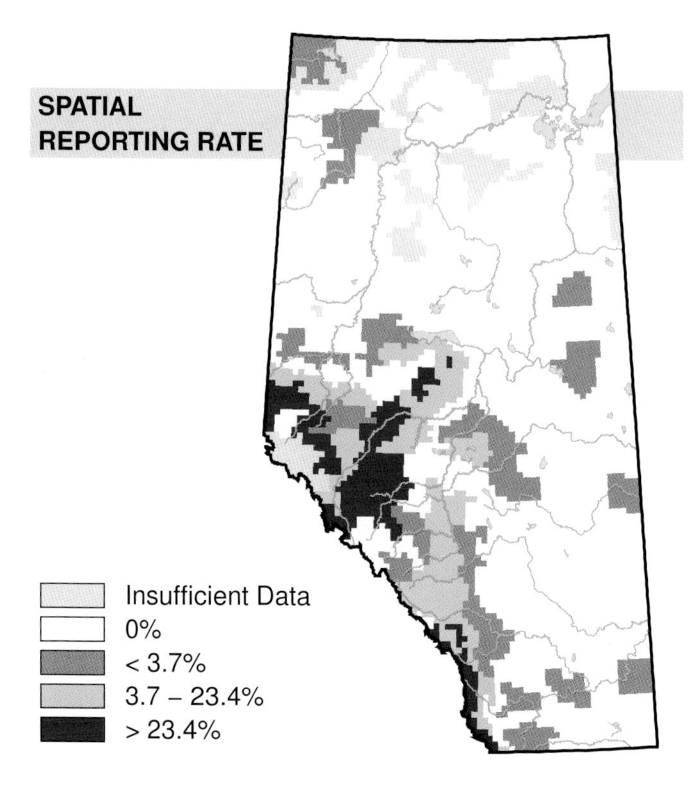

STATUS

GSWA 2000	GSWA 2005	COSEWIC Historic Rankings	COSEWIC 2007	Alberta Wildlife Act
Secure	Secure	N/A	N/A	N/A

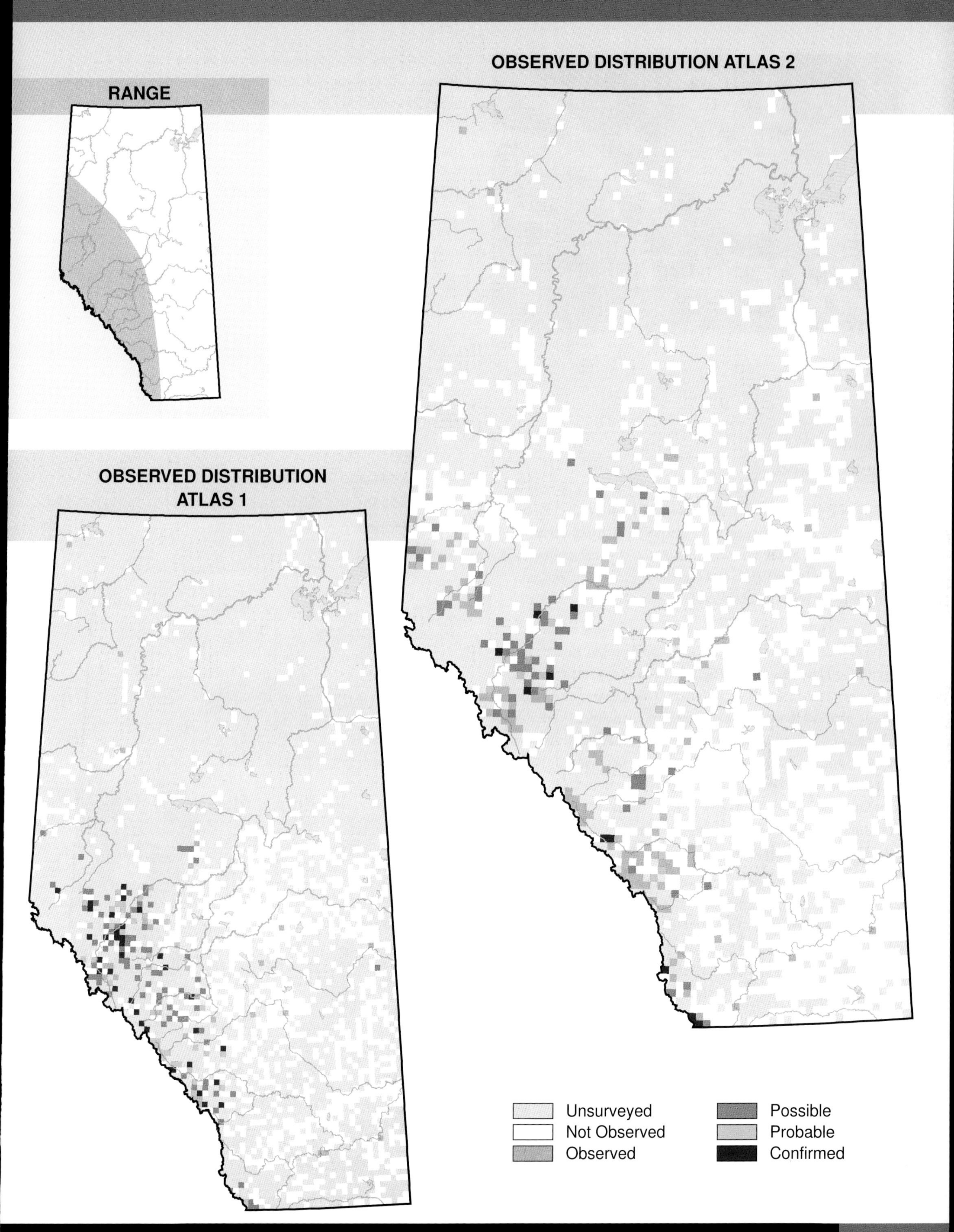
Habitat
Nest
Location
Nest
Type
Diet
RANGE
OBSERVED DISTRIBUTION ATLAS 2
OBSERVED DISTRIBUTION
ATLAS 1
Unsurveyed
Not Observed
Observed
Possible
Probable
Confirmed

Gray Catbird (*Dumetella carolinensis*)

Photo: Jim Jacobson

NESTING

CLUTCH SIZE (EGGS):	3–5
INCUBATION (DAYS):	12–13
FLEDGING (DAYS):	10
NEST HEIGHT (METRES):	<3

The Gray Catbird is found in all Natural Regions in Alberta except the Canadian Shield. The observed distribution of this species did not change between Atlas 1 and Atlas 2. However, records from the Fairview and Grande Prairie areas shifted from observed in Atlas 1 to confirmed breeding in Atlas 2.

The Gray Catbird is a representative of early-successional shrub and sapling habitat. This vocal bird prefers dense shrubs or vine tangles, along forest edges, clearings, roadsides, fencerows, and abandoned farmland. On occasion it is found in suburban areas in the densest shrubbery available. Foraging habitat is not specific and therefore does not restrict where the Gray Catbird is found. It was most often encountered in the Grassland NR, and was less commonly encountered in the Parkland and Rocky Mountain NRs. This species was least commonly observed in the Foothills and Boreal Forest NRs, where dense shrubby habitat is less common.

Relative abundance declines were detected in the Boreal Forest and Parkland NRs, and increases in relative abundance were detected in the Rocky Mountain NR. Observation rates relative to other species mirrored these changes. The Breeding Bird Survey detected an increase in Alberta for the period 1985–2005. Populations of Gray Catbird are thought to be affected positively by human activity that shifts forests towards earlier successional stages; so, the causes of the Atlas-detected declines are unclear. The elimination of shrub habitat, such as hedgerows around agricultural fields, would have a negative effect on this species. The Gray Catbird is considered Secure in Alberta.

TEMPORAL REPORTING RATE

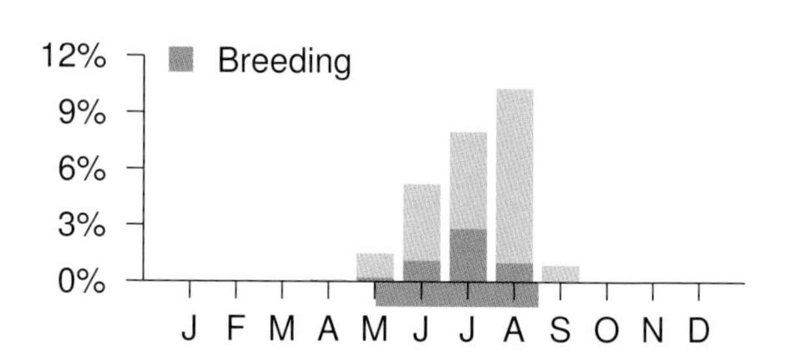

NATURAL REGIONS REPORTING RATE

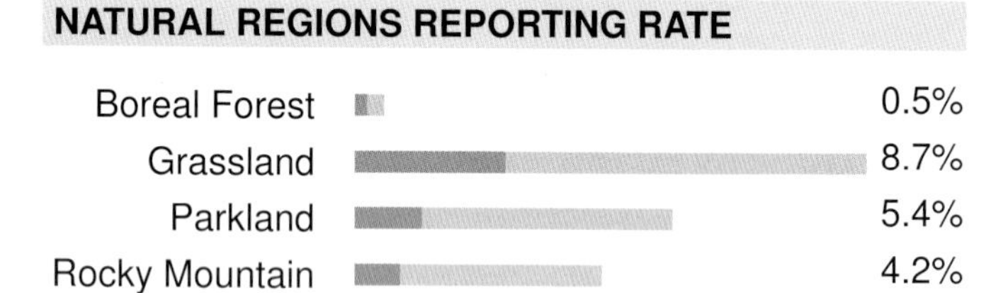

SPATIAL REPORTING RATE

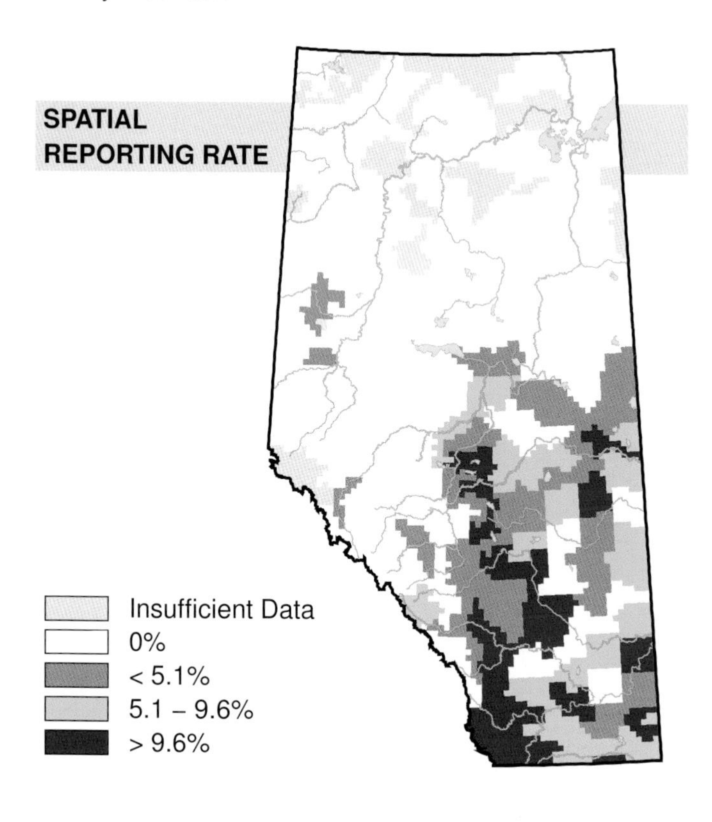

STATUS

GSWA 2000	GSWA 2005	COSEWIC Historic Rankings	COSEWIC 2007	Alberta Wildlife Act
Secure	Secure	N/A	N/A	N/A

Habitat

Nest
Location

Nest
Type

Diet

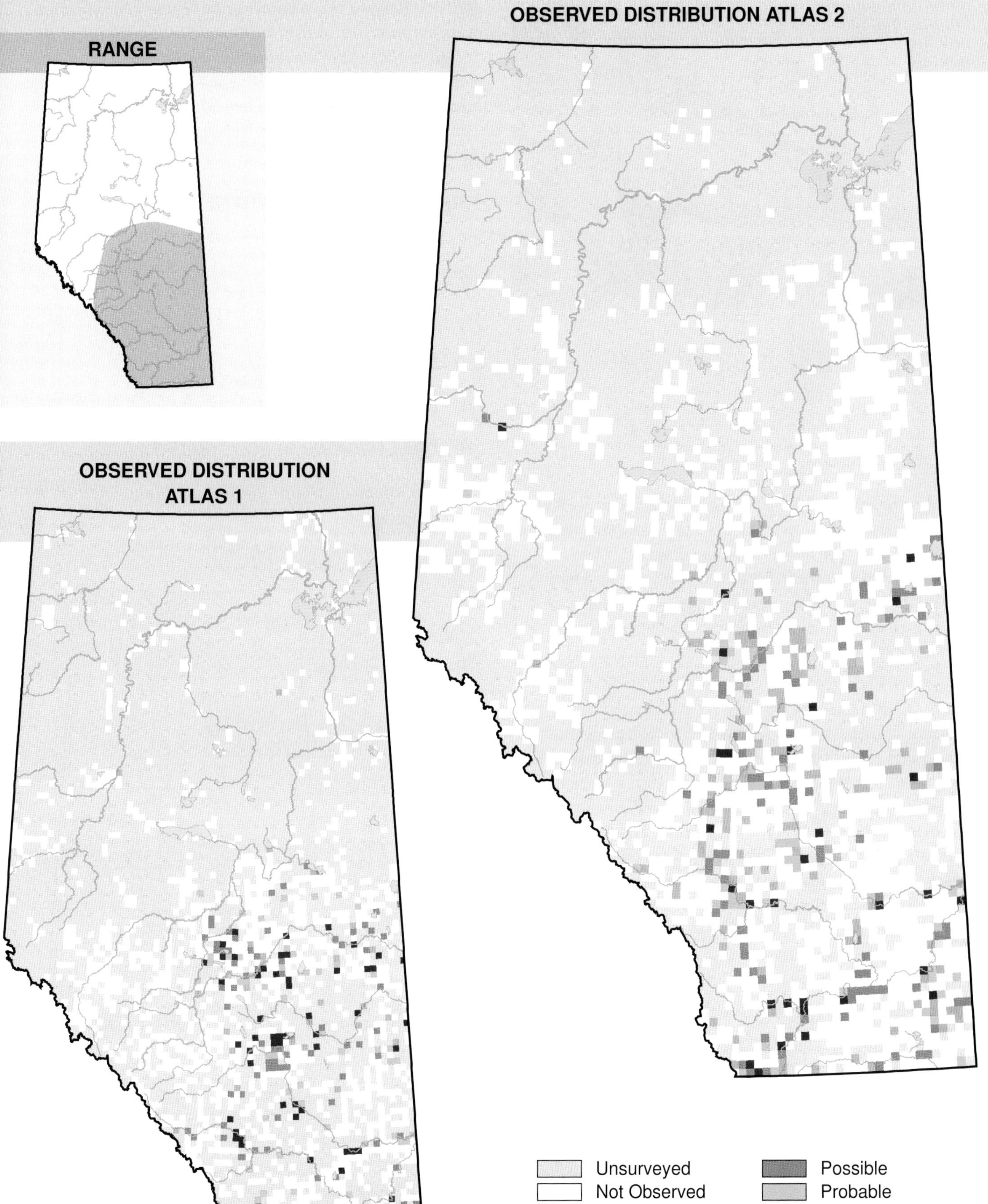
OBSERVED DISTRIBUTION ATLAS 2
RANGE
OBSERVED DISTRIBUTION
ATLAS 1
Unsurveyed
Not Observed
Observed
Possible
Probable
Confirmed

Northern Mockingbird (*Mimus polyglottos*)

Photo: Gerald Romanchuk

NESTING

CLUTCH SIZE (EGGS):	3–5
INCUBATION (DAYS):	12–13
FLEDGING (DAYS):	11–14
NEST HEIGHT (METRES):	1–4

The Northern Mockingbird is encountered infrequently in the southern portion of Alberta in the Boreal Forest, Grassland, Parkland and Rocky Mountain Natural Regions. The distribution of this species appears to be more compressed in Atlas 2 than it was during Atlas 1 because fewer Northern Mockingbird observations were reported during Atlas 2. However, records were also few in number and sparsely distributed in Atlas 1.

As its name implies, the Northern Mockingbird is a great imitator of sounds and, as a result, has one of the largest repertoires of any Alberta songbird. This species breeds near people in urban and rural settings. This species is often found in areas where land is cultivated and in second-growth habitat. It tends to avoid interior forests and is often associated with forest edges. This species was found most often in the Grassland, Parkland and Rocky Mountain NRs and was found infrequently in the Boreal Forest NR. Reporting rates for this species were low because Alberta is on the northern periphery of this species' range. Due to its conspicuous nature and tendency to live near people, it is unlikely that the presence of the Northern Mockingbird would have been missed by Atlas surveys.

There was no change in relative abundance detected in any of the NRs. Relative to other species, the Northern Mockingbird was observed more often in the Boreal Forest and Rocky Mountain NRs and less frequently in the Grassland and Parkland NRs. The Breeding Bird Survey sample size was too small to investigate abundance trends in Alberta; no trend was found on a Canada-wide scale. This species is considered Secure in Alberta.

TEMPORAL REPORTING RATE

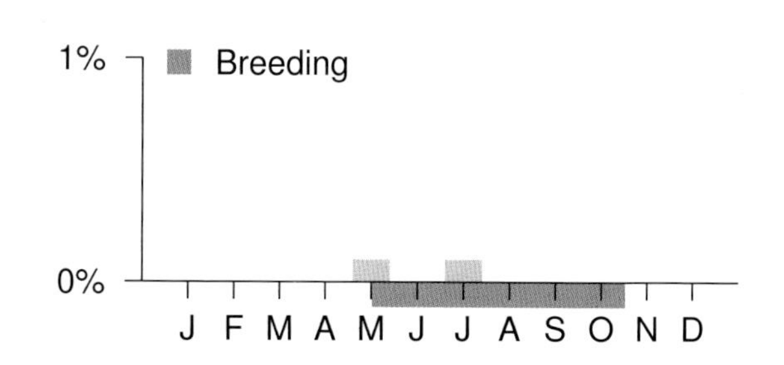

NATURAL REGIONS REPORTING RATE

SPATIAL REPORTING RATE

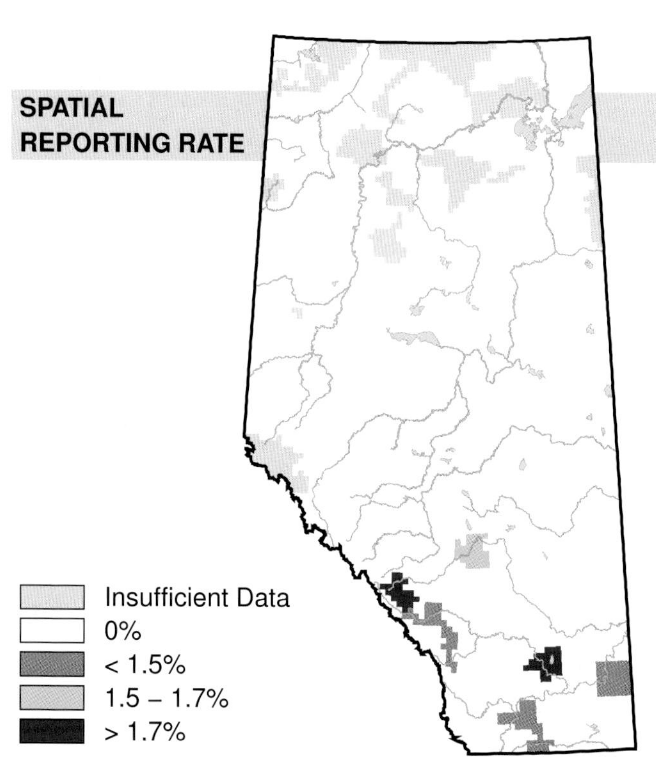

STATUS

GSWA 2000	GSWA 2005	COSEWIC Historic Rankings	COSEWIC 2007	Alberta Wildlife Act
Secure	Secure	N/A	N/A	N/A

Habitat  Nest Location  Nest Type 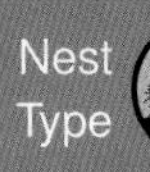 Diet 

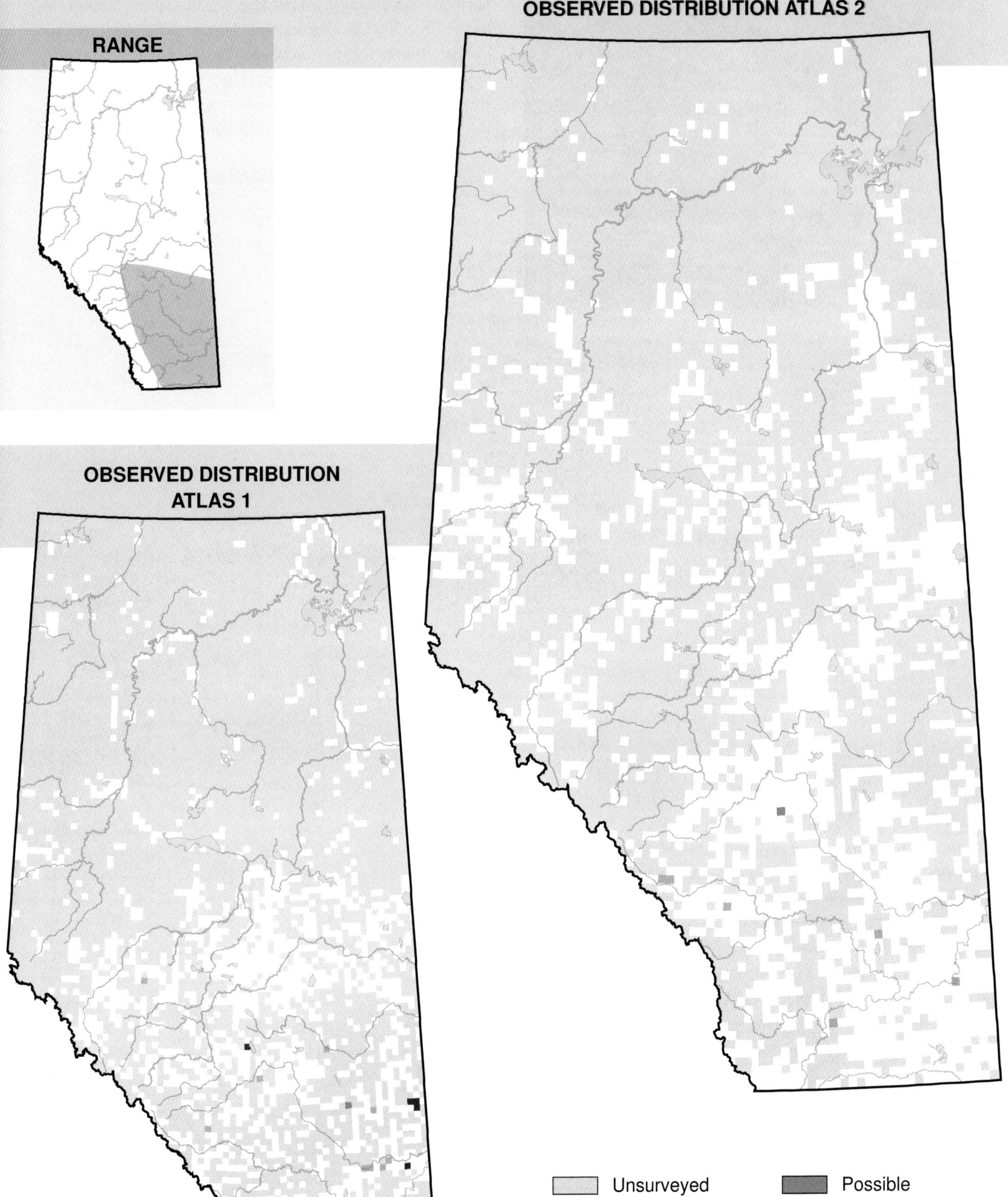

Sage Thrasher (*Oreoscoptes montanus*)

Photo: Damon Calderwood

NESTING

CLUTCH SIZE (EGGS):	4–5
INCUBATION (DAYS):	13–17
FLEDGING (DAYS):	11–14
NEST HEIGHT (METRES):	0–1.5

The Sage Thrasher is found in the southeastern corner of the province in the Grassland Natural Region. The distribution of this species did not change between Atlas 1 and Atlas 2 and reporting rates remained low for this species during Atlas 2.

As its name implies, the Sage Thrasher is closely associated with sagebrush. This species tends to build its nest in sagebrush; however, it will also occasionally build its nest in other types of shrubs, and will infrequently nest on the ground. This species breeds in arid open areas that are interspersed with shrubby vegetation. Because Alberta is on the northern periphery of the species' range and because it is associated with sagebrush, this species was found infrequently in the southern portion of the Grassland NR.

There was no change in relative abundance detected in any of the NRs, but sample sizes were too small in both Atlases to reach even our statistical threshold. Relative to other species, the Sage Thrasher was observed less frequently in the Grassland NR during Atlas 2 than during Atlas 1. The Breeding Bird Survey sample size was too small to investigate trends for the Sage Thrasher in Alberta or across Canada. Canada is on the northern periphery of this species' range, and so the Breeding Bird Survey sample sizes were small for this species. Likely the most important measure that can be taken to ensure the conservation of the Sage Thrasher is the maintenance of habitats that contain sagebrush. The status of this species is Undetermined in Alberta due to insufficient data.

TEMPORAL REPORTING RATE

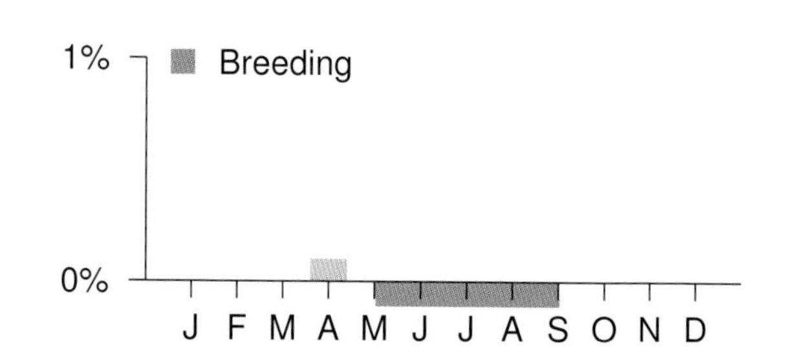

NATURAL REGIONS REPORTING RATE

SPATIAL REPORTING RATE

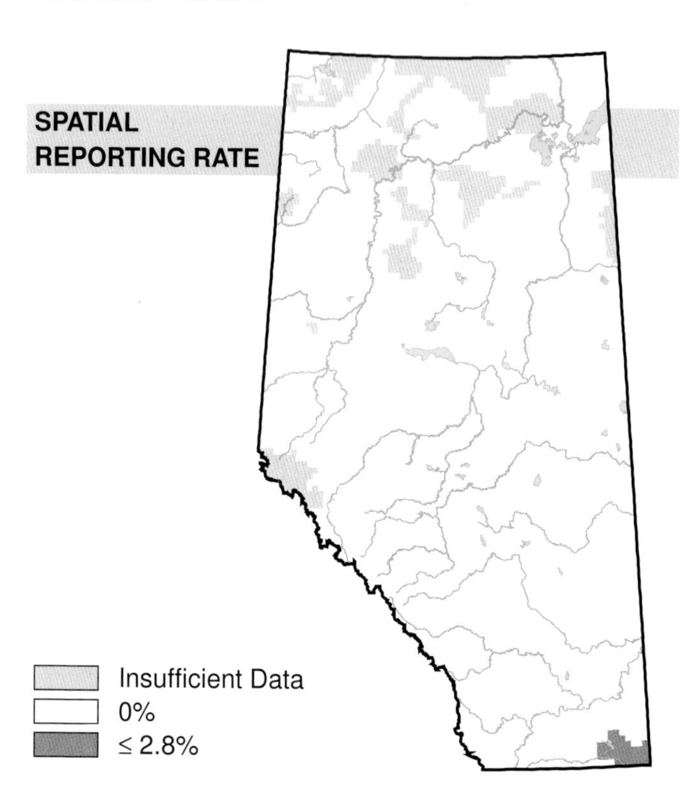

STATUS

GSWA 2000	GSWA 2005	COSEWIC Historic Rankings	COSEWIC 2007	Alberta Wildlife Act
Undetermined	Undetermined	Endangered	Endangered	N/A

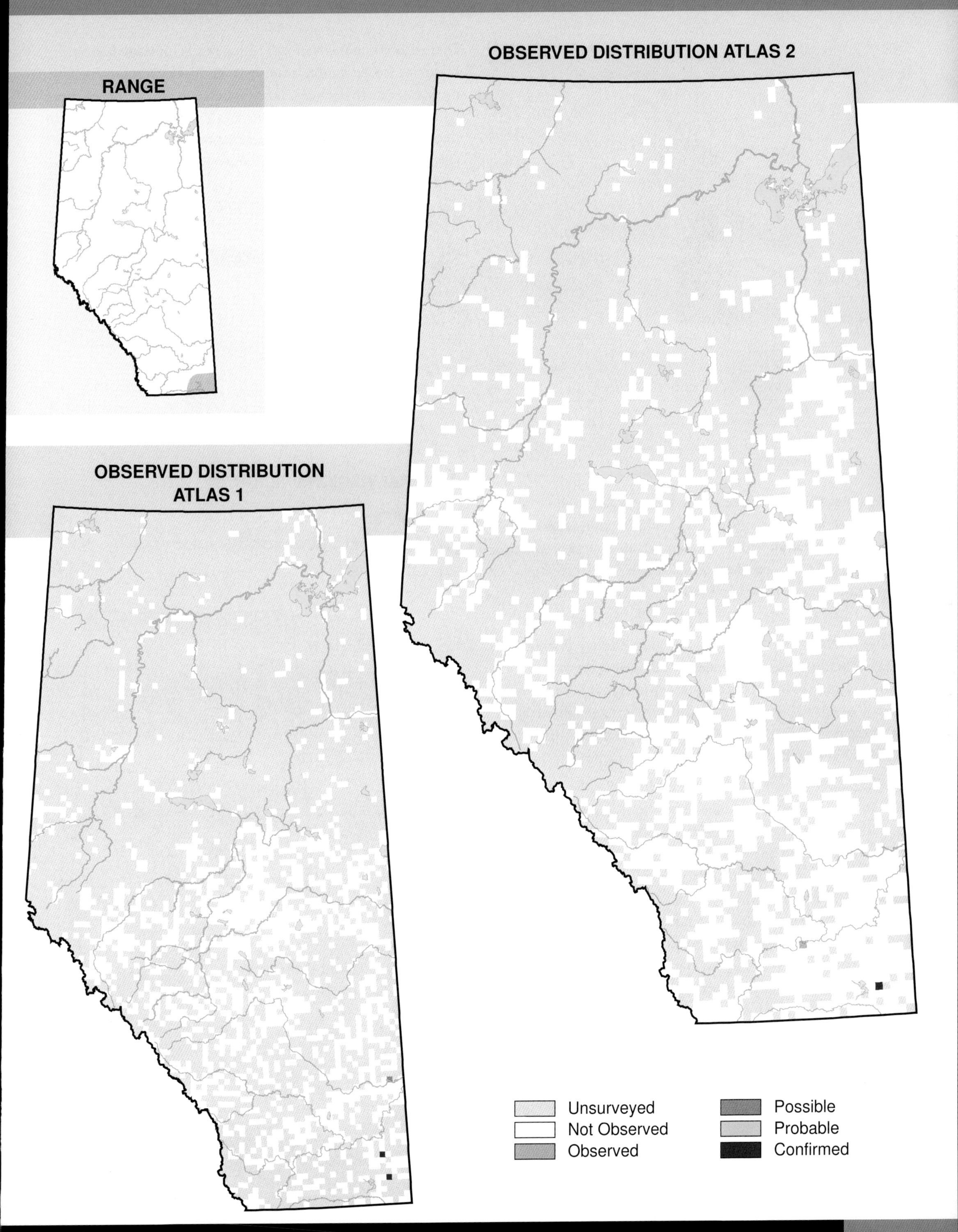
RANGE
OBSERVED DISTRIBUTION ATLAS 1
OBSERVED DISTRIBUTION ATLAS 2
Unsurveyed
Not Observed
Observed
Possible
Probable
Confirmed

Brown Thrasher (*Toxostoma rufum*)

Photo: Gerald Romanchuk

NESTING

CLUTCH SIZE (EGGS):	4–5
INCUBATION (DAYS):	10–12
FLEDGING (DAYS):	9–13
NEST HEIGHT (METRES):	<1

The Brown Thrasher breeds in the southern part of the province in the Grassland and Parkland Natural Regions. The observed distribution of this species did not change between Atlas 1 and Atlas 2, but there was a new possible breeding observation made in the Lac La Biche area during Atlas 2.

Similar to the Gray Catbird, the Brown Thrasher breeds in dense, shrubby vegetation near open areas that are often near water. This species is found usually along hedgerows and along rivers. As a consequence, this species was found mainly in the Grassland NR and was found only infrequently in the southern portion of the Parkland NR. A new possible breeding observation was reported in the Boreal Forest NR.

Declines in relative abundance were detected in the Grassland and Parkland NRs where, relative to other species, the Brown Thrasher was found less frequently in Atlas 2 than in Atlas 1. The Breeding Bird Survey did not reveal an abundance change in Alberta during the period 1985–2005, but the sample size was small. Similar to what was found by the Atlas, the Breeding Bird Survey showed a decline across Canada during the period 1985–2005. Declines could be related to changes in the abundance of shrub habitat that is available for this species. This habitat-type could be declining in riparian areas, one of the preferred breeding locations for this species in Alberta, due to the regulation of water flow by the damming of rivers.

Dunn (2005), in the National action needs for Canadian Landbird Conservation, confirms that this species is a concern primarily due to declines, possibly related to a decline in shrub habitat. The report recommends that a study of demography in important breeding locations and the maintainance of areas of shrub habitat for benefit of this and other shrub-community bird is required.

This species is considered Secure in Alberta.

TEMPORAL REPORTING RATE

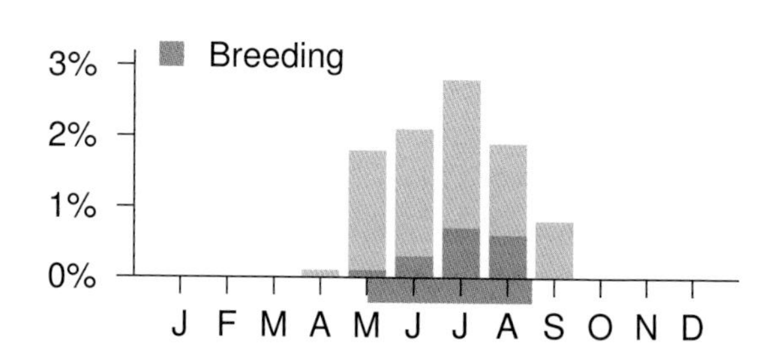

NATURAL REGIONS REPORTING RATE

Boreal Forest	< 0.1%
Grassland	6.2%
Parkland	0.6%

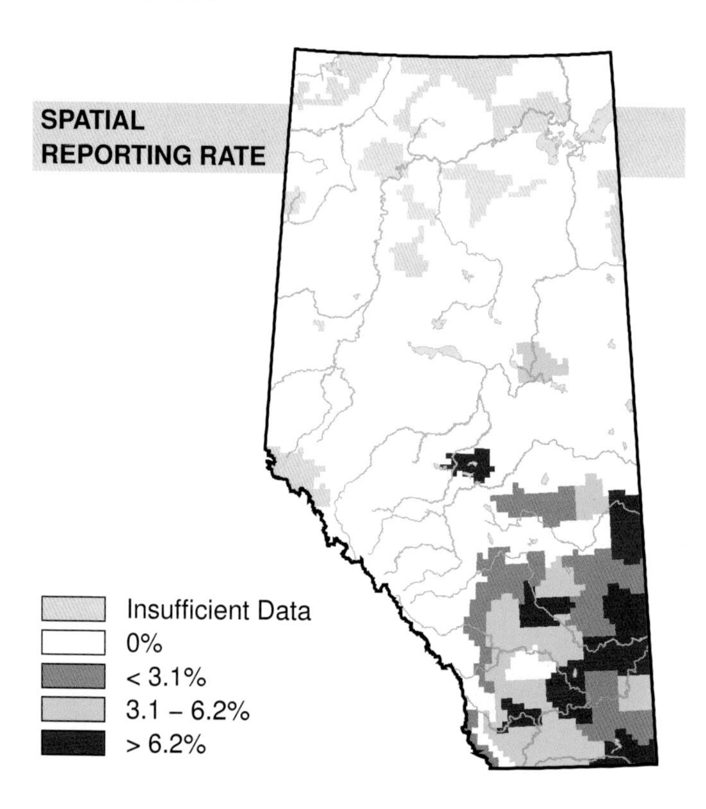

STATUS

GSWA 2000	GSWA 2005	COSEWIC Historic Rankings	COSEWIC 2007	Alberta Wildlife Act
Secure	Secure	N/A	N/A	N/A

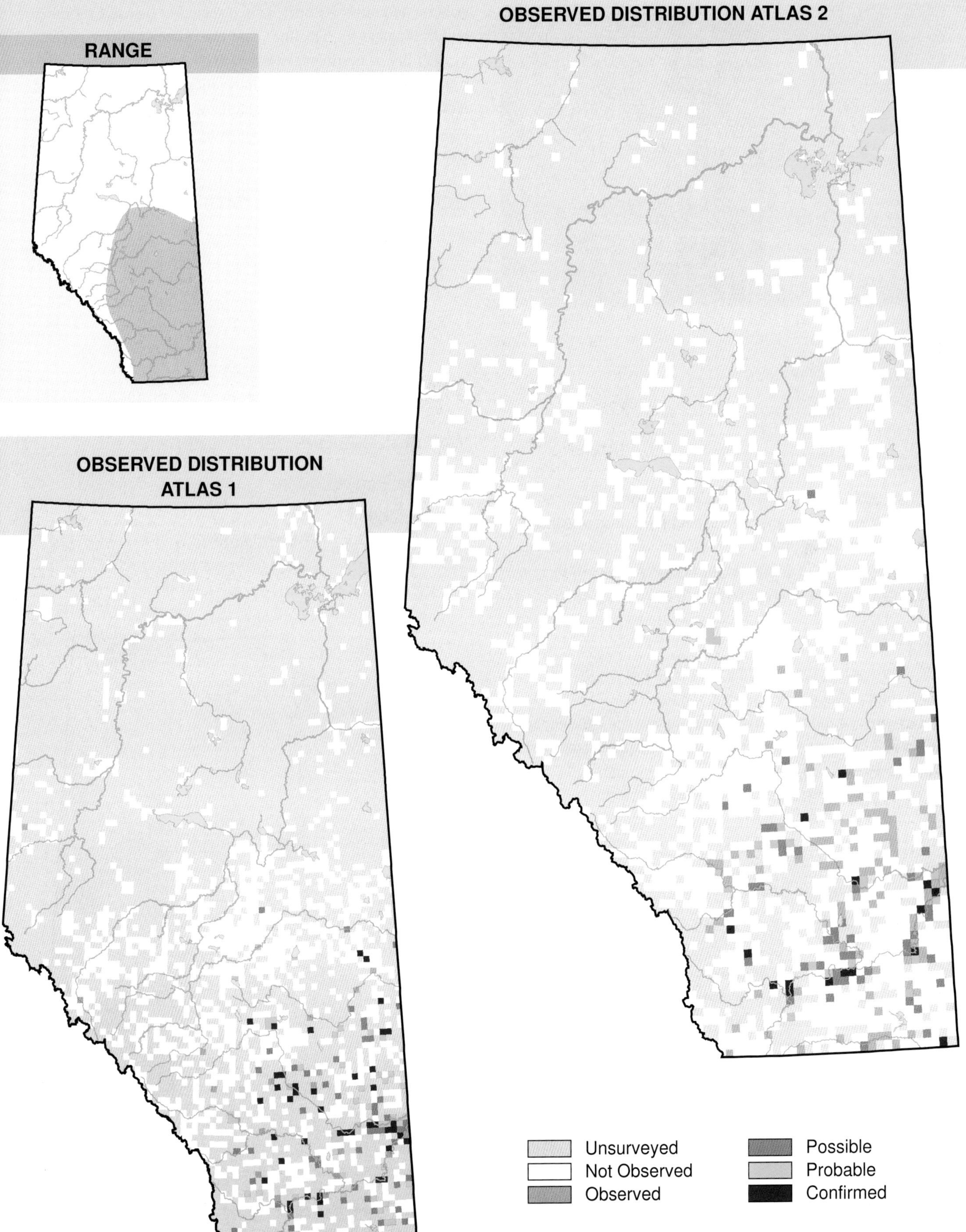
RANGE
OBSERVED DISTRIBUTION ATLAS 2
OBSERVED DISTRIBUTION ATLAS 1
Unsurveyed
Not Observed
Observed
Possible
Probable
Confirmed

European Starling (*Sturnus vulgaris*)

Photo: Gerald Romanchuk

NESTING

CLUTCH SIZE (EGGS):	5–7
INCUBATION (DAYS):	12–15
FLEDGING (DAYS):	20–22
NEST HEIGHT (METRES):	0.6–18.3

Numbering more than 200 million across North America, this introduced species is found in all Natural Regions in Alberta except the Canadian Shield. The distribution of the European Starling decreased in Atlas 2 in the northern extent of its range. No breeding records were found north of Fort McMurray and Grande Prairie in Atlas 2 but they were found there in Atlas 1. There was also a reduction in the number of confirmed breeding records from areas north and west of Edmonton.

Inhabiting a wide variety of areas, the European Starling prefers agricultural and settled areas, and avoids large expanses of wooded or forested areas. This species requires open areas where it can forage for fruits, berries, grains, and seeds, and it needs a cavity for nesting. These cavities occur in a variety of structures such as cliffs, burrows, natural cavities in trees, nest boxes, and openings in buildings. For this reason, they were often encountered in the Grassland NR, and were found less commonly in the Parkland and Rocky Mountain NRs. This species was not detected often in the Boreal Forest and Foothills NRs where forests dominate the landscape.

Declines in relative abundance were detected in the Boreal Forest, Foothills, Grassland, and Parkland NRs. Relative to other species, the European Starling was also observed less frequently in these NRs. No change was detected in the Rocky Mountain NR. Declines in abundance were also found by the Breeding Bird Survey on a Canada-wide scale during the period 1985–2005. No change in abundance was found in Alberta, although declines were detected in British Columbia and Saskatchewan. Declines could be related to decreased availability of nesting cavities, or to changes in agricultural practices such as agricultural intensification. Populations in North America seem to have stabilized, with some regions showing increases and others showing decreases. The European Starling is considered Exotic/Alien in the province.

TEMPORAL REPORTING RATE

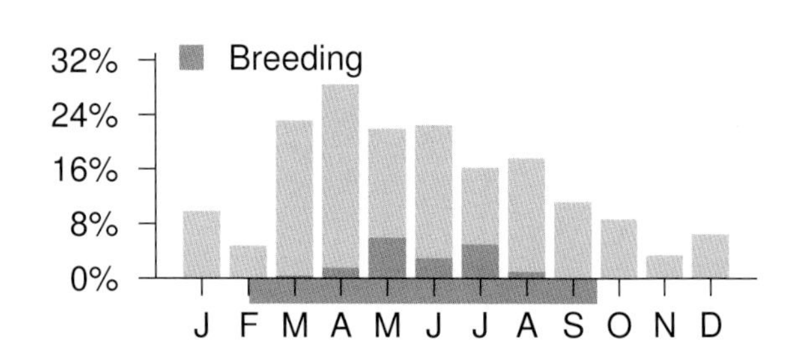

NATURAL REGIONS REPORTING RATE

SPATIAL REPORTING RATE

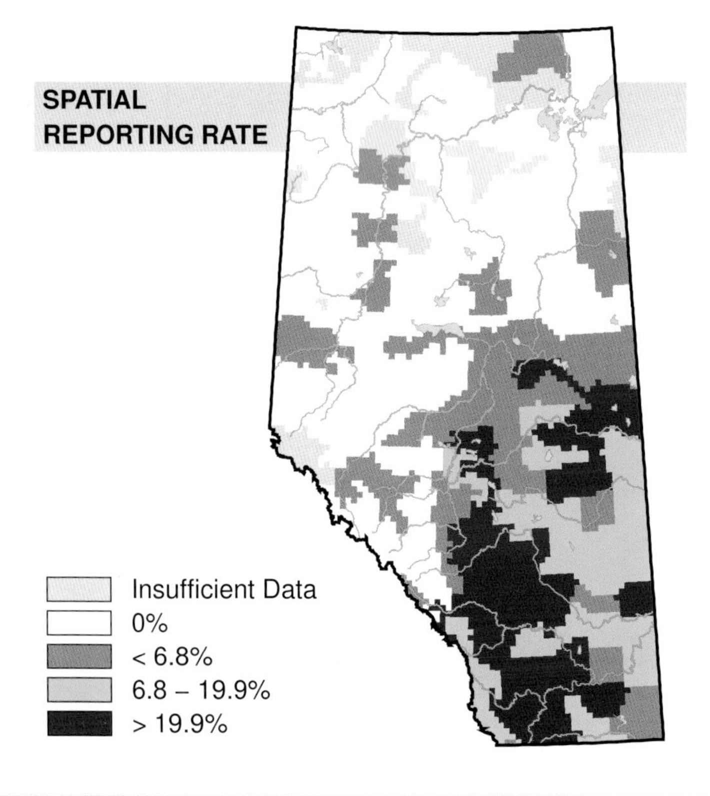

STATUS

GSWA 2000	GSWA 2005	COSEWIC Historic Rankings	COSEWIC 2007	Alberta Wildlife Act
Exotic/Alien	Exotic/Alien	N/A	N/A	N/A

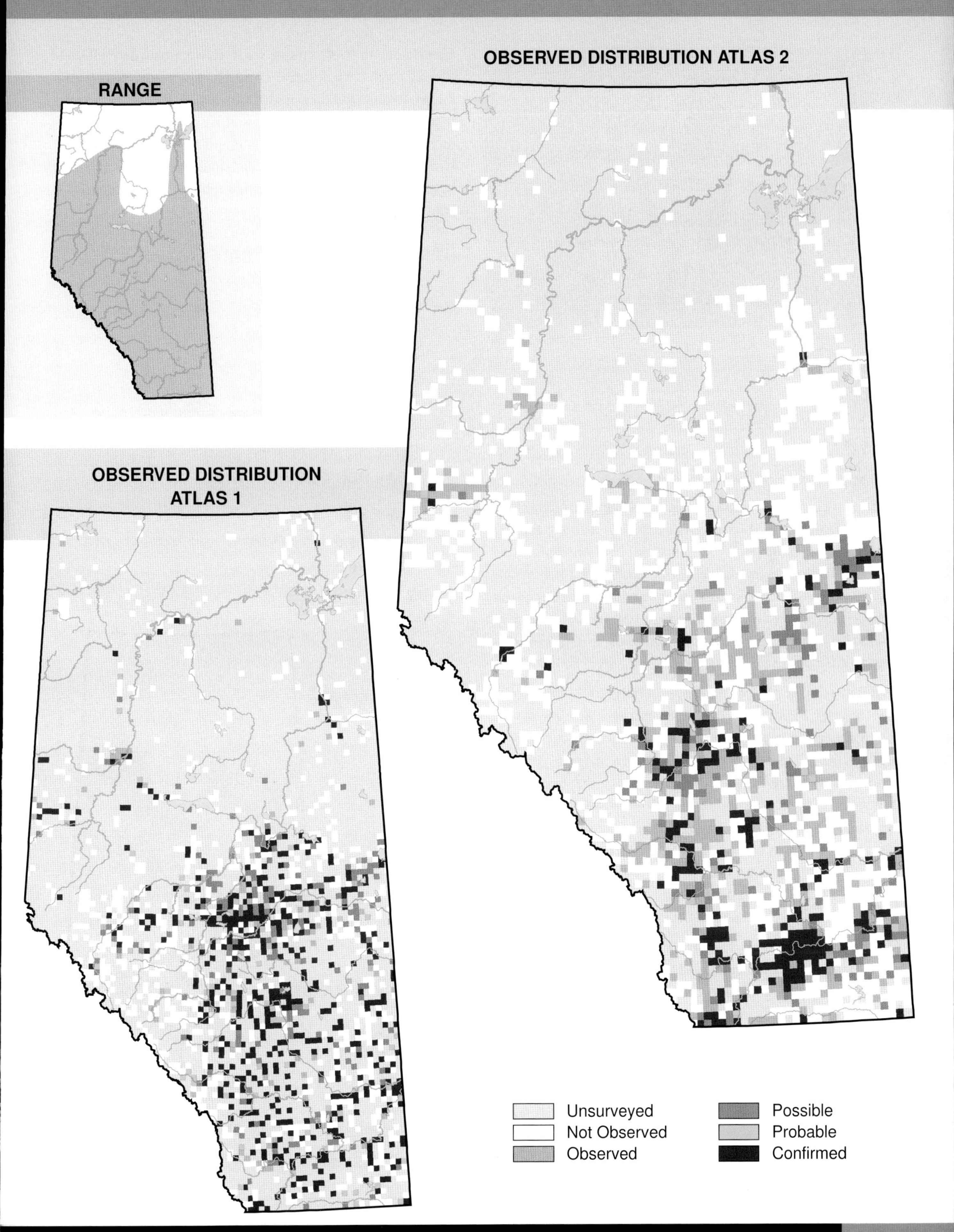
RANGE
OBSERVED DISTRIBUTION ATLAS 2
OBSERVED DISTRIBUTION
ATLAS 1
Unsurveyed
Not Observed
Observed
Possible
Probable
Confirmed

American Pipit (*Anthus rubescens*)

Photo: Gerald Romanchuk

NESTING

CLUTCH SIZE (EGGS):	4–6
INCUBATION (DAYS):	13–14
FLEDGING (DAYS):	14–16
NEST HEIGHT (METRES):	0

The American Pipit breeds in the Rocky Mountain Natural Region of Alberta. The observed distribution of this species did not change between Atlas 1 and Atlas 2. However, in Atlas 2 no breeding observations were made north of the Banff region. In Atlas 1 breeding observations were made northward through Saskatchewan River Crossing to Jasper and Grande Cache. This change is likely a function of the differences in field protocols between Atlas 1 and 2.

Found in arctic and alpine tundra regions of North America, the American Pipit prefers high-elevation subalpine meadows for breeding. Ground nests are built in wet and dry meadows, tussocks, or erosion banks, protected by overhanging vegetation. Because of this habitat preference, they were found most often in the Rocky Mountain NR. They were observed occasionally in the Boreal Forest, Grassland, and Parkland NRs; however, these records are considered to have been observations of non-breeders.

The relative abundance of the American Pipit increased in the Grassland and Parkland NRs, but these are not considered breeding observations and are likely, therefore, a function of migratory variation and not an indicator of population change. Declines in relative abundance were detected in the Rocky Mountain NR. Relative to other species, the American Pipit was observed less often in the Rocky Mountain NR. The Breeding Bird Survey has no trend data available for Alberta or Canada due to small sample sizes. Recreational activities and mining can have a negative effect on these birds, as they are known to abandon their nests if disturbed. Due to the remote nature of their breeding grounds, this species is encountered most often by people when it is migrating and during the winter. Christmas Bird Count data indicate that this species has declined throughout its wintering grounds. This species is considered Secure in the province.

TEMPORAL REPORTING RATE

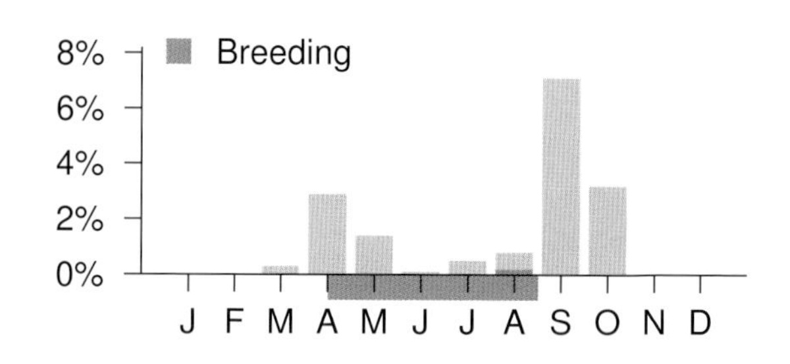

NATURAL REGIONS REPORTING RATE

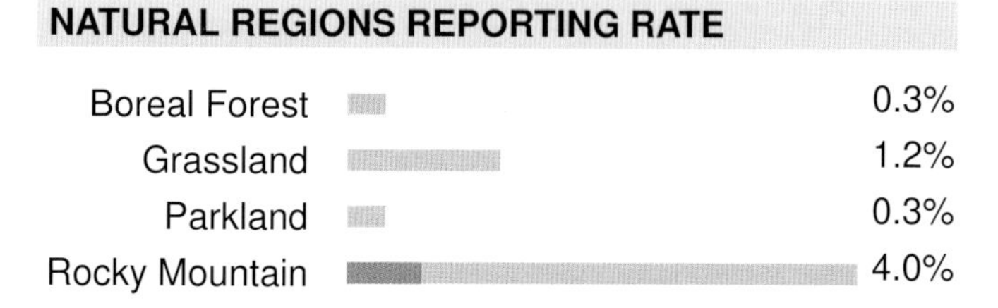

SPATIAL REPORTING RATE

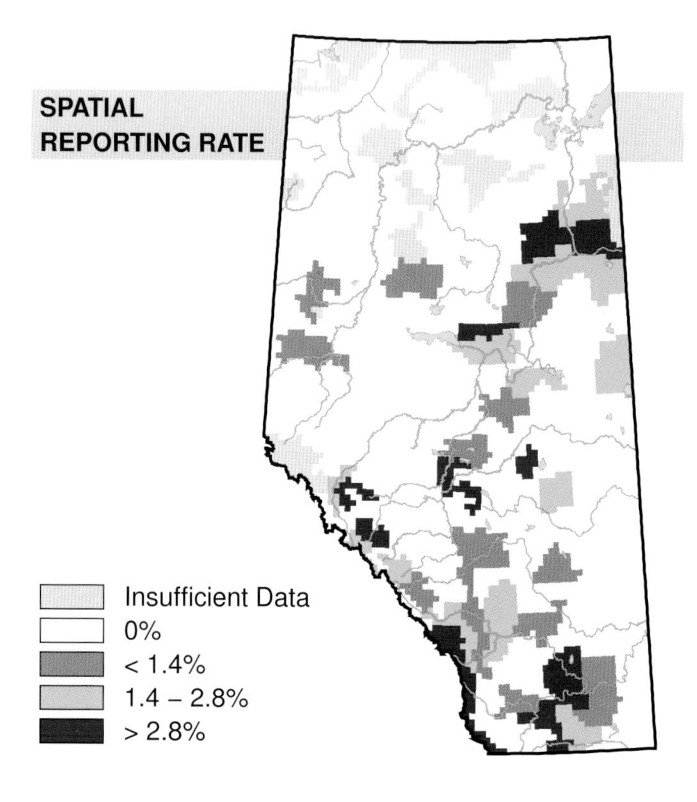

STATUS

GSWA 2000	GSWA 2005	COSEWIC Historic Rankings	COSEWIC 2007	Alberta Wildlife Act
Secure	Secure	N/A	N/A	N/A

RANGE

OBSERVED DISTRIBUTION ATLAS 2

OBSERVED DISTRIBUTION ATLAS 1

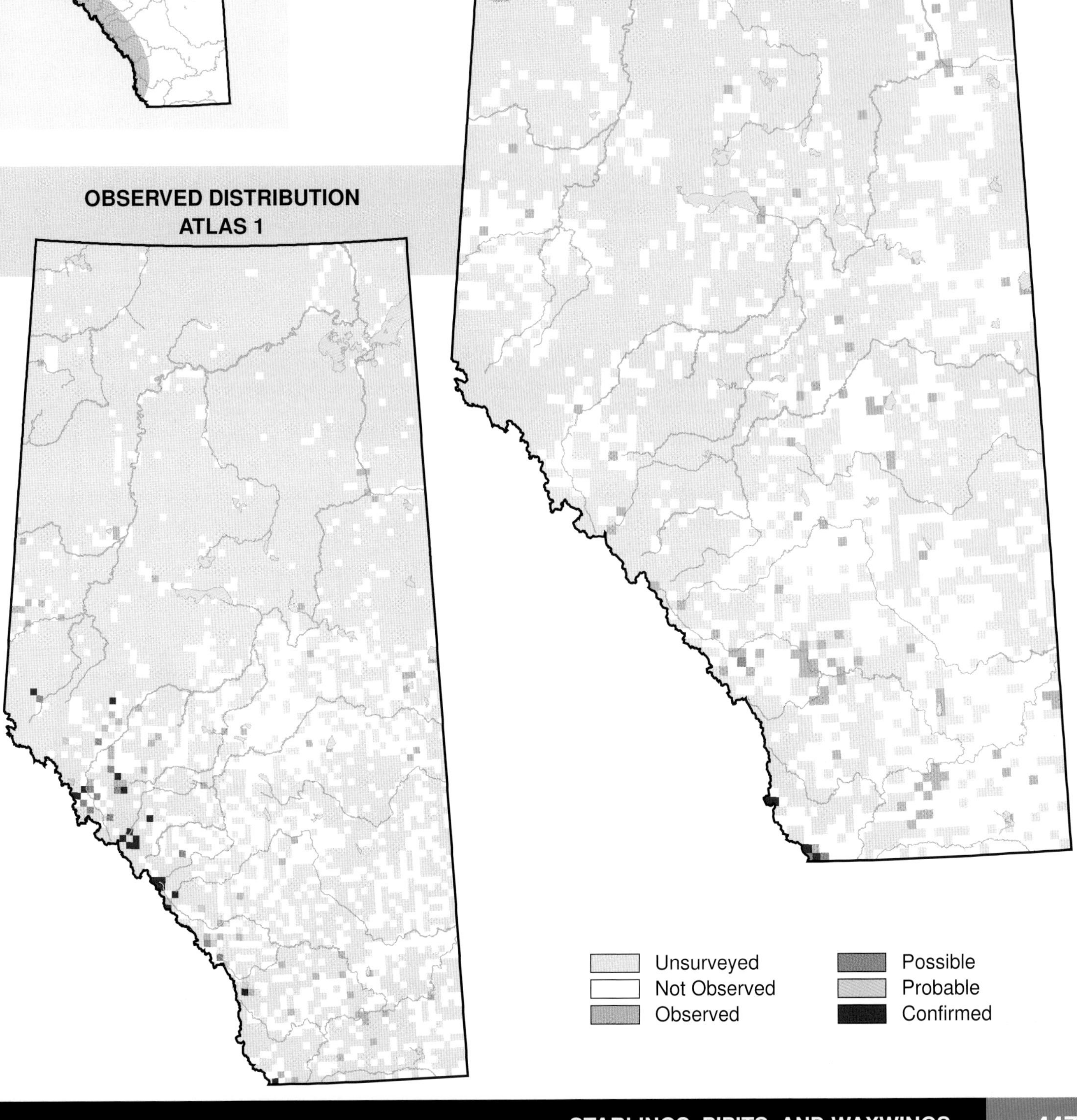

Sprague's Pipit *(Anthus spragueii)*

Photo: Gordon Court

NESTING

CLUTCH SIZE (EGGS):	3–5
INCUBATION (DAYS):	9–12
FLEDGING (DAYS):	10–11
NEST HEIGHT (METRES):	0

Sprague's Pipit occurs mainly in the Grassland Natural Region and shows no appreciable change in distribution between Atlas 1 and Atlas 2.

The highest reporting rates are from the central portion of the Grasslands NR. This is eminently consistent with their preference for large blocks of native grasses of moderate height and with few or no shrubs (Robbins and Dale, 1999). The reporting rates are lower where woody cover encroaches into grassland.

Increases in relative abundance were detected in the Grassland NR. Decreases in relative abundance were detected in the Boreal and Parkland NR. No changes in relative abundance were detected in the other NRs. Relative to other species, the frequency of occurrence data show the same pattern. Decreases in the north are consistent with the loss of 20% of native grass in the Parkland (Watmough and Schmoll, 2007). The increase in the Grassland NR may be a sampling artifact, as many of the Atlas 2 records occur in squares which were not sampled in Atlas 1 but which fall in the heart of the species' range. The Breeding Bird Survey did not detect a change provincially but specialized grassland bird surveys show declines of about 10% (Canadian Wildlife Service unpublished data). Provincial status reports from both 2000 and 2005 rate Sprague's Pipit as Sensitive; it is listed provincially as a Species of Special Concern and federally as Threatened.

Although still relatively common in areas where grassland is intact, it has shown consistent declines throughout the prairies. The species is area-sensitive (Davis, 2004). Its dependence on native grass in good-to-excellent range condition, its reluctance to use crops or planted cover (Robbins and Dale, 1999), and its reduced numbers near roads (Sutter et al., 2000) make it vulnerable to habitat loss and to habitat degradation due to exotic plant species, intensification of livestock operations, and linear disturbance.

TEMPORAL REPORTING RATE

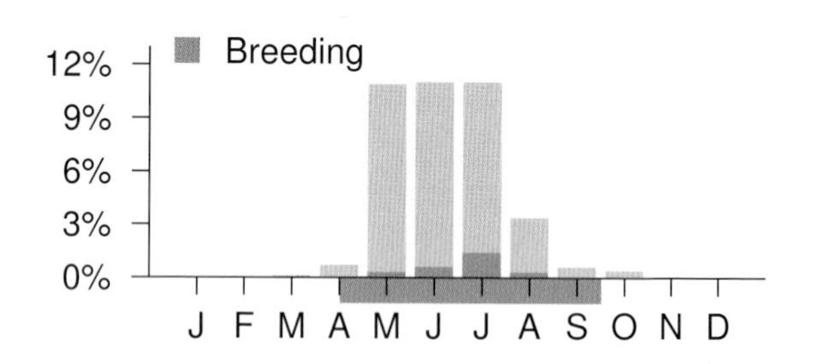

NATURAL REGIONS REPORTING RATE

SPATIAL REPORTING RATE

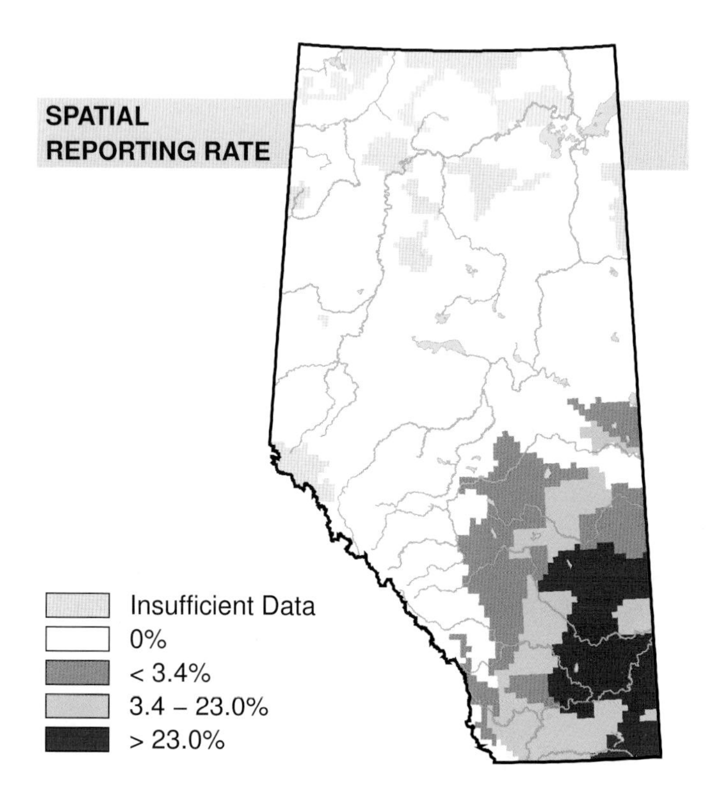

STATUS

GSWA 2000	GSWA 2005	COSEWIC Historic Rankings	COSEWIC 2007	Alberta Wildlife Act
Sensitive	Sensitive	Threatened	Threatened	N/A

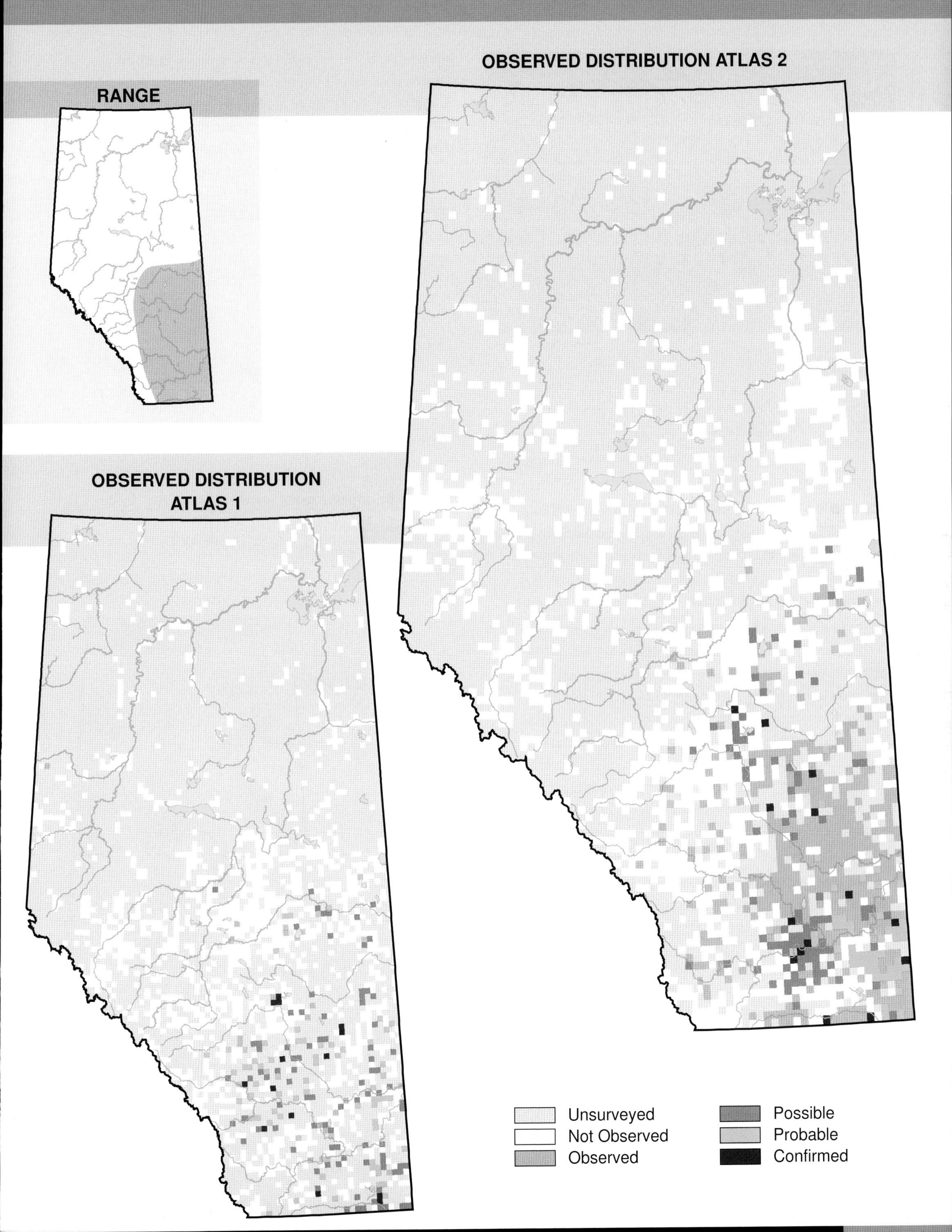

Habitat
Nest Location
Nest Type
Diet
RANGE
OBSERVED DISTRIBUTION ATLAS 2
OBSERVED DISTRIBUTION ATLAS 1
Unsurveyed
Not Observed
Observed
Possible
Probable
Confirmed

Bohemian Waxwing (*Bombycilla garrulus*)

Photo: Gerald Romanchuk

NESTING

CLUTCH SIZE (EGGS):	4–6
INCUBATION (DAYS):	13–14
FLEDGING (DAYS):	15–17
NEST HEIGHT (METRES):	1–6

Bohemian Waxwings breed in the Rocky Mountain, Foothills, and Boreal Forest Natural Regions, and at the northern limit of the Parkland NR. No change in distribution was observed between Atlas 1 and Atlas 2, although there were fewer observations in the northwestern corner of Alberta in Atlas 2.

Woodland edges in open coniferous or mixedwood forests containing birch are the preferred breeding habitat for the Bohemian Waxwing. These birds often nest in recently burned areas near water, which are usually in lowlands. Nests are made with hanging lichens on horizontal limbs near the trunk of old conifers. Habitat in second-growth or open forest is important due to the presence of a greater abundance of fruiting trees and open perches for foraging in these types of areas. They were therefore found most commonly in the Rocky Mountain, Foothills, Boreal Forest, and at the northern limit of the Parkland NRs. Grassland records were of non-breeders.

The occurrence of fruiting trees directly influences the distribution and abundance of this species. Populations of waxwings increased in the 1970s with the increased planting of ornamental fruit trees like the Mountain Ash. Declines in relative abundance were detected in the Boreal Forest NR where, relative to other species, the Bohemian Waxwing was observed less frequently in Atlas 2 than in Atlas 1. Although the biological cause for this decline is uncertain, there is no known bias to account for this change and there are many squares surveyed in both Atlases in which this bird was observed in Atlas 1 but not in Atlas 2. The Breeding Bird Survey found no change in abundance during the period 1985–2005 on a Canada-wide scale. No trends could be determined for Alberta due to the low numbers recorded. The Bohemian Waxwing is considered Secure in Alberta.

TEMPORAL REPORTING RATE

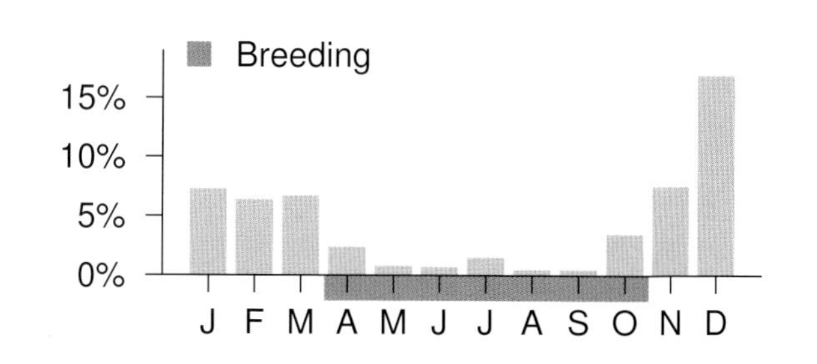

NATURAL REGIONS REPORTING RATE

Boreal Forest	0.6%
Foothills	2.1%
Grassland	0.8%
Parkland	0.7%
Rocky Mountain	2.9%

SPATIAL REPORTING RATE

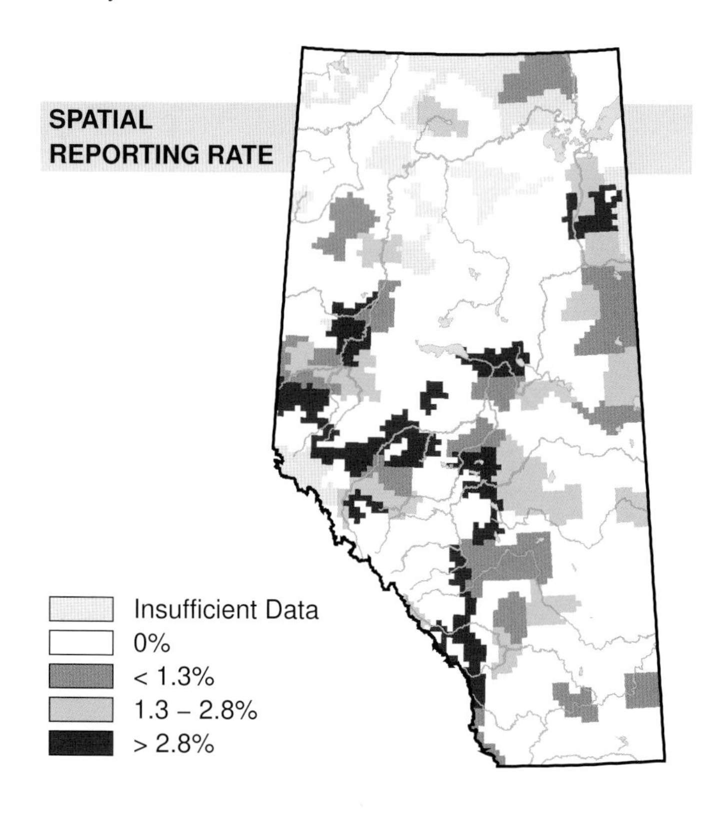

STATUS

GSWA 2000	GSWA 2005	COSEWIC Historic Rankings	COSEWIC 2007	Alberta Wildlife Act
Secure	Secure	N/A	N/A	N/A

Habitat

Nest
Location

Nest
Type

Diet

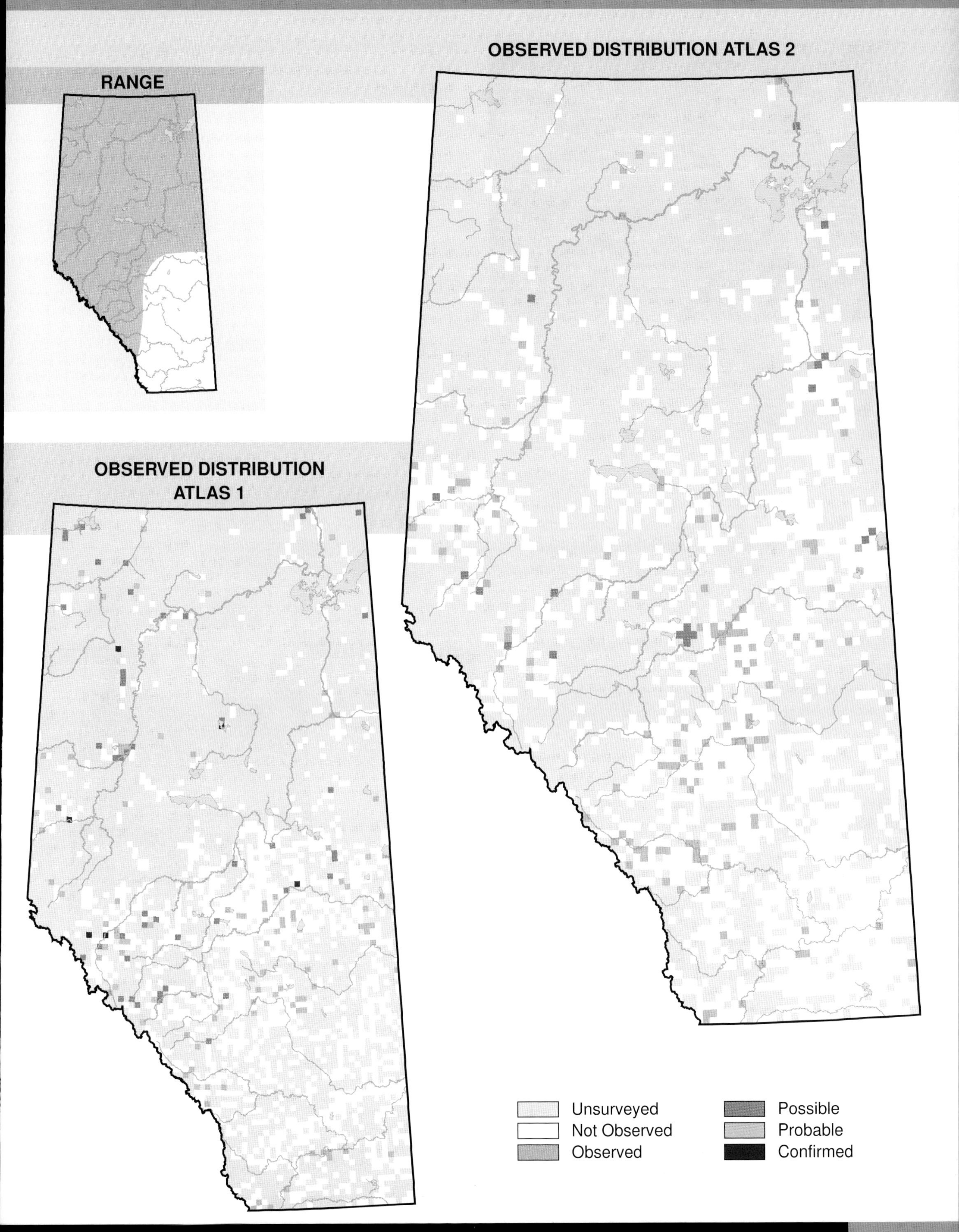
RANGE
OBSERVED DISTRIBUTION ATLAS 2
OBSERVED DISTRIBUTION
ATLAS 1
Unsurveyed
Not Observed
Observed
Possible
Probable
Confirmed

Cedar Waxwing (*Bombycilla cedrorum*)

Photo: Robert Gehlert

NESTING

CLUTCH SIZE (EGGS):	4–5
INCUBATION (DAYS):	12–14
FLEDGING (DAYS):	14–18
NEST HEIGHT (METRES):	>2

Cedar Waxwings breed in every Natural Region in Alberta. There was no change in distribution observed between Atlas 1 and Atlas 2. Fewer possible and probable breeding records were found in the northwestern part of the province during Atlas 2.

This species prefers open woodlands (deciduous, coniferous, and mixed) and old fields, and avoids interior forest habitat. Riparian areas and farms in the grasslands are also breeding sites. This species nests along edges or in open-forested regions in a variety of tree and shrub species. There is a preference for younger stands of trees in proximity to water. For this reason, Cedar Waxwings are distributed widely across the province, but are found most commonly in the Boreal Forest, Foothills, Parkland, and Rocky Mountain NRs. They are less common in the Grassland NR.

Declines in relative abundance were detected in the Boreal Forest, Grassland, and Parkland NRs where, relative to other species, Cedar Waxwings were observed less frequently in Atlas 2 than in Atlas 1. No change was detected in the other NRs. Similar to what was found by the Atlas, the Breeding Bird Survey detected a decline in abundance in Canada during the period 1985–2005. No change in abundance was found in Alberta between 1985 and 2005, but a decline was detected for the period 1995–2005. These declines are unexpected as this species is often associated with human settlement and the fruit-bearing trees associated with settlement. Further research is needed to assess the cause of these declines. The Cedar Waxwing is considered Secure in the province.

TEMPORAL REPORTING RATE

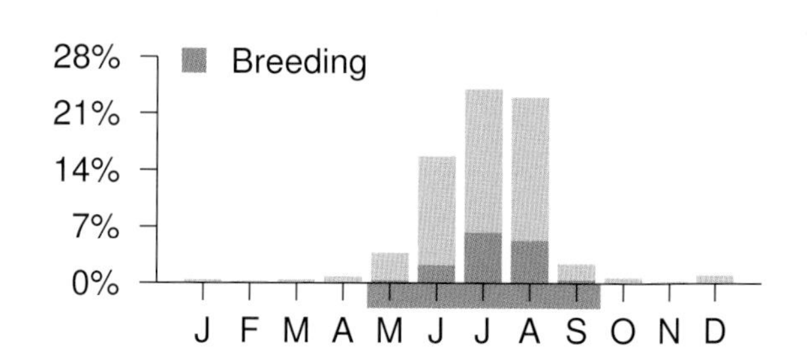

NATURAL REGIONS REPORTING RATE

SPATIAL REPORTING RATE

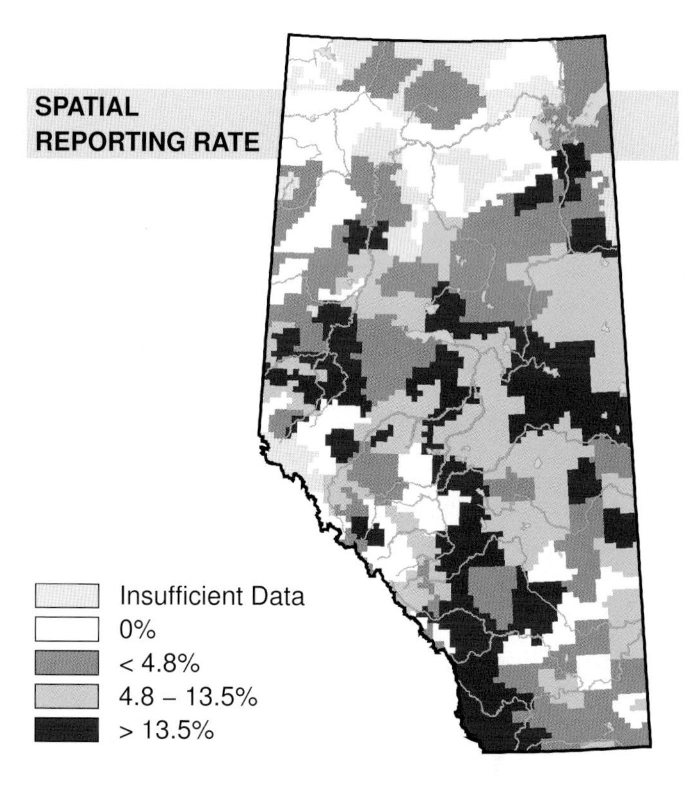

STATUS

GSWA 2000	GSWA 2005	COSEWIC Historic Rankings	COSEWIC 2007	Alberta Wildlife Act
Secure	Secure	N/A	N/A	N/A

Habitat
Nest
Location

Nest
Type

Diet

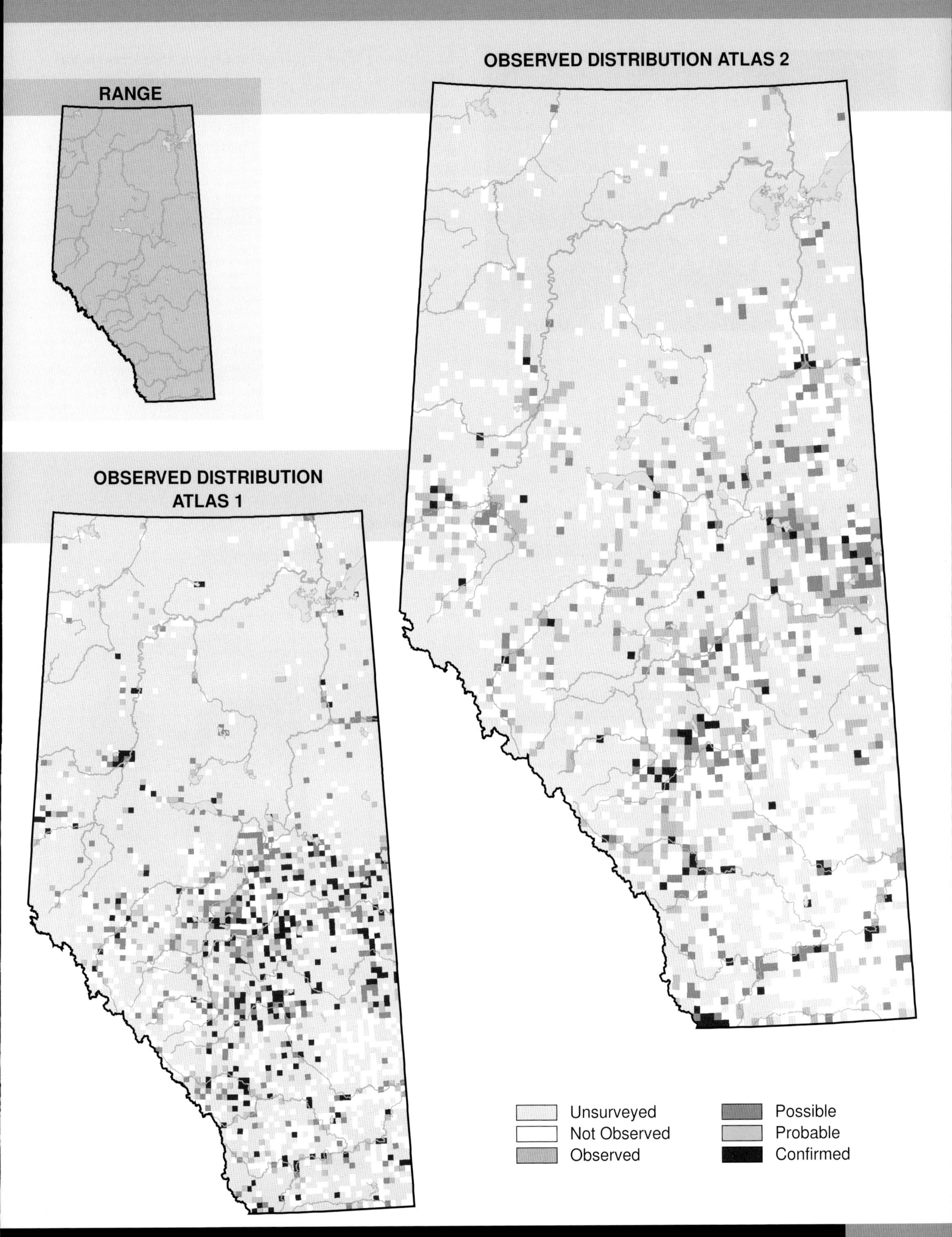
RANGE
OBSERVED DISTRIBUTION ATLAS 2
OBSERVED DISTRIBUTION
ATLAS 1
Unsurveyed
Not Observed
Observed
Possible
Probable
Confirmed

Tennessee Warbler (*Vermivora peregrina*)

Photo: Duane Boone

NESTING

CLUTCH SIZE (EGGS):	4–6
INCUBATION (DAYS):	11–12
FLEDGING (DAYS):	11–12
NEST HEIGHT (METRES):	0

One of the most common wood-warblers in Alberta, the Tennessee Warbler is found in every Natural Region. There was no change in distribution observed between Atlas 1 and Atlas 2.

The ground-nesting Tennessee Warbler breeds in deciduous, coniferous, and mixedwood forests with associated open areas containing grasses, dense shrubs, and clumps of young deciduous trees. As a result, this species was most commonly found in the Boreal Forest and Foothills NRs. It was less common in the Parkland and Rocky Mountain NRs. Records of this species from the Grassland NR were of non-breeders, except for three records of birds heard singing in suitable habitat.

Increases in relative abundance were detected in the Boreal Forest, Foothills, and Grassland NRs where, relative to other species, the Tennessee Warbler was observed more frequently in Atlas 2 than in Atlas 1. An increase was also detected in the Parkland NR where surprisingly, relative to other species, the Tennessee Warbler was observed less often in Atlas 2 than in Atlas 1. No change was detected in the Rocky Mountain NR. Since the Grassland records were made during migration, this increase is probably an artifact of the broader survey window in Atlas 2. Increases in other NRs are corroborated by the Breeding Bird Survey which found an increase in Alberta during the period 1985–2005. Abundance of this species could be increasing because it often breeds near open areas such as those that are created by forest fragmentation or forest harvesting. In addition, this species can breed in young, regenerating, post-disturbance forests. Therefore, their populations should do well in managed forests. The provincial status of this species is Secure.

TEMPORAL REPORTING RATE

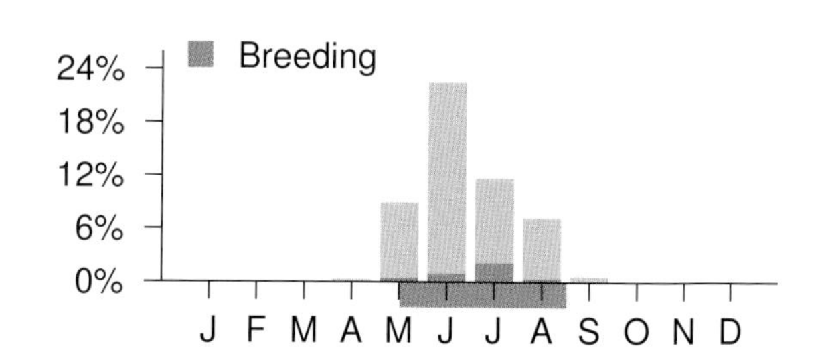

NATURAL REGIONS REPORTING RATE

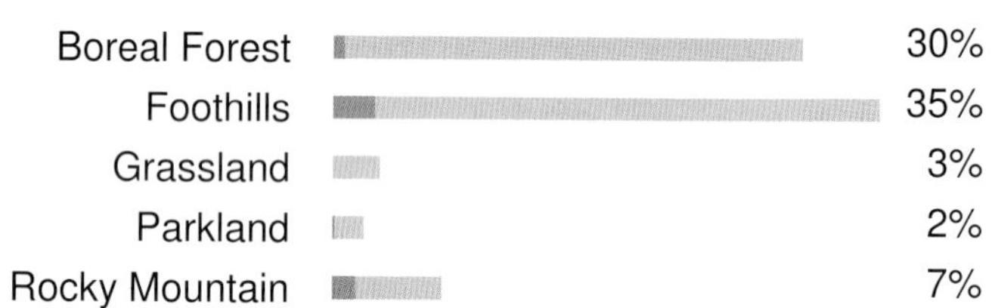

SPATIAL REPORTING RATE

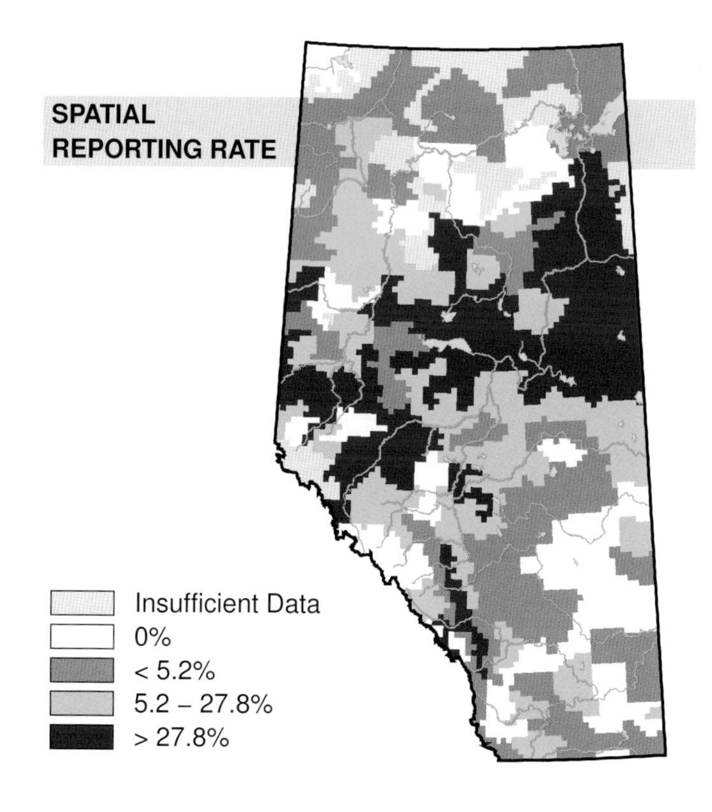

STATUS

GSWA 2000	GSWA 2005	COSEWIC Historic Rankings	COSEWIC 2007	Alberta Wildlife Act
Secure	Secure	N/A	N/A	N/A

Habitat

Nest Location

Nest Type 

Diet

RANGE

OBSERVED DISTRIBUTION ATLAS 2

OBSERVED DISTRIBUTION ATLAS 1

Unsurveyed
Not Observed
Observed
Possible
Probable
Confirmed

Orange-crowned Warbler (*Vermivora celata*)

Photo: Royal Alberta Museum

NESTING

CLUTCH SIZE (EGGS):	4–6
INCUBATION (DAYS):	12–14
FLEDGING (DAYS):	8–11
NEST HEIGHT (METRES):	0

Often numerous in suitable habitat, the nondescript Orange-crowned Warbler is found in every Natural Region of Alberta. No change in distribution was observed between Atlas 1 and Atlas 2 for this species.

The Orange-crowned Warbler breeds in a wide variety of forested habitats provided that a dense understorey is present. This species was most common in the Rocky Mountain and Foothills NRs and was less common in the Boreal Forest and Parkland NRs. The need for forested stands prevents these birds from nesting in the Grassland NR, but they may nest in the transition zone between the Grassland and Parkland NRs.

Increases in the relative abundance of this species were detected in the Boreal Forest, Grassland, and Parkland NRs. As the Grassland records were primarily non-breeding individuals, though, this increase is probably a function of migratory variation. The Breeding Bird Survey found no change in abundance in Alberta. However, the increases detected by the Atlas are mirrored by the increase in abundance in the Canada-wide Breeding Bird Survey during the period 1968–2005. As a species dependent upon secondary growth, the increases detected in the Boreal Forest and Parkland NRs are probably related to the effects of forestry as well as oil and gas activities in those NRs. The majority of BBS routes are located in the southern parts of Alberta. Thus, it seems likely that the more inclusive northern coverage by the Atlas accounts for some of the difference between the Atlas and Alberta BBS data results. The provincial status of this species is Secure in Alberta.

TEMPORAL REPORTING RATE

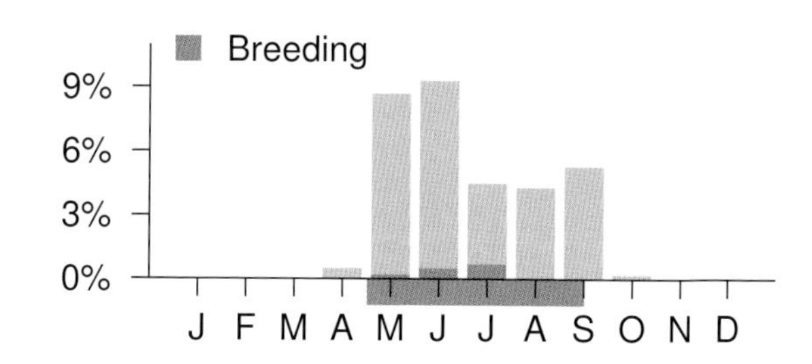

NATURAL REGIONS REPORTING RATE

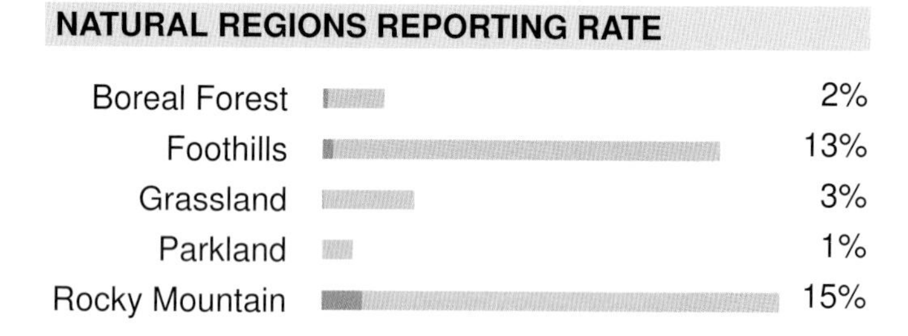

SPATIAL REPORTING RATE

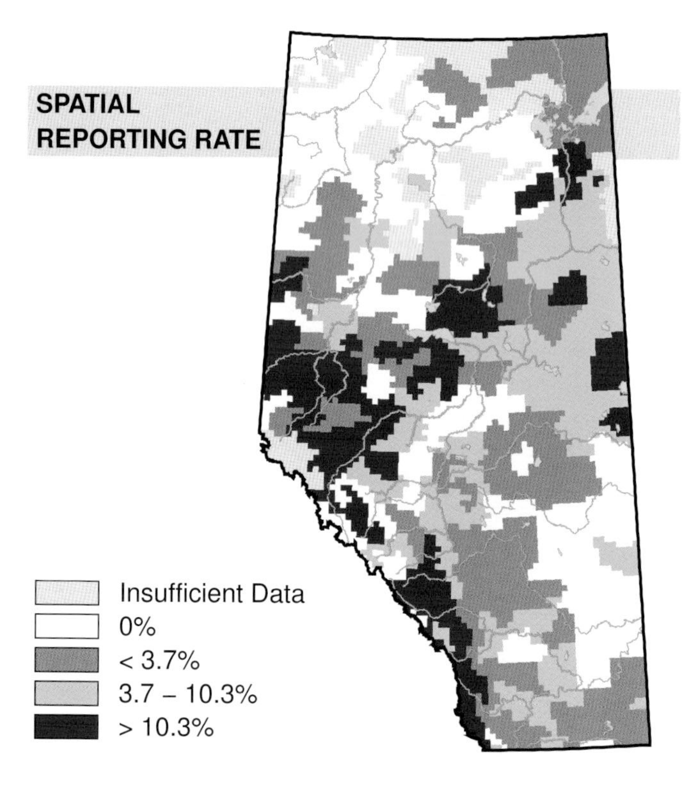

STATUS

GSWA 2000	GSWA 2005	COSEWIC Historic Rankings	COSEWIC 2007	Alberta Wildlife Act
Secure	Secure	N/A	N/A	N/A

Habitat Nest Location Nest Type Diet 

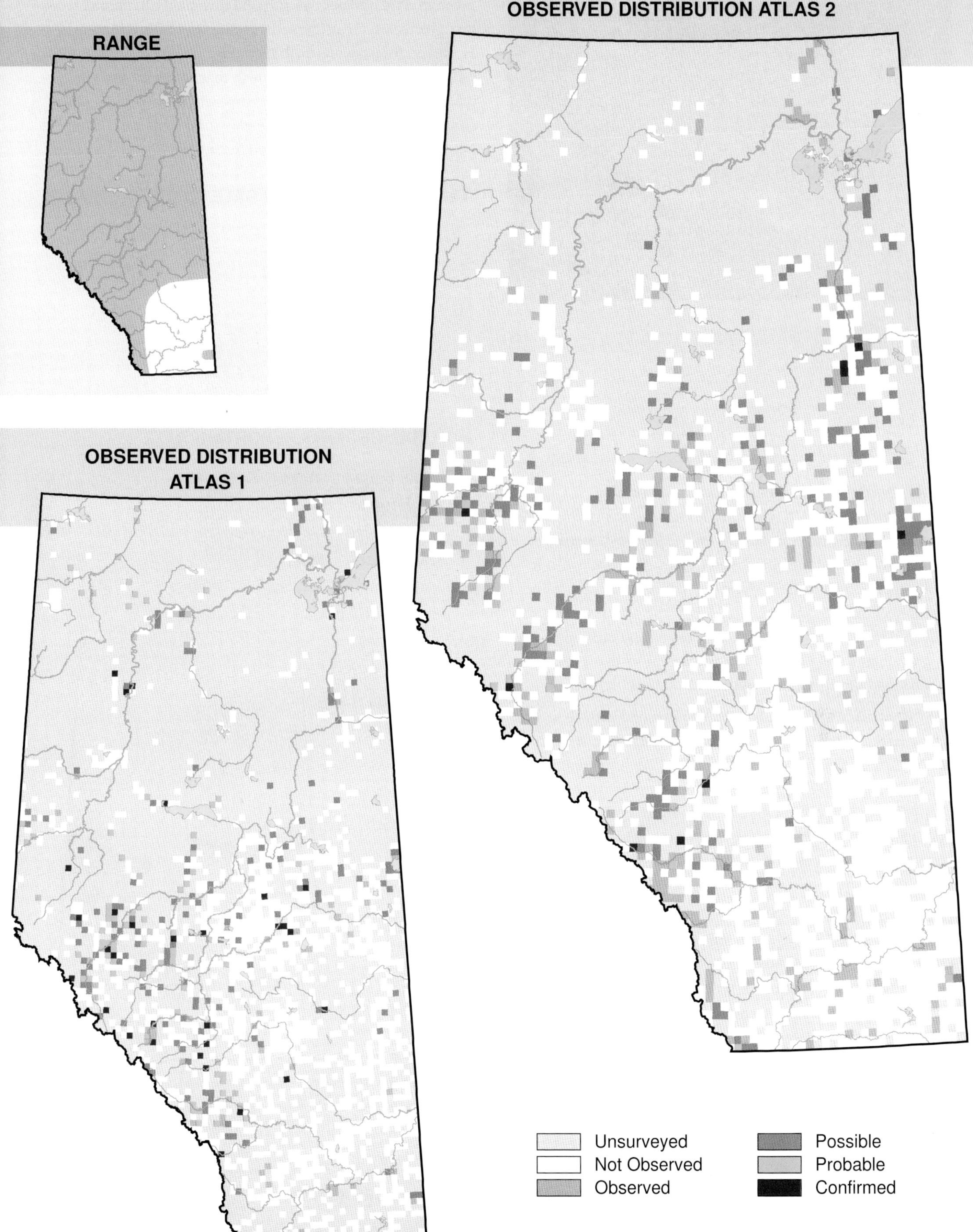

Nashville Warbler (*Vermivora ruficapilla*)

Photo: Royal Alberta Museum

NESTING

CLUTCH SIZE (EGGS):	4–5
INCUBATION (DAYS):	11–12
FLEDGING (DAYS):	11
NEST HEIGHT (METRES):	0

The Nashville Warbler is one of the rarest breeding wood-warblers found in Alberta. This species has two disjunct populations in North America. From a Canadian perspective, the eastern population generally ranges from Quebec west to Saskatchewan, while the western population breeds in southern British Columbia. Compared to Atlas 1, this species was observed more often in Atlas 2, with many new breeding records obtained in the Cold Lake area. This is likely a range extension of the eastern Nashville Warbler population.

The Nashville Warbler tends to breed in second-growth forests, open deciduous forests, or mixedwood forests. A relatively low tree density and high undergrowth density are important characteristics of the breeding habitat of these birds. Nashville Warblers were observed in the Boreal Forest, Foothills, Parkland, and Rocky Mountain NRs where these conditions were present. Records from the Grassland NR were those of non-breeders.

Changes in relative abundance were not detected for this species in any natural region in Alberta although, relative to other species, Nashville Warblers were observed more frequently in Atlas 2 than in Atlas 1. The Breeding Bird Survey sample size was too small to investigate Nashville Warbler abundance change in Alberta. However, a decline was detected across Canada during the period 1995–2005. The biological cause for the BBS report of a nation-wide decline is unclear, given that the species can respond favourably to human disturbances such as forest harvesting and fragmentation. However, forestry regrowth methods often include spraying to reduce the density of undergrowth that competes with the planted trees. It is possible that this activity is reducing the suitability of second-growth forests for this species. The Nashville Warbler is considered Secure in Alberta.

TEMPORAL REPORTING RATE

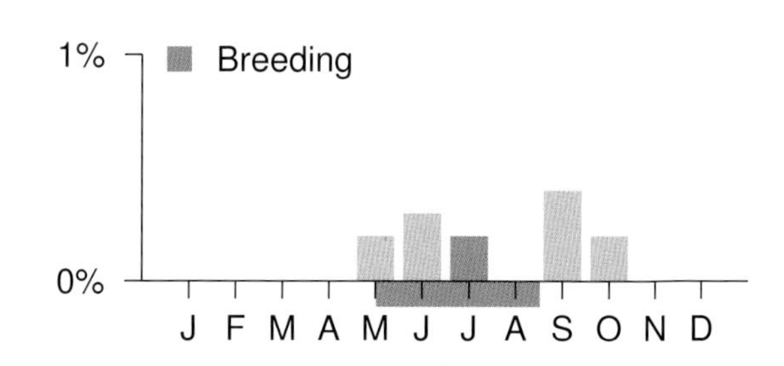

NATURAL REGIONS REPORTING RATE

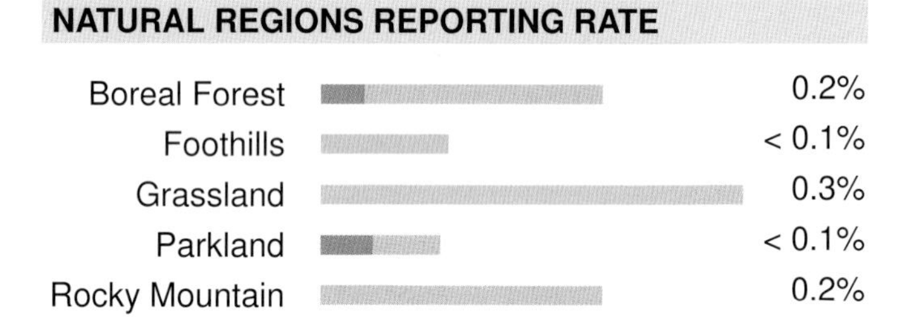

SPATIAL REPORTING RATE

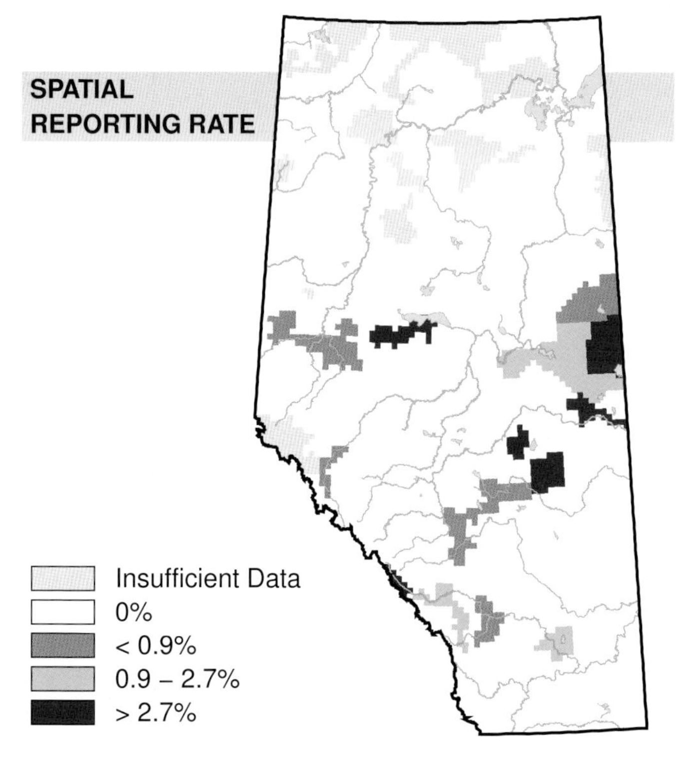

STATUS

GSWA 2000	GSWA 2005	COSEWIC Historic Rankings	COSEWIC 2007	Alberta Wildlife Act
Secure	Secure	N/A	N/A	N/A

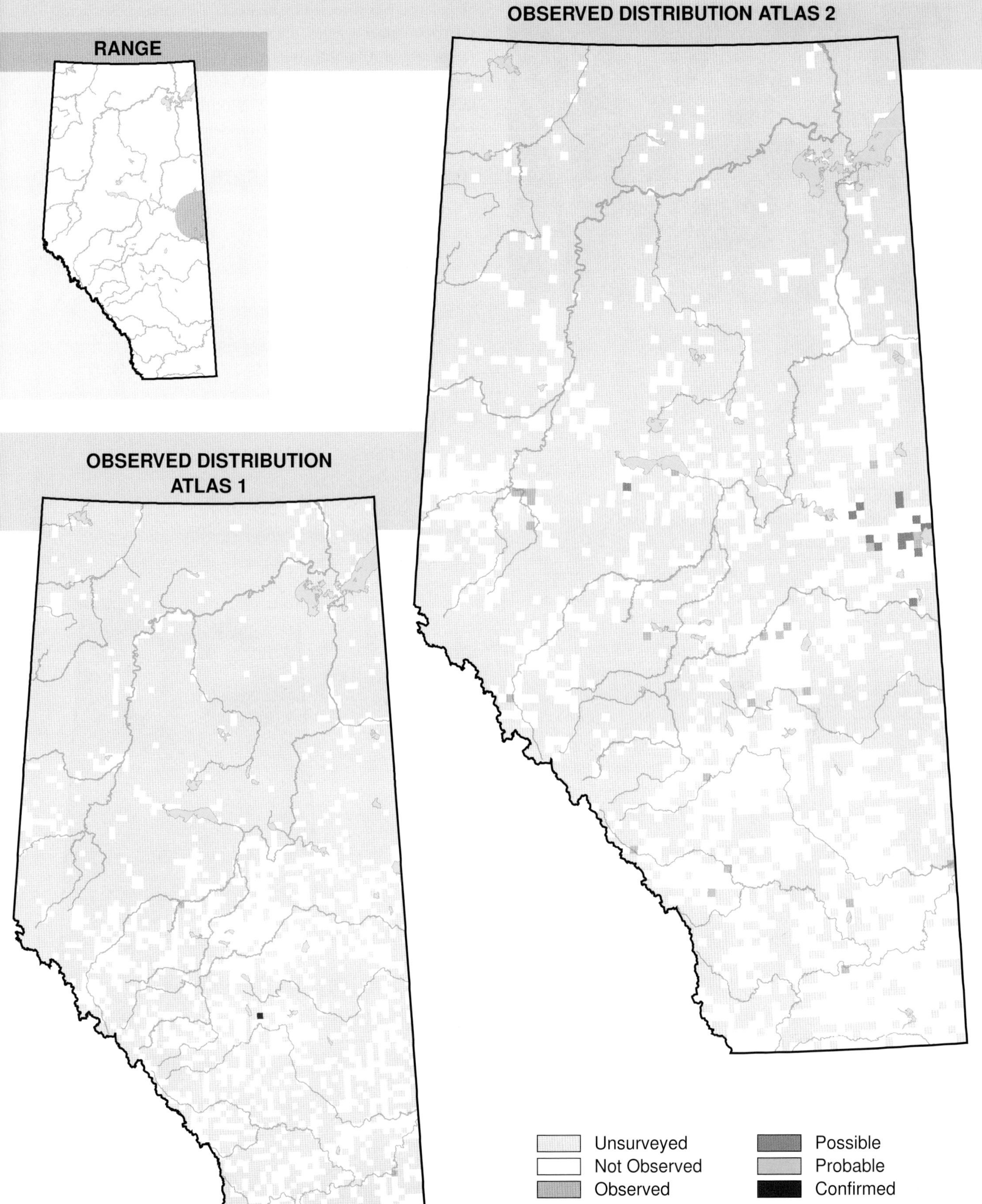
OBSERVED DISTRIBUTION ATLAS 2
RANGE
OBSERVED DISTRIBUTION ATLAS 1
Unsurveyed
Not Observed
Observed
Possible
Probable
Confirmed

Yellow Warbler (*Dendroica petechia*)

Photo: Robert Gehlert

NESTING

CLUTCH SIZE (EGGS):	4–5
INCUBATION (DAYS):	10–11
FLEDGING (DAYS):	9
NEST HEIGHT (METRES):	1–3

The Yellow Warbler breeds in every Natural Region in Alberta. The observed distribution of this species did not change between Atlas 1 and Atlas 2.

Unlike many other wood warblers that tend to breed in the interior of old forests, the Yellow Warbler is often associated with early successional habitats, especially among willows, that are near riparian areas. Due to its preference for deciduous habitats, this species was most often observed in the Parkland NR; however, it was also fairly abundant in the Boreal Forest, Foothills, Grassland, and Rocky Mountain NRs. The distribution of this species was sometimes patchy because of its preference to nest along riparian areas that are bordered with shrubs. This pattern was most apparent in the Grassland NR which tends to be a fairly dry natural region.

Declines in relative abundance were detected in the Boreal Forest, Grassland, and Parkland NRs where, relative to other species, the Yellow Warbler was observed less frequently in Atlas 2 than in Atlas 1. The Breeding Bird Survey found an abundance decrease in Alberta during the periods 1985–2005 and 1995–2005. The Boreal Forest and Parkland NRs were drier during Atlas 2 than during Atlas 1. Declines could be related to reductions in the amount of vegetation, especially shrubs such as willow, that is retained along riparian areas. Increased conversion of land to agricultural use may also have contributed to the observed declines. An increase in relative abundance was detected in the Rocky Mountain NR where, relative to other species, this bird was observed more frequently in Atlas 2 than in Atlas 1. The cause of this increase is not known, but may be a result of differences in coverage in this NR between Atlas 1 and 2. No change was detected in the Foothills NR. This species is considered Secure in Alberta.

TEMPORAL REPORTING RATE

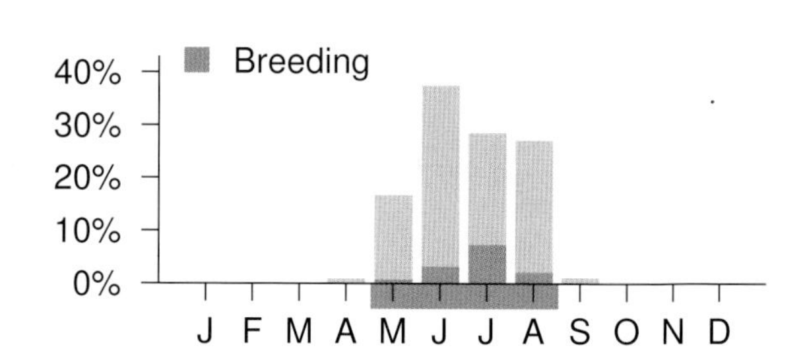

NATURAL REGIONS REPORTING RATE

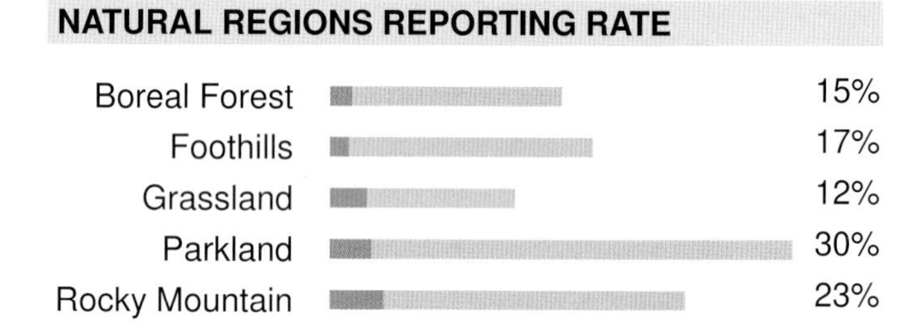

SPATIAL REPORTING RATE

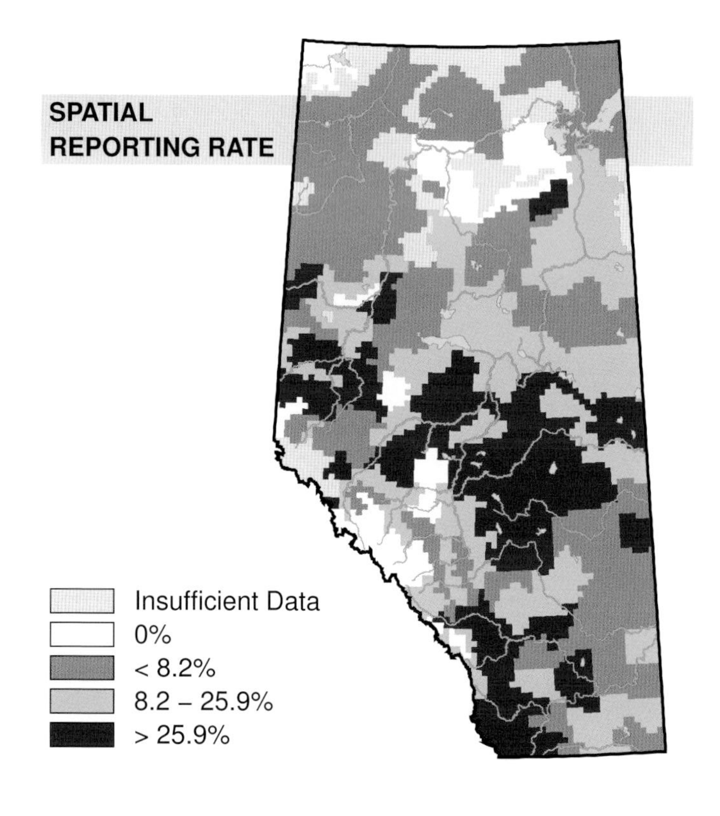

STATUS

GSWA 2000	GSWA 2005	COSEWIC Historic Rankings	COSEWIC 2007	Alberta Wildlife Act
Secure	Secure	N/A	N/A	N/A

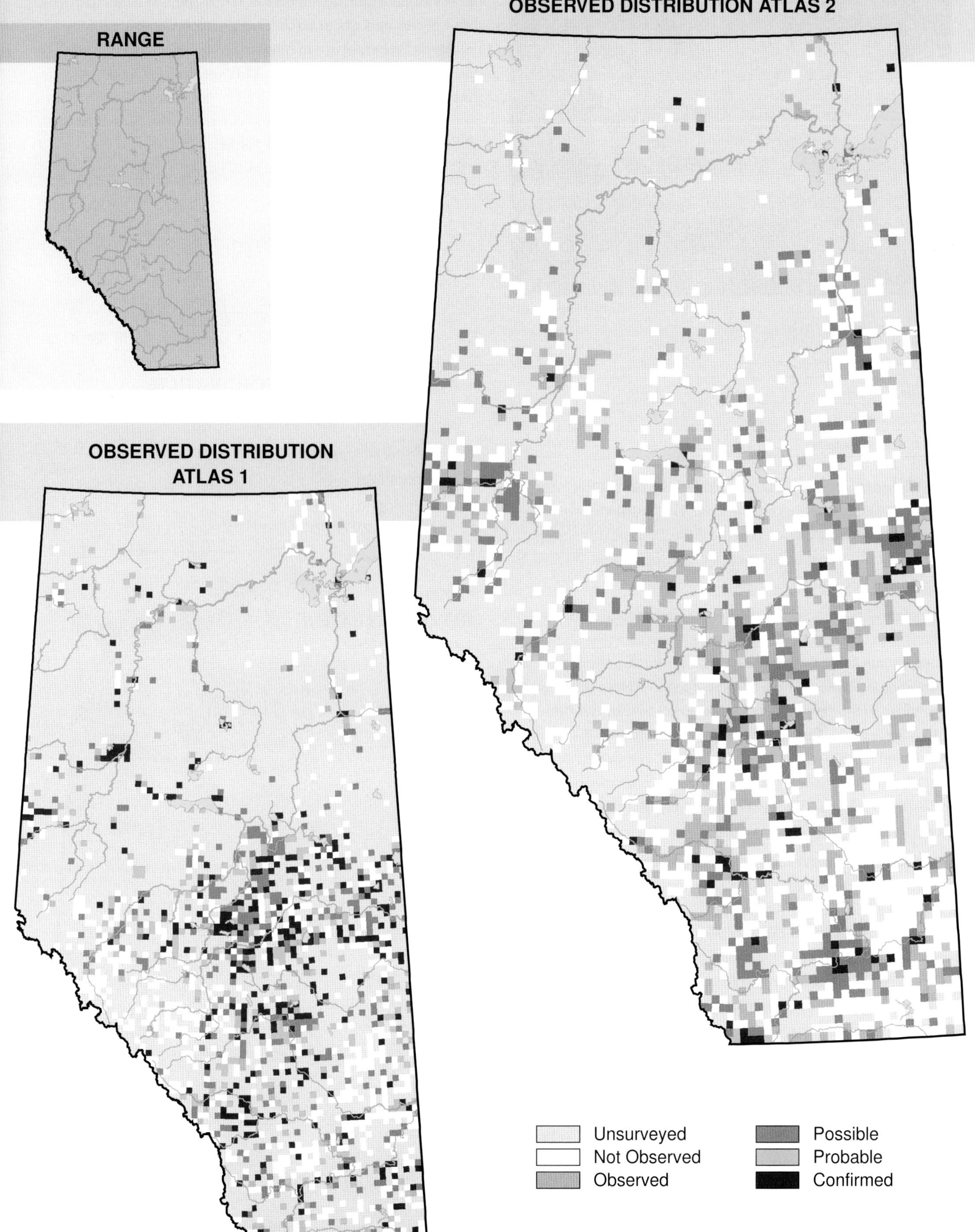
Habitat
Nest Location
Nest Type
Diet
RANGE
OBSERVED DISTRIBUTION ATLAS 2
OBSERVED DISTRIBUTION ATLAS 1
Unsurveyed
Not Observed
Observed
Possible
Probable
Confirmed

Chestnut-sided Warbler (*Dendroica pensylvanica*)

Photo: Royal Alberta Museum

NESTING

CLUTCH SIZE (EGGS):	3–5
INCUBATION (DAYS):	12–13
FLEDGING (DAYS):	10–12
NEST HEIGHT (METRES):	<1

The Chestnut-sided Warbler is found in the Boreal Forest and Foothills Natural Regions in Alberta. The distribution of this species expanded westward during Atlas 2. Typically thought of as more of an eastern species, the Chestnut-sided Warbler is becoming more common in Alberta.

Unlike most other wood warblers, the Chestnut-sided Warbler breeds in second-growth forests and along edges. As such, this species tends to be found in regenerating forests after fires and harvesting activities. The Chestnut-sided Warbler prefers to nest close to the ground in deciduous trees and shrubs but will also occasionally use coniferous trees. As a result, this species was found mainly in the Boreal Forest NR and in the northern part of the Foothills NR in areas where the preferred habitat is prevalent. Records from the Grassland and Parkland NRs were non-breeders.

An increase in relative abundance was detected in the Boreal Forest NR where, relative to other species, this warbler was observed more frequently in Atlas 2 than in Atlas 1. This increase likely resulted from a range expansion in Alberta. No change was detected in other NRs. The bird was not detected often enough during Breeding Bird Surveys to determine abundance trends for Alberta. However, a decline was detected across Canada during the period 1995–2005. It is unclear why this species appears to be declining nationally given that it responds favourably to human disturbances such as forest harvesting and fragmentation. This species is considered Secure in Alberta.

TEMPORAL REPORTING RATE

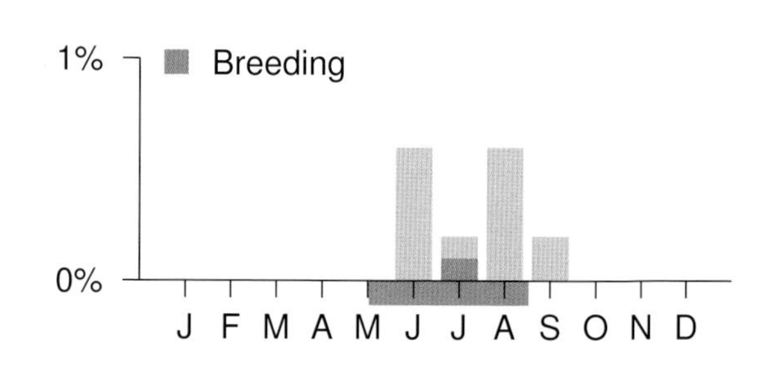

NATURAL REGIONS REPORTING RATE

SPATIAL REPORTING RATE

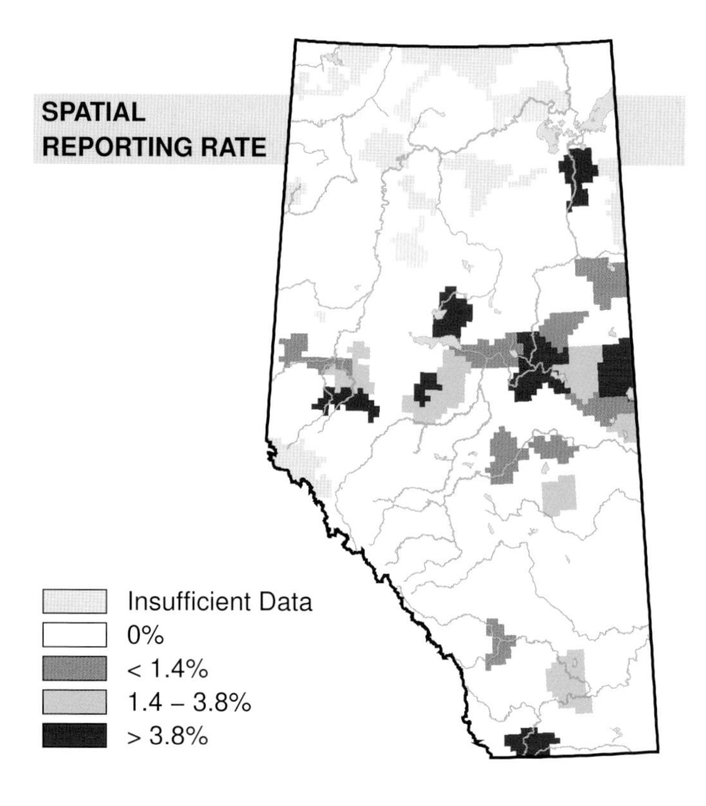

STATUS

GSWA 2000	GSWA 2005	COSEWIC Historic Rankings	COSEWIC 2007	Alberta Wildlife Act
Secure	Secure	N/A	N/A	N/A

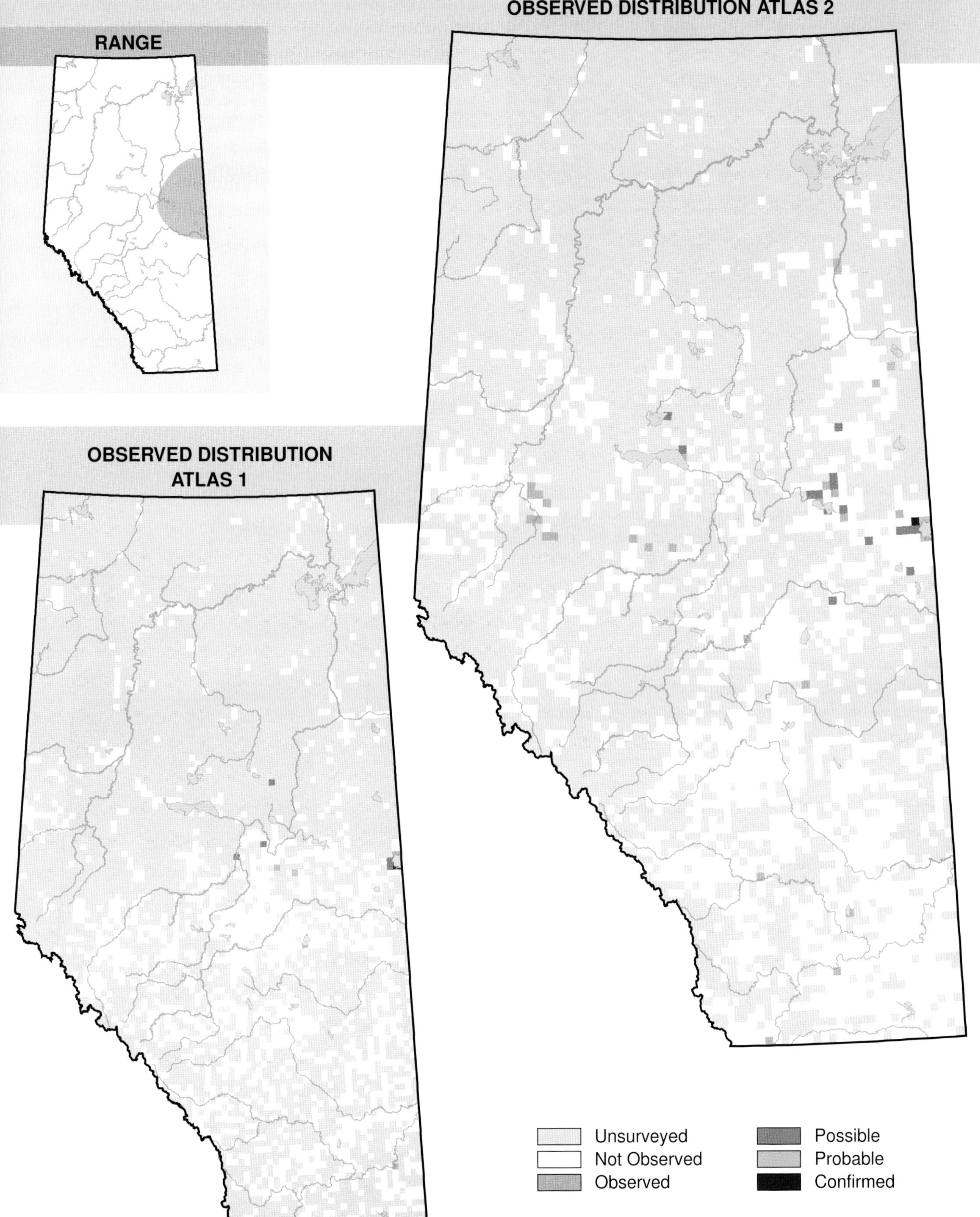
OBSERVED DISTRIBUTION ATLAS 2
RANGE
OBSERVED DISTRIBUTION ATLAS 1
Unsurveyed
Not Observed
Observed
Possible
Probable
Confirmed

Magnolia Warbler (*Dendroica magnolia*)

Photo: Robert Gehlert

NESTING

CLUTCH SIZE (EGGS):	3–5
INCUBATION (DAYS):	11–13
FLEDGING (DAYS):	8–10
NEST HEIGHT (METRES):	<2

The Magnolia Warbler is found in the Boreal Forest, Foothills, Parkland, and Rocky Mountain Natural Regions in Alberta. During Atlas 2 this species was observed farther south in the Rocky Mountain NR than during Atlas 1. This does not necessarily indicate that the range of this species has expanded because there were only a few records observed in this new area.

The Magnolia Warbler breeds in dense conifer forests or mixedwood forests. Young second-growth conifer forest habitats are often used during the breeding season, but mature stands can also be occupied provided the understorey vegetation is dense. For this reason, the Magnolia Warbler is one of the first wood-warblers to re-occupy regenerating coniferous forests after disturbances such as fires and forest harvesting. This species was found mainly in the Boreal Forest, Foothills, and Rocky Mountain NRs and was rare in the Parkland NR north of Grande Prairie. Records from the Grassland NR were non-breeders.

Increases in relative abundance were detected in the Boreal Forest and Foothills NRs where, relative to other species, Magnolia Warblers were observed more frequently in Atlas 2 than in Atlas 1. No changes were detected in other NRs. No change was detected in Alberta using Breeding Bird Survey data for this species. Nonetheless, an increase was detected on the national scale during the period 1985–2005. Increases in the relative abundance of this species are likely the result of an increased availability of young second-growth forests, its preferred breeding habitat. This species is considered Secure in Alberta.

TEMPORAL REPORTING RATE

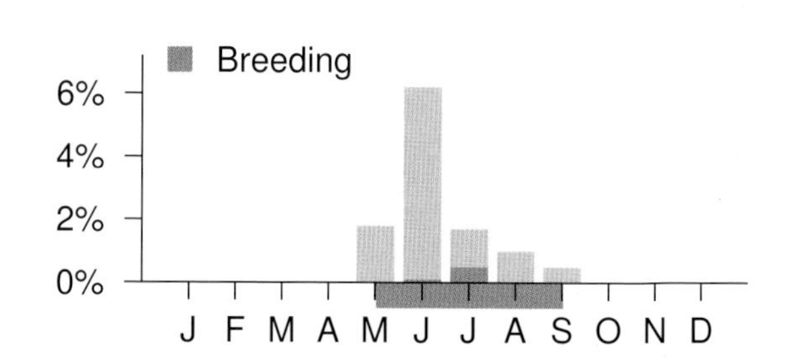

NATURAL REGIONS REPORTING RATE

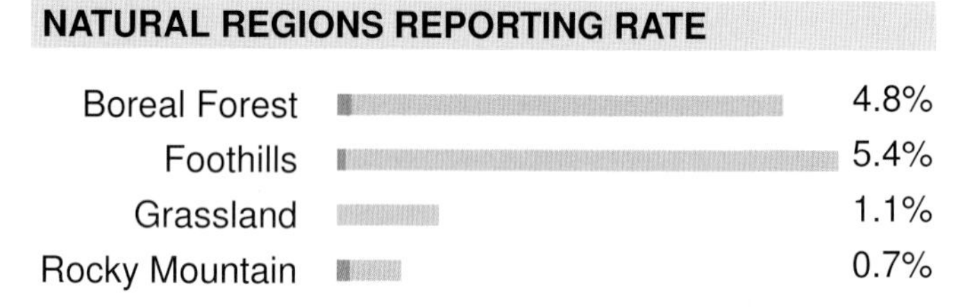

SPATIAL REPORTING RATE

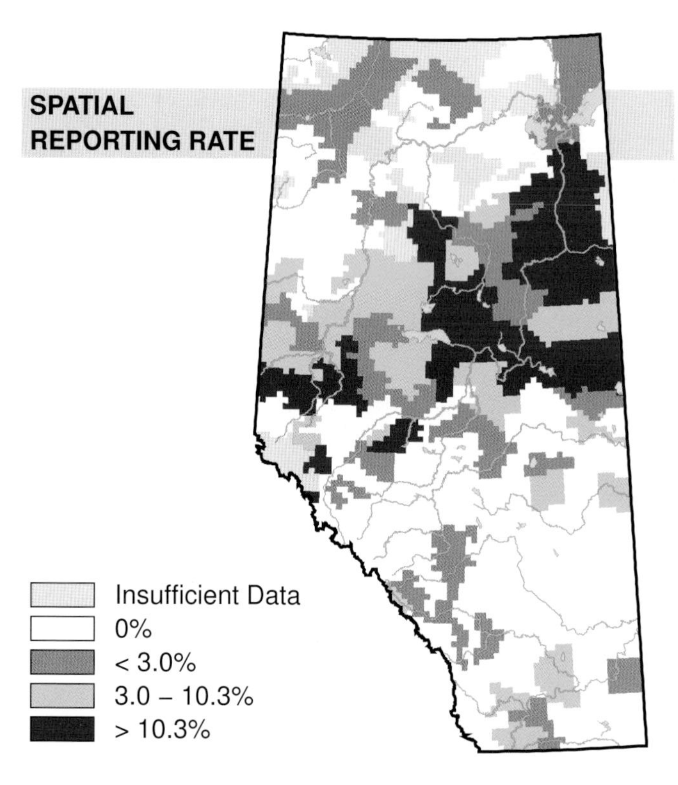

STATUS

GSWA 2000	GSWA 2005	COSEWIC Historic Rankings	COSEWIC 2007	Alberta Wildlife Act
Secure	Secure	N/A	N/A	N/A

Habitat
Nest Location
Nest Type
Diet

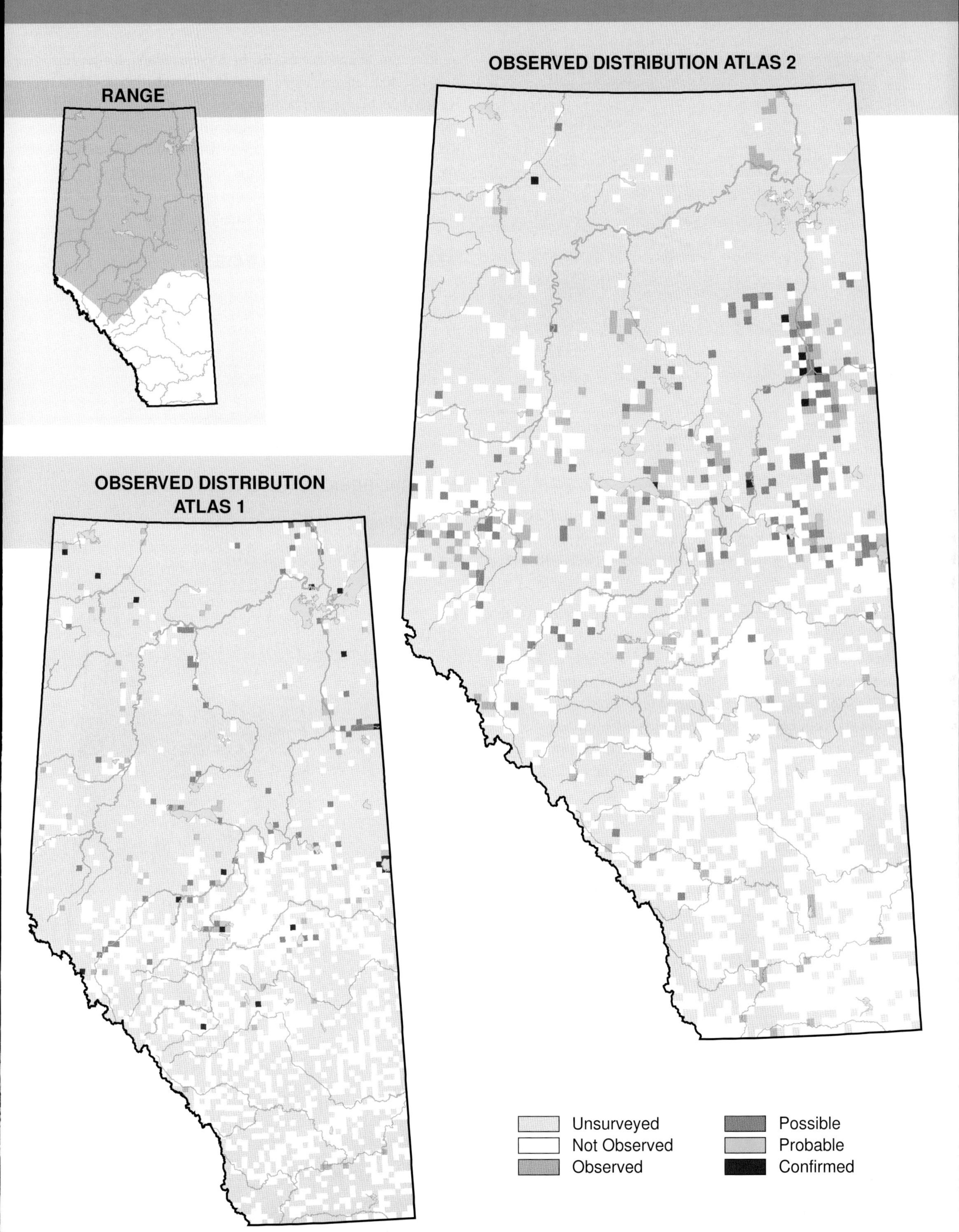

RANGE
OBSERVED DISTRIBUTION ATLAS 1
OBSERVED DISTRIBUTION ATLAS 2
Unsurveyed
Not Observed
Observed
Possible
Probable
Confirmed

Cape May Warbler (*Dendroica tigrina*)

Photo: Gerald Romanchuk

NESTING

CLUTCH SIZE (EGGS):	5–9
INCUBATION (DAYS):	unknown
FLEDGING (DAYS):	unknown
NEST HEIGHT (METRES):	3.4–18.3

The Cape May Warbler is found in the Boreal Forest, Parkland, and Foothills Natural Regions in Alberta. No change in the distribution of this species was observed between Atlas 1 and Atlas 2.

The Cape May Warbler breeds mainly in mature to old-growth coniferous forests. Tree density does not seem to limit the presence of breeding Cape May Warblers. This species has a close association with populations of one of its main food sources, the Spruce Budworm. Cape May Warbler populations expand and contract based on the availability of Spruce Budworms. As a result, this species was found mainly in the Boreal Forest NR and in the northern part of the Foothills NR, but was found infrequently in the Parkland NR around Grande Prairie. Records from the Grassland NR were non-breeders.

An increase in relative abundance was detected in the Boreal Forest NR where, relative to other species, this species was observed more frequently in Atlas 2 than in Atlas 1. This increase was somewhat surprising as this bird prefers old-growth forests, which are in decline in Alberta. This preference, together with the more complete coverage of the north in Atlas 2, and the targeted efforts in the north by expert birders, suggests that this change is artifactual. This species was not detected often enough during Breeding Bird Surveys to determine abundance trends for Alberta, although a decline was detected across Canada during the period 1985–2005. National declines could be a consequence of attempts to control Spruce Budworm outbreaks with spraying programs, as well as habitat loss. No change was detected in the Foothills and Parkland NRs. This species is considered Sensitive in Alberta due to its preference for old-growth forests during the breeding season, and to loss of wintering habitat.

TEMPORAL REPORTING RATE

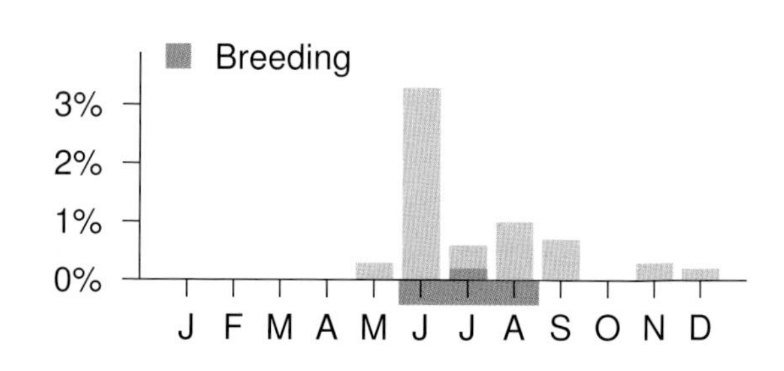

NATURAL REGIONS REPORTING RATE

SPATIAL REPORTING RATE

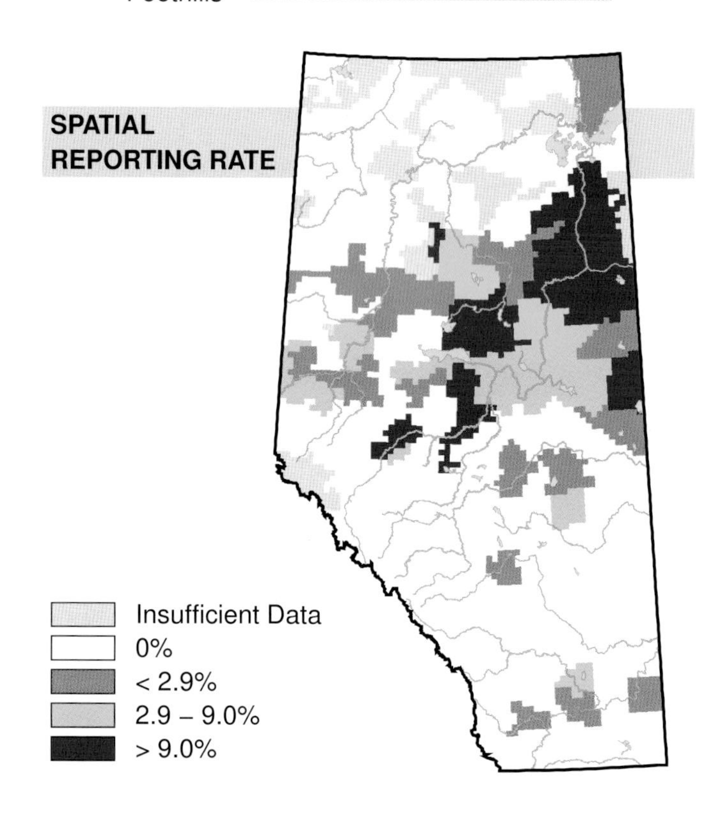

STATUS

GSWA 2000	GSWA 2005	COSEWIC Historic Rankings	COSEWIC 2007	Alberta Wildlife Act
Sensitive	Sensitive	N/A	N/A	N/A

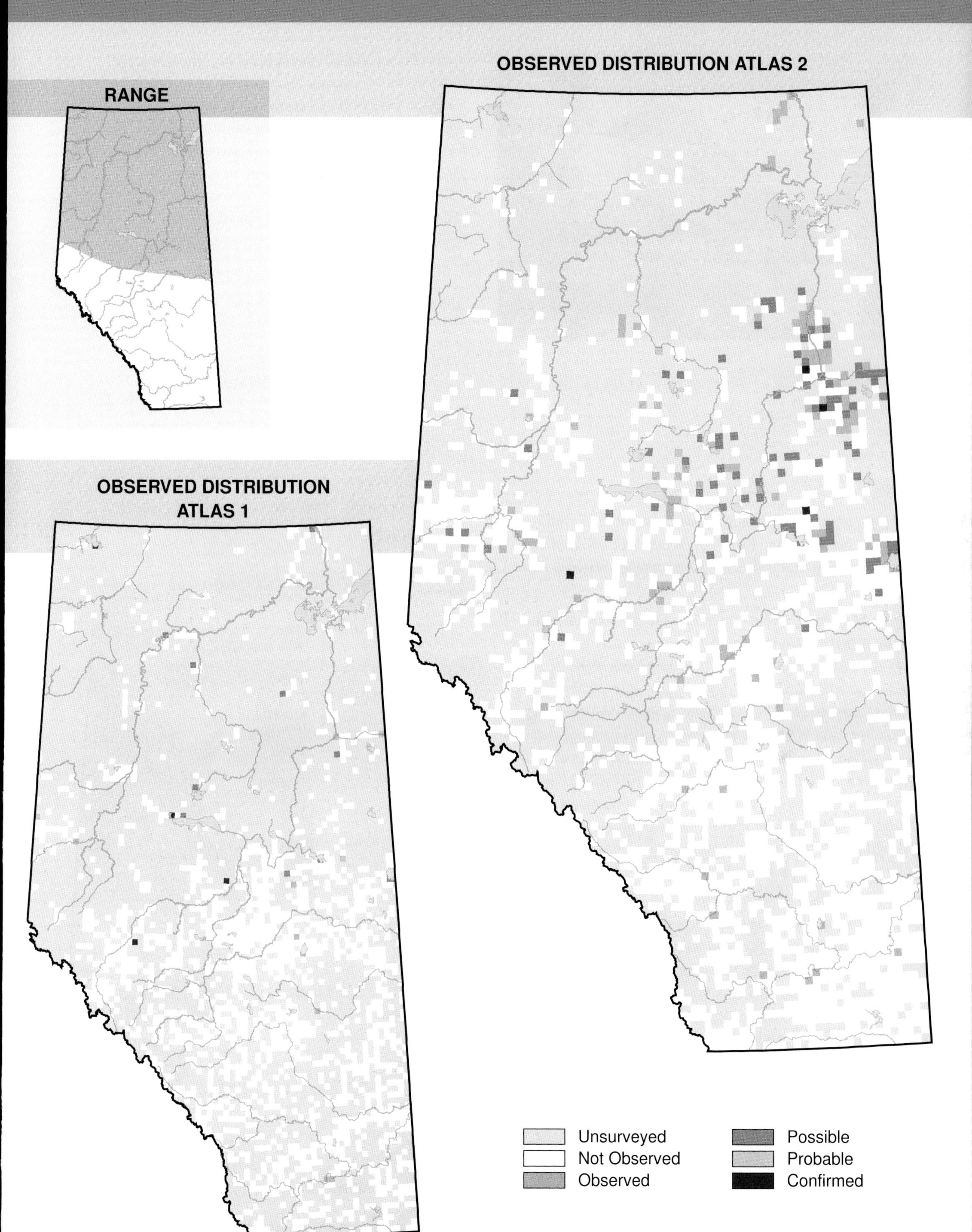
Habitat
Nest Location
Nest Type
Diet
OBSERVED DISTRIBUTION ATLAS 2
RANGE
OBSERVED DISTRIBUTION ATLAS 1
Unsurveyed
Not Observed
Observed
Possible
Probable
Confirmed

Yellow-rumped Warbler (*Dendroica coronata*)

Photo: Randy Jensen

NESTING

CLUTCH SIZE (EGGS):	4–5
INCUBATION (DAYS):	12–13
FLEDGING (DAYS):	12–14
NEST HEIGHT (METRES):	3–5

The Yellow-rumped Warbler breeds in the Boreal Forest, Foothills, Parkland, and Rocky Mountain Natural Regions. The distribution of this species did not change between Atlas 1 and Atlas 2.

The Yellow-rumped Warbler tends to breed in mature confer forests or mixedwood forests that contain mature conifer trees. Yet, despite their preference for mature conifer trees, this species can breed in forests of varying density and it tends not to avoid fragmented forests. Unlike many wood-warblers, the Yellow-rumped Warbler feeds in a variety of habitats, is able to use a variety of feeding techniques, and can have a varied diet. This species was fairly common in the Boreal Forest, Foothills, and Rocky Mountain NRs, likely because of high conifer tree abundance in those natural regions. The species was less frequently encountered in the Parkland and Grassland NRs, and records there tended to be those of non-breeders, likely because conifer trees are not common in those natural regions.

Increases in relative abundance were detected in the Boreal Forest, Grassland, and Parkland NRs where, relative to other species, the Yellow-rumped Warbler was observed more frequently in Atlas 2 than in Atlas 1. No change was detected in the Foothills and Rocky Mountain NRs. The Breeding Bird Survey did not detect any change in abundance in Alberta for the period 1985–2005, but did detect an abundance increase in Alberta during the period 1968–2005. The increases detected by the Atlas and the Alberta Breeding Bird Survey are mirrored by an increase in abundance in a Canada-wide Breeding Bird Survey during the period 1968–2005. Increases could be the result of the ability of this species to tolerate forest fragmentation, and also of its generalist feeding habits. This species is considered Secure in Alberta.

TEMPORAL REPORTING RATE

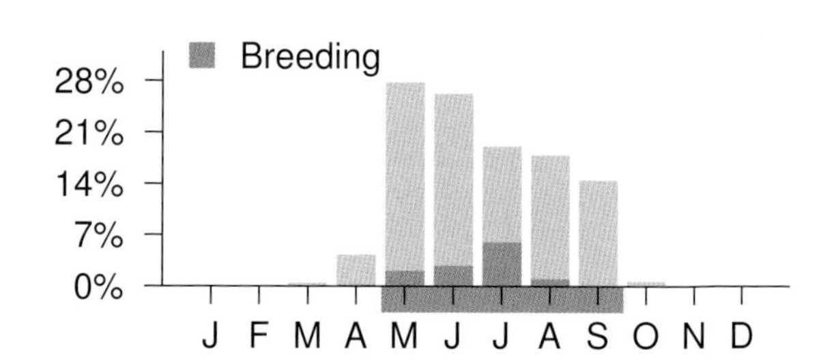

NATURAL REGIONS REPORTING RATE

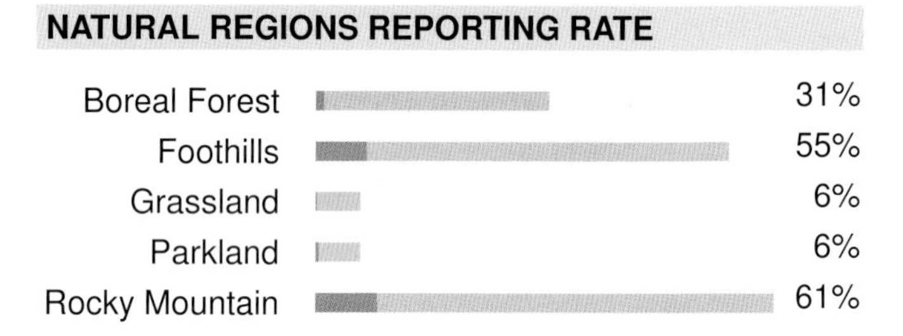

SPATIAL REPORTING RATE

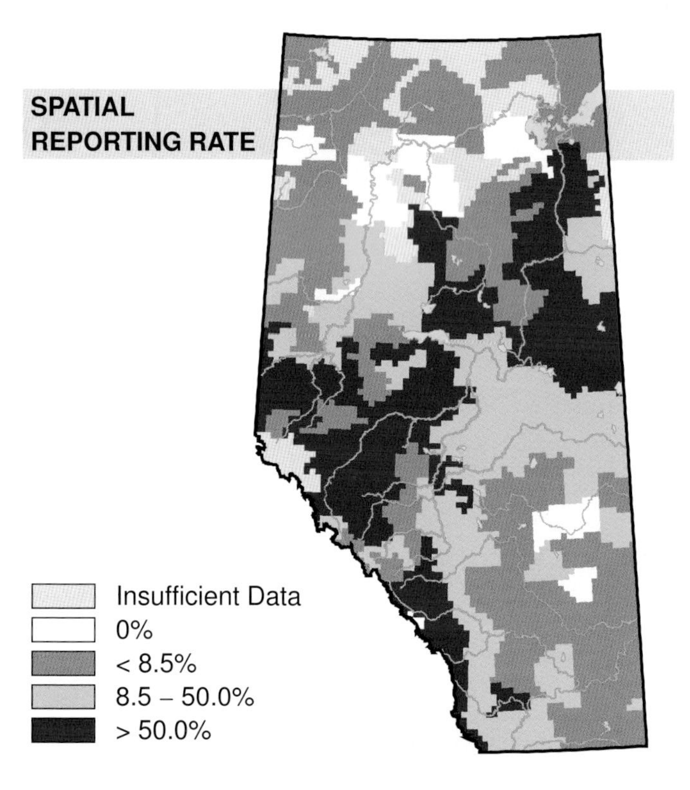

STATUS

GSWA 2000	GSWA 2005	COSEWIC Historic Rankings	COSEWIC 2007	Alberta Wildlife Act
Secure	Secure	N/A	N/A	N/A

Habitat  Nest Location  Nest Type  Diet 

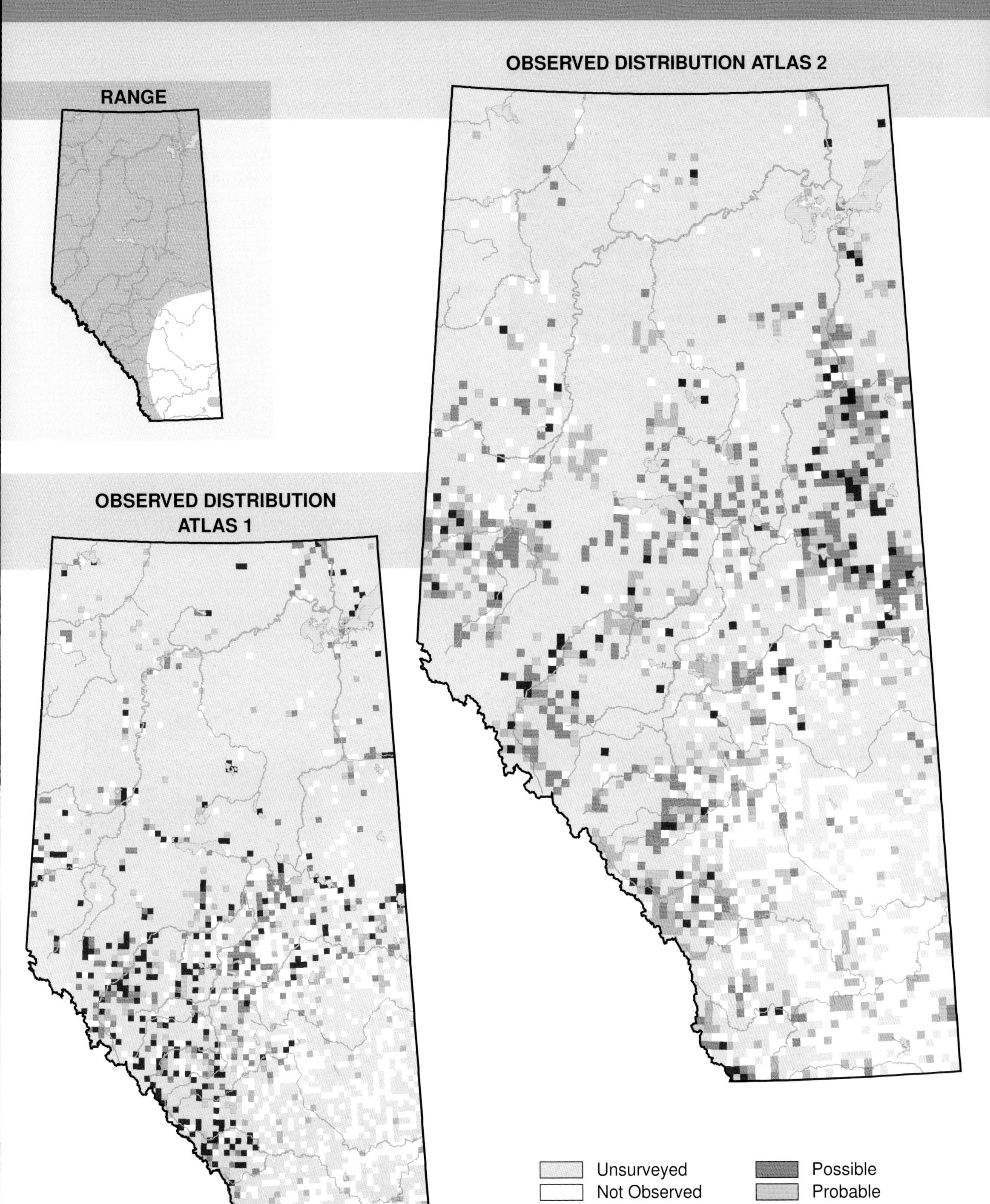

Black-throated Green Warbler (*Dendroica virens*)

Photo: Royal Alberta Museum

NESTING

CLUTCH SIZE (EGGS):	4–5
INCUBATION (DAYS):	12–13
FLEDGING (DAYS):	8–10
NEST HEIGHT (METRES):	6–11

The Black-throated Green Warbler is found in the Boreal Forest, Foothills, and Parkland Natural Regions in Alberta. No change in the distribution of this species was observed between Atlas 1 and Atlas 2.

The Black-throated Green Warbler breeds in a variety of habitats including coniferous, deciduous, and mixedwood forests, but it tends to prefer mature coniferous forests and to avoid fragmented forests. As a result, this species was found mainly in the Boreal Forest NR and the northern part of the Foothills NR, and was rare in the Parkland NR around Grande Prairie. Records from the Grassland, Parkland, and Rocky Mountain NRs were non-breeders.

Increases in relative abundance were detected in the Boreal Forest and Foothills NRs where, relative to other species, it was observed more frequently in Atlas 2 than in Atlas 1. Given this species' habitat requirements, and the extent of human activities in these areas, these increases are surprising. Rather than being related to an actual population change, these increases were likely a function of more complete coverage of the north in Atlas 2. No change was detected in the Parkland NR. The Breeding Bird Survey sample size for this species was too small to investigate abundance change in Alberta, and yet a decrease in abundance during the period 1995–2005 was detected across Canada. National declines are likely the result of increased forest fragmentation, a condition that this species generally avoids during the breeding season. The Breeding Bird Survey might not have detected changes in Alberta because northern coverage during this survey was limited and, as a road-based survey, the BBS is not likely to produce a very complete sampling of a species sensitive to forest fragmentation. The Black-throated Green Warbler is considered Sensitive and a Species of Special Concern in Alberta due to its preference for old-growth forests.

TEMPORAL REPORTING RATE

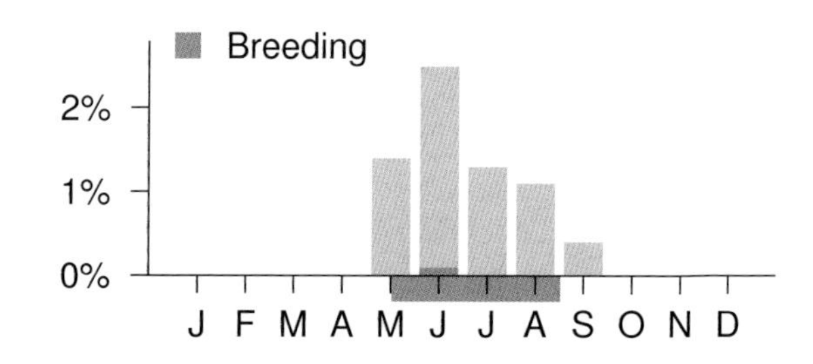

NATURAL REGIONS REPORTING RATE

SPATIAL REPORTING RATE

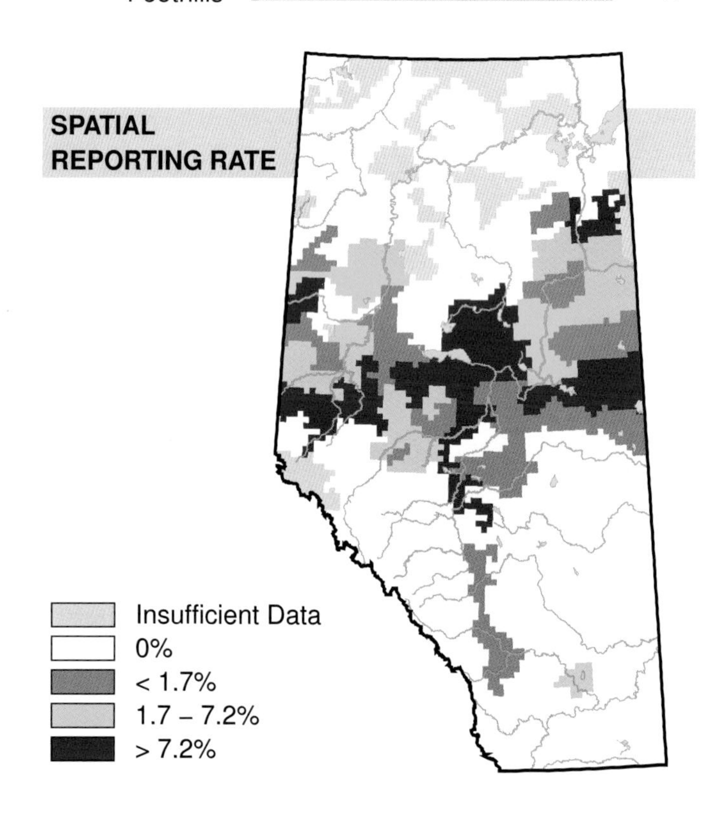

STATUS

GSWA 2000	GSWA 2005	COSEWIC Historic Rankings	COSEWIC 2007	Alberta Wildlife Act
Sensitive	Sensitive	N/A	N/A	N/A

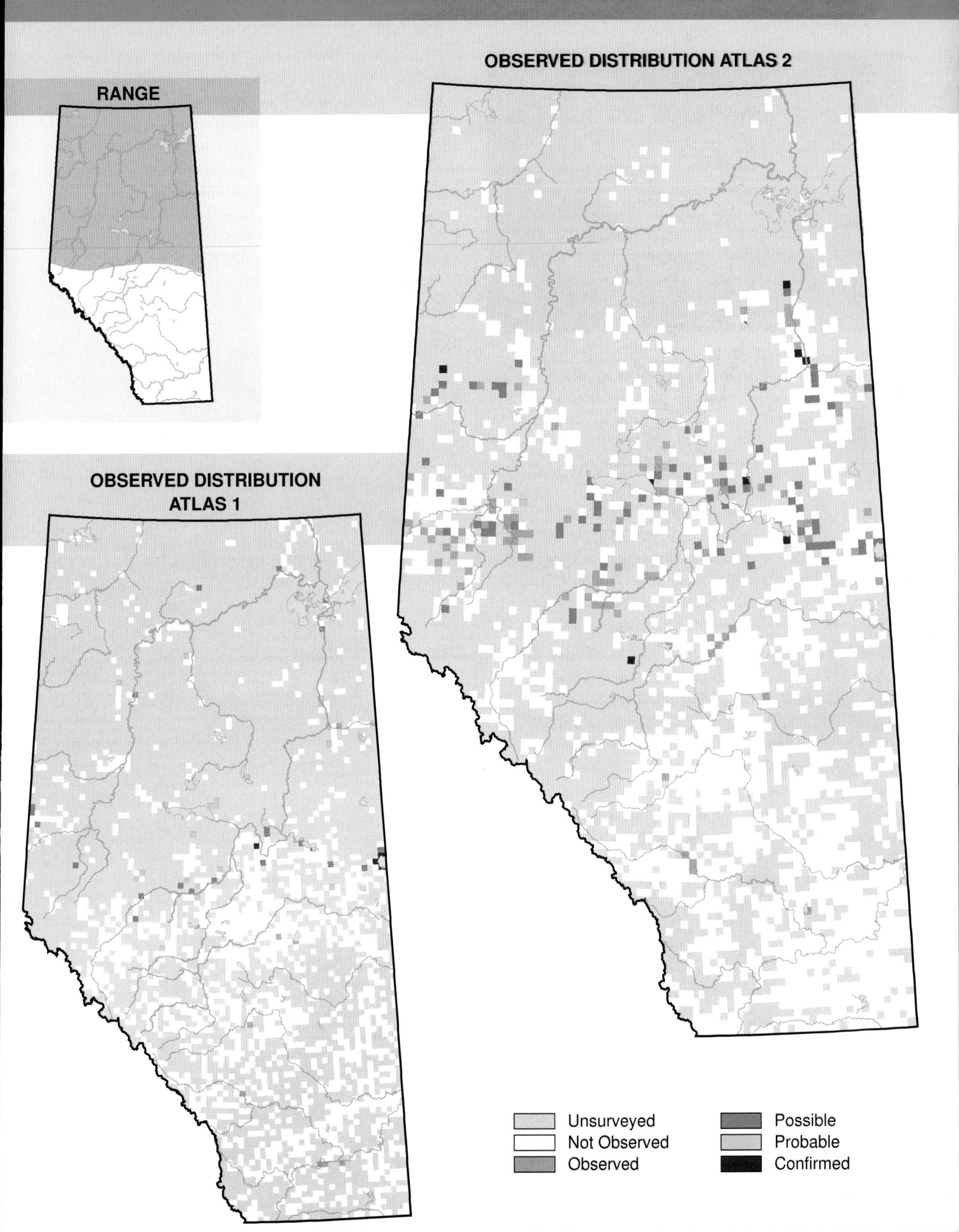
Habitat
Nest Location
Nest Type
Diet
RANGE
OBSERVED DISTRIBUTION ATLAS 2
OBSERVED DISTRIBUTION ATLAS 1
Unsurveyed
Not Observed
Observed
Possible
Probable
Confirmed

Townsend's Warbler (*Dendroica townsendi*)

Photo: Royal Alberta Museum

NESTING

CLUTCH SIZE (EGGS):	3–5
INCUBATION (DAYS):	12
FLEDGING (DAYS):	8–10
NEST HEIGHT (METRES):	2–5

Townsend's Warbler is found in Alberta in the Rocky Mountain and Foothills Natural Regions. It is common in the Banff and Kananaskis areas and uncommon north of these two parks. This species, which is usually thought of as a western species, appears to have a range that is expanding eastward. New probable breeding evidence was found near Grande Cache and between Cochrane and Turner Valley during Atlas 2.

Townsend's Warbler breeds mainly in coniferous and mixedwood forests provided there is a large coniferous component. This species generally prefers old-growth and mature forests with a thick canopy and nearby water. It does not breed in forests that are in an early successional stage; therefore, they are sensitive to forestry activities if short rotations are used. As a consequence of the habitat requirements and eastward expansion of this species' range, this bird was found mainly in the Rocky Mountain NR and only occasionally in the Foothills NR. Records from the Grassland NR were of non-breeders.

This bird was observed more frequently in the Rocky Mountain and Grassland NRs during Atlas 2.

An increase in relative abundance was detected in the Rocky Mountain NR. This is likely the result of the westward expansion of this species' range. No change was detected in the Foothills NR. The Breeding Bird Survey sample size was too small to investigate abundance changes for this species in Alberta, but an increase in abundance was detected across Canada during the period 1968–2005. It is unclear why an increase was found nationally given the sensitivity of this species to the harvesting of old-growth and mature forests. However, these increases could be related to a range expansion. This species is considered Secure in Alberta.

TEMPORAL REPORTING RATE

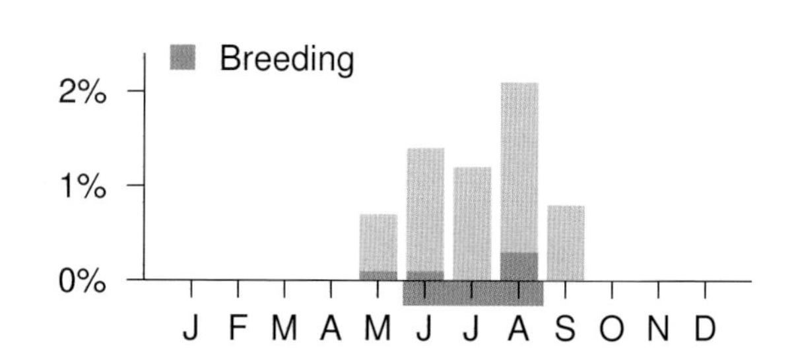

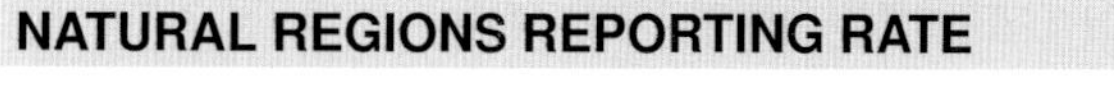

NATURAL REGIONS REPORTING RATE

SPATIAL REPORTING RATE

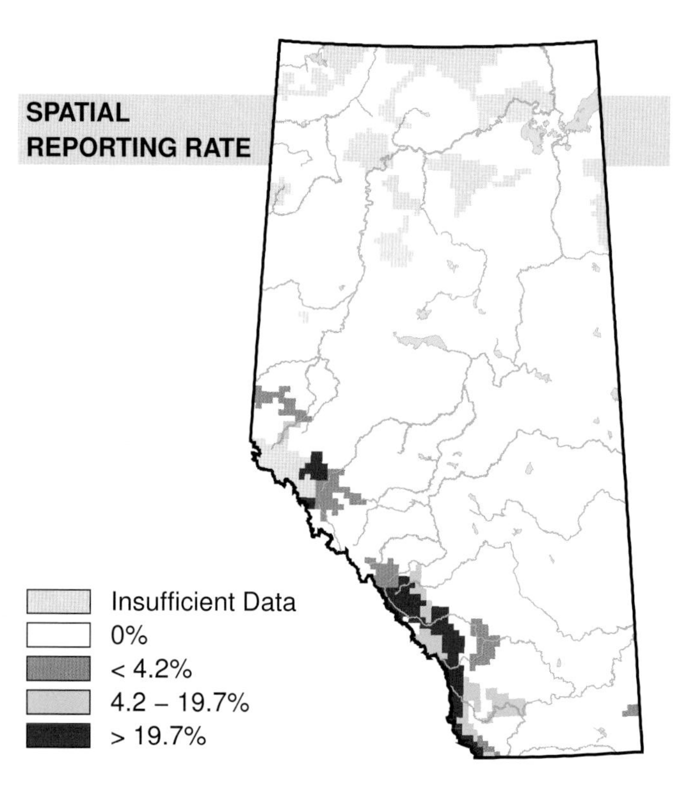

STATUS

GSWA 2000	GSWA 2005	COSEWIC Historic Rankings	COSEWIC 2007	Alberta Wildlife Act
Secure	Secure	N/A	N/A	N/A

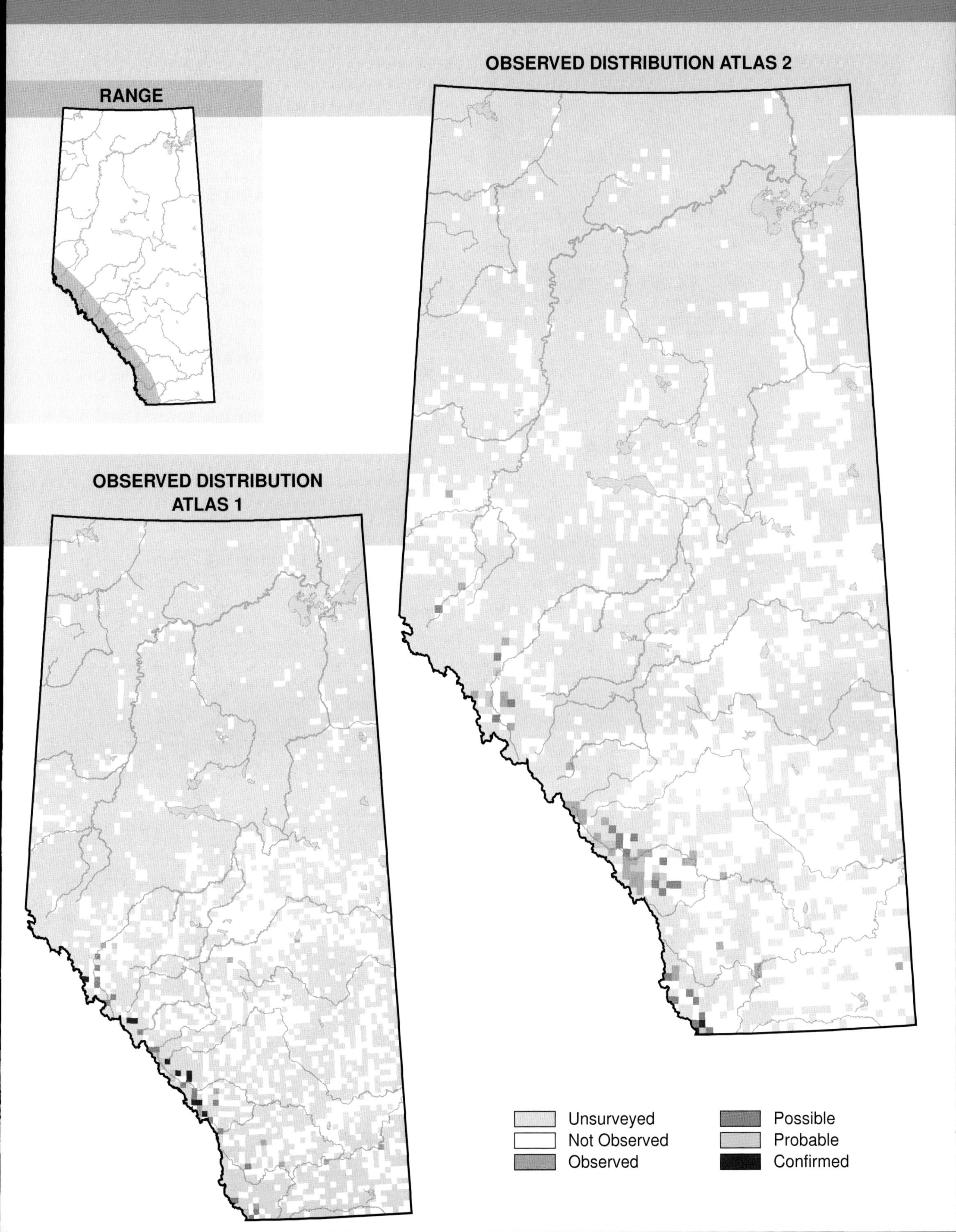
RANGE
OBSERVED DISTRIBUTION ATLAS 2
OBSERVED DISTRIBUTION ATLAS 1
Unsurveyed
Not Observed
Observed
Possible
Probable
Confirmed

Blackburnian Warbler (*Dendroica fusca*)

Photo: Barth Schorre / Cornell Laboratory of Ornithology

NESTING

CLUTCH SIZE (EGGS):	4–5
INCUBATION (DAYS):	11–13
FLEDGING (DAYS):	unknown
NEST HEIGHT (METRES):	2–26

The Blackburnian Warbler, the only wood-warbler with an orange throat, was found in the Boreal Forest Natural Region, the northern part of the Foothills NR, and in the Rocky Mountain NR. Grassland and Parkland NR records were observations of non-breeding birds. There was a change in the observed distribution of this species between Atlas 1 and Atlas 2. Atlas 2 garnered new records in the west-central part of Alberta around Grande Prairie, and in northeastern Alberta around Fort McMurray.

Associated with mature and old-growth coniferous and coniferous-deciduous mixed forests, Blackburnian Warblers nest in the tops of conifer trees. Therefore, they are found most commonly in the Boreal Forest and northern Foothills NRs where this habitat type is prevalent. The number of observations was quite low, most likely due to their low abundance and also to their particular habits: they exploit the very tops of trees for foraging; they sing quiet songs; and they spend little time singing.

No changes in relative abundance were detected for this species although, relative to other species, they were observed more frequently in Atlas 2 than in Atlas 1. This species was not detected often enough on Breeding Bird Surveys to determine trends for Alberta. However, a decline was detected across Canada during the period 1995–2005. National declines are likely the result of reductions in the amount of old-growth forest that is available, and of increases in the amount of forest fragmentation. This species is considered Secure in Alberta.

TEMPORAL REPORTING RATE

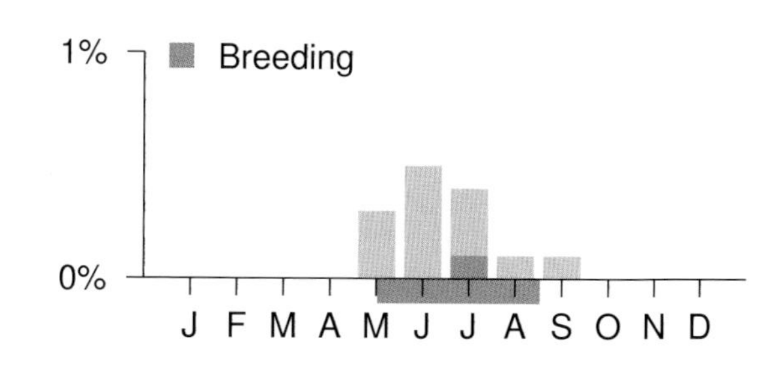

NATURAL REGIONS REPORTING RATE

SPATIAL REPORTING RATE

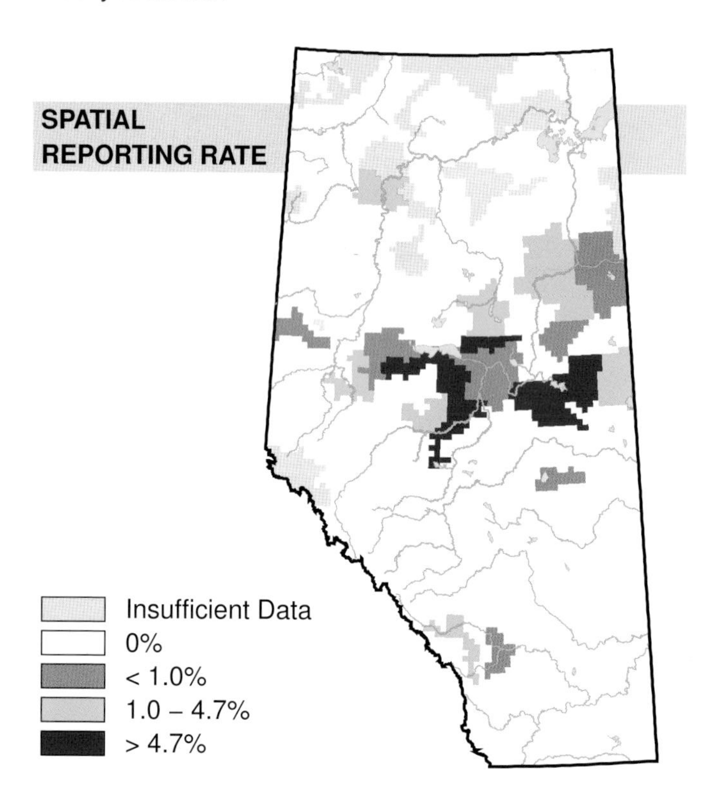

STATUS

GSWA 2000	GSWA 2005	COSEWIC Historic Rankings	COSEWIC 2007	Alberta Wildlife Act
Sensitive	Sensitive	N/A	N/A	N/A

Habitat  Nest Location  Nest Type  Diet  

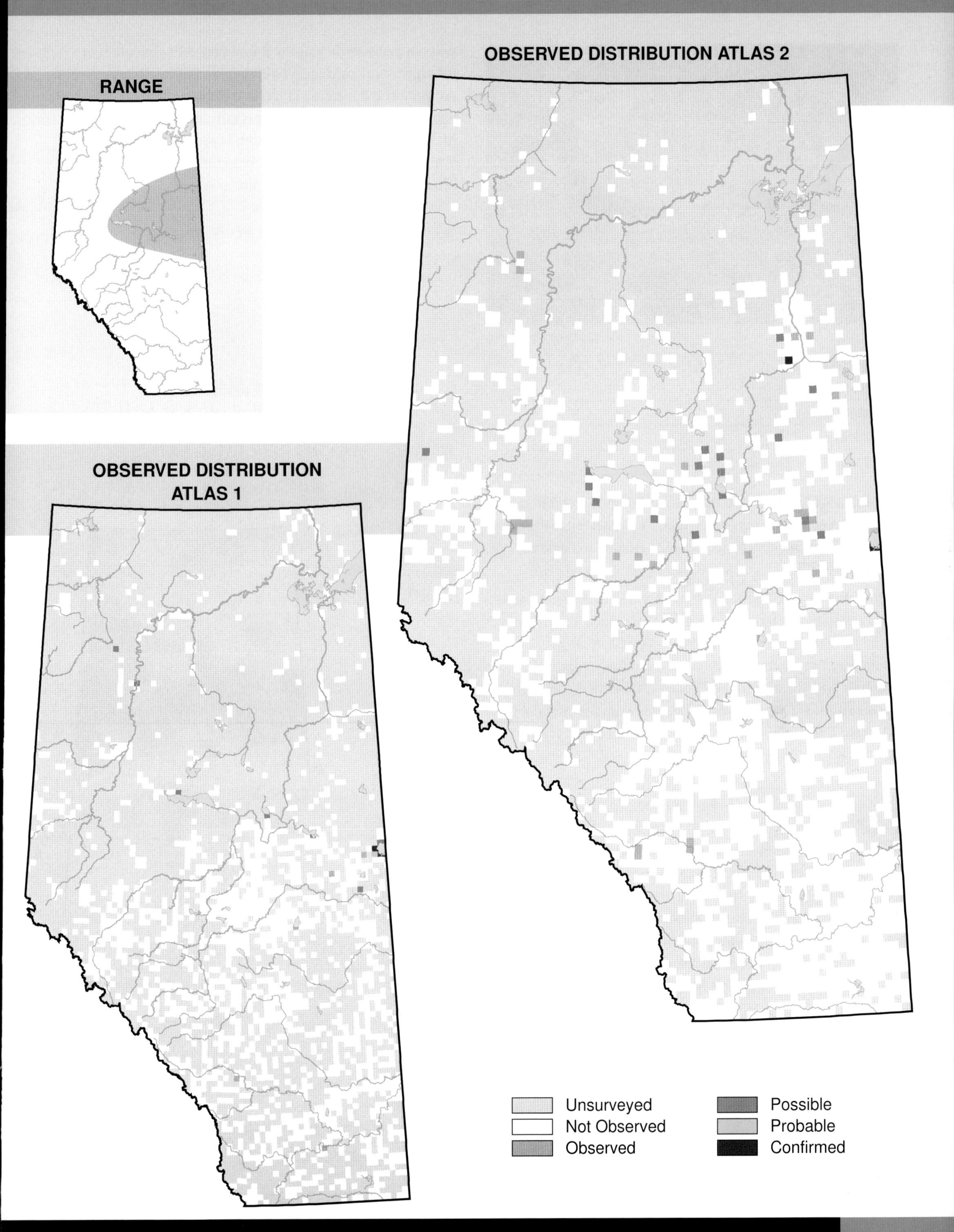

Palm Warbler (*Dendroica palmarum*)

Photo: Robert Gehlert

NESTING

CLUTCH SIZE (EGGS):	4–5
INCUBATION (DAYS):	12
FLEDGING (DAYS):	12
NEST HEIGHT (METRES):	0

The Palm Warbler is found in the Boreal Forest, Foothills, and Parkland Natural Regions in Alberta. The observed distribution of this species did not change between Atlas 1 and Atlas 2.

The Palm Warbler generally breeds in wet areas such as bogs in low-density coniferous forests with thick undergrowth and open canopies. As a result, this species was most often found in the Boreal Forest NR, where this type of habitat is commonly found, and was encountered infrequently in the Foothills NR. Records in the Parkland NR were obtained from the area around Grande Prairie. Records from the Grassland NR were from observations of non-breeders.

An increase in relative abundance was detected in the Boreal Forest NR where, relative to other species, the Palm Warbler was observed more frequently in Atlas 2 than in Atlas 1. The Breeding Bird Survey sample size was too small to investigate abundance change for this species in Alberta as most surveys are conducted south of this species' known range. No abundance change was detected by the Breeding Bird Survey across Canada. Current regrowth strategies after forest harvesting sometimes involve chemical spraying to reduce competing undergrowth; other forest disturbances, however, will not necessarily employ this strategy. Examination of differences in coverage would seem to indicate that the Atlas-detected increase is, at least in part, a function of increased survey coverage, especially around the Fort McMurray area. Further research is needed to properly evaluate the Atlas-detected increase in the Boreal Forest NR. This species is considered Secure in Alberta.

TEMPORAL REPORTING RATE

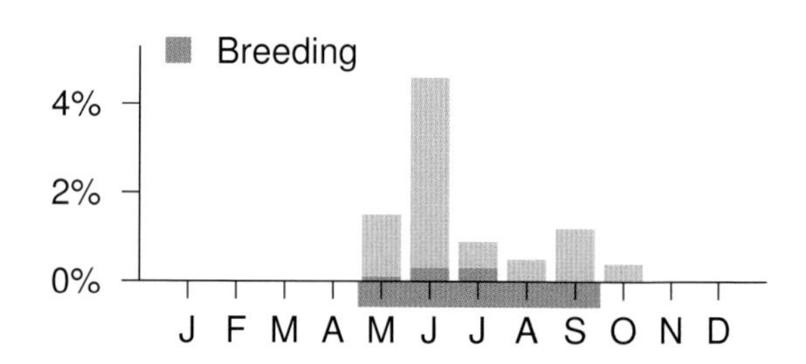

NATURAL REGIONS REPORTING RATE

Boreal Forest

1.9%

SPATIAL REPORTING RATE

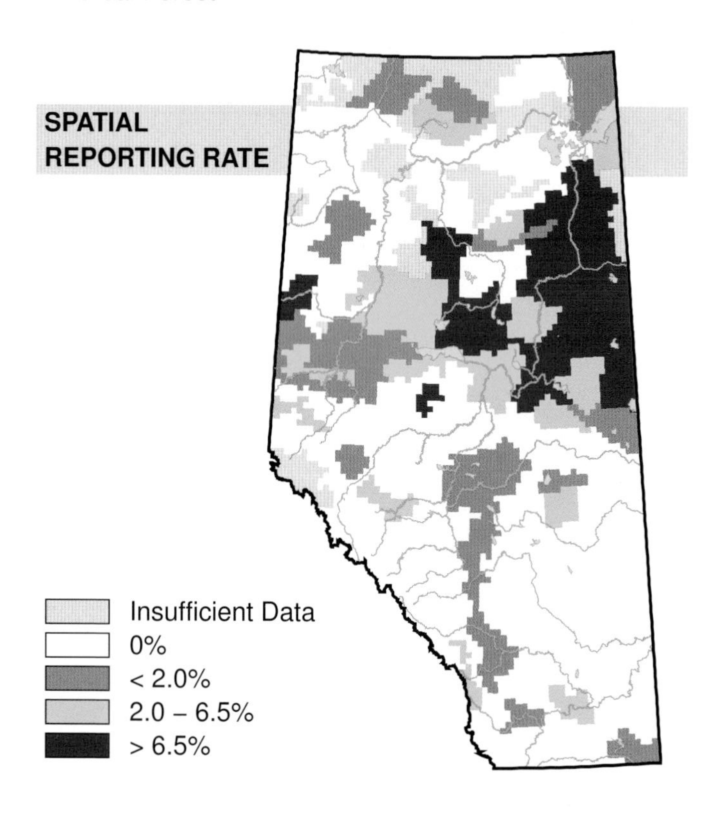

STATUS

GSWA 2000	GSWA 2005	COSEWIC Historic Rankings	COSEWIC 2007	Alberta Wildlife Act
Secure	Secure	N/A	N/A	N/A

RANGE

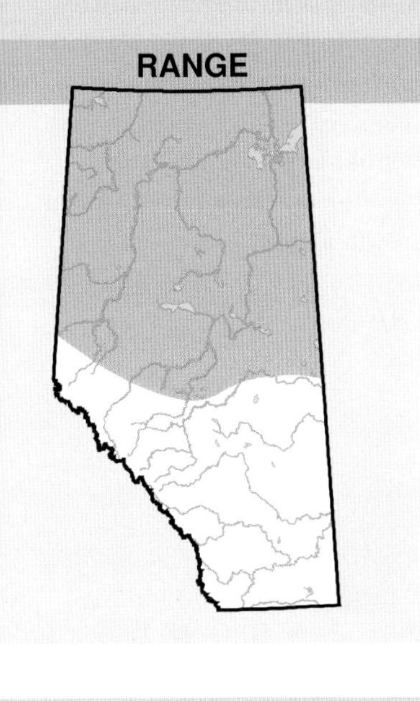

OBSERVED DISTRIBUTION ATLAS 2

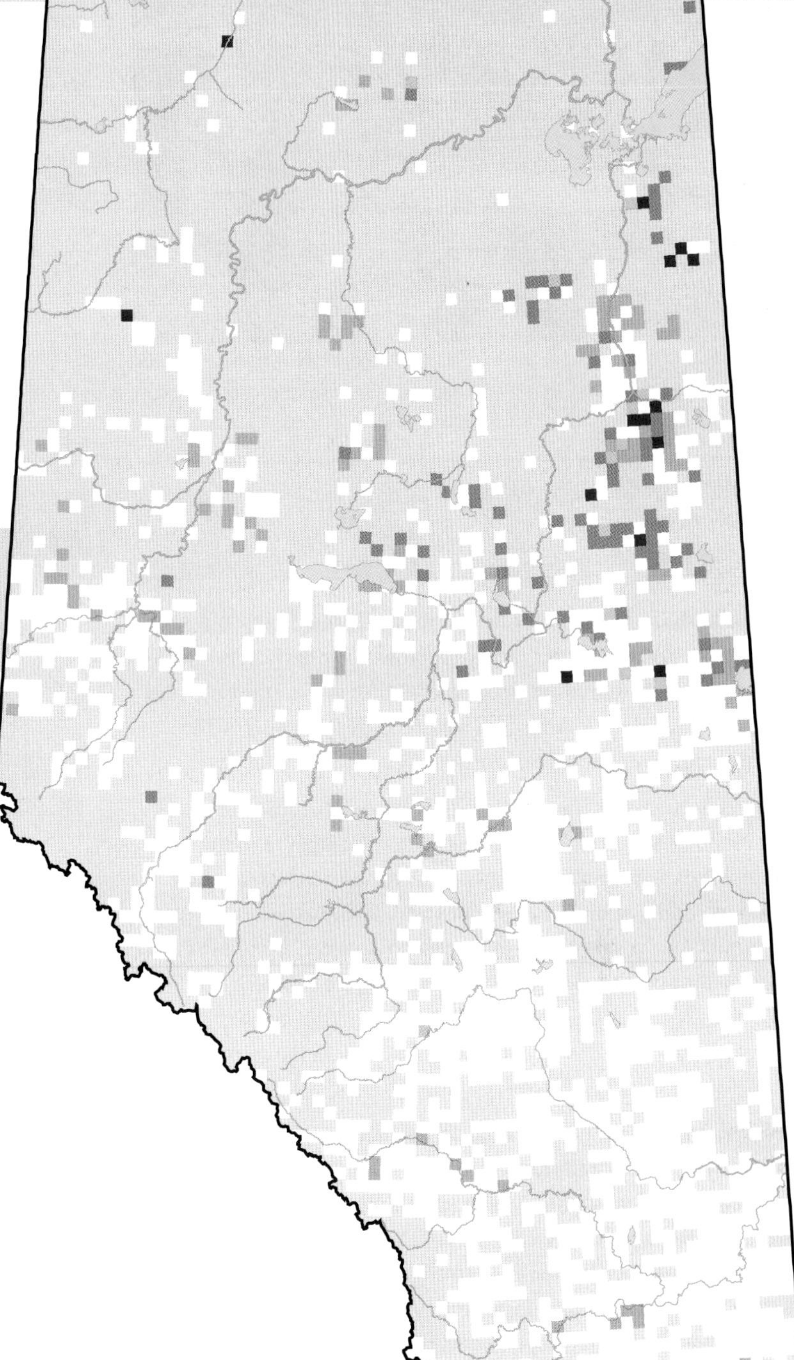

OBSERVED DISTRIBUTION ATLAS 1

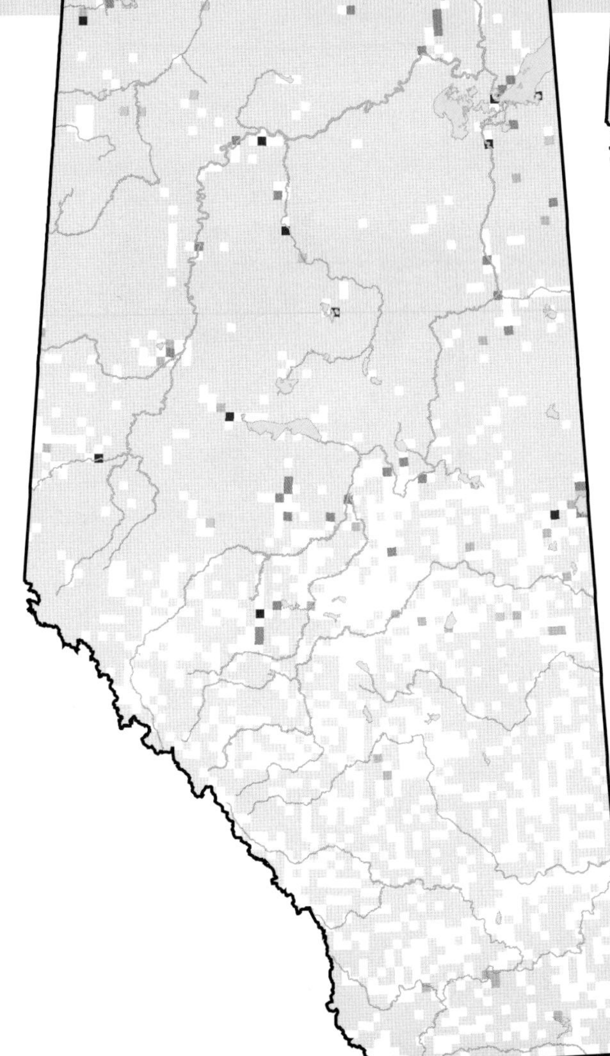

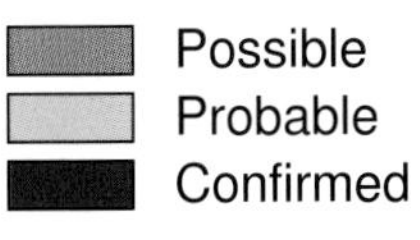

Bay-breasted Warbler (*Dendroica castanea*)

Photo: Royal Alberta Museum

NESTING

CLUTCH SIZE (EGGS):	4–6
INCUBATION (DAYS):	12
FLEDGING (DAYS):	11
NEST HEIGHT (METRES):	1–6

The Bay-breasted Warbler breeds in the Boreal Forest, Parkland, and Foothills Natural Regions in Alberta. This species was observed more often and farther west during Atlas 2 than it was during Atlas 1. This could be the result of the continued westward expansion of this species' range.

The Bay-breasted Warbler breeds in dense, mature coniferous forests that are often located near water. Compared to the closely related Blackpoll Warbler, the Bay-breasted Warbler tends to breed at higher elevations and higher above the ground. As a result, this species was distributed fairly evenly across the southern and central Boreal Forest NR between Fort MacKay, Cold Lake and Grande Prairie and the northern Foothill NR in the Swan Hills, and was rare in the Parkland NR around Grande Prairie.

An increase in relative abundance was detected in the Boreal Forest NR where, relative to other species, it was observed more frequently in Atlas 2 than in Atlas 1. This is likely the result of the westward expansion of this species' range. However, the more extensive coverage in northeastern Alberta in Atlas 2 must be considered as a contributing cause to this increase. No change was detected in the Foothills and Parkland NRs. The Breeding Bird Survey, with limited northern coverage, had a sample size too small to investigate relative abundance changes in Alberta for this species. Nevertheless, the BBS did detect a decline in abundance across Canada during the period 1985–2005. National declines could be the result of diminished quantities of old-growth forest habitat across Canada. This species is considered Sensitive in Alberta due to its dependence on old-growth forests.

TEMPORAL REPORTING RATE

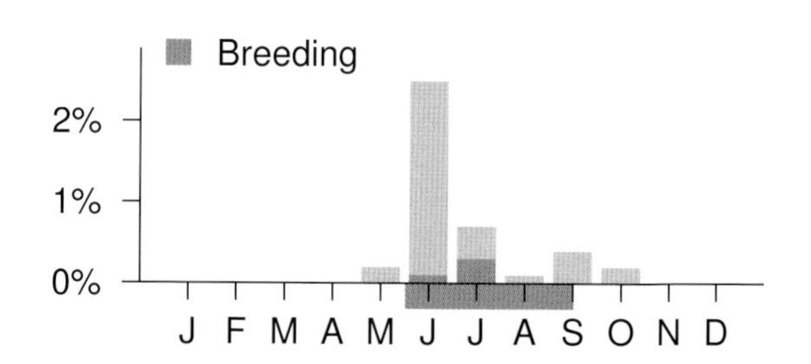

NATURAL REGIONS REPORTING RATE

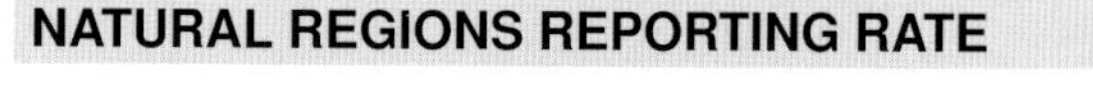

SPATIAL REPORTING RATE

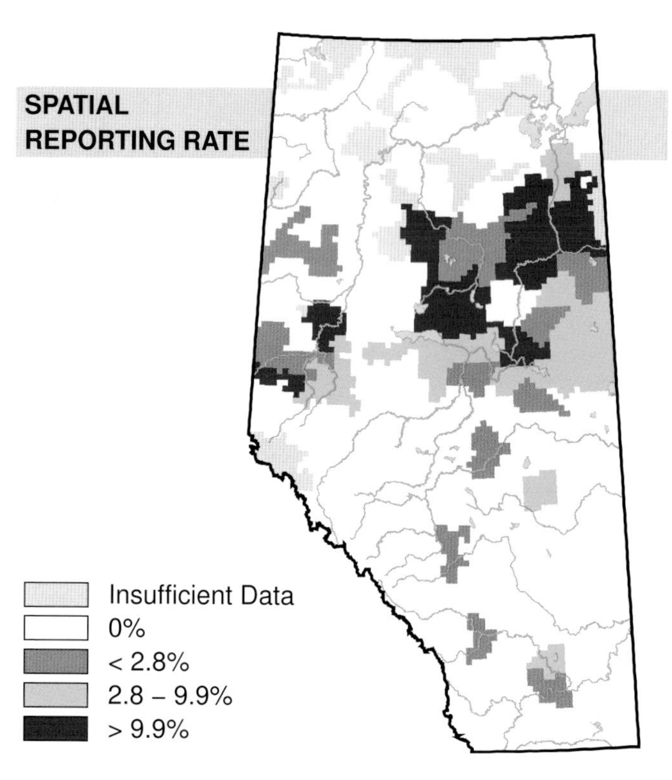

STATUS

GSWA 2000	GSWA 2005	COSEWIC Historic Rankings	COSEWIC 2007	Alberta Wildlife Act
Sensitive	Sensitive	N/A	N/A	N/A

Habitat

Nest
Location

Nest
Type

Diet

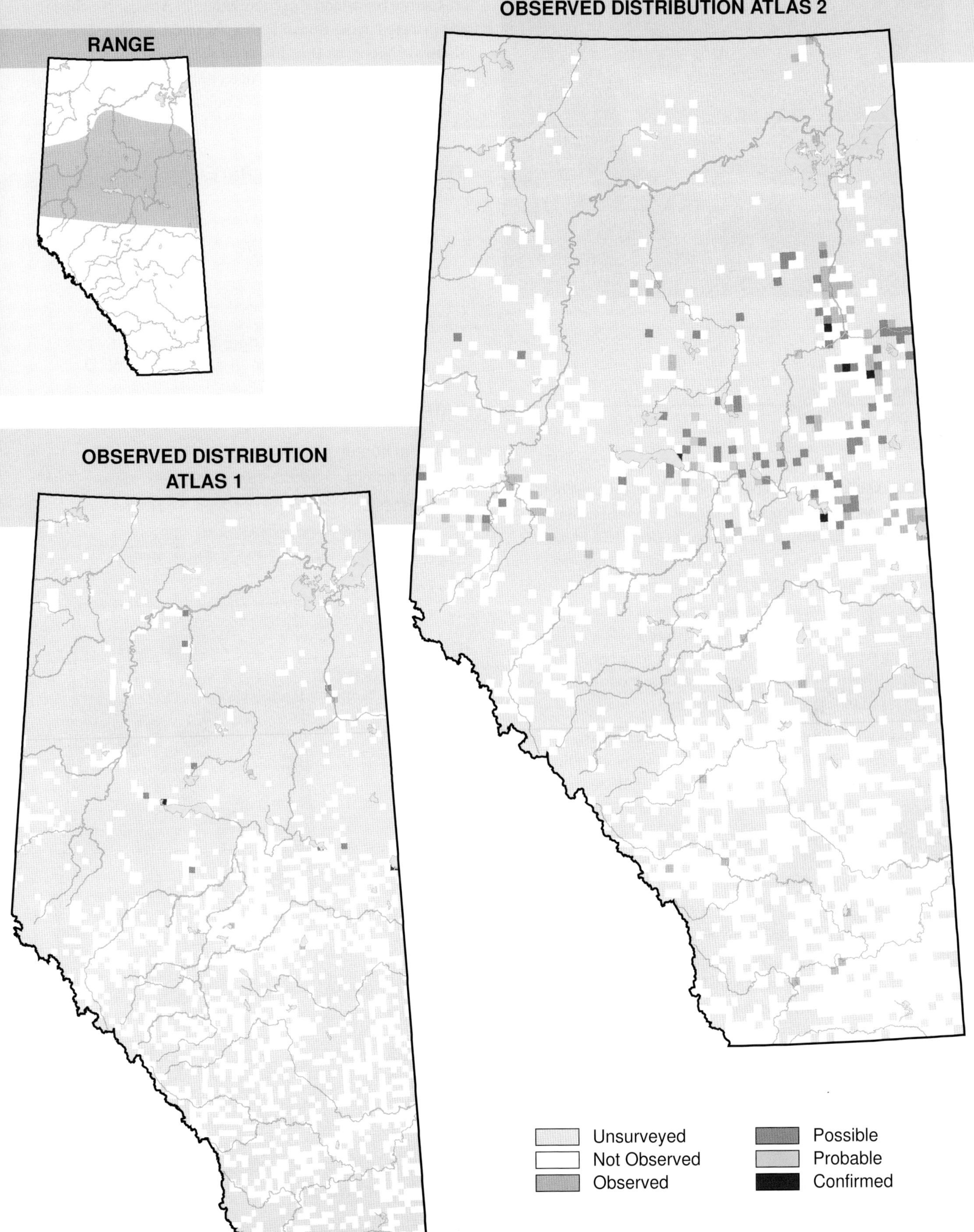
OBSERVED DISTRIBUTION ATLAS 2
RANGE
OBSERVED DISTRIBUTION
ATLAS 1
Unsurveyed
Not Observed
Observed
Possible
Probable
Confirmed

Blackpoll Warbler *(Dendroica striata)*

Photo: Royal Alberta Museum

NESTING

CLUTCH SIZE (EGGS):	4–5
INCUBATION (DAYS):	11–12
FLEDGING (DAYS):	10–12
NEST HEIGHT (METRES):	<2

The Blackpoll Warbler is found in the Boreal Forest, Foothills, Parkland, and Rocky Mountain Natural Regions in Alberta. The observed distribution of this species decreased in Atlas 2 in the northwestern part of the province around High Level and Fort Vermilion. It is unclear whether this represents an actual range contraction or whether the presence of Blackpoll Warblers in the northwest was missed during Atlas 2. The latter alternative is possible because there were not many Blackpoll Warblers recorded in thc northwest during Atlas 1 and those found were sparsely distributed.

The Blackpoll Warbler often breeds in wet areas such, as black spruce bogs, or in wet areas that contain clumps of willow thickets. Coniferous trees are usually chosen for nesting. For this reason, this species was found mainly in the Boreal Forest, Foothills, and Rocky Mountain NRs where this habitat is often found, and was rare in the Parkland NR around Grande Prairie. Records from the Grassland NR were non-breeders.

An increase in relative abundance was detected in the Grassland and Rocky Mountain NRs where, relative to other species, it was observed more frequently in Atlas 2 than in Atlas 1. The amount of suitable habitat in the Rocky Mountain NR has not changed substantially since Atlas 1 and all records from the Grassland NR were of non-breeders. Therefore, these increases are likely related to migratory variation or to the broader survey window in Atlas 2 than in Atlas 1. No change was detected in the Boreal Forest, Parkland, and Rocky Mountain NRs. The Breeding Bird Survey did not detect an abundance change in Alberta, but a decrease during the period 1968–2005 was detected across Canada. This species is considered Secure in Alberta.

TEMPORAL REPORTING RATE

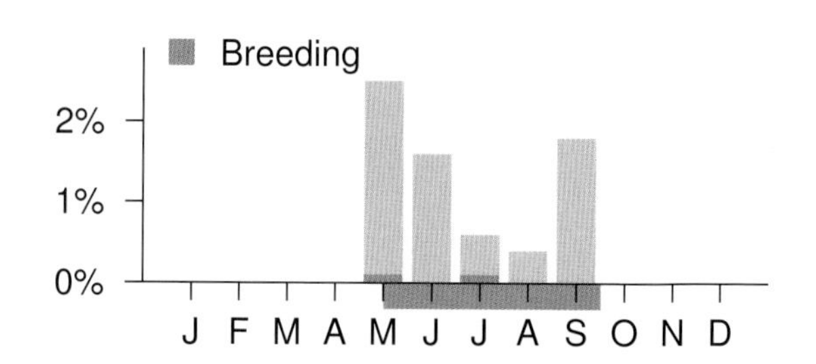

NATURAL REGIONS REPORTING RATE

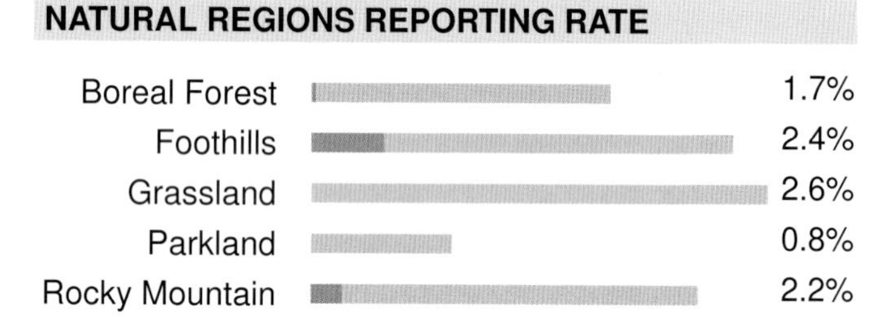

SPATIAL REPORTING RATE

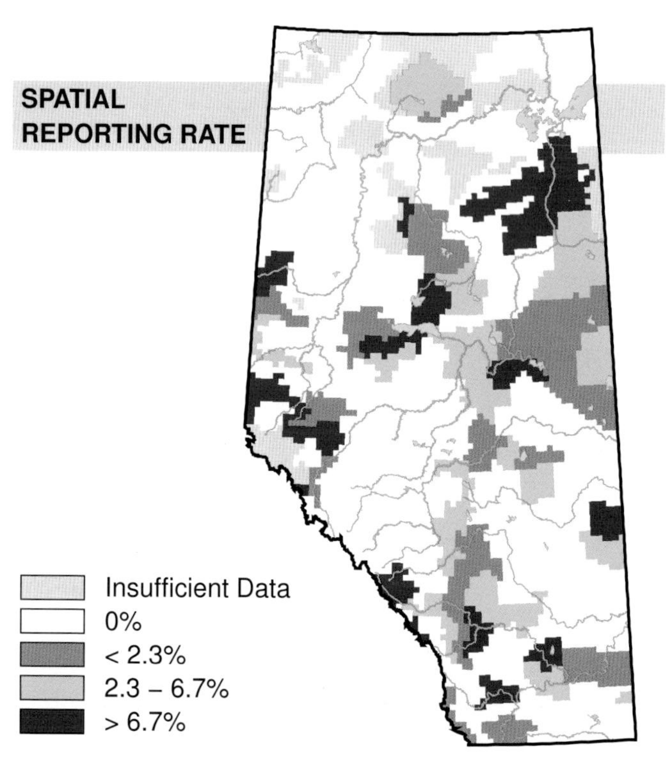

STATUS

GSWA 2000	GSWA 2005	COSEWIC Historic Rankings	COSEWIC 2007	Alberta Wildlife Act
Secure	Secure	N/A	N/A	N/A

Habitat  Nest Location  Nest Type  Diet 

RANGE

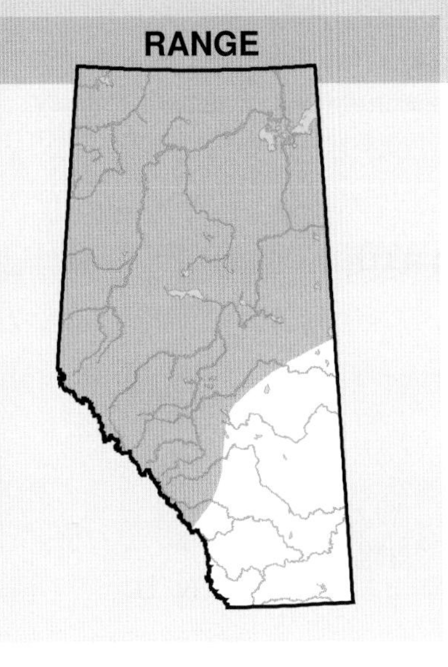

OBSERVED DISTRIBUTION ATLAS 2

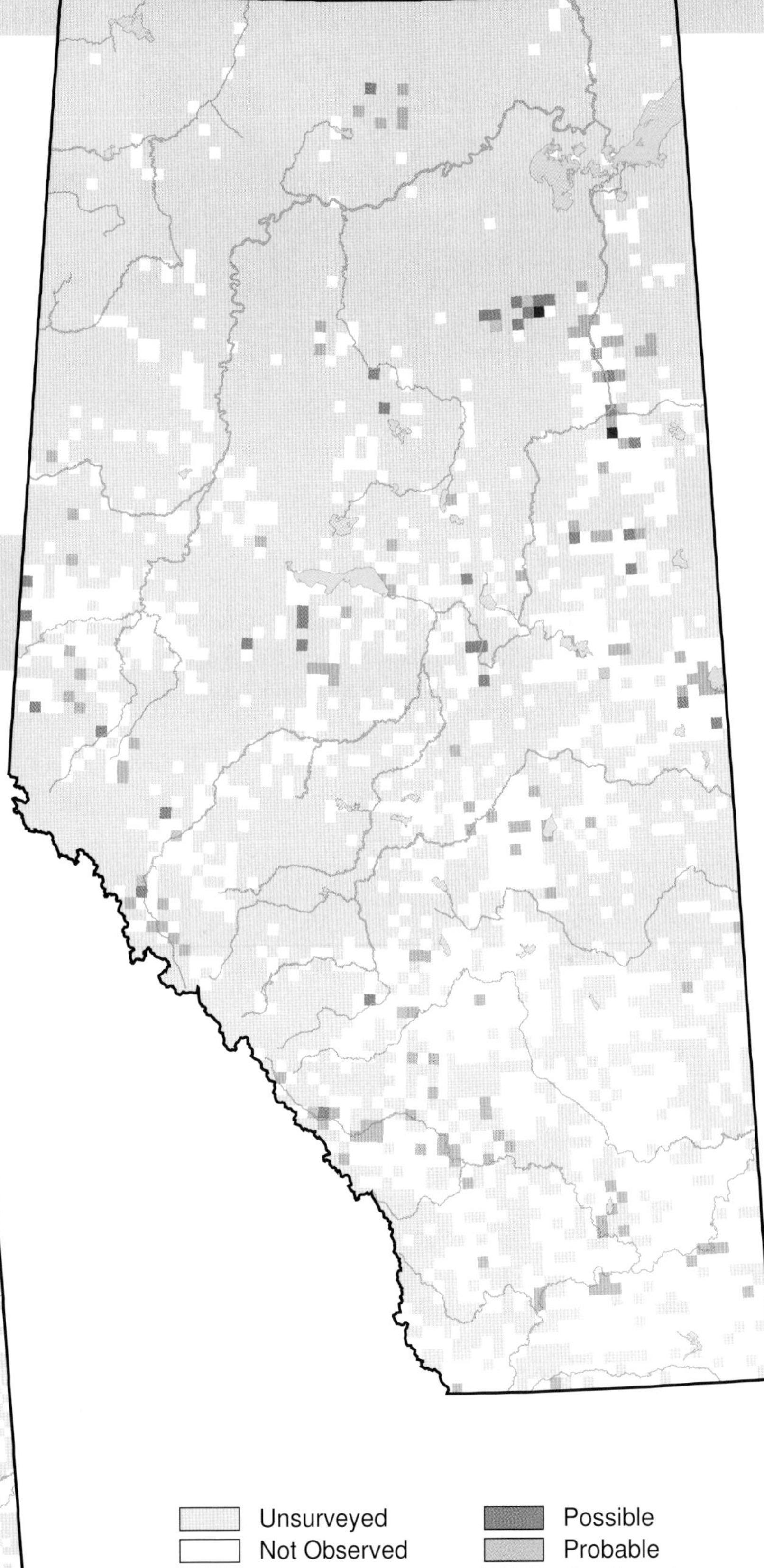

OBSERVED DISTRIBUTION ATLAS 1

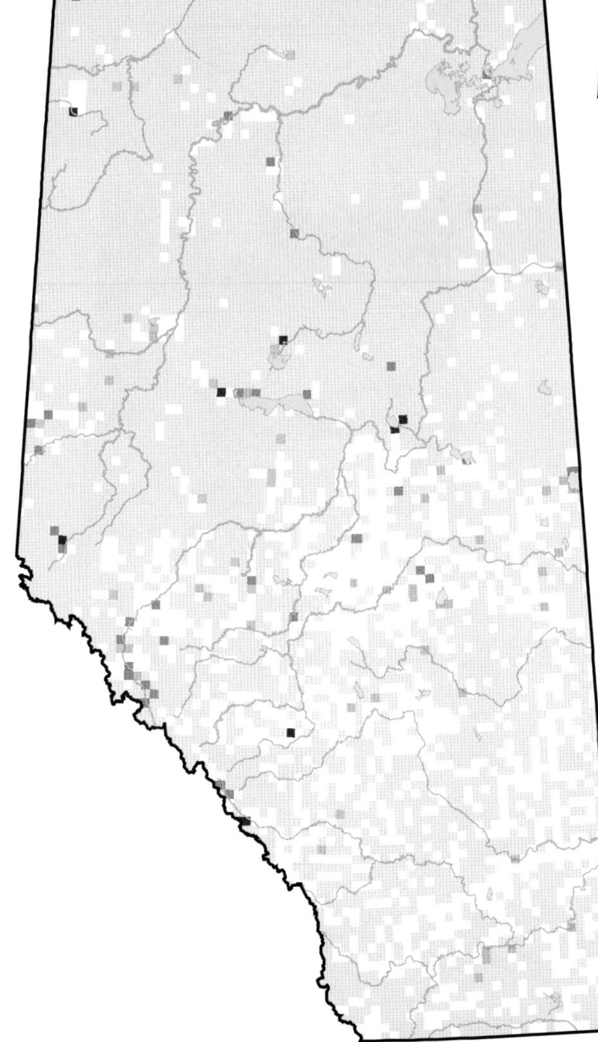

Unsurveyed
Not Observed
Observed
Possible
Probable
Confirmed

Black-and-white Warbler (*Mniotilta varia*)

Photo: Royal Alberta Museum

NESTING

CLUTCH SIZE (EGGS):	4–5
INCUBATION (DAYS):	10–13
FLEDGING (DAYS):	8–12
NEST HEIGHT (METRES):	0

The Black-and-white Warbler is found in the Boreal Forest, Foothills, Parkland, and Rocky Mountain Natural Regions in Alberta. No change in the distribution of this species was observed between Atlas 1 and Atlas 2.

The Black-and-white Warbler breeds in a variety of habitats including second-growth, mature, deciduous and mixedwood forests. However, they tend to prefer mature deciduous forests and generally avoid fragmented forests during the breeding season. For this reason, this species was found mainly in the Boreal Forest NR and in the northern part of the Foothills NR, and it was found infrequently along the periphery of the Parkland and Rocky Mountain NRs. Records from the Grasslands NR were non-breeders.

Increases in relative abundance were detected in the Boreal Forest and Foothills NRs where, relative to other species, this warbler was observed more frequently in Atlas 2 than in Atlas 1. There is no known or probable biological cause for these increases. Rather than being related to an actual population change, it is likely that more complete coverage of the north in Atlas 2 was the cause of the detected increase. No changes were detected in other NRs. The Breeding Bird Survey detected no change in abundance in Alberta, but did detect a decrease in abundance during the period 1995–2005 across Canada. National declines are likely the result of increased forest fragmentation, a condition that this species generally avoids during the breeding season. The Breeding Bird Survey might not have detected changes in Alberta because this survey has low northern coverage. This species is considered Secure in Alberta.

TEMPORAL REPORTING RATE

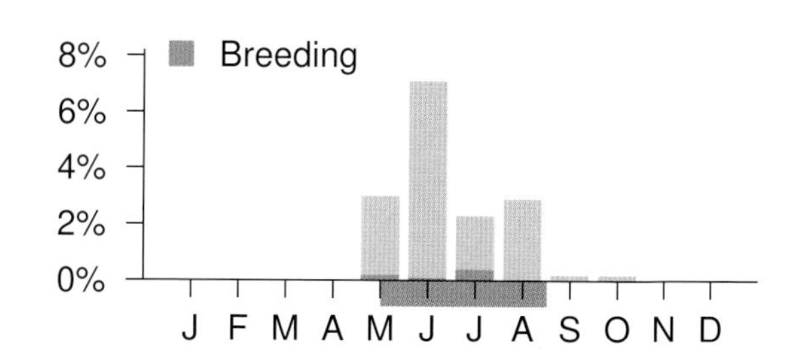

NATURAL REGIONS REPORTING RATE

SPATIAL REPORTING RATE

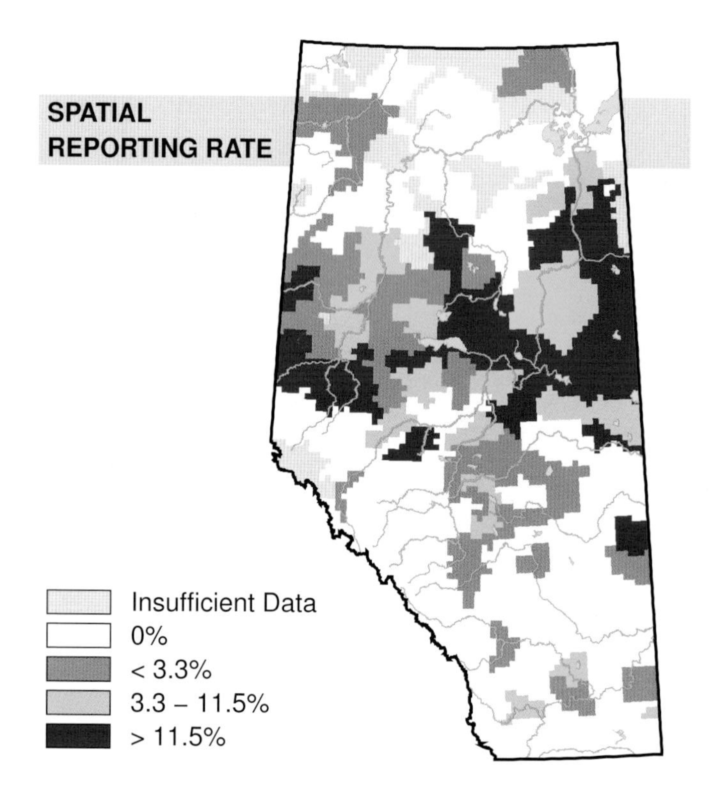

STATUS

GSWA 2000	GSWA 2005	COSEWIC Historic Rankings	COSEWIC 2007	Alberta Wildlife Act
Secure	Secure	N/A	N/A	N/A

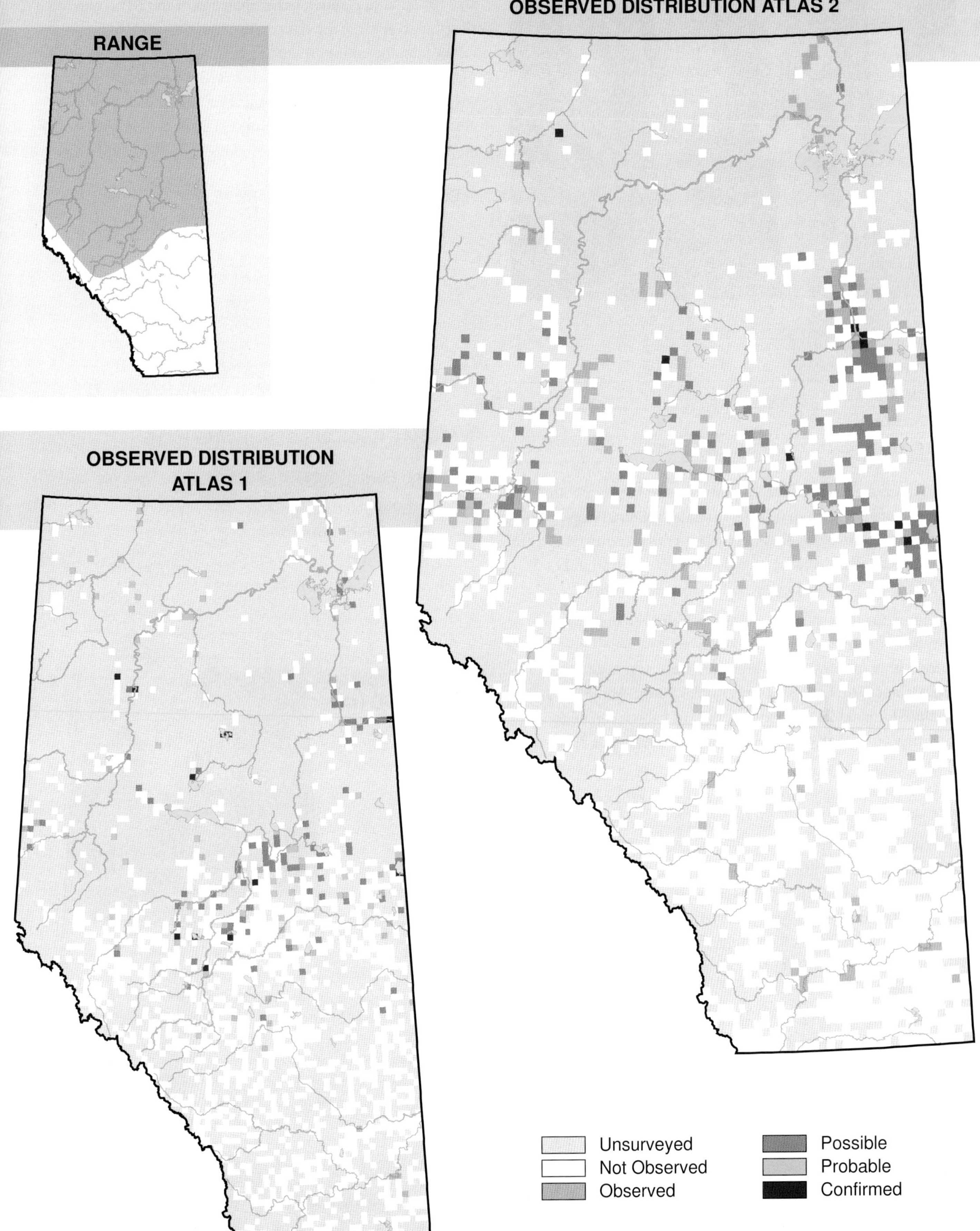
RANGE
OBSERVED DISTRIBUTION ATLAS 2
OBSERVED DISTRIBUTION ATLAS 1
Unsurveyed
Not Observed
Observed
Possible
Probable
Confirmed

American Redstart (*Setophaga ruticilla*)

Photo: Robert Gehlert

NESTING

CLUTCH SIZE (EGGS):	3–5
INCUBATION (DAYS):	12
FLEDGING (DAYS):	8–10
NEST HEIGHT (METRES):	2–8

The American Redstart breeds in the Boreal Forest, Foothills, Grassland, Parkland, and Rocky Mountain Natural Regions in Alberta. The observed distributions in Atlas 1 and Atlas 2 were similar, although breeding records were recorded slightly farther south and east into the Grasslands NR in Atlas 2.

The American Redstart is usually found in wet areas with dense deciduous habitat such as early successional, second-growth forests, or willow thickets. As a result, this species was found mainly in the Boreal Forest, Foothills, and Rocky Mountain NRs where this type of habitat can be found. Distribution in the Grassland and Parkland NRs was sparse.

Increases in relative abundance were detected in the Foothills and Rocky Mountain NRs where, relative to other species, it was observed more frequently in Atlas 2 than in Atlas 1. These increases could reflect the fact that development activities and prescribed burns in these regions have shifted the average age of forests towards earlier successional stages. An increase in relative abundance was also detected in the Grassland NR; the cause for this change is likely migratory variation, as the majority of the records for this NR were observations from the beginning and end of the breeding season and probably, therefore, of migrating birds. No change was detected in the Boreal Forest and Parkland NRs. The Breeding Bird Survey did not find any change in abundance in Alberta or across Canada. The disagreement in the results is likely a function of spatial scale and field protocol differences. The Atlas sample size for Alberta is much larger than that of the BBS, making it possible to analyze the data on the finer, Natural Region scale, rather than the entire province where regional variability will make detecting changes more difficult. Also, the BBS is a road-based, point count-style survey limiting the coverage of appropriate habitat for this species. The American Redstart is considered Secure in Alberta.

TEMPORAL REPORTING RATE

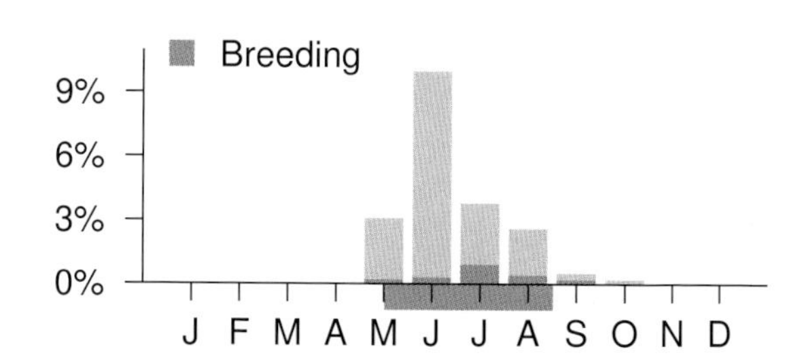

NATURAL REGIONS REPORTING RATE

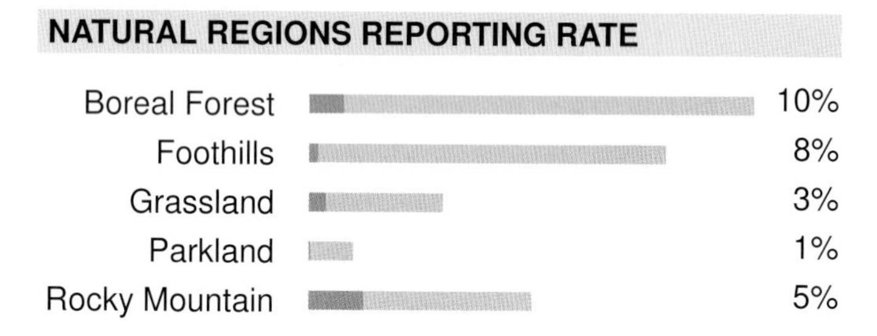

SPATIAL REPORTING RATE

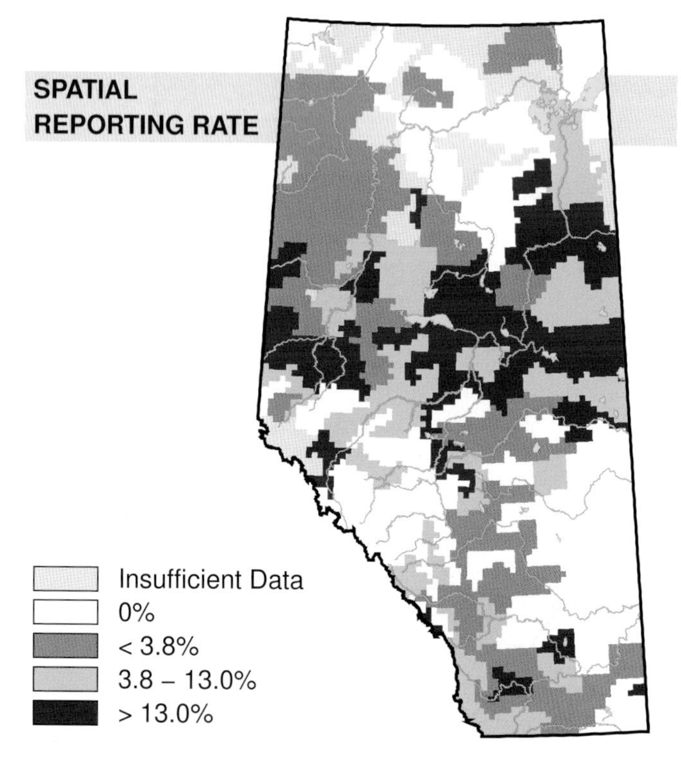

STATUS

GSWA 2000	GSWA 2005	COSEWIC Historic Rankings	COSEWIC 2007	Alberta Wildlife Act
Secure	Secure	N/A	N/A	N/A

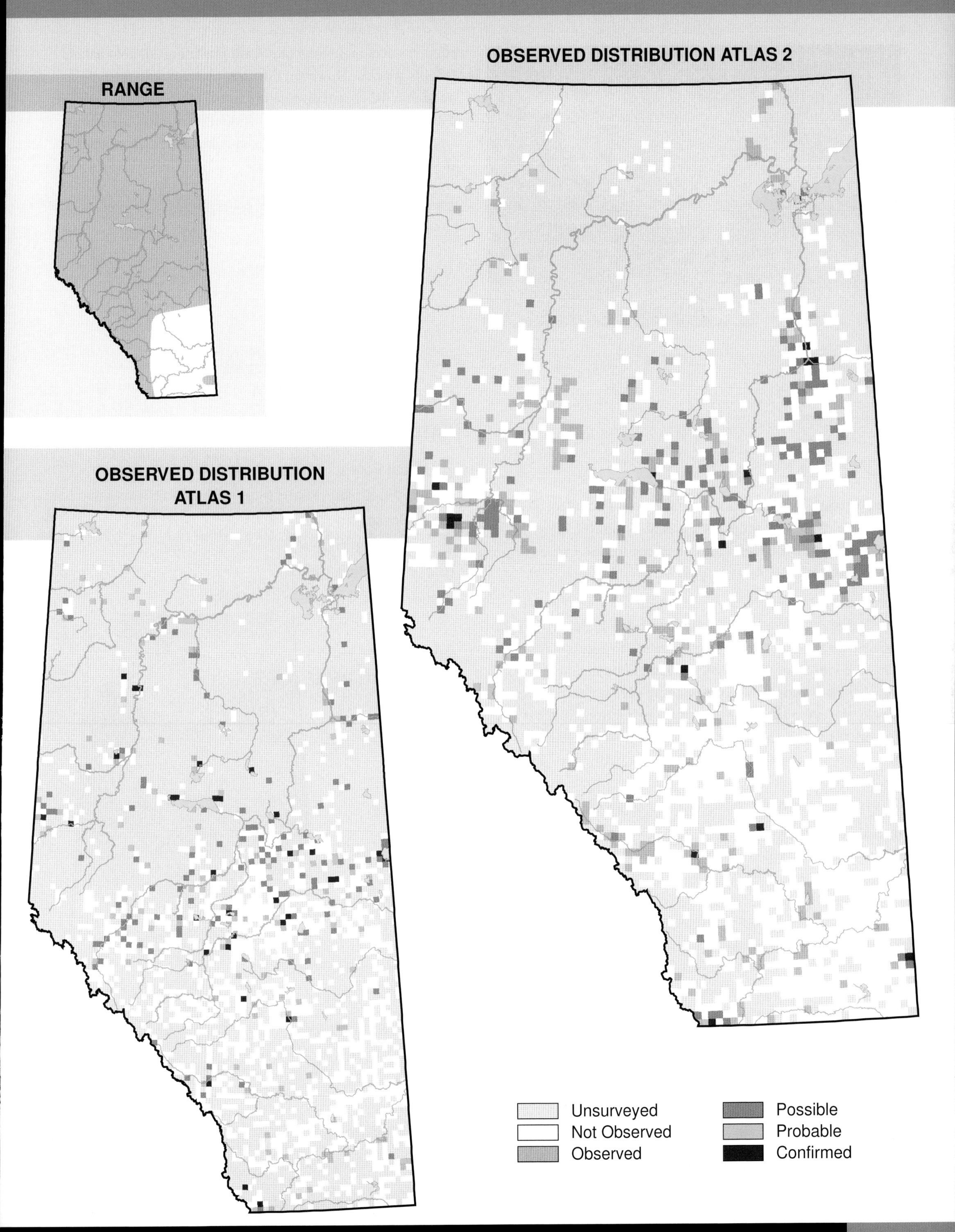
RANGE
OBSERVED DISTRIBUTION ATLAS 2
OBSERVED DISTRIBUTION ATLAS 1
Unsurveyed
Not Observed
Observed
Possible
Probable
Confirmed

Ovenbird (*Seiurus aurocapilla*)

Photo: Royal Alberta Museum

NESTING

CLUTCH SIZE (EGGS):	3–6
INCUBATION (DAYS):	11–14
FLEDGING (DAYS):	8–10
NEST HEIGHT (METRES):	0

The Ovenbird is found in the Boreal Forest, Foothills, Parkland, and Rocky Mountain Natural Regions in Alberta. The observed distribution of this species did not change between Atlas 1 and Atlas 2.

The Ovenbird generally breeds in deciduous or mixedwood forests that include a large proportion of deciduous trees. This species generally prefers to nest in mature forests, but it will also breed in young forests. Nests are built on the ground in areas with a low density of understorey vegetation. For this reason, this species was found mainly in the Boreal Forest and Foothills NRs where this habitat can be prevalent. This bird was found less commonly in the Parkland and Rocky Mountain NRs. Records from the Grassland NR were non-breeders with the exception of records on the periphery of its range.

Increases in relative abundance were detected in the Boreal Forest, Foothills, and Rocky Mountain NRs where, relative to other species, Ovenbirds were observed more frequently in Atlas 2 than in Atlas 1. No change was detected in the Parkland NR. The Breeding Bird Survey did not detect a change in Alberta, although a decline was detected across Canada during the period 1968–2005. The Atlas-detected increases are confusing considering the loss of mature trees in these NRs. However, as this species will also nest in younger forests with low density understorey vegetation, it is possible that forest regrowth methods involving chemical spraying to suppress deciduous vegetation are providing suitable habitat for this species. However, the disagreement with the national trend and the speculative nature of the biological cause of the detected increase indicate that further research is needed to assess the status of this species in Alberta. The provincial status of this species is Secure.

TEMPORAL REPORTING RATE

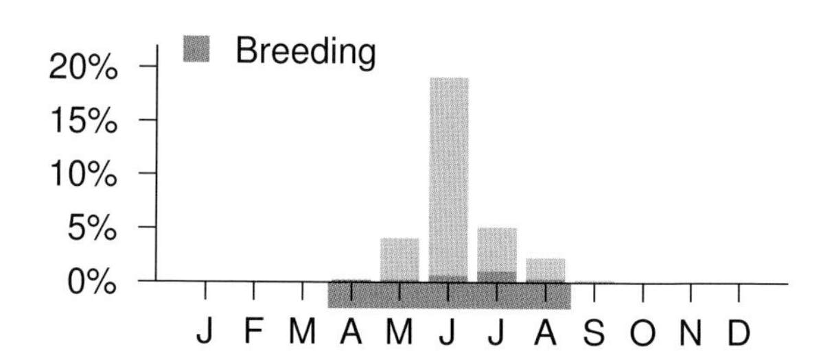

NATURAL REGIONS REPORTING RATE

Boreal Forest	13%
Foothills	13%
Grassland	1%
Parkland	1%
Rocky Mountain	2%

SPATIAL REPORTING RATE

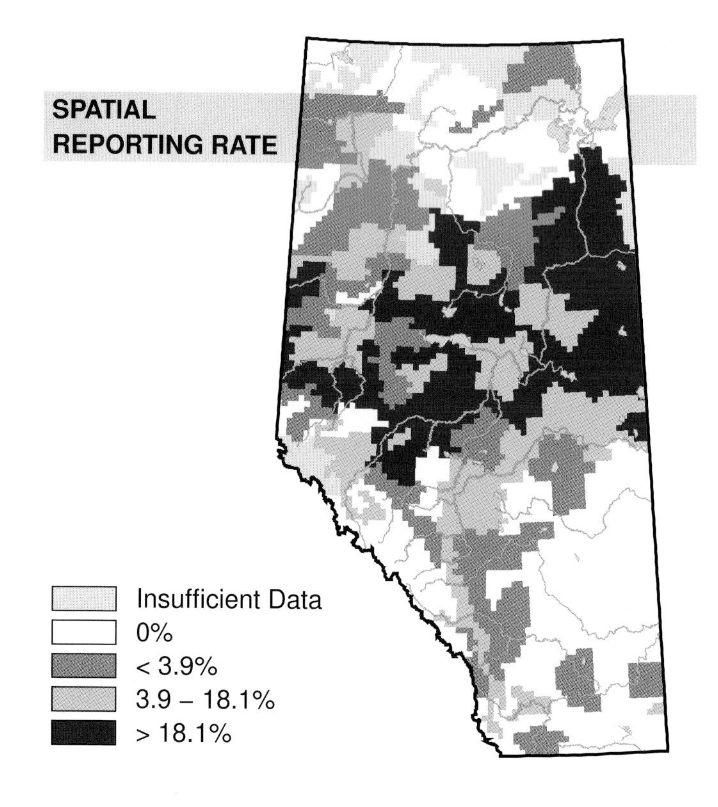

STATUS

GSWA 2000	GSWA 2005	COSEWIC Historic Rankings	COSEWIC 2007	Alberta Wildlife Act
Secure	Secure	N/A	N/A	N/A

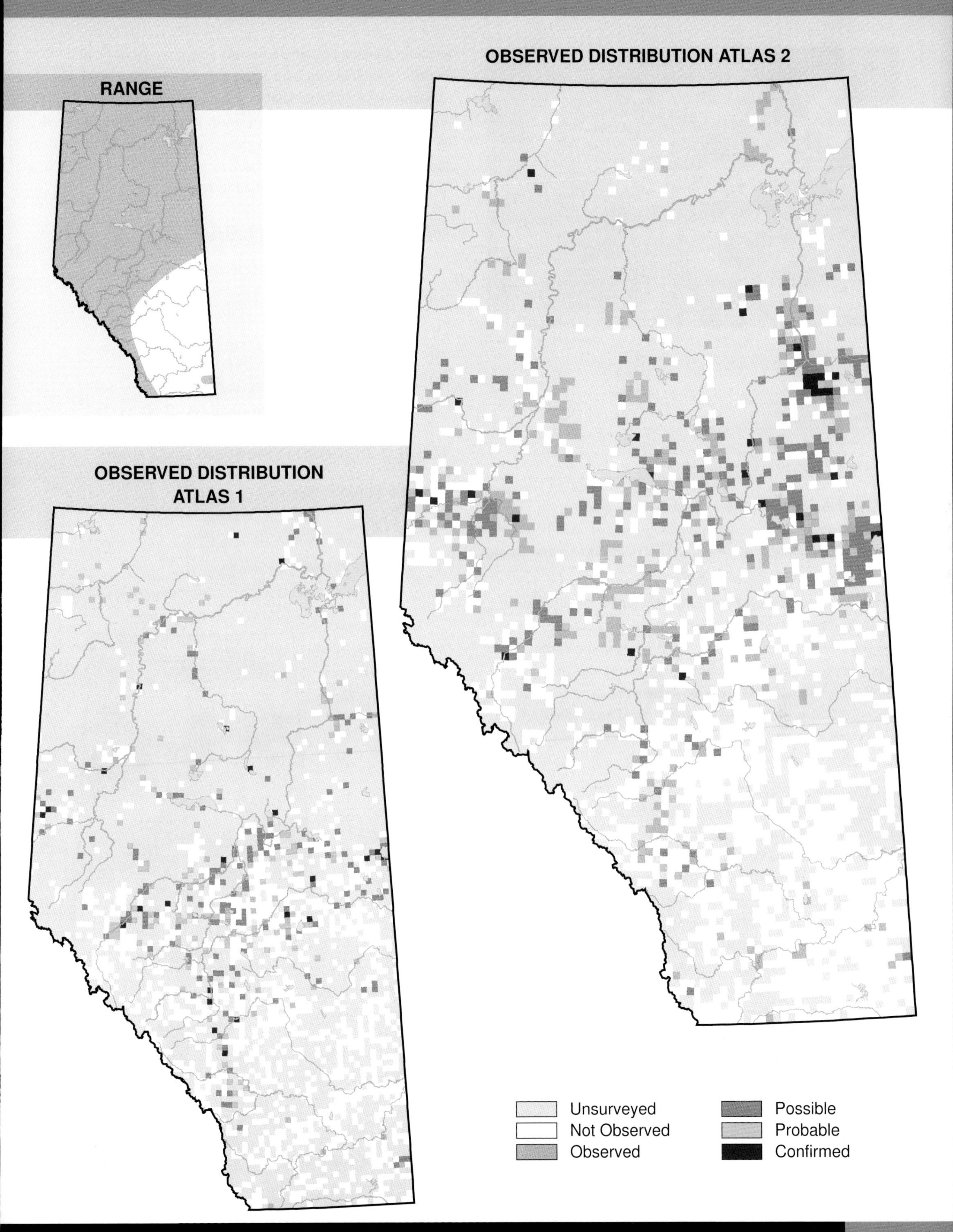
Habitat
Nest Location
Nest Type
Diet
OBSERVED DISTRIBUTION ATLAS 2
RANGE
OBSERVED DISTRIBUTION ATLAS 1
Unsurveyed
Not Observed
Observed
Possible
Probable
Confirmed

Northern Waterthrush (*Seiurus noveboracensis*)

Photo: Gerald Romanchuk

NESTING

CLUTCH SIZE (EGGS):	4–5
INCUBATION (DAYS):	12–13
FLEDGING (DAYS):	10
NEST HEIGHT (METRES):	0

The Northern Waterthrush is found in the Boreal Forest, Foothills, and Rocky Mountain Natural Regions in Alberta. No change in distribution was observed between Atlas 1 and Atlas 2 for this species.

As its name implies, the Northern Waterthrush breeds in wet forested areas such as bogs, along lake edges, and along rivers that have dense deciduous growth. As a result, the Northern Waterthrush was found mainly in the Boreal Forest, Foothills, and Rocky Mountain NRs in areas where this type of habitat was present. Records in the Grassland and Parkland NRs were mostly non-breeders, except for a few records on the periphery of this species' range.

Increases in relative abundance were detected in the Boreal Forest, Foothills, and Rocky Mountain NRs where, relative to other species, the Northern Waterthrush was observed more frequently in Atlas 2 than in Atlas 1. The causes for these increases are unclear. Increased deciduous growth is common after forest harvesting, but more current regrowth methods usually include chemical spraying that reduces deciduous growth and assists the establishment of planted conifers. Prescribed burns may have contributed to increases found in the Rocky Mountain NR, as the burned areas are usually allowed to regrow naturally. Increased survey coverage in northern Alberta may have contributed to the increase detected in the Boreal Forest NR. The Breeding Bird Survey did not detect an abundance change for this species in Alberta or Canada. The Breeding Bird Survey sample size for Northern Waterthrush was fairly small in Alberta because the breeding range of this species lies farther north than most Breeding Bird Survey routes. Lack of corroboration and probable biological cause necessitates further research to assess the Atlas findings. This species is considered Secure in Alberta.

TEMPORAL REPORTING RATE

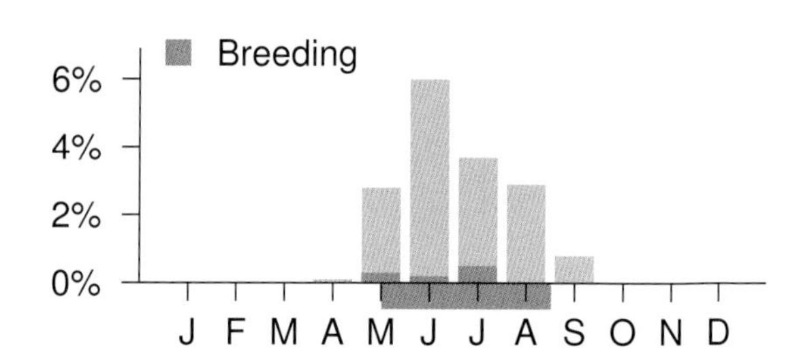

NATURAL REGIONS REPORTING RATE

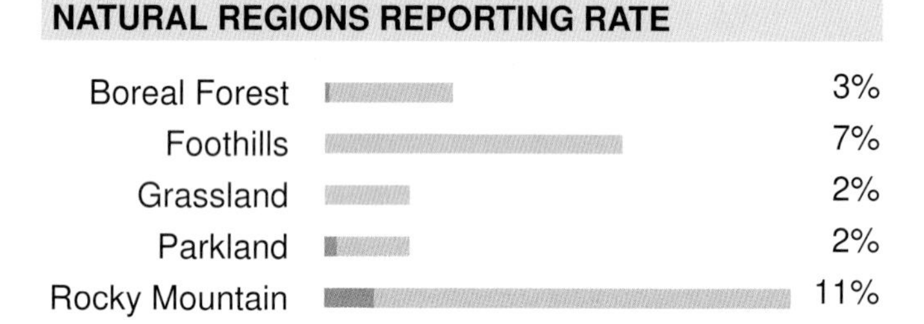

SPATIAL REPORTING RATE

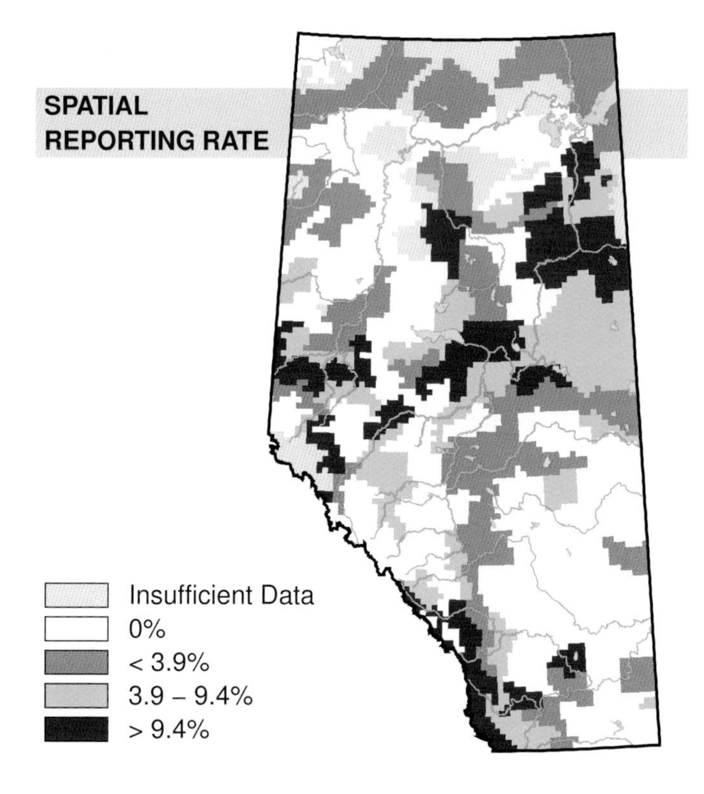

STATUS

GSWA 2000	GSWA 2005	COSEWIC Historic Rankings	COSEWIC 2007	Alberta Wildlife Act
Secure	Secure	N/A	N/A	N/A

Habitat

Nest Location

Nest Type

Diet

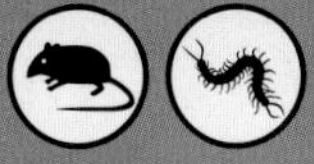

RANGE

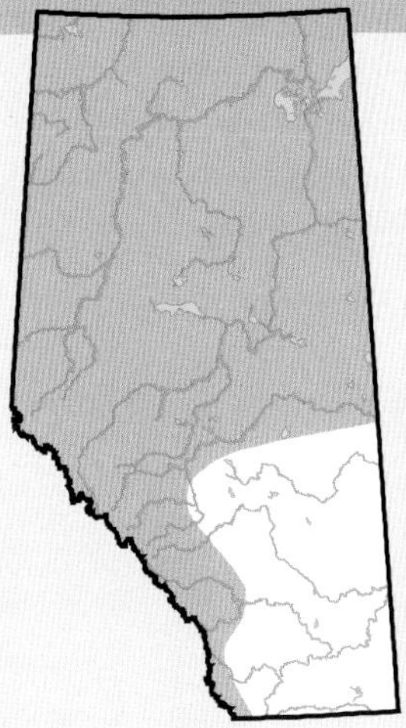

OBSERVED DISTRIBUTION ATLAS 2

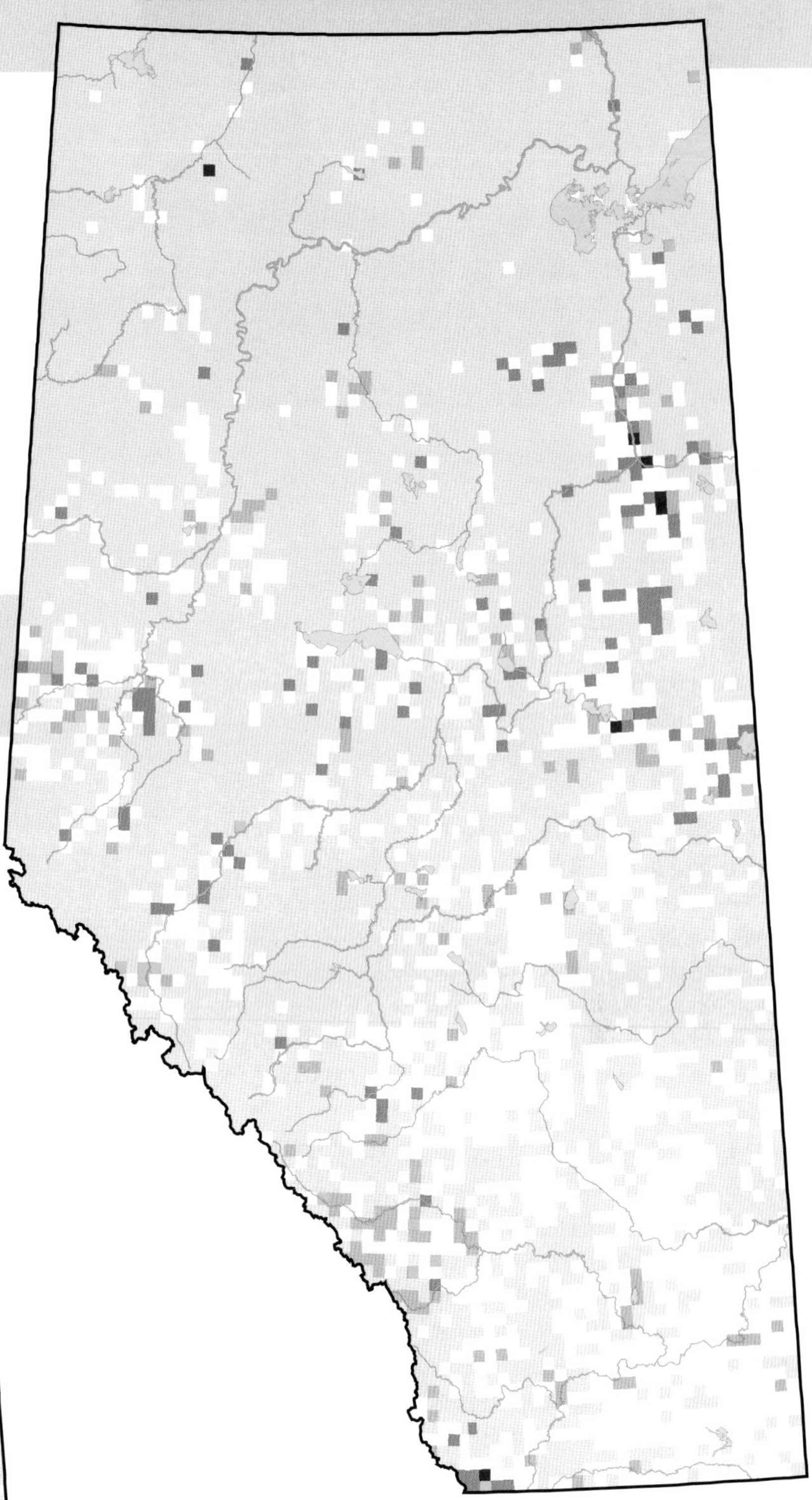

OBSERVED DISTRIBUTION ATLAS 1

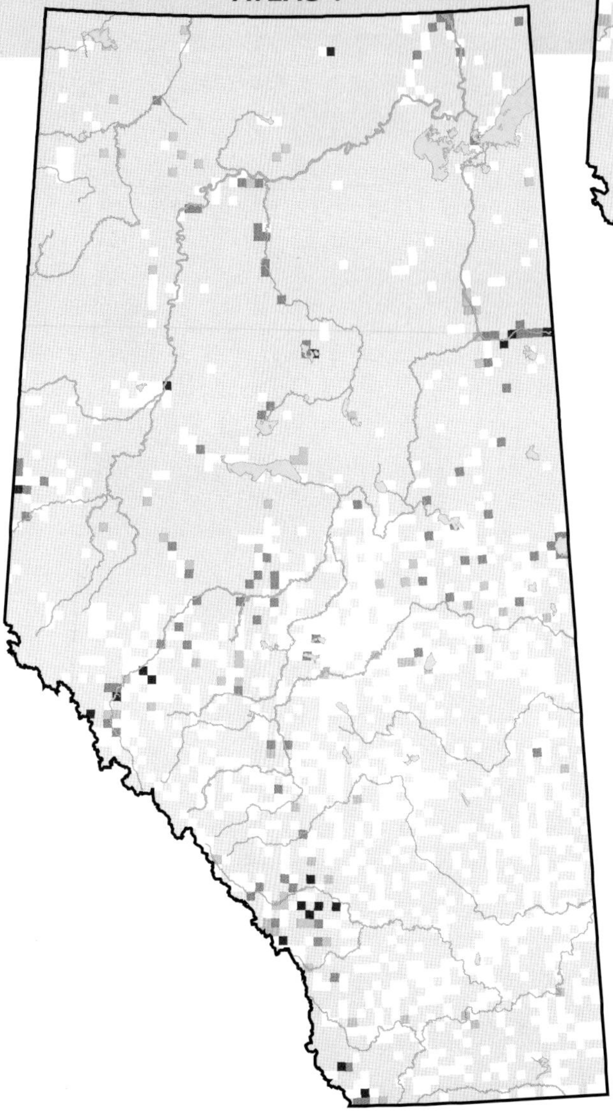

Unsurveyed
Not Observed
Observed
Possible
Probable
Confirmed

Connecticut Warbler (*Oporornis agilis*)

Photo: Royal Alberta Museum

NESTING

CLUTCH SIZE (EGGS):	4–5
INCUBATION (DAYS):	unknown
FLEDGING (DAYS):	unknown
NEST HEIGHT (METRES):	0

The Connecticut Warbler is found in the Boreal Forest, Foothills, and Parkland Natural Regions in Alberta. No change in the distribution of this species was observed between Atlas 1 and Atlas 2.

The Connecticut Warbler breeds in wet, open forests. Breeding sites can be located within deciduous, coniferous or mixedwood forests provided tree density is fairly low. The Connecticut Warbler can nest in early successional forests after logging has occurred and it can nest along forest edges. As a result, this species was found mainly in the Boreal Forest NR and in the northern part of the Foothills NR where these types of habitat can be found. Records from the Grassland NR were of non-breeders.

An increase in relative abundance was detected in the Boreal Forest NR where, relative to other species, this warbler was observed more frequently in Atlas 2 than in Atlas 1. The population may be increasing in that area because human developments are changing forest structures and providing the habitat preferred by the Connecticut Warbler. This species tends to favour open-forest habitats and can breed in regenerating and fragmented forests. No change was detected in other NRs. The Breeding Bird Survey did not find any abundance change in Alberta or Canada. The discrepancy between the Atlas and BBS findings is likely due to the much more extensive coverage of northern Alberta by the Atlas. The BBS covers little of the Canadian breeding range of the Connecticut Warbler and has low detection rates (Dunn, 2005).

Its secretive behaviour and preference for breeding habitat in remote areas has made the Connecticut Warbler very difficult to study (Pitocchelli et al., 1997). He reported that there are conflicting data on population trends for this species as the formal census data indicate stable or increasing numbers while the informal counts of migrants suggest declines.

This species is considered Secure in Alberta.

TEMPORAL REPORTING RATE

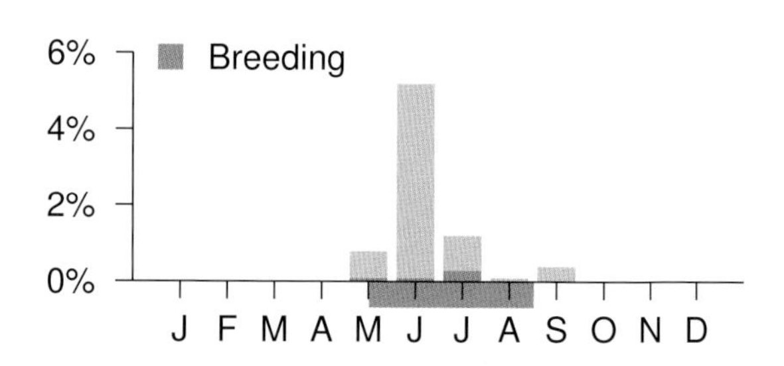

NATURAL REGIONS REPORTING RATE

SPATIAL REPORTING RATE

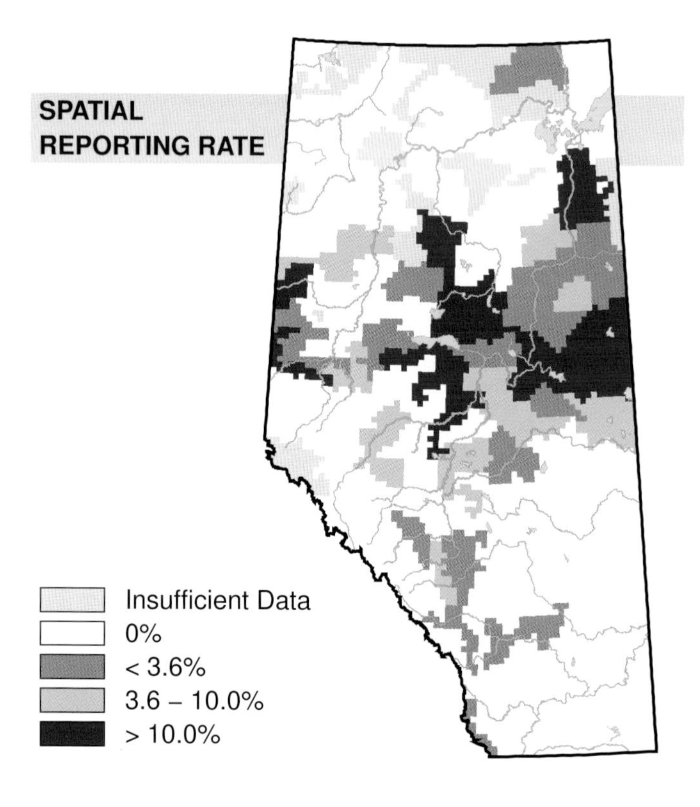

STATUS

GSWA 2000	GSWA 2005	COSEWIC Historic Rankings	COSEWIC 2007	Alberta Wildlife Act
Secure	Secure	N/A	N/A	N/A

Habitat
Nest Location
Nest Type
Diet

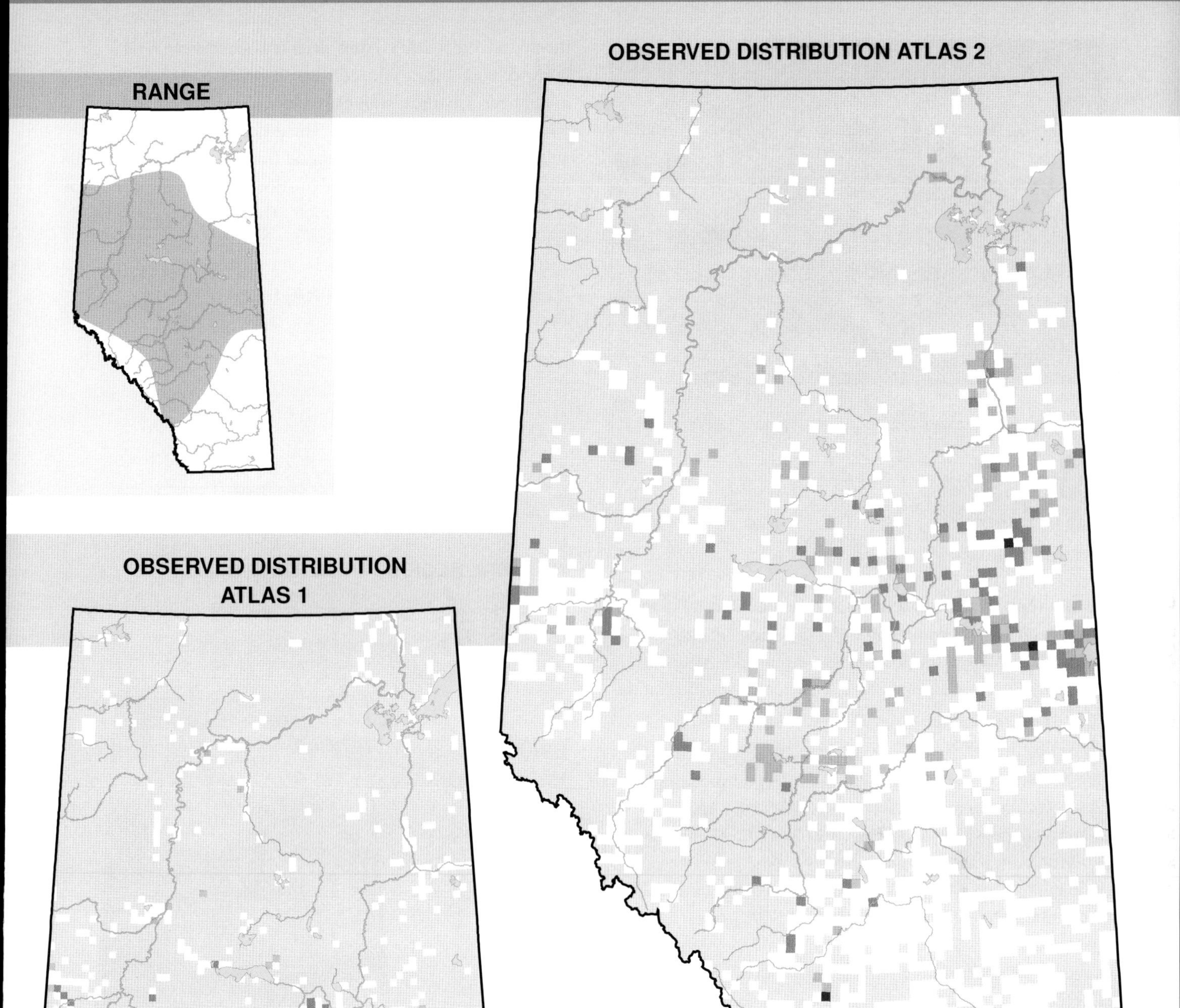

RANGE
OBSERVED DISTRIBUTION ATLAS 2
OBSERVED DISTRIBUTION ATLAS 1
Unsurveyed
Not Observed
Observed
Possible
Probable
Confirmed

Mourning Warbler (*Oporornis philadelphia*)

Photo: Alan MacKeigan

NESTING

CLUTCH SIZE (EGGS):	3–5
INCUBATION (DAYS):	12–13
FLEDGING (DAYS):	9–14
NEST HEIGHT (METRES):	0–0.8

The Mourning Warbler is found in the Boreal Forest, Foothills, Parkland, and Rocky Mountain Natural Regions in Alberta. Compared to Atlas 1, this species was observed farther west, in the Grande Prairie area, in Atlas 2. This could indicate that the range of this species, generally perceived as eastern, is expanding westward.

The Mourning Warbler tends to breed in second-growth forests and forests with relatively low tree densities but high undergrowth densities. Mourning Warblers do well in areas after disturbance when deciduous trees are the main recolonizing tree species. As a result, this species was found mainly in the Boreal Forest and Foothills NRs, as well as in the Parkland and Rocky Mountain NRs where these habitat conditions exist. Records from the Grassland NR were non-breeders.

Increases in relative abundance were detected in the Boreal Forest and Foothills NRs where, relative to other species, Mourning Warblers were observed more frequently in Atlas 2 than in Atlas 1. No changes were detected in the other NRs. The Breeding Bird Survey did not detect a change in abundance in Alberta for the period 1985–2005, although an increase was detected for the period 1968–2005. Nationally, the BBS reported a decrease in abundance for this species for the period 1985–2005. Atlas-detected increases are not unexpected in these NRs, considering the extent of forestry and oil and gas activities, which are increasing the availability of young second-growth forests. The differences between Atlas and BBS results can be understood from the viewpoint that the majority of BBS data gathered in Alberta comes from south of this species' observed distribution. This species is considered Secure in Alberta.

TEMPORAL REPORTING RATE

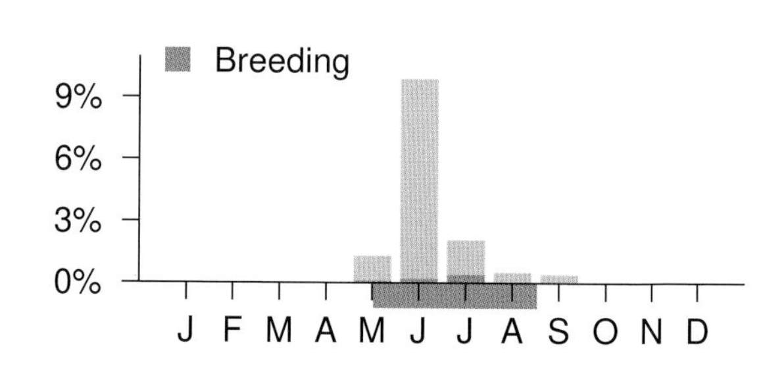

NATURAL REGIONS REPORTING RATE

SPATIAL REPORTING RATE

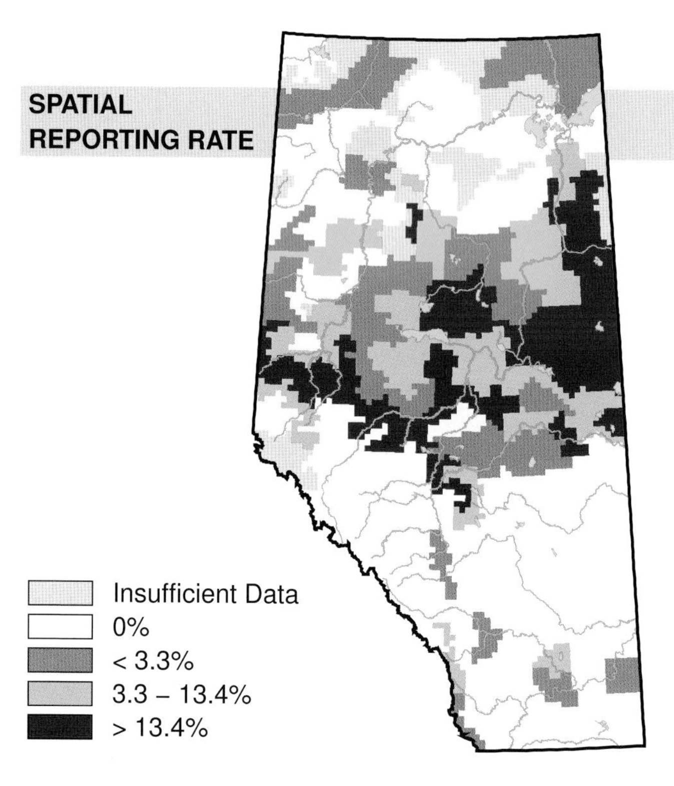

STATUS

GSWA 2000	GSWA 2005	COSEWIC Historic Rankings	COSEWIC 2007	Alberta Wildlife Act
Secure	Secure	N/A	N/A	N/A

OBSERVED DISTRIBUTION ATLAS 2

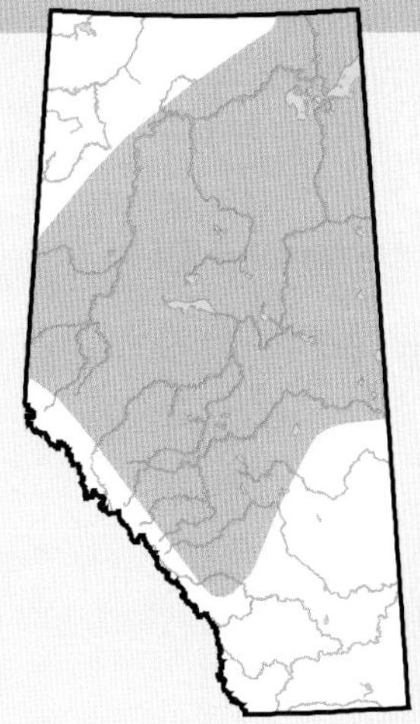

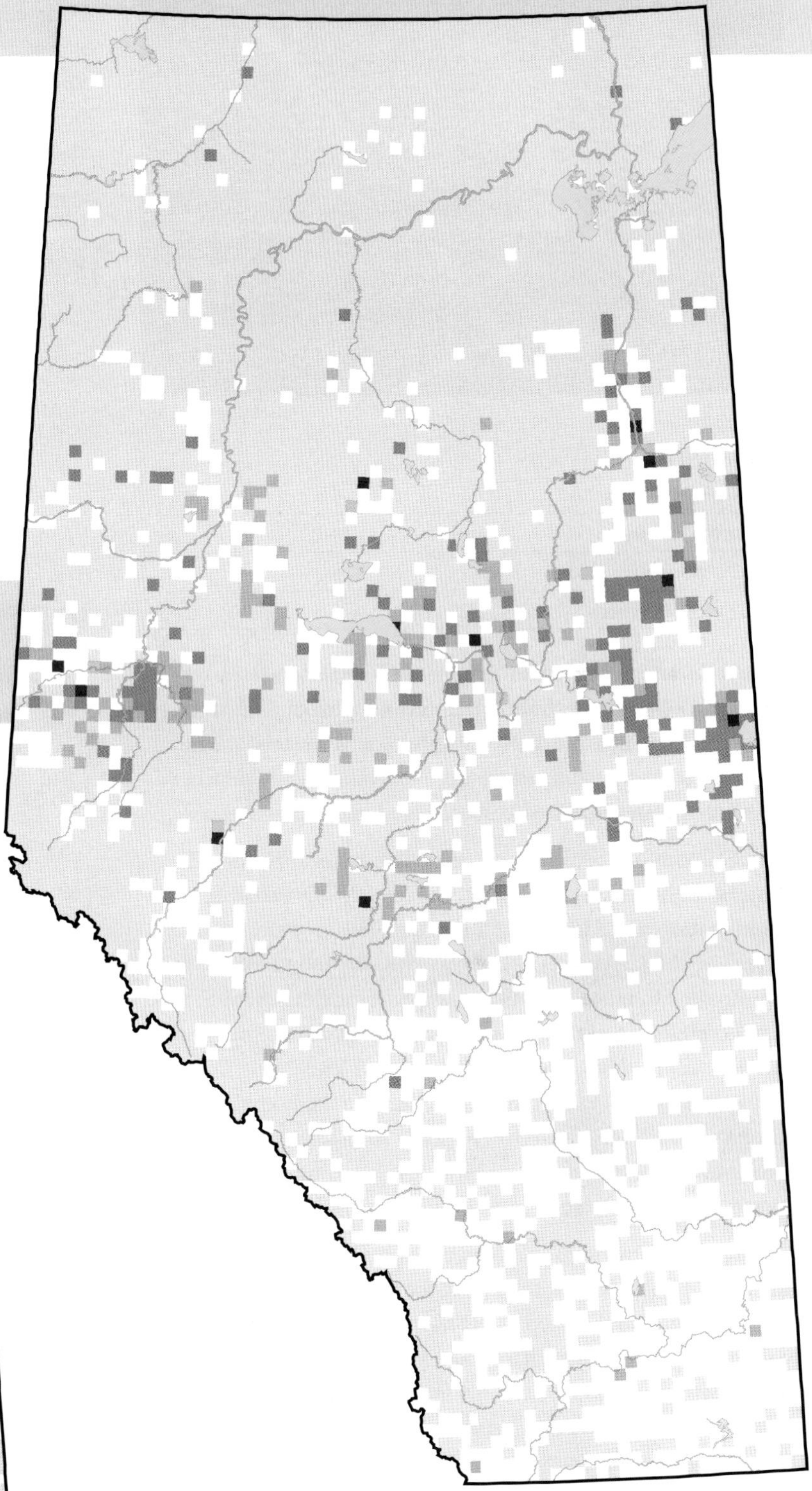

RANGE

OBSERVED DISTRIBUTION ATLAS 1

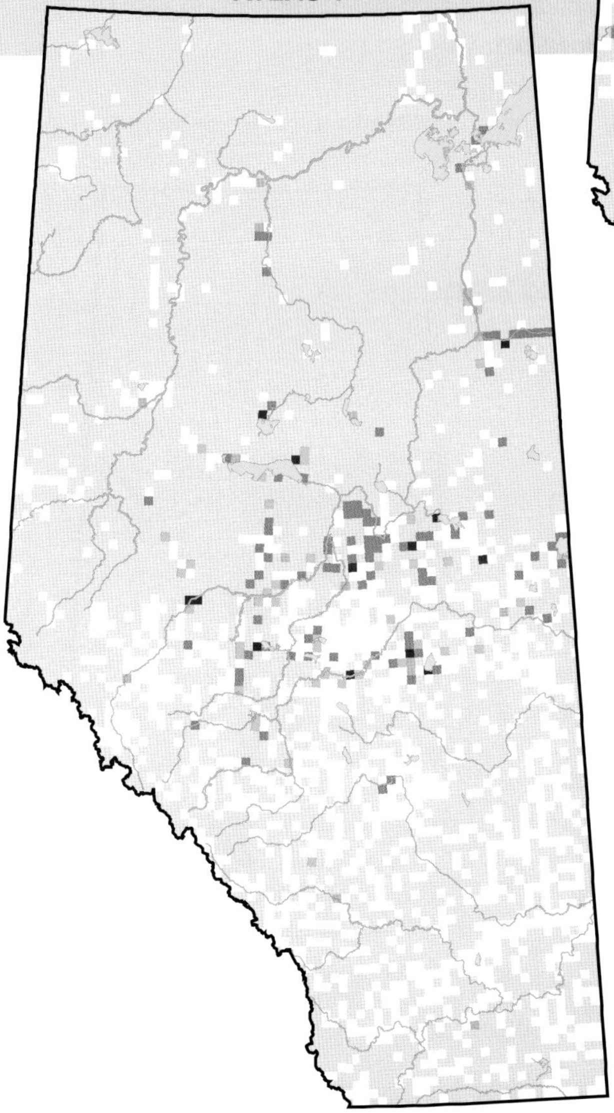

Unsurveyed
Not Observed
Observed
Possible
Probable
Confirmed

MacGillivray's Warbler (*Oporornis tolmiei*)

Photo: Gerald Romanchuk

NESTING

CLUTCH SIZE (EGGS):	3–5
INCUBATION (DAYS):	11–13
FLEDGING (DAYS):	8–9
NEST HEIGHT (METRES):	<6

MacGillivray's Warbler is found in the Boreal Forest, Foothills, and Rocky Mountain Natural Regions in Alberta. The range of this species, which is usually thought of as a western species, appears to be expanding eastward. New breeding evidence was found in the area south of Lesser Slave Lake during Atlas 2.

MacGillivray's Warbler breeds in deciduous or mixedwood forests and forests that are in process of post-disturbance regeneration. Breeding habitat is often close to water such as a marsh or creek. As a result, this species was found mainly in the Rocky Mountain NR and occasionally in the Boreal Forest and Foothills NRs where this type of habitat can be found. Some records with associated breeding evidence were found in the Parkland NR in the transition zone from the Foothills NR, near Rocky Mountain House. Records from the Grassland NR were mostly of non-breeders, but a few breeding records were noted from immediately north of the Cypress Hills.

An increase in relative abundance was detected in the Grassland NR. Most records from the Grassland NR were of non-breeders, so this increase is likely a function of migratory variation. The Breeding Bird Survey sample size was too small to investigate abundance change in Alberta, and it did not detect any abundance change in British Columbia or on a Canada-wide scale for this species. Forestry activities in Alberta's Boreal Forest and Foothills NRs will likely benefit this species as long as wetlands and watercourses are left intact. Similarly, prescribed burnings in the Rocky Mountain NR are likely to benefit this species.

Many companies have begun replacing spruce forests with unproductive pine forests; such unproductive habitat may have commercial value, but massive planting efforts may cause long-term problems for boreal forest species such as MacGillivray's Warbler (Pitocchelli et al., 1997).

MacGillivray's Warbler is considered Secure in Alberta.

TEMPORAL REPORTING RATE

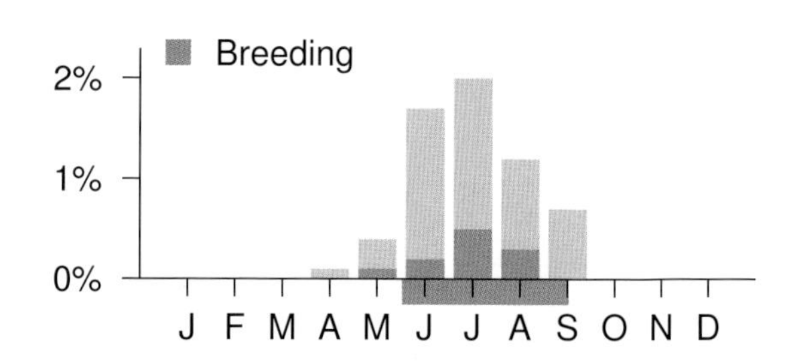

NATURAL REGIONS REPORTING RATE

SPATIAL REPORTING RATE

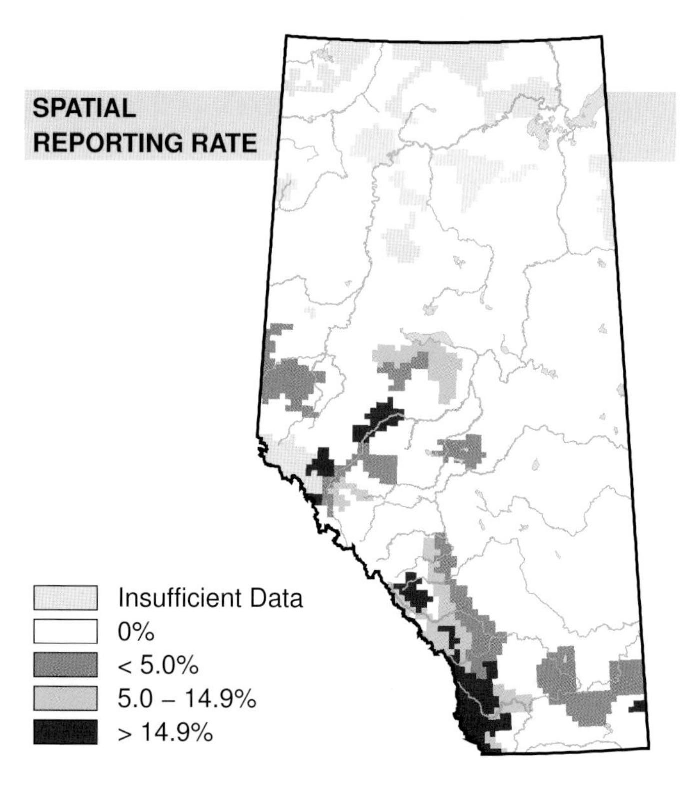

STATUS

GSWA 2000	GSWA 2005	COSEWIC Historic Rankings	COSEWIC 2007	Alberta Wildlife Act
Secure	Secure	N/A	N/A	N/A

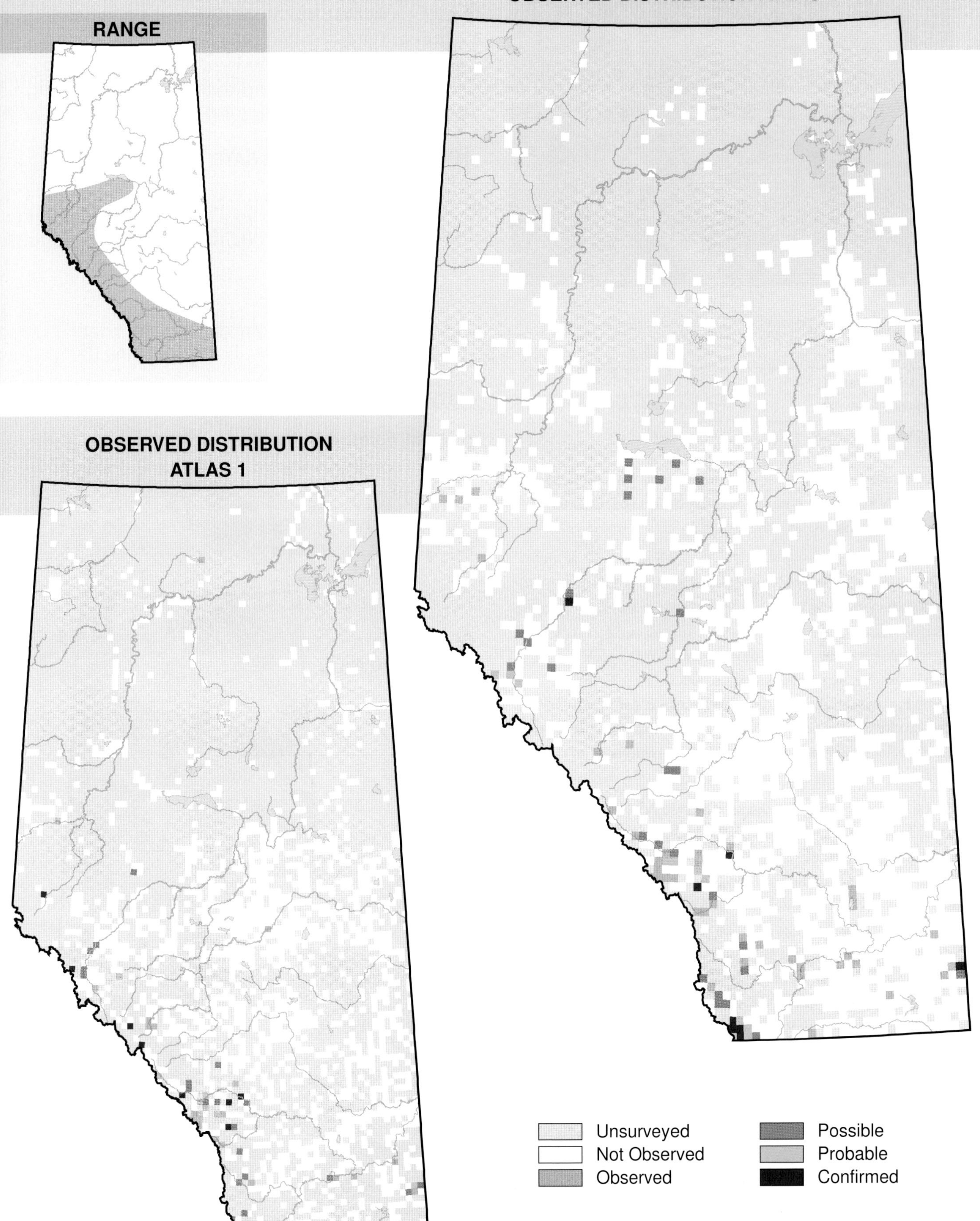
Habitat
Nest Location
Nest Type
Diet
RANGE
OBSERVED DISTRIBUTION ATLAS 2
OBSERVED DISTRIBUTION ATLAS 1
Unsurveyed
Not Observed
Observed
Possible
Probable
Confirmed

Common Yellowthroat (*Geothlypis trichas*)

Photo: Gerald Romanchuk

NESTING

CLUTCH SIZE (EGGS):	3–6
INCUBATION (DAYS):	12
FLEDGING (DAYS):	9–10
NEST HEIGHT (METRES):	0

The Common Yellowthroat is found in every Natural Region in Alberta. No change in the distribution of this species was observed between Atlas 1 and Atlas 2.

Unlike most other wood warblers, the Common Yellowthroat does not breed exclusively in forested habitats. This species tends to nest in areas that have high densities of low vegetation such as wetlands, early successional forests, and forests with high densities of undergrowth vegetation. This species sometimes nests in areas that are regenerating after fire or forest harvesting. As a result of this warbler's ability to breed in a variety of areas, the distribution of this species was fairly uniform across the province in all natural regions.

Decreases in relative abundance were detected in the Boreal Forest, Foothills, Grassland, and Parkland NRs where, relative to other species, the Common Yellowthroat was observed less frequently in Atlas 2 than in Atlas 1. No change was detected in the Rocky Mountain NR. The Breeding Bird Survey detected a decline in abundance in Alberta during the period 1985–2005. The declines detected for this species by the Atlas and the Alberta Breeding Bird Survey are mirrored in a Canada-wide decline detected by the Breeding Bird Survey during the period 1985–2005. Although the results are corroborated by the BBS, the biological cause for the decline in the Boreal Forest, Foothills, and Parkland NRs is unclear, as this species is known to nest in recently disturbed habitats. Similarly in the Grassland NR, above-normal precipitation in that region in the latter half of Atlas 2 bestowed no detectable benefit for this species. The Provincial status of the Common Yellowthroat was changed from Secure in 2000 to Sensitive in 2005, due to threats to breeding habitat.

TEMPORAL REPORTING RATE

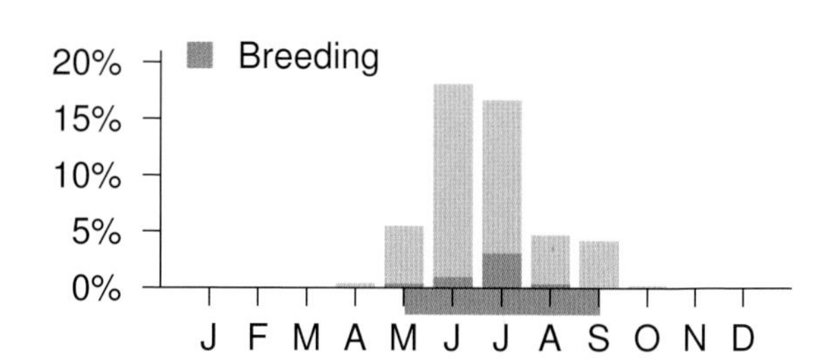

NATURAL REGIONS REPORTING RATE

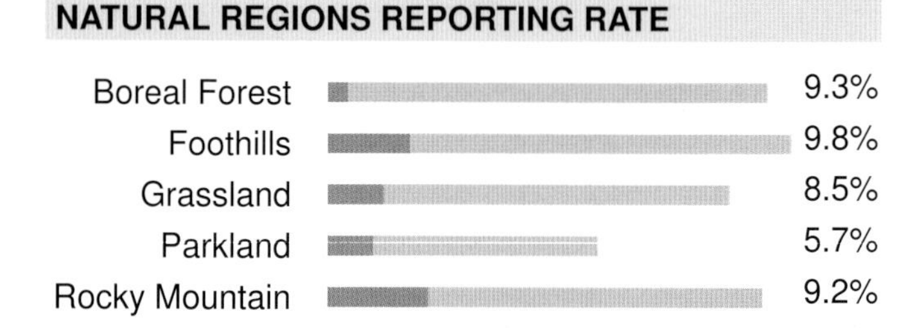

SPATIAL REPORTING RATE

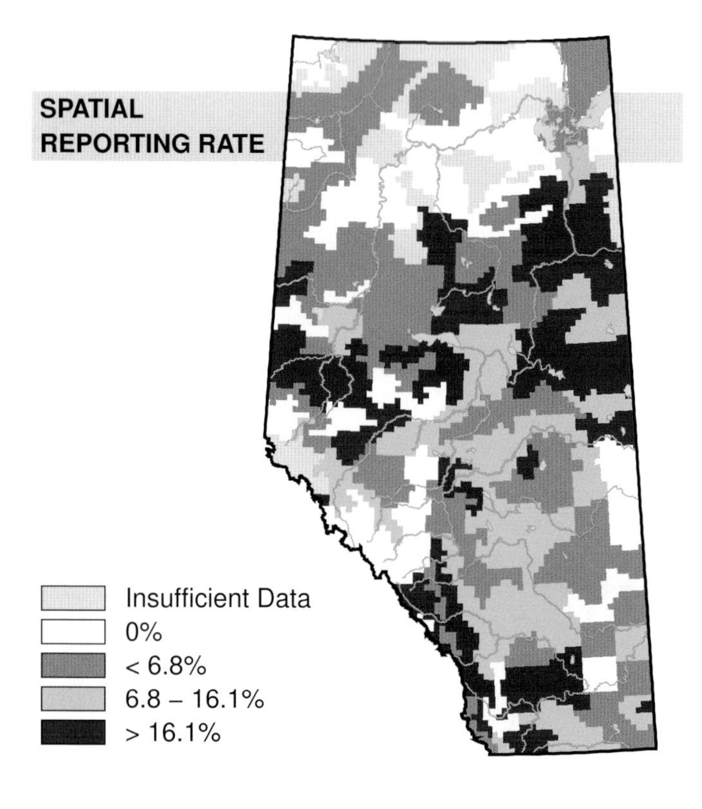

STATUS

GSWA 2000	GSWA 2005	COSEWIC Historic Rankings	COSEWIC 2007	Alberta Wildlife Act
Secure	Sensitive	N/A	N/A	N/A

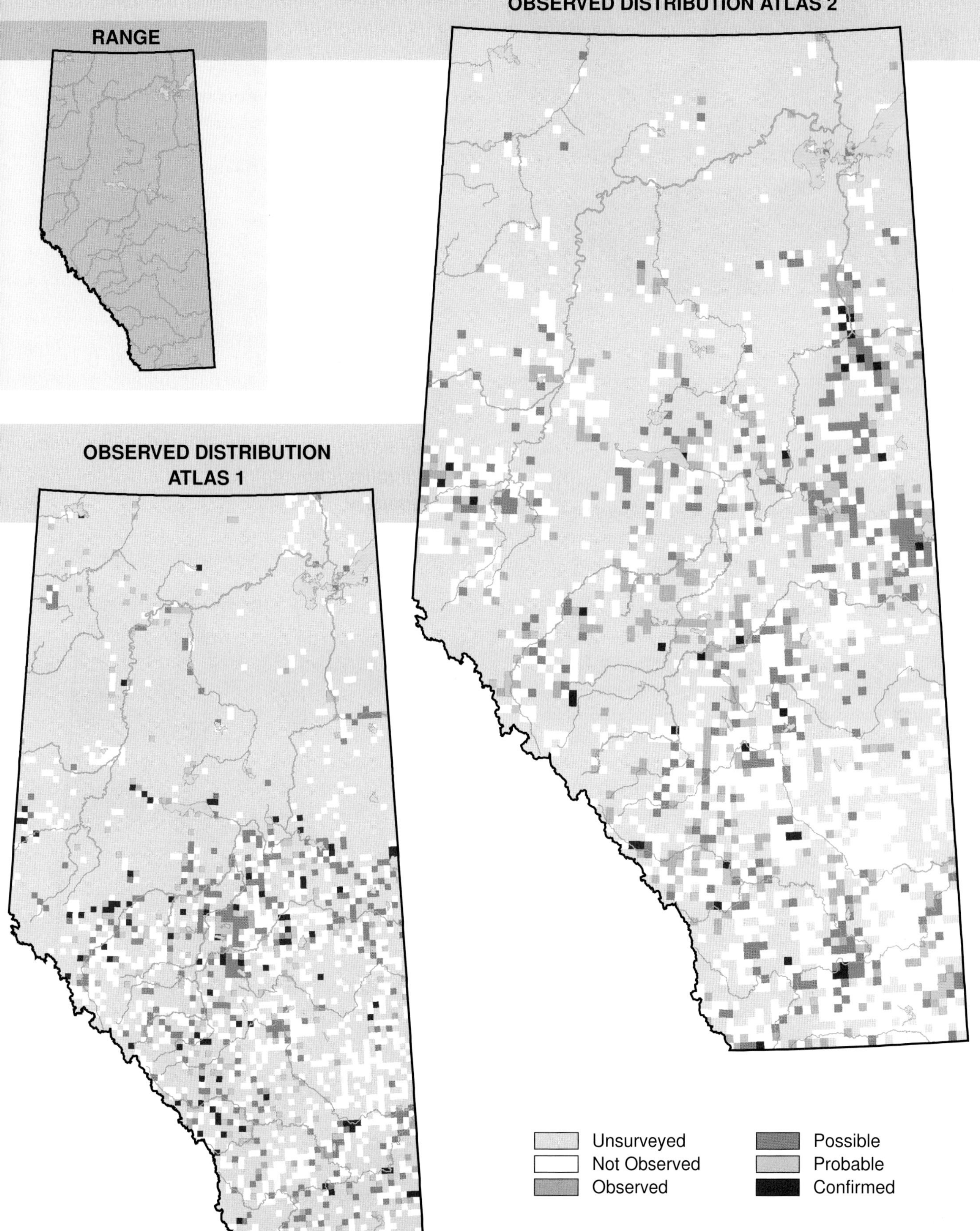
RANGE
OBSERVED DISTRIBUTION ATLAS 2
OBSERVED DISTRIBUTION ATLAS 1
Unsurveyed
Not Observed
Observed
Possible
Probable
Confirmed

Wilson's Warbler (*Wilsonia pusilla*)

Photo: Debbie Godkin

NESTING

CLUTCH SIZE (EGGS):	4–6
INCUBATION (DAYS):	10–13
FLEDGING (DAYS):	10–11
NEST HEIGHT (METRES):	<1

Wilson's Warbler breeds in the Boreal Forest, Foothills, Parkland, and Rocky Mountain Natural Regions in Alberta. This species was not observed in the northwestern part of the province during Atlas 2. It is unclear whether this represents an actual range contraction or whether the presence of this species in the northwest was missed during Atlas 2. It may have been missed because there were not many records obtained in the northwest during Atlas 1, and those few were sparsely distributed.

Wilson's Warbler is closely associated with riparian habitats. This species breeds mainly in shrubs along the edges of ponds, lakes, rivers, streams, and bogs. Regenerating forests are sometimes used during breeding as long as there is water nearby. As a result, this species was observed most often in the Foothills and Rocky Mountain NRs. Records from the Parkland NR were mostly of non-breeders, with a few scatted breeding records along the periphery of this NR. Grassland NR records were of non-breeders.

Increases in relative abundance were detected in the Boreal Forest and Grassland NRs. Rather than being related to an actual population change, the increase that was detected in the Boreal Forest NR was likely a function of broader coverage of the north in Atlas 2. Grassland NR records were of non-breeders; therefore, changes that were detected could be a function of migratory variability and may not indicate actual change in abundance. The Breeding Bird Survey did not find an abundance change in Alberta; however, a decline was found across Canada during the period 1985–2005. This decline was likely related to a reduction in the amount of riparian habitat. This species is considered Secure in Alberta.

TEMPORAL REPORTING RATE

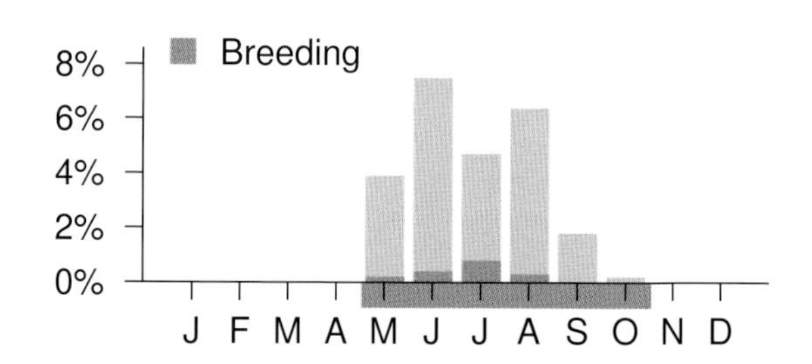

NATURAL REGIONS REPORTING RATE

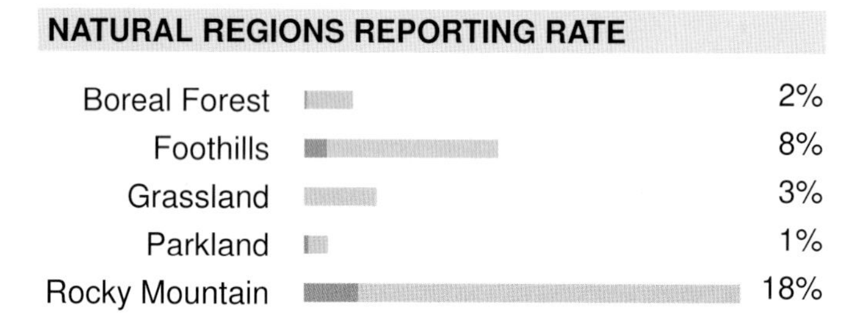

SPATIAL REPORTING RATE

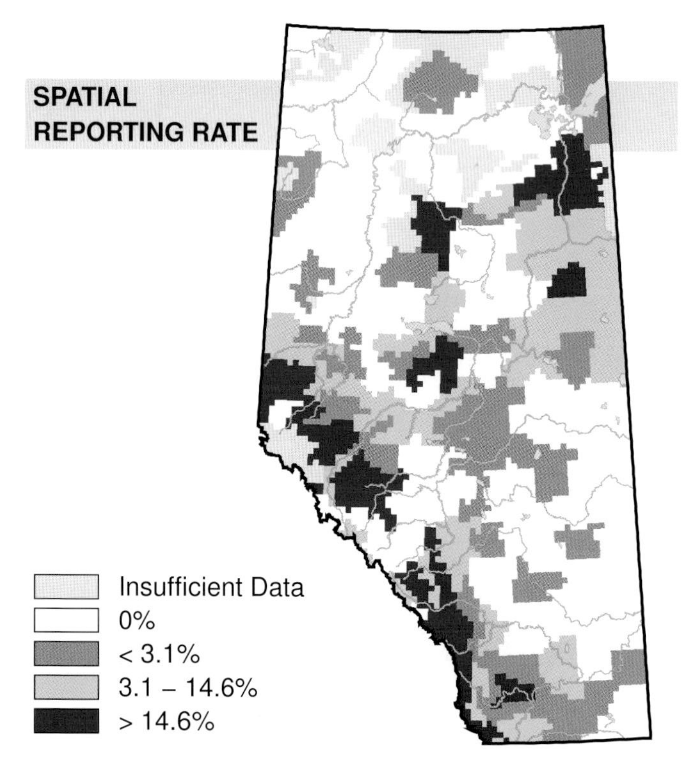

STATUS

GSWA 2000	GSWA 2005	COSEWIC Historic Rankings	COSEWIC 2007	Alberta Wildlife Act
Secure	Secure	N/A	N/A	N/A

Habitat

Nest
Location

Nest
Type

Diet

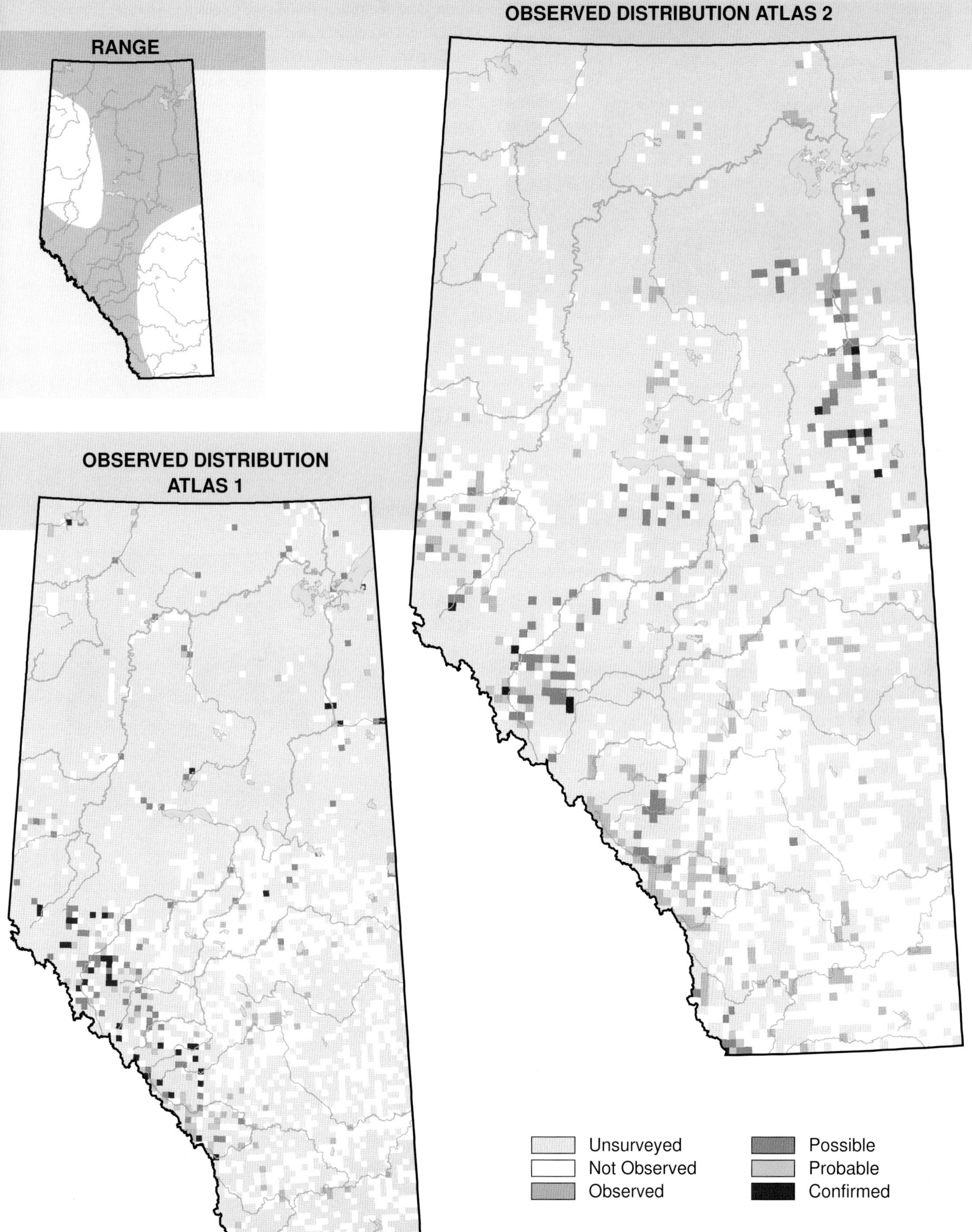
OBSERVED DISTRIBUTION ATLAS 2
RANGE
OBSERVED DISTRIBUTION
ATLAS 1
Unsurveyed
Not Observed
Observed
Possible
Probable
Confirmed

Canada Warbler (*Wilsonia canadensis*)

Photo: Royal Alberta Museum

NESTING

CLUTCH SIZE (EGGS):	3–5
INCUBATION (DAYS):	12
FLEDGING (DAYS):	8–10
NEST HEIGHT (METRES):	0

The bright yellow-breasted and black-necklaced Canada Warbler is found in the Boreal Forest, Foothills, and Parkland Natural Regions. Observations from the Grassland NR were not breeding records. There was no change in this species' distribution between Atlas 1 and Atlas 2.

More abundant in older mixed forests near water, the Canada Warbler was observed most commonly in the Boreal Forest and Foothills NRs, and less commonly in the Parkland NR in the Grande Prairie area. Possible breeding records were observed as far south as Red Deer and west to the Rocky Mountain House area. No breeding records were obtained in either the Rocky Mountain or Grassland NRs during Atlas 1 or Atlas 2.

Increases in relative abundance were detected in the Boreal Forest and Foothills NRs where, relative to other species, this warbler was observed more frequently in Atlas 2 than in Atlas 1. Considering the current level of industrial development in these NRs, this increase was unexpected as this species is known to prefer continuous rather than fragmented forests. Rather than being related to an actual population change, though, this increase was likely a function of more thorough coverage of the north in Atlas 2. There were no changes in relative abundance detected in other natural regions. This species was not detected often enough during Breeding Bird Surveys to determine abundance trends for Alberta. However, a decline was detected across Canada during the period 1985–2005. National declines could be related to increased forest fragmentation. This species is considered Sensitive in Alberta due to declines in its abundance and to its sensitivity to habitat deterioration.

TEMPORAL REPORTING RATE

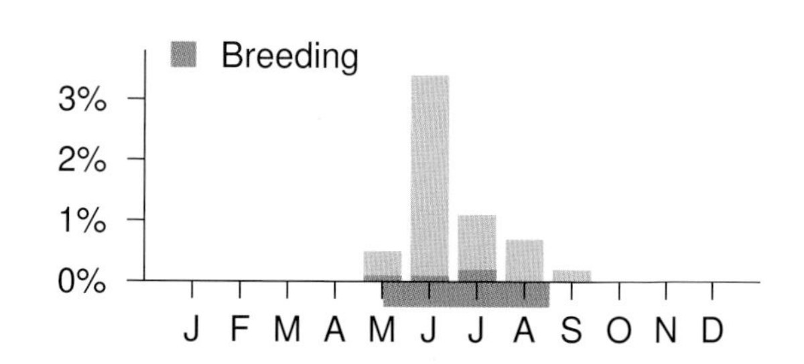

NATURAL REGIONS REPORTING RATE

SPATIAL REPORTING RATE

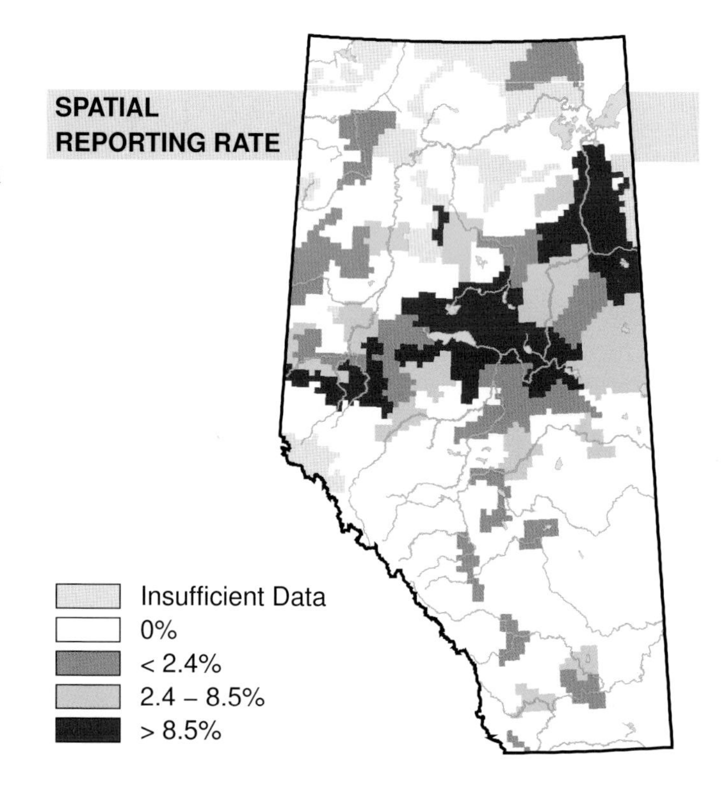

STATUS

GSWA 2000	GSWA 2005	COSEWIC Historic Rankings	COSEWIC 2007	Alberta Wildlife Act
Sensitive	Sensitive	N/A	N/A	N/A

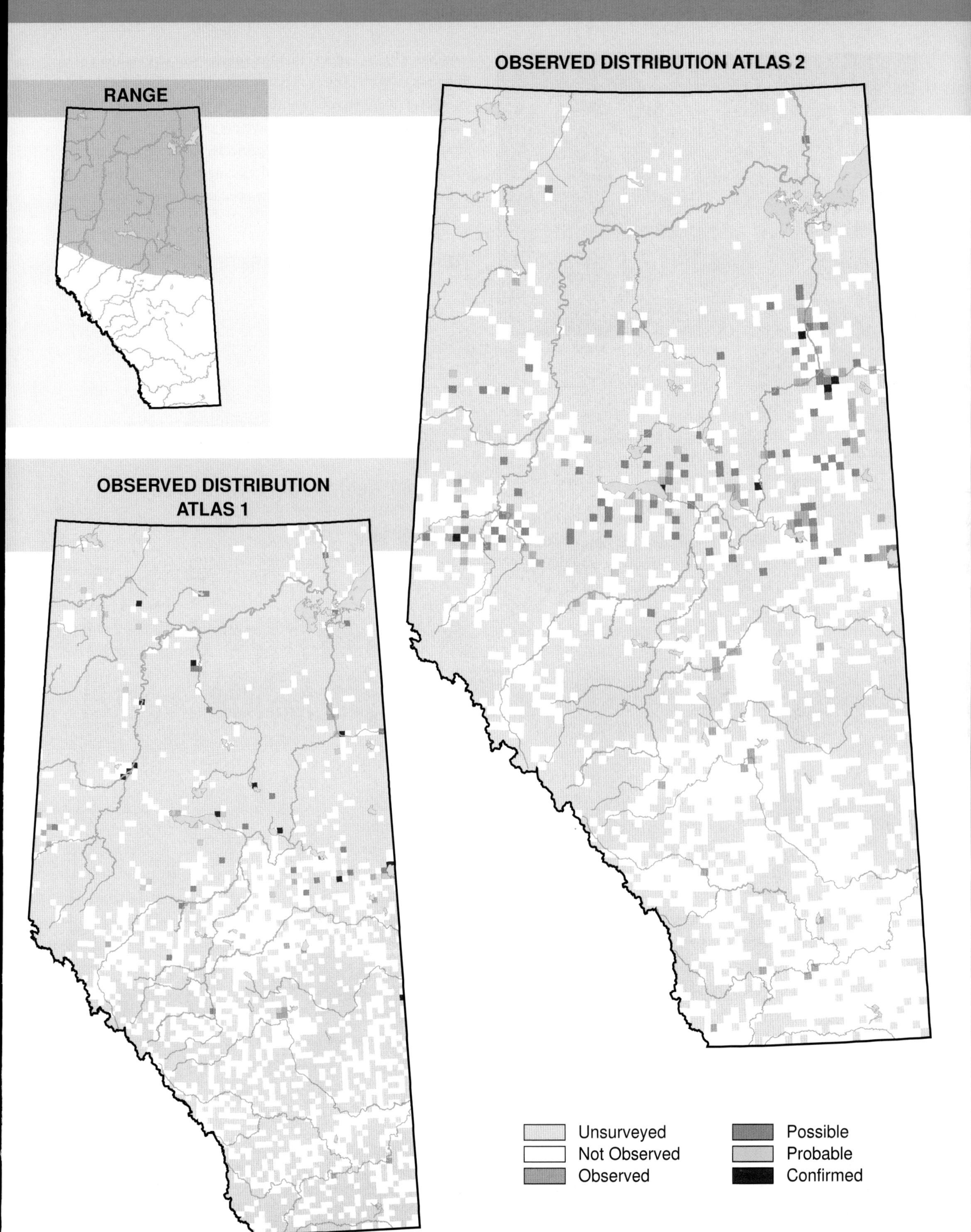
RANGE
OBSERVED DISTRIBUTION ATLAS 2
OBSERVED DISTRIBUTION
ATLAS 1
Unsurveyed
Not Observed
Observed
Possible
Probable
Confirmed

Yellow-breasted Chat (*Icteria virens*)

Photo: Alan MacKeigan

NESTING

CLUTCH SIZE (EGGS):	3–5
INCUBATION (DAYS):	11
FLEDGING (DAYS):	8
NEST HEIGHT (METRES):	1–2

Historically, the breeding range of the Yellow-breasted Chat included southeastern Alberta bounded on the north by the Red Deer River from Empress to Trochu and southwestward through Beiseker and Lethbridge to the International Boundary. Within this area it was restricted to the Milk, South Saskatchewan, Rosebud, and Red Deer river valleys (particularly the lower reaches) and adjacent coulees, as well as drainages on the southern slopes of the Cypress Hills. This range was confirmed during Atlas 1. There was no change in the observed distribution of this species between Atlas 1 and 2.

This species nests in dense shrubbery in the understorey of riparian poplar forests of major river valleys, or in dense shrubbery of smaller coulees and drainages that lack tree cover. Shaded thickets of thorny buffaloberry, hawthorn, and rose are preferred. A bulky cup-shaped nest is usually located in dense foliage 1–2 metres above the ground. Loose colonies may exist in prime habitat, although individual territories are defended.

No change in relative abundance was detected between Atlas 1 and 2. The Yellow-breasted Chat was first recorded in Alberta in 1941. The paucity of records is probably a reflection of the lack of ornithological investigations in the range of this species, rather than a range expansion into the province. Salt and Salt (1976) suggest that the species was expanding its range in Alberta. If this is happening, it was not detected by the Atlas; there were fewer records in Atlas 2 than in Atlas 1. Habitat conditions probably have not changed that dramatically between the two projects. Instream dam projects, water withdrawal projects, and grazing on riparian areas remain as potential threats. The species is considered to be Secure in Alberta.

TEMPORAL REPORTING RATE

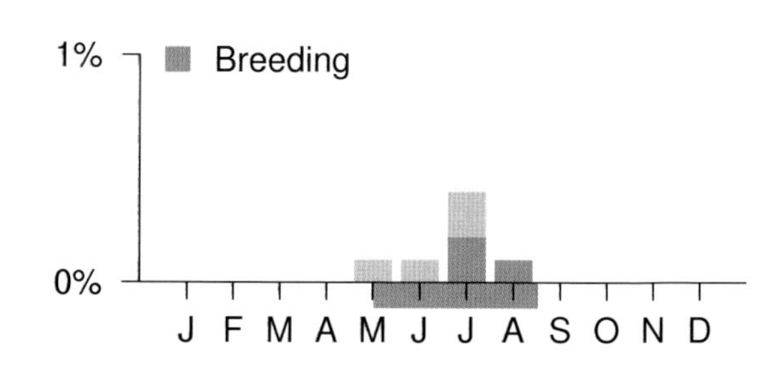

NATURAL REGIONS REPORTING RATE

SPATIAL REPORTING RATE

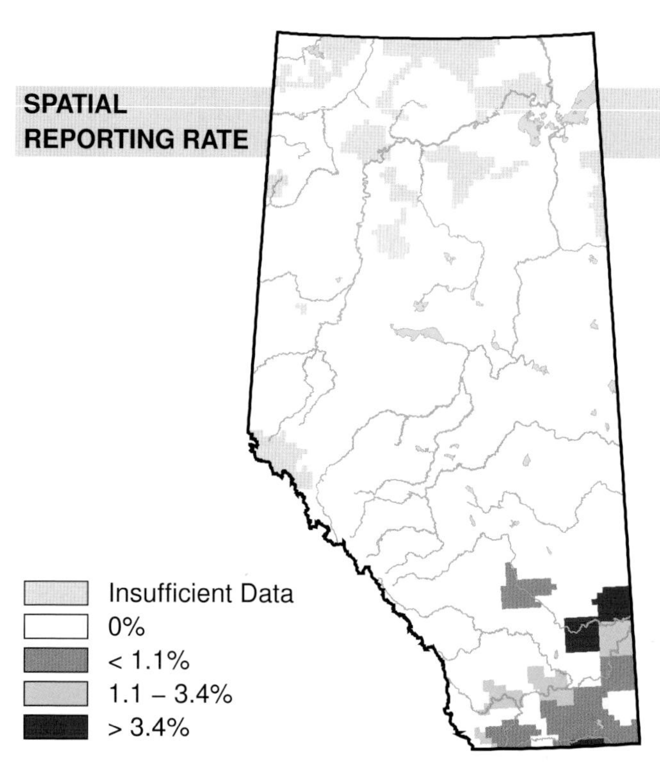

STATUS

GSWA 2000	GSWA 2005	COSEWIC Historic Rankings	COSEWIC 2007	Alberta Wildlife Act
Secure	Secure	Not At Risk	Not At Risk	N/A

RANGE

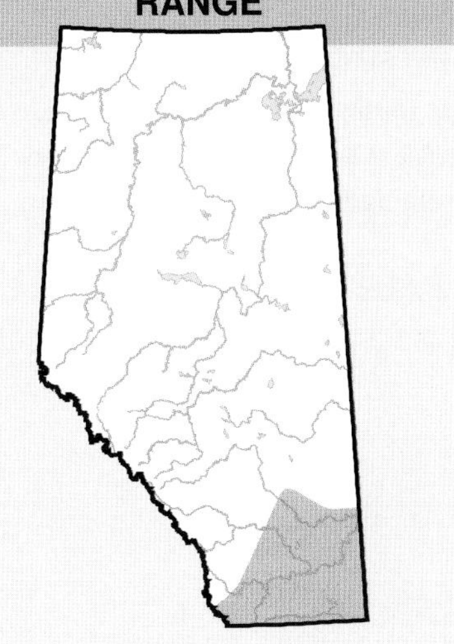

OBSERVED DISTRIBUTION ATLAS 2

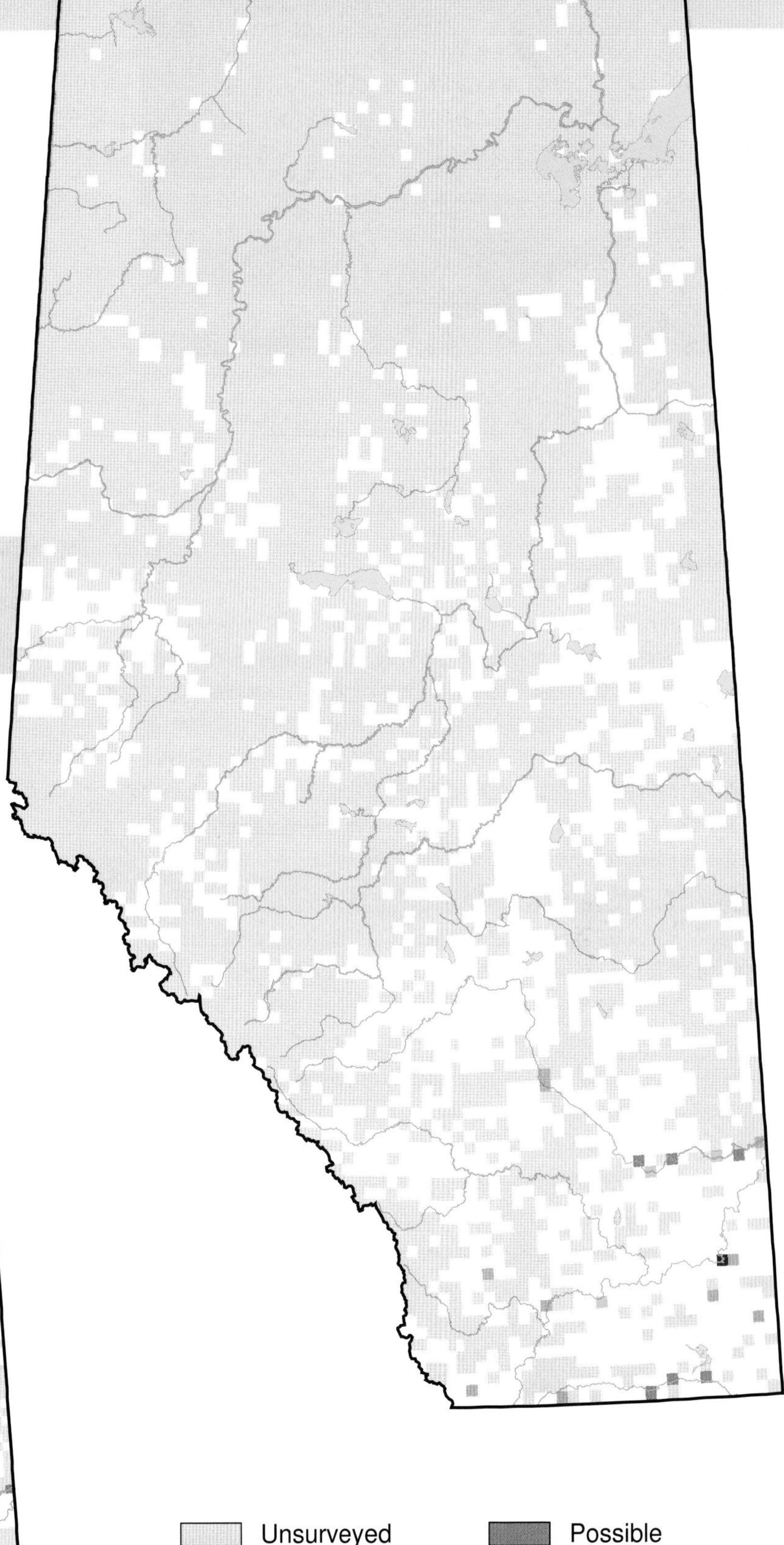

OBSERVED DISTRIBUTION ATLAS 1

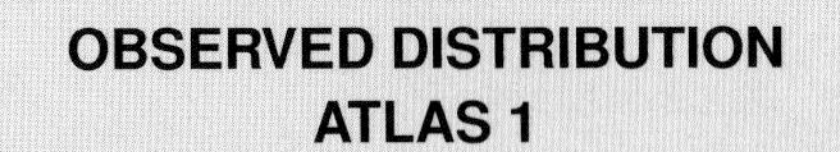

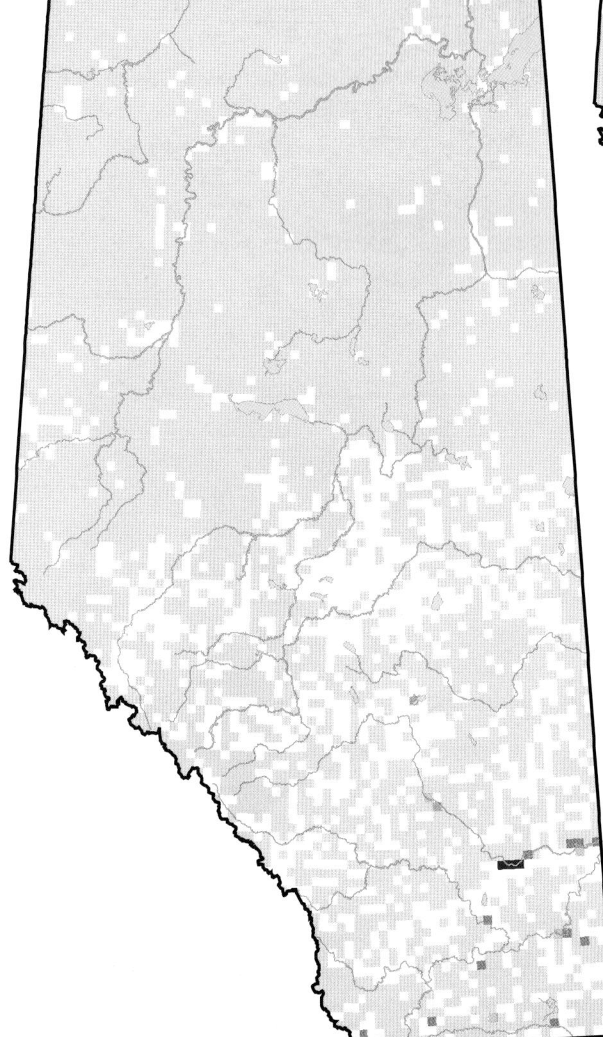

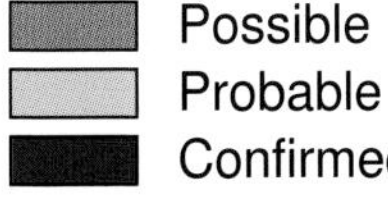

Western Tanager (*Piranga ludoviciana*)

Photo: Debbie Godkin

NESTING

CLUTCH SIZE (EGGS):	3–5
INCUBATION (DAYS):	13
FLEDGING (DAYS):	13–15
NEST HEIGHT (METRES):	<15

The Western Tanager can be found in all Natural Regions of the province for at least part of the year. A neotropical migrant from wintering grounds in Mexico and central America, this tanager spends only about a third of the year (May–August) in Alberta. The observed distribution of Western Tanagers did not change from Atlas 1, although they turned up in more squares in the Grassland NR, where the species occurs transiently, and in the northern half of the province in Atlas 2.

A species of older mixedwood forests in Alberta, the Western Tanager was observed most frequently in the Rocky Mountain and Foothills NRs, but concentrations were highest in the Boreal Forest NR north of Fort McMurray and in the Peace River country. Concentrations of birds were also high in the southwestern corner of the Rocky Mountain NR.

Increases in relative abundance were detected in the Boreal Forest, Grassland, and Rocky Mountain NRs and a decline in the Parkland NR. Observation rates relative to other species mirrored these changes. At least in the Boreal Forest NR, improved access to undisturbed patches of habitat and mapping by the Remote Areas Program probably have more to do with the reported increase than an actual increase in numbers of Western Tanagers. The changes in the Grassland and Parkland NRs may be due to migratory variability as breeding is infrequent in, or limited to, the periphery of these NRs. No change was detected in the Foothills NR.

The Breeding Bird Survey did not detect any temporal change in Western Tanager abundance in Alberta for the period 1985–2005. The species is ranked Sensitive provincially, largely because of its predilection for old coniferous and mixedwood forests, and because losses or deterioration of these habitats in the near future are anticipated as a result of changes in land use, mainly timber harvesting, in areas of occurrence.

TEMPORAL REPORTING RATE

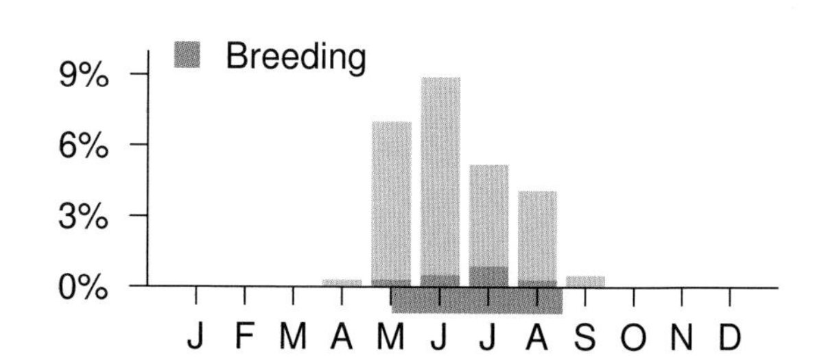

NATURAL REGIONS REPORTING RATE

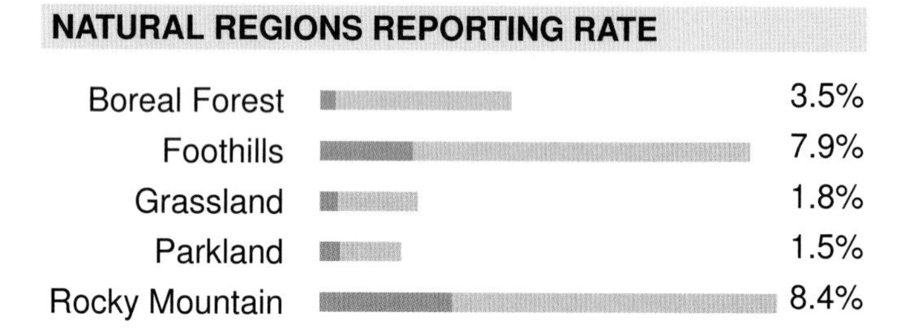

SPATIAL REPORTING RATE

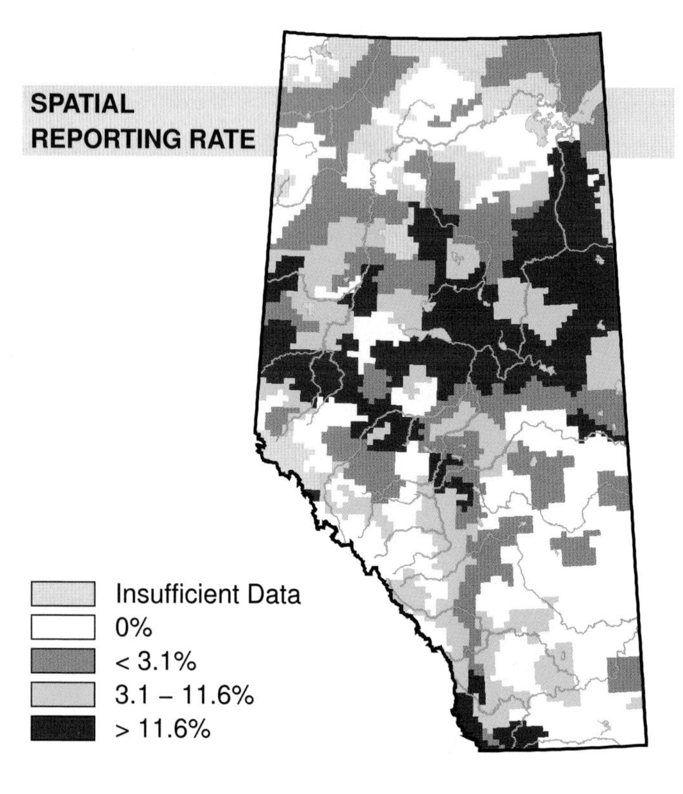

STATUS

GSWA 2000	GSWA 2005	COSEWIC Historic Rankings	COSEWIC 2007	Alberta Wildlife Act
Sensitive	Sensitive	N/A	N/A	N/A

Habitat
Nest Location
Nest Type
Diet 

OBSERVED DISTRIBUTION ATLAS 2

RANGE

OBSERVED DISTRIBUTION ATLAS 1

- Unsurveyed
- Not Observed
- Observed
- Possible
- Probable
- Confirmed

Spotted Towhee (*Pipilo maculatus*)

Photo: Gerald Romanchuk

NESTING

CLUTCH SIZE (EGGS):	2–6
INCUBATION (DAYS):	12–13
FLEDGING (DAYS):	10–12
NEST HEIGHT (METRES):	0–5.5

The Spotted Towhee is found in the southern part of the province, in the Grassland Natural Region and in the southern part of the Parkland NR. The distribution of this species did not change between Atlas 1 and Atlas 2.

The Spotted Towhee nests in dense shrubby vegetation. This species can sometimes be easily detected because it tends to make a lot of noise when it forages for food on the ground. It searches for invertebrates among the leaf litter by jumping and scraping the ground with its feet. The dense vegetation that it seems to prefer is thought to provide protection from predators for this loud forager. Nests are often built on the ground; however, these birds will also occasionally nest in vegetation above the ground. Southern Alberta is the northern extent of this species' continental range. As a result, Spotted Towhees were found mainly in shrubby vegetation along the edges of waterways in the Grassland NR, and was infrequently found along the southern periphery of the Parkland NR.

There was no change in relative abundance detected in any of the NRs. Relative to other species, the Spotted Towhee was observed less frequently in the Grassland and Parkland NRs during Atlas 2 than during Atlas 1. The Breeding Bird Survey sample size was too small to investigate trends for the Spotted Towhee in Alberta. Similar to results achieved by the Atlas, no trend was found by the Breeding Bird Survey on a Canada-wide scale. The maintenance of dense shrub habitat, especially along riparian areas in the Grassland NR, will be important for the conservation of the Spotted Towhee. The status of this species is Secure in Alberta.

TEMPORAL REPORTING RATE

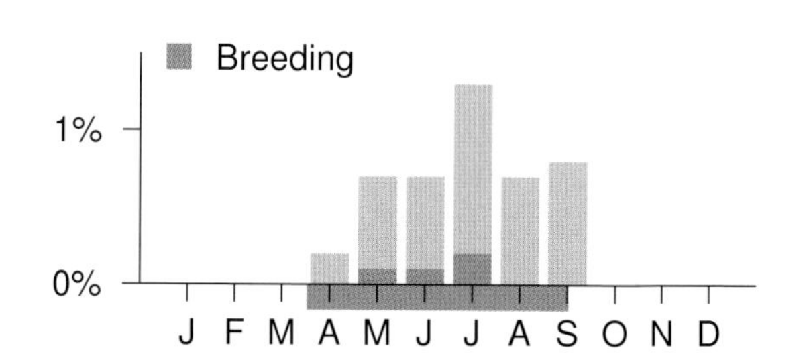

NATURAL REGIONS REPORTING RATE

Grassland 2.4%
Parkland 0.1%

SPATIAL REPORTING RATE

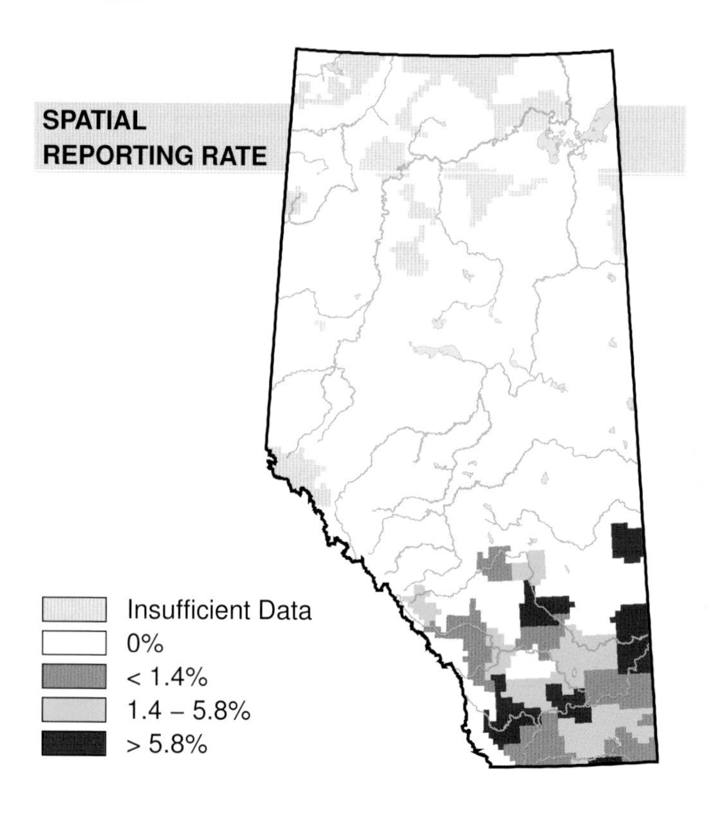

STATUS

GSWA 2000	GSWA 2005	COSEWIC Historic Rankings	COSEWIC 2007	Alberta Wildlife Act
Secure	Secure	N/A	N/A	N/A

Habitat

Nest
Location

Nest
Type
Diet
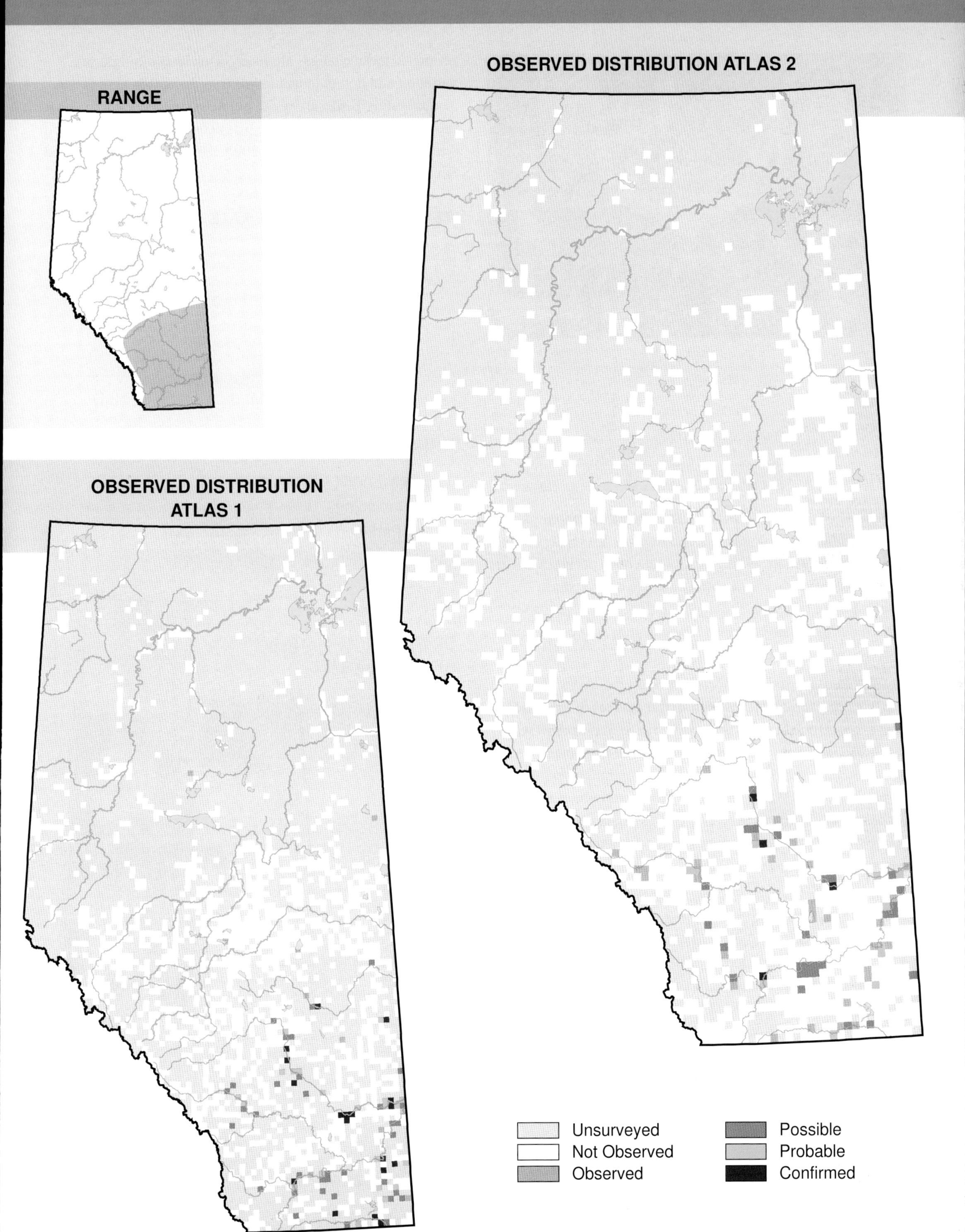
RANGE
OBSERVED DISTRIBUTION ATLAS 2
OBSERVED DISTRIBUTION
ATLAS 1
Unsurveyed
Not Observed
Observed
Possible
Probable
Confirmed

American Tree Sparrow (*Spizella arborea*)

Photo: Debbie Godkin

NESTING

CLUTCH SIZE (EGGS):	3–6
INCUBATION (DAYS):	12–13
FLEDGING (DAYS):	14–16
NEST HEIGHT (METRES):	0

The American Tree Sparrow breeds in the north-central part of Alberta in the Boreal Forest Natural Region. There were few breeding observations during Atlas 1 and none during Atlas 2. It is unlikely that this represents change to the breeding distribution of this species; rather, this species was probably missed during Atlas 2 because it breeds in a remote part of the province where survey coverage was low.

Despite its name, the American Tree Sparrow breeds in the north in open shrubby areas with sparse patches of small trees. Due to the remoteness of its breeding range, this species was most often observed during migration and in the winter in Alberta. Highest reporting rates occurred from October to December and in March and April. Observation rates were highest in the Rocky Mountain and Grassland NRs but this species was also quite commonly observed in the Parkland and Boreal Forest NRs.

A relative abundance increase was detected in the Grassland NR and, relative to other species, the American Tree Sparrow was observed more frequently in Atlas 2 than in Atlas 1 in this NR. A decline was detected in the Parkland and, relative to other species, it was observed less frequently in Atlas 2 than in Atlas 1 in this NR. No change was detected in the Boreal Forest, Foothills, or Rocky Mountain NRs. As the majority of records for this species were either migrants or winter residents, these changes are likely a function of migratory variation and not an indication of true changes in abundance. The Breeding Bird Survey did not detect an abundance change in Alberta; however, a decline was detected on a Canada-wide scale during the period 1985–2005. This species is considered Secure in Alberta.

TEMPORAL REPORTING RATE

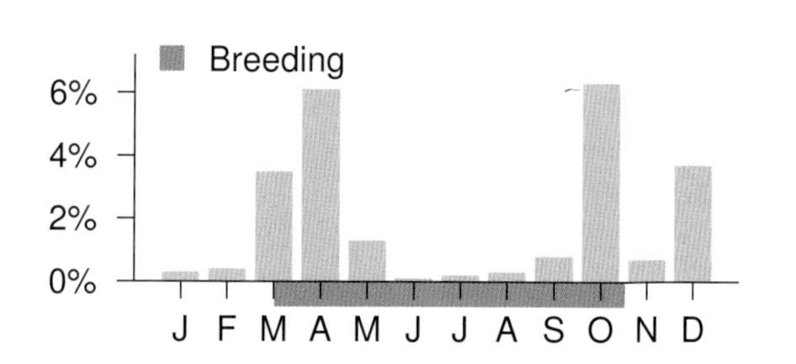

NATURAL REGIONS REPORTING RATE

SPATIAL REPORTING RATE

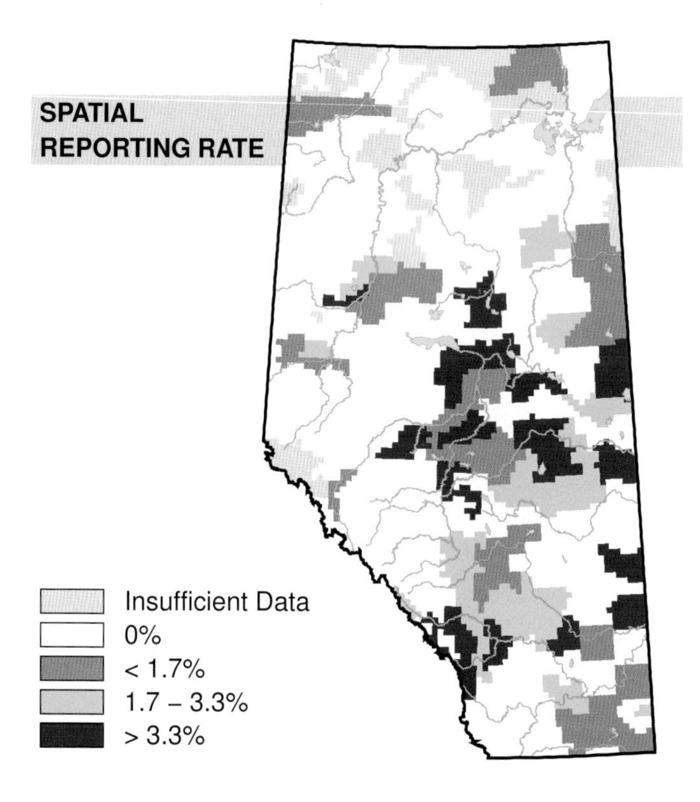

STATUS

GSWA 2000	GSWA 2005	COSEWIC Historic Rankings	COSEWIC 2007	Alberta Wildlife Act
Secure	Secure	N/A	N/A	N/A

Habitat Nest Location Nest Type Diet  

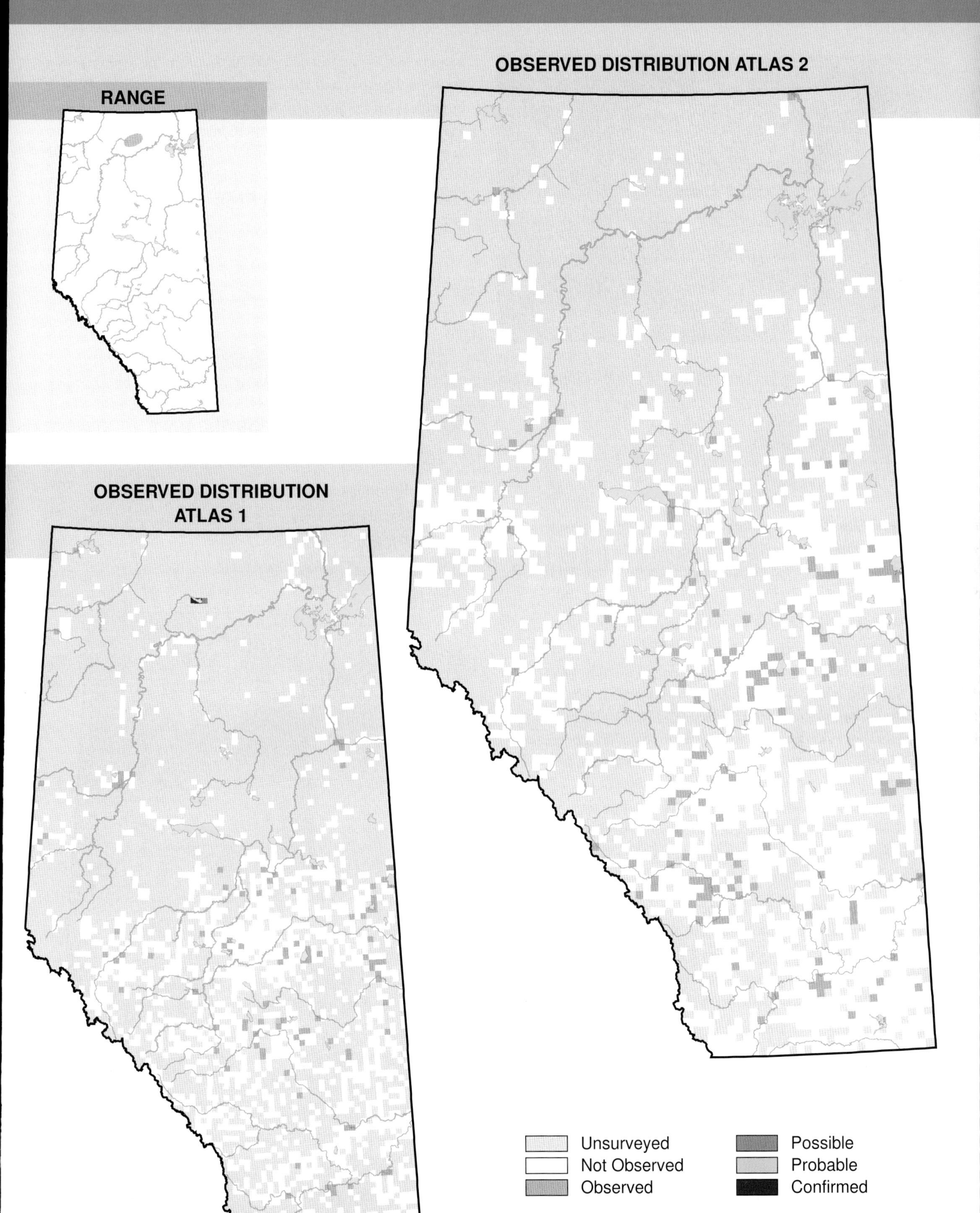

Chipping Sparrow (*Spizella passerina*)

Photo: Duane Boone

NESTING

CLUTCH SIZE (EGGS):	3–5
INCUBATION (DAYS):	11–13
FLEDGING (DAYS):	14
NEST HEIGHT (METRES):	3

One of Alberta's most common sparrows, the Chipping Sparrow breeds in every Natural Region in the province. The distribution of this species did not change between Atlas 1 and Atlas 2.

Chipping Sparrows use a wide variety of habitats during the breeding season. Generally they prefer habitats that contain open areas and forest or shrub patches. As a result, this species is quite commonly found along forest edges that may occur naturally or have resulted from human activity. This species is also common in urban areas. The presence of conifer trees seems to be an important characteristic of this species' breeding habitat. This species was most often observed in the Boreal Forest, Foothills, and Rocky Mountain NRs. Observation frequencies were lower in the Parkland and Grassland NRs, likely because coniferous trees are less common in those NRs.

A decrease in relative abundance was detected in the Parkland NR where, relative to other species, the Chipping Sparrow was observed less frequently in Atlas 2 than in Atlas 1. It is unclear why this decline was detected given that this species does well in human-modified environments. No change was detected in the Boreal Forest, Foothills, Grassland, and Rocky Mountain NRs. The Breeding Bird Survey did not detect any change in abundance for the period of 1985–2005, but did detect an increase for the period of 1995–2005. Given the discrepancy between the Atlas and BBS findings, and with no probable biological cause known for the Atlas-detected decrease in the Parkland NR, this result appears to be artifactual. This species is considered Secure in Alberta.

TEMPORAL REPORTING RATE

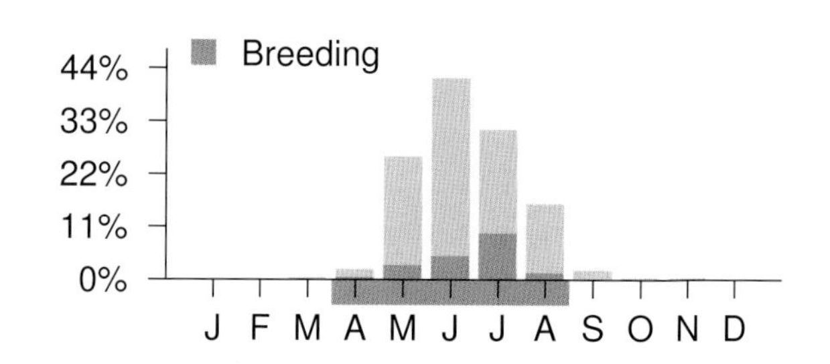

NATURAL REGIONS REPORTING RATE

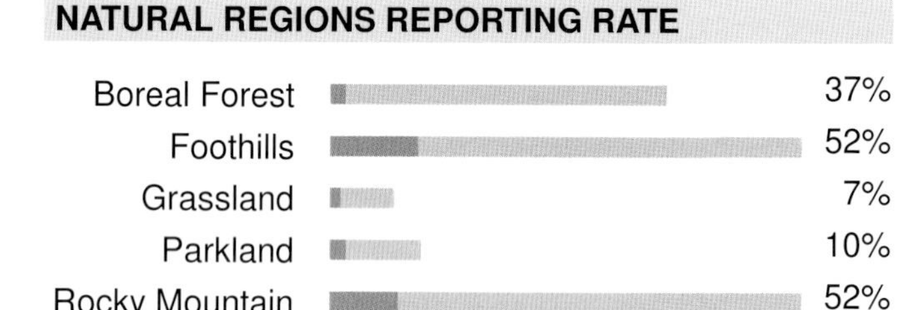

SPATIAL REPORTING RATE

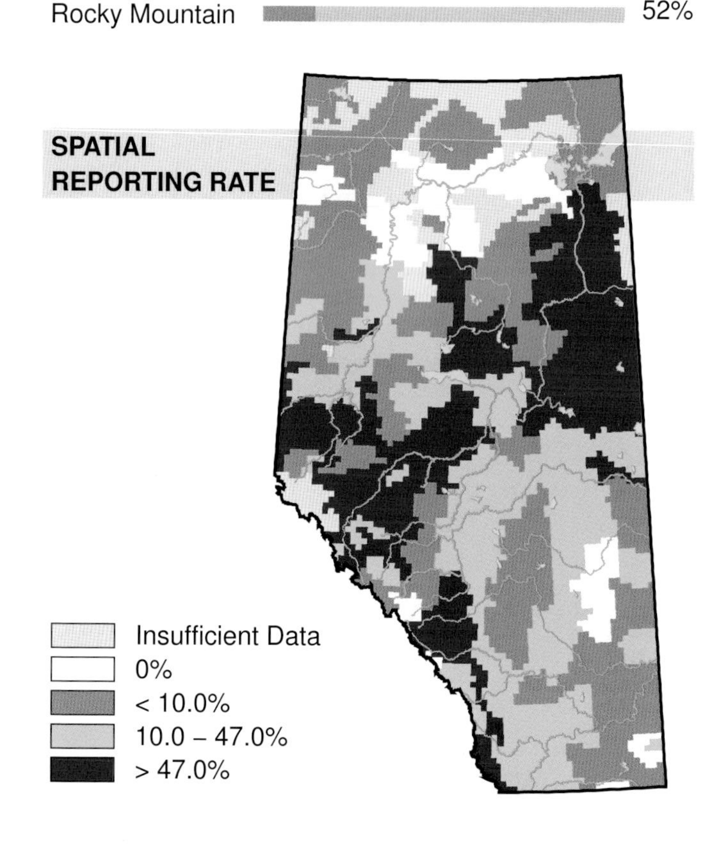

STATUS

GSWA 2000	GSWA 2005	COSEWIC Historic Rankings	COSEWIC 2007	Alberta Wildlife Act
Secure	Secure	N/A	N/A	N/A

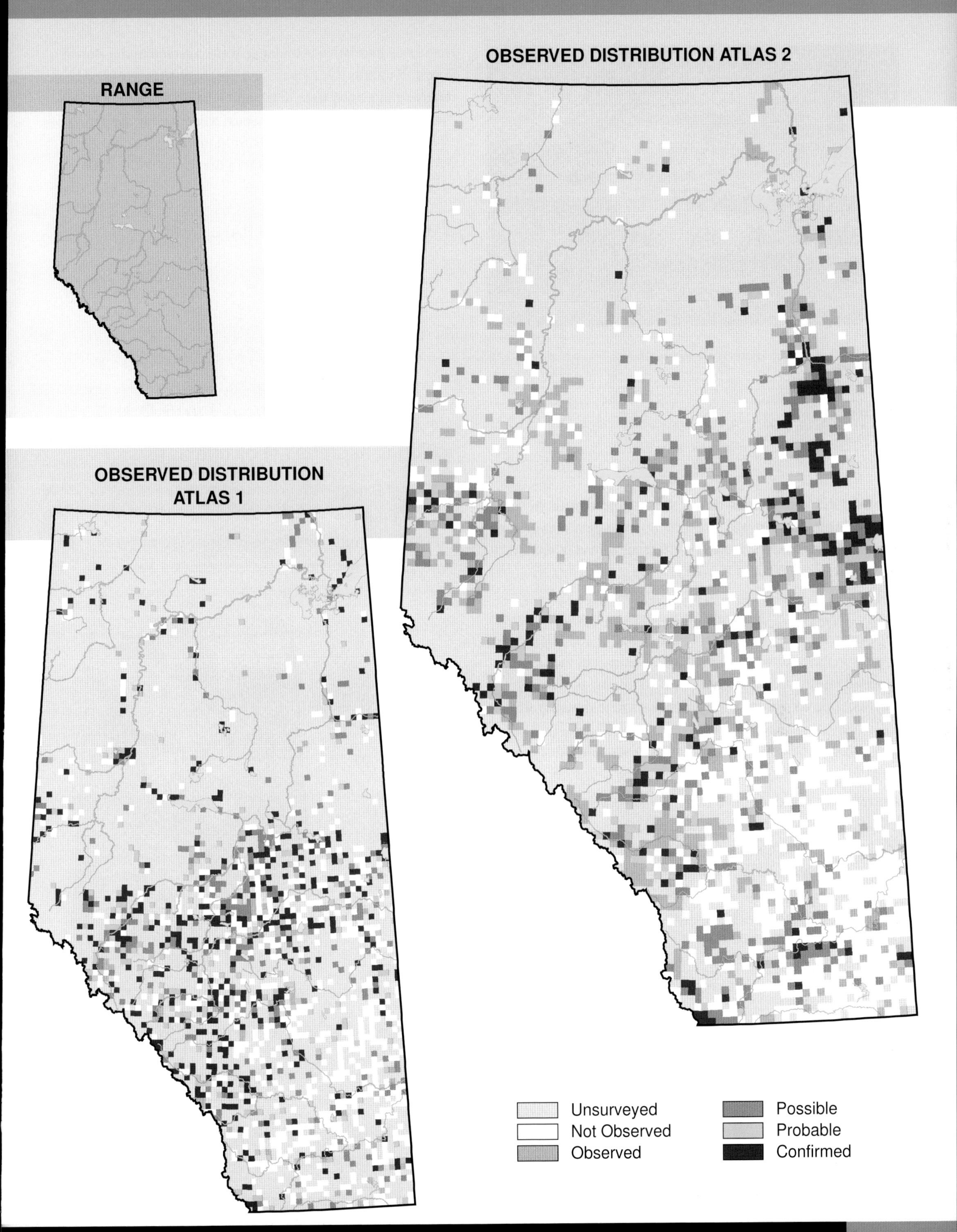

RANGE
OBSERVED DISTRIBUTION ATLAS 1
OBSERVED DISTRIBUTION ATLAS 2
Unsurveyed
Not Observed
Observed
Possible
Probable
Confirmed

Clay-colored Sparrow (*Spizella pallida*)

Photo: Stan Gosche

NESTING

CLUTCH SIZE (EGGS):	3–5
INCUBATION (DAYS):	10–11
FLEDGING (DAYS):	14–15
NEST HEIGHT (METRES):	<2

The Clay-colored Sparrow occurs throughout Alberta. In pre-settlement times it was a bird of the mixedgrass prairie but agriculture and forestry created early successional habitat in the Boreal Forest Natural Region and fire suppression has made some of the Parkland NR less suitable due to greater tree cover (Knapton, 1994). Detecting changes in distribution of a widespread species is difficult. In the Boreal Forest NR, fewer squares in the northwest yielded observations, but more squares between Cold Lake and Fort McMurray produced records. This is a shrub-obligate species and in forested areas it will occur only in early successional sites. So it may shift about in response to natural or anthropogenic disturbance that creates a mix of shrubby and open sites for short periods of time.

The highest reporting rates are in the Parkland and Grassland NRs. Clay-colored sparrows need a mix of grass for feeding and low shrub cover for nesting; they prefer snowberry where available (Knapton, 1994). The species is unlikely to occur where shrub is sparse or tall or where grass or other open habitats are not available for feeding. Parkland and the wetter parts of the mixedgrass prairie are ideal, and irrigation districts also contain much suitable habitat because the alfalfa in some planted cover is structurally very similar to shrubs. Unfortunately, birds that nest in hayfields are subject to extremely high rates of nest loss during harvest with all above-ground nests destroyed during haying activities (Frawley, 1989).

Decreases in relative abundance were detected in the Boreal Forest, Foothills, Grassland, and Parkland NRs. No changes in relative abundance were detected in the Rocky Mountain NR. The Breeding Bird Survey detected a decline for the period 1995–2005 as did Grassland Bird Monitoring (unpublished Canadian Wildlife Service data) in the period 1996–2004. Its status in Alberta is Secure.

TEMPORAL REPORTING RATE

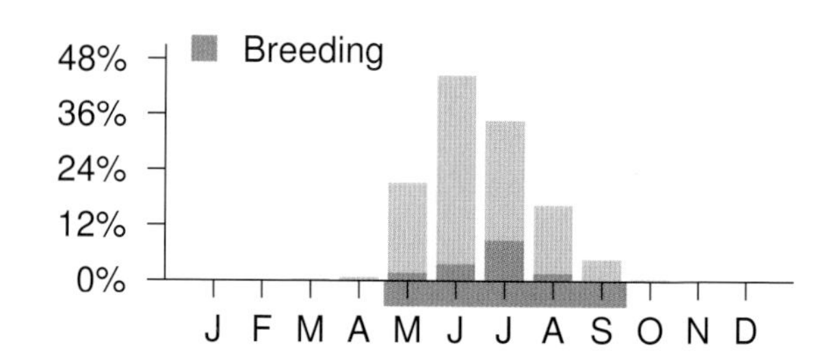

NATURAL REGIONS REPORTING RATE

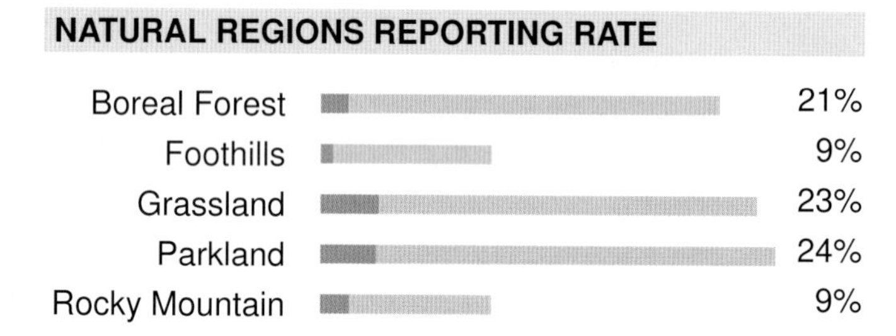

SPATIAL REPORTING RATE

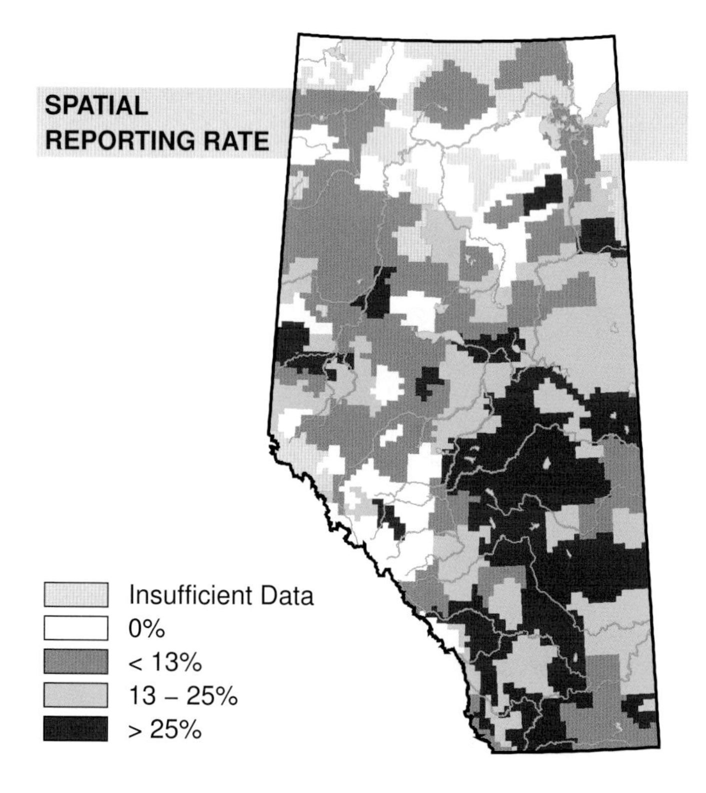

STATUS

GSWA 2000	GSWA 2005	COSEWIC Historic Rankings	COSEWIC 2007	Alberta Wildlife Act
Secure	Secure	N/A	N/A	N/A

Habitat  Nest Location  Nest Type  Diet  

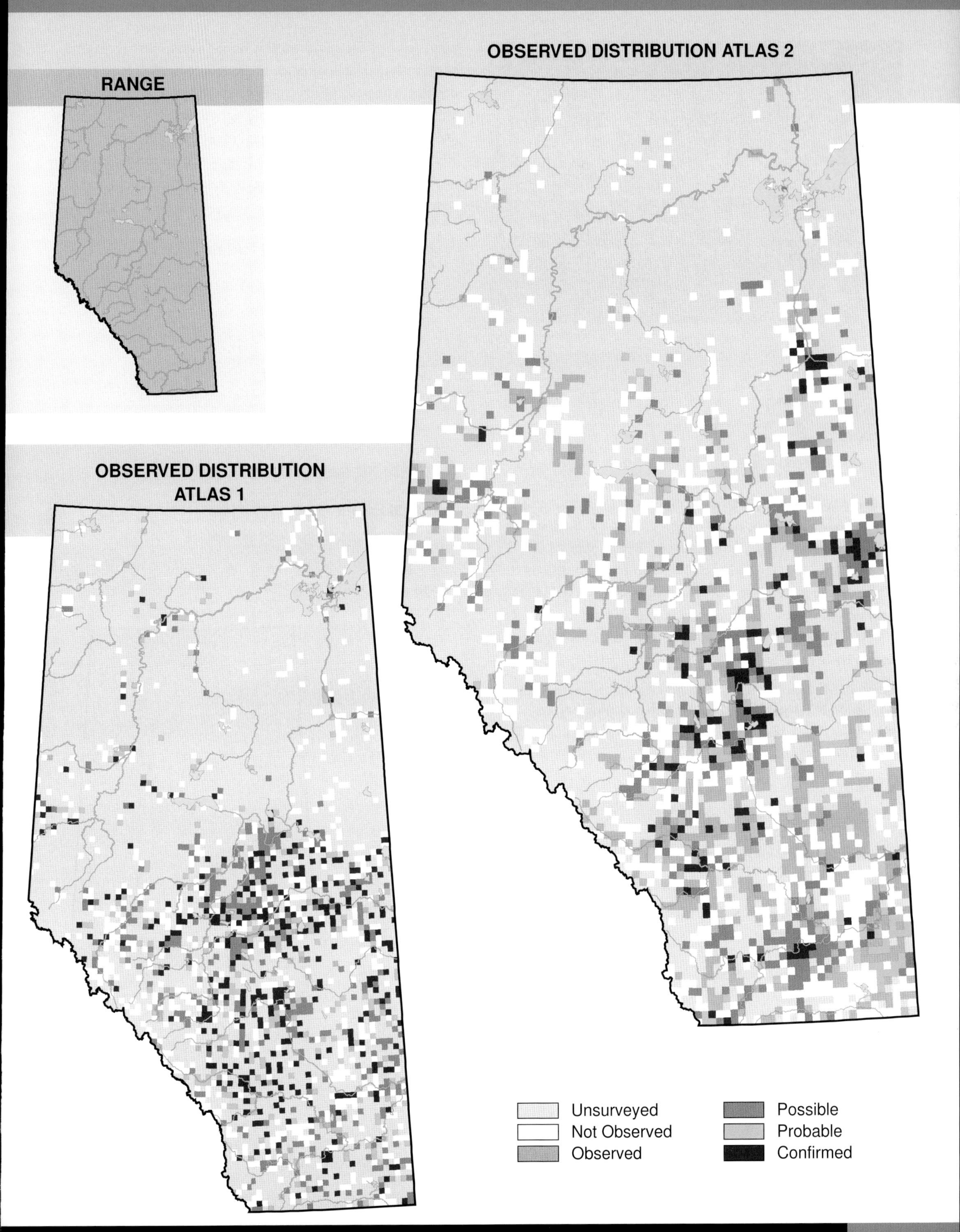

Brewer's Sparrow (*Spizella breweri*)

Photo: Damon Calderwood

NESTING

CLUTCH SIZE (EGGS):	3–5
INCUBATION (DAYS):	11–13
FLEDGING (DAYS):	8–9
NEST HEIGHT (METRES):	0–1.2

There are two subspecies of Brewer's Sparrow breeding in Alberta and they are separated by geography and habitat preference, as well as by biological and behavioural characteristics (Semenchuk, 1992). *S. b. breweri* breeds in the extreme south and southeast of the province in the Grassland Natural Region and *S. b. taverneri* is found in the Rocky Mountain NR. *S. b. breweri* was recorded farther north to Buffalo Lake in Atlas 2.

The prairie subspecies uses the semi-arid plains where short grass and low shrubs, mainly sagebrush, are found. The mountain subspecies prefers meadows that contain thickets of dwarf birch and willow.

No changes in relative abundance were detected in the Grassland, Parkland, and Rocky Mountain Natural Regions. The Breeding Bird Survey (1985–2005) has no data for Alberta and no change was detected across Canada.

Dunn (2005) indicates that the subspecies *S. b. breweri* is experiencing a population decline because it is sensitive to fragmentation and loss of its sagebrush habitat. The subspecies *S. b. taverneri*, limited to western Canada, is very poorly known. The status of each of these subspecies should be evaluated separately.

The Brewer's Sparrow is listed as Sensitive in 2005. The 2005 Alberta Status document qualifies the listing as being the result of steep population declines in Alberta since 1994. The prairie population of the species relies on the availability of natural sagebrush and such reliance is thought to be the cause of the decline. The pace of loss of this habitat has accelerated due to the complex interactions among agriculture, livestock grazing, and the invasion of exotic annual plants (Rotenberry et al., 1999). In southern Alberta the fragmentation of this habitat by increased levels of petroleum exploration and development accelerate the loss.

TEMPORAL REPORTING RATE

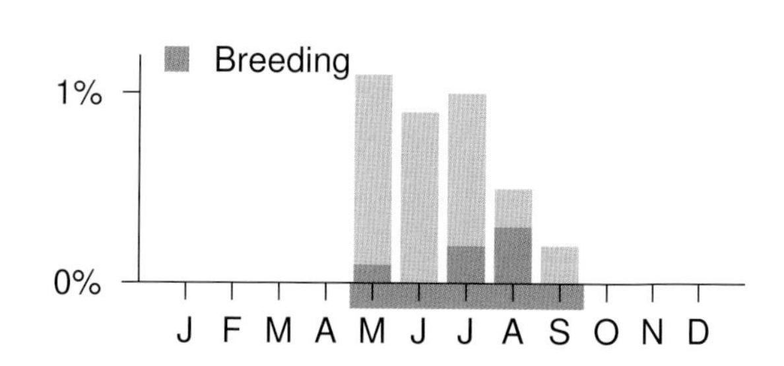

NATURAL REGIONS REPORTING RATE

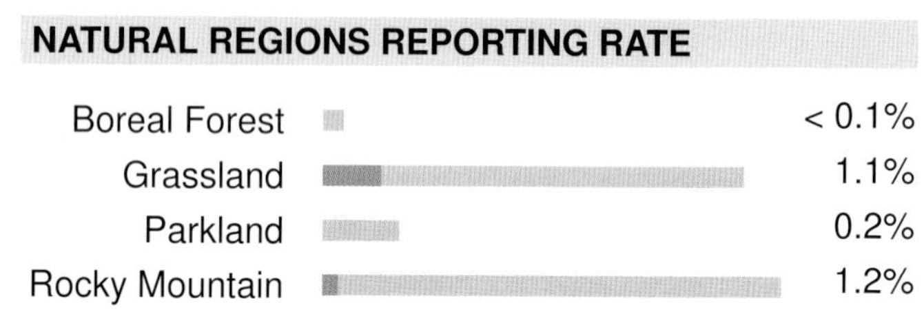

SPATIAL REPORTING RATE

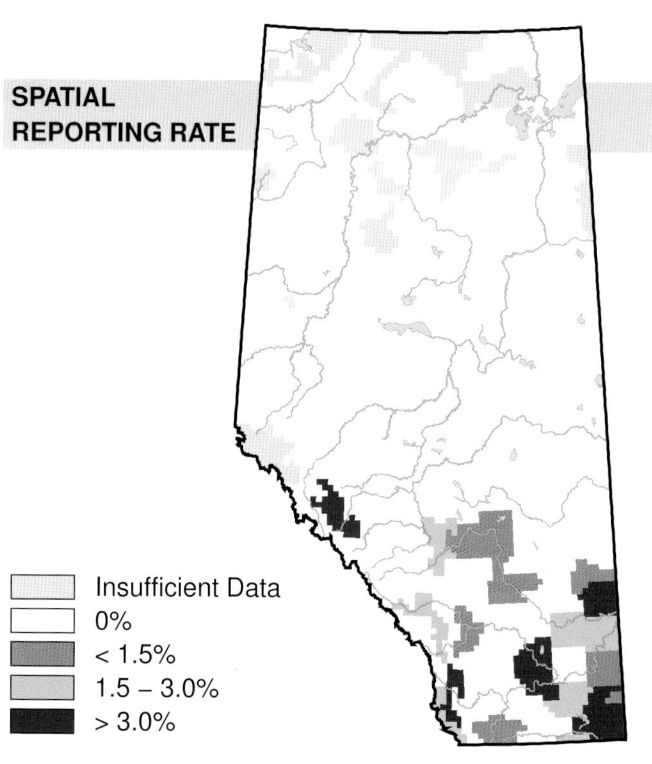

STATUS

GSWA 2000	GSWA 2005	COSEWIC Historic Rankings	COSEWIC 2007	Alberta Wildlife Act
Sensitive	Sensitive	N/A	N/A	N/A

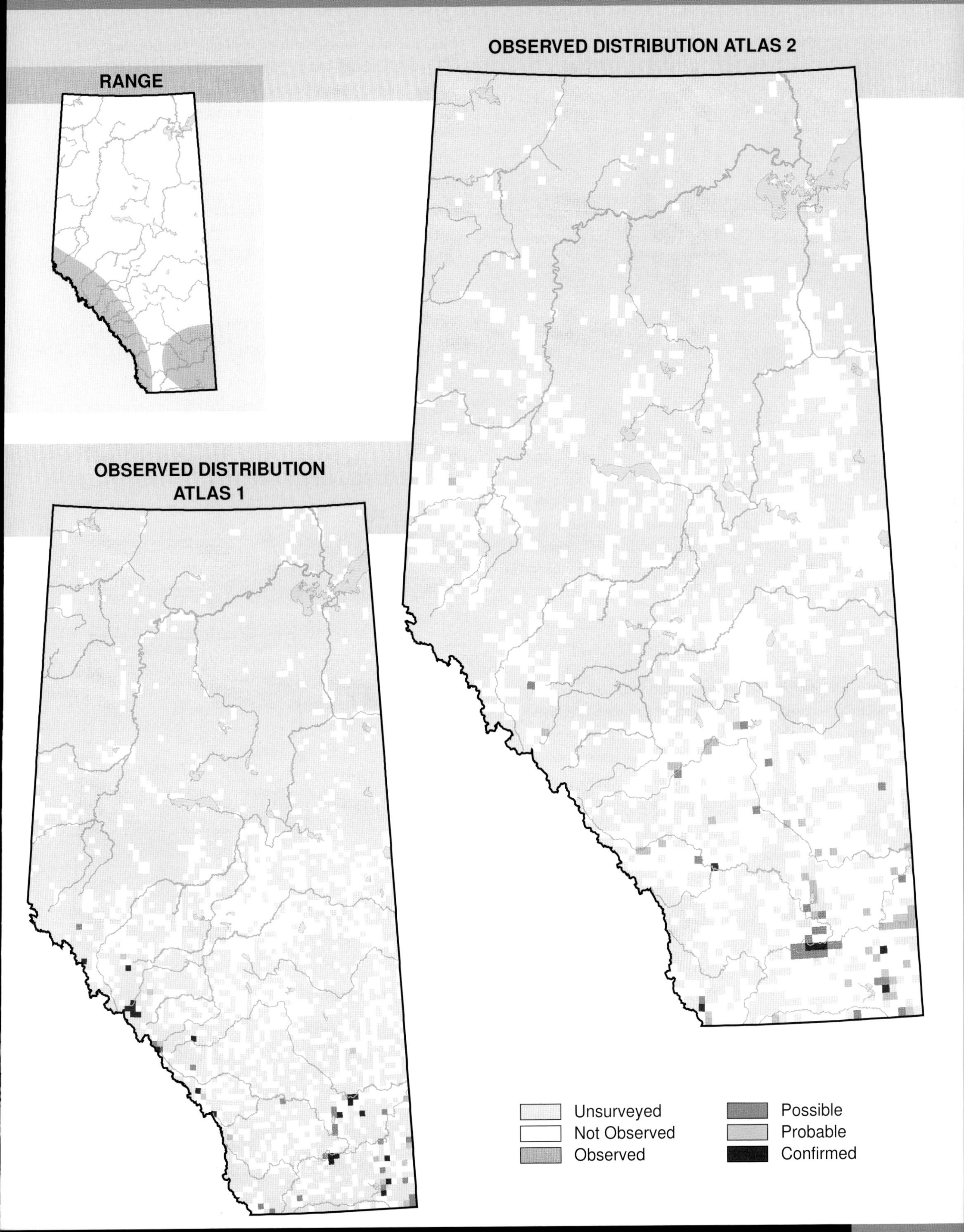
OBSERVED DISTRIBUTION ATLAS 2
RANGE
OBSERVED DISTRIBUTION ATLAS 1
Unsurveyed
Not Observed
Observed
Possible
Probable
Confirmed

Vesper Sparrow (*Pooecetes gramineus*)

Photo: Randy Jensen

NESTING

CLUTCH SIZE (EGGS):	3–5
INCUBATION (DAYS):	12–13
FLEDGING (DAYS):	20–22
NEST HEIGHT (METRES):	0

The Vesper Sparrow breeds throughout the province, but it is scarce in the extreme north. It was encountered most often in the Grassland Natural Region and less so in the Parkland, Foothills, and Rocky Mountain NRs. There was no change in distribution in Atlas 2.

This sparrow uses open, weedy, fairly dry situations such as grassy margins along roads, railways, fencelines, pastures, and grassy coulee slopes (Semenchuk, 1992). It requires song perches, such as fences, shrubs, crop residue, tall weeds, or woodlands bordering fields (Jones and Cornely, 2002). Within the aspen parkland of Alberta it is present in planted non-native grass 1–4 years after planting, but is absent from cropland with spring-seeded wheat (Prescott and Murphy, 1999).

Decreases in relative abundance were detected in the Boreal Forest and Parkland NRs. These regions were drier in Atlas 2 and Jones and Cornely (2002) record that the Vesper Sparrow moves around in response to annual rainfall; breeding success may vary greatly from year to year depending on how weather affects its habitat. No changes in relative abundance were detected in the Foothills, Grassland, and Rocky Mountain NRs. The Breeding Bird Survey (1985–2005) detected no change in abundance in Alberta or across Canada.

Dunn (2005) reports that this species is experiencing a possible decline. Although it is common and widespread, with a stable population in the western core of its range, it has experienced significant declines in Canada in the 1970s and 1990s.

Changes in farming practices, including farming that uses chemicals and large-scale tillage, have been implicated in declines of this species rangewide, as well as being a major cause of nest loss in agricultural areas (Jones and Cornely, 2002). Jones and Cornely (2002) recommend that this species will benefit from grassland easements and other management programs to preserve and restore grasslands.

The Vesper Sparrow is listed as Secure in Alberta.

TEMPORAL REPORTING RATE

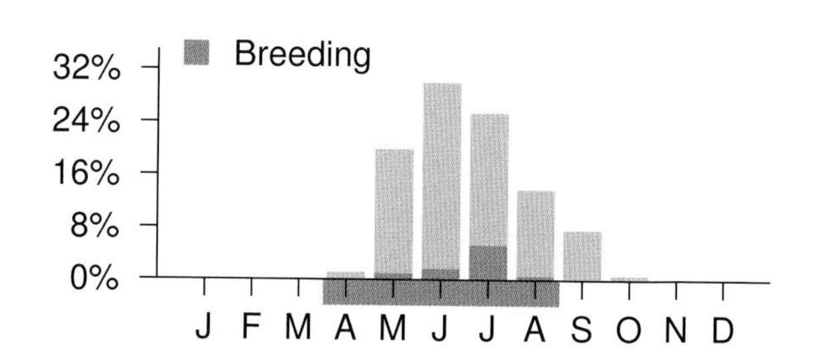

NATURAL REGIONS REPORTING RATE

SPATIAL REPORTING RATE

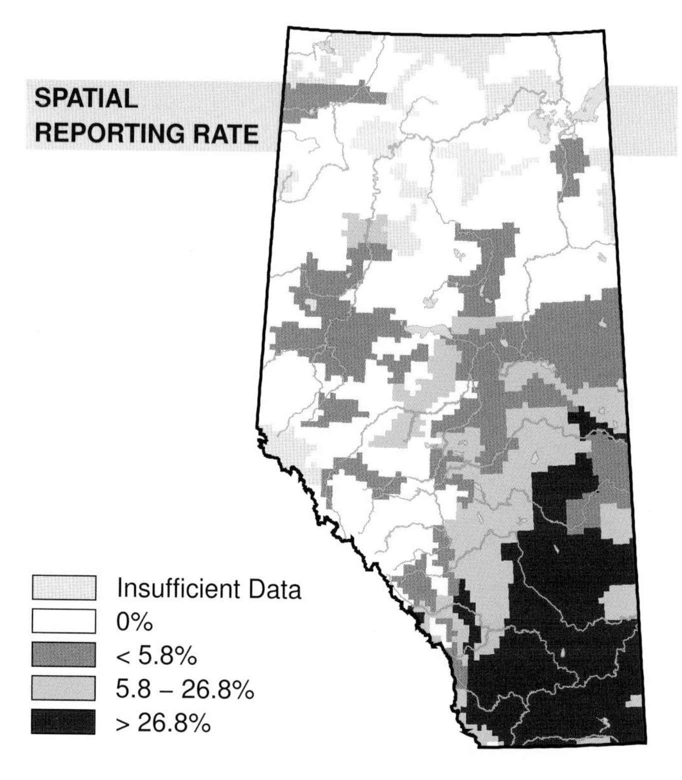

STATUS

GSWA 2000	GSWA 2005	COSEWIC Historic Rankings	COSEWIC 2007	Alberta Wildlife Act
Secure	Secure	N/A	N/A	N/A

RANGE

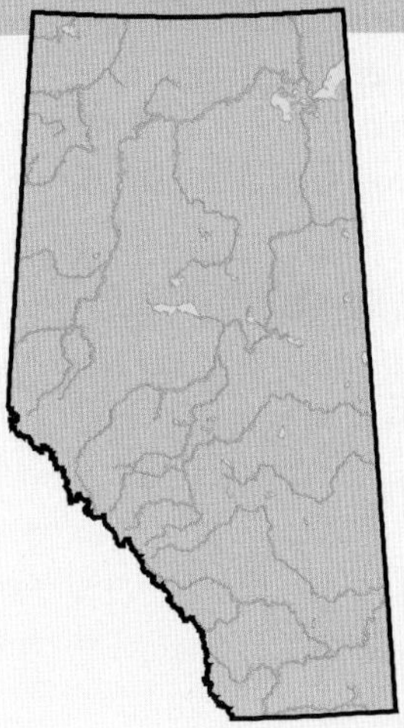

OBSERVED DISTRIBUTION ATLAS 2

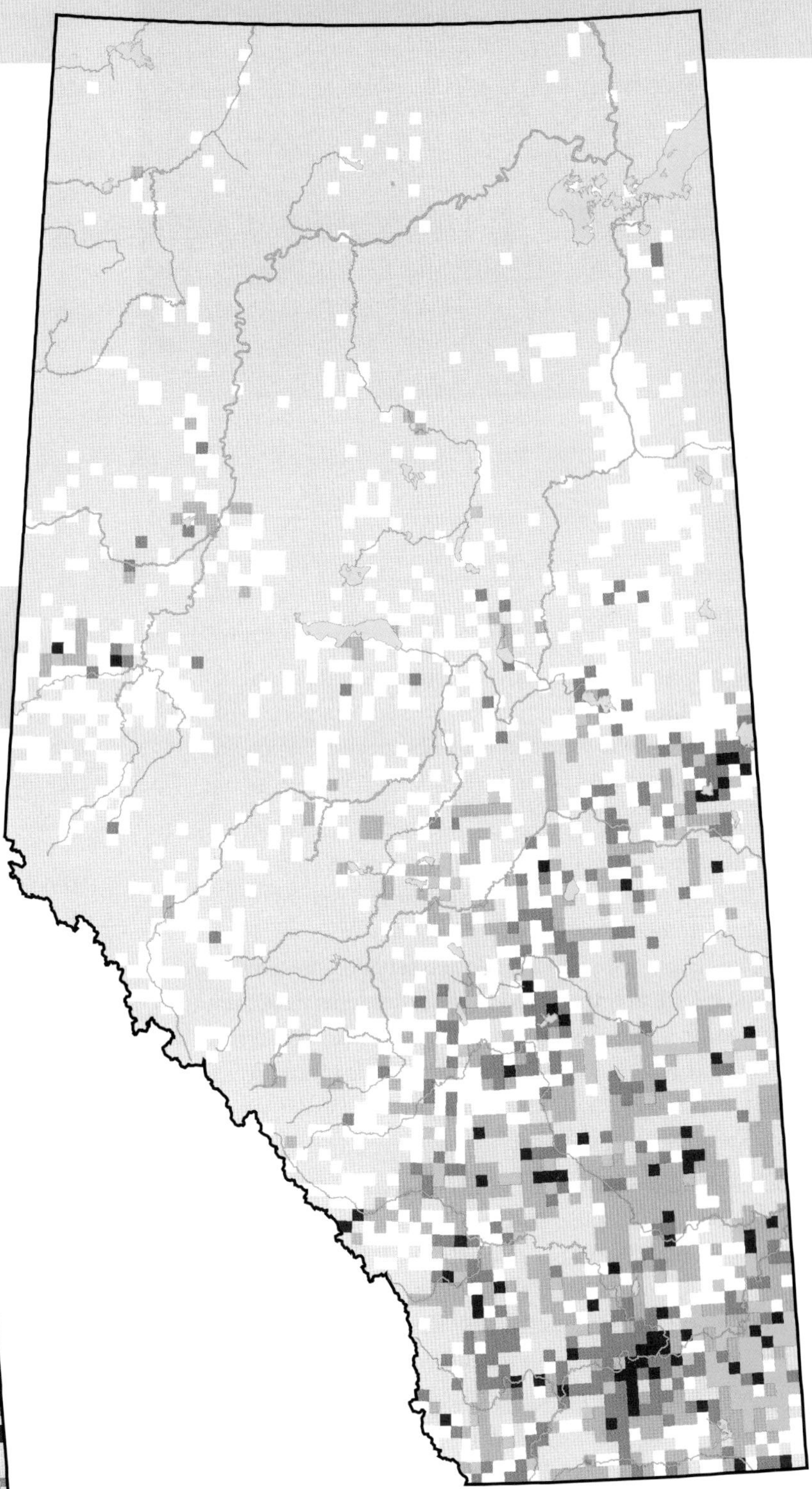

OBSERVED DISTRIBUTION ATLAS 1

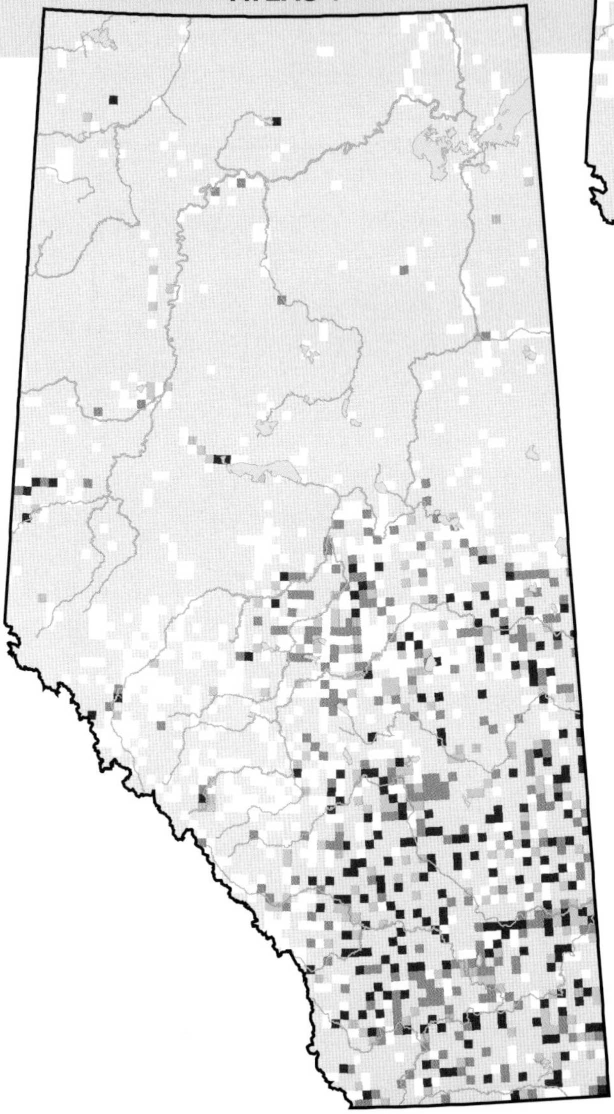

Unsurveyed
Not Observed
Observed
Possible
Probable
Confirmed

Lark Sparrow (*Chondestes grammacus*)

Photo: Alan MacKeigan

NESTING

CLUTCH SIZE (EGGS):	3–6
INCUBATION (DAYS):	11–13
FLEDGING (DAYS):	9–10
NEST HEIGHT (METRES):	0

The Lark Sparrow is found in southern Alberta in the Grassland Natural Region and in the southern portion of the Parkland NR. The distribution of this species did not change between Atlas 1 and Atlas 2.

The Lark Sparrow is an edge-associated species that prefers to breed in open areas that are interspersed with clumps of vegetation such as tree or shrub patches. Nests are built either on bare ground or off the ground in a tree or shrub. This species can breed in both cultivated habitats and natural prairie habitat. Lark Sparrows tend to prefer sites that have been grazed, and they can breed in areas that have been overgrazed. Due to its preference for open habitats, this species was found mainly in the Grassland NR and was infrequently found in the Parkland NR.

A relative abundance decline was detected in the Grassland NR where, relative to other species, the Lark Sparrow was detected less frequently in Atlas 2 than in Atlas 1. The decline could have been caused in part by long-term fire suppression because fire suppression reduces the edge habitat that is preferred by this species. No change was detected in the Parkland NR. The Breeding Bird Survey sample size was too small to investigate abundance trends for the period 1985–2005 in Alberta. However, no abundance change was found across Canada during this time period. Lacking corroboration, further research is needed to evaluate the Atlas-detected decline in the Grassland NR.

Dunn (2005) indicates that this is a species that has shown a large decline across its range. It is recommended that this species be monitored periodically to assess its status in Canada. Although the population appears stable in Canada, a small change in trend could lead to concern for the small Canadian population (Dunn, 2005).

This species is considered Secure in Alberta.

TEMPORAL REPORTING RATE

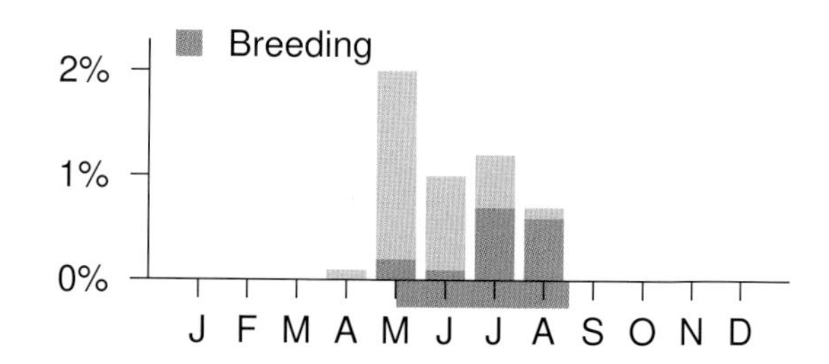

NATURAL REGIONS REPORTING RATE

SPATIAL REPORTING RATE

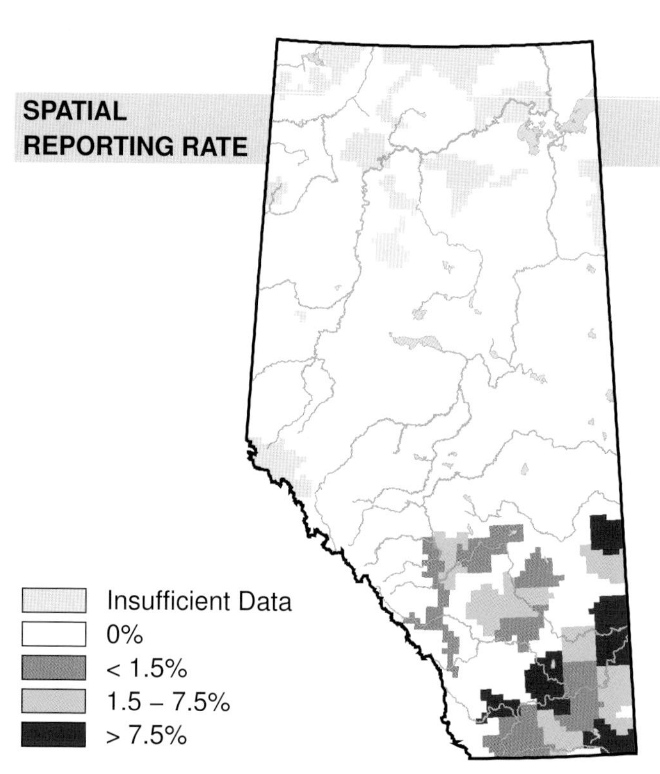

STATUS

GSWA 2000	GSWA 2005	COSEWIC Historic Rankings	COSEWIC 2007	Alberta Wildlife Act
Secure	Secure	N/A	N/A	N/A

RANGE

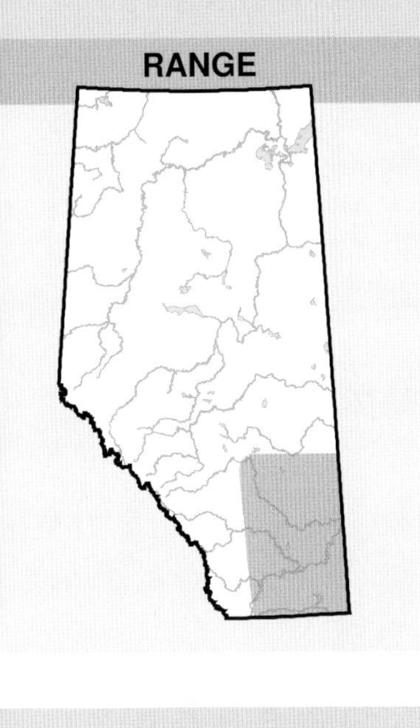

OBSERVED DISTRIBUTION ATLAS 2

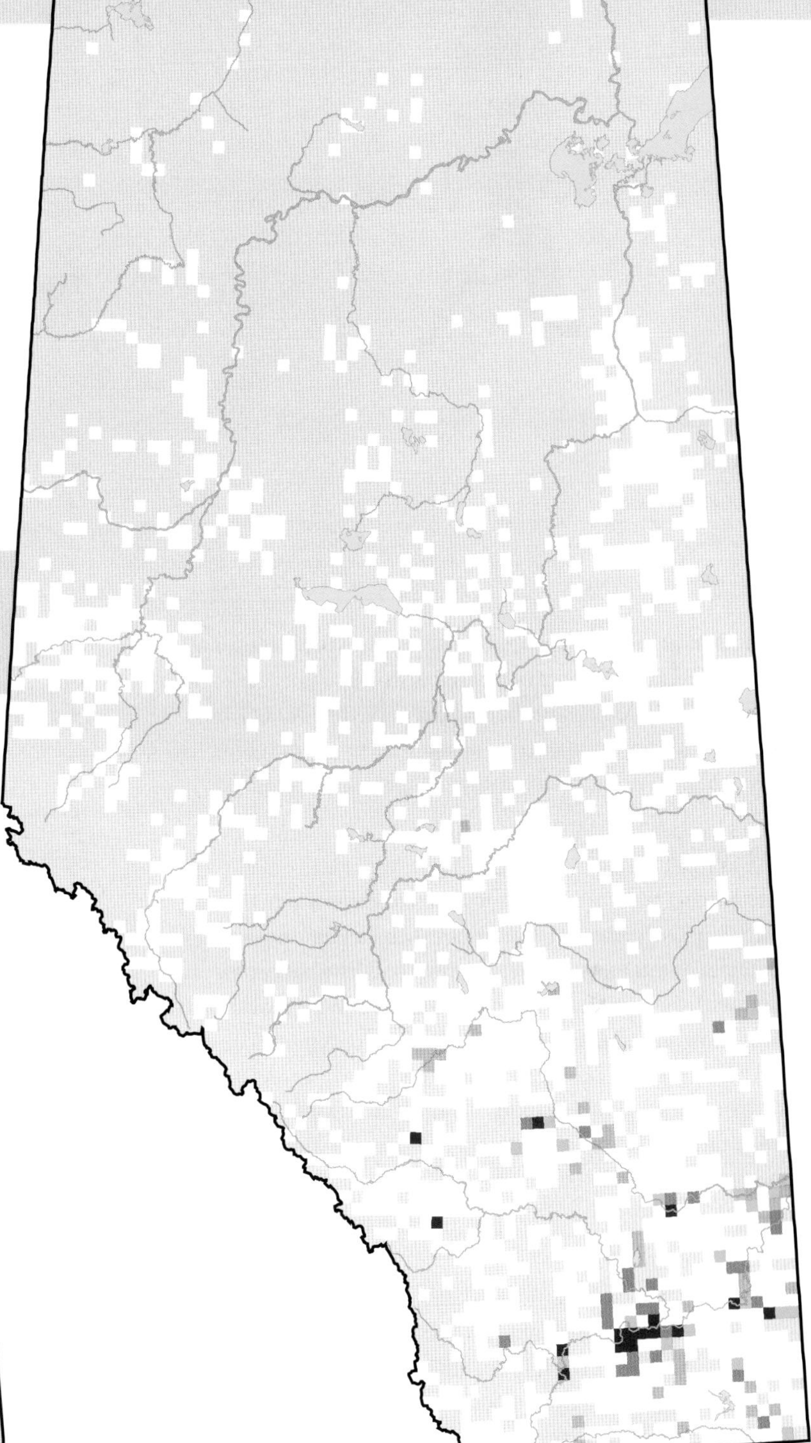

OBSERVED DISTRIBUTION ATLAS 1

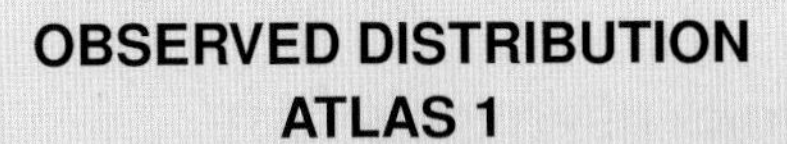

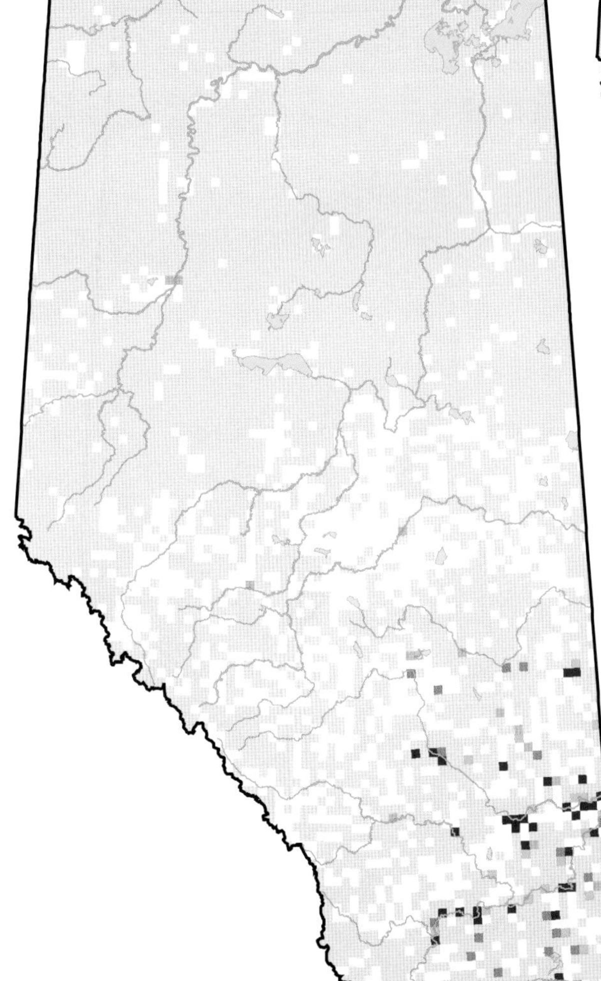

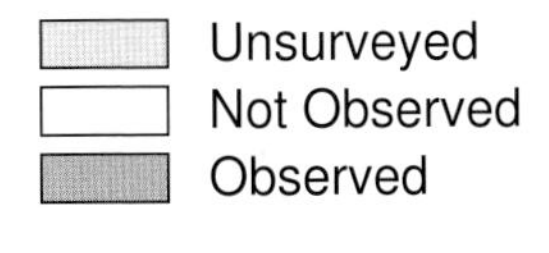

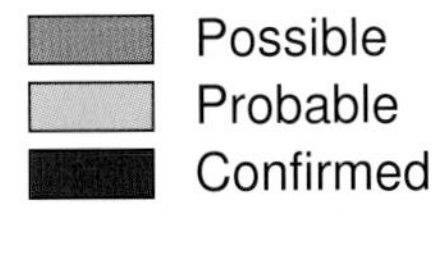

Lark Bunting (*Calamospiza melanocorys*)

Photo: Gordon Court

NESTING

CLUTCH SIZE (EGGS):	3–7
INCUBATION (DAYS):	12
FLEDGING (DAYS):	8–9
NEST HEIGHT (METRES):	0

Lark Bunting is almost entirely restricted to the Grassland Natural Region. It appears that the distribution in Atlas 2 may be slightly less extensive than that observed in Atlas 1 but some of that apparent change is due to the selection of squares revisited. There are fewer occupied squares that reached confirmed breeding status, so the distribution in Atlas 2 is more muted in its presentation.

The highest reporting rates for Lark Buntings are in the eastern Grassland NR (except the Cypress Hills) and an adjoining area extending west to around Vauxhall. This distribution fits very well with their described habitat characteristics of grassland with shrub or the alternative of tall dense agricultural cover like hayfields (Shane, 2000). They are particularly associated with sage but will use other shrubs and may treat alfalfa as a shrub substitute. Nests are placed under shrubs and are oriented to maximize sun early in the day and to provide shade and cooling breezes the rest of the time (Shane, 2000). The eastern portions of the province have sage or other shrubs and the planted cover in the irrigation district may also meet their needs.

Decreases in relative abundance were detected in the Grassland NR. No changes in relative abundance were detected in the Boreal Forest and Parkland NRs. Lark Buntings present a real challenge for trend determination at any scale. The Breeding Bird Survey detected no trend but Grassland Bird Monitoring (unpublished Canadian Wildlife Service data) found a positive trend in the period 1996–2004. This species is nomadic and moves around dramatically within its North American range from year to year in response to precipitation conditions. It can literally go from being scarcely seen in appropriate habitat in one year to common the next year, and all this variation makes detecting and interpreting trends very difficult. Provincial status reports in 2000 and 2005 rated it Sensitive and Secure, respectively.

TEMPORAL REPORTING RATE

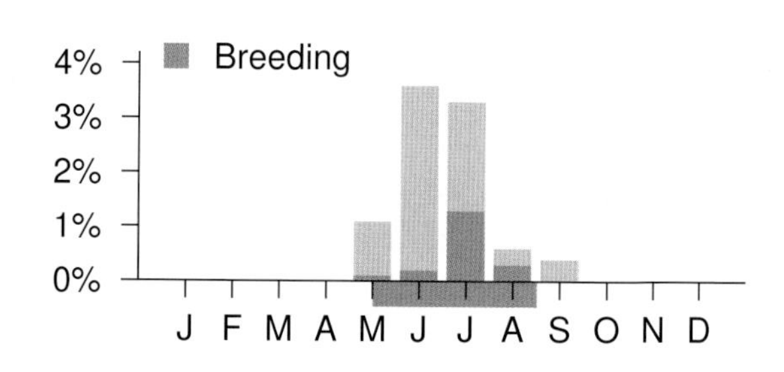

NATURAL REGIONS REPORTING RATE

Grassland 5.5%

SPATIAL REPORTING RATE

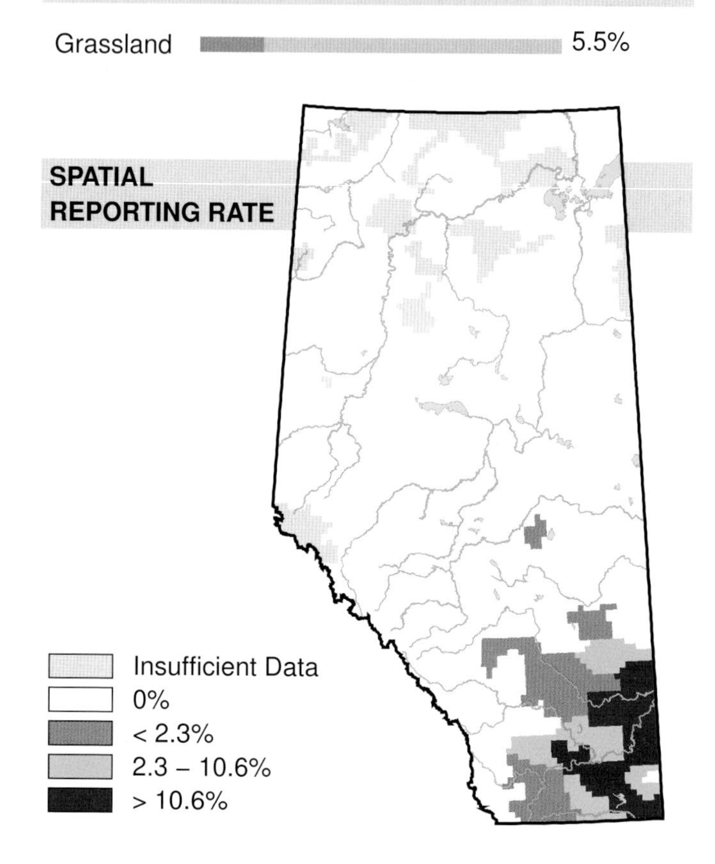

STATUS

GSWA 2000	GSWA 2005	COSEWIC Historic Rankings	COSEWIC 2007	Alberta Wildlife Act
Sensitive	Secure	N/A	N/A	N/A

Habitat Nest Location Nest Type Diet

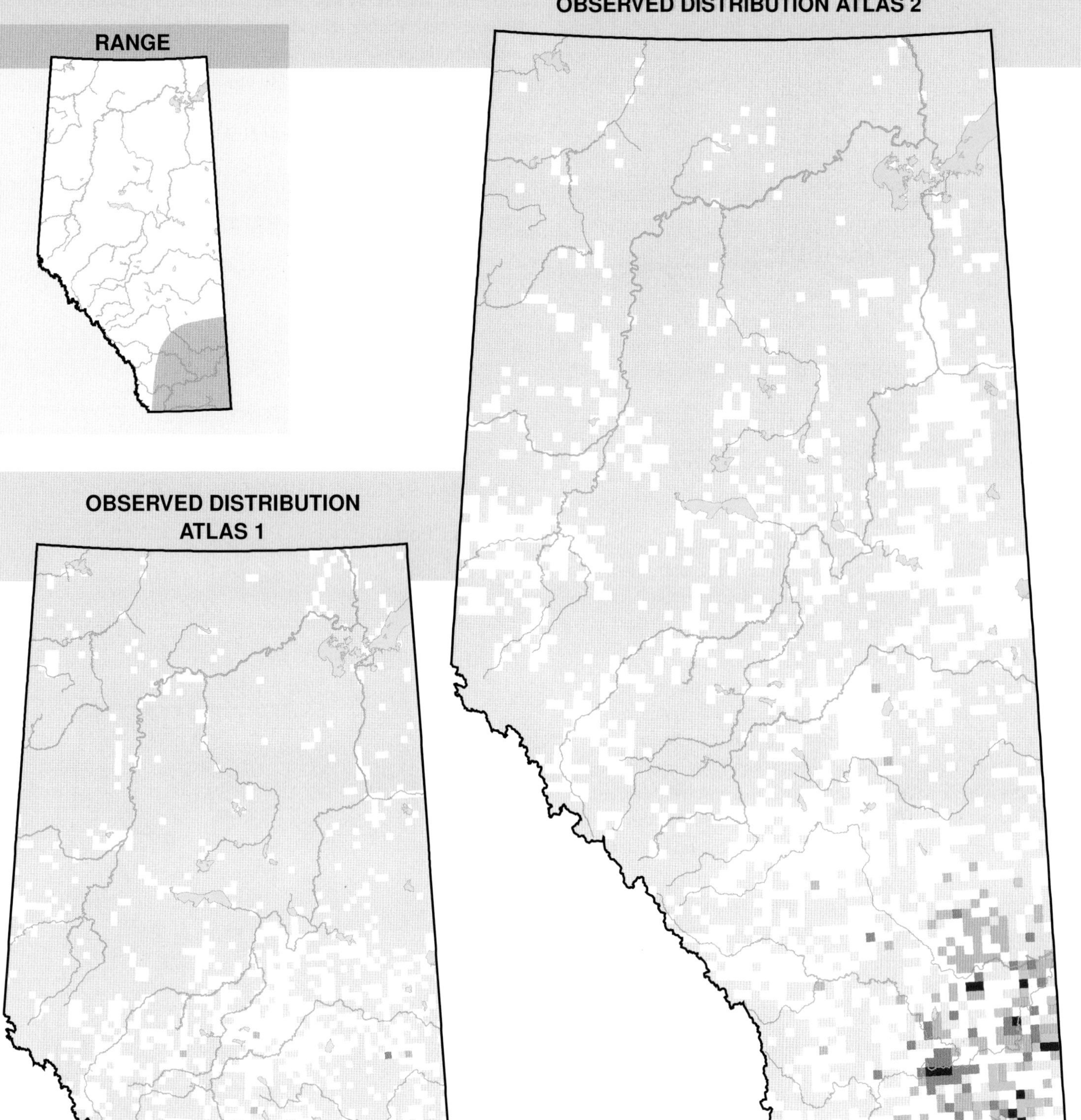

Savannah Sparrow (*Passerculus sandwichensis*)

Photo: Randy Jensen

NESTING

CLUTCH SIZE (EGGS):	4–5
INCUBATION (DAYS):	12
FLEDGING (DAYS):	7–10
NEST HEIGHT (METRES):	0

The Savannah Sparrow breeds in all Natural Regions in Alberta but was found most often in the Grassland and Parkland Natural Regions. There was no change in distribution between Atlas 1 and Atlas 2.

Dense ground vegetation, especially grasses, and moist microhabitats are the major requirements for this species (Wheelwright and Rising, 1993). In central and southern Alberta it prefers edges of prairie sloughs, marshes, moist grasslands, hayfields, and any damp, low-lying area with dense vegetation (Semenchuk, 1992). This sparrow avoids areas of extensive forest cover and areas of short grass. In northern areas, it favours sedge meadows, bogs, burns, and clearcuts.

Increases in relative abundance were detected in the Grassland NR. The increase in the Grassland NR may be due to an increase in planted pastures and hayfields and to the accelerated invasion of native prairie by exotic grasses that are more attractive to the Savannah Sparrow than grazed native grassland (Prescott and Murphy, 1996). Decreases in relative abundance were detected in the Boreal Forest, Foothills, and Parkland NRs. The drier conditions in the Boreal Forest and Parkland NRs during Atlas 2 probably affected the availability of preferred breeding habitat. No changes in relative abundance were detected in the Rocky Mountain NR. The Breeding Bird Surveys (1995–2005) detected a decrease in abundance in Alberta and across Canada.

Population size fluctuates from year to year but the changes are apparently not closely tied to events on breeding grounds in the previous year. Rather, they are probably affected most during migration by events such as predation and storms, or on wintering grounds by availability of suitable habitat (Wheelwright and Rising, 1993). As with many passerines, annual mortality usually exceeds 50%.

The Savannah Sparrow is rated as Secure in Alberta.

TEMPORAL REPORTING RATE

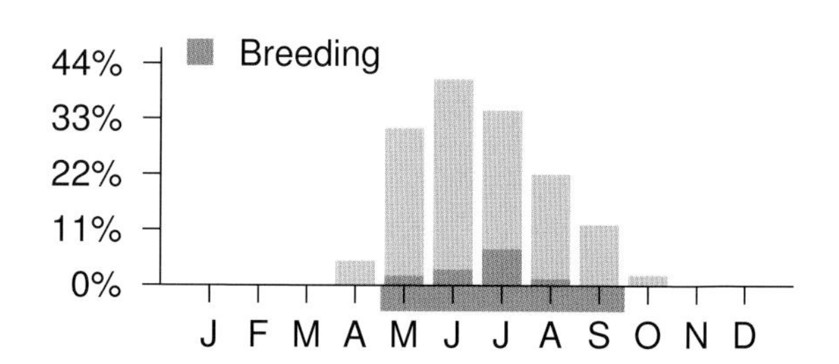

NATURAL REGIONS REPORTING RATE

Boreal Forest	6%
Foothills	4%
Grassland	37%
Parkland	23%
Rocky Mountain	14%

SPATIAL REPORTING RATE

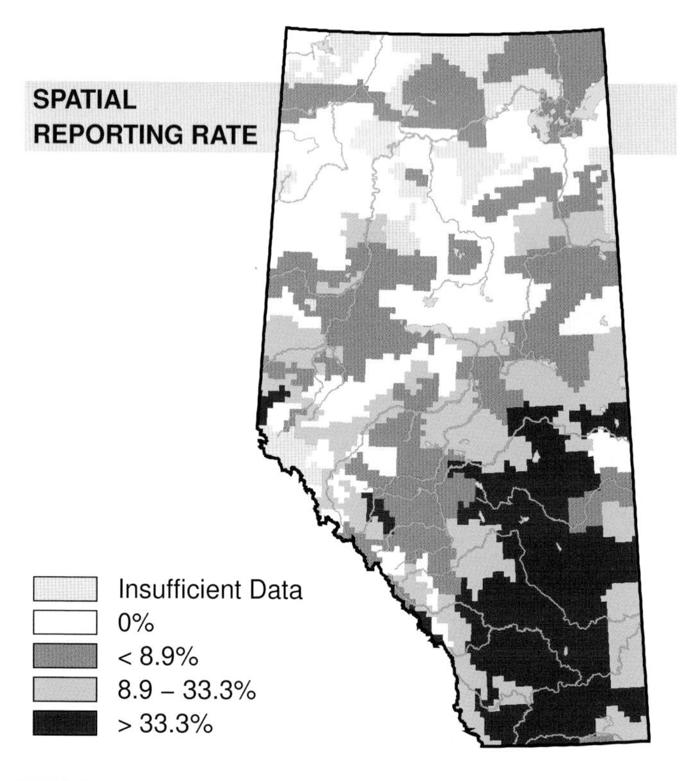

STATUS

GSWA 2000	GSWA 2005	COSEWIC Historic Rankings	COSEWIC 2007	Alberta Wildlife Act
Secure	Secure	N/A	N/A	N/A

Habitat

Nest
Location

Nest
Type

Diet

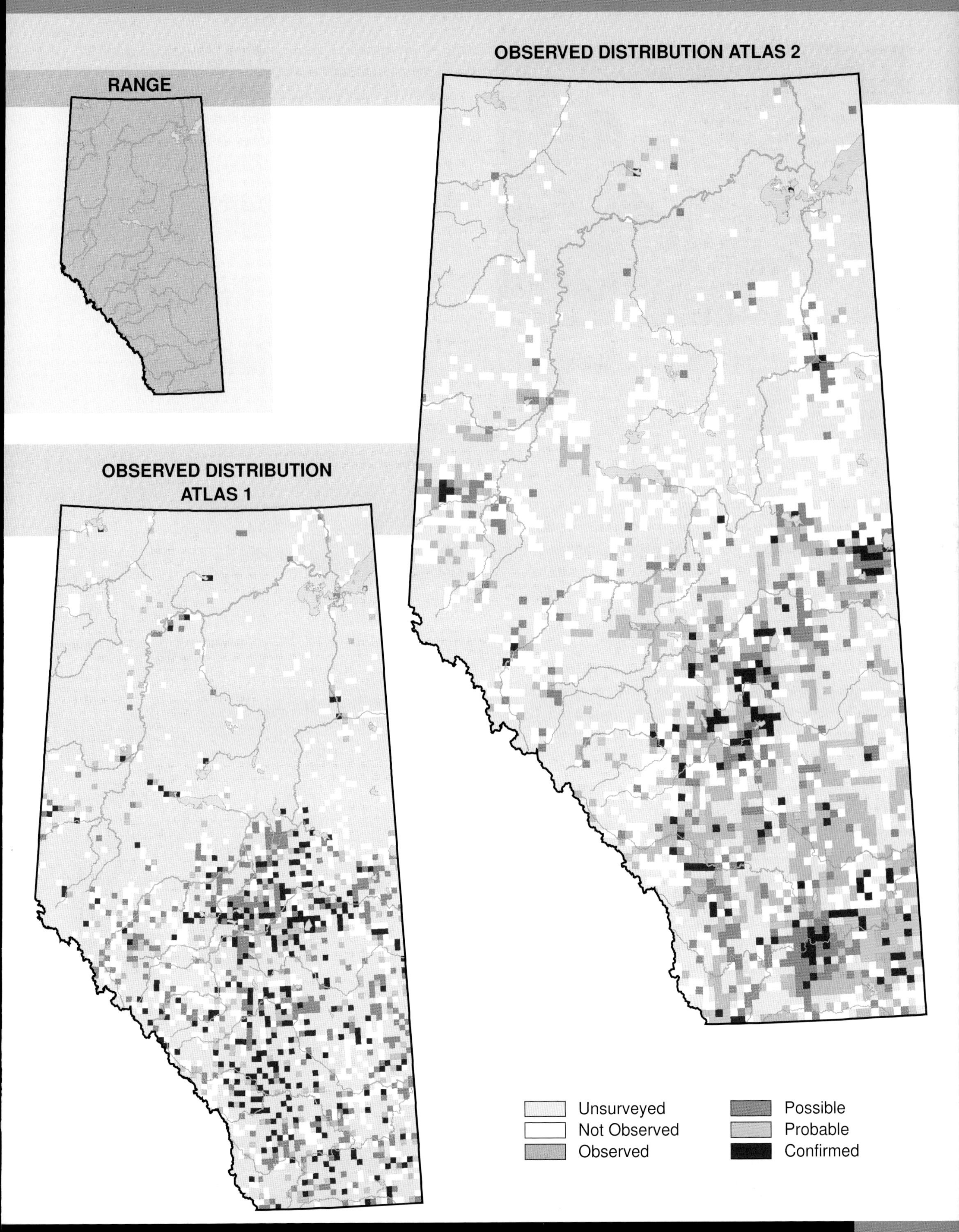
RANGE
OBSERVED DISTRIBUTION ATLAS 2
OBSERVED DISTRIBUTION
ATLAS 1
Unsurveyed
Not Observed
Observed
Possible
Probable
Confirmed

Grasshopper Sparrow (*Ammodramus savannarum*)

Photo: Robert Gehlert

NESTING

CLUTCH SIZE (EGGS):	3–6
INCUBATION (DAYS):	11–12
FLEDGING (DAYS):	9
NEST HEIGHT (METRES):	0

The Grasshopper Sparrow is almost completely confined to the Grassland Natural Region with limited occurrences in the Boreal Forest and Parkland NRs. There is no change in apparent distribution between Atlas 1 and Atlas 2 as many of the squares in the northern Grassland and southern Parkland NRs–occupied in Atlas 2–were not surveyed in Atlas 1.

The highest reporting rates are in the Sandhills and other poor soil areas near the eastern boundary of the province. Grasslands in these areas offer the right mix of standing tufts of grass with patchy bare ground and a lack of shrub cover that is the preferred habitat of Grasshopper Sparrows (Vickery, 1996).

No changes in relative abundance were detected in the Parkland NR. Increases in relative abundance were detected in the Grassland NR, but coverage in this NR was more extensive in Atlas 2. The Breeding Bird Survey did not detect any provincial trend but sample sizes are extremely small for this species. Given the variable trends from various sources and analysis types, it appears that sample sizes are inadequate to reach firm conclusions. If the apparent increase in the Grassland NR is real, it could be due to an increase in planted cover in the Grassland NR (Watmough and Schmoll, 2007), as this is one grassland species that tolerates seeded grassland well (McMaster and Davis, 2001) and it may benefit from programs like the federal Permanent Cover Program (now Green Cover). Provincial status reports from 2000 and 2005 both rate the Grasshopper Sparrow as Sensitive as it is thought to be vulnerable to the conversion of grassland to crops.

TEMPORAL REPORTING RATE

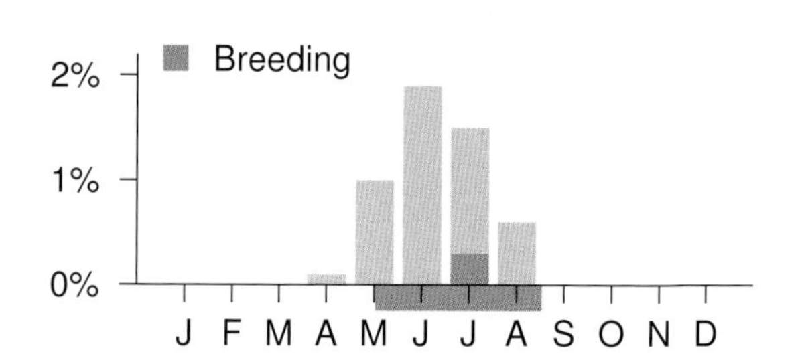

NATURAL REGIONS REPORTING RATE

SPATIAL REPORTING RATE

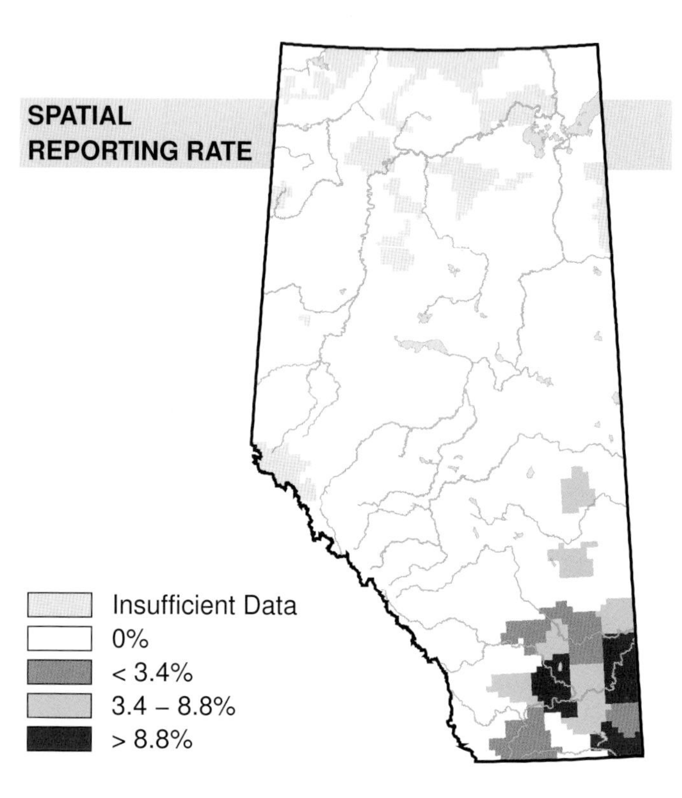

STATUS

GSWA 2000	GSWA 2005	COSEWIC Historic Rankings	COSEWIC 2007	Alberta Wildlife Act
Sensitive	Sensitive	N/A	N/A	N/A

RANGE

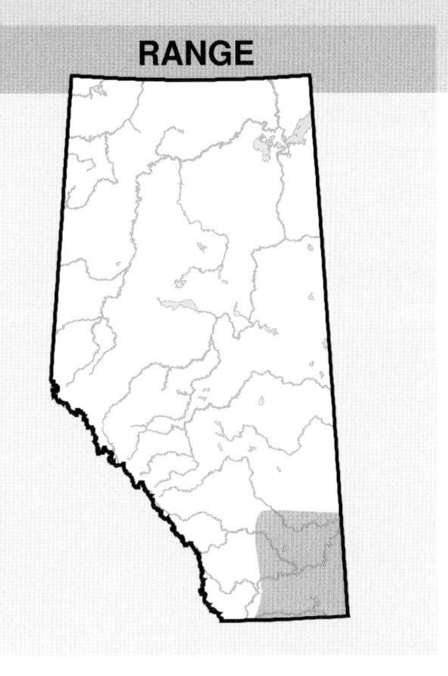

OBSERVED DISTRIBUTION ATLAS 2

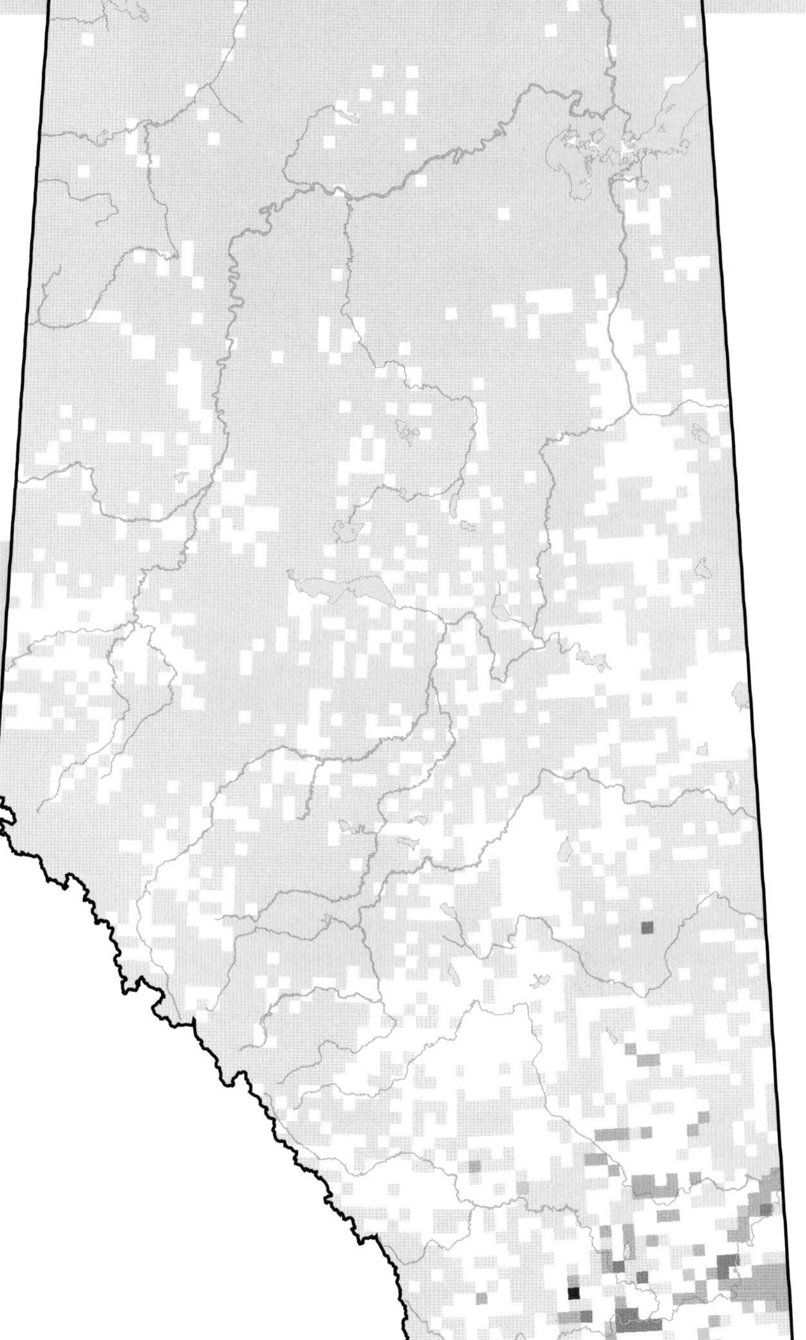

OBSERVED DISTRIBUTION ATLAS 1

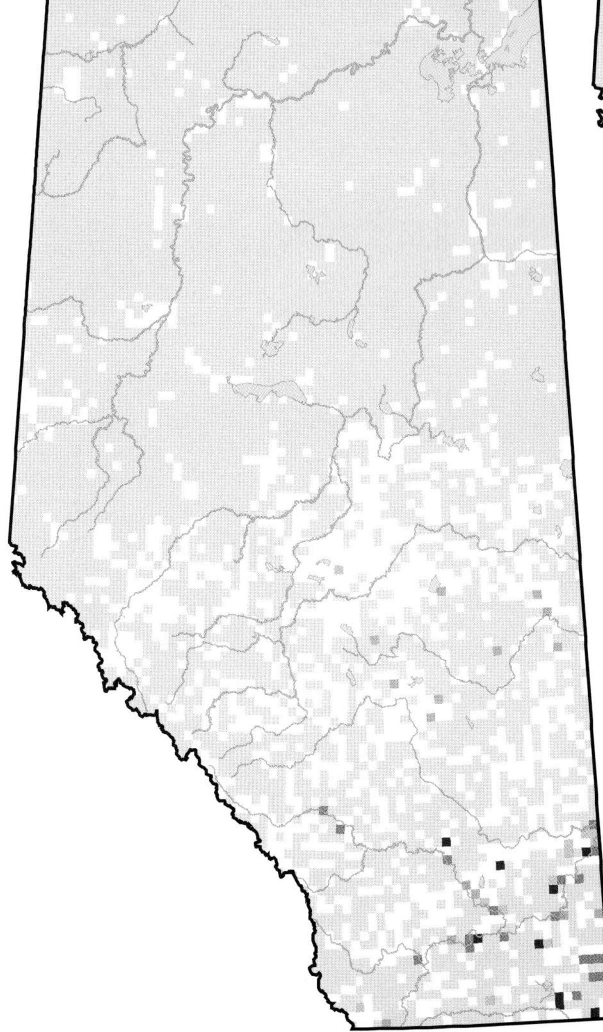

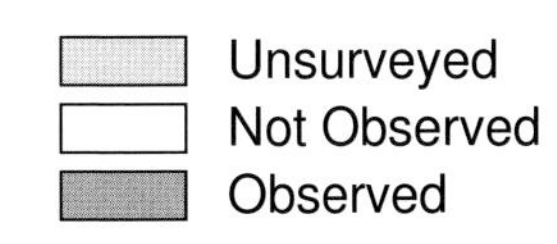

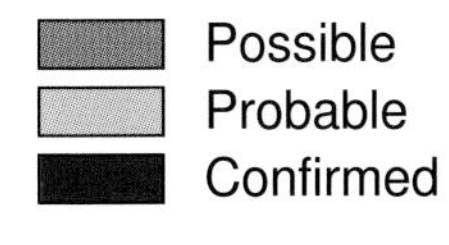

Baird's Sparrow *(Ammodramus bairdii)*

Photo: Alan MacKeigan

NESTING

CLUTCH SIZE (EGGS):	3–6
INCUBATION (DAYS):	11–12
FLEDGING (DAYS):	9
NEST HEIGHT (METRES):	0

The general distribution of Baird's Sparrow includes the Grassland and Parkland Natural Regions. There was no observable change in distribution between Atlas 1 and Atlas 2.

The highest reporting rates for Baird's Sparrow were recorded in the eastern portion of the Grassland NR, as Alberta contains the northern and western extremes of the species' range. Lower reporting rates on sandy soils and in the Parkland NR reflect the species' avoidance of anything more than scattered shrubs within grassland (Green et al., 2002).

Decreases in relative abundance were detected in the Grassland NR. Habitat monitoring shows decreases in grassland cover (varying from 7.5% in Mixed Grass to 10% in Fescue) in the Grassland NR from 1985–1999 (Watmough and Schmoll, 2007). No change in relative abundance of Baird's Sparrow was detected in the Parkland NR but the reporting rate was very low in this NR. The Breeding Bird Survey detected no change in a 20-year period but did find a decline of 8.5% for the province in the period 1995–2005. The decline appears real, as specialized grassland bird surveys confirm the BBS trend (Canadian Wildlife Service unpublished data), and the trend was detected by the Atlas. Provincial status reports from 2000 and 2005 rank Baird's Sparrow as Sensitive and May be at Risk, respectively. It was listed federally as Threatened in 1989, but was downgraded to Not at Risk in 1996. Originally it was thought to be a native-grass specialist but it will tolerate invasion by, or planted cover of, some other species so long as the management (such as grazing) creates a cover that is structurally similar to native grassland (Green et al., 2002). It is area-sensitive (Davis, 2004) and does not tolerate grazing as well as some other grassland birds, particularly in drier sites (Dale, 1983). Thus, it is vulnerable to conversion to crops or grazing that is inappropriate for the soil type.

TEMPORAL REPORTING RATE

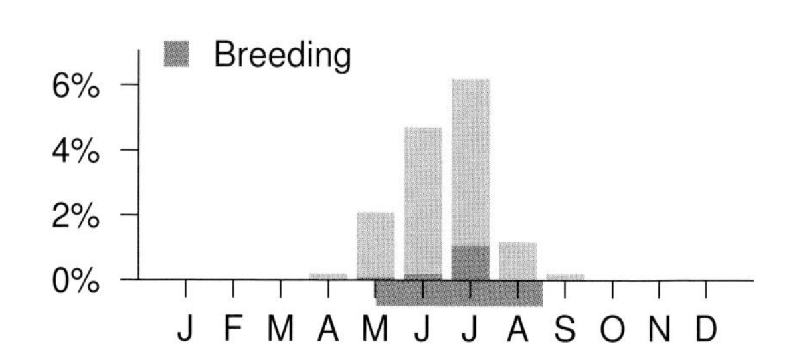

NATURAL REGIONS REPORTING RATE

Grassland	6.5%
Parkland	0.2%

SPATIAL REPORTING RATE

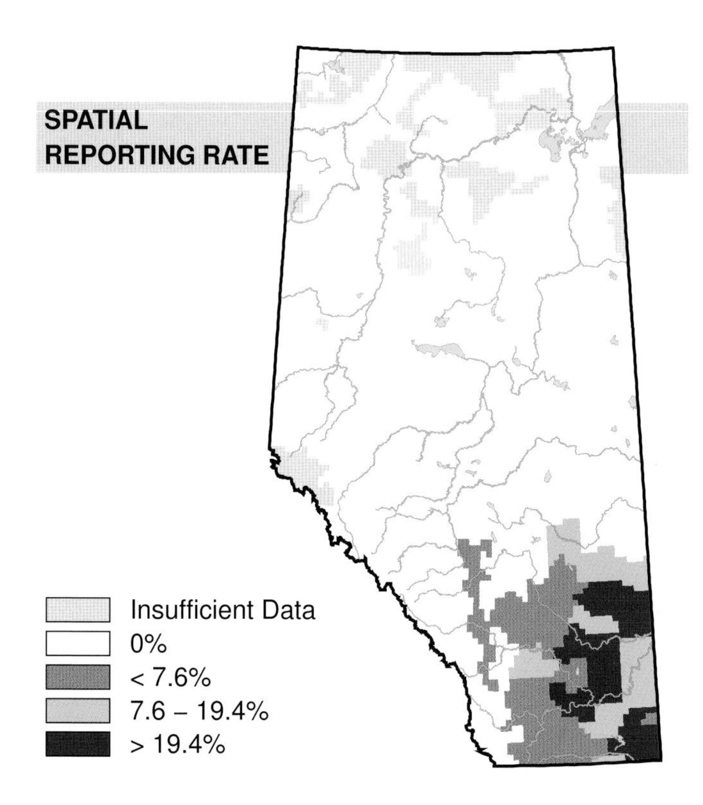

STATUS

GSWA 2000	GSWA 2005	COSEWIC Historic Rankings	COSEWIC 2007	Alberta Wildlife Act
Sensitive	May Be at Risk	Not At Risk	Not At Risk	N/A

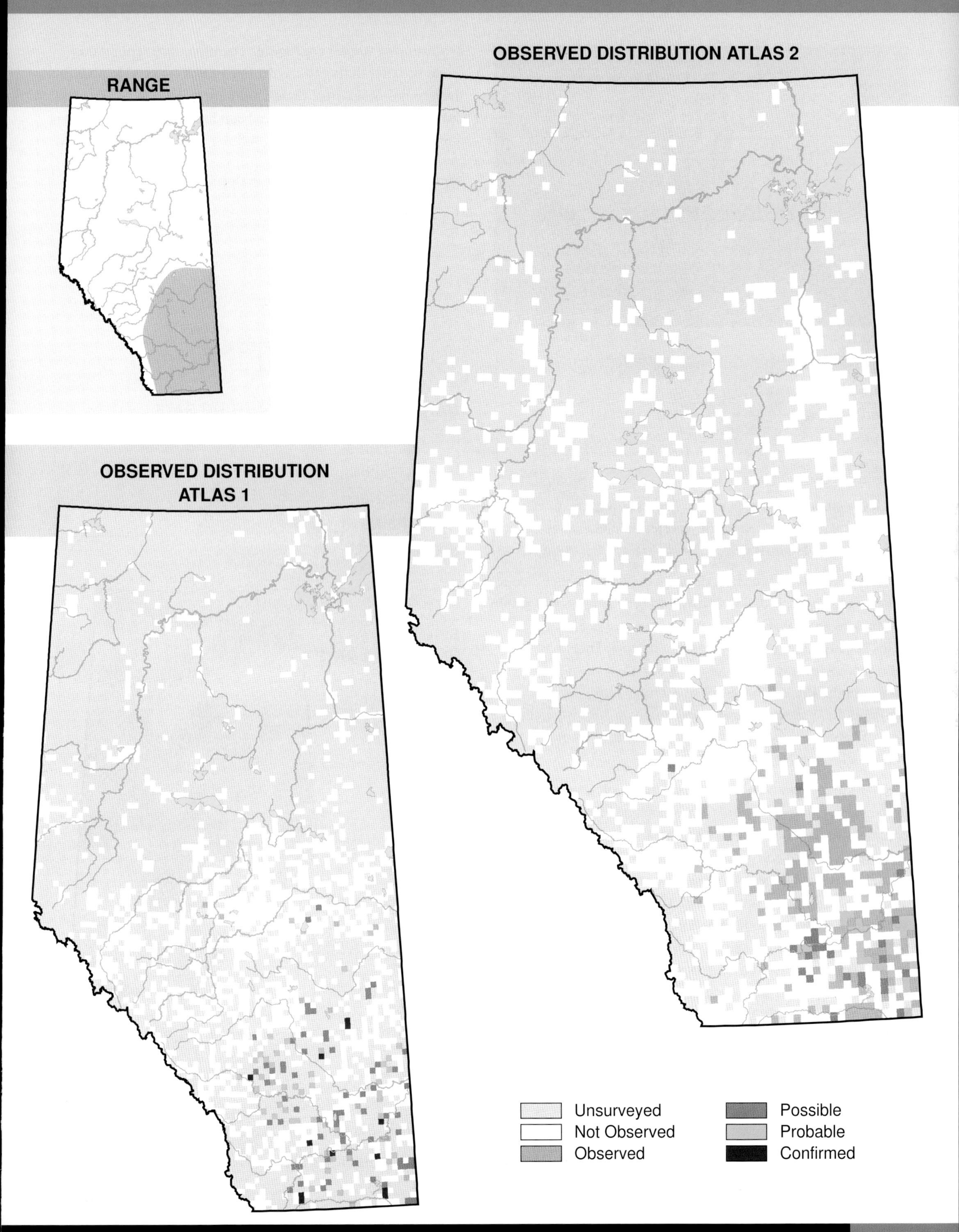

Habitat
Nest Location
Nest Type
Diet
OBSERVED DISTRIBUTION ATLAS 2
RANGE
OBSERVED DISTRIBUTION ATLAS 1
Unsurveyed
Not Observed
Observed
Possible
Probable
Confirmed

Le Conte's Sparrow (*Ammodramus leconteii*)

Photo: Gerald Romanchuk

NESTING

CLUTCH SIZE (EGGS):	4–5
INCUBATION (DAYS):	11–13
FLEDGING (DAYS):	unknown
NEST HEIGHT (METRES):	0

Le Conte's Sparrow breeds in every Natural Region in Alberta. The observed distribution of this species did not change between Atlas 1 and Atlas 2.

Le Conte's Sparrow breeds in open areas such as marshy sedge meadows, near bogs, and in marshes where tall grass and shrub patches are present. This species does not nest in cultivated fields because it requires tall grass for nesting. Due to its preference to nest in marshy bogs, this species was found mainly in the Boreal Forest NR. It was also fairly common in the Foothills, Parkland, and Rocky Mountain NRs where this type of habitat can be found. Observations made in the Grassland NR were mostly non-breeders.

Declines in the relative abundance of this species were detected in the Boreal Forest, Foothills, Grassland, and Parkland NRs and, relative to other species, Le Conte's Sparrow was observed less frequently in Atlas 2 than in Atlas 1 in these NRs. No change was detected in the Rocky Mountain NR. The Breeding Bird Survey found a decline in the abundance of this species during the period 1985–2005. Similar to the findings of the Atlas and Alberta Breeding Bird Survey, the Breeding Bird Survey detected a decline on a Canada-wide scale during the period 1995–2005. Fire is thought to have a positive effect on this species because fire stimulates new grass growth and limits expansion of woody vegetation. Therefore, fire suppression could be contributing to the decline of this species. In addition, this species is sensitive to agricultural practices and generally does not nest in fields that have been harvested. This species is considered Secure in Alberta.

TEMPORAL REPORTING RATE

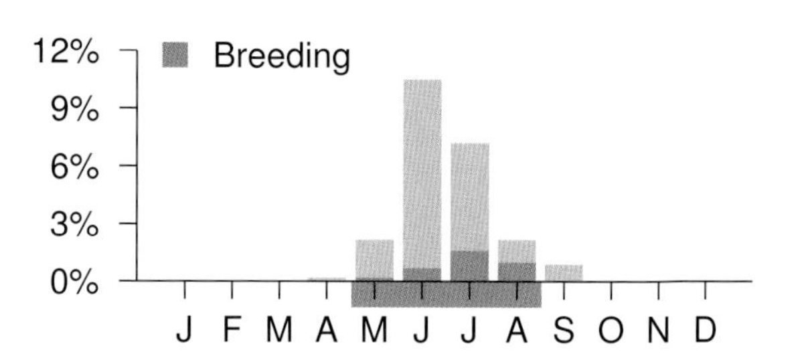

NATURAL REGIONS REPORTING RATE

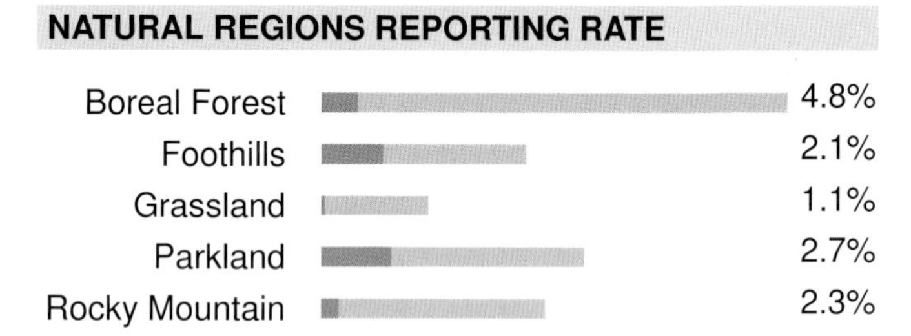

SPATIAL REPORTING RATE

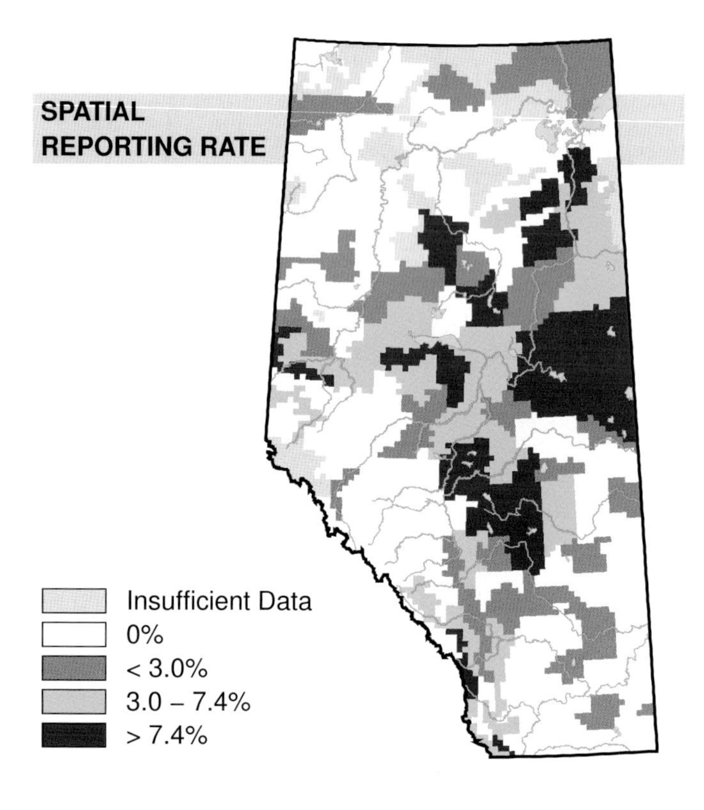

STATUS

GSWA 2000	GSWA 2005	COSEWIC Historic Rankings	COSEWIC 2007	Alberta Wildlife Act
Secure	Secure	N/A	N/A	N/A

Habitat

Nest Location

Nest Type
Diet

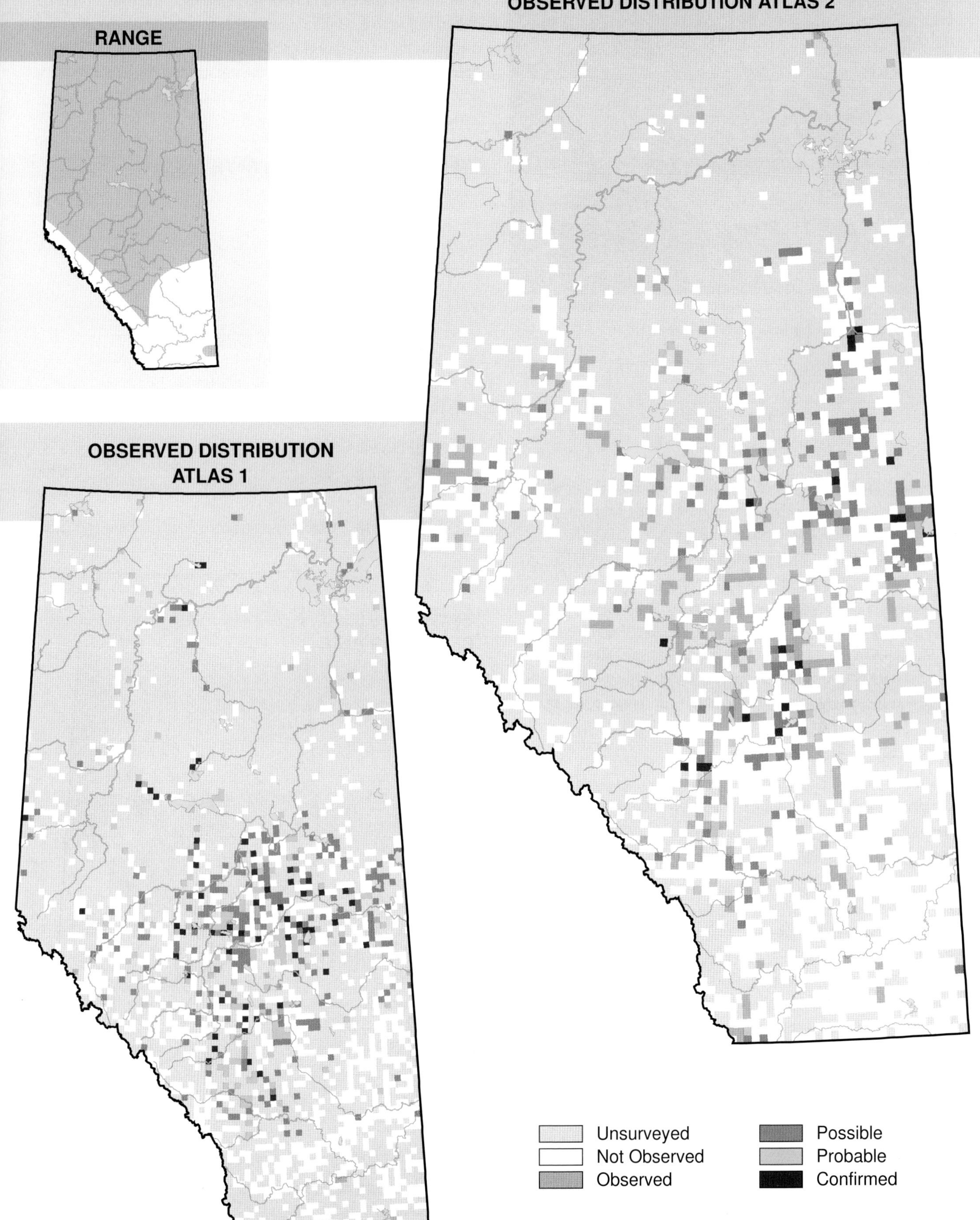
OBSERVED DISTRIBUTION ATLAS 2
RANGE
OBSERVED DISTRIBUTION ATLAS 1
Unsurveyed
Not Observed
Observed
Possible
Probable
Confirmed

Nelson's Sharp-tailed Sparrow (*Ammodramus nelsoni*)

Photo: Royal Alberta Museum

NESTING

CLUTCH SIZE (EGGS):	4–6
INCUBATION (DAYS):	11
FLEDGING (DAYS):	10
NEST HEIGHT (METRES):	0

Nelson's Sharp-tailed Sparrow breeds in the Boreal Forest, Foothills, and Parkland Natural Regions. The observed distribution of this species contracted in Atlas 2 in the northwestern part of the province, around High Level. Because there is limited survey effort in this part of Alberta, and because this species has low detectability, it is unclear whether this is a real change in distribution or whether the presence of this species in the northwest was missed during Atlas 2. In the southern parts of this bird's range, more records with breeding evidence included were obtained in the Grassland NR in Atlas 2.

The secretive Nelson's Sharp-tailed Sparrow breeds in meadows that are near marshes. This species was sparsely distributed across the province but was most often encountered in the Parkland NR. Reporting rates were fairly low in the Boreal Forest, Foothills, and Grassland NRs.

No changes in relative abundance were detected for this species in any natural region. Relative to other species, Nelson's Sharp-tailed Sparrow was observed less frequently in the Boreal Forest NR and more often in other NRs in Atlas 2 than in Atlas 1. The Breeding Bird Survey did not find a change in the relative abundance of this species in Alberta. However, an increase was detected on a Canada-wide scale during the period 1985–2005. Likely one of the main factors that limits the detectability of this species is the time of day when it is active. Unlike many other songbirds that are most active in the early morning, this species is most active in the late evening and early part of the night. Detection rates would increase if surveys were conducted at this time of day. This species is considered Secure in Alberta.

TEMPORAL REPORTING RATE

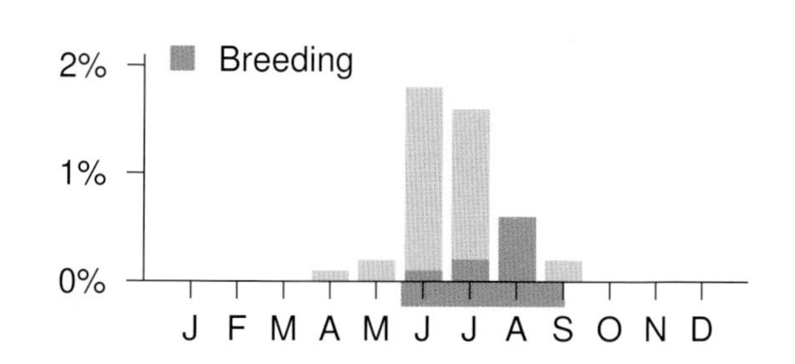

NATURAL REGIONS REPORTING RATE

SPATIAL REPORTING RATE

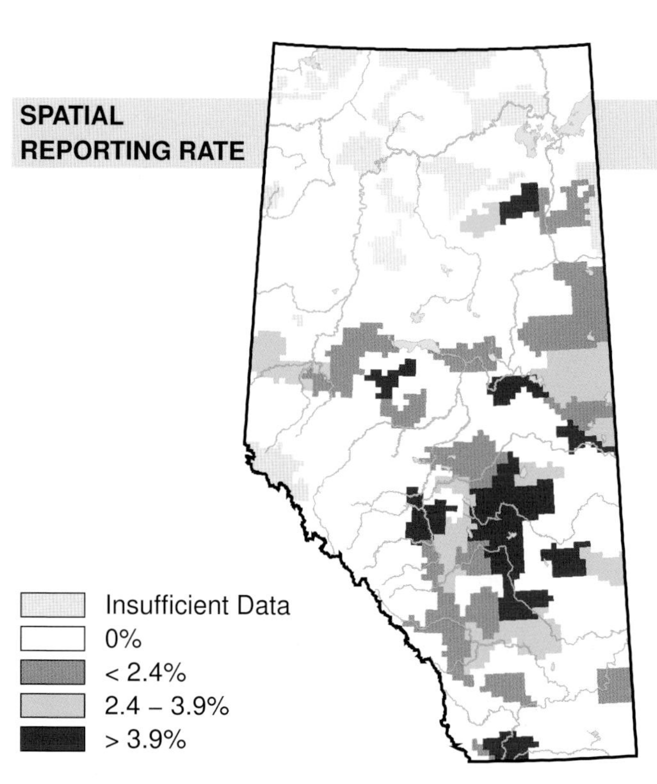

STATUS

GSWA 2000	GSWA 2005	COSEWIC Historic Rankings	COSEWIC 2007	Alberta Wildlife Act
Secure	Secure	Not At Risk	Not At Risk	N/A

Habitat

Nest Location

Nest Type

Diet

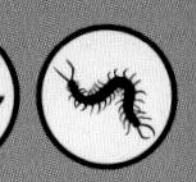

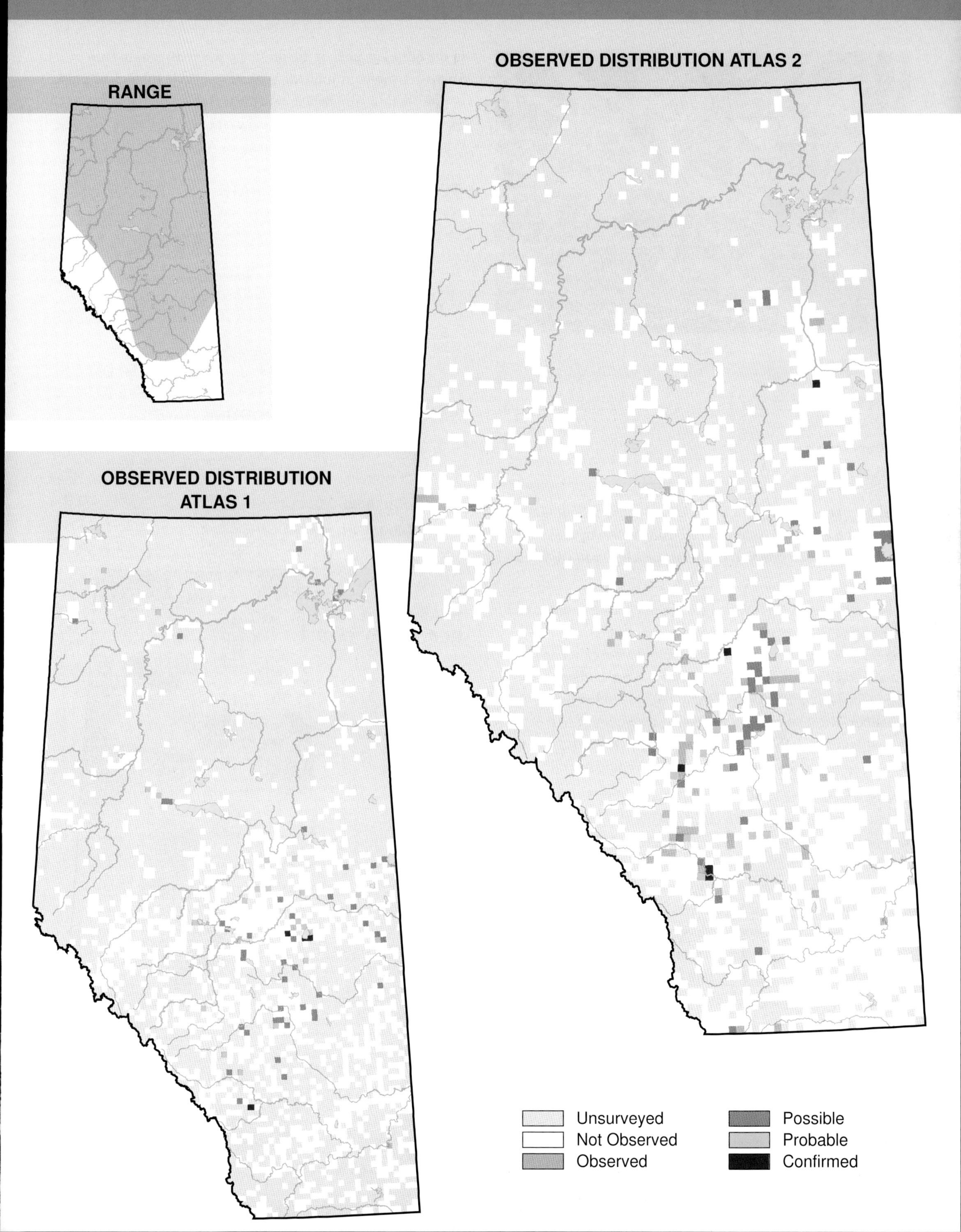

Fox Sparrow (*Passerella iliaca*)

Photo: Damon Calderwood

NESTING

CLUTCH SIZE (EGGS):	3–5
INCUBATION (DAYS):	12–14
FLEDGING (DAYS):	9–11
NEST HEIGHT (METRES):	0

The Fox Sparrow breeds in the Rocky Mountain, Foothills, Boreal, and Parkland Natural Regions. No change was observed in the distribution of this species between Atlas 1 and Atlas 2.

This sparrow prefers alder and willow thickets in spruce and fir forests, and is less common in continuous coniferous forest. Fox Sparrows are found commonly in thick cover, along woodland edges, grown up fields, and in areas bordering wet boggy areas. It was, therefore, found mainly in the Rocky Mountain NR. Observation frequencies were lower in the Boreal Forest, Foothills, and Parkland NRs. Records from the Grassland NR are not considered to be breeding records.

Declines in relative abundance were detected in the Boreal Forest NR where, relative to other species, the Fox Sparrow was observed less frequently in Atlas 2 than Atlas 1. No changes in relative abundance were detected for Foothills, Grassland, Parkland, and Rocky Mountain NRs. Reasons for the decline in the Boreal Forest NR are unknown, considering that this species prefers open coniferous forests, a habitat that is increasing in this NR due to anthropogenic disturbance. The Breeding Bird Survey did not find changes in abundance in Alberta or across Canada; however, the Atlas detected change in the north, an area of the province where Breeding Bird Survey coverage is limited. Without probable biological cause or corroborating evidence, more research will be required to assess the Atlas-detected decline in the Boreal Forest NR. The Fox Sparrow is considered Secure in Alberta.

TEMPORAL REPORTING RATE

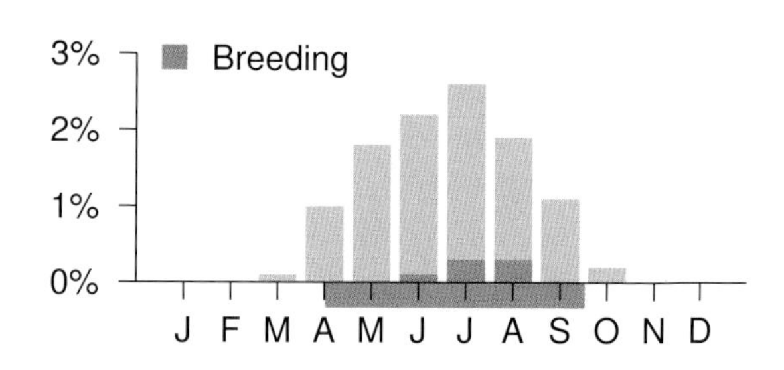

NATURAL REGIONS REPORTING RATE

Boreal Forest	0.4%
Foothills	1.2%
Parkland	0.9%
Rocky Mountain	7.3%

SPATIAL REPORTING RATE

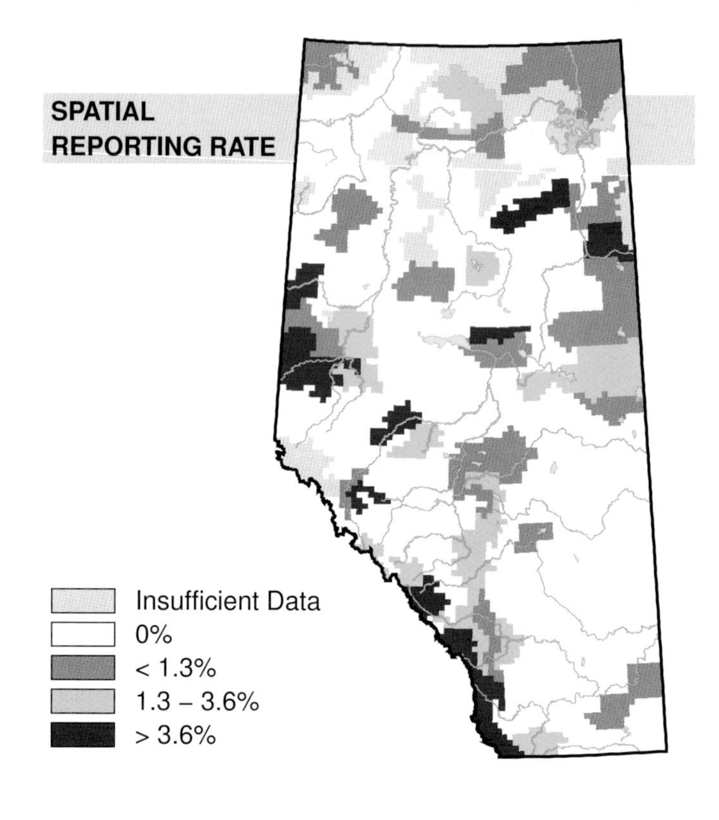

STATUS

GSWA 2000	GSWA 2005	COSEWIC Historic Rankings	COSEWIC 2007	Alberta Wildlife Act
Secure	Secure	N/A	N/A	N/A

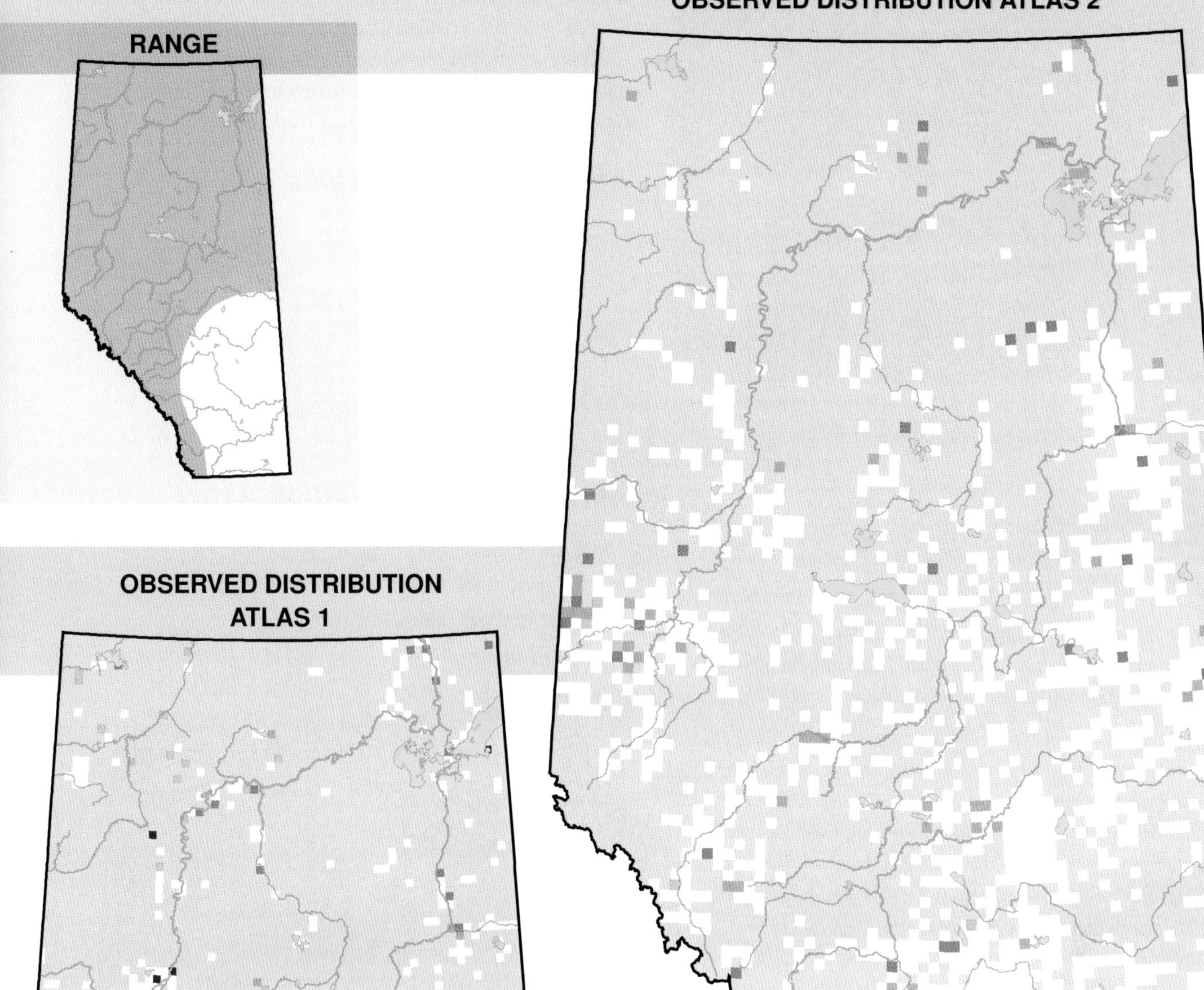

OBSERVED DISTRIBUTION ATLAS 1

Unsurveyed
Not Observed
Observed
Possible
Probable
Confirmed

Song Sparrow (*Melospiza melodia*)

Photo: Alan MacKeigan

NESTING

CLUTCH SIZE (EGGS):	3–6
INCUBATION (DAYS):	12–13
FLEDGING (DAYS):	17
NEST HEIGHT (METRES):	0

The Song Sparrow breeds in every Natural Region in Alberta. The observed distribution of this species did not change between Atlas 1 and Atlas 2.

The Song Sparrow has a widespread distribution. This species breeds in forests and shrubs that are near water. It can also breed in areas recovering from disturbance such as fires or forest harvesting, and it will sometimes nest in urban areas provided there is riparian habitat nearby. As a result of this flexibility, Song Sparrow records were fairly evenly distributed across the province. The reporting rate was lowest in the Grassland NR likely because that is the driest NR in the province and this bird nests near water.

Declines in relative abundance were detected in the Boreal Forest and Parkland NRs and, relative to other species, the Song Sparrow was observed less frequently in these NRs in Atlas 2 than in Atlas 1. The Breeding Bird Survey detected declines in abundance in Alberta and across Canada for the periods 1985–2005 and 1995–2005. These NRs were drier in Atlas 2 than in Atlas 1. The changes in precipitation, together with wetland drainage, would have reduced the amount of suitable habitat for this species in these NRs. An increase was detected in the Rocky Mountain NR where, relative to other species, the Song Sparrow was observed more frequently in that NR during Atlas 2 than during Atlas 1. Prescribed burns in this NR may have contributed to increased habitat availability for this species, but the cause for this increase is not well understood. No change was detected in the Foothills and Grassland NRs. This species is considered Secure in Alberta.

TEMPORAL REPORTING RATE

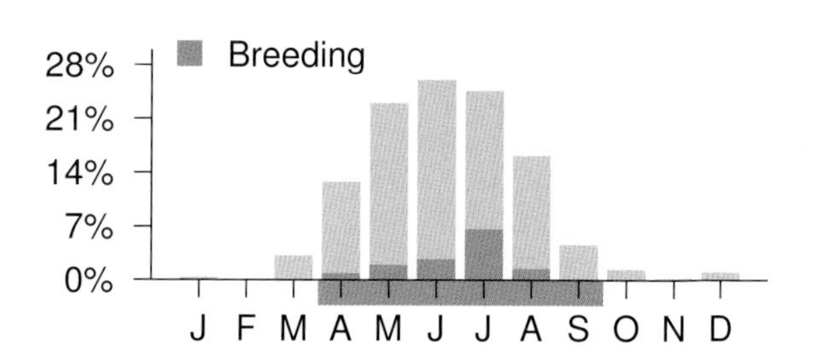

NATURAL REGIONS REPORTING RATE

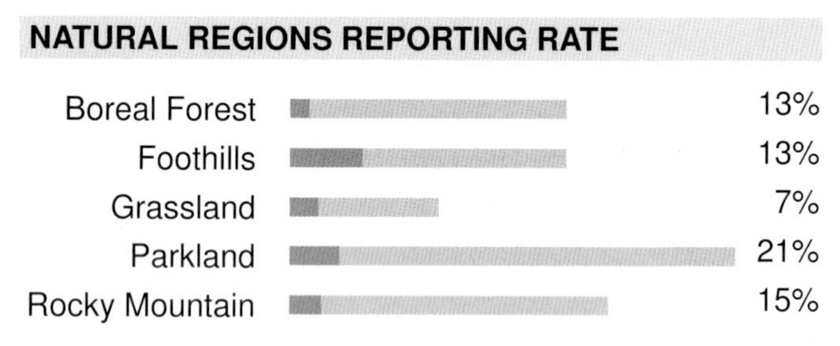

SPATIAL REPORTING RATE

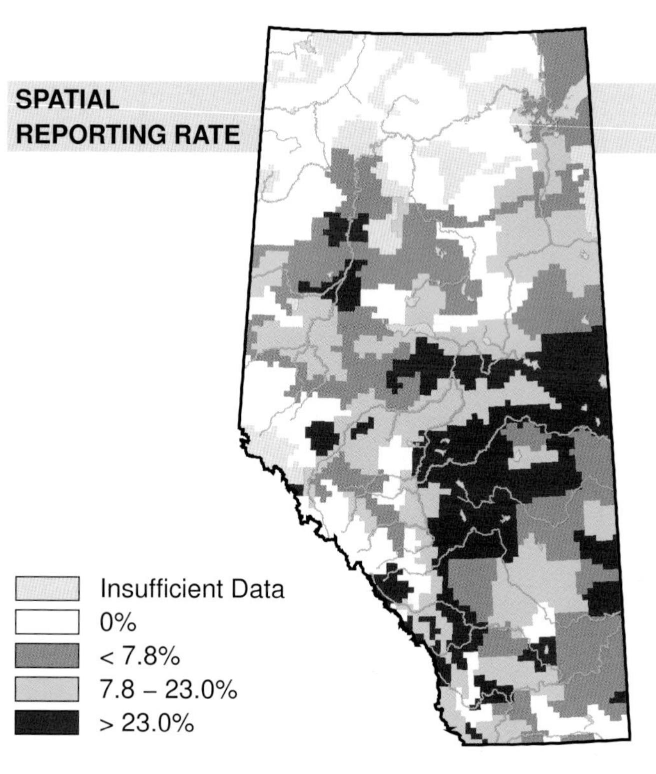

STATUS

GSWA 2000	GSWA 2005	COSEWIC Historic Rankings	COSEWIC 2007	Alberta Wildlife Act
Secure	Secure	N/A	N/A	N/A

Habitat  Nest Location  Nest Type  Diet 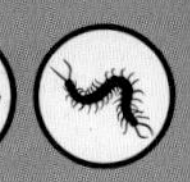

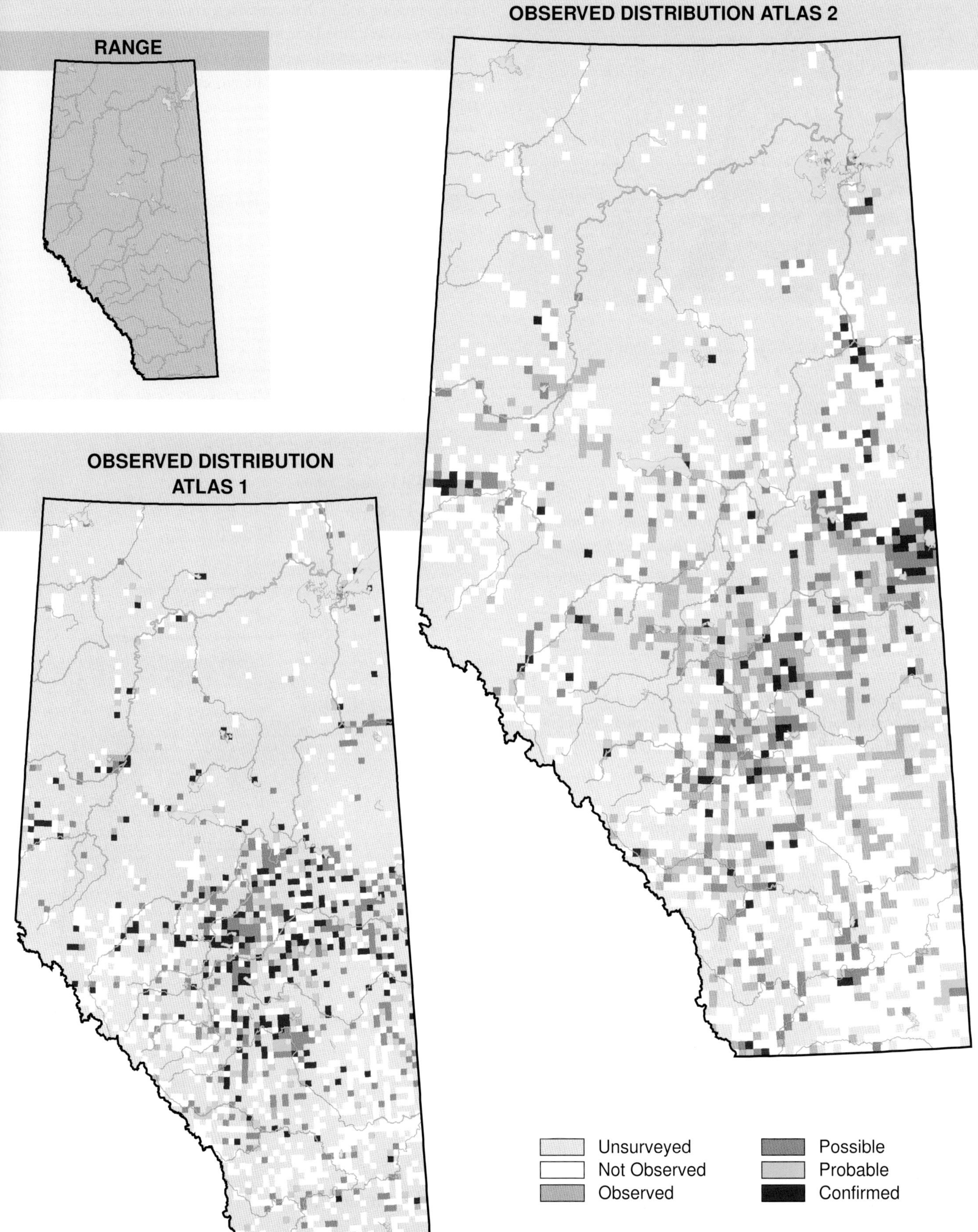

Lincoln's Sparrow (*Melospiza lincolnii*)

Photo: Randy Jensen

NESTING

CLUTCH SIZE (EGGS):	4–5
INCUBATION (DAYS):	13
FLEDGING (DAYS):	9–12
NEST HEIGHT (METRES):	0

Lincoln's Sparrow breeds in the Boreal Forest, Foothills, Parkland, and Rocky Mountain Natural Regions. The range of this species did not change between Atlas 1 and Atlas 2.

Known for its secretive nature, Lincoln's Sparrow breeds in dense shrub habitat, such as willow thickets, that is often located adjacent to a riparian area. Mixedwood forests and black spruce bogs can be used for nesting provided vegetation cover is thick and water is nearby. Due to its habitat preferences, this species was found most often in the Boreal Forest, Foothills, and Rocky Mountain NRs, and was found less often in the Grassland and Parkland NRs.

A decline in relative abundance was detected in the Boreal Forest NR even though, relative to other species, observation rates for this sparrow were similar between Atlas 2 and Atlas 1 in that NR. An increase was detected in the Grassland NR and, relative to other species, Lincoln's Sparrow was observed more frequently in Atlas 2 than in Atlas 1 in that NR. No changes were detected in the Foothills, Parkland, or Rocky Mountain NRs. However, relative to other species, Lincoln's Sparrow was observed less frequently in the Foothills and Parkland NRs, and more frequently in the Rocky Mountain NR, in Atlas 2 than in Atlas 1. The Breeding Bird Survey found an abundance decline in Alberta during the period 1995–2005. The discrepancy between what the Atlas and the Alberta Breeding Bird Survey found could be explained by differences in the temporal scales that were used to make comparisons. The time-scale of the Breeding Bird Survey (1995–2005) is narrower than the Atlas time-scale (1987–2005). Declines could be related to wetland drainage, drought, or reductions in the amount of vegetation that is retained along the boundaries with the riparian areas. This species is considered Secure in Alberta.

TEMPORAL REPORTING RATE

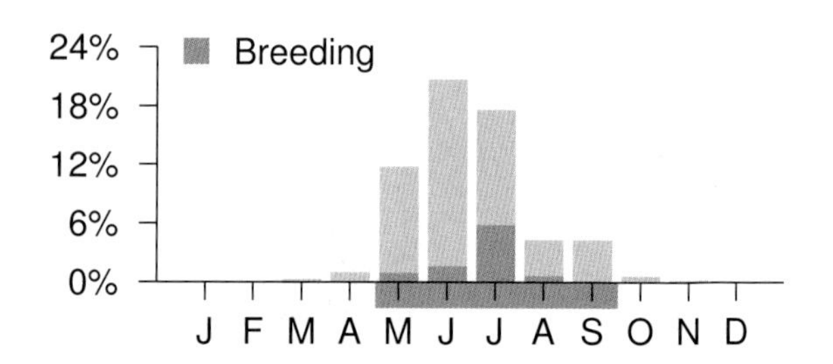

NATURAL REGIONS REPORTING RATE

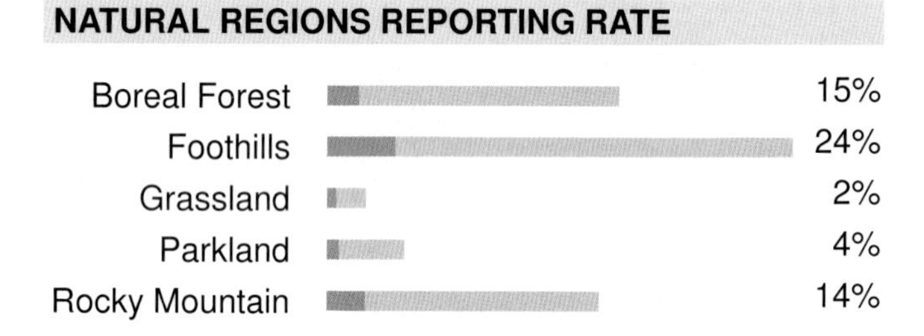

SPATIAL REPORTING RATE

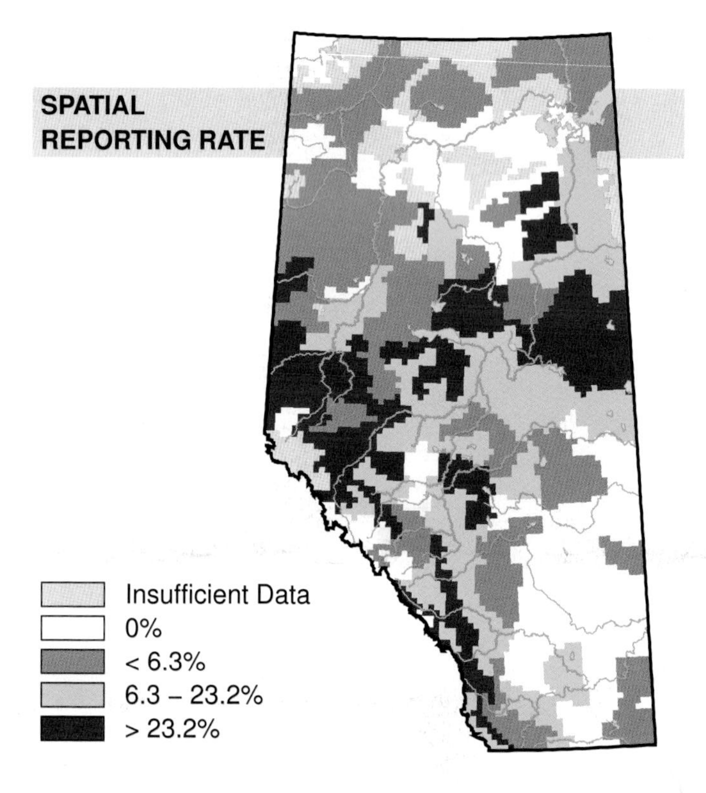

STATUS

GSWA 2000	GSWA 2005	COSEWIC Historic Rankings	COSEWIC 2007	Alberta Wildlife Act
Secure	Secure	N/A	N/A	N/A

Habitat

Nest Location

Nest Type

Diet

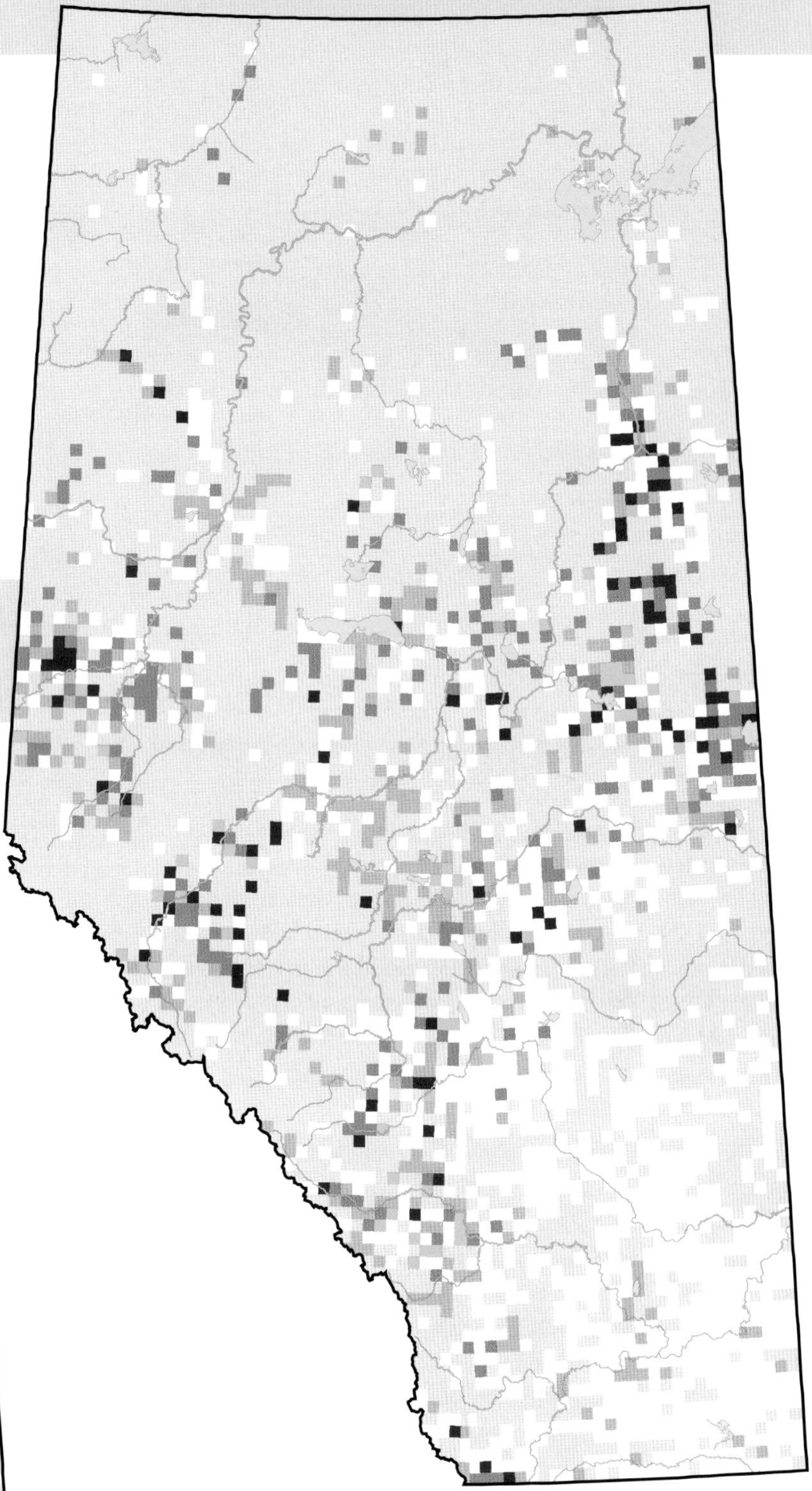

RANGE

OBSERVED DISTRIBUTION ATLAS 2

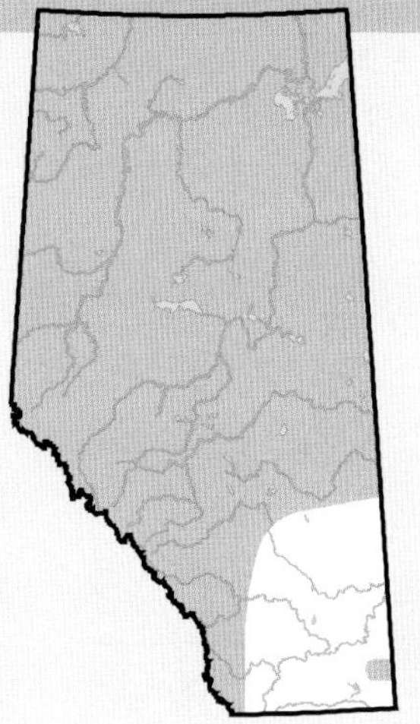

OBSERVED DISTRIBUTION ATLAS 1

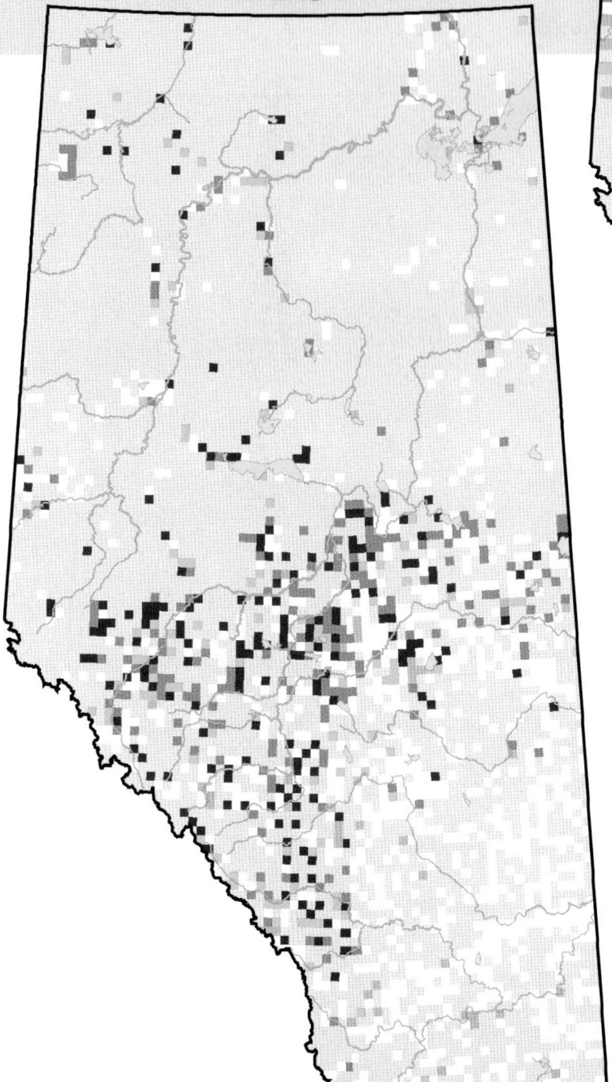

Unsurveyed
Not Observed
Observed
Possible
Probable
Confirmed

Swamp Sparrow (*Melospiza georgiana*)

Photo: Stan Gosche

NESTING

CLUTCH SIZE (EGGS):	3–5
INCUBATION (DAYS):	13
FLEDGING (DAYS):	9–10
NEST HEIGHT (METRES):	0

The Swamp Sparrow breeds in the Boreal Forest, Foothills, Parkland, and Rocky Mountain Natural Regions. The observed distribution of this species did not change between Atlas 1 and Atlas 2.

As its name suggests, the Swamp Sparrow breeds near water. This species requires the presence of shallow standing water and low dense cover, as well as tall vantage points from which it can sing, such as trees or shrubs. This species had a patchy distribution and was most often encountered near water in the Boreal Forest, Foothills, and Rocky Mountain NRs. Observations in the Parkland NR were records of non-breeders. The arid nature of the Grassland NR probably limits the presence of this species.

An increase in relative abundance was detected in the Boreal Forest NR where, relative to other species, the Swamp Sparrow was observed more frequently in Atlas 2 than in Atlas 1. This increase is unexpected as this NR was drier during Atlas 2 than during Atlas 1. The Breeding Bird Survey did not detect an abundance change in Alberta or across Canada. However, the BBS did detect an increase in the Boreal Taiga Plains Bird Conservation Region, which includes the Boreal Forest NR in Alberta, during the period 1995–2005. Lacking probable biological cause, the detected increase in the Boreal Forest NR must be interpreted cautiously and further research is needed to assess its validity and establish a cause. The Atlas detected no change in relative abundance in the Foothills, Grassland, Parkland, and Rocky Mountain NRs. This species is considered Secure in Alberta.

TEMPORAL REPORTING RATE

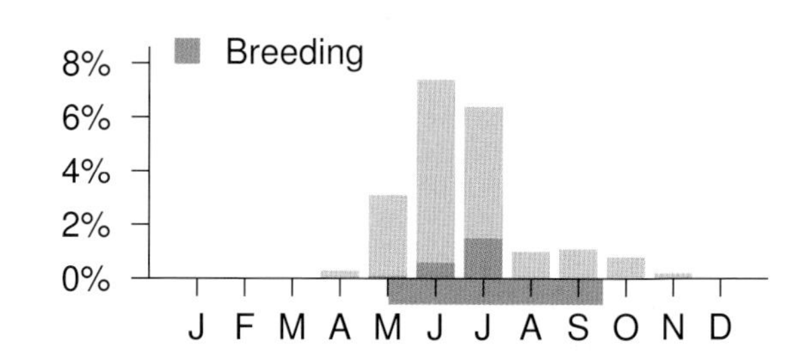

NATURAL REGIONS REPORTING RATE

Boreal Forest 2.9%
Foothills 2.9%
Parkland 0.7%

SPATIAL REPORTING RATE

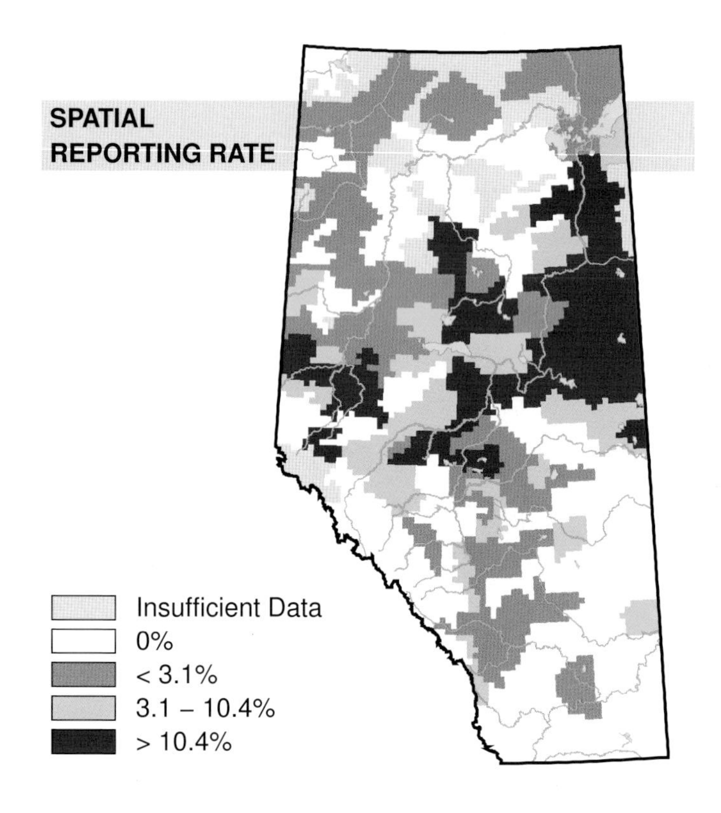

STATUS

GSWA 2000	GSWA 2005	COSEWIC Historic Rankings	COSEWIC 2007	Alberta Wildlife Act
Secure	Secure	N/A	N/A	N/A

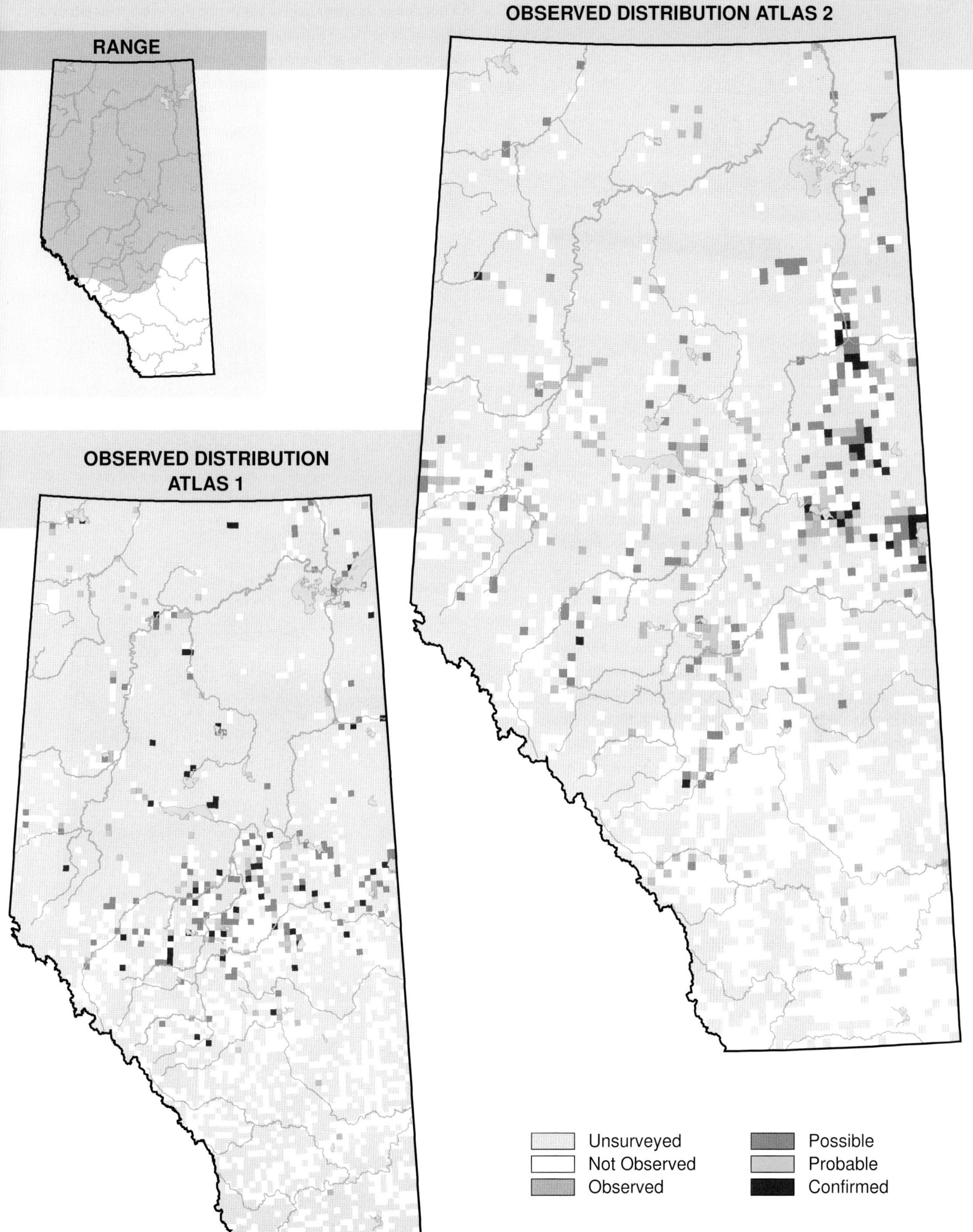
OBSERVED DISTRIBUTION ATLAS 2
RANGE
OBSERVED DISTRIBUTION ATLAS 1
Unsurveyed
Not Observed
Observed
Possible
Probable
Confirmed

White-throated Sparrow (*Zonotrichia albicollis*)

Photo: Gordon Court

NESTING

CLUTCH SIZE (EGGS):	3–5
INCUBATION (DAYS):	11–14
FLEDGING (DAYS):	9–14
NEST HEIGHT (METRES):	0.9

The White-throated Sparrow breeds in the Boreal Forest, Foothills, Parkland, and Rocky Mountain Natural Regions. The observed distribution of this species did not change between Atlas 1 and Atlas 2.

The White-throated Sparrow breeds in mixedwood forests. It tends to prefer edge habitats and densely vegetated second-growth forests after disturbances such as fire and forest harvesting, provided the forests are permitted to regenerate. Due to its preference for mixed forests, this species was commonly found in the Boreal Forest, Foothills, and Rocky Mountain NRs. It was less common in the Parkland NR. Observations made in the Grassland NR were of non-breeders, except for a few records along the periphery adjacent to the Parkland NR.

Increases in relative abundance were detected in the Foothills, Grassland, and Rocky Mountain NRs where, relative to other species, the White-throated Sparrow was observed more frequently in Atlas 2 than in Atlas 1. This species prefers edge habitat; therefore, increased forest fragmentation could be responsible for the increases that were detected. A decline was detected in the Parkland NR where, relative to other species, this species was observed less frequently in Atlas 2 than in Atlas 1. White-throated Sparrows could have declined in the Parkland as a consequence of the conversion of forests for agricultural purposes or acreage developments. No change was detected in the Boreal Forest NR. The Breeding Bird Survey did not detect an abundance change for this species in Alberta. Differences between the results of the Atlas and Alberta BBS analyses could be attributed to the fact that the data were being analyzed at different spatial scales. BBS data were analyzed at the provincial scale; whereas, given sufficient sample size, it was possible to analyze Atlas data at the smaller, Natural Region scale. The species is considered Secure in Alberta.

TEMPORAL REPORTING RATE

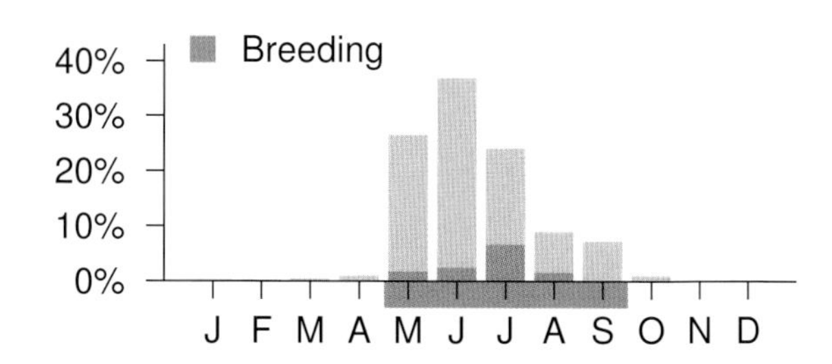

NATURAL REGIONS REPORTING RATE

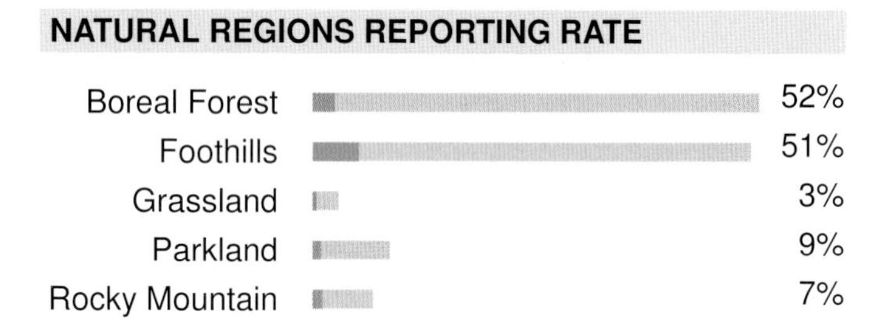

SPATIAL REPORTING RATE

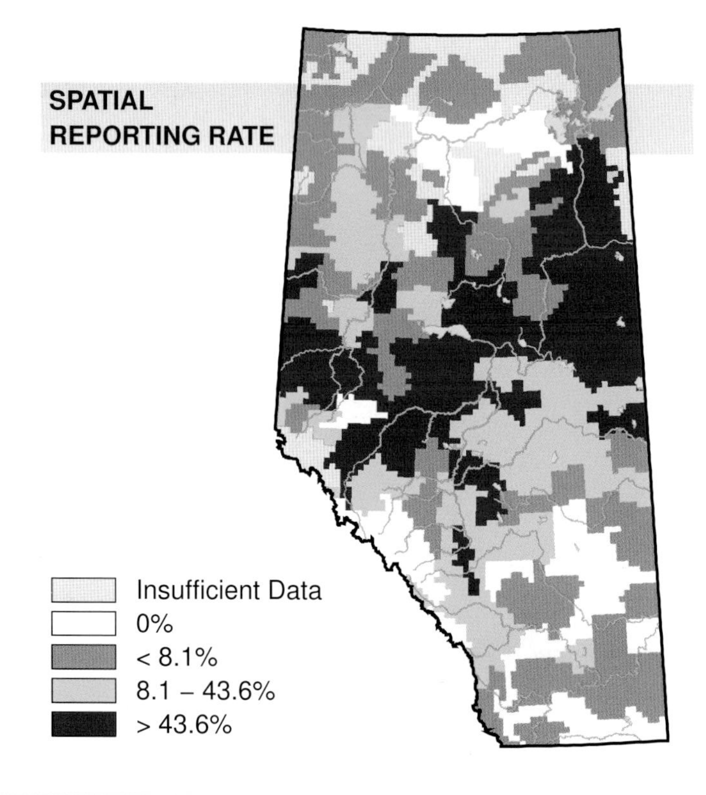

STATUS

GSWA 2000	GSWA 2005	COSEWIC Historic Rankings	COSEWIC 2007	Alberta Wildlife Act
Secure	Secure	N/A	N/A	N/A

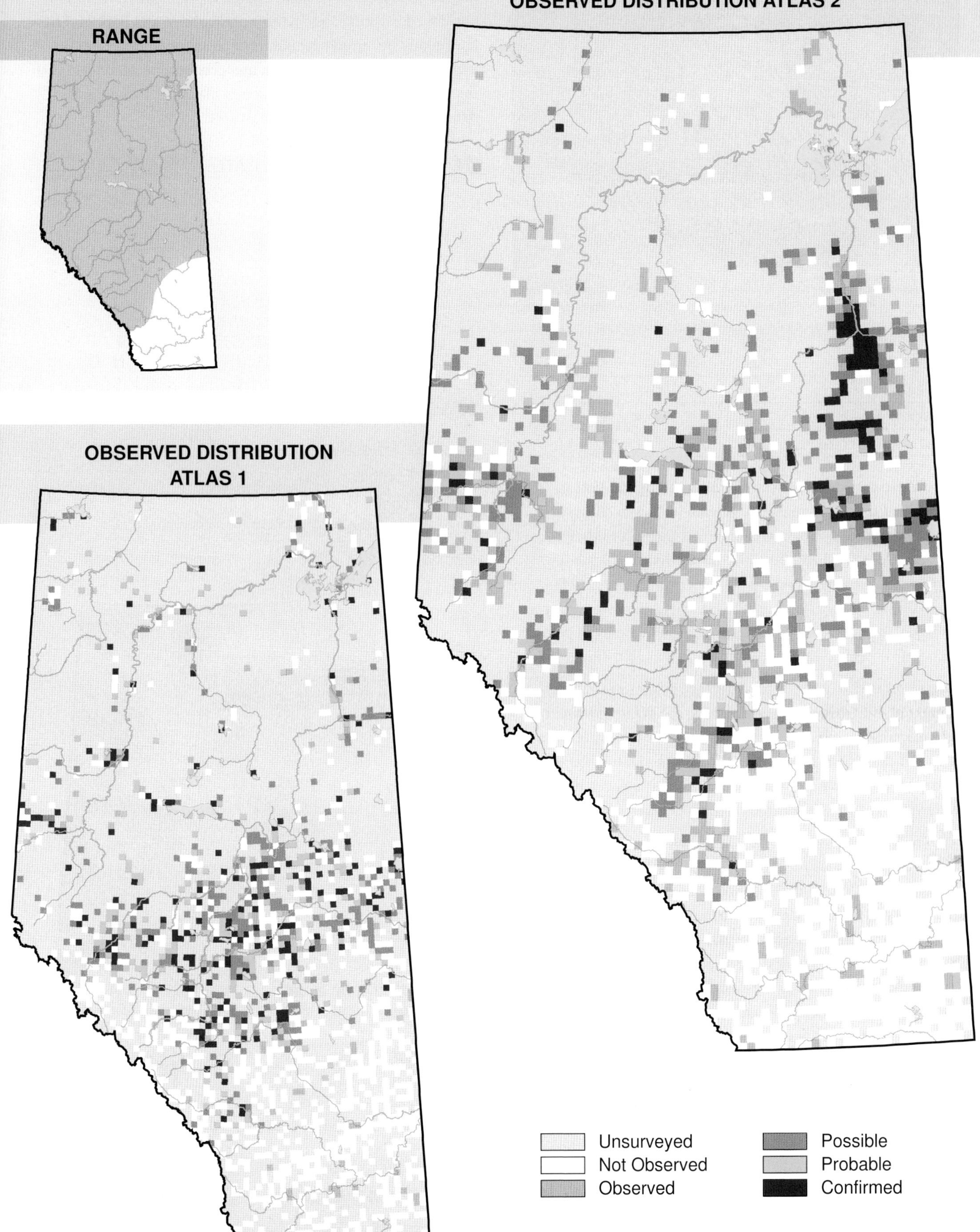
OBSERVED DISTRIBUTION ATLAS 2
RANGE
OBSERVED DISTRIBUTION
ATLAS 1
Unsurveyed
Not Observed
Observed
Possible
Probable
Confirmed

White-crowned Sparrow (*Zonotrichia leucophrys*)

Photo: Randy Jensen

NESTING

CLUTCH SIZE (EGGS):	4–6
INCUBATION (DAYS):	12–14
FLEDGING (DAYS):	25–30
NEST HEIGHT (METRES):	0

The White-crowned Sparrow breeds in the Boreal Forest, Foothills, Parkland, and Rocky Mountain Natural Regions. New evidence of possible breeding was found for this species between Cold Lake and Fort McMurray during Atlas 2. However, these new records do not necessarily indicate that the range of this species has expanded because there were only a few records and, even though there was no breeding evidence found, this species was also observed in that area during Atlas 1.

Similar to the White-throated Sparrow, the White-crowned Sparrow tends to breed in dense second-growth vegetation and it prefers edges. Suitable breeding territories should include grassy areas, bare ground, dense shrubs or young conifers, water, and tall coniferous trees. This species was most often encountered in the Rocky Mountain NR but it was also found in the Boreal Forest, Foothills, Grassland, and Parkland NRs.

Declines in relative abundance were detected in the Foothills and Rocky Mountain NRs and increases were detected in the Boreal Forest, Grassland, and Parkland NRs. The Breeding Bird Survey detected no change in abundance in Alberta or nationally for the period 1985–2005. Atlas-detected declines in the Foothills and Rocky Mountain NRs are probably related to coverage differences in these NRs between Atlas 1 and 2 and may not indicate actual changes in species abundance. Increases in the Grasslands NR are likely due in part to migratory variation, as many records in this NR were of non-breeders. Increases in the Boreal Forest and Parkland NRs have likely occurred because this species benefits from increased forest fragmentation and does well in second-growth forests post-disturbance. The species is listed as Secure in Alberta.

TEMPORAL REPORTING RATE

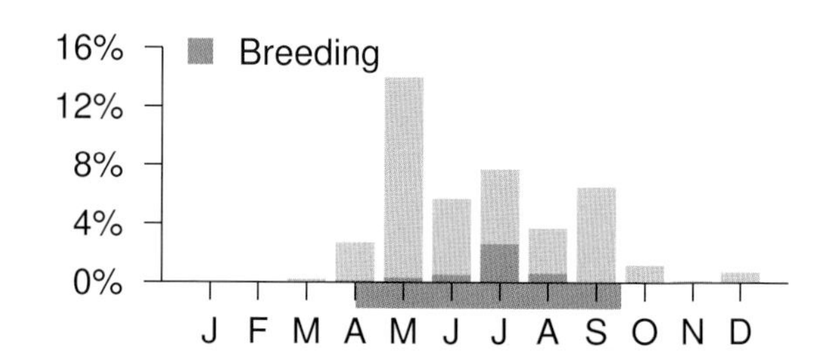

NATURAL REGIONS REPORTING RATE

Region	Rate
Boreal Forest	1%
Foothills	3%
Grassland	4%
Parkland	4%
Rocky Mountain	26%

SPATIAL REPORTING RATE

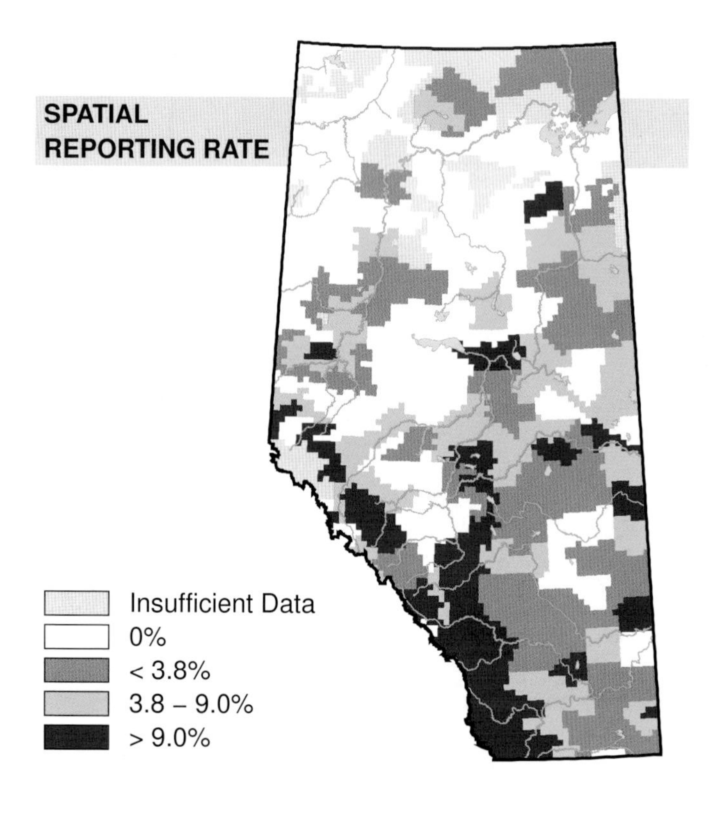

STATUS

GSWA 2000	GSWA 2005	COSEWIC Historic Rankings	COSEWIC 2007	Alberta Wildlife Act
Secure	Secure	N/A	N/A	N/A

Golden-crowned Sparrow (*Zonotrichia atricapilla*)

Photo: Royal Alberta Museum

NESTING

CLUTCH SIZE (EGGS):	4–5
INCUBATION (DAYS):	11–13
FLEDGING (DAYS):	9–11
NEST HEIGHT (METRES):	0–0.8

The Golden-crowned Sparrow breeds in the Rocky Mountain Natural Region in Alberta. It breeds in Jasper and Banff national parks but there are no breeding records for Waterton Lakes National Park. The distribution of this species did not change between Atlas 1 and Atlas 2.

The Golden-crowned Sparrow breeds in the shrubby habitat that is found at treeline in mountainous areas. For this reason, this species was found most often in the Rocky Mountain NR. Observations made in the Grassland and Parkland NR were of non-breeders.

Changes in relative abundance were not detected for this species in any natural region. However, these results should be interpreted with caution because the sample size was small for this species. Relative to other species, Golden-crowned Sparrows were observed more frequently in the Grassland NR and less frequently in the Parkland and Rocky Mountain NRs in Atlas 2 than in Atlas 1. There were not enough Golden-crowned Sparrow records to detect population changes using Breeding Bird Survey data in Alberta, and no change was detected across Canada. Due to the specificity of this bird's habitat requirements and the remote distribution of this habitat, the species is not effectively monitored by large, wide-spread, multi-species surveys such as the Atlas or the Breeding Bird Survey. Surveys specific to Golden-crowned Sparrows would be needed to determine adequately the population trends for this species.

Dunn (2005) reports that the Golden-crowned Sparrow is a species that was reviewed because of its uncertain population status and restricted winter range, but it is apparently increasing and there are no obvious threats.

Because this species breeds only in relatively isolated parts of Alaska and western Canada, it is unlikely that human activities on its breeding grounds will influence its status in the foreseeable future (Norment et al., 1998).

This species is considered Secure in Alberta.

TEMPORAL REPORTING RATE

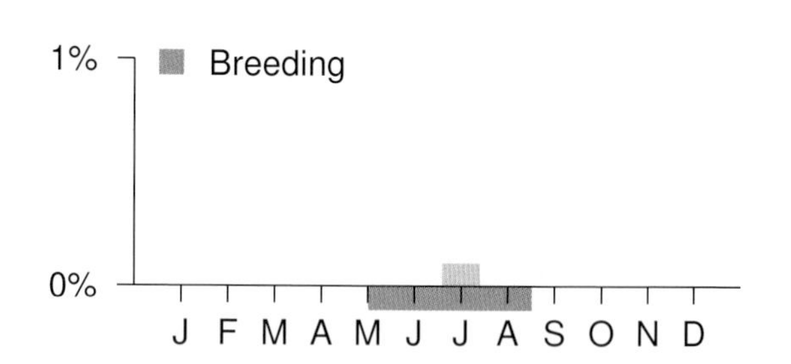

NATURAL REGIONS REPORTING RATE

Rocky Mountain 0.5%

SPATIAL REPORTING RATE

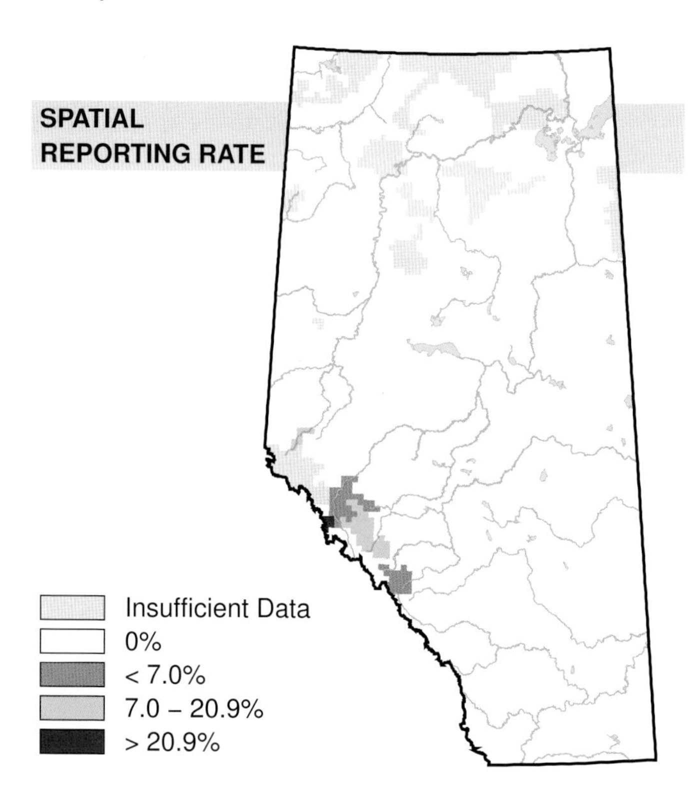

STATUS

GSWA 2000	GSWA 2005	COSEWIC Historic Rankings	COSEWIC 2007	Alberta Wildlife Act
Secure	Secure	N/A	N/A	N/A

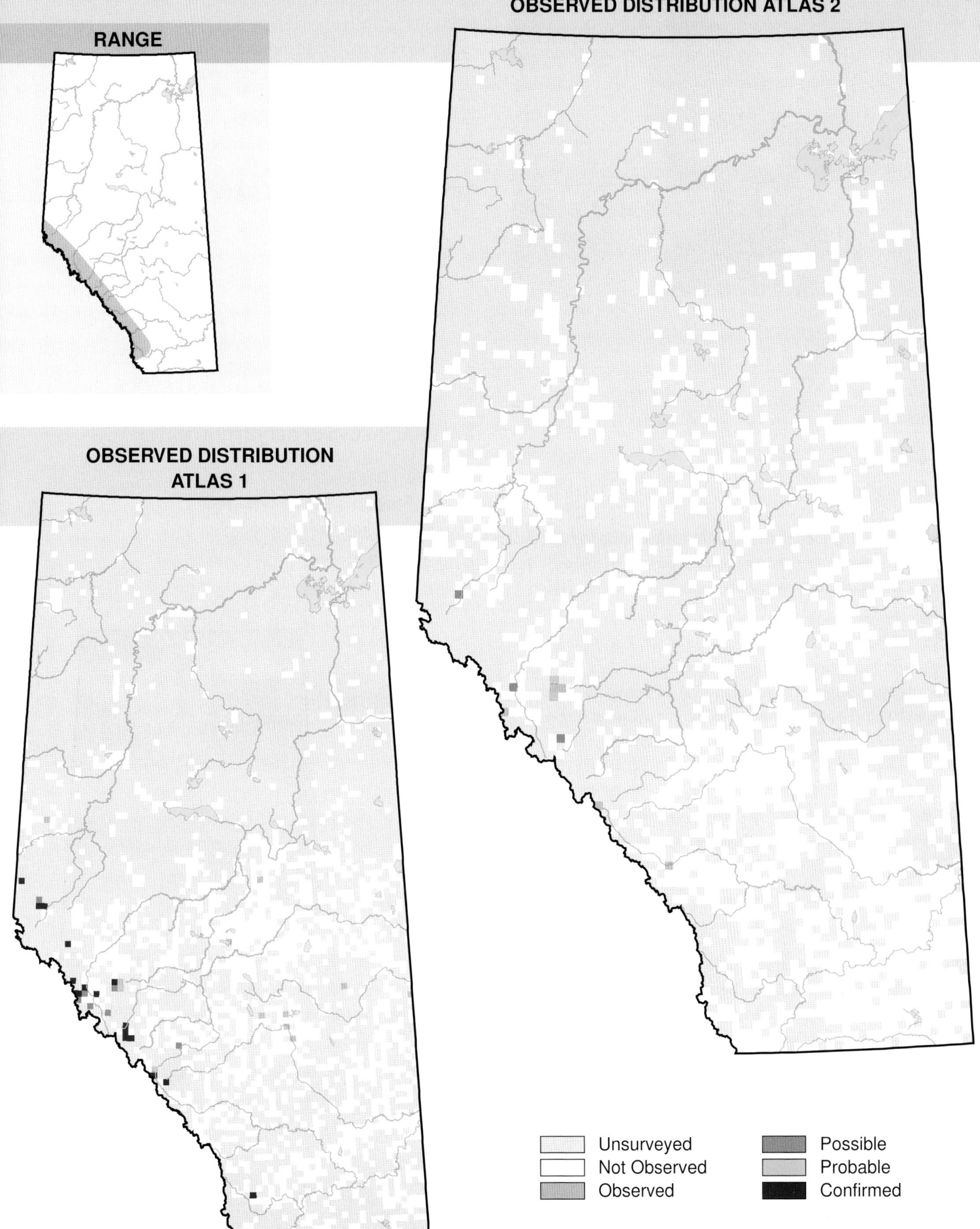
RANGE
OBSERVED DISTRIBUTION ATLAS 2
OBSERVED DISTRIBUTION ATLAS 1
Unsurveyed
Not Observed
Observed
Possible
Probable
Confirmed

Dark-eyed Junco (*Junco hyemalis*)

Photo: Gerald Romanchuk

NESTING

CLUTCH SIZE (EGGS):	3–6
INCUBATION (DAYS):	11–13
FLEDGING (DAYS):	9–13
NEST HEIGHT (METRES):	0

The Dark-eyed Junco breeds in the Boreal Forest, Foothills, Parkland, and Rocky Mountain Natural Regions in Alberta. The observed distribution of this species decreased in Atlas 2 in the northwestern part of the province around High Level. It is unclear whether this represents an actual range contraction or whether the presence of Dark-eyed Juncos in the northwest was missed during Atlas 2.

Dark-eyed Junco distribution is widespread and the birds breed in a variety of habitats. It nests in forest habitats with dense ground cover. Therefore, forests in late successional stages, with closed canopies, are less suitable than those in earlier successional stages. Conifer trees are usually present in the forests where Dark-eyed Juncos breed. This species was most often observed in the Boreal Forest, Foothills, and Rocky Mountain NRs. Observation frequencies were lower in the Parkland NR, likely because conifer trees are less common there. Observations made in the Grassland NR were non-breeders with the exception of a few isolated breeding records.

Declines in relative abundance were detected in the Boreal Forest, Foothills, and Parkland NRs and, relative to other species, Dark-eyed Juncos were observed less frequently in Atlas 2 in these NRs. No change was detected in the Grassland and Rocky Mountain NRs. The Breeding Bird Survey did not detect an abundance change for this species in Alberta. The declines detected by the Atlas were mirrored in the Canada-wide decline detected by the BBS during the period 1985–2005. The cause of the declines is unclear, as it prefers earlier successional stage forests. In Alberta's forests, industrial development is reducing mean forest ages through harvesting, roads, and cut lines. Further study is needed to understand the cause of this species' apparent decline. The provincial status of this species is Secure.

TEMPORAL REPORTING RATE

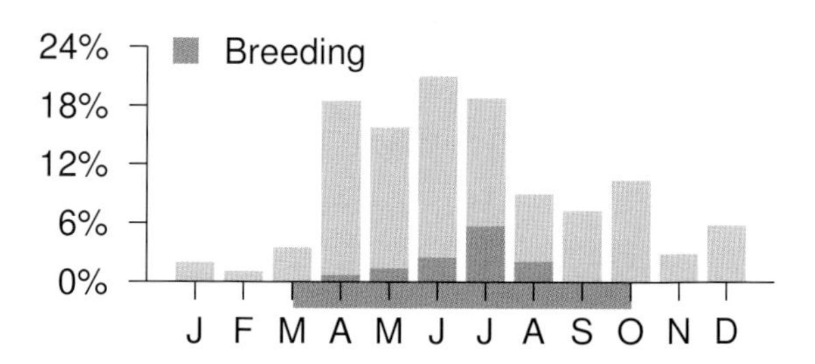

NATURAL REGIONS REPORTING RATE

Boreal Forest	10%
Foothills	39%
Grassland	2%
Parkland	4%
Rocky Mountain	53%

SPATIAL REPORTING RATE

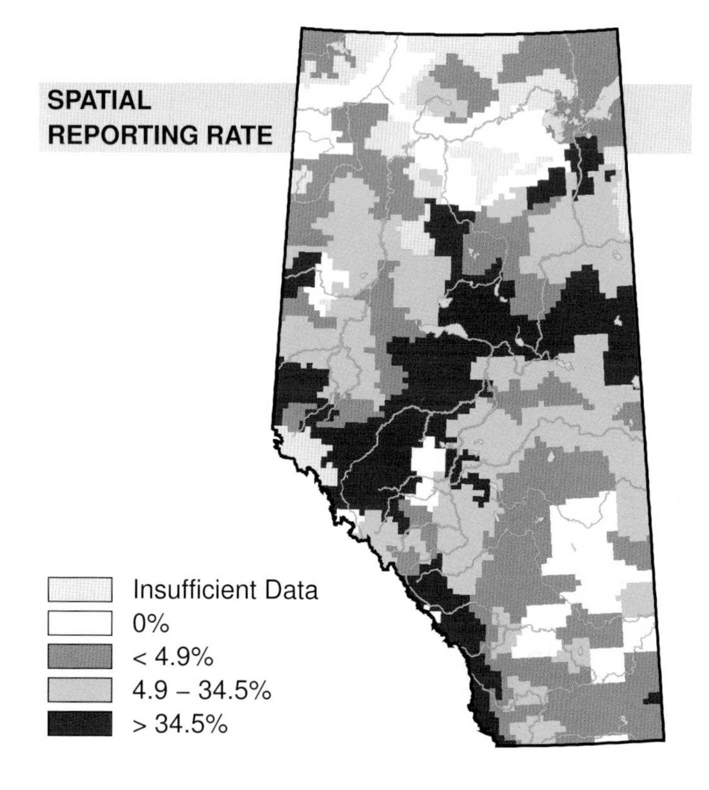

STATUS

GSWA 2000	GSWA 2005	COSEWIC Historic Rankings	COSEWIC 2007	Alberta Wildlife Act
Secure	Secure	N/A	N/A	N/A

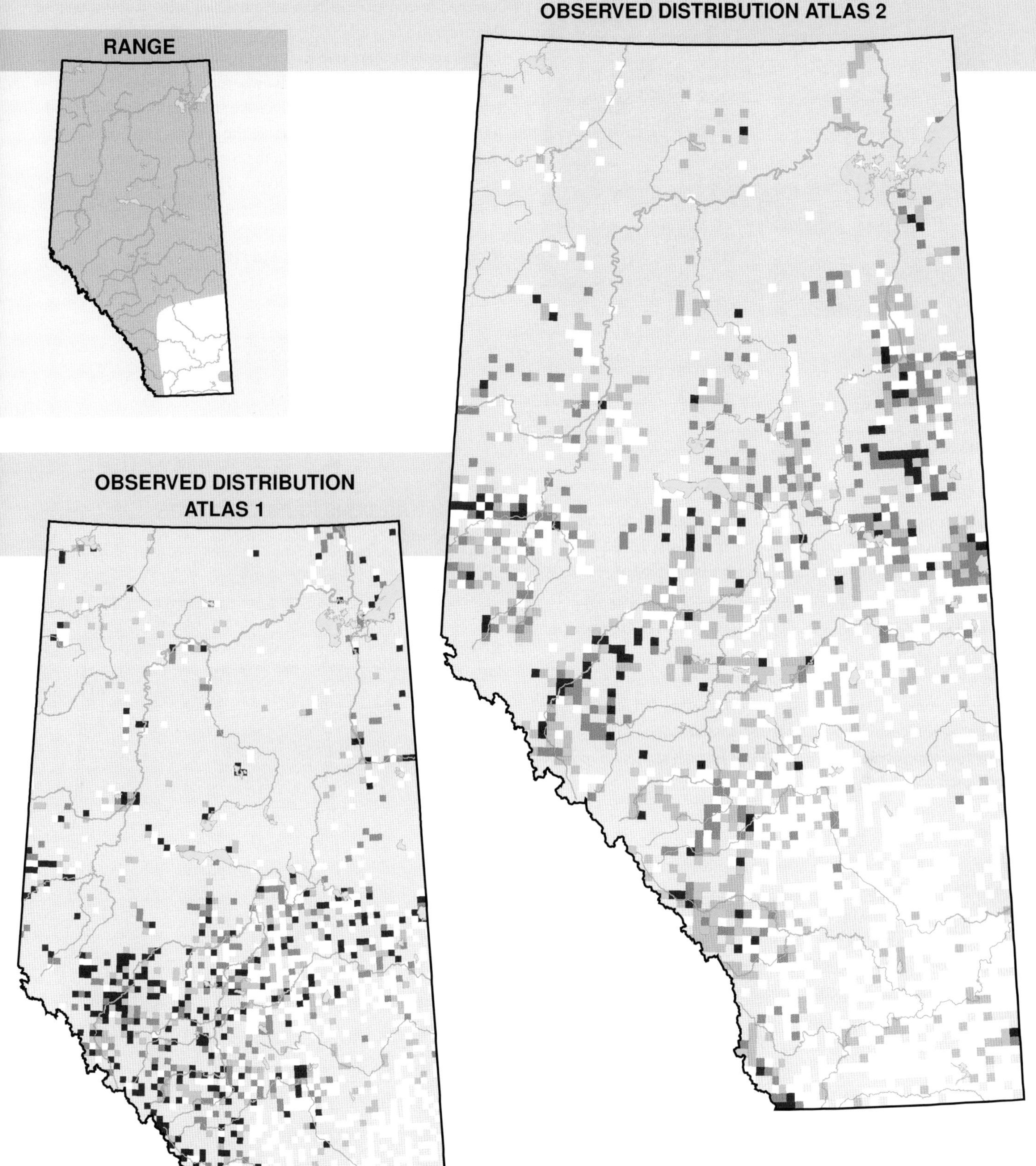

RANGE
OBSERVED DISTRIBUTION ATLAS 2
OBSERVED DISTRIBUTION ATLAS 1
Unsurveyed
Not Observed
Observed
Possible
Probable
Confirmed

McCown's Longspur (*Calcarius mccownii*)

Photo: Royal Alberta Museum

NESTING

CLUTCH SIZE (EGGS):	3–4
INCUBATION (DAYS):	12
FLEDGING (DAYS):	14
NEST HEIGHT (METRES):	0

McCown's Longspur is restricted to the Grassland Natural Region. Its distribution does not appear to have changed between Atlas 1 and Atlas 2.

The highest reporting rates observed were in a strip down the centre of the Grassland NR. This is reasonably consistent with this bird's known habitat preference for open (lacking shrub) and sparsely vegetated sites (With, 1994). Areas with greater precipitation are likely to have cover that is too tall and/or shrubby, while sandy soils along the eastern edge of the province and sagebrush areas in the southeast are likely to have too much woody cover to suit this species.

Increases in relative abundance were detected in the Grassland NR. The Breeding Bird Survey did not detect a trend and neither did Grassland Bird Monitoring. An increase seems unlikely for a number of reasons. Native grassland cover has been reduced in all subregions of the Grassland NR (Watmough and Schmoll, 2007). Even their secondary habitat of crop is less available now than in the past because some has been converted to planted pastures or hayfields which have far too high a cover to be suitable for a species that prefers bare spaces. A study near Lethbridge found that McCown's Longspur does not respond well to many of the new agricultural practices (conservation tillage, fall-sown crops) designed to conserve or take advantage of limited soil moisture (Martin and Forsyth, 2003); this species prefers fields where the soil remains bare due to frequent tillage. Ironically, those birds that accept the crop fields, where summer cover is maintained, are more productive. Although both the 2000 and 2005 Provincial status reports rate the species as Secure, it was evaluated by the Committee on the Status of Endangered Wildlife in Canada and was accorded a status of Special Concern in 2006.

TEMPORAL REPORTING RATE

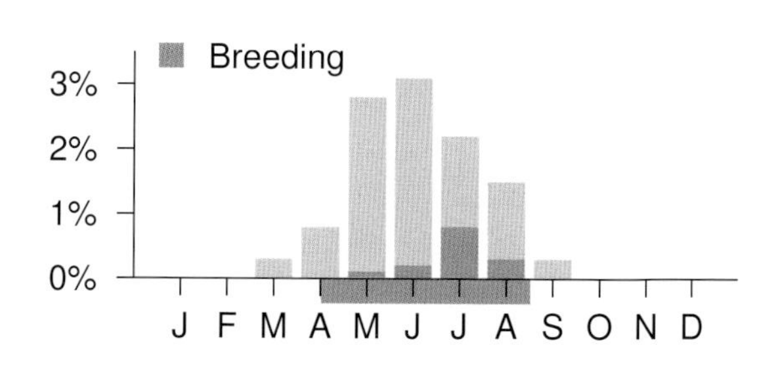

NATURAL REGIONS REPORTING RATE

Grassland 3.8%

SPATIAL REPORTING RATE

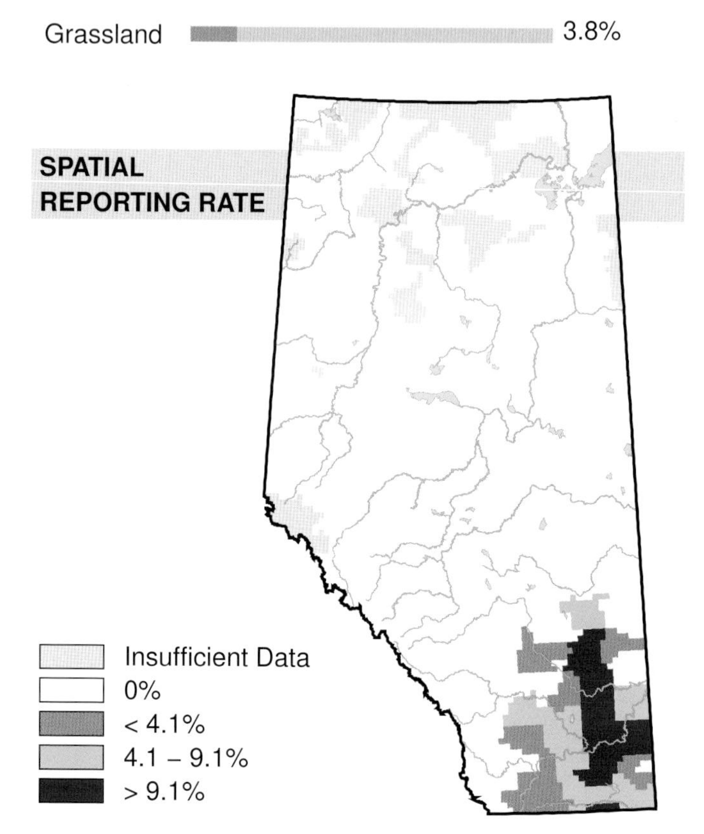

STATUS

GSWA 2000	GSWA 2005	COSEWIC Historic Rankings	COSEWIC 2007	Alberta Wildlife Act
Secure	Secure	N/A	Special Concern	N/A

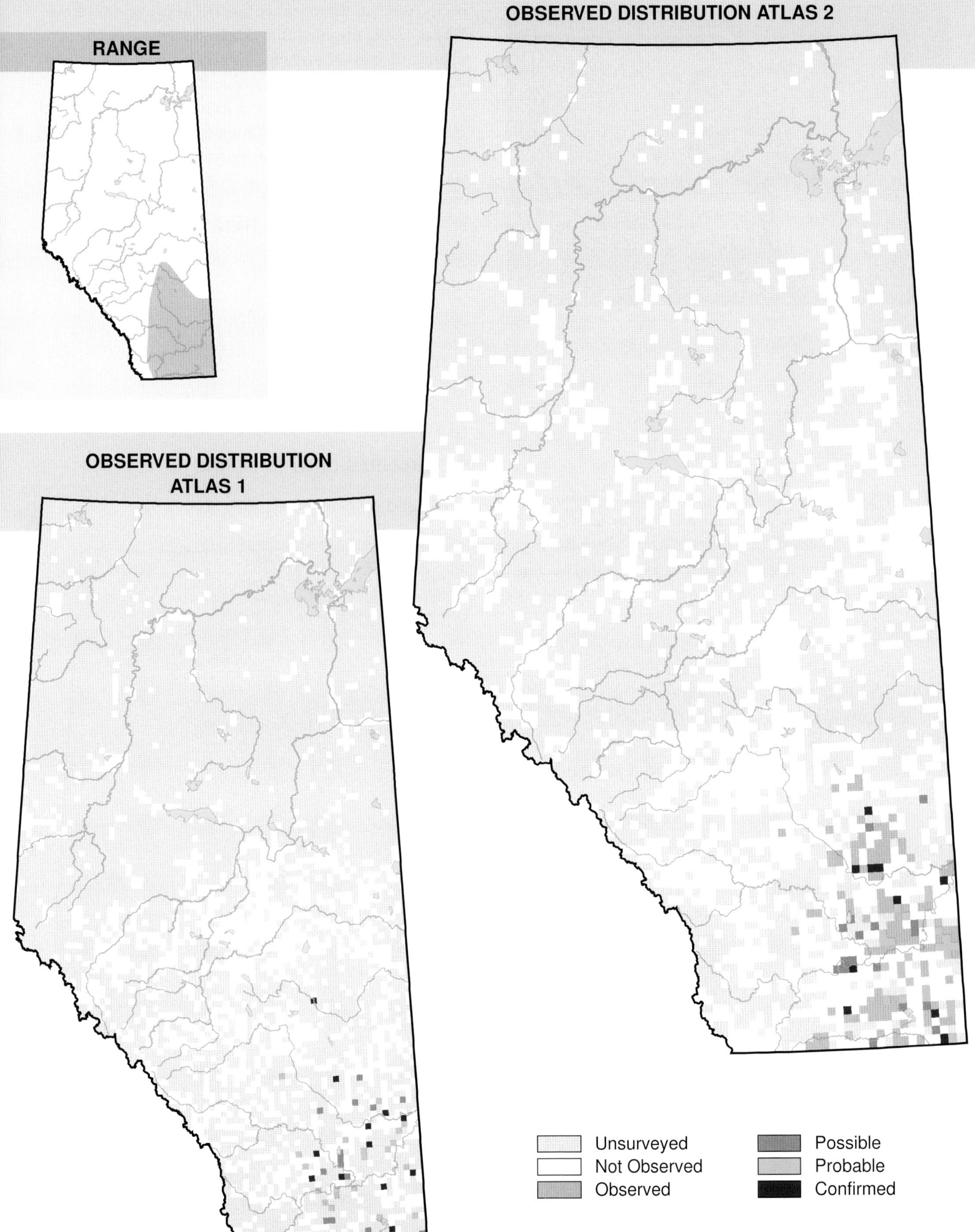
OBSERVED DISTRIBUTION ATLAS 2
RANGE
OBSERVED DISTRIBUTION ATLAS 1
Unsurveyed
Not Observed
Observed
Possible
Probable
Confirmed

Chestnut-collared Longspur (*Calcarius ornatus*)

Photo: Alan MacKeigan

NESTING

CLUTCH SIZE (EGGS):	3–6
INCUBATION (DAYS):	10–12
FLEDGING (DAYS):	24
NEST HEIGHT (METRES):	0

The Chestnut-collared Longspur is almost entirely confined to the Grassland Natural Region. Its distribution in Atlas 2 appears to have contracted slightly within that NR from the limits in Atlas 1. The contraction may be exaggerated because some squares where the bird was observed in Atlas 1 were not visited in Atlas 2, but there are also squares from the western and northern portions of the Atlas 1 distribution which were checked in Atlas 2 but no observations of the species were obtained. Wetter conditions in these areas of more productive soils may have encouraged woody growth or created cover that was too tall.

The highest reporting rates are in the centre of the Grassland NR. Moisture conditions to the west may create cover that is too lush and the sandy soils in the east support more shrub. Areas occupied by Chestnut-collared Longspurs usually have grassy cover less than 20 or 30 cm tall and only the occasional shrub (Hill and Gould, 1997). Small isolated shrubs are useful as perches from which to launch into the air and float down with wide-spread tail while singing but, like most grassland birds of the Great Plains, it avoids areas with extensive shrub cover.

No changes in relative abundance were detected in the Grassland and Parkland NRs although, relative to other species, this bird was observed less frequently in Atlas 2 than in Atlas 1. The Breeding Bird Survey detected a decline in Alberta and across Canada for the period 1985–2005. Other evidence suggesting a decline derives from the somewhat reduced distribution and losses of native grassland since 1985 throughout the Grassland NR (Watmough and Schmoll, 2007). Provincial status reports for the Chestnut-collared Longspur rank it as Secure.

TEMPORAL REPORTING RATE

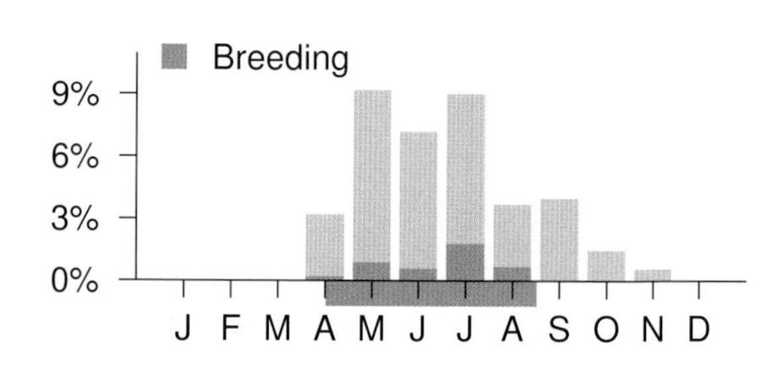

NATURAL REGIONS REPORTING RATE

Grassland 15%

SPATIAL REPORTING RATE

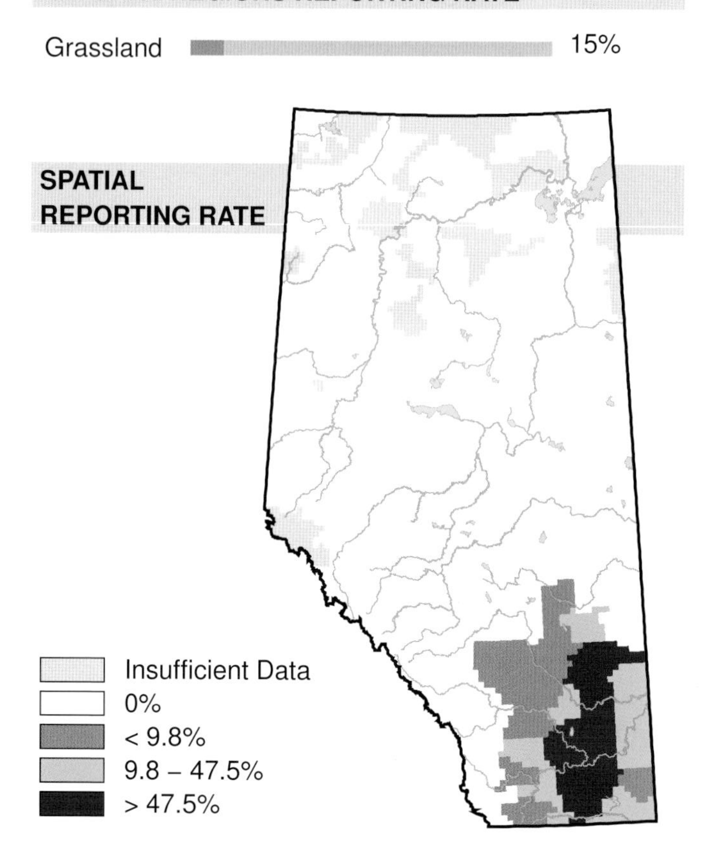

STATUS

GSWA 2000	GSWA 2005	COSEWIC Historic Rankings	COSEWIC 2007	Alberta Wildlife Act
Secure	Secure	N/A	N/A	N/A

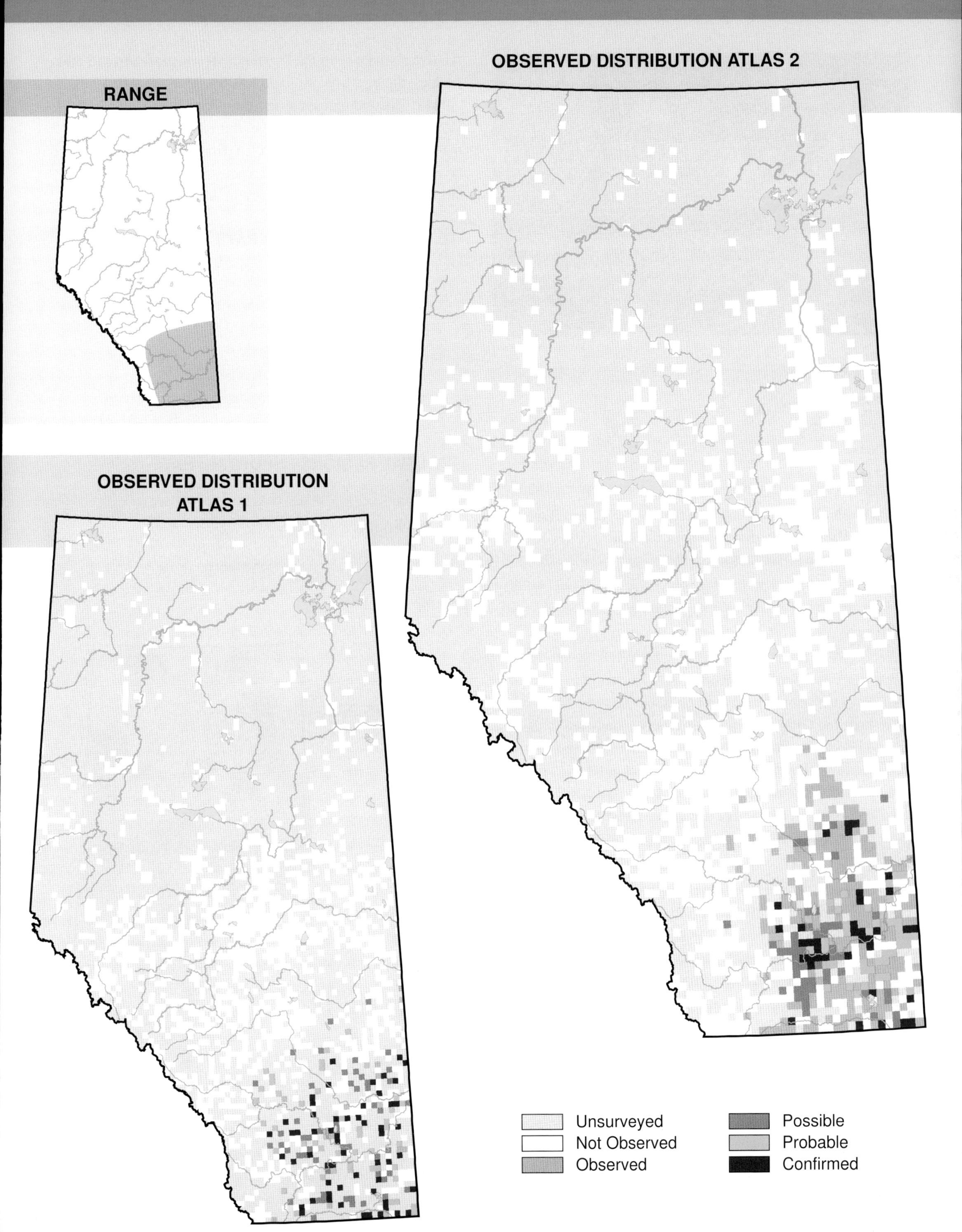
Habitat
Nest Location
Nest Type
Diet
RANGE
OBSERVED DISTRIBUTION ATLAS 2
OBSERVED DISTRIBUTION ATLAS 1
Unsurveyed
Not Observed
Observed
Possible
Probable
Confirmed

Rose-breasted Grosbeak (*Pheucticus ludovicianus*)

Photo: Robert Gehlert

NESTING

CLUTCH SIZE (EGGS):	3–5
INCUBATION (DAYS):	12–13
FLEDGING (DAYS):	9–12
NEST HEIGHT (METRES):	1.5–4.5

The Rose-breasted Grosbeak breeds in every Natural Region in Alberta. The observed distribution of this species did not change between Atlas 1 and Atlas 2.

This grosbeak breeds in deciduous and mixedwood forests and is rarely found in coniferous dominated forests. This species tends to be found in second-growth, and often breeds near forest edges. It does not seem to be negatively affected by forest fragmentation. The breeding area is usually near wetlands such as creeks, rivers, ponds, or marshes. The Rose-breasted Grosbeak was most often encountered in the Boreal Forest and Foothills NRs. However, it was also quite common in the Parkland and Rocky Mountain NRs in areas where deciduous forests occur. In the Grassland NR this species was found in tracts of deciduous forest along rivers.

Increases in relative abundance were detected in the Boreal Forest, Foothills, Grassland, and Rocky Mountain NRs where, relative to other species, the Rose-breasted Grosbeak was observed more frequently in Atlas 2. Similar to the Atlas, the Breeding Bird Survey revealed an abundance increase in Alberta during the periods 1985–2005 and 1995–2005. Increases are likely the result of forest management that has increased the quantity of second-growth stands and has shifted forests towards younger age-classes. Most of the records from the Grassland NR were non-breeders; so, the increase may reflect migratory variability. The Atlas detected a decline in relative abundance in the Parkland NR where, relative to other species, the bird was observed less frequently in Atlas 2 than in Atlas 1. The biological cause for this decrease is unknown. However, with greater coverage of the Parkland NR in Atlas 2 than in Atlas 1, the Atlas-detected decrease does not appear to be a function of coverage differences. Rose-breasted Grosbeaks are considered Secure in Alberta.

TEMPORAL REPORTING RATE

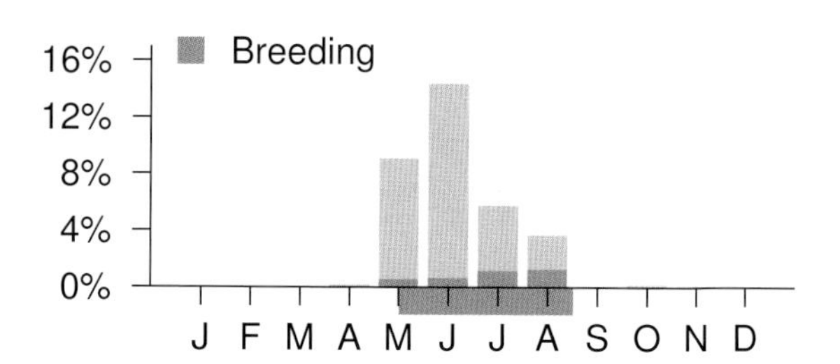

NATURAL REGIONS REPORTING RATE

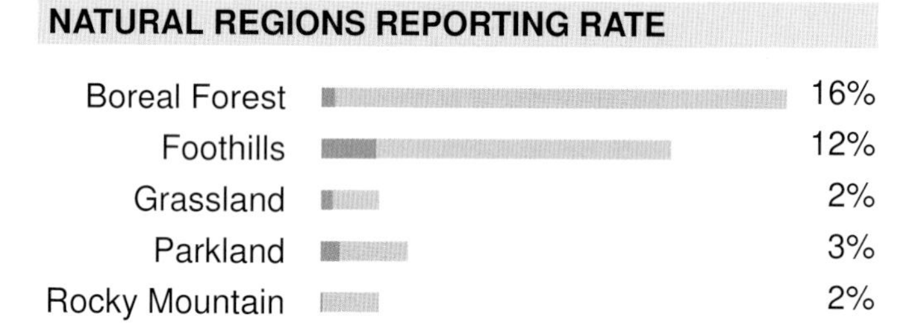

SPATIAL REPORTING RATE

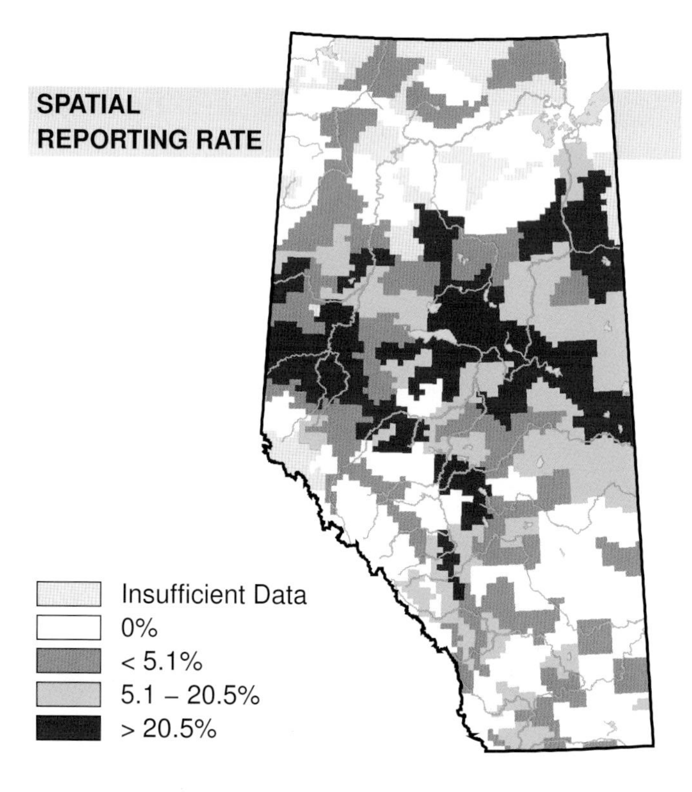

STATUS

GSWA 2000	GSWA 2005	COSEWIC Historic Rankings	COSEWIC 2007	Alberta Wildlife Act
Secure	Secure	N/A	N/A	N/A

Habitat  Nest Location  Nest Type  Diet 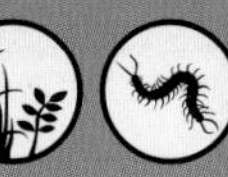

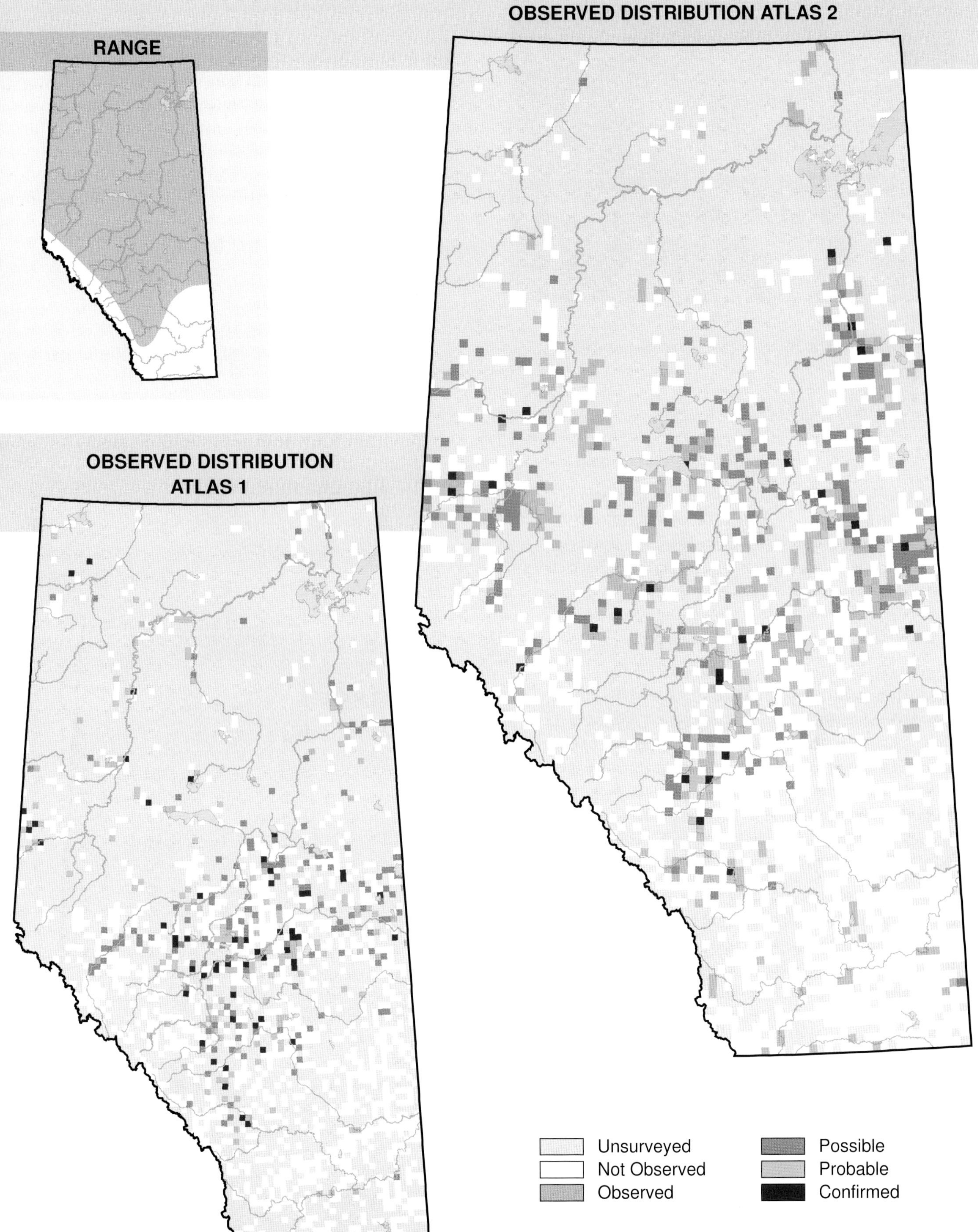

Black-headed Grosbeak (*Pheucticus melanocephalus*)

Photo: Royal Alberta Museum

NESTING

CLUTCH SIZE (EGGS):	3–4
INCUBATION (DAYS):	12–13
FLEDGING (DAYS):	12
NEST HEIGHT (METRES):	1–6

The Black-headed Grosbeak breeds in the very southern parts of the province and, as in Atlas 1, the majority of records came from Waterton Lakes National Park. There was breeding recorded in the Foothills Natural Region in Atlas 2, which was not recorded in Atlas 1. Also in Atlas 2, there were extra-limital observations in the Wainwright area, a fair distance north and east of its expected range.

It was found most frequently in the Rocky Mountain NR, and less frequently in the Foothills and Grassland NRs. This species occupies diverse habitats, but prefers cottonwood groves and other riparian habitats in dry grassland, openings in mature pine forest and aspen groves, especially in mountain valleys. This habitat is most prevalent in southern Alberta, from Waterton Lakes National Park east to the Cypress Hills. The few records found in the Parkland NR are not considered to be breeding records.

No changes in relative abundance of the Black-headed Grosbeak were detected, but relative to other species, Black-headed Grosbeaks were observed more frequently in Atlas 2 than in Atlas 1. The Breeding Bird Survey lacked sufficient records to assess changes in abundance in Alberta, and it detected no changes in abundance from 1985–2005 in British Columbia or Canada, although Dunn (2005) reports that there has been a significant range-wide increase in population affecting the northern edge of the range in Canada as well.

This species is relatively tolerant of human disturbance and will breed in yards and gardens, if there is adequate cover for nesting, and trees and understorey shrubs for gleaning insects. Black-headed Grosbeaks benefit from irrigation of arid areas and the opening of dense forests. Indirect evidence suggests that suitable breeding habitat may limit reproductive opportunities for both males and females and hence regulate populations (Hill, 1995). This species is considered Secure in Alberta.

TEMPORAL REPORTING RATE

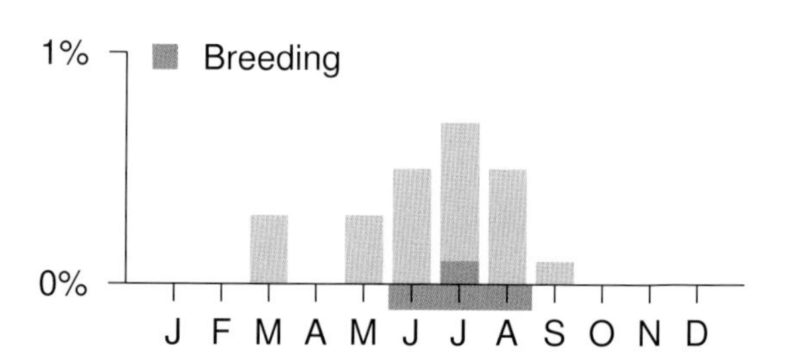

NATURAL REGIONS REPORTING RATE

Grassland	0.6%
Parkland	0.8%
Rocky Mountain	3.5%

SPATIAL REPORTING RATE

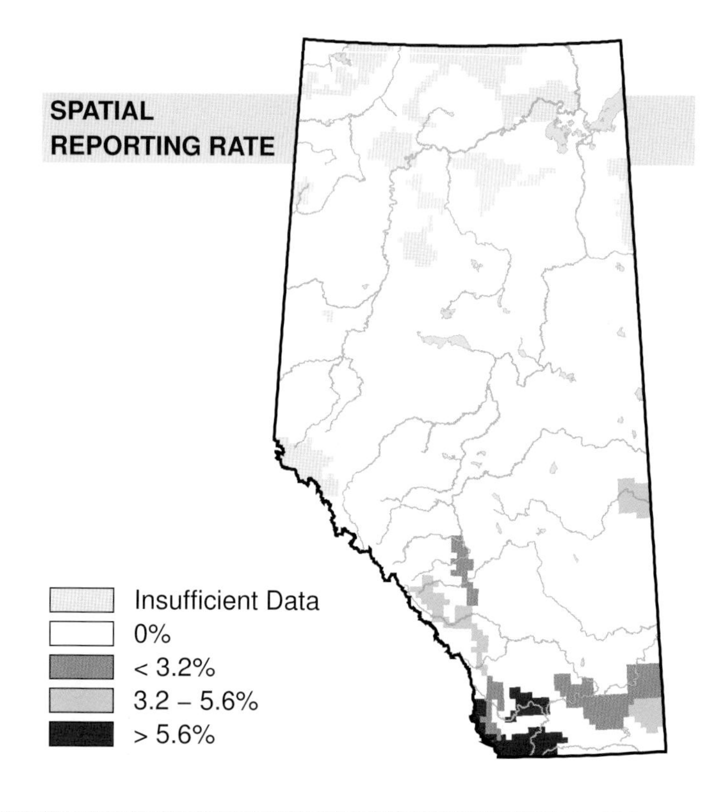

STATUS

GSWA 2000	GSWA 2005	COSEWIC Historic Rankings	COSEWIC 2007	Alberta Wildlife Act
Secure	Secure	N/A	N/A	N/A

Habitat
Nest Location
Nest Type
Diet

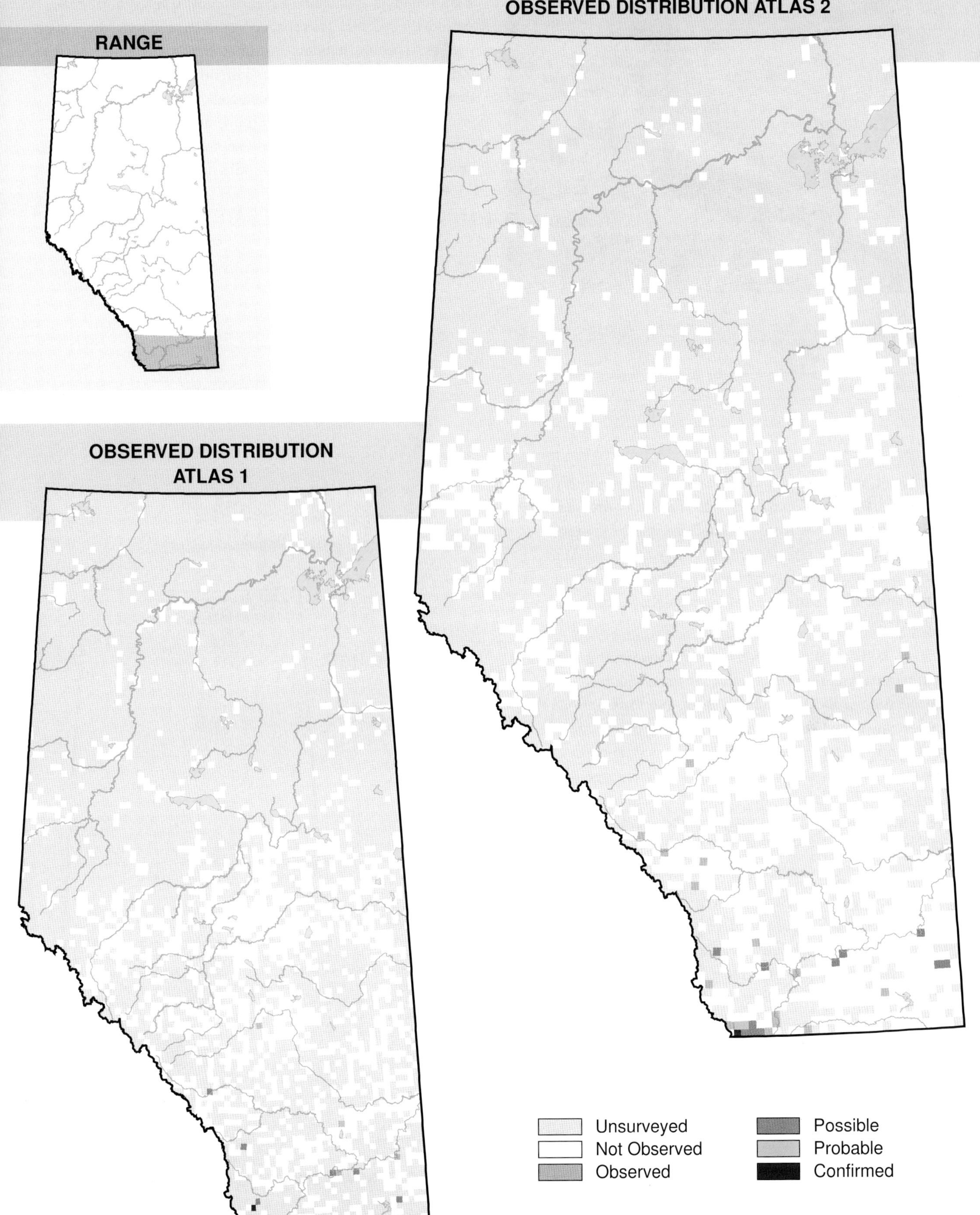

RANGE
OBSERVED DISTRIBUTION ATLAS 1
OBSERVED DISTRIBUTION ATLAS 2
Unsurveyed
Not Observed
Observed
Possible
Probable
Confirmed

Lazuli Bunting (*Passerina amoena*)

Photo: Randy Jensen

NESTING

CLUTCH SIZE (EGGS):	3–4
INCUBATION (DAYS):	12
FLEDGING (DAYS):	10–12
NEST HEIGHT (METRES):	<3

The Lazuli Bunting is found in the southern part of the province in the Grassland, Parkland, and Rocky Mountain Natural Regions. The distribution of this species appears to have become more compressed between Atlas 1 and Atlas 2. There were fewer observations made in the northwestern part of the range in the area north of the Bow River during Atlas 2.

The Lazuli Bunting tends to be associated with brushy habitat such as willow or alder thickets. For this reason, this species often nests along riparian areas and in areas that are regenerating post-disturbance, such as after fires. Despite their tendency to occupy areas post-disturbance, they do not commonly move into forests that are regenerating after harvesting has occurred. The Lazuli Bunting's range reaches its northernmost limit in British Columbia and trickles into the western portion of Alberta; therefore, this species was found mainly in the Rocky Mountain NR and was found only infrequently in the Grassland NR. Observations made in the Parkland NR were of non-breeders.

An increase in relative abundance was detected in the Rocky Mountain NR where, relative to other species, the Lazuli Bunting was observed more frequently during Atlas 2 than during Atlas 1. The Breeding Bird Survey sample size was too small to investigate abundance trends in Alberta. Similar to what was found by the Atlas in the Rocky Mountain NR, the BBS found an abundance increase across Canada during the period 1968–2005. This species could be on the rise because it prefers to occupy habitats post-disturbance and there have been quite a few fires in recent years in British Columbia. The increased relative abundance detected by the Atlas is reasonable given that the BBS found an increase across Canada. More individuals could be moving into Alberta. This species is considered Secure in Alberta.

TEMPORAL REPORTING RATE

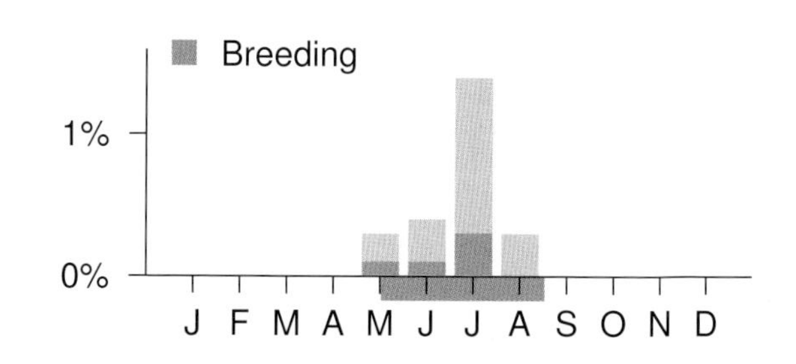

NATURAL REGIONS REPORTING RATE

SPATIAL REPORTING RATE

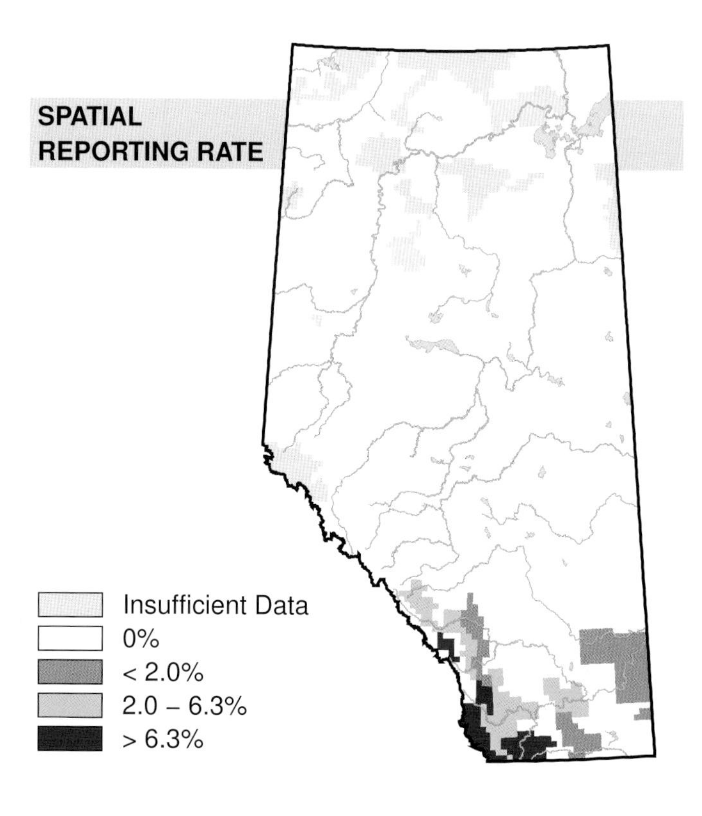

STATUS

GSWA 2000	GSWA 2005	COSEWIC Historic Rankings	COSEWIC 2007	Alberta Wildlife Act
Secure	Secure	N/A	N/A	N/A

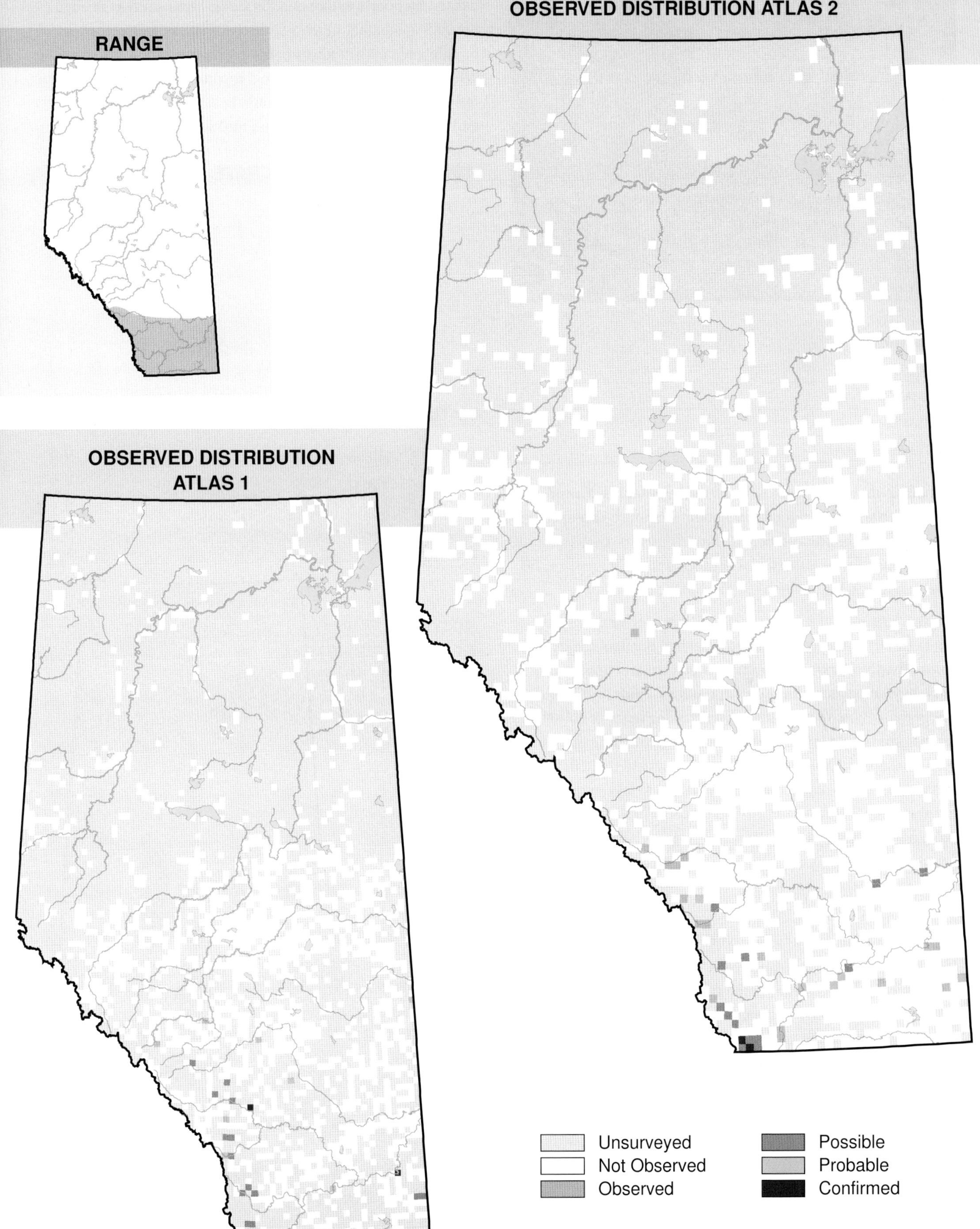
RANGE
OBSERVED DISTRIBUTION ATLAS 2
OBSERVED DISTRIBUTION ATLAS 1
Unsurveyed
Not Observed
Observed
Possible
Probable
Confirmed

Bobolink (*Dolichonyx oryzivorus*)

Photo: Royal Alberta Museum

NESTING

CLUTCH SIZE (EGGS):	4–7
INCUBATION (DAYS):	10–13
FLEDGING (DAYS):	10–14
NEST HEIGHT (METRES):	0

The polygynous Bobolink is found in the Foothills, Grassland, Parkland, and Rocky Mountain Natural Regions. This species' observed distribution contracted from Atlas 1 to Atlas 2. In the northern portion of this species' expected range there were far fewer observations and none in the north-east.

Historically, this species nested in tall-grass or mixed-grass prairie, until intensive agriculture changed the landscape. Currently, the Bobolink is found in open country and prefers large hayfields, moist meadows and weedy fields dominated by tall grasses. This species builds well-hidden nests in hollows on the ground at the base of tall plants. Although rare in all NRs, this species was most frequently observed in the Rocky Mountain, Grassland, and Foothills NRs, and less frequently in the Parkland NR.

There was no change in relative abundance detected in any of the NRs, and relative to other species, the Bobolink was observed less often in the Grassland and Parkland NRs. The most pronounced changes in the observed distribution were in the Parkland NR, where its rarity placed it below our statistical threshold. The Breeding Bird Survey detected a decline in abundance across Canada for the period 1985–2005, but there was not a large enough sample size to determine trends in Alberta. Conservation issues for this species include predation of eggs and nestlings by generalist predators and nest damage caused by poor weather and flooding. More study is needed to determine whether human-related activities, such as earlier haying and nest-site disturbance by humans and cattle, are affecting the productivity of this species. The Bobolink's status is considered Sensitive in Alberta because of its low abundance and its dependence on tall-grass meadows.

TEMPORAL REPORTING RATE

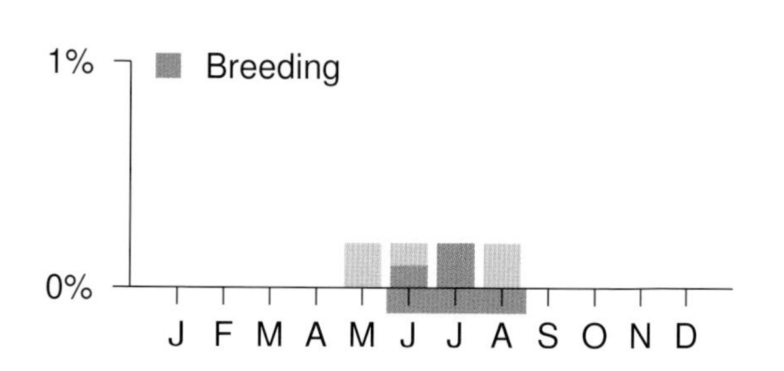

NATURAL REGIONS REPORTING RATE

SPATIAL REPORTING RATE

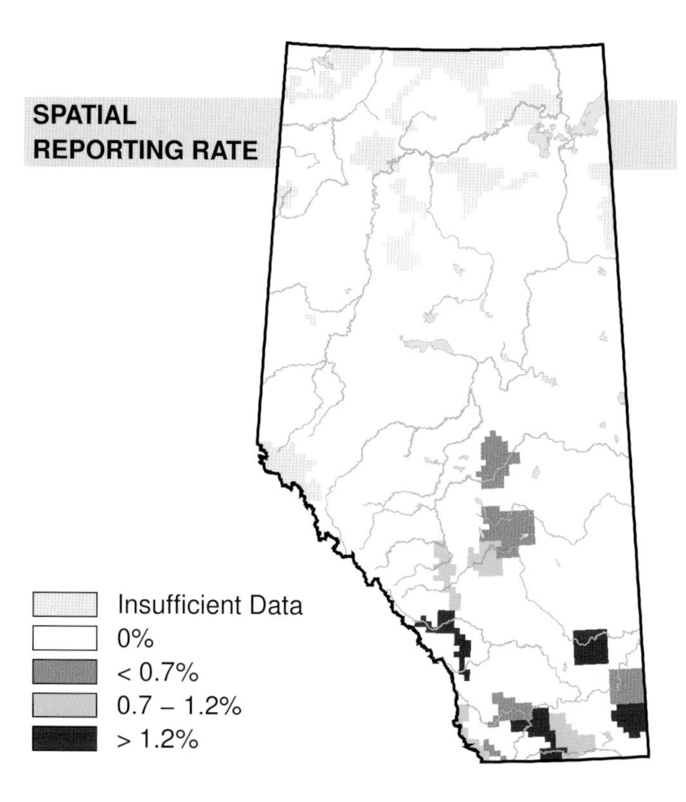

STATUS

GSWA 2000	GSWA 2005	COSEWIC Historic Rankings	COSEWIC 2007	Alberta Wildlife Act
Sensitive	Sensitive	N/A	N/A	N/A

Habitat

Nest
Location

Nest
Type

Diet

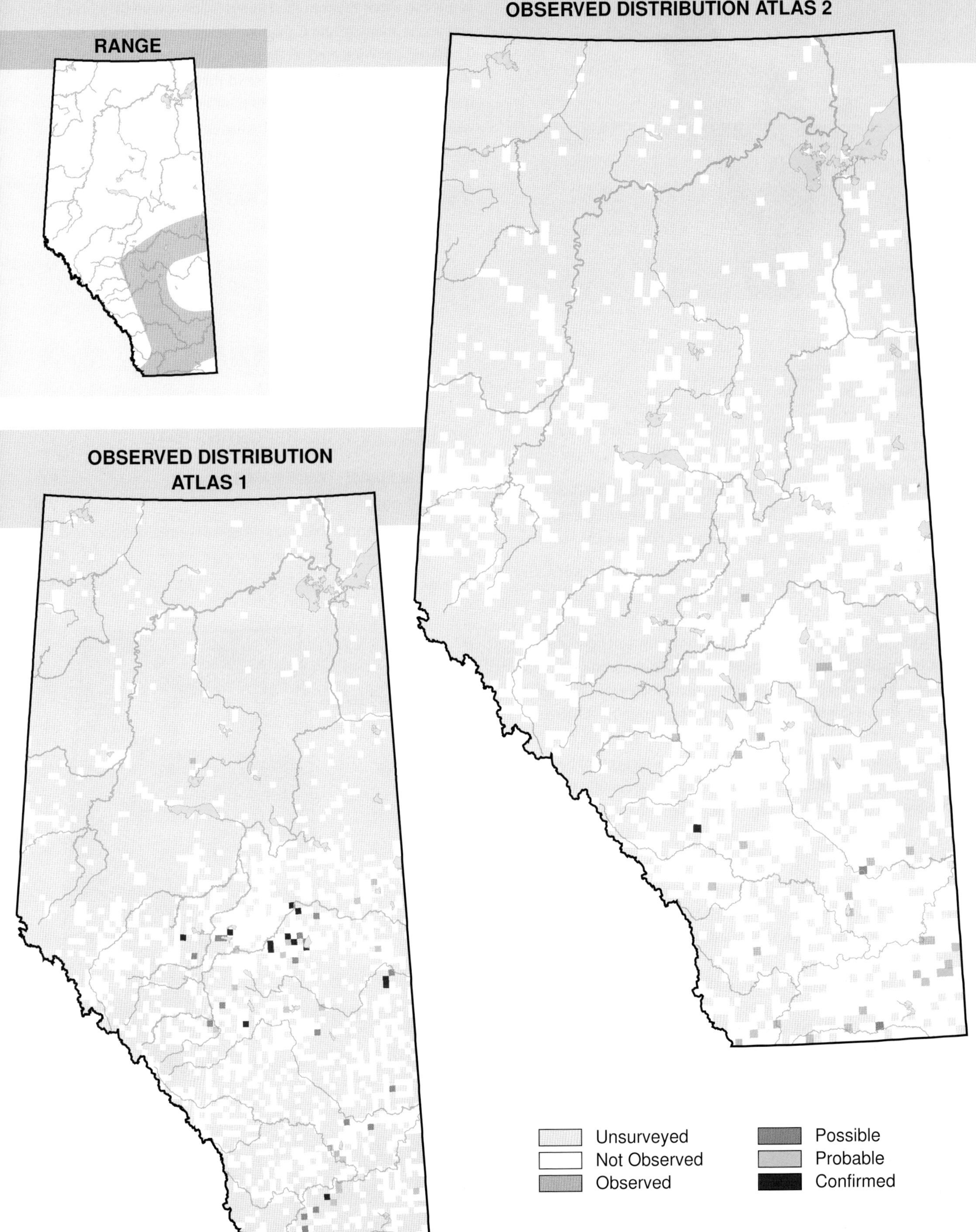
RANGE
OBSERVED DISTRIBUTION ATLAS 2
OBSERVED DISTRIBUTION
ATLAS 1
Unsurveyed
Not Observed
Observed
Possible
Probable
Confirmed

Red-winged Blackbird (*Agelaius phoeniceus*)

Photo: Gerald Romanchuk

NESTING

CLUTCH SIZE (EGGS):	4
INCUBATION (DAYS):	10–12
FLEDGING (DAYS):	10–14
NEST HEIGHT (METRES):	0.2–6.1

The familiar Red-winged Blackbird is found throughout Alberta in every Natural Region. The distribution of this species did not change between Atlas 1 and Atlas 2.

The Red-winged Blackbird can breed in a variety of wetlands. In Alberta, it is usually found breeding among the cattails along the perimeter of shallow wetlands and lakes. However, they can also nest in upland habitat in willow thickets, along the edges of forests, and in fields where agricultural plants are grown. Due to its generalist nature, this species was found commonly right across the province. Reporting rates were highest in the Grassland and Parkland NRs. This was likely the result of the greater potential for reports to be sent in, because more people live in those regions; as well, the habitat there is more open and, consequently, observations are easier to make. This species was also encountered commonly in the Boreal Forest, Foothills, and Rocky Mountain NRs.

Declines in relative abundance were detected in the Boreal Forest, Foothills, and Parkland NRs where, relative to other species, the Red-winged Blackbird was observed less frequently in Atlas 2 than in Atlas 1. The Breeding Bird Survey found a decline in Alberta and across Canada during the period 1985–2005. Declines could be related to drought, wetland drainage, or the failure to retain enough buffer habitat around the perimeter of wetlands. In addition, Red-winged Blackbirds are considered agricultural pests because they damage crops and are loud when they roost in large groups. It is legal to shoot them in the United States where they breed and where many of the Canadian breeders spend the winter. The Atlas also detected an increase in relative abundance in the Rocky Mountain NR. The cause of this increase is not understood, but may be a function of differences in survey coverage between Atlas 1 and Atlas 2. This species is considered Secure in Alberta.

TEMPORAL REPORTING RATE

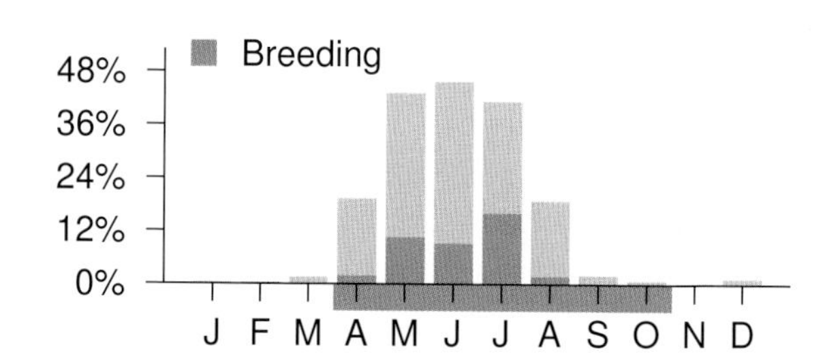

NATURAL REGIONS REPORTING RATE

SPATIAL REPORTING RATE

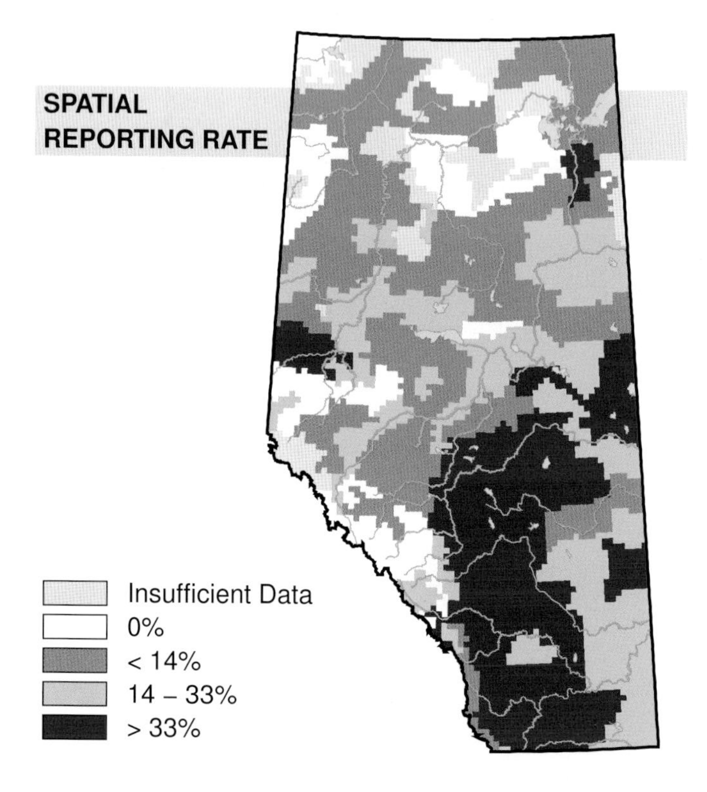

STATUS

GSWA 2000	GSWA 2005	COSEWIC Historic Rankings	COSEWIC 2007	Alberta Wildlife Act
Secure	Secure	N/A	N/A	N/A

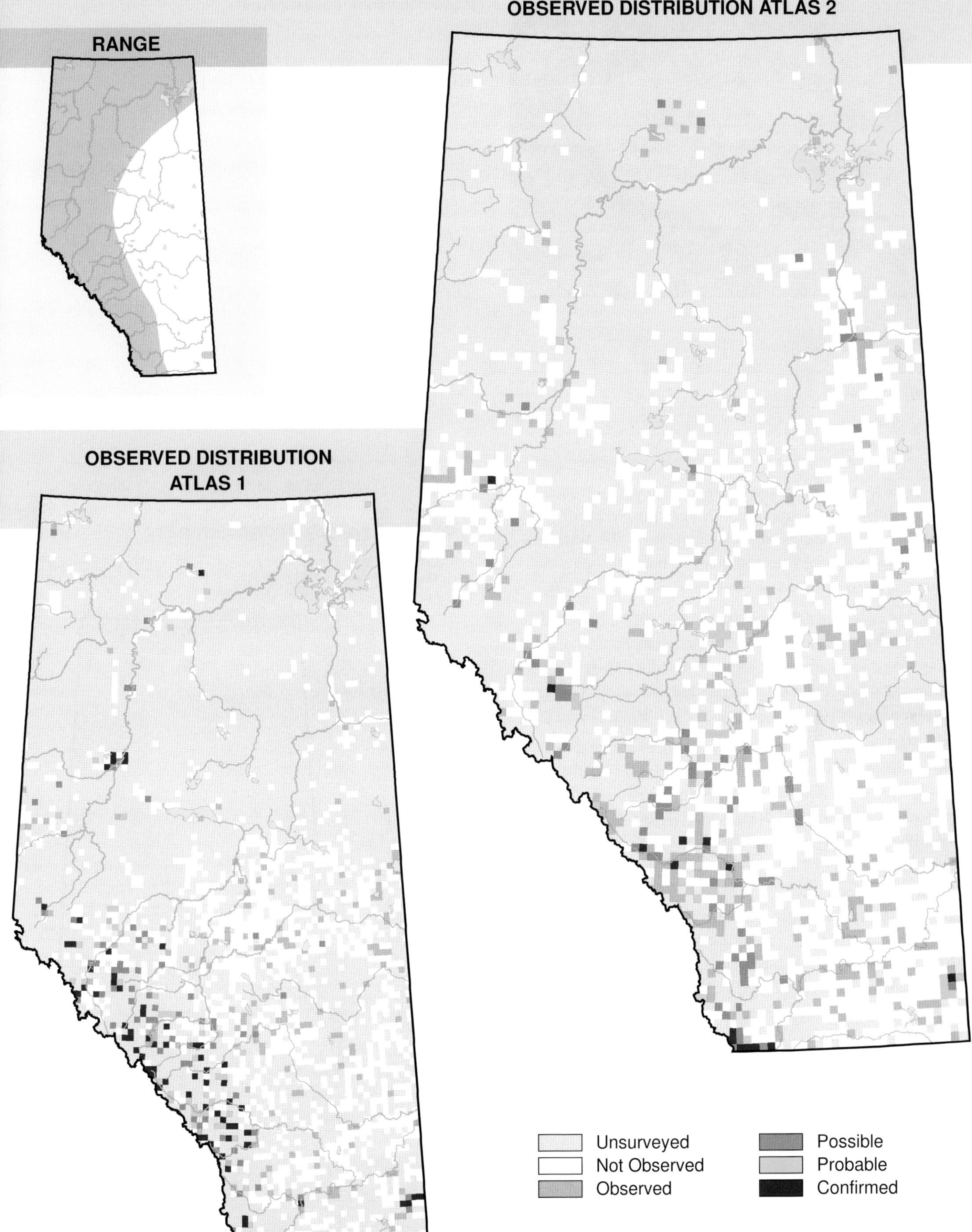
Habitat
Nest Location
Nest Type
Diet
RANGE
OBSERVED DISTRIBUTION ATLAS 2
OBSERVED DISTRIBUTION ATLAS 1
Unsurveyed
Not Observed
Observed
Possible
Probable
Confirmed

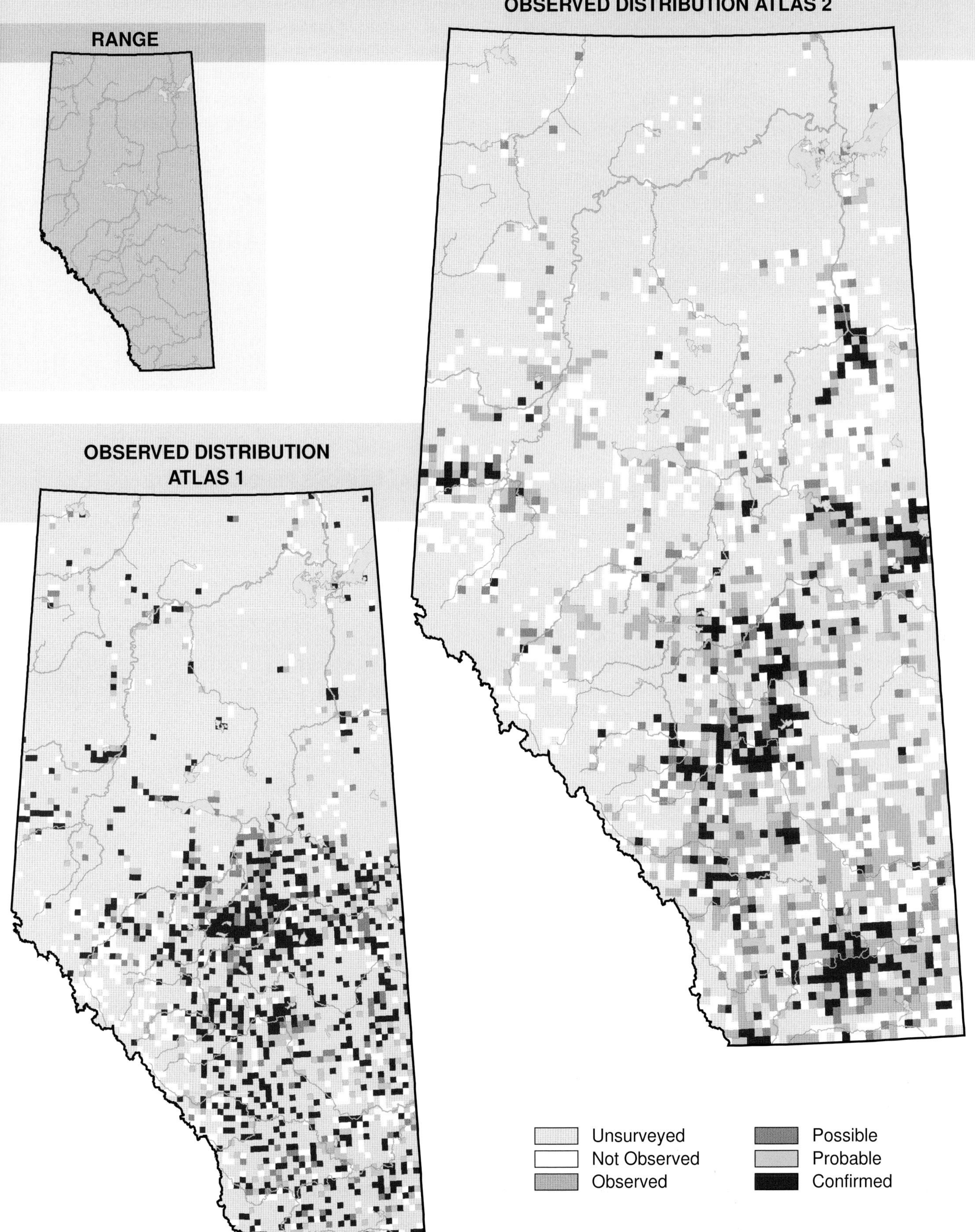
RANGE
OBSERVED DISTRIBUTION ATLAS 2
OBSERVED DISTRIBUTION ATLAS 1
Unsurveyed
Not Observed
Observed
Possible
Probable
Confirmed

Western Meadowlark (*Sturnella neglecta*)

Photo: Jim Jacobson

NESTING

CLUTCH SIZE (EGGS):	3–7
INCUBATION (DAYS):	13–15
FLEDGING (DAYS):	26
NEST HEIGHT (METRES):	0

Western Meadowlark is largely confined to the Grassland and Parkland Natural Regions. There is no apparent distribution change between Atlas 1 and Atlas 2. Historically, there have been distributional changes as this species was limited by forest edge (Lanyon, 1994); they avoid heavily wooded cover. Human-induced change since settlement has altered the distribution of wooded cover with expansion of aspens through fire suppression and openings in the forest brought about by logging and agricultural expansion.

The highest reporting rates occur in the Grassland NR (but not Cypress Hills) and in the area around Calgary. Western Meadowlark is widespread in open grassy or grass like habitats. It is most common in native grassland but also occurs in hayfields and the edges of crop fields (Lanyon, 1994). The Cypress Hills has a lot of woody cover and the area around Calgary is now largely developed.

Increases in relative abundance were detected in the Grassland and Rocky Mountain NRs (Cypress Hills). Decreases in relative abundance were detected in the Boreal and Parkland NRs. No changes in relative abundance were detected in the Foothills NR. The Breeding Bird Survey detected an increase in Alberta for the period 1995–2005 but reported a prairie-wide decrease for the period 1985–2005. Grassland Bird Monitoring (unpublished Canadian Wildlife Service data) didn't detect a trend from 1996 through 2004. An increase in population seems less likely given conflicting trends from several sources with good sample sizes, and declines in native grassland within the Grassland and Parkland NRs in the period 1985–2000 (Watmough and Schmoll, 2007). The loss of native cover is partially offset by the conversion of some crop to planted pasture or hay but most species nesting in hayfields are subject to catastrophic nest losses due to haying operations (Frawley, 1989). Provincial status reports from 2000 and 2005 both rank the species as Secure.

TEMPORAL REPORTING RATE

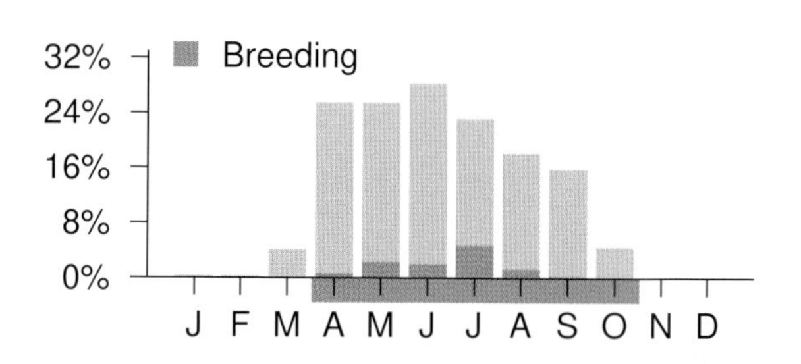

NATURAL REGIONS REPORTING RATE

Region	Rate
Grassland	52%
Parkland	5%
Rocky Mountain	3%

SPATIAL REPORTING RATE

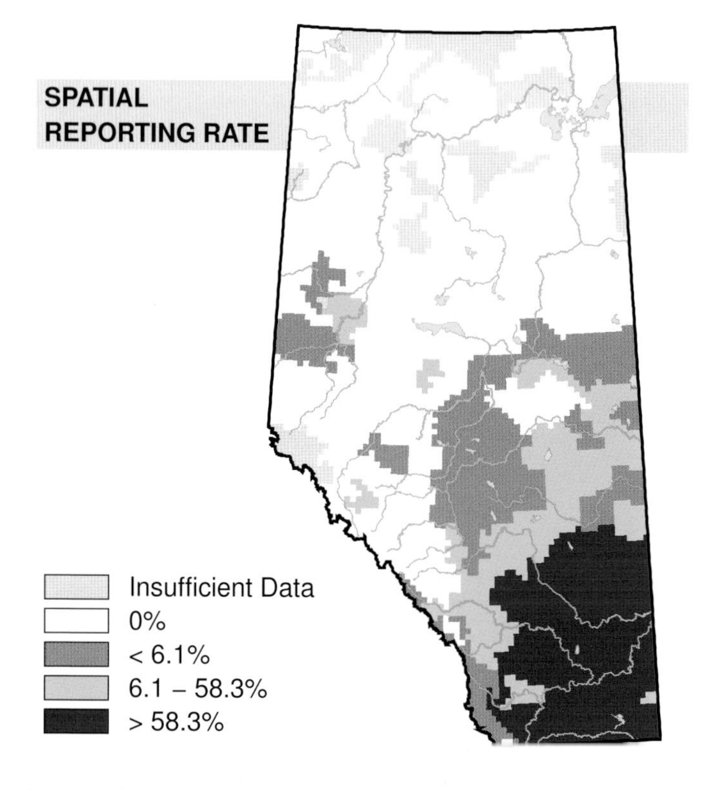

STATUS

GSWA 2000	GSWA 2005	COSEWIC Historic Rankings	COSEWIC 2007	Alberta Wildlife Act
Secure	Secure	N/A	N/A	N/A

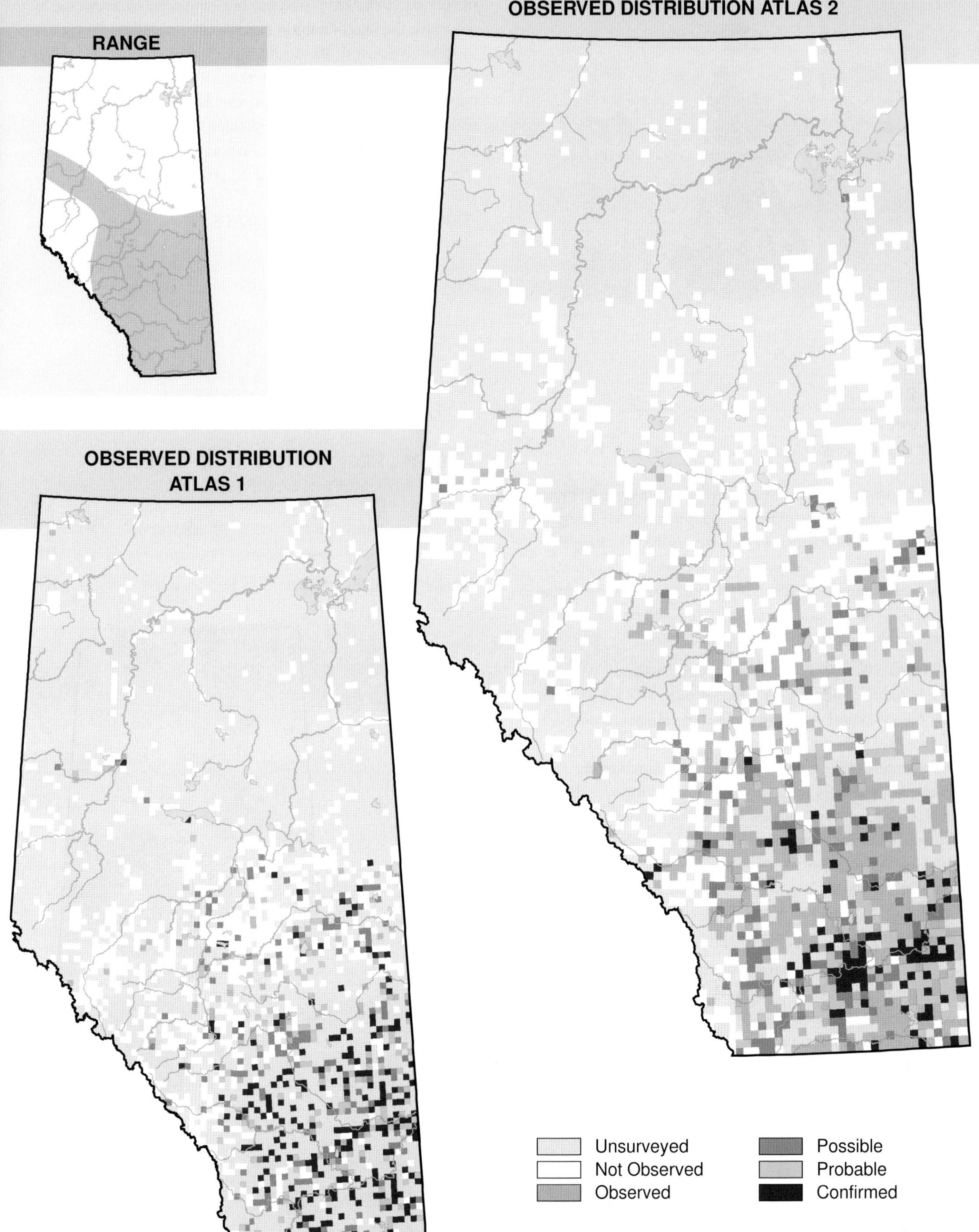
RANGE
OBSERVED DISTRIBUTION ATLAS 2
OBSERVED DISTRIBUTION
ATLAS 1
Unsurveyed
Not Observed
Observed
Possible
Probable
Confirmed

Yellow-headed Blackbird (*Xanthocephalus xanthocephalus*)

Photo: Gerald Romanchuk

NESTING

CLUTCH SIZE (EGGS):	3–4
INCUBATION (DAYS):	11–14
FLEDGING (DAYS):	20
NEST HEIGHT (METRES):	0.2–0.9

The Yellow-headed Blackbird is found in the Boreal Forest, Grassland, Parkland, and Rocky Mountain Natural Regions. Although, the distribution of this species did not change between Atlas 1 and Atlas 2, fewer probable and confirmed breeding records were found in the northwest portion of the province. There was also one possible breeding record farther north, in Wood Buffalo National Park.

Yellow-headed Blackbirds prefer prairie wetlands, where they nest in the emergent vegetation. However, they are also associated with trembling aspen and mountain meadows. Unlike its cousin the Red-winged Blackbird, which is usually found in areas of shallower waters, the Yellow-headed Blackbird is usually found in cattails and bulrushes where there is deeper water. However, there is considerable overlap between the habitat preferences of these two species and, therefore, they are sometimes found nesting in the same area. The Yellow-headed Blackbird was most commonly observed in the Grassland and Parkland NRs and small numbers were reported in the Rocky Mountain and Boreal Forest NRs.

Declines in relative abundance of this species were detected in the Boreal Forest, Grassland, and Parkland NRs and increases were detected in the Rocky Mountain NR. Differences in observation frequency relative to other species between Atlas 1 and 2 mirrored these changes. The Breeding Bird Survey found abundance declines in Alberta and Canada-wide during the period 1985–2005. Because breeding sites are restricted to wetlands, the loss of habitat due to draining and drought is the most likely cause of the declines that have been detected. The Rocky Mountain NR is mostly protected by National Parks, so therefore the opposite trend that was detected there is reasonable. Populations will fluctuate with wetland conditions, and can recover quite quickly if conditions are favourable. This species is considered Secure in the province.

TEMPORAL REPORTING RATE

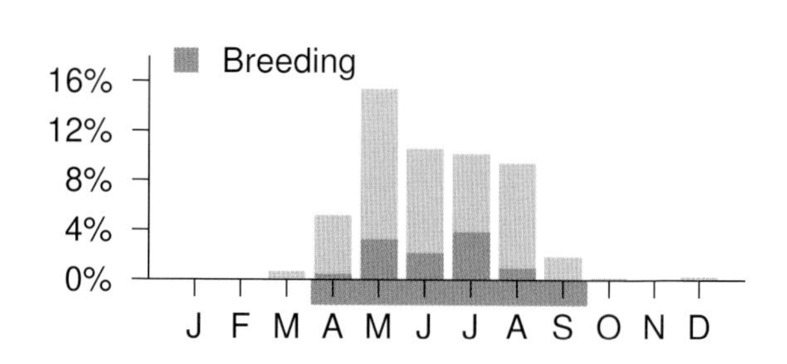

NATURAL REGIONS REPORTING RATE

SPATIAL REPORTING RATE

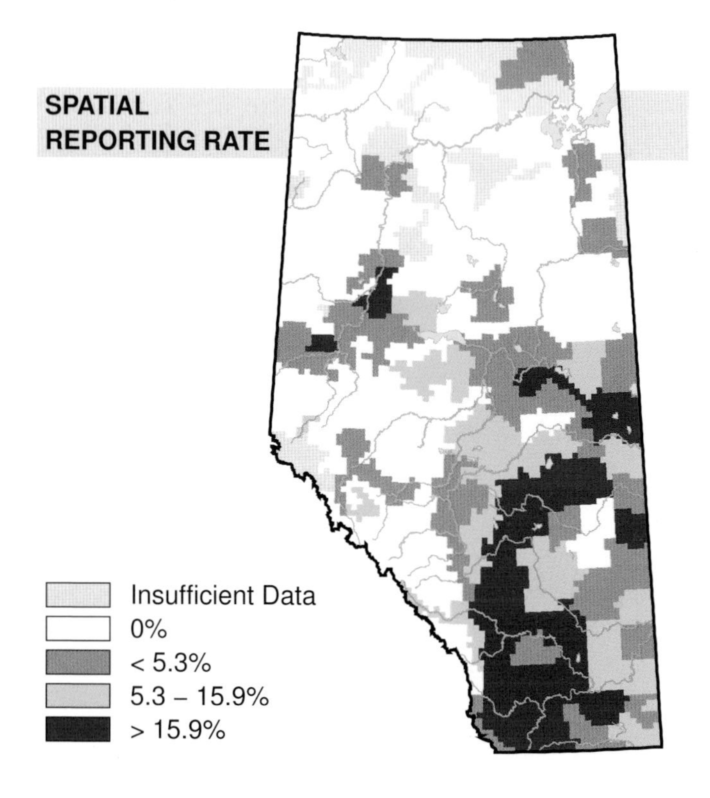

STATUS

GSWA 2000	GSWA 2005	COSEWIC Historic Rankings	COSEWIC 2007	Alberta Wildlife Act
Secure	Secure	N/A	N/A	N/A

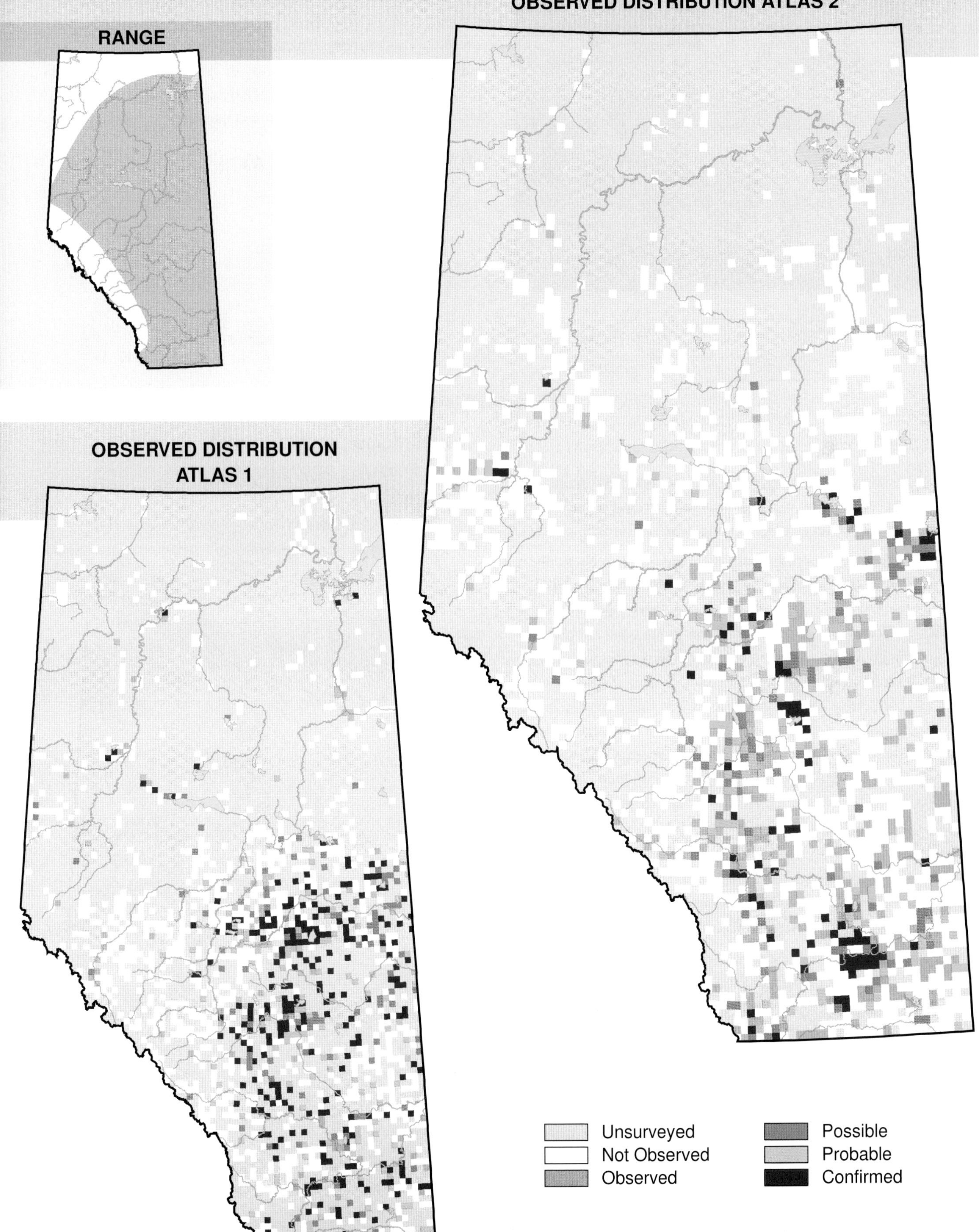
RANGE
OBSERVED DISTRIBUTION ATLAS 2
OBSERVED DISTRIBUTION ATLAS 1
Unsurveyed
Not Observed
Observed
Possible
Probable
Confirmed

Rusty Blackbird (*Euphagus carolinus*)

Photo: Royal Alberta Museum

NESTING

CLUTCH SIZE (EGGS):	4–5
INCUBATION (DAYS):	14
FLEDGING (DAYS):	13
NEST HEIGHT (METRES):	<3

The Rusty Blackbird breeds in the Boreal Forest, Parkland, and Foothills Natural Regions. The observed distribution of this species did not change between Atlas 1 and Atlas 2.

One of Alberta's least frequently encountered blackbirds, the Rusty Blackbird is found in wet forest habitat. This species often breeds in bogs, muskeg swamps, and along the perimeter of beaver ponds. Nests are built off the ground and are often located in coniferous trees such as black spruce, but can also be found in willow, birch, and tamarack. Due to its preference for wet coniferous forests, this species was found mainly in the Boreal Forest, Foothills, and Parkland NRs. Due to its preference for wet coniferous forests, this species was found mainly in the Boreal Forest, Foothills, and Parkland NRs. Lower reporting rates in the Boreal Forest NR may be related to the more limited coverage and greater habitat diversity in this NR. This species is probably more abundant in the Boreal Forest and Foothills NRs than in the Parkland NR because more suitable habitat is available in the first two NRs. Records from the Grassland and Rocky Mountain NRs were those of non-breeders.

A decline in relative abundance was detected in the Boreal Forest where, relative to other species, the Rusty Blackbird was observed less frequently in Atlas 2 than in Atlas 1. The Breeding Bird Survey (1985–2005) lacked a sufficient sample size to estimate population trends in Alberta, but did detect a Canada-wide decline. No changes in relative abundance were detected in other NRs. The Alberta Government changed the status of this species from Secure in 2000 to Sensitive in 2005 due to the declines that have been detected, and to this bird's close association with a threatened habitat.

TEMPORAL REPORTING RATE

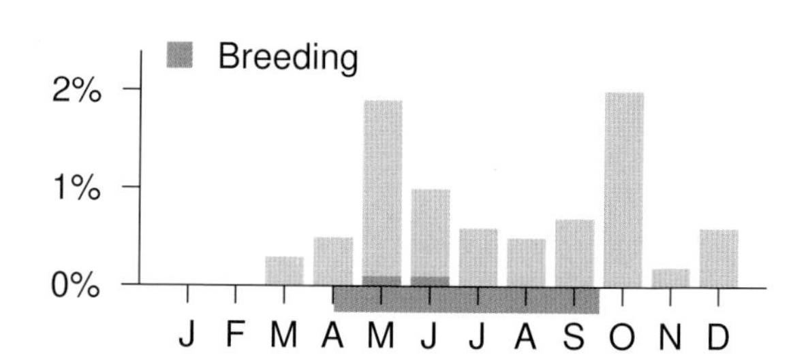

NATURAL REGIONS REPORTING RATE

SPATIAL REPORTING RATE

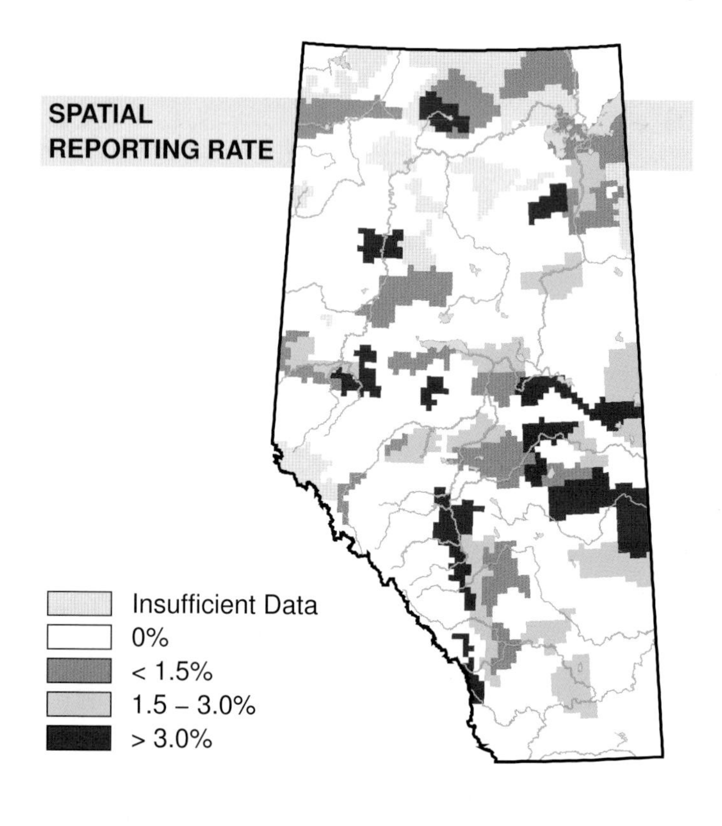

STATUS

GSWA 2000	GSWA 2005	COSEWIC Historic Rankings	COSEWIC 2007	Alberta Wildlife Act
Secure	Sensitive	N/A	Special Concern	N/A

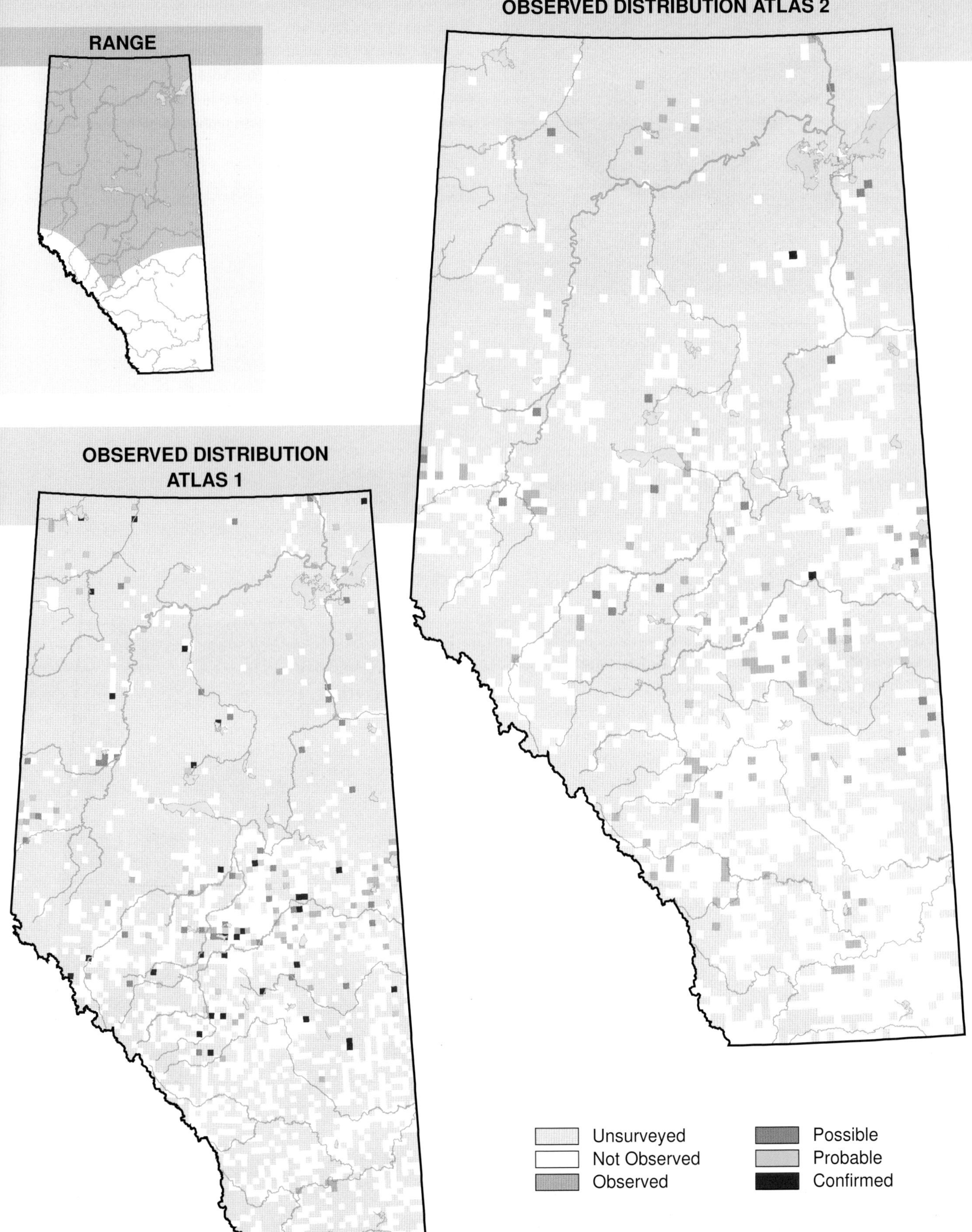
RANGE
OBSERVED DISTRIBUTION ATLAS 2
OBSERVED DISTRIBUTION
ATLAS 1
Unsurveyed
Not Observed
Observed
Possible
Probable
Confirmed

Brewer's Blackbird (*Euphagus cyanocephalus*)

Photo: Dave McKenzie

NESTING

CLUTCH SIZE (EGGS):	5–6
INCUBATION (DAYS):	12–14
FLEDGING (DAYS):	11–13
NEST HEIGHT (METRES):	0

In Alberta, Brewer's Blackbird is found in every Natural Region. There were fewer records of this species in northern Alberta in Atlas 2. In Atlas 1, this species had been observed to expand northward using modified areas and roadways (Semenchuk, 1992). The change in Atlas 2 may represent a contraction of the bird's distribution to historical limits from before Atlas 1.

Brewer's Blackbird tends to breed in human-modified habitats such as along roads, in forest clearcuts, and in open urban parks. This habitat contains a combination of open areas and patches of trees or shrubs, often near water. Because of its preference for open areas and human-modified environments, this species was most common in the Grassland and Parkland NRs and was more sparsely distributed in the Boreal Forest, Foothills, and Rocky Mountain NRs.

Declines in relative abundance were detected in the Boreal Forest, Foothills, Grassland, and Parkland NRs where, relative to other species, Brewer's Blackbird was observed less frequently in Atlas 2 than in Atlas 1. Given this species' preference for human-modified environments, there is no apparent biological cause for these declines. The range contraction in northern Alberta, however, would have contributed to the detected decline in the Boreal Forest NR. An increase was detected in the Rocky Mountain NR. The Breeding Bird Survey did not find an abundance change for this species in Alberta or Canada. The Breeding Bird Survey has adequate survey coverage for this species because Breeding Bird Surveys are generally conducted along roads where Brewer's Blackbirds often breed. While there is no known probable biological cause for the Atlas-detected changes, there are also no apparent sources of bias in the Atlas data. Further research is needed to assess these findings and the true status of this species. This species is considered Secure in Alberta.

TEMPORAL REPORTING RATE

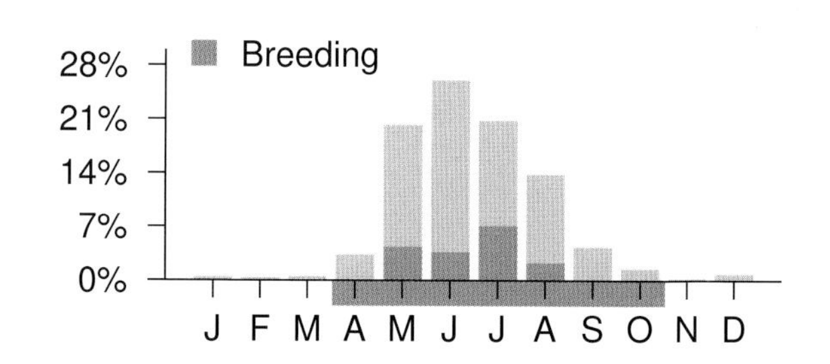

NATURAL REGIONS REPORTING RATE

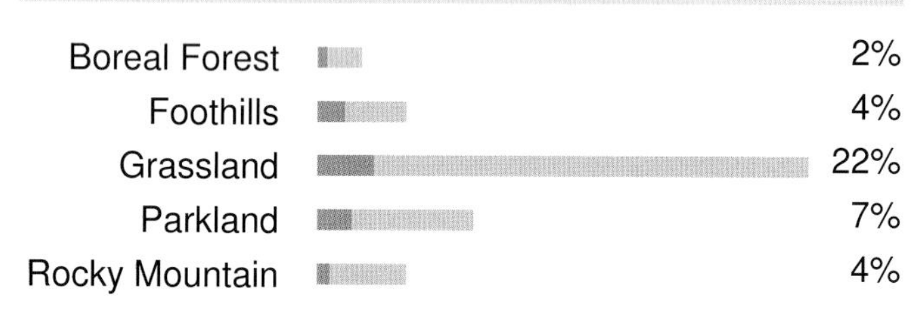

SPATIAL REPORTING RATE

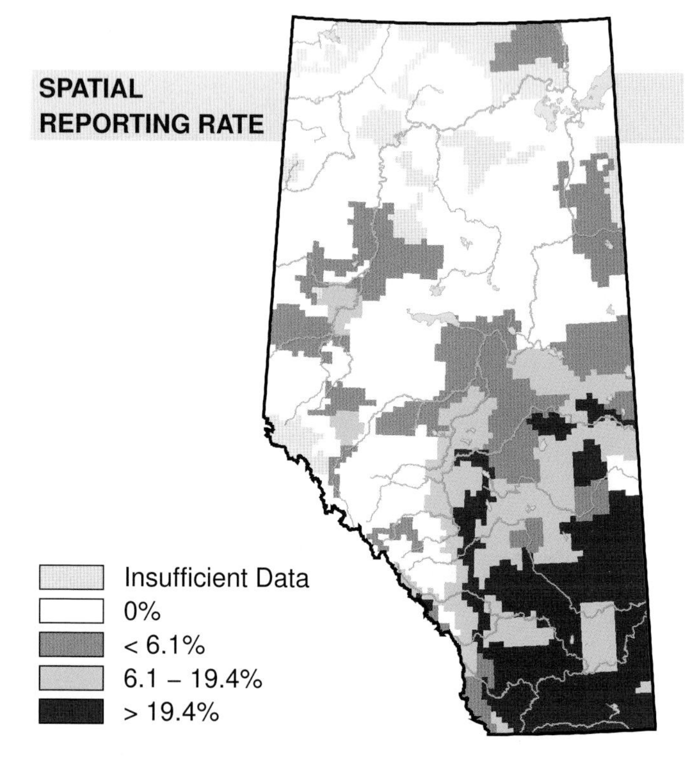

STATUS

GSWA 2000	GSWA 2005	COSEWIC Historic Rankings	COSEWIC 2007	Alberta Wildlife Act
Secure	Secure	N/A	N/A	N/A

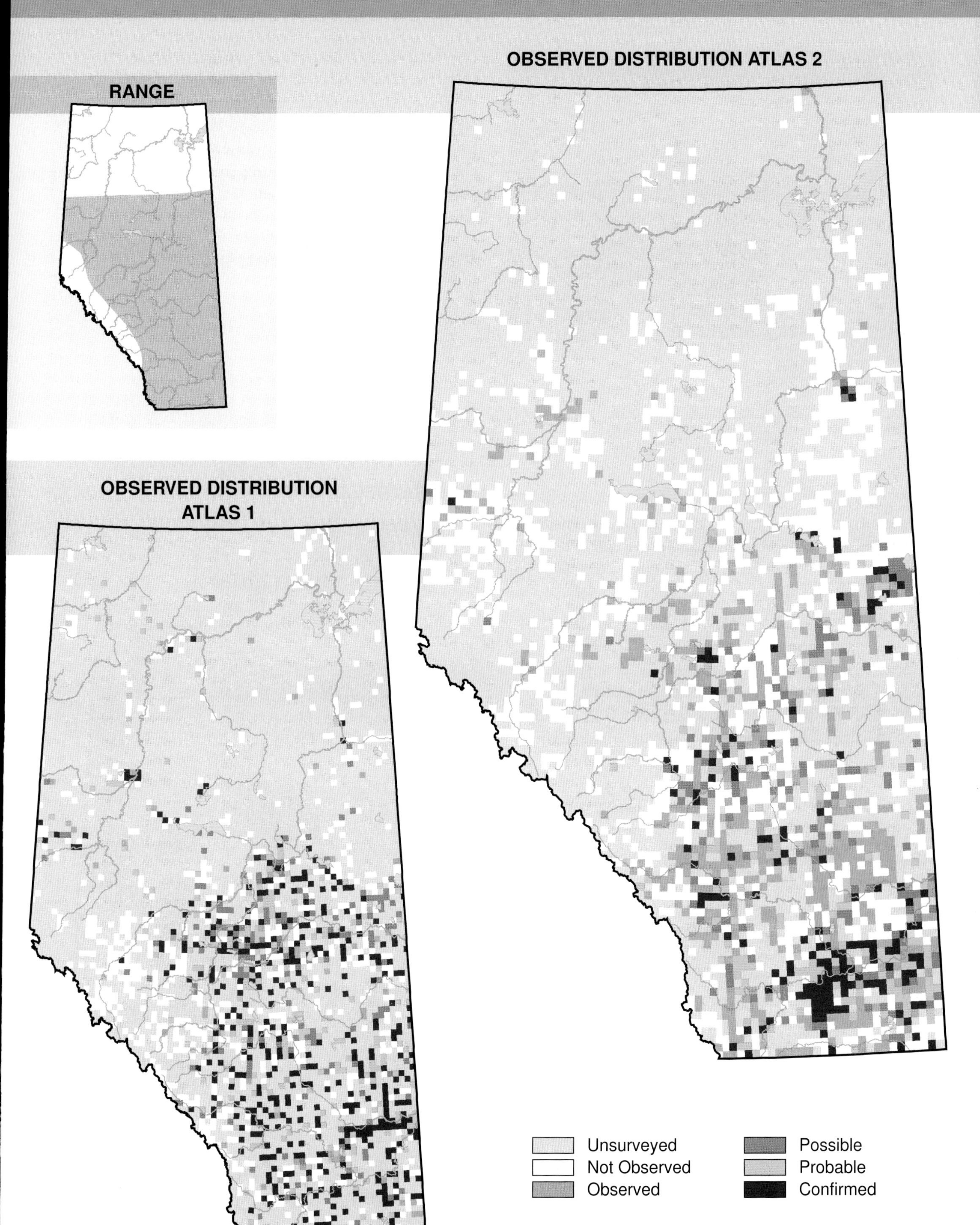
OBSERVED DISTRIBUTION ATLAS 2
RANGE
OBSERVED DISTRIBUTION
ATLAS 1
Unsurveyed
Not Observed
Observed
Possible
Probable
Confirmed

Common Grackle (*Quiscalus quiscula*)

Photo: Gordon Court

NESTING

CLUTCH SIZE (EGGS):	4–6
INCUBATION (DAYS):	12–14
FLEDGING (DAYS):	12–15
NEST HEIGHT (METRES):	0.6–30.5

The Common Grackle is found in the Boreal Forest, Foothills, Grassland, and Parkland Natural Regions. Records of this species were sparser in Atlas 2, but the observed distribution of this species did not change between Atlas 1 and Atlas 2.

The Common Grackle is found mainly in open areas that contain patches of trees or shrubs. This species can also be found along forest edges, along the edge of marshes, and around human settlements. This species can be common in agricultural areas where open areas and shelter belts provide ideal habitat and where their main food items—invertebrates and grain—can be plentiful. As a consequence, Common Grackles were most often encountered in the Grassland and Parkland NRs and were encountered less often in the Boreal Forest NR.

A decline in relative abundance was detected in the Boreal Forest NR where, relative to other species, the Common Grackle was observed less often in Atlas 2 than in Atlas 1. An increase in relative abundance was detected in the Grassland NR where, relative to other species, it was observed more frequently in Atlas 2 than in Atlas 1. The Breeding Bird Survey found no change in Alberta for the period 1985–2005. The net effect of the Atlas results matches this. However, the BBS did detect a decline in Alberta for the period 1968–2005 and across Canada during the period 1985–2005. It is surprising that declines have been found considering this species is thought to benefit from expanding agriculture, and it breeds in human-modified environments. However, because they are considered pests in some parts of their range, Common Grackles are killed because they eat crops such as corn. These management practices could be contributing to the declines that have been detected. This species is considered Secure in Alberta.

TEMPORAL REPORTING RATE

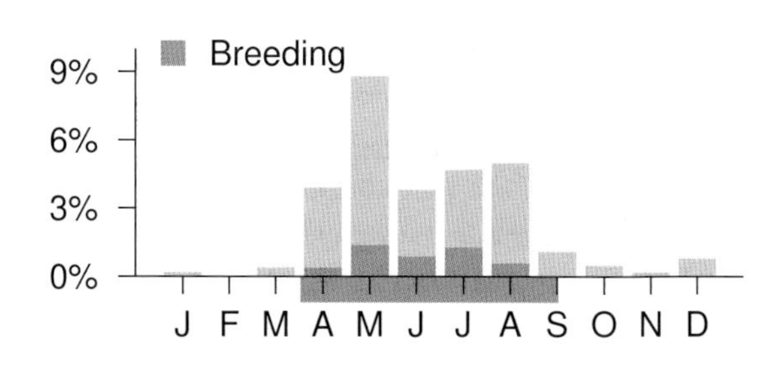

NATURAL REGIONS REPORTING RATE

SPATIAL REPORTING RATE

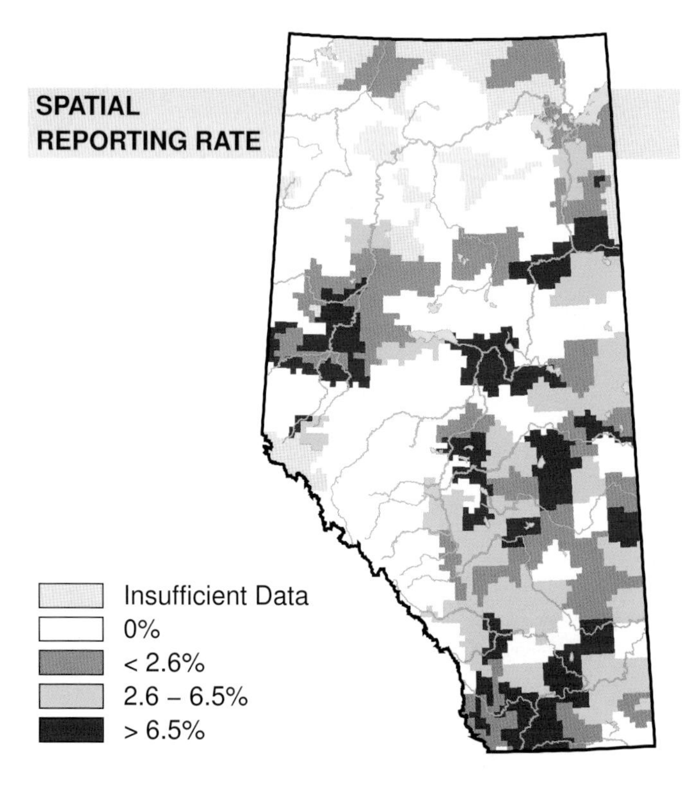

STATUS

GSWA 2000	GSWA 2005	COSEWIC Historic Rankings	COSEWIC 2007	Alberta Wildlife Act
Secure	Secure	N/A	N/A	N/A

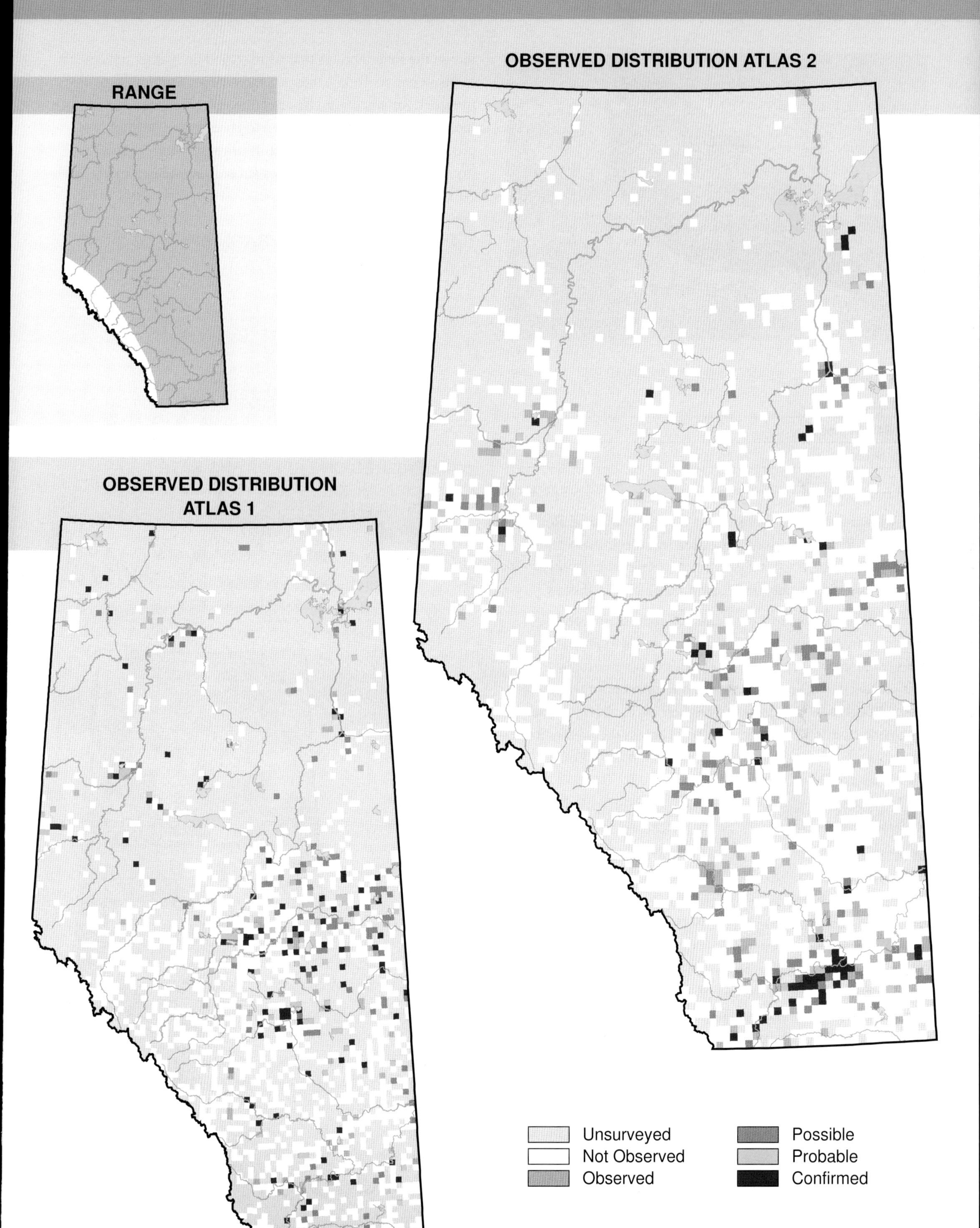
Habitat
Nest Location
Nest Type
Diet
OBSERVED DISTRIBUTION ATLAS 2
RANGE
OBSERVED DISTRIBUTION ATLAS 1
Unsurveyed
Not Observed
Observed
Possible
Probable
Confirmed

Brown-headed Cowbird (*Molothrus ater*)

Photo: Gary Kurtz

NESTING

CLUTCH SIZE (EGGS):	25–50 in total
INCUBATION (DAYS):	11–12
FLEDGING (DAYS):	10
NEST HEIGHT (METRES):	host species

The Brown-headed Cowbird is found in every Natural Region in the province. There were fewer records for this species in northern Alberta in Atlas 2, but it is not clear whether this represents a true change in the distribution of this species.

Brown-headed Cowbirds are best known for their breeding strategy of nest parasitism. This species tends to prefer patchy forests that are fragmented and avoids contiguous forests. For this reason, expansion of human development has favoured this species because it has gained access to new areas via forest fragmentation. This species was found across the province but the highest reporting rates were in the Foothills, Grassland, Parkland, and Rocky Mountain NRs.

Declines in relative abundance were detected in the Boreal Forest, Foothills, Grassland, and Parkland NRs where, relative to other species, the Brown-headed Cowbird was observed less frequently in Atlas 2 than in Atlas 1. Similarly, the Breeding Bird Survey found a decline in abundance in Alberta and across Canada during the period 1985–2005. Additionally, provincial-scale declines were found in British Columbia, Saskatchewan, Manitoba, Ontario, Quebec, New Brunswick, and Nova Scotia during the period 1985–2005. Declines are surprising considering that this species is generally thought to benefit from human developments and increased forest fragmentation. Declines may be a reflection of declines in nest host populations, which have been suffering from the effects of forest fragmentation. Historically, Brown-headed Cowbirds fed on the invertebrates that were stirred up by the movements of bison herds. Today, they are more dependent on domestic livestock for stirring up invertebrate prey. Declines in this bird's numbers could be related to changes in livestock production—specifically the shift towards using feedlots instead of grazing. This species is considered Secure in Alberta.

TEMPORAL REPORTING RATE

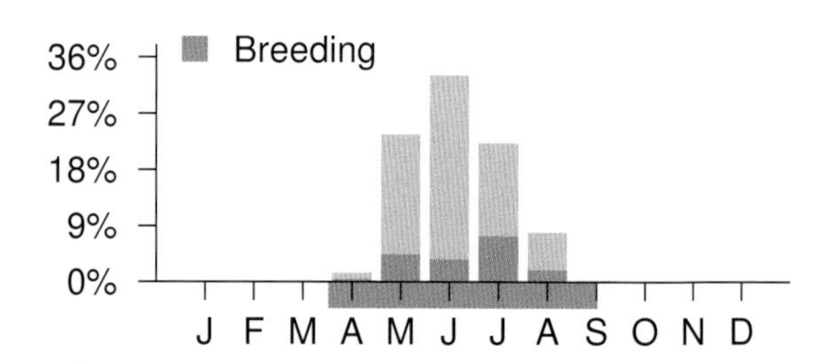

NATURAL REGIONS REPORTING RATE

Region	Rate
Boreal Forest	9%
Foothills	10%
Grassland	19%
Parkland	11%
Rocky Mountain	23%

SPATIAL REPORTING RATE

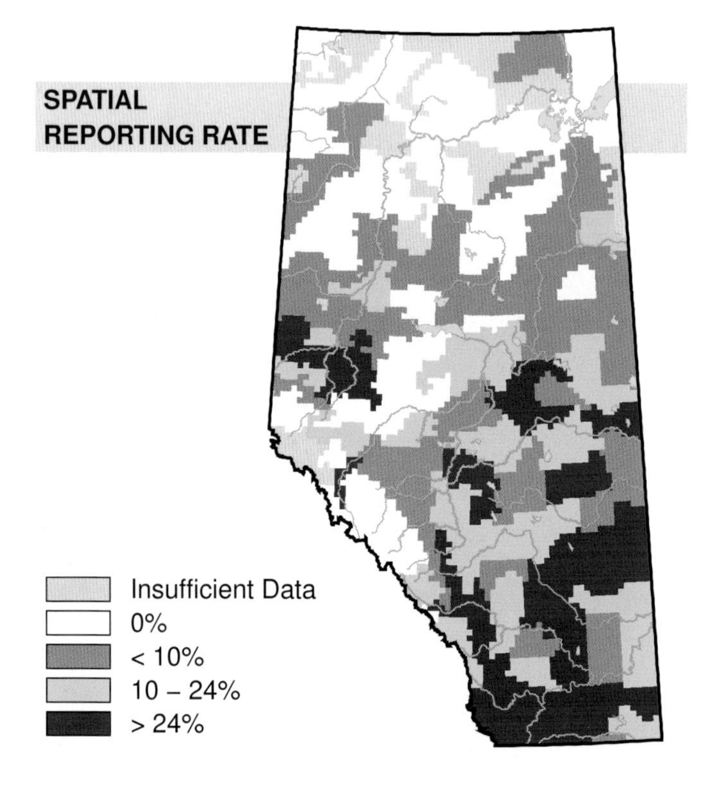

STATUS

GSWA 2000	GSWA 2005	COSEWIC Historic Rankings	COSEWIC 2007	Alberta Wildlife Act
Secure	Secure	N/A	N/A	N/A

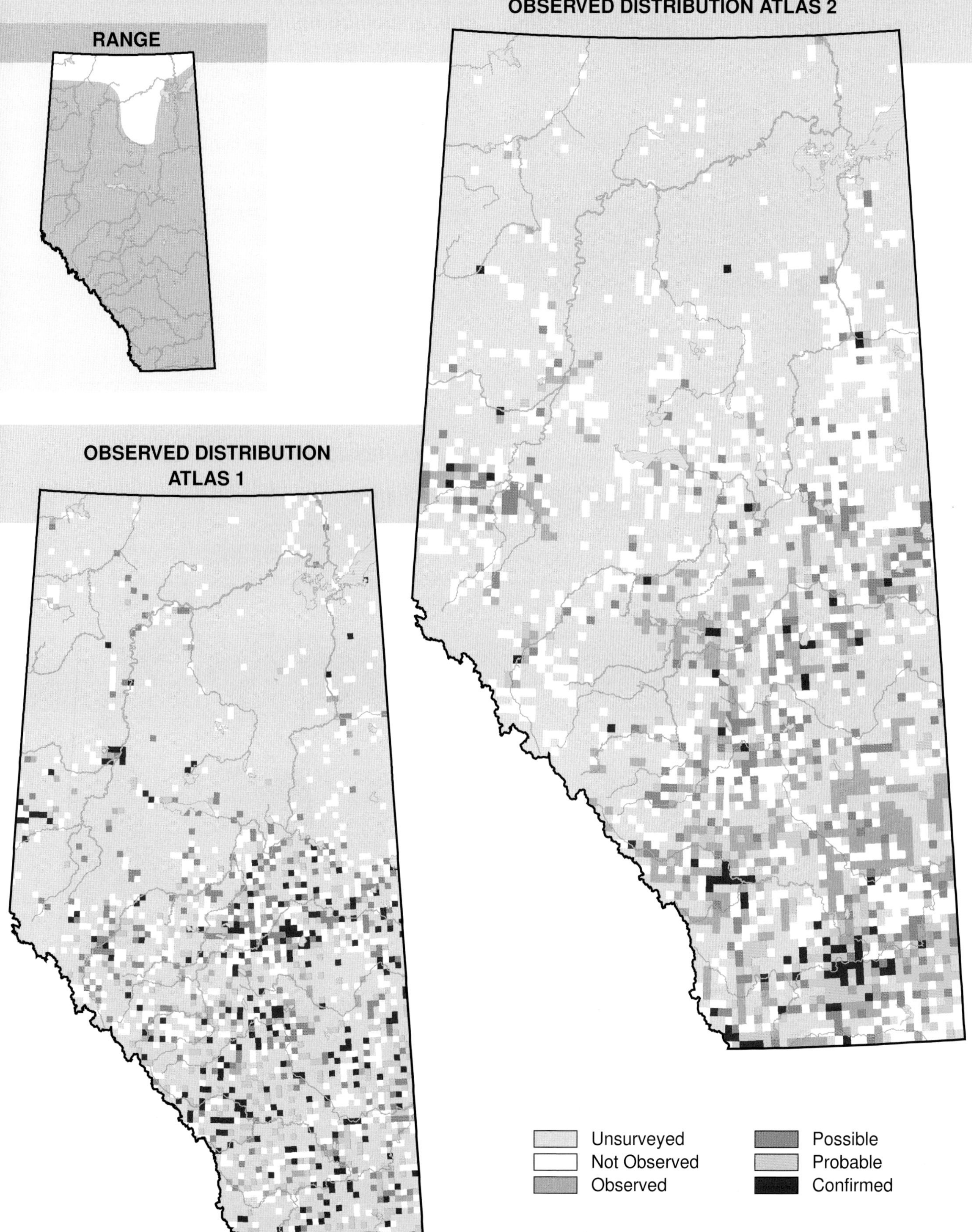
RANGE
OBSERVED DISTRIBUTION ATLAS 2
OBSERVED DISTRIBUTION ATLAS 1
Unsurveyed
Not Observed
Observed
Possible
Probable
Confirmed

Baltimore Oriole (*Icterus galbula*)

Photo: Gerald Romanchuk

NESTING

CLUTCH SIZE (EGGS):	4–5
INCUBATION (DAYS):	12–15
FLEDGING (DAYS):	12–14
NEST HEIGHT (METRES):	7–10

The Baltimore Oriole is found in every Natural Region in Alberta except for the Canadian Shield. The observed distribution of this species decreased in Atlas 2 in the northwestern part of the province around High Level and Fort Vermilion. It is unclear whether this represents an actual range contraction or whether the presence of Baltimore Orioles in the northwest was missed during Atlas 2. The new probable breeding evidence found south of Fort Chipewyan could represent a range expansion in that area. Since Atlas 1 the Northern Oriole has been split into two species, the Baltimore Oriole and Bullock's Oriole. The ranges of these two species overlap in southern Alberta.

The Baltimore Oriole usually breeds in forests that are dominated by deciduous trees. Nests are often built near edges, in areas with low tree density, and near water. Due to their preference for this habitat type, this species was found most often in the Parkland NR and in the Grassland NR along water courses such as the South Saskatchewan and Bow rivers. Reporting rates in the Boreal Forest and Rocky Mountain NRs were lower likely because these NRs are dominated by coniferous forests.

Declines in relative abundance were detected in the Boreal Forest, Foothills, Grassland, and Parkland NRs where, relative to other species, Baltimore Orioles were observed less often in Atlas 2 than in Atlas 1. The splitting of the Northern Oriole into two separate species has not likely had a large effect on the trends that were detected by the Atlas because the Bullock's Oriole is rare in Alberta. No change was detected in the Rocky Mountain NR. Similar to what was found by the Atlas in most NRs, the Breeding Bird Survey found a decline in Alberta and across Canada during the period 1985–2005. The Alberta Government changed the status of this species from Secure in 2000 to Sensitive in 2005, due to the observed declines and to cultivation that has threatened parkland habitat.

TEMPORAL REPORTING RATE

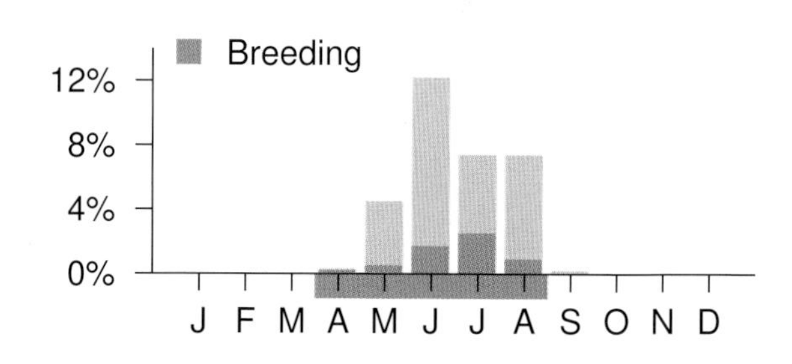

NATURAL REGIONS REPORTING RATE

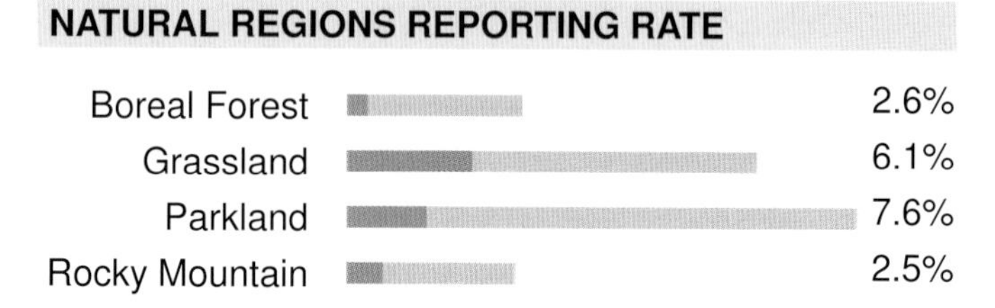

SPATIAL REPORTING RATE

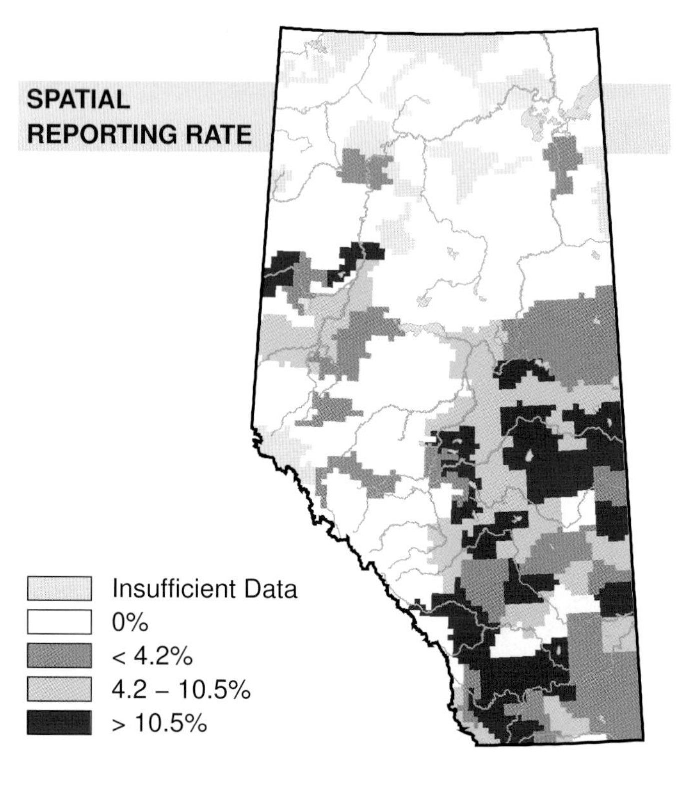

STATUS

GSWA 2000	GSWA 2005	COSEWIC Historic Rankings	COSEWIC 2007	Alberta Wildlife Act
Secure	Sensitive	N/A	N/A	N/A

Habitat  Nest Location  Nest Type 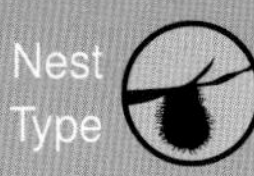 Diet  

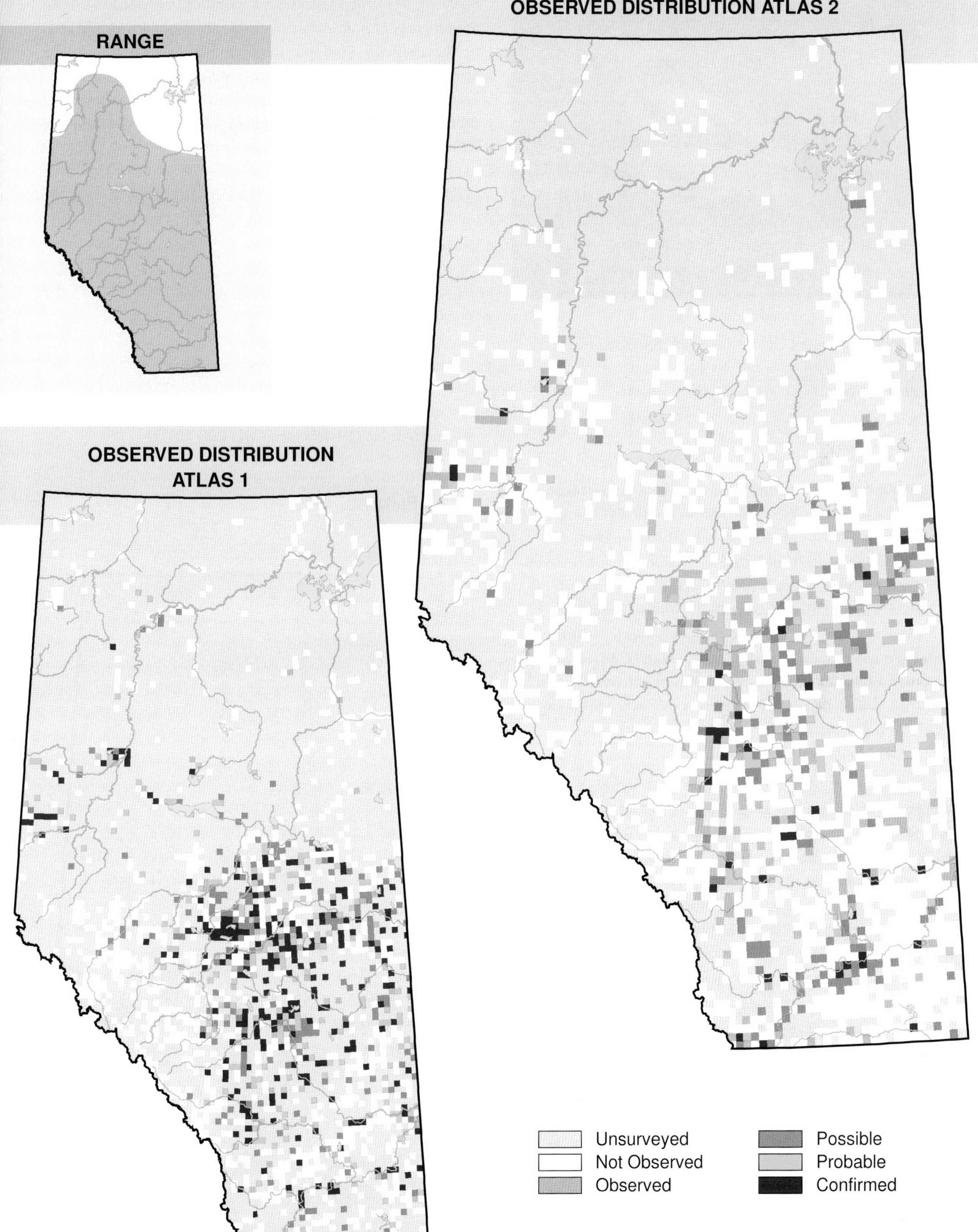

Bullock's Oriole (*Icterus bullockii*)

Photo: Royal Alberta Museum

NESTING

CLUTCH SIZE (EGGS):	4–5
INCUBATION (DAYS):	12–15
FLEDGING (DAYS):	12–14
NEST HEIGHT (METRES):	7–10

After the completion of Atlas 1, the Northern Oriole taxon was split into two species: the Baltimore and Bullock's orioles. As records from Atlas 1 did not distinguish between the two species, comparisons would need to have been made between the Northern Oriole records in Atlas 1 and the combined records of the Baltimore and Bullock's orioles from Atlas 2. However, as the known range of Bullock's Oriole is contained with the range of the Baltimore Oriole and is relatively rare in the province, it is not possible to determine if the range of this new species has changed. The difference between these two species' ranges and the relative rarity of Bullock's Oriole in comparison with the Baltimore Oriole is very clearly reflected in the distribution maps for this species.

Similar to the Baltimore Oriole, Bullock's Oriole usually breeds in deciduous forests. These forests are usually fragmented and have a low tree density. This species often nests near water. Due to its habitat preferences and southern range, Bullock's Oriole was found mainly in the Grassland NR along rivers such as the South Saskatchewan River and Red Deer River.

It is not possible to investigate whether there are relative abundance trends for this species because it was not recognized as a unique species during Atlas 1. The Breeding Bird Survey, however, has kept a separate track of these two species during the progress of both Atlases because the Northern Oriole had been classified as two species previously. Despite this advantage, the BBS did not report trends in Alberta for this species due to small sample size. However, declines were found in British Columbia and across Canada during the period 1985–2005. Bullock's Oriole is considered Secure in Alberta.

TEMPORAL REPORTING RATE

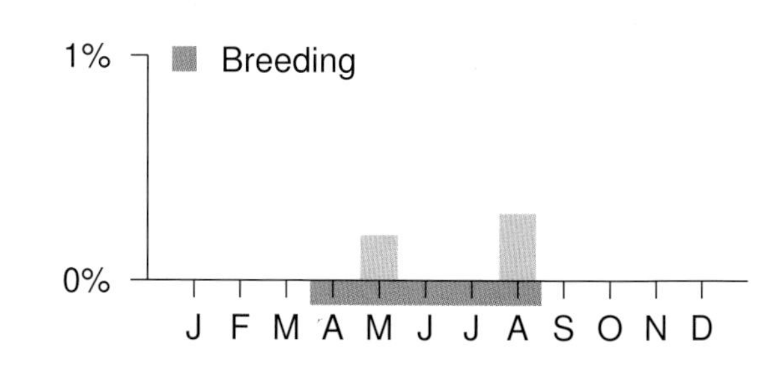

NATURAL REGIONS REPORTING RATE

SPATIAL REPORTING RATE

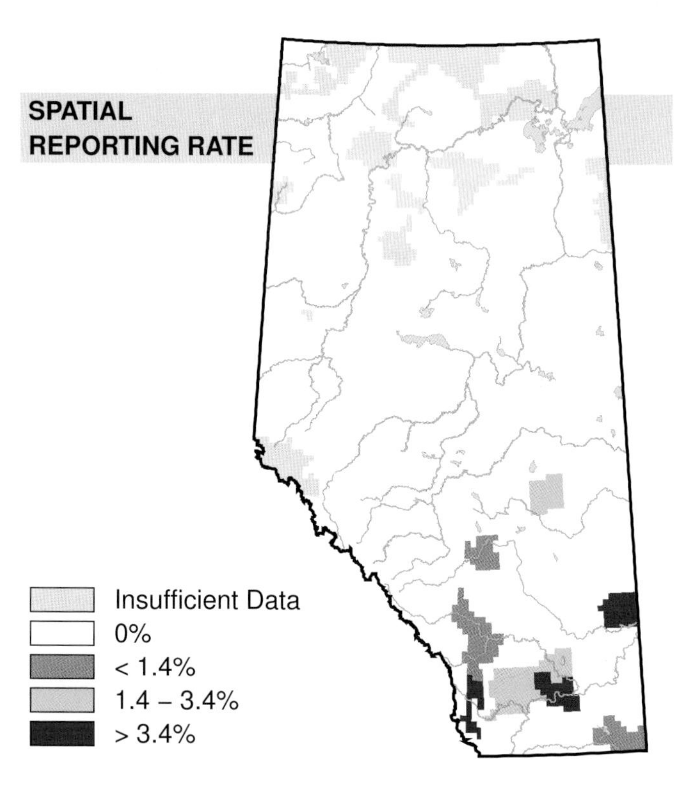

STATUS

GSWA 2000	GSWA 2005	COSEWIC Historic Rankings	COSEWIC 2007	Alberta Wildlife Act
Secure	Secure	N/A	N/A	N/A

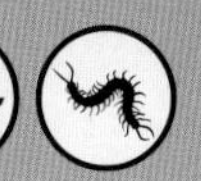

Habitat
Nest Location
Nest Type
Diet

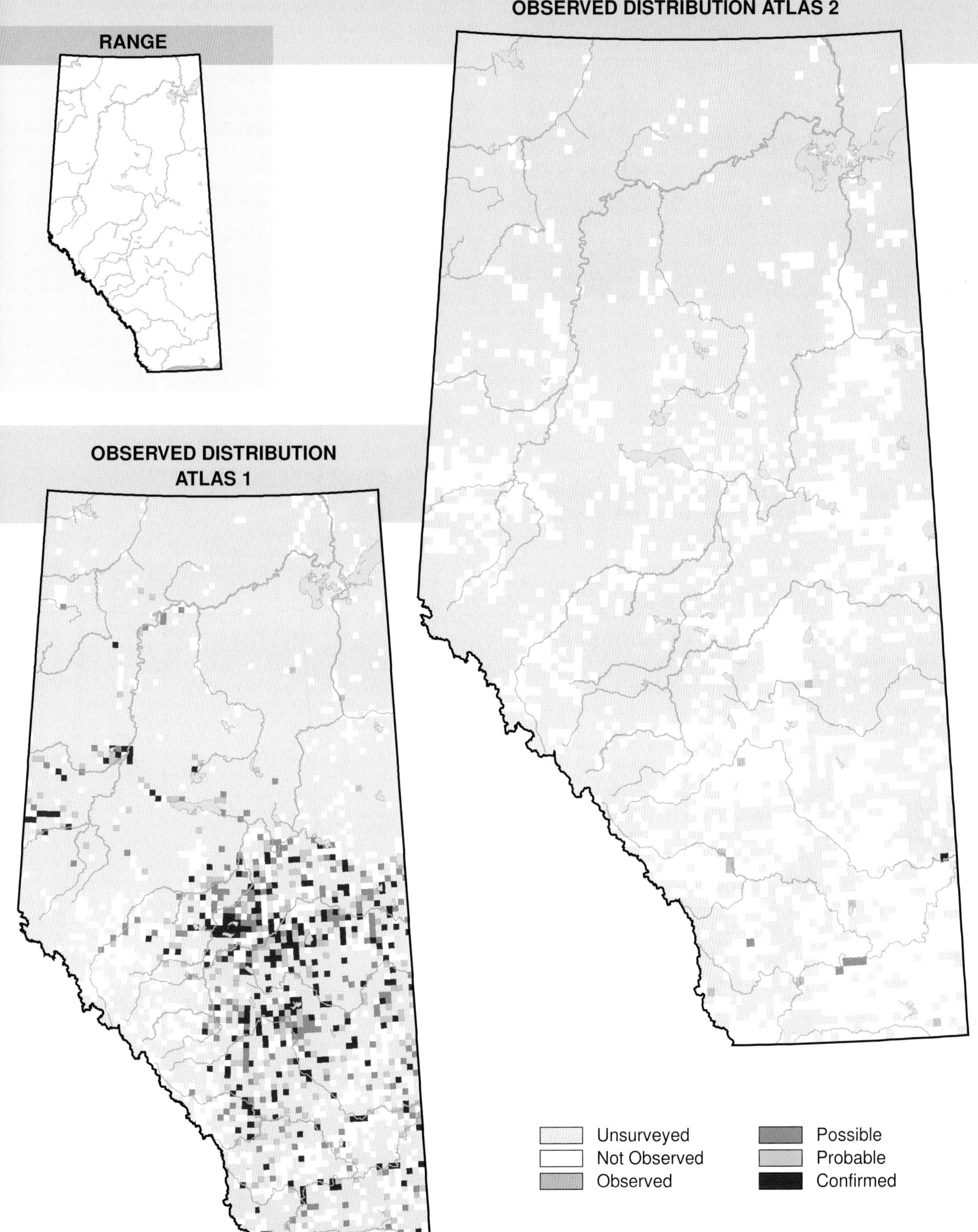

RANGE
OBSERVED DISTRIBUTION ATLAS 1
OBSERVED DISTRIBUTION ATLAS 2
Unsurveyed
Not Observed
Observed
Possible
Probable
Confirmed

Gray-crowned Rosy-Finch (*Leucosticte tephrocotis*)

Photo: Gerald Romanchuk

NESTING

CLUTCH SIZE (EGGS):	4–5
INCUBATION (DAYS):	12–14
FLEDGING (DAYS):	32–34
NEST HEIGHT (METRES):	0–7.6

The Gray-crowned Rosy-Finch breeds in the Rocky Mountain Natural Region in Alberta. The observed distribution of this species did not change between Atlas 1 and Atlas 2.

The Gray-crowned Rosy-Finch breeds at high altitudes in rocky alpine areas such as talus slopes, rock piles, and cliffs. Breeding habitat is usually above treeline and is often near glaciers. For this reason, this species was found most often in the Rocky Mountain NR. Observations made in the Boreal Forest, Foothills, Grassland, and Parkland NRs were of non-breeders.

Changes in relative abundance were not detected for this species in any natural region. These results should be interpreted with caution, though, because the sample size was small for this species. Relative to other species, the Gray-crowned Rosy-Finch was observed less frequently in the Boreal Forest, Foothills, Parkland, and Rocky Mountain NRs and more frequently in the Grassland NR in Atlas 2 than in Atlas 1. There were not enough Gray-crowned Rosy-Finch records to detect population changes in Alberta or Canada using Breeding Bird Survey data. Due to the specificity of this species' habitat requirements and the remote distribution of its habitat, it is not effectively monitored by large, wide-spread, multi-species surveys such as the Atlas or the Breeding Bird Survey. As all species of rosy-finch are mixed together in winter counts (Dunn, 2005), surveys designed specifically to monitor the Gray-crowned Rosy-Finch are needed to determine adequately the population trends for this species.

Dunn (2005) reports that little is known about the status of this species. There is an apparent population decline, although most of the montane breeding habitat is not under threat, and the species appears to be tolerant of human-impacted landscapes in winter.

This species is considered Secure in Alberta.

TEMPORAL REPORTING RATE

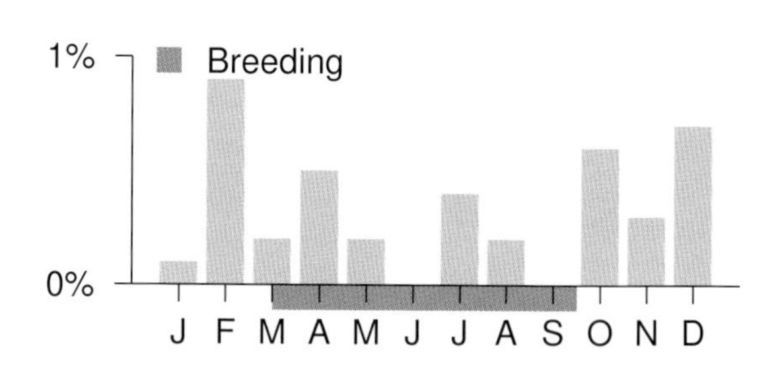

NATURAL REGIONS REPORTING RATE

Foothills	0.2%
Parkland	< 0.1%
Rocky Mountain	2.3%

SPATIAL REPORTING RATE

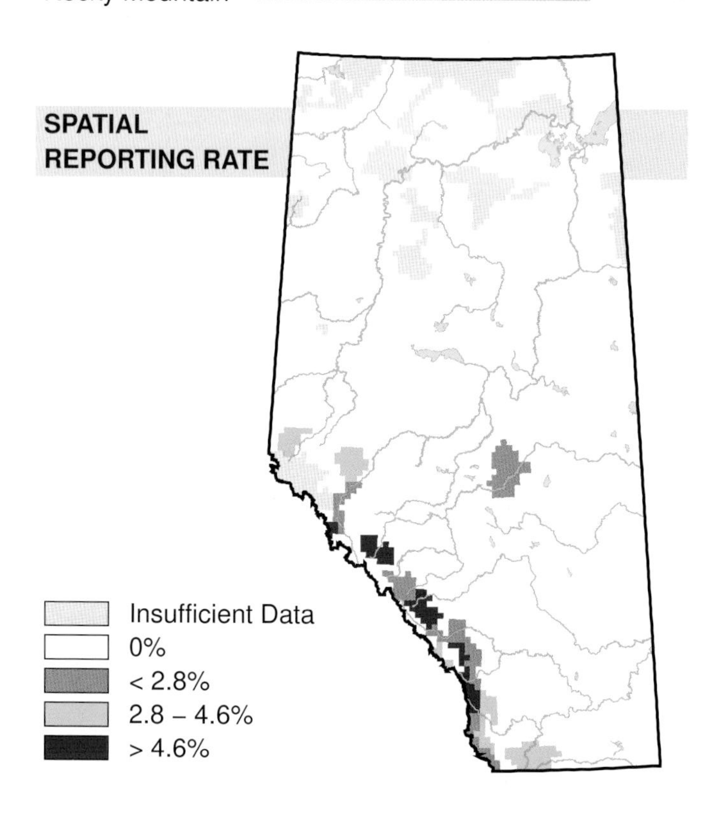

STATUS

GSWA 2000	GSWA 2005	COSEWIC Historic Rankings	COSEWIC 2007	Alberta Wildlife Act
Secure	Secure	N/A	N/A	N/A

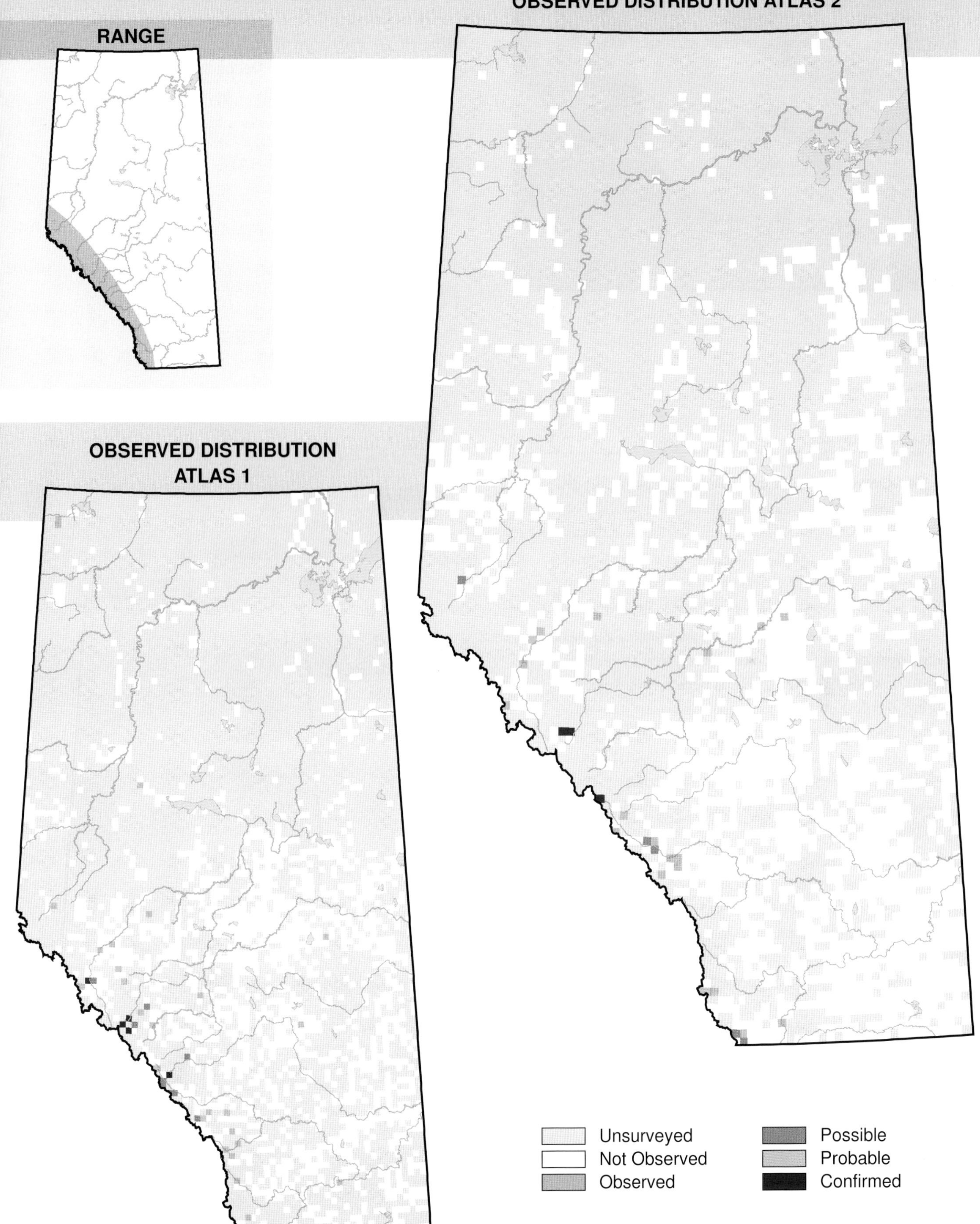
RANGE
OBSERVED DISTRIBUTION ATLAS 2
OBSERVED DISTRIBUTION ATLAS 1
Unsurveyed
Not Observed
Observed
Possible
Probable
Confirmed

Pine Grosbeak (*Pinicola enucleator*)

Photo: Gerald Romanchuk

NESTING

CLUTCH SIZE (EGGS):	4–5
INCUBATION (DAYS):	13–14
FLEDGING (DAYS):	13–20
NEST HEIGHT (METRES):	3–4.5

The Pine Grosbeak breeds in the Boreal Forest, Foothills, Parkland, and Rocky Mountain Natural Regions. The observed distribution of this species decreased in Atlas 2 in the northern part of the province, but it is unclear whether this represents an actual range contraction or whether the presence of Pine Grosbeaks was missed in the north during Atlas 2.

The Pine Grosbeak breeds in coniferous forests. This species prefers open forests and is not found in forests with high stand densities. As such, this species can benefit from selective logging or forest thinning provided coniferous trees are retained. Due to its close association with coniferous trees, this species was most often encountered in the Boreal Forest, Foothills, and Rocky Mountain NRs. It was infrequently encountered in the Parkland. Records from the Grassland NR were non-breeders.

An increase in relative abundance was detected in the Rocky Mountain NR where, relative to other species, the Pine Grosbeak was observed more frequently in Atlas 2 than in Atlas 1. An abundance increase in this NR is possible because most of the NR is within parks where the coniferous forests that this species requires are not harvested. No change was detected in the Boreal Forest, Foothills, Grassland, and Parkland NRs. However, relative to other species, Pine Grosbeaks were observed less frequently in Atlas 2 than in Atlas 1 in these NRs. Breeding Bird Survey sample size was too small to investigate Pine Grosbeak abundance changes in Alberta, but an abundance decline was detected across Canada during the period 1985–2005. Declines could be related to the harvesting of coniferous forests. This species is considered Secure in Alberta.

TEMPORAL REPORTING RATE

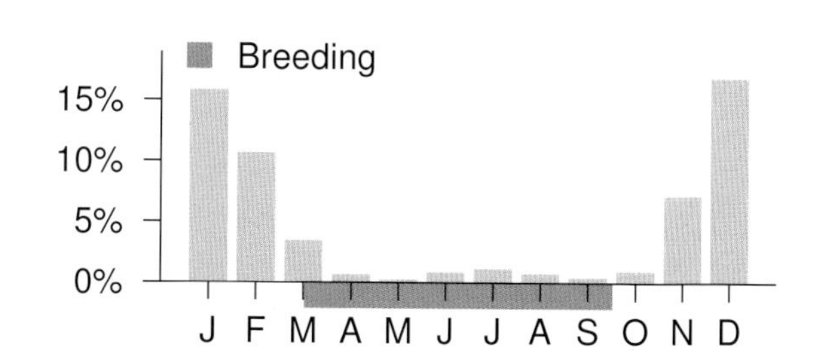

NATURAL REGIONS REPORTING RATE

SPATIAL REPORTING RATE

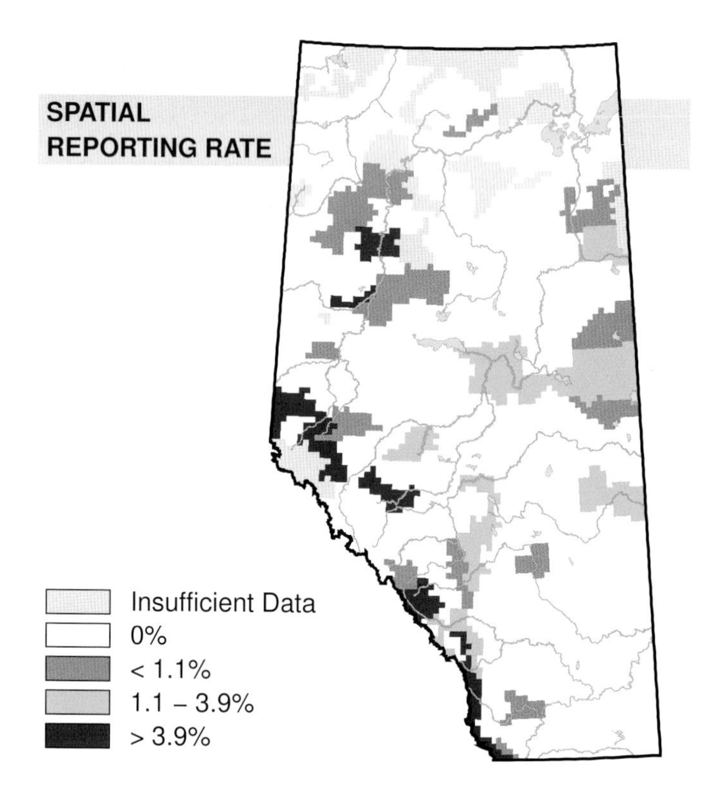

STATUS

GSWA 2000	GSWA 2005	COSEWIC Historic Rankings	COSEWIC 2007	Alberta Wildlife Act
Secure	Secure	N/A	N/A	N/A

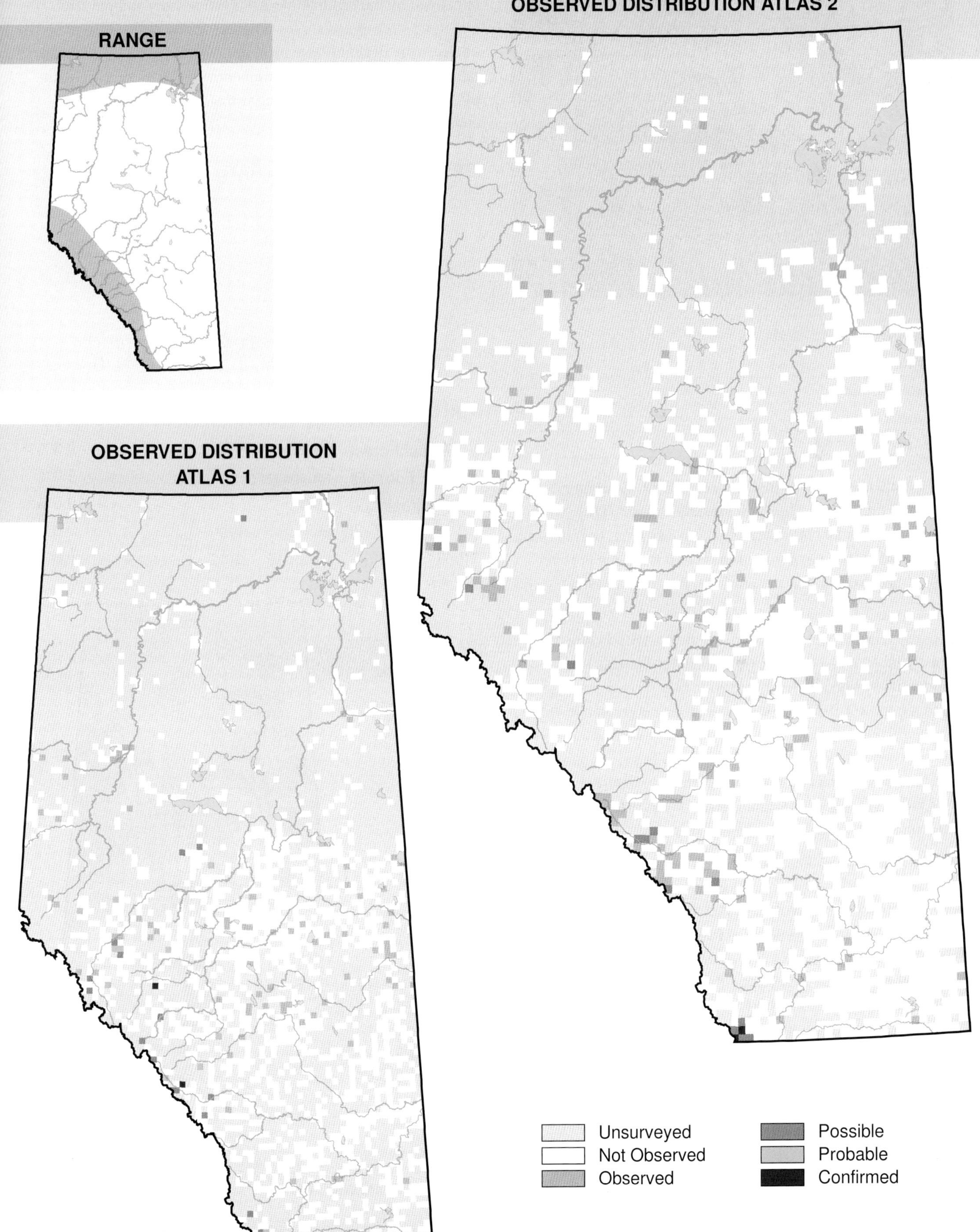
OBSERVED DISTRIBUTION ATLAS 2
RANGE
OBSERVED DISTRIBUTION ATLAS 1
Unsurveyed
Not Observed
Observed
Possible
Probable
Confirmed

Purple Finch (*Carpodacus purpureus*)

Photo: Robert Gehlert

NESTING

CLUTCH SIZE (EGGS):	4–5
INCUBATION (DAYS):	13
FLEDGING (DAYS):	14
NEST HEIGHT (METRES):	18

The Purple Finch breeds in the Boreal Forest, Foothills, Parkland, and Rocky Mountain Natural Regions. There were fewer records in the northwest during Atlas 2 but, overall, the observed distribution of this species did not change between Atlas 1 and Atlas 2.

The Purple Finch can nest in a variety of forests, but coniferous trees are usually present in areas where this species breeds. This species can have irruptive populations because their productivity is tied to yearly variation in conifer cone production. In years when cone production is greater, Purple Finches have greater reproductive output. Due to its association with conifers, this species was found primarily in the Foothills NR and it was quite common in the Boreal Forest, Parkland, and Rocky Mountain NRs. Observations made in the Grassland NR were records of non-breeders.

Declines in relative abundance were detected in the Boreal Forest and Parkland NR where, relative to other species, the Purple Finch was observed less frequently in Atlas 2 than in Atlas 1. No change was detected in the Foothills, Grassland, and Rocky Mountain NRs. The Breeding Bird Survey did not reveal an abundance change in Alberta. However, similar to what was found for some NRs by the Atlas, a decline was found on a Canada-wide scale during the period 1985–2005. The reason that the Atlas found a negative trend while the Breeding Bird Survey found no change in Alberta could be explained by the fact that the Atlas had a larger sample size and more widespread survey coverage in the north. It is unclear why declines have been found, although they could be related to a reduction in the number of conifers; these trees are often targeted during forest harvesting. This species is considered Secure in Alberta.

TEMPORAL REPORTING RATE

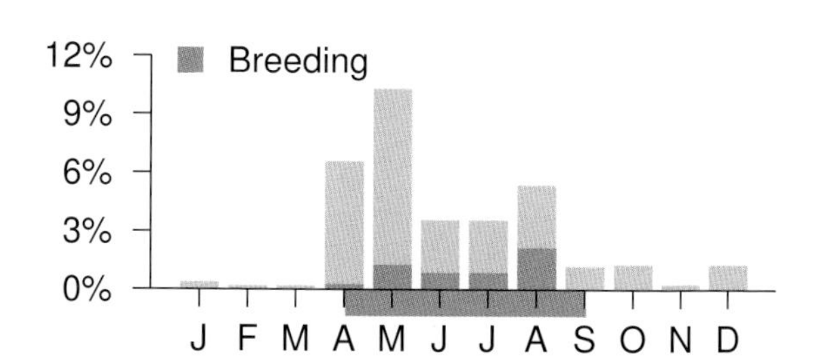

NATURAL REGIONS REPORTING RATE

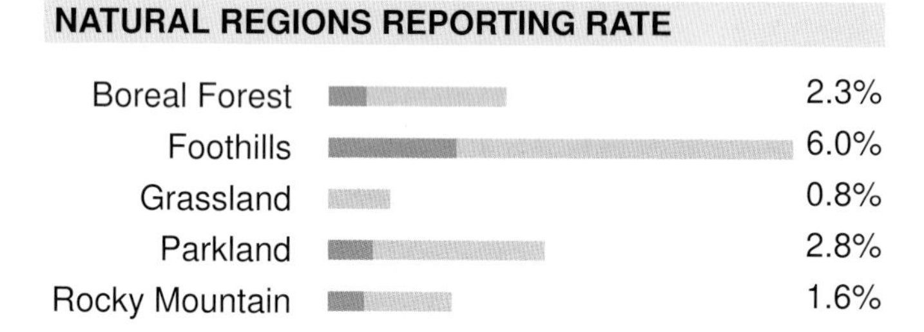

SPATIAL REPORTING RATE

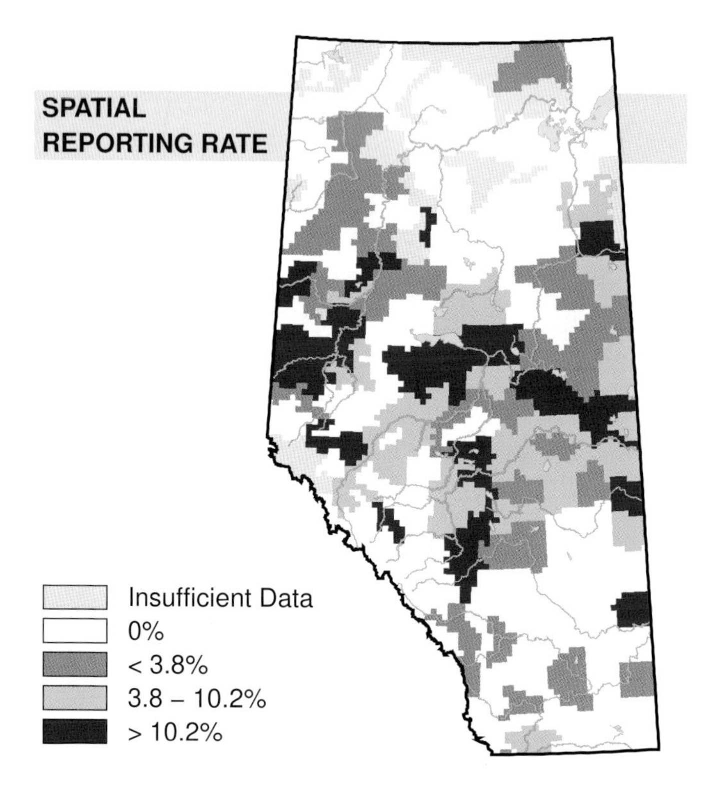

STATUS

GSWA 2000	GSWA 2005	COSEWIC Historic Rankings	COSEWIC 2007	Alberta Wildlife Act
Secure	Secure	N/A	N/A	N/A

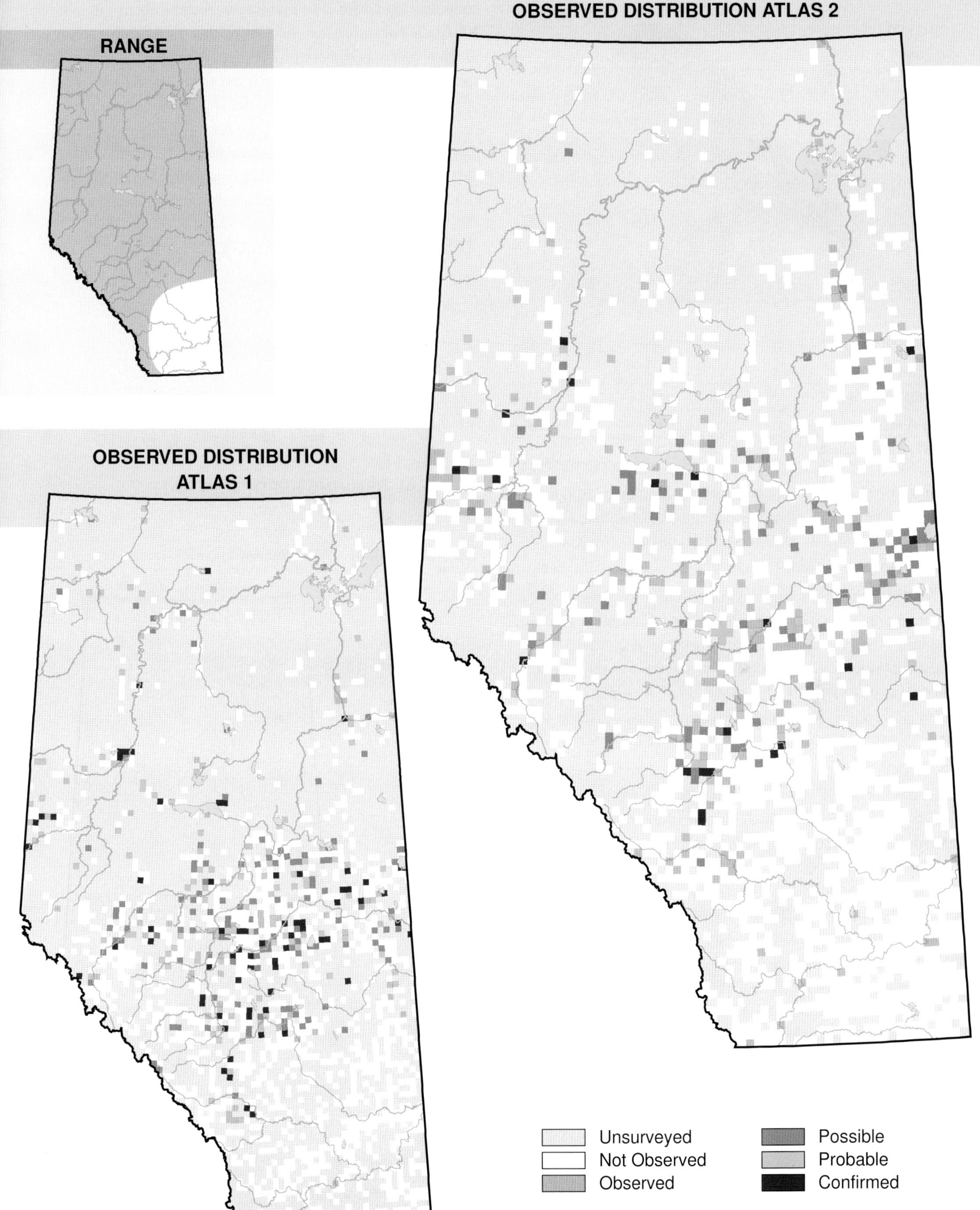
RANGE
OBSERVED DISTRIBUTION ATLAS 2
OBSERVED DISTRIBUTION
ATLAS 1
Unsurveyed
Not Observed
Observed
Possible
Probable
Confirmed

Cassin's Finch (*Carpodacus cassinii*)

Photo: Raymond Toal

NESTING

CLUTCH SIZE (EGGS):	4–5
INCUBATION (DAYS):	12–14
FLEDGING (DAYS):	14
NEST HEIGHT (METRES):	3–25

One of Alberta's rarest finches, Cassin's Finch breeds in the southwest part of the province in the Rocky Mountain Natural Region. The distribution of this species did not change between Atlas 1 and Atlas 2.

Cassin's Finch prefers to breed in coniferous forests in mountainous areas. Generally thought of as a western species, Cassin's Finch can be observed in the extreme southwestern corner of Alberta. Consequently, this species was most often encountered in the Rocky Mountain NR. It was infrequently recorded in the Parkland and Grassland NRs where there was no breeding evidence associated with any of the observations of this species.

Changes in relative abundance were not detected for this species in any natural region. However, these results should be interpreted with caution because the sample size for this species was small. Relative to other species, Cassin's Finch was observed more frequently in the Rocky Mountain NR in Atlas 2 than in Atlas 1. There were not enough Cassin's Finch records to detect population changes using Breeding Bird Survey data in Alberta, but a decline was detected in British Columbia and across Canada during the period 1985–2005. Further declines are expected in Canada, at least in the short term, because the Mountain Pine Beetle has killed many coniferous trees in British Columbia and the range of the beetle continues to expand. The inability to detect a change in Alberta is not surprising considering that, in Alberta, Cassin's Finch is on the periphery of its range. Along the periphery of a species' range, high variability is expected, even if the population is stable. Therefore, tracking changes within its core range, rather than on the periphery, is likely to give the more reliable measure of its general status. In Alberta, this means referring to BBS and other data sources from British Columbia. This species is considered Secure in Alberta.

TEMPORAL REPORTING RATE

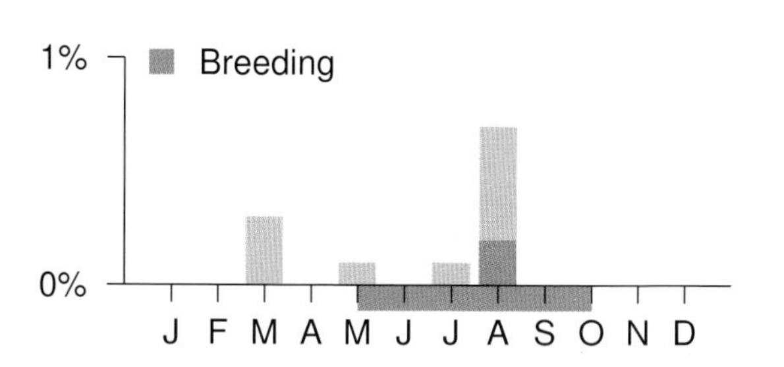

NATURAL REGIONS REPORTING RATE

SPATIAL REPORTING RATE

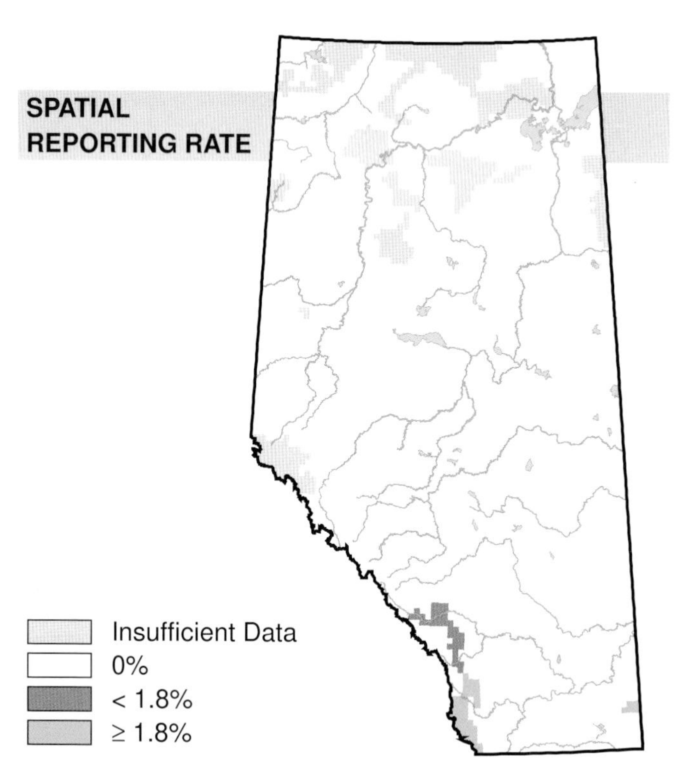

STATUS

GSWA 2000	GSWA 2005	COSEWIC Historic Rankings	COSEWIC 2007	Alberta Wildlife Act
Secure	Secure	N/A	N/A	N/A

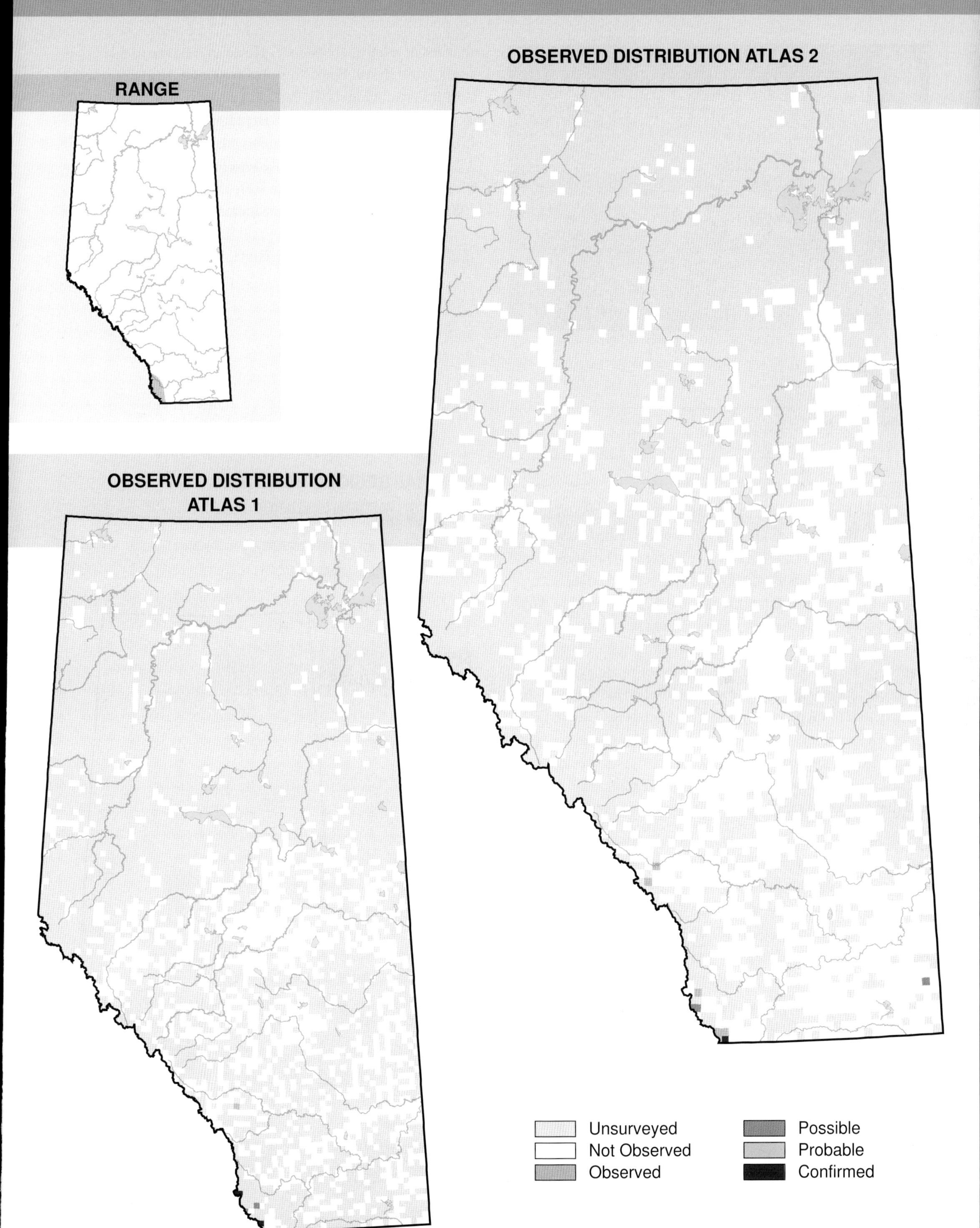
Habitat
Nest Location
Nest Type
Diet
RANGE
OBSERVED DISTRIBUTION ATLAS 2
OBSERVED DISTRIBUTION ATLAS 1
Unsurveyed
Not Observed
Observed
Possible
Probable
Confirmed

House Finch (*Carpodacus mexicanus*)

Photo: Randy Jensen

NESTING

CLUTCH SIZE (EGGS):	4–5
INCUBATION (DAYS):	12–14
FLEDGING (DAYS):	11–19
NEST HEIGHT (METRES):	1.5–10.7

The House Finch is found across the southern portion of the province. The distribution of this species expanded between Atlas 1 and Atlas 2. During Atlas 1, House Finches were found infrequently and there was only one notice of breeding evidence (possible breeding) found in the Jasper area. In Atlas 2, many more observations were reported and multiple breeding observations were made, including quite a few confirmed breeding records. This pattern of range expansion has been documented across the range of the House Finch. An introduction of this species on Long Island, New York in the mid-1900s facilitated its range expansion throughout the eastern part of the continent. In recent years, this species has become more common along the northern extent of its range and the edge of its range appears to be shifting northward.

The House Finch tends to be found near people in cities and towns, especially in older neighbourhoods that have established trees and shrub patches. This species will also be found occasionally in rural areas near farms and human settlements. They are found infrequently along the edges of sparsely treed coniferous forests. This species tends to avoid continuous or densely treed habitats. Alberta is on the northern periphery of this species' range; therefore, it was found most often in the Grassland NR and less often in the Parkland NR. It was found infrequently in the Boreal Forest NR.

With so few records in Atlas 1 it was not possible to evaluate changes in relative abundance. However, relative to other species, the House Finch was observed more frequently in the Boreal Forest, Grassland, Parkland, and Rocky Mountain NRs in Atlas 2 than in Atlas 1. The Breeding Bird Survey sample size was too small to investigate trends for this species in Alberta. An abundance increase was found by the Breeding Bird Survey on a Canada-wide scale during the period 1985–2005. This species is considered Secure in Alberta.

TEMPORAL REPORTING RATE

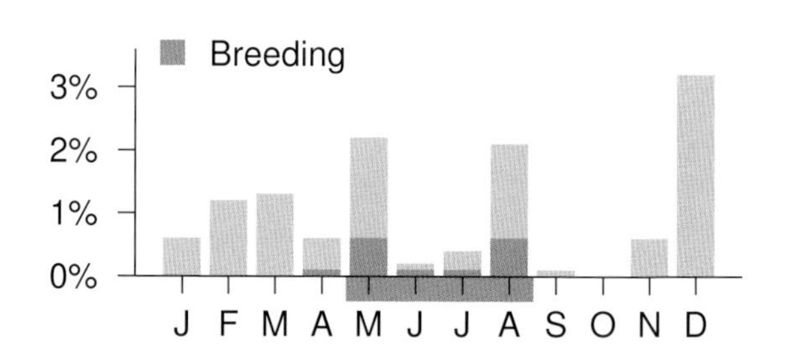

NATURAL REGIONS REPORTING RATE

SPATIAL REPORTING RATE

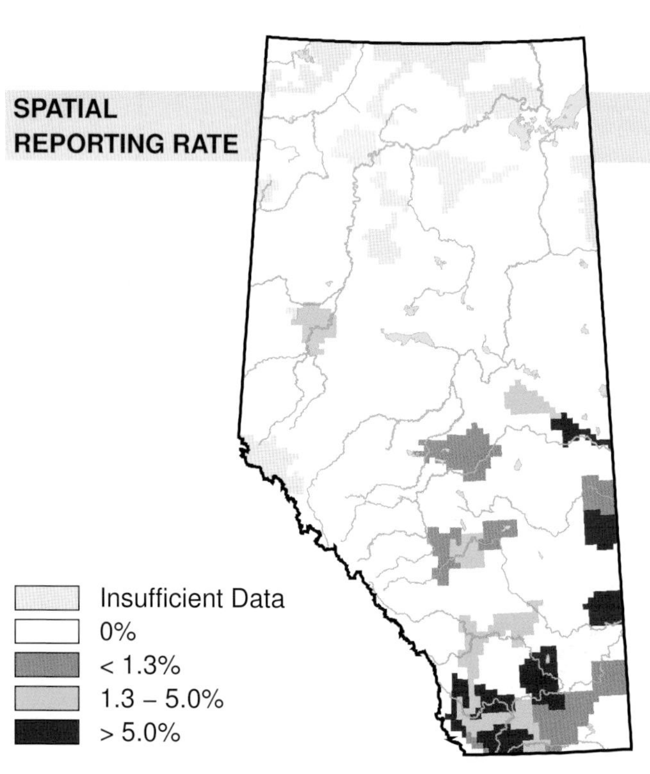

STATUS

GSWA 2000	GSWA 2005	COSEWIC Historic Rankings	COSEWIC 2007	Alberta Wildlife Act
Secure	Secure	N/A	N/A	N/A

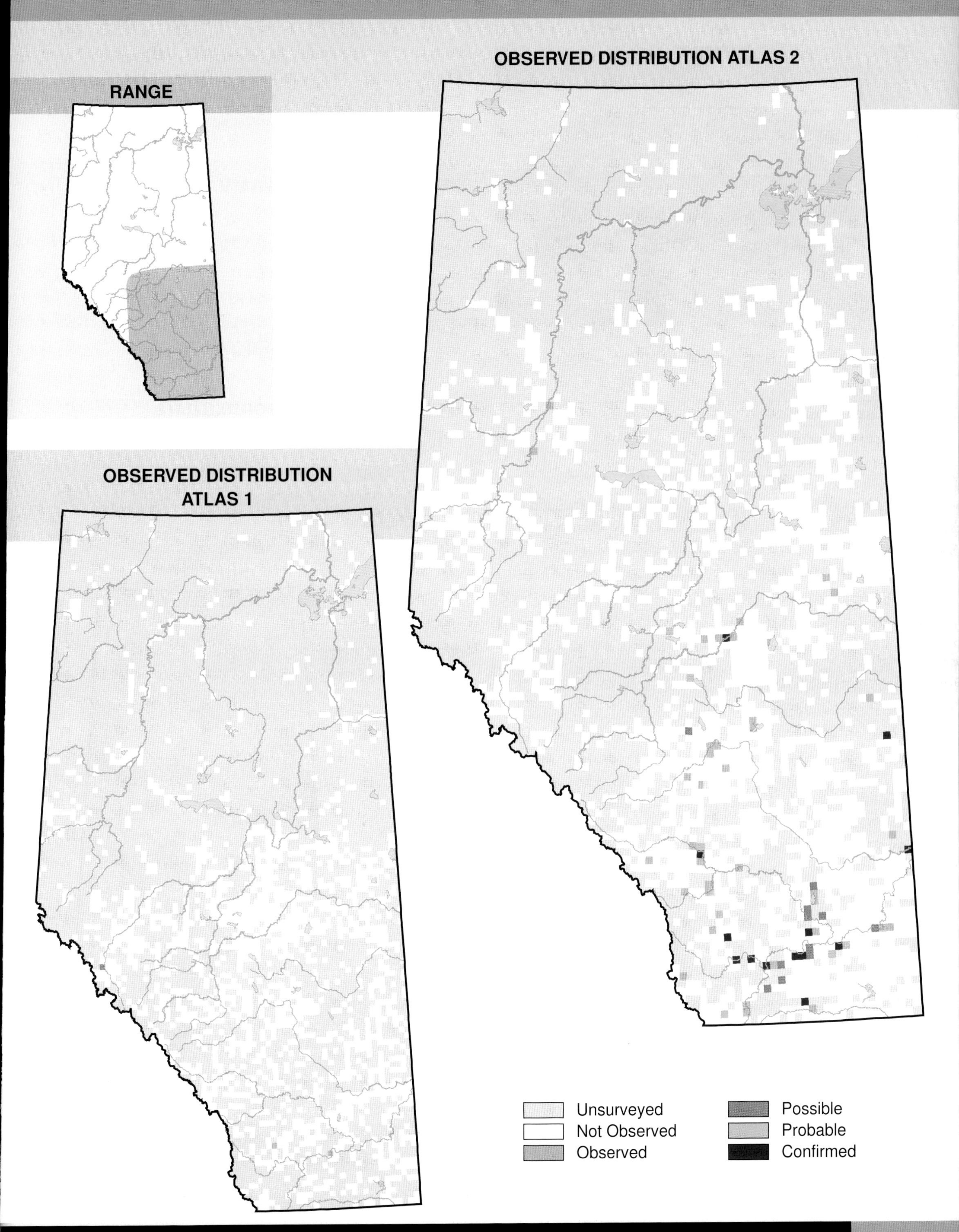
RANGE
OBSERVED DISTRIBUTION ATLAS 2
OBSERVED DISTRIBUTION ATLAS 1
Unsurveyed
Not Observed
Observed
Possible
Probable
Confirmed

Red Crossbill (*Loxia curvirostra*)

Photo: Gerald Romanchuk

NESTING

CLUTCH SIZE (EGGS):	3–5
INCUBATION (DAYS):	12–15
FLEDGING (DAYS):	17–22
NEST HEIGHT (METRES):	1–18

The Red Crossbill is rare in Alberta, but it can be found in the Boreal Forest, Foothills, Parkland, and Rocky Mountain Natural Regions. The observed distribution of this species did not change between Atlas 1 and Atlas 2.

The Red Crossbill is closely associated with conifer trees. The crossed bill of this species is specialized for prying open conifer cones so that the seeds that are contained within can be extracted efficiently. The timing of breeding is closely tied to the availability of cones. Therefore, unlike many other types of birds, crossbills can breed throughout most of the year. Due to their close association with conifers, and because their range encompasses more of the western part of the province, Red Crossbills were found mainly in the Foothills and Rocky Mountain NRs. This species was rare in the Boreal Forest and Parkland NRs. Observations made in the Grassland NR were records of non-breeders.

Relative abundance changes were not detected in any NR. However, this result should be interpreted with caution: the sample size for this species was small because Red Crossbills are sparsely distributed across the province. Relative to other species, they were observed more frequently in Atlas 2 than in Atlas 1 in all NRs. The Breeding Bird Survey did not detect an abundance change in Alberta; yet, a decline was detected on a Canada-wide scale during the period 1985–2005. It is unclear why this decline was found, but it could be related to a reduction in the number of conifers because these trees are often targeted during forest harvesting. This species is considered Secure in Alberta.

TEMPORAL REPORTING RATE

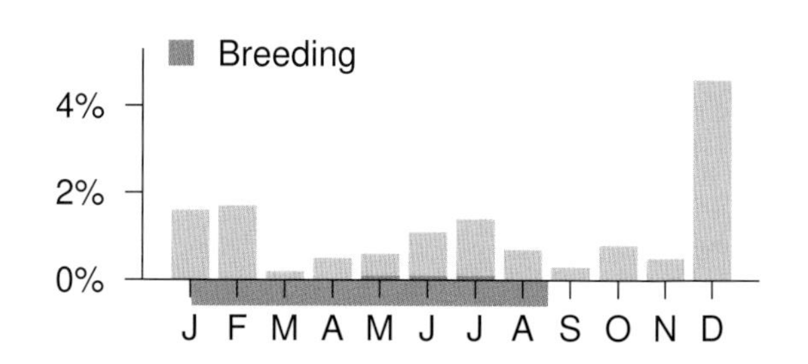

NATURAL REGIONS REPORTING RATE

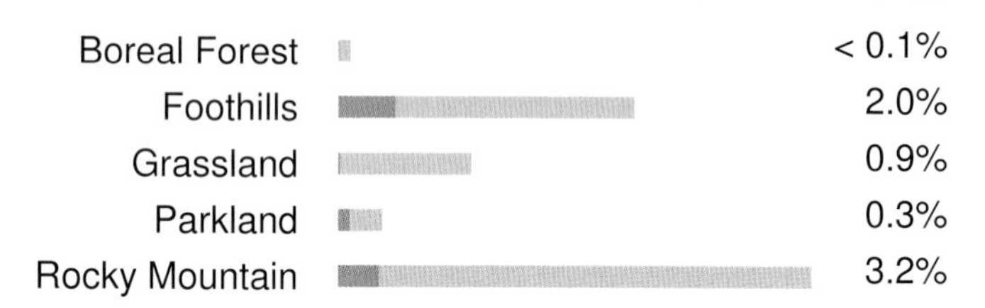

SPATIAL REPORTING RATE

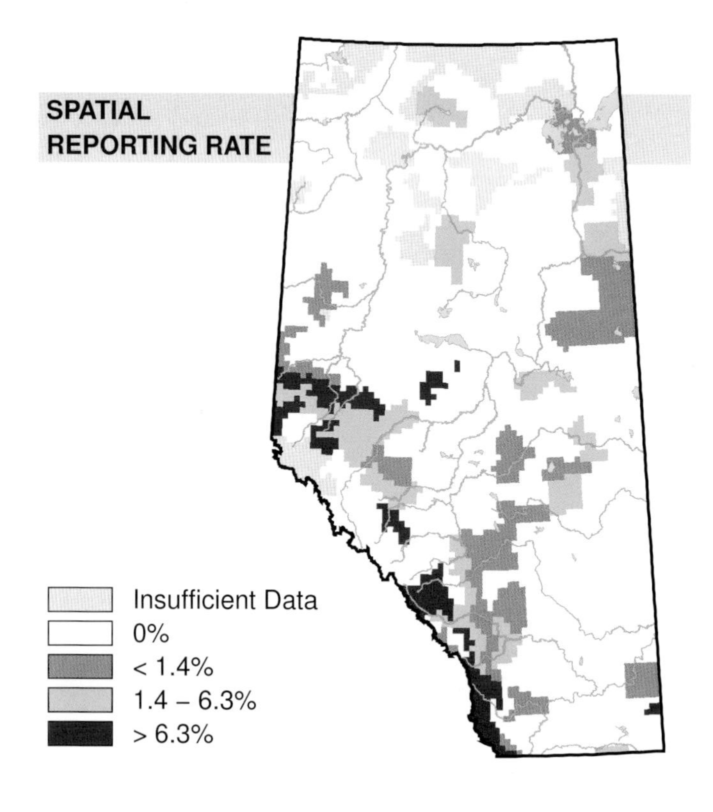

STATUS

GSWA 2000	GSWA 2005	COSEWIC Historic Rankings	COSEWIC 2007	Alberta Wildlife Act
Secure	Secure	N/A	N/A	N/A

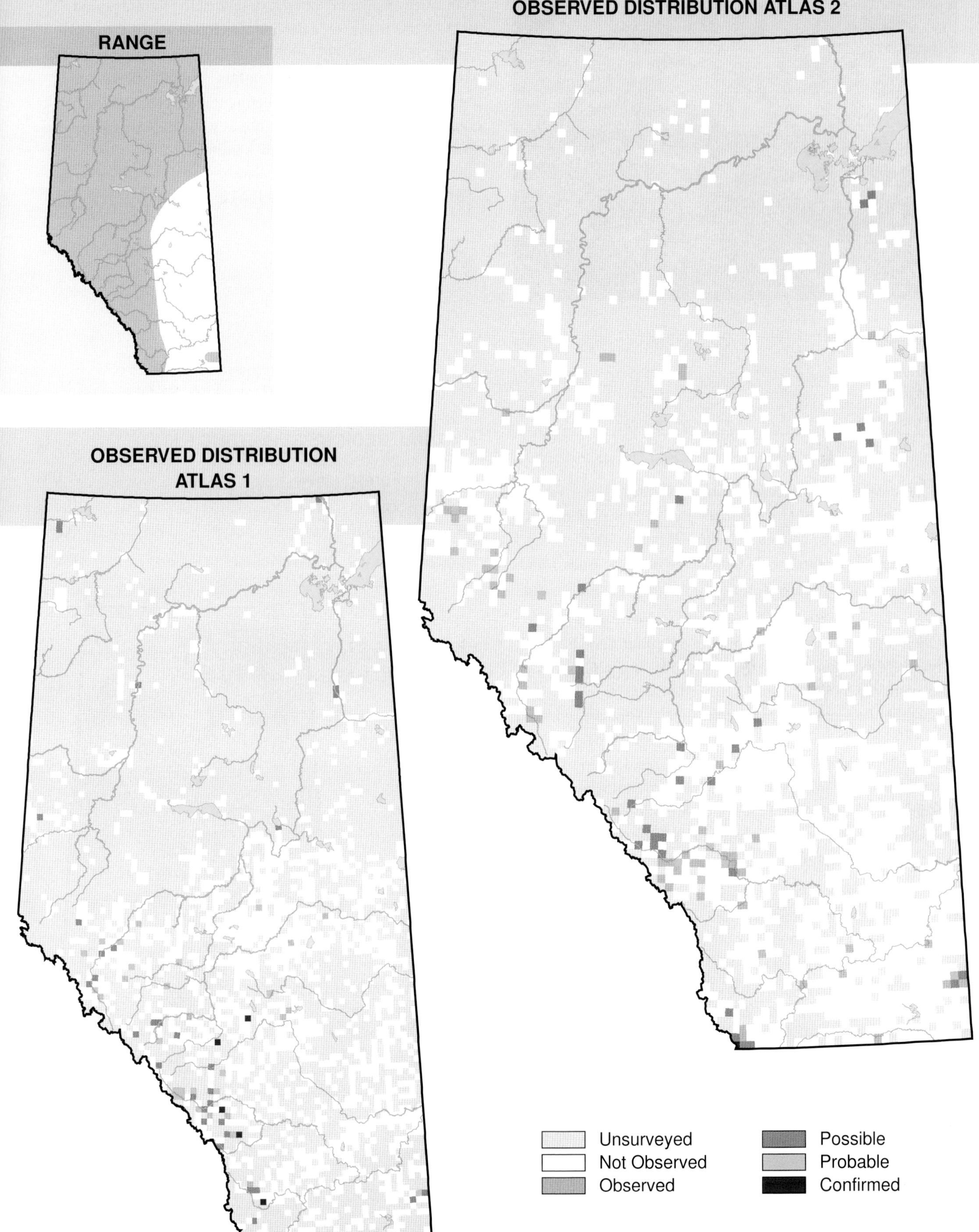
OBSERVED DISTRIBUTION ATLAS 2
RANGE
OBSERVED DISTRIBUTION ATLAS 1
Unsurveyed
Not Observed
Observed
Possible
Probable
Confirmed

White-winged Crossbill (*Loxia leucoptera*)

Photo: Gordon Court

NESTING

CLUTCH SIZE (EGGS):	3–4
INCUBATION (DAYS):	12–14
FLEDGING (DAYS):	unknown
NEST HEIGHT (METRES):	1–20

The more common of Alberta's two crossbills, the White-winged Crossbill breeds in the Boreal Forest, Foothills, Parkland, and Rocky Mountain Natural Regions. The distribution of this species did not change between Atlas 1 and Atlas 2.

White-winged Crossbills are considered nomadic because the location of their breeding territories can be quite varied between years depending on the availability of conifer seeds, their preferred food. This bird's crossed bill enables it to extract the seeds contained in conifer cones. Similar to the Red Crossbill, the White-winged Crossbill can breed throughout most of the year, if sufficient food is available. Due to its close association with coniferous forests, this species was most often encountered in the Boreal Forest, Foothills, and Rocky Mountain NRs. Observations made in the Grasslands were of non-breeders except for a few records along the periphery of this NR.

Increases in relative abundance were detected in the Boreal Forest and Rocky Mountain NRs where, relative to other species, the White-winged Crossbill was observed more frequently in Atlas 2 than in Atlas 1. Being nomadic, it is difficult to assess if the detected changes are related to annual variation in breeding locations or whether they are related to actual changes in population size. No change was detected in the other NRs. The Breeding Bird Survey did not detect an abundance change in Alberta, or nationally, for the period 1985–2005. However, a decline was found across Canada during the period 1995–2005. This decline could be related to a reduction in the number of conifers. In addition, shorter logging rotations could be affecting this species because conifers tend to produce more cones when they are older. This species is considered Secure in Alberta.

TEMPORAL REPORTING RATE

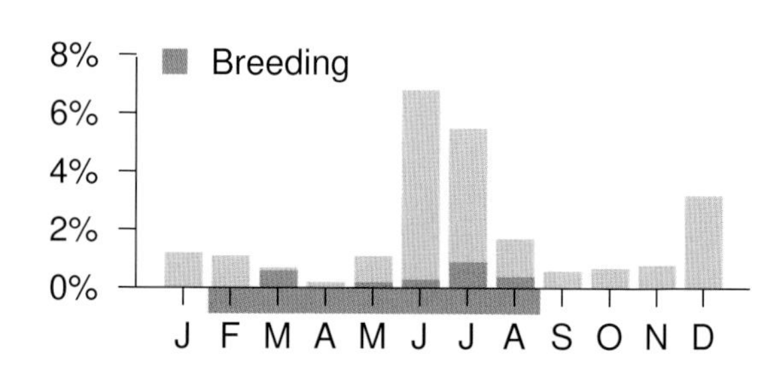

NATURAL REGIONS REPORTING RATE

SPATIAL REPORTING RATE

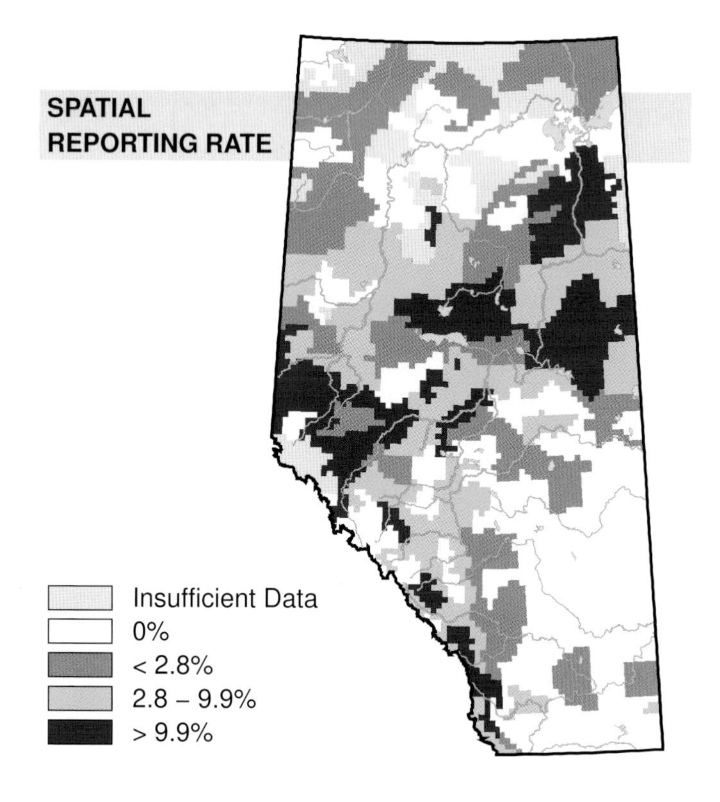

STATUS

GSWA 2000	GSWA 2005	COSEWIC Historic Rankings	COSEWIC 2007	Alberta Wildlife Act
Secure	Secure	N/A	N/A	N/A

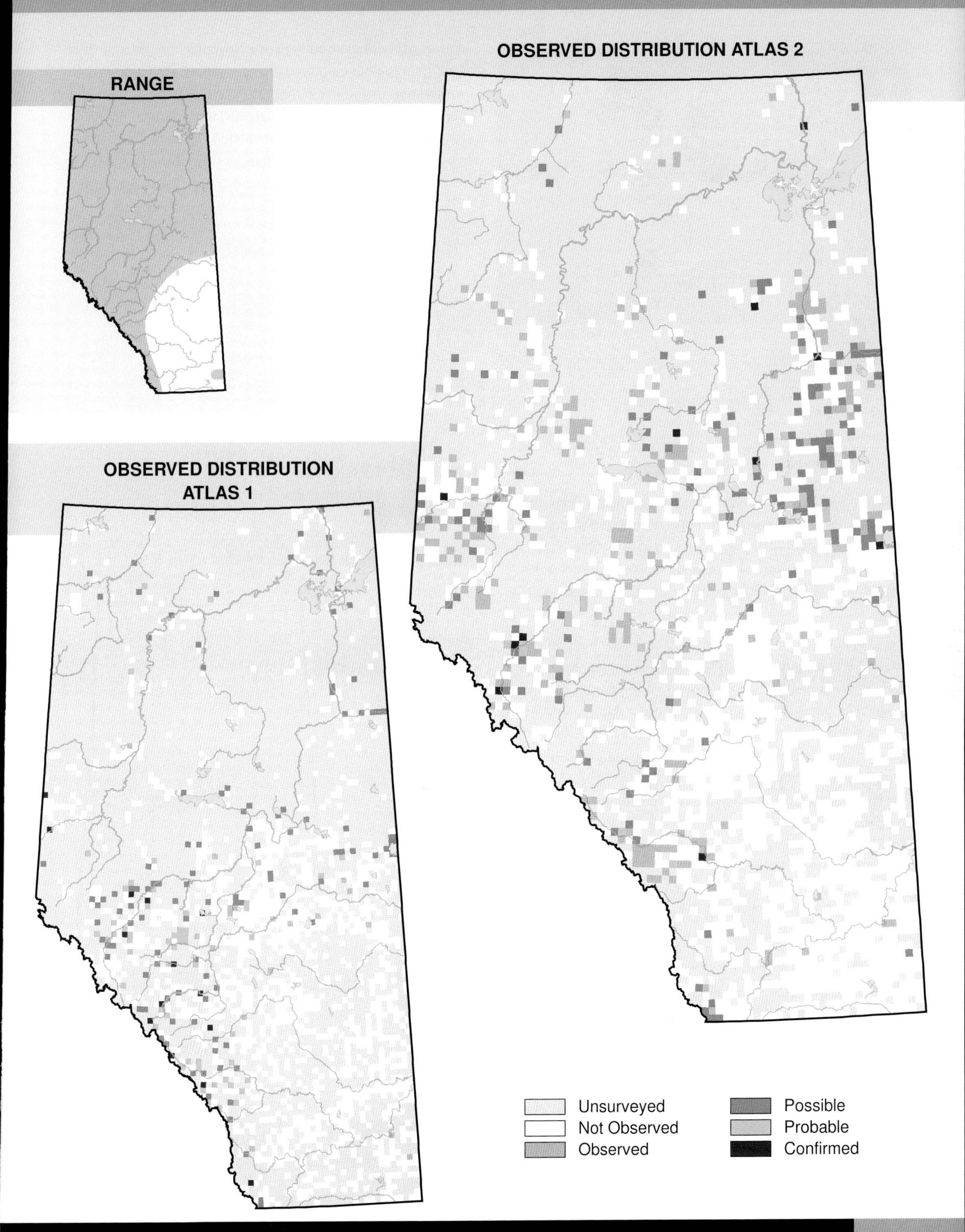
OBSERVED DISTRIBUTION ATLAS 2
RANGE
OBSERVED DISTRIBUTION ATLAS 1
Unsurveyed
Not Observed
Observed
Possible
Probable
Confirmed

Common Redpoll (*Carduelis flammea*)

Photo: Gerald Romanchuk

NESTING

CLUTCH SIZE (EGGS):	4–6
INCUBATION (DAYS):	10–11
FLEDGING (DAYS):	12
NEST HEIGHT (METRES):	1–3

The Common Redpoll is a rare breeder in Alberta. This species is most often observed in the province during the winter when it ventures south in search of food. The distribution of this species remained patchy; however, none of the observations made during Atlas 2 had breeding evidence associated with them.

The Common Redpoll breeds in open areas that contain patches of coniferous trees or shrubs. This species tends to avoid both densely forested areas and wide-open areas. The presence of vegetation is important because this species prefers to build its nest off the ground; however, it will occasionally use rocky ledges. In Alberta, this species would most likely be found breeding in the far northeast part of the province where this species' preferred type of habitat can be found. Common Redpolls were most often encountered in the Boreal Forest and Rocky Mountain NRs but were also fairly common in the Grassland and Parkland NRs. However, none of the observations made during Atlas 2 had breeding evidence associated with them. Therefore, observation rates from the natural regions do not reflect a breeding distribution.

Increases in relative abundance were detected in the Foothills, Grassland, and Rocky Mountain NRs where, relative to other species, Common Redpolls were observed more frequently in Atlas 2 than in Atlas 1. The majority of the observations made during Atlas 2 occurred between November and March. It is important to note that these changes reflect change to the wintering populations, rather than changes to breeding populations. No change was detected in the Boreal Forest and Parkland NRs. There were not enough Common Redpoll records to detect population changes using Breeding Bird Survey data in Alberta and no change was detected across Canada. This species is considered Secure in Alberta.

TEMPORAL REPORTING RATE

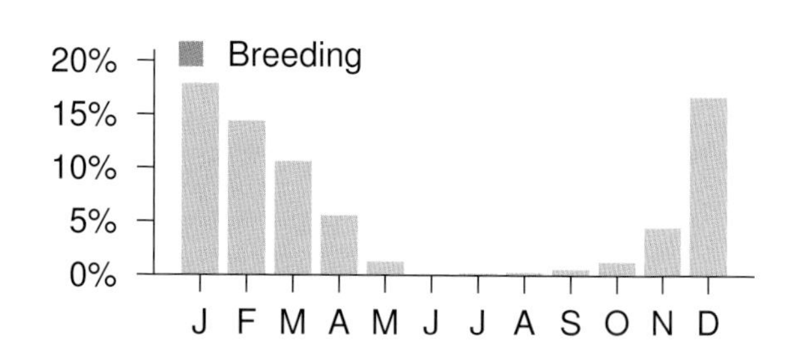

NATURAL REGIONS REPORTING RATE

SPATIAL REPORTING RATE

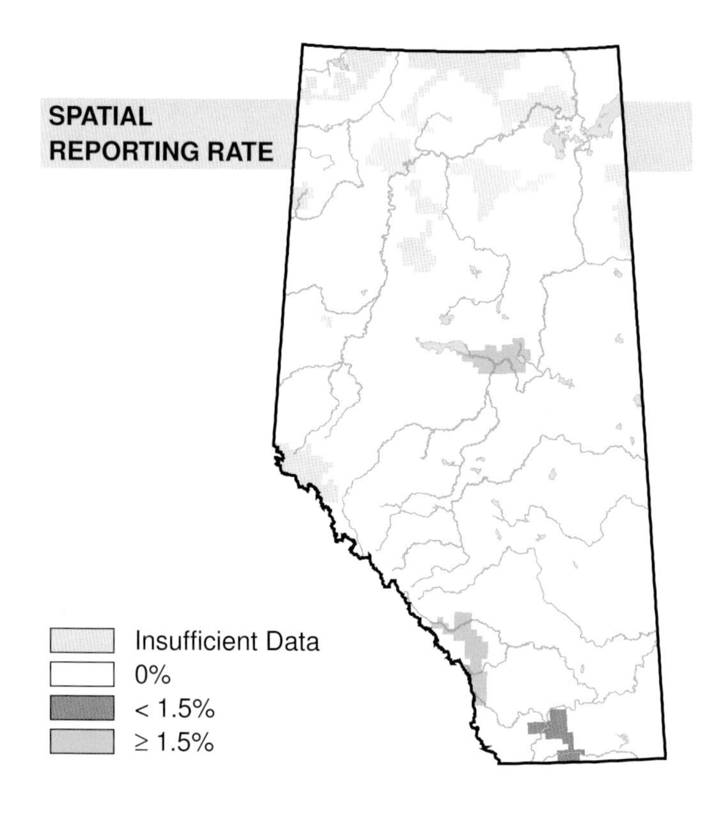

STATUS

GSWA 2000	GSWA 2005	COSEWIC Historic Rankings	COSEWIC 2007	Alberta Wildlife Act
Secure	Secure	N/A	N/A	N/A

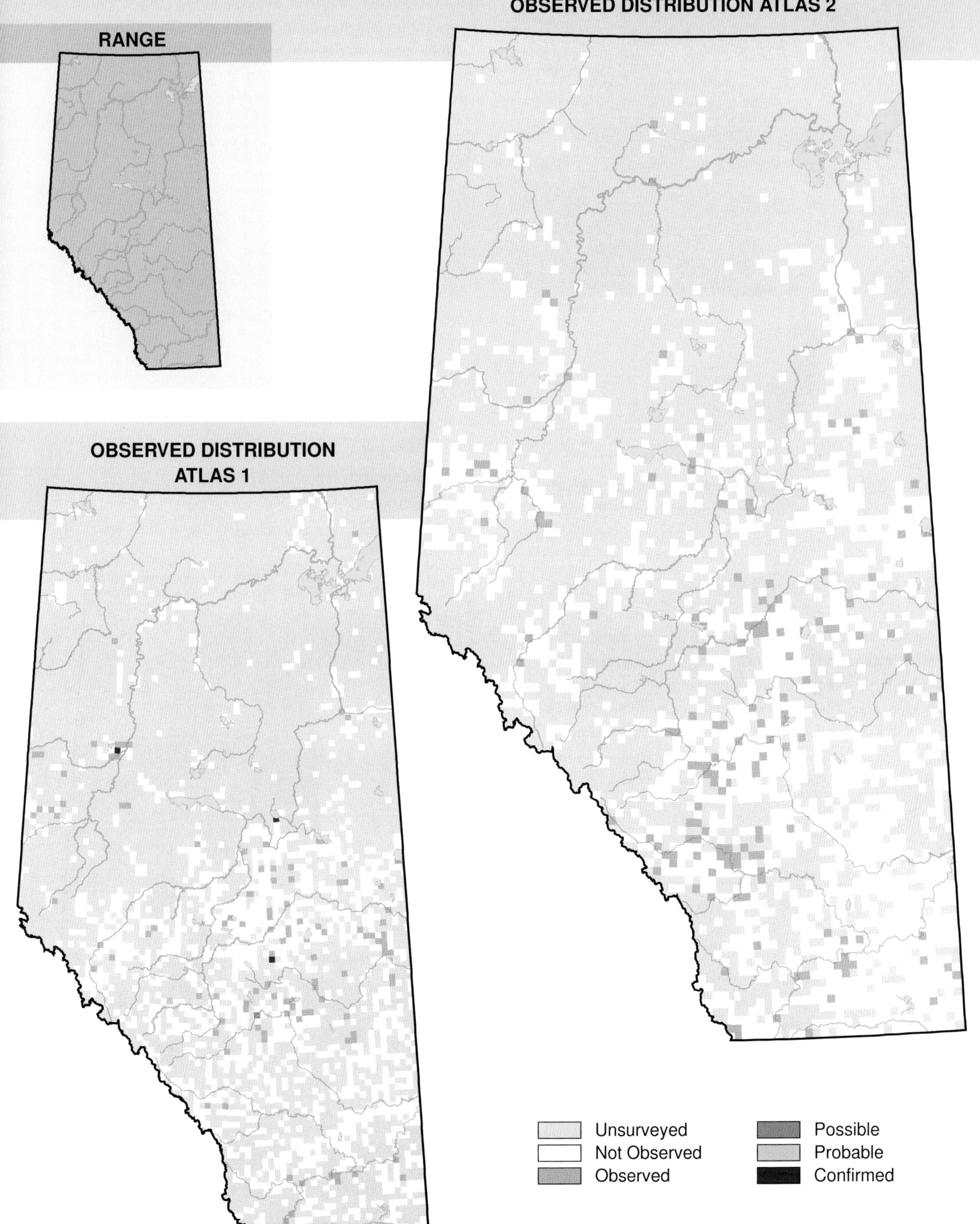

OBSERVED DISTRIBUTION ATLAS 2
RANGE
OBSERVED DISTRIBUTION ATLAS 1
Unsurveyed
Not Observed
Observed
Possible
Probable
Confirmed

Pine Siskin (*Carduelis pinus*)

Photo: Raymond Toal

NESTING

CLUTCH SIZE (EGGS):	3–6
INCUBATION (DAYS):	7–9
FLEDGING (DAYS):	15–15
NEST HEIGHT (METRES):	2–12

The Pine Siskin breeds in the Boreal Forest, Foothills, Parkland, and Rocky Mountain Natural Regions. The observed distribution of this species did not change between Atlas 1 and Atlas 2.

The Pine Siskin is most often found in coniferous forests; however, mixedwood forests are also occasionally used. Nests are usually built in conifer trees. This species tends to eat conifer seeds, although their diet can be quite varied and sometimes includes: seeds from grasses and deciduous trees; deciduous tree buds; and invertebrates. Due to their close association with conifers, this species was found mainly in the Boreal Forest, Foothills, and Rocky Mountain NRs and was only occasionally found in the Parkland and Grassland NRs.

Increases in relative abundance were detected in the Boreal Forest, Grassland, and Parkland NRs. Relative to other species, the Pine Siskin was observed more frequently in the Boreal Forest and Grassland NRs, and less frequently in the Parkland NR, during Atlas 2 than during Atlas 1. No change was detected in the Foothills and Rocky Mountain NRs. The Breeding Bird Survey found a decline in abundance in Alberta and across Canada during the periods 1985–2005 and 1995–2005. In light of the BBS results and the lack of apparent biological cause, the Atlas-detected increases were likely a function of more inclusive coverage during Atlas 2, rather than an indication of actual population change. The BBS-detected declines could be related to a reduction in the number of conifers because these trees are often targeted during forest harvesting. This species is considered Secure in Alberta.

TEMPORAL REPORTING RATE

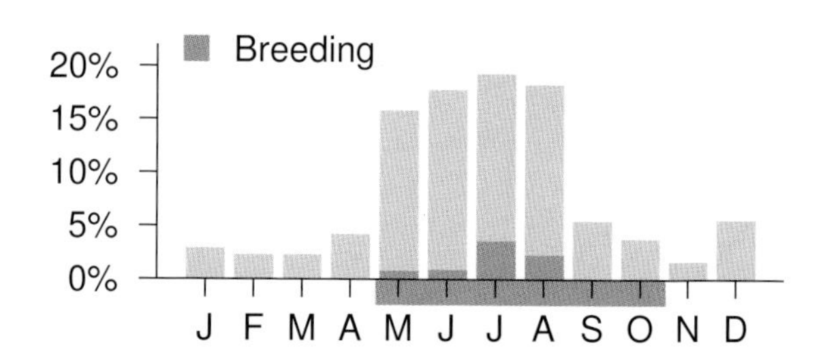

NATURAL REGIONS REPORTING RATE

Region	Rate
Boreal Forest	9%
Foothills	38%
Grassland	3%
Parkland	5%
Rocky Mountain	47%

SPATIAL REPORTING RATE

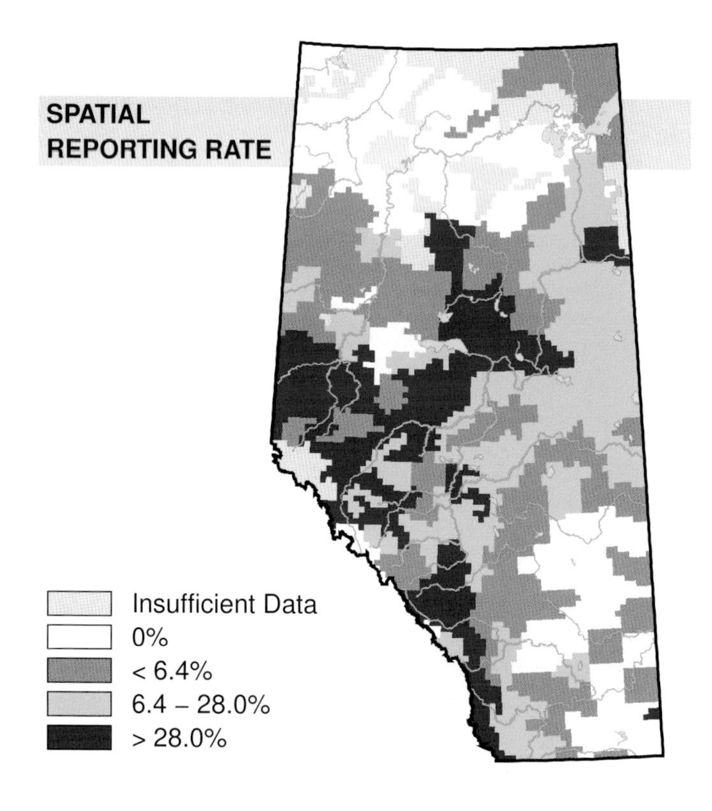

STATUS

GSWA 2000	GSWA 2005	COSEWIC Historic Rankings	COSEWIC 2007	Alberta Wildlife Act
Secure	Secure	N/A	N/A	N/A

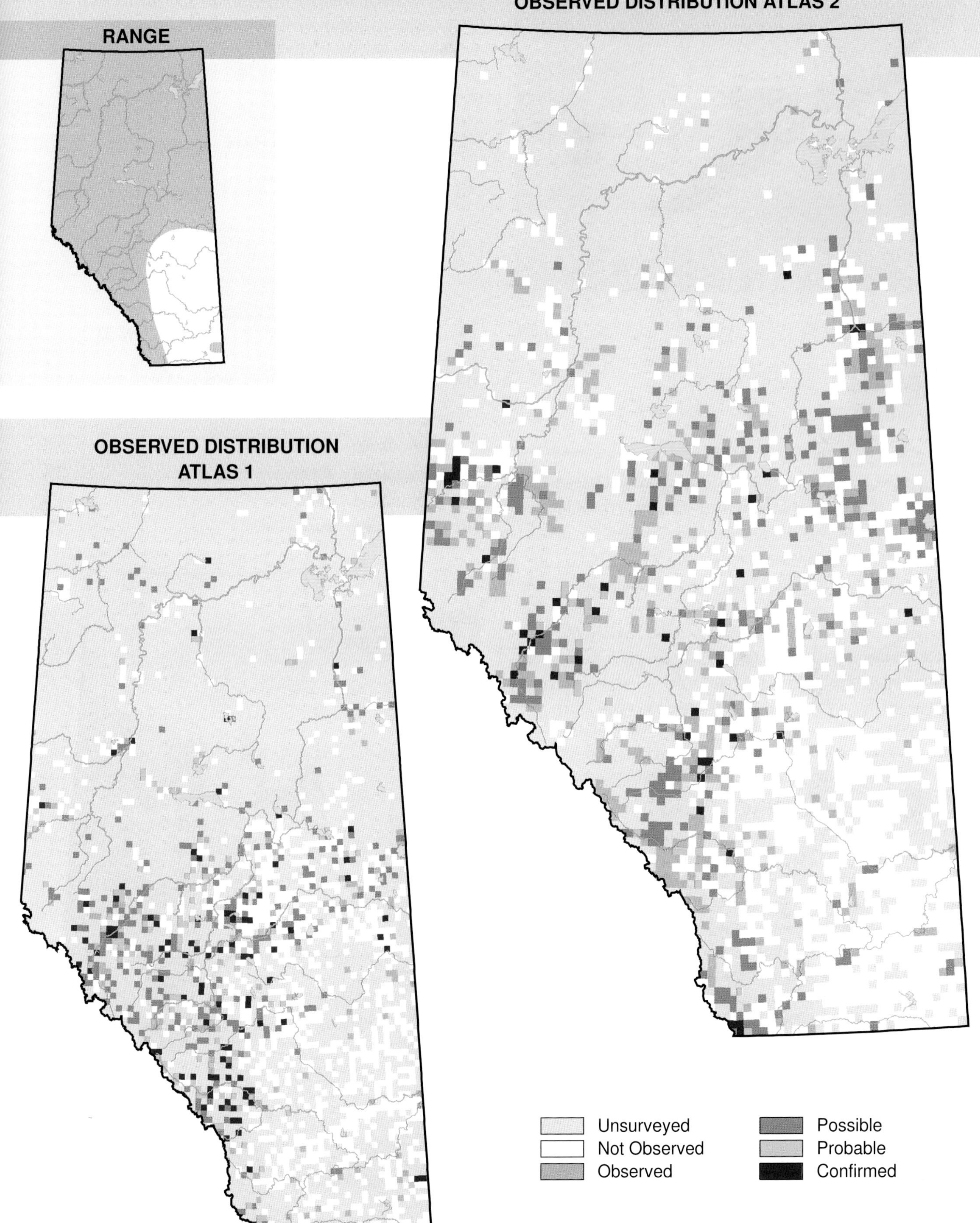
Habitat
Nest Location
Nest Type
Diet
OBSERVED DISTRIBUTION ATLAS 2
RANGE
OBSERVED DISTRIBUTION ATLAS 1
Unsurveyed
Not Observed
Observed
Possible
Probable
Confirmed

American Goldfinch (*Carduelis tristis*)

Photo: Debbie Godkin

NESTING

CLUTCH SIZE (EGGS):	4–6
INCUBATION (DAYS):	12–14
FLEDGING (DAYS):	11–17
NEST HEIGHT (METRES):	6

The American Goldfinch is found in every Natural Region in the province. The observed distribution of this species did not change between Atlas 1 and Atlas 2.

The American Goldfinch avoids mature forests and is usually found in dense early successional forests near open areas. This species is often found along the edge of cultivated fields, hedgerows, roadsides, and gardens. Because of its preference for the combination of open areas and dense shrub patches, this species often benefits from the conversion of forests for agriculture. As a result, this species was most often encountered in the Grassland and Parkland NRs, but they were also relatively common in the southern part of the Boreal Forest NR and in the Foothills and Rocky Mountain NRs.

Declines in relative abundance were detected in the Boreal Forest, Grassland, and Parkland NRs and, relative to other species, American Goldfinch was observed less frequently in Atlas 2 than in Atlas 1 in these NRs. No change was detected in the Foothills and Rocky Mountain NRs. The Breeding Bird Survey found an abundance decline in Alberta during the period 1985–2005. The declines found by the Atlas and the Alberta Breeding Bird Survey are mirrored in a Canada-wide Breeding Bird Survey decline during the period 1985–2005. Trends detected by the Breeding Bird Survey corroborate those detected by the Atlas. However, given that this species is thought to benefit from the expansion of agriculture, it is unclear why declines were found. Perhaps changes to agricultural practices have negatively affected this species. Further research is needed. This species is considered Secure in Alberta.

TEMPORAL REPORTING RATE

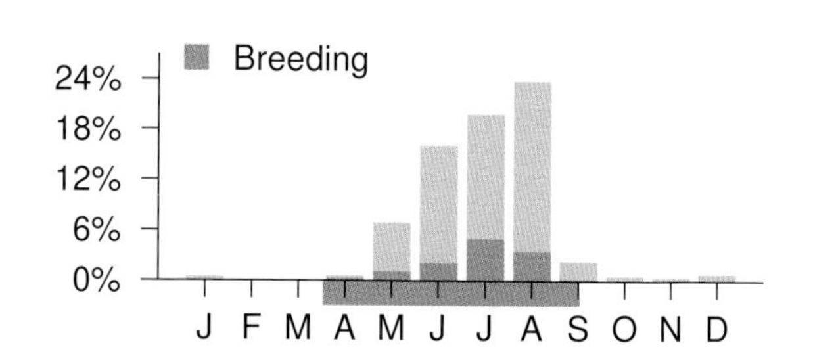

NATURAL REGIONS REPORTING RATE

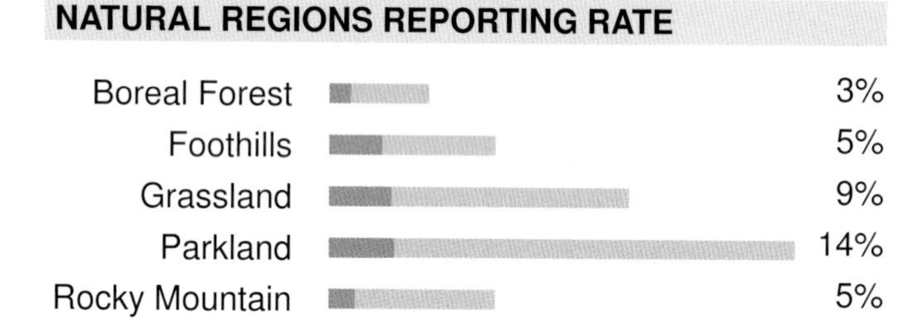

SPATIAL REPORTING RATE

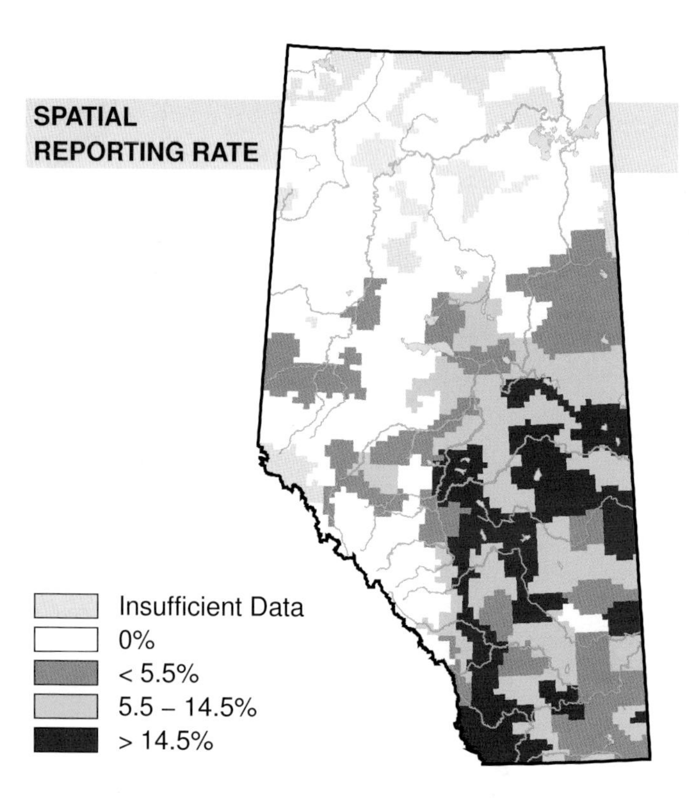

STATUS

GSWA 2000	GSWA 2005	COSEWIC Historic Rankings	COSEWIC 2007	Alberta Wildlife Act
Secure	Secure	N/A	N/A	N/A

Habitat   Nest Location  Nest Type  Diet

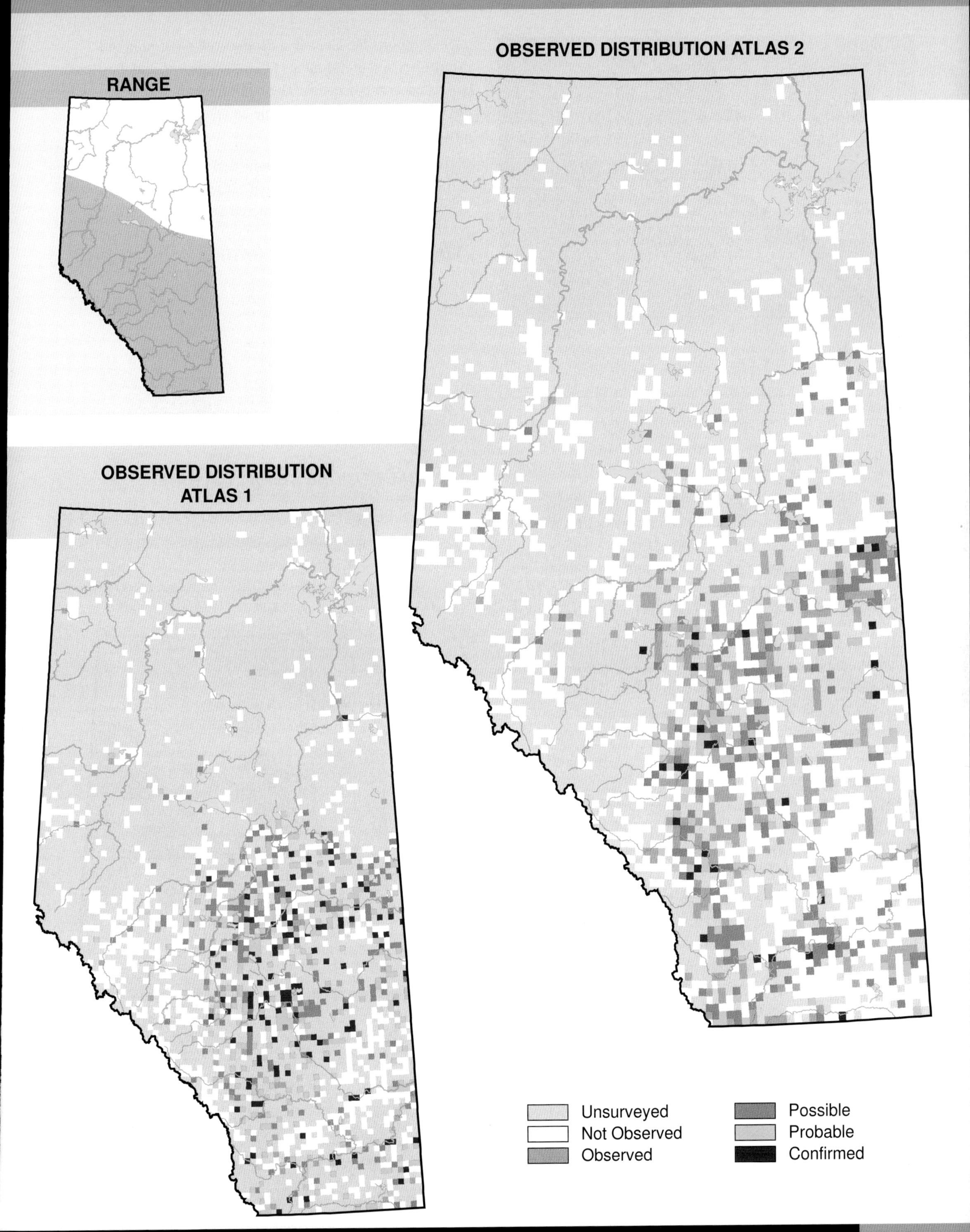

Evening Grosbeak (*Coccothraustes vespertinus*)

Photo: Gerald Romanchuk

NESTING

CLUTCH SIZE (EGGS):	3–5
INCUBATION (DAYS):	12–14
FLEDGING (DAYS):	13–14
NEST HEIGHT (METRES):	18

The Evening Grosbeak breeds in northern, central, and southwestern Alberta in the Boreal Forest, Foothills, Parkland, and Rocky Mountain Natural Regions. It is a permanent resident throughout much of central and northern Alberta, but in winter it ranges widely and is sporadic in occurrence. The distribution of this species did not change between Atlas 1 and Atlas 2.

It breeds mainly in coniferous or mixedwood forests that contain a large proportion of coniferous trees. This species prefers more open forests rather than dense stands. Despite having a varied diet, populations of this species expand during outbreaks of forest insect populations, such as that of the Spruce Budworm. Due to the bird's preference for coniferous trees, this species was most often encountered in the Foothills and Rocky Mountain NRs and was common in the Boreal Forest and Parkland NRs. Observations made in the Grassland NR were mostly of non-breeders.

Declines in relative abundance were detected in the Boreal Forest and Parkland NRs where, relative to other species, the Evening Grosbeak was observed less frequently in Atlas 2 than in Atlas 1. No changes were detected in the other NRs. The Breeding Bird Survey did not find an abundance change in Alberta, but a decline was detected across Canada during the period 1985–2005. The nation-wide decline can be explained partially by the decline in Spruce Budworm populations brought about through improved control measures. Dunn (2005) reports the Canadian declines are greatest in the Maritimes and the Montane Cordillera region, and they have persisted over 3 decades. The report recommends research to investigate the causes of decline and the effect of land-use practices on demography.

In Alberta, habitat loss from industrial activities and expanding agriculture is likely the principal cause of the observed declines. This species is considered Secure in Alberta.

TEMPORAL REPORTING RATE

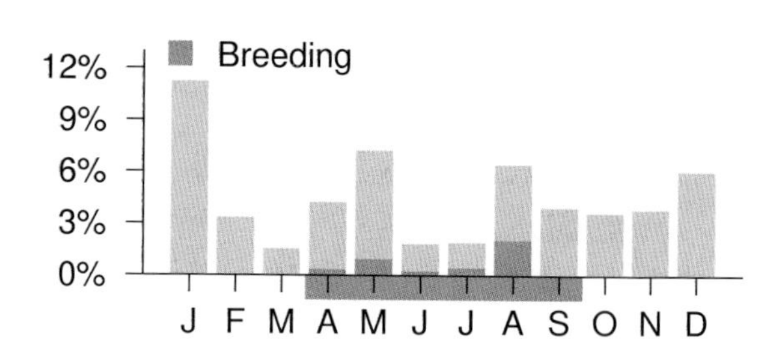

NATURAL REGIONS REPORTING RATE

SPATIAL REPORTING RATE

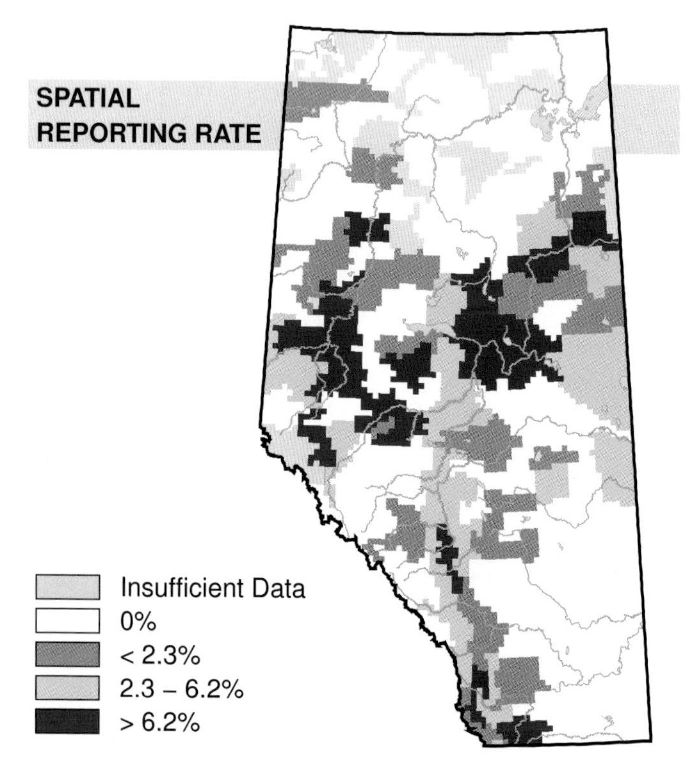

STATUS

GSWA 2000	GSWA 2005	COSEWIC Historic Rankings	COSEWIC 2007	Alberta Wildlife Act
Secure	Secure	N/A	N/A	N/A

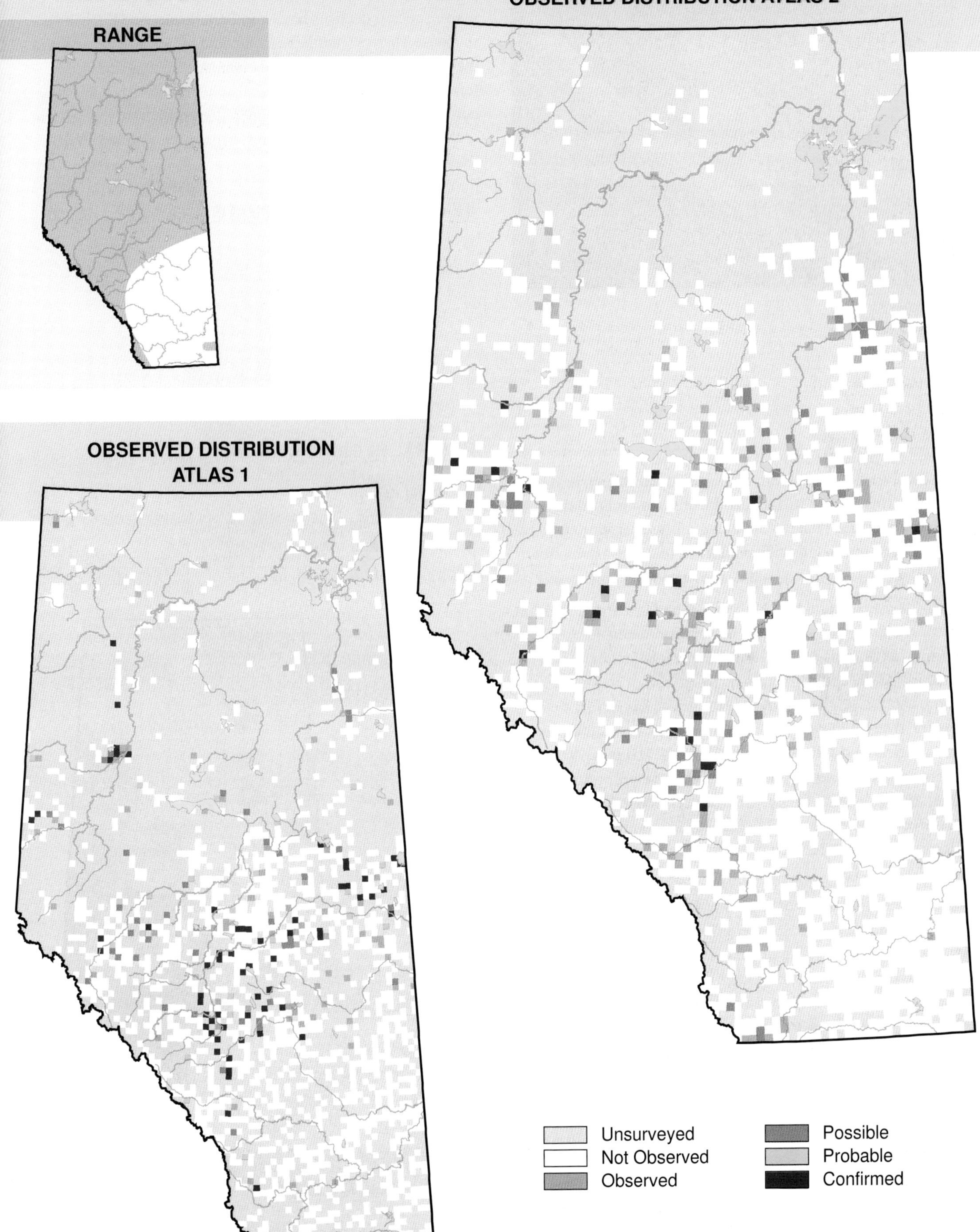

OBSERVED DISTRIBUTION ATLAS 2
RANGE
OBSERVED DISTRIBUTION ATLAS 1
Unsurveyed
Not Observed
Observed
Possible
Probable
Confirmed

House Sparrow (*Passer domesticus*)

Photo: Gerald Romanchuk

NESTING

CLUTCH SIZE (EGGS):	3–7
INCUBATION (DAYS):	11–14
FLEDGING (DAYS):	12–18
NEST HEIGHT (METRES):	<12.2

The House Sparrow is found in every Natural Region in Alberta. This species was found farther north during Atlas 1 than it was during Atlas 2. The range of this species could have contracted because, given its preference for close association with people, it is unlikely that its presence would be missed.

The House Sparrow is one of the most familiar birds in Alberta because it is closely associated with human-modified environments. This species nests near people in urban and rural areas where it is often found along hedges and in dense shrub habitat. The distribution of this species is similar to the general distribution of people in Alberta. As a consequence, this species was more common in the south and was encountered less often farther north. This species was most often encountered in the Grassland and Parkland NRs and was less frequently found in the Boreal Forest, Foothills, and Rocky Mountain NRs.

Declines in relative abundance were detected in the Boreal Forest, Foothills, Grassland, and Parkland NRs where, relative to other species, House Sparrows were observed less frequently in Atlas 2 than in Atlas 1. An increase was detected in the Rocky Mountain NR where, relative to other species, they were observed more often in Atlas 2 than in Atlas 1. The Breeding Bird Survey found a decline in abundance in Alberta during the period 1968–2005. The declines found by the Atlas and the Alberta Breeding Bird Survey are mirrored in a decline detected by the Canada-wide Breeding Bird Survey during the period 1985–2005. In Europe, where House Sparrows are a native species, declines have also been documented. Declines in House Sparrow numbers are thought to be tied to changes in the types of gardens that people maintain in urban areas; important in this case is the fact that hedges have become less common. In rural areas, farm operations have become larger in scale, with fewer farm dwellings. This species is considered Exotic/Alien in Alberta.

TEMPORAL REPORTING RATE

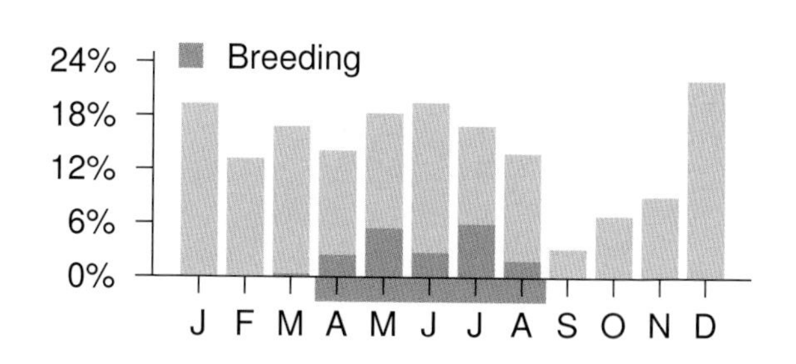

NATURAL REGIONS REPORTING RATE

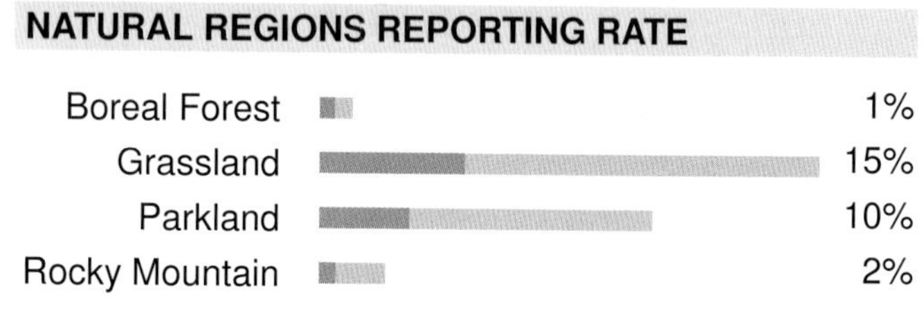

SPATIAL REPORTING RATE

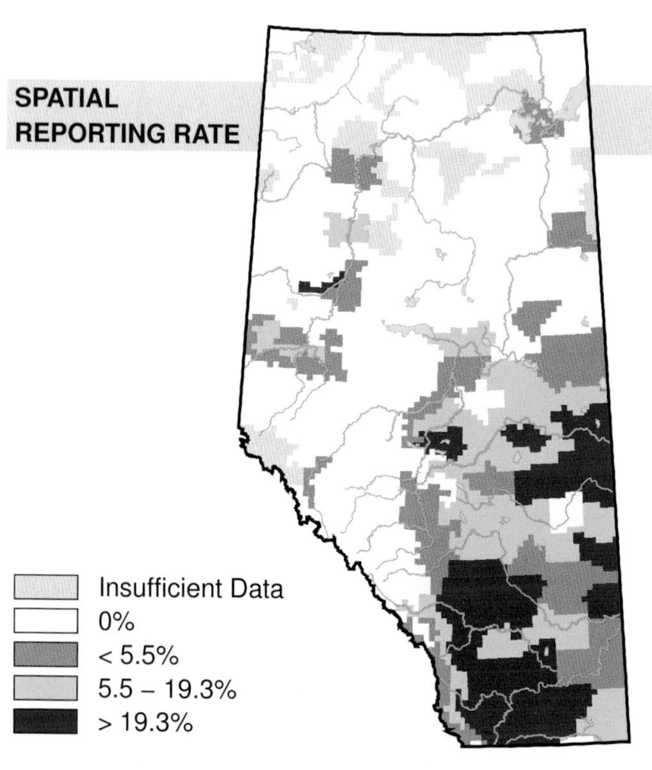

STATUS

GSWA 2000	GSWA 2005	COSEWIC Historic Rankings	COSEWIC 2007	Alberta Wildlife Act
Exotic/Alien	Exotic/Alien	N/A	N/A	N/A

RANGE

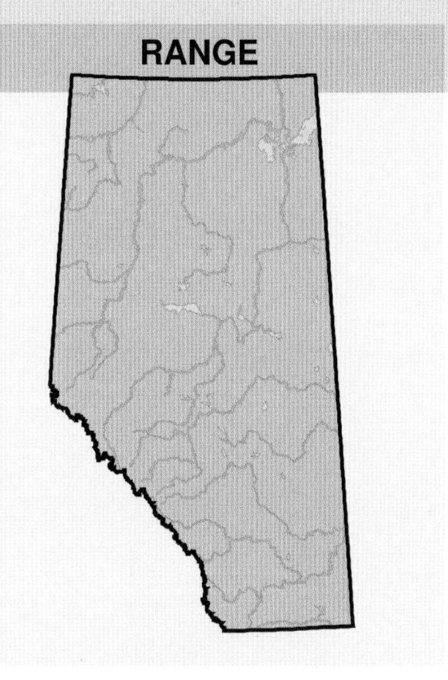

OBSERVED DISTRIBUTION ATLAS 2

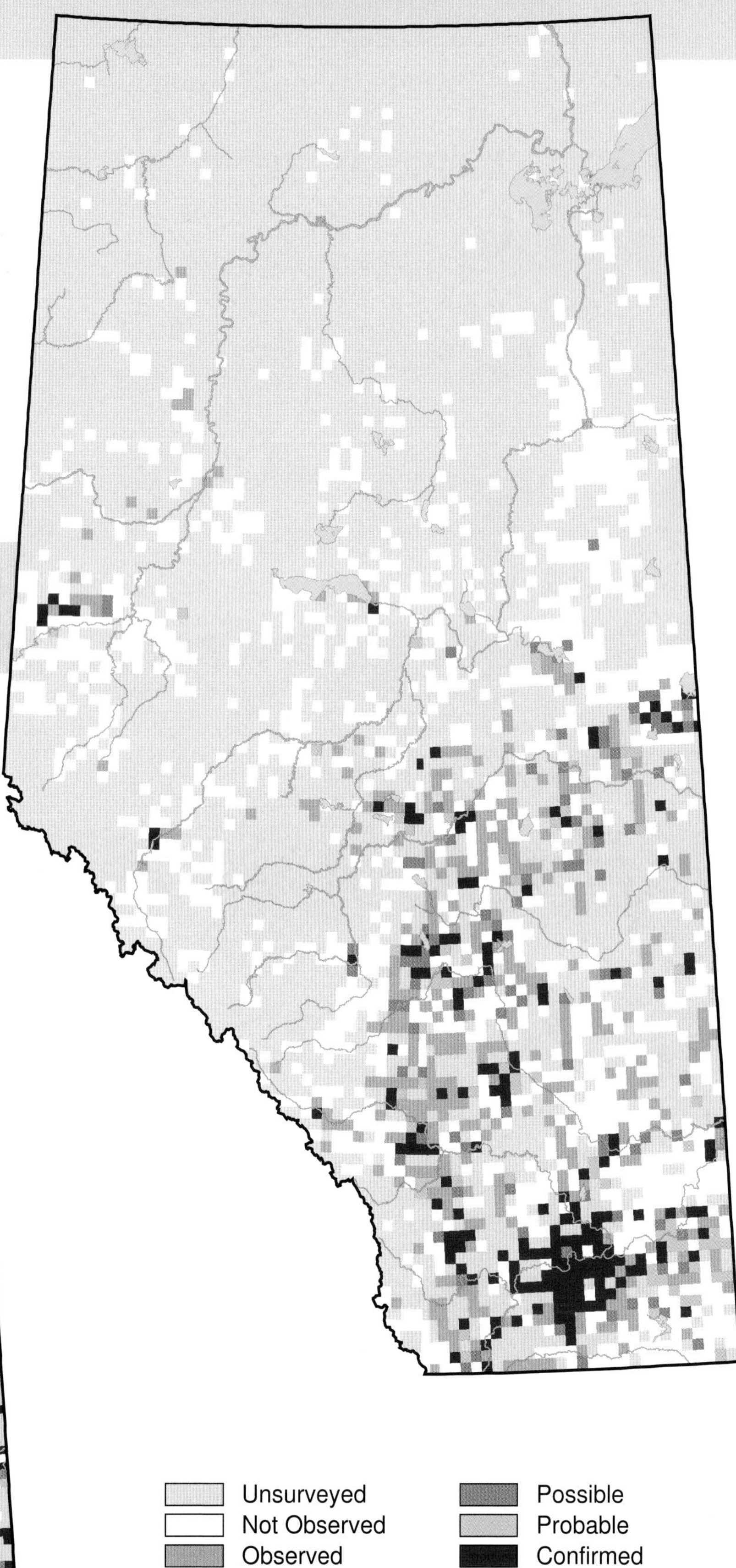

OBSERVED DISTRIBUTION ATLAS 1

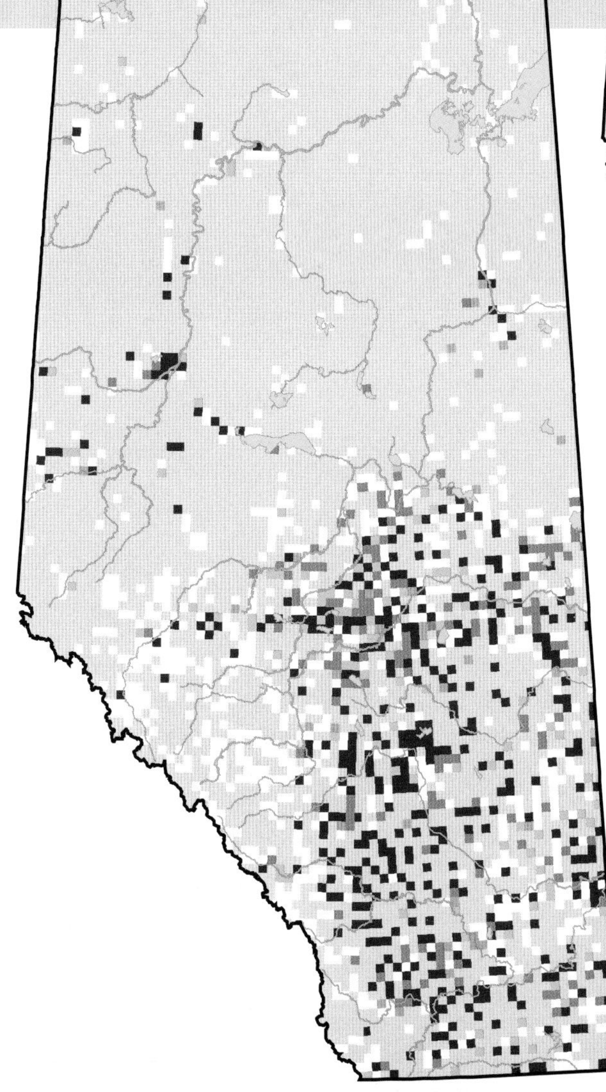

Unsurveyed
Not Observed
Observed
Possible
Probable
Confirmed

Alberta Species List

Description

This is version 1.03 of the Official List of the Birds of Alberta, published on the 4th of August 2006 by The Alberta Bird Record Committee. This list is continually updated and can be viewed at www.royalalbertamuseum.ca/natural/birds/birdlist/taxon.htm.

Birds are listed in taxonomic order. Next to each species name are codes indicating the species status. For accidental and vagrant species, the first year this species was confirmed in the province is provided in parentheses.

Status Codes

Br A regular breeder in the province. These species do not stay year-round but rather come to the province in the spring to breed, leaving again in the fall. Abundance varies from abundant to scarce or local but all are findable when sought in the right habitat and season. Species with abundance modifiers are usually more difficult to find.

PR A permanent resident. These species stay in the province year-round. They breed in the province and all or most individuals remain in the province for the winter months. Most of these species move only short distances from their breeding grounds.

Mig A species that only passes through the province. They do not breed or winter regularly in the province. Some species only pass through the province during the spring or fall migration, or abundance may differ between the two seasons. In this case, the following qualifiers apply:

SMig Spring migration

FMig Fall migration

F Fall

WV A winter visitor. Individuals occur during the winter period but do not necessarily remain all winter.

WR A winter resident species. Individuals arrive in the winter and remain until they leave for the north in the spring.

Species in which the majority of individuals leave the province for the winter (though, on the basis that some individuals remain, could be considered permanent residents) are coded as follows:

W A species wintering in the province in some numbers. They are usually fewer than during the breeding season.

w Very few individuals winter in the province. The species is irregular in winter, or does not winter on an annual basis. It should be noted that more species are attempting to winter in the province.

Acc Accidental. These species have been documented on fewer than ten occasions in the province. Many of these species are not likely to reoccur, or at best do only infrequently (Br indicates that the species has bred in the province, however).

V Vagrant. These species have been documented more than 10 times in the province, but fewer than 50 times. These species are of very irregular occurrence but most are likely to be reported again in the future. Br indicates that the species has bred in the province.

Acc/V A species that, though it may have been reported more than 10 times (often by reliable observers), is not well supported by documentation or material evidence. For accidental and vagrant species, we provide the year when that species' occurrence in the province was first confirmed.

Ext Extirpated. These are species which no longer occur in the province.

Int Introduced. These are species introduced and now well established in the province.

* Sight records by single observers that receive four favourable votes and no more than one dissenting vote. For record adjudication purposes, such a record is acceptable, but it does not pass the more stringent requirements for inclusion in the Official Provincial List.

Abundance Qualifiers

vr Very rare. The species occurs in the province on a near-annual basis but is extremely difficult to locate due to very low abundance, localized distribution or secretive nature. Extensive effort and luck are required for an observer to find one of these species in the province.

r Rare. The species occurs annually and is difficult to find, but numbers are higher than **vr** or it is less local in distribution.

s Scarce. The species may breed in low numbers, is difficult to locate due to its habits (secretive, nocturnal), or occupies habitats where it is difficult to observe or find.

l Local. This is a species which occupies only a very small percentage of available or suitable habitat and which is often known from only a few sites in the province.

u Uncommon.

Erratic This is a species whose local abundance fluctuates widely from year to year.

Decl This is a species known to be declining in abundance (question mark when it is suspected).

Incr This is a species known to be increasing in abundance (a question mark when it is suspected).

Official List of the Birds of Alberta

Order Anseriformes

Family Anatidae

1 Greater White-fronted Goose .... Mig
2 Snow Goose .... Mig
3 Ross' Goose .... Mig
4 Brant .... Acc/V (1957)
5 Cackling Goose .... Mig
6 Canada Goose .... Br/w
7 Trumpeter Swan .... rBr/w
8 Tundra Swan .... Mig/w
9 Wood Duck .... rBr/w
10 Gadwall .... Br/w
11 Eurasian Wigeon .... rSMig/vrFMig (1959)
12 American Wigeon .... Br/W
13 American Black Duck .... vrBr/rMig (1920s)
14 Mallard .... Br/W
15 Blue-winged Teal .... Br
16 Cinnamon Teal .... Br incr
17 Northern Shoveler .... Br
18 Northern Pintail .... Br/w
19 Garganey .... Acc (1977)
20 Green-winged Teal .... Br/w
21 Canvasback .... Br/w
22 Redhead .... Br/w
23 Ring-necked Duck .... Br/w
24 Tufted Duck .... Acc (1992)
25 Greater Scaup .... rSMig/vrFMig/w
26 Lesser Scaup .... Br/w
27 King Eider .... Acc (1894)
28 Common Eider .... Acc (1993)
29 Harlequin Duck .... Br/w
30 Surf Scoter .... sBr/uMig
31 White-winged Scoter .... Br
32 Black Scoter .... V (1975)
33 Long-tailed Duck .... rMig/w
34 Bufflehead .... Br/w
35 Common Goldeneye .... Br/W
36 Barrow's Goldeneye .... Br/w
37 Hooded Merganser .... rBr/uMig/w
38 Common Merganser .... Br/w
39 Red-breasted Merganser .... Br
40 Ruddy Duck .... Br

Order Galliformes

Family Phasianidae

41 Gray Partridge .... Int (1908) PR
42 Ring-necked Pheasant .... Int (1908) PR
43 Ruffed Grouse .... PR
44 Greater Sage-Grouse .... PR decl
45 Spruce Grouse .... sPR
46 Willow Ptarmigan .... rBr/sWV
47 White-tailed Ptarmigan .... sPR
48 Dusky Grouse .... sPR
49 Sharp-tailed Grouse .... PR decl?
50 Greater Prairie-Chicken .... Ext
51 Wild Turkey .... Int (1962) PR(l)

Order Gaviiformes

Family Gaviidae

52 Red-throated Loon .... vrBr/Mig
53 Pacific Loon .... vrBr/SMig, sFMig, w
54 Common Loon .... Br

55 Yellow-billed Loon . . . . . V (1975)

Order Podicipediformes

Family Podicipedidae

56 Pied-billed Grebe . . . . . Br
57 Horned Grebe . . . . . Br
58 Red-necked Grebe . . . . . Br
59 Eared Grebe . . . . . Br
60 Western Grebe . . . . . Br
61 Clark's Grebe . . . . . vrBr(l)

Order Pelecaniformes

Family Pelecanidae

62 American White Pelican . . . . . Br

Family Phalacrocoracidae

63 Double-crested Cormorant . . . . . Br/w

Order Ciconiiformes

Family Ardeidae

64 American Bittern . . . . . Br decl?
65 Great Blue Heron . . . . . Br/w
66 Great Egret . . . . . V (1954)
67 Snowy Egret . . . . . Acc/V (1901)
68 Little Blue Heron . . . . . Acc (1991)
69 Tricolored Heron . . . . . Acc (1981)
70 Cattle Egret . . . . . Acc/V (1964)
71 Green Heron . . . . . Acc/V (1975) incr?
72 Black-crowned Night-Heron . . . . . Br/w
73 Yellow-crowned Night-Heron . . . . . Acc (1999)

Family Threskiornithidae

74 White-faced Ibis . . . . . rBr incr

Family Cathartidae

75 Turkey Vulture . . . . . sBr(l)

Order Falconiformes

Family Accipitridae

76 Osprey . . . . . Br
77 Bald Eagle . . . . . Br/w
78 Northern Harrier . . . . . Br/w
79 Sharp-shinned Hawk . . . . . Br/w
80 Cooper's Hawk . . . . . Br (decl?), w
81 Northern Goshawk . . . . . Br/w decl?
82 Broad-winged Hawk . . . . . Br
83 Swainson's Hawk . . . . . Br
84 Red-tailed Hawk . . . . . Br/w
85 Ferruginous Hawk . . . . . Br
86 Rough-legged Hawk . . . . . Mig/WV
87 Golden Eagle . . . . . Br/w

Family Falconidae

88 American Kestrel . . . . . Br/w
89 Merlin . . . . . Br/w
90 Gyrfalcon . . . . . rWV/WR
91 Peregrine Falcon . . . . . Br
92 Prairie Falcon . . . . . Br/w

Order Gruiformes

Family Rallidae

93 Yellow Rail . . . . . r-sBr(l)
94 Virginia Rail . . . . . r-sBr(l)
95 Sora . . . . . Br
96 American Coot . . . . . Br/w

Family Gruidae

97 Sandhill Crane . . . . . Br
98 Common Crane . . . . . Acc (1957)
99 Whooping Crane . . . . . rMig, vrBr

Order Charadriiformes

Family Charadriidae

100 Black-bellied Plover . . . . . Mig
101 American Golden-Plover . . . . . Mig
102 Pacific Golden-Plover . . . . . Acc (1925)
103 Lesser Sand-Plover . . . . . Acc (1984)
104 Snowy Plover . . . . . Acc (1990)
105 Semipalmated Plover . . . . . Mig, rBr
106 Piping Plover . . . . . rBr(l)
107 Killdeer . . . . . Br/w
108 Mountain Plover . . . . . vrBr (1941)

Family Recurvirostridae

109 Black-necked Stilt . . . . . rBr (1970) incr
110 American Avocet . . . . . Br

Family Scolopacidae

111 Spotted Sandpiper . . . . . Br
112 Solitary Sandpiper . . . . . Br
113 Wandering Tattler . . . . . Acc (1938)
114 Spotted Redshank . . . . . Acc (1987)
115 Greater Yellowlegs . . . . . Br
116 Willet . . . . . Br
117 Lesser Yellowlegs . . . . . Br
118 Upland Sandpiper . . . . . Br decl?
119 Eskimo Curlew . . . . . Extinct?
120 Whimbrel . . . . . r-sSMig/vrFMig
121 Long-billed Curlew . . . . . Br decl
122 Hudsonian Godwit . . . . . sSMig/rFMig
123 Marbled Godwit . . . . . Br
124 Ruddy Turnstone . . . . . sSMig/rFMig
125 Black Turnstone . . . . . Acc (1998)
126 Surfbird . . . . . Acc (1975)
127 Red Knot . . . . . sSMig/rFMig
128 Sanderling . . . . . Mig
129 Semipalmated Sandpiper . . . . . Mig
130 Western Sandpiper . . . . . vr-rMig (1972)
131 Red-necked Stint . . . . . Acc (1995)
132 Little Stint . . . . . Acc (2000)
133 Least Sandpiper . . . . . Mig
134 White-rumped Sandpiper . . . . . sSMig/vrFMig

135 Baird's Sandpiper .... Mig
136 Pectoral Sandpiper .... Mig
137 Sharp-tailed Sandpiper .... Acc/V (1975)
138 Dunlin .... sSMig/rFMig
139 Curlew Sandpiper .... Acc (1975)
140 Stilt Sandpiper .... Mig
141 Spoon-billed Sandpiper .... Acc (1984)
142 Buff-breasted Sandpiper .... sSMig/rFMig
143 Ruff .... Acc/V (1967)
144 Short-billed Dowitcher .... Br
145 Long-billed Dowitcher .... Mig
146 Wilson's Snipe .... Br/w
147 Wilson's Phalarope .... Br
148 Red-necked Phalarope .... Mig, vrBr
149 Red Phalarope .... vrFMig

Family Stercorariidae

150 Pomarine Jaeger .... Acc (1996)
151 Parasitic Jaeger .... vrSMig/rFMig
152 Long-tailed Jaeger .... Acc (1932)

Family Laridae

153 Franklin's Gull .... Br
154 Little Gull .... Acc (1985)
155 Bonaparte's Gull .... Br
156 Mew Gull .... sBr/Mig
157 Ring-billed Gull .... Br
158 California Gull .... Br
159 Herring Gull .... Br
160 Thayer's Gull .... vrSMig/rFMig (1928)
161 Iceland Gull .... Acc/V (1975)
162 Lesser Black-backed Gull .... Acc (1989)
163 Slaty-backed Gull .... Acc (2000)
164 Glaucous-winged Gull .... Acc/V (1936)
165 Glaucous Gull .... rMig/WV (1915)
166 Great Black-backed Gull .... Acc (1986)
167 Sabine's Gull .... rMig
168 Black-legged Kittiwake .... Acc (1976)
169 Caspian Tern .... rBr/Mig incr?
170 Black Tern .... Br
171 Common Tern .... Br
172 Arctic Tern .... vrBr/Mig
173 Forster's Tern .... Br

Family Alcidae

174 Black Guillemot .... Acc (1988)
175 Long-billed Murrelet .... Acc (1994)
176 Ancient Murrelet .... Acc (1975)

Order Columbiformes

Family Columbidae

177 Rock Pigeon .... Int PR
178 Band-tailed Pigeon .... V (1967)
179 Eurasian Collared-Dove .... Acc (2002)
180 White-winged Dove .... Acc (1997)
181 Mourning Dove .... Br
182 Passenger Pigeon .... Extinct

Order Cuculiformes

Family Cuculidae

183 Yellow-billed Cuckoo .... Acc (1968)
184 Black-billed Cuckoo .... r-sBr Erratic

Order Strigiformes

Family Tytonidae

185 Barn Owl .... Acc (1967)

Family Strigidae

186 Flammulated Owl .... Acc (2000)
187 Western Screech-Owl .... Acc (1897)
188 Eastern Screech-Owl .... Acc (1930)
189 Great Horned Owl .... PR
190 Snowy Owl .... Mig/WV
191 Northern Hawk Owl .... sPR, W Erratic
192 Northern Pygmy-Owl .... sPR(l)
193 Burrowing Owl .... Br decl
194 Barred Owl .... sPR(l)
195 Great Gray Owl .... sPR(l), W Erratic
196 Long-eared Owl .... sBr
197 Short-eared Owl .... Br/w decl?
198 Boreal Owl .... sBr/w
199 Northern Saw-whet Owl .... Br/w

Order Caprimulgiformes

Family Caprimulgidae

200 Common Nighthawk .... Br decl?
201 Common Poorwill .... V (1945), vrBr

Order Apodiformes

Family Apodidae

202 Black Swift .... rBr(l)
203 White-throated Swift .... Acc (1996)

Family Trochilidae

204 Green Violet-ear .... Acc (1994)
205 Ruby-throated Hummingbird .... Br
206 Black-chinned Hummingbird .... Acc (1979)
207 Anna's Hummingbird .... Acc (1976)
208 Costa's Hummingbird .... Acc (1988)
209 Calliope Hummingbird .... Br
210 Rufous Hummingbird .... Br

Order Coraciiformes

Family Alcedinidae

211 Belted Kingfisher .... Br/w

Order Piciformes

Family Picidae

212 Lewis's Woodpecker .... V, vrBr
213 Red-headed Woodpecker .... V, Br?

214 Williamson's Sapsucker .... Acc (1992)
215 Yellow-bellied Sapsucker .... Br
216 Red-naped Sapsucker .... Br
217 Red-breasted Sapsucker .... Acc (1994)
218 Downy Woodpecker .... PR
219 Hairy Woodpecker .... PR
220 American Three-toed Woodpecker .... PR
221 Black-backed Woodpecker .... PR
222 Northern Flicker .... Br/w
223 Pileated Woodpecker .... PR

Order Passeriformes

Family Tyrannidae

224 Olive-sided Flycatcher .... Br decl?
225 Western Wood-Pewee .... Br
226 Yellow-bellied Flycatcher .... sBr
227 Alder Flycatcher .... Br
228 Willow Flycatcher .... Br(l)
229 Least Flycatcher .... Br
230 Hammond's Flycatcher .... sBr(l)
231 Gray Flycatcher .... Acc (1999)
232 Dusky Flycatcher .... Br
233 "Western" Flycatcher .... Br
234 Eastern Phoebe .... Br
235 Say's Phoebe .... Br
236 Great Crested Flycatcher .... sBr(l)
237 Western Kingbird .... Br
238 Eastern Kingbird .... Br
239 Scissor-tailed Flycatcher .... Acc (1943)

Family Laniidae

240 Loggerhead Shrike .... Br decl
241 Northern Shrike .... uMig/WV, vrBr

Family Vireonidae

242 Yellow-throated Vireo .... Acc (2003)
243 Cassin's Vireo .... sBr
244 Blue-headed Vireo .... Br
245 Warbling Vireo .... Br
246 Philadelphia Vireo .... Br
247 Red-eyed Vireo .... Br

Family Corvidae

248 Gray Jay .... PR
249 Steller's Jay .... PR(l)
250 Blue Jay .... PR
251 Clark's Nutcracker .... PR
252 Black-billed Magpie .... PR
253 American Crow .... Br/w
254 Common Raven .... PR incr

Family Alaudidae

255 Horned Lark .... Br/Mig/w

Family Hirundinidae

256 Purple Martin .... Br
257 Tree Swallow .... Br
258 Violet-green Swallow .... Br
259 Northern Rough-winged Swallow .... Br
260 Bank Swallow .... Br
261 Cliff Swallow .... Br
262 Barn Swallow .... Br

Family Paridae

263 Black-capped Chickadee .... PR
264 Mountain Chickadee .... PR
265 Chestnut-backed Chickadee .... Acc/V (1979)
266 Boreal Chickadee .... PR

Family Sittidae

267 Red-breasted Nuthatch .... Br, w Erratic
268 White-breasted Nuthatch .... PR

Family Certhiidae

269 Brown Creeper .... Br/w

Family Troglodytidae

270 Rock Wren .... Br
271 Carolina Wren .... Acc (1987)
272 House Wren .... Br
273 Winter Wren .... Br
274 Sedge Wren .... rBr(l) Erratic
275 Marsh Wren .... Br

Family Cinclidae

276 American Dipper .... Br/w

Family Sylviidae

277 Golden-crowned Kinglet .... Br/w
278 Ruby-crowned Kinglet .... Br
279 Blue-gray Gnatcatcher .... Acc (1987)

Family Turdidae

280 Northern Wheatear .... Acc (1989)
281 Eastern Bluebird .... vrBr (1977) incr?
282 Western Bluebird .... vrBr (1984)
283 Mountain Bluebird .... Br
284 Townsend's Solitaire .... Br/w
285 Veery .... Br
286 Gray-cheeked Thrush .... vrBr(l)/r-sMig
287 Swainson's Thrush .... Br
288 Hermit Thrush .... Br
289 Wood Thrush .... Acc (1980)
290 American Robin .... Br/w
291 Varied Thrush .... Br/w

Family Mimidae

292 Gray Catbird .... Br
293 Northern Mockingbird .... V (1928), vrBr
294 Sage Thrasher .... V (1924), vrBr
295 Brown Thrasher .... Br
296 Bendire's Thrasher .... Acc (1988)
297 Curve-billed Thrasher .... Acc (1998)

Family Sturnidae

298 European Starling .... Int Br/w

Family Prunellidae

299 Siberian Accentor .... Acc (2002)

Family Motacillidae
300 American Pipit .... Br/Mig
301 Sprague's Pipit.................................. Br decl?
Family Bombycillidae
302 Bohemian Waxwing .... Br/W
302 Cedar Waxwing .... Br
Family Parulidae
304 Blue-winged Warbler .... Acc (1996)
305 Golden-winged Warbler..................... Acc (1985)
306 Tennessee Warbler.................................. Br
307 Orange-crowned Warbler........................... Br
308 Nashville Warbler .... vrBr/Mig (1974)
309 Northern Parula .... Acc (1958)
310 Yellow Warbler .... Br
311 Chestnut-sided Warbler .... rBr(l)
312 Magnolia Warbler .... Br
313 Cape May Warbler.................................. sBr
314 Black-throated Blue Warbler................... V (1917)
315 Yellow-rumped Warbler............................ Br
316 Black-throated Gray Warbler .... Acc (2001)
317 Black-throated Green Warbler .... Br
318 Townsend's Warbler .... Br
319 Hermit Warbler .... Acc (2002)
320 Blackburnian Warbler.............................. sBr(l)
321 Pine Warbler................................ Acc (1924)
322 Palm Warbler .... Br
323 Bay-breasted Warbler.............................. Br
324 Blackpoll Warbler .... Br
325 Black-and-white Warbler........................... Br
326 American Redstart.................................. Br
327 Ovenbird .... Br
328 Northern Waterthrush .... Br
329 Kentucky Warbler .... Acc (1988)
330 Connecticut Warbler................................ Br
331 Mourning Warbler................................... Br
332 MacGillivray's Warbler .... Br
333 Common Yellowthroat .... Br
334 Hooded Warbler............................. Acc (1991)
335 Wilson's Warbler.................................... Br
336 Canada Warbler .... Br
337 Yellow-breasted Chat .... Br(l)
Family Thraupidae
338 Summer Tanager .... Acc (1995)
339 Scarlet Tanager .... Acc/V (1964)
340 Western Tanager.................................... Br
Family Emberizidae
341 Green-tailed Towhee........................ Acc (1996)
342 Spotted Towhee .... Br
343 Eastern Towhee .... Acc (1994)
344 Cassin's Sparrow .... Acc (1986)
345 American Tree Sparrow .... Mig/rWV, vrBr
346 Chipping Sparrow .... Br
347 Clay-colored Sparrow............................... Br
348 Brewer's Sparrow .... Br
349 Field Sparrow................................ Acc (2000)
350 Vesper Sparrow .... Br
351 Lark Sparrow .... Br
352 Black-throated Sparrow .... Acc (1993)
353 Lark Bunting................................ Br Erratic
354 Savannah Sparrow................................... Br
355 Grasshopper Sparrow .... sBr decl?
356 Baird's Sparrow .... sBr, decl
357 Le Conte's Sparrow................................. Br
358 Nelson's Sharp-tailed Sparrow .... sBr(l)
359 Fox Sparrow .... Br/w
360 Song Sparrow....................................... Br/w
361 Lincoln's Sparrow .... Br
362 Swamp Sparrow .... Br/w
363 White-throated Sparrow .... Br/w
364 Harris's Sparrow.................................. Mig/w
365 White-crowned Sparrow .... Br/w
366 Golden-crowned Sparrow .... sBr(l)
367 Dark-eyed Junco .... Br/w
368 McCown's Longspur................................. Br
369 Lapland Longspur .... Mig/w
370 Smith's Longspur .... rSMig/vrFMig
371 Chestnut-collared Longspur .... Br
372 Snow Bunting.................................. Mig/WV
Family Cardinalidae
373 Northern Cardinal .... Acc/V (1987)
374 Rose-breasted Grosbeak............................ Br
375 Black-headed Grosbeak .... sBr(l) incr
376 Lazuli Bunting .... Br
377 Indigo Bunting.......................... V (1926), vrBr
378 Painted Bunting .... Acc (2000)
379 Dickcissel .... Acc/V (1940)
Family Icteridae
380 Bobolink .... sBr decl?
381 Red-winged Blackbird .... Br/w
382 Eastern Meadowlark .... Acc (1989)
383 Western Meadowlark .... Br/w
384 Yellow-headed Blackbird .... Br/w
385 Rusty Blackbird................................... Br/w
386 Brewer's Blackbird .... Br/w
387 Common Grackle.................................... Br
388 Brown-headed Cowbird .... Br
389 Baltimore Oriole .... Br
390 Bullock's Oriole.................................... sBr(l)
Family Fringillidae
391 Brambling .... Acc (1989)
392 Gray-crowned Rosy-Finch .... Br/W
393 Pine Grosbeak .... Br/WV Erratic
394 Purple Finch....................................... Br/w
395 Cassin's Finch .... sBr(l)

396 House Finch . . . . sBr/W (1944) incr
397 Red Crossbill . . . . PR Erratic
398 White-winged Crossbill . . . . PR Erratic
399 Common Redpoll . . . . Mig/WV, vrBr
400 Hoary Redpoll . . . . rWV
401 Pine Siskin . . . . Br/w
402 American Goldfinch . . . . Br/w
403 Evening Grosbeak . . . . PR Erratic decr

Family Passeridae

404 House Sparrow . . . . Int PR

Appendix

Ivory Gull . . . . *

Bibliography

Aldridge, C.L. and Brigham, R.M. 2001. Nesting and reproductive activities of Greater Sage-Grouse in a declining northern fringe population. Condor, volume 103: 537–543.

Altman, B. and Sallabanks, R. 2000. Olive-sided Flycatcher (*Contopus cooperi*). *In* Poole, A. and Gill, F. (Editors), The Birds of North America, 502. The Birds of North America, Inc., Philadelphia, PA.

Austin, J.E., Custer, C.M., and Afton, A.D. 1998. Lesser Scaup (*Aythya affinis*). *In* Poole, A. and Gill, F. (Editors), The Birds of North America, 338. The Birds of North America, Inc., Philadelphia, PA.

Austin, J.E. and Miller, M.R. 1995. Northern Pintail (*Anas acuta*). *In* Poole, A. and Gill, F. (Editors), The Birds of North America, 163. The Birds of North America, Inc., Philadelphia, PA.

Barr, J.F., Eberl, C., and McIntyre, J.W. 2000. Red-throated Loon (*Gavia stellata*). *In* Poole, A. and Gill, F. (Editors), The Birds of North America, 513. The Birds of North America, Inc., Philadelphia, PA.

Barrett, G., Silcocks, A., Barry, S., Cunningham, R., and Poulter, R. 2003. The New Atlas of Australian Birds (1998–2001). Birds Australia, Melbourne.

Bart, J., Burnham, K.P., Dunn, E., Francis, C.M., and Ralph, C.J. 2004. Goals and strategies for estimating trends in landbird abundance. Journal of Wildlife Management, volume 68(3): 611–626.

Bart, J. and Klosiewski, S.P. 1989. Use of presence-absence to measure changes in avian density. Journal of Wildlife Management, volume 53(3): 847–852.

Beason, R.C. 1995. Horned Lark (*Eremophila alpestris*). *In* Poole, A. and Gill, F. (Editors), The Birds of North America, 195. The Birds of North America, Inc., Philadelphia, PA.

Beauchamp, W.D., Koford, R.R., Nudds, T.D., Clark, R.G., and Johnson, D.H. 1996. Long-term declines in nest success of prairie ducks. Journal of Wildlife Management, volume 60: 247–257.

Beckingham, J.D. and Archibald, J.H. 1996. Field Guide to Ecosites of Northern Alberta. Special Report 5, Natural Resources Canada, Canadian Forest Service, Northern Forestry Centre.

Bemis, C. and Rising, J.D. 1999. Western Wood-Pewee (*Contopus sordidulus*). *In* Poole, A. and Gill, F. (Editors), The Birds of North America, 451. The Birds of North America, Inc., Philadelphia, PA.

Bethke, R.W. and Nudds, T. 1995. Effects of climate change and land use on duck abundance in Canadian Prairie-Parklands. Ecological Applications, volume 5: 588–600.

Bibby, C.J., Burgess, N.D., and Hill, D.A. 2000. Bird Census Techniques. Academic Press, London, UK.

Bird, D., Bildstein, K., Ardia, D., Steenhof, K., Smallwood, J., Mason, J., Causey, M., Mossop, D., Dibernardo, A., Lindsay, R., McCartney, D., and Hendrickson, J. 2004. Are American Kestrel Populations in a State of Decline in North America? *In* Raptor Research Foundation Annual Meeting.

Boag, D.A. and Schroeder, M.A. 1992. Spruce Grouse (*Falcipennis canadensis*). *In* Poole, A. and Gill, F. (Editors), The Birds of North America, 5. The Birds of North America, Inc., Philadelphia, PA.

Bookhout, T.A. 1995. Yellow Rail (*Coturnicops noveboracensis*). *In* Poole, A. and Gill, F. (Editors), The Birds of North America, 139. The Birds of North America, Inc., Philadelphia, PA.

Bowen, R.V. 1997. Townsend's Solitaire (*Myadestes townsendi*). *In* Poole, A. and Gill, F. (Editors), The Birds of North America, 269. The Birds of North America, Inc., Philadelphia, PA.

Braun, C.E., Martin, K., and Robb, L.A. 1993. White-tailed Ptarmigan (*Lagopus leucura*). *In* Poole, A. and Gill, F. (Editors), The Birds of North America, 68. The Birds of North America, Inc., Philadelphia, PA.

Brown, P.W. and Fredrickson, L.H. 1997. White-winged Scoter (*Melanitta fusca*). *In* Poole, A. and Gill, F. (Editors), The

Birds of North America, 274. The Birds of North America, Inc., Philadelphia, PA.

Brua, R.B. 2001. Ruddy Duck (*Oxyura jamaicensis*). *In* Poole, A. and Gill, F. (Editors), The Birds of North America, 696. The Birds of North America, Inc., Philadelphia, PA.

Buehler, D. 2000. Bald Eagle (*Haliaeetus leucocephalus*). *In* Poole, A. and Gill, F. (Editors), The Birds of North America, 506. The Birds of North America, Inc., Philadelphia, PA.

Cadman, M.D., Eagles, P.F.J., and Helleiner, F. (Editors). 1990. Atlas of the Breeding Birds of Ontario. University of Waterloo Press, Waterloo, Ontario, Canada.

Canada Centre for Remote Sensing 2000. Land Cover Map of Canada. atlas.nrcan.gc.ca.

Carroll, J.P. 1993. Gray Partidge (*Perdix perdix*). *In* Poole, A. and Gill, F. (Editors), The Birds of North America, 58. The Birds of North America, Inc., Philadelphia, PA.

Chamberlain, D.E. and Fuller, R. 2001. Contrasting patterns of change in the distribution and abundance of farmland birds in relation to farming system in lowland Britain. Global Ecology and Biogeography, volume 10: 399–409.

Chapman, B.A., Goossen, J.P., and Ohanjanian, I. 1985. Occurrences of Black-necked Stilts, *Himantopus mexicanus*, in Western Canada. Canadian Field-Naturalist, volume 99: 254–257.

Cyr, A. and Larivée, J. 1993. A checklist approach for monitoring neotropical migrant birds: Twenty-year trends in birds of Québec using ÉPOQ. *In* Finch, D.M. and Stangel, P. (Editors), Status and management of Neotropical migratory birds, volume General technical report RM-229, 229–236. U.S. Forest Service. Rocky Mountain Forest ocl Range Experiment Station, Fort Collins, Colorado, USA.

Cyr, A. and Larivée, J. 1995. Atlas saisonnier des oiseaux du Québec. Presses de l'Université de Sherbrooke et Société de loisir ornithologique de l'Estrie, Sherbrooke, Québec, Canada.

Dale, B. 1983. Habitat relationships of seven species of passerine birds at Last Mountain Lake. Master's thesis, University of Regina, Regina, Saskatchewan.

Davis, S. 2004. Area sensitivity in grassland passerines: effects of patch size, patch shape, and vegetation structure on bird abundance and occurrence in southern Saskatchewan. Auk, volume 121: 1130–1145.

Donald, P.F. and Fuller, R.J. 1998. Ornithological atlas data: a review of uses and limitations. Bird Study, volume 45: 129–145.

Downing, D.J. and Pettapiece, W.W. 2006. Natural Regions and Subregions of Alberta. Technical Report T/852, Government of Alberta.

Drilling, N., Titman, R., and McKinney, F. 2002. Mallard (*Anas platyrhynchos*). *In* Poole, A. and Gill, F. (Editors), The Birds of North America, 658. The Birds of North America, Inc., Philadelphia, PA.

Droege, S., Cyr, A., and Larivée, J. 1998. Checklists: An under-used tool for the inventory and monitoring of plants and animals. Conservation Biology, volume 12(5): 1134–1138.

Dugger, B.D. and Dugger, K.M. 2002. Long-billed Curlew (*Numenius americanus*). *In* Poole, A. and Gill, F. (Editors), The Birds of North America, 628. The Birds of North America, Inc., Philadelphia, PA.

Dugger, B.D., Dugger, K.M., and Fredrickson, L.H. 1994. Hooded Merganser (*Lophodytes cucullatus*). *In* Poole, A. and Gill, F. (Editors), The Birds of North America, 98. The Birds of North America, Inc., Philadelphia, PA.

Dunn, E. 2005. National action needs for Canadian Landbird Conservation. Special report 1, Canadian Wildlife Service Landbird Committee, Ottawa, Canada.

Dunn, E.H., Francis, C.M., Blancher, P.J., Drennan, S.R., Howe, M.A., LePage, D., Robbins, C.S., Rosenberg, K.V., Sauer, J.R., and Smith, K.G. 2005. Enhancing the scientific value of the Christmas Bird Count. The Auk, volume 122(1): 338–346.

Dunn, E.H., Larivée, J., and Cyr, A. 1996. Can checklist programs be used to monitor populations of birds recorded during the migration season? Wilson Bulletin, volume 108(3): 540–549.

Dunn, E.H., Larivée, J., and Cyr, A. 2001. Site-specific observation in breeding season improves the ability of checklist data to track population trends. Journal of Field Ornithology, volume 72(2): 547–555.

Eadie, J.M., Mallory, M.L., and Lumsden, H.G. 1995. Common Goldeneye (*Bucephala clangula*). *In* Poole, A. and Gill, F. (Editors), The Birds of North America, 170. The Birds of North America, Inc., Philadelphia, PA.

Eadie, J.M. and Savard, J.P.L. 2000. Barrow's Goldeneye (*Bucephala islandica*). *In* Poole, A. and Gill, F. (Editors), The Birds of North America, 548. The Birds of North America, Inc., Philadelphia, PA.

Eaton, S.W. 1992. Wild Turkey (*Meleagris gallopavo*). *In* Poole, A. and Gill, F. (Editors), The Birds of North America, 22. The Birds of North America, Inc., Philadelphia, PA.

Erskine, A.J. 1992. Atlas of Breeding Birds of the Maritime Provinces. Nimbus Publishing Ltd. and the Nova Scotia Museum, Halifax, Nova Scotia, Canada.

Ficken, M.S., McLaren, M.A., and Hailman, J.P. 1996. Boreal Chickadee (*Parus hudsonicus*). *In* Poole, A. and Gill, F. (Editors), The Birds of North America, 254. The Birds of North America, Inc., Philadelphia, PA.

Francis, C.M., Bart, J., Dunn, E.H., Burnham, K.P., and Ralph, C.J. 2005. Enhancing the value of the breeding bird survey: reply to Sauer et al. (2005). Journal of Wildlife Management, volume 69(4): 1327–1332.

Frawley, B. 1989. The dynamics of nongame bird breeding ecology in Iowa alfalfa fields. Master's thesis, Iowa State University, Ames, Iowa.

Gammonley, J.H. 1996. Cinnamon Teal (*Anas cyanoptera*). *In* Poole, A. and Gill, F. (Editors), The Birds of North America, 209. The Birds of North America, Inc., Philadelphia, PA.

Gaston, K.J. 1999. Implications of interspecific and intraspecific abundance-occupancy relationships. Oikos, volume 86: 195–207.

Gaston, K.J., Blackburn, T.M., Greenwood, J.J.D., Gregory, R.D., Quinn, R.M., and Lawton, J.H. 2000. Abundance-occupancy relationships. Journal of Applied Ecology, volume 37(Suppl. 1): 39–59.

Gauthier, G. 1993. Bufflehead (*Bucephala albeola*). *In* Poole, A. and Gill, F. (Editors), The Birds of North America, 67. The Birds of North America, Inc., Philadelphia, PA.

Giudice, J.H. and Ratti, J.T. 2001. Ring-necked Pheasant (*Phasianus colchicus*). *In* Poole, A. and Gill, F. (Editors), The Birds of North America, 572. The Birds of North America, Inc., Philadelphia, PA.

Gratto-Trevor, C.L. 2000. Marbled Godwit (*Limosa fedoa*). *In* Poole, A. and Gill, F. (Editors), The Birds of North America, 492. The Birds of North America, Inc., Philadelphia, PA.

Green, M.T., Lowther, P., Jones, S., Davis, S., and Dale, B. 2002. Baird's Sparrow (*Ammodramus bairdii*). *In* Poole, A. and Gill, F. (Editors), The Birds of North America, 638. The Birds of North America, Inc., Philadelphia, PA.

Greene, E., Davison, W., and Muehter, V.R. 1998. Steller's Jay (*Cyanocitta stelleri*). *In* Poole, A. and Gill, F. (Editors), The Birds of North America, 343. The Birds of North America, Inc., Philadelphia, PA.

Greenwood, J.J.D. 2003. The monitoring of British breeding birds: a success story for conservation science? The Science of the Total Environment, volume 310: 221–230.

Gregoire, P. 2000. Harlequin Duck surveys on the eastern slopes of Alberta, preliminary results 1998, 1999. *In* Proceedings of the Fifth Harlequin Duck Symposium. Washington Department of Fish and Wildlife, Blaine, WA.

Group, E.S.W. 1995. A National Ecological Framework for Canada. Map at 1: 7,500,000, Agriculture and Agri-Food Canada, Research Branch, Centre for Land and Biological Resource Research and Environment Canada, State of the Environment Directorate, Ecozone Analysis Branch, Ottawa/Hull. 125 pp.

Hannon, S.J., Eason, P.K., and Martin, K. 1998. Willow Ptarmigan (*Lagopus lagopus*). *In* Poole, A. and Gill, F. (Editors), The Birds of North America, 369. The Birds of North America, Inc., Philadelphia, PA.

Harrison, J.A., Allan, D.G., Underhill, L.G., Herremans, M., Tree, A.J., Parker, V., and Brown, C.J. (Editors). 1997. The Atlas of Southern African Birds. BirdLife South Africa, Johannesburg, South Africa.

Hatch, J.J. and Weseloh, D.V. 1999. Double-crested Cormorant (*Phalacrocorax auritus*). *In* Poole, A. and Gill, F. (Editors), The Birds of North America, 441. The Birds of North America, Inc., Philadelphia, PA.

Hejl, S.J., Newlon, K.R., McFadzen, M.E., Young, J.S., and Ghalambor, C.K. 2002. Brown Creeper (*Certhia americana*). *In* Poole, A. and Gill, F. (Editors), The Birds of North America, 669. The Birds of North America, Inc., Philadelphia, PA.

Hepp, G.R. and Bellrose, F.C. 1995. Wood Duck (*Aix sponsa*). *In* Poole, A. and Gill, F. (Editors), The Birds of North America, 169. The Birds of North America, Inc., Philadelphia, PA.

Hill, D.P. and Gould, L.K. 1997. Chestnut-collared Longspur (*Calcarius ornatus*). *In* Poole, A. and Gill, F. (Editors), The Birds of North America, 288. The Birds of North America, Inc., Philadelphia, PA.

Hill, G.E. 1995. Black-headed Grosbeak (*Pheucticus melanocephalus*). *In* Poole, A. and Gill, F. (Editors), The Birds of North America, 143. The Birds of North America, Inc., Philadelphia, PA.

Hoffman, R.W. 2006. White-tailed Ptarmigan (*Lagopus leucura*): a technical conservation assessment. Online 2.1, USDA Forest Service, Rocky Mountain Region.

Hohman, W.L. and Eberhardt, R.T. 1998. Ring-necked Duck (*Aythya collaris*). *In* Poole, A. and Gill, F. (Editors), The Birds of North America, 329. The Birds of North America, Inc., Philadelphia, PA.

Holt, D.W. and Petersen, J.L. 2000. Northern Pygmy-Owl (*Glaucidium gnoma*). *In* Poole, A. and Gill, F. (Editors), The Birds of North America, 494. The Birds of North America, Inc., Philadelphia, PA.

Houston, S.C. and Bowen, D.E., Jr 2001. Sprague's Pipit (*Anthus spragueii*). *In* Poole, A. and Gill, F. (Editors), The Birds of North America, 580. The Birds of North America, Inc., Philadelphia, PA.

Hunt, L. 1993. Diet and habitat use of nesting Prairie Falcons (*Falco mexicanus*) in an agricultural landscape in southern Alberta. Master's thesis, University of Alberta, Edmonton.

Ingold, J.L. and Wallace, G.E. 1994. Ruby-crowned Kinglet (*Regulus calendula*). *In* Poole, A. and Gill, F. (Editors), The Birds of North America, 119. The Birds of North America, Inc., Philadelphia, PA.

Jenny, H. 1941. Factors of Soil Formation: A System of Quantitative Pedology. McGraw-Hill, New York.

Johnson, K. 1995. Green-winged Teal (*Anas crecca*). *In* Poole, A. and Gill, F. (Editors), The Birds of North America, 193. The Birds of North America, Inc., Philadelphia, PA.

Jones, S.L. and Cornely, J.E. 2002. Vesper Sparrow (*Pooecetes gramineus*). *In* Poole, A. and Gill, F. (Editors), The Birds of North America, 624. The Birds of North America, Inc., Philadelphia, PA.

Kachigan, S.K. 1986. Statistical Analyis: An interdisciplinary introduction to univariate and multivariate methods. Radius Press, New York, USA.

Knapton, R.W. 1994. Clay-colored Sparrow (*Spizella pallida*). *In* Poole, A. and Gill, F. (Editors), The Birds of North America, 120. The Birds of North America, Inc., Philadelphia, PA.

LaDeau, S., Kilpatrick, A.M., and Marra, P.P. 2007. West Nile virus emergence and large-scale declines of North American bird populations. Online, Nature International Weekly Journal of Science.

Lanyon, W.E. 1994. Western Meadowlark (*Sturnella neglecta*). *In* Poole, A. and Gill, F. (Editors), The Birds of North America, 104. The Birds of North America, Inc., Philadelphia, PA.

Lanyon, W.E. 1997. Great Crested Flycatcher (*Myiarchus crinitus*). *In* Poole, A. and Gill, F. (Editors), The Birds of North America, 300. The Birds of North America, Inc., Philadelphia, PA.

LeSchack, C.R., McKnight, S.K., and Hepp, G.R. 1997. Gadwall (*Anas strepera*). *In* Poole, A. and Gill, F. (Editors), The Birds of North America, 283. The Birds of North America, Inc., Philadelphia, PA.

Lowther, P.E., Douglas, H.D., and Gratto-Trevor, C.L. 2001. Willet (*Tringa semipalmata*). *In* Poole, A. and Gill, F. (Editors), The Birds of North America, 579. The Birds of North America, Inc., Philadelphia, PA.

Major, J. 1951. A Functional, Factorial Approach to Plant Ecology. Ecology, volume 32: 392–412.

Mallory, M. and Metz, K. 1999. Common Merganser (*Mergus merganser*). *In* Poole, A. and Gill, F. (Editors), The Birds of North America, 442. The Birds of North America, Inc., Philadelphia, PA.

Martin, P. and Forsyth, D. 2003. Occurrence and productivity of songbirds in prairie farmland under conventional versus minimum tillage regimes. Agriculture, Ecosystems and Environment, volume 96: 107–117.

McCallum, D.A., Grundel, R., and Dahlsten, D.L. 1999. Mountain Chickadee (*Poecile gambeli*). *In* Poole, A. and Gill, F. (Editors), The Birds of North America, 453. The Birds of North America, Inc., Philadelphia, PA.

McGeoch, M.A. and Gaston, K.J. 2002. Occupancy frequency distributions: patterns, artefacts and mechanisms. Biological Reviews, volume 77: 311–331.

McMaster, D. and Davis, S.K. 2001. An evaluation of Canada's Permanent Cover Program: habitat for grassland birds? Journal of Field Ornithology, volume 72: 195–210.

Mitchell, C.D. 1994. Trumpeter Swan (*Cygnus buccinator*). *In* Poole, A. and Gill, F. (Editors), The Birds of North America, 105. The Birds of North America, Inc., Philadelphia, PA.

Mowbray, T. 1999. American Wigeon (*Anas americana*). *In* Poole, A. and Gill, F. (Editors), The Birds of North America, 401. The Birds of North America, Inc., Philadelphia, PA.

Mowbray, T.B. 2002. Canvasback (*Aythya valisineria*). *In* Poole, A. and Gill, F. (Editors), The Birds of North America, 659. The Birds of North America, Inc., Philadelphia, PA.

Mowbray, T.B., Ely, C.R., Sedinger, J.S., and Trost, R.E. 2002. Canada Goose (*Branta canadensis*). *In* Poole, A. and Gill, F. (Editors), The Birds of North America, 682. The Birds of North America, Inc., Philadelphia, PA.

Murphy, M.T. 1996. Eastern Kingbird (*Tyrannus tyrannus*). *In* Poole, A. and Gill, F. (Editors), The Birds of North America, 253. The Birds of North America, Inc., Philadelphia, PA.

Norment, C.J., Hendricks, P., and Santonocito, R. 1998. Golden-crowned Sparrow (*Zonotrichia atricapilla*). *In* Poole, A. and Gill, F. (Editors), The Birds of North America, 352. The Birds of North America, Inc., Philadelphia, PA.

Pierotti, R.J. and Good, T.P. 1994. Herring Gull (*Larus argentatus*). *In* Poole, A. and Gill, F. (Editors), The Birds of North America, 128. The Birds of North America, Inc., Philadelphia, PA.

Pinel, H.W., Smith, W.W., and Wershler, C.R. 1991. Alberta Birds, 1971–1980, Volume 1. Non-Passerines. Occasional Paper 13, Natural History Section, Provincial Museum of Alberta, Edmonton, Alberta.

Pitocchelli, J., Bouchie, J., and Jones, D. 1997. Connecticut Warbler (*Oporornis agilis*). *In* Poole, A. and Gill, F. (Editors), The Birds of North America, 320. The Birds of North America, Inc., Philadelphia, PA.

Poole, A., Stettenheim, P., and Gill, F. (Editors). 1992. Birds of North America. The American Ornithologists' Union, The Academy of Natural Sciences, Washington, D.C.

Poulin, R.G., Grindal, S.D., and Brigham, R.M. 1996. Common Nighthawk (*Chordeiles minor*). *In* Poole, A. and Gill, F. (Editors), The Birds of North America, 213. The Birds of North America, Inc., Philadelphia, PA.

Prescott, D.R.C. and Murphy, A.J. 1996. Habitat associations of grassland birds on native and tame pastures of the Aspen Parkland of Alberta. Technical report, Alberta Centre. NAWMP-021. Edmonton, Alberta.

Prescott, D.R.C. and Murphy, A.J. 1999. Bird populations of seeded grasslands in the aspen parkland of Alberta. Studies in Avian Biology, volume 19: 203–210.

Radford, J.Q. and Bennett, A.F. 2005. Terrestrial avifauna of the Gippsland Plain and Strzelecki Ranges, Victoria, Australia: insights from Atlas data. Wildlife Research, volume 32: 531–555.

Reed, J.M., Causey, D., Hatch, J.J., Cooke, F., and Crowder, L. 2003. Review of the Double-crested Cormorant Management Plan 2003: Final Report of the AOU Conservation Committee's Panel. Technical report, AOU Conservation Committee.

Robbins, C.S. and Geissler, P.H. 1990. Sampling methods and mapping grids. *In* Smith, C.R. (Editor), Handbook for Atlassing North American Breeding Birds. Vermont Institute of Natural Science, Woodstock, Vermont. www.bsc-eoc.org/norac/atlascont.htm.

Robbins, M.B. and Dale, B.C. 1999. Sprague's Pipit (*Anthus spragueii*). *In* Poole, A. and Gill, F. (Editors), The Birds of North America, 439. The Birds of North America, Inc., Philadelphia, PA.

Roberts, R.L., Donald, P.F., and Fisher, I.J. 2005. Worldbirds: developing a web-based data collection system for global monitoring of bird distribution and abundance. Biodiversity and Conservation, volume 14: 2807–2820.

Robinson, J.A., Oring, L.W., Skorupa, J.P., and Boettcher, R. 1997. American Avocet (*Recurvirostra americana*). *In* Poole, A. and Gill, F. (Editors), The Birds of North America, 275. The Birds of North America, Inc., Philadelphia, PA.

Robinson, J.A., Reed, J.M., Skorupa, J.P., and Oring, L.W. 1999. Black-necked Stilt (*Himantopus mexicanus*). *In* Poole, A. and Gill, F. (Editors), The Birds of North America, 449. The Birds of North America, Inc., Philadelphia, PA.

Root, T.L. 1988. Atlas of Wintering North American Birds: An analysis of Christmas Bird Count data. University of Chicago Press, Chicago, USA.

Rotenberry, J.T., Patten, M.A., and Preston, K.L. 1999. Brewer's Sparrow (*Spizella breweri*). *In* Poole, A. and Gill, F. (Editors), The Birds of North America, 275. The Birds of North America, Inc., Philadelphia, PA.

Rowher, F.C., Johnson, W.P., and Loos, E.R. 2002. Blue-winged Teal (*Anas discors*). *In* Poole, A. and Gill, F. (Editors), The Birds of North America, 625. The Birds of North America, Inc., Philadelphia, PA.

Rusch, D.H., DeStefano, S., Reynolds, M.C., and Lauten, D. 2000. Ruffed Grouse (*Bonasa umbellus*). *In* Poole, A. and Gill, F. (Editors), The Birds of North America, 515. The Birds of North America, Inc., Philadelphia, PA.

Ryan, M., Renken, R., and Dinsmore, J. 1984. Marbled Godwit habitat selection in the northern prairie region. Journal of Wildlife Management, volume 48: 1206–1218.

Ryder, R.A. and Manry, D.E. 1994. White-faced Ibis (*Plegadis chihi*). *In* Poole, A. and Gill, F. (Editors), The Birds of North America, 130. The Birds of North America, Inc., Philadelphia, PA.

Salisbury, C.D.C. and Salisbury, L.D. 1989. Successful breeding of Black-necked Stilts in Saskatchewan. Blue Jay, volume 47: 154–156.

Salt, W.R. and Salt, J.R. 1976. The Birds of Alberta. Hurtig Publishers, Edmonton, Alberta.

Sauer, J.R., Hines, J.E., and Fallon, J. 2005a. The North American Breeding Bird Survey, results and analysis 1966–2005. Technical report, USGS Patuxent Wildlife Research Center, Laurel, MD. Version 6.2.2006.

Sauer, J.R., Link, W.A., Nichols, J.D., and Royle, J.A. 2005b. Using the North American Breeding Bird Survey as a tool for conservation: A critique of Bart et al. (2004). Journal of Wildlife Management, volume 69(4): 1321–1326.

Sauer, J.R., Link, W.A., and Royle, J.A. 2004. Estimating population trends with a linear model: technical comments. The Condor, volume 106: 435–440.

Savard, J.P.L., Bordage, D., and Reed, A. 1998. Surf Scoter (*Melanitta perspicillata*). *In* Poole, A. and Gill, F. (Editors), The Birds of North America, 363. The Birds of North America, Inc., Philadelphia, PA.

Schieck, J., Stuart-Smith, K., and Norton, M. 2000. Bird communities are affected by amount and dispersion of vegetation retained in mixedwood boreal forest harvest areas. Forest and Ecology Management, volume 126: 239–254.

Schroeder, M.A., Young, J.R., and Braun, C.E. 1999. Sage Grouse (*Centrocercus urophasianus*). *In* Poole, A. and Gill, F. (Editors), The Birds of North America, 425. The Birds of North America, Inc., Philadelphia, PA.

Schukman, J.M. and Wolf, B.O. 1998. Say's Phoebe (*Sayornis saya*). *In* Poole, A. and Gill, F. (Editors), The Birds of North America, 374. The Birds of North America, Inc., Philadelphia, PA.

Scott, G. 1995. Canada's Vegetation: A World Perspective. McGill-Queen's University Press, Montreal, Canada.

Semenchuk, G.P. (Editor). 1992. The Atlas of Breeding Birds of Alberta. The Federation of Alberta Naturalists, Edmonton, Alberta.

Shane, T.G. 2000. Lark Bunting (*Calamospiza melanocorys*). *In* Poole, A. and Gill, F. (Editors), The Birds of North America, 542. The Birds of North America, Inc., Philadelphia, PA.

Smallwood, J.A. and Bird, D.M. 2002. American Kestrel (*Falco sparverius*). *In* Poole, A. and Gill, F. (Editors), The Birds of North America, 602. The Birds of North America, Inc., Philadelphia, PA.

Smith, A.R. 1996. Atlas of Saskatchewan Birds. Saskatchewan Nat. Hist. Society, Regina, Saskatchewan.

Smith, C.R. (Editor). 1990. Handbook for Atlassing North American Breeding Birds. Vermont Institute of Natural Science, Woodstock, Vermont. www.bsc-eoc.org/norac/atlascont.htm.

Smith, S.M. 1993. Black-capped Chickadee (*Poecile atricapillus*). *In* Poole, A., Stettenheim, P., and Gill, F. (Editors), The Birds of North America, 39. The Birds of North America, Inc., Philadelphia, PA.

Storer, R.W. and Nuechterlein, G.L. 1992. Western Grebe and Clark's Grebe. *In* Poole, A., Stettenheim, P., and Gill, F. (Editors), The Birds of North America, 26. The Birds of North America, Inc., Philadelphia, PA.

Sutter, G.C., Davis, S.K., and Duncan, D.C. 2000. Grassland songbird abundance along roads and trails in southern Saskatchewan. Journal of Field Ornithology, (71): 110–116.

Takats, D.L., Francis, C.M., Holroyd, G.L., Duncan, J.R., Mazur, K.M., Cannings, R.J., Harris, W., and Holt, D. 2001. Guidelines for nocturnal owl monitoring in North America. Technical report, Beaverhill Bird Observatory and Bird Studies Canada, Edmonton, Alberta. 32 pp.

Team, A.P.F.R. 2005. Alberta Peregrine Falcon Recovery Plan 2004–2010. Alberta Species at Risk Recovery Plan 3, Alberta Sustainable Resource Development, Fish and Wildlife Division, Edmonton, Alberta. 16 pp.

Temple, S.A. and Cary, J.R. 1990. Description of the Wisconsin checklist project. *In* Sauer, J. and Droege, S. (Editors), Survey designs and statistical methods for the evaluation of avian population trends, Biological Report, volume 90(1), 14–17. U.S. Fish and Wildlife Service, Wisconsin. 166 pp.

van Belle, G. 2002. Statistical Rules of Thumb. John Wiley & Sons Canada.

Vickery, P.D. 1996. Grasshopper Sparrow (*Ammodramus savannarum*). *In* Poole, A. and Gill, F. (Editors), The Birds of North America, 239. The Birds of North America, Inc., Philadelphia, PA.

Wagner, S.F. 1992. Introduction to Statistics. Harper Collins, New York, USA.

Wallis, C. and Wershler, C. 1984. Kazan Upland Resource Assessment for Ecological Reserves Planning in Alberta. Technical report, Prepared by Cottonwood Consultants Ltd. Calgary, Alberta for Alberta Energy and Natural Resources, Public Lands Division.

Walter, H. 1979. Vegetation of the Earth and Ecological Systems of the Geo-Biosphere. Springer-Verlag, New York, second edition.

Watmough, M.D. and Schmoll, M.J. 2007. Environment Canada's Prairie and Northern Habitat Monitoring Program Phase II: Recent habitat trends in the PHJV. Technical Report Draft, Canadian Wildlife Service.

Wedgwood, J.A. and Taylor, P.S. 1988. Black-necked Stilt in Saskatchewan. Blue Jay, volume 46: 80–83.

Weeks, H.P., Jr 1994. Eastern Phoebe (*Sayornis phoebe*). *In* Poole, A. and Gill, F. (Editors), The Birds of North America, 94. The Birds of North America, Inc., Philadelphia, PA.

Wheelwright, N.T. and Rising, J.D. 1993. Savannah Sparrow (*Passerculus sandwichensis*). *In* Poole, A. and Gill, F. (Editors), The Birds of North America, 45. The Birds of North America, Inc., Philadelphia, PA.

With, K.A. 1994. McCown's Longspur (*Calcarius mccownii*). *In* Poole, A. and Gill, F. (Editors), The Birds of North America, 96. The Birds of North America, Inc., Philadelphia, PA.

Woodin, M.C. and Michot, T.C. 2002. Redhead (*Aythya americana*). *In* Poole, A. and Gill, F. (Editors), The Birds of North America, 695. The Birds of North America, Inc., Philadelphia, PA.

Zwickel, F.C. and Bendell, J.F. 2005. Blue Grouse (*Dendragapus obscurus*). *In* Poole, A. and Gill, F. (Editors), The Birds of North America, 15. The Birds of North America, Inc., Philadelphia, PA.

Index

C

D

E

F

G

N

O

P

Q

R

S

T

V

W

X

Y

Z